Infectious Forest Diseases

FSC
www.fsc.org
MIX
Paper from
responsible sources
FSC® C013604

Dedicated to Giovanni Nicolotti

Infectious Forest Diseases

Edited by

Paolo Gonthier

Department of Agricultural, Forest and Food Sciences
University of Torino, Grugliasco (TO), Italy

and

Giovanni Nicolotti†

Formerly of Department of Exploitation and Protection of Agricultural and Forestry Resources
University of Torino, Grugliasco (TO), Italy

With the assistance of Luana Giordano

www.cabi.org

CABI is a trading name of CAB International

CABI
Nosworthy Way
Wallingford
Oxfordshire OX10 8DE
UK

Tel: +44 (0)1491 832111
Fax: +44 (0)1491 833508
E-mail: info@cabi.org
Website: www.cabi.org

CABI
38 Chauncey Street
Suite 1002
Boston, MA 02111
USA

Tel: +1 800 552 3083 (toll free)
Tel: +1 (0)617 395 4051
E-mail: cabi-nao@cabi.org

A catalogue record for this book is available from the British Library, London, UK.

Library of Congress Cataloging-in-Publication Data

Infectious forest diseases / edited by Paolo Gonthier and Giovanni Nicolotti.
p. cm.
Includes bibliographical references and index.
ISBN 978-1-78064-040-2 (alk. paper)
1. Trees--Diseases and pests. 2. Forest health. 3. Forest protection. I. Gonthier, Paolo. II. Nicolotti, Giovanni.
SB761.I54 2013
582.16--dc23

2012044403

ISBN-13: 978 1 78064 040 2

Commissioning editor: Rachel Cutts
Editorial assistant: Alexandra Lainsbury
Production editor: Lauren Povey

Typeset by SPi, Pondicherry, India.
Printed and bound in the UK by CPI Group (UK) Ltd, Croydon, CR0 4YY.

Contents

Contributors ix

Foreword xiii
S. Woodward

Preface xv

Contributors' Acknowledgements xvii

List of Acronyms xix

Introduction to Infectious Forest Diseases xxi

PART I – INTRODUCTORY CONCEPTS

1 **Concepts of Epidemiology of Forest Diseases** 1
J. Oliva, J.B. Boberg, A.J.M. Hopkins and J. Stenlid

2 **General Strategies of Forest Disease Management** 29
R.L. Edmonds

PART II – NON-FUNGAL INFECTIOUS FOREST DISEASES

3 **Forest Diseases Caused by Viruses** 50
C. Büttner, S. von Bargen, M. Bandte and H.-P. Mühlbach

4 **Forest Diseases Caused by Prokaryotes: Phytoplasmal and Bacterial Diseases** 76
H.M. Griffiths

5 **Forest Diseases Caused by Higher Parasitic Plants: Mistletoes** 97
D.C. Shaw and R.L. Mathiasen

6 **Pine Wilt Disease and Other Nematode Diseases** 115
N. Kamata and Y. Takeuchi

PART III – FOREST DISEASES CAUSED BY FUNGI AND FUNGAL-LIKE ORGANISMS

ROOT AND BUTT ROTS

7 Annosus Root and Butt Rots 128
P. Gonthier and M. Thor

8 Armillaria Root Rots 159
J.-J. Guillaumin and P. Legrand

9 Laminated and Tomentosus Root Rots 178
K.J. Lewis

STEM ROTS

10 Heart Rots, Sap Rots and Canker Rots 197
R. Vasaitis

VASCULAR DISEASES

11 Ceratocystis Diseases 230
T.C. Harrington

12 Dutch Elm Disease and Other Ophiostoma Diseases 256
T. Kirisits

13 Blackstain Root Disease and Other Leptographium Diseases 283
L.G. Eckhardt

CANKER DISEASES

14 Cankers and Other Diseases Caused by the *Botryosphaeriaceae* 298
J.W.M. Mehl, B. Slippers, J. Roux and M.J. Wingfield

15 Chestnut Blight 318
S. Prospero and D. Rigling

16 Other Cankers Caused by *Cryphonectria* and Sibling Species 340
M. Gryzenhout

17 Cypress Canker 359
R. Danti, G. Della Rocca and A. Panconesi

18 Pitch Canker 376
T.R. Gordon

19 Larch Canker 392
T.L. Cech

20 Hypoxylon Canker 407
M.E. Ostry

BRANCH AND TIP BLIGHTS

21 Branch and Tip Blights 420
P. Capretti, A. Santini and H. Solheim

FOLIAR DISEASES

22 Dothistroma Needle Blight 436
L.S. Bulman, M.A. Dick, R.J. Ganley, R.L. McDougal, A. Schwelm and R.E. Bradshaw

23 Other Foliar Diseases of Coniferous Trees 458
M. Bednářová, M. Dvořák, J. Janoušek and L. Jankovský

24 Foliar Diseases of Broadleaved Trees 488
T. Kowalski

OOMYCETE DISEASES

25 Oomycete Diseases 519
K.J. Hayden, G.E.St.J. Hardy and M. Garbelotto

RUST DISEASES

26 Tree Rusts 547
R.C. Hamelin

PART IV – NURSERY DISEASES

27 Seed, Seedling and Nursery Diseases 567
A. Lilja and M. Poteri

PART V – INTRODUCED PATHOGENS

28 Responding to Diseases Caused by Exotic Tree Pathogens 592
M. Pautasso

Index 613

Colour plate section can be found following p. 296

Contributors

Martina Bandte, Division of Phytomedicine, Faculty of Agriculture and Horticulture, Humboldt-Universität zu Berlin, Berlin, D-14195, Germany. E-mail: martina.bandte@agrar.hu-berlin.de

Miroslava Bednářová, Department of Forest Pathology and Wildlife Management, Faculty of Forestry and Wood Technology, Mendel University, Brno, 613 00, Czech Republic. E-mail: svezi.mirka@email.cz

Johanna B. Boberg, Department of Forest Mycology and Plant Pathology, Swedish University of Agricultural Sciences, Uppsala, SE-750 07, Sweden. E-mail: johanna.boberg@slu.se

Rosie E. Bradshaw, Bio-Protection Research Centre, Massey University, Institute of Fundamental Sciences (formerly Institute of Molecular BioSciences), Massey University, Palmerston North, 4442, New Zealand. E-mail: r.e.bradshaw@massey.ac.nz

Lindsay S. Bulman, Scion, Rotorua, 3010, New Zealand. E-mail: lindsay.bulman@scionresearch.com

Carmen Büttner, Division of Phytomedicine, Faculty of Agriculture and Horticulture, Humboldt-Universität zu Berlin, Berlin, D-14195, Germany. E-mail: carmen.buettner@agrar.hu-berlin.de

Paolo Capretti, Section of Plant Protection, Department of Agricultural Biotechnology, University of Firenze, Florence, 50144, Italy. E-mail: paolo.capretti@unifi.it

Thomas L. Cech, Institute of Forest Protection, Federal Research and Training Centre for Forests, Natural Hazards and Landscape, Vienna, 1131, Austria. E-mail: thomas.cech@bfw.gv.at

Roberto Danti, Plant Protection Institute, National Research Council (CNR), Sesto Fiorentino, Florence, I-50019, Italy. E-mail: danti@ipp.cnr.it

Gianni Della Rocca, Plant Protection Institute, National Research Council (CNR), Sesto Fiorentino, Florence, I-50019, Italy. E-mail: g.dellarocca@ipp.cnr.it

Margaret A. Dick, Scion, Rotorua, 3010, New Zealand. E-mail: margaret.dick@scionresearch.com

Miloň Dvořák, Department of Forest Pathology and Wildlife Management, Faculty of Forestry and Wood Technology, Mendel University, Brno, 613 00, Czech Republic. E-mail: milon.dvorak@seznam.cz

Lori G. Eckhardt, Forest Health Dynamics Laboratory, School of Forestry and Wildlife Sciences, Auburn University, Auburn, AL 36849, USA. E-mail: ECKHALG@auburn.edu

Robert L. Edmonds, School of Forest Resources, University of Washington, Seattle, WA 98195, USA. E-mail: bobe@u.washington.edu

Rebecca J. Ganley, Scion, Rotorua, 3010, New Zealand. E-mail: rebecca.ganley@scionresearch.com

Matteo Garbelotto, Department of Environmental Science, Policy and Management, University of California, Berkeley, CA 94720, USA. E-mail: matteog@berkeley.edu

Paolo Gonthier, Department of Agricultural, Forest and Food Sciences, University of Torino, Grugliasco, I-10095, Italy. E-mail: paolo.gonthier@unito.it

Thomas R. Gordon, Department of Plant Pathology, University of California, Davis, CA 95616, USA. E-mail: trgordon@ucdavis.edu

Helen M. Griffiths, Department of Plant Pathology and Plant Microbe Biology, Cornell University, Ithaca, NY 14853, USA. E-mail: hmg1@cornell.edu

Marieka Gryzenhout, Department of Plant Sciences, University of the Free State, Bloemfontein, 9300, South Africa. E-mail: GryzenhoutM@ufs.ac.za

Jean-Jacques Guillaumin, retired from National Institute for Agricultural Research (INRA), France. E-mail: jj.guillaumin@gmail.com

Richard C. Hamelin, Laurentian Forestry Centre, Natural Resources Canada, Quebec, QC, G1V 4C7, Canada, and Department of Forest Sciences, The University of British Columbia, Vancouver, BC, V6T 1Z4, Canada. E-mail: richard.hamelin@ubc.ca

Giles E.St.J. Hardy, School of Biological Sciences and Biotechnology, Murdoch University, Murdoch, WA 6150, Australia. E-mail: G.Hardy@murdoch.edu.au

Thomas C. Harrington, Department of Plant Pathology and Microbiology, Iowa State University, Ames, IA 50011, USA. E-mail: tcharrin@iastate.edu

Katherine J. Hayden, Department of Environmental Science, Policy and Management, University of California, Berkeley, CA 94720, USA. E-mail: khayden@berkeley.edu

Anna J.M. Hopkins, Department of Forest Mycology and Plant Pathology, Swedish University of Agricultural Sciences, Uppsala, SE-750 07, Sweden. E-mail: anna.hopkins@slu.se

Libor Jankovský, Department of Forest Pathology and Wildlife Management, Faculty of Forestry and Wood Technology, Mendel University, Brno, 613 00, Czech Republic. E-mail: jankov@mendelu.cz

Joseph Janoušek, Department of Forest Pathology and Wildlife Management, Faculty of Forestry and Wood Technology, Mendel University, Brno, 613 00, Czech Republic. E-mail: janousek.jose@gmail.com

Naoto Kamata, Graduate School of Agricultural and Life Sciences, The University of Tokyo, Tokyo, 113-8657, Japan. E-mail: kamatan@uf.a.u-tokyo.ac.jp

Thomas Kirisits, Institute of Forest Entomology, Forest Pathology and Forest Protection, Department of Forest and Soil Sciences, University of Natural Resources and Life Sciences, Vienna (BOKU), Vienna, A-1190, Austria. E-mail: thomas.kirisits@boku.ac.at

Tadeusz Kowalski, Department of Forest Pathology, University of Agriculture in Krakow, Krakow, 31-425, Poland. E-mail: rltkowal@cyf-kr.edu.pl

Philippe Legrand, formerly of Forest Health Service, France. E-mail: phlegrand.lempdes@laposte.net

Kathy J. Lewis, University of Northern British Columbia, Prince George, BC, V2N 4Z9, Canada. E-mail: lewis@unbc.ca

Arja Lilja, Vantaa Research Unit, Finnish Forest Research Institute, Vantaa, FI-01301, Finland. E-mail: arja.lilja@metla.fi

Robert L. Mathiasen, School of Forestry, Northern Arizona University, Flagstaff, AZ 86011-5018, USA. E-mail: robert.mathiasen@nau.edu

Rebecca L. McDougal, Scion, Rotorua, 3010, New Zealand. E-mail: rebecca.mcDougal@scionresearch.com

James W.M. Mehl, DST/NRF (Department of Science and Technology/National Research Foundation) Centre of Excellence in Tree Health Biotechnology (CTHB), Forestry and Agricultural Biotechnology Institute (FABI), University of Pretoria, Pretoria, 0028, South Africa. E-mail: james.mehl@fabi.up.ac.za

Hans-Peter Mühlbach, Biocentre Klein Flottbek and Botanical Garden Ohnhorststrasse, University of Hamburg, Hamburg, D-22609, Germany. E-mail: muehlbach@botanik.uni-hamburg.de

Jonàs Oliva, Department of Forest Mycology and Plant Pathology, Swedish University of Agricultural Sciences, Uppsala, SE-750 07, Sweden. E-mail: jonas.oliva@slu.se

Michael E. Ostry, Northern Research Station, US Department of Agriculture Forest Service, St Paul, MN 55108, USA. E-mail: mostry@fs.fed.us

Alberto Panconesi, retired from Plant Protection Institute, National Research Council (CNR), Sesto Fiorentino, Florence, I-50019, Italy. E-mail: a.panconesi@ipp.cnr.it

Marco Pautasso, Centre d'Ecologie Fonctionnelle et Evolutive (CEFE), National Centre for Scientific Research (CNRS), Montpellier, 34293, France. E-mail: marpauta@gmail.com

Marja Poteri, Suonenjoki Research Unit, Finnish Forest Research Institute, Suonenjoki, FI-77600, Finland. E-mail: marja.poteri@metla.fi

Simone Prospero, Swiss Federal Institute for Forest, Snow and Landscape Research (WSL), Birmensdorf, CH-8903, Switzerland. E-mail: simone.prospero@wsl.ch

Daniel Rigling, Swiss Federal Institute for Forest, Snow and Landscape Research (WSL), Birmensdorf, CH-8903, Switzerland. E-mail: daniel.rigling@wsl.ch

Jolanda Roux, DST/NRF (Department of Science and Technology/National Research Foundation) Centre of Excellence in Tree Health Biotechnology (CTHB), Forestry and Agricultural Biotechnology Institute (FABI), University of Pretoria, Pretoria, 0028, South Africa. E-mail: jolanda.roux@up.ac.za

Alberto Santini, Plant Protection Institute, National Research Council (CNR), Sesto Fiorentino, 50019, Italy. E-mail: a.santini@ipp.cnr.it

Arne Schwelm, formerly of Institute of Molecular BioSciences, Massey University, Palmerston North, 4442, New Zealand. Now at: Department of Plant Biology and Forest Genetics, Uppsala BioCenter, Swedish University of Agricultural Sciences, Uppsala, PO Box 7080, SE-750 07 Uppsala, Sweden. E-mail: Arne.Schwelm@slu.se

David C. Shaw, Department of Forest Engineering, Resources, and Management, Oregon State University, Corvallis, OR 97331, USA. E-mail: dave.shaw@oregonstate.edu

Bernard Slippers, DST/NRF (Department of Science and Technology/National Research Foundation) Centre of Excellence in Tree Health Biotechnology (CTHB), Forestry and Agricultural Biotechnology Institute (FABI), University of Pretoria, Pretoria, 0028, South Africa. E-mail: bernard.slippers@up.ac.za

Halvor Solheim, The Norwegian Forest and Landscape Institute, Ås, 1431, Norway. E-mail: halvor.solheim@skogoglandskap.no

Jan Stenlid, Department of Forest Mycology and Plant Pathology, Swedish University of Agricultural Sciences, Uppsala, SE-750 07, Sweden. E-mail: jan.stenlid@slu.se

Yuko Takeuchi, Graduate School of Agriculture, Kyoto University, Kyoto, 606-8502, Japan. E-mail: yuuko@kais.kyoto-u.ac.jp

Magnus Thor, Skogforsk, The Forest Research Institute of Sweden, Uppsala Science Park, Uppsala, SE-751 83, Sweden. E-mail: magnus.thor@skogforsk.se

Rimvydas Vasaitis, Department of Forest Mycology and Plant Pathology, Swedish University of Agricultural Sciences, Uppsala, SE-750 07, Sweden. E-mail: rimvys.vasaitis@slu.se

Susanne von Bargen, Division of Phytomedicine, Faculty of Agriculture and Horticulture, Humboldt-Universität zu Berlin, Berlin, D-14195, Germany. E-mail: susanne.von.bargen@agrar.hu-berlin.de

Michael J. Wingfield, DST/NRF (Department of Science and Technology/National Research Foundation) Centre of Excellence in Tree Health Biotechnology (CTHB), Forestry and Agricultural Biotechnology Institute (FABI), University of Pretoria, Pretoria, 0028, South Africa. E-mail: mike.wingfield@up.ac.za

Foreword

Although arguably first recorded in the writing of Plinius over 2000 years ago, pathogens attacking trees are receiving more attention now than at any time in the last 140 years, when the 'father of forest pathology', Robert Hartig, published his foundation treatise *Important Diseases of Forest Trees* in 1874. Over recent years, a rise in numbers of known damaging pathogens, usually invasive species associated with serious dieback of forest ecosystems and the death of urban trees has led to increased public perception of pathogens as a force shaping forest ecosystems and having a great impact on our use of trees in all landscape and productive situations.

The publication of this book is very timely. Recent general texts on forest and shade tree pathology are lacking, despite there being a number of books covering individual diseases or groups of pathogens. The science is moving on very rapidly, paralleling the huge leaps forward in understanding made possible by technological advances. Hence, there is a great deal of new information to be summarized and made available in this condensed form, both to interested parties – including undergraduate and postgraduate students of forest pathology, forestry, ecology and conservation sciences, and to practitioners.

The editors of this book have called on a range of contributors with suitable expertise to present their reviews and thoughts on the diseases affecting trees. Following the opening two chapters of Part I, which cover concepts of epidemiology and the management of disease in forests, the book is divided into four further parts. Part II has four chapters on non-fungal infectious diseases, and Part III – the largest section of the book – has 20 chapters on diseases caused by fungi or fungal-like organisms, which are logically arranged on the basis of the tree tissues attacked: root and butt tots, stem rots, vascular diseases, cankers, branch and tip blights, foliar diseases, and then the highly important oomycete and rust diseases. Part IV, with one chapter, covers diseases in forest nurseries. Finally, Part V, also with one chapter, focuses on invasive alien species – introduced pathogens, which figure among the greatest threats to the integrity of forest ecosystems worldwide.

Steve Woodward
University of Aberdeen
Editor-in-Chief, *Forest Pathology*

Preface

Forest pathology combines fields as diverse as botany, plant physiology, microbiology, climatology, soil science, forestry and ecology, but was born an applied science. As a multidisciplinary science with an inevitable applied component, forest pathology not only needs to keep up with the advances made in each field, but is also required to answer the needs of a modern and changing society, whose concept of ecosystem services provided by forests has broadened to encompass not only timber production, but also hydrogeological protection, biodiversity, recreational use and the preservation of home environments for native people the world over.

The development and the extensive use of new analytical approaches, including molecular methods and techniques based in geographic information systems (GIS), has rapidly advanced the knowledge on the infection biology and epidemiology of forest pathogens. These approaches have led to the accelerated discovery of emergent diseases, caused both by known and novel pathogen species. They have helped to clarify the taxonomy of many microbes and their associations with hosts, and have provided much insight into the epidemiology of many diseases, both newly discovered and long known, but poorly understood. A thorough knowledge of the biology of pathogens and of the epidemiology of any disease remains pivotal in the design of modern, effective, and sustainable control strategies, including integrated disease management programmes.

This book aims to provide a review of the most serious infectious diseases in both forest and urban environments, and in landscape settings, and provides practical guidelines for their management based upon both basic and applied scientific research. It was conceived as a comprehensive reference manual for biotic forest diseases, dealing not only with the better known fungal and fungal-like pathogens, but also with viruses, phytoplasmas, bacteria, higher parasitic plants and nematodes.

My co-editor, colleague and friend Giovanni Nicolotti passed away before reading the chapters that make up this book. He was proud of this ambitious project and this book is dedicated to him. I am grateful to all contributors and authors who have made the book possible. I would also like to acknowledge several people who have been of invaluable assistance during the preparation of this book: Matteo Garbelotto and Michael J. Wingfield, for exciting and useful discussions on the project; Steve Woodward for writing an insightful foreword; and Luana Giordano, whose competent and patient editorial effort has been instrumental at all stages of the preparation of the book.

While our intended audience includes forest managers, researchers and extension specialists, professional arborists and urban foresters, the book is really meant as a resource for anybody who cares about and cares for forest and tree health. Finally, it was our intent to produce a useful aid for all those teaching and studying forest pathology and ecology, in the hope that it may positively contribute to shape those who will be the future stewards of our forests. We hope we have succeeded in this final task, and we humbly apologize for all the inevitable shortcomings and omissions.

For Readers

The following sources were generally adopted for the biological nomenclature:

- Index Fungorum for fungi and fungal-like organisms, version 2012. Available at: http://www.indexfungorum.org
- US Department of Agriculture, Natural Resources Conservation Service PLANTS Database, version 2012. Available at: http://plants.usda.gov
- Fauna Europaea, version 2012. Available at: http://www.faunaeur.org

Paolo Gonthier
University of Torino
Grugliasco
Italy

Contributors' Acknowledgements

Chapter 1

Jonàs Oliva, Johanna B. Boberg and Anna J.M. Hopkins were funded by Future Forests, a multidisciplinary research programme supported by the Foundation for Strategic Environmental Research (MISTRA), the Swedish Forestry Industry, the Swedish University of Agricultural Sciences (SLU), Umeå University and the Forestry Research Institute of Sweden. Jonàs Oliva was also funded by the postdoctoral grant 2008 BP A 00013 from the Agency for Administration of University and Research Grants (AGAUR) of the Autonomous Government of Catalonia. Anna J.M. Hopkins was also funded by the European Union (EU) Seventh Framework Programme FP7 2007–2013 (KBBE 2009-3) under grant agreement 245268 ISEFOR (Increasing Sustainability of European Forests – Modelling for Security Against Invasive Pests and Pathogens under Climate Change).

Jonàs Oliva, Johanna B. Boberg, Anna J.M. Hopkins and Jan Stenlid gratefully acknowledge comments received on earlier versions of this manuscript from Jonathan Yuen (SLU) and the editors.

Chapter 4

Helen M. Griffiths sincerely thanks Wayne A. Sinclair for training in phytoplasmas, Rasa Jomantiene and James Walla who provided a critical and valuable review of the manuscript, and Carol Fisher for assistance with formatting.

Chapter 7

Paolo Gonthier and Magnus Thor gratefully acknowledge Matteo Garbelotto for critically reviewing the manuscript.

Chapter 10

Rimvydas Vasaitis thanks Jan Stenlid for constructive discussions and Audrius Menkis for technical assistance. The work was supported by the Swedish Research Council for Environment, Agricultural Sciences and Spatial Planning (FORMAS) and by the Carl Tryggers Foundation for Scientific Research.

Chapter 12

Thomas Kirisits dedicates the chapter to his teacher, mentor and friend Edwin Donaubauer (1932–2012).

Chapter 14

James W.M. Mehl, Bernard Slippers, Jolanda Roux and Michael J. Wingfield thank the Department of Science and Technology (DST)/National Research Foundation (NRF) Centre of Excellence in Tree Health Biotechnology (CTHB) and members of the Tree Protection Co-operative Program (TPCP), South Africa, for financial support.

Chapter 15

Simone Prospero and Daniel Rigling thank Ursula Heiniger for having introduced them to the fascinating world of chestnut blight, and Esther Jung and Hélène Blauenstein for excellent technical support of their ongoing research on chestnut blight. They are grateful to Silvia Dingwall for the English revision of the manuscript. Funding for much of the authors' work cited in the chapter was from the Swiss National Science Foundations (Scopes projects IB73A0-111089 and IZ73Z0-12792), the Competence Center Environment and Sustainability (CCES) of the ETH (Swiss Federal Institute of Technology Zurich) Domain (project GEDIHAP – The role of genetic diversity in host-pathogen interactions in dynamic environments), and the Cantons of Lucerne, St Gallen, Vaud, Valais and Zug.

Chapter 16

Marieka Gryzenhout is grateful for the valuable insights on the manuscript provided by Charles S. Hodges, North Carolina State University, USA.

Chapter 28

Marco Pautasso would like to thank Victoria Chadfield, Matteo Garbelotto, Tom Harwood, Ottmar Holdenrieder, Mike Jeger, Mathieu Moslonka-Lefebvre, Mike Shaw, Joan Webber and Xiangming Xu for insights and discussions, and Ottmar Holdenrieder for helpful comments on a previous draft. The review was partly funded by the Rural Economy and Land Use Programme (RELU), UK and the French Foundation for Research on Biodiversity (FRB).

List of Acronyms

a.s.l.	above sea level
ELISA	enzyme-linked immunosorbent assays
PCR	polymerase chain reaction
s.l.	*sensu lato*
s.s.	*sensu stricto*

Introduction to Infectious Forest Diseases

Paolo Gonthier

Although considerable research efforts are still needed to precisely disentangle the role and quantify the impact of all factors contributing to forest decline, there is little doubt that the overwhelming majority of timber losses are caused by diseases, ranking well ahead of insects, fire, adverse weather and animals (Tainter and Baker, 1996). Interestingly, fire and insects probably rank first as agents of forest destruction and decline in the eyes of the public and of the regulators and policy makers. One has to wonder whether the underestimation of the role played by forest diseases may be the result of the human inability to visually recognize the microscopic agents that cause these diseases, as opposed to the ability to see insects and to witness destructive events such as firestorms or hurricanes.

Diseases of forest trees are responsible not only for tree death, but also for reductions in growth and growth abnormalities, and can be caused either by abiotic agents, e.g. climatic and edaphic factors, chemical pollutants, or by biotic infectious agents. It should be noted that abiotic factors may sometimes simply cause injuries without having a true pathogenic effect on the affected plant, but they can also be true disease agents with prolonged effects on the physiology of the tree and causing induced active plant responses. Although pathogenic effects on trees may also be occasionally associated with insect attacks, the overwhelming majority of tree diseases are caused by viruses, phytoplasmas, bacteria, fungi, fungal-like organisms (e.g. oomycetes), parasitic plants and nematodes. Biotic disease agents are infectious, whether on their own or thanks to a vector, and the diseases they cause are thus transmissible from diseased to healthy trees. As expected, the number of pages of forest pathology textbooks (Boyce, 1961; Lanier *et al.*, 1976; Tainter and Baker, 1996; Capretti and Ragazzi, 2009; Edmonds *et al.*, 2011) dedicated to infectious diseases is one order of magnitude larger than that dedicated to non-infectious diseases.

Infectious disease agents vary in their epidemiology depending on their biology and/or on the biology of their associated vectors. In general, infectious forest pathogens have a moderate impact in native balanced ecosystems where they play more of an ecological rather than a strict pathological role (Garbelotto, 2004). The key role assigned to tree diseases as drivers of forest regeneration and succession, and as supporters of biodiversity, may in some circumstances justify the artificial inoculation of forest pathogens, e.g. for the promotion of wildlife habitats (Filip *et al.*, 2011; see also Vasaitis, Chapter 10, this volume).

Outbreaks of disease with destructive effects are associated either with the emergence of native pathogens due, for instance, to intensive forest management or fire suppression (see Edmonds, Chapter 2, this volume), or can occur when pathogen or host introductions lead to novel host–pathogen interactions. Tremendous epidemics have been attributed to the lack of co-evolution between hosts and pathogens (see e.g. Oliva *et al.*, Chapter 1, this volume; Kirisits, Chapter 12, this volume; Prospero and Rigling, Chapter 15, this volume). Climate change may also be responsible for an increase of outbreaks, especially of necrotrophic pathogens (Manning and Vontiedemann, 1995; Jactel *et al.*, 2011; Sturrock *et al.*, 2011; Pautasso *et al.*, 2012).

The direct economic losses caused by infectious forest pathogens, whether introduced or not, are incalculable. In addition, forest pathogens can have severe and permanent effects on entire ecosystems (Loo, 2009). Although the effects of pathogens on the level and quality of ecosystem services provided by forests may be difficult to evaluate in a strictly monetary sense, their negative impact is undeniable and often more far-reaching than simple economic damage.

Costs and investments required to control infectious forest diseases might be too high to be affordable when managing forests. Therefore, there is a need to optimize resources by adopting modern, effective and sustainable control measures based on a deep knowledge of the infection biology and epidemiology of pathogens. Theoretical control strategies should be translated into integrated disease management (IDM) tactics, which could be implemented locally and vary from site to site depending on the pathogen, the host and the environment.

Forests are the dominant and most productive terrestrial ecosystems on our planet, and despite all of the cumulative knowledge on the forest diseases that threaten them, every year a number of new 'pathosystems' are discovered as a result of the introduction of exotic pathogens, host shifts and jumps, hybridization and recombination among pathogens, etc. Additionally, current predictions on trends caused by climate change suggest that the incidence of some diseases will rise, while some may decline (Pautasso *et al.*, 2012); but as climate becomes less optimal for forests, it is inevitable that we will witness an emergence of diseases that currently are regarded as insignificant or secondary. This emergence is a clear indication for the need of an increased effort by forest pathologists to protect extant forests and to regenerate forests where they have been destroyed by excessive urbanization, exploitation, or disease and pest outbreaks.

References

Boyce, J.S. (1961) *Forest Pathology,* 3rd edn. McGraw-Hill, New York.

Capretti, P. and Ragazzi, A. (2009) *Elementi di Patologia Forestale.* Pàtron Editore, Bologna, Italy.

Edmonds, R.L., Agee, J.K. and Gara, R.I. (2011) *Forest Health and Protection,* 2nd edn. Waveland Press, Long Grove, Illinois.

Filip, G., Chadwick, K., Zambino, P., Omdal, D., Ramsey-Kroll, A., Schmitt, C., Maffei, H., Saavedra, A., Rall, W. and Parks, C. (2011) *Seven- to 14-Year Effects of Artificially Inoculating Living Conifers to Promote Stem Decay and Subsequent Wildlife Use in Oregon and Washington Forests.* US Department of Agriculture, Forest Service, Forest Health Protection, Pacific Northwest Region, Portland, Oregon.

Garbelotto, M. (2004) Root and butt rot diseases. In: Burley, J., Evans, J. and Youngquist, J.A. (eds) *Encyclopedia of Forest Sciences: Volume 2.* Elsevier, Oxford, UK, pp. 750–758.

Jactel, H., Petit, J., Desprez-Loustau, M.-L., Delzon, S., Piou, D., Battisti, A. and Koricheva, J. (2011) Drought effects on damage by forest insects and pathogens: a meta-analysis. *Global Change Biology* 18, 267–276.

Lanier, L., Joly, P., Bondoux, P. and Bellemère, A. (1976) *Mycologie et Pathologie Forestière. Tome II. Pathologie Forestière.* Masson, Paris.

Loo, J.A. (2009) Ecological impacts of non-indigenous invasive fungi as forest pathogens. *Biological Invasions* 11, 81–96.

Manning, W.J. and Vontiedemann, A. (1995) Climate-change: potential effects of increased atmospheric carbon-dioxide (CO_2), ozone (O_3), and ultraviolet-b (UV-B) radiation on plant diseases. *Environmental Pollution* 88, 219–245.
Pautasso, M., Döring, T.F., Garbelotto, M., Pellis, L. and Jeger, M.J. (2012) Impacts of climate change on plant diseases – opinions and trends. *European Journal of Plant Pathology* 133, 295–313.
Sturrock, R.N., Frankel, S.J., Brown, A.V., Hennon, P.E., Kliejunas, J.T., Lewis, K.J., Worrall, J.J. and Woods, A.J. (2011) Climate change and forest diseases. Review. *Plant Pathology* 60, 133–149.
Tainter, F.H. and Baker, F.A. (1996) *Principles of Forest Pathology*. John Wiley, Hoboken, New Jersey.

1 Concepts of Epidemiology of Forest Diseases

Jonàs Oliva,* Johanna B. Boberg, Anna J.M. Hopkins and Jan Stenlid
Swedish University of Agricultural Sciences, Uppsala, Sweden

1.1 The Definition of Epidemiology

Epidemiology refers to the study of the occurrence and cause of infectious diseases, their origin, and their distribution in space and time within a population (Schumann, 1991). Its literary meaning refers to diseases among a human population, but the term is also widely used for the spread of diseases in plant populations. The term epidemic is often misused to describe only rapid and widespread diseases, but it should refer to the development of any disease, irrespective of its speed and extent (Tainter and Baker, 1996). Questions typically dealt with in epidemiological studies in an attempt to understand disease development are the following (Schumann, 1991):

- What kind of pathogen is causing the disease?
- Where did it come from and how was it dispersed?
- How will the disease spread, seen both from a short time (within a year) and a longer time (between years) perspective?

In forestry, infectious diseases are caused by a diverse range of organisms, including fungi, oomycetes, bacteria, phytoplasmas, parasitic higher plants, viruses and nematodes (Tainter and Baker, 1996). Among these, the fungi are the largest group causing disease in forest trees (Sinclair and Lyon, 2005). Different *Phytophthora* spp. belonging to the oomycetes group are also important as the causal agent of several devastating diseases on trees.

1.2 The Disease Triangle/Pyramid

Forest disease epidemics develop as the ultimate consequence of the interaction between a susceptible host plant, a virulent pathogen and a favourable climate. The interaction between these three elements of plant diseases, i.e. the host plant, the infectious agent and the environment, can be visualized and described by the disease triangle (Fig. 1.1) (Agrios, 2005). This conceptual model was first formalized by McNew (1960) and can be used to understand how epidemics develop and how they may be predicted and limited (Scholthof, 2007).

1.2.1 Pathogen factors

A pathogen has a basal set of pathogenicity factors that make it able to cause disease.

* E-mail: jonas.oliva@slu.se

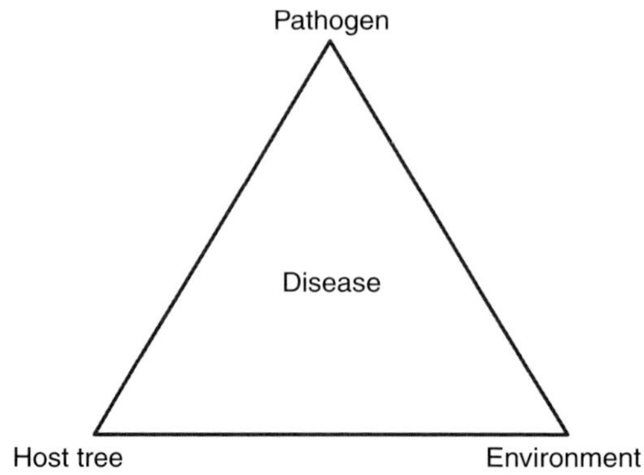

Examples of factors that affect disease epidemics:

Pathogen factors	Host tree factors	Environmental factors
• Virulence	• Host resistance	• Humidity
• Inoculum type	• Host age	• Temperature
• Dispersal potential	• Seasonal development	• Drought
• Inoculum potential	• Host uniformity	• Frost
• Host range		• Biotic environment
• Alternation between hosts		• Competition
		• Silviculture

Fig. 1.1. The disease triangle illustrates the interaction between pathogen, host and the environment as a prerequisite for disease to occur. The triangle may serve as a conceptual model describing the factors that influence the development of an epidemic. Also, included are examples of pathogen, host and environmental factors that may affect their interaction and favour disease and, thereby, the extent of disease epidemics.

It can have high or low aggressiveness depending on how efficiently it will cause disease. An aggressive pathogen can rapidly infect the host and produce large quantities of inoculum. Thus the aggressiveness of a pathogen is related to several factors; these include the quantity of inoculum produced, the mode of dispersal of the pathogen and the pathogen reproductive cycle. The form of the inoculum is also important as this can influence the speed with which the host is colonized.

The quantity of inoculum (bacteria, fungal spores, nematode eggs, etc.) is important for infection as high amounts increase the likelihood of a reproductive propagule successfully reaching and infecting a potential host. Knowledge about the initial amount of inoculum present (e.g. spores) is also important when describing disease spread using modelling or simulations. However, the potential for averting disease is not only dependent on the amount of spores. The concept of inoculum potential (Eqn 1), as defined by Garrett (1956a), acknowledges this and attempts to quantify the energy that is carried by the pathogen at the point of infection and is available to overcome host defences and initiate disease. Garrett stated that the inoculum potential may be increased either by increasing the quantity of inoculum (e.g. spores, individual mycelium) per unit area of host surface or by increasing the energy status of each unit. In his experiments using rhizomorph-producing *Armillaria* spp., Garrett (1956b) showed that the speed of infection by the rhizomorphs increased with the amount of inoculum and decreased with the distance between the inoculum source and the host.

Subsequently, when the inoculum potential of the fungus was low, the infections were inhibited by host resistance.

$$\text{Inoculum potential} = \text{Quantity of inoculum} \times \text{Energy reserves/inoculum} \quad (1)$$

Similarly, the closer the source of the inoculum is to suitable hosts, the higher the probability that the propagules will reach the host, which will also increase the chance of an epidemic. For example, spread from an infected tree to neighbouring trees in a forest is more likely than spread from the same tree to trees in neighbouring forests. The quantity of infective propagules that reach a specific host is known as the inoculum loading. When the inoculum loading, and thus also the inoculum potential, is very high hosts previously considered resistant to a pathogen may become susceptible. Foliar infections of *Dothistroma* spp. have been observed on the atypical host Norway spruce (*Picea abies* (L.) Karst.) for example, as a result of high inoculum loading from adjacent severely infected pines (*Pinus* spp.) (Lang, 1987).

In pathogenic fungi, several different types of inoculum are often produced. Spores, both asexual (conidia) and sexual, can be emitted in large amounts; they are often microscopic in size and are effectively spread in air, water or soil (Tainter and Baker, 1996), either directly or by vectors. Not only the nature of the propagule, but the way in which the reproductive propagules are produced, determines the speed and severity of epidemics. For example, spores produced on the outside of the host (such as on the bark of a tree or on its leaves) can be easily and effectively dispersed to new hosts. Other pathogens may use the host seed for reproduction and dispersal. Some pathogens produce propagules within the host plant material, so it is difficult for them to spread without the help of vectors. An example is *Ophiostoma novo-ulmi* Brasier, the causal agent of Dutch elm disease, which produces its spores within the wood of host elms (*Ulmus* spp.) and relies on bark beetles to collect and disperse the spores to new hosts (Brasier, 1991). Other pathogens produce inoculum and infective structures such as mycelia or rhizomorphs on infected plant parts within the soil, where they have a limited capacity of spread.

The mode of dispersal and the number of disease cycles per season are important factors in epidemics. Some pathogens have short reproductive cycles that allow several generations per year, while others reproduce less frequently. Propagules can be airborne, soil-borne or waterborne, or dispersed by host seeds or by vectors. Pathogens with airborne propagules are possibly the most important group as they can spread easily and cause sudden and widespread epidemics. The second most important group of pathogens have airborne vectors, e.g. *O. novo-ulmi*, and the pine wilt nematode *Bursaphelenchus xylophilus* Steiner & Buhrer, which are both spread by insect vectors. Pathogens spread by windblown rain can cause severe but localized epidemics, while when pathogens are seed dispersed, such as can occur for the pine pitch canker pathogen *Gibberella circinata* Nirenberg & O'Donnell ex Britz, T.A. Cout., M.J. Wingf. & Marasas (Storer *et al.*, 1998), or spread in soil (like many *Phytophthora* spp.), they tend to cause localized, slow-spreading and severe diseases. In reality, many pathogens produce more than one type of propagule and so can use several different modes of dispersal, thereby increasing the possibility for epidemics. While spores are important for long distance dispersal of fungal pathogens, for example, other sources of inoculum, such as rhizomorphs and mycelia, may be very important locally. Spread via rhizomorphs and mycelia has a significant influence on the development of disease caused by decay fungi such as species of *Armillaria* and *Heterobasidion* (Woodward *et al.*, 1998). Humans can also play a role in the dispersal of pathogens.

Following dispersal, there are a number of distinct steps that lead to the development and perpetuation of disease in a host. Together, these steps make up the disease cycle. The disease cycle can closely align with the life cycle of the pathogen, but it actually refers to the disease in the host as a function of the pathogen. The disease cycle can be broken up into four main phases: pre-entry, entry, colonization and dissemination. The pre-entry phase starts with inoculation:

the first contact between propagules of the pathogen and the host. Following inoculation, the propagules start to grow on the surface of the new host substrate towards a penetration point. In the case of spores, the growth is divided into two stages: germination of the spores and the production of a germ tube. Favourable environmental conditions (such as suitable temperature and moisture) are required for these growth stages to occur. During the subsequent entry phase, the pathogen may enter the host by directly penetrating cells, by penetrating through natural openings (such as stomata) or by penetrating through wounds. The pathogen may form specialized structures such as appressoria or infection pegs to penetrate the host substrate.

During the colonization phase, the host is colonized by the pathogen, followed by the appearance of the first symptoms of disease, and then the production of spores and their dispersal. Colonization of the host can be intracellular or intercellular. Pathogens typically utilize one of two different strategies when colonizing their host: they act as either biotrophs or necrotrophs. Biotrophs obtain their nutrients from living cells, so keep their host's cells viable during the colonization process. Typically, biotrophs produce infection structures and specialized organs for nutrient uptake and metabolite exchange, e.g. haustoria. Biotrophs often cause limited damage to the host, target fast-growing plants and have a limited host range. Rusts and powdery mildews are examples of biotrophs. In contrast, necrotrophs obtain their nutrients from dead cells. This means that they must kill their host cells before colonization. Unlike biotrophs, some necrotrophs have no infection structures and instead enter the host via wounds. They can cause severe symptoms and target weakened plants; they usually have a wide host range. The widespread forest pathogen *Phytophthora cinnamomi* Rands is an example of a necrotroph. It is also possible for pathogens to have a significant lag time between entry into the host and the development of disease symptoms, i.e. an incubation period. Frequently this occurs because the pathogen enters a dormant or latent phase; it then remains inside the host in a dormant stage, neither actively growing nor causing disease, until conditions within the host are suitable.

Disease cycles can be classified as either continuous (where the pathogen is found on one or one of several hosts throughout the entire disease cycle) or discontinuous (where the cycle is broken by a resting or non-pathogenic phase in the pathogen).

In continuous disease cycles, the unbroken disease cycle can take place on a single (primary) host, although sometimes part of the disease cycle can take place on an alternative or alternate host. An alternative host acts as an alternative if the primary host is not present. The alternative host is often closely related to the primary host and can show few or no disease symptoms. In contrast, disease cycles with an alternate host usually involve the pathogen completing part of its life cycle on that alternate host. This is usually an obligate part of the disease cycle or life cycle of the pathogen, and the alternate host is usually unrelated to the primary host. Many species of macrocyclic rusts, such as *Cronartium ribicola* J.C. Fisch., the causal agent of white pine blister rust, require the presence of the alternate host, in this case *Ribes* spp., to reinfect the primary host and cause widespread disease (Cobb *et al.*, 2010). Some disease cycles include survival or overwintering of the pathogen when the primary host is not present or unsuitable for new inoculation and colonization. This is a critical step in the life cycle of the pathogen; therefore, if this phase is present in the disease cycle, it is often manipulated for control or management of the disease.

In discontinuous disease cycles, the pathogen can survive outside the primary host using one of a number of different mechanisms. One such mechanism is to form resting stages or survival structures. Many *Phytophthora* spp., for example, are able to form thick-walled chlamydospores that can survive for many years in the soil, waiting for suitable environmental conditions to germinate and infect a new host. Other examples of resting structures include oospores, teliospores, sclerotia and cleistothecia. Pathogens with a discontinuous disease cycle can also survive in a saprophytic phase in which the pathogen obtains nutrients from dead organic matter. Many leaf pathogens can survive as saprophytes on debris on the forest floor, and spores produced on the debris can then infect new hosts when environmental and host conditions are suitable. Finally, some pathogens survive outside their primary host by forming an epiphytic phase, i.e. they live on the surface of their host for some time without causing disease.

Most pathogens have a narrow host range, i.e. they are specialists that are specific to host species within the same genus, e.g. *O. novo-ulmi*, which causes disease only in elm trees (Brasier, 1991). Other pathogens are generalists and have a much broader host range; these can be exemplified by *Phytophthora ramorum* Werres, De Cock & Man in't Veld and *P. cinnamomi*, which both infect plant and tree species belonging to a broad range of genera (Hansen *et al.*, 2005).

1.2.2 Host factors

The host response to an invading pathogen contributes to determining the impact and severity of the resulting disease, and can range from extreme susceptibility to complete resistance. The susceptibility or resistance of the plant host can be age related (ontogenetic) (Develey-Rivière and Galiana, 2007) or organ specific (Blodgett *et al.*, 2005). Resistance can be expressed constitutively or induced (Eyles *et al.*, 2010) and can also be expressed systemically or locally. The resistance can be specific for a given pathogen (race-specific or vertical resistance) or general (basal or horizontal resistance) (Király *et al.*, 2007).

Ontogenic resistance is related to the developmental stage of the host and is often observed as a decrease of susceptibility with host age. As an example, the susceptibility of Monterey pine (*Pinus radiata* D. Don) to needle blight caused by *Dothistroma* spp. in New Zealand is highest in young stands, up to 12 years old and then decreases (Watt *et al.*, 2011b). The susceptibility of certain developmental stages to disease can also be related to environmental factors. Scots pine (*Pinus sylvestris* L.), for example, is only susceptible to snow blight caused by *Phacidium infestans* P. Karst. when it is very young, when branches and needles may be covered by snow during the winter (Roll-Hansen *et al.*, 1992). This last example also illustrates that it can be hard to disentangle the effect of developmental stage of the individual host from environmental influences; in this case, a young tree is also a small tree. The opposite situation can be observed for *Porodaedalea pini* (Brot.) Murrill infecting pine trees. Young trees can be carriers of fungal propagules, but the pathogen requires the heartwood formation that occurs in older trees for heart rot development to begin (Haddow, 1938).

Constitutive resistance includes all those factors that are present in the host before challenge by a pathogen. These can be structural, including mechanical barriers such as bark, cuticle or wax layers on leaves. Preformed components of chemical resistance can be macromolecules, e.g. lignin, or extractable low molecular weight metabolites, e.g. the stilbene pinosylvin in pines. Host phenology can also be a form of resistance, for example, it can match or mismatch the pathogen life cycle.

When the host is exposed to a pathogen, several responses are triggered. The pathogen can be recognized through innate immunity, when signals are transmitted via signalling pathways that include salicylic acid, jasmonate or ethylene. The downstream responses include the activation of hypersensitive cell death and the formation of reactive oxygen species, phenolic substances and resin production. Subsequently, new polyphenol-rich cells, traumatic resin ducts and, later on, periderm are formed. Previous exposure to a pathogen can also affect the level of resistance to future attacks. Induced resistance is an activation of the host plant's defence mechanism, which enhances the protection against later attack (Eyles *et al.*, 2010). The resistance may be induced by both pathogenic and non-pathogenic microorganisms and herbivores, as well as through abiotic mechanisms. Such acquired resistance can either be expressed locally or systemically.

At the population level, resistance within the host varies depending on the 'pathosystem' studied and its evolutionary history.

In the case that the host plant will be infected, the development of the epidemic will depend on the level of resistance in the population and on the environmental factors influencing it (Agrios, 2005). At the population level, two types of resistance emerge: (i) a basal or horizontal resistance, which is a non-specific general resistance of the host that is effective against a broad range of pathogens; and (ii) a race-specific or vertical resistance, which is a specific differential resistance that is effective against some races of pathogens but ineffective against others (Agrios, 2005). While horizontal resistance is characterized by the absence of genetic interaction between the host genotype and the pathogen genotype, race-specific vertical resistance has been developed through gene-for-gene relationships (Leonard and Czochor, 1980). Susceptible hosts lack the genes for resistance and will, in the presence of a virulent pathogen and favourable conditions, develop the disease.

Trees might be able to tolerate disease, meaning that they are infected but the disease is not causing a major phenotypic disadvantage in terms of vitality or growth capacity. Resistance can be costly, and tolerance can provide an evolutionarily stable strategy to handle a moderate pathogen pressure. Another advantage of a tree being tolerant could be that the tolerant host becomes an infection source for less tolerant trees that, if infected, will have to spend energy on resistance. In that case, tolerant hosts can win the growth competition for light and resources. Besides its biological advantages, tolerance can also have positive effects from an economic perspective. Trees investing in resistance will decrease their growth in trying to cope with the pathogen, and this will ultimately affect ring width and timber quality. An example of possible tolerant–resistant responses was observed by Cruickshank *et al.* (2011) when they studied the effects of *Armillaria ostoyae* (Romagn.) Herink infection on Douglas-fir (*Pseudotsuga menziesii* (Mirb.) Franco) trees. In this study, trees were observed to shift (decrease) their growth immediately after infection, and this growth reduction would accumulate over the years, regardless of the amount of root system infected. The growth shift was not uniform: on the one hand, Cruickshank *et al.* (2011) observed trees with one lesion in a primary root that was already showing growth reductions; on the other hand, they observed that a number of *A. ostoyae*-infected trees were able to maintain similar growth to that of the disease-free trees in the same plot. This latter group could represent trees with a disease-tolerant strategy; these would not decrease their secondary growth as quickly as the highly disease-responsive trees that would invest a lot of resources in building up defence structures.

1.2.3 Environmental factors

Disease is ultimately a synchronization of different biological events affected by environmental factors. The environment includes climatic factors, such as humidity and temperature, as well as other abiotic factors, such as soil nutrient status and pH. Biotic factors are also very important for disease development, and these include the interaction of the host and the pathogen with other organisms. Conducive conditions are the combination of environmental factors favourable for the development of disease. Environmental factors affect the plant host and the pathogen in different ways. Adverse conditions stress the host and leave less energy to allocate to resistance, which can make it more susceptible to necrotrophic fungal attacks. For example, drought-stressed pines are more susceptible to *Diplodia pinea* (Desm.) J. Kickx f. than unstressed trees (Blaschke and Cech, 2007). Stress not only increases susceptibility during the period in which it occurs, it can also predispose the host even if the stress condition has been alleviated when the tree is challenged with a pathogen (Desprez-Loustau *et al.*, 2006). Stress does not always favour disease development though; for example, biotrophic organisms, such as rusts and powdery mildews, normally cause less extensive disease on stressed hosts.

The atmospheric environment can be viewed at three scales: climate, weather and microclimate. Climate, which describes the long-term weather conditions of a certain region, affects the likelihood of both host and pathogen species being able to establish in a particular area. Climatic factors may limit the distribution of certain pathogens even though potential hosts are present. The hemiparasitic plant *Viscum album* L. is common in southern and central Europe, yet rare in Scandinavia where susceptible pines are present but winter temperatures are too low for pathogen survival (Skre, 1979). Weather factors influence pathogens at their source by controlling their reproduction and release, and also their arrival at the host by their effects on transport, deposition and infection. Microclimatic conditions – such as high needle surface moisture – can also play an important role in the development of disease epidemics. The pine pathogen *Mycosphaerella pini* Rostr. ex Munk requires 10 h of constant needle wetness for infection by conidia and is thus favoured by dense stands with high humidity (Gilmour, 1981). Similarly, the widespread pathogen *P. cinnamomi* can be extremely sensitive to variation in temperature and soil water content within a stand (Shearer *et al.*, 2010; Meadows *et al.*, 2011).

Soil nutrient status and pH can also influence disease development, sometimes in complex ways. High nutrient levels can be both beneficial and detrimental to disease. The incidence of pitch canker disease on slash pine (*Pinus elliottii* Engelm.) was found to be spatially correlated with high levels of nitrogen deposited from a poultry operation (Lopez-Zamora *et al.*, 2007). Scots pines growing in more fertile sites were more frequently infected by the pine twisting rust *Melampsora populnea* f.sp. *pinitorqua* Boerema & Verh., possibly owing to the higher nutritional value and longer shoots of the host (Mattila *et al.*, 2001). Adding nutrients to mountain hemlocks (*Tsuga mertensiana* (Bong.) Carrière), however, significantly increased the resistance of the trees against infection by the laminated root rot fungus *Phellinus sulphurascens* Pilát in the laboratory (Matson and Waring, 1984). High pH and liming predispose trees to annosus root rot, probably by facilitating ectotrophic growth of the causal fungus, *Heterobasidion annosum* (Fr.) Bref. *s.l.*, on roots (Stenlid and Redfern, 1998). Light conditions may also have a role in disease susceptibility. Shading, for example, significantly increased the susceptibility of mountain hemlocks to root rot regardless of whether nutrients were added (Matson and Waring, 1984).

In addition to the physical environment, the biological environment can also play a role in disease development. This includes both the internal and the external biological environment of both the host and the pathogen. The presence of other organisms, such as fungal or bacterial endophytes, may significantly influence the development of disease in the host. Endophyte-mediated resistance was demonstrated in western white pine (*Pinus monticola* Douglas ex D. Don) against the white pine blister rust pathogen (Ganley *et al.*, 2008). Similarly, the presence of mycorrhizal symbionts has been shown to increase host resistance against pathogens by increasing host growth and vigour (Lei *et al.*, 1995). Other organisms in the environment may also reduce the probability of disease by establishing or outcompeting the pathogen in its chosen niche. The biological control agent *Phlebiopsis gigantea* (Fr.) Jülich most likely prevents infection of Norway spruce stumps by *H. annosum s.l.* by outcompeting it during stump colonization (Holdenrieder and Greig, 1998). For some pathogens, disease development is influenced by the availability of alternative or alternating hosts in the vicinity of the original host. The planting of host species in monocultures can also influence disease levels in other ways, and this is discussed further later on in this chapter.

Uniformity of the population and the forest stand may influence disease development. Genetic uniformity within a population of trees may make them vulnerable to major epidemics, but can also decrease disease if the population generally is resistant against a certain pathogen. The effect of tree species diversity on the frequency of diseases is not straightforward, and both positive and negative results have been observed in association with both monocultures and forests with a higher tree species diversity (Pautasso *et al.*, 2005). Uniformity can also be applicable to age structure. In Denmark, outbreaks of *Gremmeniella abietina* (Lagerb.) M. Morelet on Austrian pine (*Pinus nigra* Arnold) partly depend on the presence of dense and widespread stands aged between 30 and 40 years (Thomsen, 2009).

1.2.4 Temporal and human factors

Time is another important factor in the development of disease or epidemics as all three components of the disease triangle must interact within a given period of time (Schumann, 1991). Also, the specific time when a certain disease event takes place, as well as the length of that time, will affect the disease severity (Agrios, 2005). For example, the sporulation of conifer cone pathogens may be out of phase with the production of new cones in the potential host. The lag between these events can vary between years, thus allowing certain overlaps. This sort of equilibrium is especially sensitive to (future) changes in climate. Time is sometimes added to the disease triangle to form a disease pyramid (Schumann, 1991; Agrios, 2005), but it can also be viewed as an integral part of the different components of the disease triangle (Scholthof, 2007).

The influence of humans on forests is another factor that affects forest disease epidemics. Through different forest management practices, humans affect the number and type of host species, the genetic resistance or susceptibility of the planted material and the degree of genetic and structural uniformity in a forest stand (both within and between host species). Management practices performed in a growing forest stand may also increase disease frequency. For example, *H. annosum s.l.* has increased in frequency in managed forests compared with natural forests owing to its ability to utilize newly created stumps as an infection court (Redfern and Stenlid, 1998; Berglund *et al.*, 2011). Humans also affect the prevalence of pathogens by increasing pathogen dispersal globally through trade and travel, thereby introducing pathogens to new potentially susceptible host populations (Levine and D'Antonio, 2003). Nurseries, in particular, are not only a frequent source of host dispersal but could also be an ideal site for the creation of new species where pathogens from different hosts meet in a common place (Olson and Stenlid, 2002).

1.3 Disease in Managed and Natural Forests

1.3.1 Disease in managed forests

Forest management activities have direct effects on the environment, the host and the pathogen. Managed forests differ from natural forests in many ways, such as between and within tree species diversity, tree age distribution, understory diversity, canopy structure, amount of dead wood, fungal community and soil fertility. As an example, thinning operations increase the chances of damage by root rot pathogens while they reduce the chance of other pathogens attacking (Table 1.1).

Managed forests are subjected to a different rotation schedule than are natural forests. One of the main differences is the younger age of the trees and the occurrence of more even-aged stands. Several species of heart rot decay fungi, e.g. *Porodaedalea pini, Phellinus chrysoloma* (Fr.) Donk and *Inonotus leporinus* (Fr.) Gilb. & Ryvarden, are disadvantaged by the younger tree age of managed forests. However, the uniformity of tree age found in managed forests will favour the spread of disease between trees with similar susceptibility within a stand or region. For example, the sapling stage of Scots pine trees is highly susceptible to needle cast disease caused by *Lophodermium seditiosum* Minter, Staley & Millar or by *Lophodermella sulcigena* (Link) Tubeuf, while older trees are more resistant. Likewise, susceptible young Monterey pines are grown over large areas in New Zealand and this has exaggerated outbreaks of *Dothistroma* spp. Moreover, because there is a constant harvest of biomass from managed forests, there will consequently be lower amounts of coarse woody debris. This has a significant negative effect on the diversity of saprotrophic wood decay fungi that might be competing with pathogenic fungi during certain phases of the life cycle.

The wounding of trees is a negative side effect of a range of silvicultural practices such

Table 1.1. Types of pathogens favoured (+) and not favoured (–) by management operations (including thinning, precommercial thinning, selective fellings and final felling).

Effects	Decay pathogens	Root rot pathogens	Wilt pathogens	Canker pathogens	Foliar pathogens
Creation of fresh stump surfaces: airborne infection		+ + +			
Creation of substrate available for colonization: soil-borne infection		+ +			
Reduction of number of tree-to-tree root contacts		–			
Promotion of tree growth	–	+			+
Tree wounds	+ + +	+	+	+	
Slash production				+	
Cull pieces		+			
Slash burning		+			
Increased light and temperature			–	–	–
Creation of fresh branch surfaces by pruning	+			+ + +	–

as pruning or tree felling and timber extraction. Bark wounds will inevitably result in a potential infection court for fungi. Studies of Norway spruce have shown that small wounds (<15 cm^2) rarely resulted in subsequent rot, while infection rates in larger wounds were more frequent and related to wound size (Isomäki and Kallio, 1974; Vasiliauskas *et al.*, 1996). The pruning of green branches – done in an attempt to reduce the spread of and incidence of *Dothistroma* spp. in New Zealand (Gadgil, 1984), may create entry points for canker fungi such as *Diplodia pinea* (Chou and MacKenzie, 1988). Similarly, the pruning of conifers in autumn frequently promotes the spread of *Phacidium coniferarum* (G.G. Hahn) DiCosmo, Nag Raj & W.B. Kendr. (Uotila, 1990).

The biotic environment influences the capacity of pathogens to cause damage as they have to compete with an already established microbial community in either living tissues, e.g. endophytes, or in the soil, as in the case of soil suppressiveness. One interesting example is provided by the much higher susceptibility to *H. annosum s.l.* of forests planted on former agricultural land compared with plantations on previously forested sites (Oliva *et al.*, 2010). Although former forest sites tend to get infected at an earlier age, once the pathogen gets introduced in a first rotation stand on a former agricultural site, it will spread more freely, presumably as a result of the lack of competition from wood-inhabiting fungi in the agricultural soil.

Plantation forests are often seen as a very simplified version of a forest ecosystem. They are usually monospecific, even-aged, and they may have a different genetic origin from the local natural populations. As long as the uniformity of the stand is coupled with resistance it may favour tree health, but if the resistance to disease is broken, uniformity in these three characteristics can allow for increased disease levels. The link between diversity and disease resistance is not always straightforward, and every case has to be carefully inspected (Koricheva *et al.*, 2006). High densities in plantation forests are also known to increase disease spread, such as in the case of oak wilt caused by *Ceratocystis fagacearum* (Bretz) J. Hunt on mixed northern red oak (*Quercus rubra* L.) forests (Fig. 1.2). However, when the planted species is an early colonizer, the plantation may well resemble the natural habitat of that colonizer. One example of broken resistance is provided by plantations of non-native tree species (Wingfield *et al.*, 2001). Non-native tree species escape pathogens in their original habitats and initially display a low level of disease. Often, with time, these plantations are attacked by pathogens making host jumps from native tree species, or by the introduction of pathogens from the original habitat of the planted

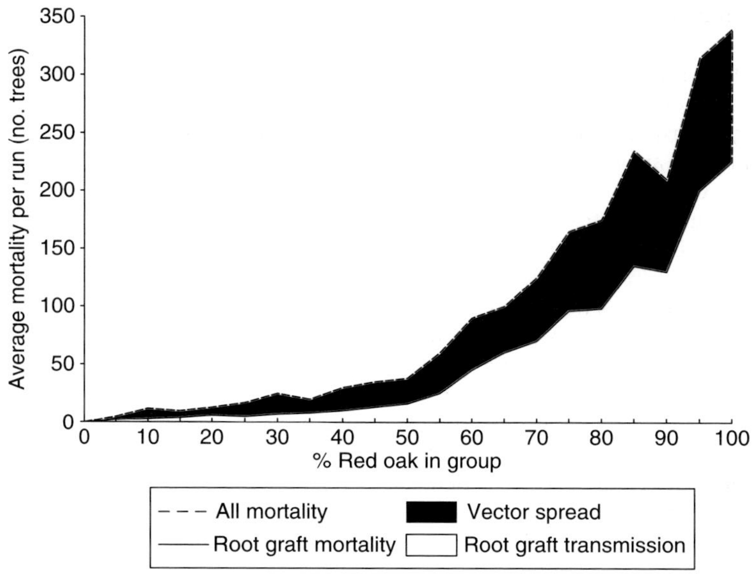

Fig. 1.2. A simulation of the spread of *Ceratocystis fagacearum* (Bretz) J. Hunt in red oak (*Quercus rubra* L.) after 40 years showing how the proportion of hosts in a stand will affect the rate of disease spread by either vectors or root grafts. There is a clear increase in the mortality when the amount of oak trees in a stand exceeds 50%. There is also a change in how much vectors and root grafts contribute to the total mortality when the oak density increases. In stands with less than 50% oak trees, a higher proportion of the spread, and thus mortality, is related to the activity of different vectors, e.g. beetles. In stands with an oak density above this threshold, transmission through root grafts is the main means of spread and related mortality. The figure is modified from Menges and Loucks (1984), with permission.

tree species or elsewhere. When disease outbreaks happen, the spread rate can be high owing to the uniformity of the stands.

1.3.2 Disease in natural forests

In natural habitats, forest structure and dynamics are more complex than in plantation forestry. Disease can be viewed as a biological disturbance affecting the structure and composition of the forest. This fact is well illustrated in tropical forest ecosystems where no climatic limitation exists, and where the high tree diversity is thought to be due to density-dependent mortality. Factors causing density-dependent mortality will predominantly affect dominant tree species, so allowing the emergence of rare species. The density-dependent concept first observed in tropical environments was later expanded by Hille Ris Lambers *et al.* (2002), who concluded that density-dependent mortality was probably operating at every latitude. The proportion of density-dependent mortality due to disease both affects and is influenced by host density in several ways (Gilbert, 2002): (i) the number of available hosts in time and space; (ii) the distance between hosts, hence the probability of transmission; (iii) the competition between individuals, hence host vigour; (iv) frequency-dependent effects on vector behaviour and herbivore damage; (v) density and composition of the associated plant community; and (vi) the physical environment for host–pathogen interaction.

One conceptual model that explains the density dependence of mortality is the Janzen–Connell hypothesis (reviewed by Gilbert, 2002), which is based on the assumptions that there is a diversity of pathogens that are either host specialists or have differential effects on local host species, and that both host and pathogen dispersal are spatially constrained. In the model, the majority of tree seeds do not disperse far from the parent tree, while the parent tree also acts as a reservoir of specialist pathogens. Thus, seedlings closer to the parent tree will be exposed to the most pathogens, leading to high mortality, while seeds landing under the canopy of an adult tree of a different species would be in a more advantageous position. The result is that the spatial distribution of adult trees is less clumped than would be expected from a random distribution pattern. This dependence was well shown by Bell *et al.* (2006) when they experimentally manipulated the pathogen population of a tropical forest site with fungicide targeting damping-off pathogens. Untreated plots with a high density of *Sebastiana longicuspis* Standl. had three times less survival than low-density plots. Fungicide application increased seedling survival, and plots with both a high and low density ended up with the same survival, supporting the idea that density-dependent mortality was due to host-specific pathogens.

Pathogen disturbance can also operate at a larger scale than the individual tree. *A. ostoyae* infection centres can cover several hectares. In mixed conifer forests in north-western USA and British Columbia, *A. ostoyae* infections will influence tree species composition (Hansen and Goheen, 2000). Some pathogens are adapted to large-scale disturbances such as forest fire. The resting spores of *Rhizina undulata* Fr. are activated by heat shock of more than 22°C (Jalaluddin, 1967), so after a fire, the fungus is ready to kill emerging pine regeneration or parts of the remaining stand (Lygis *et al.*, 2005).

1.3.3 Baseline mortality

There is always a certain level of disease in a forest that maintains the structure of the forest with time; this can be expressed by baseline mortality. In natural forests, mature trees are the result of a selection process from a high number of seedlings. Thus, at a regional level, trees of small diameter appear at a higher number than larger trees. This density decrease along the life of the forest follows an exponential decay function called de Liocourt's law. Under this law, the level of mortality needed for trees to move from one diameter class to the next is constant. Manion and Griffin (2001) used this concept to differentiate 'healthy' mortality from disease levels that would compromise the present forest structure. According to them, the level of baseline mortality in a tree species could be easily calculated from a diameter distribution plot. When plotting the logarithm of the density against diameter, the log plot appears as a straight decreasing line with a certain slope 'a'. The baseline mortality 'm' can then be calculated using Eqn 2:

$$m = 1 - e^{ax} \qquad (2)$$

where 'x' represents an increase in diameter class.

Using Eqn 2, information such as 'an increase of 2.5 cm correlates with a 20% decrease in the number of trees' can be obtained. The baseline mortality is then compared with observed mortality due to fellings and to biotic and abiotic disturbances. Comparison of the baseline mortality with the observed mortality can be used to assess whether a tree population is stable in the long term. An example is provided in Fig. 1.3.

1.4 Forest Decline Concepts

Recurrent forest disease outbreaks during the second half of the 20th century could not be satisfactorily explained by classical forest pathology concepts, e.g. Koch's postulates, one pathogen–one disease, or the disease triangle. These diseases of complex aetiology were included within the concept of forest decline. Several theories have tried to explain forest decline from new perspectives, such as it being the result of a repeated stress alteration affecting the tree, thereby making it susceptible to secondary pathogens (Houston, 1992). Others have explained decline as a result of the demographic structure of tree populations, in which even-aged stands would exaggerate the effects of allelopathy, nutrient depletion and natural senescence that would make certain sites appear unsuitable (Mueller-Dombois, 1992). Decline has also been related to environmental pollution and global climate change (Auclair *et al.*, 1992).

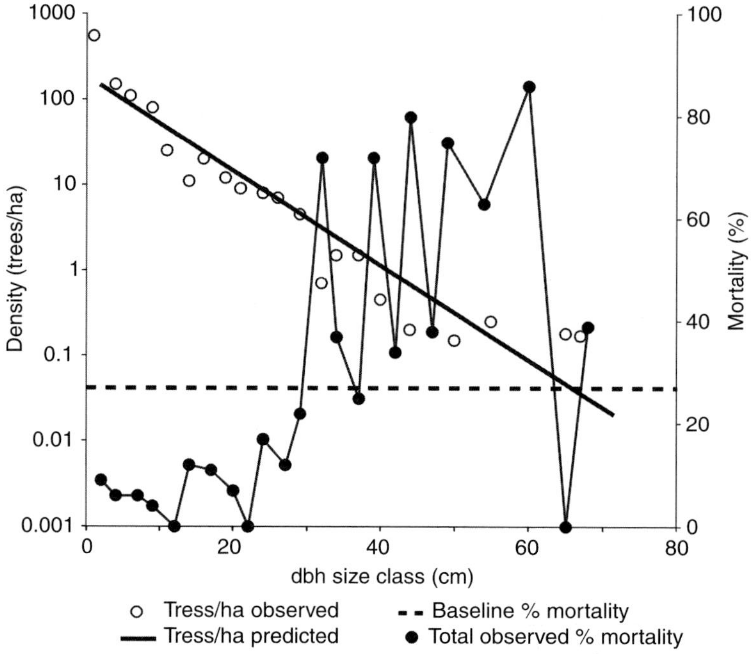

Fig. 1.3. Observed tree density, predicted tree density (least squares best-fit regression line), predicted baseline mortality and total observed mortality in relation to diameter at breast height (dbh) size class of a population of American beech (*Fagus grandifolia* Ehrh.) in New York state. The baseline mortality represents the 'natural' mortality in each dbh class that can be expected in the development of a healthy forest. The baseline mortality is calculated from the slope of the regression line fitted to the observed tree density. In other words, 27% of the trees in each size class should die in order to maintain the current forest structure. However, the observed % mortality in this population shows the effect of beech bark disease, which mainly causes mortality in trees of dbh larger than 30 cm. This will result in small beech trees being more common and larger trees being less common. Taken together, the mean observed mortality will exceed the baseline mortality, suggesting an unsustainable development of the forest stand. The figure is redrawn from Manion and Griffin (2001), with permission.

The most accepted theory was the spiral of decline proposed by Manion (1981). He defined the decline syndrome as 'caused by the interaction of interchangeable, specifically ordered abiotic and biotic factors that produce a gradual general deterioration, often ending in the death of trees'. According to the spiral, decline involves three or more factors. The first set of factors includes non- and slow-changing factors such as climate, site or an unsuitable genetic background, which cause a permanent stress to the trees. This stress predisposes the trees to the action of a second set of factors. This second set of factors is of short duration and causes a drastic injury such as defoliation by insects, drought or frost. These factors incite the tree into the decline process. Recovery is hampered by the stress caused by the predisposing conditions. At this point, a third group of factors, typically biotic, contributes to further decline of the trees, and ultimately causes their death. Contributing factors include secondary pest and pathogens such as bark beetles, canker fungi and root rots that would otherwise not be able to cause significant damage.

Not all decline syndromes fit into the same model. For example, Sinclair and Hudler (1988) listed four main types of decline: (i) decline caused primarily by perennial or continual irritation by one factor; (ii) decline caused by drastic injury plus secondary stress; (iii) interchangeable predisposing, inciting and contributing factors, i.e. the Manion spiral; and (iv) synchronous cohort senescence, i.e. the Mueller-Dombois model. Forest decline as a concept has also received some criticism such as that of Ostry *et al.* (2011). Some of the arguments against forest decline are that: (i) many 'simple' diseases have 'decline' symptoms; (ii) all diseases have a complex aetiology, not just decline diseases; (iii) the disease triangle adequately describes the expression of decline symptoms; (iv) a decline diagnosis does not forecast the fate of a tree; (v) tree decline is a normal progression in the development of forest stands; and (vi) diagnosis of decline disease does not improve tree management.

1.5 Co-evolution and Genetic Determinants of Pathogenicity and Virulence

1.5.1 Co-evolution

One important aspect of the risk of epidemic outbreaks of a certain disease is the degree to which host and pathogen are adapted to each other, often referred to as co-evolution. Over evolutionary time, this can be seen as an arms race between host and pathogen populations. It is not in the long-term interest (often expressed as an evolutionarily stable strategy) of the pathogen to completely destroy the host – kill it – and by this means exhaust the habitat and nutrient base that is needed for long-term survival. This is often expressed as an evolutionarily stable strategy for not maximizing aggressiveness. For the host, it is less obvious what the benefits from hosting a pathogen could be that allow for an evolutionarily stable coexistence with a pathogen. Still, at least three types of benefits can be envisioned. Firstly, the costs, either in terms of the resource spending required to completely deny a pathogen access to the tissue it needs, or in terms of autotoxicity, when the production of excessive amounts of toxic secondary metabolites can be too high. The costs of both of these factors can prevent trees from successfully competing with others of the same or a different species that allocate fewer resources to defence. Second, the pathogen might be providing a relatively cheap way of sensing environmental stress. Third, when they are moderately susceptible, trees might promote disease in competing members of the population, and thereby increase the fitness of genotypes that can tolerate disease.

Co-evolution can act at a regional scale, leading to host populations with differences in their average susceptibility to the pathogen. This is well illustrated by an experiment conducted by Hunt and Sickle (1984), in which seed lots were collected from lodgepole pine (*Pinus contorta* Douglas ex Loudon) trees growing in areas with or without sweetgale (*Myrica gale* L.) and hence with or without the rust fungus *Cronartium comptoniae* Arthur. Sweetgale is an obligate alternate host in the life cycle of the rust fungus and basidiospore spread is limited in distance. Hence, lodgepole pine populations in habitats with sweetgale had experienced the rust in their local population history, while populations from habitats lacking sweetgale were naive (Fig. 1.4). This history is reflected in highly rust-resistant seed lots from the rust environment and seed lots differing in infection rates from zero to 40% from rust-naive habitats when challenged with the rust in a common garden experiment.

Following the same logic, in the pathogen population, one should expect to find highly aggressive individuals in situations where epidemics are in their initial phase compared with the later stages when co-evolutionary forces have had time to affect virulence. Such lack of resistance/high virulence is typically observed following pathogen introduction. For example, highly aggressive clones of *O. novo-ulmi* have frequently been isolated from the epidemic front of Dutch elm disease in Europe (Brasier, 1988).

1.5.2 Genetic determinants of pathogenicity and virulence

The interactions between hosts and pathogens are mediated through recognition and signalling processes. The host tree reacts to

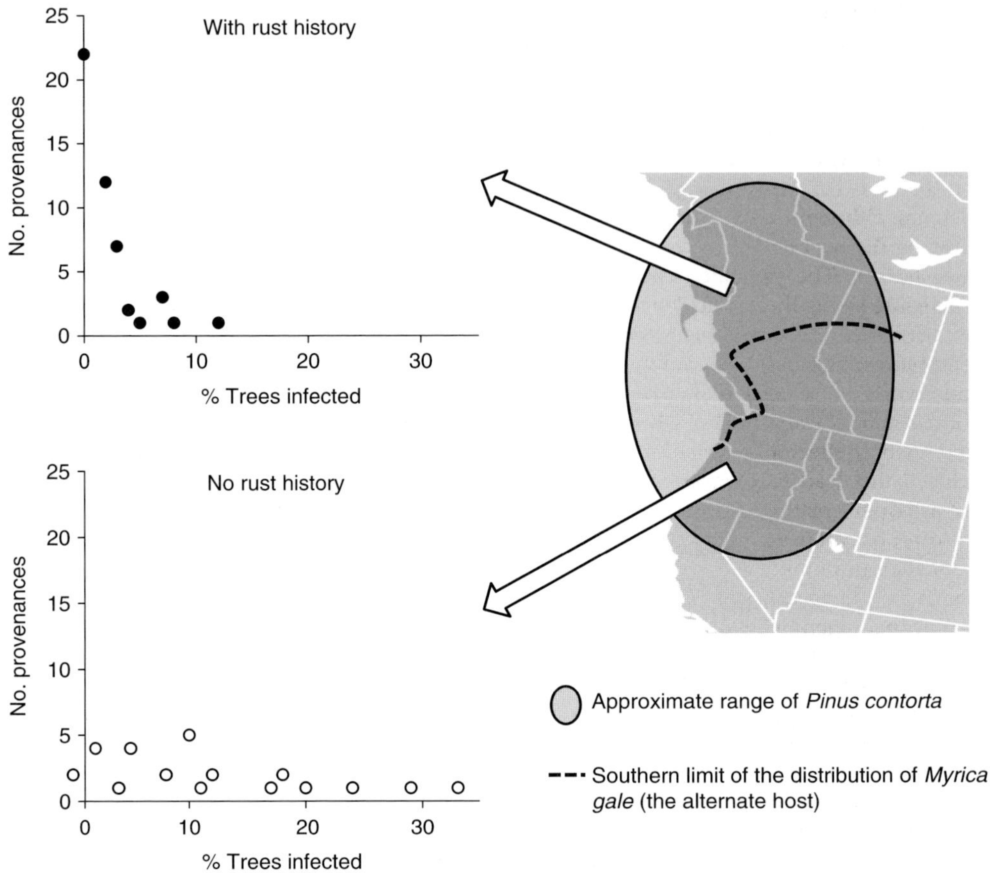

Fig. 1.4. The potential effect of co-evolution on disease severity. The map illustrates the range of lodgepole pine (*Pinus contorta* Douglas ex Loudon) (dark grey shading) in western North America. West and north of the dashed line is the range of sweetgale (*Myrica gale* L.), the obligate alternate host of the rust fungus *Cronartium comptoniae* Arthur. Seed lots were collected from *P. contorta* trees growing in areas with or without the presence of *M. gale* and, hence, the rust fungus *C. comptoniae*. Provenances of *P. contorta* from source locations with *M. gale* were less infected when challenged with the pathogen in a common garden experiment. Pine provenances from locations lacking the alternate host were naive and thus more susceptible to the rust fungus. The high proportion of resistant provenances from the pine population within the range of the alternate host is an illustration of the co-evolutionary forces that drive the development of host resistance. Figure adapted from Hunt and Sickle (1984), with permission.

the presence of a potential pathogen by recognizing pathogen associated molecular patterns (PAMPs). PAMPs can be generic structures common across many pathogens, e.g. chitin from fungal cell walls. PAMP recognition puts the tree into an alert state so that it can rapidly initiate defence responses such as the production of secondary metabolites and reactive oxygen species, as well as programmed cell death. Necrotrophic and hemi-necrotrophic pathogens can benefit from these cell death responses and live on the dying tissue. In contrast, biotrophic pathogens depend on living tissue and will be controlled by such responses. From an epidemiological point of view, the consequences of the different pathogen lifestyles is that under stress conditions trees are normally more susceptible to necrotrophs, while biotrophs can thrive better on vigorously growing cells of fast-growing trees.

Pathogens have effector molecules that are designed to act in the host cells to block recognition or destroy defence reactions. These molecules typically have a low molecular weight and one of several highly conserved amino acid motifs that act to facilitate penetration of the molecule across host cell membranes. Firstly, there is a basic recognition of PAMPs. This can be suppressed by pathogen effectors, although the effectors can be recognized by the host and then function as avirulence factors. The pathogen can continue to evolve into losing these avirulence factors, and the arms race then continues. Biotrophs and hemi-biotrophs are able to deliver these molecules within host cells to block recognition or to affect gene expression. If the host recognizes the biotroph effector, a series of reactions are induced that lead to programmed cell death, i.e. a hypersensitive response. In contrast, necrotrophs can use other types of effectors to induce programmed cell death, thus giving them access to dead tissue. For example, necrotrophic pathogens can produce host-specific toxins that possibly have the PAMP recognition molecules as targets (Friesen *et al.*, 2008).

The specific patterns of recognition by the host and response from the pathogen result in co-evolution and population changes. Gene-for-gene situations result when the host recognizes a pathogen and begins a defence reaction that will trigger the pathogen to react and avoid the host response by altering the recognized target. This arms race, typically seen in biotrophs, can lead to a genetic dominance for resistance in the host population and to avirulence in the pathogen population. In contrast, toxic effectors produced by necrotrophs result in a genetic dominance for susceptibility in the host population and in virulence in the pathogen population.

A key aspect of both pathogenicity in the fungus and resistance in the host is the importance of sexual reproduction and recombination for diversification. Sexual recombination in a pathogen population helps to spread genes for favourable traits into various genetic backgrounds. This can be, for instance, toxin production or the ability to avoid recognition by the host. Conversely, once a highly fit pathogen genotype has been achieved, asexual reproduction can help to propagate this, so making infections more efficient. In the host, sexual recombination can promote a wide distribution in the population of factors associated with resistance, e.g. receptors with specificity for the recognition of pathogen traits. Resistant genotypes that have been favoured by natural selection or by breeding can be multiplied through clonal propagation, thereby offering an opportunity to proliferate host genotypes with low susceptibility to certain pathogens. Temporarily, this can be a highly efficient way to manage a disease situation. However, history has taught us that almost all resistance genes have been overcome by rapid evolution in the pathogen populations. The unequal generation times for trees (tens or hundreds of years) and microorganisms (which can have several generations a year) frequently give the pathogen the upper hand in the arms race. Pathogens with the greatest chance of breaking down resistance genes are those that have a mixed reproduction system; they have at least one sexual cycle a growing season, combined with asexual reproduction which can multiply propagules of high pathogenic capacity, and a high potential for gene flow. Resistance breakdown can be counteracted by taking breeding approaches based on several resistance genes or several types of resistance, including structural and chemical defences as well as recognition responses. It has also been argued that planting mixes of host genotypes, or promoting diverse forests with a varied species composition, would help to reduce the selection of aggressive pathogens.

1.6 Models to Describe Epidemics

The temporal dynamics and spatial patterns of an epidemic are, as previously described, a result of the interaction between a pathosystem and certain environmental conditions. Mathematical and statistical models can be used to understand the spatio-temporal dynamics of diseases. In epidemiology, modelling is generally used to understand and predict the invasion and persistence of diseases in order to develop sustainable strategies to manage them (van Maanen and Xu, 2003). Disease can be assessed at different levels (leaf, branch, tree or stand) and is often measured as incidence or severity. Seem (1984) defined incidence as the proportion of disease entities within the sampling unit, e.g. the proportion of infected trees on a site, or the proportion of leaves infected on a branch. Severity was defined as the quantity of disease affecting units within the sampling unit, e.g. the number of spots per leaf or branch. How and at which level we measure disease will affect the way we model it. As an example, most models are based on the assumption that the amount of disease is proportional to the number of inocula; that is, no synergy or antagonism exists between inocula (van der Plank, 1975). However, disease measurements, typically as the number of diseased trees or leaves, do not correspond with the number of lesions and therefore the amount of inocula, because trees or leaves may be infected more than once. If lesions were scattered at random, 357 lesions, i.e. 357 spores, for example, would be needed to infect 30% of 1000 plants. Of those, 300 lesions would be the first lesion, and the remaining 57 would correspond to the second, third, or the *n*th lesion in some plants. The disconnection between the amount of inoculum and the measured disease may affect how realistic our model is.

1.6.1 Disease progress curves

A disease epidemic can be defined as any change in disease level in a host population. Disease increase will vary depending on which temporal or spatial scale is used, and on the type of disease measure. The trend in disease with time can be described by a disease progress curve where, for example, the proportion of diseased hosts (x) is plotted against time (t). The increase in the number of diseased hosts (Δx) per unit of time (Δt), i.e. dx/dt, when Δt tends towards zero, can be expressed mathematically, and can occur biologically in two ways (van der Plank, 1963).

- Simple interest: the amount of disease (x) increases linearly with time, and the amount of inoculum remains constant (Q). Depending on the interaction between host and environment and pathogen, a similar amount of inoculum can result in more or less disease. This is depicted by R, the basic infection rate, which measures the rate of increase per unit of infectious tissue per unit of time. In Eqn 3, the disease increase with time (t) is constant (K): $dx/dt = K = QR$. This is the case for the early stages of diseases caused by many root rot fungi which have resting spores in the soil.

$$x = QRt \tag{3}$$

- Compound interest: the disease increases exponentially (see Eqn 4) because the infected tissues increase the amount of inoculum: $dx/dt = Kx$. In the case of true exponential increase, the amount of increase relative to the amount of infected tissue (r) remains constant with time (e.g. the population increases 5% every day), which is typical in the early stages (x <0.05) of disease outbreaks caused by foliar pathogens.

$$x = x_0\, e^{rt} \tag{4}$$

The nature of the pathogen and the host will determine the shape of the disease progress curve. The main difference between linear growth and exponential growth is whether the disease increase depends on the previous disease quantity or not; in other words, whether or not the infected hosts contribute to further spread of the disease. Hosts or host tissues can contribute to infection by being the source of extra inoculum or by bridging infection with healthy hosts or host cells. Inoculum build-up will be influenced by the type of reproduction of the pathogen. In contrast with monocyclic diseases, polycyclic diseases complete more than one life cycle during the life of the host. Foliar diseases will typically spread exponentially owing to their capacity to reproduce asexually from newly infected tissue. In contrast, root rot diseases such as *Armillaria mellea* (Vahl) P. Kumm. *s.l.* will typically spread linearly, because disease progression depends on the number of infected stumps from the previous rotation (i.e. the initial inoculum) and newly infected saplings take a long time to become infectious (Fig. 1.5B versus 1.5A).

In reality, disease progress curves hardly follow either one of the categories depicted in

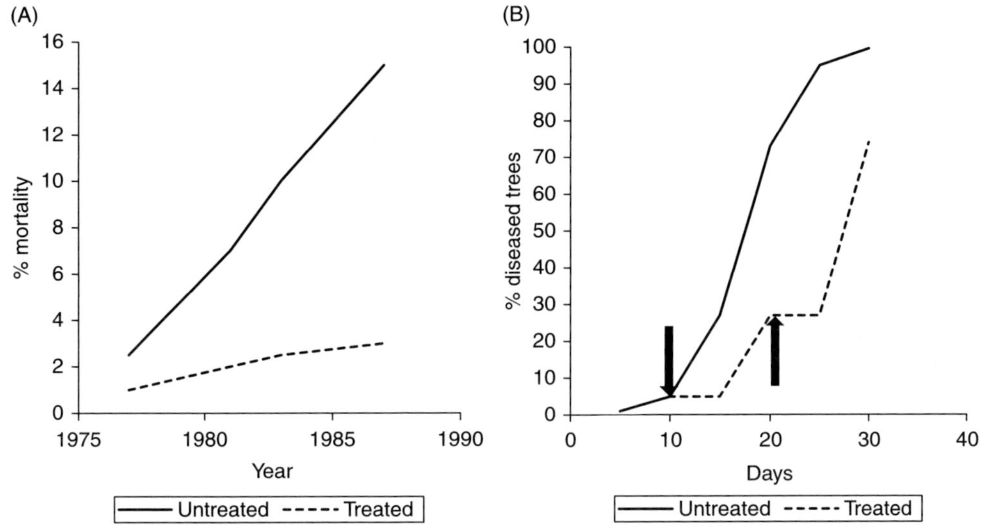

Fig. 1.5. Examples of disease progress curves. (A) Simple interest disease curve: the cumulative percentage mortality of Douglas-fir (*Pseudotsuga menziesii* (Mirb.) Franco) seedlings caused by *Armillaria ostoyae* (Romagn.) Herink in untreated forest blocks (stumps retained from the previous rotation) and blocks treated by stumping and root raking to remove any existing infected material before planting. (B) Compound interest disease curve: a theoretical study of disease progress of a foliage disease in red Christmas tree plantations. The upper curves show progress of the disease under untreated conditions, while the lower curves show the disease progress after two applications of a short-lived fungicide on days 10 and 20 (arrowed). Figure adapted from Morrison *et al.* (1988), with permission, and from Manion (1981), with permission of Pearson Education, Inc., Upper Saddle River, New Jersey.

Eqns 3 and 4, because: (i) disease increases intermittently, for example as a result of variable weather conditions; (ii) there is a limit to the number of hosts that can be infected, i.e. the higher the amount of disease the more difficult to find a healthy host; (iii) newly infected tissues are not immediately infectious; and (iv) disease tends to occur in foci. In the case of a simple interest disease, the effects of decreasing numbers of hosts are incorporated into Eqn 3 by applying a correction factor '$1 - x$': $dx/dt = QR\ (1 - x)$ (Eqn 5). Whether the disease follows such a model can then be assessed by transforming the disease data x by $\ln(1/1 + x)$ and plotting it against time. If linear, the slope of the adjusted line will correspond to QR.

$$x = QRt\ (1 - x) \qquad (5)$$

In the case of compound interest diseases, the same correction factor, $dx/dt = rx\ (1 - x)$, will be applied, converting Eqn 4 into a logistic curve (Eqn 6) (Fig. 1.5B). In Eqn 6, r is referred as the apparent infection rate. This type of curve depicts a disease that progresses slowly to start with owing to a lack of infectious tissue, then accelerates until 50% of the population is infected and then slows down after this point as a result of the increasing lack of susceptible tissue. This curve implies that the logit of the proportion of infected units increases linearly with time. In order to know whether a disease follows a logistic curve, the logit of the amount of disease x can be plotted against time and assessed for linearity. If the logit values fit on to a line, its slope will be equal to r.

$$\text{In}\ (x/1 - x) = x_0 + rt \qquad (6)$$

The latency period (p) represents the time it takes for infected tissues to become infectious. In a compound interest disease, when assessing the disease level at a time (t), we are describing the disease created from the infectious tissue at time ($t - p$). Because disease assessments consider infected tissue and not infectious tissue, the apparent infection rate (r), and not the basic infection rate

(R) is usually obtained. The latency period is important from a disease management perspective, because, during that time, compound interest diseases behave as simple interest diseases, so epidemics are less explosive, i.e. a lower r, assuming an equal R. The apparent infection rate, r, depends on p and R as shown in Eqn 7.

$$R/r = xt/xt - p \quad (7)$$

Disease progress curves often fit well into a logistic curve (e.g. Fig. 1.5B), implying that r remains constant (Eqn 6) during the whole epidemic. Under the assumption of a constant latency period, logistic growth implies a decrease of R. A decrease in R during the epidemic could be due to the fact that (i) most susceptible tissues/hosts are infected first, (ii) the disease itself changes the conditions that favour reproduction, e.g. defoliation produces increased light under the canopy, and (iii) lesions become older and produce less inoculum.

1.6.2 Applications of disease curves

Disease progress curves can be used to compare different diseases. Two types of variables are often used: R and r (Eqns 3 and 4). These variables represent a combination of the pathogen aggressiveness, the host susceptibility and the environmental conditions, and how they affect disease levels. R is used for simple interest diseases (Eqn 3), i.e. monocyclic diseases. If the diseased tissue does not act as secondary inoculum, the disease will depend on the initial amount of infectious inoculum, Q. Note that the same disease progress can be attained with a higher amount of initial inoculum and a lower rate of spread. Thus, an independent measure of the initial quantity of inoculum is needed in order to be able to compare different environments. In the case of diseases with exponential growth (Eqn 4), the amount of infected host is critical to the progress of the disease. The increase at any time will depend on the proportion of infected hosts (x_0) and on the capacity of the inoculum to infect new hosts, the host susceptibility and the environmental conditions, represented by r. Note that although R and r are related, R (Eqn 7) is not directly extracted from field observations.

Disease progress curves can be used for designing control strategies by assessing the impact of changes in t, R or r and Q or x_0. In both simple and compound interest diseases (Eqns 3 and 4), the amount of disease can be decreased by reducing the time of the epidemic, e.g. reducing the rotation time for root rot pathogens. In the case of simple interest diseases, control can be achieved by decreasing the quantity of the initial inoculum Q, e.g. by stump removal in the case of *A. ostoyae* (Fig. 1.5A), or by the application of *Phlebiopsis gigantea* to stumps, which reduces the amount of *H. annosum s.l.* (Q) that establishes after thinning. R can be changed by reducing the capacity of the pathogen to cause disease, for example by increasing horizontal resistance (general resistance) in the stand. In the case of diseases that have a logarithmic increase, changing x_0 implies reducing the number of diseased hosts, e.g. removing trees infected with *O. novo-ulmi* or reducing the quantity of overwintering structures; r can be modified by reducing the spread of the pathogen, e.g. by spraying with fungicide, so preventing new infections (Fig. 1.5B). Spraying fungicide on healthy hosts can prevent the germination of spores on healthy tissues; r will then be zero as long as the fungicide is effective (Fig. 1.5B). Equations 3 and 4 can be used to model the effects of resistance on disease progress. Assuming a majority of avirulent inoculum, the initial amount of infection can be reduced by increasing vertical resistance (race-specific resistance), thereby preventing the establishment of avirulent races of the pathogen. Note that increasing vertical resistance will not affect r, because virulent races will reproduce normally, but they will start from fewer foci owing to their lower frequency in the population. Instead, r can be reduced by increasing horizontal resistance, thus all races of the pathogen will be equally affected by either increasing their latency period or by reducing their capacity of infection (R) (Eqn 7).

The optimal combination of inoculum reduction and infection rate will have to be evaluated in each case. Generally speaking though, diseases that are better controlled by reductions in r and R are not well controlled by measures targeting x_0 and Q. In the equation for simple interest diseases, all variables are linear, therefore proportional changes in each of them will result in similar disease changes. Reductions of Q are normally easier to achieve than reductions in R, i.e. it is technically easier to remove 50% of the stumps in a plantation than to introduce 50% resistance in a tree species. In compound interest diseases, decreases in x_0 will delay the disease, which, if combined with a lowered r, can make the disease imperceptible until the end of the rotation. However, the longer the rotation and the higher the infection rate, the more efficient the control measures based on reductions in r will be, as opposed to control measures targeting x_0.

1.6.3 Dispersal and spread in space and time

Pathogens differ in the way they disperse and spread in space and time. Models can be used to unravel the mechanisms operating in different types of spread, and to predict the extent of the damage in the future. Invasive forest pathogens provide good examples of different paces of spread across continents (Mundt *et al.*, 2009) (Fig. 1.6). As an example, *Cryphonectria parasitica* (Murrill) M.E. Barr was first detected in New York in 1904. By 1950, it had spread throughout the native population of American chestnut (*Castanea dentata* (Marsh.) Borkh.) in the eastern USA. Other pathogens, such as *Cronartium ribicola*, although also facing a naive host population, did not spread so dramatically. It is easy to observe that *C. ribicola* kept a more or less constant pace of spread over the years, while the rate of *C. parasitica* spread was increasing with time. The spatial spread of a disease can be viewed as an expanding wave. The height of the wave can be regarded as a constant quantity of disease, e.g. 30% of the infected hosts. If two points are placed at a known distance, the time it takes to move from the first point to the second point can be regarded as a velocity measure. The velocity of propagation can be used to compare different diseases and different control methods, e.g. using a mixture of susceptible and resistant hosts.

The most simple models of disease spread assume a constant speed while the more complex refer to an increasing speed. The first type of disease spreads via so-called travelling waves. The second type of disease follows so-called accelerating or dispersive waves. Whether the spread will increase or will remain steady will depend on the type of dispersal. Spatially limited dispersal will result in constant waves. If the dispersal is less restrictive, inoculum can disperse over long distances and there will be a certain amount of disease transmitted well in advance of the main wave, so the disease front will travel faster and faster as the epidemic builds up.

The dispersal of a pathogen can be basically described by two types of function: exponential and power-law equations (Fitt *et al.*, 1987). These two types of equations focus on different physical aspects of the dispersal. Exponential dispersal curves focus on deposition, and assume a constant decrease of the amount of spores with distance. Power-law equations focus on spatial diffusion due, for example, to turbulence in the canopy. The power-law equation fits well with the dispersal of fungi with small spores, and it typically yields fat-tailed distribution curves. In general, both models fit reasonably well when trying to predict dispersal at short distances from the focus, but as they differ in the tail of the distribution they result in significant differences at large scales.

In order to generate disease, there is need for not only dispersal but also the presence of a suitable tissue. This type of function is called a contact function. As with disease progress curves, at low disease incidences an exponential gradient will occur, but at higher disease incidences the probability of finding a suitable host will be difficult as most of the hosts will be already infected; a correction will therefore be needed, e.g. a logistic correction. Two types of spatio-temporal models emerge, logistic and power logistic, linked, respectively, with exponential and power-law dispersal. The logistic model will provide a constant expansion of the trailing edge, while the power-logistic model will provide an accelerating wave (Madden *et al.*, 2007).

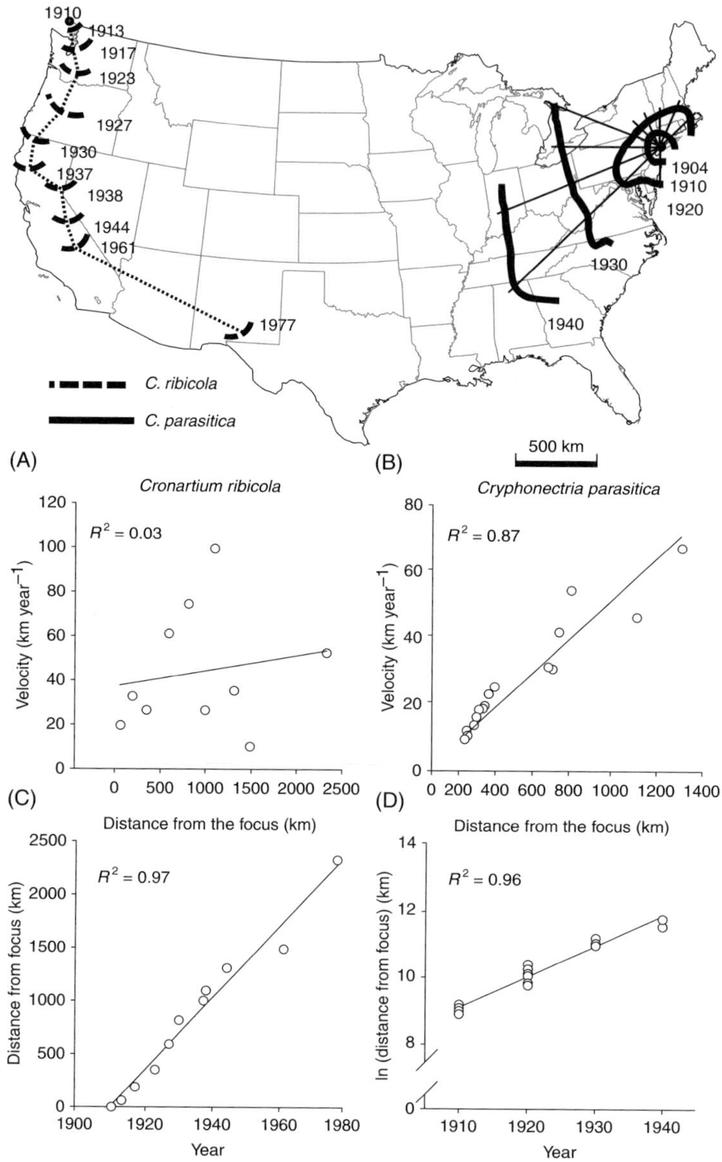

Fig. 1.6. Spatio-temporal dynamics of *Cronartium ribicola* J.C. Fisch. and *Cryphonectria parasitica* (Murrill) M.E. Barr epidemics in the USA. Continental spread can occur at a constant rate, thus the velocity of the epidemic front has a constant speed as in the case of *C. ribicola* (A). In the case of *C. parasitica*, the velocity of the epidemic front increased with the distance (B). The advance of the epidemic waves occurs at an exponential rate with time, thus the epidemic front is found at longer distances every time (D). This does not occur when the velocity is constant; in that case, the waves advance linearly (C). Accelerating waves normally arise from pathogens whose dispersal follows a fat-tailed distribution such as a power-law function. On the contrary, pathogens whose wave speed is constant may have a spatially limited dispersal, and the tail of their dispersal distribution declines faster or as fast as that of an exponential function, i.e. it is exponentially bounded. Velocity calculated as in Mundt *et al.* (2009) using dispersal data for *C. parasitica* from Gravatt (1949) (courtesy of the FAO) and for *C. ribicola* (courtesy of the USDA Forest Service, Pacific Southwest Research Station, see www.fs.fed.us/psw/programs/ifg/disease.shtml).

Both logistic and power-logistic models only assume that dispersal occurs from a single point. More complex spatio-temporal models include dispersal from the front and the trailing edge of the wave, so they allow dispersal towards both positive and negative distance values. The contact functions are normal, double exponential, root and modified power law. When trying to predict whether a pathogen will spread as a dispersive wave or as a travelling wave, one can look to see whether the dispersal gradient is exponentially bounded. An exponentially bounded contact function would imply that the contact gradient decreases with distance as fast, or faster, than with an exponential function, i.e. normal and double exponential distributions. If the decrease occurs more slowly than under an exponential function, e.g. following a modified power-law dispersal, an accelerating spread wave could be expected (Frantzen and van den Bosch, 2000).

1.6.4 Dynamic and static models

Based on our understanding of the disease cycle, we can use static models to predict disease severity based upon constant factors. Depending on the temporal and spatial scale at which we want to make predictions, these factors will have different influences. As the scale narrows, new factors emerge as being significant; these are called emerging properties. For example, at a regional level, temperature can govern the decay incidence of *H. annosum s.l.* (Thor *et al.*, 2005) and can, therefore, be used to make a prediction at a regional level. At the stand level, the history of the stand and its management will be critical (Oliva *et al.*, 2010); thus, an approach taking into account the time and place of primary and secondary infections is more relevant. Several diseases can be satisfactorily explained with these static models. The probability of pine twisting rust damage caused by *M. populnea* f.sp. *pinitorqua* is very dependent on the presence of nearby aspen stands (the alternate host) (Mattila, 2005). However, sometimes, in order to capture the key mechanisms of the epidemic, we have to consider the development of the disease in relation to time. This type of model is called a dynamic model and includes the variable time.

1.6.5 Spatially explicit models

The so-called spatially explicit models of tree disease can take into account the spatial spread of the pathogen and the spatial distribution of the host. This type of model is especially useful when modelling diseases that can be spatially constrained, such as root rot diseases. After a thinning operation, the closer the trees are to an infected stump (primary spread) the earlier they will become infected. Once trees become infected they also act as a source of inoculum (secondary spread). Root diseases are not the only ones to have primary and secondary spread stages with some spatial constraints. Wilt diseases, such as those caused by *O. novo-ulmi*, *Ceratocystis platani* (J.M. Walter) Engelbr. & T.C. Harr., and *C. fagacearum* also spread in the canopy and through root grafts. Typically, primary and secondary spread responds to different biological processes, e.g. spore infection by bark beetle activity versus mycelial spread through root-to-root grafts. Depending on the initial distribution of the host and inoculum, there will be a delay between these two processes that typically results in a high variability of the outcomes. It is possible to build up simple models to study the probability of infection of a single tree from either primary or secondary infections. The second step is to apply models for individual hosts to a spatially explicit model for the whole population, such as cellular automata. Cellular automata represent cells in a grid, which show different states, such as susceptible, latent and infectious. By using a spatially explicit procedure, Kleczkowski *et al.* (1997) satisfactorily reproduced epidemics of *Rhizoctonia* spp. on radish, and captured the mean disease progress curve as well as the increasing variance with time, i.e. the longer the time the higher the variability of the outcome.

Spatially explicit models have been useful to simulate forest root rot diseases. In this type of disease, primary infections come from former disease centres or from spore infections. Secondary infections are due to transfer between trees. The Western Root Disease Model (WRDM) simulates root diseases caused by *A. ostoyae*, *H. annosum s.l.*, and *P. sulphurascens* (Frankel, 1998). Another model, Rotstand, followed the same principle and was adapted to European conditions (Fig. 1.7) (Pukkala *et al.*, 2005).

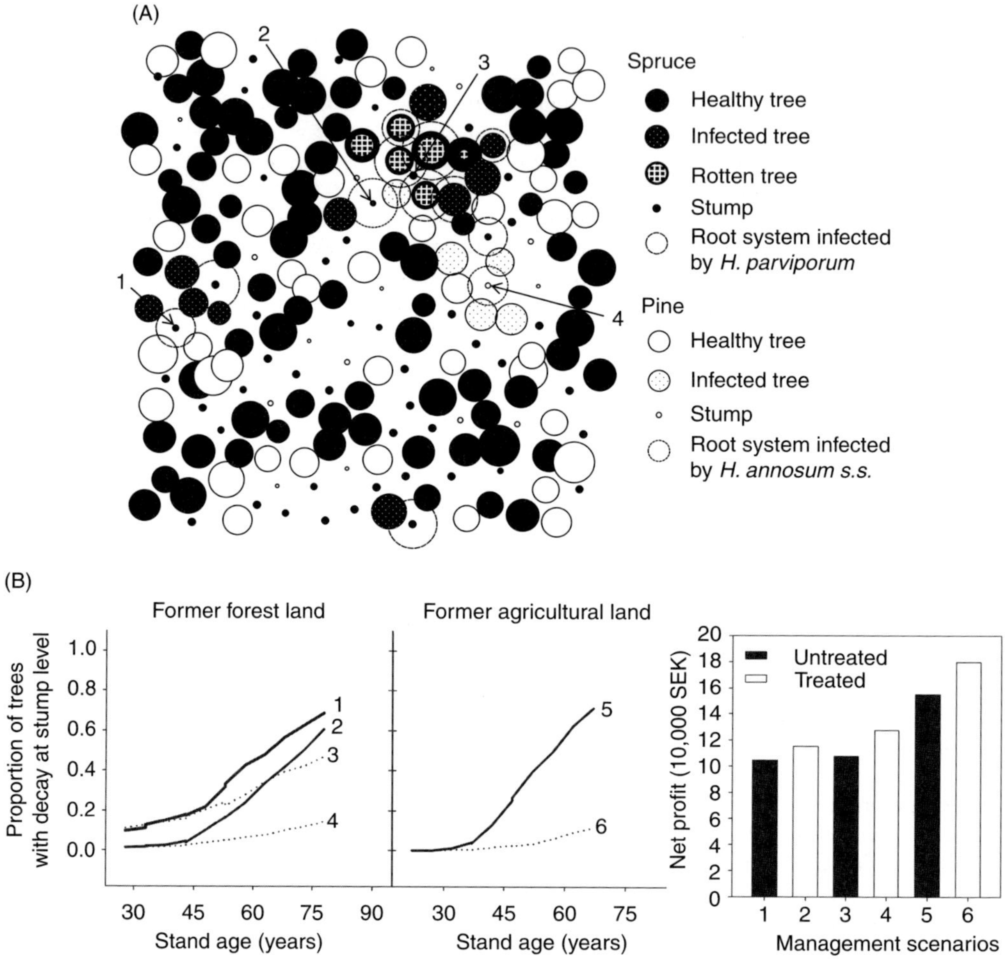

Fig. 1.7. Dynamic modeling of annosus root rot with the Rotstand model. (A) Graphical interface of the status of a mixed pine–spruce forest 5 years after the first commercial thinning (age = 35 years). The program simulates the spread of infection by *Heterobasidion* spp. on unprotected thinning stumps (1, 2 and 4) and carried over from infected stumps from the previous generation (3). In 1, 2 and 4, infection occurred recently (5 years), thus neighbouring trees have only their root systems infected. In 3, the trees were infected before thinning, thus they have already developed decay columns and have themselves spread the disease to their neighbours. The host specificity of *H. annosum s.s.* (Fr.) Bref. (for Scots pine, *Pinus sylvestris* L.) and *H. parviporum* Niemelä & Korhonen (for Norway spruce, *Picea abies* (L.) Karst.) is indicated. In 1, spruce stump infection with *H. parviporum* does not spread towards the pine trees. The size of the dotted areas indicates the proportion of decay in the section. (B) Dynamic modelling can be used to assess disease development and the economic outcome of different management scenarios, such as in the case of *Picea abies* planted on former forest land with and without carry-over from the previous rotation, either with (management scenarios 2, 4) or without stump treatment (1, 3). The same situation can be simulated in stands planted on former agricultural land with and without stump treatment (6 and 5, respectively). Note that on agricultural land, a faster development of decay is predicted than on forest land, as is observed in reality. Data from Thor *et al.* (2006), with permission.

However, while the WRDM focuses on tree mortality, Rotstand puts special emphasis on the development of decay in the stem caused by *H. annosum s.l.* Furthermore, Rotstand simulates the development of different European tree species under different forest management regimes. It calculates the value of the stand based on the assortment of different timber products. In addition, it incorporates elements of the management that are critical for the establishment of the pathogen, such as stump treatment during thinning and the season of harvest, as well as the tree species and decay incidence in the previous generation. This comprehensive model can produce realistic results (Oliva and Stenlid, 2011), which can be used for management decisions. The model has been used to illustrate the need for stump treatment even in situations where a substantial amount of decay is carried over from the infected stumps of the previous rotation (Thor *et al.*, 2006).

1.6.6 Climatic envelopes and species distribution

Recently, a stronger focus has been placed on predicting forest disease based on climatic conditions. On the one hand, climate matching can be used to predict whether an area can be potentially suitable for a pathogen to establish and cause disease. This typically applies to plantations of exotic tree species in areas where the pathogen of interest is not yet present, e.g. Monterey pine and *Gibberella circinata* in Australia and New Zealand (Ganley *et al.*, 2011). On the other hand, the global change in climate is predicted to alter current patterns of temperature, precipitation and frequency of extreme events. These changes can modify existing host–pathogen equilibria, for example, by increasing the number of days conducive to disease development (Fig. 1.8), or by increasing host susceptibility through stress (Desprez-Loustau *et al.*, 2006). At a geographic scale, a shift in the climate may enable the establishment of pathogens in areas that were not suitable earlier due to climatic constraints. The range of suitable climatic conditions for a specific species that prevail in a particular region is often referred to as the 'climate envelope' of a species and can be used to predict future expansions or reductions in range based on current and future climate scenarios. Climatic data are comparable across regions and can be interpolated at a regional scale. This facilitates matching with distribution maps.

Changes in species distribution patterns in relation to climate have primarily used two different modelling approaches: statistical or machine learning models, and mechanistic models. Climatic envelope models are typically 'statistical' models (e.g. BIOCLIM, DOMAIN, GAM, MAXENT) based on empirical relationships between species occurrence data and environmental variables such as climate (Elith *et al.*, 2006). They do not attempt to describe the 'cause and effect' between model parameters and response (Hijmans and Graham, 2006); instead they describe the correlation between the two. Species distributions can also be modelled using mechanistic models, which are based on knowledge of the ecophysiology of the modelled species. These models provide a mechanistic understanding of the underlying relationship between environmental conditions and species distribution. A major drawback is that the required physiological data for the species modelled are often not available.

CLIMEX is another type of climatic envelope model, which uses a hybrid statistical/mechanistic approach to predict potential geographical distribution and seasonal abundance of a species (Sutherst *et al.*, 2007). CLIMEX includes a simple population biology model, which is parameterized to describe the response of the organism to moisture, temperature and light (if applicable). The climatic requirements and their interactions are based on the known geographical distribution of the species. The model needs to be tested with an independent data set. CLIMEX has been widely used to predict the potential spread of invasive species in quarantine pest risk assessments, as well as to study changes in species distribution as affected by a changing climate, e.g. in the case of *P. cinnamomi* (Brasier and Scott, 1994), *Dothistroma* spp. and *G. circinata* (Watt *et al.*, 2011a).

The use of climatic envelopes alone has been the focus of criticism because factors other than climate determine the current

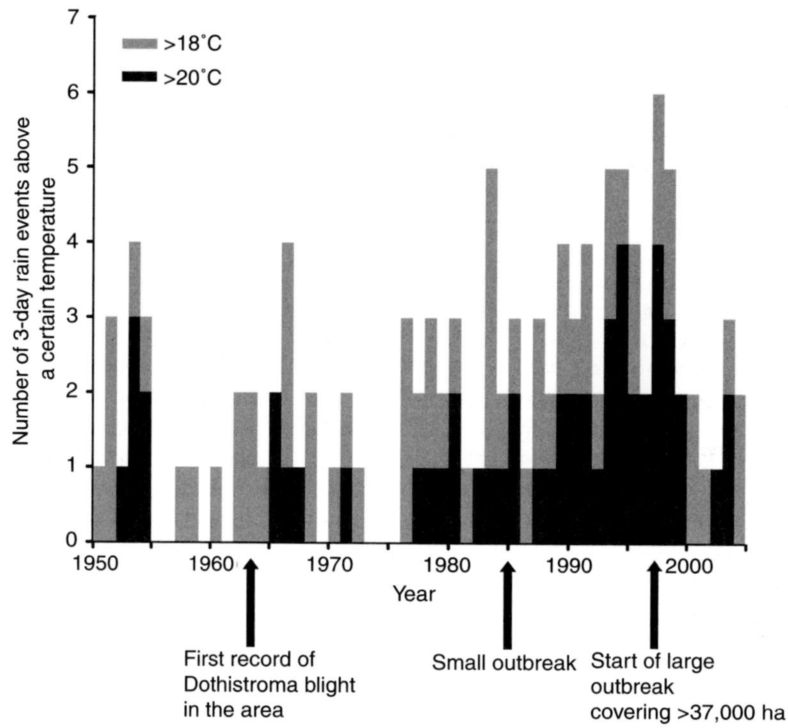

Fig. 1.8. Number of 3-day rain events above temperature thresholds of 18 and 20°C at the weather station 'Smithers' in British Columbia, Canada, between the years 1950 and 2004. The figure illustrates the importance of environmental factors in causing conditions conducive for the development of epidemics. Dothistroma needle blight was first observed in the area at the beginning of the 1960s. A small outbreak in the mid 1980s was followed by a large outbreak in plantations of *Pinus contorta* Douglas ex Loudon some 12 years later, which coincided with an increased frequency of warm summer rain events during the 1990s. The figure is modified from Woods *et al.* (2005), with permission.

range of a species (Hulme, 2003) and the dynamics of distribution changes. In terms of forest pathogens, biotic interactions such as competition, host distribution and host resistance variability have a large impact on distribution patterns (Kearney and Porter, 2009). These factors are not modelled in a direct way but may be taken into account indirectly because climatic envelope models predict the realized (observed) niche rather than the fundamental niche (Hijmans and Graham, 2006). In terms of potential range expansion, the means of spread and limiting factors that act as barriers – such as oceans, mountains and a fragmented landscape – are not taken into account. These models are also based on a number of assumptions, such as species niche constancy, genetic homogeneity among populations and equilibrium of the species population with the environment (Elith *et al.*, 2006). These assumptions are very often violated for pathogens and especially for invading pathogens that are introduced into a new environment. For example, *Neonectria fuckeliana* (C. Booth) Castl. & Rossman is native to northern Europe on Norway spruce but has been found causing cankers under much warmer conditions on Monterey pine in New Zealand (Crane *et al.*, 2009). Non-native pathogens can also be spread via sink–source dynamics. Thus, species observed in a new area may be repeatedly introduced from somewhere else.

In the case of some pathogens, extreme weather conditions are responsible for disease outbreaks. These extreme events are difficult to capture in climate niche models because they are based on climatic variables, i.e. weather averaged over long time periods. An alternative approach can be to model the probability of disease-conducive conditions, and this may provide better predictions than those based on climatic envelopes alone. Models of this kind have been developed for *P. cinnamomi* in France (Bergot *et al.*, 2004), for Dothistroma needle blight in the western USA (Woods *et al.*, 2005) and for *G. abietina* in Denmark (Thomsen, 2009).

References

Agrios, G.N. (2005) *Plant Pathology*, 5th edn. Academic Press, New York.

Auclair, A.N.D., Worrest, R.C., Lachance, D. and Martin, H.C. (1992) Climatic perturbation as a general mechanism of forest dieback. In: Manion, P.D. and Lachance, D. (eds) *Forest Decline Concepts*. APS Press, St Paul, Minnesota, pp. 38–58.

Bell, T., Freckleton, R.P. and Lewis, O.T. (2006) Plant pathogens drive density-dependent seedling mortality in a tropical tree. *Ecology Letters* 9, 569–574.

Berglund, H., Hottola, J., Penttilä, R. and Siitonen, J. (2011) Linking substrate and habitat requirements of wood-inhabiting fungi to their regional extinction vulnerability. *Ecography* 34, 864–875.

Bergot, M., Cloppet, E., Pérarnaud, V., Déqué, M., Marçais, B. and Desprez-Loustau, M.-L. (2004) Simulation of potential range expansion of oak disease caused by *Phytophthora cinnamomi* under climate change. *Global Change Biology* 10, 1539–1552.

Blaschke, M. and Cech, T.L. (2007) Absterbende Weisskiefern – eine langfristige Folge des Trockenjahres 2003? [Declining Scots pines: a consequence of the drought in 2003?]. *Forstschutz Aktuell* 40, 32–34.

Blodgett, J.T., Herms, D.A. and Bonello, P. (2005) Effects of fertilization on red pine defense chemistry and resistance to *Sphaeropsis sapinea*. *Forest Ecology and Management* 208, 373–382.

Brasier, C.M. (1988) Rapid changes in genetic structure of epidemic populations of *Ophiostoma ulmi*. *Nature* 332, 538–541.

Brasier, C.M. (1991) *Ophiostoma novo-ulmi* sp. nov., causative agent of current Dutch elm disease pandemics. *Mycopathologia* 115, 151–161.

Brasier, C.M. and Scott, J.K. (1994) European oak declines and global warming: a theoretical assessment with special reference to the activity of *Phytophthora cinnamomi*. *EPPO Bulletin* 24, 221–232.

Chou, C.K.S. and MacKenzie, M. (1988) Effect of pruning intensity and season on *Diplodia pinea* infection of *Pinus radiata* stem through pruning wounds. *European Journal of Forest Pathology* 18, 437–444.

Cobb, R.C., Meentemeyer, R.K. and Rizzo, D.M. (2010) Apparent competition in canopy trees determined by pathogen transmission rather than susceptibility. *Ecology* 91, 327–333.

Crane, P.E., Hopkins, A.J.M., Dick, M.A. and Bulman, L.S. (2009) Behaviour of *Neonectria fuckeliana* causing a pine canker disease in New Zealand. *Canadian Journal of Forest Research* 39, 2119–2128.

Cruickshank, M.G., Morrison, D.J. and Lalumière, A. (2011) Site, plot, and individual tree yield reduction of interior Douglas-fir associated with non-lethal infection by *Armillaria* root disease in southern British Columbia. *Forest Ecology and Management* 261, 297–307.

Desprez-Loustau, M.-L., Marçais, B., Nageleisen, L.-M., Piou, D. and Vannini, A. (2006) Interactive effects of drought and pathogens in forest trees. *Annals of Forest Science* 63, 597–612.

Develey-Rivière, M.-P. and Galiana, E. (2007) Resistance to pathogens and host developmental stage: a multifaceted relationship within the plant kingdom. *New Phytologist* 175, 405–416.

Elith, J. *et al.* (2006) Novel methods improve prediction of species' distributions from occurrence data. *Ecography* 29, 129–151.

Eyles, A., Bonello, P., Ganley, R. and Mohammed, C. (2010) Induced resistance to pests and pathogens in trees. *New Phytologist* 185, 893–908.

Fitt, B.D.L., Gregory, P.H., Todd, A.D., McCartney, H.A. and Macdonald, O.C. (1987) Spore dispersal and plant disease gradients; a comparison between two empirical models. *Journal of Phytopathology* 118, 227–242.

Frankel, S. (1998) *User's Guide to the Western Root Disease Model, Version 3.0*. General Technical Report No. PSW-GTR-165, US Department of Agriculture, Forest Service, Pacific Southwest Research Station, Berkeley, California.

Frantzen, J. and van den Bosch, F. (2000) Spread of organisms: can travelling and dispersive waves be distinguished? *Basic and Applied Ecology* 1, 83–92.

Friesen, T.L., Faris, J.D., Solomon, P.S. and Oliver, R.P. (2008) Host-specific toxins: effectors of necrotrophic pathogenicity. *Cellular Microbiology* 10, 1421–1428.

Gadgil, P.D. (1984) Dothistroma needle blight. In: Gadgil, P.D. (ed.) *Forest Pathology in New Zealand*. New Zealand Forest Research Institute, Rotorua, New Zealand.

Ganley, R.J., Sniezko, R.A. and Newcombe, G. (2008) Endophyte-mediated resistance against white pine blister rust in *Pinus monticola*. *Forest Ecology and Management* 255, 2751–2760.

Ganley, R.J., Watt, M.S., Kriticos, D.J., Hopkins, A.J.M. and Manning, L.K. (2011) Increased risk of pitch canker to Australasia under climate change. *Australasian Plant Pathology* 40, 228–237.

Garrett, S.D. (1956a) *Biology of Root-Infecting Fungi*. Cambridge University Press, Cambridge, UK.

Garrett, S.D. (1956b) Rhizomorph behaviour in *Armillaria mellea* (Vahl) Quél. II. Logistics of infection. *Annals of Botany* 20, 193–209.

Gilbert, G.S. (2002) Evolutionary ecology of plant diseases in natural ecosystems. *Annual Review of Phytopathology* 40, 13–43.

Gilmour, J.W. (1981) The effect of season on infection of *Pinus radiata* by *Dothistroma pini*. *European Journal of Forest Pathology* 11, 265–269.

Gravatt, F. (1949) Chestnut blight in Asia and North America. *Unasylva* 3, 3–7.

Haddow, W.R. (1938) The disease caused by *Trametes pini* (Thore) Fries in white pine (*Pinus strobus* L.). *Transactions of the Royal Canadian Institute* 22, 21–80.

Hansen, E.M. and Goheen, E.M. (2000) *Phellinus weirii* and other native root pathogens as determinants of forest structure and process in western North America. *Annual Review of Phytopathology* 38, 515–539.

Hansen, E.M., Parke, J.L. and Sutton, W. (2005) Susceptibility of Oregon forest trees and shrubs to *Phytophthora ramorum*: a comparison of artificial inoculation and natural infection. *Plant Disease* 89, 63–70.

Hijmans, R.J. and Graham, C.H. (2006) The ability of climate envelope models to predict the effect of climate change on species distributions. *Global Change Biology* 12, 2272–2281.

Hille Ris Lambers, J., Clark, J.S. and Beckage, B. (2002) Density-dependent mortality and the latitudinal gradient in species diversity. *Nature* 417, 732–735.

Holdenrieder, O. and Greig, B.J.W. (1998) Biological methods of control. In: Woodward, S., Stenlid, J., Karjalainen, R. and Hüttermann, A. (eds) Heterobasidion annosum: *Biology, Ecology, Impact and Control*. CAB International, Wallingford, UK, pp. 235–258.

Houston, D.R. (1992) A host-stress–saprogen model for forest dieback–decline diseases. In: Manion, P.D. and Lachance, D. (eds) *Forest Decline Concepts*. APS Press, St Paul, Minnesota, pp. 3–25.

Hulme, P.E. (2003) Biological invasions: winning the science battles but losing the conservation war? *Oryx* 37, 178–193.

Hunt, R.S. and Sickle, G.A.V. (1984) Variation in susceptibility to sweet fern rust among *Pinus contorta* and *P. banksiana*. *Canadian Journal of Forest Research* 14, 672–675.

Isomäki, A. and Kallio, T. (1974) Consequences of injury caused by timber harvesting machines on the growth and decay of spruce (*Picea abies* (L.) Karst.). *Acta Forestalia Fennica* 136, 1–24.

Jalaluddin, M. (1967) Studies on *Rhizina undulata*: I. Mycelial growth and ascospore germination. *Transactions of the British Mycological Society* 50, 449–459.

Kearney, M. and Porter, W. (2009) Mechanistic niche modelling: combining physiological and spatial data to predict species' ranges. *Ecology Letters* 12, 334–350.

Király, L., Barna, B. and Király, Z. (2007) Plant resistance to pathogen infection: forms and mechanisms of innate and acquired resistance. *Journal of Phytopathology* 155, 385–396.

Kleczkowski, A., Gilligan, C.A. and Bailey, D.J. (1997) Scaling and spatial dynamics in plant–pathogen systems: from individuals to populations. *Biological Sciences* 264, 979–984.

Koricheva, J., Vehvilainen, H., Riihimaki, J., Ruohomaki, K., Kaitaniemi, P. and Ranta, H. (2006) Diversification of tree stands as a means to manage pests and diseases in boreal forests: myth or reality? *Canadian Journal of Forest Research* 36, 324–336.

Lang, K.J. (1987) *Dothistroma pini* on young Norway spruce (*Picea abies*). *European Journal of Forest Pathology* 17, 316–317.

Lei, Z.P., Jin, J.R and Wang, C.W. (1995) Antagonism between ectomycorrhizal fungi and plant pathogens. In: Brundett, M., Dell, B., Malajczuk, N. and Gong, M.Q. (eds) *Mycorrhizas for Plantation Forestry in Asia. ACIAR Proceedings Series* 62, 77–81.

Leonard, K.J. and Czochor, R.J. (1980) Theory of genetic interactions among populations of plants and their pathogens. *Annual Review of Phytopathology* 18, 237–258.

Levine, J.M. and D'Antonio, C.M. (2003) Forecasting biological invasions with increasing international trade. *Conservation Biology* 17, 322–326.

Lopez-Zamora, I., Bliss, C., Jokela, E.J., Comerford, N.B., Grunwald, S., Barnard, E. and Vasquez, G.M. (2007) Spatial relationships between nitrogen status and pitch canker disease in slash pine planted adjacent to a poultry operation. *Environmental Pollution* 147, 101–111.

Lygis, V., Vasiliauskas, R. and Stenlid, J. (2005) Clonality in the postfire root rot ascomycete *Rhizina undulata*. *Mycologia* 97, 788–792.

Madden, L.V., Hughes, G. and van den Bosch, F. (2007) *The Study of Plant Disease Epidemics*. APS Press, St Paul, Minnesota.

Manion, P.D. (1981) *Tree Disease Concepts*. Prentice Hall, Englewood Cliffs, New Jersey.

Manion, P.D. and Griffin, D.H. (2001) Large landscape scale analysis of tree death in the Adirondack Park, New York. *Forest Science* 47, 542–549.

Matson, P.A. and Waring, R.H. (1984) Effects of nutrient and light limitation on mountain hemlock: susceptibility to laminated root rot. *Ecology* 65, 1517–1524.

Mattila, U. (2005) Probability models for pine twisting rust (*Melampsora pinitorqua*) damage in Scots pine (*Pinus sylvestris*) stands in Finland. *Forest Pathology* 35, 9–21.

Mattila, U., Jalkanen, R. and Nikula, A. (2001) The effects of forest structure and site characteristics on probability of pine twisting rust damage in young Scots pine stands. *Forest Ecology and Management* 142, 89–97.

McNew, L. (1960) The nature, origin, and evolution of parasitism. In: Horsfall, J.G. and Dimond, A.E. (eds) *Plant Pathology – An Advanced Treatise*. Academic Press, New York, pp. 19–69.

Meadows, I.M., Zwart, D.C., Jeffers, S.N., Waldrop, T.A. and Bridges, W.C. (2011) Effects of fuel reduction treatments on incidence of *Phytophthora* species in soil of a southern Appalachian mountain forest. *Plant Disease* 95, 811–820.

Menges, E.S. and Loucks, O.L. (1984) Modeling a disease-caused patch disturbance: oak wilt in the midwestern United States. *Ecology* 65, 487–498.

Morrison, D.J., Wallis, G.W. and Weir, L.C. (1988) *Control of Armillaria and Phellinus Root Diseases: 20-year Results from the Skimikin Stump Removal Experiment*. Information Report BC-X-302, Forestry Canada, Pacific and Yukon Region, Victoria, BC, Canada.

Mueller-Dombois, D. (1992) A natural dieback theory, cohort senescence as an alternative to the decline disease theory. In: Manion, P.D. and Lachance, D. (eds) *Forest Decline Concepts*. APS Press, St Paul, Minnesota, pp. 26–37.

Mundt, C.C., Sackett, K.E., Wallace, L.D., Cowger, C. and Dudley, J.P. (2009) Long-distance dispersal and accelerating waves of disease: empirical relationships. *American Naturalist* 173, 456–466.

Oliva, J. and Stenlid, J. (2011) Validation of the Rotstand model for simulating *Heterobasidion annosum* root rot in *Picea abies* stands. *Forest Ecology and Management* 261, 1841–1851.

Oliva, J., Thor, M. and Stenlid, J. (2010) Long term effects of mechanized stump treatment against *Heterobasidion annosum s.l.* root rot in *Picea abies*. *Canadian Journal of Forest Research* 40, 1020–1033.

Olson, A. and Stenlid, J. (2002) Pathogenic fungal species hybrids infecting plants. *Microbes and Infection* 4, 1353–1359.

Ostry, M.E., Venette, R.C. and Juzwik, J. (2011) Decline as a disease category: is it helpful? *Phytopathology* 101, 404–409.

Pautasso, M., Holdenrieder, O. and Stenlid, J. (2005) Susceptibility to fungal pathogens of forests differing in tree diversity. In: Scherer-Lorenzen, M., Körner, C. and Schulze, E.D. (eds) *Forest Diversity and Function: Temperate and Boreal Systems. Ecological Studies* 176, Springer, Berlin, pp. 263–289.

Pukkala, T., Möykkynen, T., Thor, M., Rönnberg, J. and Stenlid, J. (2005) Modeling infection and spread of *Heterobasidion annosum* in even-aged Fennoscandian conifer stands. *Canadian Journal of Forest Research* 35, 74–84.

Redfern, D.B. and Stenlid, J. (1998) Spore dispersal and infection. In: Woodward, S., Stenlid, J., Karjalainen, R. and Hüttermann, A. (eds) Heterobasidion annosum: *Biology, Ecology, Impact and Control*. CAB International, Wallingford, UK, pp. 105–124.

Roll-Hansen, F., Roll-Hansen, H. and Skroppa, T. (1992) *Gremmeniella abietina, Phacidium infestans*, and other causes of damage in alpine, young pine plantations in Norway. *European Journal of Forest Pathology* 22, 77–94.

Scholthof, K.B.G. (2007) The disease triangle: pathogens, the environment and society. *Nature Reviews Microbiology* 5, 152–156.

Schumann, G.L. (1991) *Plant Diseases: Their Biology and Social Impact*. APS Press, St Paul, Minnesota.

Seem, R.C. (1984) Disease incidence and severity relationships. *Annual Review of Phytopathology* 22, 133–150.

Shearer, B.L., Dillon, M.J., Kinal, J. and Buehrig, R.M. (2010) Temporal and spatial soil inoculum dynamics following *Phytophthora cinnamomi* invasion of *Banksia* woodland and *Eucalyptus marginata* forest biomes of south-western Australia. *Australasian Plant Pathology* 39, 293–311.

Sinclair, W.A. and Hudler, G.W. (1988) Tree declines: four concepts of causality. *Journal of Arboriculture* 14, 29–35.

Sinclair, W.A. and Lyon, H.H. (2005) *Diseases of Trees and Shrubs*. Cornell University Press, New York.

Skre, O. (1979) The regional distribution of vascular plants in Scandinavia with requirements for high summer temperatures. *Norwegian Journal of Botany* 26, 295–318.

Stenlid, J. and Redfern, D.B. (1998) Spread within the tree and stand. In: Woodward, S., Stenlid, J., Karjalainen, R. and Hüttermann, A. (eds) Heterobasidion annosum: *Biology, Ecology, Impact and Control*. CAB International, Wallingford, UK, pp. 125–141.

Storer, A.J., Gordon, T.R. and Clark, S.L. (1998) Association of the pitch canker fungus, *Fusarium subglutinans* f.sp. *pini*, with Monterey pine seeds and seedlings in California. *Plant Pathology* 47, 649–656.

Sutherst, R.W., Maywald, G.F. and Kriticos, D.J. (2007) CLIMEX version 3: user's guide. Hearne Scientific Software Pty Ltd., Melbourne, Australia. Available from: www.hearne.com.au (accessed 19 November 2012).

Tainter, F.H. and Baker, F.A. (1996) *Principles of Forest Pathology*. John Wiley, Hoboken, New Jersey.

Thomsen, I.M. (2009) Precipitation and temperature as factors in *Gremmeniella abietina* epidemics. *Forest Pathology* 39, 56–72.

Thor, M., Ståhl, G. and Stenlid, J. (2005) Modelling root rot incidence in Sweden using tree, site and stand variables. *Scandinavian Journal of Forest Research* 20, 165–176.

Thor, M., Arlinger, J.D. and Stenlid, J. (2006) *Heterobasidion annosum* root rot in *Picea abies*: modelling economic outcomes of stump treatment in Scandinavian coniferous forests. *Scandinavian Journal of Forest Research* 21, 414–423.

Uotila, A. (1990) Infection of pruning wounds in Scots pine by *Phacidium coniferarum* and selection of pruning season. *Acta Forestalia Fennica* 215, 1–36.

van der Plank, J.E. (1963) *Plant Diseases: Epidemics and Control*. Academic Press, New York.

van der Plank, J.E. (1975) *Principles of Plant Infection*. Academic Press, New York.

van Maanen, A. and Xu, X.-M. (2003) Modelling plant disease epidemics. *European Journal of Plant Pathology* 109, 669–682.

Vasiliauskas, R., Stenlid, J. and Johansson, M. (1996) Fungi in bark peeling wounds of *Picea abies* in central Sweden. *European Journal of Forest Pathology* 26, 285–296.

Watt, M.S., Ganley, R.J., Kriticos, D.J. and Manning, L.K. (2011a) Dothistroma needle blight and pitch canker: the current and future potential distribution of two important diseases of *Pinus* species. *Canadian Journal of Forest Research* 41, 412–424.

Watt, M.S., Palmer, D.J. and Bulman, L.S. (2011b) Predicting the severity of *Dothistroma* on *Pinus radiata* under current climate in New Zealand. *Forest Ecology and Management* 261, 1792–1798.

Wingfield, M.J., Slippers, B., Roux, J. and Wingfield, B.D. (2001) Worldwide movement of exotic forest fungi, especially in the tropics and the southern hemisphere. *BioScience* 51, 134–140.

Woods, A., Coates, K.D. and Hamann, A. (2005) Is an unprecedented Dothistroma needle blight epidemic related to climate change? *BioScience* 55, 761–769.

Woodward, S., Stenlid, J., Karjalainen, R. and Hüttermann, A. (eds) (1998) Heterobasidion annosum: *Biology, Ecology, Impact and Control*. CAB International, Wallingford, UK.

2 General Strategies of Forest Disease Management

Robert L. Edmonds*
University of Washington, Seattle, USA

2.1 Introduction to Forest Diseases

Pathogens occur in all the world's forests: tropical, temperate and boreal. All tree species are susceptible to pathogens to varying degrees, and all tree components, such as foliage, twigs, branches, stems and roots can be diseased (Tainter and Baker, 1996). However, not all tree species are susceptible to the same pathogens and there is a considerable degree of between and within host variation in resistance. In natural ecosystems, pathogens have typically co-evolved with their hosts and host resistance mechanisms have developed. In contrast, introduced pathogens have not evolved with their new hosts and so new host trees are particularly susceptible to exotic pathogens. Introduced pathogens have caused millions of tree deaths around the world and there is always the threat of new introductions. Human influences, including forest management, can also cause native pathogens to be more aggressive (Slaughter and Rizzo, 1999).

Tree diseases cause considerable mortality, growth loss and disturbance in the world's forest ecosystems. In unmanaged pristine forests they strongly influence forest structure, including the presence of ecologically important coarse woody debris (logs and standing dead trees), tree species succession, fire and nutrient cycling. In managed forests, tree diseases can reduce timber yield and biomass for fuel as well as influencing other ecosystem services such as water, wildlife, biodiversity, aesthetics and recreation. Thus there is a need to manage diseases. Diseases are not the only disturbances in forests though. Insects, fire, drought, flooding, erosion, air pollution and animal damage are also present and commonly interact with diseases to change the landscape (Edmonds *et al.*, 2011). Moreover, future climate change is expected to influence these disturbances as well (Dukes *et al.*, 2009; Woods *et al.*, 2010; Sturrock *et al.*, 2011). Therefore, forest disease management should not focus just on pathogens if we want healthy and sustainable forests in the future.

Humans have attempted to manage plant diseases for centuries with the hope of controlling or even eradicating them, at least in the case of agricultural diseases (Whetzel, 1929). Following in the agricultural tradition, forest pathologists initially emphasized the diseases of the important timber tree species, and expressed the damage found in terms of money or timber volume lost to mortality or growth reduction. Forest diseases were rarely considered in the context of whole

* E-mail: bobe@u.washington.edu

ecosystems, although considerable information exists about the ecology of many forest pathogens, e.g. *Heterobasidion* and *Armillaria* spp., and the effects of environmental factors on plant diseases. Furthermore, information about the ecology of the major tree species has rarely been related to the ecology of pathogens. Forests, however, are different from agricultural crops and fall along a gradient from natural forests with little or no disease management, where natural processes are allowed to proceed, to intensively managed plantations. Forest disease organisms can play both negative and beneficial roles. If we want to understand their full impact we must study diseases in the ecosystem context. Only in this way can we make intelligent management decisions for the future.

2.2 Disease Detection, Surveying and Monitoring

The first step in disease management is disease detection, followed by surveying and monitoring. Disease detection is practised not only at the individual tree or stand level but now commonly also at the landscape level using aerial photography, satellite imagery and lidar (light detection and ranging). Commonly, annual surveys of forest conditions from light aircraft are used to describe changes in disease conditions from year to year.

In addition, the recent expansion of molecular techniques has greatly improved our knowledge of forest pathosystems and has given forest pathologists new insights into the management of forest pathogens (Lundquist and Hamelin, 2005). Now, a precise identification of pathogen species, races and individuals can be made even without isolation by using PCR-based molecular tools (Kim *et al.*, 2005; Glaeser and Lindner, 2011). Pathogens can be identified directly by extracting DNA from soil, tree boles, branches and leaves without the need for culturing and microscopic examination. We can now differentiate between pathogenic and non-pathogenic species or races. Molecular techniques are particularly important in the identification of newly introduced pathogens, such as *Phytophthora ramorum* Werres, De Cock & Man in't Veld, the cause of sudden oak death (SOD), which is a problem in North America and Europe (Kliejunas, 2010). Breeding programmes for host disease resistance have also benefited from the use of molecular techniques to examine the structure of resistant host and pathogen populations and determine the genes for disease resistance.

Other new tools, such as global positioning systems (GPS) and geographic information systems (GIS) are now used in forest pathology as well. The precise location of a sample can be determined using GPS, and a combination of molecular marker technology and GIS can be used to manage diseases at the landscape level. The geographic distribution and genetic structure of pathogens and host tree populations can readily be related to site attributes including soil properties, fire history, insect activity, temperature and moisture, topography and management practices.

2.3 Importance of Knowledge of Infection Biology for Disease Management

Causative agents of forest diseases include viruses, phytoplasmas, bacteria, fungi, fungal-like organisms, nematodes and protozoans (Tainter and Baker, 1996; Agrios, 2005). Fungi cause the majority of forest diseases (Edmonds *et al.*, 2011). The major concern with these organisms lies with their ability to spread and infect new hosts. Disease-causing organisms can be dispersed in the soil, through water movement and in the air, with airborne dispersal being the most rapid means of spread at both local and long distance scales. Propagules can also spread long distances as a result of human movement (via infected plants or seeds), through transportation (via aeroplanes, boats, cars and trucks, or even on feet), and by birds, animals and insects. The spread of disease organisms, however, cannot be understood simply by understanding dispersion processes. It is also necessary to study the production, release, dispersal, deposition and viability of inoculum, as well as the effects of environmental factors on these processes, in order to fully understand infection biology (Edmonds, 1996). Many viable spores may be released, but unless they are able to arrive alive and the conditions are right to enable the infection of a susceptible host, disease will not spread.

2.4 Epidemiological Aspects for Disease Management

2.4.1 Epidemiology

To manage plant diseases one has to understand epidemiology or epiphytology. Technically, epidemiology, or the study of epidemics and the factors that influence them, applies only to humans as 'epidemic' is Greek for 'on the people'; in contrast, 'epiphytotic' is Greek for 'on the plants'. Nevertheless, the two terms are commonly used interchangeably. Epidemiology is the basis for the development of pathogen control strategies (Agrios, 2005). The essence of epidemiology is the disease triangle (susceptible host plants, a virulent pathogen and a favourable environment), but in the study of plant disease epidemics this is generally expanded to include time and human effects. Time has a large effect on disease development because factors such as the duration and frequency of favourable temperature and moisture conditions vary during the year. The pathogen infection cycle is also a function of time. The specific host factors that affect the development of epidemics are: host susceptibility and level of genetic resistance, degree of uniformity of host plants, type of crop, plant density, planting time and age of host plants. Pathogen factors that affect epidemic development are: levels of virulence, quantity of inoculum near hosts, type of reproduction of the pathogen, pathogen ecology and mode of pathogen spread. Temperature and moisture are the most important environmental factors that influence the development of epidemics. Human activities of influence include site selection and preparation, propagation techniques, cultural practices, disease control measures and, of course, the introduction of new pathogens. In contrast to agricultural crops, most forest disease epidemics, such as Dutch elm disease, chestnut blight and white pine blister rust, take years to develop. However, in forest nurseries epidemics can develop in a growing season.

2.4.2 Use of models in managing forest diseases and developing risk maps

Epidemics are dynamic processes and managers are particularly interested in the conditions that bring about epidemics, their rate of increase and their demise. Modelling is used to predict the rate of spread of native and introduced diseases and the effects of climate change, as well as to determine strategies to be used by managers (Tainter and Baker, 1996; Rosso and Hansen, 2003; Meentemeyer *et al.*, 2008; Dukes *et al.*, 2009). There are several approaches to modelling diseases: simple and multiple regression models; simulation models that take into account site, environmental and silvicultural factors; and spatially explicit models that use satellite imagery, GPS coordinates and sophisticated mathematical algorithms to model the risk of spread of diseases – particularly for invasive pathogens like *P. ramorum* (Kelley *et al.*, 2007).

Regression models have been widely used to predict disease spread, the amount of decay in trees and the spread of dwarf mistletoes (*Arceuthobium* spp.). An example of a multiple regression model is the prediction of the spread of western dwarf mistletoe (*Arceuthobium campylopodum* Engelm.) in ponderosa pine (*Pinus ponderosa* Lawson & C. Lawson) in Arizona and California, which is based on the six-class dwarf mistletoe rating of infected trees, stand density and infection age (Dixon and Hawksworth, 1979). Regression models have also been developed for the management of foliage diseases such as Swiss needle cast (Shaw *et al.*, 2011). For example, Rosso and Hansen (2003) found that needle cast severity in Douglas-fir (*Pseudotsuga menziesii* (Mirb.) Franco) stands in coastal Oregon was related to fog occurrence, precipitation, temperature, elevation and slope aspect. Manter *et al.* (2005) further refined these findings and found that winter temperature, together with spring and summer precipitation, were key independent variables.

Simulation models are more universally applicable than regression models and have more utility for ecosystem management. An example of a simulation model is the Western Root Disease Model (WRDM; Frankel, 1998) (Fig. 2.1), which is a sub-model of the tree growth model Forest Vegetation Simulator (FVS) for the western USA. Input to the model is the type of root disease (laminated root rot, annosus root and butt rot, or Armillaria root disease), the proportion of stand area that is inside the root disease centre and the density of infected trees. Data from aerial photographs, stand inventories and GIS can be utilized, and model outputs include graphs, charts, spreadsheets and landscape visualizations showing changes in trees and root disease through time. The Landscape Visualization System (LVS) of Oliver and McCarter (1995) is linked to the WRDM. This model can be run under a number of management scenarios that allow managers to make decisions on root disease treatment, and has been used to assess root disease impacts in mixed conifer stands in southwest Oregon (Maffei *et al.*, 2008). It has also been used to analyse how root disease affects stand structure, mortality, fuel accumulation, and the development and retention of key structures for northern spotted owl

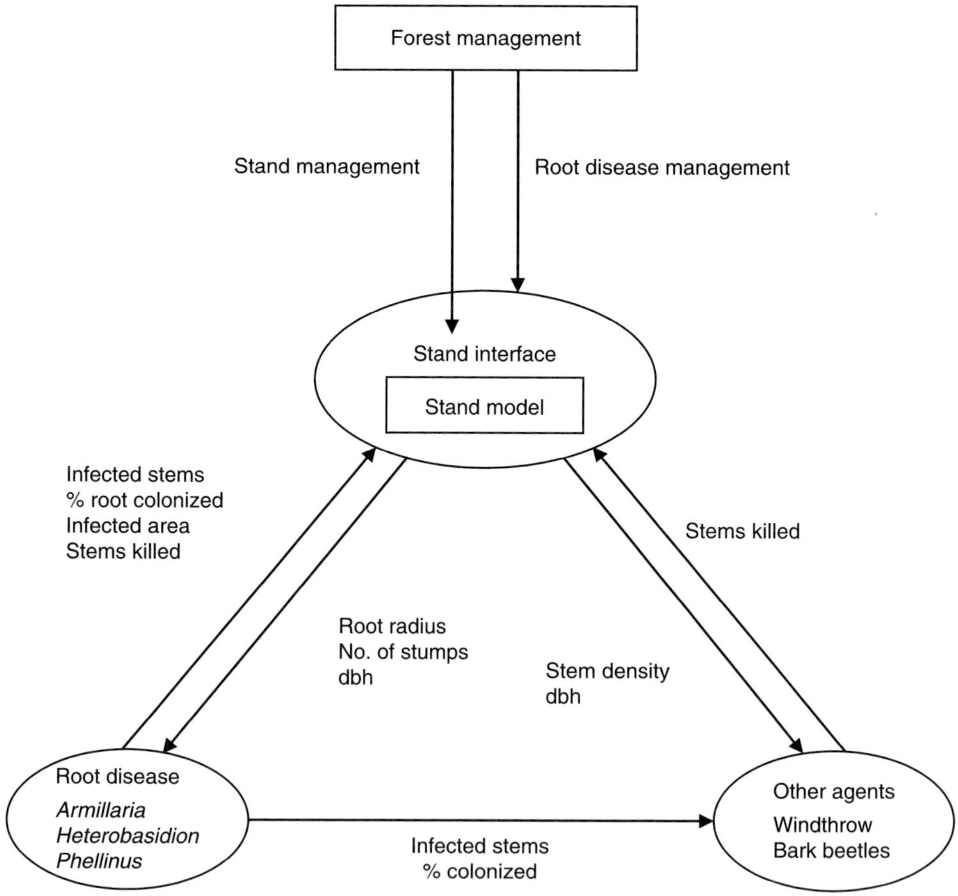

Fig. 2.1. Structure of the Forest Vegetation Simulator (FVS) stand growth and yield model for the interior western USA and its two sub-models: the root disease model (Western Root Disease Model or WRDM); and the 'Other agents' model (which covers windthrow and bark beetles) (Frankel, 1998). dbh, diameter at breast height.

habitat under various management scenarios in late successional reserves.

Other examples of simulation models involve prediction of the spread of *Phellinus sulphurascens* Pilát, *Armillaria* spp. and *Onnia tomentosa* (Fr.) P. Karst. in coastal areas in western Canada (Peet and Hunt, 2005), modelling the infection and spread of *Heterobasidion annosum* (Fr.) Bref. *s.l.* in managed Norway spruce (*Picea abies* (L.) Karst.) and Scots pine (*Pinus sylvestris* L.) stands in Finland, Sweden and Norway (Pukkala *et al.*, 2005), and the prediction of the life history, ballistics and contagion of dwarf mistletoes in the western USA (Robinson and Geils, 2006). The latter model is linked to two individual tree growth models, FVS and Tree and Stand Simulator (TASS), and provides a practical tool for assessing the long-term effects of disease and forest management in dwarf mistletoe-infected stands.

There is also considerable interest in modelling the spread risk of invasive disease organisms such as *P. ramorum*. Kelley *et al.* (2007) used topographic and climatic data along with digital host vegetation data to predict the spread of *P. ramorum* within California and to other areas where it currently does not occur in the continental USA. They used a new approach involving five different spatial environmental niche models: Expert-driven Rule-based, Logistic Regression, Classification and Regression Trees, Genetic Algorithms and Support Vector Machines (SVM). Final risk maps were produced based on model agreement.

Rule-based models have used research data and expert input rather than statistical inference to determine the importance of predictor variables. All models have predicted some risk of SOD in coastal California, Oregon and Washington, in the northern foothills of the Sierra Nevada Mountains in California, and in parts of Oklahoma, Arkansas, Tennessee, Kentucky, Mississippi, Alabama, Georgia, South Carolina, North Carolina, Virginia, Delaware and Maryland. Total precipitation, frost days, and average maximum and minimum temperatures were the most important predictors of the niche for *P. ramorum*. A process model (CLIMEX) has been used to predict the potential range for Dothistroma needle blight for susceptible pine and other tree species throughout the world (Watt *et al.*, 2009).

In addition to these modelling techniques, economic models can also be used for decision making. For example, an economic analysis was done using present net worth to predict the return on investment in stumping to control *P. sulphurascens* in Douglas-fir stands (Russell *et al.*, 1986). Costs of site preparation and stump removal, planting, thinning, vegetation control and fertilization were included, and volume/acre was predicted with and without laminated root rot caused by *P. sulphurascens*.

To adequately understand and manage forest health problems it is necessary to determine where on the landscape the highest risk areas of unhealthy forests will be. Thus, the US Department of Agriculture (USDA) Forest Service and partners have developed a national risk assessment process that maps the potential future risk of mortality due to diseases and insects (Krist *et al.*, 2007). Here, risk is defined as the expectation that 25% or more of the standing live biomass of trees greater than 2.5 cm in diameter will die over the next 15 years as a result of insect and disease activity. In the assessment conducted in 2006, it was calculated that more than 23 million ha of the 297 million ha of forest land in the USA are at risk. Most of this was assigned to 11 risk agents, including western conifer bark beetles (mountain and western pine beetle, pine and fir engravers, and Douglas-fir and spruce beetles), southern pine beetle, gypsy moth, oak decline on red oaks, hardwood decline and root diseases. The National Insect and Disease Risk Map (NIDRM) represents an integration of 188 individual risk models. In addition, an early warning system for forest health threats in the USA has been developed.

2.5 Management Strategies and Tactics

There are seven general strategies or tactics for managing forest diseases: (i) avoidance, (ii) exclusion, (iii) eradication, (iv) protection, (v) resistance, (vi) therapy and (vii) integrated disease management (Fig. 2.2).

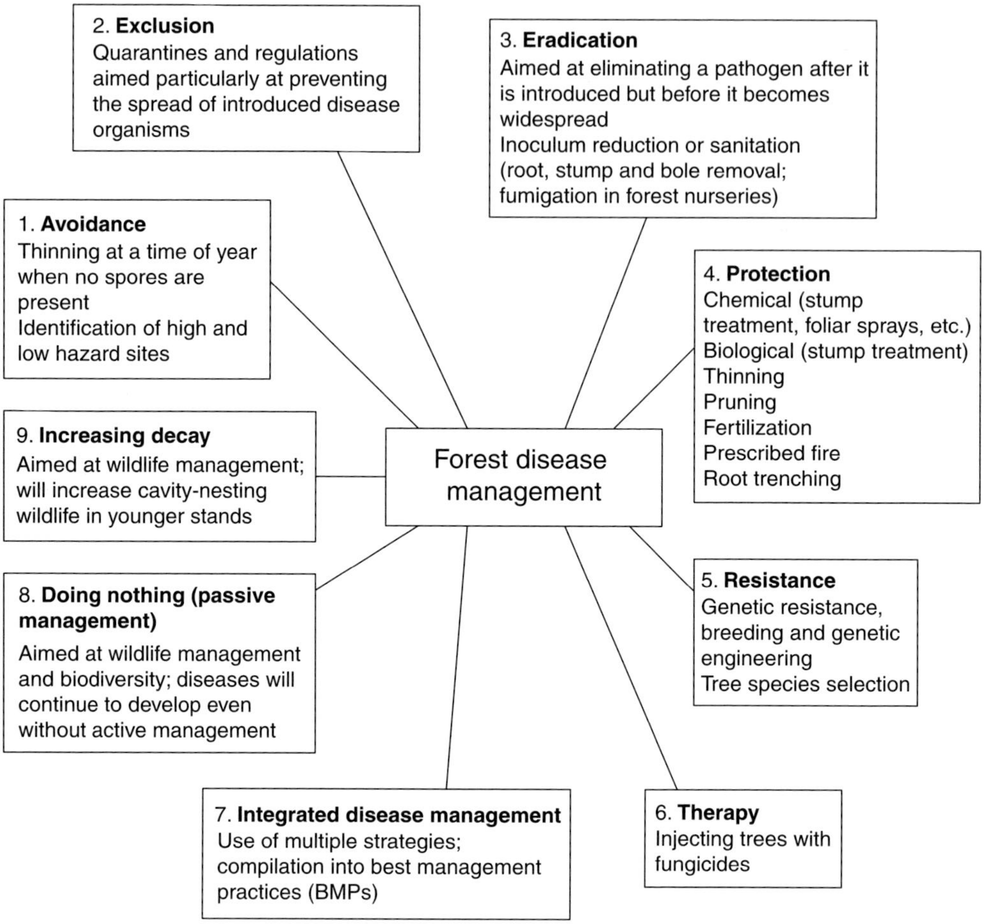

Fig. 2.2. Strategies or tools available for the management of forest diseases.

These strategies and tactics are discussed below relative to forest disease management. Many forest managers though may decide not to manage diseases actively at all. This is termed passive management (Fig. 2.2), and it may be appropriate if the impact of a disease is anticipated to be low or if creation of wildlife habitat is a favoured management objective. For example, dead and dying trees in root disease pockets caused by *P. sulphurascens* may provide suitable habitats for cavity-nesting birds (e.g. woodpeckers), mammals and even reptiles. During harvesting operations, leaving small patches of forest that are infected but in which the fungus is spreading slowly, promotes biodiversity and wildlife populations (Thies and Sturrock, 1995). Passive management might be an option if land is to be managed primarily for wildlife and biodiversity, and habitat is needed for cavity-nesting birds (Bull *et al.*, 1997), although land owners do need to recognize that diseases will continue to develop even without active management. When infection centres are large and expanding rapidly, then site productivity can be greatly impacted for many years. If a landowner owns only a small, heavily infected area, then active management to reduce inoculum and change species is more appropriate.

Because many forests are grown on relatively short rotations, many stem decay fungi do not have time to develop, which is good for timber production. However, if wildlife protection is an ecosystem management goal, then increasing decay for cavity-nesting birds may have to be practised (Fig. 2.2). Management guidelines for federal lands in the western USA now dictate that a certain number of standing dead and green trees be left for wildlife after logging. Some of the remaining green trees are inoculated with decay fungi to enhance habitat for woodpeckers, which typically excavate cavities in dead trees in areas softened by decay (Filip *et al.*, 2004). Because they remain alive for a long time, inoculated green trees are safer during logging operations than standing dead trees. Also, because they remain alive, they provide wildlife habitat for longer time periods than dead trees, which soon fall. The decay fungi that have been suggested for inoculation are: *Porodaedalea pini* (Brot.) Murrill, *Fomitopsis cajanderi* (P. Karst.) Kotl. & Pouzar and *F. pinicola* (Sw.) P. Karst. for all conifers; *Hericium abietis* (Weir ex Hubert) K.A. Harrison for western hemlock (*Tsuga heterophylla* (Raf.) Sarg.); *Phellinus hartigii* (Allesch. & Schnabl) Pat. for western hemlock and Douglas-fir; and *Wolfiporia extensa* (Peck) Ginns for larch (*Larix* spp.).

2.5.1 Avoidance

Avoidance aims to avoid conditions or environments that are conducive to the disease. This may be as simple as thinning at times of the year when spores are not present or will not be able to infect the host. For example, high and low temperatures may prevent spore production and/or stump infection by *H. annosum s.l.* (Woodward *et al.*, 1998). In the south-eastern USA, pine thinning can be conducted in the warm summer months because stump temperatures are high and *H. annosum s.l.* is outcompeted by another fungus, *Trichoderma* spp., which is favoured by high temperatures (Driver and Ginns, 1969). Conversely, in Scandinavia, thinning can be practised during cold winter months when spores are not released and the possibility of stump infection is very low.

The identification of high and low hazard sites has been employed for white pine blister rust (*Cronartium ribicola* J.C. Fisch) and root diseases. In Wisconsin, areas close to lake shores have low hazard because airflows carrying rust basidiospores from *Ribes* spp. go below the foliage of the white pines (*Pinus* spp.). Areas on ridges between lakes, in contrast, are designated as high-hazard areas. Here, resistant white pines should be planted.

Soil characteristics have been used to estimate hazard to annosus root and butt rots and littleleaf disease. In the south-eastern USA, incidence of *H. annosum s.l.* is highest in deep sandy soils with low organic matter. In the UK, evaluation of site hazard for controlling *H. annosum s.l.* was adopted as Forestry Commission policy in 2007 (Redfern *et al.*, 2010). Littleleaf disease, caused by *Phytophthora cinnamomi* Rands, is worse on eroded soils with hardpans and poor drainage (Hansen and Lewis, 1997).

P. sulphurascens tends to be more prevalent on mid and upper slopes (Kastner *et al.*, 1994). Sites with high soil pH and low soil aluminium may favour *Armillaria ostoyae* (Romagn.) Herink (Browning and Edmonds, 1993). In the northern Rocky Mountains, it appears that the prevalence of *Armillaria* spp. can be predicted by habitat type (McDonald *et al.*, 1987). Hot-dry, cold-dry and frost-pocket habitat types appeared not to support *Armillaria* spp. Incidence was higher in cool-dry/moderate, cool-moderate/wet and warm-moderate/wet habitat types. In undisturbed sites, the probability of infection by *Armillaria* spp. decreased as productivity increased. High-productivity sites that had been disturbed, however, had more *Armillaria* spp. than undisturbed high-productivity plots. Recently burned sites may favour *Rhizina undulata* Fr. (Tainter and Baker, 1996).

2.5.2 Exclusion

Exclusion involves quarantine regulations and best management practices (BMPs) to prevent pathogen introduction. Quarantines restrain the transport of materials that may harbour pests and involve legal regulations and penalties. Most countries have regulations that impose restrictions on live trees and forest products being imported and on the movement of disease organisms. There are many concerns about the introduction of new diseases in Europe (Stenlid *et al.*, 2011), the USA and other countries, and organizations have been established to counter new introductions. For example, the European and Mediterranean Plant Protection Organization (EPPO), an intergovernmental organization with 50 members, is responsible for European cooperation in plant health, and the International Plant Protection Convention (IPPC) represents an international agreement on plant health with 177 current signatories.

In the USA, the Animal and Plant Health Inspection Service (APHIS) of the USDA has a mission that includes protecting natural resources. There is considerable concern about the introduction of new diseases and pests on logs imported from Asia, New Zealand and Chile (Filip and Morrell, 1996). Relying solely on inspection at the time of importation is not acceptable for addressing new plant disease threats. A detailed pest risk assessment was requested by APHIS in the 1990s and proposed regulations were developed with specific importation requirements.

Quarantine is most effective against exotic introductions, particularly for root diseases caused by *Phytophthora* spp., even after pathogen introduction. Because *Phytophthora* spp. spread by swimming spores in the soil, it is important that inoculum is not spread in soil particles on boots, equipment and machinery. A simple washing or treatment in Clorox will reduce or eliminate the potential for spread. In areas with susceptible ornamental trees, surface water flows should be managed to prevent spore spread and the creation of wet soil conditions. In Western Australia, quarantine and hygiene procedures are practised because *P. cinnamomi* has the capacity to build up to a very high population level from just a few propagules. Marked reductions in the rate of extension of the disease can be achieved with even moderate attempts at hygiene. For example, in the past, gravel from dead and dying forest areas was used for roads. When this practice was stopped, there was a marked reduction in disease extension. Contamination from forest nurseries also was reduced by the imposition of stringent hygiene measures involving fungicidal treatment of vehicles, tools, boots, and so on, that could carry contaminated soil from nurseries to the forest. The same principles apply for Port Orford cedar (*Chamaecyparis lawsoniana* (A. Murray bis) Parl.) in the western USA.

In the USA, *P. ramorum* has been managed by regulations and quarantine enforced by APHIS. Movement of plants from California to other states is banned unless they have been inspected and treated (Kliejunas, 2010). There are even restrictions on the movement of host material at the county level in California, where 14 counties are currently under quarantine. Movement from national forests is restricted as well. For example, the policy on the Mendocino National Forest is that firewood or other host material can be moved into and between any of the quarantine counties, but it cannot be moved out of quarantine counties. Canada also prohibits the import of nursery stock, logs and mulch from areas where *P. ramorum* has been found.

2.5.3 Eradication

Eradication is aimed at reducing or eliminating a pathogen after it is introduced, but before it becomes widespread. In actual practice, pathogens are never fully eliminated, so the focus tends to be on inoculum reduction and sanitation.

Inoculum reduction or sanitation

Inoculum reduction or sanitation can be effective for the management of a variety of forest disease types, including vascular wilt diseases, such as Dutch elm disease and oak wilt, and dwarf mistletoes, foliage diseases, cankers and root diseases. This can be accomplished by chemical treatment or by the physical removal of infected substrates. Perhaps the most cited example of sanitation is the removal and burning of insect brood material to prevent the spread of Dutch elm disease (caused by *Ophiostoma* spp.) (Tainter and Baker, 1996) and of wood infected with the oak wilt fungus, *Ceratocystis fagacearum* (Bretz) J. Hunt, (Koch *et al*., 2010). Movement of infected firewood is strongly discouraged (Koch *et al*., 2010), and public education on the role of firewood movement in the spread of insects and diseases is now practised in many states of the USA. In southwest Oregon, state and federal agencies have attempted to eradicate *P. ramorum* from infested sites by cutting and burning all infected host materials, creating a buffer of apparently uninfected plants, since 2001 (Václavík *et al*., 2010).

Root and stump removal has been utilized for structural root diseases such as Armillaria root disease, laminated root rot and annosus root and butt rot (Morrison *et al*., 1991; Thies and Sturrock, 1995; Vasaitis *et al*., 2008). There are three approaches to inoculum reduction: stump removal, tree pushing and fumigation. In Canada, tree pushing using a backhoe with a thumb to harvest whole trees has also been employed for the control of *P. sulphurascens* (Thies and Sturrock, 1995). This tends to remove most of the infected root material and no stumps remain to harbour inoculum. Fumigation is mentioned later on in this section in relation to the chemical treatment of stumps, and the latter is also discussed in Section 2.5.4 under '*Chemical control*'. Soil trenching to reduce inoculum and the spread of *Armillaria* spp. and *Phytophthora* spp. in urban areas or orchards is possible, but this is generally impractical in forested areas.

Most of the research on stump removal has been in the coniferous forests of western North America, Europe and New Zealand (Vasaitis *et al*., 2008). In the western USA and Canada, stump removal has been utilized to aggressively manage killing strains of *A. ostoyae*, *P. sulphurascens*, and *O. tomentosa* (Morrison *et al*., 1991; Vasaitis *et al*., 2008). Stump removal has also been effectively used for the control of *H. annosum s.l.* in Europe (Woodward *et al*., 1998; Vasaitis *et al*., 2008).

Large tractors with toothed blades or backhoes are typically used to remove stumps. It is generally not necessary to completely remove them from the site; just pushing the stumps out of the ground and upturning them suffices. This dries out the wood and reduces the inoculum over time, especially for the fungi that tend to spread mostly by vegetative means. Leaving upturned stumps may reduce the inoculum of *A. ostoyae*, *P. sulphurascens* and *O. tomentosa*, but may not be effective for *H. annosum s.l.* because spores can be produced from fruiting bodies on the stumps (Vasaitis *et al*., 2008). Thus, collecting and removing *H. annosum s.l.* infected stumps and roots from harvested areas is advisable. Stumps removed for root disease treatment could be used for biofuel production along with healthy stumps. Bioenergy production and the sanitation of forest land to clear root disease could go hand in hand.

Vasaitis *et al*. (2008) concluded that stump removal from clear-felled forest areas has, in most cases, been successful in controlling root disease as well as being economically feasible. It has resulted in: (i) reduction of root rot in the next rotation; (ii) improved seedling establishment; and (iii) increased tree productivity. Disturbances caused by stumping (including soil disturbance, erosion and compaction) have been generally acceptable, although this does largely depend on soil type. Stumping on moist, high bulk density soils results in severe compaction and poor seedling performance, and is best practised on flat land or on relatively mild slopes with light (relatively low clay content) soils.

Clear-cutting (or clear-felling) will reduce the incidence of obligate parasites such as dwarf mistletoes, particularly if all infected trees are removed, including unmerchantable trees (Hawksworth and Wiens, 1996). This practice has reduced the incidence of western hemlock dwarf mistletoe (*Arceuthobium tsugense* (Rosend.) G.N. Jones) in coastal Oregon and Washington, because the spread of dwarf mistletoes is mostly local and slow. Clear-cutting and short rotations can be used to manage decay and canker fungi that take a considerable time to develop, but rust diseases are not readily managed by clear-cutting. In this case, spores of the rust fungi are capable of spreading long distances, and in the case of white pine blister rust and other long-cycle rusts, populations of the alternate hosts are not reduced by clear-cutting. Also, clear-cutting alone will also not reduce the incidence of root diseases because the fungi involved, such as *Armillaria mellea* (Vahl ex Fr.) P. Kumm. *s.l.*, *H. annosum s.l.* and *P. sulphurascens*, can survive as saprophytes in stumps and roots, and have the capacity for attacking planted or naturally regenerated trees on the site as they develop (Hadfield *et al.*, 1986). *Phytophthora* spp. can survive as resting spores in the soil after clear-cutting.

Fumigant chemicals have also been used to reduce *P. sulphurascens*, *A. ostoyae* and *H. annosum s.l.* inoculum in woody substrates, as well as to control *Phytophthora* spp. in soil, but mostly in nursery or ornamental situations. Placing liquid fumigant (allyl alcohol, chloropicrin, metam sodium, or ipconazole) in holes drilled in Douglas-fir stumps infected with *P. sulphurascens* can reduce inoculum in high-hazard areas such as campgrounds (Thies and Sturrock, 1995). Fumigation has also been tested on infected living Douglas-fir trees in Oregon, with some success (Harrington and Thies, 2007). Soil and woody residues containing *A. ostoyae* have been fumigated with methyl bromide.

The reduction of pathogen populations in bare-root forest nurseries is particularly important, and soil fumigation for root diseases, parasitic nematodes and weeds is commonly practised (Aldhous and Mason, 1994; Wang *et al.*, 2005). Most nursery root diseases are caused by fungi in the genera *Fusarium*, *Cylindrocarpon* and *Rhizoctonia*, and the fungal-like genera *Pythium* and *Phytophthora*. Typically, gaseous fumigants such as methyl bromide/chloropicrin or metam sodium are used to manage these pathogens. However, once a fumigation cycle is started, beds usually have to be re-fumigated every few years because pathogens can readily build up their populations and cannot easily be eradicated. Many fungicides also decrease ectomycorrhizal development.

Although methyl bromide is extremely effective, it causes environmental problems because bromine resulting from the breakdown of the compound enters the atmosphere and contributes to depleting the ozone that protects life on earth from harmful UV radiation (Wang *et al.*, 2005). Methyl bromide was scheduled to be phased out in 2005, but this has not yet happened (US EPA, 2009). Yearly exemptions are still being granted for certain crops, including use in forest tree seedling nurseries. Alternative chemicals being investigated include metam sodium, methyl iodide and dazomet (which breaks down into methyl isothiocyanate). The use of new tarping material (e.g. virtually impermeable plastic film, VIF), and soil pasteurization by solar irradiation in warm climates and steam/hot water treatments, are also being investigated.

Hazard/danger tree management

Hazard or danger trees can be managed through tree removal, which, concomitantly, will result in reduction of any inoculum. Trees are large objects and once they pass the sapling stage can cause considerable damage if they fall. To qualify as a hazard or danger tree, a tree must have a defect or be potentially unstable and a target (people or objects of value to people). Thus, the majority of trees falling in forests do no damage because there are no human-related targets. Factors contributing to hazard or danger trees are tree location, the presence of defects, insects and diseases, and weather conditions (Toupin *et al.*, 2008). Targets include forest workers, campers, hikers, people driving in the forest, and buildings. Root and butt rots, heart rots, sap rots, dwarf mistletoes and fungal cankers can all contribute to tree failure. Root and butt rots, in particular, are large contributors to tree failure in forests worldwide, particularly at recreation sites. Wind storms and root diseases strongly interact in bringing trees down. All of the major structural root rots, including laminated root rot, Armillaria root disease, annosus root and butt rot, schweinitzii root and butt rot (caused by *Phaeolus schweinitzii* (Fr.) Pat.), and tomentosus root rot, contribute to danger tree or hazard tree problems (Toupin *et al.*, 2008). In general, the best guideline is that the longer people are exposed to a tree the greater the chance of exposure to danger. High traffic areas have a greater chance of exposure than low traffic areas. Trees should be removed if there is any danger of failure.

2.5.4 Protection

Protection involves the provision of chemical, biological, physical, temporal or spatial barriers to prevent pathogen infection. Silvicultural management techniques are designed to maintain or increase the overall health and vigour of trees, and include the use of thinning, pruning, fertilization and prescribed fire. The philosophy here is to prevent diseases from occurring or to slow the rate of spread of established diseases. Chemical and biological control, including the use of mycorrhizas, and the influence of thinning, pruning, fertilizing and prescribed fire are separately discussed below.

Chemical control

Chemicals have been used widely to prevent infection by disease-causing agents. They include seed treatment chemicals, chemical sprays for foliage diseases and chemical stump treatments for root diseases. Many of these chemicals inhibit spore germination. Foliar sprays for controlling foliage diseases on ornamental and Christmas trees and in Monterey pine (*Pinus radiata* D. Don) plantations in New Zealand have also been used successfully (Hansen and Lewis, 1997). Stem injections of phosphite have been used to control disease front extension of *P. cinnamomi* in jarrah (*Eucalyptus marginata* Donn ex Sm.) forests in Western Australia (Shearer *et al.*, 2006) and phosphonate (= phosphite) has promise for control of SOD (Garbelotto and Schmidt, 2009). As discussed earlier, chemicals can also be used to eradicate soil-borne diseases, or at least to reduce inoculum levels, e.g. the use of fumigants in bare-root tree seedling nurseries. In container nurseries, as well as damping-off, other diseases occur, particularly grey mould caused by *Botrytis cinerea* Pers. and powdery mildews. Seedlings are commonly treated with fungicides (Lilja *et al.*, 2010; see also Lilja and Poteri, Chapter 27, this volume).

A variety of chemicals have been employed to prevent infection by root disease fungi. Because *H. annosum s.l.* spreads by airborne spores, chemical stump treatment has been widely practised (Woodward *et al.*, 1998). Stump treatments to control spread were first suggested in the 1950s in the UK by Rishbeth (1959a,b). Creosote was found to be effective but not practical to use in the forest. A search for better chemicals resulted in the testing of urea, sodium nitrate and powdered borax. Urea and sodium nitrate applied to stump surfaces promote other organisms that outcompete *H. annosum s.l.* In North America, the infection of stumps by *H. annosum s.l.* can be prevented by treatment with granular sodium tetraborate decahydrate or disodium tetrahydrate. Two borate products (trade names Sporax® and Timbor®) are registered (Asiegbu *et al.*, 2005). Borate stump treatment is used operationally in the south-eastern USA and in high-hazard areas such as campgrounds in the western USA, especially on large stumps of susceptible pine, fir (*Abies* spp.) and hemlock species where it can be applied carefully. It is not effective on trees that are already infected.

Biological control

Biological control is an alternative to the chemical control of root diseases, and many fungi, including *Phlebiopsis gigantea* (Fr.) Jülich, *Bjerkandera adusta* (Willd.) P. Karst., *F. pinicola*, *Resinicium bicolor* (Alb. & Schwein.) Parmasto, *Hypholoma* spp., *Trichoderma* spp. and *Scytalidium* spp. have been tested on stumps against *H. annosum s.l.* (Asiegbu *et al.*, 2005). *P. gigantea* has given the best results and commercial products of this saprotrophic fungus are available under the trademarks of Rotstop® in Fennoscandanavia, PG Suspension in the UK and PG IBL in Poland (Asiegbu *et al.*, 2005). These products can either be applied manually to stumps or automatically by spraying during logging operations. They are not yet used in North America because of ecological concerns, and Vainio (2008) recommends that local biocontrol strains should be used in order to prevent the global spread of *P. gigantea* and hybrid formation between geographically isolated populations. Rotstop® treatment in Finland, however, does not cause a significant reduction in the species diversity of wood-inhabiting fungi (Vainio, 2008). *Trichoderma* spp. have been tested on stumps containing *P. sulphurascens* in Oregon, with some success (Thies and Sturrock, 1995).

Biological controls are also being tested in forest nurseries, including commercially available biocontrols such as Companion®

(*Bacillus subtilis*), Actinovate® (*Streptomyces lydicus*), Soil Guard® (*Gliocladium virens* J.H. Mill., Giddens & A.A. Foster) and Root Shield® (*Trichoderma harzianum* Rifai). The use of plants, such as mustard (*Brassica* spp.), that produce natural fungicides, has also been proposed. An integrated approach to managing nursery diseases is preferable to a single approach such as methyl bromide fumigation.

Ectomycorrhizal fungi on conifer roots also act as biological control agents, especially in forest nurseries. There is substantial evidence from field studies that considerable potential exists for improving tree survival and production by inoculating seedlings with specific ectomycorrhizal fungi (Robson *et al.*, 1994; Smith and Read, 2008). Southern pines inoculated with *Pisolithus arhizus* (Scop.) Rauschert performed better in the nursery and in outplanting studies than non-inoculated control seedlings (Cordell, 1985). The survival of conifer seedlings inoculated with *P. arhizus* is particularly enhanced on eroded and degraded sites such as mine spoils. *P. arhizus* does not universally give a positive growth response, and other fungi may give better responses because they are adapted to that environment. For example, in dry southwestern Oregon, *Rhizopogon* spp. are better suited to the soils and environment and are more effective in promoting early growth than *P. arhizus*. Responses to inoculation with ectomycorrhizal fungi are not confined to harsh sites though, and some good responses have been observed where ectomycorrhizal inoculum is abundant. For example, inoculation of fast-growing eucalyptus trees with specific fungi increased early tree growth. More than 500 specific fungal-host tree associations have now been studied, and many appear to have a positive effect on growth of the host, while other fungi have a neutral effect and yet others cause growth loss and may be considered parasitic.

Although nursery seedling roots may be well inoculated with mycorrhizal fungi that are capable of enhancing growth, nursery mycorrhizal fungi are usually replaced by natural mycorrhizal fungi when seedlings are planted in the field. The ability to identify sites where inoculated trees are likely to succeed is of great consequence. Two factors are very important: (i) soil nutrient status (particularly for phosphorus) and (ii) the abundance of inoculum. Responses are likely to be greatest where there are low populations of indigenous fungi. The largest effects of inoculation are expressed when there is high nutrient requirement; that is, when trees are young. Inoculated Douglas-fir seedlings in Oregon and eucalyptus seedlings in southwestern Australia had significantly increased growth on recently clear-cut sites (Robson *et al.*, 1994).

An injected biological control agent to induce resistance to Dutch elm disease has also been developed, which consists of a suspension of live spores of the fungus *Verticillium dahliae* Kleb.

Thinning

Thinning overstocked stands to improve forest health and restore ecosystems is now commonly used, particularly in western North America, in forests where fire suppression has been practised for more than a century. Thinning has variable effects on root diseases, but the consequences may not be apparent for decades. Thinning non-symptomatic trees around a laminated root rot pocket in Douglas-fir stands in a zone 15 m from the last tree showing symptoms reduces the spread rate of *P. sulphurascens* (Thies and Sturrock, 1995) because the fungus does not spread as rapidly along or through dead root systems as through living root systems. Thinning has also been used in stands that have a high incidence of *H. annosum s.l.* in grand fir (*Abies grandis* (Douglas ex D. Don) Lindl.) in Washington to favour ponderosa pine and Douglas-fir, and reduce wildfire fire hazard.

However, thinning may increase the incidence of root diseases (Maloney *et al.*, 2008). The rate of spread by *A. ostoyae* may increase as a result of thinning because it can occupy the stumps and roots of thinned trees, especially of large trees (Morrison *et al.*, 1991; Slaughter and Rizzo, 1999). *H. annosum s.l.* may infect trees remaining after thinning through spore infection of stumps, resulting in the need for stump treatment. Wounds created during thinning operations are also an important source of infection. Avoidance of wounding, particularly when thinning non-resinous species, reduces the incidence of infection by *H. annosum s.l.* Generally, the larger the wound, the greater the chance of infection. Thinning should be avoided in the spring when the bark is loose and the chance of damage to residual trees is high, especially in thin-bark species.

The effects of precommercial thinning are also variable. This may increase the incidence of blackstain root disease caused by *Leptographium wageneri* (W.B. Kendr.) M.J. Wingf. because the insects that spread the disease may be attracted to slash and fresh stumps (Hansen and Lewis, 1997), although it will not increase the spread of *A. ostoyae* or *H. annosum s.l.* because the stumps are too small. Also, cutting high stumps reduces the chance of residual tree infection in western hemlock stands because the fungus is replaced or outcompeted by other organisms.

Thinning has also been used to increase the health of a stands infected with dwarf mistletoes in western North America. Trees that are heavily infected with dwarf mistletoes can be removed, leaving the stand in a healthier condition. However, unless the stand is re-thinned after the initial thinning to remove mistletoe-infected trees that were not noticed at the time of the first thinning, the value of the original thinning may be lost.

The incidence of foliage diseases in conifer plantations may also be reduced by thinning because it increases airflow and reduces the humid conditions that favour disease development. Then again, thinning may increase the local incidence of a disease; for example, the incidence of white pine blister rust can increase because populations of its alternate host *Ribes* spp. may also increase (Maloney *et al.*, 2008).

Pruning

Pruning can be used to reduce the incidence of dwarf mistletoes, branch cankers, rust diseases and vascular wilts such as Dutch elm disease, but other diseases may increase as a result of improper pruning. For example, decay fungi can enter through the pruning wounds, especially on hardwoods, and in Monterey pine in New Zealand cankers caused by *Diplodia pinea* (Desm.) J. Kickx f. (syn. *Sphaeropsis sapinea* (Fr.) Dyko & B. Sutton) can develop (Hansen and Lewis, 1997). Pruning of the branches of young white pine trees infected by *C. ribicola* in western Washington and Idaho has been successful in preventing the development of white pine blister rust but requires frequent monitoring.

Fertilization

Fertilization is used in forestry to increase tree growth. Three types of fertilizer are generally used: nitrogen (urea or ammonium nitrate), phosphorus (superphosphate or rock phosphate) and trace elements (copper, boron and zinc). Potassium fertilization also has been used to a limited extent. Urea fertilization may increase the impact of foliage diseases and influence decay fungi by increasing diameter growth, thereby decreasing the percentage loss to decay when fungi are compartmentalized.

Losses due to root diseases may also be reduced by fertilization. For example, Douglas-fir stands in Idaho showed less loss to Armillaria root disease when fertilized with urea and potassium (Mandzak and Moore, 1994). Potassium fertilization, though, had little effect on laminated root rot in western Washington (Thies *et al.*, 2006). Also, long-term trials on stumped sites did not reveal any effect of ammonium nitrate fertilization on the occurrence of *P. sulphurascens* in replanted Douglas-fir (Thies and Westlind, 2005). Filip *et al.* (2002) also found no significant fertilizer influence of urea or combined nitrogen, phosphorus, potassium and sulfur treatments on grand fir mortality due to *A. ostoyae* in north-eastern Oregon. It is not known how fertilization influences *H. annosum s.l.*

Feeder root diseases appear to be reduced by fertilization. The effects of *P. cinnamomi* on shortleaf pine (*Pinus echinata* Mill.) in the south-eastern USA can be alleviated by the application of nitrogen fertilizer. Similarly, phosphate application to infected Monterey pine stands in New Zealand also reduced the disease. Attempts to control *P. cinnamomi* in eucalyptus trees in Australia using fertilizers, however, have failed.

Prescribed fire

With the current emphasis on managing forests as ecosystems, there has been considerable interest in using prescribed fire as a management tool. Prescribed fire has been used to control brown spot needle disease (*Scirrhia acicola* Dearn., syn. *Mycosphaerella dearnessii* M.E. Barr) on longleaf pine (*Pinus palustris* Mill.) in the south-eastern USA, and to reduce dwarf mistletoe inoculum by killing infected suppressed trees in eastern Washington and Oregon (Edmonds *et al.*, 2011).

In forest ecosystems that typically have high fire frequencies, long-term fire suppression has resulted in an increased incidence of root disease and insect problems. Thus, the reintroduction of fire should result in a healthier forest with fewer root disease and insect problems.

Prescribed fire may have negative consequences for diseases too. Maloney *et al.* (2008) found that levels of Armillaria root disease and annosus root and butt rot, as well as bark beetle mortality, were very high in sugar pine (*P. lambertiana* Douglas) and Jeffrey pine (*P. jeffreyi* Balf.) after prescribed burning treatments in an old-growth mixed conifer forest in the Sierra Nevada (California), especially in the larger trees. Most slash burns do not get hot enough to sterilize the soil, and although stumps and roots may burn, the inoculum of root disease fungi is generally not eliminated. Root disease activity needs to be monitored in the years following thinning to see if there is a continuing lag effect. In general, slash burning after harvesting in the western USA has generally had little or no impact on root fungi such as *A. ostoyae*, *H. annosum s.l.* and *P. sulphurascens* (Hadfield *et al.*, 1986). Prescribed fire also increases the incidence of the root disease caused by *R. undulata* because this fire-loving fungus requires fire and soil heating for spore germination.

In Australia, prescribed fire is used for hazard reduction and regeneration of desirable tree species. Low-intensity hazard reduction burns are used in dry sclerophyll forests at 7–10 year intervals, although periodic removal of the litter layer for the purpose of hazard reduction could significantly increase the susceptibility of these forests to *P. cinnamomi*. Removal of the litter layer in jarrah forests in Western Australia increases soil temperatures in spring, thereby prolonging the period when the soil moisture and temperature regimes are suitable for spore production. Hazard reduction burning also changes the understorey to favour bull banksia (*Banksia grandis* Willd.), which harbours the fungus. However, the use of fire may reduce inoculum in dead bull banksia trunks (Collins *et al.*, 2012).

Other protection practices

Other protection practices include the use of physical or spatial barriers, such as soil trenching and buried plastic tarps to prevent the spread of *Armillaria* spp. and the Dutch elm disease fungus through soil and roots. Temporal barriers, such as the time of year of thinning for prevention of *H. annosum s.l.* infection of cut stumps (Tainter and Baker, 1996), are also used. Trenching is used as a treatment option to prevent the spread of oak wilt through oak (*Quercus* spp.) root grafts (Koch *et al.*, 2010).

2.5.5 Resistance

Resistance employs hosts that are resistant to diseases as well as cultivars or varieties of the same host that may have been bred or genetically modified for resistance. Disease resistance has been employed successfully to manage forest diseases and falls into two categories: genetic resistance within species involving breeding for resistance and genetic engineering; and tree species selection.

Genetic resistance, breeding for resistance and genetic engineering

Genetic resistance is an important factor in disease management (Hansen and Lewis, 1997). Sniezko (2006) noted successes in resistance breeding in four tree species in

North America against non-native pathogens: western white pine (*P. monticola* Douglas ex D. Don) and white pine blister rust; Port Orford cedar and Phytophthora root disease; American chestnut (*Castanea dentata* (Marsh) Borkh.) and chestnut blight (*Cryphonectria parasitica* (Murrill) M.E. Barr); and American elm (*Ulmus americana* L.) and Dutch elm disease. Breeding has also been successful in loblolly pine (*P. taeda* L.) against fusiform rust (*Cronartium fusiforme* Hedg. & Hunt ex Cumm.), and clones of poplars (*Populus* spp.) resistant to poplar rust (*Melampsora* spp.) have been developed. Trees resistant to fungi causing structural root diseases, such as *P. sulphurascens* and *H. annosum s.l.*, have been noted, but to date, genetic resistance has not been used to manage these diseases. There is considerable interest in the genetic engineering of trees, particularly against introduced diseases (Adams *et al.*, 2002), but there is also great concern about the potential effects of introducing genetically modified plants into landscapes.

Tree species selection

In some areas, tree species selection has great promise for disease management because most disease organisms have specific hosts or a range of susceptible hosts (Hansen and Lewis, 1997). Selection of tree species is particularly important for managing root diseases. For example, western redcedar (*Thuja plicata* Donn ex D. Don) or Douglas-fir could be planted as an alternative species to Port Orford cedar in areas with Port Orford cedar root disease. In areas with laminated root rot, red alder (*Alnus rubra* Bong., which is immune), western white pine, or western redcedar could be planted as alternatives to Douglas-fir, while recognizing that western white pine is susceptible to white pine blister rust and that western redcedar is prone to deer browsing in many areas. Table 2.1 shows the relative susceptibility of western North American conifers and hardwoods to the four most common root diseases.

Alternative species are not very effective against Armillaria root rot because its host range is so wide, although western larch (*Larix occidentalis* Nutt.) is seldom damaged (Table 2.1). The use of combinations of species has been successful in Canada; fewer and smaller diseased patches were observed when rows of susceptible Douglas-fir were alternated with less susceptible western redcedar or paper birch (*Betula papyrifera* Marsh.) than were found in pure Douglas-fir. Alternative species can also be used in the control of annosus root and butt rot and blackstain root disease (Table 2.1). Tables like this can be compiled for the relative susceptibility of tree species in other areas of the world.

2.5.6 Therapy

Some pathogens can be managed after they have infected a host by using chemotherapy. Injecting elms with fungicides to eliminate the Dutch elm disease fungus is a good example of the use of chemotherapy in forestry. Several chemicals are commonly used, including thiabendazole and propiconazole (Hubbes, 2004). Thiabendazole can also be used for oak wilt. The effectiveness of these compounds depends on how well they distribute in the crown, the rate at which they are applied, their concentration in plant tissues and how well they persist in the tree. Injection is only effective in trees with <10% wilt and only when beetles introduce the disease. Also, it is usually done only for highly valued trees because it is extremely expensive; it has not been practised in a forest situation. Systemic induced resistance (SIR) can also be acquired using biological agents such as *V. dahliae*, which has been used to manage Dutch elm disease. *V. dahliae* and *Diplodia scrobiculata* J. de Wet, Slippers & M.J. Wingf. have been shown to induce SIR in Aleppo pine (*P. halepensis* Mill.) against *D. pinea* infections (Muñoz *et al.*, 2008).

Branch pruning could also be considered a therapeutic technique for Dutch elm disease before it reaches the main stem, and for dwarf mistletoes and white pine blister rust. Root pruning has also been practised for removing roots infected with *A. ostoyae*.

Table 2.1. Relative susceptibility of western North American conifers to damage by Armillaria root disease, annosus root and butt rot, laminated root rot and blackstain root disease (adapted from Hadfield *et al.*, 1986).

Tree species	Armillaria root disease	Annosus root and butt rot	Laminated root rot	Blackstain root disease
Abies amabilis (Douglas ex Loudon) Douglas ex Forbes	2[a]	1	2	4
Abies concolor (Gord. & Glend.) Lindl. ex Hildebr.	1	1	1	4
Abies grandis (Douglas ex D. Don) Lindl.	1	1	1	4
Abies lasiocarpa (Hook.) Nutt.	2	2	2	4
Abies magnifica A. Murray bis	2	2	2	4
Abies procera Rehder	2	2	2	4
Calocedrus spp.	3	3	4	4
Chamaecyparis lawsoniana (A. Murray bis) Parl.	3	3	4	4
Larix occidentalis Nutt.	3	3	2	4
Picea engelmannii Parry ex Engelm.	2	3	2	4
Picea sitchensis (Bong.) Carrière	2	3	3	4
Pinus contorta Douglas ex Loudon	2	2	3	3
Pinus lambertiana Douglas	2	3	3	4
Pinus monticola Douglas ex D. Don	2	3	3	4
Pinus ponderosa Lawson & C. Lawson	2	2	3	1
Pseudotsuga menziesii (Mirb.) Franco west of Cascade crest	2	3	1	1
P. menziesii (Mirb.) Franco east of Cascade crest	1	3	1	3
Thuja plicata Donn ex D. Don	2	3	4	4
Tsuga heterophylla (Raf.) Sarg.	2	2	2	3
Tsuga mertensiana (Bong.) Carrière	2	1	1	3
Hardwoods	2	3	4	4

[a]Susceptibility: 1, severely damaged; 2, moderately damaged; 3, seldom damaged; 4, not damaged (immune).

2.5.7 Integrated disease management

More than one disease management strategy might be employed at a time, and successful management often employs multiple strategies. Many disease management practices are most effective when applied as a system rather than individually, a concept known as integrated disease management (IDM) or integrated pest management (IPM). Silviculture is an important component of IDM and can be used effectively to manage introduced diseases (Waring and O'Hara, 2005); it has been effectively used for white pine blister rust (Schwandt *et al.*, 2010; Zeglen *et al.*, 2010). Cost plays an important role in decisions about how to manage forest diseases because of the relatively low value of forest products.

IDM practices are usually compiled as BMPs. A good example of this are the BMPs compiled to prevent or reduce the spread of *P. ramorum* (Kliejunas, 2010). The California Oak Mortality Task Force has compiled guidelines for user groups in wildlands or urban–forest interface zones for forestry, recreational users, forest collectors, arborists, homeowners, etc. BMPs for woodlands, parks and gardens in the UK are also available.

As mentioned previously, diseases do not always occur alone on the landscape. In fact, in large areas of the world, forests are affected by combinations of diseases, insects and fire. Thus, the management of disease, insect and fire need to be integrated when possible. An example of this involves the management of national forests in Oregon and Washington, where recommendations are different by region. East of the Cascade crest and in south-west Oregon, where forests are dry, and insect and disease problems and fire danger are high, the focus is on thinning to improve forest health and reduce tree stress and fire danger (Table 2.2) (Filip and Schmitt, 1990; Campbell and Liegel, 1996). In comparison, west of the Cascade crest, where forests are wet and fire danger and insect problems are lower, the focus is on increasing the diversity of tree species away from single-species plantations (Table 2.2).

IDM is also practised in forest nurseries. Management usually involves an understanding of the life cycles of disease organisms and of how environmental factors affect their development. Regulating irrigation, fertilization and cultural regimes so that they do not favour diseases, using sanitation and exclusion practices, establishing disease-monitoring schemes and using chemicals judiciously usually results in the production of healthy seedlings (Dumroese and James, 2005). Not allowing movement of infected plant material or contaminated equipment within or among nurseries, weed control, removing infected seedlings, checking water supplies periodically for pathogens, and removing diseased trees and alternative hosts for rusts from areas around nurseries are particularly effective.

In conclusion, there is no doubt that forest nurseries and forest ecosystems are greatly affected by diseases. They are common disturbance agents in natural ecosystems and change successional patterns, promote biodiversity and provide habitat for birds and other animals by changing the structure of the forest. However, they also cause great losses of timber production in managed forests. Forest managers now have many tools for detecting, monitoring, modelling and risk rating forest landscapes for disease. They also have many management strategies or tools, including avoidance, exclusion (through quarantine and regulation), eradication, protection, resistance, therapy and integrated disease management. In some areas, just doing nothing or even increasing decay for wildlife may be appropriate, while in other areas, particularly where diseases interact with insects and fire, active ecosystem management will

Table 2.2. Integrated management recommendations for maintaining healthy forests in Oregon and Washington (USA) in dry and moist forest ecosystems (From Filip and Schmitt, 1990; Campbell and Liegel, 1996).

East of the Cascade crest and in south-west Oregon (dry forests)
- Thin stands to reduce competition, stress and mortality from bark beetles.
- Harvest certain species, such as lodgepole pine (*Pinus contorta* Douglas ex Loudon) to create a mosaic of age classes across the landscape to prevent bark beetle outbreaks.
- Keep in mind the effects that certain activities (such as thinning, harvesting or replanting with certain species) will have on root diseases and dwarf mistletoes.
- Reduce forest floor fuel loads to prevent destructive stand-replacement fires.
- Introduce prescribed fire to lighten fuel loads and remove Douglas-fir (*Pseudotsuga menziesii* (Mirb.) Franco) and true firs (*Abies* spp.). Use fire to regenerate species such as larch (*Larix* spp.) or quaking aspen (*Populus tremuloides* Michx.) that depend on fire for seedbeds or to stimulate root sprouting.
- Design site-specific regeneration (natural or planted) to promote desired species composition and structure.

West of the Cascade crest (moist forests)
- Shift from single species to multiple species to reduce insect outbreaks and proliferation of diseases. This involves trade-offs between maximizing timber production and minimizing insect and disease damage.
- Replant harvested areas with seedlings grown from local seed sources or use natural regeneration. The severity of diseases such as Swiss needle cast (*Phaeocryptopus gaeumannii* (T. Rohde) Petr.) of Douglas-fir is much less when trees are adapted to the site.
- Maintain a mix of species and age classes, thereby preventing the landscape from being dominated by uniform, highly susceptible stands.

be necessary to restore ecosystems to a healthy condition. A clear set of objectives for land management is needed in order to choose the appropriate management strategy. Introduced disease organisms will still continue to be a threat, and climate change is likely to increase stress levels for trees and perhaps make root and foliage diseases more aggressive.

References

Adams, J.M., Piovesan, G., Strauss, S. and Brown, S. (2002) The case for genetic engineering of native and landscape trees against introduced pests and diseases. *Conservation Biology* 16, 874–879.

Agrios, G.N. (2005) *Plant Pathology*, 5th edn. Academic Press, New York.

Aldhous, J.R. and Mason, W.L. (1994) *Forest Nursery Practice*. Forestry Commission Bulletin 111, Her Majesty's Stationery Office, London.

Asiegbu, F.O., Adomas, A. and Stenlid, J. (2005) Conifer root and butt rot caused by *Heterobasidion annosum* (Fr.) Bref. *s.l. Molecular Plant Pathology* 6, 395–409.

Browning, J.E. and Edmonds, R.L. (1993) Influence of soil aluminum and pH on Armillaria root rot in Douglas-fir in western Washington. *Northwest Science* 67, 37–43.

Bull, E.L., Parks, C.G. and Torgersen, T.R. (1997) *Trees and Logs Important to Wildlife in the Interior Columbia River Basin*. General Technical Report PNW-GTR-391, US Department of Agriculture, Forest Service, Pacific Northwest Station, Portland, Oregon.

Campbell, S. and Liegel, L. (1996) *Disturbance and Forest Health in Oregon and Washington*. General Technical Report PNW-GTR-381, US Department of Agriculture, Forest Service, Pacific Northwest Station, Portland, Oregon.

Collins, S., McComb, J.A., Howard, K., Shearer, B.L., Colquhoun, I.J. and Hardy, G.E.St.J. (2012) The long-term survival of *Phytophthora cinnamomi* in mature *Banksia grandis* killed by the pathogen. *Forest Pathology* 42, 28–36.

Cordell, C.E. (1985) The application of *Pisolithus tinctorius* ectomycorrhizae in forest land management. In: Molina, R. (ed.) *Proceedings of the Sixth North American Conference on Mycorrhizae, 25–29 June 1984, Bend, Oregon*. Forest Research Laboratory, Oregon State University, Corvallis, Oregon, pp. 69–71.

Dixon, G.E. and Hawksworth, F.G. (1979) A spread intensification model for southwestern dwarf mistletoe in ponderosa pine. *Forest Science* 25, 43–52.

Driver, C.H. and Ginns, J.H. Jr (1969) Ecology of slash pine stumps: fungal colonization and infection by *Fomes annosus*. *Forest Science* 15, 1–10.

Dukes, J.S. *et al.* (2009) Response of insect pests, pathogens, and invasive plant species to climate change in the forests of northeastern North America: what can we predict? *Canadian Journal of Forest Research* 39, 231–248.

Dumroese, R.K. and James, R.L. (2005) Root diseases in bare-root and container nurseries of the Pacific Northwest: epidemiology, management, and effects on outplanting performance. *New Forests* 30, 185–202.

Edmonds, R.L. (1996) Inoculum. In: Tainter, F.H. and Baker, F.A. (eds) *Principles of Forest Pathology*. John Wiley, Hoboken, New Jersey, pp. 164–184.

Edmonds, R.L., Agee, J.K. and Gara, R.I. (2011) *Forest Health and Protection*, 2nd edn. Waveland Press, Long Grove, Illinois.

Filip, G.M. and Morrell, J.J. (1996) Importing Pacific Rim wood: pest risks to domestic resources. *Journal of Forestry* 94, 22–26.

Filip, G.M. and Schmitt, C.L. (1990) *R_x for* Abies: *Silvicultural Options for Diseased Firs in Oregon and Washington*. General Technical Report PNW-GTR-252, US Department of Agriculture, Forest Service, Pacific Northwest Station, Portland, Oregon.

Filip, G.M., Ganio, L., Moester, P.T., Mason, R.R. and Wickman, B.E. (2002) Ten-year effect of fertilization on tree growth and mortality associated with Armillaria root disease, fir engravers, dwarf mistletoe, and western spruce budworm in northeastern Oregon. *Western Journal of Applied Forestry* 17, 122–128.

Filip, G.M., Parks, C.C., Baker, F.A. and Daniels, S.E. (2004) Artificial inoculation of decay fungi into Douglas-fir with rifle or shotgun to produce wildlife trees in western Oregon. *Western Journal of Applied Forestry* 19, 211–215.

Frankel, S.J. (1998) *User's Guide to the Western Root Disease Model, Version 3.0*. General Technical Report PSW-GTR-165, US Department of Agriculture, Forest Service, Pacific Southwest Station, Berkeley, California.

Garbelotto, M. and Schmidt, D.J. (2009) Phosphonate controls sudden oak death pathogen for up to 2 years. *California Agriculture* 63, 1–17.

Glaeser, J.A. and Lindner, D.L. (2011) Use of fungal biosystematics and molecular genetics in detection and identification of wood-decay fungi for improved forest management. *Forest Pathology* 41, 341–348.

Hadfield, J.S., Goheen, D.J., Filip, G.M., Schmitt, C.L. and Harvey, R.D. (1986) *Root Diseases in Oregon and Washington Conifers.* Report R6-FPM-250–286, US Department of Agriculture, Forest Service, State and Private Forestry, Forest Pest Management, Portland, Oregon.

Hansen, E.M. and Lewis, K.J. (1997) *Compendium of Conifer Diseases.* APS Press, St Paul, Minnesota.

Harrington, C.A. and Thies, W.G. (2007) Laminated root rot and fumigant injection affect survival and growth of Douglas-fir. *Western Journal of Applied Forestry* 22, 220–227.

Hawksworth, F.G. and Wiens, D. (1996) *Dwarf Mistletoes: Biology, Pathology, and Systematics.* Agricultural Handbook 709, US Department of Agriculture, Forest Service, Washington, DC.

Hubbes, M. (2004) Induced resistance for control of Dutch elm disease. *Forest Systems* 13, 185–196.

Kastner, W.W., Goheen, D.J. and Edmonds, R.L. (1994) Relationship between occurrence of laminated root rot and site characteristics in Douglas-fir forests of the northern Oregon Coast Range. *Western Journal of Applied Forestry* 9, 14–17.

Kelley, M., Guo, Q., Liu, D. and Shaari, D. (2007) Modeling the risk for a new invasive forest disease in the United States: an evaluation of five environmental niche models. *Computers, Environment and Urban Systems* 31, 689–710.

Kim, M.-S., Klopfenstein, N.B. and Hamelin, R.C. (2005) Application of molecular genetic tools to studies of forest pathosystems. In: Lundquist, J.E. and Hamelin, R.C. (eds) *Forest Pathology: From Genes to Landscapes*. APS Press, St Paul, Minnesota, pp. 9–20.

Kliejunas, J.T. (2010) *Sudden Oak Death and* Phytophthora ramorum*: A Summary of the Literature.* General Technical Report PSW-GTR-234, US Department of Agriculture, Forest Service, Pacific Southwest Station, Albany, California.

Koch, K.A., Quiram, G.L. and Venette, R.C. (2010) A review of oak wilt management: a summary of treatment options. *Urban Forestry and Greening* 9, 1–8.

Krist, F.J. Jr, Sapio, F.J. and Tkacz, B.M. (2007) *Mapping Risk from Forest Insects and Diseases.* Report FHTET 2007-06, US Department of Agriculture, Forest Service, Forest Health Protection, Forest Health Technology Enterprise Team, Fort Collins, Colorado.

Lilja, A., Poteri, M., Petäistö, R.-J., Rikala, R., Kurkela, T. and Kasanen, R. (2010) Fungal diseases in forest nurseries in Finland. *Silva Fennica* 44, 525–545.

Lundquist, J.E. and Hamelin, R.C. (2005) *Forest Pathology: From Genes to Landscapes.* APS Press, St Paul, Minnesota.

Maffei, H.M., Filip, G.M., Chadwick, K.L. and David, L. (2008) Western root disease model simulation versus plot remeasurement: 11 years of change in stand structure and density induced by Armillaria root disease on central Oregon. In: Havis, R.N. and Crookston, N.L. (eds) *Third Forest Vegetation Simulator Conference, Fort Collins, Colorado, February 13–15, 2007.* Proceedings RMRS-P-54, US Department of Agriculture, Forest Service, Rocky Mountain Research Station, Fort Collins, Colorado, pp. 55–67.

Maloney, P.E., Smith, T.F., Jensen, C.E., Innes, J., Rizzo, D.M. and North, M.P. (2008) Initial tree mortality and insect pathogen response to fire and thinning restoration treatments in an old-growth mixed-conifer forest of the Sierra Nevada, California. *Canadian Journal of Forest Research* 38, 3011–3020.

Mandzak, J.M. and Moore, J.A. (1994) The role of nutrition in the health of inland western forests. In: Sampson, R.N. and Adams, D.L. (eds) *Assessing Forest Ecosystem Health in the Inland West.* Haworth Press, Binghamton, New York, pp. 191–210.

Manter, D.K., Reeser, P.W. and Stone, J.K. (2005) A climate-based model for predicting geographic variation in Swiss needle cast severity in the Oregon coast range. *Phytopathology* 95, 1256–1265.

McDonald, G.I., Martin, N.E. and Harvey, A.E. (1987) Armillaria *in the Northern Rockies: Pathogenicity and Susceptibility on Pristine and Disturbed Sites.* Research Note INT-371, US Department of Agriculture, Forest Service, Intermountain Research Station, Ogden, Utah.

Meentemeyer, R.K., Rank, N.E., Shoemaker, D.A., Oneal, C.B., Wickland, A.C., Frangioso, K.M. and Rizzo, D.M. (2008) Impact of sudden oak death on tree mortality in the Big Sur ecoregion of California. *Biological Invasions* 10, 1243–1255.

Morrison, D.J., Merler, H. and Norris, D. (1991) *Detection, Recognition and Management of Armillaria and Phellinus Root Diseases in the Southern Interior of British Columbia.* FRDA Report 179, Joint Publication of Forestry Canada and the British Columbia Ministry of Forests, Forestry Canada, Victoria, BC, Canada.

Muñoz, Z., Moret, A. and Garcés, S. (2008) The use of *Verticillium dahliae* and *Diplodia scrobiculata* to induce resistance in *Pinus halepensis* against *Diplodia pinea* infection. *European Journal of Plant Pathology* 120, 331–337.

Oliver, C.D. and McCarter, J.B. (1995) Developments in decision support for landscape management. In: *Proceedings of the Ninth Symposium on Geographic Information Systems held March 27–30, Vancouver, British Columbia, Canada*. Also available as: Oliver, C.D. and McCarter, J.B. (1996) Developments in decision support for landscape management. In: Heit, M., Parker, H.D. and Shortreid, A. (eds) *GIS Applications in Natural Resources 2*. GIS World Books, Fort Collins, Colorado, pp. 501–509.

Peet, F.G. and Hunt, R.S. (2005) A new simulator for the spread of forest root diseases by individual root contacts. *Forest Science* 51, 425–437.

Pukkala, T., Möykkyen, T., Thor, M., Rönnberg, J. and Stenlid, J. (2005) Modeling infection and spread of *Heterobasidion annosum* in even-aged Fennoscandian conifer stands. *Canadian Journal of Forest Research* 35, 74–84.

Redfern, D.B., Pratt, J.E., Hendry, S.J. and Low, J.D. (2010) Development of a policy and strategy for controlling infection by *Heterobasidion annosum* in British forests: a review of supporting research. *Forestry* 83, 207–218.

Rishbeth, J. (1959a) Stump protection against *Fomes annosus*. I. Treatment with creosote. *Annals of Applied Biology* 47, 519–528.

Rishbeth, J. (1959b) Stump protection against *Fomes annosus*. II. Treatment with substances other than creosote. *Annals of Applied Biology* 47, 529–541.

Robinson, D.C.E. and Geils, B.W. (2006) Modelling dwarf mistletoe at three scales: life history, ballistics and contagion. *Ecological Modelling* 199, 23–38.

Robson, A.D., Abbott, L.K. and Malajczuk, N. (1994) *Management of Mycorrhizas in Agriculture, Horticulture and Forestry*. Developments in Plant and Soil Sciences, Vol. 56, Kluwer Academic Publishers, Dordrecht, The Netherlands.

Rosso, P.H. and Hansen, E.M. (2003) Predicting Swiss needle cast disease distribution and severity in young Douglas-fir plantations in coastal Oregon. *Phytopathology* 93, 790–798.

Russell, K., Johnsey, R. and Edmonds, R.L. (1986) Disease and insect management. In: Oliver, C.D., Hanley, D.P. and Johnson, J.A. (eds) *Douglas-fir: Stand Management for the Future*. Contribution 55, Institute of Forest Products, College of Forest Resources, University of Washington, Seattle, Washington, pp. 189–207.

Schwandt, J.W., Lockman, I.B., Kliejunas, J.T. and Muir, J.A. (2010) Current health issues and management strategies for white pines in the western United States and Canada. *Forest Pathology* 40, 226–250.

Shaw, D.C., Filip, G.M., Kanaskie, A., Maguire, D.A. and Littke, W.A. (2011) Managing an epidemic of Swiss needle cast in the Douglas-fir region of Oregon: the role of the Swiss needle cast cooperative. *Journal of Forestry* 109, 109–119.

Shearer, B.L., Fairman, R.G. and Grant, M.J. (2006) Effective concentration of phosphite in controlling *Phytophthora cinnamomi* following stem injection of *Banskia* species and *Eucalyptus marginata*. *Forest Pathology* 36,119–135.

Slaughter, G.W. and Rizzo, D.M. (1999) Past forest management promoted root disease in Yosemite Valley. *California Agriculture* 53, 17–24.

Smith, S.E. and Read, D.J. (2008) *Mycorrhizal Symbiosis*, 3rd edn. Academic Press, New York.

Sniezko, R.A. (2006) Resistance breeding against nonnative pathogens in forest trees – current successes in North America. *Canadian Journal of Plant Pathology* 29, S270–S279.

Stenlid, J., Oliva, J., Boberg, J.B. and Hopkins, A.J.M. (2011) Emerging diseases in European forest ecosystems and responses in society. *Forests* 2, 486–504.

Sturrock, R.N., Frankel, S.J., Brown, A.V., Hennon, P.E., Kliejunas, J.T., Lewis, K.J., Worrall, J.J. and Woods, A.J. (2011) Climate change and forest diseases. *Plant Pathology* 60, 133–149.

Tainter, F.H. and Baker, F.A. (1996) *Principles of Forest Pathology*. John Wiley, Hoboken, New Jersey.

Thies, W.G. and Sturrock, R.N. (1995) *Laminated Root Rot in Western North America*. Resource Bulletin/ General Technical Report PNW-GTR-349, US Department of Agriculture, Forest Service, Pacific Northwest Research Station Portland, Oregon.

Thies, W.G. and Westlind, D.J. (2005) Stump removal and fertilization of five *Phellinus weirii*-infested stands in Washington and Oregon affect mortality and growth of planted Douglas-fir 25 years after treatment. *Forest Ecology and Management* 219, 242–258.

Thies, W.G., Kelsey, R.G., Westlind, D.J. and Madsen, J. (2006) Potassium fertilizer applied immediately after planting had no impact on Douglas-fir seedling mortality caused by laminated root rot on a forested site in Washington State. *Forest Ecology and Management* 229, 195–201.

Toupin, R., Filip, G., Erkert, T. and Barger, M. (2008) *Field Guide for Danger Tree Identification and Response.* Report R-6-NR-FP-PR-01-08, US Department of Agriculture, Forest Service, Pacific Northwest Region, Portland, Oregon.

US EPA (2009) The Phaseout of Methyl Bromide. US Environmental Protection Agency, Washington, DC. Available at: www.epa.gov/ozone/mbr (accessed 27 February 2012).

Václavík, T., Kanaskie, A., Hansen, E.M., Ohmann, J.L. and Meentemeyer, R.K. (2010) Predicting potential and actual distribution of sudden oak death in Oregon: prioritizing landscape contexts for early detection and eradication of disease outbreaks. *Forest Ecology and Management* 260, 1026–1035.

Vainio, E.J. (2008) Ecological impacts of *Phlebiopsis gigantea* biocontrol against *Heterobasidion* spp. as revealed by fungal community profiling and population analysis. PhD thesis, University of Helsinki, Helsinki, Finland.

Vasaitis, R., Stenlid, J., Thomsen, I.M., Barklund, P. and Dahlberg, A. (2008) Stump removal to control root rot in forest stands. A literature study. *Silva Fennica* 42, 457–483.

Wang, D., Juzwik, J., Fraedrich, S.W., Spokas, K., Zhang, Y. and Koskinen, W.C. (2005) Atmospheric emissions of methyl isothiocyanate and chloropicrin following soil fumigation and surface containment treatment in bare-root forest nurseries. *Canadian Journal of Forest Research* 35, 1202–1212.

Waring, K.M. and O'Hara, K.L. (2005) Silvicultural strategies in forest ecosystems affected by introduced pests. *Forest Ecology and Management* 209, 27–41.

Watt, M.S., Kriticos, D.J., Alcaraz, S., Brown, A.V. and Leriche, A. (2009) The hosts and potential geographic range of Dothistroma needle blight. *Forest Ecology and Management* 257, 1505–1519.

Whetzel, H.H. (1929) The terminology of phytopathology. *Proceedings of International Congress of Plant Sciences, First Congress, Ithaca, New York, 1926, Vol.* 2, pp. 1204–1215.

Woods, A.J., Heppner, D., Kope, H.H., Burleigh, J. and Maclauchlan, L. (2010) Forest health and climate change: a British Columbia perspective. *Forestry Chronicle* 86, 412–422.

Woodward, S., Stenlid, J., Karjalainen, R. and Hüttermann, A. (1998) Heterobasidion annosum: *Biology, Ecology Impact and Control.* CAB International, Wallingford, UK.

Zeglen, S., Pronos, J. and Merler, H. (2010) Silvicultural management of white pines in western North America. *Forest Pathology* 40, 347–368.

3 Forest Diseases Caused by Viruses

Carmen Büttner,[1]* Susanne von Bargen,[1] Martina Bandte[1] and Hans-Peter Mühlbach[2]
[1]*Humboldt-Universität zu Berlin, Germany;*
[2]*University of Hamburg, Germany*

3.1 Pathogens, Significance and Distribution

A lack of knowledge of the presence and frequency of occurrence of viral diseases in forest trees leads to the impression that they are rare and therefore not important. The opposite is true. Viruses are responsible for far greater economic losses than generally recognized. This is because the symptoms are very different from those attributed to bacteria and fungi, and the losses are often more insidious and less conspicuous, so go unnoticed and untreated.

Plant viruses play a central role in the plant health status of forest trees because the high degree of disease that they cause leads to extensive tissue damage (Nienhaus, 1985). Viral pathogens are present in plants in every ecosystem and induce substantial losses worldwide in agriculture, horticulture and forestry (King *et al.*, 2011). There is an urgent need for reliable methods for virus detection and identification in forest trees – as well as a need for tools for disease management. Knowledge on virus characteristics followed by knowledge of their epidemiology are the first steps in developing appropriate phytosanitary strategies to produce virus-free plants and to keep tree seedlings free of plant viruses (Jones, 2004). The mode of transmission has to be considered as an important factor affecting the spread and impact of a virus infection within a forest.

Viruses have been recovered and identified from many deciduous fruit trees and a scattered number of broadleaved and coniferous forest trees. Conversely, other viruses, such as maple mosaic virus and oak ringspot virus are still not classified. In this chapter we focus on important plant genera cultivated in forests, park areas and on roadsides (Table 3.1). To date, viruses from 17 different genera have been identified. The latest, *European mountain ash ringspot-associated virus* (EMARaV) has recently been classified by the International Committee on Taxonomy of Viruses (ICTV) (Mühlbach and Mielke-Ehret, 2011). Interestingly, some plant genera, such as *Carpinus*, are susceptible to only one or two virus species while others, such as *Betula* and *Fraxinus*, are hosts to multiple virus species. Because of the economic and ecological importance of forest trees for industry, for fuel and in the future (to offset the impact of climate change), there is a need to increase research on plant viruses.

Obviously, viruses had infected trees and caused diseases for centuries before they were

* E-mail: carmen.buettner@agrar.hu-berlin.de

Table 3.1. Economically important viruses detected in forest and roadside trees.

Host genus	Virus species	Particle morphology[a]	Genus[b]	References
Abies	Not identified	Isometric		Flachmann *et al.*, 1990
Acer	*Arabis mosaic virus* (ArMV)		*Nepovirus*	Thomas, 1970
	Maple mosaic virus	Not visualized		Szirmai, 1972
	Maple leaf perforation virus	Not visualized		Šubíková, 1973
	Not identified	Rod-shaped		Lana *et al.*, 1980; Führling and Büttner, 1998
	ArMV		*Nepovirus*	Erdiller, 1986
	Cucumber mosaic virus (CMV)		*Cucumovirus*	
	Sowbane mosaic virus (SoMV)		*Sobemovirus*	
Aesculus	*Apple chlorotic leaf spot virus* (ACLSV)		*Trichovirus*	Hentsch *et al.*, 1997
	Strawberry latent ringspot virus (SLRSV)		*Nepovirus*	
	Apple mosaic virus (ApMV)		*Ilarvirus*	Sweet and Barbara, 1979; Polák and Zieglerová, 1997
	Cherry leaf roll virus (CLRV)		*Nepovirus*	Sweet and Campbell, 1976
Betula	ArMV, CLRV, *Tomato ringspot virus* (ToRSV)		*Nepovirus*	Cooper and Massalski, 1984
	Tobacco necrosis virus (TNV)		*Necrovirus*	
	Prunus necrotic ringspot virus (PNRSV)		*Ilarvirus*	
	Prune dwarf virus (PDV)		*Ilarvirus*	
	CLRV		*Nepovirus*	Cooper and Atkinson, 1975; Jones *et al.*, 1990; Jalkanen *et al.*, 2007; von Bargen *et al.*, 2009; Büttner *et al.*, 2011
	ApMV		*Ilarvirus*	Gotlieb and Berbee, 1973; Hardcastle and Gotlieb, 1980; Polák and Procházková, 1996; Polák and Zieglerová, 1997
Carpinus	ApMV		*Ilarvirus*	Polák and Zieglerová, 1997
Fagus	CLRV		*Nepovirus*	Nienhaus *et al.*, 1985; Winter and Nienhaus, 1989
	Brome mosaic virus (BMV)		*Bromovirus*	
	Bean yellow mosaic virus (BYMV)		*Potyvirus*	
	Not identified	Filamentous	–	
	CLRV		*Nepovirus*	Jones *et al.*, 1990
	Tomato black ring virus (ToBRV)		*Nepovirus*	Schmelzer *et al.*, 1966
	TNV		*Necrovirus*	Büttner and Nienhaus, 1989a
Fraxinus	ArMV		*Nepovirus*	Cooper, 1975; Cooper and Sweet, 1976; Cooper *et al.*, 1983

Continued

Table 3.1. Continued.

Host genus	Virus species	Particle morphology[a]	Genus[b]	References
	CLRV		*Nepovirus*	Hamacher and Quadt, 1991; Giersiepen, 1993
	TNV		*Necrovirus*	Ciferri *et al.*, 1961; Casalicchio, 1965
	Tobacco mosaic virus (TMV)		*Tobamovirus*	Lana and Agrios, 1974a,b
	TMV		*Tobamovirus*	Ferris *et al.*, 1989
	Tobacco rattle virus (TRV)		*Tobravirus*	
	ToRSV		*Nepovirus*	
	TMV		*Tobamovirus*	Castello *et al.*, 1984; Shiel and Castello, 1985
	Tobacco ringspot virus (TRSV)		*Nepovirus*	
	Not identified	Isometric		Bertioli *et al.*, 1993
	TRSV		*Nepovirus*	Hibben and Walker, 1971; Hibben and Bozarth, 1972
Picea	Not identified	Rod-shaped		Cech *et al.*, 1961
	Tomato mosaic virus (ToMV)		*Tobamovirus*	Jacobi and Castello, 1992; Castello *et al.*, 1995; Bachand and Castello, 1998, 2001
	Not identified	Filamentous		Castello *et al.*, 2000
	Not identified	Rod-shaped		Biddle and Tinsley, 1968
	Not identified	Rod-shaped		Soikkeli, 1983
Pinus	Not identified		Unassigned genus within *Partitiviridae*	Veliceasa *et al.*, 2006
	Not identified	Rod-shaped		Soikkeli, 1983
	Scots pine mosaic virus	Not visualized		Schmelzer *et al.*, 1966
	Scots pine bushy stunt virus	Not visualized		Jančařík and Blattný, 1966
	Not identified	Rod shaped		Biddle and Tinsley, 1968
	TNV		*Necrovirus*	Büttner and Nienhaus, 1989a
Populus	*Poplar mosaic virus* (PopMV)		*Carlavirus*	Schmelzer, 1966; Navrátil and Boyer, 1968; Cooper and Edwards, 1981; Cooper *et al.*, 1986; Smith and Campbell, 2004; Smith *et al.*, 2004
	ArMV, ToBRV		*Nepovirus*	Cooper and Sweet, 1976
	TRV		*Tobravirus*	
	TNV		*Necrovirus*	Hibben *et al.*, 1979
Prunus	*Plum pox virus* (PPV)		*Potyvirus*	Schimanski and Funk, 1968;
	Prune dwarf virus (PDV)		*Ilarvirus*	Schimanski *et al.*, 1975; Németh *et al.*, 2010
	CLRV		*Nepovirus*	
	Apple stem pitting virus (ASPV)		*Foveavirus*	
	CLRV		*Nepovirus*	Löw, 1995

	Tomato bushy stunt virus (TBSV)		*Tombusvirus*	Allen and Davidson, 1967
	ACLSV		*Trichovirus*	Rana *et al.*, 2008
Quercus	TMV		*Tobamovirus*	Yarwood and Hecht-Poinar, 1970; Yarwood, 1971; Nienhaus and Yarwood, 1972; Yarwood and Hecht-Poinar, 1973; Horvath *et al.*, 1975; Nienhaus, 1975
	TNV		*Necrovirus*	Büttner and Nienhaus, 1989a
	Not identified			Nienhaus, 1985
	Not identified	Not visualized		Kim and Fulton, 1973; Büttner and Führling, 1993, 1996
Robinia	*Peanut stunt virus,* PSV (formerly *Robinia mosaic virus*)		*Cucumovirus*	Atanasoff, 1935
	SLRSV		*Nepovirus*	Bang *et al.*, 2006; Borodynko *et al.*, 2007
Salix	ToMV		*Tobamovirus*	Kopp *et al.*, 1999
	TNV		*Necrovirus*	
	BMV		*Bromovirus*	
Sambucus	CLRV (*Golden elderberry virus,* GEV)		*Nepovirus*	Hansen and Stace-Smith, 1971; Ellis *et al.*, 1992
	Tomato black ring virus (TBRV)		*Nepovirus*	Pospieszny *et al.*, 2004
	Elderberry latent virus (EILDV)		*Carmovirus*	Jones, 1972, 1974; Ellis *et al.*, 1992
Sorbus	ApMV		*Nepovirus*	Polák and Zieglerová, 1997
	CLRV		*Nepovirus*	Rebenstorf *et al.*, 2006
	ACLSV		*Trichovirus*	Sweet, 1980
	European mountain ash ringspot-associated virus (EMARaV)		*Emaravirus*	Ebrahim-Nesbat and Izadpanah, 1992; Mielke *et al.*, 2008; Kallinen *et al.*, 2009; Mielke-Ehret *et al.*, 2010; Valkonen and Rännäli, 2010; Mühlbach and Mielke-Ehret, 2011
Ulmus	*Elm mottle virus* (EMoV)		*Ilarvirus*	Schmelzer *et al.*, 1966; Schmelzer, 1969; Jones and Mayo, 1973; Jones, 1974
	CLRV (formerly *Elm mosaic virus*)		*Nepovirus*	Swingle *et al.*, 1941, 1943; Ford *et al.*, 1972; Schmelzer, 1972
	ToRSV		*Nepovirus*	Varney and Moore, 1952
	TBSV		*Tombusvirus*	Novák and Lanzová, 1980
	Not identified	Filamentous		Bandte *et al.*, 2004

[a]Particle morphology is only accounted for viruses that are not yet identified; those that are not visualized in the electron microscope are termed 'not visualized'.
[b]The virus genus is given for all classified virus species, which are written in italic letters, as is the genus in which it is classified.

first detected and demonstrated to be the causal agents of those diseases. As stated above, virus symptoms often go unrecognized; this is because they can produce visible symptoms in some varieties but remain latent in others. Conversely, some viruses produce distinctive symptoms, whereas others are more difficult to detect visually. As an example, a wide distribution of *Cherry leaf roll virus* (CLRV) has been confirmed in Finnish birch species (*Betula* spp.) (Jalkanen *et al.*, 2007; von Bargen *et al.*, 2009) where they cause significant reductions in tree vitality and yield, whereas CLRV infection in olive trees is symptomless and does not cause significant damage (Langer *et al.*, 2010). Furthermore, symptoms caused by mineral deficiency, ozone or drought can be confused with chlorotic and necrotic discolorations induced by fungal or viral pathogens and sucking insects. Grafting and budding failures, which still occur frequently in fruit and ornamental trees, were often considered to be the result of incompatibility problems associated with a particular rootstock/scion cultivar combination. In reality, viruses are frequently responsible for these complete or partial incompatibilities and bud-take problems. This problem is often observed in fruit trees affected by viruses (Németh, 1986).

It is known that CLRV in birch leads to degeneration and in many other tree species to decline (Table 3.2). Similar effects are known from decline-affected European mountain ash (*Sorbus aucuparia* L.) infected with EMARaV, which produces chlorotic ringspots, mottling and staghead (Table 3.3, Plate 1(H)). Other symptoms, such as characteristic virus-like symptoms in oak, associated with degenerated twigs and a distinct loss of vigour have not yet been correlated with a viral pathogen (Table 3.4).

Losses from virus-diseased forest trees are difficult to measure unless trees are visibly damaged or deficiencies in wood quality are noticed. Furthermore, variations in damage by one particular pathogen or by interacting pathogens can, over time, cause different degrees of losses between growing regions or even forest stands. In addition, different management practices, cultivar and species differences, differences in virulence of the virus isolate or strain, and climate conditions can hamper an accurate assessment of virus-induced losses. However, knowledge and information on losses are essential in order to determine economic thresholds for effective control measures.

Viruses are pathogens of small size. They can only be observed by electron microscopy. The structure of a virus is given by its coat proteins, which surround the viral genome. Over 50% of known plant viruses are rod shaped and between 300 and 500 nm in length and 15 and 20 nm in diameter. The second most common structure is formed by isometric particles 40–50 nm in diameter. In addition, a very small number of plant viruses have a lipid envelope around the viral capsid. Viruses are obligate parasites and have no energy metabolism of their own; consequently they require living cells to replicate. They may induce metabolic and structural disarrays in plant cells to varying degrees depending on the viral species and plant susceptibility. Once entry into the cell is achieved, nucleic acids, amino acids and enzymes of the host are taken over by viruses for their replication, placing additional demands on host metabolism. This host–pathogen interaction shows how viruses alter plant vitality.

As shown in fruit trees, virus infections can affect the size, shape and quality of wood and seeds, and can induce coloured patterns on and malformations of leaves. For instance, *Citrus tristeza virus* (CTV), which represents one of the major threats to citrus production worldwide, can cause three different syndromes, depending on the virus strain and on scion cultivar rootstock combinations. These are: (i) decline; (ii) yellowing and stunting; and (iii) stem pitting (Moreno *et al.*, 2008). The stem pitting is accompanied by lower yield and diminished fruit quality, independently of the rootstock. Another devastating viral disease, Little cherry disease (LChD), caused by members of the *Closteroviridae* family, is reported in *Prunus* spp. (Ludvíková and Suchá, 2011). Fruits of sensitive cultivars do not mature up to picking time, and therefore most cherries of diseased trees are pointed in shape, imperfectly coloured, small and insipid in taste; they are

Table 3.2. Main characteristics of *Cherry leaf roll virus* (CLRV).

Significance and distribution	Numerous plant species have been recorded that are affected by CLRV (Jones, 1985). Natural host range comprises at least 18 genera of broadleaved trees and shrubs as well as a variety of herbaceous plants. The virus is widespread in forest tree species, for example throughout Europe and North America (Büttner *et al.*, 2011). Incidence of CLRV has been reported mostly from tree species of the temperate regions, but it was shown by von Bargen *et al.* (2009) that the virus was also detectable in broadleaved trees north of the Arctic Circle. CLRV is most commonly detected in birch (*Betula* spp.), black elderberry (*Sambucus nigra* L.), European mountain ash (*Sorbus aucuparia* L.) and European ash (*Fraxinus excelsior* L.) (Cooper and Massalski, 1984; Führling and Büttner, 1997; Rebenstorf *et al.*, 2006; Jalkanen *et al.*, 2007).
Diagnosis	Most commonly, CLRV induces leaf roll, chlorotic ringspots and vein-banding symptoms in leaves. However, symptoms differ depending on plant species, virus strain, location and season. Usually, CLRV can be detected by mechanical transmission to herbaceous indicator plants. Another possibility is transmission by grafting to woody plants in their early physiological stages, which has been shown for diverse birch species, sweet cherry (*Prunus avium* (L.) L.) and blackberry (*Rubus* spp.). However, a long latent period has to be expected until the appearance of symptoms, as summarized in Büttner *et al.* (2011). Electron microscopy has been applied for detection of CLRV-like particles in infected birch species (Hamacher and Giersiepen, 1989). Detection and identification of viruses in infected plants by immunosorbent electron microscopy (ISEM) was reported to be as sensitive as ELISA (Lesemann, 1982). However, the most sensitive detection of CLRV from woody hosts is by immunocapture-reverse transcription-PCR (IC-RT-PCR), as established by Werner *et al.* (1997). This technique has been refined in order to allow the routine testing of large sample numbers of woody hosts (Gentkow *et al.*, 2007) as well as the characterization of CLRV variants by restriction fragment length polymorphism (RFLP) analyses (Buchhop *et al.*, 2009).
Biology, epidemiology and management	The genome of CLRV contains two (+) single-stranded (ss)RNA segments, which are separately encapsidated in isometric particles, with a diameter of approximately 28 nm (Büttner *et al.*, 2011; von Bargen *et al.*, 2012). Both particles are required to establish an infection in host plants (Jones and Duncan, 1980). CLRV is a seed- and pollen-borne virus (Mink, 1993; Johansen *et al.*, 1994) that infects seedlings of woody plants. Dissemination of CLRV by infected pollen is considered to be an important natural mode of transmission in forest tree species (Card *et al.*, 2007). The virus can be horizontally transmitted between adjacent trees, as well as transmitted vertically from mother plants to progenies by fertilization with CLRV-carrying pollen. CLRV is also readily transmitted by water (Bandte *et al.*, 2007) and root connections, and might therefore be soil transmittable (Büttner *et al.*, 2011). It is not considered to be transmitted by nematodes. Insects may contribute to the infection of plants with CLRV in an indirect manner through wounds via virus-contaminated pollen, or pollinating insects may transfer infected pollen to healthy plants. It is of major importance to eliminate CLRV-infected trees as main virus reservoirs. However, large scale eradication is not advisable owing to ecological and economic constraints. As the impact of insect vectors in the dissemination of CLRV within and between host plant species is still unknown, combating putative vectors in forest stands cannot be recommended either. The lack of knowledge of key epidemiological aspects of virus dispersal illustrates that effective control of CLRV in forested areas is not possible. Hence, an integrated strategy relying on prophylactic control measures has to be established to prevent the spread of CLRV in forested areas. This must include the use of virus-free certified planting material, monitoring and early detection of the virus in symptomatic trees, followed by eradication of infected plants.

Table 3.3. Main characteristics of *European mountain ash ringspot-associated virus* (EMARaV).

Significance and distribution	EMARaV is a novel plant RNA virus with a multipartite genome of negative polarity, which infects European mountain ash (*Sorbus aucuparia* L.) in many parts of Europe, from the mountainous areas of Austria (Tyrol) in the south to Finland and Russia in northern Europe (Benthack *et al.*, 2005; Mielke *et al.*, 2008; Kallinen *et al.*, 2009).
Diagnosis	The characteristic symptoms are chlorotic ringspots and chlorotic mottling on leaves. In addition, affected trees often show reduced growth and gradual decay over years. The serological detection of EMARaV would require undertaking the laborious procedure of preparing nucleocapsid fractions from European mountain ash leaf extracts, followed by protein analyses by polyacrylamide gel electrophoresis and Western blotting. This technique is not well suited to the routine processing of large quantities of samples. In contrast, RNA preparation for reverse transcription-PCR (RT-PCR) is done by quick and simple standard procedures (i.e. methods based on the silica binding of RNA). Leaves, inner bark and vegetative buds can all be used as source of RNA, which allows diagnosis through all seasons of the year.
Biology, epidemiology and management	The means of natural transmission of EMARaV are still unclear. The agent has been shown to be transmissible by grafting (Führling and Büttner, 1995), but the first efforts to transmit the virus mechanically on to healthy European mountain ash seedlings or indicator plants by means of abrasion were only partially successful. Furthermore, the vertical distribution of EMARaV via seeds or pollen has not yet been reported. However, recent studies have revealed that an eriophyid mite, *Phytoptus pyri* Pagenstecher, might be a candidate vector of EMARaV (Mielke-Ehret *et al.*, 2010). Whenever European mountain ash seedlings produced in nurseries are used for new plantations, it is mandatory to test them for EMARaV infection before planting. Only virus-free seedlings, carefully tested by RT-PCR, must be used. However, trees in the wild are naturally dispersed via seeds or by the suckers of older trees, and while the transmission of EMARaV via seed has not yet been reported, vegetative transmission by suckers – a very common means of natural propagation of trees – is possible. In the case that the putative vector transmission of EMARaV by the eriophyid mite *P. pyri* is confirmed, the population density of these gall mites must be controlled. However, adult-stage gall mites escape from the galls and are distributed by the wind, therefore can cover large distances (Schliesske, 1995). This feature would make any forest phytosanitary measures that use chemical or mechanical treatments an extremely difficult task.

not fit for the fresh fruit market (Bajet *et al.*, 2008). Severe damage has occurred in particular in North America. Thus, in the Kooltenay region (British Columbia, Canada), where the disease was first observed in 1933, sweet cherry (*Prunus avium* (L.) L.) production dropped by roughly 90% within 30 years (1949–1979). Just recently, *Little sweet cherry virus* was detected in flowering and sweet cherry trees in China (Rao *et al.*, 2011).

Baur (1907) was the first to publish observations of an infectious chlorosis on *Fraxinus* spp. and *Sorbus* spp., though the nature of the causal agent was not known and even remained disputed for another century. In the case of *Sorbus* spp., Mielke and Mühlbach (2007) described the new virus EMARaV (Table 3.3). Kegler (1960) and Schmelzer (1966) carried out initial studies on virus infection by visually observing symptoms on the leaves and habit of forest trees. Overall, various investigations have significantly expanded our knowledge of viruses in forest ecosystems, and these are thoroughly summarized in an extensive review by Nienhaus and Castello (1989). Continuing surveys of forested areas, public greens and young seedlings in nurseries seem to confirm,

Table 3.4. Main characteristics of the graft-transmissible chlorotic ringspots of *Quercus robur* associated with a virus.

Significance and distribution	Surveys in nurseries, forest stands, park areas and roadsides have led to the registration of many English oaks (*Quercus robur* L.). In Northern Europe, seedlings and trees with characteristic virus-like symptoms such as chlorotic ringspots, chlorotic spots and mottling were observed (Büttner and Führling, 1996). Some of these plants also exhibited twig degeneration and suffered from a distinct loss of vigour. Over time, the proportion of those diseased oak seedlings in the total number of cultivated oak seedlings increases (Bandte *et al.*, 2002). Similar observations have been made in an oak gene conservation seed orchard established in 1992. Earlier studies have also observed these virus-like symptoms (Nienhaus, 1975; Nienhaus and Castello, 1989), but neither the aetiology nor the epidemiology of a viral agent has been described so far.
Diagnosis	Although mechanical transmission to indicator plants failed, the graft-transmissible properties of the causal agent give a prominent pointer to an infectious disease (Büttner and Führling, 1996). Infections with *Tobacco mosaic virus* (TMV), *Tobacco necrosis virus* (TNV), *Cherry leaf roll virus* (CLRV) and *Brome mosaic virus* (BMV) were ruled out by applying serological and molecular assays. The difficulty in purifying and visualizing virus particles prompted an attempt at double-stranded (ds)RNA detection.
Biology and epidemiology	Small dsRNA could be isolated from leaf, bud and bark tissue of trees independent of symptom development, geographic location and sampling date (Büttner and Führling, 1996). The physical characterization of these fragments suggested an infection with a non-symptomatic latent cryptic virus. This assumption is supported by the partial characterization of dsRNA fragments via reverse transcription-degenerate oligonucleotide primed-PCR (RT-DOP-PCR) and complementary (c)DNA cloning (Hahn, 2006). Sequence analysis revealed strong similarities to the RNA-dependent RNA polymerase (RdRp) associated with the family *Partitiviridae*. Additionally, a fragment of approximately 10 kb appeared in some samples. The size and its occurrence in different tissues and developmental stages pointed to the *Endornaviridae*, a new family of plant RNA viruses. These viruses share common properties, such as symptom-free infection of their hosts, transmission only via ovaries and pollen (vertical transmission), and the lack of formation of typical virions. In particular, they are not known to have an effect on their hosts. The observed characteristic chlorotic ringspots and mottling may neither be related to the presence of cryptic plant viruses nor to the presence of endornaviruses.

step by step, the dispersal of viruses in many deciduous trees.

In the Ninth Report of the ICTV, some 2285 virus and viroid species have now been listed (King *et al.*, 2011). Most of them are well-characterized viruses, but only a few have been detected in trees and shrubs. Investigations on viruses in the forest ecosystem are extremely rare compared with those on viruses of the agricultural and horticultural environments.

3.2 Diagnosis

The diagnosis of plant viruses involves more than just the detection of a disease, because detection is also the basis of the major means of control, i.e. effective quarantine, and thereby the means of prevention of disease establishment. The early and accurate diagnosis of plant diseases is a crucial component of any tree management system. Virus diseases can be managed most effectively when control measures are introduced in the early stages of plant growth. For example, there are effective tests for virus detection in seeds and mother plants, which are conducted before the initiation of cuttings. Further tools for control are mentioned later in the chapter in Section 3.4 on management strategies.

Many of the symptoms observable in virus-diseased trees suggest that alterations of endogenous levels of hormones or alterations in the metabolism of plant growth substances may be involved. Changes in endogenous hormone levels are the cause of symptoms associated with virus infection such as growth inhibition, tumour formation, and hyperplastic and hypoplastic changes in leaves. Virus infections can usually be recognized by mosaic-like leaf patterns, ringspots, lines and mottling of light and dark green. The infection often spreads over the whole leaf blade, having begun at one focal point. Young growing leaves infected during early development are often deformed or involute. Leaf areas that are yellow or slightly faded are considered chlorotic and develop around the primary site of infection. Areas with brown or dead tissue are considered necrotic. Chlorosis is caused by a breakdown of chlorophyll, resulting in a decreased rate of photosynthesis. Heavy infections are characterized by complete loss of chlorophyll in infected tissues, which are typically yellow-like due to the presence of only carotenoids.

Interactions between a host and a virus may result in visible symptoms. Infection causes changes in choloroplast morphology and cell/tissue metabolism, which adversely affect photosynthesis and, consequently, plant growth (Hamacher, 1994). Chloroplast abnormalities occur, such as changes in external morphology and size, disorganization or impaired development of lamellae, membrane vesiculation, increases in size and number of plastoglobuli, and an accumulation of phytoferritin and starch grains. Not all abnormalities are necessarily observed in one situation. They are also dependent on the severity of the symptoms induced by a particular strain of virus and on the type of tissue affected.

Virus infection also leads to a reduction in transpiration rate, which is often correlated with reduced leaf stomatal aperture. It has also been shown that the turgidity of a leaf both on and immediately after inoculation can influence the creation or development of infection sites and the apparent susceptibility of a plant to virus infection.

Reliance on symptoms alone is not adequate for proper identification of viruses. However, a thorough knowledge of symptomatology is absolutely essential to recognize virus-induced symptoms in the field (Plate 1).

Virus infections are often confused with symptoms caused by other biotic and abiotic factors, and these have to be distinguished by reliable diagnostic methods (Büttner and Führling, 1993). Advances in the development of modern methods for virus detection provide specific and sensitive tools that are needed for the detection of viruses in early stages of infection. However, it has to be understood that the detection of viruses in woody plants is much more difficult than in herbaceous hosts. The high amount of phenolic compounds, the irregular distribution of viruses in trees and the often low virus concentrations make detection complicated. These factors demand the development of techniques and methods that are adaptable to woody plants. In addition, complications in detection can develop when the trees are infected simultaneously by several viruses, a common phenomenon in nature. The complex nature of virus infections in trees requires development in the future of more refined diagnostic techniques for both the determination of known viruses and the identification of possibly unknown viruses.

As a general tool to detect viruses in forest trees is not available, we suggest a variety of techniques that have to be carried out in combination (Fig. 3.1). The methods suggested include: biological indexing, electron microscopic observations, antibody-based methods, double-stranded (ds)RNA-analysis, molecular hybridization and PCR. DNA microarrays have to be discussed as possible future tools. When using these diagnostic tools, it is important to have standard controls whenever tissue from healthy plants and cultivars is available.

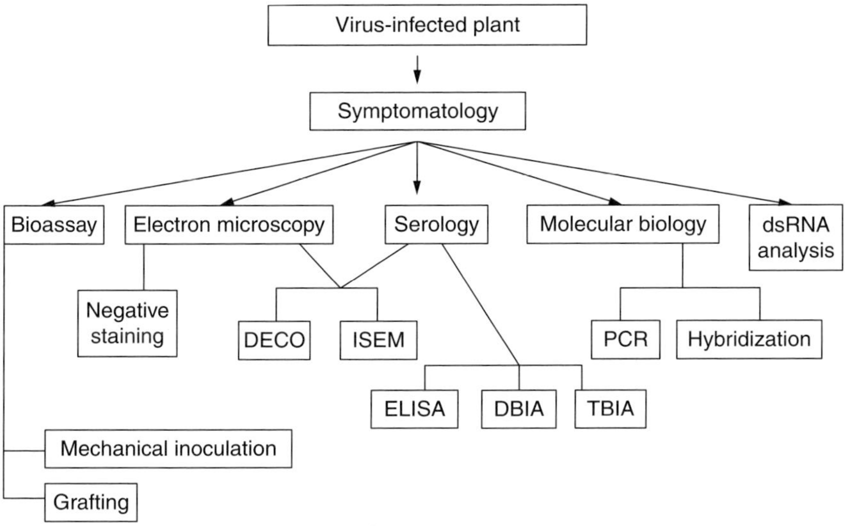

Fig. 3.1. Flow-chart of diagnostic methods and tools for the detection of viruses in trees. DBIA, dot-blot immunoassay; DECO, method of decoration of virus particles by antibodies (antibody binding); dsRNA, double-stranded RNA; ISEM, immunosorbent electron microscopy; TBIA, tissue-blot immunoassay.

3.2.1 Biological indexing

Biological indexing is a time-consuming, labour-intensive technique that requires greenhouse or orchard space. It is the method of choice for the detection of graft-transmissible pathogens for which no precise information is available. It enables the transmission of viruses to healthy plants and the demonstration of both known and unknown or uncharacterized agents, and confirms their infectious character. Biological indexing comprises two general methodologies: (i) the herbaceous host bioassay, which involves inoculation to a range of sensitive indicator cultivars; and (ii) the woody host bioassay, which relies on budding and grafting techniques. The first method requires that sample tissue is macerated in suitable buffer and rubbed on to the leaves of herbaceous plants, which are then incubated and observed for symptom development over time. Efforts to transmit a virus from a woody to a herbaceous plant can often be difficult due to phenolic compounds in the tissue. Nepoviruses such as CLRV can be reliably transmitted (Table 3.2), but other virus genera might be more difficult to transmit or may fail to induce symptoms and thus can escape detection. The techniques used in the woody host bioassay are well known from traditional techniques used to propagate cultivars. All viruses can be transmitted by grafting.

3.2.2 Electron microscopy

Electron microscopy is used to identify particle morphology as a complement to biological indexing or to confirm the viral origin. The sensitivity of this method is poor for detection purposes. It is not applicable to large-scale analyses and is not a suitable tool for detecting viruses directly in plant material from trees, because of the irregular distribution and the low concentration of the pathogens. Some improvement has been made to the method based on the use of antibodies. Combination of the high resolution of the electron microscopy with the specificity of the antibodies has a significant advantage in the detection process. The principle is the selective trapping of plant virus particles on grids that are pre-coated with specific antibodies (Roberts and Harrison, 1979; Russo *et al.*, 1980). The technique has also been coupled with the use of protein A or colloidal gold for a better decoration of the virus particles (Kalashjan and Lipartia, 1986). Immune electron microscopy is highly reliable when samples contain virus particles, but its application remains limited by the low number of samples that can be processed. These methods have been used to detect/locate CLRV on and in pollen from birch and walnut (*Juglans* spp.) trees (Massalski and Cooper, 1984; Massalski *et al.*, 1988).

3.2.3 ELISA

ELISA, first established by Clark and Adams (1977), is still an inexpensive and reliable assay for routine testing for the presence of viruses in leaves, stems, bark, flowers and roots. However, the quality of the test results depends on the quality and availability of the pathogen-specific polyclonal and monoclonal antibodies. Various formats have been developed and assayed based on the use of specific antibodies, as reported by Koenig and Paul (1982) and Cambra *et al.* (1991, 1997). Grüntzig *et al.* (1994) evaluated a sampling schedule for the best virus detection by ELISA considering their irregular distribution within the sample tree and demonstrated that the sampling point in the tree canopy has an influence on the reliability of the diagnosis.

The main versions of ELISA used are the double antibody sandwich (DAS) and triple antibody sandwich (TAS). Lateral flow devices and sticks have also been developed and applied. Their use is limited to the rapid diagnosis of materials showing virus-typical associated symptoms. According to Cambra *et al.* (2006), ELISA will remain the primary method of choice for the universal detection and routine screening of viruses in woody plants, using either a single monoclonal antibody or mixtures of monoclonal antibodies. However, the assays lack the sensitivity required for the detection of woody plant viruses, which are usually present in low titres in their hosts.

3.2.4 Hybridization and PCR

Hybridization and PCR are sensitive diagnostic techniques and enable virus detection in the smallest amounts. Specific DNA or complementary (c)DNA sequences are amplified *in vitro* from trace amounts in a complex mixture of templates. RNA viruses require the additional step of reverse transcription (RT) to convert their sequences from RNA to DNA before the amplification process begins.

The dot–blot hybridization is the most common molecular hybridization. Nucleic acid solution is directly applied on to a membrane such as nitrocellulose or nylon, followed by specific probes (Mühlbach *et al.*, 2003). The technique is based on the specific interactions between complementary purine and pyrimidine bases forming A–T and G–C base pairs. This interaction results in a stable hybrid formed by part of the nucleic acid sequence of the target molecule and the labelled complementary sequence. The dot–blot hybridization can be applied to detect particular plant RNA viruses in those cases where a genome specific probe is available. By varying the hybridization conditions, a sequence identity of 60–70% between the specific probe and the target RNA is sufficient for the detection. The dot–blot hybridization is well qualified for routine diagnostic testing of large-scale samples.

Werner *et al.* (1997) evaluated a method for detecting CLRV in seeds of birch and concluded that the use of specific antibodies for immune capture (IC), followed by RT-PCR, is the most sensitive technique for detecting viral RNAs. The method provides a cheaper and more reliable system for routine use. PCR is more specific than the hybridization and, moreover, enables the differentiation of virus strains.

3.2.5 DsRNAs detection

The tools for detecting dsRNAs led to a step forward when searching for an unknown single-stranded (ss)RNA virus. DsRNAs are formed during virus replication, and therefore consist of full-length genomes. After separation on gels, the size and patterns of these virus-specific dsRNAs can be useful in virus characterization and provide useful information on the type of virus (Valverde *et al.*, 1990). This technique is based on the premise that healthy plants not infected with a virus do not contain dsRNAs. Multiple dsRNA bands on gels after gel electrophoresis may indicate the detection of a monopartite virus that produces sub-genomic RNAs during its replication cycle. Such multiple bands can also indicate that the plant is infected by at least two viruses or by a replicating multipartite virus (Jelkmann, 1995). The method described by Morris and Dodds (1979) involves total nucleic acid extraction, followed by binding of dsRNA to cellulose and then its elution. The protocol of Tzanetakis and Martin (2008) has been adapted to woody plant material and been further modified to isolate dsRNA successfully from leaves of diseased English oaks (*Quercus robur* L.). The leaves show chlorotic ringspots caused by an unknown graft-transmissible agent (Table 3.4). The quality of extracted dsRNA is dependent on the amount of plant compounds such as glycoside, polysaccharides and polyphenols that are present and can interfere with electrophoretic mobility of nucleic acids. Alternatively, monoclonal antibodies to dsRNA are applicable to detect infection in plants by RNA viruses in crude nucleic acids (Lukács, 1994).

3.2.6 Tissue print

Tissue printing of plant material directly on to nitrocellulose membranes or filter paper is a suitable procedure for sample collection and storage off-site, prior to subsequent processing in the laboratory. Direct immunoprinting ELISA using monoclonal or recombinant antibodies without the need for extract preparation is the officially recommended protocol in the European Union (EU) for the detection of CTV (López *et al.*, 2003). Tissue print can also be coupled with amplification methods based on RT-PCR. The so-called print-capture procedure described by Olmos *et al.* (1996) for the detection of *Plum pox virus* (PPV) in peach and apricot seedlings provides similar sensitivity to that of immune capture-RT (IC-RT-PCR) applying PPV-specific antibodies.

3.2.7 Multiplex RT-PCR

A multiplex RT-PCR was reported by Saade *et al.* (2000), enabling the detection of *Prunus necrotic ringspot virus* (PNRSV), *Prune dwarf virus* (PDV) and *Apple mosaic virus* (ApMV), which are the three most important ilarviruses affecting stone fruits and forest trees (Table 3.1). Similarly, Menzel *et al.* (2002) established a multiplex RT-PCR for the reliable detection of four economically important apple viruses, including an internal control by co-amplification of the mitochondrial *nad*5 gene from plant mRNA. The assay allowed the reliable detection of different virus isolates from different geographic regions all year round. Coupling the amplification of multiple RNA target sequences with a colorimetric detection system called RT-PCR-ELISA (Menzel *et al.*, 2003) revealed that this technique was more reliable for the detection of *Apple chlorotic leaf spot virus* (ACLSV), *Apple stem grooving virus* (ASGV), *Apple stem pitting virus* (ASPV) and ApMV than indexing by woody indicators. Sánchez-Navarro *et al.* (2008) pushed multiplexing by RT-PCR to its limits by the development of an assay for the simultaneous identification of eight important viruses affecting stone fruit trees. The method included the detection of ApMV, PNRSV, PDV and ACLSV, which are also frequently found in important forest tree species (Table 3.1).

Apart from PCR, a protocol based on the loop-mediated isothermal amplification (LAMP) has been developed as a very sensitive diagnostic tool for the reliable detection of different strains of PPV in nectarines and peach trees (Varga and James, 2006). The major advantage of the recently reported one-step RT-LAMP for the fast detection of PPV in woody hosts is that it has the same sensitivity as RT-PCR paired with very high specificity and low artefact susceptibility, especially when combined with a simplified and standardized virus extraction method from woody plants (Hadersdorfer *et al.*, 2011).

Although this technique is not applied for the detection of viruses that are relevant in forest trees, it may be a promising low-cost alternative to the widely applied PCR in such instances.

3.2.8 Microarrays

There are only few published reports describing the use of microarrays for the detection of viruses relevant to woody hosts. The detection and genotyping of PPV using an oligonucleotide array has been employed by Pasquini *et al.* (2008), while Lenz *et al.* (2008) described an array for simultaneous detection of RT-PCR-amplified target sequences of several viruses from fruit tree samples. Abdullahi and Rott (2009) developed a microarray immunoassay by immobilizing specific antibodies on to glass slides followed by colorimetric detection of target antigens following the DAS procedure. The antibody microarray was suitable for the detection of several grapevine and fruit tree viruses, with sensitivities and specificities comparable to those of microtitre plate-based ELISA. Recently, Engel *et al.* (2010) and Abdullahi *et al.* (2011) reported the development of diagnostic oligonucleotide microarrays. These were designed for the simultaneous detection of up to 13 different grapevine viruses, including the nepoviruses *Arabis mosaic virus* (ArMV), and *Tomato ringspot virus* (ToRSV), which have also been reported in forest trees (Table 3.1). Although there are no DNA arrays available for the detection of the all-important viruses of forest trees, the multiplexing capacity of microarrays for use with samples from woody plants has been demonstrated by the above-mentioned examples. The method has to be considered as a powerful tool for the simultaneous detection of viruses in woody host species suitable, for instance, for high throughput plant certification purposes.

3.3 Infection Biology and Epidemiology

Plant viruses are distributed worldwide and they have been detected in different environments, such as woody and herbaceous plants, soil, surface waters, glacier ice, seawater and clouds. The mode of transmission has to be considered as an important factor influencing spread and impact of a virus infection, regardless of the plant involved (Büttner and Bandte, 2001). Vector or non-vector transmission, however, has to be determined. The great majority of plant viruses are dependent for their spread upon efficient transmission from infected to healthy plants by vectors. Prominent specific vector organisms for plant viruses are plant-feeding arthropods, nematodes and plant parasitic fungi. About 94% of known plant virus vectors are arthropods, with 55% of them being aphids (Ng and Falk, 2006). Whiteflies, thrips, mealy bugs, plant hoppers, grasshoppers, scales and a few beetles also serve as vectors for certain viruses. Some viruses may persist for weeks or months and even replicate in their insect vectors; others are carried for less than an hour. The vector transmission is a very specific process. A specific virus can only be transmitted by a single vector type. Conversely, each vector species can transmit one or more different viruses. Furthermore, some viruses are transmitted by seeds and pollen, and/or by soil and water. Mechanical transmission has to be regarded as an important manner, which often occurs in the form of human handling of plant material during production, or by animals carrying the virus on their body parts and/or mouthparts, and even after passage through the human alimentary tract (Tomlinson *et al.*, 1982).

3.3.1 Mechanical transmission

Most viruses of forest trees and urban greens spread from diseased plants by contact with each other, by infected propagative material or through natural root grafting to neighbouring trees. They are distributed inadvertently by man on a massive scale because of vegetative propagation via rooting cuttings during propagation, e.g. *Poplar mosaic virus* (PopMV) in poplar (*Populus* spp.). Budding and grafting, in which infected mother plants perpetuate the viruses, are not common for these types of trees. Viruses can be spread by direct transfer of sap through the contact of sap of the wounded plant with a healthy plant (Nienhaus *et al*., 1990). Such contact may occur during normal forestry management, such as damage caused by contact with human tools, or naturally by animals feeding on the plant.

3.3.2 Transmission through water and soil

Büttner and Nienhaus (1989b) detected plant viruses in water samples collected from forest districts in Germany. In this study, mechanically transmissible tobamoviruses, potexviruses and tombusviruses were isolated from water samples taken from creeks, ponds and drainage ditches. Several studies on the water transmission of viruses in different parts of the world have confirmed a relatively high concentration of several plant viruses in surface waters such as ditches, lakes, rivers and streams (Koenig, 1986; Rohwer *et al*., 2009). The significance of water transmission on the epidemiology and ecology of plant viruses has not been well studied. Our own investigations and those of other scientists indicate that water transmission of viruses occurs more frequently than was once thought (Büttner and Koenig, 2013). During a survey of plant viruses in small ponds and creeks in a forested area in Germany, several viruses were isolated. The *Carnation Italian ringspot virus* (CIRV) was identified; this had previously been isolated only twice from carnations originating from Italy and USA, and infected carnations only with difficulty (Büttner *et al*., 1987). The role of this tombusvirus in the forest ecosystem is still unknown.

In the early 1980s, *Tomato bushy stunt virus* (TBSV) was used to demonstrate that humans can act as carriers of plant pathogens by consuming infected plants and shedding infective viral particles in their faeces (Tomlinson *et al*., 1982). It was suggested that plant viruses with no known vectors, such as most tobamoviruses and tombusviruses, may have a certain 'alimentary resistance' and stay intact after passing through the alimentary tract (Tomlinson *et al*., 1982; Zhang *et al*., 2006), which enables humans and other animals to act as carriers of these viruses in all ecosystems, including forested areas. In a metagenomic study of the RNA viral communities in human faeces, sequence portions of more than 30 different plant viruses, especially of tobamoviruses, tymoviruses, marafiviruses, maculaviruses, carmoviruses and necroviruses were detected (Rohwer *et al*., 2009). These findings are important for those working on virus epidemiology and involved in forest tree health because this form of transmission has an impact on plant protection services, quarantine stations, breeders, nurserymen and research laboratories.

Analysis of soils from forest stands in Germany using the bait-plant technique confirmed the presence of potexviruses, tobamoviruses, necroviruses and potyviruses (Büttner and Nienhaus, 1989a).

The need and use of water and soil for crop production makes it necessary to be fully aware of the importance that these media have as a source and vehicle of virus movement. The viruses of highest significance in the water-borne and soil-borne transmission of infectious diseases are essentially those that have the ability to maintain their infectivity in this environment even though viruses cannot replicate outside the cells of hosts. Most viruses that are passively transmitted in this manner are released from disturbed plant roots, are very stable, reach high concentrations in plants, and infect a wide range of plant species and genera. Virus particles are known to be absorbed to clay particles and organic plant debris (Kegler *et al*., 1995). The stability of the coat protein explains their longevity in water and soils, and their stability in extreme environments (Koenig, 1986; Tomlinson, 1987). Viruses can be transported by water and soil from infected to healthy plants, where they can cause an infection after entering the plant passively through their roots (Schwarz *et al*., 2010).

Furthermore, it is remarkable that viruses have been found in ancient glacial ice (Castello *et al*., 1999). *Tomato mosaic virus* (ToMV), a very stable plant virus with a wide host range that includes spruce, was detected in this environment (Castello *et al*., 1995; Bachand and Castello, 1998). ToMV-infected red spruce (*Picea rubens* Sarg.) seedlings are affected by a reduction in seedling height, weight, root volume, bud break and mean shoot length when compared with control seedlings (Bachand *et al.*, 1996). Surprisingly, needles of infected seedlings were less susceptible to freezing damage than those of control seedlings. Fillhart *et al.* (1997) postulated that the atmospheric spread of infectious plant viruses without invertebrate vectors represents a potentially long-distance transport mechanism for stable plant viruses like ToMV. The authors detected ToMV in more than half of the investigated cloud samples collected from the summit of a mountain in New York and in collection sites along the coast of Maine, USA. Perhaps the virus becomes airborne as a result of ToMV contaminated soil particles that serve as cloud condensation nuclei.

3.3.3 Transmission through vectors

Vectors are of major importance in the dynamics of virus populations and they have a strong influence on virus aetiology. Many species of invertebrates, nematodes, fungi and fungal-like organisms are involved in virus transmission.

Insects

When all known plant viruses are considered, approximately 70% are insect transmitted. Aphids constitute the most important group of vectors, because of their natural abundance in the ecosystem and due to their feeding behaviour. Vectors are classified based on the properties of the relationship established, including length of time required for acquisition, latency and retention. Two major categories of transmission are distinguished, circulative and non-circulative, based on the sites of retention and routes of movement through the vector (Matthews, 1991). Circulative transmission can be classified into: (i) non-propagative, when the virus does not replicate in the vector although it reaches the digestive tract and haemolymph, and from there the salivary glands, from which it is inoculated during the subsequent feeding; and (ii) propagative, when the virus replicates inside cells of the vector and has a long-term association with the vector. These categories of plant viruses encode genes that are differentially expressed in their infection cycle during the infection of plants or insects. They belong to the families *Bunyaviridae*, *Reoviridae* and *Rhabdoviridae*, and to the genera *Marafivirus* and *Tenuivirus*.

Numerous experiments on vector transmission of the above-mentioned viruses have been carried out with herbaceous plants or fruit trees and grapevines, but not with common forest trees. Therefore, any information obtained on the potential transmission of viruses within forest trees is more a logical consequence than a proof of experiment. There are only a few studies on vectors within the forest context.

Nematodes

Only a small number of genera of nematodes are involved in plant virus transmission, but they are important and widespread. Some soil-borne viruses are transmitted by these plant parasitic soil-borne nematodes to horticultural plants (Brown *et al.*, 1995). Their movement is limited, and their activity greatly depends on the texture and water content of the soil. ArMV, a member of the nepoviruses and a well-known pathogen of a number of woody hosts, is transmitted by species in the two genera of *Xiphinema* and *Longidorus*. Members of the tobraviruses are transmitted by species of *Trichodorus*. Once acquired, viruses may persist in transmissible form in nematodes for up to a year or more, depending on the species. The viruses do not replicate in the nematode vector. They are bound reversibly and in a specific manner to the upper regions in the digestive tract of the nematodes, from which they are released when the nematode's saliva is pumped into the host cells during the feeding process. Tobraviruses, as well as nepoviruses, have wide host ranges, which aids their survival in a given region.

Fungi and fungal-like organisms

Some plant viruses in several genera are transmitted by plant pathogenic fungi, the majority to herbaceous host plants such as grain crops and sugarbeet. A successful transmission is dependent on the way in which the fungus carries the virus (Adams, 1991).

Furoviruses, Pecluviruses, Pomoviruses, Benyviruses and bymoviruses are transmitted by species of *Polymyxa* or *Spongospora*, which were previously classified as fungi but have now been assigned to the protozoa. Viruses are carried in zoospores, which are released into the surrounding aqueous medium and carry the viruses inside the protoplast. These organisms deposit viruses into root cells when infecting the roots of a new host plant. The virus may also remain infectious in resting spores in air-dried soil for many years. Viruses transmitted by *Olpidium brassicae* (Woronin) P.A. Dang and *O. bornovanus* (Sahtiy.) Karling, such as *Tobacco necrosis virus* (TNV), are carried on the surface of the spores. Viruses transmitted by *Olpidium* spp. do not remain infectious in air-dried soil and are characteristic in having a wide host range.

Knowledge of the importance of the fungal transmission of viruses to forest tree species is for the most part non-existent and urgently requires research input. About 40 years ago, particles resembling *Tobacco mosaic virus* (TMV) were found in preparations of five out of 12 rust (*Pucciniales*) and two out of seven powdery mildew (*Erysiphales*) species (Yarwood and Hecht-Poinar, 1973). The strains were transmissible when asexual spores (conidia) were dusted on to leaves of *Chenopodium quinoa* Willd. The recovery of infective virus from oaks was reported formerly by Yarwood and Hecht-Poinar (1970). Nienhaus (1971) confirmed the hypothesis of fungal transmissibility by isolating TMV from oaks (*Quercus agrifolia* Née, *Q. phellos* L.) as well as from powdery mildew (*Sphaerotheca lanestis* Harkn.) from the same diseased oaks.

3.3.4 Seed transmission

Forest viruses may be responsible for a wide range of economic losses due to their impact on tree seed survival and growth, and also because of the long distance spread of seeds. For example, the germination of seeds from CLRV-infected birch trees is strongly reduced (Cooper and Atkinson, 1975). Presumably the embryo is invaded during the transmission of CLRV by seeds.

Two general types of seed transmission can be distinguished. First, transmission may be the result of contamination of the seed coat with virus, resulting in subsequent infection of the germinating seedling by mechanical means. The external virus can be inactivated by certain treatments, thus eliminating almost all seed-borne infection. In the second and more common type of seed transmission, the virus is found within the tissues of the embryo, which may become infected through the ovary or via the pollen. Most seed-transmitted viruses are also transmitted through the pollen from infected plants, and all known pollen-transmitted viruses are seed transmissible. The mechanisms by which pollen transmits viruses are either through their presence within the sperm cell nucleus or cytoplasm or, alternatively, their presence on the exine of the pollen grains. The germ tubes growing from such pollen may then pick up virus particles and actively carry them to the ovule.

Viral infection of the embryo axis has been associated with the transmission by seed of numerous viruses that are important to forest trees and shrubs, such as Ilarviruses and Nepoviruses, e.g. ArMV, *Strawberry latent ringspot virus* (SLRSV) and *Tobacco ringspot virus* (TRSV). Virus replication within the embryo is presumably reinitiated when the metabolism of infected cells increases at germination (Johansen *et al.*, 1994). Cooper *et al.* (1986) detected Ilarviruses in sweet cherry seed, and PopMV in large-scale surveys of poplar germplasm. Cucumoviruses are transmitted through seeds and are very commonly found in cultivated plants and also in woody hosts, e.g. *cucumber mosaic virus* (CMV) in *Berberis thunbergii* DC. (Cooper, 1993). Tobamoviruses are known to be transmitted by contaminated seed coats and thereby infect the seedling at germination; these viruses have been detected in English oak and maple (*Acer* spp.) trees (Führling and Büttner, 1998).

3.4 Management Strategies and Tactics

Once infected with a virus, perennial woody trees cannot be cured. Therefore, forest tree virology plays a major role in conserving the economic value of a forest stand (Hubbes, 1993). There is substantial evidence that shows the potential impact of virus disease on forest trees.

The two requirements for effective control of viruses are the identification of the causal pathogen and the determination of the possible mechanisms of transmission; this was shown for CLRV by Hamacher and Quadt (1994). A strong basic knowledge of virus characteristics, experience in symptomatology and access to reliable and sensitive virus detection are prerequisites for the proper diagnosis of forest tree viruses and for their effective management. Virus control in forests is based on prevention, and so strategies have to be adapted to the origin of the planting material. Knowledge of the status of gene banks, nurseries, forest stands and trees in urban areas, such as roadside trees, adds supporting information for developing control plans.

With the exception of the elimination of infected and unthrifty trees, it is usually impractical, if not impossible, to prevent the spread of viruses between trees in the field once infected trees have been transplanted into the field or are growing under natural conditions. Attention, inevitably, has to be focused on the nursery production of planting stock, where critical examination and assessment of vigour can be practised routinely. Prevention is only possible when the pathogen is detected and its general properties, including its mechanisms of natural dispersal, are known. Further promising approaches to prevent the transmission of viruses within nurseries are the use of stringent hygienic measures, the use of surfactants to avoid the spread of viruses by tools and containers or in set-up areas, and the control of potential water or vector transmission of viruses. Plant nurseries are typically placed on open-textured soil that might potentially harbour soil-borne vectors and stable viruses. Healthy plants growing in this soil are commonly infected with viruses (Cooper, 1993). So it is recommended to leave wide headlands and to avoid planting seedlings in recently cleared areas that contained infected plants for at least 3 years, unless the soil is partially sterilized. Alternatively, trees should be propagated in steam-sterilized horticultural soil mixtures.

Virus-infected trees in forest stands cannot be eliminated on a large scale. Moreover, they have to be registered and the cause of virus emergence has to be analysed. A programme for the choice of tree species for future planting has to be developed, taking into consideration not only seedling stage but also plant health over a tree's lifetime, as well as the complication of natural succession forests. As mentioned previously, management strategies are prevention tools. They can be classified in three categories: (i) removal or avoidance of source of infection; (ii) control or avoidance of vectors; and (iii) protection of the plant from systemic disease. This requires breeding for resistance, the propagation of virus-free plants, the use of virus-free seeds and practices designed to reduce virus spread by vectors.

A continuous effort to breed virus-resistant plant varieties, especially for renewable resources (e.g. poplar species) should be a major future concern. Breeding, however, does not provide an advantage in all cases. One has to be aware that viruses continually mutate in the field with respect to both virulence and the range of plants they can infect. For instance Atreya *et al.* (1990) were the first to confirm the hypothesis that a coat protein mutation affects the aphid transmissibility of a potyvirus, and thus plays a role in determining aphid transmissibility. Evolutionary constraints on the emergence of plant RNA viruses are summarized and reviewed by Elena *et al.* (2011). Thereby, viral emergence results from an evolutionary process in which the main players are ecological factors, virus genetic plasticity and host factors. Hence, breeding for resistance or the development of transgenic plants are unlikely to give a permanent reliable solution.

When testing seeds, there are two protocols for the preparation of the samples: (i) direct preparation of the seeds, when possible; and (ii) using leaf tissue after the seeds have germinated. In all cases, the seed coat has to be removed before sample preparation. When viruses are transmitted through seed, major efforts should be made to control stock plants and seeds from nurseries and breeding programmes. Certification schemes would lead to access to virus-free seeds and mother plants, as has been achieved in the successful certification schemes in fruit crop production. It should be remembered that the development of high-quality certification in fruit trees took over 20 years within the EU. The time invested in developing certification for trees is a secondary consideration when fighting for a sustainable virus-free and healthy forest.

Using fruit trees grown in temperate areas that are affected by many viruses as an example, it should be possible to develop efficient strategies for virus elimination in forest trees over time. This can be achieved using a number of different techniques: (i) meristem tissue culture; (ii) thermotherapy *in vivo* or *in vitro*; (iii) a combination of *in vitro* thermotherapy and meristem tissue culture; and (iv) *in vitro* micrografting (Barba *et al.*, 1992). In all cases, pathogen elimination is a host–pathogen dependent process. Simplification of recommendations for certification should be avoided. However, tools are available that allow logical steps to be taken to obtain virus-free propagation material in forest trees. Worldwide plant health is a major concern to forestry phytosanitary agencies, simply because, as already noted, no effective cure exists to treat trees with established infections.

Virus testing in nurseries requires good diagnosis, assisted by information on symptoms, mode of transmission and mode of dispersal and host range. The detection of the progress of the epidemic has to be determined using data from: (i) plant material; (ii) age of the plant; (iii) season when samples are taken; and (iv) information on virus species. From our experience, leaves, stems and bark should always be tested in parallel to achieve optimum test results in a survey. The application of indicator plants for grafting is also recommended to confirm replication of a viral agent after transmission from diseased to healthy plants. This type of indexing is a common tool for the confirmation of virus detection for some viruses in fruit trees, such as ApMV, ACLSV and PPV.

References

Abdullahi, I. and Rott, M. (2009) Microarray immunoassay for the detection of grapevine and tree fruit viruses. *Journal of Virological Methods* 160, 90–100.

Abdullahi, I., Gryshan, Y. and Rott, M. (2011) Amplification-free detection of grapevine viruses using an oligonucleotide microarray. *Journal of Virological Methods* 178, 1–15.

Adams, M.J. (1991) Transmission of plant viruses by fungi. *Annals of Applied Biology* 118, 479–492.

Allen, W.R. and Davidson, T.R. (1967) *Tomato bushy stunt virus* from *Prunus avium* L.: I. Field studies and virus characterization. *Canadian Journal of Botany* 45, 2375–2383.

Atanasoff, D. (1935) Old and new viruses of trees and shrubs. *Phytopathologische Zeitschrift* 8, 197–223.

Atreya, C.D., Raccah, R. and Pirone, T.P. (1990) A point mutation in the coat protein abolishes aphid transmissibility of a *potyvirus*. *Virology* 178, 161–165.

Bachand, G.D. and Castello, J.D. (1998) Seasonal pattern of *tomato mosaic tobamovirus* infection and concentration in red spruce seedlings. *Applied and Environmental Microbiology* 64, 1436–1441.

Bachand, G.D. and Castello, J.D. (2001) Immunolocalization of *tomato mosaic tobamovirus* in roots of red spruce seedlings. *Journal of Phytopathology* 149, 415–419.

Bachand, G.D., Castello, J.D., Schaedle, M., Stehman, S.V. and Livingston, W.H. (1996) Effects of *tomato mosaic tobamovirus* infection on red spruce seedlings. *Canadian Journal of Forest Research* 26, 973–981.

Bajet, N.B., Unruh, T.R., Druffel, K.L. and Eastwell, K.C. (2008) Occurrence of two little cherry viruses in sweet cherry in Washington State. *Plant Disease* 92, 234–238.

Bandte, M., Schraudner, M. and Büttner, C. (2002) Viruserkrankungen an Jungpflanzen aus Baumschulen [Viral diseases of seedlings in nurseries]. In: Dujesiefken, D. and Kockerbeck, P. (eds) *Jahrbuch der Baumpflege 2002*. Thalacker, Braunschweig, Germany, pp. 196–202.

Bandte, M., Essing, M., Obermeier, C. and Büttner, C. (2004) Virus-diseased *Ulmus laevis* in eastern Germany. *Investigación Agraria: Sistemas y Recursos Forestales* 13, 65–69.

Bandte, M., Eschevarria-Laza, H.-J., Paschek, U., Ulrichs, C., Pestemer, W., Schwarz, D. and Büttner, C. (2007) Transmission of plant viruses by water. In: Fischer, G., Magnitskiy, S., Flórez, L.E., Miranda, D. and Medina, A. (eds) *Congresso Columbiana de Horticultura. Colombia Horticola: Retos y Oportunidades.*

Memorias, Sociedad Colombiana de Ciencias Hortícolas (SCCH), Bogotá. Produmedios, Bogota, Colombia, pp. 31–43.

Bang, J.H., Choi, J.K. and Lee, S.Y. (2006) Characterization of *Peanut stunt virus* isolated from black locust tree (*Robinia pseudoacacia* L.). *Plant Pathology Journal* 22, 125–130.

Barba, M., Martino, L. and Lauretti, F. (1992) Comparison of different methods to produce virus free stone fruits. *Acta Horticulturae* 309, 385–392.

Baur, F. (1907) Über infektiöse Chlorosen bei *Ligustrum, Laburnum, Fraxinus, Sorbus* und *Ptelea* [About infectious chlorosis in *Ligustrum, Laburnum, Fraxinus, Sorbus* and *Ptelea*]. *Berichte der Deutschen Botanischen Gesellschaft* 25, 410–413.

Benthack, W., Mielke, N., Büttner, C. and Mühlbach, H.-P. (2005) Double-stranded RNA pattern and partial sequence data indicate plant virus infection associated with ringspot disease of European mountain ash (*Sorbus aucuparia* L.). *Archives of Virology* 150, 37–52.

Bertioli, D.J., Hayle, A. and Cooper, J.I. (1993) A new virus isolated from an ash tree with dieback. *Journal of Phytopathology* 139, 367–372.

Biddle, P.G. and Tinsley, T.W. (1968) Virus diseases in conifers in Great Britain. *Nature* 219, 1387–1388.

Borodynko, N., Hasiów, B., Figlerowicz, M. and Pospieszny, H. (2007) Identification of the new strain of *Strawberry latent ringspot virus* isolated from black locust (*Robinia pseudoacacia* L.). *Journal of Phytopathology* 155, 738–742.

Brown, D.J.F., Robertson, W.M. and Trudgill, D.L. (1995) Transmission of viruses by plant nematodes. *Annual Review of Phytopathology* 33, 223–249.

Buchhop, J., von Bargen, S. and Büttner, C. (2009) Differentiation of *Cherry leaf roll virus* isolates from various host plants by immunocapture-reverse transcription-polymerase chain reaction-restriction fragment length polymorphism according to phylogenetic relations. *Journal of Virological Methods* 157, 147–154.

Büttner, C. and Bandte, M. (2001) Übertragungsmöglichkeiten von Forstvirosen und deren Bekämpfung [Transmission facilities and their control of forest viruses]. In: Dujesiefken, D. and Kockerbeck, P. (eds) *Jahrbuch der Baumpflege*. Thalacker-Medien, Braunschweig, Germany, pp. 230–233.

Büttner, C. and Führling, M. (1993) Beobachtungen zu virusbedingten Symptomen an Stieleichen (*Quercus robur* L.) – eine Abgrenzung zu ähnlichen nicht virusbedingten Krankheitsbildern [Studies on virus-induced symptoms of diseased oak trees (*Quercus robur* L.) – a delimitation to similar symptoms not caused by viruses]. *Nachrichtenblatt Deutscher Pflanzenschutzdienst* 45, 110–115.

Büttner, C. and Führling, M. (1996) Studies on virus infection of diseased *Quercus robur* (L.) from forest stands in northern Germany. *Annales des Sciences Forestières* 53, 383–388.

Büttner, C. and Koenig, R. (2013) Viruses. In: Hong, C., Moormann, G., Wohanka, W. and Büttner, C. (eds) *Biology, Detection and Management of Plant Pathogens in Irrigation Water*. APS Press, St Paul, Minnesota.

Büttner, C. and Nienhaus, F. (1989a) Virus contamination of soils in forest ecosystems of the Federal Republic of Germany. *European Journal of Forest Pathology* 19, 47–53.

Büttner, C. and Nienhaus, F. (1989b) Virus contamination of waters in two forest districts of the Rhineland area (FRG). *European Journal of Forest Pathology* 19, 206–211.

Büttner, C., Jacobi, V. and Koenig, R. (1987) Isolation of *carnation Italian ringspot virus* from a creek in a forested area south west of Bonn. *Journal of Phytopathology* 118, 131–134.

Büttner, C., von Bargen, S., Bandte, M. and Myrta, A. (2011) *Cherry leaf roll virus*. In: Hadidi, A., Barba, M., Candresse, T. and Jelkmann, W. (eds) *Virus and Virus-Like Diseases of Pome and Stone Fruits*. APS Press, St Paul, Minnesota, pp. 119–125.

Cambra, M., Camarasa, E., Gorris, M.T., Garnsey, S.M. and Carbonell, E. (1991) Comparison of different immunosorbent assays for *Citrus tristeza virus* (CTV) using CTV-specific monoclonal and polyclonal antibodies. In: Briansky, R.H., Lee, R.F. and Timmer, L.W. (eds) *Proceedings of the Eleventh Conference of the International Organization of Citrus Virologists, November 6–10, 1989, Orlando, Florida*. IOCV, University of California, Riverside, California, pp. 38–45.

Cambra, M., Olmos, A., Gorris, M.T., Duran, N., Roman, M.P., Camasara, E. and Dasi, M.A. (1997) Sensitive detection of plant pathogens by using immobilized targets in tissue imprinted membranes. In: Dehne, H.-W., Adam, G., Diekmann, M., Mauler-Machnik, A. and van Halteren, P. (eds) *Diagnosis and Identification of Plant Pathogens*. Kluwer Academic Publishers, Dordrecht, The Netherlands, pp. 95–98.

Cambra, M., Boscia, D., Myrta, A., Palkovics, L., Navrátil, M., Barba, M., Gorris, M.T. and Capote, N. (2006) Detection and characterization of *Plum pox virus*: serological methods. *EPPO Bulletin* 36, 254–261.

Card, S.D., Pearson, M.N. and Clover, G.R.G. (2007) Plant pathogens transmitted by pollen. *Australasian Plant Pathology* 36, 455–461.

Casalicchio, G. (1965) La picchiettatura cloro-necrotica del frassino maggiore [Punctation in chloro-necrotic ash]. *Monti e Boschi* 16, 39–46.

Castello, J.D., Amico, L.A. and O'Shea, M.T. (1984) Detection of tobacco mosaic and tobacco ringspot viruses in white ash trees by enzyme-linked immunosorbent assay. *Plant Disease* 68, 787–790.

Castello, J.D., Wargo, P.M., Jacobi, V., Bachand, G.D., Tobi, D.R. and Rogers, M.A.M. (1995) *Tomato mosaic virus* infection of red spruce on Whiteface Mountain, New York: prevalence and potential impact. *Canadian Journal of Forest Research* 25, 1340–1345.

Castello, J.D., Rogers, S.O., Starmer, W.T., Catranis, C.M., Ma, L., Bachand, G.D., Zhao, Y. and Smith, J.E. (1999) Detection of *tomato mosaic tobamovirus* RNA in ancient glacial ice. *Polar Biology* 22, 207–212.

Castello, J.D., Rogers, S.O., Bachand, G.D., Fillhart, R.C., Murray, J.S., Weidemann, K., Bachand, M. and Almond, M.A. (2000) Detection and partial characterization of tenuiviruses from black spruce. *Plant Disease* 84, 143–147.

Cech, M., Kralik, O. and Blattny, C. (1961) Rod-shaped particles associated with virosis of spruce. *Phytopathology* 51, 183–185.

Ciferri, R., Corte, A. and Rui, D. (1961) Due virosi del frassino: l'arricciamento fogliare con necrosi e la marmoreggiatura fogliare [Two viruses of ash: leaf curl with necrosis and leaf marbling]. *Rivista di Patologia Vegetale* 3, 241–250.

Clark, N.F. and Adams, A.N. (1977) Characteristics of the microplate method of enzyme-linked immunosorbent assay for the detection of plant viruses. *Journal of General Virology* 34, 475–483.

Cooper, J.I. (1975) *Arabis mosaic virus,* a cause of chlorotic symptoms in leaves of *Fraxinus excelsior* L. *Plant Pathology* 24, 114–116.

Cooper, J.I. (1993) *Virus Diseases of Trees and Shrubs,* 2nd edn. Chapman and Hall, London.

Cooper, J.I. and Atkinson, M.A. (1975) *Cherry leaf roll virus* causing a disease of *Betula* spp. in the United Kingdom. *Forestry* 48, 193–203.

Cooper, J.I. and Edwards, M.L. (1981) The distribution of *poplar mosaic virus* in hybrid poplars and virus detection by ELISA. *Annals of Applied Biology* 99, 53–61.

Cooper, J.I. and Massalski, P.R. (1984) Viruses and virus-like diseases affecting *Betula* spp. *Proceedings of the Royal Society of Edinburgh* 85B, 183–195.

Cooper, J.I. and Sweet, J.B. (1976) The detection of viruses with nematode vectors in six woody hosts. *Forestry* 49, 73–78.

Cooper, J.I., Edwards, M.L., Arnold, M.K. and Massalski, P.R. (1983) A tobravirus that invades *Fraxinus mariesii* in the United Kingdom. *Plant Pathology* 32, 469–472.

Cooper, J.I., Edwards, M.L. and Siwecki, R. (1986) The detection of *poplar mosaic virus* and its occurrence in a range of clones in England and Poland. *European Journal of Forest Pathology* 16, 116–125.

Ebrahim-Nesbat, F. and Izadpanah, K. (1992) Virus-like particles associated with ringfleck mosaic of mountain ash and a mosaic disease of raspberry in the Bavarian Forest. *European Journal of Forest Pathology* 22, 1–10.

Elena, S.F., Carrera, J. and Rodrigo, G. (2011) A systems biology approach to the evolution of plant–virus interactions. *Current Opinion in Plant Biology* 14, 372–377.

Ellis, P.J., Converse, R.H. and Stace-Smith, R. (1992) Viruses of *Sambucus canadensis* in North America. *Acta Horticulturae* 308, 69–80.

Engel, E.A., Escoar, P.F., Rojas L.A., Rivera, P.A., Fiore, N. and Valenzuela, P.D.T. (2010) A diagnostic oligonucleotide microarray for simultaneous detection of grapevine viruses. *Journal of Virological Methods* 163, 445–451.

Erdiller, G. (1986) *Acer* virus diseases in Turkey. *Journal of Turkish Phytopathology* 15, 46–59.

Ferris, M.A., Castello, J.D. and Sinclair, W.A. (1989) Effects of virus and mycoplasmalike organism infection on green and white ash. *Phytopathology* 79, 579–583.

Fillhart, R.C., Bachand, G.D. and Castello, J.D. (1997) Airborne transmission of *tomato mosaic tobamovirus* and its occurrence in red spruce in the northeastern United States. *Canadian Journal of Forest Research* 27, 1176–1181.

Flachmann, M., Lesemann, D.-E., Frenzel, B. and Koenig, R. (1990) Isometric virus-like particles in *Abies alba* Mill. and other *Abies* species: partial purification and improved detection by means of immunoelectron microscopy. *Journal of Phytopathology* 129, 193–202.

Ford, R.E., Moline, H.E., McDaniel, G.L., Mayhew, D.E. and Epstein, A.H. (1972) Discovery and characterization of *elm mosaic virus* in Iowa. *Phytopathology* 62, 987–992.

Führling, M. and Büttner, C. (1995) Transmission experiments of viruses to woody seedlings (*Quercus robur* L. and *Sorbus aucuparia* L.) by grafting and mechanical inoculation. *European Journal of Forest Pathology* 25, 129–135.

Führling, M. and Büttner, C. (1997) Viruserkrankungen an Laubgehölzen – zu beachten in Baumschulen, im Garten und Landschaftsbau sowie im Forst [Viral diseases of deciduous trees – to be observed in nurseries, gardening, landscaping and forestry]. In: Dujesiefken, D. and Kockerbeck, P. (eds) *Jahrbuch der Deutschen Baumpflege 1997*. Thalacker, Braunschweig, Germany, pp. 184–188.

Führling, M. and Büttner, C. (1998) Nachweis von Tobamo-Viren in Bergahorn (*Acer pseudoplatanus* L.) mit Scheckung und Blattdeformation [Detection of tobamoviruses in sycamore maple (*Acer pseudoplatanus* L.) with spotting and leaf deformation]. *Forstwissenschaftliches Centralblatt* 117, 92–97.

Gentkow, J., von Bargen, S., Petrik, K. and Büttner, C. (2007) Möglichkeiten zur Detektion des Kirschenblattrollvirus (CLRV) in Gehölzen durch serologische und molekulare Methoden [Opportunities for the detection of *Cherry leaf roll virus* (CLRV) in woody plants by serological and molecular methods]. In: Dujesiefken, D. and Kockerbeck, P. (eds) *Jahrbuch der Baumpflege 2007*. Thalacker, Braunschweig, Germany, pp. 279–284.

Giersiepen, R. (1993) Untersuchungen zur Charakterisierung des Kirschenblattrollvirus (CLRV) als Ursache einer Eschenvirose [Studies on the characterization of *Cherry leaf roll virus* (CLRV) as a cause of ash virosis]. Dissertation, Rheinische Friedrich-Wilhelms-Universität zu Bonn, Bonn, Germany.

Gotlieb, A.R. and Berbee, J.G. (1973) Line pattern of birch caused by *apple mosaic virus*. *Phytopathology* 63, 1470–1477.

Grüntzig, M., Fuchs, E., Pfeilstetter, E., Al Kai, B. and Otto, F. (1994) Untersuchungen zur Verteilung mechanisch übertragbarer Viren in Bäumen des kern- und Steinobstes [Studies on the distribution of mechanically transmissible virus in pome and stone fruit trees]. *Archives of Phytopathology and Plant Protection* 29, 217–248.

Hadersdorfer, J., Neumüller, M., Treutter, D. and Fischer, T.C. (2011) Fast and reliable detection of *Plum pox virus* in woody host plants using the Blue LAMP protocol. *Annals of Applied Biology* 159, 456–466.

Hahn, S. (2006) Virologische Untersuchungen an Stieleichen (*Quercus robur* L.) zum verursachenden Pathogen der pfropfübertragbaren chlorotischen Ringflecken [Virological studies of oak (*Quercus robur* L.) for the causative pathogen of graft-transmissible chlorotic ringspot]. Available at: www.openthesis.org/documents/Virologische-Untersuchungen-Stieleichen-Quercus-robur-523673.html (accessed 28 November 2011).

Hamacher, J. (1994) *Licht- und Elektronenmikroskopische Untersuchungen an Laubbäumen unter Verschiedenen Stresseinwirkungen* [*Light and electron microscope studies on deciduous trees under different stress conditions*]. Verlag Kovac, Hamburg, Germany.

Hamacher, J. and Giersiepen, R. (1989) Histologisch–cytologische Veränderungen in Kirschenblattrollvirus-infizierten und durch Luftschadstoffe gestressten Betula-Arten [Histological–cytological changes in cherry leaf roll virus-infected and air pollutant stressed *Betula* spp.]. *Nachrichtenblatt Deutscher Pflanzenschutzdienst* 41, 124–130.

Hamacher, J. and Quadt, A. (1991) Light- and electron microscopic studies of *Cherry leaf roll virus* (CLRV) on European ash (*Fraxinus excelsior* L.). *Journal of Phytopathology* 131, 215–226.

Hamacher, J. and Quadt, A. (1994) Isolation of *cherry leaf roll* and *brome mosaic viruses* from European beech and transmission to beech seedlings. *Plant Disease* 78, 849–853.

Hansen, A.J. and Stace-Smith, R. (1971) Properties of virus isolated from golden elderberry, *Sambucus nigra aurea*. *Phytopathology* 61, 1222–1229.

Hardcastle, T. and Gotlieb, A.R. (1980) An enzyme-linked immunosorbent assay for the detection of *apple mosaic virus* in yellow birch. *Canadian Journal of Forest Research* 10, 278–283.

Hentsch, T., Fuchs, E. and Grüntzig, M. (1997) Das Strawberry latent ringspot (?) nepovirus (SLRSV) an *Aesculus hippocastanum* L. [Strawberry latent ringspot (?) nepovirus (SLRSV) on *Aesculus hippocastanum* L.]. *Archives of Phytopathology and Plant Protection* 31, 13–23.

Hibben, C.R. and Bozarth, B.F. (1972) Identification of an ash strain of *Tobacco ringspot virus*. *Phytopathology* 62, 1023–1029.

Hibben, C.R. and Walker, J.T. (1971) Nematode transmission of the ash strain of *tobacco ringspot virus*. *Plant Disease Reporter* 55, 475–478.

Hibben, C.R., Bozarth, R.F. and Reese, J. (1979) Identification of *tobacco necrosis virus* in deteriorating clones of aspen. *Forest Science* 25, 557–567.

Horvath, J., Eke, I., Gal, T. and Dezsery, M. (1975) Demonstration of virus-like particles in sweet chestnut and oak with leaf deformations in Hungary. *Zeitschrift Pflanzenkrankheiten und Pflanzenschutz* 82, 498–502.

Hubbes, M. (1993) Impact of molecular biology on forest pathology: a literature review. *European Journal of Forest Pathology* 23, 201–217.
Jacobi, V. and Castello, J.D. (1992) Infection of red spruce, black spruce, and balsam fir seedlings with *tomato mosaic virus*. *Canadian Journal of Forest Research* 22, 919–924.
Jalkanen, R., Büttner, C. and von Bargen, S. (2007) *Cherry leaf roll virus* abundant on *Betula pubescens* in Finland. *Silva Fennica* 41, 755–762.
Jančařík, V. and Blattný, C. (1966) Brooming in Scots pine (*Pinus sylvestris*) in Czechoslovakia. *Lesnícky Časopis* 12, 507–520.
Jelkmann, W. (1995) *Cherry virus* A: cDNA cloning of dsRNA, nucleotide sequence analysis and serology reveal a new plant capillovirus in sweet cherry. *Journal of General Virology* 76, 2015–2024.
Johansen, E., Edwards, M.C. and Hampton, R.O. (1994) Seed transmission of viruses: current perspectives. *Annual Review of Phytopathology* 32, 363–386.
Jones, A.T. (1972) Purification and properties of *elderberry latent virus*, one of four sap-transmissible viruses obtained from American elder (*Sambucus canadensis* L.). *Annals of Applied Biology* 70, 49–58.
Jones, A.T. (1974) *Elm mottle virus. CMI/AAB Descriptions of Plant Viruses No. 139*, Commonwealth Mycological Institute, Kew, Surrey, UK, and Association of Applied Biologists, Wellsbourne, Warwick, UK.
Jones, A.T. (1985) *Cherry leaf roll virus. CMI/AAB Description of Plant Viruses No. 306.* Commonwealth Mycological Institute, Kew, Surrey, UK, and Association of Applied Biologists, Wellsbourne, Warwick, UK.
Jones, A.T. and Duncan, G.H. (1980) The distribution of some genetic determinants in the two nucleoprotein particles of *cherry leaf roll virus*. *Journal of General Virology* 50, 269–277.
Jones, A.T. and Mayo, M.A. (1973) Purification and properties of *elm mottle virus*. *Annals of Applied Biology* 74, 211–217.
Jones, A.T., Koenig, R., Lesemann, D.-E., Hamacher, J., Nienhaus, F. and Winter, S. (1990) Serological comparison of isolates of *cherry leaf roll virus* from diseased beech and birch trees in a forest decline area in Germany with other isolates of the virus. *Journal of Phytopathology* 129, 339–344.
Jones, R.A.C. (2004) Using epidemiological information to develop effective integrated virus disease management strategies. *Virus Research* 100, 5–30.
Kalashjan, J.A. and Lipartia, Z.M. (1986) *Apple chlorotic leaf spot virus* (ACLSV) in affected tissues and possibility of its detection by immune electron microscopy (IEM). *Acta Horticulturae* 193, 311–318.
Kallinen, A.K., Lindberg, I.L., Tugume, A.K. and Valkonen, J.P. (2009) Detection, distribution, and genetic variability of *European mountain ash ringspot-associated virus*. *Phytopathology* 99, 344–352.
Kegler, H. (1960) Das Ringfleckenmosaik der Eberesche (*Sorbus aucuparia* L.) [Ring spot mosaic of mountain ash (*Sorbus aucuparia* L.)]. *Phytopathologische Zeitschrift* 37, 214–216.
Kegler, H., Fuchs, E., Spaar, D. and Kegler, J. (1995) Viren in Boden und Grundwasser [Viruses in soil and groundwater]. *Archives of Phytopathology and Plant Protection* 29, 349–371.
Kim, K.S. and Fulton, J.P. (1973) Association of virus-like particles with a ringspot disease of oak. *Plant Disease Reporter* 57, 1029–1031.
King, A.M.Q., Adams, M.J., Carstens, E.B. and Lefkowitz, E.J. (eds) (2011) *Virus Taxonomy: Ninth Report of The International Committee on Taxonomy of Viruses*. Elsevier Academic Press, San Diego, California.
Koenig, R. (1986) Plant viruses in rivers and lakes. *Advances in Virus Research* 31, 321–333.
Koenig, R. and Paul, H.L. (1982) Detection and differentiation of plant viruses by various ELISA procedures. *Acta Horticulturae* 127, 147–158.
Kopp, R.F., Castello, J.D. and Abrahamson, L.P. (1999) Viruses in *Salix* grown for bioenergy. *European Journal of Forest Pathology* 29, 117–122.
Lana, A.O. and Agrios, G.N. (1974a) Transmission of a mosaic disease of white ash to woody and herbaceous hosts. *Plant Disease Reporter* 58, 536–540.
Lana, A.O. and Agrios, G.N. (1974b) Properties of a strain of *tobacco mosaic virus* isolated from white ash trees. *Phytopathology* 64, 1490–1495.
Lana, A.F., Thomas, O.T. and Peterson, J.F. (1980) A virus isolated from sugar maple. *Phytopathologische Zeitschrift* 97, 214–218.
Langer, J., von Bargen, S., Bandte, M., Hamacher, J. and Büttner, C. (2010) Auftreten und Bedeutung des *Cherry leaf roll virus* in Laubbäumen am Beispiel von Steinfrüchten [Occurrence and significance of *Cherry leaf roll virus* in deciduous trees on the example of stone fruits]. In: Dujesiefken, D. (ed.) *Jahrbuch der Baumpflege 2010*. Haymarket Media, Braunschweig, Germany, pp. 237–244.
Lenz, O., Petrzik, K. and Spak, J. (2008) Investigating the sensitivity of a fluorescence-based microarray for the detection of fruit-tree viruses. *Journal of Virological Methods* 148, 96–105.

Lesemann, D.-E. (1982) Advances in virus identification using immunosorbent electron microscopy. *Acta Horticulturae* 127, 159–171.

López, M.M., Bertolini, E., Olmos, A., Caruso, P., Gorris, M.T., Llop, P., Penyalver, R. and Cambra, M. (2003) Innovative tools for detection of plant pathogenic viruses and bacteria. *International Microbiology* 6, 233–243.

Löw, A. (1995) Untersuchungen zum ganzjährigen serologischen und immuncytologischen Nachweis von *Cherry leaf roll virus* (CLRV) und *Prune dwarf virus* (PDV) in Wildkirschen (*Prunus avium*) [Serological tests for year-round and immunocytological detection of *Cherry leaf roll virus* (CLRV) and *Prune dwarf virus* (PDV) in wild cherry (*Prunus avium*)]. Dissertation, Rheinische Friedrich-Wilhelms, Universität Bonn, Bonn, Germany.

Ludvíková, H. and Suchá, J. (2011) First report of little cherry disease from sweet cherry (*Prunus avium*) and sour cherry (*P. cerasus*) in the Czech Republic. *Plant Disease* 95, 1197.

Lukács, N. (1994) Detection of virus infection in plants and differentiation between coexisting viruses by monoclonal antibodies to double-stranded RNA. *Journal of Virological Methods* 47, 255–272.

Massalski, P.R. and Cooper, J.I. (1984) The location of virus-like particles in the male gametophyte of birch, walnut and cherry naturally infected with *cherry leaf roll virus* and its relevance to vertical transmission of the virus. *Plant Pathology* 33, 255–262.

Massalski, P.R., Cooper, J.I., Hellen, C.U.T. and Edwards, M.L. (1988) The effect of *cherry leaf roll virus* infection on the performance of birch pollen and studies on virus replication in germinating pollen. *Annals of Applied Biology* 112, 415–425.

Matthews, R.E.F. (1991) *Plant Virology*, 3rd edn. Academic Press, New York.

Menzel, W., Jelkmann, W. and Maiss, E. (2002) Detection of four apple viruses by multiplex RT-PCR assays with coamplification of plant mRNA as internal control. *Journal of Virological Methods* 99, 81–92.

Menzel, W., Zahn, V. and Maiss, E. (2003) Multiplex RT-PCR-ELISA compared with bioassay for the detection of four apple viruses. *Journal of Virological Methods* 110, 153–157.

Mielke, N. and Mühlbach, H.-P. (2007) A novel, multipartite, negative-strand RNA virus is associated with the ringspot disease of European mountain ash (*Sorbus aucuparia* L.). *Journal of General Virology* 88, 1337–1346.

Mielke, N., Weber, M., Khan, S. and Mühlbach, H.-P. (2008) Detection of *European mountain ash ringspot-associated virus* (EMARaV) in *Sorbus aucuparia* L. by a specific antiserum and reverse transcription-PCR. *Forest Pathology* 38, 371–380.

Mielke-Ehret, N., Thoma, J., Schlatermund, N. and Mühlbach, H.-P. (2010) Detection of *European mountain ash ringspot-associated virus*-specific RNA and protein P3 in the pear leaf blister mite *Phytoptus pyri* (*Eriophyidae*). *Archives of Virology* 155, 987–991.

Mink, G.I. (1993) Pollen- and seed-transmitted viruses and viroids. *Annual Review of Phytopathology* 31, 375–402.

Moreno, P., Ambrós, S., Albiach-Marti, M.R., Guerri, J. and Peña, L. (2008) *Citrus tristeza virus*: a pathogen that changed the course of the citrus industry. *Molecular Plant Pathology* 9, 251–268.

Morris, T.J. and Dodds, J.A. (1979) Isolation and analysis of double-stranded RNA from virus-infected plants and fungal tissue. *Phytopathology* 69, 854–858.

Mühlbach, H.-P. and Mielke-Ehret, N. (2011) Emaravirus. In: King, A., Lefkowitz, E., Adams, M.J. and Carstens, E.B. (eds) *Virus Taxonomy: Ninth Report of the International Committee on Taxonomy of Viruses*. Elsevier Academic Press, San Diego, California, pp. 767–770.

Mühlbach, H.-P., Weber, U., Gomez, G., Pallas, V., Duran-Vila, N. and Hadidi, A. (2003) Molecular hybridization. In: Hadidi, A., Flores, R., Randles, J.W. and Semancik, J.S. (eds) *Viroids*. CSIRO Publishing, Collingwood, Victoria, Australia, pp. 103–114.

Navrátil, S. and Boyer, M.G. (1968) The identification of *poplar mosaic virus* in Canada. *Canadian Journal of Botany* 46, 722–723.

Németh, M. (1986) *Virus, Mycoplasma and Rickettsia Diseases of Fruit Trees*. Akadémiai Kiado, Budapest.

Németh, M., Nyerges, K., Hangyál, R. and Kósa, G. (2010) Surveying viruses on ornamental trees and shrubs in two Hungarian botanical gardens and an arboretum. *Julius-Kühn-Archiv* 427, 293–299.

Ng, J.C.K. and Falk, B.W. (2006) Virus–vector interactions mediating nonpersistent and semi-persistent transmission of plant viruses. *Annual Review of Phytopathology* 44, 183–212.

Nienhaus, F. (1971) *Tobacco mosaic virus* strains extracted from conidia of powdery mildews. *Virology* 46, 504–505.

Nienhaus, F. (1975) Viren und virusverdächtige Erkrankungen in Eichen (*Quercus robur* und *Quercus sessiliflora*) [Viruses and suspected viruses in diseased oaks (*Quercus robur* and *Quercus sessiliflora*)]. *Zeitschrift Pflanzenkrankheiten und Pflanzenschutz* 85, 739–749.

Nienhaus, F. (1985) Infectious diseases in forest trees caused by viruses, mycoplasma-like organisms and primitive bacteria. *Cellular and Molecular Life Sciences* 41, 597–603.

Nienhaus, F. and Castello, J.D. (1989) Viruses in forest trees. *Annual Review of Phytopathology* 27, 165–186.

Nienhaus, F. and Yarwood, C.E. (1972) Transmission of virus from oak leaves fractionated with Sephadex. *Phytopathology* 62, 313–315.

Nienhaus, F., Ebrahim-Nesbat, F., Fricke, M., Büttner, C. and Welter, K. (1985) Investigations on viruses from declining beech trees (*Fagus sylvatica* L.) in Rhineland and Westfalia, Federal Republic of Germany. *European Journal of Forest Pathology* 15, 402–411.

Nienhaus, F., Büttner, C. and Hamacher, J. (1990) Virus infection of forest trees by mechanical transmission. *Journal of Phytopathology* 129, 141–150.

Novák, J.B. and Lanzová, J. (1980) Demonstration of *tomato bushy stunt virus* in some forest tree species and plants. *Lesnictví* 26, 1009–1016.

Olmos, A., Dasí, M.A., Candresse, T. and Cambra, M. (1996) Print-capture PCR: a simple and highly sensitive method for the detection of *plum pox virus* (PPV) in plant tissues. *Nucleic Acids Research* 24, 2192–2193.

Pasquini, G. *et al.* (2008) Oligonucleotide microarray based detection and genotyping of *Plum pox virus*. *Journal of Virological Methods* 147, 118–126.

Polák, Z. and Procházková, Z. (1996) *Apple mosaic virus* associated with decline of silver birch. *Ochrana Rostlin – UZPI* 32, 15–18.

Polák, Z. and Zieglerová, J. (1997) Spontaneous occurrence of *apple mosaic virus* in some forest and ornamental woody species. In: Cagan, L. and Praslicka, J. (eds) *Proceedings of Fourteenth Slovak and Czech Plant Protection Conference (Slovak Republic), 3–4 September 1997*. Slovak University of Agriculture, Nitra, Slovakia, pp. 87–88.

Pospieszny, H., Borodynko, N. and Jończyk, M. (2004) First report of *Tomato black ring virus* (TBRV) in the natural infection of *Sambucus nigra* in Poland. *Journal of Plant Protection Research* 44, 373–376.

Rana, T., Chandel, V., Hallan, V. and Zaidi, V.V. (2008) Characterisation of *Apple chlorotic leaf spot virus* infecting almonds in India. *Australasian Plant Disease Notes* 3, 65–67.

Rao, W.-L., Li, F., Zuo, R.J. and Li, R. (2011) First report of *Little cherry virus* 2 in flowering and sweet cherry trees in China. *Plant Disease* 95, 1484.

Rebenstorf, K., Candresse, T., Dulucq, M.J., Büttner, C. and Obermeier, C. (2006) Host species-dependent population structure of a pollen-borne plant virus, *Cherry leaf roll virus*. *Journal of Virology* 80, 2453–2462.

Roberts, I.M. and Harrison, B.D. (1979) Detection of *potato leafroll* and *potato mop-top viruses* by immunosorbent electron microscopy. *Annals of Applied Biology* 93, 289–297.

Rohwer, F., Prangishvili, D. and Lindell, D. (2009) Roles of viruses in the environment. *Environmental Microbiology* 11, 2771–2774.

Russo, M., Martelli, G.P. and Savino, V. (1980) Immunosorbent electron microscopy for detection of sap-transmissible viruses of grapevine. In: McGinnis, A.J. (ed.) *Proceedings of the Seventh Meeting of the International Council for the Study of Virus and Virus-like Diseases of the Grapevine*. Niagara Falls, Ontario, Canada. ICVG, Genolier, Switzerland, pp. 251–257.

Saade, M., Aparicio, F., Sánchez-Navarro, J.A., Herranz, M.C., Myrta, A., Di Terlizzi, B. and Pallás, V. (2000) Simultaneous detection of three ilarviruses affecting stone fruit trees by nonisotopic molecular hybridization and multiplex reverse-transcription polymerase chain reaction. *Phytopathology* 90, 1330–1336.

Sánchez-Navarro, J.A., Aparicio, F., Herranz, M.C., Minafra, A., Myrta, A. and Pallás, V. (2008) Simultaneous detection and identification of eight stone fruit viruses by one-step RT-PCR. *European Journal of Plant Pathology* 111, 77–84.

Schimanski, H.-H. and Funk, T. (1968) Untersuchungen über das Ausmaβ der Übertragung von Kirschenringflecken-Viren durch den Samen von Vogelkirschen (*Prunus avium* L.) [Studies on the extent of transfer of *Cherry ringspot virus* by the seeds of wild cherry (*Prunus avium* L.)]. *Archiv Gartenbau* 16, 143–150.

Schimanski, H.-H., Schmelzer, K., Kegler, H. and Albrecht, H.-J. (1975) Wild growing *Prunus* species of the subgenera *Prunophora* and *Padus* as natural host plants for cherry ringspot viruses. *Zentralblatt für Bakteriologie, Parasitenkunde, Infektionskrankheiten und Hygiene (Abteilung 2)* 130, 109–120.

Schliesske, J. (1995) *Gallmilben an Obstgewächsen: Morphologie und Symptomatologie* [Gall mites on fruit plants: morphology and symptomatology]. Schriftenreihe der Deutschen Phytomedizinischen Gesellschaft, UlmerVerlag Hohenheim, Germany.

Schmelzer, K. (1966) Untersuchungen an Viren der Zier- und Wildgehölze. 5. Mitteilung: Virosen an *Populus* und *Sambucus* [Studies on viruses of ornamental and wild shrubs. 5. Viruses of *Populus* and *Sambucus*]. *Phytopathologische Zeitschrift* 55, 311–351.
Schmelzer, K. (1969) Das Ulmenscheckungs-Virus [The *Elm mottle virus*]. *Phytopathologische Zeitschrift* 64, 39–67.
Schmelzer, K. (1972) Nachweis der Verwandtschaft zwischen Herkünften des Kirschenblattroll-Virus (*cherry leaf roll virus*) und dem Ulmenmosaik-Virus (*Elm mosaic virus*) [Proof of relationship between the origins of *Cherry leaf roll virus* and *Elm mosaic virus*]. *Zentralblatt für Bakteriologie, Parasitenkunde, Infektionskrankheiten und Hygiene (Abteilung 2)* 127, 140–144.
Schmelzer, K., Schmelzer, E.S. and Schmidt, H. (1966) Viruskrankheiten und virusverdächtige Erscheinungen an Forstgehölzen [Symptoms of virus diseases and suspect viruses in forest trees]. *Archiv für Forstwesen* 5, 107–120.
Schwarz, D., Beuch, U., Bandte, M., Fakhro, A., Büttner, C. and Obermeier, C. (2010) Spread and interaction of *Pepino mosaic virus* (PepMV) and *Pythium aphanidermatum* in a closed nutrient solution recirculation system: effects on tomato growth and yield. *Plant Pathology* 59, 443–452.
Shiel, P.J. and Castello, J.D. (1985) Detection of tobacco mosaic and tobacco ringspot viruses in herbaceous and woody plants near virus-infected white ash trees in central New York. *Plant Disease* 69, 791–795.
Smith, C.M. and Campbell, M.M. (2004) *Populus* genotypes differ in infection and systemic spread of *poplar mosaic virus* (PopMV). *Plant Pathology* 53, 780–787.
Smith, C.M., Rodriguez-Buey, M., Karlsson, J. and Campbell, M.M. (2004) The response of the poplar transcriptome to wounding and subsequent infection by a viral pathogen. *New Phytologist* 164, 123–136.
Soikkeli, S. (1983) Viruses in conifer needles in Finland: description of visible symptoms and ultrastructural aberrations of mesophyll tissue. *Communicationes – Instituti Forestalis Fenniae* 116, 77–83.
Šubíková, V. (1973) The mechanical transmission of *Euonymus mosaic virus*, maple leaf perforation by leaf extract of leaf nucleic acid to herbaceous plants. *Biologia Plantarum* 15, 166–170.
Sweet, J.B. (1980) Fruit tree virus infections of woody exotic and indigenous plants in Britain. *Acta Phytopathologica Academiae Scientiarum Hungaricae* 15, 231–238.
Sweet, J.B. and Barbara, D.J. (1979) A yellow mosaic disease of horse chestnut (*Aesculus* spp.) caused by *apple mosaic virus*. *Annals of Applied Biology* 92, 335–341.
Sweet, J.B. and Campbell, A.I. (1976) Pome fruit virus infections of some woody ornamental and indigenous species of *Rosaceae*. *Journal of Horticultural Science* 51, 91–97.
Swingle, R.U., Tilford, P.E. and Irish, D.F. (1941) A transmissible mosaic of elm. *Phytopathology* 13, 22 [abstr.].
Swingle, R.U., Tilford, P.E. and Irish, D.F. (1943) A graft transmissible mosaic of American elm. *Phytopathology* 33, 1196–1200.
Szirmai, J. (1972) An *Acer* virus disease in maple trees planted in avenues. *Acta Phytopathologica Academiae Scientiarum Hungaricae* 7, 197–207.
Thomas, P.R. (1970) Host status of some plants for *Xiphinema diversicaudatum* (Micol.) and their susceptibility to viruses transmitted by this species. *Annals of Applied Biology* 65, 169–178.
Tomlinson, J.A. (1987) Epidemiology and control of virus diseases of vegetables. *Annals of Applied Biology* 110, 661–681.
Tomlinson, J.A., Faithfull, E., Flewett, T.H. and Beards, G. (1982) Isolation of infective *tomato bushy stunt virus* after passage through the human alimentary tract. *Nature* 300, 637–638.
Tzanetakis, I.E. and Martin, R.R. (2008) A new method for extraction of double-stranded RNA from plants. *Journal of Virological Methods* 149, 167–170.
Valkonen, J.P.T. and Rännäli, M. (2010) First report of *European mountain ash ringspot-associated virus* (EMARaV) in *Sorbus aucuparia* from eastern Karelia, Russia. *Plant Disease* 94, 921.
Valverde, R.A., Nameth, S.T. and Jordan, R.L. (1990) Analysis of double-stranded RNA for plant virus diagnosis. *Plant Disease* 74, 255–258.
Varga, A. and James, D. (2006) Use of reverse transcription loop-mediated isothermal amplification for the detection of *Plum pox virus*. *Journal of Virological Methods* 138, 184–190.
Varney, E.H. and Moore, J.D. (1952) Strain of *tomato ringspot virus* from American elm. *Phytopathology* 42, 476–477.
Veliceasa, D., Enünlü, N., Kós, P.B., Köster, S., Beuther, E., Morgun, B., Deshmukh, S.D. and Lukács, N. (2006) Searching for a new putative cryptic virus in *Pinus sylvestris* L. *Virus Genes* 32, 177–186.
von Bargen, S., Grubits, E., Jalkanen, R. and Büttner, C. (2009) *Cherry leaf roll virus* – an emerging virus in Finland? *Silva Fennica* 43, 727–738.

von Bargen, S., Langer, J., Robel, J., Rumbou, A. and Büttner, C. (2012) Complete nucleotide sequence of *Cherry leaf roll virus* (CLRV), a subgroup C nepovirus. *Virus Research* 163, 678–683.
Werner, R., Mühlbach, H.-P. and Büttner, C. (1997) Detection of *cherry leaf roll nepovirus* (CLRV) in birch, beech and petunia by immuno-capture RT-PCR using a conserved primer pair. *European Journal of Forest Pathology* 27, 309–318.
Winter, S. and Nienhaus, F. (1989) Identification of viruses from European beech (*Fagus sylvatica* L.) of declining forests in Northrhine-Westfalia (FRG). *European Journal of Forest Pathology* 19, 111–118.
Yarwood, C.E. (1971) *Erysiphaceae* transmit virus to *Chenopodium*. *Plant Disease Reporter* 55, 342–344.
Yarwood, C.E. and Hecht-Poinar, E. (1970) A virus resembling *tobacco mosaic virus* in oak. *Phytopathology* 60, 1320.
Yarwood, C.E. and Hecht-Poinar, E. (1973) Viruses from rusts and mildews. *Phytopathology* 63, 1111–1115.
Zhang, T., Breitbart, M., Lee, W.H., Run, J.-Q., Wei, C.L., Soh, S.W.L., Hibberd, M.L., Liu, E.T., Rohwer, F. and Ruan, Y. (2006) RNA viral community in human feces: prevalence of plant pathogenic viruses. *PLoS Biology* 4(1): e3. doi: 10.1371/journal.pbio.0040003.

4 Forest Diseases Caused by Prokaryotes: Phytoplasmal and Bacterial Diseases

Helen M. Griffiths*
Cornell University, Ithaca, New York, USA

4.1 Pathogens, Significance and Distribution

4.1.1 Phytoplasmas

Historically, of the prokaryotes, phytoplasmas have caused the greatest economic damage to forest trees. Phytoplasmas are associated with the diseases of several hundred plant species, including a number of forest tree species, in many places in the world (Maramorosch and Raychaudhuri, 1981; McCoy *et al.*, 1989; Lee *et al.*, 2000). Most phytoplasmas have been detected in angiosperms, although recently phytoplasmas have also been confirmed in gymnosperms (Paltrinieri *et al.*, 1998; Schneider *et al.*, 2005; Śliwa *et al.*, 2008; Davis *et al.*, 2010; Valiunas *et al.*, 2010). The phytoplasmas that have been documented to be of concern to temperate forest tree species are listed in Table 4.1.

Because of the inability to culture phytoplasmas outside the plant or insect host, obtaining information about these organisms has been difficult. However, with the advent of molecular biology, significant advances have been made in understanding these pathogens. Based on housekeeping genes such as 16S rRNA, phytoplasmas have been shown to belong to the class *Mollicutes* (Lim and Sears, 1989; Gundersen *et al.*, 1994; Lee *et al.*, 2000, 2006; Martini *et al.*, 2007; Hodgetts *et al.*, 2008). Mycoplasmas, acholeplasmas and spiroplasmas belong to the same class, but phytoplasmas are considered to be more closely related to *Bacillus*, *Clostridium* and *Streptococcus* (Woese *et al.*, 1980; Gundersen *et al.*, 1994). For the identification, differentiation and preliminary classification of phytoplasmas, primer pairs based on the highly conserved 16S rRNA gene sequence have been developed. Using the restriction fragment length polymorphism (RFLP) analysis of the 16S rRNA gene, Lee *et al.* (1993) constructed a classification scheme for phytoplasmas. This scheme, with periodic updates (Lee *et al.*, 1998, 2000, 2004a,b), has provided reliable markers for the identification and classification of many phytoplasmas into groups and subgroups. In addition, the sequence information from the 16S rRNA genes has served as a baseline for the delineation of '*Candidatus* Phytoplasma' candidate species as recommended by the Phytoplasma Taxonomy Group of the International Research Program on Comparative Mycoplasmology (IRPCM Phytoplasma/Spiroplasma Working Team – Phytoplasma Taxonomy Group, 2004). Table 4.1 shows the relationship between the two phytoplasmal classification systems.

* E-mail: hmg1@cornell.edu

Table 4.1. Phytoplasmas commonly detected in temperate forest tree species.

Plant host	Common disease name	'*Candidatus* Phytoplasma' species	16S rDNA group (subgroup)	Country of occurrence	References
Alnus glutinosa (L.) Gaertn., *A. incana* (L.) Moench, *A. rubra* Bong.	Alder yellows	'*Ca.* Phytoplasma ulmi'	16SrV-C or AldY	Austria, Bosnia and Herzegovina, France, Germany, Hungary, Italy, Lithuania, North America, Switzerland	Lederer and Seemüller, 1991; Marcone *et al.*, 1994; Valiunas *et al.*, 2001; Berges and Seemüller, 2002; Lee *et al.*, 2004a
Castanea crenata Siebold & Zucc.	Chestnut witches' broom	'*Ca.* Phytoplasma castaneae'	16SrVI or CnWB	Kyongnam and Chonbuk Provinces, South Korea	Jung *et al.*, 2002
Cupressus spp.	X-disease	'*Ca.* Phytoplasma pruni' related	16SrIII	Italy	Paltrinieri *et al.*, 1998
Fraxinus americana L., *F. anomala* Torr. ex S. Watson, *F. excelsior* L., *F. pennsylvanica* Marsh., *F. uhdei* (Wenzig) Lingelsh., *F. velutina* Torr.	Ash yellows	'*Ca.* Phytoplasma fraxini', '*Ca.* Phytoplasma asteris'	16SrVII or AshY 16SrI-B	North America (Canada and USA), South America (Colombia), Poland	Sinclair and Griffiths, 1994; Walla *et al.*, 2000; Filgueira *et al.*, 2004; Kamiñska and Berniak, 2009
Juglans nigra L.	Walnut witches' broom	'*Ca.* Phytoplasma pruni' related	16SrIII-G/A or WWB	Georgia, USA	Chen *et al.*, 1992; Zhao *et al.*, 2009
Juniperus occidentalis Hook.	Juniper witches' broom	'*Ca.* Phytoplasma phoenicium' related	16SrIX-E or JunWB	Oregon, USA	Davis *et al.*, 2010
Larix sp.	Larch dwarfed needle proliferation	'*Ca.* Phytoplasma asteris'	16SrI-B	Ukraine	Jomantiene *et al.*, 2011
Pinus halepensis Mill., *P. sylvestris* L.	Phytoplasma disease of pine	'*Ca.* Phytoplasma pini'	16SrXXI-A	Germany, Lithuania, Poland, Spain	Schneider *et al.*, 2005; Śliwa *et al.*, 2008; Valiunas *et al.*, 2010
Populus alba L., *P. nigra* L., *P. tremula* L.	Aster yellows	'*Ca.* Phytoplasma asteris'	16SrI-P or AY	Croatia, France, Germany, Hungary	Berges *et al.*, 1997; Šeruga *et al.*, 2003; Musić *et al.*, 2008
Quercus robur L.	Oak decline	'*Ca.* Phytoplasma mali' related	16SrX or AP	Germany	Ahrens and Seemüller, 1994; Berges *et al.*, 2000
Salix babylonica L.	Yellows-type disease	'*Ca.* Phytoplasma asteris'	16SrI-C	Shaanxi Province, China	Wei *et al.*, 2009
Ulmus alata Michx., *U. americana* L., *U. chenmoui* W.C. Cheng, *U. crassifolia* Nutt., *U. laevis* Pall., *U. procera* Salisb., *U. pumila* L., *U. rubra* Muhl., *U. serotina* Sarg.	Elm yellows	'*Ca.* Phytoplasma ulmi'	16SrV-A	France, Germany, Italy, North America, Serbia	Mäurer *et al.*, 1993; Lee *et al.*, 1995, 2004a; Sfalanga *et al.*, 2002; Jović *et al.*, 2008, 2011; Romanazzi and Murolo, 2008a
Zelkova serrata (Thunb.) Makino (*Ulmaceae*)			16SrV-A		Romanazzi and Murolo, 2008a

Ash yellows disease caused by the phytoplasma AshY causes slow growth and decline in many ash species (*Fraxinus* spp.), and occurs widely in North America (Canada and the USA) and in Colombia (South America) (Sinclair and Griffiths, 1994; Sinclair *et al.*, 1994; Walla *et al.*, 2000; Filgueira *et al.*, 2004). From the 16 AshY strains obtained from the known range of ash yellows in North America, three different RFLP profiles were observed, indicating there is heterogeneity within the AshY group (Griffiths *et al.*, 1999b). By sequencing PCR products generated using 16S rRNA gene primers, AshY strains were shown to cluster as a discrete group, and this was given the 16S rRNA gene RFLP classification of 16SrVII (Lee *et al.*, 1998). Further, the proposed name is '*Candidatus* (*Ca.*) Phytoplasma fraxini' (Griffiths *et al.*, 1999b). In Poland, European ash (*F. excelsior* L.) trees were observed to have symptoms similar to those described by Sinclair *et al.* (1996). However, the phytoplasma was shown to be a member of the 16SrI group. This was the first record of a member of '*Ca.* Phytoplasma asteris' occurring in ash trees (Kamiñska and Berniak, 2009). Variation in strain virulence has been demonstrated for several phytoplasmas, including AshY. Sinclair and Griffiths (2000) showed that different strains of this pathogen caused significantly different degrees of growth suppression and loss of foliar greenness in green ash (*F. pennsylvanica* Marsh.).

The phytoplasma EY, the causal agent of elm yellows, is important in the decline of American elm (*Ulmus americana* L.) trees in North America, and of Eurasian and hybrid elms in Europe (Sinclair *et al.*, 1976; Marcone *et al.*, 1997; Griffiths *et al.*, 1999a). The EY phytoplasma group represents one of the largest. It includes phytoplasmal strains that would appear to share no ecological niche in that they affect different plant taxa, for example alder (*Alnus*), grape vine (*Vitis*), Virginia creeper (*Parthenocissus quinquefolia* (L.) Planch.), peach and cherry (*Prunus* spp.), and have been identified on plants in many parts of the world (Lee *et al.*, 2004a). Based on analysis of the 16S rDNA sequences, the elm yellows group was shown to form a discrete group, 16SrV (Lee *et al.*, 2004a). RFLP analysis of 16S rRNA gene sequences differentiated them into five subgroups in which all EY strains, whether from Europe or North America, were in the same subgroup. As all the EY strains from elm trees considered at the time fulfilled the requirements for '*Ca.* Phytoplasma' as proposed by IRPCM Phytoplasma/Spiroplasma Working Team – Phytoplasma Taxonomy Group (2004); phytoplasmas in 16SrV-A were designated members of the taxon '*Ca.* Phytoplasma ulmi', with $EY1^T$ the type strain (Lee *et al.*, 2004a).

However, Jacobs *et al.* (2003) determined that there was a phytoplasma associated with the decline of the elm trees in the urban area near Chicago (Illinois) and that it belonged to the clover proliferation group (16SrVI). The symptoms observed were not entirely consistent with those described for elm yellows; for example, the characteristic phloem discoloration and aroma of wintergreen oil (methyl salicylate) was not always experienced. In addition, some trees showing dieback were positive for the aster yellows (AY) phytoplasma (Jacobs *et al.*, 2003). It appears then that in North America there are at least three different phytoplasmas that are involved in the formation of symptoms similar to those characteristic of elm yellows. On further analysis using *secY* gene sequences, one of the Italian EY strains could be distinguished from EY1 from North America (Lee *et al.*, 2004a). Examination of 25 isolates from elm trees in different geographic regions of Serbia using four different genetic markers showed that these Serbian isolates could be assigned to four different subtypes of '*Ca.* Phytoplasma ulmi' (Jović *et al.*, 2011). Strain variation within the 16SrV-A group appeared to be significant, and these findings raise the question as to whether similar variation would be found in other '*Ca.* Phytoplasma' species if more isolates were examined using the genes utilized in the Serbian study.

Alder yellows phytoplasma (AldY) infects alder species (*Alnus* spp.) in North America and Europe (Lederer and Seemüller, 1991; Mäurer *et al.*, 1993; Cvrković *et al.*, 2008). The phytoplasma belongs to the 16SrV-C group and is closely related to some strains of phytoplasma Flavescence Dorée (FD), which is a quarantine pathogen of grapevine; it is, therefore, of significant concern (Davis and Dally,

2001; Arnaud *et al.*, 2007). Similar to AshY, AldY can have varying impact on alder trees depending upon the virulence of the strain. Avirulent and low virulence strains have been identified that do not result in disease symptoms (Berges and Seemüller, 2002).

Using molecular tools, an unspecified cypress species, western juniper (*Juniperus occidentalis* Hook.), Scots pine (*Pinus sylvestris* L.) and Aleppo pine (*Pinus halepensis* Mill.) (Paltrinieri *et al.*, 1998; Schneider *et al.*, 2005; Śliwa *et al.*, 2008; Davis *et al.*, 2010; Valiunas *et al.*, 2010) were all shown to be infected with phytoplasmas. These infected trees were identified in Italy, the western USA, Germany, Lithuania, Poland and Spain. The phytoplasmal strains were identified as members of group 16SrIII (X-disease), 16SrIX-E ('*Ca.* Phytoplasma phoenicium') and 16SrXXI ('*Ca.* Phytoplasma pini'). Recently, Jomantiene *et al.* (2011) identified a phytoplasmal strain belonging to group 16SrI subgroup B ('*Ca.* Phytoplasma asteris') in larch (*Larix* spp.) in Ukraine. As larch is widely distributed throughout Eurasia and is highly valuable for timber, this finding could have significant economic impact on the timber industry (Lin *et al.*, 2010). As the western juniper (Davis *et al.*, 2010) and the trees in the study reported by Schneider *et al.* (2005) showed symptoms that often result in tree mortality, these phytoplasmas are likely to be important gymnosperm pathogens.

4.1.2 Bacteria

Of the myriad of walled bacteria involved in plant diseases only a few have been found to be associated either directly or indirectly with forest tree diseases in woods and forestry settings.

Bacterial wetwood is common in a number of tree species and is found in many geographic locations. Several bacterial species have been associated with this condition, but none has been identified as the causal agent. There have been as many as 14 species of bacteria isolated from American elms showing symptoms characteristic of wetwood. *Enterobacter* and *Klebsiella* spp. are the most frequently isolated, but they are considered to be associated with, rather than the cause of, the symptoms (Murdoch and Campana, 1983). Boards cut from trees with wetwood often warp, and when they dry they may split, thus devaluing the lumber (timber). As sawdust and shavings are used extensively for bedding on dairy farms, this is an important source of *Klebsiella* spp. involved in mastitis, a serious disease of dairy cows that results in milk loss and culling on dairy farms in many parts of North America (Zadoks *et al.*, 2011).

Bacterial leaf scorch (BLS) caused by *Xylella fastidiosa* has been detected in a number of forest tree species from the mid-western states of North America, as far north as New York State, Ottawa, Ontario and through the Atlantic seaboard states, and in a number of the southern states (Sherald and Kostka, 1992). Locations have mainly been urban, except for Delaware and New Jersey where the disease has been detected in oak species (*Quercus* spp.) in several state forests.

X. fastidiosa is a xylem-inhabiting Gram-negative, aerobic, non-flagellated bacterium that has a large host range. It was cultured and identified as the cause of Pierce's disease (PD) in 1987 (Wells *et al.*, 1987). The whole genome – initially of a citrus strain of *X. fastidiosa* – was sequenced, and the sequencing has now been extended to three strains, including almond (*Prunus dulcis* (Mill.) D.A. Webb) and oleander (*Nerium oleander* L.) (Simpson *et al.*, 2000; Van Sluys *et al.*, 2003). Some phylogenetic studies have been performed and show that neither the elm nor oak inhabiting strains belonged to the same cluster as the grape-isolated strains (Hendson *et al.*, 2001; Rodrigues *et al.*, 2003).

For many years, *X. fastidiosa* has been of major concern in orchards and urban trees. It has been observed mainly in North America, but also in Brazil in citrus (Chang *et al.*, 1993), in Kosovo in grape (Berisha *et al.*, 1998) and in Taiwan in pear (Leu and Su, 1993). There are many wild plant species that carry the pathogen and they are often symptomless. It has only been since the beginning of 2000 that trees in woodlands and forested areas were observed dying from BLS, and, to date, only in the forests of New Jersey and Delaware have oaks been reported to be dying in areas where monitoring at several of their state forests is ongoing. In Kentucky, mature pin oaks (*Q. palustris* Münchh.) lining streets have been observed dying from BLS (Sherald and Kostka, 1992; Gould and Lashomb, 2007; Hartman, 2007). Elm was first found to be affected by the disease in the south-eastern states of North America as early as 1957 (Wester and Jylkka, 1959). At the time, the disease was thought to be due to a virus. More recently, leaf scorch of elm in urban settings has been observed in Maryland and New Orleans (Louisiana). Elm trees were reported to be infected with *X. fastidiosa* in Ontario and in New York State (Goodwin and Zhang, 1997; Purcell and Almeida, 2010). Decline of the American sycamore (*Platanus occidentalis* L.) caused by *X. fastidiosa* has been reported in Washington (DC), Lousiana, North Carolina, South Carolina, Texas and Florida. Red maple (*Acer rubrum* L.) appears to be less affected, and infected trees have only been reported in northern Virginia.

Bacterial galls on trees are common, but there have only been a few reports of bacteria causing galls in forest tree species that result in economic levels of damage. Galls caused by two *Pantoea* (syn. *Erwinia*) species were identified on Douglas-fir (*Pseudotsuga menziesii* (Mirb.) Franco) in British Columbia (BC), though the frequency of occurrence was low (DeYoung *et al.*, 1998). Bacterial galls on Douglas-fir have been observed from southern Arizona to BC and Alberta (Garrard, 1960; Sinclair and Lyon, 2005). There have been many reports of galls formed, probably by bacteria, but no details of characterization of the organisms have been reported. Douglas-fir growing in south-western BC were observed with galls on the branches of mature trees. The galls were of two distinct types and resulted in the isolation of two *Pantoea* strains. They differed in colony morphology, temperature sensitivity, growth rate, antibiotic resistance, serological response and host range. Molecular tools were not used to characterize these strains, so they were confirmed to be *Pantoea* spp. using standard bacteriological methods, but only one strain could be identified to group. This identification was based on the fatty acid analysis and identified the isolate that gave morphologically atypical galls as *Pantoea agglomerans* (DeYoung *et al.*, 1998).

4.2 Diagnosis

4.2.1 Phytoplasmas

Phytoplasmas are Gram-positive eubacteria with DNA characterized by low guanine and cytosine (G+C) content; they lack cell walls and currently cannot be cultured *in vitro*. The fact that all attempts to culture these organisms in artificial media has failed has made identification difficult. The symptoms of diseased plants vary with the phytoplasma, host, stage of disease development, age of host and environment at the time of infection (McCoy *et al.*, 1989; Lee *et al.*, 2000; Seemüller *et al.*, 2002). In woody plants, symptoms may be non-specific and are sometimes similar to those observed in virus-infected plants. These symptoms include yellowing or reddening of the foliage, small leaves, leaf curl, vein necrosis and, often, premature fall coloration. Premature defoliation, poor terminal growth, dieback and decline in plant health are also often reported as being associated with phytoplasmal infection. White ash (*F. americana* L.), green ash and velvet ash (*F. velutina* Torr.) are the only ash species that have been observed sufficiently to allow comment on phytoplasmal infection symptoms. Detailed descriptions of symptoms in these tree species at various ages is presented in a review by Sinclair and Griffiths (1994) and in a feature article by Sinclair *et al.* (1996). In brief, symptoms include slow growth, diminished apical dominance, suppressed root development, precocious shoot growth, witches' brooms, and subnormal greenness of foliage with early foliage colour change often observed in the autumn.

Some of the early efforts made to determine the cause of this wide range of specific and non-specific symptoms of phytoplasmal infection used electron and light microscopy. In light microscopy, two stains, Dienes' (Deeley *et al.*, 1979) and 4′-6-diamidino-2-phenylindole (DAPI) (Seemüller, 1976), allow non-specific detection of the *Mollicutes* in the phloem. DAPI staining of longitudinal sections of root can be performed rapidly and provides reliable results. However, there is need for a microscope equipped for fluorescence work. Dienes' staining can be used with bright field microscopy but can give ambiguous results. Immunofluorescence microscopy is also an option, but monoclonal antibodies need to be available for the phytoplasma of interest. Guo and Chen (1993) produced a monoclonal antibody to strain AshY2 and this was used to detect AshY phytoplasma in four ash and lilac (*Syringa*) species in nine locations (Griffiths *et al.*, 1994b). With the development in the 1990s of PCR and various additional procedures, these microscopy methods have in general been replaced or are used as a first step in diagnosis.

Phytoplasmas are considered to be primarily phloem limiting in infected plants, and vectoring is mainly by the phloem-feeding insects of the order Hemiptera (Weintraub and Beanland, 2006). The success of molecular analysis depends upon obtaining nucleic acid preparations of good quality. This can be difficult, particularly with woody plants where the concentration of the pathogen is low and often unevenly distributed in the phloem; this also makes it challenging to obtain reproducible results. There are many protocols and DNA extraction kits available, but those that will generate suitable DNA for PCR and further analysis usually need to include a phytoplasma enrichment step, and a step to remove the polyphenolics and polysaccharides present in woody plants, which can inhibit the Taq DNA polymerase used in PCR (Marzachí *et al.*, 2004). It is important to understand the biology of the system under investigation, and thus harvest tissues for nucleic acid extraction that are likely to have the highest concentration of phytoplasmas for the time in the growing season. In most cases, diagnosis requires a multiple step detection procedure. After an initial PCR amplification with universal primers, it is common for nested PCR to be performed with either a different pair of phytoplasma universal primers followed by RFLP analysis, or with group-specific primers (Griffiths *et al.*, 1999a; Sinclair and Griffiths, 2000; Bertaccini, 2007). In either case sequence analysis will usually be performed on the final PCR products.

A virtual classification technique called '*i*PhyClassifier' based on computer-generated RFLP patterns from PCR sequence data was recently developed that allows identification and differentiation of phytoplasma strains (Zhao *et al.*, 2009, 2010). This user-friendly program can greatly speed up the detection and classification of a phytoplasma. It allows the generation of sequence similarity scores, RFLP pattern similarities coefficients, virtual gel images, tentative assignment to a 16Sr group and also '*Ca.* Phytoplasma' species assignment without the need to perform RFLP digestions and gel electrophoresis. Real-time PCR is in many areas of diagnosis replacing traditional PCR, and has the advantage of being faster, more sensitive and more suited for mass screening. Galetto and Marzachí (2010) reviewed the real-time PCR system and its role in phytoplasma detection and quantification. The use of microarrays is also another option, and has been proposed by some researchers (Hadidi *et al.*, 2004; Nicolaisen and Bertaccini, 2007). All these methods have the potential to increase the number of samples evaluated in a given time frame and improve sensitivity. However, there is still the requirement for good quality nucleic acid, and less laborious methods that are reliable and can be performed at a cheaper rate per sample. Many aspects of the detection, characterization, classification, biology and epidemiology of phytoplasmas are presented in an interactive web page from the Phytoplasma Resource Center (2011).

4.2.2 Bacteria

Pathological wetwood, which is observed in elm, along with quaking aspen (*Populus tremuloides* Michx.), eastern cottonwood (*Populus deltoides* Bartram ex Marsh.) and willow (*Salix* spp.) in the western and northern USA, and in northern red oak (*Q. rubra* L.) in the southern USA (Murdoch and Campana, 1983; Xu *et al.*, 2001; Jacobi, 2009), involves dead or dying tree cells and is sometimes called slime flux. Symptoms include yellowing–browning of the central core of the tree and the oozing of a foul-smelling slime flux, which is forced out by the high gas pressure produced by bacterial fermentation in the xylem. The ooze is toxic to new shoots and to surrounding herbaceous plants. Therefore, if new shoots and

probably transmit in European white elm. It is yet to be determined which vectors are involved in EY disease spread from Japanese zelkova.

Alder yellows is more common in Europe than in North America. The AldY phytoplasma belongs to 16SrV-C subgroup and is closely related to the FD phytoplasma, which is of economic importance to vineyards in Europe (Lee *et al.*, 2004a). The leafhopper *Oncopsis alni* Schrank (Maixner and Reinert, 1999) is responsible for the transmission of AldY, which primarily infects *Alnus* spp. but occasionally infects grapevine (Arnaud *et al.*, 2007). Some AldY strains induce severe symptoms in alder, but avirulent strains also occur as many plants in which phytoplasmas were detected showed little in the way of symptoms (Berges and Seemüller, 2002). These latent infections of alder by AldY are potentially important inoculum reservoirs, and removing alder trees from areas close to vineyards would be advised as a recent hypothesis is that some strains of FD-related phytoplasmas may have originated from insects feeding on AldY-infected alders (Arnaud *et al.*, 2007; Cvrković *et al.*, 2008).

It is now clear that gymnosperms can be hosts for phytoplasmas (Paltrinieri *et al.*, 1998; Schneider *et al.*, 2005; Śliwa *et al.*, 2008; Davis *et al.*, 2010; Valiunas *et al.*, 2010; Jomantiene *et al.*, 2011). The phytoplasmas are present at such low titres in these trees that they could not be detected by direct PCR or by electron microscopy; it was necessary instead to use nested PCR (Śliwa *et al.*, 2008). Phytoplasmas belonging to 16SrXXI subgroup A (Śliwa *et al.*, 2008), 16SrIII (Paltrinieri *et al.*, 1998), 16SrIX subgroup E (Davis *et al.*, 2010) and 16SrI (Jomantiene *et al.*, 2011) have been detected in gymnosperms. Except for 16SrIX subgroup E, which was detected in western juniper in Oregon (Davis *et al.*, 2010), the detection of phytoplasmas in gymnosperms has, to date, occurred in Europe. It will be interesting to learn the geographic range, whether additional 16SrRNA groups and subgroups are identified, whether more families within the gymnosperms are hosts and which insect vectors are involved in transmission. Gymnosperms do not appear to be suitable hosts for phytoplasmas in that the pore sizes in the sieve cells are small (Esau, 1969) and could be too small to allow entry of the phytoplasmas, which are in the size range of 200–800 nm diameter. This could indicate that phytoplasmal spread and colonization occurs in a different manner in gymnosperms than in dicotyledons. Some phytoplasmas can, under certain situations, spread via root grafts (e.g. EY; Sinclair and Lyon, 2005), but whether this could be a method of spread in gymnosperms is unknown.

4.4.2 Bacteria

The epidemiology of bacterial wetwood is not understood. Wetwood typically develops in trees that are several years old, though trees that have recently been transplanted may show symptoms if the root system is not well established. Any activity that causes damage or stress to the roots and stems of trees can result in a wetwood problem. It is thought that the wetwood sap is forced from infected xylem through wounds by gas pressure, which is caused by bacterial fermentation in the wetwood (Zeikus and Ward, 1974; Jacobi, 2009). A number of bacterial species, which include *Klebsiella oxytoca* and *Pseudomonas* spp., and also yeasts, appear to be involved in producing the symptoms associated with wetwood in American elm (Murdoch and Campana, 1983). In some situations, *Clostridium* spp. have also been isolated from elm wetwood (Schink *et al.*, 1981). The variation in bacterial populations found in the wetwood is likely to be related to the microbial populations of the surrounding soil; this is considered to be the initial source of the inoculum, with root infections through wounds considered to be secondary. Where oozing occurs, the bacteria could be transferred to a new stem or branch via the ooze. Insects are attracted to the oozing liquid, but there is no research to confirm their involvement in disease spread.

BLS of shade and forest trees caused by *X. fastidiosa* not only infects a number of forest tree species, but also infects many common landscape plants, trees and weeds which may not show symptoms but act as an inoculum source (Gould and Lashomb, 2007). From studies performed on economically important crops (e.g. grapevine), spread has been determined to occur primarily by xylem feeding sharpshooters, leafhoppers and also spittlebugs (Janse and Obradovic, 2010). The insect(s) that vector BLS have not been identified, although insects known as vectors of other diseases caused by *X. fastidiosa* have been observed at BLS oak sites. It is thought that the manner in which the BLS vector(s) acquire and transmit the bacterium will be similar to that for vectors of PD, in which the xylem-feeding insects acquire the bacterium by feeding on infected host tissues. Once the xylem fluid is drawn into the insect foregut, the bacteria multiply and can transmit the pathogen to a new host within an hour of its acquisition. An adult insect is able to transmit the bacterium for the remainder of its life, whereas nymphs only transmit until the next moult (Gould and Lashomb, 2007). Leafhoppers can feed on several different plants, and can overwinter on the alternative host plants, which can provide substantial inoculum. It is thought that root grafts are an unlikely means of disease spread, as diseased trees are usually observed randomly within a stand. In hosts where leaf scorch is the primary symptom, the bacteria are in highest numbers in the veins and petioles of symptomatic leaves, but overwinter in the roots (Henneberger *et al.*, 2004). As leaf scorch commonly affects elm, oak and sycamore growing close to each other, it was suspected that the trees were probably infected by the same strain of *X. fastidiosa*. However, Sherald (1993) showed that the strains from elm and sycamore were only pathogenic in their respective hosts.

X. fastidiosa is sensitive to cool temperatures, but the strain responsible for BLS is more cold tolerant than the strains involved with PD. The BLS strain involved with oak scorch appears to be more cold tolerant than the strains affecting sycamore and elm, although Goodwin and Zhang (1997) detected *X. fastidiosa* in three American elm trees in the Niagara Peninsula of southern Ontario (Canada). Researchers using sycamore BLS strains from two locations in Georgia (USA) were unable to show that low temperature (5°C) had a negative influence on bacterial viability (Henneberger *et al.*, 2004). Vector movement and ecological niche is probably as much involved with geographic range of the diseases as with sensitivity of the bacterial isolates to temperature. Trees infected by BLS are predisposed to attacks by other organisms and insects.

The epidemiology of the *Pantoea* spp. involved in gall formation on Douglas-fir has not been investigated. There is speculation as to disease spread in that the original galls were within the current year's growth, which suggests the involvement of insect vectors (DeYoung *et al.*, 1998). In the inoculated trees, galls formed at locations other than the inoculation sites, which would suggest systemic movement or, possibly, insect transmission (DeYoung *et al.*, 1998).

Vector identification is extremely important for all the host–pathogen systems discussed. Once the vectors are identified for each plant–pathogen–vector system a start can be made on developing effective management procedures.

A number of reviews have been published on the implications of climate change on forest diseases (Sturrock *et al.*, 2011) and plant health (Pautasso *et al.*, 2010). Foissac and Wilson (2010) made some predictions as to how climate change might affect phytoplasma diseases. It is thought that an increase in temperature is likely to be experienced early in the season and could shorten the epidemiological cycle for phytoplasmas. Symptoms on infected plants may occur earlier and could be more severe in some situations as there are likely to be higher populations of phytoplasmal organisms in the plants. Some plant species, though, harbour high concentrations of phytoplasmas and show little in the way of symptoms (e.g. alder) (Lederer and Seemüller, 1991). As phytoplasmal infections depend upon insect vectors, higher temperatures earlier in the season and possible changes in rainfall patterns could alter insect dynamics and may not necessarily favour phloem feeders. Disease range is likely to change as temperatures increase in northern locations; this could have an impact on the geographic range of phytoplasmas as well as BLS. Even though BLS is not considered to be as temperature sensitive as other diseases caused by *X. fastidiosa*, it does appear to have a boundary that is influenced by winter temperature; whether this is related to survival of the insects involved in transmission or to the survival of the bacteria is unclear. However, with climate change, both would be affected, which could result in BLS occurring a lot more frequently in forest trees in more northerly locations. Our understanding of the epidemiology of bacterial wetwood is limited. Hence the implications of climate change are not clear, although stress on the trees does appear to exacerbate the symptoms, and therefore increase in temperatures and changes in rainfall could increase the occurrence of symptoms.

4.5 Management Strategies and Tactics

Even though integrated management regimes are emerging for the control of phytoplasmas in field crops (Olivier *et al.*, 2009), extending these to forests is unlikely to occur in the near future as a result of logistical issues. In some forest stands, if disease is not advanced and the trees have economic value, harvesting may be the best course of action, followed by replanting with either a less susceptible cultivar or a line with 'cross protection' (see below), if available, or with another species. For most phytoplasmas detected in forest tree species, alternative plant hosts have rarely been detected.

From phytoplasmal disease systems that have been well characterized, it is apparent that pathogen transmission occurs quickly and that using insecticides in the conventional manner, even frequently, would not act sufficiently quickly to have any impact on the epidemic. Treating crop borders with systemic insecticides was helpful in controlling vector populations in cases of lucerne yellows in Australia (Pilkington *et al.*, 2004). Unfortunately, extending such a protocol to a forest situation would be difficult.

Controlling the insects involved in disease spread is normally not practical in forests as, in most situations, insecticides can only be applied to the understorey. Aerial applications would be needed to provide the necessary overstorey coverage, for which the cost involved would be high; furthermore, unless the forests were very isolated, there would be too many concerns related to chemical drift and human health and safety.

The control of weeds that act as alternative hosts for phytoplasmas is a potential method for disease control. In North America, there are currently no major phytoplasmal diseases of forest trees for which this would be a useful method, but in Poland, where AY (the aster yellows phytoplasma) appears to be involved in the appearance of ash yellows type symptoms in ash trees, controlling alternative hosts of AY, which may include weed species, may be a useful option (Kamiñska and Berniak, 2009).

As both phytoplasmas and *X. fastidiosa* affect the ability of infected trees to take up water, providing water or mulching around the trees, and thus reducing evaporation during times of water stress, may increase the longevity of the trees. Reduction in bacterial wetwood may also be achieved by reducing stress, particularly by providing irrigation to recently transplanted trees. Different therapeutic treatments were recently evaluated in oak by Hartman *et al.* (2010) with the goal of identifying a practical control method for BLS. Injections of the antibiotic oxytetracycline were shown to be the best treatment; they delayed the onset of scorch symptoms in the late summer by about 2 weeks, but were not able to cure trees. Timing of the injections appeared to be important, with injections given about 3 weeks after full expansion of the first leaves providing the best results. This, again, is only practical for trees growing in areas such as golf courses, or growing as street trees or as private landscape trees.

Breeding resistant varieties is the most desirable approach for controlling forest tree diseases. With forest trees, considering all the constraints related to the production system, it is the only method that would be of value for foresters. However, breeding trees with the desired attributes of disease resistance and silvicultural characters is an expensive, long-term project. Breeding for resistance of elm to Dutch elm disease (a fungal disease) has generated a number of cultivars which have been released in Europe and North America (Townsend, 2000). Six of these cultivars, which represented Eurasian species and elm hybrids, were evaluated for their response to a North American strain of EY (Sinclair *et al.*, 2000). Cultivars Frontier, Pathfinder and Patriot all appeared to be tolerant. In the Homestead trees, no phytoplasmas were detected, which could indicate that this is resistant to EY. Nursery performance was reported for some Dutch elm disease-tolerant elm, but only Patriot was included in this study, and it performed well (Kuser and Polanin, 2001).

Cross protection using different phytoplasmas has been demonstrated in a number of plants. Sinclair and Griffiths (2000) conducted studies in periwinkle (*Catharanthus roseus* (L.) G. Don) plants with two strains of AshY that differed greatly in aggressiveness. When either strain was inoculated 11 weeks before the other only the initial strain could be detected in periwinkle after a further 12 weeks. Similar success was achieved by other researchers in cross protection studies with phytoplasmas from two other 16SrDNA groups (Freitag, 1964; Kison and Seemüller, 2001), which indicates that this method may have significant implications for phytoplasmal disease control, even in a forest tree situation. AshY could then be potentially controlled by introducing a non-aggressive strain of AshY into the forest stand. This could be performed by either graft-inoculating already established trees, or by planting trees that were already infected. As these trees would be infected with the non-aggressive strain of AshY (which will not kill the host), over time it is possible that this could become the dominant AshY strain in the tree stand. Moreover, insects will feed on these trees and spread the non-aggressive strain, thus protecting ash trees from more aggressive strains. Currently, with the EAB situation in ash, there is little interest in breeding ash trees specifically for resistance to ash yellows, although there are some ash tree breeding projects underway in North America to generate resistance to EAB. It is not immediately apparent that the germplasm used in these studies has known tolerance to AshY, and therefore the cross protection could be useful. It would certainly be ideal to use germplasm for these EAB breeding projects that demonstrate some tolerance to AshY. From what has been observed to date, velvet ash and also single leaf ash (*Fraxinus anomala* Torr. ex S. Wats.) appear to have a degree of resistance that is not evident in American and green ashes (Sinclair *et al.*, 1993, 1994), and these could be useful sources of resistance in future breeding programmes.

Physical and chemical methods are procedures used extensively in plant pathology to eliminate pathogens from plants, particularly to generate 'virus-free' plants (Griffiths *et al.*, 1990). Recently, researchers have been evaluating such methods for eliminating phytoplasmas from grape (Osler *et al.*, 2003; Romanazzi and Murolo, 2008b), and also '*Ca.* Phytoplasma solani' from periwinkle (Ćurković Perica, 2008). Even though these methods may result in clean plants, the plants can easily become reinfected unless methods of protection are put into place.

It would be ideal to utilize integrated pest management (IPM) methods for disease control in forest trees. In annual crops, or even in orchard situations, IPM has been used successfully for many years for many disease and pest situations. However, as has been highlighted in earlier parts of this chapter, this has not, as yet, been readily implemented for trees in forest settings. Preferably, IPM would require the growing of trees with at least some disease resistance, or cross protection, and application of chemicals (insecticides and/or bactericides) with good levels of efficacy but with low mammalian toxicity, or the use of biological control agents. It will probably be a number of years before any of these tools are available for controlling the prokaryotic diseases of forest trees.

As is apparent from this chapter, there are some large gaps in our knowledge about the prokaryotes and their associated forest tree diseases, thus making it difficult to implement good management practices. Nevertheless, governments and forest managers need to be aware that the future of our forests can be jeopardized by our actions and that efforts should be made to limit the occurrence of such diseases.

References

Ahrens, U. and Seemüller, E. (1994) Detection of mycoplasmalike organisms in declining oaks by polymerase chain reaction. *European Journal of Forest Pathology* 24, 55–63.

Arnaud, G., Malembic-Maher, S., Salar, P., Bonnet, P., Maixner, M., Marcone, C., Boudon-Padieu, E. and Foissac, X. (2007) Multilocus sequence typing confirms the close genetic interrelatedness of three distinct flavescence dorée phytoplasma strain clusters and group 16SrV phytoplasmas infecting grapevine and alder in Europe. *Applied and Environmental Microbiology* 73, 4001–4010.

Bai, X., Zhang, J., Ewing, A., Miller, S.A., Radek, A.J., Shevchenko, D.V., Tsukerman, K., Walunas, T., Lapidus, A., Campbell, J.W. and Hogenhout, S.A. (2006) Living with genome instability: the adaptation of phytoplasmas to diverse environments of their insect and plant host. *Journal of Bacteriology* 188, 3682–3696.

Bai, X., Ammar, E.D. and Hogenhout, S.A. (2007) A secreted effector protein of AY-WB phytoplasma accumulates in nuclei and alters gene expression of host plant cells, and is detected in various tissues of the leafhopper *Macrosteles quadrilineatus*. *Bulletin of Insectology* 60, 217–218.

Bai, X., Correa, V.R., Toruño, T.Y., Ammar, E.D., Kamoun, S. and Hogenhout, S.A. (2009) AY-WB phytoplasma secretes a protein that targets plant cell nuclei. *Molecular Plant–Microbe Interactions* 22, 18–30.

Berges, R. and Seemüller, E. (2002) Impact of phytoplasma infection of common alder (*Alnus glutinosa*) depends on strain virulence. *Forest Pathology* 32, 357–363.

Berges, R., Cousin, M.-T., Roux, J., Mäurer, R. and Seemüller, E. (1997) Detection of phytoplasma infections in declining *Populus nigra* 'Italica' trees and molecular differentiation of the aster yellows phytoplasma identified in various *Populus* species. *European Journal of Forest Pathology* 27, 33–43.

Berges, R., Rott, M. and Seemüller, E. (2000) Range of phytoplasma concentrations in various plant hosts as determined by competitive polymerase chain reaction. *Phytopathology* 90, 1145–1152.

Berisha, B., Chen, Y.D., Zhang, G.Y., Xu, B.Y. and Chen, T.A. (1998) Isolation of Pierce's disease bacteria from grapevines in Europe. *European Journal of Plant Pathology* 104, 427–433.

Bertaccini, A. (2007) Phytoplasmas: diversity, taxonomy, and epidemiology. *Frontiers in Bioscience* 12, 673–689.

Braun, E.J. and Sinclair, W.A. (1976) Histopathology of phloem necrosis in *Ulmus americana*. *Phytopathology* 66, 598–607.

Carraro, L., Ferrini, F., Ermacora, P., Loi, N., Martini, M. and Osler, R. (2004) *Macropsis mendax* as a vector of elm yellows phytoplasma of *Ulmus* species. *Plant Pathology* 53, 90–95.

Chang, C.J., Garnier, M., Zreik, L., Rossetti, V. and Bové, J.M. (1993) Culture and serological detection of xylem-limited bacterium causing citrus variegated chlorosis and its identification as a strain of *Xylella fastidiosa*. *Current Microbiology* 27, 137–142.

Chen, J., Chang, C.J. and Jarret, R.L. (1992) DNA probes as molecular markers to monitor the seasonal occurrence of walnut witches'-broom mycoplasmalike organism. *Plant Disease* 76, 1116–1119.

Ćurković Perica, M. (2008) Auxin-treatment induces recovery of phytoplasma-infected periwinkle. *Journal of Applied Microbiology* 105, 1826–1834.

Cvrković, T., Jović, J., Mitrović, M., Petrović, A., Krnjajić, S., Malembic-Maher, S. and Toševski, I. (2008) First report of alder yellows phytoplasma on common alder (*Alnus glutinosa*) in Serbia. *Plant Pathology* 57, 773.

Davis, M.J., Purcell, A.H. and Thomson, S.V. (1978) Pierce's disease of grapevines: isolation of the causal bacterium. *Science* 199, 75–77.

Davis, R.E. and Dally, E.L. (2001) Revised subgroup classification of group 16SrV phytoplasmas and placement of *flavescence dorée*-associated phytoplasmas in two distinct subgroups. *Plant Disease* 85, 790–797.

Davis, R.E., Dally, E.L., Zhao, Y., Lee, I.-M., Jomantiene, R., Detweiler, A.J. and Putnam, M.L. (2010) First report of a new subgroup 16SrIX-E ('*Candidatus* Phytoplasma phoenicium'-related) phytoplasma associated with juniper witches' broom disease in Oregon, USA. *Plant Pathology* 59, 1161.

Deeley, J.A., Stevens, W.A. and Fox, R.T.V. (1979) Use of Dienes' stain to detect plant diseases induced by mycoplasmalike organisms. *Phytopathology* 69, 1169–1171.

DeYoung, R.M., Copeman, R.J. and Hunt, R.S. (1998) Two strains in the genus *Erwinia* cause galls on Douglas-fir in southwestern British Columbia. *Canadian Journal of Plant Pathology* 20, 194–200.

Drooz, A. (1985) *Insects of Eastern Forests*. Miscellaneous Publication 1426, US Department of Agriculture, Forest Service, Washington, DC.

Esau, K. (1969) The phloem. In: Zimmermann, W., Ozenda, P. and Wulff, H.D. (eds) *Encyclopedia of Plant Anatomy*, Vol. V, Part 2. Gebrüder Borntraeger, Berlin, pp. 479–504.
Feeley, C.J., Hart, E.R., Thompson, J.R. and Harrington, T.C. (2001) Occurrence, associated symptoms, and potential insect vectors of the ash yellows phytoplasma in Iowa, U.S. *Journal of Arboriculture* 27, 331–339.
Filgueira, J.J., Franco-Lara, L., Salcedo, J.E., Gaitan, S.L. and Boa, E.R. (2004) Urapan (*Fraxinus udhei*) dieback, a new disease associated with a phytoplasma in Colombia. *Plant Pathology* 53, 520.
Firrao, G. and Bazzi, C. (1994) Specific identification of *Xylella fastidiosa* using the polymerase chain reaction. *Phytopathologia Mediterranea* 33, 90–92.
Foissac, X. and Wilson, M.R. (2010) Current and possible future distributions of phytoplasma diseases and their vectors. In: Weintraub, P.G. and Jones, P. (eds) *Phytoplasmas: Genomes, Plant Hosts and Vectors*. CAB International, Wallingford, UK, pp. 309–324.
Franc, G.D. (1994) Bacterial wetwood slim flux. [Bulletin] B-994, Cooperative Extension Service, University of Wyoming, Laramie, Wyoming. Available at: www.wyomingextension.org/agpubs/pubs/B994.pdf (accessed 30 June 2011).
Freitag, J.H. (1964) Interaction and mutual suppression among three strains of aster yellows virus. *Virology* 24, 401–413.
Galetto, L. and Marzachí, C. (2010) Real-time PCR diagnosis and quantification of phytoplasmas. In: Weintraub, P.G. and Jones, P. (eds) *Phytoplasmas: Genomes, Plant Hosts and Vectors*. CAB International, Wallingford, UK, pp. 1–18.
Garrard, E.H. (1960) *Bacterial Diseases of Plants*. Ontario Department of Agriculture, Toronto, Ontario, Canada.
Goodwin, P.H. and Zhang, S. (1997) Distribution of *Xylella fastidiosa* in southern Ontario as determined by the polymerase chain reaction. *Canadian Journal of Plant Pathology* 19, 13–18.
Gould, A.B. and Lashomb, J.H. (2007) Bacterial leaf scorch (BLS) of shade trees. *The Plant Health Instructor*, doi: 10.1094/PHI-1-2007-0403-07. Available at: http://www.apsnet.org/edcenter/intropp/lessons/prokaryotes/Pages/BacterialLeafScorch.aspx (accessed 22 November 2012).
Griffiths, H.M., Slack, S.A. and Dodds, J.H. (1990) Effect of chemical and heat therapy on virus concentrations in *in vitro* potato plantlets. *Canadian Journal of Botany* 68, 1515–1521.
Griffiths, H.M., Gundersen, D.E., Sinclair, W.A., Lee, I.-M. and Davis, R.E. (1994a) Mycoplasmalike organisms from milkweed, goldenrod and spirea represent two new 16S rRNA subgroups and three new strain subclusters related to peach X-disease MLOs. *Canadian Journal of Plant Pathology* 16, 255–260.
Griffiths, H.M., Sinclair, W.A., Davis, R.E., Lee, I.-M., Dally, E.L., Guo, Y.-H., Chen, T.A. and Hibben, C.R. (1994b) Characterization of mycoplasmalike organisms from *Fraxinus*, *Syringa* and associated plants from geographically diverse sites. *Phytopathology* 84, 119–126.
Griffiths, H.M., Sinclair, W.A., Boudon-Padieu, E., Daire, X., Lee, I.-M., Sfalanga, A. and Bertaccini, A. (1999a) Phytoplasmas associated with elm yellows: molecular variability and differentiation from related organisms. *Plant Disease* 83, 1101–1104.
Griffiths, H.M., Sinclair, W.A., Smart, C.D. and Davis, R.E. (1999b) The phytoplasma associated with ash yellows and lilac witches' broom: '*Candidatus* Phytoplasma fraxini'. *International Journal of Systematic Bacteriology* 49, 1605–1614.
Gundersen, D.E., Lee, I-M., Rehner, S.A., Davis, R.E. and Kingsbury, D.T. (1994) Phylogeny of mycoplasmalike organisms (phytoplasmas), a basis for their classification. *Journal of Bacteriology* 176, 5244–5254.
Guo, Y.H. and Chen, T.A (1993) Monoclonal antibodies against ash yellows agent. *Phytopathology* 93, 243–244.
Hadidi, A., Czosnek, H. and Barba, M. (2004) DNA microarrays and their potential applications for the detection of plant viruses, viroids, and phytoplasmas. *Journal of Plant Pathology* 86, 97–104.
Hansen, H.N. and Smith, R.E (1937) A bacterial gall disease of Douglas-fir. *Pseudotsuga taxifolia*. *Hilgardia* 10, 569–577.
Harper, S.J., Ward, L.I. and Clover, G.R.G. (2010) Development of LAMP and real-time PCR methods for the rapid detection of *Xylella fastidiosa* for quarantine and field applications. *Phytopathology* 100, 1282–1288.
Hartman, J. (2007) *Bacterial Leaf Scorch*. Plant Pathology Fact Sheet PPFS-OR-W-12, Co-operative Extension Service, University of Kentucky.
Hartman, J., Dixon, E. and Bernick, S. (2010) Evaluation of therapeutic treatments to manage oak bacterial leaf scorch. *Arboriculture and Urban Forestry* 36, 140–146.

Hendson, M., Purcell, A.H., Chen, D., Smart, C., Guilhabert, M. and Kirkpatrick, B. (2001) Genetic diversity of Pierce's disease strains and other pathotypes of *Xylella fastidiosa. Applied and Environmental Microbiology* 67, 895–903.

Henneberger, T.S.M., Stevenson, K.L., Britton, K.O. and Chang, C.J. (2004) Distribution of *Xylella fastidiosa* in sycamore associated with low temperature and host resistance. *Plant Disease* 88, 951–958.

Hill, G.T. and Sinclair, W.A. (2000) Taxa of leafhoppers carrying phytoplasmas at sites of ash yellows occurrence in New York State. *Plant Disease* 84, 134–138.

Hodgetts, J., Boonham, N., Mumford, R., Harrison, N. and Dickinson, M. (2008) Phytoplasma phylogenetics based on analysis of *secA* and 23S rRNA gene sequences for improved resolution of candidate species of '*Candidatus* Phytoplasma'. *International Journal of Systematic and Evolutionary Microbiology* 58, 1826–1837.

Hogenhout, S.A. and Musić, M.Š. (2010) Phytoplasma genomics, from sequencing to comparative and functional genomics – what have we learnt? In: Weintraub, P.G. and Jones, P. (eds) *Phytoplasmas: Genomes, Plant Hosts and Vectors*. CAB International, Wallingford, UK, pp. 19–36.

Hopkins, D.L. and Purcell, A.H. (2002) *Xylella fastidiosa*: cause of Pierce's disease of grapevine and other emergent diseases. *Plant Disease* 86, 1056–1066.

Huang, Q. (2009) Specific detection and identification of *Xylella fastidiosa* strains causing oleander leaf scorch using polymerase chain reaction. *Current Microbiology* 58, 393–398.

IRPCM Phytoplasma/Spiroplasma Working Team – Phytoplasma Taxonomy Group (2004) '*Candidatus* Phytoplasma', a taxon for the wall-less, non-helical prokaryotes that colonize plant phloem and insects. *International Journal of Systematic Evolutionary Microbiology* 54, 1243–1255.

Jacobi, W.R. (2009) *Bacterial Wetwood*. Fact Sheet No. 2.910, Colorado State University Extension, Fort Collins, Colorado. Available at: http://www.ext.colostate.edu/pubs/garden/02910.pdf (accessed 22 November 2012).

Jacobs, K.A., Lee, I.-M., Griffiths, H.M., Miller, F.D. and Bottner, K.D. (2003) A new member of the clover proliferation phytoplasma group (16SrVI) associated with elm yellows in Illinois. *Plant Disease* 87, 241–246.

Janse, J.D. and Obradovic, A. (2010) *Xylella fastidiosa*: its biology, diagnosis, control and risks. *Journal of Plant Pathology* 92, S1.35–S1.48.

Jomantiene, R., Valiunas, D., Ivanauskas, A., Urbanavieiene, L., Staniulis, J. and Davis, R.E. (2011) Larch is a new host for a group 16SrI, subgroup B, phytoplasma in Ukraine. *Bulletin of Insectology* 64, S101–S102.

Jović, J., Cvrković, T., Mitrović, M., Petrović, A., Krnjajić, S. and Toševski, I. (2008) New strain of '*Candidatus* Phytoplasma ulmi' infecting *Ulmus minor* and *U. laevis* in Serbia. *Plant Pathology* 57, 1174.

Jović, J., Cvrković, T., Mitrović, M., Petrović, A., Krstić, O., Krnjajić, S. and Toševski, I. (2011) Multigene sequence data and genetic diversity among '*Candidatus* Phytoplasma ulmi' strains infecting *Ulmus* spp. in Serbia. *Plant Pathology* 60, 356–368.

Jung, H.-Y., Sawayanagi, T., Kakizawa, S., Nishigawa, H., Miyata, S., Oshima, K., Ugaki, M., Lee, J.-T., Hibi, T. and Namba, S. (2002) '*Candidatus* Phytoplasma castaneae', a novel phytoplasma taxon associated with chestnut witches' broom disease. *International Journal of Systematic and Evolutionary Microbiology* 52, 1543–1549.

Kakizawa, S., Oshima, K., Nighigawa, H., Jung, H.-Y., Wei, W., Suzuki, S., Tanaka, M., Miyata, S., Ugaki, M. and Namba, S. (2004) Secretion of immunodominant membrane protein from onion yellows phytoplasma through the Sec protein – translocation system in *Escherichia coli*. *Microbiology* 150, 135–142.

Kakizawa, S., Oshima, K., Jung, H.-Y., Suzuki, S., Nighigawa, H., Arashida, R., Miyata, S., Ugaki, M., Kishino, H. and Namba, S. (2006) Positive selection acting on a surface membrane protein of the plant-pathogenic phytoplasmas. *Journal of Bacteriology* 188, 3424–3428.

Kamiñska, M. and Berniak, H. (2009) '*Candidatus* Phytoplasma asteris' in *Fraxinus excelsior* and its association with ash yellows newly reported in Poland. *Plant Pathology* 58, 788.

Kashian, D.M. and Witter, J.A. (2011) Assessing the potential for ash canopy tree replacement via current regeneration following emerald ash borer-caused mortality on southeastern Michigan landscapes. *Forest Ecology and Management* 26, 480–488.

Kison, H. and Seemüller, E. (2001) Differences in strain virulence of the European stone fruit yellows phytoplasma and susceptibility of stone fruit trees on various rootstocks to the pathogen. *Journal of Phytopathology* 149, 533–541.

Kube, M., Schneider, B., Kuhl, H., Dandekar, T., Heitmann, K., Migdoll, A.M., Reinhardt, R. and Seemüller, E. (2008) The linear chromosome of the plant-pathogenic mycoplasma '*Ca.* P. mali'. *BMC Genomics* 9, 306.

Kuser, J. and Polanin, N. (2001) Comparative test of six DED-tolerant elms: a preliminary report on nursery performance. *Journal of Arboriculture* 27, 281–282.

Lederer, W. and Seemüller, E. (1991) Occurrence of mycoplasma-like organisms in diseased and non-symptomatic alder trees (*Alnus* spp.). *European Journal of Forest Pathology* 21, 90–96.

Lee, I.-M., Hammond, R.W., Davis, R.E. and Gundersen, D.E. (1993) Universal amplification and analysis of pathogen 16S rDNA for classification and identification of mycoplasmalike organisms. *Phytopathology* 83, 834–842.

Lee, I.-M., Bertaccini, A., Vibio, M., Gundersen, D.E., Davis, R.E., Mittempergher, L., Conti, M. and Gennari, F. (1995) Detection and characterization of phytoplasmas associated with disease in *Ulmus* and *Rubus* in northern and central Italy. *Phytopathologia Mediterranea* 34, 174–183.

Lee, I.-M., Gundersen-Rindal, D.E., Davis, R.E. and Bartoszyk, I.M. (1998) Revised classification scheme of phytoplasmas based on RFLP analyses of 16S rRNA and ribosomal protein gene sequences. *International Journal of Systematic Bacteriology* 48, 1153–1169.

Lee, I.-M., Davis, R.E. and Gundersen-Rindal, D.E. (2000) Phytoplasma: phytopathogenic mollicutes. *Annual Review of Microbiology* 54, 221–255.

Lee, I.-M., Martini, M., Marcone, C. and Zhu, S.F. (2004a) Classification of phytoplasma strains in the elm yellows group (16SrV) and proposal of '*Candidatus* Phytoplasma ulmi' for the phytoplasma associated with elm yellows. *International Journal of Systematic and Evolutionary Microbiology* 54, 337–347.

Lee, I.-M., Gundersen, D.E., Davis, R.E., Bottner, K.D., Marcone, C. and Seemüller, E. (2004b) '*Candidatus* Phytoplasma asteris', a novel phyoplasma taxon associated with aster yellows and related diseases. *International Journal of Systematic and Evolutionary Microbiology* 54, 1037–1048.

Lee, I.-M., Zhao, Y. and Bottner, K.D. (2006) *SecY* gene sequence analysis for finer differentiation of diverse strains in the aster yellows phytoplasma group. *Molecular and Cellular Probes* 20, 87–91.

Leu, L.S. and Su, C.C. (1993) Isolation, cultivation, and pathogenicity of *Xylella fastidiosa*, the causal bacterium of pear leaf scorch disease in Taiwan. *Plant Disease* 77, 642–646.

Li, Y., Hao, G., Galvani, C.D., Meng, Y., De La Fuente, L., Hoch, H.C. and Burr, T.J. (2007) Type I and type II pili of *Xylella fastidiosa* affect twitching motility, biofilm formation and cell–cell aggregation. *Microbiology* 153, 719–726.

Lim, P.O. and Sears, B.B. (1989) 16S rRNA sequence indicates that plant-pathogenic mycoplasmalike organisms are evolutionarily distinct from animal mycoplasmas. *Journal of Bacteriology* 171, 5901–5906.

Lin, C.-P., Huang, J.-P., Wu, C.-S., Hsu, C.-Y. and Chaw, S.-M. (2010) Comparative chloroplast genomics reveals the evolution of *Pinaceae* genera and subfamilies. *Genome Biology and Evolution* 2, 504–517.

Maixner, M. and Reinert, W. (1999) *Oncopsis alni* (Schrank) (Auchenorrhyncha: Cicadellidae) as a vector of the alder yellows phytoplasma of *Alnus glutinosa* (L.) Gaertn. *European Journal of Plant Pathology* 105, 87–94.

Maramorosch, K. and Raychaudhuri, S.P. (1981) *Mycoplasma Diseases of Trees and Shrubs*. Academic Press, New York.

Marcone, C., Firrao, G., Ragozzino, A. and Locci, R. (1994) Detection of MLOs in declining alder trees in southern Italy and their characterization by RFLP analysis. *European Journal of Forest Pathology* 24, 217–228.

Marcone, C., Ragozzino, A. and Seemüller, E. (1997) Identification and characterization of the phytoplasma associated with elm yellows in southern Italy and its relatedness to other phytoplasma of the elm yellows group. *European Journal of Forest Pathology* 27, 45–54.

Martini, M., Lee, I.-M., Bottner, K.D., Zhao, Y., Botti, S., Bertaccini, A., Harrison, N.A., Carraro, L., Marcone, C., Khan, A.J. and Osler, R. (2007) Ribosomal protein gene-based phylogeny for finer differentiation and classification of phytoplasmas. *International Journal of Systematic and Evolutionary Microbiology* 57, 2037–2051.

Marzachí, C., Milne, R.G. and Bosco, D. (2004) Phytoplasma–plant–vector relationships. In: Pandalai, S.G. and Gayathri, A. (eds) *Recent Research Development in Plant Pathology, Vol.3*. Research Signpost, Trovandrum, Kerala, India, pp. 211–241.

Matteoni, J.A. and Sinclair, W.A. (1988) Elm yellows and ash yellows. In: Hiruki, C. (ed.) *Tree Mycoplasma Diseases and Epidemiology*. University of Alberta, Edmonton, Alberta, Canada.

Mäurer, R., Seemüller, E. and Sinclair, W.A. (1993) Genetic relatedness of mycoplasmalike organisms affecting elm, alder and ash in Europe and North America. *Phytopathology* 83, 971–976.

McCoy, R.E. *et al.* (1989) Plant diseases associated with mycoplasma-like organisms. In: Whitcomb, R.F. and Tully, J.G. (eds) *The Mycoplasmas, Vol. V. Spiroplasmas, Acholeplasmas and Mycoplasmas of Plants and Arthropods*. Academic Press, New York, pp. 545–640.

Mehta, A. and Rosato, Y.B. (2001) Phylogenetic relationships of *Xylella fastidiosa* strains from different hosts, based on 16S rDNA and 16S-23S intergenic spacer sequences. *International Journal of Systematic and Evolutionary Microbiology* 51, 311–318.

Murdoch, C.W. and Campana, R.J. (1983) Bacterial species associated with wetwood of elm. *Phytopathology* 73, 1270–1273.

Musić, M.S., Novokmet, M., Novak, R., Krajačić, M. and Škorić, D. (2008) Distribution of phytoplasma diseases in the Lombardy poplar tree population of Zagreb urban area. *Acta Botanica Croatia* 67, 119–130.

Nault, L.R. and Rodriguez, J.G. (1985) *The Leafhoppers and Planthoppers*. John Wiley, New York.

Newman, K.L., Almeida, R.P., Purcell, A.H. and Lindow, S.E. (2004) Cell–cell signaling controls *Xylella fastidiosa* interactions with both insects and plants. *Proceedings of the National Academy of Sciences of the United States of America* 101, 1737–1742.

Nicolaisen, M. and Bertaccini, A. (2007) An oligonucleotide microarray-based assay for identification of phytoplasma 16S ribosomal groups. *Plant Pathology* 56, 332–336.

Olivier, C.Y., Lowery, D.T. and Stobbs, L.W. (2009) Phytoplasma diseases and their relationships with insects and plant hosts in Canadian horticultural and field crops. *Canadian Entomologist* 141, 425–462.

Oshima, K., Kakizawa, S., Nishigawa, H., Jung, H.-Y., Wei, W., Suzuki, S., Arashida, R., Nakata, D., Miyata, S.-I., Ugaki, M. and Namba, S. (2004) Reductive evolution suggested from the complete genome sequence of a plant-pathogenic phytoplasma. *Nature Genetics* 36, 27–29.

Osler, R., Carraro, L., Ermacora, P., Ferrini, F., Loi, N., Loschi, A., Martini, M., Muton, P.B. and Refatti, E. (2003) Roguing: a controversial practice to eradicate grape yellows caused by phytoplasmas. In: *Extended Abstracts, Fourteenth Meeting of the International Council for the Study of Virus and Virus-Like Diseases of Grapevine (ICVG), 12–17th September, 2003, Locorotondo, Italy*, p. 68.

Paltrinieri, S., Martini, M. and Bertaccini, A. (1998) X-disease related phytoplasmas in ornamental trees and shrubs with witches' broom and malformation symptoms. *Journal of Plant Pathology* 80, 261.

Pautasso, M., Dehnen-Schmutz, K., Holdenrieder, O., Pietravalle, S., Salama, N., Jeger, M.J., Lange, E. and Hehl-Lange, S. (2010) Plant health and global change – some implications for landscape management. *Biological Reviews* 85, 729–755.

Phytoplasma Resource Center (2011) *i*PhyClassifier. Phytoplasma Resource Center, US Department of Agriculture, Agricultural Research Service, Beltsville Agricultural Research Center, Beltsville, Maryland. Available at: http://plantpathology.ba.ars.usda.gov/cgi-bin/resource/iphyclassifier.cgi (accessed 14 September 2011).

Pilkington, L.J., Gurr, G.M., Fletcher, M.J., Elliott, E., Nikandrow, A. and Nicol, H.L. (2004) Reducing the immigration of suspected leafhopper vectors and severity of Australian lucerne yellows disease. *Australian Journal of Experimental Agriculture* 44, 983–992.

Purcell, A. and Almeida, R. (2010) *Xylella fastidiosa*: a scientific and community Internet resource on plant diseases caused by the bacterium *Xylella fastidiosa*. Available at: http://nature.berkeley.edu/xylella/diseases/diseases.html (accessed 30 June 2011).

Rodrigues, J.L.M., Silvia-Stenico, M.E., Gomes, J.E., Lopes, J.R.S. and Tsai, S.M. (2003) Detection and diversity assessment of *Xylella fastidiosa* in field-collected plant and insect samples by using 16S rRNA and *gyrB* sequences. *Applied and Environmental Microbiology* 7, 4249–4255.

Romanazzi, G. and Murolo, S. (2008a) '*Candidatus* Phytoplasma ulmi' causing yellows in *Zelkova serrata* newly reported in Italy. *Plant Pathology* 57, 1174.

Romanazzi, G. and Murolo, S. (2008b) Partial uprooting and pulling to induce recovery in bois noir-infected grapevines. *Journal of Phytopathology* 156, 747–750.

Roper, M.C., Greve, L.C., Warren, J.G., Labavitch, J.M. and Kirkpatrick, B.C. (2007a) *Xylella fastidiosa* requires polygalacturonase for colonization and pathogenicity in *Vitis vinifera* grapevines. *Molecular Plant–Microbe Interactions* 20, 411–419.

Roper, M.C., Greve, L.C., Labavitch, J.M. and Kirkpatrick, B.C. (2007b) Detection and visualization of an exopolysaccharide produced by *Xylella fastidiosa in vitro* and *in planta*. *Applied and Environmental Microbiology* 73, 7252–7258.

Schaper, U. and Seemüller, E. (1984) Recolonization of the stem of apple proliferation and pear decline-diseased trees by the causal organisms in spring. *Zeitschrift für Pflanzenkrankheiten und Pflanzenschutz* 91, 608–613.

Schink, B., Ward, J.C. and Zeikus, J.G. (1981) Microbiology of wetwood: importance of pectin degradation and *Clostridium* species in living trees. *Applied and Environmental Microbiology* 42, 526–532.

Schneider, B., Torres, E., Martín, M., Schröder, M., Behnke, H.D. and Seemüller, E. (2005) '*Candidatus* Phytoplasma pini', a novel taxon from *Pinus sylvestris* and *Pinus halepensis*. *International Journal of Systematic and Evolutionary Microbiology* 55, 303–307.

Seemüller, E. (1976) Investigations to demonstrate mycoplasma-like organisms in diseased plants by fluorescence microscopy. *Acta Horticulturae* 67, 109–112.

Seemüller, E., Garnier, M. and Schneider, B. (2002) Mycoplasmas of plants and insects. In: Razin, S. and Hermann, R. (eds) *Molecular Biology and Pathogenicity of Mycoplasmas*. Kluwer Academic/Plenum Publishers, New York, pp. 91–115.

Šeruga, M., Škorić, D., Botti, S., Paltrinieri, S., Juretić, N. and Bertaccini, A. (2003) Molecular characterization of a phytoplasma from the aster yellows (16SrI) group naturally infecting *Populus nigra* L. 'Italica' trees in Croatia. *Forest Pathology* 33, 113–125.

Sfalanga, A., Martini, M., Surico, G. and Bertaccini, A. (2002) Involvement of phytoplasmas in decline of *Ulmus chenmoui* in Central Italy. *Forest Pathology* 32, 265–275.

Sherald, J.L. (1993) Pathogenicity of *Xylella fastidiosa* in American elm and failure of reciprocal transmission between strains from elm and sycamore. *Plant Disease* 77, 190–193.

Sherald, J.L. and Kostka, S.J. (1992) Bacterial leaf scorch of landscape trees caused by *Xylella fastidiosa*. *Journal of Arboriculture* 18, 57–63.

Simpson, A.J. *et al.* (2000) The genome sequence of the plant pathogen *Xylella fastidiosa*. *Nature* 406, 151–159.

Sinclair, W.A. (2000) Elm yellows in North America. In: Dunn, C.P. (ed.) *The Elms: Breeding, Conservation, and Disease Management*. Kluwer Academic Publishers, Norwell, Massachusetts, pp. 121–136.

Sinclair, W.A. and Griffiths, H.M. (1994) Ash yellows and its relationship to dieback and decline of ash. *Annual Review of Phytopathology* 32, 49–60.

Sinclair, W.A. and Griffiths, H.M. (1995) Epidemiology of a slow-decline phytoplasmal disease: ash yellows on old-field sites in New York State. *Phytopathology* 85, 123–128.

Sinclair, W.A. and Griffiths, H.M. (2000) Variation in aggressiveness of ash yellows phytoplasmas. *Plant Disease* 84, 282–288.

Sinclair, W.A. and Lyon, H. (2005) *Diseases of Trees and Shrubs*. Comstock Publishing Associated, New York.

Sinclair, W.A., Braun, E.J. and Larsen, A.O. (1976) Update on phloem necrosis of elms. *Journal of Arboriculture* 2, 106–113.

Sinclair, W.A., Griffiths, H.M. and Treshow, M. (1993) Impact of ash yellows mycoplasmalike organisms on radial growth of naturally infected white, green, and velvet ash. *Canadian Journal of Forest Research* 23, 2467–2472.

Sinclair, W.A., Griffiths, H.M. and Treshow, M. (1994) Ash yellows in velvet ash in Zion National Park, Utah: high incidence but low impact. *Plant Disease* 78, 486–490.

Sinclair, W.A., Griffiths, H.M. and Davis, R.E. (1996) Ash yellows and lilac witch broom: phytoplasmal diseases of concern to forestry and horticulture. *Plant Disease* 80, 468–475.

Sinclair, W.A., Townsend, A.M., Griffiths, H.M. and Whitlow, T.H. (2000) Reponses of six Eurasian *Ulmus* cultivars to North American elm yellows phytoplasma. *Plant Disease* 84, 1266–1270.

Śliwa, H., Kamińska, M., Korszun, S. and Adler, P. (2008) Detection of '*Candidatus* Phytoplasma pini' in *Pinus sylvestris* trees in Poland. *Journal of Phytopathology* 156, 88–92.

Sturrock, R.N., Frankel, S.J., Brown, A.V., Hennon, P.E., Kliejunas, J.T., Lewis, K.J., Worrall, J.J. and Woods, A.J. (2011) Climate change and forest diseases. *Plant Pathology* 60, 133–149.

Townsend, A.M. (2000) USDA genetic research on elms. In: Dunn, C.P. (ed.) *The Elms: Breeding, Conservation, and Disease Management*. Kluwer Academic Publishers, Norwell, Massachusetts, pp. 271–278.

Tran-Nguyen, L.T.T., Kube, M., Schneider, B., Reinhardt, R. and Gibb, K.S. (2008) Comparative genome analysis of '*Candidatus* Phytoplasma australiense' (Subgroup *tuf*-Australia I; *rp*-A) and '*Ca.* Phytoplasma asteris' strains OY-M and AY-WB. *Journal of Bacteriology* 190, 3979–3991.

Valiunas, D., Alminaite, A., Staniulis, J., Jomantiene, R. and Davis, R.E. (2001) First report of alder yellows phytoplasma in the eastern Baltic Region. *Plant Disease* 85, 1120.

Valiunas, D., Jomantiene, R., Ivanauskas, A., Sneideris, D., Staniulis, J. and Davis, R.E. (2010) A possible threat to the timber industry: '*Candidatus* Phytoplasma pini' in Scots pine (*Pinus sylvestris* L.) in Lithuania. In: Bertaccini, A., Laviña, A. and Torres, E. (eds) *Current Status and Perspectives of Phytoplasma Disease Research and Management*. Abstract book of the combined meeting of WG 1-4, Stiges, Spain on 1–2 February 2010. COST action FA 0807: Integrated Management of Phytoplasma Epidemics in Different Crop Systems, pp. 38. Available from: http://costphytoplasma.eu/ (accessed 22 November 2012).

Van Sluys, M.A. *et al.* (2003) Comparative analyses of the complete genome sequences of Pierce's disease and citrus variegated chlorosis strains of *Xylella fastidiosa*. *Journal of Bacteriology* 185, 1018–1026.

Walla, J.A., Jacobi, W.R., Tisserat, N.A., Harrell, M.O., Ball, J.J., Neill, G.B., Reynard, D.A., Guo, Y.H. and Spiegel, L. (2000) Condition of green ash, incidence of ash yellows phytoplasmas, and their association in the Great Plains and Rocky Mountain Regions of North America. *Plant Disease* 84, 268–274.

Wei, T., Wu, Y.F., Wu, K.K., Hou, W. and Li, Y.R. (2009) First report of a 16SrI-C group phytoplasma associated with yellows-type disease affecting willow plants in China. *Plant Disease* 93, 197.

Weintraub, P.G. and Beanland, L. (2006) Insect vectors of phytoplasmas. *Annual Review of Entomology* 51, 91–111.

Weintraub, P.G. and Wilson, M.R. (2010) Control of phytoplasma diseases and vectors. In: Weintraub, P.G. and Jones, P. (eds) *Phytoplasmas: Genomes, Plant Hosts and Vectors*. CAB International, Wallingford, UK, pp. 233–249.

Wells, J.M., Raju, B.C., Hung, H.-Y., Weisburg, W.G., Mandelco-Paul, L. and Brenner, D.J. (1987) *Xylella fastidiosa* new-genus new species Gram-negative xylem-limited fastidious plant bacteria related to *Xanthomonas* spp. *International Journal of Systematic Bacteriology* 37, 136–143.

Wester, H.V. and Jylkka, E.W. (1959) Elm leaf scorch, graft transmissible virus of American elm. *Plant Disease Reporter* 43, 519.

Woese, C.R., Maniloff, J. and Zablen, L.B. (1980) Phylogenetic analysis of the mycoplasmas. *Proceedings of the National Academy of Sciences of the United States of America* 77, 494–498.

Xu, Z.C., Leininger, T.D., Lee, A.W.C. and Tainter, F.H. (2001) Physical, mechanical, and drying properties associated with bacterial wetwood in red oaks. *Forest Products Journal* 51, 79–84.

Zadoks, R.N., Griffiths, H.M., Munoz, M.A., Ahlstrom, C., Bennett, G.J., Thomas, E. and Schukken, Y.H. (2011) Sources of *Klebsiella* and *Raoultella* species on dairy farms: be careful where you walk. *Journal of Dairy Science* 94, 1045–1051.

Zeikus, J.G. and Ward, J.C. (1974) Methane formation in living trees: a microbial origin. *Science* 184, 1181–1183.

Zhao, Y., Wei, W., Lee, I.-M., Shao, J., Suo, X. and Davis, R.E. (2009) Construction of an interactive online phytoplasma classification tool, *i*PhyClassifier, and its application in analysis of the peach X-disease phytoplasma group (16SrIII). *International Journal of Systematic and Evolutionary Microbiology* 59, 2582–2593.

Zhao, Y., Wei, W., Davis, R.E. and Lee, I.-M. (2010) Recent advances in 16S rRNA gene-based phytoplasma differentiation, classification and taxonomy. In: Weintraub, P.G. and Jones, P. (eds) *Phytoplasmas, Genomes, Plant Hosts and Vectors*. CAB International, Wallingford, UK, pp. 64–92.

5 Forest Diseases Caused by Higher Parasitic Plants: Mistletoes

David C. Shaw[1]* and Robert L. Mathiasen[2]
[1]*Oregon State University, Corvallis, Oregon, USA;* [2]*Northern Arizona University, Flagstaff, Arizona, USA*

5.1 Pathogens, Significance and Distribution

We define a mistletoe as a parasitic flowering plant from sandalwood order (*Santalales*) that attaches to the above-ground stem of another plant (primarily gymnosperms and angiosperms). While mistletoes are widespread and ecologically important components of forests worldwide, some of them are also damaging parasites of commercially important trees. The five families of plants that include mistletoes are: *Loranthaceae, Viscaceae, Misodendraceae, Eremolepidaceae* and *Santalaceae*. However, not all species within these families are mistletoes. Even within the *Loranthaceae*, the family with the most genera of mistletoes, three genera attach only to roots. By our definition, these root parasites are not considered to be mistletoes. In addition, other angiosperm genera, such as *Cuscuta* (*Convolvulaceae*), *Striga* (*Orobanchaceae*), and *Cassytha* (*Lauraceae*) that also attach to host stems or roots are not considered mistletoes and will not be included in our discussion. Parker and Riches (1993) provide a good review of the biology and control of these non-mistletoe parasitic weeds.

Some mistletoes parasitize a very large number of hosts in different families. Two notable examples are *Viscum album* subsp. *album* L., with over 450 host species (Barney *et al.*, 1998), and *Amyema miquelli* (Lehm. ex Miq.) Tiegh., which parasitizes hosts in 17 plant families (Downey, 1998). In contrast, a few mistletoes only parasitize one host species, for example some dwarf mistletoes (*Arceuthobium* spp.) (Hawksworth *et al.*, 2002). Although host specificity is commonly mentioned in the mistletoe literature, and many authors have pointed to its pathological importance, few investigations have examined the mechanisms that control mistletoe–host compatibility (Scharpf, 1984; Norton and Carpenter, 1998). A better understanding of the mechanisms related to host specificity would have practical applications to mistletoe management, particularly in developing greater host resistance (Shamoun and DeWald, 2002). How the mechanisms of host–mistletoe compatibility or incompatibility function remains one of the most fascinating and challenging areas of mistletoe biology and pathology yet to be understood.

Mistletoes are distributed all over the world and occur on all of the continents except for Antarctica. Although several species occur in temperate zones, they are most common and abundant in the tropical forests of Asia,

* E-mail: dave.shaw@oregonstate.edu

Central America, South America and Africa (Kuijt, 1969; Calder and Bernhardt, 1983; Polhill and Wiens, 1998). The loranthaceous mistletoes, which represent by far the most species, are primarily distributed in the southern hemisphere, but several genera, such as *Loranthus*, *Psittacanthus* and *Struthanthus*, also occur north of the equator. In contrast, viscaceous mistletoes are distributed primarily in the northern hemisphere, with notable exceptions such as *Viscum*, *Korthalsella* and *Notothixos* occurring south of the equator. Although many mistletoes affect tropical trees, the most damaging and economically important species occur in the coniferous forests of North America and, to some extent, Europe. Mistletoes also parasitize a variety of fruit- and nut-producing trees throughout the world and cause economic losses by reducing the productivity of infected trees (Table 5.1) (Pundir, 2009). As common and sometimes abundant parasitic plants in many forest ecosystems of the world, the management of mistletoes is often necessary, but they may also play critical roles in the function and structure of the forests they inhabit (Watson, 2001).

Mistletoes have been associated with losses in food production, rubber production and wood and fibre production (Table 5.1). However, quantitative data for the amount of economic losses are lacking for most mistletoe–host associations (Knutson, 1983; Pundir, 2009), except for the dwarf mistletoes, which are common and widespread pathogens of commercially valuable conifers (Hawksworth and Wiens, 1996; Geils *et al.*, 2002).

Because mistletoes reduce the growth of commercially valuable timber species,

Table 5.1. Pathogenic genera of mistletoes, the host groups seriously affected and the regions where economic losses are associated with mistletoe parasitism. Modified from Knutson (1983).

Genera	Hosts affected	Region
Family: *Loranthaceae*		
Amyema	*Acacia*, *Eucalyptus*	Australia
Dendropemon	Citrus (*Citrus*)	Caribbean
Dendrophthoe	Citrus, fig (*Ficus*), guava (*Psidium guajava* L.), mulberry (*Morus*)	India
Macrosolen	Citrus	India
Oryctanthus	Cacao (*Theobroma cacao* L.)	Central America
Phthirusa	Avocado (*Persea americana* Mill.), cacao, citrus, coffee (*Coffea*), *Erythrina*, mango (*Mangifera indica* L.), rubber (*Hevea*)	Central and South America
Psittacanthus	Citrus, pine (*Pinus*)	Mexico and Central America
Scurrula	Citrus	Philippines, Indonesia
Struthanthus	Citrus, mango, pine	Central America
Tapinanthus	Cacao	Africa
Tolypanthus	Citrus	India
Family: *Mesodendraceae*		
Misodendrum spp.	*Nothofagus* spp.	South America
Family: *Viscaceae*		
Arceuthobium	Conifers in the *Pinaceae*	North America and Asia
Dendrophthora	Avocado, cacao, mango, rubber	South America
Korthalsella	*Acacia*, *Eucalyptus*	Australia, Hawaii
Phoradendron	Avocado, cacao, citrus, coffee, *Erythrina*, fir (*Abies*), oak (*Quercus*), pecan (*Carya illinoinensis* (Wangenh.) K. Koch), walnut (*Juglans*)	North, Central, and South America
Viscum	Almond (*Prunus*), apple (*Malus*), fig, fir, olive (*Olea*), peach (*Prunus*), pear (*Pyrus*), persimmon (*Diospyros*), pine, prune (*Prunus*), rubber, walnut	Asia, Europe, and Africa (introduced into California)

their pathological effects are usually evaluated by quantifying the percentage loss compared with healthy trees. For example, the effects of dwarf mistletoes on the growth of their conifer hosts has been assessed by investigators, who have reported reductions in radial diameter growth or estimated reduced volume growth for individual trees, stands of trees or large geographic regions (Hawksworth and Scharpf, 1978; Mathiasen, 1996; Shaw *et al.*, 2004). Economic losses from dwarf mistletoes amount to billions of US dollars annually, but no detailed economic analysis has been published recently (Hawksworth and Wiens, 1996; Geils and Hawksworth, 2002). Pundir (2009) has worked extensively in the western Himalayas and recently summarized mistletoe damage on hosts while including documentation of the impacts on fruit and seed production of trees infected by mistletoe.

Reports of mistletoe damage often use only a qualitative ranking system of low, moderate or severe infection levels; exceptions are the severity rating systems developed for the dwarf mistletoes. While many rating systems have been published (Dooling, 1978; Parker and Mathiasen, 2004), the Hawksworth six-class system (Hawksworth, 1977) remains the standard disease severity rating system (Fig. 5.1) used for dwarf mistletoes throughout North America, and has been adapted for use when rating the severity of infection for other mistletoes (Daugherty and Mathiasen, 2003; Howell and Mathiasen, 2004). However, Pundir (2009) has developed a rating system for hardwoods that is used extensively in north-western India (Fig. 5.2).

Divide live crown into vertical thirds:
upper, middle, lower.

Rate each third separately as 0,1 or 2, where:

0 = no visible infections

1 = less than 1/2 of branches infected

2 = more than 1/2 of branches infected.

Add the ratings of each third for a total of 0 to 6.

Example:

Upper with <50% of branches infected = 1

Middle with <50% of branches infected = 1

Lower with >50% of branches infected = 2

Total = 1 + 1 + 2 = 4.

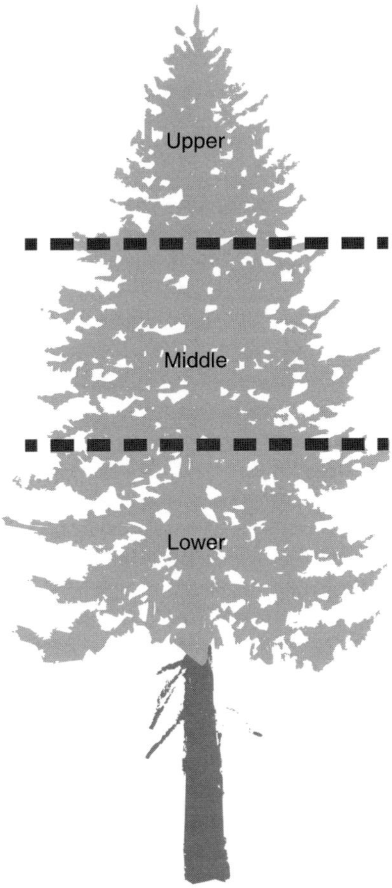

Fig. 5.1. The six-class dwarf mistletoe (*Arceuthobium* spp.) rating system for conifers. From Hawksworth and Wiens (1996).

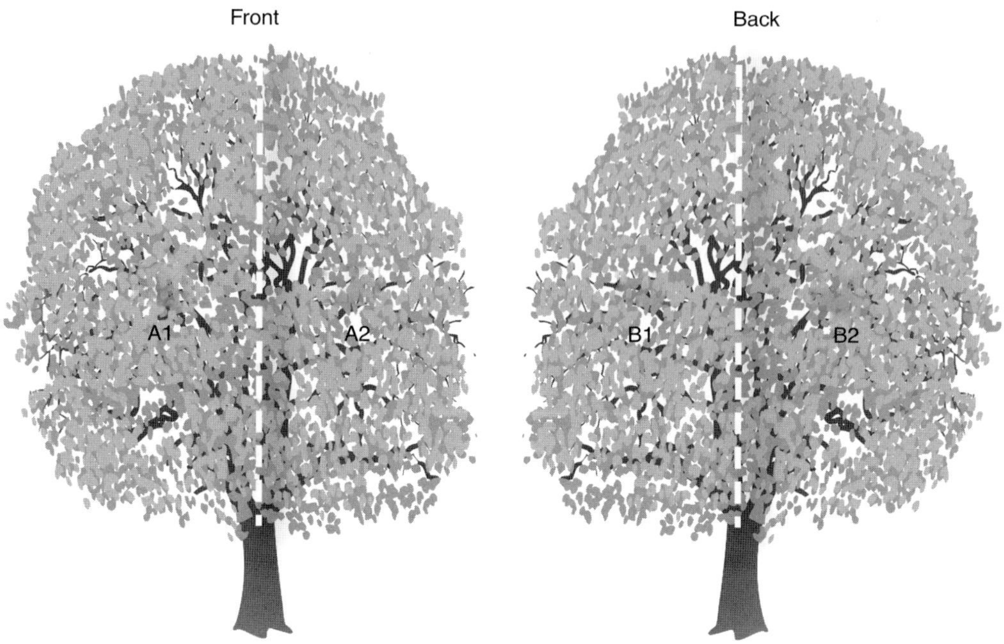

Divide crown into two parts with the broader side designated as the front, A, and the back side as B.

Divide the front into A1 and A2 compartments.

Divide the back into B1 and B2 compartments.

Estimate percentage of the crown that is infected for each compartment (0–100%) separately.

Sum A1, A2, B1 and B2, then divide by 4.

Example:

A1 = 20%, A2 = 80%, B1 = 20%, B2 = 40%

Rating = 20 + 80 + 20 + 40 = 160/4 = 40%

Fig. 5.2. The quadrat rating system for hardwoods. From Pundir (2009).

5.2 Diagnosis

The effects of mistletoes on their hosts include reductions in growth, vigour, fruiting and seed production. Infections by mistletoes may also cause the hypertrophy of infected branches, dieback of infected limbs, death of the primary leader (spike tops) and the formation of witches' brooms (Knutson, 1983). Severe infection by mistletoes is often associated with premature mortality of host trees, particularly trees infected by dwarf mistletoes. Dead tops are commonly observed on conifers severely infected with species of *Arceuthobium*, *Phoradendron* and *Viscum* (Hawksworth and Wiens, 1996). Nutrients and water absorbed by the host's roots are diverted to supply the mistletoe infections occurring between the roots and the tree top, thereby depriving the topmost branches of needed resources. Eventually, as the number of mistletoe infections increases, a point is reached where the top of the tree can no longer survive, and a dead top develops as a symptom of severe mistletoe infection.

The most common disease symptom associated with mistletoe infection is branch dieback (Knutson, 1983). Over a period of time, the branch distal to the point of attachment by the mistletoe dies, whereas the branch segment proximal to the mistletoe remains alive and continues to supply water and nutrients to the parasite. In many cases, the distal end of the branch eventually falls away, leaving a live branch supporting a large mistletoe plant at its end (Kuijt, 1969; Howell and Mathiasen, 2004). Branch dieback has been reported for many mistletoe–host combinations, but its pathological effects on hosts have not been adequately investigated (Knutson, 1983). Dieback of branches is particularly prevalent during droughts when the host is taxed by lack of water, but the mistletoe continues its demand for the scarce resource (Lei, 2001). During droughts, mistletoe plants distal to others often die as well.

Many mistletoes cause localized hypertrophy of host tissues at the site of infection; this is primarily a consequence of the disruption of normal tissue development (Kuijt, 1969; Knutson, 1983). The swelling of host tissue may be associated with either an increase in host cambial activity, resulting in the formation of secondary xylem cells, or the displacement of host xylem and phloem tissue (Kuijt, 1969). Large swellings on a host branch are also associated with profusely branched extraxylary absorptive structures produced by the mistletoe, such as with some species of *Phoradendron* (Thoday, 1957).

A few mistletoes, most notably dwarf mistletoes, alter host phytohormone balance resulting in the formation of dense masses of branches called witches' brooms (Plate 2) (Livingston *et al.*, 1984; Hawksworth and Wiens, 1996). Branches from witches' brooms exhibit features atypical of uninfected host tissue, such as: increased longevity (Weir, 1916), the elimination or reduction of seed and cone production (Kuijt, 1969), increased branch elongation (on some hosts) (Tinnin and Knutson, 1980), increased biomass compared with uninfected branches of the same age (Tinnin and Knutson, 1980), and decreased number, length and mass of needles (Broshot and Tinnin, 1986; Reblin *et al.*, 2006).

Signs of mistletoe infection are, obviously, the plants produced on infected branches and trunks of their host plants. Many mistletoes produce relatively large plants, sometimes with leaves that are readily observed, but some produce small plants that may be overlooked without careful observation. In deciduous trees, non-dwarf mistletoes are often only readily observed during leafless periods. Some mistletoes may also produce large, showy flowers that attract their bird pollinators, while others have undergone extreme reduction in floral morphology, as in all *Viscaceae*, *Misodendraceae* and some *Loranthaceae* (Gill and Hawksworth, 1961; Kuijt, 1969). Leaf size also varies greatly among mistletoes; some species have very large leaves, while others may be squamate, i.e. with leaves that are reduced to very small scales (Kuijt, 1969).

Because mistletoes are distributed worldwide, their identification requires the use of a wide array of literature (taxonomic monographs, regional floras, agriculture handbooks and refereed papers) (e.g. Hawksworth and Wiens, 1996; Polhill and Wiens, 1998; Kuijt, 2003).

5.3 Infection Biology

Host infection has been described in detail for some groups of mistletoes but not for others, but is considered to be similar for all mistletoes (Kuijt, 1969; Calder and Bernhardt, 1983; Hawksworth and Wiens, 1996; Kolb, 2002). Upon germination, seeds form a hypocotyl that elongates until it forms a holdfast which attaches firmly to the host branch. As for other flowering plants, seed germination is influenced by temperature, moisture and light (Lamont, 1983). The seeds of mistletoes in the family *Viscaceae* have a chlorophyllous endosperm and embryo, and so are capable of producing simple sugars as an energy source after germination (Lamont, 1983). A penetration peg develops on the lower surface of the holdfast that mechanically penetrates the epidermis or bark of the host, eventually contacting the host's phloem and/or xylem. Penetration of host tissue is primarily by mechanical means, but there is some evidence that pectinases may also be involved (Thoday, 1951; Knutson, 1983). Once the mistletoe has entered host tissue it develops a haustorium (Kuijt, 1969; Calvin and Wilson, 1998), and then aerial plants. Many tropical mistletoes begin forming plants soon after they establish their connection to their host, while the dwarf mistletoes may take 2–6 years to form aerial plants.

Competition for water and nutrients is the most obvious explanation for the detrimental effects that mistletoes have on trees. High transpiration rates by mistletoes cause reduced xylem water potentials in host branches, which reduces the net photosynthetic rates of the host (Stewart and Press, 1990; Kolb, 2002). Mistletoes generally have higher leaf transpiration and stomatal conductance than their hosts. The accumulation of osmotically active solutes in mistletoe tissue also contributes to a lower xylem water potential in their tissue compared with that in the host, thus further facilitating the absorption of water and solutes. Mistletoes can also decrease xylem hydraulic conductivity of host branches distal to the point of infection. This can cause the end of the branch to die, but the mistletoe remains living, drawing water and nutrients from the infected branch (Hull and Leonard, 1964; Tennakoon and Pate, 1996). High concentrations of minerals in mistletoes demonstrate that they are also efficient parasites of these nutrients (Glatzel, 1983; Lamont, 1983; Ehleringer and Schulze, 1985).

Several studies have now clearly demonstrated that many mistletoes previously thought to be only water parasites actually derive some or most of their carbon requirements from their hosts as dissolved compounds in the host xylem sap (Marshall and Ehleringer, 1990; Richter *et al.*, 1995). It has now been estimated that some mistletoes absorb low amounts of carbon (5–20% of their requirements) from their hosts, while others absorb as much as 80% of their carbon requirements (Kolb, 2002). Recent experimental evidence suggests that mistletoes are most robust on the most vigorous host trees (Bickford *et al.*, 2005), a concept that has long been assumed (Watson, 2009), but lacked substantiating data.

5.4 Epidemiology

Mistletoes disperse their seeds by three distinct mechanisms: (i) animal dispersal (primarily birds); (ii) wind dispersal; and (iii) explosive discharge of seed. Management systems for mistletoes depend on these dispersal mechanisms, as they are the foundation of mistletoe spread and intensification in managed forests and orchards.

5.4.1 Bird dispersal

The co-evolution of mistletoes with their avian vectors has resulted in attractive and nutritious fruits that provide valuable food for many bird species throughout the world (Reid, 1991; Ladley and Kelly, 1996; Martínez del Rio *et al.*, 1996). The mature fruits of mistletoes are brightly coloured (usually white, yellow, red, blue or purple) and their seeds are coated with a natural 'glue', termed viscin (Kuijt, 1969; Reid, 1991). Birds either swallow mistletoe fruits whole, peel off the outer exocarp and ingest the seed and viscin, or eat only the viscin coating around the seed (Kuijt, 1969; Snow and Snow, 1988). Once the bird has eaten the seed, it is either regurgitated or defaecated, but the seed is still covered with some of its viscin coat, which allows it to adhere to potential hosts. In many instances, seeds adhering to a bird's beak, legs or feathers are rubbed off on to a branch of a potential host.

Approximately 90 bird species from ten families are considered to be mistletoe fruit specialists, and these exhibit a range of behavioural and morphological adaptations to their narrow diet. Most of these groups are represented by four or fewer species, except for the flowerpeckers (Dicaeidae, a family of passerines) of Asia (44 species) and the euphonias (Carduelinae, Fringillidae) of Latin America (33 species). While most discussion of mistletoe dispersal is typically restricted to these dietary specialists, a wide range of other avian species also disperse mistletoe seeds (Watson, 2001), accounting for all dispersal in Europe and most regions of North America. Within the USA, vectors of *Phoradendron* spp. are fairly well known (Mallams and Mathiasen, 2010), but only a few studies have examined in any detail the relationships between birds and *Phoradendron* (Aukema and Martínez del Rio, 2002; Aukema, 2004).

For many mistletoes, particularly those in Central and South America, the key vectors have not been investigated to any large extent (Martínez del Rio *et al.*, 1996; Restrepo *et al.*, 2002; Medel *et al.*, 2004). The control of economically damaging mistletoes in managed areas is often confounded by their reintroduction by birds (Gill and Hawksworth, 1961). In South and North America, animals other than birds have also been implicated in the dispersal of mistletoe seeds. In South America, a marsupial disperses seeds (Amico and Aizen, 2000), and in North America, squirrels and other mammals have been shown to rarely disperse dwarf mistletoe seeds adhering to their fur (Hawksworth and Wiens, 1996; Mathiasen, 1996).

Host distribution, size, or sex may influence mistletoe distribution and abundance (Aukema, 2003, 2004). For example, when hosts are widely scattered, their mistletoe parasites may be less common and widely distributed as well. Birds that disseminate mistletoes often perch at the tops of the larger trees, thus depositing mistletoe seeds high in the canopy. For dioecious tree species, bird visitation may be biased in favour of fruiting plants, thereby influencing overall mistletoe distribution (Carlo and Aukema, 2005). Furthermore, the consistent availability of mistletoe fruits can attract birds which also feed on the host's fruits (van Ommeren and Whitham, 2002). This favours the spread of seeds from infected hosts over non-infected hosts during years when susceptible trees have not produced an abundance of fruits. It has been suggested that when the seeds of infected trees are also spread by mistletoe vectors and this is correlated with greater tree regeneration, the relationship between the mistletoe and host tree approaches mutualism (van Ommeren and Whitham, 2002).

5.4.2 Wind-dispersed seeds (feathery mistletoes, *Misodendrum* spp.)

The *Misodendraceae* of temperate South America, which are endemic to South American temperate forests where they commonly infect *Nothofagus* spp., is the only group of mistletoes that have wind-dispersed seeds. The female flower has three setae inserted in furrows on the ovary (Heide-Jørgensen, 2008). These setae become plumose and elongate with maturation of the nutlet, and then serve in wind transport of the seed and perhaps as mechanisms for entwining the branches of host trees. Although we are not aware of published accounts of dispersal of *Misodendrum* spp., we assume that in theory, stand structure and landscape topography that influences wind patterns would also influence the pattern of mistletoe spread. Wind patterns during the dispersal period should control where seed flies and which trees become infected, and non-host size and distribution would also influence where seed can be dispersed by wind. Perhaps in some locations in the mountainous terrain of South America the wind is not unidirectional and may spread seed in multiple directions; but if wind is primarily from one direction, then a pattern of spread downwind from a seed source is likely. Hypothetically, mistletoe that occurred on ridge tops would spread further and wider than mistletoes in valleys or from wind-protected sites.

5.4.3 Explosive discharge of seeds (dwarf mistletoes)

The dwarf mistletoes are primarily disseminated by an 'explosive fruit' system involving both hydrostatic and mechanical mechanisms (Hawksworth and Wiens, 1996; Hawksworth *et al.*, 2002). Dwarf mistletoe seeds are expelled from fruits at initial velocities of about 24 m s^{-1} and may fly 10 m or more. These seeds, like bird-distributed seeds, are also coated with viscin, which allows them to adhere to potential host surfaces. Factors affecting the spread and intensification of dwarf mistletoes associated with both the explosively disseminated seed mechanism and random spread by seeds sticking to animal vectors have been summarized by several investigators (Parmeter, 1978; Hawksworth and Wiens, 1996; Mathiasen, 1996). These two mechanisms contribute differently to the spread of dwarf mistletoes, the former producing primarily localized intensification, and the latter contributing to the occasional establishment of new infection centres. In general, however, forest stand structure and species composition are the primary controls on the spread and intensification of dwarf mistletoes in conifer stands.

5.4.4 Pollination of mistletoes

Pollination of mistletoe flowers by biotic agents (primarily birds and insects) as well as by wind, is essential for reproduction and therefore it is relevant to epidemiology. Many tropical and subtropical mistletoes in the *Loranthaceae* have large, colourful flowers that are borne in groups that produce large amounts of sugar-rich nectar that attract avian pollinators. Elaborate pollination and seed dispersal mechanisms involving birds have evolved in some of these loranthaceous species (Freehan, 1985; Ladley *et al.*, 1997; Kirkup, 1998). For example, birds pry open the fused corollas to reach their nectar reward, upon which the pollen 'explodes' on to the bird's head. These mistletoes are often dichogamous (protandrous), and after the birds have visited flowers in the male phase, they eventually visit flowers in the female phase, thereby effecting pollination. The co-evolutionary relationships between mistletoes and their bird pollinators are so closely linked that disruption of this association could have long-term negative consequences for both interacting organisms and, possibly, the entire ecosystem (Reid *et al.*, 1995; Watson, 2001). However, many bird-pollinated mistletoes are serviced by a broad range of species, and no bird can be considered a mistletoe pollen specialist (Watson, 2001). In Mexico, Central America and South America, hummingbirds are key pollinators of mistletoes with large, showy red or yellow flowers (Azpeitia and Lara, 2006). A variety of insects are the key pollinators of mistletoes in the *Viscaceae* and *Loranthaceae* (Kuijt, 1969; Hawksworth and Wiens, 1996).

5.5 Management Strategies and Tactics

Management of mistletoes is dependent on the nature of the mistletoe and the situation (Reid and Shamoun, 2009). Whereas the dwarf mistletoes in North America (Geils *et al.*, 2002) and all mistletoes in the western Himalayas (Pundir, 2009) may be managed as destructive forest and orchard pathogens, in New Zealand, several loranths are considered endangered species and are managed for preservation (de Lange and Norton, 1997). The European mistletoe (*V. album* L.) is used in the pharmaceutical industry and is cultivated commercially (Büssing, 2000), while the famous horticulturist Luther Burbank intentionally established the species on fruit trees for the commercial Christmas ornament industry in California (USA) about 100 years ago (Hawksworth and Scharpf, 1986). The European mistletoe is also used extensively in the UK in the Christmas ornament industry (Biggs, 2010).

Some mistletoes are being managed to benefit wildlife in Australia and North America (Shaw *et al.*, 2004). For example, in Oregon and Washington (USA), witches' brooms caused by Douglas-fir dwarf mistletoe (*A. douglasii* Engelm.) are important nesting sites for the endangered spotted owl (*Strix occidentalis* Xantus de Vesey) (Marshall *et al.*, 2003), while hemlock dwarf mistletoe (*A. tsugense* (Rosend.) G.N. Jones) is known to provide nesting platforms for the endangered marbled murrelet (*Brachyramphus marmoratus* Gmelin) near the coast of the Pacific Ocean

(Hamer and Nelson, 1995). No longer, then, are mistletoes considered simply tree pathogens (Watson, 2001). This has led to a more nuanced and ecological approach to managing them in many situations.

5.5.1 Avoidance

One way to avoid mistletoes is to plant the host outside the known range of the mistletoe. In temperate regions, mistletoe distribution is often limited by temperature. Pacific mistletoe (*Phoradendron villosum* (Nutt.) Nutt.) in Oregon does not occur north of the Columbia River or east in the Columbia River Gorge, so that Oregon white oak (*Quercus garryana* Douglas ex Hook.) is free of mistletoe in these regions. Hemlock dwarf mistletoe is restricted to the lower elevations and does not occur throughout the range of its primary host, western hemlock (*Tsuga heterophylla* (Raf.) Sarg.) in Alaska (USA). Although this situation is a limited possibility for management, it is relevant for hosts that may be especially affected at a local scale.

Many mistletoes occur in distinct plant communities where they are more successful or occur at higher densities than in other vegetation types. It may be possible to avoid mistletoes by managing hosts in forest stands where the mistletoes are unlikely to be encountered, but the mistletoe could easily grow if present. Forest pathologists have sampled diseases across habitat types and found that certain plant communities have more intense and common occurrences of dwarf mistletoes (Mathiasen and Blake, 1984; Merrill *et al.*, 1987).

Although we are not aware of research that has tested these ideas, it is likely that since *Misodendrum* spp. is dispersed by wind, planting susceptible trees (*Nothofagus* spp.) upwind of known infection areas may avoid exposure to seed. If known infection areas are along ridge tops, it would be preferable to avoid planting stands below and downwind of these areas. It may be possible to block spread with taller trees that could protect small stature stands, although this strategy is untested.

Dwarf mistletoes disperse mostly by explosive discharge of seeds (see Section 5.4.3), which can reach a maximum distance of about 10 m. In areas of large burns where dwarf mistletoe has been eradicated locally, it is possible to plant trees that will be likely to avoid exposure to seed. Dwarf mistletoes are patchily distributed, and it is also possible to find areas where they are not present. These areas may be successfully planted to avoid exposure. Many authors recommend using physical features of the landscape, such as rivers, roads and non-forested areas to avoid and/or block exposure to dwarf mistletoe seed from nearby infection centres (Hawksworth and Wiens, 1996; Hawksworth *et al.*, 2002). In addition, a key management recommendation for the avoidance of dwarf mistletoe seed exposure is to protect a plantation with a buffer of non-hosts that will block the spread of seeds from adjacent infected stands. This may also be considered to be exclusion (see Section 5.5.2).

The majority of mistletoes are dispersed by birds and therefore it is quite difficult to avoid exposure of a particular area to seed without ecological knowledge of the local dispersal agents and their ecology. Some data indicate that birds prefer larger trees to small and more densely managed trees (Aukema, 2004). So it may be possible to use trap trees, i.e. to leave larger trees that may attract birds, thereby avoiding exposure to seeds of trees in regenerating and young plantations.

5.5.2 Exclusion

An often discussed management approach, yet rarely attempted for obvious reasons, is to eradicate the vectors of mistletoe seeds from a localized area. Although this could exclude the seeds of mistletoes from a plantation, it has very negative conservation biology implications, and is likely to fail owing to the rapid recolonization of sites by birds (Pundir, 2009). Watson (2011) details accounts of work in Australia that attempted to prevent mistletoes from intensifying by killing the avian vectors, which was apparently a common practice for some time during the early 1900s. He notes that the vector birds recolonize forests quickly, are specialists on mistletoe fruit, and are rarely found in areas without mistletoe.

Therefore, they are unlikely to vector the seed into areas of no mistletoe.

The primary method of exclusion for dwarf mistletoes is to eradicate the disease from the site, by clear-cutting and possibly burning the site, and then replanting with uninfected trees (Hawksworth and Wiens, 1996; Muir and Geils, 2002). To exclude dwarf mistletoes from the replanted site, it may be necessary to plant a buffer of non-hosts around the plantation. However, Muir and Geils (2002) note that this often does not work operationally, as the host trees often naturally seed into these planting areas and may grow faster than the non-hosts. Muir and Geils (2002) note several strategies for preventing the spread of dwarf mistletoes into cut blocks, which include: (i) placing the cutting boundaries in non-infested portions of the stand and using natural features of the landscape, such as rivers, roads, or non-forested areas to prevent seed dispersal into the plantation; (ii) minimizing the length of the cut block border that has dwarf mistletoe along it (e.g. larger circular cuts have less border than long narrow cut blocks); (iii) planting buffer strips only where this has been demonstrated to work; (iv) avoiding the leaving of single infected trees or small clumps of infected trees in the cut block; and (v) choosing residual trees that are uninfected when regenerating stands with shelterwood or seed tree systems.

An alternative to planting a buffer strip of non-hosts is to plant host trees very densely because dwarf mistletoe spreads more slowly in dense, fast-growing stands of trees than through widely spaced and slow-growing trees (Parmeter, 1978; Muir and Geils, 2002). However, this is only the case when there is no infested overstorey raining seeds down on to the planted trees.

The dynamics of wind-dispersed seed is complex, and buffers and other mechanisms that work for *Arceuthobium* spp. are likely not to work for feathery mistletoes.

5.5.3 Eradication

The primary control of mistletoes remains the pruning of infected branches and removing infected trees (Butin, 1995; Geils *et al.*, 2002; Pundir, 2009; Mallams and Mathiasen, 2010). Depending on dispersal mechanisms, eradication may only be a temporary solution.

The three subspecies of *V. album* in Europe (Dobbertin and Rigling, 2006; Biggs, 2010) are the primary mistletoes of concern for forest, orchard and ornamental trees. Pruning and tree removal are the primary control for *V. album* (Butin, 1995). Because its seeds are spread by birds, it is difficult to prevent reintroductions and therefore pruning must be practised continuously. *V. album* subsp. *abietis* (Wiesb.) Abromeit. is an important parasite of the silver fir (*Abies alba* Mill.) and is managed by replacing this species with other conifers, and selectively thinning infected trees. However, there are now conservation concerns that silver fir populations are being reduced to unacceptable levels and that selective thinning is favouring the growth of *V. album* by increasing light levels in thinned forests (Noetzli *et al.*, 2003).

Loranthaceous mistletoes, primarily *Amyema* spp., are of particular concern in south-eastern Australia because they can intensify on isolated and clumped paddock trees in the farming landscape (Reid and Shamoun, 2009). These trees are valued for shade, biodiversity and aesthetics. Although there is a long history of experimental herbicide injections, the principle control of these mistletoes is pruning. In Australia, another possible option is pollarding, where the small diameter branches are removed from the main large diameter trunk and major lateral branches (New South Wales Landcare, 2008). In many cases, pollarding can be successfully used to remove mistletoe from a tree without killing it.

Pundir (2009) has extensively reviewed the experiences of plantation managers in India and elsewhere in the world who have utilized pruning (physical removal) to control mistletoe for over 100 years. In most locations, pruning must be continually done to maintain low populations of mistletoes.

The interaction of mistletoes and prescribed and wildland fires is currently an important area of fire behaviour research, particularly because of the increased frequency and intensity of wildfires in the western USA and Australia (Parker *et al.*, 2006;

Hoffman *et al.*, 2007). Wildfire controls dwarf mistletoes by killing infected trees, reducing stand density, killing infected regeneration and scorch pruning infected branches (Alexander and Hawksworth, 1976).

As a management tool, prescribed fire (Plate 3) can directly control dwarf mistletoes by scorch pruning lower infected branches and brooms, killing severely infected trees and causing the dehiscence of mistletoe plants exposed to smoke and heat. Prescribed fire has been shown to reduce dwarf mistletoe populations, but the intensity of fire used is critical to the effectiveness of this management practice (Koonce and Roth, 1980; Harrington and Hawksworth, 1990; Conklin and Geils, 2008). Prescribed fire can also reduce the susceptibility of a forest to spread of dwarf mistletoe, especially if combined with thinning, by killing the young regeneration that would become infected from overstorey trees, and maintaining wider spacing between trees (Alexander and Hawksworth, 1976).

5.5.4 Protection

The protection of host trees from mistletoes is an undeveloped science. In theory, it may be possible to protect trees from mistletoe by using a physical barrier on to the bark of young twigs so that seed cannot come in contact with the bark. Applying a systemic herbicide by injecting trees may result in protection of the tree by influencing the success of mistletoe seedlings.

The use of shading to reduce the impacts of mistletoes on crop trees is possible because mistletoes do best in high light environments (Parker and Riches, 1993). In Africa, control of African mistletoe (*Tapinanthus bangwensis* (Engler & K. Krause) Danser) on cacao (*Theobroma cacao* L.) by growing the crop trees under the canopies of shade trees has been suggested by Room (1973). However, in many situations, the use of shade to control mistletoes may be very difficult, and it is not always possible to experimentally determine the impact of shade on the growth of the hosts and the mistletoe (Parker and Riches, 1993). Some trees can outgrow and shade out mistletoes on branches, so fertilizing has been suggested for situations where tree vigour and growth can be improved (Parker and Riches, 1993).

5.5.5 Resistance

There is robust evidence that plant breeding could succeed in developing varieties of forest and crop trees that are resistant to loranthaceous and viscaceous mistletoes (Roth, 1974; Hariri *et al.*, 1991; Shamoun and DeWald, 2002), but the expense of breeding programmes apparently has limited its application. Shamoun and DeWald (2002) suggested that the low cost and ease with which mistletoes have been controlled silviculturally has prevented investment in developing resistant varieties of commercially valuable trees. Information on the resistance of western conifers to dwarf mistletoes has been reviewed by Roth (1978), Scharpf (1984) and Shamoun and DeWald (2002). Thus far, resistance to dwarf mistletoes has not been applied to any large extent on an operational basis and the mechanisms of resistance remain largely unknown.

Roth (1974) examined the use of resistant trees to manage western dwarf mistletoe on ponderosa pine (*Pinus ponderosa* Lawson & C. Lawson) in the western USA. He demonstrated that seedlings of ponderosa pine from putatively resistant parents had less infection and grew more rapidly than similar seedlings from less resistant parents. Some of the most definitive work on the possibilities of using resistance to manage dwarf mistletoes has been conducted by Robert Scharpf (Scharpf, 1987), who reported resistance to western dwarf mistletoe in low elevation Jeffrey pines (*P. jeffreyi* Balf.) from Placer County, California. Seedlings grown from these pines and planted near infected overstorey trees demonstrated lower levels of infection 13 years after their establishment than trees grown from susceptible trees.

Another example of resistance to dwarf mistletoe infection was reported by Smith *et al.* (1993) for western hemlock. They found substantial differences in resistance between artificially inoculated trees grown from grafts of putatively resistant and susceptible western hemlocks in British Columbia (Canada). However, dwarf mistletoe-resistant western hemlocks have not been propagated to any extent and are not available for planting out at this time.

Although it has now been shown that there are dwarf mistletoe-resistant trees in nature, the propagation of resistant stock for widespread use has not progressed to any extent.

5.5.6 Therapy

Cultural methods

The primary cultural therapy for mistletoe infection is pruning the branch with the mistletoe plant below its point of attachment, or killing the tree (see Section 5.5.3, Eradication, which reviews this approach).

The removal of mistletoe plants from infected branches does not kill the mistletoe and resprouting from the haustorial/endophyte system often occurs. This has led to the application of sunlight exclusion. For example, black plastic or aluminium foil can be wrapped around the infected portion of the branch after removal of the aerial shoots to prevent resprouting (Parker and Riches, 1993), although this is not always effective. Another option tested by Watson and Martinez-Trinidad (2006) for Christmas mistletoe (*Phoradendron tomentosum* (DC.) Engelm. ex A. Gray) on cedar elm (*Ulmus crassifolia* Nutt.) was to apply caulk to the mistletoe stump surface and surrounding bark after removal of the aerial portion of the mistletoe; this reduced mistletoe resprouting by 57% after 29 months. Watson and Martinez-Trinidad (2006) also tested black paint in combination with a growth regulator applied to the stump surface, but this was not as effective as the caulk treatment after 8 months.

Chemical methods

Chemical control of mistletoes has been investigated since before the 1930s (Parker and Riches, 1993; Shamoun and DeWald, 2002; Pundir, 2009; Reid and Shamoun, 2009), although the use of chemicals is not particularly common in mistletoe control today due to mixed results, environmental considerations and potential harm to the host. Reid and Shamoun (2009) noted that there continues to be a search for inexpensive chemicals that selectively kill mistletoe without harming the host tree. This has made the use of chemicals difficult for practitioners owing to the uncertainty of results unless there has already been a study that directly relates to a specific product, mistletoe, host and geographic area, which can then be applied with a relative degree of certainty.

Before the 1980s, in North America and Canada the research emphasis was on herbicides that were mostly variations on phenoxy compounds such as 2,4-D (2,4-dichlorophenoxy acetic acid), or 2,4,5-T (2,4,5-trichlorophenoxy butyric acid), either sprayed directly on to plants or injected into the bole (Graser, 1952; Quick, 1964), and often in combination with other chemicals, such as dicamba (3,6-dichloro-2-methoxybenzoic acid). These herbicides have not been effective and often injured the host tree. Currently no herbicides are recommended for the control of *Arceuthobium* or *Phoradendron* spp. in the USA, although Harris (1992) has recommended cutting *Phoradendron* and leaving a stump 45–90 mm long to which a 2,4-D and dicamba solution is applied. Harris felt that the treatment was effective, but 2,4-D is not currently registered for this use. Wood and Reilly (2004) have recently tested several herbicides and a growth regulator for the control of *P. flavescens* (Pursh) Nutt. ex A. Gray. in pecan trees in the USA. Ethephon, glyphosate (N-(phosphonomethyl) glycine), paraquat dicholoride (N,N′-dimethyl-4,4′-bipyridinium dichloride), and polyborate (boron compound) had little long-term effect. The dimethylamine salt of 2,4-D at 1.2 to 2.4 g l^{-1}, sprayed 2 weeks before tree bud break, was the most effective treatment, and the authors felt that spraying mistletoe plants with 2,4-D offered a practical way to eradicate mistletoe from pecan trees. Although some progress has been made with these herbicides, it is our understanding that no herbicides (apart from ethephon) are currently registered for use in controlling mistletoes in the USA, thus making it illegal to apply herbicides for such use, unless for research and development purposes.

In the 1980s, interest increased in the plant growth regulator ethephon for the control of *Arceuthobium* spp. and *Phoradendron* spp. in the USA (Hawksworth and Wiens, 1996). Although ethephon was shown to be effective in causing aerial shoots to dehisce, it did not kill the endophytic portion of the plant, and resprouting generally occurred. Currently, ethephon is the only chemical registered in the USA for the control of mistletoes. Adams *et al.* (1993) have summarized the cautions to consider in using ethephon for the control of both *Arceuthobium* and *Phoradendron* spp. The application of ethephon to *Arceuthobium* spp. shoots has consistently caused dehiscence, but the endophytic system of the dwarf mistletoe has not been killed. Ethephon application has been of most utility on high-value ornamental trees where it could be applied once every few years and pruning was not acceptable (Shamoun and DeWald, 2002). Adams *et al.* (1993) noted that ethephon was not effective when sprayed on to the mistletoe plants of *Phoradendron* spp., and the most effective use of ethephon on *Phoradendron* was to treat the cut stub of the mistletoe plant with ethephon after removal, though the product is not registered for this use.

Reid and Shamoun (2009) summarized the potential for herbicides to provide economic control of mistletoes in south-east Australia. During the 1950s and 1960s, a number of studies were conducted in Australia using 2,4-D for mistletoe control on eucalyptus (Brown and Greenham, 1965). The most common recommendations on the use of chemicals in Australia have been to inject dilute variations of 2,4-D into the tree. Minko and Fagg (1989) followed this up with tests of stem injections of four additional herbicides, and concluded that these herbicides had promise. Reid and Shamoun (2009) have also noted that spraying mistletoe plants with a 0.5 label dose of 2,4-D (1.38 g l^{-1}) killed about 42% of mistletoe plants (*Amyema* spp.) without harming the host eucalyptus. They felt that this recent research on chemical control of mistletoes of south-east Australia indicated that there may be a role for herbicides in long term economic control, although the most effective herbicide (2,4-D Na) is not registered for this application in New South Wales (New South Wales Landcare, 2008). Because stem injections are labour intensive for large numbers of trees, and cannot be used on severely infected trees, the aerial (helicopter) application of 2,4-D Na has also been suggested (Reid and Shamoun, 2009).

Pundir (2009) summarized work on the use of chemicals for the control of mistletoe around the world, but particularly for India. In India, a wide range of chemicals, but especially focused on 2,4-D mixtures and combinations, either injected into the tree or sprayed on to the mistletoe plants, have been tested over the years. A wide variety of other chemicals has also been tested, including a refined diesel oil emulsion (powerine added to a bar soap and water mixture) that was sprayed directly on to the mistletoe plant. The emulsion was cheap and easy to make, and quite effective. However, a primary interest in India has been to avoid aerial spraying and exposure of people and animals to chemicals, so the injection of herbicides into the tree has been preferred.

In Europe, the emphasis during the 1950s to 1990s was on the control of European mistletoe species by injecting herbicides into the tree, or spraying them on to the plants (Delabraze and Lanier, 1972; Baillon *et al.*, 1988; Pundir, 2009). The focus was on the herbicide 2,4-D and variations. However, it is not known to what degree the operational use of herbicides for the control of European mistletoe continues today in Europe, as the public has a high acceptance of mistletoes and a low acceptance of herbicides.

Biological methods

Biological control with fungi has been studied and is still being developed for some dwarf mistletoes (Shamoun and DeWald, 2002; Reid and Shamoun, 2009), but there have been no reports of the control of non-dwarf mistletoes with fungi or insects (Parker and Riches, 1993). Pundir (2009) has also synthesized the existing knowledge on biological control of mistletoes and reviewed insects, fungi and epiparasites that occur on mistletoes. There is potential for development of biological control given the number of different biotic organisms reported to have an impact on mistletoes, yet there are no operational controls in existence.

Shamoun and DeWald (2002) reported on attempts to develop mycoherbicides for application as an inundative treatment for the localized control of dwarf mistletoes. Two fungi in particular, *Glomerella cingulata* (Stoneman) Spauld. & H. Schrenk and *Neonectria neomacrospora* (C. Booth & Samuels) Mantiri & Samuels, have been shown to be specific to dwarf mistletoes. Reid and Shamoun (2009) noted that this type of treatment required the integration of mycoherbicide applications with production forest management practices and planning, as well as market research to determine the commercial feasibility of the production of mycoherbicides for dwarf mistletoes.

5.5.7 Integrated disease management

Integrated disease management (IDM) tactics follow the principles of integrated pest management (IPM) (Harris *et al.* 2004). The foundation for management has been knowledge of the distribution and impacts of the pest in question through survey and impact assessments. Economic analyses are also needed to determine thresholds to specify when the disease is causing damage at a level that requires action. Prevention and control techniques can then be implemented, and the long-term management of disease agents can be dealt with through monitoring and research that aids in understanding the specific factors that relate to the epidemiology of the disease. Management of mistletoes as plant pathogens has required an added dimension owing to its positive impacts on wildlife and biodiversity in many settings, even in orchards and forest plantations. Therefore, integrated management of mistletoes has required an examination of the ecological context of the problem, its relationship to regional biodiversity and the benefits that mistletoes might bring.

IDM tactics will only solve the problem in the context of a long-term management plan that incorporates the ecological context, distribution and impacts of problem mistletoes, the setting of thresholds for management action, the implementation of prevention and control techniques, and monitoring. Resistance and genetic improvement research and testing may also be important for long-term management of disease, while research into and development of new management techniques and new knowledge of mistletoe biology can be incorporated into long-term plans. Reid and Shamoun (2009) have contrasted research approaches to managing mistletoes in the commercial forests of Canada versus wooded pastures in Australia. In Canada, the impacts of dwarf mistletoes are estimated to cost millions of dollars in losses annually, and therefore the development of an expensive research and development programme for mycoherbicides is defensible. In contrast, in Australia, the impacts on trees may be profound, but they are happening in private farm paddocks rather than across large surfaces of commercial forest, and a research and development programme for new control techniques for mistletoe is not economically viable.

References

Adams, D.H., Frankel, S.J. and Lichter, J.M. (1993) Considerations when using ethephon for suppressing dwarf and leafy mistletoe infestations in ornamental landscapes. *Journal of Arboriculture* 19, 351–357.

Alexander, M.E. and Hawksworth, F.G. (1976) Fire and dwarf mistletoes in North American coniferous forests. *Journal of Forestry* 74, 446–449.

Amico, G. and Aizen, M.A. (2000) Mistletoe seed dispersal by a marsupial. *Nature* 408, 929–930.

Aukema, J.E. (2003) Vectors, viscin, and *Viscaceae*: mistletoes as parasites, mutualists, and resources. *Frontiers in Ecology and the Environment* 1, 212–219.

Aukema, J.E. (2004) Distribution and dispersal of desert mistletoe is scale-dependent, hierarchically nested. *Ecography* 27, 137–144.

Aukema, J.E. and Martínez del Rio, C. (2002) Where does a fruit-eating bird deposit mistletoe seeds? Seed deposition patterns and an experiment. *Ecology* 83, 3489–3496.

Azpeitia, F. and Lara, C. (2006) Reproductive biology and pollination of the parasitic plant *Psittacanthus calyculatus* (*Loranthaceae*) in central Mexico. *Journal of the Torrey Botanical Society* 133, 429–438.
Baillon, F., Chamel, A., Fer, A., Frochot, H., Gambonnet, B. and Manzato, M.C. (1988) Lutte chimique contre le gui (*Viscum album* L.). Pénétration, transport, efficacité de deux herbicides phloème-mobiles (2,4-DB et glyphosate) [Chemical control against mistletoes (*Viscum album* L.). Penetration, transport, and effectiveness of two herbicides phloem-mobile (2,4-DB and glycophosate)]. *Annales des Sciences Forestières* 45, 1–16.
Barney, C.W., Hawksworth, F.G. and Geils, B.W. (1998) Hosts of *Viscum album*. *European Journal of Forest Pathology* 28, 187–208.
Bickford, C.P., Kolb, T.E. and Geils, B.W. (2005) Host physiological condition regulates parasitic plant performance: *Arceuthobium vaginatum* subsp. *cryptopodum* on *Pinus ponderosa*. *Oecologia* 146, 179–189.
Biggs, J. (2010) *A Little Book About Mistletoes*. Potamogeton Press, Oxford, UK.
Broshot, N.E. and Tinnin, R.O. (1986) The effect of dwarf mistletoe on starch concentrations in the twigs and needles of lodgepole pine. *Canadian Journal of Forest Research* 16, 658–660.
Brown, A.G. and Greenham, C.G. (1965) Further investigations in the control of mistletoe by trunk injections. *Australian Journal of Experimental Agriculture and Animal Husbandry* 5, 305–309.
Butin, H. (1995) *Tree Diseases and Disorders*. Oxford University Press, New York.
Büssing, A. (2000) *Mistletoe. The genus* Viscum. Harwood Academic Publishers, Amsterdam, The Netherlands.
Calder, M. and Bernhardt, P. (1983) *The Biology of Mistletoes*. Academic Press, San Diego, California.
Calvin, C.L. and Wilson, C.A. (1998) The haustorial system in African *Loranthaceae*. In: Polhill, R. and Wiens, D. (eds) *The Mistletoes of Africa*. Royal Botanic Gardens, Kew, UK, pp. 17–36.
Carlo, T.A. and Aukema, J.E. (2005) Female-directed dispersal and facilitation between a tropical mistletoe and a dioecious host. *Ecology* 86, 3245–3251.
Conklin, D.A. and Geils, B.W. (2008) Survival and sanitation of dwarf mistletoe-infected ponderosa pine following prescribed underburning. *Western Journal of Applied Forestry* 23, 216–222.
Daugherty, C.M. and Mathiasen, R.L. (2003) Incidence of mistletoes in the pinyon–juniper woodlands of the Coconino National Forest, Arizona. *Western North American Naturalist* 63, 382–390.
de Lange, P.J. and Norton, D.A. (1997) *Ecology and Conservation of New Zealand's Loranthaceous Mistletoes*. Department of Conservation, Wellington, New Zealand.
Delabraze, P.P. and Lanier, L. (1972) Contribution à la lutte chimique contre le gui (*Viscum album* L.) [Contribution to the chemical control of mistletoes (*Viscum album* L.)]. *European Journal of Forest Pathology* 2, 95–103.
Dobbertin, M. and Rigling, A. (2006) Pine mistletoe (*Viscum album* ssp. *austriacum*) contributes to Scots pine (*Pinus sylvestris*) mortality in the Rhone valley of Switzerland. *Forest Pathology* 36, 309–322.
Dooling, O.J. (1978) Survey methods to determine the distribution and intensity of dwarf mistletoes. In: Scharpf, R.F. and Parmeter, J.R. Jr (eds) *Proceedings of the Symposium on Dwarf Mistletoe Control Through Forest Management, April 11–13, 1978*. General Technical Report PSW-31, US Department of Agriculture, Forest Service, Pacific Southwest Forest and Range Experiment Station, Berkeley, California, pp. 36–44.
Downey, P.O. (1998) An inventory of host species for each aerial mistletoe species (*Loranthaceae* and *Viscaceae*) in Australia. *Cunninghamia* 5, 685–720.
Ehleringer, J.R. and Schulze, E.-D. (1985) Mineral concentrations in an autoparasitic *Phoradendron californicum* growing on a parasitic *P. californicum* and its host, *Cercidium floridum*. *American Journal of Botany* 72, 568–571.
Freehan, J. (1985) Explosive flower opening in ornithophily: a study of pollination mechanisms in some central African *Loranthaceae*. *Botanical Journal of the Linnean Society* 90, 129–144.
Geils, B.W. and Hawksworth, F.G. (2002) Damage, effects, and importance of dwarf mistletoes. In: Geils, B.W., Cibrián-Tovar, J. and Moody, B. (eds) *Mistletoes of North American Conifers*. General Technical Report RMRS-GTR-98, US Department of Agriculture, Forest Service, Rocky Mountain Research Station, Ogden, Utah, pp. 57–65.
Geils, B.W., Cibrián-Tovar, J. and Moody, B. (2002) *Mistletoes of North American Conifers*. General Technical Report RMRS-GTR-98, US Department of Agriculture, Forest Service, Rocky Mountain Research Station, Ogden, Utah.
Gill, L.S. and Hawksworth, F.G. (1961) *The Mistletoes: A Literature Review*. Technical Bulletin 1242, US Department of Agriculture, Washington, DC.
Glatzel, G. (1983) Mineral nutrition and water relations of hemiparasitic mistletoes: a question of partitioning. Experiments with *Loranthus europaeus* and *Quercus petraea* and *Quercus robur*. *Oecologia* 56, 193–201.
Graser, H.L. (1952) A progress report on mistletoe control in walnuts. *Western Fruit Grower* 6, 23–24.

Hamer, T.E. and Nelson, S.K. (1995) Characteristics of marbled murrelet nest trees and nesting stands. In: Ralph, C.J., Hunt, G.L. Jr, Raphael, M.G. and Piatt, J.F. (eds) *Ecology and Conservation of the Marbled Murrelet*. General Technical Report PSW-152, US Department of Agriculture, Forest Service, Pacific Southwest Forest and Range Experiment Station, Albany, California, pp. 68–92.
Hariri, B., Salle, G. and Andary, C. (1991) Involvement of flavonoids in the resistance of two poplar cultivars to mistletoe (*Viscum album* L.). *Protoplasma* 162, 20–26.
Harrington, M.G. and Hawksworth, F.G. (1990) Interactions of fire and dwarf mistletoe on mortality of southwestern ponderosa pine. In: Krammes, J.S. (ed.) *Proceedings of the Symposium on Effects of Fire Management of Southwestern Natural Resources, November 15–17, 1988, Tucson, Arizona*. General Technical Report RM-191, US Department of Agriculture, Forest Service, Rocky Mountain Research Station, Fort Collins, Colorado, pp. 234–240.
Harris, R.W. (1992) *Arboriculture: Integrated Management of Landscape Trees, Shrubs, and Vines*, 2nd edn. Prentice Hall, Upper Saddle River, New Jersey.
Harris, R.W., Clark, J.R. and Matheny, N.P. (2004) *Arboriculture: Integrated Management of Landscape Trees, Shrubs, and Vines*, 4th edn. Prentice Hall, Upper Saddle River, New Jersey.
Hawksworth, F.G. (1977) *The 6-class Dwarf Mistletoe Rating System*. General Technical Report RM-48, US Department of Agriculture, Forest Service, Rocky Mountain Forest and Range Experiment Station, Fort Collins, Colorado.
Hawksworth, F.G. and Scharpf, R.F. (1978) Literature on the dwarf mistletoes: damage and control. In: Scharpf, R.F. and Parmeter, J.R. (eds) *Proceedings of the Symposium on Dwarf Mistletoe Control Through Forest Management, April 11–13, 1978*. General Technical Report PSW-31, US Department of Agriculture, Forest Service, Pacific Southwest Forest and Range Experiment Station, Berkeley, California, pp. 178–184.
Hawksworth, F.G. and Scharpf, R.F. (1986) Spread of European mistletoe (*Viscum album*) in California, USA. *European Journal of Forest Pathology* 16, 1–5.
Hawksworth, F.G. and Wiens, D. (1996) *Dwarf Mistletoes: Biology, Pathology, and Systematics*. Agriculture Handbook 709, US Department of Agriculture, Forest Service, Washington, DC.
Hawksworth, F.G., Wiens, D. and Geils, B.W. (2002) *Arceuthobium* in North America. In: Geils, B.W., Cibrián-Tovar, J. and Moody, B. (eds) *Mistletoes of North American Conifers*. General Technical Report RMRS-GTR-98, US Department of Agriculture, Forest Service, Rocky Mountain Research Station, Ogden, Utah, pp. 29–56.
Heide-Jørgensen, H.S. (2008) *Parasitic Flowering Plants*. Brill, Leiden, The Netherlands.
Hoffman, C., Mathiasen, R. and Sieg, C. (2007) Dwarf mistletoe effects on fuel loadings in ponderosa pine forests in northern Arizona. *Canadian Journal of Forest Research* 37, 662–670.
Howell, B.E. and Mathiasen, R.L. (2004) Growth impacts of *Psittacanthus angustifolius* Kuijt on *Pinus oocarpa* Schiede in Honduras. *Forest Ecology and Management* 198, 75–88.
Hull, R.J. and Leonard, O.A. (1964) Physiological aspects of parasitism in mistletoes (*Arceuthobium* and *Phoradendron*). I. The carbohydrate nutrition of mistletoe. *Plant Physiology* 39, 996–1007.
Kirkup, D. (1998) Pollination mechanisms in African *Loranthaceae*. In: Polhill, R. and Wiens, D. (eds) *The Mistletoes of Africa*. Royal Botanic Gardens, Kew, UK, pp. 37–60.
Knutson, D.M. (1983) Physiology of mistletoe parasitism and disease responses in the host. In: Calder, M. and Bernhardt, P. (eds) *The Biology of Mistletoes*. Academic Press, San Diego, California, pp. 295–316.
Kolb, T.E. (2002) Ecophysiology of parasitism in the plant kingdom. In: López-Sáez, J.A., Catalán, P. and Sáez, L. (eds) *Parasitas de la Péninsula Ibérica e Islas Baleares*. Mundi-Prensa, Madrid, pp. 57–85.
Koonce, A.L. and Roth, L.F. (1980) The effects of prescribed burning on dwarf mistletoe in ponderosa pine. In: Martin, R.E. (ed.) *Proceedings of the Sixth Conference on Fire and Forest Meteorology, April 22–24, 1980, Seattle, Washington*. Society of American Foresters, Washington, DC, pp. 197–203.
Kuijt, J. (1969) *The Biology of Parasitic Flowering Plants*. University of California Press, Berkeley, California.
Kuijt, J. (2003) *Monograph of* Phoradendron (Viscaceae). Systematic Botany Monographs 66, American Society of Plant Taxonomists, Ann Arbor, Michigan.
Ladley, J.J. and Kelly, D. (1996) Dispersal, germination and survival of New Zealand mistletoes (*Loranthaceae*): dependence on birds. *New Zealand Journal of Ecology* 20, 69–79.
Ladley, J.J., Kelly, D. and Robertson, A.W. (1997) Explosive flowering, nectar production, breeding systems, and pollinators of New Zealand mistletoes (*Loranthaceae*). *New Zealand Journal of Botany* 35, 345–360.
Lamont, B. (1983) Mineral nutrition of mistletoes. In: Calder, M. and Bernhardt, P. (eds) *The Biology of Mistletoes*. Academic Press, San Diego, California, pp. 185–201.
Lei, S. (2001) Survival and development of *Phoradendron californicum* and *Acacia reggii* during a drought. *Western North American Naturalist* 61, 78–84.

Livingston, W.H., Brenner, M.L. and Blanchette, R.A. (1984) Altered concentrations of abscisic acid, indole-3-acetic acid, and zeatin riboside associated with eastern dwarf mistletoe infections on black spruce. In: Hawksworth, F.G. and Scharpf, R.F. (eds) *Proceedings of the Symposium of Biology of Dwarf Mistletoes, August 8, 1984, Fort Collins, Colorado*. General Technical Report RM-111, US Department of Agriculture, Forest Service, Rocky Mountain Forest and Range Experiment Station, Fort Collins, Colorado, pp. 53–61.

Mallams, K.M. and Mathiasen, R.L. (2010) *Mistletoes on Hardwoods in the United States*. Forest Insect and Disease Leaflet 147, US Department of Agriculture, Forest Service, Pacific Northwest Region, Portland, Oregon.

Marshall, J.D. and Ehleringer, J.R. (1990) Are xylem-tapping mistletoes partially heterotrophic? *Oecologia* 84, 244–248.

Marshall, K., Mamone, M. and Barclay, R. (2003) A survey of Douglas-fir dwarf mistletoe brooms used for nests by northern spotted owls on the Applegate Ranger District, Ashland Resource Area in southwest Oregon. *Western Journal of Applied Forestry* 18, 115–117.

Martínez del Rio, C., Silva, A., Medel, R. and Hourdequin, M. (1996) Seed dispersers as disease vectors: bird transmission of mistletoe seeds to plant hosts. *Ecology* 77, 912–921.

Mathiasen, R.L. (1996) Dwarf mistletoes in forest canopies. *Northwest Science* 7, 61–72.

Mathiasen, R.L. and Blake, E.A. (1984) Relationships between dwarf mistletoes and habitat types in western coniferous forests. In: Hawksworth, F.G. and Scharpf, R.F. (eds) *Proceedings of the Symposium of Biology of Dwarf Mistletoes, August 8, 1984, Fort Collins, Colorado*. General Technical Report RM-111, US Department of Agriculture, Forest Service, Rocky Mountain Forest and Range Experiment Station, Fort Collins, Colorado, pp. 111–116.

Medel, R., Vergara, E., Silva, A. and Kalin-Arroyo, M. (2004) Effects of vector behavior and host resistance on mistletoe aggregation. *Ecology* 85, 120–126.

Merrill, L.M., Hawksworth, F.G. and Jacobi, W.R. (1987) Frequency and severity of ponderosa pine dwarf mistletoe in relation to habitat type and topography in Colorado. *Plant Disease* 71, 342–344.

Minko, G. and Fagg, P.C. (1989) Control of some mistletoe species on eucalypts by trunk injection with herbicides. *Australian Forestry* 52, 94–102.

Muir, J.A. and Geils, B.W. (2002) Management strategies for dwarf mistletoes: silviculture. In: Geils, B.W., Cibrián-Tovar, J. and Moody, B. (eds) *Mistletoes of North American Conifers*. General Technical Report RMRS-GTR-98, US Department of Agriculture, Forest Service, Rocky Mountain Research Station, Ogden, Utah, pp. 83–94.

New South Wales Landcare (2008) *Methods to Manage Mistletoe. A Landholder's Guide*. Southern New England Landcare, Armidale, Australia.

Noetzli, K.P., Müller, B. and Sieber, T.N. (2003) Impact of population dynamics of white mistletoe (*Viscum album* spp. *abietis*) on European silver fir (*Abies alba*). *Annals of Forest Science* 60, 773–779.

Norton, D.A. and Carpenter, M.A. (1998) Mistletoes as parasites: host specificity and speciation. *Trends in Ecology and Evolution* 13, 101–105.

Parker, C. and Riches, C.R. (1993) *Parasitic Weeds of the World: Biology and Control*. CAB International, Wallingford, UK.

Parker, T.E. and Mathiasen, R.L. (2004) A comparison of rating systems for dwarf mistletoe induced witches' brooms in ponderosa pine. *Western Journal of Applied Forestry* 19, 54–59.

Parker, T.E., Clancy, K.M. and Mathiasen, R.L. (2006) Interactions among fire, insects, and pathogens in coniferous forests of the interior western United States and Canada. *Agricultural and Forest Entomology* 8, 167–189.

Parmeter, J.R. Jr (1978) Forest stand dynamics and ecological factors in relation to dwarf mistletoe spread, impact, and control. In: Scharpf, R.F. and Parmeter, J.R. (eds) *Proceedings of the Symposium on Dwarf Mistletoe Control Through Forest Management, April 11–13, 1978*. General Technical Report PSW-31, US Department of Agriculture, Forest Service, Pacific Southwest Forest and Range Experiment Station, Berkeley, California, pp. 16–30.

Polhill, R. and Wiens, D. (1998) *The Mistletoes of Africa*. Royal Botanic Gardens, Kew, UK.

Pundir, Y.P.S. (2009) *Mistletoes of Western Himalayas: Biology, Damage, and Control*. Bishen Singh Mahendra Pal Singh, Dehra Dun, India.

Quick, C.R. (1964) *Experimental Herbicidal Control of Dwarf Mistletoe on Some California Conifers*. Research Note PSW-47, US Department of Agriculture, Forest Service, Rocky Mountain Forest and Range Experiment Station, Berkeley, California.

Reblin, J., Logan, B. and Tissue, D. (2006) Impact of dwarf mistletoe (*Arceuthobium pusillum*) infection on needles of red spruce (*Picea rubens*) and white spruce (*Picea glauca*): oxygen exchange, morphology, and composition. *Tree Physiology* 26, 1325–1332.

Reid, N. (1991) Coevolution of mistletoes and frugivorous birds? *Australian Journal of Ecology* 16, 457–469.

Reid, N. and Shamoun, S.F. (2009) Contrasting research approaches to managing mistletoes in commercial forests and wooded pastures. *Botany* 87, 1–9.

Reid, N., Stafford Smith, M. and Yan, Z. (1995) Ecology and population biology of mistletoes. In: Lowman, M.D. and Nadkarni, N.M. (eds) *Forest Canopies*. Academic Press, San Diego, California, pp. 285–310.

Restrepo, C., Sargent, S., Levey, D.J. and Watson, D.M. (2002) The role of vertebrates in the diversification of the New World mistletoes. In: Levey, D.J., Silva, W.R. and Galetti, M. (eds) *Seed Dispersal and Frugivory: Ecology and Conservation*. CAB International, Wallingford, UK, pp. 83–98.

Richter, A., Popp, M., Mensen, R., Stewart, G.R. and von Willert, D.J. (1995) Heterotrophic carbon gain of the parasitic angiosperm *Tapinanthes oleifolius*. *Australian Journal of Plant Physiology* 22, 537–544.

Room, P.M. (1973) Ecology of the mistletoe *Tapinanthus bangwensis* growing on cocoa in Ghana. *Journal of Ecology* 61, 729–742.

Roth, L.F. (1974) Resistance of ponderosa pine to dwarf mistletoe. *Silvae Genetica* 23, 116–120.

Roth, L.F. (1978) Genetic control of dwarf mistletoe. In: Scharpf, R.F. and Parmeter, J.R. (eds) *Proceedings of the Symposium on Dwarf Mistletoe Control Through Forest Management, April 11–13, 1978*. General Technical Report PSW-31, US Department of Agriculture, Forest Service, Pacific Southwest Forest and Range Experiment Station, Berkeley, California, pp. 16–30.

Scharpf, R.F. (1984) Host resistance to dwarf mistletoes. In: Hawksworth, F.G. and Scharpf, R.F. (eds) *Proceedings of the Symposium of Biology of Dwarf Mistletoes, August 8, 1984, Fort Collins, Colorado*. General Technical Report RM-111, US Department of Agriculture, Forest Service, Rocky Mountain Forest and Range Experiment Station, Fort Collins, Colorado, pp. 70–76.

Scharpf, R.F. (1987) Resistance of Jeffrey pine to dwarf mistletoes. In: Weber, H.C. and Forstreuter, W. (eds) *Proceedings of the Fourth International Symposium of Parasitic Flowering Plants, 1987, August 2–7, Marburg, Germany*. Philipps Universität, Marburg, Germany, pp. 745–753.

Shamoun, S.F. and DeWald, L.E. (2002) Management strategies for dwarf mistletoes: biological, chemical, and genetic approaches. In: Geils, B.W., Cibrián-Tovar, J. and Moody, B. (eds) *Mistletoes of North American Conifers*. General Technical Report RMRS-GTR-98, US Department of Agriculture, Forest Service, Rocky Mountain Research Station, Ogden, Utah, pp. 75–82.

Shaw, D.C., Watson, D.M. and Mathiasen, R.L. (2004) Comparison of dwarf mistletoes (*Arceuthobium* spp., *Viscaceae*) in the western United States with mistletoes (*Amyema* spp., *Loranthaceae*) in Australia – ecological analogs and reciprocal models for ecosystem management. *Australian Journal of Botany* 52, 481–498.

Smith, R.B., Wass, E.F. and Meagher, M.D. (1993) Evidence of resistance to hemlock dwarf mistletoe (*Arceuthobium tsugense*) in western hemlock (*Tsuga heterophylla*) clones. *European Journal of Forest Pathology* 23, 163–170.

Snow, B.K. and Snow, D.W. (1988) *Birds and Berries: A Study of an Ecological Interaction*. T. and A.D. Poyser, Calton, UK.

Stewart, G.R. and Press, M.C. (1990) The physiology and biochemistry of parasitic angiosperms. *Annual Review of Plant Physiology and Plant Molecular Biology* 41, 127–151.

Tennakoon, K.U. and Pate, J.S. (1996) Effects of parasitism by a mistletoe on the structure and functioning of branches of its host. *Plant, Cell and Environment* 19, 517–528.

Thoday, D. (1951) The endophytic system of *Viscum album*. *Journal of Experimental Botany* 2, 1–19.

Thoday, D. (1957) Modes of union and interaction between parasite and host in the *Loranthaceae*. II. *Phoradendreae*. *Proceedings of the Royal Society of London, Series B, Biological Sciences* 146, 320–413.

Tinnin, R.O. and Knutson, D.M. (1980) Growth characteristics of the brooms on Douglas-fir caused by *Arceuthobium douglasii*. *Forest Science* 26, 149–158.

van Ommeren, R.J. and Whitham, T.G. (2002) Changes in interactions between juniper and mistletoe mediated by shared avian frugivores: parasitism to potential mutualism. *Oecologia* 130, 281–288.

Watson, D.M. (2001) Mistletoe – a keystone resource in forests and woodlands worldwide. *Annual Review of Ecology and Systematics* 32, 219–250.

Watson, D.M. (2009) Determination of parasitic plant distribution: the role of host quality. *Canadian Journal of Botany* 87, 16–21.

Watson, D.M. (2011) *Mistletoes of Southern Australia*. CSIRO Publishing, Collingwood, Victoria, Australia.

Watson, W.T. and Martinez-Trinidad, T. (2006) Strategies and treatments for leafy mistletoe (*Phoradendron tomentosum* (DC.) Engelm. ex Gray) suppression on cedar elm (*Ulmus crassifolia* Nutt.). *Arboriculture and Urban Forestry* 32, 265–270.

Weir, J.R. (1916) *Mistletoe Injury to Conifers in the Northwest*. Bulletin 360, US Department of Agriculture, Washington, DC.

Wood, B.W. and Reilly, C.C. (2004) Control of mistletoe in pecan trees. *HortScience* 39, 110–114.

6 Pine Wilt Disease and Other Nematode Diseases

Naoto Kamata[1]* and Yuko Takeuchi[2]
[1]*The University of Tokyo, Tokyo, Japan;* [2]*Kyoto University, Kyoto, Japan*

6.1 Pathogens, Significance and Distribution

Nematodes, generally termed roundworms, are invertebrate animals belonging to the phylum Nematoda. About 10% of nematode species are considered plant parasitic and often cause serious damage to the host woody plants, mostly on their roots; for example, fig trees, coffee trees and maple trees may be damaged by root-knot nematodes, and many fruit trees can be hosts for root-lesion nematodes, dagger nematodes, burrowing nematodes or ring nematodes. Among the most important infectious tree diseases caused by nematode pathogens are pine wilt disease and red ring disease. These share several similarities: a relatively short history of one century, both pathogens belong to the genus *Bursaphelenchus,* which comprises over 100 mycophagous species, and both diseases are transmitted by specific insect vectors. This chapter mostly deals with these two nematode-caused diseases.

Pine wilt disease was first found in 1905 in Nagasaki City on Kyushu Island, which is the most western mainland island in Japan. It had been speculated that mortality was caused by wood borers and/or phloem feeders because various species of these insects emerged from the diseased and dead pine trees. In 1971, the pine wood nematode *Bursaphelenchus xylophilus* (Steiner & Buhrer) was identified as the pathogen that killed healthy pine trees (Kiyohara and Tokushige, 1971). This was the first example in which a nematode had been proved to kill healthy woody plants. Results of several genetic studies support the origin of *B. xylophilus* from North America. To date, the pine wilt disease is spreading in wide areas of eastern Asia and within a well-defined area in Europe, namely Portugal and Spain.

B. xylophilus (Fig. 6.1) was first described as the new species *Aphelenchoides xylophilus* by Steiner and Buhrer (1934), and transferred to the genus *Bursaphelenchus* by Nickle (1970). While looking for the pathogen of pine wilt, *Bursaphelenchus* sp. was found in the wood of pine trees killed by the disease. Kiyohara and Tokushige (1971) inoculated this *Bursaphelenchus* sp. into 25 year old pine trees and proved its pathogenicity. The nematode species was described as *B. lignicolus* (Mamiya and Kiyohara, 1972) and then synonymized as *B. xylophilus* (Nickle *et al.*, 1981). In Japan, *B. xylophilus* is transmitted almost solely by the Japanese pine sawyer *Monochamus alternatus* (Hope) (Plate 4), although it can

* E-mail: kamatan@uf.a.u-tokyo.ac.jp

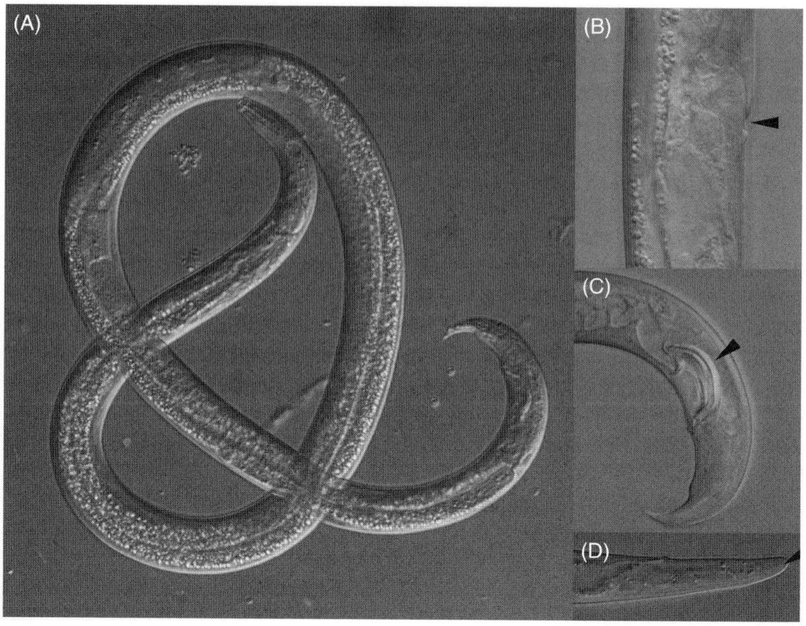

Fig. 6.1. The pine wood nematode *Bursaphelenchus xylophilus* (Steiner & Buhrer): (A) adult male; (B) vulval area of adult female with an overlapping flap at anterior vulval lip; (C) lateral view of male tail showing unique spicule; and (D) non-mucronated tail of adult female. Credits: Natsumi Kanzaki.

arbitrarily find a suitable vector belonging to the (same) genus *Monochamus*. Pine species that originated from eastern North America are generally resistant, probably because *B. xylophilus* is of North American origin, but many Eurasian pine species are susceptible. Taxonomically, pine species belonging to the subsections *Australes* and *Contortae* tend to be resistant to *B. xylophilus*, while those in the subsections *Ponderosae* and *Oocarpae* are generally susceptible. Species of *Pinaceae* other than those belonging to the genus *Pinus* can also be hosts of *B. xylophilus*.

Red ring disease was first found in 1905 in the Cedros peninsula, Trinidad. It was initially considered to be a symptom of bud rot of coconut palm, being attributed to the pathogenic fungus *Botryodiplodia* sp. The association between the red ring disease and a nematode was reported only in 1918 (Griffith, 1987; Giblin-Davis, 1990). Now red ring disease is one of the most common and most serious diseases recorded in coconut and oil palms in the neotropics, where it causes 10–15% loss annually (Chinchilla, 1988). To date, it is limited in its distribution to Latin America, including Central America, south-eastern Mexico, the Caribbean and parts of South America.

The red ring nematode, or coconut palm nematode, *B. cocophilus* (Cobb) Baujard, causes red ring disease of trees of the family *Palmae*, including the coconut palm (*Cocos nucifera* L.), African oil palm (*Elaeis guineensis* Jacq.) and date palm (*Phoenix dactylifera* L.). It is vectored primarily by adults of the American palm weevil *Rhynchophorus palmarum* (Linnaeus), but also by the sugarcane weevils *Dynamis borassi* (Fabricius) and *Metamasius hemipterus* (Linnaeus). Palm weevils alone can cause damage to palm trees in Ecuador, but the level of the nematode population is low and such intense weevil attacks without red ring disease are rare in other countries (Griffith, 1987).

B. cocophilus is a very long, slender nematode; females and males are 60–139 and 65–179 times longer than wide, respectively, with the greatest body width being less than 15.5 μm and the total length ranging from 775 to 1369 μm (Giblin-Davis, 1993). It is an obligate phytoparasite: it does not survive long when separated from its palm or weevil hosts, and it shows little growth and reproduction on monoxenic cultures of fungi, such as *Alternaria alternata* (Fr.) Keissl. (Goberdhan, 1962), *Botrytis cinerea* Pers. and *Monilinia fructicola* (G. Winter) Honey, all good hosts for several species of *Bursaphelenchus* (Giblin-Davis *et al.*, 1989).

The red ring disease agent is considered of Latin American origin, while the genera of both *Rhynchophorus* (the vector) and *Cocos* (the host) originated in Indo-Pacific region. The world distribution of *B. cocophilus* includes: Belize, Brazil, Colombia, Costa Rica, Ecuador, El Salvador, French Guiana, Grenada, Guatemala, Guyana, Honduras, Mexico, Nicaragua, Panama, Peru, St Vincent and the Grenadines, Suriname, Trinidad and Tobago and Venezuela. Although the nematode has been reported to be present in Barbados, Dominica and Jamaica, these reports have not been confirmed or refuted (Giblin-Davis *et al.*, 2010). The vector insects, including potential ones, have a wider distribution than *B. cocophilus*, and therefore quarantine measures are strongly recommended to avoid disease introduction into new countries which have other *Rhynchophorus* spp. and susceptible hosts.

6.2 Diagnosis

Because temperature greatly influences the development of symptoms, mortality from the pine wilt disease generally occurs from late summer (under warm conditions) to the next spring following infection under cool conditions. A reddish-brown crown is a characteristic symptom of pine wilt-diseased trees, although, the reduction and cessation of oleoresin exudation is the first detectable internal symptom, so that a simple diagnosis to detect pine wilt-diseased trees is the measurement of oleoresin exudation flow after mechanical wounding, even when no external symptoms are observable. Direct detection of *B. xylophilus* is also good evidence for the disease. A Baermann funnel technique has commonly been used for extracting nematodes, although the nematode then requires to be identified.

According to a recent taxonomic system, the genus *Bursaphelenchus* is divided into 'groups', which are roughly equivalent to a subgenus. Groups are defined mainly by spicule morphology. *B. xylophilus* is grouped with *B. mucronatus* Mamiya & Enda, *B. fraudulentus* Rühm and others; these are distinctive from all other *Bursaphelenchus* spp. by the presence in the female of a vulval flap of a characteristic shape (Fig. 6.1B) and by the shape of the male spicules (Fig. 6.1C). *B. xylophilus* has a broad tail with rounded tip, which is a characteristic that distinguishes it from the other species in the group (Fig. 6.1D). However, identification by morphological features requires well-prepared microscope slides, a high-quality microscope and considerable experience in nematode taxonomy. In contrast to such difficulty in morphological identification, molecular biological techniques can be applied to identify the nematode that need much less technical experience and very little nematological training, although these used to require expensive equipment. Several methods, mostly based on PCR techniques, have been proposed for identifying and/or discriminating *B. xylophilus* from numbers of closely related *Bursaphelenchus* spp. A newly developed product, the '*Bursaphelenchus xylophilus* Detection Kit', based on a diagnostic method by loop-mediated isothermal amplification (LAMP) (Kikuchi *et al.*, 2009) is available, which enables a rapid and precise diagnosis at low costs.

In red ring disease, the affected palm tree is characterized by a discrete brick to brownish-red circular band that appears on the cut surface, 2–6 cm in width and 2–6 cm from the exterior, within the cortex of affected organs, i.e. roots, stem and petioles. The red ring develops internally before the appearance of external symptoms. External symptoms are confined to the leaves, which become short and deformed, turn yellow bronze, then deep red or brown. Colour change generally begins at the leaf tip, starting with older, lower leaves, and progressing upwards until eventually the tree wilts and dies. The precise diagnosis of the disease is quite difficult because the symptoms vary widely according to palm type and age, and biotic/abiotic environmental conditions. For example, in African oil palms 'little leaf syndrome' is often manifested in stunted growth of the fronds, while leaf colour remains normal. In a plantation where both 'hartrot' (fatal wilt) and red ring disease occur, these two diseases are not readily distinguishable by external symptoms; in general, hartrot seems to be red ring disease without the red ring (Kastelein *et al.*, 1985). In coconut palm, the disease can be confirmed by examining stem tissue extracted with a coring device for evidence of discoloured tissue and red ring nematodes. In African oil palm, however, stem coring for diagnostic purposes is not recommended because of the high probability of false negative samplings (Chinchilla, 1988).

Red ring nematodes are most often recovered from necrotic lesions in the middle and distal parts of symptomatic leaves. Adults, propagative juveniles and eggs are usually recovered from the stem areas where discoloured lesions have not yet coalesced into the red ring, while dauer juveniles are most abundant in the red ring tissues of stem and petioles (Griffith, 1987; Giblin-Davis, 1993). The nematodes isolated are usually identified microscopically, as unlike the case of *B. xylophilus*, PCR-based identification is not common for *B. cocophilus*.

6.3 Infection Biology

Although the transmission of *B. xylophilus* through root grafts among pine trees has been reported (Tanaka and Gyokusen, 2004), most pine wilt infections are vectored by cerambycid beetles belonging to the genus *Monochamus*: *M. carolinensis* (Olivier), *M. mutator* LeConte, *M. scutellatus* (Say), *M. titillator* (Fabricius) in North America, *M. alternatus* and *M. saltuarius* (Gebler) in Japan, and *M. galloprovincialis* (Olivier) in Portugal and Spain.

During infection, *B. xylophilus* assumes a special dauer stage for transportation (Fig. 6.2); this is the fourth-stage dispersal juveniles (J_{IV}). The vector *Monochamus* beetle harbouring the J_{IV} of *B. xylophilus* under its elytra (i.e. its wing cases) and in the tracheae (i.e. its breathing

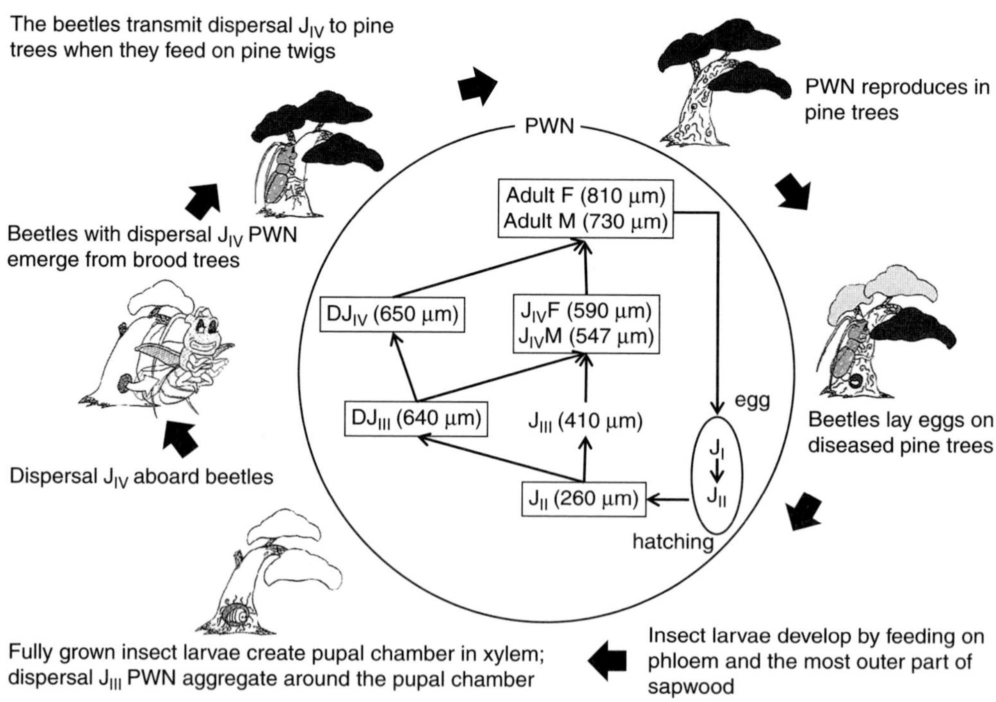

Fig. 6.2. Interrelationships between the pine wood nematode (PWN) *Bursaphelenchus xylophilus* (Steiner & Buhrer) and its insect vector *Monochamus* spp. DJ, dispersal juvenile; J, juvenile.

tubes), flies to healthy pine trees and feeds on twig bark (maturation feeding). At this time, *B. xylophilus* migrates into the tree through the feeding wounds. An alternative is via oviposition slits made by the beetles, as happens mainly with the close relative *B. mucronatus*. Transmission during the maturation feeding of beetles is the major pathway of infection and initiates the phytophagous phase of the nematode, which has the greatest importance for the potential development of pine wilt disease. In a susceptible host tree species and under favourable climatic conditions, the nematodes switch into the propagative mode and grow vigorously, feeding on plant tissues and moving around from the cambium into the xylem. The generation time is 6 days at 20°C and 3 days at 30°C. Affected trees often become symptomatic as a result of a water conductance anomaly (cavitation), through the xylem, and wilt to death. The trees thus damaged are available for oviposition by *Monochamus* females; therefore, the pine wilt-diseased trees possess both the pathogen *B. xylophilus* and the vector *Monochamus* beetle inside. In dead trees, the nematodes feed on fungi, in particular on blue-stain fungi such as *Ceratocystis*, *Ophiostoma*, *Leptographium* and *Gliocladium* spp. The fungal flora in the wood, especially the presence of favourable fungi, is important for the reproduction, survival and initial load of *B. xylophilus* carried by the emerging beetles (Mamiya, 1984; Linit, 1988; Evans *et al.*, 1996; Maehara *et al.*, 2005). As *Monochamus* spp. larvae develop during winter, the J_{III} ('predauer', third-stage dispersal juveniles) of *B. xylophilus* aggregate around pupal chambers made by the mature beetle larvae. Between late spring and early summer, when the beetles pupate or eclose, the J_{III} juveniles develop into J_{IV} juveniles and become ready to be carried by the beetles.

The dauer stage juveniles (J_{III}) of the red ring nematode are harboured internally in the adult females of the weevil vector, either in the proliferated tracheal sacs, from where they move down to the ovipositor, or in the haemocoel among the loops of the gut tract, from where they enter the ovipositor. The J_{III} nematodes are probably deposited during weevil oviposition (rather than during feeding, as in the case of *B. xylophilus*) into the leaf bases or wounds in a susceptible palm tree. The low densities of nematodes that persist in the soil (Blair, 1970), or contaminate the faces or bodies of the weevil (Hagley, 1963), seem to be consistent with this idea. Inside the tree, the nematode feeds, grows and reproduces on host parenchyma cells in the stems, petioles and root cortex, causing a lethal wilt in as little as 2–4 months. The nematode has a generation time of 9–10 days in coconut tissue at 25°C. Once the weevil egg hatches, as many as 10,000 immature nematodes enter and infest the weevil larva. The nematodes remain inside the weevil through metamorphosis, without moulting or reproducing. The mature weevils leave their cocoons in the rotted palm, carrying with them new batches of J_{III} nematodes ready to infest a new tree (Griffith, 1987; Giblin-Davis, 1993).

6.4 Epidemiology

Not all pine trees are diseased by the presence of *B. xylophilus*. The balance between pathogen (virulence and abundance of *B. xylophilus*) and host (susceptibility) determines the onset of the disease, which is also affected by environmental factors, such as temperature and soil moisture. Furthermore, symptomless or latent carrier trees can survive for many years and serve as a source of infection (Bergdahl and Halik, 1999).

The spread rate of pine wilt disease depends critically on the dispersal kernel of vector insects (Takasu *et al.*, 2000). *Monochamus* spp. are generally good flyers. In Japan, the longest distance of one flight of *M. alternatus* was estimated at 3.1 km from information on the natural invasion of the disease into the islands (Kawabata, 1979). A laboratory experiment with a flight mill recorded a maximum distance of 3.3 km or a maximum duration of 58 min of continuous flight (Enda, 1985). An average travel distance obtained by the mark–recapture technique was 1820 m for long-distance dispersal (Takasu *et al.*, 2000) and 14 m for short-distance dispersal (Shibata, 1986). The disease expansion rate depends largely on the proportion of the beetles that become long-distance dispersers. With short-distance dispersal, the speed was at most 5 m a year. However, if a proportion of about 10% of the population are long-distance dispersers, this is sufficient to produce a travelling wave that advances at 4 km a year. Estimating further from the same kernel, say 33.3% of the long-distance dispersers move over 2 km and 19.2% move over 3 km. Therefore, under the assumption of the presence of 10% of the population as long-distance dispersers, the probability of dispersal is 3.3% for 2 km and 1.9% for 3 km. These estimates support the tradition that widths of the barrier zone are usually 2 km.

Monochamus spp. are attracted to α-pinene released by wounded or weakened pine trees (Ikeda *et al.*, 1980). Males of *M. alternatus* produce 2-(undecyloxy)-ethanol, a major component of aggregation pheromone synergized by both ethanol and α-pinene that attracts both sexes (Teale *et al.*, 2011).

Depending on temperature, *M. alternatus* has one or two year life cycles in Japan and undergoes one to three generations a year in the southern China. The mean longevity of *M. alternatus* adults is approximately 70 days in outdoor cages (Togashi and Magira, 1981), though they can live longer than half a year in the laboratory (Zang and Linit, 1998). The seasonal occurrence of adult emergence and the temporal pattern of J_{VI} transmission into pine trees in relation to the age of beetles are critical for determining timing of spraying insecticide against flying beetles. The J_{VI} transmission into pine trees peaked from 10 to 30 days after emergence for *M. alternatus* (Jikumaru and Togashi, 2000), in the third to fourth week for *M. carolinensis* (Linit, 1990) and in second to sixth week for *M. galloprovincialis* (Naves et al., 2007).

The adults of the vector weevils of *B. cocophilus* are strong flyers, averaging about 6 m s^{-1} in the field, and may travel up to 1.6 km a day (Griffith, 1987). *Rhynchophorus* spp. are attracted to semiochemicals released by palms that are physically wounded or by those with diverse types of rot. In addition, they have the sex-specific pheromone 'rhynchophorol' (6-methyl-2-hept-2-en-4-ol) as an aggregation pheromone, and this is attractive to adults of both sexes (Rochat *et al.*, 1991a,b). Host plant volatiles may serve as short-distance attractants or arrestants, while the weevil pheromone coupled with host volatiles probably acts as a long-range attractant (Giblin-Davis, 1993; Jaffé *et al.*, 1993). The adult weevil population fluctuates during the year. It shows a peak in the dry season, while the percentage of the weevils carrying the nematodes is lowest in this season and gradually increases towards the end of the rainy season (Chinchilla and Escobar, 2007).

Currently, the literature does not indicate palm seeds as a pathway for infestation by *B. cocophilus*. A possibility of soil-borne infection via roots was suggested by Warwick and Bezerra (1992), but recent work on the nematode distribution inside host palms suggested that the chances of transmission of the disease by means of the roots from one plant to another are minimal (Duarte *et al.*, 2008).

6.5 Management Strategies and Tactics

6.5.1 Avoidance

Temperature greatly affects the onset of pine wilt disease. In susceptible pines, the disease only occurs where the duration of mean daily temperatures of >20°C last for a long period. Those susceptible pines can grow disease free at high elevations or at high latitudes. Some local populations of susceptible pines that are endangered by pine wilt epidemics are transplanted to cooler regions to avoid their extinction and maintain their genetic resources. These individuals should be transplanted back to the original location before starting to make flowers to avoid genetic disturbance in the refuge.

6.5.2 Exclusion

Human transportation of unprocessed wood, packaging and wood products is a high-risk pathway for the movement of *B. xylophilus*. It is apparent that the invasions into Asian countries and Portugal were not caused by insect dispersal but by human transportation.

Restrictions on the movement and importation of wood products, and a rigorous inspection system for wood materials, are needed internationally and domestically. The International Standards for Phytosanitary Measures (ISPM) describe heat treatment that achieves a minimum wood core temperature of 56°C for a minimum of 30 min and methyl bromide fumigation for eliminating *B. xylophilus* from wood packaging materials (IPPC, 2006).

Local movement of *B. cocophilus* occurs via its weevil vector, but wider dissemination can only occur by the transport of infested palms. Red ring disease usually goes undetected until the damage is very severe, and *B. cocophilus* and its vector weevils conceal themselves within the stems of their hosts. This makes it difficult to detect these pests during port-of-entry inspection. Now, many countries are toughening up border controls on palm or palm product imports. *R. palmarum* was added to the European and Mediterranean Plant Protection Organization (EPPO) A1 list in 2005, and in EPPO member countries palm plants for planting should be imported from pest-free areas. African countries, in which *R. palmarum*, *R. ferrugineus* (Olivier) and *B. cocophilus* do not exist so far, regard them all as high-risk pests. The governments of Asian countries, including India and Thailand, require a phytosanitary certificate issued by an appropriate authority at the country of origin when oil palms are imported. In the USA, the Federal Order that prevents the introduction and dissemination of *R. palmarum*, *R. ferrugineus* and *B. cocophilus* from foreign countries was implemented in February 2010.

6.5.3 Eradication

Pine wilt disease

Physical and chemical controls are major tactics after felling pine trees infested by *B. xylophilus*. One of the most effective measures for controlling damage due to pine wilt disease is to kill the *Monochamus* spp. vectors before they emerge from brood trees. Practices that are important for such eradication, but are apt to be neglected are: (i) the complete treatment of small branches larger than 2 cm in diameter, as well as trunks, because larval density tends to be higher in those parts of the tree with thinner bark, such as the upper portions of the main trunk and branches; and (ii) the completion of eradication before the adult insects emerge.

CUT AND CHEMICAL APPLICATION. Approximately 100% mortality of the insect vector is possible by spraying insecticide in the autumn when the larvae are living beneath the bark or are constructing pupal chambers. For both winter and spring treatments, when the entrance of the pupal chamber is plugged with excelsior-like shavings produced by the larvae, an oil solution of chemicals with a high permeability should be used because the chemicals cannot reach the pupal chambers easily. In contrast, almost 100% pine sawyer mortality is possible by fumigation, with the fumigant killing both the nematode and its insect vector. The procedure is to first put pruned branches on the forest floor after removing the humus layer. Pine wilt-affected logs are piled on the branches and treated with the chemicals and the pile is covered with a polyvinyl chloride (PVC) sheet. Soil is placed around the edge of the PVC sheet to seal it. Fumigation should last at least 7 days.

CUT AND CRUSH. A regulation of the Japanese Government specifies two sizes of chips that should result from the crushing of diseased trees:, <6 mm using an ordinary chipper at a factory and <15 mm using a mobile chipper in the field. The former is based on the size of the pupal chamber of *M. alternatus* (6–9 mm wide). The latter is based on the results of field experiments that caused 100% mortality for overwintering larvae when the maximum size of chips was 80 × 60 × 16 mm.

CUT, BURN AND USE FOR MAKING CHARCOAL. Complete mortality of overwintering *M. alternatus* larvae can be achieved by charring the outer 1 cm of the wood, even after formation of the pupal chambers. Portable incinerators are recommended for burning and making the charcoal to avoid infections by the pine root rot fungal pathogen *Rhizina undulata* Fr., the spores of which are germinated by high temperature.

CUT AND BURY OR SOAK IN WATER. A soil covering can completely prevent the emergence of sawyer beetle adults, which is possible for coastal pine stands on a sandy soil. In contrast, the 'cut and soak in water option' is not practical because it takes over 100 days to kill beetles in infested logs by soaking in seawater.

Natural enemies, such as bird and insect predators, parasitoids and entomopathogenic microbes have been studied as biological control agents against pine wilt disease. To date, products using the entomopathogenic fungus *Beauveria bassiana* (Bals.-Criv.) Vuill. are commercially supplied in Japan. The non-woven fabric strips containing the asexual spores (conidia) of the fungus can be applied to the bark of dead pine trunks and branches in various ways and stapled in place where new adult beetles may emerge.

In inaccessible locations, such as cliffs, aerial spot spraying on to a pine wilt-killed tree and adjacent healthy pines is efficient.

Red ring disease

Phytosanitation is currently the best method for the management of red ring disease. This strategy targets both the vector population and the sources of nematode inocula. Treatment with an appropriate herbicide, which produces a rapid death of the palm and prevents insect breeding in the dying tissues, should be promptly conducted as soon as the tree is found to be symptomatic. Sprays of methomyl (an insecticide) and trunk application of monosodium acid methanearsonate (MSMA; a herbicide) are recommended. Palms that are heavily infested with vector weevils should be cut and treated with an insecticide, or burned. The way the affected palm is eliminated may be important for the subsequent incidence of the disease, as pointed out by Chinchilla and Escobar (2007); sawdust made when felling the affected palm and the remaining stem may both become the source of several generations of infected weevils, which will propagate the disease.

There have been several research studies that attempted to reduce the adult population of the vector weevil. A mass trapping-based strategy, which is usually adopted with other strategies, will be described in Section 6.5.7.

The use of natural enemies against the weevil may be possible. For example, entomopathogenic nematodes in the families *Steinernematidae* and *Heterorhavditidae*, the bacterium *Micrococcus roseus*, the tachinid parasite *Parabillaea rhynchophorae* Blanchard, a *Cytoplasmic polyhedrosis virus* (CPV) and human activities may be effective against the insect vector. De Foliart (1993) and Cerda *et al.* (2001) described the use of larvae of *R. palmarum* as food by a local community in Venezuela, providing nutritional information and discussing the potential of the larvae both as a nutritional food and as a source of cash income for the community.

6.5.4 Protection

The two major methods of protection against pine wilt disease are preventive spraying of insecticide to kill vector beetles and injection of nematicide into tree trunks to kill *B. xylophilus* and/or depress the nematode population after infection.

Sawyer beetle adults prefer young branches and shoots for their maturation feeding, which occurs mainly in the upper forest canopy, indicating the importance of applying insecticide to this part of the trees. Because the transmission of *B. xylophilus* peaks early in the sawyer's adult life, insecticide should be applied at the beginning of adult emergence, although there is a major problem in estimating the timing of this as it differs greatly among years depending on the temperature. Also, preventive sprays can only reduce the probability of infection. Previously, aerial spraying was common because insecticide can be sprayed from the top of tree canopies, although recently the practice has decreased because of its negative impact on ecosystems and human health. In contrast, spraying insecticide from the ground has been increasing because drift of the insecticide outside the pine stand is less than for aerial spraying. However, the effect of the ground spraying tends to be lower as tree height increases because the insecticide cannot reach high positions. Ground spraying is also more time-consuming and needs more insecticide than aerial spraying.

The trunk injection of nematicides is the only practical control method for *B. xylophilus*. Chemicals similar to anthelmintics, such as morantel tartrate, are used for this purpose. To avoid chemical injury, it is important not to let the chemicals into contact with cambium during injection and to crust the drilled hole afterwards by filling with vinyl acetate emulsion. This method has generated excellent results in disease prevention, with few adverse environmental effects and long-lasting efficacy for the entire feeding period of adults (Kamata, 2008). The efficacy period has extended to 6 years (Pfizer, 2012a,b). The main drawback of the technique is its high cost. Previously, the method was mostly used for important or irreplaceable trees growing at famous temples, shrines, golf courses and the like; its use in forests was not common, but has been increasing, especially in coastal pine stands.

Another method of protection against pine wilt disease is the use of a barrier zone in which (putative) host pine trees are removed to reduce probability of invasion of the vector beetles into areas to be protected. This 'barrier zone' method has been applied in the north-eastern region of Japan, in Shenzhen City, China, and along the Portugal–Spain border. According to the distance of beetle dispersal, the recommended width of the barrier zone is >2 km.

For red ring disease, preventive treatment with nematicides and/or insecticides is not always effective and poses environmental risks. Blair (1970) suggested instead a protective approach that uses mechanical sealing of the stem cracks in the crown area of healthy palms, in order to prevent the palm weevil from infesting the trees.

6.5.5 Resistance

Since 1978, to cope with the serious damage resulting from pine wilt, a breeding project for resistance to this disease had been carried out under the leadership of the Forest Tree Breeding Center, Forestry and Forest Products Research Institute, Japan (Nose and Shiraishi, 2008). Surviving pine trees were selected from heavily pine wilt-affected stands, and grafts from them were inoculated artificially with *B. xylophilus* (pre-screening). The candidates that showed higher survival were again grafted and challenged with *B. xylophilus* at the nursery (second screening). The pines that were selected at this stage were then used as mother trees in seed orchards; the open pollinated seeds were collected and, resistant clones were registered after careful screening in a similar manner to that of the first generation. Such breeding programmes are being conducted in several regions in Japan and also in Anhui Province in China. The number of resistant trees that have resulted is still not enough though, and the demand for more resistant clones is increasing.

Hirao *et al.* (2012) has described the difference between resistant and susceptible pine clones in response to *B. xylophilus* infection at the transcript level. The day will soon come when an efficient and reliable DNA marker is available for resistance pine screening.

Promising tolerance against *B. cocophilus* has been identified in some coconut ecotypes, but has not yet been utilized (Harrison and Jones, 2003).

6.5.6 Therapy

If pine wilt disease develops fully after infection, there is no possibility of recovering from the disease. However, the probability of the disease development can be decreased by reducing drought stress. Hence, irrigation is useful for reducing the mortality rate of the disease, especially in hot and dry summers.

To control *B. cocophilus*, injections of systemic nematicides such as fenamiphos, oxamyl or carbofuran into little leaf symptomatic palms has caused apparent recovery of the palms in some of the trees tested. Another chemical, aldicarb, was partially effective in African oil palm in Honduras when it was applied directly to the whorl, but was not effective when it was applied on the soil, trunk or axil (Chinchilla, 1988). Nematicide treatment for palms with the classical symptoms of red ring is not feasible, irrespective of palm type, because such trees have been internally damaged in the vascular system and die anyway. Thus, nematicide treatment of red ring-affected palm is not recommended.

6.5.7 Integrated disease management

At the first stage of its invasion, complete eradication of *B. xylophilus* can be achieved, and there are a few examples of success in eliminating a *B. xylophilus* population after its establishment, albeit restricted to isolated islands. In contrast, once *B. xylophilus* has established its population over a wide area it is only possible to eliminate it at a local level, and it becomes impossible to completely prevent its reinvasion. In these cases, the management strategy should aim to reduce the incidence of pine wilt and keep the disease at an endemic level. To keep the incidence low enough to maintain pine stands, a combination of tactics is important. In coastal Japanese black pine (*P. thunbergii* Parl.) plantations, mortality from pine wilt disease is tolerable if the plantation can persist (Plate 5). On the contrary, no mortality, or a minimum mortality, is allowable in gardens, urban parks, golf courses and like places. The strategy should differ according to the situation.

The Kenrokuen Garden is known as one of the three most beautiful Japanese Gardens in Japan. Among approximately 600 pine trees, 15 are especially important elements of its scenery. No pine tree mortality is tolerated in the garden. The key points of the control strategy are as follows (Kamata, 2008). The first is to decrease the probability of a new invasion from outside: a *B. xylophilus*–pine sawyer management strategy is needed for areas including surrounding neighbourhoods. Incidence of the disease in the neighbourhood should also be reduced to decrease the probability of introduction of *B. xylophilus*, and so trunk injection by nematicides is carried out for pine trees in neighbouring urban parks as well as inside the Kenrokuen Garden itself. The second strategy point is complete eradication of the disease – because so many people visit the garden every day, it is unlikely that diseased pine trees go unnoticed; unhealthy trees that are noticed can then be treated. The third strategy point is prevention. Trunk injection by nematicides is employed for the pine trees other than the 15 trees of special interest because chemical injury is possible. Preventive sprays also have been conducted for nematicide-injected trees because such trees could act as 'stepping stones' for pine sawyers if they are not sprayed.

In forests or plantations, trunk injection of nematicides cannot be a major tactic in the pine wilt management strategy because the cost would be quite high. Eradication and preventive sprays should be used instead. Complete eradication of diseased trees from forests or plantations is not easy, but is effective in reducing the disease. Therefore, intensive surveys on diseased trees should be carried out from the ground and air before the emergence of adult beetles. As previously mentioned, irrigation is also effective in summer for decreasing water stress.

However, despite intensive efforts in removing dead trees from the stands, new dead trees continue to appear in the vicinity of the stumps of trees killed in the previous year, and wilting recurs in the same pine stand every year because of asymptomatic carriers spreading the disease to surrounding pine trees (Futai, 2003). Infection by *B. xylophilus* through root grafts among individual pine trees has also been reported. In Korea, the removal of all pine trees, both symptomatic and asymptomatic, within 10–50 m of diseased trees has succeeded in controlling pine wilt disease (Kwon *et al.*, 2011).

The incidence of red ring disease can be kept at a low level if an integrated management approach is followed. This includes the elimination of sources of inocula (diseased palms), destruction of the weevil's breeding places and lowering the adult population of the weevil by mass trapping. Among these strategies, mass trapping of the weevil may hold the key. Attraction by certain volatile attractants, often coupled with poisoned bait, has been used to capture adults in diverse types of traps (Jaffé *et al.*, 1993). Early studies used pieces of stem from several plant species as the emitter of semiochemicals, but the costs of this approach are high because the stem pieces must be normally replaced every week. When red ring incidence and weevil population are both high, these traps may not be enough to control the disease. The use of the aggregation pheromone rhynchophorol has drastically improved captures in traps, and rhynchophorol has been proven efficient in practical use. In Costa Rica, mass trapping by means of rhynchophorol coupled with insecticide-laced sugarcane was carried out in combination with a continuous removal of diseased palms, resulting in reduction of disease incidence in oil palm plantations in Costa Rica (Oehlschlager *et al.*, 1995, 2002). Captured insects need then be killed with insecticides or by other means. Rhynchophorol can be obtained both by commercial synthesis and by filling a trap with males of *R. palmarum* and activating them with ethyl acetate odours to induce production of the pheromone (Sánchez and Cerda, 1993; EPPO, 2005). Future study needs to determine whether a similar strategy would be effective in other palm species.

References

Bergdahl, D. and Halik, S. (1999) Inoculated *Pinus sylvestris* serve as long term hosts for *Bursaphelenchus xylophilus*. In: Futai, K., Togashi, K. and Ikeda, T. (eds) *Sustainability of Pine Forests in Relation to Pine Wilt and Decline. Proceedings of International Symposium, Tokyo, 27–28 October, 1998*. Shokado, Kyoto, Japan, pp. 73–78.

Blair, G.P. (1970) Studies on red ring disease of the coconut palm. *Oléagineux* 25, 19–22.

Cerda, H., Martinez, R., Briceno, N., Pizzoferrato, L., Manzi, P., Tommaseo Ponzetta, M., Marin, O. and Paoletti, M.G. (2001) Palm worm (*Rhynchophorus palmarum*) traditional food in Amazonas, Venezuela – nutritional composition, small scale production and tourist palatability. *Ecology of Food and Nutrition* 40, 13–32.

Chinchilla, C. (1988) El Sindrome del anillo rojo-hoja pequeña en palma aceitera y cocotero. *Boletín Técnico* (Oil Palm Operations, Chiquita Brands, Costa Rica) No. 2, 113–136 [in Spanish].

Chinchilla, C. and Escobar, R. (2007) Red ring and other diseases of the oil palm in Central and South America. *ASD Oil Palm Papers* (San José, Costa Rica) No. 30, 1–27.

DeFoliart, G.R. (1993) Hypothesizing about palm weevil and palm *Rhinoceros* beetle larvae as traditional cuisine, tropical waste recycling, and pest and disease control on coconut and other palms – can they be integrated? *Principes* 37, 42–47.

Duarte, A.G., de Lima, I.S., Araújo, J.V., Duarte, A.G., de Albuquerque, A.L.S. and Cruz, M.M. (2008) Disposition of the nematode *Bursaphelenchus cocophilus* (Cobb) Baujard, in coconut palm trees with the red ring disease. *Revista Brasileira de Fruticultura* 30, 622–627 [in Portuguese, with English summary].

Enda, N. (1985) Flight performance of *Monochamus alternatus* adults investigated by flight mill technique. In: *Transactions of the 96th Meeting of Japanese Forestry Society*. Japanese Forestry Society, Tokyo, pp. 517–518 [in Japanese].

EPPO (2005) Data sheets on quarantine pests: *Rhynchophorus palmarum*. *EPPO Bulletin* 35, 468–471.

Evans, H.F., McNamara, D.G., Braasch, H., Chadoeuf, J. and Magnusson, C. (1996) Pest risk analysis (PRA) for the territories of the European Union (as PRA area) on *Bursaphelenchus xylophilus* and its vectors in the genus *Monochamus*. *EPPO Bulletin* 26, 199–249.

Futai, K. (2003) Role of asymptomatic carrier trees in epidemic spread of pine wilt disease. *Journal of Forest Research* 8, 253–260.

Giblin-Davis, R.M. (1990) *The Red Ring Nematode and Its Vectors*. Nematology Circular No. 181, Florida Department of Agriculture and Consumer Services, Division of Plant Industry, Gainesville, Florida.

Giblin-Davis, R.M. (1993) Interactions of nematodes with insects. In: Khan, M.W. (ed.) *Nematode Interactions*. Chapman and Hall, London, pp. 302–344.

Giblin-Davis, R.M., Gerber, K. and Griffith, R. (1989) *In vivo* and *in vitro* culture of the red ring nematode, *Rhadinaphelenchus cocophilus*. *Nematropica* 19, 135–142.

Giblin-Davis, R.M., Inserra, R.N., Stanley, J.D. and Hodges, G.S. (2010) *The Red Ring Nematode, Bursaphelenchus cocophilus (Cobb, 1919) Baujard, 1989 (Nematoda: Tylenchida)*. DACS-P-01748, Pest Alert, Florida Department of Agriculture and Consumer Services, Division of Plant Industry, Gainesville, Florida.

Goberdhan, L.C. (1962) The '*in vitro*' culture of *Rhadinaphelenchus cocophilus* (Cobb, 1919) Goodey, 1960. In: *Annual Report of the Trinidad and Tobago Coconut Research Ltd for 1961, Trinidad*, pp. B1–B7.

Griffith, R. (1987) Red ring disease of coconut palm. *Plant Disease* 71, 193–196.

Hagley, E.A.C. (1963) The role of the palm weevil, *Rhynchophorus palmarum*, as a vector of red ring disease of coconuts. I. Results of preliminary investigations. *Journal of Economic Entomology* 56, 375–380.

Harrison, N.A. and Jones, P. (2003) Diseases of coconut. In: Ploetz, R.C. (ed.) *Diseases of Tropical Fruit Crops*. CAB International, Wallingford, UK, pp. 197–225.

Hirao, T., Fukatsu, E. and Watanabe, A. (2012) Characterization of resistance to pine wood nematode infection in *Pinus thunbergii* using suppression subtractive hybridization. *BMC Plant Biology* 12: 13, doi: 10.1186/1471-2229-12-13.

Ikeda, T., Enda, N., Yamane, A., Oda, K. and Toyoda, T. (1980) Attractants for the Japanese pine sawyer, *Monochamus alternatus* Hope (Coleoptera: Cerambycidae). *Applied Entomology and Zoology* 15, 358–361.

IPPC (2006) *International Standards for Phytosanitary Measures, 1 to 24*, 2005 edn. Secretariat of the International Plant Protection Convention, FAO, Rome, Italy.

Jaffé, K., Sánchez, P., Cerda, H., Hernández, J.V., Jaffé, R., Urdaneta, N., Guerra, G., Martínez, R. and Miras, B. (1993) Chemical ecology of the palm weevil *Rhynchophorus palmarum* (L.) (Coleoptera: Curculionidae): attraction to host plants and to a male-produced aggregation pheromone. *Journal of Chemical Ecology* 19, 1703–1720.

Jikumaru, S. and Togashi, K. (2000) Temperature effects on the transmission of *Bursaphelenchus xylophilus* (Nemata: Aphelenchoididae) by *Monochamus alternatus* (Coleoptera: Cerambycidae). *Journal of Nematology* 32, 110–116.

Kamata, N. (2008) Integrated pest management of pine wilt disease in Japan: tactics and strategies. In: Zhao, B.G., Futai, K., Sutherland, J.R. and Takeuchi, Y. (eds) *Pine Wilt Desease*. Springer, Tokyo, Japan, pp. 304–322.

Kastelein, P., Sanchit-Bekker, M.L. and Dipotaroeno, M.S. (1985) Confusion in the recognition of 'hartrot' and red ring disease in coconut. *De Surinaamse Landbouw* 33, 65–66.

Kawabata, K. (1979) Dispersal of the Japanese pine sawyer, *Monochamus alternatus*, among small islands. *Transactions of the 32nd Annual Meeting of Kyushu Branch of the Japanese Forestry Society*, pp. 281–282 [in Japanese].

Kikuchi, T., Aikawa, T., Oeda, Y., Karim, N. and Kanzaki, N. (2009) A rapid and precise diagnostic method for detecting the pinewood nematode *Bursaphelenchus xylophilus* by loop-mediated isothermal amplification. *Phytopathology* 99, 1365–1369.

Kiyohara, T. and Tokushige, Y. (1971) Inoculation experiments of a nematode, *Bursaphelenchus* sp., onto pine trees. *Journal of Japanese Forestry Society* 53, 210–218 [in Japanese, with English summary].

Kwon, T.-S., Shin, J.H., Lim, J.-H., Kim, Y.-K. and Lee, E.J. (2011) Management of pine wilt disease in Korea through preventative silvicultural control. *Forest Ecology and Management* 261, 562–569.

Linit, M.J. (1988) Nematode–vector relationships in the pine wilt disease system. *Journal of Nematology* 20, 227–235.

Linit, M.J. (1990) Transmission of pinewood nematode through feeding wounds of *Monochamus carolinensis* (Coleoptera: Cerambycidae). *Journal of Nematology* 22, 231–236.

Maehara, N., Hata, K. and Futai, K. (2005) Effect of bluestain fungi on the number of *Bursaphelenchus xylophilus* (Nematoda: Aphelenchoididae) carried by *Monochamus alternatus* (Coleoptera: Cerambycidae). *Nematology* 7, 161–167.

Mamiya, Y. (1984) The pine wood nematode. In: Nickle, W.R. (ed.) *Plant and Insect Nematodes*. Marcel Dekker, New York, pp. 589–626.

Mamiya, Y. and Kiyohara, T. (1972) Description of *Bursaphelenchus lignicolus* n.sp. (Nematoda: Aphelenchoididae) from pine wood and histopathology of nematode-infested trees. *Nematologica* 18, 120–124.

Naves, P., Camacho, S., De Sousa, E. and Quartau, J. (2007) Transmission of the pine wood nematode *Bursaphelenchus xylophilus* through feeding activity of *Monochamus galloprovincialis* (Coleoptera: Cerambycidae). *Journal of Applied Entomology* 131, 21–25.

Nickle, W.R. (1970) A taxonomic review of the genera of Aphelenchoidea (Fuchs, 1937) Thorne, 1949 (Nematoda: Tylenchida). *Journal of Nematology* 2, 375–392.

Nickle, W.R., Golden, A.M., Mamiya, Y. and Wergin, W.P. (1981) On the taxonomy and morphology of the pine wood nematode, *Bursaphelenchus xylophilus* (Steiner & Burher, 1934) Nickle 1970. *Journal of Nematology* 13, 385–392.

Nose, M. and Shiraishi, S. (2008) Breeding for resistance to pine wilt disease. In: Zhao, B.G., Futai, K., Sutherland, J.R. and Takeuchi, Y. (eds) *Pine Wilt Disease*. Springer, Tokyo, pp. 334–350.

Oehlschlager, A.C., McDonald, R.S., Chinchilla, C. and Patschke, S.N. (1995) Influence of a pheromone-based mass-trapping system on the distribution of *Rhynchophorus palmarum* (Coleoptera: Curculionidae) in oil palm. *Environmental Entomology* 24, 1005–1012.

Oehlschlager, A.C., Chinchilla, C., Castillo, G. and Gonzalez, L. (2002) Control of red ring disease by mass trapping of *Rhynchophorus palmarum* (Coleoptera: Curculionidae). *Florida Entomologist* 85, 507–513.

Pfizer (2012a) Effect of six years: Greenguard® NEO. Available at: www.greenguard.jp/common/pdf/neo.pdf (accessed 6 May 2012) [in Japanese].

Pfizer (2012b) Greenguard®: Greenguard® Eight. Available at: www.greenguard.jp/common/pdf/eight.pdf (accessed 6 May 2012) [in Japanese].

Rochat, D., González, A.V., Mariau, D., Villanueva, A.G. and Zagatti, P. (1991a) Evidence for male-produced aggregation pheromone in American palm weevil, *Rhynchophorus palmarum* (L.) (Coleoptera: Curculionidae). *Journal of Chemical Ecology* 17, 1221–1230.

Rochat, D., Malosse, C., Lettere, M., Ducrot, P.-H., Zagatti, P., Renou, M. and Descoins, C. (1991b) Male-produced aggregation pheromone of the American palm weevil, *Rhynchophorus palmarum* (L.) (Coleoptera, Curculionidae): collection, identification, electrophysiological activity, and laboratory bioassay. *Journal of Chemical Ecology* 17, 2127–2141.

Sánchez, P. and Cerda, H. (1993) The complex *Rhynchophorus palmarum/Bursaphelenchus cocophilus* in palms. *Boletín di Entomología Venezolana* 8, 1–18 [in Spanish].

Shibata, E. (1986) Dispersal movement of the adult Japanese pine sawyer, *Monochamus alternatus* Hope (Coleoptera: Cerambycidae) in a young pine forest. *Applied Entomology and Zoology* 21, 184–186.

Steiner, G. and Buhrer, E.M. (1934) *Aphelenchoides xylophilus*, n.sp., a nematode associated with bluestain and other fungi in timber. *Journal of Agriculture Research* 48, 949–951.

Takasu, F., Yamamoto, N., Kawasaki, K., Togashi, K., Kishi, Y. and Shigesada, N. (2000) Modeling the expansion of an introduced tree disease. *Biological Invasions* 2, 141–150.

Tanaka, H. and Gyokusen, K. (2004) Infection of pine wilt disease through root grafts among *Pinus thunbergii* trees. *Kyushu Journal of Forest Research* 57, 241–242 [in Japanase].

Teale, S.A., Wickham, J.D., Zhang, F., Su, J., Chen, Y., Xiao, W., Hanks, L.M. and Millar, J.G. (2011) A male-produced aggregation pheromone of *Monochamus alternatus* (Coleoptera: Cerambycidae), a major vector of pine wood nematode. *Journal of Economic Entomology* 104, 1592–1598.

Togashi, K. and Magira, H. (1981) Age-specific survival rate and fecundity of the adult Japanese pine sawyer, *Monochamus alternatus* Hope (Coleoptera: Cerambycidae), at different emergence times. *Applied Entomology and Zoology* 16, 351–361.

Warwick, D.R.N. and Bezerra, A.P.T. (1992) Possible root transmission of the red ring nematode (*Rhadinaphelenchus cocophilus*) to coconut palms. *Plant Disease* 76, 809–811.

Zang, X. and Linit, M.J. (1998) Comparison of oviposition and longevity of *Monochamus alternatus* and *M. carolinensis* (Coleoptera: Cerambycidae) under laboratory conditions. *Journal of Nematology* 27, 36–41.

7 Annosus Root and Butt Rots

Paolo Gonthier[1]* and Magnus Thor[2]
[1]*University of Torino, Grugliasco, Italy;* [2]*Skogforsk (Forest Research Institute of Sweden), Uppsala, Sweden*

7.1 Pathogens, Significance and Distribution

The root and butt rots caused by the *Heterobasidion annosum* (Fr.) Bref. species complex, hereafter referred as *H. annosum s.l.*, are among the most destructive diseases of conifers. This fungus, which belongs to the *Basidiomycota*, is widely distributed in the coniferous forests of the northern hemisphere, especially in Europe, North America (i.e. Mexico, the USA, western and, marginally, eastern Canada), southern Russia, China (i.e. the Himalayan region and Manchuria) and Japan (Korhonen *et al.*, 1998a; Dai *et al.*, 2003; Worrall *et al.*, 2010).

The fungus had long been regarded as a single species until mating experiments revealed the occurrence of intersterile groups within *H. annosum s.l.* (Korhonen, 1978; Chase and Ullrich, 1988; Capretti *et al.*, 1990) – three in Eurasia and two in North America. All the intersterile groups obtained formal description. The Eurasian groups were named *H. annosum s.s.*, *H. abietinum* Niemelä & Korhonen and *H. parviporum* Niemelä & Korhonen (Niemelä and Korhonen, 1998), while the North American groups were described as *H. irregulare* (Underw.) Garbel. & Otrosina and *H. occidentale* Otrosina & Garbel. (Otrosina and Garbelotto, 2010). Although defined on the basis of partial reproductive isolation and morphology, further supported by phylogenetic analyses based on a range of markers (Otrosina *et al.*, 1993; Harrington *et al.*, 1998; Linzer *et al.*, 2008; Dalman *et al.*, 2010), these species are also characterized by a distinct host preference. *H. annosum s.s.* is mostly associated with pines (*Pinus* spp.), especially Scots pine (*P. sylvestris* L.), but attacks several other conifers and even some broadleaved tree species. *H. parviporum* shows a relatively strict specialization for Norway spruce (*Picea abies* (L.) Karst.), while *H. abietinum* is commonly associated with silver fir (*Abies alba* Mill.) and other species of the genus *Abies*. In North America, *H. irregulare* attacks pines, junipers (*Juniperus* spp.) and incense cedar (*Calocedrus decurrens* (Torr.) Florin), while *H. occidentale* shows a broader host range, being mainly associated with the genera *Abies*, *Picea*, *Tsuga*, *Pseudotsuga* and *Sequoiadendron*.

The distribution of the species of *H. annosum s.l.* reflects that of the forests of their main hosts. *H. annosum s.s.* is found all over Europe, except on sites north of 66°N, and its distribution area expands east at least in the Altai region of southern Siberia

* E-mail: paolo.gonthier@unito.it

(Korhonen *et al.*, 1998a; Dai *et al.*, 2003; Asiegbu *et al.*, 2005). *H. parviporum* occurs from northern Fennoscandia to the southern Alps and from western Europe to southern Siberia, northern China and Japan (Korhonen *et al.*, 1998a; Dai *et al.*, 2003; Tokuda *et al.*, 2011). *H. abietinum* is distributed in central and southern Europe, and in the Mediterranean Basin. The fungus has been reported from Poland in the north to Spain, Italy, Greece and western Turkey in the south (Korhonen *et al.*, 1998a; Kowalski and Łakomy, 1998; Doğmuş-Lehtijärvi *et al.*, 2006; Sánchez *et al.*, 2007). *H. occidentale* is found only in western North America, from Alaska in the north to southern Mexico in the south (Filip and Morrison, 1998; Garbelotto and Chapela, 2000), while *H. irregulare* is reported both in eastern and western forests, from Quebec to Florida and from British Columbia to Mexico, respectively (Filip and Morrison, 1998; Otrosina and Garbelotto, 2010); the fungus is less common in the central part of North America, although it can be found along with *H. occidentale* in the western USA (Worrall *et al.*, 2010), and by itself in many Midwestern states (Linzer *et al.*, 2008). *H. irregulare* was introduced into central Italy during World War II (Gonthier *et al.*, 2004), and subsequently became invasive by spreading in Italian stone pine (*P. pinea* L.) stands (Gonthier *et al.*, 2007). As a consequence, the current distribution area of *H. irregulare* also includes an area of approximately 100 km along the Tyrrhenian coast around Rome (D'Amico *et al.*, 2007; Gonthier *et al.*, 2007). Interestingly, in the invasion area in Italy, *H. irregulare* was also found in pure oak (*Quercus* spp.) stands, a habitat unavailable to the native *H. annosum s.s.* (Gonthier *et al.*, 2012b).

The biogeography and evolutionary history of species within *H. annosum s.l.* were recently elucidated by Linzer *et al.* (2008) for *H. irregulare* and *H. annosum s.s.*, and by Dalman *et al.* (2010) for the whole species complex. Speciation within the species complex began about 60 million years ago (mya), well after the radiation of host genera. The last emerging species was *H. abietinum*, which arose 14–31 mya (Dalman *et al.*, 2010). Although *H. annosum s.l.* species are characterized by partial interfertility in laboratory experiments, and by overlaps in their distribution areas, hybrids among species have been found very rarely between the sympatric *H. irregulare* and *H. occidentale* in North America (Garbelotto *et al.*, 1996; Lockman *et al.*, 2011). Nevertheless, a massive and almost unidirectional introgression of alleles has been reported from the native *H. annosum s.s.* to the invasive *H. irregulare* in Italy (Gonthier and Garbelotto, 2011).

H. annosum s.l. is the most common wood decay agent of conifers. It is a major concern in several areas of the world, especially where intensive forest management is practised. Despite a remarkable regional variation in the frequency of root and butt rots, the incidence and losses caused by *H. annosum s.l.* are generally high. In the UK, an incidence of decay as high as 68% was recorded in Sitka spruce (*Picea sitchensis* (Bong.) Carrière) stands (Pratt, 1979a), corresponding to a loss in value of 43% (Pratt, 1979b). In alpine Norway spruce and mixed stands, disease incidence attained 71% locally, and the direct financial losses were estimated as between 18% and 34%, depending on the assortment criteria (Gonthier *et al.*, 2012a). At a greater geographic scale, financial losses caused by the species complex in the European Union (EU) were estimated at 790 million euros a year (Woodward *et al.*, 1998). This estimate includes losses due to decay and overall reduction in diameter growth of infected trees (Bendz-Hellgren and Stenlid, 1995), but does not include the windthrow and reduction of resistance of stands to storm damage that are caused by the fungus, which may be locally significant.

Like other wood decay fungi, *H. annosum s.l.* affects species composition, stand structure, stand density, and the direction and rate of forest succession (Goheen and Otrosina, 1998; Garbelotto, 2004). When tree mortality occurs, gaps appear in the forest canopy, and these change the light, moisture and temperature regimes in the forest. In addition, as a facultative pathogen being able to survive saprotrophically on dead wood, the fungus is expected to contribute significantly to nutrient recycling.

7.2 Diagnosis

The detection of *H. annosum s.l.* in the forest is based on the examination of symptoms and finding fruiting bodies (Greig, 1998). Two kinds of symptoms may be associated with *H. annosum s.l.* attacks, depending on tree species and age: root rot and heart rot.

In general, in species characterized by a resinous heartwood (e.g. *Pinus* spp., *Juniperus* spp.), both young and adult trees are susceptible to a white root rot and may be killed by the fungus. Mortality occurs in a short period of time (i.e. one season) in young trees, and proceeds more slowly in older trees. Symptoms before death may include the decline of annual shoot growth and the shading of old needles, resulting in the so-called lion-tailing phenomenon and in a thinner crown. It should be noted that up to two-thirds of a root system may be killed by *H. annosum s.l.* before symptoms appear in the crown (Greig, 1998). Gaps in the forest resulting from tree mortality are frequently colonized by broadleaves: this trait is characteristic of the disease.

In species characterized by a non-resinous heartwood (e.g. *Picea* spp., *Abies* spp.), although mortality has been observed in saplings and in very young trees, the fungus is generally responsible for extensive heart rots in the roots, the bole (butt) and the stem. Often the decay is confined to the butt. However, in Norway spruce, decay columns may extend several metres into the stem. When heart rot develops, eventually resulting in cavities, external symptoms are rarely visible and mortality does not occur. Occasionally, there may be an exudation of resin at the base of the tree or on surface roots. In cases of advanced decay in older trees, foliage yellowing may appear. Furthermore, in Norway spruce butt swelling may indicate the presence of decay within the stem (Greig, 1998).

Few exceptions to the above symptoms have been reported. Provenances of silver fir from southern Italy and pinsapo fir (*A. pinsapo* Boiss) from southern Spain are susceptible to root rot and mortality instead of heart rot (Puddu *et al.*, 2003; Sánchez *et al.*, 2007). Mortality has also been observed in white fir (*A. concolor* (Gord. & Glend.) Lindl. ex Hildebr.) forests in California, southern Utah and near the California–Nevada border (Garbelotto *et al.*, 1999; Worrall *et al.*, 2010). Conversely, the closely related Swiss stone pine (*P. cembra* L.) and Siberian pine (*P. sibirica* Du Tour) have been reported to bear heart rots (Gonthier *et al.*, 2003). A few reports of Scots pine heart rots caused by *H. annosum s.l.* in UK and in the Alps have also been published (Greig, 1995; Gonthier *et al.*, 2012a).

Fruiting bodies of *H. annosum s.l.* most commonly develop at ground level, at the base of stumps or dead trees, and are often partially covered by litter. Characteristically, they may envelop litter debris. They may also form inside the internal cavities of old decayed stumps, on the underside of the root plates of windthrown trees and, occasionally, at the collar of living trees in which butt rot is at a very advanced stage. In dry forests of western USA, fruiting bodies have been found under the intact top surface of pine stumps (Otrosina and Garbelotto, 2010); the epidemiological role played by fruiting bodies produced in such sealed habitats is unclear, although their lifespan extends through several years, potentially lasting until the stump surface may crack. The fruiting body is 1–40 cm across, perennial, typically in the form of brackets or resupinate, and rubbery in texture. The top surface, if present, is reddish or dark brown (tobacco brown) in colour and becomes darker with age. The margin is distinct and white. The lower surface is characterized by numerous small pores and is white or cream coloured. Incipient fruiting bodies may form under the bark. These are smaller in size (0.5–1 cm in diameter) and are referred to as 'popcorns' because of their appearance (Otrosina and Garbelotto, 2010). Slight differences among *H. annosum s.l.* species occur in the macromorphology and micromorphology of fruiting bodies, especially in the size of pores, in the length of brown hairs on the upper margin of basidiomata (basidiocarps) and, to a lesser extent, in the size of spores (Mitchelson and Korhonen, 1998; Otrosina and Garbelotto, 2010). Unfortunately, differences are small and morphological traits are partially overlapping among species, which often leads to erroneous field identifications.

In the absence of fruiting bodies, *H. annosum s.l.* may not be easily distinguished from other root and butt rot agents. When the fungus behaves as a root rot agent, by infecting the cambium and then spreading into the root sapwood, it produces sheets of a white, paper-thin mycelium beneath the bark of infected roots or at the tree collar (Greig, 1998). Although this cannot be regarded as a reliable diagnostic feature of the disease, it is generally sufficient to discriminate between annosus root rots and Armillaria root rots, whose signs consist of a thicker mycelium and/or in the presence of rhizomorphs (see Guillaumin and Legrand, Chapter 8, this volume). Heart rot is a white pocket rot. Incipient decay in most trees is pale. Later it becomes stained in various colours, depending on the hosts. In advanced decay, *H. annosum s.l.* rot is soft, fibrous and stringy in texture, dry in nature and containing numerous small black flecks running parallel to the grain. Eventually, decayed portions are replaced by a hollow. Decay may be detected in standing trees by extracting wood from roots and from the tree collar using an increment borer.

For an accurate diagnosis of *H. annosum s.l.*, putatively infected wood samples should be collected and incubated at room temperature under moist conditions. After 5–10 days of incubation, infected samples will result in the growth of the anamorphic stage of the fungus (*Spiniger meineckellus* (A.J. Olson) Stalpers), which is easily recognizable using a 10× lens or under a dissecting microscope. The conidiophores appear as a mass of whitish 'pinheads' on stalks. The fungus can be isolated and cultured on a range of media, including selective media (Kuhlman and Hendrix, 1962). Abundant conidiophores and asexual spores (conidia) are produced in culture (Stalpers, 1978), making it easy to distinguish *H. annosum s.l.* from other fungi. Nevertheless, the macromorphological and micromorphological characters of colonies within *H. annosum s.l.* are not sufficiently divergent to allow a reliable identification of species within the complex (Mitchelson and Korhonen, 1998).

The identification of species within *H. annosum s.l.* was traditionally based on the use of mating tests, i.e. sexual compatibility tests, through which an unknown isolate is paired on agar medium with homokaryotic tester strains of the different species. Details on how to perform mating tests had been provided earlier by Stenlid (1985) and Mitchelson and Korhonen (1998). The partial *in vitro* interfertility among species may occasionally lead to ambiguous diagnosis. A number of molecular diagnostics assays based on both nuclear and mitochondrial DNA have been developed and are now available for the identification of species within *H. annosum s.l.* (Mitchelson and Korhonen, 1998). For practical purposes, PCR with taxon-specific primers should be preferred as this may allow identification from both pure cultures and environmental samples, i.e. wood samples. A summary of taxon-specific priming PCRs developed for the identification of annosus root and butt rot pathogens has been provided by Nicolotti *et al.* (2010), and is reported in Table 7.1.

7.3 Infection Biology

Early studies by Rishbeth (1950, 1951a,b) elucidated the general mechanisms of *H. annosum s.l.* infection in managed forest stands. The fungal pathogen is known as having a mixed infection biology. Infections from airborne propagules (primary infection) can originate only from freshly exposed wood surfaces: stumps or wounds in the stem or roots. Once primary mycelia have established and eventually formed secondary mycelia, the fungus can infect uninjured trees by vegetative growth of mycelium through root contacts or grafts (secondary infection) (Fig. 7.1).

The fungus produces both sexual spores and conidia and they are both infective in experimental inoculations (Redfern and Stenlid, 1998). None the less, it is largely accepted that the overwhelming majority of airborne infections are caused by sexual spores rather than by conidia (Redfern and Stenlid, 1998). Although conidia have been reported to be present in the air (Hsiang *et al.*, 1989), conidiophores are apparently rare and delicate in nature compared with fruiting bodies; hence, primary infections are mostly operated by sexual propagules. It has been suggested that there is an active role of conidia only in short-distance transmission in substrates or when they are accidentally vectored by root-feeding insects (Kadlec *et al.*, 1992). However, no population studies have been able to identify identical individual mycelia

Table 7.1. Summary of taxon-specific PCR-based methods for the detection of species within *Heterobasidion annosum s.l.* Adapted from Nicolotti *et al.* (2010), with kind permission of Springer Science+Business Media.

Primer combination and sequence (5′–3′)			
Forward	Reverse	Specificity/amplicon size	Reference/s
ITS1F[a] (cttggtcatttagaggaagtaa)	ITS4[b] (tcctccgcttattgatatgc)	Fungi	Garbelotto *et al.*, 1996
ITS P1 (gtcggtcgggttcttttgatc)		*H. irregulare*/518 bp	
ITS S1 (gccgcgtcttctcacaaact)		*H. occidentale*/486 bp	
MLF (taaaaatttaaattagccataa)	Mito7 (gccaatttattttgctacc)	*H. annosum s.s.*/230 bp	Garbelotto *et al.*, 1998; Gonthier *et al.*, 2001, 2003
MLS (aaattagccatattttaaaag)	Mito5 (taagaccgctatawaccagac)	*H. abietinum*/195 bp	
		H. parviporum/185 bp	
EfaHaFor (ctatgtcgcggtacagcttg)	EfaHaRev (gcgaggayaagaagtaatcagca)	*H. annosum s.l.*/169 bp	Gonthier *et al.*, 2007
EfaNAPFor (gtacatggtcactgtacgtagatgc)		*H. irregulare*/71 bp	
EfaEuPFor (atggtcactgtacgtagatcatgc)		*H. annosum s.s.*/69 bp	
MJ-F (ggtcctgtctggctttgc)	MJ-R (ctgaagcacaccttgcca)	*H. annosum s.s.*/100 bp	Hantula and Vainio, 2003
KJ-F (ccattaacggaaccgacgtg)	KJ-R (gtgcggctcattctacgctatc)	*H. parviporum*/350 bp	
HET-7 (cttctcacaaactcttcg)	HET-8 (caggtcccccacaatcg)	*H. annosum s.l.*/400 bp	Bahnweg *et al.*, 2002

[a]Universal primer from Gardes and Bruns (1993).
[b]Universal primer from White *et al.* (1990).

Fig. 7.1. Infection cycle of *Heterobasidion annosum s.l.* Infections from airborne spores occur through freshly cut stumps or wounds in the stem or roots. Wounds might be more important as infection courts in true firs (*Abies* spp.) than in pines (*Pinus* spp.). Once established in the root systems of stumps or trees, the mycelium may spread through root contacts or grafts, thus resulting in the infection of neighbouring healthy trees. The two trees on the left are Norway spruce (*Picea abies* (L.) Karst), the third is a Scots pine (*Pinus sylvestris* L.) and the fourth a silver fir (*Abies alba* Mill.).

(genets) at significant distances, as would be expected if they had been generated by conidia (e.g. Swedjemark and Stenlid, 1993; Garbelotto *et al.*, 1999; Gonthier *et al.*, 2003). Every time an identical genet was found, it was better explained by secondary tree-to-tree infection.

Population studies of *H. annosum s.l.* in diseased stands have pointed to the importance of stump infection in the epidemiology of the disease (Chase and Ullrich, 1983; Swedjemark and Stenlid, 1993). The fungus, by exploiting its ability to infect living cells, is a primary stump colonist, taking early possession of the stump surface before the arrival of competitive microorganisms. The period of susceptibility of stumps to infection is variable, but it rarely exceeds a month after cutting (felling) (Redfern and Stenlid, 1998). While susceptibility has been reported to be maximum on the day of felling on Norway spruce and Sitka spruce, it has been noted to increase in the first few days after felling, before subsequently declining, on some pine species in North America (Redfern and Stenlid, 1998; Woods *et al.*, 2000). The decrease in susceptibility with time is probably the result of colonization of the stumps by microorganisms that compete with and replace *H. annosum s.l.* Once established, *H. annosum s.l.* colonies penetrate rapidly to form vertical columns that can enlarge to occupy large portions of stumps. In general, as colonization proceeds downwards to the root system, colonies may die back from the upper portions of stumps. Growth rate depends on both host species and climate, and has been reported to attain up to 20 cm a month (Hunt *et al.*, 1976). In most species, e.g. pine species, Norway spruce, Douglas-fir (*Pseudotsuga menziesii* (Mirb.) Franco), sapwood is more readily colonized than heartwood. There are examples, though, which show that either the opposite may occur or, in some other species, that the entire stump surface may be colonized; colonization may even be restricted to the boundary between sapwood and heartwood (Redfern and Stenlid, 1998). Wood moisture content has been suggested as a major factor affecting colonization (Redfern, 1993); locally or temporarily, heartwood may be too dry or sapwood too wet for a suitable colonization to occur.

In general, both the incidence of colonized stumps and the surface area colonized increase with increasing stump diameter (Morrison and Johnson, 1999). Although small diameter stumps can harbour airborne colonies of *H. annosum s.l.*, most of these colonies seem to disappear later (Schönar, 1975), with no obvious consequences for disease epidemiology.

Primary infection may also occur through wounded roots or through stem wounds. While infection of wounded Scots pine roots, although possible, is not likely to provide a significant means of disease entry into stands, the probability that Norway spruce roots may be infected by spores seems to be greater (Redfern and Stenlid, 1998). Such different patterns between these two species could be related to their different ability to inhibit infection or the first stages of colonization through resinosis. There is evidence that the amount of infection is related to the severity of wounding, with wounds exposing transverse sections including heartwood being the most conducive for infection. While there is no evidence that wounds on pine stems may become infected naturally, western hemlock (*Tsuga heterophylla* (Raf.) Sarg.) and, to a lesser extent, Sitka spruce, have been reported to be susceptible to wound infection (Redfern and Stenlid, 1998). In a population genetics study performed in white fir forests in California, 86% of fungal genotypes were found only within a single tree or stump, with no stumps nearby, while 14% had spread to adjacent trees (Garbelotto *et al.*, 1999). The authors concluded that live trees, and therefore wounds rather than stumps, provide the most common infection courts for *H. occidentale* in Californian white fir stands. A similar conclusion was drawn in another study focused on grand fir (*Abies grandis* (Douglas ex D. Don) Lindl.) (Lockman, 1993). The vulnerability of fir trees to wound infections may reflect their relatively simple constitutive defence systems, which are based on resin blisters, as opposed to the more efficient secretory structures present in other tree genera, i.e. resin passages and resin ducts in *Picea* and *Pinus* trees, respectively (McGarvey and Croteau, 1995). Variable levels of stem wound infections have been reported in Norway spruce (Redfern and Stenlid, 1998). Locally, stem wounds may have a significant role in the life cycle of the pathogen, especially in unthinned stands.

H. annosum s.l. is incapable of freely growing in the soil, so a close contact between a woody inoculum and living host roots is pivotal for secondary infection to occur. Interestingly, root grafts have been shown to be less effective than contacts for disease transmission because grafted stump roots tend to remain alive, thus confining the fungus to the inner wood (Otrosina and Cobb, 1989). Depending on the soil properties, the fungus may either colonize the bark of live roots directly (in low pH soils) or grow ectotrophically under the outer bark scales (in high pH soils), thereby reducing the effects of active host resistance (Stenlid and Redfern, 1998). Subsequently, colonization involves either the phloem and the cambium in host species susceptible of root rot, or the central cylinder in species susceptible to heart rots.

In order to select the most appropriate control measures, it might be necessary to determine whether stumps have been infected secondarily through root contacts, indicating that the disease was already present in the root systems of trees, or primarily by spores landing on their surfaces. In addition to what previously reported in the diagnosis section, the following indication may have a practical significance: decay is visible on recently felled stumps only if infection has occurred on the standing trees, either through root contacts or through wounds.

7.4 Epidemiology

7.4.1 Patterns and rate of spread

The dynamics of spread of *H. annosum s.l.* in space and time have been elucidated by studies on the size and distribution of genets in forest stands, which are based on the use of both traditional mycological methods (i.e. somatic incompatibility tests, identification of mating alleles) and molecular fingerprinting markers (e.g. microsatellite and RAPD (random amplification of polymorphic DNA) analyses).

In general, in new afforestations and in first rotation stands, numerous small genets can be found, indicating that all the inoculum for disease has originated from spore-infected stumps (Swedjemark and Stenlid, 1993). In old forest sites and in second or later rotation stands, relatively large genets may develop as the result of the vegetative spread of mycelium through root contacts. The largest genet reported so far belonged to *H. abietinum*; it was 57 m long and had colonized ten pinsapo fir trees in southern Spain (Sánchez *et al*., 2007). *H. parviporum* was also reported to develop genets extending over 50 m in diameter and colonizing up to 15 trees in a Todo fir (*Abies sachalinensis* (Fr. Schmidt) Mast.) plantation in Japan (Tokuda *et al*., 2011). In most cases though, the largest diameter of single *H. annosum s.l.* genets is less than 30 m – a figure that is based on 14 investigations of *H. annosum s.l.* genet size that are summarized by Queloz and Holdenrieder (2005), and the size difference among the various species within *H. annosum s.l.* is small (Piri *et al*., 1990; Queloz and Holdenrieder, 2005). Although several trees can be colonized by the same genet, in general, a single genet colonizes one to three trees even at the end of the rotation (Stenlid, 1985; Piri *et al*., 1990; Garbelotto *et al*., 1999; Piri and Korhonen, 2001, 2008; Gonthier *et al*., 2003).

The transfer of the disease and of the same genet between forest generations has been demonstrated (Stenlid, 1987; Piri, 1996), indicating that a build-up of inoculum is likely to occur with time. However, incidence of decay in the preceding rotations is not always a good predictor of the incidence in the next rotations, suggesting that locally the build-up of inoculum may be due to spore-infected stumps (Stenlid and Redfern, 1998). In a long-term study of canopy gap expansion in California, it was shown that canopy gaps associated with *H. annosum s.l.* may completely arrest their expansion (Rizzo *et al*., 2000); new primary infections are, therefore, required to maintain a certain level of new infections and the pathogen cannot rely simply on carry-over from previous rotations.

Bendz-Hellgren and Stenlid (1998) investigated the relative susceptibility of different stump types. Thinning stumps were most susceptible to spore infection, probably as a result of a favourable microclimate and because they were large enough to provide suitable nutrition and water conditions. Stumps created from a final felling were slightly less susceptible than stumps created from thinning, due to more extreme microclimatic conditions. Stumps created in precommercial thinning were too small to provide a good substrate for *H. annosum s.l.*, and dried out before the infection could spread to neighbouring root systems; this is supported by results obtained by Vollbrecht *et al*. (1995). These small stumps provide a small target area for spore infection and, in general, decompose quickly, thus providing inoculum only for a short period of time. Furthermore, in small stumps there will be a lower probability of infection of two sexually compatible spores and hence a lower probability of the formation of heterokaryotic mycelia (Stenlid and Redfern, 1998). Although homokaryotic mycelia may be as virulent as heterokaryotic mycelia (Garbelotto *et al*., 1997), these latter are deemed to be more viable in the long term (Korhonen and Piri, 1994). The susceptibility of stumps created from a final felling to spore infections is sufficient to provide important inoculum for the next generation (Stenlid, 1987; Piri, 1996; Vollbrecht and Stenlid, 1999; Rönnberg and Jørgensen, 2000), and the fungus may survive in old stumps for several decades (Hodges, 1969). Greig and Pratt (1976) reported that *H. annosum s.l.* remained active in stumps for up to 62 years. From the south-eastern USA a more rapid decay of stumps is reported owing to high temperature and humidity (Kuhlman and Ross, 1970).

Some differences among species of *H. annosum s.l.* may exist in ability to spread from stumps to trees. In a Norway spruce stand, Oliva *et al*. (2011a) found that *H. parviporum* spread more frequently to the neighbouring trees from stumps inoculated 15 years previously than did *H. annosum s.s.* Furthermore, while the probability of stump-to-tree spread increased with the diameter of the stumps in the case of *H. parviporum*, no such association was found between the probability of stump-to-tree spread and the diameter of the stumps of *H. annosum s.s.* (Oliva *et al.*, 2011a).

The expansion of infection centres (i.e. mortality centres when species susceptible to root rot are involved) has been reported to occur at a variable rate and up to 2.1 m year^{-1} (Stenlid and Redfern, 1998; Rizzo *et al.*, 2000). As previously mentioned, at a certain time infection centres may stop enlarging (Rizzo *et al.*, 2000). The reason for this is still unclear.

Thinnings have been suggested to promote not only the primary infection but also the tree-to-tree spread of the pathogen in infected Norway spruce stands (Bendz-Hellgren *et al.*, 1999; Piri and Korhonen, 2008). The growth rate of the fungus in the roots increases after the felling of infected trees. In living Norway spruce roots, *H. annosum s.l.* is typically confined to dead heartwood. Hence, the transfer of the fungus between living trees may be limited to functional root grafts that enable the fungus to grow from the xylem of infected roots into the xylem of healthy roots. After the tree is cut, *H. annosum s.l.* expands outwards from the centre of the root, and it may be able to spread to the surrounding trees more easily through non-grafted root contacts.

In Italy, the invasion of *H. irregulare* provided a unique opportunity for an estimate of the overall rate of spread of *H. annosum s.l.* inclusive of within-stand stump-to-tree or tree-to-tree contagion through root contacts and of spread via airborne spores. Based on the distance between the point of introduction and the furthermost site where the fungus was detected, and the time elapsed since introduction, the overall rate of spread was estimated at 1.3 km year^{-1} (Gonthier *et al.*, 2007). Such a rate is likely to be an underestimate of the potential rate of spread because of the high fragmentation of the pine forests in the invasion area and also because the time necessary for the fungus to establish itself after introduction (which is currently unknown) could not be excluded from the calculation. It should be noted, however, that the ability of *H. irregulare* to colonize oak stands in the invasion area may imply that the spread of the fungus is not limited by the fragmented distribution of pine woodlands in central Italy (Gonthier *et al.*, 2012b).

7.4.2 Potential dispersal range of spores and temporal patterns of primary infection

Although spores of *H. annosum s.l.* have been reported to travel long distances, i.e. hundreds of kilometres (Rishbeth, 1959; Kallio, 1970), there is increasing evidence of a geographically limited range of dispersal for *H. annosum s.l.* spores. Based on data from Kallio (1970), Stenlid (1994) fitted a model describing the dispersal gradient of *H. annosum s.l.* The majority of spores are deposited on a scale of a few metres from the fruiting body: only 3% of spores travel 10 m and 0.1% of spores 100 m. In a study in eastern Finland, gradients were steeper around *H. annosum s.s.* infection centres than around *H. parviporum* infection centres (Möykkynen *et al.*, 1997). The background deposition level was reached at distances of 98 m and 1255 m, respectively, from the *H. annosum s.s.* and *H. parviporum* infection centres. Indirect evidence of a geographically limited range of influence of *H. abietinum* and *H. parviporum* spores has been obtained in the Alps by sampling nearby located silver fir and Norway spruce forests (Gonthier *et al.*, 2001). The authors were unable to detect significant spore flow between these two forests, which were located only a few kilometres apart. In a recently completed study in central Italy, the probability of finding *H. annosum s.s.* spores dropped to under 50% at a distance of 70 m from patches of pine, to 20% at 200 m and 1% 500 m from the patches of pine (Gonthier *et al.*, 2012b).

This information is important from an epidemiological point of view and supports the belief that during thinning operations or fellings, the presence of basidiospore-producing fruiting bodies increases the risk of stump infection within but not between forests. Low levels of spore movement among forests may be relevant only when dealing with the colonization of areas where *H. annosum s.l.* may not yet be present (Gonthier *et al.*, 2001).

Temporal patterns of stump infection have been found to vary largely among different sampling periods in the same stands, and especially among forests in different climatic zones. In northern Europe, infections follow a bell-shaped curve, with a maximum probability of infection during the summer and with no infection during the winter (Brandtberg *et al.*, 1996). In the UK, stump infection occurs during most of the year (Rishbeth, 1957; Meredith, 1959). In the Alps, but also in other areas of central Europe (Sylvestre-Guinot and Delatour, 1978), the airborne inoculum of *H. annosum s.l.*, although present starting in February at most sites, is higher in August to October, reaching a peak in September (Gonthier *et al.*, 2005). A relative peak, lower than the late summer one, is observed in late spring, around May (Gonthier *et al.*, 2005). Such a trend might be common in continental climates, characterized by hot dry weather in the summer combined with snow and low temperatures in the winter, which may restrict significant spore production to the autumn. Temporal patterns of spore deposition have been found to differ slightly among the European species of *H. annosum s.l.*, with deposition of *H. parviporum* increasing significantly during the summer, and deposition of the other two species showing only small fluctuations throughout the year (Gonthier *et al.*, 2005). There is a general lack of knowledge of the aerobiology of *H. annosum s.l.* in the Mediterranean area. However, in central Italy, Garbelotto *et al.* (2010) detected high levels of spore deposition of *H. annosum s.s.* in the winter and very low levels in the summer. Interestingly, in the same area, spores of the invasive *H. irregulare* were consistently detected both in the winter and in the summer (Garbelotto *et al.*, 2010), indicating that the risk of infection by the exotic species might be significant throughout the year. In south-eastern USA, high summer temperatures reduce sporulation (Driver and Ginns, 1969; Ross, 1973) and also result in high stump temperatures that prevent infection (Ross, 1973).

In addition to the susceptibility of stumps, the probability of stump infection depends upon the availability and abundance of viable airborne inoculum landing on stumps, which may be related to either seasonal effects or weather effects. It should be noted that the contribution of seasonal effects versus weather effects in determining inoculum availability is hard to estimate. There is good evidence that spore production is primarily affected by temperature and humidity (Redfern and Stenlid, 1998). The average minimum air temperature of a 4 week period has been identified as a suitable predictor variable for modelling *H. annosum s.l.* spore deposition (Gonthier *et al.*, 2005).

Little is known about the minimum number of spores required for successful stump infection. Punter (1970) suggested that infection is rare if the deposition rate is lower than 100 spores m^{-2} h^{-1}. Redfern *et al.* (1997) found that 50,000 m^{-2} viable spores applied to stumps significantly raised the incidence of infection compared with the natural background level. Based on the probability of two compatible homokaryon meeting on a stump, 10 spores m^{-2} h^{-1} has been identified as the threshold above which considerable risk of stump infection occurs during fellings (Möykkynen and Kontiokari, 2001).

7.4.3 Factors affecting disease incidence in a stand

In natural forest ecosystems, *H. annosum s.l.* plays a more subordinate role than in managed forests (Shaw *et al.*, 1994). In intensively managed forests, characterized by monocultures and, perhaps, establishment on former agricultural land, *H. annosum s.l.* greatly influences the outcome for the forest manager (Korhonen *et al.*, 1998b).

The disease dynamics imply that disease incidence increases with stand age (Thor *et al.*, 2005a). Depending on the host species, the management technique adopted may also play an important role. For instance, selection felling, small unit clear-felling up to 1.0 ha and introducing multistoried forest have been suggested to increase the incidence of *H. parviporum* in Todo fir plantations in Japan, as they may enhance the chance for the transfer of the fungus through root contacts (Tokuda *et al.*, 2011).

Site location, soil and site history have all been reported to affect disease incidence. While contrasting results have been obtained by relating disease incidence to site elevation, slope and aspect (Korhonen and Stenlid, 1998; Thor *et al.*, 2005a; Mattila and Nuutinen, 2007; Gonthier *et al.*, 2012a), the effects of soil type on the spread of the fungus are more clear and consistent. Damage caused by *H. annosum s.l.* is generally greater on fertile soils and on sandy soils with a low organic matter content (Korhonen and Stenlid, 1998). Furthermore, high calcium content and pH greatly influence the occurrence of the disease by reducing the activity of antagonistic fungi in the soil. Results from long-term experiments in Sitka spruce stands in the UK clearly indicate that the rate of infection is much lower on deep peat soils than on mineral soils (Redfern *et al.*, 2010). Damage caused by the pathogen is generally greater on former agricultural lands and on pasture lands than on forest soils (Korhonen and Stenlid, 1998).

Little is known of the effects of forest fires on the occurrence of *H. annosum s.l.* According to recent observations, fires significantly reduced the occurrence of *H. annosum s.s.* in mugo pine (*P. mugo* Turra) stands compared to non-burned sites (Lygis *et al.*, 2010). Nevertheless, the fungus persisted in the large majority of burned sites.

7.4.4 Estimating and predicting disease incidence

The frequency of diseased trees in a stand is an important piece of information from a planning perspective. The accurate assessment of disease incidence is problematic, especially in forest stands comprising tree species susceptible to heart rot, as the detection of decay from external symptoms is practically impossible (Vollbrecht and Agestam, 1995). This makes assessment based on remote sensing unreliable. Surveys can be carried out by the following methods: (i) using an increment borer to detect decayed wood from bore core samples – in a study by Stenlid and Wästerlund (1986), the accuracy of this method at stump height was 69–77%; (ii) visually inspecting freshly cut stumps in thinned or clear-felled areas; (iii) in mechanized thinning, programming the harvester computer to calculate the frequency of decayed trees based on assortments assigned by the operator (only the severely decayed butt logs will affect the result); and (iv) using a device for the detection of decay, e.g. Rotfinder®, which has an accuracy of more than 80% (Oliva *et al.*, 2011b). Surveys based on the observation of fruiting bodies rarely allow the accurate assessment of disease incidence. In fact, it has been reported that fruiting bodies may be rare or even absent in stands characterized by high levels of disease (Tokuda *et al.*, 2011; Gonthier *et al.*, 2012a), and are therefore unsuitable predictors of disease incidence (Gonthier *et al.*, 2012a).

Disease incidence may be predicted by using models, which have been developed for several different forest types in Europe and North America (Table 7.2). A comprehensive modelling effort was the Western Root Disease Model (WRDM), which describes the infection process of *H. annosum s.l.*, *Armillaria* spp. and *Phellinus sulphurascens* Pilát in even-aged stands with one or several tree species in western North America (McNamee *et al.*, 1989; Frankel, 1998). The EU project 'Modelling of *Heterobasidion annosum* in European forests' (MOHIEF) aimed to develop a model for *H. annosum s.l.* under European conditions that considered variations in soil, forest conditions, climate and hosts (Woodward *et al.*, 2003; Thor *et al.*, 2005b). A Fennoscandian model within MOHIEF, named Rotstand, has also been developed (Pukkala *et al.*, 2005). Oliva and Stenlid (2011) validated the Rotstand model for Swedish conditions, and concluded that Rotstand can predict decay originating from stumps resulting from thinning if correctly parameterized.

Thor *et al.* (2005a) produced models describing the probability of decay in individual trees of Norway spruce (Fig. 7.2). The model described by Thor *et al.* (2005a), as well as the model created for alpine forests (Gonthier *et al.*, 2012a), were developed to incorporate only data that it is possible to derive from stand records or easy and cheap to collect in the field. Consequently, these models could be used in strategic,

Table 7.2. Models available for the prediction of *Heterobasidion annosum s.l.* disease parameters.

Region, subregion; name of model	Pathogen(s)	Host(s) and forest type(s)	Output variables	References
Europe; MOHIEF	*H. annosum s.l.*	*Abies alba* Mill., *Picea abies* (L.) Karst., *Pinus sylvestris* L., several types of forests	Incidence of heart rot Incidence of mortality Disease development	Woodward *et al.*, 2003
Europe, Alps	*H. annosum s.l.*	*A. alba*, *Larix decidua* Mill., *P. abies*, *Pinus cembra* L., *P. sylvestris*, mixed even-aged, uneven-aged and irregular forests	Incidence of heart rot Direct financial losses	Gonthier *et al.*, 2012a
Europe, Denmark	*H. annosum s.l.*, *Armillaria mellea s.l.*	*P. abies*, even-aged forest	Incidence of heart rot	Vollbrecht and Jørgensen, 1995
Europe, Fennoscandia; Rotstand	*H. annosum s.l.*	*P. abies*, *P. sylvestris*, even-aged forests	Incidence of heart rot Incidence of mortality Disease development Decay volume	Pukkala *et al.*, 2005
Europe, Finland	*H. annosum s.l.*	*P. abies*, even-aged forests	Incidence of heart rot Disease development	Möykkynen *et al.*, 1998
Europe, Finland	*H. annosum s.l.*	*P. abies*, even-aged forests	Incidence of heart rot Disease development	Möykkynen *et al.*, 2000
Europe, Finland	*H. annosum s.l.*	*P. abies*, even-aged forests	Incidence of heart rot Disease development	Möykkynen and Miina, 2002
Europe, Finland	*H. annosum s.l.*	*P. abies*, even-aged forests	Incidence of heart rot	Mattila and Nuutinen, 2007
Europe, Sweden	*H. annosum s.l.*	*P. abies*, even-aged forests	Tree-level infection risk Disease development at the stand level	Stenlid, 1987
Europe, Sweden	*H. annosum s.l.*, *A. mellea s.l.*	*P. abies*, even-aged forest	Incidence of heart rot	Vollbrecht and Agestam, 1995
Europe, Sweden	*H. annosum s.l.*	*P. abies*, pure and mixed even-aged forests	Incidence of heart rot	Lindén and Vollbrecht, 2002
Europe, Sweden	*H. annosum s.l.*	*P. abies*, even-aged forests	Incidence of heart rot	Thor *et al.*, 2005a
Europe, UK	*H. annosum s.s.*	*Picea sitchensis* (Bong.) Carrière, even-aged forests	Disease hazard rating of site	Pratt *et al.*, 1989
Europe, UK	*H. annosum s.s.*	*P. sitchensis*, even-aged forests	Risk of disease Disease development	Redfern, 2001

Continued

Table 7.2. Continued.

Region, subregion; name of model	Pathogen(s)	Host(s) and forest type(s)	Output variables	References
North America	Any forest root disease with tree-to-tree spread	Any forest	Incidence of disease Incidence of mortality Disease development Yield losses	Peet and Hunt, 2005
North America, USA, California	*H. annosum s.l.*	*Juniperus occidentalis* Hook., *Pinus jeffreyi* Balf., *Pinus ponderosa* Lawson & C. Lawson, even-aged forests	Incidence of mortality Disease development	Preisler and Slaughter, 1997
North America, USA, Oregon	*H. occidentale*	*Abies procera* Rehder	Decay volume	Sullivan *et al.*, 2001
North America, south-eastern USA	*H. irregulare*	*Pinus elliottii* Engelm., *Pinus taeda* L., even-aged forests	Disease hazard rating of site	Baker *et al.*, 1993
North America, western USA and Canada; Western Root Disease Model (WRDM)	*H. annosum s.l.*, *Armillaria* spp., *Phellinus sulphurascens* (*P. weirii*)	Several conifers, even-aged forests	Incidence of mortality Disease development	McNamee *et al.*, 1989; Frankel, 1998
Russia	*H. annosum s.l.*	*P. sylvestris*, even-aged forests	Disease incidence	Davidenko *et al.*, 1984

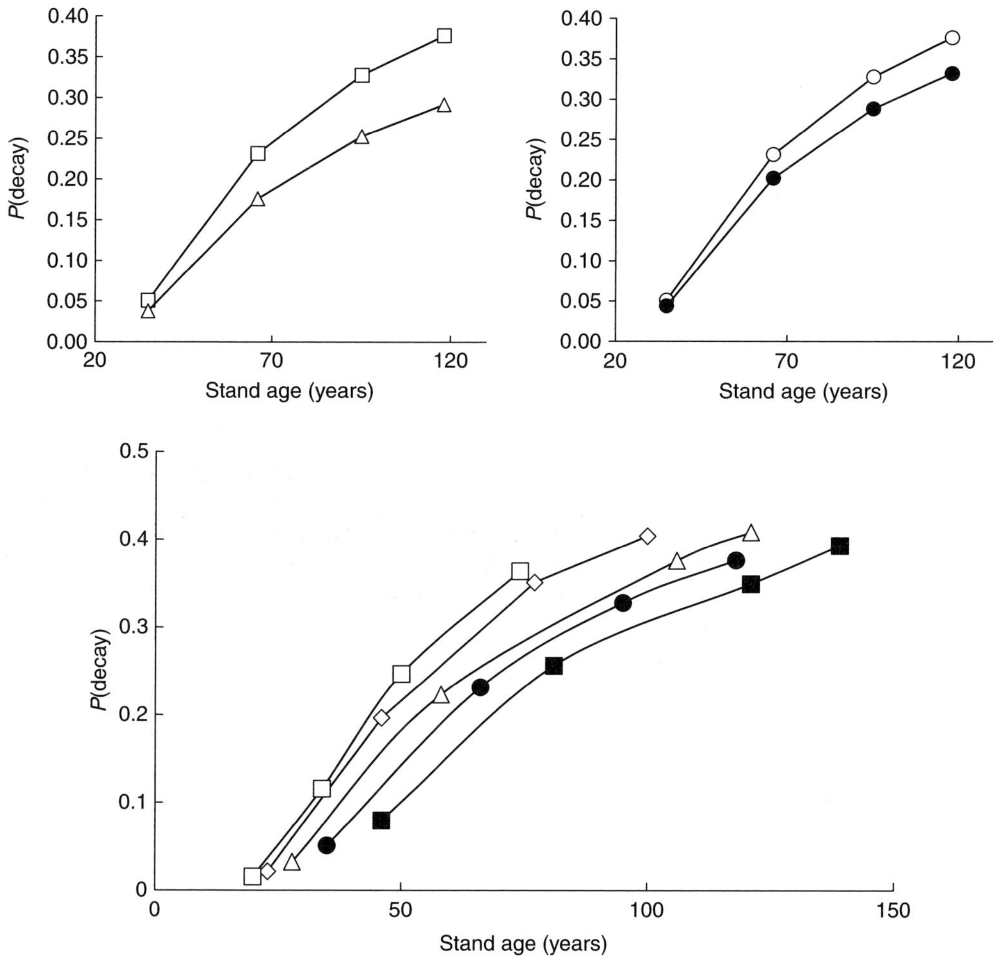

Fig. 7.2. Examples of estimates of the probability (*P*) of decay at stump height depending on the proportion of *Picea abies* (L.) Karst. in the stand (left top), the soil texture class (right top) and the site index (bottom). At 100% *P. abies* (white squares), *P*(decay) is higher than at 10% *P. abies* (white triangles). *P. abies* trees growing in sandy–silty till/sand (black circles) have lower *P*(decay) than trees in other soil texture classes (white circles). The probability of decay varies depending on site class index, SI: SI = 36, white squares; SI = 32, white diamonds; SI = 28, white triangles; SI = 24, black circles; SI = 20, black squares. Models from Thor *et al.* (2005a).

tactical and operational planning systems in practical forestry.

Usually, models are most useful in comparative analyses. By simulating different scenarios, most models can be used to determine the most appropriate management strategies to control the disease, following an integrated disease management (IDM) approach. Caution should be taken when assessing absolute levels of disease or value losses.

7.5 Management Strategies and Tactics

7.5.1 Avoidance

Logging favours the spread of disease because a great number of infection routes are created in the form of stumps and wounds. Therefore, logging operations should be allocated to periods in which the risk of spore infections is low. For instance, in northern Europe, this will imply logging during the wintertime, when temperatures are below 0°C (Brandtberg *et al.*, 1996). In the Alps, logging during late summer and autumn should be avoided (Gonthier *et al.*, 2005); in this area, although winter fellings may not be feasible at all sites, *H. annosum s.l.* could be successfully controlled by planning logging and thinning in most of the spring and in the early summer (Gonthier *et al.*, 2005; Gonthier, 2010). In order to minimize the risk of stump infection, in the south-eastern USA logging should be carried out during the summer, when temperatures can reach 40°C, thus inhibiting spore germination (Driver and Ginns, 1969; Ross, 1973).

It should be noted though that the rapid expansion of old infection centres after a thinning may significantly contribute to the occurrence of root rot in the residual stand, independently on whether thinning is performed in periods characterized by a low risk of stump infection or not (Piri and Korhonen, 2008).

Keeping down the number of thinning operations will minimize the disease on first-generation forest sites, whereas on sites with second or later rotations, the amount of disease present in the root systems will play a role in disease development. Hence, an unthinned stand with many root contacts could provide as high *H. annosum s.l.* incidence as a stand subjected to thinning (Oliva *et al.*, 2010). Based on current knowledge on the infection biology of the fungus, the avoidance of wounding may be particularly important in the prevention of infection of western hemlock and fir species.

Site characteristics can be used to estimate the hazard for annosus root and butt rot. When planning a new planting, forest managers should consider the following hazard factors: high soil pH and high soil lime content, high soil fertility, a fluctuating groundwater level, temporary drought, former use as arable land, former tree generation infected by *H. annosum s.l.* and neighbouring forests heavily infected by the pathogen (Korhonen *et al.*, 1998b). Conversely, sites characterized by wet soils, peat soils, or very dry and infertile soils are mostly safe (Korhonen *et al.*, 1998b).

7.5.2 Exclusion

H. annosum s.l. has long been regarded as an unlikely exotic pathogen because of the short lifespan of its airborne spores and its inability to freely grow in the soil. However, the introduction of *H. irregulare* genotypes from south-eastern North America into central Italy during World War II (Gonthier *et al.*, 2004) and the subsequent invasion process (Gonthier *et al.*, 2007) clearly indicate that the pathogen may become established and spread when inoculum is transported into a conducive site. In this case, it has been hypothesized that the introduction occurred through the movement of crates, pallets or other military equipment made of untreated lumber (rough timber) from infected trees (Gonthier *et al.*, 2004). The transfer of *H. annosum s.l.* species over oceans should be prevented, as well as their introduction into the southern hemisphere, e.g. into the conifer plantations of New Zealand and Australia, where the fungus does not occur. The fact that one of the most destructive tree pathogens may undergo successful introductions and, once established, threaten forests as did *H. irregulare* in Italy, suggests that quarantine regulations may be needed. The movement of woody inoculum into a site, usually in the form of fence posts, should also be prevented, especially if the site has not yet been colonized by *H. annosum s.l.*

7.5.3 Eradication

Experience and the literature clearly show that there is little chance of eradicating the fungus totally once it is established in a stand. Laflamme *et al.* (1998) reported on an eradication trial in which *H. annosum s.l.* (*H. irregulare*) was detected in Quebec (Canada) in a 50 year old red pine (*P. resinosa* Aiton) plantation. Trees in root rot centres were removed together with apparently healthy trees surrounding the centres, and the top soil and organic matter was scraped from the surface. Post treatment, infected seedlings as well as fruiting bodies of the fungus were found at the limit of the treated areas.

Removal of stumps is practised routinely after the clear-felling of heavily infected stands on alkaline soils in the UK (Greig, 1984). Complete stump removal, including careful removal of all roots, has proved to be efficient on an experimental scale in Sweden (Stenlid, 1987), although the method is not practised at all for the eradication of *H. annosum s.l.*

Stump removal is an expensive, time-consuming control method that requires the use of machines (Korhonen *et al.*, 1998b). In a study by Omdal *et al.* (2001), which aimed to compare the efficiency of different harvesting machines in terms of the amount of root biomass they left in the ground, a Vibro-Driver/Extractor was found to be the most powerful machine. All the machines tested removed at least 83% of the below-ground coarse root biomass. In general, following the treatments, residual biomass was not deemed to represent a significant source of root disease as nearly 90% of broken roots remaining were smaller than 5 cm in diameter and predicted to decompose rapidly (Omdal *et al.*, 2001). The goal to be achieved is rarely the total eradication of inoculum, but rather a significant reduction in inoculum volume.

Stump extraction for the recovery of forest energy might also contribute to reducing the risk of disease (Vasaitis *et al.*, 2008; Walmsley and Godbold, 2010), although it will not result in its eradication. In a recent study, the long-term efficacy of stump removal for *H. annosum s.l.* ranged from 20% to 53% depending on the site (Cleary *et al.*, 2013).

Environmental impacts related to stump harvesting for bioenergy have been recently reviewed (Walmsley and Godbold, 2010). These include: removal of soil organic matter inputs, increased soil erosion, increased soil compaction, depletion of soil nutrient stocks, unknown impacts on future productivity, loss of valuable habitat for many organisms and an increase in non-forest vegetation. Best practice guidelines are needed in order to minimize the environmental impacts of stump removal (Walmsley and Godbold, 2010).

While stump removal is rarely used in mountain forests or after selective fellings (Gonthier, 2010), and although it is less effective in reducing damage caused by *H. annosum s.l.* than by *P. sulphurascens* and *Armillaria ostoyae* (Romagn.) Herink (Cleary *et al.*, 2013), this treatment is deemed the only means by which adequate control can be accomplished in certain sites (Gibbs *et al.*, 2002).

7.5.4 Protection

Stump treatment

When logging occurs during the period when spore infection is expected, stump surfaces can be treated with a control agent (Rishbeth, 1957; Pratt *et al.*, 1998). To be fully effective and acceptable, materials for stump treatment need to be efficacious in a wide range of conditions, cheap, readily available, non-toxic to the user and the environment, easy and safe to handle, and approved for this use by the regulatory authorities (Pratt, 1999; Pratt and Thor, 2001). Among the many studied compounds, urea, disodium octaborate tetrahydrate (DOT) and sodium tetraborate decahydrate (borax), together with the biological control agent *Phlebiopsis gigantea* (Fr.) Jülich, have proved both effective and possible to use (Korhonen *et al.*, 1994; Nicolotti and Gonthier, 2005; Oliva *et al.*, 2010). In many cases, in northern Europe the effects are as good as winter logging, i.e. 95–100% control, although there are also reports on the poor efficacy of urea (Pratt, 1994) and *P. gigantea* (Berglund and Rönnberg, 2004).

The effectiveness of urea is dose dependent, with water solutions containing over 20% urea (w/v) being more effective than solutions with lower urea concentrations (Nicolotti and Gonthier, 2005). The failure of urea to protect stumps of Sitka spruce (Pratt, 1994) is probably due to its mode of action. Urea applied to stumps is hydrolysed by enzymes in the living tissue (Johansson *et al.*, 2002), resulting in an increase in pH, which is toxic to *H. annosum s.l.*; this hydrolysis occurs more readily in pine than in spruce stumps, where the urea may even act as a fertilizer and enhance pathogen growth (Pratt, 2003). Under Scandinavian conditions, Norway spruce is often infected in the sapwood (living tissue), whereas in Sitka spruce in the UK, many infections occur in the heartwood (dead tissue) (Redfern, 1993;

Bendz-Hellgren and Stenlid, 1998). Urea has not been registered as a pesticide in several countries and, consequently, cannot be used for stump protection in large-scale forestry; in high concentrations, such as seen in spillage near the stump during mechanized application, it may be toxic to plants (Thor *et al.*, 1997; Westlund and Nohrstedt, 2000). Borates such as DOT and borax, registered as Timbor® and Sporax® or Polybor®, respectively, have the same advantages as urea in price and handling and storage possibilities. Borates are efficient against *H. annosum s.l.* infection as long as the concentration is at least 5% (Pratt and Lloyd, 1996; Pratt *et al.*, 1998). Boron at high concentration is toxic to plants and could harm ground vegetation near treated stumps (Thor *et al.*, 1997; Westlund and Nohrstedt, 2000). The primary mode of action of borates is on the metabolism of basidiomycetes (Lloyd, 1997); they are not registered as pesticides in most European countries and, consequently, they cannot be used everywhere in practical forestry.

P. gigantea is a wood decaying fungus with saprotrophic rather than pathogenic behaviour. It is used for stump treatment in three different formulations across Europe: PG suspension in the UK, PG IBL in Poland and Rotstop® in Finland (Pratt *et al.*, 2000); of these, PG suspension and Rotstop® are most suited for mechanized stump treatment. Rotstop® is based on a Finnish isolate with a high ability for colonizing Norway spruce stumps (Korhonen *et al.*, 1994); from 2004 though, Rotstop® contains a formulation of a Swedish strain of *P. gigantea* (Berglund *et al.*, 2005). The products used elsewhere in Europe are derived from isolates growing preferably on pine stumps. Being a biological substance, *P. gigantea* is sensitive to handling, e.g. to temperature and pressure, which is of particular importance when application is mechanized. Recently, Rotstop® has also been formulated as a gel, thus enabling easier handling in mechanized applications.

The impact of stump treatments on non-target microorganisms has been investigated with contrasting results. In one study, stump treatment decreased species richness in both *P. gigantea*-treated and urea-treated stumps of Norway spruce. However, stumps treated with *P. gigantea* were colonized mainly by the same fungi as those occurring naturally in untreated stumps, whereas urea-treated stumps were colonized to a greater extent by *Ascomycota* and anamorphic fungi rather than by *Basidiomycota* (Vasiliauskas *et al.*, 2004). Based on the results of another study aimed at comparing the side effects of treatments on Norway spruce stumps, while the effects of urea on the naturally occurring mycoflora were only transient, those of *P. gigantea* and, especially, of borax were more significant and persistent (Varese *et al.*, 2003).

Across Europe, stumps are treated on more than 200,000 ha $year^{-1}$, with an average cost of 1.2 euros m^{-3} from thinning and 0.4 euros m^{-3} from the final felling (Thor, 2003). In most countries, more than 95% of the area is treated by mechanized methods, but in the UK and in Poland, respectively, there is 20% and 100% manual treatment. In Sweden, where the degree of mechanization in stump treatment is close to 100%, approximately 35,000 ha $year^{-1}$ are treated, which is estimated to be half of the need in thinnings (Thor, 2003).

In North America, different stump treatments are recommended depending on the region (Filip and Morrison, 1998). Stump treatments with borax are recommended in coastal British Columbia, the western USA, south-eastern Canada and north-eastern USA. In the pine plantations of south-eastern USA, treatment with borax or urea is recommended as a complement to summer thinning when temperatures are not high enough to prevent spore germination on stumps (Robbins, 1984). Given the peculiar infection biology of *H. occidentale* on fir trees, treatment of true fir (*Abies* spp.) stumps may be unnecessary and is not always recommended (Woodruff, 2008). Rather, it is recommended that wounding is minimized during logging operations.

The control agent is applied to the stump surface at the same moment as the rest of the tree is severed from the stump. Treatment must remain effective for at least 7 days, although if stump treatment is not applied immediately after felling, it could still have some protective value for up to 7 days later (Woods *et al.*, 2000). Standards for manual and mechanized stump treatments (Plates 6 and 7) have been published previously (Pratt and Thor, 2001). For mechanized stump treatments, the harvester requires a tank containing the liquid, a pump, hoses and valves, and a spraying device. There are two types of spraying devices: the through-the-bar sprayer, in which the liquid control agent is sprayed through a row of holes on the underside of the bar, and the under-bar sprayer, in which a spray nozzle is attached to the bar bracket, and the liquid is sprayed towards the underside of the bar (Fig. 7.3). The drilled guide bars could provide a good coverage of stumps of all sizes, all of which would be possible to handle with the harvester head, although the spillage could be high on small stumps. The special guide bars are about 50% more expensive than conventional bars. In practice, through-the-bar systems dominate in most applications. To keep costs down, it is of vital importance to decrease the machine downtime to an absolute minimum. This implies rational handling of water and stump protectant, and a rational supply of water. Recently, a mix pump enabling online mixing of Rotstop® gel and water has been developed, which is an improvement.

Trenching

Digging isolation trenches around diseased trees may prevent the vegetative spread of

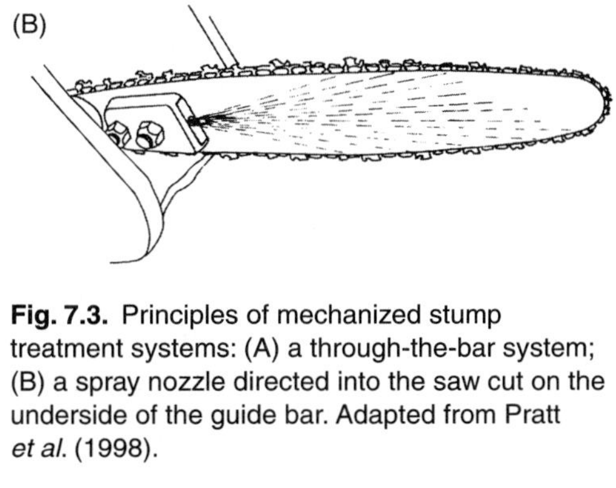

Fig. 7.3. Principles of mechanized stump treatment systems: (A) a through-the-bar system; (B) a spray nozzle directed into the saw cut on the underside of the guide bar. Adapted from Pratt *et al.* (1998).

H. annosum s.l. (Korhonen *et al.*, 1998b; Kliejunas *et al.*, 2005; Eyles *et al.*, 2008). The trench should be 40–70 cm deep. Instead of an open trench, it could be more practical to bury a plastic sheet in a vertical position into the soil (Korhonen *et al.*, 1998b). The effectiveness of trenching is largely dependent on the regular maintenance of the trenches to prevent the re-establishment of root contacts, and on the proper siting of the trenches to ensure that all infected trees are isolated (Eyles *et al.*, 2008). Trenching needs to include at least one row of completely healthy trees (Kliejunas *et al.*, 2005). However, diagnosis targeted to the root systems is normally difficult, and hence infected trees may remain undetected for long periods of time. Furthermore, observational data suggest that instead of controlling the disease, trenching could promote the spread of the fungus, because by breaking and injuring roots it can make the tree susceptible to airborne infections (Korhonen *et al.*, 1998b). Thus, trenching is a very expensive and impractical control method requiring careful field operations, and might be advisable only when pathogen inoculum is very local, e.g. in case of a new establishment of the pathogen, or when protecting high value properties. Digging 150 cm deep trenches, combined with the uprooting of all symptomatic pine trees and at least one row of healthy trees at the edges of trenches, was effective against annosus root rots and not prohibitive in terms of costs in a recreation site in southern California (Kliejunas *et al.*, 2005).

Other silvicultural methods of protection

Based on knowledge of the infection biology and epidemiology of *H. annosum s.l.*, a number of silvicultural methods aimed at protecting trees have been suggested (Korhonen *et al.*, 1998b). These include: (i) adopting the widest possible spacing among trees in new plantations and reforestations so as to decrease the probability of root contacts; (ii) delaying thinnings; (iii) reducing rotation length so that there is premature clear-felling in heavily infected stands. Decisions on the above silvicultural strategies should be adopted within an IDM programme, for which modelling might be needed. Precommercial thinning is deemed to be safe in many species (Vollbrecht *et al.*, 1995; Morrison and Johnson, 1999) as small-diameter stumps are of minor importance to the spread of the fungus, and independent of stand density and the strength of the thinning (Vollbrecht *et al.*, 1995). A diameter threshold of 10 cm under which precommercial thinnings are mostly safe has been identified and is recommended in North America (Morrison and Johnson, 1999). As shown in western hemlock stands, felling high stumps may also reduce the chance of residual tree infection in stands because the fungus is replaced by or outcompeted by other organisms (see Edmonds, Chapter 2, this volume).

7.5.5 Resistance

Inoculation assays and field studies indicate that there is a potential for improving resistance against annosus root and butt rots, at least in Norway spruce (Swedgemark and Stenlid, 1997; Karlsson and Swedjemark, 2006).

Changes in tree species composition that turn a susceptible forest into a more tolerant one are an important control method against root rot diseases, including annosus root and butt rots. Broadleaves in general are almost resistant to *H. annosum s.l.*, and certain conifers are tolerant to certain *H. annosum s.l.* species despite their susceptibility to others (Table 7.3). In heavily infected forests, planting or favouring a tree species tolerant to the *H. annosum s.l.* spp. present in the stand can be effective in controlling the disease. For instance, in areas with only *H. parviporum* present, e.g. in the northern part of Sweden or in western Finland, there is a possibility of replacing Norway spruce with Scots pine. In the Alps, the control of *H. abietinum* in heavily infected silver fir forests could be achieved either by promoting Scots pine or Norway spruce trees (Gonthier, 2010). In severely damaged subalpine Norway spruce stands, the disease can be reduced by favouring the more resistant European larch (*Larix decidua* Mill.) (Gonthier, 2010). In North America, the apparent lack of overlap in host range between *H. irregulare* and *H. occidentale* suggests that substitution of tree species may be a useful management approach (Worrall *et al.*, 2010). For example, disease caused by *H. occidentale* in stands of white fir could be managed by stand conversion to ponderosa pine (*Pinus ponderosa* Lawson & C. Lawson) (Worrall *et al.*, 2010). The species requirements on a site destined for high volume production may differ though. Furthermore, the choice of which species to plant depends on species suitability for the specific site conditions as well as on the management objectives. Often, the only practical solution is to continue to grow a susceptible species, and to accept a certain amount of disease.

It has been suggested that a rotation of a resistant tree species can clean an infected site of the pathogen inoculum, thus combining resistance management methods with eradication methods. However, although broadleaved tree species are not very susceptible to *H. annosum s.l.*, significant levels of infection by *H. annosum s.s.* were recorded on European beech (*Fagus sylvatica* L.) after pine rotations (Łakomy and Cieślak, 2008), indicating that a carry-over of inoculum may occur even when tolerant species are used. Furthermore, it has been demonstrated in Finland and Lithuania that birch (*Betula* spp.) planting does not fully prevent attack in the subsequent generation of pine (Lygis *et al.*, 2004).

Changes in tree species can be achieved in a relatively short period of time only through clear-felling followed by artificial planting. In naturally regenerated forests, shifts in tree species composition are more difficult to obtain and they should be driven by appropriate silvicultural practices. Obviously, in such conditions, for a reduction of the pathogen inoculum to occur, several decades, or even centuries, may be necessary.

Admixture of tree species normally decreases the incidence of disease by reducing the number of root contacts between susceptible trees (Korhonen *et al.*, 1998b; Lindén and Vollbrecht, 2002), although contrasting results have been reported on the effects of admixtures against the pathogen (Korhonen *et al.*, 1998b). The susceptibility of each tree species in general, and the site conditions, both of which affect the tree vitality, will set

Table 7.3. Tentative susceptibility of some forest tree species to species of *Heterobasidion annosum s.l.* based on the authors' own experience and previously published data (Delatour *et al.*, 1998; Filip and Morrison, 1998; Korhonen *et al.*, 1998a; Gonthier *et al.*, 2007, 2012a,b; Otrosina and Garbelotto, 2010; Worrall *et al.*, 2010).

	H. abietinum	*H. annosum s.s.*	*H. parviporum*	*H. irregulare*	*H. occidentale*
Eurasia					
Abies alba Mill.	+++[a]	+	0	–	–
Larix decidua Mill.	+	+++	++	–	–
Picea abies (L.) Karst.	0	++	++++	–	–
Pinus cembra L./*P. sibirica* Du Tour	0	+	++	–	–
Pinus pinea L.	0	++	0	++[b]	–
Pinus sylvestris L.	0	+++	0	–	–
North America					
A. concolor (Gord. & Glend.) Lindl. ex Hildebr.	–	–	–	0	+++
A. grandis (Douglas ex D. Don) Lindl.	–	–	–	0	+++
Calocedrus decurrens (Torr.) Florin	–	–	–	+++	+
Juniperus spp.	–	–	–	++++	++
Picea sitchensis (Bong.) Carrière	–	++++[b]	–	0	++++
Pinus elliottii Engelm.	–	–	–	++++	+
Pinus jeffreyi Balf.	–	–	–	++(+)	+
Pinus ponderosa Lawson & C. Lawson	–	–	–	+++	+
Pinus radiata D. Don	–	–	–	+	+
Pinus resinosa Aiton	–	–	–	+++	+
Pinus taeda L.	–	–	–	++++	+
Pseudotsuga menziesii (Mirb.) Franco	+[b]	+++[b]	–	+	++
Sequoiadendron giganteum (Lindl.) J. Buchholz	–	–	–	0	++
Thuja plicata Donn ex D. Don	–	–	–	0	++
Tsuga heterophylla (Raf.) Sarg.	–	–	–	0	++++
Broadleaves	0	+[c]	0	+?[b]	0

[a]Susceptibility: ++++, severely damaged; +++, moderately damaged; ++, seldom damaged; +, rarely reported; 0, not damaged (immune).
[b]Allopatric host–pathogen combination.
[c]Mostly *Betula* spp., *Fagus* spp., *Populus* spp.

the scene for the host–pathogen interaction in a forest with mixed tree species. Models indicate that the planting of Scots pines in clusters around colonized stumps markedly decreases the butt rot of Norway spruce trees at the age of 20 years and at clear-felling (Möykkynen and Pukkala, 2011).

7.5.6 Therapy

Although the infection biology of *H. annosum s.l.* was not revealed until Rishbeth's work, forest pathology research was early of the opinion that little could be done once the disease had got a grip of a stand (Hütterman and

Woodward, 1998). Consequently, prevention of the disease should be the aim rather than cure. Therapy measures against *H. annosum s.l.* are still not considered relevant.

The idea of using mycoviruses as control agents for the development of therapeutic measures has been tested by Ihrmark (2001). She found that partitivirus-like double-stranded (ds)RNA elements could not be used as biocontrol agents against *H. annosum s.l.* because they did not cause hypovirulence in this species. The results did though indicate the potential promise of deleterious mycoviruses as biocontrol agents. The search for partitiviruses able to infect pathogenic fungi such as *H. annosum s.l.* is therefore still in progress (Vainio *et al.*, 2010).

7.5.7 Integrated disease management

Systems combining different management approaches to reduce parasites are generally more effective and even cheaper than single control methods. Current knowledge on the infection biology and epidemiology of *H. annosum s.l.* has allowed to the development of IDM systems for annosus root and butt rots. IDM systems should fit and meet the requirements of the general forest management system of the area (Tainter and Baker, 1996), and are generally based on local epidemiological features (e.g. disease incidence, patterns of infection, etc.) making them useful in tactical rather than strategic disease management.

IDM may be as simple as treating stumps exclusively if they exceed a certain diameter threshold in forest sites with known disease potential. For instance, in the western USA, the recommended minimum stump treatment size is 30 cm when using a chainsaw for pine in dry pine stands and for true fir everywhere. In high elevation true fir stumps, and where mechanical shearers are used, the minimum diameter should be reduced to 20 cm (Schmitt *et al.*, 2000). In the UK, the need for stump treatment can be deduced based on a combination of soil type, climatic zone and management options (Pratt, 2003; Redfern *et al.*, 2010). An IDM system has been designed to be used in the western Italian Alps (Gonthier, 2010); this is based on stump treatment with urea at 30% concentration combined with an appropriate timing of thinning and logging operations to minimize the risk of stump infections, and with practices aimed at promoting tolerant tree species (Fig. 7.4). While planning the timing of thinnings, priority is given to the most

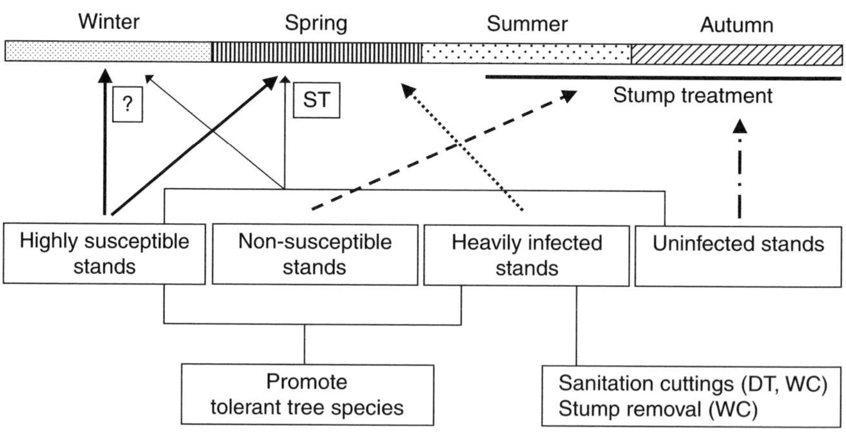

Fig. 7.4. Diagram of the integrated disease management (IDM) system developed to fight annosus root and butt rots in the western Italian Alps. Arrows indicate the appropriate timing of thinning. Stump treatment is performed with urea at 30% concentration. Key: ?, thinning where possible; DT, dead trees; ST, stump treatment; WC, where convenient.

susceptible and heavily infected stands, which are preferably thinned in spring and early summer when the risk of stump infection in the area is still limited. Less susceptible and uninfected stands are thinned in the summer or autumn. Stump treatment is necessary during summer and autumn thinnings, and is strongly recommended whenever dealing with uninfected or susceptible stands.

Models such as those described above can be used to simulate and compare the economic outcomes of management and control strategies under varying conditions. The Rotstand model has been used for analyses of optimized management of decayed stands of Norway spruce (Möykkynen and Pukkala, 2009) and mixed stands of Norway spruce and Scots pine in Finland (Möykkynen and Pukkala, 2010). On sites with old stumps that are heavily infected, a mixture of species, including less susceptible tree species, in combination with postponed thinnings and shorter rotations, improved the outcome. The economic outcomes of stump treatment have been modelled by Thor *et al.* (2006). The results suggested that stump treatment should be carried out in all logging operations of Norway spruce stands of at least site index (SI) 26 (an SI of 26 indicates that the dominant height at 100 years of age is 26 m), irrespective of the stand being established on forest soil or former agricultural soil. The same is implied for mixed stands in southern Sweden that contain at least 50% Norway spruce. In mixed stands (maximum 50% Norway spruce) in northern Sweden, or if *H. annosum s.s.* is not expected, stump treatment is not economically justified, at least not in the final felling.

As an example of IDM tactics, the model case study at the Swedish State Forest Company Sveaskog based on the methodology previously described by Thor *et al.* (2006) is briefly reported here. Model stands were defined as comprising Norway spruce stands on poor, medium and highly productive sites (SI = 20, 26 and 32). In addition, for the medium SI, the proportion of Norway spruce (the most susceptible tree species) was varied from 20% up to 100%, the other species being Scots pine. Three treatment programmes – from no protection to full stump protection – were simulated with the Rotstand model (Pukkala *et al.*, 2005). The root rot incidence increased with soil fertility (Fig. 7.5) and the proportion of susceptible tree species. The length of rotation also affected the outcome. At the time for final felling of the more productive stands, the root rot incidence was as high as about 65%. When applying a full treatment programme, the simulated root rot incidence was 5–10% at the end of the rotation. The simulations showed losses in volume production of the magnitude of 3–5% due to growth losses and mortality.

The economic outcome of the simulations implied that on the sites of SI 26 and 32 a full stump treatment programme was the most profitable option, whereas on the poorer site (SI = 20) the treatment was not profitable. The proportion of susceptible tree species in the stand also affected the profitability of stump treatment. When the proportion of the susceptible tree species exceeded 50%, stump treatment in all operations was the most profitable option, whereas at lower proportions of susceptible species, stump treatment was the most profitable option only in thinnings.

Based on the results of the simulations, literature reports and experience, a set of recommendations were stated: (i) in the southern regions (more fertile soils, greater difference in price between sound timber and decayed fuelwood, longer period with risk of spore infection, etc.) stumps of Norway spruce should be treated in all operations during temperatures exceeding 0°C. However, on sites with low site indexes, treatment is not recommended. The proportion of Norway spruce should be 50% or more. On sites with mixed forests (Norway spruce proportion of 20–50%) stump protection is recommended in thinning operations only; (ii) in the northern regions, stump treatment is recommended in all operations on fertile soils. At the assumed levels of interest rate (2.56%) and wood price, treatment was not profitable; (iii) the maximum coverage of protective agent should be striven for. Follow-up results showing 90% coverage on 90% of the stumps indicate a good result from mechanized stump treatment; (iv) for stump treatment in heavily infected stands, if the decay

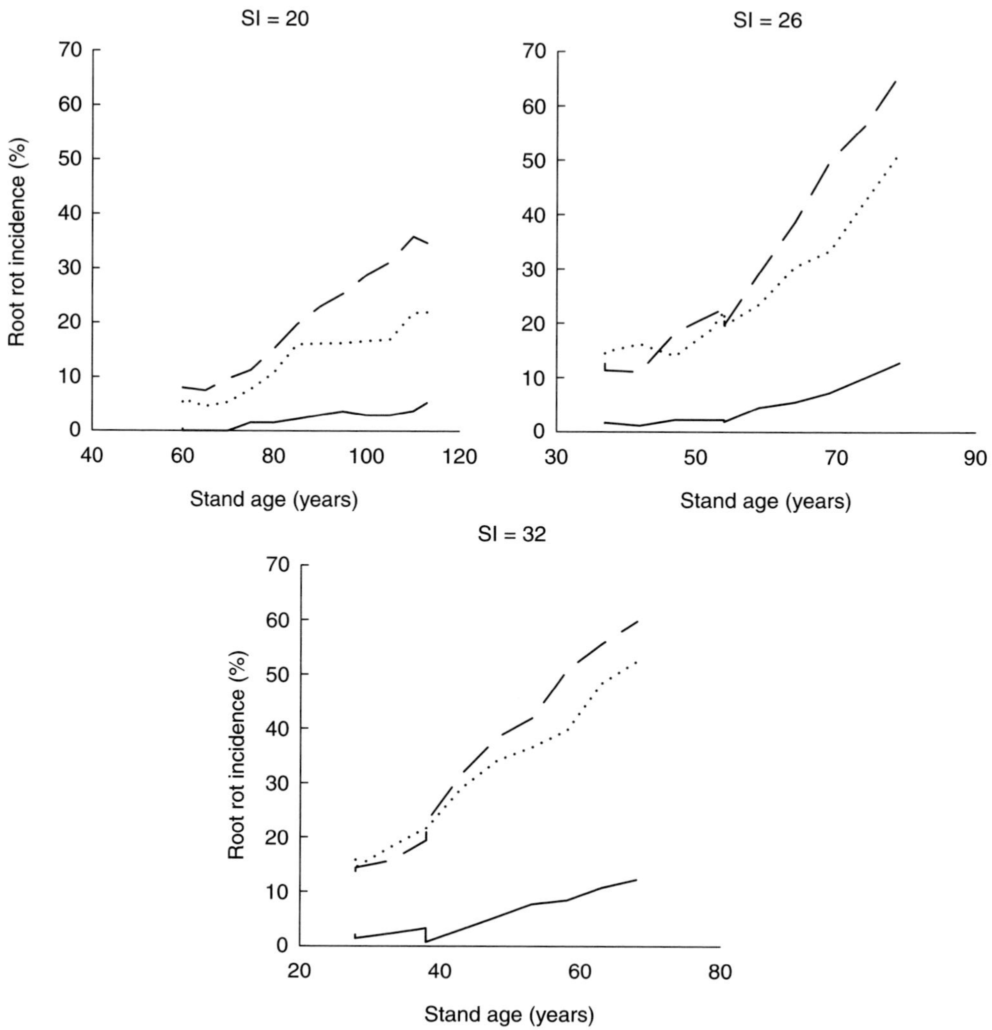

Fig. 7.5. Simulation of root rot development in three protection strategies in three Swedish stand types of site index (SI) 20, 26 and 32. The incidence of root rot increases with age, fertility of the soil and the lack of stump treatment. ———, stump treatment at all operations; ········, stump treatment in thinnings only; – – – –, no stump treatment.

incidence turns out to be very high, the treatment can be interrupted; (v) pre-cleaning of Norway spruce-dominated thinning stands with the use of brush saws should be carried out during the winter; (vi) stands of Scots pine on high risk sites (*H. annosum s.s.* present, fluctuating water table, first generation stand, sand or fine material soil, alkaline soils, etc.) should also be subject to stump treatment (Rönnberg *et al.*, 2006).

Today's knowledge on the infection biology and epidemiology of *H. annosum s.l.* is sufficient to successfully fight against annosus root and butt rots, although there is a whole system of people – consisting of policy makers, managers, researchers, engineers, trainers, forest owners, contractors and machine operators – who need to collaborate in this (Pratt and Thor, 2001). In addition, for foresters to make the right decisions, there is a need for integrated planning and decision support systems, including root rot aspects. Hence, it is important to keep *H. annosum s.l.* in mind when working out specifications for these systems.

References

Asiegbu, F.O., Adomas, A. and Stenlid, J. (2005) Conifer root and butt rot caused by *Heterobasidion annosum* (Fr.) Bref. *s.l. Molecular Plant Pathology* 6, 395–409.

Bahnweg, G., Möller, E.M., Anegg, S., Langebartels, C., Wienhaus, O. and Sandermann, H. Jr (2002) Detection of *Heterobasidion annosum s.l.* [(Fr.) Bref.] in Norway Spruce by polymerase chain reaction. *Journal of Phytopathology* 150, 382–389.

Baker, F.A., Verbyla, D.L., Hodges, C.S. Jr and Ross, E.W. (1993) Classification and regression tree analysis for assessing hazard of pine mortality caused by *Heterobasidion annosum*. *Plant Disease* 77, 136–139.

Bendz-Hellgren, M. and Stenlid, J. (1995) Long-term reduction in the diameter growth of butt rot affected Norway spruce, *Picea abies*. *Forest Ecology and Management* 74, 239–243.

Bendz-Hellgren, M. and Stenlid, J. (1998) Effects of clear-cutting, thinning, and wood moisture content on the susceptibility of Norway spruce stumps to *Heterobasidion annosum*. *Canadian Journal of Forest Research* 28, 759–765.

Bendz-Hellgren, M., Brandtberg, P.O., Johansson, M., Swedjemark, G. and Stenlid, J. (1999) Growth rate of *Heterobasidion annosum* in *Picea abies* established on forest land and arable land. *Scandinavian Journal of Forest Research* 14, 402–407.

Berglund, M. and Rönnberg, J. (2004) Effectiveness of treatment of Norway spruce stumps with *Phlebiopsis gigantea* at different rates of coverage for the control of *Heterobasidion*. *Forest Pathology* 34, 233–243.

Berglund, M., Rönnberg, J., Holmer, L. and Stenlid, J. (2005) Comparison of five strains of *Phlebiopsis gigantea* and two *Trichoderma* formulations for treatment against natural *Heterobasidion* spore infections on Norway spruce stumps. *Scandinavian Journal of Forest Research* 20, 12–17.

Brandtberg, P.O., Johansson, M. and Seeger, P. (1996) Effects of season and urea treatment on infection of stumps of *Picea abies* by *Heterobasidion annosum* in stands on former arable land. *Scandinavian Journal of Forest Research* 11, 261–268.

Capretti, P., Korhonen, K., Mugnai, L. and Romagnoli, C. (1990) An intersterility group of *Heterobasidion annosum* specialized to *Abies alba*. *European Journal of Forest Pathology* 20, 231–240.

Chase, T.E. and Ullrich, R.C. (1983) Sexuality, distribution, and dispersal of *Heterobasidion annosum* in pine plantations of Vermont. *Mycologia* 75, 825–831.

Chase, T.E. and Ullrich, R.C. (1988) *Heterobasidion annosum* root and butt rot of trees. *Advances in Plant Pathology* 6, 501–510.

Cleary, M.R., Arhipova, N., Morrison, D.J., Thomsen, I.M., Sturrock, R.N., Vasaitis, R., Gaitnieks, T. and Stenlid, J. (2013) Stump removal to control root disease in Canada and Scandinavia: a synthesis of results from long-term trials. *Forest Ecology and Management* 290, 5–14.

Dai, Y.-C., Vainio, E.J., Hantula, J., Niemelä, T. and Korhonen, K. (2003) Investigations on *Heterobasidion annosum s. lat.* in central and eastern Asia with the aid of mating tests and DNA fingerprinting. *Forest Pathology* 33, 269–286.

Dalman, K., Olson, Å. and Stenlid, J. (2010) Evolutionary history of the conifer root rot fungus *Heterobasidion annosum sensu lato*. *Molecular Ecology* 19, 4979–4993.

D'Amico, L., Motta, E., Annesi, T., Sciré, M., Luchi, N., Hantula, J., Korhonen, K. and Capretti, P. (2007) The North American P group of *Heterobasidion annosum s.l.* is widely distributed in *Pinus pinea* forests of the western coast of central Italy. *Forest Pathology* 37, 303–320.

Davidenko, M.V., Kulikov, A.I. and Storozhenko, V.G. (1984) Forecasting foci of *Heterobasidion annosum* in Scots pine plantations in Buzuluk forest. *Lesnoe Khozyaistvo* 9, 66–67.

Delatour, C., von Weissenberg, K. and Dimitri, L. (1998) Host resistance. In: Woodward, S., Stenlid, J., Karjalainen, R. and Hüttermann, A. (eds) Heterobasidion annosum, *Biology, Ecology, Impact and Control*. CAB International, Wallingford, UK, pp. 143–166.

Doğmuş-Lehtijärvi, H.T., Lehtijärvi, A. and Korhonen, K. (2006) *Heterobasidion abietinum* on *Abies* species in western Turkey. *Forest Pathology* 36, 280–286.

Driver, C.H. and Ginns, J.H. (1969) Ecology of slash pine stumps: fungal colonisation and infection by *Fomes annosus*. *Forest Science* 15, 2–10.

Eyles, A., Beadle, C., Barry, K., Francis, A., Glen, M. and Mohammed, C. (2008) Management of fungal root-rot pathogens in tropical *Acacia mangium* plantations. *Forest Pathology* 38, 332–355.

Filip, G.M. and Morrison, D.J. (1998) North America. In: Woodward, S., Stenlid, J., Karjalainen, R. and Hüttermann, A. (eds) Heterobasidion annosum, *Biology, Ecology, Impact and Control*. CAB International, Wallingford, UK, pp. 405–427.

Frankel, S. (1998) *User's Guide to the Western Root Disease Model, Version 3.0*. General Technical Report PSW-GTR-165, US Department of Agriculture, Forest Service, Pacific Southwest Station, Berkeley, California.
Garbelotto, M. (2004) Root and butt rot diseases. In: Burley, J., Evans, J. and Youngquist, J.A. (eds) *Encyclopedia of Forest Sciences: Volume 2*. Elsevier, Oxford, UK, pp. 750–758.
Garbelotto, M. and Chapela, I. (2000) First report of *Heterobasidion annosum* on the endemic *Abies hickeli* of southern Mexico. *Plant Disease* 84, 1047.
Garbelotto, M., Ratcliff, A., Bruns, T.D., Cobb, F.W. Jr and Otrosina, W.J. (1996) Use of taxon-specific competitive-priming PCR to study host specificity, hybridization, and intergroup gene flow in intersterility groups of *Heterobasidion annosum*. *Phytopathology* 86, 543–551.
Garbelotto, M.M., Lee, H.K., Slaughter, G., Popenuck, T., Cobb, F.W. Jr and Bruns, T.D. (1997) Heterokaryosis is not required for virulence of *Heterobasidion annosum*. *Mycologia* 89, 92–102.
Garbelotto, M., Otrosina, W.J., Cobb, F.W. Jr and Bruns, T.D. (1998) The European S and F intersterility groups of *Heterobasidion annosum* may represent sympatric protospecies. *Canadian Journal of Botany* 76, 397–409.
Garbelotto, M., Cobb, F.W. Jr, Bruns, T.D., Otrosina, W.J., Popenuck, T. and Slaughter, G. (1999) Genetic structure of *Heterobasidion annosum* in white fir mortality centers in California. *Phytopathology* 89, 546–554.
Garbelotto, M., Linzer, R., Nicolotti, G. and Gonthier, P. (2010) Comparing the influences of ecological and evolutionary factors on the successful invasion of a fungal forest pathogen. *Biological Invasions* 12, 943–957.
Gardes, M. and Bruns, T.D. (1993) ITS primers with enhanced specificity for basidiomycetes – application to the identification of mycorrhizae and rusts. *Molecular Ecology* 2, 113–118.
Gibbs, J.N., Greig, B.J.W. and Pratt, J.E. (2002) Fomes root rot in Thetford Forest, East Anglia: past, present and future. *Forestry* 75, 191–202.
Goheen, D.J. and Otrosina, W.J. (1998) Characteristics and consequences of root diseases in forests of western North America. In: Frankel, S. (ed.) *User's Guide to the Western Root Disease Model, Version 3.0*. General Technical Report PSW-GTR-165, US Department of Agriculture, Forest Service, Pacific Southwest Research Station, Berkeley, California, pp. 3–8.
Gonthier, P. (2010) Controlling root and butt rot diseases in alpine European forests. In: Arya, A. and Perelló, A.E. (eds) *Management of Fungal Plant Pathogens*. CAB International, Wallingford, UK, pp. 345–361.
Gonthier, P. and Garbelotto, M. (2011) Amplified fragment length polymorphism and sequence analyses reveal massive gene introgression from the European fungal pathogen *Heterobasidion annosum* into its introduced congener *H. irregulare*. *Molecular Ecology* 20, 2756–2770.
Gonthier, P., Garbelotto, M., Varese, G.C. and Nicolotti, G. (2001) Relative abundance and potential dispersal range of intersterility groups of *Heterobasidion annosum* in pure and mixed forests. *Canadian Journal of Botany* 79, 1057–1065.
Gonthier, P., Garbelotto, M. and Nicolotti, G. (2003) Swiss stone pine trees and spruce stumps represent an important habitat for *Heterobasidion* spp. in subalpine forests. *Forest Pathology* 33, 191–203.
Gonthier, P., Warner, R., Nicolotti, G., Mazzaglia, A. and Garbelotto, M. (2004) Pathogen introduction as a collateral effect of military activity. *Mycological Research* 108, 468–470.
Gonthier, P., Garbelotto, M. and Nicolotti, G. (2005) Seasonal patterns of spore deposition of *Heterobasidion* species in four forests of the western Alps. *Phytopathology* 95, 759–767.
Gonthier, P., Nicolotti, G., Linzer, R., Guglielmo, F. and Garbelotto, M. (2007) Invasion of European pine stands by a North American forest pathogen and its hybridization with a native interfertile taxon. *Molecular Ecology* 16, 1389–1400.
Gonthier, P., Brun, F., Lione, G. and Nicolotti, G. (2012a) Modelling the incidence of *Heterobasidion annosum* butt rots and related economic losses in alpine mixed naturally-regenerated forests of northern Italy. *Forest Pathology* 42, 57–68.
Gonthier, P., Lione, G., Giordano, L. and Garbelotto, M. (2012b) The American forest pathogen *Heterobasidion irregulare* colonizes unexpected habitats after its introduction in Italy. *Ecological Applications* 22, 2135–2143.
Greig, B.J.W. (1984) Management of east England pine plantations affected by *Heterobasidion annosum* root rot. *European Journal of Forest Pathology* 14, 392–397.
Greig, B.J.W. (1995) Butt-rot of Scots pine in Thetford Forest caused by *Heterobasidion annosum*: a local phenomenon. *European Journal of Forest Pathology* 25, 95–99.
Greig, B.J.W. (1998) Field recognition and diagnosis of *Heterobasidion annosum*. In: Woodward, S., Stenlid, J., Karjalainen, R. and Hüttermann, A. (eds) Heterobasidion annosum, *Biology, Ecology, Impact and Control*. CAB International, Wallingford, UK, pp. 35–41.

Greig, B.J.W. and Pratt, J.E. (1976) Some observations on the longevity of *Fomes annosus* in conifer stumps. *European Journal of Forest Pathology* 6, 250–253.

Hantula, J. and Vainio, E. (2003) Specific primers for the differentiation of *Heterobasidion annosum* (*s.str.*) and *H. parviporum* infected stumps in Northern Europe. *Silva Fennica* 37, 181–187.

Harrington, T.C., Stenlid, J. and Korhonen, K. (1998) Evolution in the genus *Heterobasidion*. In: Delatour, C., Guillaumin, J.J., Lung-Escarmant, B. and Marçais, B. (eds) *Root and Butt Rots of Forest Trees. Proceedings of the Ninth International Conference on Root and Butt Rots of Forest Trees.* Les Colloques No. 89, INRA Editions, Paris, pp. 63–74.

Hodges, C.S. (1969) Modes of infection and spread of *Fomes annosus*. *Annual Review of Phytopathology* 7, 247–266.

Hsiang, T., Edmonds, R.L. and Driver, C.H. (1989) Conidia of *Heterobasidion annosum* from *Tsuga heterophylla* forests in western Washington. *Canadian Journal of Botany* 67, 1262–1266.

Hunt, R.S., Cobb, F.W. Jr and Parmeter, J.R. (1976) *Fomes annosus* stump colonization and fungus development in the California mixed-conifer type. *Canadian Journal of Forest Research* 6, 159–165.

Hüttermann, A. and Woodward, S. (1998) Historical aspects. In: Woodward, S., Stenlid, J., Karjalainen, R. and Hüttermann, A. (eds) Heterobasidion annosum, *Biology, Ecology, Impact and Control.* CAB International, Wallingford, UK, pp. 1–25.

Ihrmark, K. (2001) Double-stranded RNA elements in the root rot fungus *Heterobasidion annosum*. Doctoral thesis, Swedish University of Agricultural Sciences, Uppsala 2001. *Acta Universitatis Agriculturae Sueciae, Silvestria 210.*

Johansson, M., Pratt, J.E. and Asiegbu, F.O. (2002) Treatment of Norway spruce and Scots pine stumps with urea against the root and butt rot fungus *Heterobasidion annosum*: possible modes of action. *Forest Ecology and Management* 157, 87–100.

Kadlec, Z., Starý, P. and Zumr, V. (1992) Field evidence for the large pine weevil, *Hylobius abietis* as a vector of *Heterobasidion annosum*. *European Journal of Forest Pathology* 22, 316–318.

Kallio, T. (1970) Aerial distribution of the root-rot fungus *Fomes annosus* (Fr.) Cooke in Finland. *Acta Forestalia Fennica* 107, 1–55.

Karlsson, B. and Swedjemark, G. (2006) Genotypic variation in natural infection frequency of *Heterobasidion* spp. in a *Picea abies* clone trial in southern Sweden. *Scandinavian Journal of Forest Research* 21, 108–114.

Kliejunas, J.T., Otrosina, W.J. and Allison, J.R. (2005) Uprooting and trenching to control annosus root disease in a developed recreation site: 12-year results. *Western Journal of Applied Forestry* 20, 154–159.

Korhonen, K. (1978) Intersterility groups of *Heterobasidion annosum*. *Communicationes Instituti Forestalis Fenniae* 94, 1–25.

Korhonen, K. and Piri, T. (1994) The main hosts and distribution of the S and P groups of *Heterobasidion annosum* in Finland. In: Johansson, M. and Stenlid, J. (eds) *Proceedings of the Eighth International Conference on Root and Butt Rots, Wik, Sweden and Haikko, Finland, August 9–16, 1993.* Swedish University of Agricultural Sciences, Uppsala, Sweden, pp. 260–267.

Korhonen, K. and Stenlid, J. (1998) Biology of *Heterobasidion annosum*. In: Woodward, S., Stenlid, J., Karjalainen, R. and Hüttermann, A. (eds) Heterobasidion annosum, *Biology, Ecology, Impact and Control.* CAB International, Wallingford, UK, pp. 43–70.

Korhonen, K., Lipponen, K., Bendz, M., Johansson, M., Ryen, I., Venn, K., Seiskari, P. and Niemi, M. (1994) Control of *Heterobasidion annosum* by stump treatment with 'Rotstop', a new commercial formulation of *Phlebiopsis gigantea*. In: Johansson, M. and Stenlid, J. (eds) *Proceedings of the Eighth International Conference on Root and Butt Rots, Wik, Sweden and Haikko, Finland, August 9–16, 1993.* Swedish University of Agricultural Sciences, Uppsala, Sweden, pp. 675–685.

Korhonen, K., Capretti, P., Karjalainen, R. and Stenlid, J. (1998a) Distribution of *Heterobasidion annosum* intersterility groups in Europe. In: Woodward, S., Stenlid, J., Karjalainen, R. and Hüttermann, A. (eds) Heterobasidion annosum, *Biology, Ecology, Impact and Control.* CAB International, Wallingford, UK, pp. 93–104.

Korhonen, K., Delatour, C., Greig, B.J.W. and Schönar, S. (1998b) Silvicultural control. In: Woodward, S., Stenlid, J., Karjalainen, R. and Hüttermann, A. (eds) Heterobasidion annosum, *Biology, Ecology, Impact and Control.* CAB International, Wallingford, UK, pp. 283–313.

Kowalski, T. and Łakomy, P. (1998) A new record of *Heterobasidion annosum* (Fr.) Bref. F group occurrence in Poland in connection with interesting mycological findings. *Phytopathologia Polonica* 15, 49–55.

Kuhlman, E.G. and Hendrix, F.F. Jr (1962) A selective medium for the isolation of *Fomes annosus*. *Phytopathology* 52, 1310–1312.

Kuhlman, E.G. and Ross, E.W. (1970) Regeneration of pine on *Fomes annosus*-infested sites in the southeastern United States. In: Hodges, C.S., Rishbeth, J. and Yde-Andersen, A. (eds) *Proceedings of the Third International Conference on* Fomes annosus, *Denmark, July–August 1968*. US Department of Agriculture, Forest Service, Washington, DC, pp. 71–76.

Laflamme, G., Blais, R. and Bussières, G. (1998) Eradication trial of annosus root rot in red pine plantations. In: Delatour, C., Guillaumin, J.J., Lung-Escarmant, B. and Marçais, B. (eds) *Root and Butt Rot of Forest Trees. Proceedings of the Ninth International Conference on Root and Butt Rot of Forest Trees.* Les Colloques No. 89, INRA Editions, Paris, pp. 375–380.

Łakomy, P. and Cieślak, R. (2008) Early infection of *Fagus sylvatica* by *Heterobasidion annosum sensu stricto. Forest Pathology* 38, 314–319.

Lindén, M. and Vollbrecht, G. (2002) Sensitivity of *Picea abies* to butt rot in pure stands and in mixed stands with *Pinus sylvestris* in southern Sweden. *Silva Fennica* 36, 767–778.

Linzer, R.E., Otrosina, W.J., Gonthier, P., Bruhn, J., Laflamme, G., Bussières, G. and Garbelotto, M. (2008) Inferences on the phylogeography of the fungal pathogen *Heterobasidion annosum*, including evidence of interspecific horizontal genetic transfer and of human-mediated, long range dispersal. *Molecular Phylogenetics and Evolution* 46, 844–862.

Lloyd, J.D. (1997) Borates and their biological applications. In: *Proceedings of the Second International Conference on Wood Protection with Diffusible Preservatives and Pesticides.* Forest Products Society, Madison, Wisconsin, pp. 45–54.

Lockman, I.B. (1993) Population structure and incidence of *Heterobasidion annosum* in grand fir and Douglas-fir on the Nez Perce National Forest, Idaho. MSc thesis, Oregon State University, Corvallis, Oregon.

Lockman, I.B., Mascheretti, S. and Garbelotto, M. (2011) A first generation hybrid discovered in *Larix lyalli* in Montana. In: *Book of Abstracts of the Thirteenth International Conference on Root and Butt Rots.* University of Florence, Florence, Italy, pp. 30–31.

Lygis, V., Vasiliauskas, R. and Stenlid, J. (2004) Planting *Betula pendula* on pine sites infested by *Heterobasidion annosum*: disease transfer, silvicultural evaluation, and community of wood-inhabiting fungi. *Canadian Journal of Forest Research* 34, 120–130.

Lygis, V., Vasiliauskas, I., Stenlid, J. and Vasaitis, R. (2010) Impact of forest fire on occurrence of *Heterobasidion s.s.* root rot and other wood-inhabiting fungi in roots of *Pinus mugo*. *Forestry* 83, 83–92.

Mattila, U. and Nuutinen, T. (2007) Assessing the incidence of butt rot in Norway spruce in southern Finland. *Silva Fennica* 41, 29–43.

McGarvey, D.J. and Croteau, R. (1995) Terpenoid metabolism. *Plant Cell* 7, 1015–1026.

McNamee, P. *et al.* (1989) Description of a multi-species root disease model developed for silvicultural planning and management in coniferous forests of western North America. In: Morrison D.J. (ed.) *Proceedings of the Seventh International Conference on Root and Butt Rots, Vernon and Victoria, British Columbia, Canada, August 9–16, 1988.* Forestry Canada, Pacific Forestry Centre, Victoria, British Columbia, Canada, pp. 320–335.

Meredith, D.S. (1959) The infection of pine stumps by *Fomes annosus* and other fungi. *Annals of Botany, New Series* 24, 63–78.

Mitchelson, K. and Korhonen, K. (1998) Diagnosis and differentiation of intersterility groups. In: Woodward, S., Stenlid, J., Karjalainen, R. and Hüttermann, A. (eds) Heterobasidion annosum, *Biology, Ecology, Impact and Control.* CAB International, Wallingford, UK, pp. 71–92.

Morrison, D.J. and Johnson, A.L.S. (1999) Incidence of *Heterobasidion annosum* in precommercial thinning stumps in coastal British Columbia. *European Journal of Forest Pathology* 29, 1–16.

Möykkynen, T. and Kontiokari, J. (2001) Spore deposition of *Heterobasidion annosum* coll. in *Picea abies* stands of North Karelia, eastern Finland. *Forest Pathology* 31, 107–114.

Möykkynen, T. and Miina, J. (2002) Optimizing the management of a butt-rotted *Picea abies* stand infected by *Heterobasidion annosum* from the previous rotation. *Scandinavian Journal of Forest Research* 17, 47–52.

Möykkynen, T. and Pukkala, T. (2009) Optimizing the management of a Norway spruce stand on a site infected by *Heterobasidion* coll. *Scandinavian Journal of Forest Research* 24, 149–159.

Möykkynen, T. and Pukkala, T. (2010) Optimizing the management of Norway spruce and Scots pine mixtures on a site infected by *Heterobasidion* coll. *Scandinavian Journal of Forest Research* 25, 127–137.

Möykkynen, T. and Pukkala, T. (2011) Effect of planting Scots pine around Norway spruce stumps on the spread of *Heterobasidion* coll. *Forest Pathology* 41, 212–220.

Möykkynen, T., von Weissenberg, K. and Pappinen, A. (1997) Estimation of dispersal gradients of S- and P-type basidiospores of *Heterobasidion annosum*. *European Journal of Forest Pathology* 27, 291–300.

Möykkynen, T., Miina, J., Pukkala, T. and von Weissenberg, K. (1998) Modelling the spread of butt rot in a *Picea abies* stand in Finland to evaluate the profitability of stump protection against *Heterobasidion annosum*. *Forest Ecology and Management* 106, 247–257.

Möykkynen, T., Miina, J. and Pukkala, T. (2000) Optimizing the management of a *Picea abies* stand under risk of butt rot. *Forest Pathology* 30, 65–76.

Nicolotti, G. and Gonthier, P. (2005) Stump treatment against *Heterobasidion* with *Phlebiopsis gigantea* and some chemicals in *Picea abies* stands in the western Alps. *Forest Pathology* 35, 365–374.

Nicolotti, G., Gonthier, P. and Guglielmo, F. (2010) Advances in detection and identification of wood rotting fungi in timber and standing trees. In: Gherbawy, Y. and Voigt, K. (eds) *Molecular Identification of Fungi*. Springer, Berlin, pp. 251–276.

Niemelä, T. and Korhonen, K. (1998) Taxonomy of the genus *Heterobasidion*. In: Woodward, S., Stenlid, J., Karjalainen, R. and Hüttermann, A. (eds) Heterobasidion annosum, *Biology, Ecology, Impact and Control*. CAB International, Wallingford, UK, pp. 27–33.

Oliva, J. and Stenlid, J. (2011) Validation of the Rotstand model for simulating *Heterobasidion annosum* root rot in *Picea abies* stands. *Forest Ecology and Management* 261, 1841–1851.

Oliva, J., Thor, M. and Stenlid, J. (2010) Long-term effects of mechanized stump treatment against *Heterobasidion annosum* root rot in *Picea abies*. *Canadian Journal of Forest Research* 40, 1020–1033.

Oliva, J., Bendz-Hellgren, M. and Stenlid, J. (2011a) Spread of *Heterobasidion annosum s.s.* and *Heterobasidion parviporum* in *Picea abies* 15 years after stump inoculation. *FEMS Microbiology Ecology* 75, 414–429.

Oliva, J., Romeralo, C. and Stenlid, J. (2011b) Accuracy of the Rotfinder instrument in detecting decay on Norway spruce (*Picea abies*) trees. *Forest Ecology and Management* 262, 1378–1386.

Omdal, D.W., Shaw, G.C. and Jacobi, W.R. (2001) Evaluation of three machines to remove armillaria- and annosum-infected stumps. *Western Journal of Applied Forestry* 16, 22–25.

Otrosina, W.J. and Cobb, F.W. Jr (1989) Biology, ecology, and epidemiology of *Heterobasidion annosum*. In: Otrosina, W.J. and Scharpf, R.F. (eds) *Proceedings of the Symposium on Research and Management of Annosus Root Disease (*Heterobasidion annosum*) in Western North America. April 18–21, 1989, Monterey, California*. General Technical Report PSW-GTR-116, US Department of Agriculture, Forest Service, Pacific Southwest Forest and Range Experiment Station, Berkeley, California, pp. 26–34.

Otrosina, W.J. and Garbelotto, M. (2010) *Heterobasidion occidentale* sp. nov. and *Heterobasidion irregulare* nom. nov.: a disposition of North American *Heterobasidion* biological species. *Fungal Biology* 114, 16–25.

Otrosina, W.J., Chase, T.E., Cobb, F.W. Jr and Korhonen, K. (1993) Population structure of *Heterobasidion annosum* from North America and Europe. *Canadian Journal of Botany* 71, 1064–1071.

Peet, F.G. and Hunt, R.S. (2005) A new simulator for the spread of forest root diseases by individual root contacts. *Forest Science* 51, 425–437.

Piri, T. (1996) The spreading of the S type of *Heterobasidion annosum* from Norway spruce stumps to the subsequent tree stand. *European Journal of Forest Pathology* 26, 193–204.

Piri, T. and Korhonen, K. (2001) Spatial distribution and persistence of *Heterobasidion parviporum* genets on a Norway spruce site. *Forest Pathology* 37, 1–8.

Piri, T. and Korhonen, K. (2008) The effect of winter thinning on the spread of *Heterobasidion parviporum* in Norway spruce stands. *Canadian Journal of Forest Research* 38, 2589–2595.

Piri, T., Korhonen, K. and Sairanen, A. (1990) Occurrence of *Heterobasidion annosum* in pure and mixed spruce stands in southern Finland. *Scandinavian Journal of Forest Research* 5, 113–125.

Pratt, J.E. (1979a) *Fomes annosus* butt rot of Sitka spruce: I. Observations on the development of butt rot in individual trees and in stands. *Forestry* 52, 11–29.

Pratt, J.E. (1979b) *Fomes annosus* butt rot of Sitka spruce: III. Losses in yield and value of timber in diseased trees and stands. *Forestry* 52, 113–127.

Pratt, J.E. (1994) Some experiments with borates and with urea to control stump infection by *H. annosum* in Britain. In: Johansson, M. and Stenlid, J. (eds) *Proceedings of the Eighth International Conference on Root and Butt Rots, Wik, Sweden and Haikko, Finland, August 9–16, 1993*. Swedish University of Agricultural Sciences, Uppsala, Sweden, pp. 662–667.

Pratt, J.E. (1999) Registration of *Phlebiopsis gigantea* as a forest biocontrol agent in the UK: recent experience. *Biocontrol Science and Technology* 9, 113–118.

Pratt, J.E. (2003) Stump treatment against *Fomes*. In: *Forest Research: Annual Report and Accounts 2001–2002*. Forest Research: An Agency of the Forestry Commission, Stationery Office, Edinburgh, UK, pp. 76–85.

Pratt, J.E. and Lloyd, J.D. (1996) The use of disodium octaborate tetrahydrate to control conifer butt rot caused by *Heterobasidion annosum*. In: *Proceedings: Crop Protection in Northern Britain 1996, University of Dundee*, pp. 207–212.

Pratt, J.E. and Thor, M. (2001) Improving mechanised stump protection against *Fomes* root rot in Europe. *Quarterly Journal of Forestry* 95, 119–127.

Pratt, J.E., Redfern, D.B. and Burnand, A.C. (1989) Modelling the spread of *Heterobasidion annosum* in Sitka spruce plantations in Britain. In: Morrison, D.J. (ed.) *Proceedings of the Seventh International Conference on Root and Butt Rots, Vernon and Victoria, British Columbia, Canada, August 9–16, 1988*. Forestry Canada, Pacific Forestry Centre, Victoria, British Columbia, Canada, pp. 308–319.

Pratt, J.E., Johansson, M. and Hüttermann, A. (1998) Chemical control of *Heterobasidion annosum*. In: Woodward, S., Stenlid, J., Karjalainen, R. and Hüttermann, A. (eds) Heterobasidion annosum, *Biology, Ecology, Impact and Control*. CAB International, Wallingford, UK, pp. 259–282.

Pratt, J.E., Niemi, M. and Sierota, Z.H. (2000) Comparison of three products based on *Phlebiopsis gigantea* for the control of *Heterobasidion annosum* in Europe. *Biocontrol Science and Technology* 10, 469–479.

Preisler, H.K. and Slaughter, G.W. (1997) A stochastic model for tree survival in stands affected by annosum root disease. *Forest Science* 43, 78–86.

Puddu, A., Luisi, N., Capretti, P. and Santini, A. (2003) Environmental factors related to damage by *Heterobasidion abietinum* in *Abies alba* forests in Southern Europe. *Forest Ecology and Management* 180, 37–44.

Pukkala, T., Möykkynen, T., Thor, M., Rönnberg, J. and Stenlid, J. (2005) Modeling infection and spread of *Heterobasidion annosum* in even-aged Fennoscandian conifer stands. *Canadian Journal of Forest Research* 35, 74–84.

Punter, D. (1970) *Fomes annosus* in eastern Canada. In: Toussoun, T.A., Bega, R.V. and Nelson, P.E. (eds) *Root Diseases and Soil-borne Pathogens*. University of California Press, Berkeley and Los Angeles, California, pp. 156–170.

Queloz, V. and Holdenrieder, O. (2005) Wie gross wird *Heterobasidion annosum s.l.*? – Ein literaturübersicht [Clonal size of *Heterobasidion annosum s.l.*? – a literature overview]. *Schweizerische Zeitschrift für Forstwesen* 156, 395–398.

Redfern, D.B. (1993) The effect of wood moisture on infection of Sitka spruce stumps by basidiospores of *Heterobasidion annosum*. *European Journal of Forest Pathology* 23, 218–235.

Redfern, D.B. (2001) Fomes decay in Sitka spruce: new perceptions of risk. In: *Forest Research: Annual Report and Accounts 1999–2000*. Forest Research: An Agency of the Forestry Commission, Stationery Office, Edinburgh, UK, pp. 17–23.

Redfern, D.B. and Stenlid, J. (1998) Spore dispersal and infection. In: Woodward, S., Stenlid, J., Karjalainen, R. and Hüttermann, A. (eds) Heterobasidion annosum, *Biology, Ecology, Impact and Control*. CAB International, Wallingford, UK, pp. 105–124.

Redfern, D.B., Gregory, S.C. and Macaskill, G.A. (1997) Inoculum concentration and the colonization of *Picea sitchensis* stumps by basidiospores of *Heterobasidion annosum*. *Scandinavian Journal of Forest Research* 12, 41–49.

Redfern, D.B., Pratt, J.E., Hendry, S.J. and Low, J.D. (2010) Development of a policy and strategy for controlling infection by *Heterobasidion annosum* in British forests: a review of supporting research. *Forestry* 83, 207–218.

Rishbeth, J. (1950) Observations on the biology of *Fomes annosus*, with particular reference to East Anglian pine plantations. I. The outbreaks of disease and ecological state of the fungus. *Annals of Botany* 14, 365–383.

Rishbeth, J. (1951a) Observations on the biology of *Fomes annosus* with particular reference to East Anglian pine plantations. II. Spore production, stump infection, and saprophytic activity in stumps. *Annals of Botany* 15, 1–21.

Rishbeth, J. (1951b) Observations on the biology of *Fomes annosus* with particular reference to East Anglian pine plantations. III. Natural and experimental infection of pines and some factors affecting severity of the disease. *Annals of Botany* 15, 221–246.

Rishbeth, J. (1957) Some further observations on *Fomes annosus* Fr. *Forestry* 30, 69–89.

Rishbeth, J. (1959) Dispersal of *Fomes annosus* Fr. and *Peniophora gigantea* (Fr.) Massee. *Transactions of the British Mycological Society* 42, 243–260.

Rizzo, D.M., Slaughter, G.W. and Parmeter, J.R. Jr (2000) Enlargement of canopy gaps associated with a fungal pathogen in Yosemite Valley, California. *Canadian Journal of Forest Research* 30, 1501–1510.

Robbins, K. (1984) *Annosus Root Rot in Eastern Conifers*. Insect and Disease Leaflet 76, US Department of Agriculture, Forest Service, Washington, DC.

Rönnberg, J. and Jørgensen, B.B. (2000) Incidence of root and butt rot in consecutive rotations of *Picea abies*. *Scandinavian Journal of Forest Research* 15, 210–217.

Rönnberg, J., Petrylaite, E., Nilsson, G. and Pratt, J.E. (2006) Two studies to assess the risk to *Pinus sylvestris* from *Heterobasidion* spp. in southern Sweden. *Scandinavian Journal of Forest Research* 21, 405–413.

Ross, E.W. (1973) Fomes annosus *in the Southeastern United States: Relation of Environmental and Biotic Factors to Stump Colonization and Losses in the Residual Stand.* Forest Service Technical Bulletin 1459, US Department of Agriculture, Forest Service, Washington, DC.

Sánchez, M.E., Luchi, N., Jímenez, J.J., De Vita, P., Sánchez, J.E., Trapero, A. and Capretti, P. (2007) An isolated population of *Heterobasidion abietinum* on *Abies pinsapo* in Spain. *Forest Pathology* 37, 348–356.

Schmitt, C.L., Parmeter, J.R. and Kliejunas, J.T. (2000) *Annosus Root Disease of Western Conifers.* Insect and Disease Leaflet 172, US Department of Agriculture, Forest Service, Washington, DC.

Schönar, S. (1975) Zur Besiedlung frischer Stubben in Fichten-Erstaufforstungen durch *Fomes annosus* [The colonization of fresh stumps in spruce afforestation by *Fomes annosus*]. *Allgemeine Forst- und Jagdzeitung* 146, 177–179.

Shaw, D., Edmonds, R., Littke, W., Browning, J., Russell, K. and Driver, C. (1994) Influence of forest management on annosus root disease in coastal western hemlock, Washington State, USA. In: Johansson, M. and Stenlid, J. (eds) *Proceedings of the Eighth International Conference on Root and Butt Rots, Wik, Sweden and Haikko, Finland, August 9–16, 1993.* Swedish University of Agricultural Sciences, Uppsala, Sweden, pp. 646–655.

Stalpers, J.A. (1978) *Identification of Wood-Inhabiting* Aphyllophorales *in Pure Culture.* Studies in Mycology 16, Centraalbureau Voor Schimmelcultures, Baarn, The Netherlands.

Stenlid, J. (1985) Population structure of *Heterobasidion annosum* as determined by somatic incompatibility, sexual incompatibility and isoenzyme patterns. *Canadian Journal of Botany* 63, 2268–2273.

Stenlid, J. (1987) Controlling and predicting the spread of *Heterobasidion annosum* from infected stumps and trees of *Picea abies*. *Scandinavian Journal of Forest Research* 2, 187–198.

Stenlid, J. (1994) Regional differentiation in *Heterobasidion annosum*. In: Johansson, M. and Stenlid, J. (eds) *Proceedings of the Eighth International Conference on Root and Butt Rots, Wik, Sweden and Haikko, Finland, August 9–16, 1993.* Swedish University of Agricultural Sciences, Uppsala, Sweden, pp. 243–248.

Stenlid, J. and Redfern, D.B. (1998) Spread within the tree and stand. In: Woodward, S., Stenlid, J., Karjalainen, R. and Hüttermann, A. (eds) Heterobasidion annosum, *Biology, Ecology, Impact and Control.* CAB International, Wallingford, UK, pp. 125–141.

Stenlid, J. and Wästerlund, I. (1986) Estimating the frequency of stem rot in *Picea abies* using an increment borer. *Scandinavian Journal of Forest Research* 1, 303–308.

Sullivan, K.F., Filip, G.M., Arena, J.V., Fitzgerald, S.A. and Tesch, S.D. (2001) Incidence of infection and decay caused by *Heterobasidion annosum* in managed noble fir on the Warm Springs Indian Reservation, Oregon. *Western Journal of Applied Forestry* 16, 106–113.

Swedjemark, G. and Stenlid, J. (1993) Population dynamics of the root rot fungus *Heterobasidion annosum* following thinning of *Picea abies*. *Oikos* 66, 247–254.

Swedjemark, G. and Stenlid, J. (1997) Between-tree and between-isolate variation for growth of S-group *Heterobasidion annosum* in sapwood of *Picea abies* cuttings. *Canadian Journal of Forest Research* 27, 711–715.

Sylvestre-Guinot, G. and Delatour, G. (1978) Recherches sur les variations saisonnières de l'inoculum aérien du *Fomes annosus* (Fr.) Cooke dans l'est de la France [Research on seasonal variations of airborne inoculum of *Fomes annosus* (Fr.) Cooke in the east of France]. *Annales des Sciences Forestières* 35, 151–163.

Tainter, F.H. and Baker, F.A. (1996) *Principles of Forest Pathology.* John Wiley, Hoboken, New Jersey.

Thor, M. (2003) Operational stump treatment against *Heterobasidion annosum* in European forestry: current situation. In: Laflamme, G., Bérubé, J. and Bussières, G. (eds) *Proceedings of the Tenth International Conference on Root and Butt Rot of Forest Trees, Québec City, Canada, September 16–22, 2001.* Information Report LAU-X-126, Canadian Forest Service, Laurentian Forestry Centre, Sainte-Foy, Quebec, Canada, pp. 170–175.

Thor, M., Nohrstedt, H.-Ö. and Weslien, J. (1997) *Possible Environmental Effects of Stump Treatment with Borate,* Phlebiopsis gigantea *and Urea: a Literature Study.* Skogforsk Report 1, Forestry Research Institute of Sweden, Uppsala, Sweden.

Thor, M., Ståhl, G. and Stenlid, J. (2005a) Modelling root rot incidence in Sweden using tree, site and stand variables. *Scandinavian Journal of Forest Research* 20, 165–176.

Thor, M., Möykkynen, T., Pratt, J.E., Pukkala, T., Rönnberg, J., Shaw, C.G. III, Stenlid, J., Ståhl, G. and Woodward, S. (2005b) Modeling infection and spread of *Heterobasidion annosum* in coniferous forests in Europe. In: Bevers, M. and Barrett, T.M. (eds) *Systems Analysis in Forest Resources: Proceedings of the 2003 Symposium.* General Technical Report PNW-GTR-656, US Department of Agriculture, Forest Service, Pacific Northwest Research Station, Portland, Oregon, pp. 105–111.

Thor, M., Arlinger, J.D. and Stenlid, J. (2006) *Heterobasidion annosum* root rot in *Picea abies*: modelling economic outcomes of stump treatment in Scandinavian coniferous forests. *Scandinavian Journal of Forest Research* 21, 414–423.

Tokuda, S., Ota, Y., Hattori, T., Shoda-Kagaya, E. and Sotome, K. (2011) The distribution of closely related large genets of *Heterobasidion parviporum* in a Todo fir (*Abies sachalinensis*) stand in Hokkaido, Japan. *Forest Pathology* 41, 482–492.

Vainio, E.J., Korhonen, K., Tuomivirta, T. and Hantula, J. (2010) A novel putative partitivirus of the saprotrophic fungus *Heterobasidion ecrustosum* infects pathogenic species of the *Heterobasidion annosum* complex. *Fungal Biology* 114, 955–965.

Varese, G.C., Gonthier, P. and Nicolotti, G. (2003) Impact of biological and chemical treatments against *Heterobasidion annosum* on non-target micro-organisms. In: Laflamme, G., Bérubé, J. and Bussières, G. (eds) *Proceedings of the Tenth International Conference on Root and Butt Rot of Forest Trees, Québec City, Canada, September 16–22, 2001*. Information Report LAU-X-126, Canadian Forest Service, Laurentian Forestry Centre, Sainte-Foy, Quebec, Canada, pp. 145–154.

Vasaitis, R., Stenlid, J., Thomsen, I.M., Barklund, P. and Dahlberg, A. (2008) Stump removal to control root rot in forest stands. A literature study. *Silva Fennica* 42, 457–483.

Vasiliauskas, R., Lygis, V., Thor, M. and Stenlid, J. (2004) Impact of biological ('Rotstop') and chemical (urea) treatments on fungal community structure in freshly cut *Picea abies* stumps. *Biological Control* 31, 405–413.

Vollbrecht, G. and Agestam, E. (1995) Modelling incidence of root rot in *Picea abies* plantations in southern Sweden. *Scandinavian Journal of Forest Research* 10, 74–81.

Vollbrecht, G. and Jørgensen, B.B. (1995) Modelling the incidence of butt rot in plantations of *Picea abies* in Denmark. *Canadian Journal of Forest Research* 25, 1887–1896.

Vollbrecht, G. and Stenlid, J. (1999) Transfer of the P-type of *Heterobasidion annosum* from old-growth stumps of *Picea abies* to *Picea abies* and *Larix × eurolepis*. *European Journal of Forest Pathology* 29, 153–159.

Vollbrecht, G., Gemmel, P. and Pettersson, N. (1995) The effect of pre-commercial thinning on the incidence of *Heterobasidion annosum* in planted *Picea abies*. *Scandinavian Journal of Forest Research* 10, 37–41.

Walmsley, J.D. and Godbold, L. (2010) Stump harvesting for bioenergy: a review of the environmental impacts. *Forestry* 83, 17–38.

Westlund, M. and Nohrstedt, H.-Ö. (2000) Effects of stump-treatment substances for root-rot control on ground vegetation and soil properties in a *Picea abies* forest in Sweden. *Scandinavian Journal of Forest Research* 15, 550–560.

White, T.J., Bruns, T.D., Lee, S. and Taylor, J.W. (1990) Amplification and direct sequencing of fungal ribosomal RNA genes for phylogenetics. In: Innis, M.A., Gelfand, D.H., Sninsky, J.J. and White, T.J. (eds) *PCR Protocols: A Guide to Methods and Applications*. Academic Press, San Diego, California, pp. 315–322.

Woodruff, W.C. (2008) Sporax®-ing conifer stumps in the North Moonlight Salvage Sale. FHP (Forest Health Protection) Evaluation No. NE08-06, US Department of Agriculture, Forest Service, Forest Health Protection, Pacific Southwest Region, Vallejo, California. Available at: https://fs.usda.gov/Internet/FSE_DOCUMENTS/fsbdev3_046399.pdf (accessed 27 November 2012).

Woods, C.M., Woodward, S. and Redfern, D.B. (2000) Receptivity of *Picea sitchensis* stumps to infection by *Heterobasidion annosum* basidiospores. *Forestry* 73, 457–465.

Woodward, S., Stenlid, J., Karjalainen, R. and Hüttermann, A. (1998) Preface. In: Woodward, S., Stenlid, J., Karjalainen, R. and Hüttermann, A. (eds) Heterobasidion annosum, *Biology, Ecology, Impact and Control*. CAB International, Wallingford, UK, pp. xi–xii.

Woodward, S., Pratt, J.E., Pukkala, T., Spanos, K.A., Nicolotti, G., Tomiczek, C., Stenlid, J. Marçais, B. and Łakomy, P. (2003) MOHIEF: modelling of *Heterobasidion annosum* in European forests, a EU-funded research program. In: Laflamme, G., Bérubé, J. and Bussières, G. (eds) *Proceedings of the Tenth International Conference on Root and Butt Rot of Forest Trees, Québec City, Canada, September 16–22, 2001*. Information Report LAU-X-126, Canadian Forest Service, Laurentian Forestry Centre, Sainte-Foy, Quebec, Canada, pp. 423–427.

Worrall, J.J., Harrington, T.C., Blodgett, J.T., Conklin, D.A. and Fairweather, M.L. (2010) *Heterobasidion annosum* and *H. parviporum* in the southern Rocky Mountains and adjoining states. *Plant Disease* 94, 115–118.

8 Armillaria Root Rots

Jean-Jacques Guillaumin[1]* and Philippe Legrand[2]
[1]*Formerly of INRA (National Institute for Agricultural Research), France;*
[2]*Formerly of Forest Health Service, France*

8.1 Pathogens, Significance and Distribution

Armillaria is definitely the most studied fungus in the world. Several dozens of papers are now published every year that cover all aspects and facts on this genus of the *Basidiomycota*, which causes root disease of woody plants (forest trees, amenity trees, orchard trees, vines, etc.). This chapter will summarize the knowledge of the fungus in forestry and emphasize interesting facts for the forest manager; many more details are available in literature (Shaw and Kile, 1991; Fox, 2000; Guillaumin *et al.*, 2005a).

The fruiting bodies of the *Armillaria* spp. are generally cespitose, rarely solitary. The pileus is fleshy, its colour is highly variable from yellow to brown, tawny or olivaceous, according to the species and environment. It generally carries squamules darker than the ground colour, and these remain at the surface of the cap at maturity. The stipe is central, fibrous to fleshy, often becoming hollow in old samples. With the exception of *A. tabescens* (Scop.) Emel and *A. ectypa* (Fr.) Lamoure, the stipe carries an annulus or at least the remnants of the general veil in the form of an arachnoid coating (Fig. 8.1). The lamellae are adnexed, sub-decurrent or decurrent. The trama is bilateral, the basidiospores are ellipsoid, hyaline, inamyloid and weakly cyanophilic, smooth or showing inconspicuous crests that are more visible under the scanning electronic microscope (Bennell *et al.*, 1985). Masses of spores appear white to cream coloured. Two other criteria of the vegetative mycelium characterize the genus better than those that describe the fruiting body: (i) all the *Armillaria* spp. can initiate rhizomorphs (linear aggregated organs), at least in pure culture, and (ii) in all species the secondary mycelium is diploid (and not dikaryotic).

8.1.1 Main species of *Armillaria*

Until the 1970s, all the armillarias with an annulus on the stipe were included in a sole species: *A. mellea* (Vahl ex Fr.) P. Kumm. *s.l.* This vast and polymorphic taxon was regarded as widespread over all continents. However, based on sexual incompatibility, morphological differences and immunological, biochemical and molecular traits, some mycologists recognized several species in this broad complex. The use of sexual

* E-mail: jj.guillaumin@gmail.com

Fig. 8.1. Fruiting bodies of *Armillaria mellea* (Vahl ex Fr.) P. Kumm. Credits: Jean-Jacques Guillaumin.

compatibility (leading to the concept of 'biological species') was a decisive step in this process (Korhonen, 1978).

Only four species are common to all the areas of the northern hemisphere, and these are also the most important *Armillaria* spp. in their impact on woody vegetation. They differ from each other in their morphology, their morphogenesis (modalities of initiation of rhizomorphs and fruiting bodies), their ecology and also in their biological behaviour and pathogenicity (Guillaumin *et al.*, 1993; Guillaumin and Legrand, 2005).

A. tabescens has no ring on the stipe. It is a thermophilic species, common in southern Europe, the south-eastern USA, southern Japan and China. In natural ecosystems, *A. tabescens* generally behaves as a saprophyte which mostly colonizes the dead stumps of the Mediterranean oaks (*Quercus* spp.). It is rarely pathogenic in forests, but can cause heavy damage in orchards and amenity plantations, mostly in the south-eastern USA (Florida, Georgia, South Carolina). Its impact is lower in Europe and Asia.

A. mellea s.s. is a fairly thermophilic taxon, although it is a little less so than *A. tabescens* (its optimum growth temperature is close to 25°C). Molecular studies show that this species is highly variable. It consists of two subspecies: *A. mellea* subsp. *mellea*, found in Europe and North America, which is heterothallic and tetrapolar, and *A. mellea* subsp. *nipponica*, found in Asia (and introduced by humans into Africa), which is homothallic. The pathogenicity of *A. mellea*, whatever the subspecies, is generally high. On all the continents of the northern hemisphere, it is generally the species responsible for damage to fruit trees (especially *Prunus* and *Citrus* spp.) and amenity deciduous trees. It also plays an important role in forests, both on deciduous trees and resinous trees of the family *Cupressaceae*. It can also attack the *Pinaceae* in areas with a Mediterranean climate (California and the Balkans).

A. ostoyae (Romagn.) Herink is markedly less thermophilic than *A. mellea* and *A. tabescens*, with an optimum growth temperature of 22°C. This species is well characterized by the morphology of its fruiting bodies, which have dense, dark scales on the cap. It is fundamentally linked to conifers (*Pinaceae*), on which it behaves both as a saprophyte decaying the dead stumps and roots, and as a parasite causing high damage in coniferous forests and plantations in the northern hemisphere. Although the list of the hosts from which *A. ostoyae* has been reported is shorter than that for *A. mellea*, *A. ostoyae* is probably the *Armillaria* species that has been the subject of the highest number of studies since the species were separated, in 1978. *A. ostoyae* can sometimes also infect deciduous trees or shrubs, generally when they are mixed with conifers. The subterranean rhizomorphs of the species show a characteristic dichotomous pattern of branching (Morrison, 1982).

A. gallica Marxm. & Romagn. is a species that is essentially saprophytic and plays a major ecological role as a stump decomposer in natural deciduous and mixed forests. This species is characterized by its subterranean rhizomorphs, which are highly differentiated, long living, and are initiated easily and early from dead wood

(Guillaumin and Legrand, 2005); the brown or black rhizomorphs that are so common in the litter of the deciduous forests of Europe generally belong to this species (or to *A. cepistipes* Velen. at high altitudes or latitudes). These organs have a high rate of branching and their numerous branches play a foraging role in the soil, in search of new woody food bases, either dead or alive; in contrast to *A. ostoyae*, these branches are initiated according to a monopodial pattern (Morrison, 1982). *A. gallica* is clearly less pathogenic than *A. mellea* or *A. ostoyae*.

Seven *Armillaria* spp. have been reported in Australasia, all different from those of the northern hemisphere. Of these, six species have received binomial names. Only two are common to Australia and New Zealand. In New Zealand, the most important species are *A. novae-zelandiae* (G. Stev.) Boesew. and *A. limonea* (G. Stev.) Boesew. (Shaw *et al.*, 1981). The former species also exists in Australia and has also been reported from Indonesia and Malaya. In Australia, the most widespread and aggressive species is *A. luteobubalina* Watling & Kile, a significant pathogen on *Eucalyptus* spp. in eastern Australia and the sole *Armillaria* species reported from Western Australia (Shearer *et al.*, 1997). Both *A. novae-zelandiae* and *A. luteobubalina* are present in Argentina, Chile and southern Brazil. Other species described by morphologists in the 1950s and the 1960s could in fact be synonymous with either of these two species, or represent taxa that are very close to them (Pildain *et al.*, 2009).

In tropical Africa (south of the Sahara) the taxa reported are quite different. There are special cases: *A. mellea* subsp. *nipponica* has probably been introduced from Asia into east Africa by colonizers (Ota *et al.*, 2000). The taxon 'SIG III' from the highlands of Kenya appears to be unclassifiable (Mohammed, 1994). All the other armillarias from tropical Africa belong to the so-called '*heimii* group'. This group shows a series of common traits: (i) tetrapolar heterothallism is absent (some populations are homothallic, others are heterothallic bipolar; Abomo-Ndongo, 1997); (ii) the fruiting bodies are rare in natural conditions; (iii) the ring on the stipe of the fruiting bodies is provisional and sometimes absent; (iv) from the molecular point of view, the ribosomal gene 5S is in inverse position in comparison with all other armillarias. Structuring this group is difficult, particularly because of homothallism. Abomo-Ndongo (1997) considered it to be a sole species, *A. heimii* Pegler, divided into five subgroups. In contrast, the studies carried out in east and South Africa distinguish at least two distinct species: *A. heimii* and *A. fuscipes* Petch. Only the latter taxon is reported as a pathogen on American pines (*Pinus* spp.) planted in East and South Africa (Coetzee *et al.*, 2000; Gezahgne *et al.*, 2003).

8.1.2 Significance and distribution of *Armillaria* spp.

When establishing a rough typology of the *Armillaria* occurrences throughout the world, a double distinction can be made: (i) between the sub-natural forests (of natural origin, but more or less managed by man and with a low level of artificialization) and the tree plantations (involving artificial interventions and intensive management); and (ii) between deciduous and resinous forests. Deciduous trees and conifers in the northern hemisphere are not attacked by the same *Armillaria* spp. In addition, whatever the species, damage is generally higher on conifers.

Australia is the only part of the world where important damage due to *Armillaria* has been reported in natural deciduous forests. The species responsible is *A. luteobubalina*. In the eastern part of the continent, the hosts concerned are different species of *Eucalyptus* (*E. obliqua* L'Her., *E. regnans* F. Muell. and others; Kile, 1980; Kellas *et al.*, 1987). In Western Australia, the disease has been reported from various natural forest ecosystems. The fungus mainly attacks *E. wandoo* Blakely, but it also attacks trees belonging to families *Myrtaceae*, *Proteaceae* and *Mimosaceae*. The fungus plays an important role in the remodelling of the natural ecosystems (Schearer *et al.*, 1997).

In the northern hemisphere, in sub-natural deciduous forests, *Armillaria* generally plays a minor role. It has been reported mostly on oaks: for instance on the red oaks in Ozark Mountains in the highland south-central region of the USA (Bruhn *et al.*, 2000), on English oak (*Quercus robur* L.) in France (Guillaumin *et al.*, 1985) and on Mediterranean oaks in Italy (Sicoli *et al.*, 1998). The fungal species involved are *A. mellea* and *A. gallica*, and sometimes *A. tabescens*. These fungi generally are just an element of complex declines in which drought and insects also play a major role. According to the scheme of Manion (1981), armillarias appear sometimes as a predisposing factor (through numerous early infections of the root systems), and sometimes as a contributing factor giving the deathblow to enfeebled trees.

In the mountains of western North America, *A. ostoyae*, in association with the two other root rot agents *Phellinus sulphurascens* Pilát and *Heterobasidion annosum* (Fr.) Bref. *s.l.*, plays a major role in natural or sub-natural coniferous and mixed forests. In the Rocky Mountains of the USA (Montana, Idaho and Wyoming), *A. ostoyae* is, together with fire and bark beetles, one of the factors that governs the natural evolution of the forest and especially the balance between the tree species. In these multi-species stands, the most susceptible species are the grand fir (*Abies grandis* (Douglas ex D. Don) Lindl.), the subalpine fir (*Abies lasiocarpa* (Hook.) Nutt.) and the Douglas-fir (*Pseudotsuga menziesii* (Mirb.) Franco). Hagle (2008) showed that the disease appears into two major patterns: (i) in the form of distinct patches, almost circular, corresponding to the extension of one clone, where the disease advances centrifugally, while some regeneration reoccupies the centre of the patch; and (ii) as a dispersed mortality forming a continuous coverage of the sites and probably resulting from the merging of multiple clones. Hagle (2008) also points out two additional, less important, syndromes due to *A. ostoyae*: mortalities of seedlings and saplings, and butt rot. *A. ostoyae* is also responsible for heavy damage in interior British Columbia where the most risky host species are Douglas-fir and lodgepole pine (*P. contorta* Douglas ex Loudon) (Morrison and Mallett, 1996). In eastern Oregon and Washington, ponderosa pine (*P. ponderosa* Lawson & C. Lawson) is one of the most susceptible species, with grand fir and white fir (*A. concolor* (Gord. & Glend.) Lindl. ex Hildebr.) (Filip and Goheen, 1984). The impact of the fungus is generally lower in the coastal Pacific coniferous forests (western British Columbia, Oregon and Washington).

In Europe, Armillaria root disease plays some role in the resinous forests of Scandinavia and Poland, mostly on Scots pine (*P. sylvestris* L.). The disease has also been reported from the south of the continent: mostly on Greek fir (*A. cephalonica* London) and Bulgarian fir (*A. borisii-regis* Mattf.) in Greece and Albania (Tsopelas, 1999), on silver fir (*A. alba* Mill.) in the Italian Appenines, on Swiss stone pine (*P. cembra* L.) in the Alps (Anselmi and Lanata, 1989), and on mountain pine (*P. uncinata* Mill. ex Mirb.) in the French Pyrenees (Durrieu *et al.*, 1985). In Europe, as in North America, the parasite involved is generally *A. ostoyae*, although this is replaced by *A. mellea s.s.* on *Abies* spp. in the Balkans.

The role of *Armillaria* is also of importance in deciduous plantations. It seems to be crucial in plantations of quaking aspen (*Populus tremuloides* Michx.) in Wisconsin (Stanosz and Patton, 1987). In the UK, plantations of English oak and European beech (*Fagus sylvatica* L.) on ancient farm soils have been reported to be infected by *A. gallica*, and also by *A. ostoyae* (Rishbeth, 1978). In France, plantations of 'precious hardwoods', especially walnut (*Juglans regia* L.) and sweet cherry (*Prunus avium* (L.) L.) are sometimes endangered by *A. mellea*. In south-western France, *A. tabescens* is the agent of decline of English oak in areas liable to flooding. The same parasite in the same region was one of the factors that put in check attempts to introduce eucalyptus culture in the 1980s (Legrand and Lung-Escarmant, 2005).

In Africa, the frequent and heavy attacks of *Armillaria* spp. on industrial cultures (coffee, tea, cocoa, hevea, cassava, etc.) are not in the scope of this forestry-oriented account, but damage by *Armillaria* spp. is also frequent in plantations of teak (*Tectona grandis* L. f.). In Kenya, scattered plantations of hardwoods established to provide wood for cooking and building are sometimes endangered by the presence of an *Armillaria* inoculum issuing from former indigenous trees (Mwangi *et al.*, 1994). In Australia, artificial plantations of mountain ash (*E. regnans*) have been infected by *A. luteobubalina* (Podger *et al.*, 1978).

Forestry in the second half of the 20th century was characterized by frequent planting of coniferous species, often of exotic origin, within the framework of an artificial and intensive forest management regime. Most of the time, these plantations took place following the clear-cutting (clear-felling) of more or less natural forests. Under these conditions, the damage provoked by root rot fungi, and particularly by *Armillaria* spp., was generally more severe than in forests that were closer to natural management. The damage, often already significant on the first-rotation stand, generally worsens during the second rotation owing to the increase of inoculum.

The pine species originating from southern North America, such as slash pine (*P. elliottii* Engelm.), Monterey pine (*P. radiata* D. Don), loblolly pine (*P. taeda* L.) and Mexican weeping pine (*P. patula* Schiede ex Schltdl. & Cham.) were abundantly planted throughout the world, wherever the climate was mild, and they were often heavily attacked by the native *Armillaria* spp. Plantations of Monterey pine in New Zealand were laid to waste by *A. novae-zelandiae* and *A. limonea* (Hood, 1989). Slash pine in Brazil was attacked by local species close to *A. luteobubalina* (Auer *et al.*, 2003). In east and South Africa, plantations of slash pine and ocote chino (*Pinus oocarpa* Schiede ex Schltdl.) and, to a lesser degree, of Mexican weeping pine, Monterey pine and loblolly pine suffer from infection by *A. fuscipes* (Coetzee *et al.*, 2000; Gezahgne *et al.*, 2003).

In Europe, a number of coniferous species were introduced from north-western North America. Among them, grand fir appeared to be particularly susceptible to the European populations of *A. ostoyae*. In contrast, Douglas-fir, also from the same area, showed a high level of tolerance to *Armillaria* in Europe, even though it is regarded as one of the most susceptible conifer species in its native area (Legrand and Lung-Escarmant, 2005). In northern and eastern Europe, artificial plantations of indigenous species such as Scots pine (Finland, Poland) or Norway spruce (*Picea abies* (L.) Karst.) (Poland, southern Germany) are frequently attacked (Hintikka, 1974; Schönhar, 1994; Lech and Sierota, 2007). In south-western France, the forest of 'Landes de Gascogne' is a monoculture of maritime pine (*P. pinaster* Aiton) of over 1 million ha. Mostly of artificial origin, this forest is managed with intensive silvicultural methods (fertilization, weeding, etc.). Under these conditions, *A. ostoyae* causes either scattered elimination of seedlings and saplings or, in older stands, round patches of mortality (Lung-Escarmant and Taris, 1984).

In western North America, although subnatural forests are the norm, conifer plantations also exist, and they do not escape attack by *A. ostoyae* (Morrison, 1981). In the eastern part of the continent, coniferous plantations are also infected by *A. ostoyae*, particularly in Ontario (Bruhn *et al.*, 1996; Whitney *et al.*, 2002) and Newfoundland (Singh and Richardson, 1973). This is also the case in Asia for plantations of species such as Chinese pinenut (*P. koraiensis* Siebold & Zucc.) in China and Korea (Jie, 1982), and for plantations of Japanese larch (*Larix kaempferi* (Lam.) Carrière) and hinoki false cypress (*Chamaecyparis obtusa* (Siebold & Zucc.) Siebold & Zucc. ex Endl.) in Japan (Ota *et al.*, 1998); the species involved in the case of hinoki false cypress is *A. mellea* subsp. *nipponica*.

8.2 Diagnosis

The colonization of the root system by the fungus provokes local reactions of the tissues (cortex and cambium) of the root and collar, and results in more general symptoms on the whole tree, due to the decay of the roots which lose a great part of their functionalities.

Subterranean rhizomorphs can often be observed at the surface of the roots, but the most aggressive species (*A. mellea* and *A. ostoyae* in Europe) are not the most rhizomorphogenic and it is frequently the case that the *Armillaria* species that has initiated rhizomorphs around the roots is not the one responsible for the main infection. On the large or middle roots (especially in conifers), black lumps can often be observed; they correspond to latent necroses where the fungus has been shut in by the histological reactions of the root. When the fungus succeeds in colonizing the cortical and cambial tissues, white mycelial fans are easily detected by debarking the root with a knife (Plate 8). The portions of the roots that are infected are frequently blackened and their diameter is increased by a proliferation of the mycelium and/or reaction tissues. Observation of the roots can also reveal the presence of exudations around the infected zones: resin on conifers, gums on *Prunus* spp. and *Citrus* spp., and latex on *Hevea* spp.

At the level of the collar, when infection is advanced, exudations are particularly obvious. Sometimes cankers or cracks can also be seen at this level, and these are a sign of the presence of the fungus under the bark. The cracks are especially frequent under tropical conditions, where mycelial proliferation is faster. At this stage, conspicuous mycelial fans are present at the collar between the wood and the bark. These fans can grow up along the cambium of the trunk to a height of 1.5 m or more. When the mycelial fans reach the collar, symptoms generally begin to be visible on the aerial part of the tree. In the autumn, fruiting bodies sometimes appear at the base of declining trees, issuing from the subcortical fans or rhizomorphs (Plate 9).

Colonization of the roots can cause a general growth loss, and also a more visible growth decrease of the shoots, often more conspicuous on fruit and amenity trees, and a discoloration of the foliage. Foliage discoloration generally starts in the form of a chlorosis, then it evolves into yellowing or even in reddening, followed by drying up of the foliage. Sometimes only leaves of one or a few main branches show these symptoms, the branches involved being fed by the roots that are most affected by the fungus. These symptoms can evolve slowly with time, with the foliage density decreasing year by year until the death of the whole crown; in contrast, the drying up of the whole foliage can happen suddenly.

These symptoms are not specific, they just reveal the dysfunction of the roots, and can be confused with the results of the action of other biotic or abiotic agents. In the case of complex declines involving predisposing, inciting and contributing factors (Manion, 1981), it is often impossible to sort out which part of the symptoms depends on *Armillaria* spp. and which on other agents (e.g. drought).

Exacerbation of fruiting in trees is another frequent symptom, especially in conifers. It is also a classical reaction to various stressing agents (Morrison *et al.*, 1991).

The subterranean rhizomorphs of *Armillaria* spp. are easily observed in the litter or in the upper 10 cm of mineral soil. They are cylindrical and brown or black. The observation of their anatomy allows them to be distinguished from creeping angiosperm roots, fern rhizomes or rhizomorphs of other fungi. The presence of *Armillaria* spp. in soil can be detected by trapping it using woody fragments buried in the soil (Blodgett and Lundquist, 2006).

Identification of the *Armillaria* spp. present on a site can be of great practical importance because the aggressiveness and host range of the diverse species are quite different. The morphology of the fruiting bodies (when they are present) and the morphology and abundance of the rhizomorphs can give some information. Traditionally, the identification of species was based on sexual compatibility or mating tests. In this method, the unknown diploid isolate was mated with a series of haploid strains ('testers') belonging to the different *Armillaria* spp. known to exist on the continent concerned. The unknown isolate appears compatible with the testers of only one of the species and incompatible with all the others (when a black line appears at the frontier between the mycelia). The isolate (of course) belongs to the species with which it is compatible. However, using this method requires some experience and also the maintenance of the tester species in culture.

Several molecular methods were proposed to replace the mating tests method, and a routine method was designed by Harrington and Wingfield (1995). This is based on PCR amplification of the non-coding DNA fragment PCR-1, followed by cutting up using the restriction enzyme AluI (if necessary, the use of two additional enzymes, NDeI and BSMI, allows the separation of taxons with the same AluI pattern). According to Escofet Crespo *et al.* (2006), *Armillaria* spp. can be identified by molecular methods without isolation of the fungus, directly from soil samples taken close to the roots.

8.3 Infection Biology

All *Armillaria* spp. behave as saprophytes decaying wood. They can directly invade dead woody substrates and they are also able to continue to exploit wood as saprophytes after they have killed a host as parasites. The colonization of stumps and dead wood in the soil maintains an inoculum potential that threatens the roots of living trees. The fungal mycelium can be preserved in stumps over several decades (Rishbeth, 1972).

Basidiospores are unable to directly infect trees although they probably are at the origin of the contamination of the stands through colonization of dead substrates. Infection of the root by the subterranean rhizomorph has been described both in roots sampled under natural conditions (Day, 1927; Rykowski, 1975) and as a result of artificial inoculation in containers (Thomas, 1934; Guillaumin and Rykowski, 1980). The rhizomorph that is growing in the soil reaches a root, creeps over its surface and adheres to it by a polysaccharidic mucus produced by the apex. On the interface of the rhizomorph and the root, a short rhizomorphic branch, the 'infection cone', appears (Thomas, 1934), and this penetrates the external tissues of the root as a result of both mechanical and enzymatic (suberases) effects. In a second step, the tip of the cone splits into individual hyphae, which invade the cortical parenchyma, challenge the reaction tissues initiated by the root and, possibly, reach the cambium.

Armillaria spp. can also infect roots by direct contact between a sound root and a source of inoculum: a stump, a dead root or an infected but still living root (Zeller, 1926; Guillaumin and Legrand, 2005). In this case, it is generally the growth of the sound root that brings it into the vicinity of the inoculum. This mode of infection, which is easy to reproduce in artificial inoculation, is realized by the undifferentiated mycelium and it probably involves a maceration of the living tissues by fungal enzymes and/or toxins before mycelial penetration. This mode of penetration is typical of a necrotrophic parasite.

The most aggressive species, particularly *A. mellea* and *A. ostoyae*, are often poorly rhizomorphogenic under natural conditions, and for these species, penetration by direct contact with inoculum is more frequent than that involving the rhizomorphs. The frequency of the latter mode of infection may have been overestimated in the past. Conversely, for *A. gallica* and its related species, infection by rhizomorphs is the rule: the intense development of the rhizomorphic net of these species results in a high number of contacts between rhizomorphs and roots, and some of them end by being successful in spite of the low intrinsic aggressiveness of the species.

There are cases when dead wood is in close contact with living tissues, for instance in coppice trees when the roots feeding the new stems are contiguous with the dead wood resulting from the cutting of the ancient stems. A necrotrophic fungus such as *Armillaria* can colonize first the dead wood and then, from this food base, can infect the living tissues (Stanosz and Patton, 1987). This mode of infection (perthophytism – a subdivision of necrophytism in which the dead substratum is mixed with parts of the living host) is likely to be frequent for *A. gallica*.

Opinions differ as to the role of wounds in infection. Some authors (Solla *et al.*, 2002) consider that wounds on roots favour infection, at least by weakening the reactions of the tree, but most of those who have performed artificial infections observed that wounds do not increase the infection rates, or even decrease it (Guillaumin and Pierson, 1977). However, penetration can be made easier by certain root structures that constitute points of weakness, such as points of emergence of the secondary roots, the extremity of a tap root, and contact between two roots that causes their partial crushing (Desray *et al.*, 1998). Bad planting conditions or the presence of stones, which cause malformations on the roots, are also favourable to infection.

After it has crossed the cortical parenchyma, the fungus comes into contact with the cambium and phloem, tissues with a weak mechanical resistance and a high nutritional potential. At this level, the mycelium aggregates into fans, which advance both laterally and longitudinally along the cambium. The growth of the fans towards the tips of the roots constitutes the main mode of short-range dissemination of the *Armillaria*, and initiates a few rhizomorphs. In the opposite centripetal direction, the fans invade broader and broader roots until they reach the collar. From this point, the colonization of the root can start again in a centrifugal direction in the other large roots contiguous to the first root attacked. In addition, the fans girdle the collar, which makes the death of the tree certain. They also climb up along the trunk.

In certain hosts, the fans can also progress along other paths of weakness: the gap between cork and bark in the cherry tree, the medullary rays in cacao (*Theobroma cacao* L.) (Dade, 1927) and the young tap root of English walnut (*Juglans regia* L.) (Desray *et al.*, 1998). The consequences of a given penetration by *Armillaria* spp. often depend on the distance between the infection point and the collar. The attacks on the taproot are more dangerous than those on lateral roots, which reach the collar later and more uncertainly.

The mechanisms that tend to contain and isolate the fungus are often (on a tree that is not weakened) able to stop the subsequent mycelial growth along the infected root. The reaction of the cortical parenchyma can even sometimes eradicate the parasite, which is expelled at the periphery of the root, but most of the time, the mycelium remains alive in a latent lesion (a lump of decayed tissues surrounded by the histological barriers). These latent infections correspond to a balance that can last for several years (Delatour and Guillaumin, 1995), although this balance can also be upset by changes in the environment. Fungal growth can start again from the lesion if the tree is weakened by any agent. Another possible event is the death of the distal part of the root, which allows the extension of the fungus in the dead tissues, thus increasing of its inoculum potential and consequently causing a new advance of the fungus in the living tissues.

The death of the tree, by felling or another cause, is of course the most drastic breaking of the balance: from all the pre-existing latent lesions in the root system, the mycelium of *Armillaria* spp. can colonize the whole of the roots and the stumps, giving rise to an inoculum that will threaten the next stand (Legrand *et al.*, 1996). Because of the latent lesions, a site can be risky even though there was no evidence of the presence of *Armillaria* spp. in the previous woody culture.

8.4 Epidemiology

8.4.1 Inocula involved in an epidemic progress of *Armillaria* spp. in a stand

Whatever the mode of infection (by direct contact or through rhizomorphs), infection in a stand has its origin in a woody, subterranean inoculum. The development and severity of the disease largely depend on quantity, nature and distribution of this woody inoculum.

In every plant disease, epidemiology distinguishes the primary from the secondary inoculum. In the particular case of root rots, primary inoculum consists of the contaminated stumps and woody remnants remaining from the previous culture and present in soil at the beginning of the new stand. The secondary inoculum consists of the trees of the current stand infected by the fungus and which, dead or still alive, can potentially contaminate their neighbours.

It is often difficult in each particular situation to sort out the relative role of the two types of inocula. This distinction is yet crucial. Classical phytopathologists (Van der Plank, 1963) showed that, when only primary inoculum is concerned, the disease shows a linear development with time, but when a secondary inoculum is involved in addition, the development of the disease with time is consistent with a logistic curve ('interest yields interest').

Several studies have been carried out to follow the development with time of an *Armillaria* epidemic from the establishment of a young forest stand or orchard. Some of these studies have tried to determine the relative role of the two types of inocula from the relative matching of the experimental mortality curves with the theoretical curves representing the linear and logistic models, respectively. Another line of research has been based on the calculation of the distances between recently dead trees and (i) the sources of inoculum from the previous stand (the primary inoculum), and (ii) sources from the dead trees of the current stand (secondary inoculum) (Lung-Escarmant and Guyon, 2004). These studies gave varied results. In the first few years, disease development always fits the linear model, as expected. Afterwards, the relative importance of the secondary inoculum appears to differ considerably according to the situation (Swift, 1970; Bruhn *et al.*, 1996; van der Kamp and Hood, 2002; Lung-Escarmant and Guyon, 2004). In many cases though, it seems that the secondary inoculum only plays a minor role compared with primary inoculum. It appears that many situations that could seem to be critical in stands aged 20–30 years do not worsen later: the mortality rates increase steeply in the first few years, then reach a plateau. Several reasons have been proposed for this phenomenon. First of all, as shown by Rishbeth (1972), the young seedlings or saplings killed at the beginning of a stand can be too small to constitute an efficient secondary inoculum. In addition, the level of resistance to *Armillaria* spp. generally increases with tree age (van der Kamp and Hood, 2002; Lung-Escarmant and Guyon, 2004). Over 25–30 years old, many root infections are encircled by defence responses and transformed into latent lesions. Certain infected trees can even evolve as far as recovery (MacKenzie, 1987).

A heavy primary inoculum can also (statistically) screen the intervention of a secondary inoculum. An important role for the latter depends on the efficiency of the contamination between neighbouring trees, which is itself dependent not only on the tree age, but also on the density of the stand and on species susceptibility (Hagle, 2008).

8.4.2 Long-distance dissemination and the role of basidiospores

The means of establishment of *Armillaria* spp. on a new site is a problem that has not been completely solved and which questions the role of the basidiospores. In very young stands without any forest history, *Armillaria* is generally represented by numerous clones (or genets), each covering a very small area, which suggests that the site has been sown by a rain of basidiospores (Legrand *et al.*, 1996; Dodd *et al.*, 2006). The new infection foci that colonize a new plantation are all genetically different from one another (Prospero *et al.*, 2008). The haploid mycelia resulting from the germination of the basidiospores are probably short living (Guillaumin, 1986), but, by mating with other compatible haploids, they give rise to new diploid genotypes.

By analogy with another root rot fungus, *H. annosum s.l.* (Rishbeth, 1950), certain researchers have tried to obtain the germination of basidiospores at the surface of stumps. With the *Armillaria* spp. of the northern hemisphere, these attempts have always ended in failure (Sites, 1967). By contrast, in New Zealand, *A. novae-zelandiae* could be successfully established from basidiospores on wood billets (Hood *et al.*, 2002, 2008). The same team (Power *et al.*, 2008) could collect the basidiospores by trapping them in the air and showed that they are transported to some distance outside the forest sites where the fruiting bodies appear. In France, Legrand *et al.* (1996) pointed out the high number and low area of the genets of *A. ostoyae* on a plot where the stumps had been completely removed. They suggested that the basidiospores of this species germinate and mate on tiny woody fragments rather than at the stump surface. Power *et al.* (2008) showed that the germination of the spores of *A. novae-zelandiae* was better on stem pieces than on stumps.

The introduction of *Armillaria* spp. to new site can also be achieved through human activities, such as the transport of soil containing infected roots and of organic fertilizers, or by mulching or even transport of contaminated supports as well as the use of infected young seedlings.

8.5 Management Strategies and Tactics

As *Armillaria* spp. are natural compounds of each forest ecosystem, the various influences of the diversity, structure and dynamics of plant communities and forest succession in which they are found lead to different root disease situations, and different strategies for controlling them can be distinguished.

Kile *et al.* (1991) described four cases of primary disease syndromes in the world: (i) Armillaria root disease in boreal forests and western North American coastal coniferous forests; (ii) ring disease in ancient stands of mountain pine in the Pyrenees (France); (iii) root rot of mixed species coniferous forests in western North America; and (iv) root rot of dry sclerophyll eucalyptus forests in Australia. In most cases the fungus behaves as an unobtrusive wood decomposer and recycler of organic matter in the forest ecosystems. It can cause root lesions or butt rot on trees, but more often it kills only suppressed or over-matured trees. Thus, *Armillaria* spp. acts as a partial thinning agent that eliminates the less vigorous trees. In this equilibrium, the fungus causes little damage, losses are scattered or appear only from time to time, and the role of the forest manager is to maintain this natural balance by sustainable management of the forest ecosystem. Frequently, and disrupting the previous equilibrium, a complex forest decline can occur, involving several abiotic and/or biotic factors. Generally, *Armillaria* is not the primary cause of the decline, but only a consequence of a situation in which the fungus finds an opportunity to develop. In such cases, *Armillaria* can also become an important agent of mortality, but the action of the forest manager should focus on the primary causes of the decline, and improve the resilience of the ecosystem.

Controlling a root disease is technically difficult and expensive, and the control strategy will be different according to the degree of damage caused or expected in the forest stand. Assessment of the impact of the disease is required before a decision is made. When losses are particularly low, which is the most frequent case in forestry, no control is needed, but when losses are (or might become) of economic importance, control will be required.

For this purpose, two key factors must be taken into account. Firstly, host adaptation to the site, appropriate silvicultural practices and, more generally, global forest health in a sustainable ecosystem, favour biodiversity and maintain or improve tree vigour; in this case, *Armillaria* can be maintained as a wood decomposer or as a secondary weakness pathogen. Second, the amount of dead wood, and particularly the amount of stumps, determines the inoculum pressure of the fungus and its epidemiology. Each event that increases the amount of dead wood in the soil can increase the incidence of the disease, which can become a primary disease when the inoculum potential is particularly high. For example, the build-up of inoculum in stumps can allow the pathogen to invade and kill trees in the residual stand following partial felling, and in the second-growth stand following clear-felling. In contrast, the *Armillaria* inoculum can persist in stumps over 15–50 years, and allow the fungus to act as a secondary pathogen every time trees are weakened. In consequence, the combination of a high inoculum potential and of a stress that weakens the trees will encourage the disease.

8.5.1 Avoidance

In establishing new plantations, the choice of the site is important. Young saplings are particularly vulnerable when exposed to a high inoculum potential. *Armillaria* is generally absent from former agricultural lands, but can reach a high inoculum potential in sites that are to be planted on clear-felled former forest. Young healthy plants, adapted to the site, without any root deformations, preferably in containers rather than as bare-rooted seedlings (to reduce plant stress) and planted at low density are generally more tolerant.

8.5.2 Eradication

Armillaria spp. cannot be eradicated from a site but damage can be kept at tolerable levels with appropriate management. Ring barking before felling trees is supposed to decrease the carbohydrates stored within the roots and to render them unsuitable for colonization by *Armillaria* spp. It was effective and gave excellent results in tea plantations following the clear-felling of tropical forests (Leach, 1939; Swift, 1970). In temperate forests though, lesions on live roots and rhizomorphs in the soil are usually abundant, and ring barking is not effective in reducing mortality in subsequent plantations. In some cases, it has even been shown to cause more rhizomorph production (Redfern, 1968).

The use of herbicides has been recommended to kill trees or to poison stumps. The objective was to make roots unsuitable for colonization by *Armillaria* spp., but more easily colonized by other saprophytic fungi. However, the method gave the opposite results according to the experiments cited in the literature (Rishbeth, 1976), so it cannot be recommended for reducing Armillaria disease.

The use of fire in burning forest remnants before planting can reduce the density of rhizomorphs in soil, but a high inoculum potential persists in deep roots, which are not affected by fire. Burning stumps has no significant effect on mortality caused on seedlings by Armillaria root disease either (Whitney and Irwin, 2005). So fire is not efficient in reducing mortality in subsequent plantations. Neither is the fungus usually eradicated by intercalating a herbaceous crop for several years before the new plantation is established; it can survive several decades on small woody debris in soil.

Direct reduction of inoculum by the physical removal of stumps and roots during site preparation before planting has given positive and significant results in long-term trials. This method has been well studied in Monterey pine plantations in New Zealand (van der Pas and Hood, 1984) and in ponderosa pine plantations or other conifer stands in North America (Roth *et al.*, 2000; Cleary *et al.*, 2013). The method is much more efficient when not only stumps are extracted, but roots as well, for example by using an excavator in preference to a classic bulldozer. This method is particularly recommended in the case of high inoculum potential, and when economic losses due to Armillaria disease are expected, notably when the disease is well developed in the stand that is felled to be subsequently planted. It needs to be borne in mind though that stump removal may damage the soil. Finally, avoiding hazardous sites with a high inoculum of pathogenic *Armillaria* spp. can be a better solution both technically and economically.

If necessary, the removal of diseased trees, stumps and roots can be applied not only for clear-fellings, but also for partial thinnings or for regeneration cuttings in diseased forest stands. The method is more easily applied and recommended for mortality gaps or rings, and for treating and isolating a diseased zone from other healthy areas of the stand.

8.5.3 Protection

Chemical control

Pawsey and Rahman (1976), Hagle and Shaw (1991) and West (2000) have reviewed the chemical control of *Armillaria* spp. Many experiments, both in the laboratory and in the field, have studied the use of chemical treatments to protect uninfected plants, to eradicate the fungus in infected stumps and roots, or to cure infected living plants. Soil disinfection has been the most successful technique and is sometimes used as routine in orchards and vineyards (Guillaumin *et al.*, 2005b). Preventive or curative applications of systemic, non-lethal fungicides on plants (especially of compounds that can be transported by basipetal translocation) have also been tried, generally with poorer success, and their use remains at the experimental level. These techniques are, anyway, not usable in forestry because of their cost and because of the complexity of the forest ecosystem.

Biological control

The microbiological environment of the roots in the soil has been studied from two points of view. Firstly, some fungi and other microorganisms inhibit the growth of *Armillaria* mycelium or rhizomorphs, while others favour it. So it may be possible to explain the gravity of disease development by the balance between the microorganisms that are present in the soil (Mañka, 1980). Second, some microorganisms that inhibit *Armillaria* spp. may be tried as biological control agents. Those that have been most studied are different species of *Trichoderma*, which are sometimes highly efficient in the laboratory. This efficacy has not been confirmed in the field, neither in forest stands nor in nurseries, probably because of the complexity of the interactions in the soil. Other microorganisms have also been studied for use in biological control (e.g. parasitic or mycorrhizal fungi, bacteria and nematodes), but they do not have any concrete application in forestry up to now.

Another research approach concerns saprophytic fungi acting as wood decomposers, particularly in stumps and roots. Some of these are cord forming or have aggregate mycelium and occupy ecological niches very close to *Armillaria* spp. (Rayner and Boddy, 1988). These are good candidates to compete with or to inhibit *Armillaria* colonization and growth, and inoculation tests in stumps have given promising results. For example, in Australia, *Hypholoma australe* O.K. Mill. is a good antagonist for *A. luteobubalina* on karri (*Eucalyptus diversicolor* F. Muell.) stumps (Pearce *et al.*, 1995).

Even within the genus *Armillaria*, some species are much less pathogenic, acting mainly as wood decomposers, and occupying the same ecological niche as pathogenic species. These saprophytic species need to be favoured by silvicultural practices in order to reduce the development of the pathogenic species. For example, in central France, the preservation of some broadleaved trees in coniferous stands allows the maintenance of a higher fungal diversity that prevails in mixed stands, and reduces root contacts and the propagation of *A. ostoyae* between living conifer trees and stumps; and in addition, it favours the development of the saprophytic *A. gallica* and *A. cepistipes*. These two species are able to colonize conifer stumps and reduce the colonization and the development of the pathogenic *A. ostoyae* (Legrand *et al.*, 1996). Further, in their ability to colonize Norway spruce stumps, the two species *A. ostoyae* and *A. cepistipes* experience a mutual negative effect (Prospero *et al.*, 2006).

The best biological control approach then appears the natural one: that provided by appropriate silvicultural practices and allowing a great biodiversity in a well-balanced ecosystem.

Pruning

Pruning and girdling or scraping out diseased roots or the diseased collar, as in the Yangambi method used in Africa for the rubber tree (*Hevea brasiliensis* (Willd. ex A. Juss.) Müll. Arg.) (Pichel, 1956), are economically only usable on individual trees (amenity trees or orchard trees), and not in forests.

Fertilization

The use of fertilizer to improve the resistance of the trees to Armillaria disease, and liming to modify soil characteristics, has led to contradictory results in the different forestry experiments cited in the literature.

Trenching

Trenching deeply in the soil to isolate infected plants from the healthy parts of a stand was traditionally used in maritime pine stands in south-western France, but is now more profitably used in orchards. This technique has been improved by using a vertical plastic layer in the soil as a mechanical barrier to root and rhizomorph growth.

More generally, the forest manager should modify silvicultural practices to minimize the exposure of trees to *Armillaria* inoculum, most particularly in second-growth stands.

8.5.4 Resistance

In establishing new plantations, the choice of the plant is also important. Locally adapted seed sources are recommended, and when natural regeneration is used, it should favour the species that is the most tolerant to the fungus and most adapted to the site. Precommercial thinning, commercial thinning, and site regeneration offer opportunities to change species composition (Hagle, 2008).

Mixing species in forest stands, particularly broadleaved trees with conifers, reduces root contacts and the propagation of the fungus. Broadleaved trees are generally more resistant to Armillaria root disease, while conifers are more susceptible to *A. ostoyae*. So although reducing the number of broad-leaved trees present can improve conifer growth, it can also increase Armillaria root disease incidence in conifer stands (Baleshta *et al.*, 2005; Simard *et al.*, 2005).

Between conifers, differences in susceptibility to *A. ostoyae* are known too. In plantations of exotic species in Europe, grand fir is highly damaged while Douglas-fir is much more resistant. This is not the case in North America, where substantial losses can affect Douglas-fir, more particularly in monospecific plantations out of its natural range. In western North America, pines, western larch (*L. occidentalis* Nutt.), spruces, western red-cedar (*Thuja plicata* Donn ex D. Don) and hemlocks (*Tsuga* spp.) are more resistant to Armillaria root disease than are true firs (*Abies* spp.) and Douglas-fir (Hagle, 2008).

Within the same genus, some differences in susceptibility between species are known as well. In south-western France, loblolly pine, Monterey pine, Corsican pine (*P. nigra* Arnold subsp. *laricio* Maire), and lodgepole pine are less susceptible than indigenous maritime pine. In addition, within a species, some differentiated populations can show differences in susceptibility, for example, in maritime pine (Lung-Escarmant and Taris, 1988; Zas *et al.*, 2007), Norway spruce (Prospero *et al.*, 2004) and Japanese cedar (*Cryptomeria japonica* (L. f.) D. Don) (Melo *et al.*, 2004). Furthermore, it seems that genetic resistance cannot be ruled out as an option in Monterey pine affected by *A. novae-zelandiae* in New Zealand (Hood *et al.*, 2009).

8.5.5 Integrated disease management

As many forestry practices affect the incidence and severity of Armillaria root disease, control of the disease must be based on knowledge of the whole forest ecosystem, and close-to-nature silvicultural principles that allow sustainable management, the maintenance of biodiversity and the resilience of the ecosystem. Selecting the management options that are best suited to each situation depends on understanding the whole forest ecosystem, and the role of *Armillaria* and its epidemiology. Several concrete proposals based on long-term experiments and well-studied cases of Armillaria root disease in different parts of the world have recently been made by various authors. Some (non-exhaustive) examples follow.

In the Rocky Mountains, the Western Root Disease Model (WRDM) described by Stage *et al.* (1990), and Shaw *et al.* (1991), was historically the first model developed to manage forest stands affected by root diseases. The WRDM provides an analytical and dynamic representation of the spatial epidemiology of three types of root diseases: laminated root rot caused by *P. sulphurascens*, annosus root and butt rot disease caused by *H. annosum s.l.* and Armillaria root disease. It can simulate the impact of various management scenarios on the development of the disease, and has been improved all over the years by comparisons between simulation results and remeasurements from long-term experimental plots (Maffei *et al.*, 2008).

Also in North America, the Missouri Ozark Forest Ecosystem Project was initiated to examine the impacts of forest management practices on an ecosystem affected by oak decline. In this forest, *Armillaria* and oak species are important interacting components of the ecosystem, and the comprehension of the ecosystem will improve silvicultural practices (Bruhn *et al.*, 2000; Baucom, 2005).

McDonald *et al.* (2005) have developed a tool to assess the effects of fuel treatments on Armillaria root disease in the dry forests of the western USA. It indicates how some fire planning (fuel management) activities may exacerbate Armillaria disease within high-risk stands, and helps to determine an appropriate fuel management plan for reducing future damage by the disease.

In order to manage tree species that will survive on infected sites, Hagle (2008) has published an important monograph on *A. ostoyae* disease affecting conifer stands in the northern and central Rocky Mountains, killing most particularly Douglas-fir and grand fir. Detailed management options vary according to the mode of disease development, with site and local climate determining host adaptation and host susceptibility.

Cleary *et al.* (2008) have proposed a decision key to help natural resource managers to choose among several different treatments in British Columbia's southern interior forests affected by *A. ostoyae*, where it kills both natural and planted coniferous trees. These treatments are: (i) to remove the inoculum; (ii) to plant or encourage natural regeneration of resistant species or uniform mixtures containing resistant species; and (iii) to ignore disease and accept a loss of volume at rotation. In British Columbia too, Cruickshank *et al.* (2009, 2011) have estimated the impact of *A. ostoyae* root disease in Douglas-fir plantations, the importance of non-lethal root infections and the volume reduction that is attributable to disease. The widespread below-ground incidence of *A. ostoyae*, its rapid colonization of stumps and its wide host range can reduce site potential in managed stands. Trees suffer accumulating growth reduction without recovery, and root diseases affect forest productivity even without mortality. The minimization of disease impact in respect of other forest management goals is also discussed.

Hood and Kimberley (2009) have studied the long-term impact of Armillaria root disease in Monterey pine plantations in New Zealand. The effect of thinnings on the distribution and severity of the disease, and possible reductions in volume growth, have been clarified in second-rotation plantations. Filip *et al.* (2009) have determined the long-term effects of thinning in ponderosa pine plantations affected by *A. ostoyae* in central Oregon and have given recommendations for multiple thinning. Filip *et al.* (2010), in a mixed conifer forest in south-central Oregon, have compared the effect of different silvicultural treatments on tree growth loss and mortality caused by *A. ostoyae*, and on the balance between susceptible and tolerant host species. Leave-tree mortality appeared not to be enhanced by harvesting. In addition, more disease-resistant seedlings and saplings survived in the shelterwood harvest stands and group selection openings than in comparable areas that had not been harvested.

Finally, as Armillaria root disease is more severe in trees that are maladapted to climate-induced stress, forest management practices that intend to minimize the adverse impacts of diseases should also take in account more global changes in environmental conditions, such as climate change (Kim *et al.*, 2008; Klopfenstein *et al.*, 2009a,b), which can affect not only the development of the disease, but also the adaptation of the tree species and interactions between hosts and pathogens in a complex ecosystem.

References

Abomo-Ndongo, S. (1997) Morphogenèse et systèmes sexuels chez les isolats du genre *Armillaria* d'Afrique tropicale [Morphogenesis and sexual systems in isolates of the genus *Armillaria* from tropical Africa]. PhD thesis, Université Blaise-Pascal, Clermont-Ferrand, France.

Anselmi, N. and Lanata, F. (1989) Distribuzione e comportamento delle specie di *Armillaria* riscontrata nell'Italia settentrionale [Distribution and behaviour of *Armillaria* species found in northern Italy]. *Micologia Italiana* 3, 57–70.

Auer, C.G., Gomes, N.S.B. and Grigoletti, A. (2003) Novas ocorrências da armillariose em *Pinus* no Brasil [New occurrences of Armillarias on *Pinus* in Brazil]. *Summa Phytopathologica* 29, 205–207.

Baleshta, K.E., Simard, S.W., Guy, R.D. and Chanway, C.P. (2005) Reducing paper birch density increases Douglas-fir growth rate and Armillaria root disease incidence in southern interior British Columbia. *Forest Ecology and Management* 208, 1–13.

Baucom, D.L. (2005) *Armillaria* species in the Missouri Ozark forests. MSc thesis, University of Missouri, Columbia, Missouri.

Bennell, A.P., Watling, R. and Kile, G.A. (1985) Spore ornamentation in *Armillaria* (*Agaricales*). *Transactions of the British Mycological Society* 84, 447–455.

Blodgett, J.T. and Lundquist, J.G. (2006) Wood blocks as *Armillaria* traps in burned ponderosa pine stands. *Phytopathology* 96, S13.

Bruhn, J.N., Mihail, J.D. and Meyer, T.R. (1996) Using spatial and temporal patterns of Armillaria root disease to formulate management recommendations for Ontario's black spruce (*Picea mariana*) seeds orchards. *Canadian Journal of Forest Research* 26, 298–305.

Bruhn, J.N., Wetteroff, J.J., Mihail, J.D., Kabrick, J.M. and Pickens, J.B. (2000) Distribution of *Armillaria* species in upland Ozark Mountain forests with respect to site, overstory species composition and oak decline. *European Journal of Forest Pathology* 30, 43–60.

Cleary, M., van der Kamp, B. and Morrison, D.J. (2008) British Columbia's southern interior forests: Armillaria root disease stand establishment decision aid. *British Columbia Journal of Ecosystems and Management* 9, 60–65.

Cleary, M.R., Arhipova, N., Morrison, D.J., Thomsen, I.M., Sturrock, R.N., Vasaitis, R., Gaitnieks, T. and Stenlid, J. (2013) Stump removal to control root disease in Canada and Scandinavia: a synthesis of results from long-term trials. *Forest Ecology and Management* 290, 5–14.

Coetzee, M.P.A., Wingfield, B.D., Coutinho, T.A. and Wingfield, M.J. (2000) Identification of the causal agent of Armillaria root rot of *Pinus* species in South Africa. *Mycologia* 92, 777–785.

Cruickshank, M.G., Morrison, D.J. and Lalumière, A. (2009) The interaction between competition in interior Douglas-fir plantations and disease caused by *Armillaria ostoyae* in British Columbia. *Forest Ecology and Management* 257, 443–452.

Cruickshank, M.G., Morrison, D.J. and Lalumière, A. (2011) Site, plot, and individual tree yield reduction of interior Douglas-fir associated with non-lethal infection by Armillaria root disease in southern British Columbia. *Forest Ecology and Management* 261, 297–307.

Dade, H.A. (1927) 'Collar crack' of cacao (*Armillaria mellea* [Vahl] Fr.). *Gold Coast Department of Agriculture Bulletin* 5, 7–21.

Day, W.R. (1927) The parasitism of *Armillaria mellea* in relation to conifers. *Quarterly Journal of Forestry* 21, 9–21.

Delatour, C. and Guillaumin, J.-J. (1995) Role of *Armillaria* in the decline of silver fir in the Vosges and the Massif Central. In: Landmann, G. and Bonneau, M. (eds) *Forest Decline and Atmospheric Deposition Effects in the French Mountains*. Springer, Berlin, pp. 353–360.

Desray, P., Jay-Allemand, C., Fady, B. and Guillaumin, J.-J. (1998) Susceptibility to *Armillaria mellea* of different progenies of *Juglans regia* selected for timber. In: Delatour, C., Guillaumin, J.-J., Lung-Escarmant, B. and Marçais, B. (eds) *Proceedings of Ninth International Conference on Root and Butt Rots*. Les Colloques No. 89, INRA Editions, Paris, pp. 287–300.

Dodd, S.L., Shah, F.A. and Hood, I.A. (2006) Molecular tools to determine the mode of spread of the root pathogen *Armillaria* in commercial pine forests. *Phytopathology* 96, S30.

Durrieu, G., Beneteau, A. and Niocel, S. (1985) *Armillaria obscura* dans l'écosystème forestier de Cerdagne [*Armillaria obscura* in the forest ecosystem of Cerdagne]. *European Journal of Forest Pathology* 15, 350–355.

Escofet Crespo, P.E., Aguín Casal, O. and Mansilla Vázquez, J.P. (2006) Detección e identificación por técnicas moleculares de especies del género *Armillaria* a partir de muestras de suelo [Detection and identification by molecular techniques of species of the genus *Armillaria* from soil samples]. *Boletín de Sanidad Vegetal. Plagas* 32, 231–240.

Filip, G.M. and Goheen, D.J. (1984) Root diseases cause severe mortality in white and grand fir stands of the Pacific Northwest. *Forest Science* 30, 138–142.

Filip, G.M., Fitzgerald, S.A., Chadwick, K.L. and Max, T.A. (2009) Thinning ponderosa pine affected by Armillaria root disease: 40 years of growth and mortality on an infected site in Central Oregon. *Western Journal of Applied Forestry* 24, 88–94.

Filip, G.M., Maffei, H.M., Chadwick, K.L. and Max, T.A. (2010) Armillaria root disease-caused tree mortality following silvicultural treatments (shelterwood or group selection) in an Oregon mixed-conifer forest: insights from a 10-year case study. *Western Journal of Applied Forestry* 25, 136–143.

Fox, R.T.V. (2000) *Armillaria Root Rot: Biology and Control of Honey Fungus*. Intercept, Andover, UK.

Gezahgne, A., Roux, J. and Wingfield, M.J. (2003) Diseases of exotic *Eucalyptus* and *Pinus* species in Ethiopian plantations. *South African Journal of Science* 99, 29–33.

Guillaumin, J.-J. (1986) Contribution à l'étude des armillaires phytopathogènes, en particulier du groupe mellea: cycle caryologique, notion d'espèce, rôle biologique des espèces [Contributions to the study of phytopathogenic armillarias, in particular the group mellea: karyological cycle, notion of species, and biological role of species]. MSc thesis, Université Claude Bernard (Lyon 1), Lyon, France.

Guillaumin, J.-J. and Legrand, P. (2005) Cycle infectieux de l'armillaire – stratégie biologique des espèces européennes [Infectious cycle of armillaria – biological strategy of European species]. In: Guillaumin, J.-J., Legrand, P., Lung-Escarmant, B. and Botton, B. (eds) *L'Armillaire et le Pourridié-agaric des Végétaux Ligneux* [*Armillaria and Fungal Root Rot of Woody Plants*]. Editions INRA [now Editions Quae], Paris, pp. 177–201.

Guillaumin, J.-J. and Pierson, J. (1977) Inoculation de deux clones de *Prunus mahaleb* par l'armillaire. Influence du greffage et des blessures [Inoculation of two clones of *Prunus mahaleb* by armillaria. Influence of grafting and wounds]. *Annales de Phytopathologie* 9, 533.
Guillaumin, J.-J. and Rykowski, K. (1980) Study of infection of walnut (*Juglans regia* L.) by honey fungus (*Armillaria mellea* [Vahl] Quél.) in model experiments. *Folia Forestalia Polonica* 24, 192–213 [in Polish].
Guillaumin, J.-J., Bernard, C., Delatour, C. and Belgrand, M. (1985) Contribution à l'étude du dépérissement du chêne: pathologie racinaire en forêt de Tronçais [Contribution to the study of oak decline: root pathology in the forest of Tronçais]. *Annales des Sciences Forestières* 42, 1–22.
Guillaumin, J.-J. *et al.* (1993) Geographical distribution and ecology of the six *Armillaria* species in western Europe. *European Journal of Forest Pathology* 23, 321–341.
Guillaumin, J.-J., Legrand, P., Lung-Escarmant, B. and Botton, B. (2005a) *L'Armillaire et le Pourridié-agaric des Végétaux Ligneux* [*Armillaria and Fungal Root Rot of Woody Plants*]. Editions INRA [now Editions Quae], Paris.
Guillaumin, J.-J., Lung-Escarmant, B. and Dubos, B. (2005b) Lutte chimique [Chemical control]. In: Guillaumin, J.-J., Legrand, P., Lung-Escarmant, B. and Botton, B. (eds) *L'Armillaire et le Pourridié-agaric des Végétaux Ligneux* [*Armillaria and Fungal Root Rot of Woody Plants*]. Editions INRA [now Editions Quae], Paris, pp. 365–375.
Hagle, S.K. (2008) *Management Guide for Armillaria Root Disease*: Armillaria ostoyae *(Romagnesi) Herink.* US Department of Agriculture, Forest Health Protection and State Forestry Organizations. Available at: http://www.fs.usda.gov/Internet/FSE_DOCUMENTS/stelprdb5187208.pdf (accessed 28 November 2012).
Hagle, S.K. and Shaw, C.G. (1991) Avoiding and reducing losses from armillaria root disease. In: Shaw, C.G. and Kile, G.A. (eds) *Armillaria Root Disease*. Agriculture Handbook 691, US Department of Agriculture, Forest Service, Washington, DC, pp. 157–173.
Harrington, T.C. and Wingfield, B.D. (1995) A PCR-based identification method for species of *Armillaria*. *Mycologia* 87, 280–288.
Hintikka, V. (1974) Notes on the ecology of *Armillariella mellea* in Finland. *Karstenia* 14, 12–31.
Hood, I.A. (1989) Armillaria root rot disease in New Zealand forests. *New Zealand Journal of Forestry Science* 19, 180–197.
Hood, I.A. and Kimberley, M.O. (2009) Impact of armillaria root disease and the effect of thinning in a late-rotation *Pinus radiata* plantation. *Forest Pathology* 39, 415–427.
Hood, I.A., Horner, I.J., Gardner, J.F. and Sandberg, C.J. (2002) Armillaria root disease of *Pinus radiata* in New Zealand. 1. Basidiospore dispersal. *New Zealand Journal of Forestry Science* 32, 94–102.
Hood, I.A., Petrini, L.E. and Gardner, J.F. (2008) Colonisation of woody material in *Pinus radiata* plantations by *Armillaria novae-zelandiae* basidiospores. *Australasian Plant Pathology* 37, 347–352.
Hood, I.A., Kimberley, M.O. and Gardner, J.F. (2009) Susceptibility to *Armillaria novae-zelandiae* among clones of *Pinus radiata*. *Forest Pathology* 39, 405–414.
Jie, G.Z. (1982) Root rot of red pine *Armillariella mellea* (Vahl ex Fr.) Karst. In: Chinese Academy of Forestry Science (ed.) *Forest Diseases in China*. Chinese Forestry Publications, Beijing, pp. 34–36 [in Chinese].
Kellas, J.D., Kile, G.A., Jarrett, R.G. and Morgan, B.J.T. (1987) The occurrence and effects of *Armillaria luteobubalina* following partial cutting in mixed eucalypt stands in the Wombat Forest, Victoria. *Australian Forest Research* 17, 263–267.
Kile, G.A. (1980) Behaviour of an *Armillaria* in some *Eucalyptus obliqua–Eucalyptus regnans* forests in Tasmania and its role in their decline. *European Journal of Forest Pathology* 10, 278–296.
Kile, G.A., McDonald, G.I. and Byler, G.W. (1991) Ecology and disease in natural forests. In: Shaw, C.G. and Kile, G.A. (eds) *Armillaria Root Disease*. Agriculture Handbook 691, US Department of Agriculture, Forest Service, pp. 102–121.
Kim, M.-S., Klopfenstein, N.B., Richardson, B.A., Hanna, J.W. and McDonald, G.I. (2008) Toward determining the influence of climate change on armillaria root disease in western North America. *Phytopathology* 98, S192.
Klopfenstein, N.B., Kim, M.-S., Hanna, J.W., Richardson, B.A. and Lundquist, J.E. (2009a) *Approaches to Predicting Potential Impacts of Climate Change on Forest Disease: An Example with Armillaria Root Disease*. Research Paper RMRS-RP-76, US Department of Agriculture, Forest Service, Rocky Mountain Research Station, Fort Collins, Colorado.
Klopfenstein, N.B., Kim, M., Hanna, J.W., Richardson, B.A., Smith, A.L. and Maffei, H. (2009b) Predicting potential impacts of climate change on armillaria root disease in the inland northwestern USA. *Phytopathology* 99, S65.
Korhonen, K. (1978) Interfertility and clonal size in the *Armillariella mellea* complex. *Karstenia* 18, 31–42.

Leach, R. (1939) Biological control and ecology of *Armillaria mellea* (Vahl) Fr. *Transactions of the British Mycological Society* 23, 320–329.

Lech, P. and Sierota, Z. (2007) Occurrence of root diseases in Poland in relation to host species, forest site types, injuries of trees and crown conditions. *Schriften aus der Forstlichen Fakultät der Universität Göttingen und der Nordwestdeutschen Forstlichen Versuchsanstalt* 142, 187–192.

Legrand, P. and Lung-Escarmant, B. (2005) L'armillaire en plantations forestières en Europe [Armillaria in forest plantations in Europe]. In: Guillaumin, J.-J., Legrand, P., Lung-Escarmant, B. and Botton, B. (eds) *L'Armillaire et le Pourridié-agaric des Végétaux Ligneux [Armillaria and Fungal Root Rot of Woody Plants]*. Editions INRA [now Editions Quae], Paris, pp. 307–318.

Legrand, P., Ghahari, S. and Guillaumin, J.-J. (1996) Occurrence of genets of *Armillaria* spp. in four mountain forests in central France: the colonization strategy of *Armillaria ostoyae*. *New Phytologist* 133, 321–332.

Lung-Escarmant, B. and Guyon, D. (2004) Temporal and spatial dynamics of primary and secondary infection by *Armillaria ostoyae* in a *Pinus pinaster* plantation. *Phytopathology* 94, 125–131.

Lung-Escarmant, B. and Taris, B. (1984) L'armillaire, parasite du pin maritime dans les Landes de Gascogne [Armillaria, a parasite of maritime pine in Landes de Gascogne]. *Phytoma* 362, 44–47.

Lung-Escarmant, B. and Taris, B. (1988) Étude du comportement d'une gamme d'hôtes feuillus et résineux vis-à-vis d'*Armillaria obscura*: approche méthodologique et perspectives [Study of the behaviour of a range of broadleaved and coniferous hosts vis-à-vis *Armillaria obscura*: methodological approach and outlook]. In: *Compte Rendus Annales ANPP, 2me Conférence Internationale sur Les Maladies des Plantes, Bordeaux, France*, 8–10 Nov. 1988, pp. 1185–1191.

MacKenzie, M. (1987) Infection change and volume loss in a 19-year-old *Pinus radiata* stand affected by armillaria root-rot. *New Zealand Journal of Forestry Science* 17, 100–108.

Maffei, H.M., Filip, G.M., Chadwick, K.L. and David, L. (2008) Western Root Disease Model simulation versus plot remeasurement: 11 years of change in stand structure and density induced by armillaria root disease in Central Oregon. In: Havis, R.N. and Crookston, N.L. (eds) *Third Forest Vegetation Simulator Conference, 2007, February 13–15, Fort Collins, Colorado*. Proceedings RMRS-P-54, US Department of Agriculture, Forest Service, Rocky Mountain Research Station, Fort Collins, Colorado, pp. 55–67.

Manion, P.D. (1981) *Tree Disease Concepts*. Prentice Hall, New York.

Mańka, K. (1980) Soil–site conditions in relation to the *Armillariella mellea* problem. In: Dimitri, L. (ed.) *Proceedings of the Fifth International Conference on Problems of Root and Butt Rots, Kassel, Federal Republic of Germany, August 7–12, 1978*. Hessian Forstliche Laboratory, Hann-Münden, pp. 234–244.

McDonald, G.I., Tanimoto, P.D., Rice, T.M., Hall, D.E., Stewart, J.E., Zambino, P.J., Tonn, J.R., Klopfenstein, N.B. and Kim, M.-S. (2005) *Fuels Planning: Science Synthesis and Integration Environmental Consequences. Fact Sheet 13. Root Disease Analyzer –* Armillaria *response tool (ART)*. Research Note RMRS-RN-23-13-WWW, US Department of Agriculture, Forest Service, Rocky Mountain Research Station, Fort Collins, Colorado.

Melo, M.J., Machado, M.H., Moniz, M. de F., Bragança, M.H. and Sousa Santos, M.N. (2004) Resistance patterns of *Cryptomeria japonica* to *Armillaria mellea* in S. Miguel Island – Azores. Poster, 4° Congresso da Sociedade Portuguesa de Fitopatologia, Universidade do Algarve, Faro, 4–6 February 2004.

Mohammed, C. (1994) The detection and species identification of African *Armillaria*. In: Schotts, A., Dewey, F.M. and Olivier, R. (eds) *Modern Detection Assays for Plant Pathogenic Fungi*. CAB International, Wallingford, UK, pp. 141–148.

Morrison, D.J. (1981) *Armillaria Root Disease. A Guide to Disease Diagnosis, Development and Management in British Columbia*. Information Report BC-X-203, Canadian Forestry Service, Pacific Forest Research Centre, Victoria, British Columbia, Canada.

Morrison, D.J. (1982) Variation among British isolates of *Armillaria mellea*. *Transactions of the British Mycological Society* 78, 459–464.

Morrison, D.J. and Mallett, K. (1996) Silvicultural management of armillaria root disease in western Canadian forests. *Canadian Journal of Plant Pathology* 18, 194–199.

Morrison, D.J., Williams, R.E. and Whitney, R.D. (1991) Infection, disease development, diagnosis and detection. In: Shaw, C.G. and Kile, G.A. (eds) *Armillaria Root Disease*. Agriculture Handbook 691, US Department of Agriculture, Forest Service, Washington, DC, pp. 62–75.

Mwangi, L.M., Mwenje, E., Makambila, C., Chanakira-Nyahwa, F., Guillaumin, J.-J. and Mohammed, C. (1994) Ecology and pathogenicity of *Armillaria* in Kenya, Zimbabwe and the Congo. In: Johansson, M. and Stenlid, J. (eds) *Proceedings of the Eighth International Conference on Root and Butt Rots, Wik, Sweden and Haikko, Finland, August 9–16, 1993*. Swedish University of Agricultural Sciences, Uppsala, Sweden, pp. 34–44.

Ota, Y., Matsushita, N., Nagasawa, E., Terashita, T., Fukuda, K. and Suzuki, K. (1998) Biological species of *Armillaria* in Japan. *Plant Disease* 82, 537–543.

Ota, Y., Intini, M. and Hattori, T. (2000) Genetic characterization of heterothallic and non-heterothallic *Armillaria mellea sensu stricto*. *Mycological Research* 104, 1046–1054.

Pawsey, R.G. and Rahman, M.A. (1976) Chemical control of infection by honey fungus, *Armillaria mellea*: a review. *Arboricultural Journal* 2, 468–477.

Pearce, M.H., Nelson, E.E. and Malajczuk, N. (1995) Effects of the cord-forming saprotrophs *Hypholoma australe* and *Phanerochaete filamentosa* and of ammonium sulphamate on establishment of *Armillaria luteobubalina* on stumps of *Eucalyptus diversicolor*. *Mycological Research* 99, 951–956.

Pichel, R.J. (1956) *Les Pourridiés de l'Hévéa dans la Cuvette Congolaise* [Root Rots of Hevea in the Congo Basin]. Séries Techniques 49, Institut National pour l'Étude Agronomique du Congo Belge, Brussels, Belgium.

Pildain, M.B., Coetzee, M.P.A., Rajchenberg, M., Petersen, R.H., Wingfield, M.J. and Wingfield, B.D. (2009) Molecular phylogeny of *Armillaria* from the Patagonian Andes. *Mycological Progress* 8, 181–194.

Podger, F.D., Kile, G.A., Watling, R. and Fryer, J. (1978) Spread and effects of *Armillaria luteobubalina* sp. nov. in an Australian *Eucalyptus regnans* plantation. *Transactions of the British Mycological Society* 71, 77–87.

Power, M.W.P., Ramsfield, T.D. and Hood, I.A. (2008) Detection of *Armillaria* basidiospore dispersal. *New Zealand Plant Protection* 61, 35–40.

Prospero, S., Holdenrieder, O. and Rigling, D. (2004) Comparison of the virulence of *Armillaria cepistipes* and *Armillaria ostoyae* on four Norway spruce provenances. *Forest Pathology* 34, 1–14.

Prospero, S., Holdenrieder, O. and Rigling, D. (2006) Rhizomorph production and stump colonization by co-occurring *Armillaria cepistipes* and *Armillaria ostoyae*: an experimental study. *Forest Pathology* 36, 21–31.

Prospero, S., Lung-Escarmant, B. and Dutech, C. (2008) Genetic structure of an expanding Armillaria root rot fungus (*Armillaria ostoyae*) population in a managed pine forest in southwestern France. *Molecular Ecology* 17, 3366–3378.

Rayner, A.D.M. and Boddy, L. (1988) *Fungal Decomposition of Wood; its Biology and Ecology*. John Wiley, Hoboken, New Jersey.

Redfern, D.B. (1968) The ecology of *Armillaria mellea* in Britain. Biological control. *Annals of Botany* 32, 293–300.

Rishbeth, J. (1950) Observations on the biology of *Fomes annosus*, with particular reference to East Anglian pine plantations. I. The outbreaks of disease and ecological status of the fungus. *Annals of Botany* 15, 1–21.

Rishbeth, J. (1972) The production of rhizomorphs by *Armillaria mellea* from stumps. *European Journal of Forest Pathology* 2, 193–205.

Rishbeth, J. (1976) Chemical treatment and inoculation of hardwood stumps for control of *Armillaria mellea*. *Annals of Applied Biology* 82, 57–70.

Rishbeth, J. (1978) Infection foci of *Armillaria mellea* in first rotation hardwoods. *Annals of Botany* 42, 1131–1139.

Roth, L.F., Shaw, C.G. and Rolph, L. (2000) Inoculum reduction measures to control armillaria root disease in a severely infected stand of ponderosa pine in south-central Washington: 20 year results. *Western Journal of Applied Forestry* 15, 92–100.

Rykowski, K. (1975) Modalité d'infection des pins sylvestres par l'*Armillariella mellea* (Vahl) Karst. dans les cultures forestières [Mode of infection of Scots pines by *Armillariella mellea* (Vahl) Karst. in forest plantations]. *European Journal of Forest Pathology* 5, 65–82.

Schönhar, S. (1994) Der Hallimasch (*Armillaria*) als Kernfäuleerreger in Fichtenbeständen der Schwäbischen Alb [The honey fungus (*Armillaria*) as core root pathogens in spruce stands in the Swabian Alps]. *Allgemeine Forst-und Jagdzeitung* 7, 132–136.

Shaw, C.G. and Kile, G.A. (eds) (1991) *Armillaria Root Disease*. Agriculture Handbook 691, US Department of Agriculture, Forest Service, Washington, DC.

Shaw, C.G., MacKenzie, M., Toes, E.H.A. and Hood, I.A. (1981) Cultural characteristics and pathogenicity to *Pinus radiata* of *Armillaria novae-zelandiae* and *A. limonea*. *New Zealand Journal of Forest Science* 11, 65–70.

Shaw, C.G., Stage, A.R. and McNamee, P. (1991) Modeling the dynamics, behaviour and impact of Armillaria root disease. In: Shaw, C.G. and Kile, G.A. (eds) *Armillaria Root Disease*. Agriculture Handbook 691, US Department of Agriculture, Forest Service, Washington, DC, pp. 150–156.

Shearer, B.L., Byrne, A., Dillon, M. and Buehring, R. (1997) Distribution of *Armillaria luteobubalina* and its impact on community diversity and structure in *Eucalyptus wandoo* woodland of southern Western Australia. *Australian Journal of Botany* 45, 151–165.

Sicoli, G., de Gioia, T., Luisi, N. and Lerario, P. (1998) Multiples factors associated with oak decline in southern Italy. *Phytopathologia Mediterranea* 37, 1–8.

Simard, S.W., Hagerman, S.M., Sachs, D.L., Heineman, J.L. and Mather, W.J. (2005) Conifer growth, *Armillaria ostoyae* root disease, and plant diversity responses to broadleaf competition reduction in mixed forests of southern interior British Columbia. *Canadian Journal of Forest Research* 35, 843–859.

Singh, P. and Richardson, J. (1973) Armillaria root rot in seeded and planted areas in Newfoundland. *Forestry Chronicle* 49, 180–182.

Sites, W.H. (1967) The comparative colonization rates of *Fomes annosus* (Fr.) Cke. and *Armillaria mellea* (Vahl) Quélet in loblolly pine and three selected hardwoods. MSc thesis, Duke University, Durham, North Carolina.

Solla, A., Tomlinson, F. and Woodward, S. (2002) Penetration of *Picea sitchensis* rootbark by *Armillaria mellea*. *Forest Pathology* 32, 55–70.

Stage, A.R., Shaw, C.G., Marsden, M.A., Byler, J.W., Renner, D.L., Eav, B.B., McNamee, P.J., Sutherland, G.D. and Webb, T.M. (1990) *User's Manual for Western Root Disease Model*. General Technical Report INT-GTR-267, US Department of Agriculture, Forest Service, Intermountain Research Station, Ogden, Utah.

Stanosz, G.R. and Patton, R.F. (1987) Armillaria root rot in aspen stands after repeated short rotations. *Canadian Journal of Forest Research* 17, 1001–1005.

Swift, M.J. (1970) *Armillaria mellea* in Central Africa: studies on substrate colonisation relating to the mechanism of biological control by ring-barking. In: Toussoun, T.A., Bega, R.V. and Nelson, P.E. (eds) *Root Diseases and Soil Borne Pathogens*. University of California Press, Berkeley, California, pp. 150–152.

Thomas, H.E. (1934) Studies on *Armillaria mellea* (Vahl) Quél.: infection, parasitism and host resistance. *Journal of Agricultural Research* 48, 187–218.

Tsopelas, P. (1999) Distribution and ecology of *Armillaria* species in Greece. *European Journal of Forest Pathology* 29, 103–116.

van der Kamp, B.J. and Hood, I.A. (2002) Armillaria root disease of *Pinus radiata* in New-Zealand. 2. Invasion and host reaction. *New Zealand Journal of Forest Science* 32, 103–115.

van der Pas, J.B. and Hood, I.A. (1984) The effect of site preparation on the incidence of armillaria root rot on *Pinus radiata* four years after conversion from indigenous forest in Omataroa forest. In: Kile, G.A (ed.) *Proceedings of the Sixth International Conference on Root and Butt Rots, Melbourne, Victoria, and Gympie, Queensland, Australia, August 25–31, 1983*. CSIRO, Melbourne, Australia, pp. 387–397.

Van der Plank, J.E. (1963) *Plant Diseases, Epidemics and Control*. Academic Press, London.

West, J.S. (2000) Chemical control of *Armillaria*. In: Fox, R.T.V. (ed.) *Armillaria Root Rot: Biology and Control of Honey Fungus*. Intercept, Andover, UK, pp. 173–182.

Whitney, R.D. and Irwin, R.N. (2005) Comparison of armillaria root disease on burned and unburned, harvested sites in Ontario. *Forestry Chronicle* 81, 56–60.

Whitney, R.D., Fleming, R.L., Zhou, K. and Mossa, D.S. (2002) Relationship of root rot to black spruce windfall and mortality following strip clear-cutting. *Canadian Journal of Forest Research* 32, 283–294.

Zas, R., Solla, A. and Sampedro, L. (2007) Variography and kriging allow screening *Pinus pinaster* resistant to *Armillaria ostoyae* in field conditions. *Forestry* 80, 201–209.

Zeller, S.M. (1926) Observations on infections of apple and prune roots by *Armillaria mellea* Vahl. *Phytopathology* 16, 479–484.

9 Laminated and Tomentosus Root Rots

Kathy J. Lewis*
University of Northern British Columbia, Prince George, Canada

9.1 Pathogens, Significance and Distribution

Phellinus sulphurascens Pilát (syn. *P. weirii* (Murrill) Gilb.) and *Onnia tomentosa* (Fr.) P. Karst. (syn. *Inonotus tomentosus* (Fr.) Teng) are fungi that cause root rot of conifer species, and are most prevalent in the northern hemisphere. Both fungal species are members of the family *Hymenochaetaceae*, and the two genera are closely related (Jeong *et al.*, 2005).

P. sulphurascens is the causal agent of laminated root rot, which attacks a number of conifer species, with Douglas-fir (*Pseudotsuga menziesii* (Mirb.) Franco) as its most economically important host. Following Murrill's description (Murrill, 1914) of a fungus causing white rot on western redcedar (*Thuja plicata* Donn ex. D. Don) as *Poria weirii* (Murrill) Murrill, the host range was extended to include Douglas-fir by Bier and Buckland (1947). Later, Gilbertson (1974) assigned the fungus to the genus *Phellinus* based on the morphology of the fruiting body. By this time, two forms of the fungus were commonly referred to in the literature, differentiated by host preference and characters such as fruiting body periodicity (Buckland *et al.*, 1954) and cultural characteristics (Clark, 1958). Angwin and Hansen (1988) determined that the two forms were intersterile, and in 1994 they were separated into two species by Larsen *et al.* (1994); the name of the Douglas-fir form was changed to *P. sulphurascens* and that of the redcedar form retained the name *P. weirii*. The name *P. sulphurascens* has been generally adopted for the Douglas-fir form of what was formerly called *P. weirii*, and will be used in this chapter.

O. tomentosa is the causal agent of tomentosus root disease, which is found on a range of conifer species, most notably spruce (*Picea* spp.). Its taxonomic history is convoluted, and the species has been transferred to different genera multiple times. The name *O. tomentosa* has been adopted in recent studies (e.g. Germain *et al.*, 2009) and will be used in this chapter.

Laminated root rot is found in the east Asian countries of China, Japan, Russia (Siberia) and (just recently) in Turkey (Doğan and Karadelev, 2009); it is also found in North America where it occurs in the Pacific Northwest, from northern California east to Montana and north to British Columbia (BC), where it is found in the southern interior and on the south coast. Although Douglas-fir is the most economically important host, true

* E-mail: lewis@unbc.ca

firs (*Abies* spp.) and mountain hemlock (*Tsuga mertensiana* (Bong.) Carrière) are also highly susceptible, followed by spruce and western hemlock (*T. heterophylla* (Raf.) Sarg.). Pine species (*Pinus* spp.) are tolerant of infection, which means that they are infrequently infected unless growing in close association with inoculum sources, and they are rarely killed. Cedars (*Thuja* spp. and *Chamaecyparis* spp.) are resistant to attack by *P. sulphurascens* and hardwood species are immune (Thies and Sturrock, 1995).

O. tomentosa has a much larger geographic range, and is found extensively throughout the boreal, sub-boreal and maritime forests of North America (Davidson and Redmond, 1957; Robinson-Jeffrey and Loman, 1963; Aho, 1971; Whitney, 1995). The actual range of *O. tomentosa* is somewhat uncertain due to the existence of the related species, *O. circinata* (Fr.) P. Karst. This fungus has been reported from a number of pine species in the south-east USA (Boyce, 1963; Ross, 1966) and differences between the two species are subtle and may be overlooked. *O. circinata* occurs more commonly on pine species than does *O. tomentosa*, it has curved hymenial setae (compared with straight setae in *O. tomentosa*), and the *O. circinata* fruiting body is often sessile rather than stipitate. In addition to its occurrence in North America, *O. tomentosa* has been reported from other regions in the northern hemisphere, including Russia, Scandinavia, China and Europe. While *O. tomentosa* has been isolated from almost all the conifer genera found in temperate and boreal forests, spruce species are the most susceptible and the most damaged. In North America, *O. tomentosa* has caused extensive losses from mortality and growth reduction in white spruce (*P. glauca* (Moench) Voss), black spruce (*P. mariana* (Mill.) Britton, Sterns & Poggenb.) and Engelmann spruce (*P. engelmannii* Parry ex Engelm.), and their hybrids (Whitney, 1962; van Groenewoud and Whitney, 1969; Lewis, 1997; Lewis *et al.*, 2005).

P. sulphurascens and *O. tomentosa* are both facultative parasites that infect a live host, eventually killing it before entering a saprophytic phase during which the fungus lives by breaking down and utilizing wood constituents for nutrition. Both fungi are able to break down and utilize cellulose and lignin, leaving behind a characteristic pitted decay pattern, which is called a white rot owing to the bleached appearance that results from lignin removal (Blanchette, 1980). However, both fungi differ from other decay-causing fungi in that they are able to attack and kill live hosts. Of the two fungi, *P. sulphurascens* is the more aggressive pathogen on its main hosts, and can kill small trees in a few years, while larger trees take up to several decades to die (Thies and Sturrock, 1995). Surveys of infested sites have shown mortality rates in the most susceptible hosts of 20% to 40% (Hagle, 2009). Estimates of the percentage of forested area infested with *P. sulphurascens* range widely and have been shown to vary with site conditions, species composition and stand history (Byler *et al.*, 1990). During the time from infection to mortality, the fungus causes reduction in growth due to reduced nutrient and water uptake, and to the allocation of resources to defence rather than to growth. Thies (1983) estimated growth rates in trees killed by *P. sulphurascens* as 32% less than those of healthy counterparts in their last 10 years of growth. Goheen and Hansen (1993a) estimated a combined growth reduction and mortality loss of 40–70% in infested areas.

In comparison, *O. tomentosa* develops more slowly along a tree's root system and typically takes several decades to cause mortality (Whitney, 1962). Surveys for root disease caused by *O. tomentosa* have shown that the disease occurs patchily, with some areas having no disease, and others with over 80% of the spruce trees dead (Whitney, 1962; Lewis and Hansen, 1991a). In spruce, the fungus causes a root and butt rot, which weakens the root system and causes significant reductions in net volume due to cull. In pine, though, the fungus causes cambial necrosis, which results in a faster rate of mortality than in spruce (Schulting, 1987; Lewis and Hansen, 1991a). The growth of infected mature spruce trees declines slowly over time relative to healthy trees, and the reduction does not become significant until the later stages of disease progression. Growth averages 10% less than that of healthy trees over the most recent 5 year period (Whitney, 1980; Lewis, 1997). Both growth reduction and losses due to butt cull are related to the number of roots colonized by the fungus, and the subsequent reduction in net volume (up to 30%) is significant when about 75% of the root system is colonized (Lewis, 1997). Effects of the disease on spruce plantations in western BC have not been well quantified, but in eastern Canada net losses of up to 19% of the highly valued butt log were found in 65 year old white spruce in Ontario. These losses apply to decay only; losses were greater if the stained wood, which precedes decay, could not be utilized either (Whitney, 1993). Significant tree mortality due to *O. tomentosa* was reported in Quebec (Lachance, 1978) and Ontario (Whitney, 1993), where cumulative mortality over a 6 year period was over 10%, and butt cull averaged 13.8% of gross merchantable volume. Losses to *P. sulphurascens* in second-growth stands that had been either naturally or artificially regenerated with host species were estimated to be even greater, up to half of the potential harvest volume after two or more rotations of a susceptible host species (Thies and Sturrock, 1995).

Both *P. sulphurascens* and *O. tomentosa* interact with other disturbance agents, particularly bark beetles, by predisposing infected trees to attack by these secondary pest species. Douglas-fir beetle (*Dendroctonus pseudotsugae* Hopkins) often preferentially attacks Douglas-fir infected with *P. sulphurascens* (James *et al.*, 1984) and sometimes masks the contribution of the fungus to tree mortality. In contrast, the spruce beetle *Dendroctonus rufipennis* Kirby does not appear to preferentially select trees infected by *O. tomentosa*, although beetle attacks are more successful (more galleries and more brood) on infected trees during endemic population levels (Lewis and Lindgren, 2002). This difference is probably due to the greater length of time that it takes for *O. tomentosa* to kill its host compared with *P. sulphurascens*, which results in deterioration of the phloem habitat as well as reduced defence ability in the former case, and just a reduced defence response in the latter case.

Finally, with regard to the significance of these two root disease fungi, it is important to note that both of them are natural disturbance agents and, as such, influence ecosystem dynamics. They both produce openings in the forest canopy, which can cause significant changes in species composition and distribution in infested compared with non-infested areas, generally as a result of changes in resource availability, but also because of species-specific responses to infection by the root disease fungi. Changes in stand structure and composition in response to infection have been well documented for *P. sulphurascens* (e.g. Dickman and Cook, 1989; Holah *et al.*, 1993; Ingersoll *et al.*, 1996), but have not been well studied for *O. tomentosa*. More is known about *P. sulphurascens* in general because it has caused more obvious losses to an important commercial species in a region where forest harvesting has a relatively long history compared with areas affected by the more subtle *O. tomentosa*. Canopy openings caused by *P. sulphurascens* can be quite diffuse or well defined depending on the forest type affected; however, in general they tend to be much larger than those caused by *O. tomentosa*, which is limited in spread by the prevalence of resistant species (e.g. subalpine fir, *Abies lasiocarpa* (Hook.) Nutt.) in many spruce forests and the slow rate of spread of the fungus along colonized roots (Whitney and van Groenewoud, 1964). In both cases, mortality of host trees results in coarse woody debris, standing snags and greater stand complexity, all of which can enhance habitat and biodiversity values, which are increasingly important in society's view in comparison with more traditional timber values.

9.2 Diagnosis

P. sulphurascens and *O. tomentosa* cause a root rot in conifers and therefore disrupt the normal functions of the roots, including water and nutrient uptake and translocation, storage and stability. Symptoms of the diseases in standing trees therefore reflect root dysfunction and include reduced shoot growth, chlorosis and thinning of the foliage and, a few years before mortality, a stress-induced cone crop of many smaller than usual cones. Basal resinosis is an occasional symptom on spruce infected with *O. tomentosa*, but is not frequent enough to be reliable for diagnosis. When a diseased tree uproots owing to loss of structural integrity, many of the roots are no more than stubs, forming what is called a 'root ball'. This tree-level symptom is most prominent in laminated root rot. Tomentosus root disease can also result in root balls, but more commonly only a few of the roots will be stubby and rotted off, with others lifting up as more of a 'root plate' (Plate 10). This may be due to a tendency for windthrow of even healthy trees in spruce ecosystems, which typically have moist to wet soils, and to the shallow root systems of spruce compared with the deeper rooting habit of Douglas-fir and other main hosts of *P. sulphurascens*.

Trees weakened by root rot before falling often fall in a random pattern compared with the ordered pattern that results from wind-driven failure. Standing tree and windthrow mortality caused by these root pathogens result in canopy gaps that are often colonized by immune or resistant species, and this shift in species composition can persist for many years (Ingersoll *et al.*, 1996). Stand-level symptoms of these root diseases therefore include structural diversity (uneven-aged stands), patches of hardwoods or resistant conifers, and brushy patches with scattered standing dead and windthrown trees. Live trees showing crown symptoms are often at the margins of infection centres, and secondary bark beetles may also be present, particularly in areas with laminated root rot (Goheen and Hansen, 1993a).

Roots infected by *P. sulphurascens* can be identified by the presence of ectotrophic (surface) mycelium, which covers much of the colonized root. This mycelium is cream coloured and is most evident on roots with thin bark. A brown crust of mycelium can occur in patches on root surfaces below the duff (i.e. the partly decayed organic matter on the forest floor), particularly around root crotches. Thin, brown, whisker-like setal hyphae can be found in the ectotrophic mycelium as well as in the colonized wood itself (Plate 11A). The presence of these hyphae is a diagnostic feature for *P. sulphurascens* (Thies and Sturrock, 1995). Ectotrophic mycelium is not common in roots infected by *O. tomentosa*, but in roots that are less than 2–3 cm in diameter yellowish white mycelium can be found in the bark (Plate 11B), just below the outer surface (Lewis *et al.*, 1992).

The early stages of root decay appear as stain in the wood. Both stains are reddish brown in colour and can extend from the roots up the bole of the tree for several metres. Wood colonized by both the fungi has a notable pitted appearance due to the breakdown of cellulose and lignin. Decay caused by *O. tomentosa* progresses from a light pink to a dark brown stain, followed by the initiation of small pits in the wood, which grow larger as the decay progresses (Lewis *et al.*, 1992). The pattern of decay in spruce is very similar to that caused by *Porodaedalea pini* (Brot.) Murrill, which causes a true heart rot of several conifer species, including spruce, and can confuse the diagnosis of tomentosus root rot. Decay caused by *P. sulphurascens* also has a pitted appearance, but the main distinguishing feature is the separation of the annual rings resulting in the laminated appearance for which the disease is named (Plate 11C).

The fruiting bodies of *P. sulphurascens* are uncommon, particularly in interior forests due to the dry conditions (Thies and Sturrock, 1995), and they are difficult to find as they form on the underside of fallen trees and failed root systems. They are annual (perennial for *P. weirii* on cedar), flat, with a pale brown margin and a light brown pore layer, turning darker with age. The pores are circular to angular and the context (tissue constituting the fruiting body) is yellowish (Gilbertson and Ryvarden, 1987). The fruiting bodies of *O. tomentosa* are commonly produced, except during dry summers. They are also annual, but with a stipe up to 3.5 cm long. The cap is often centrally depressed, circular and up to 11 cm in diameter. The upper surface is yellowish brown in colour and slightly fuzzy. The pore layer is pale brown, becoming darker with age, and the pores are angular (Gilbertson and Ryvarden, 1987). The fruiting bodies of *O. tomentosa* form on the ground, directly above infected roots, but due to annual and seasonal variation in their production, are not reliable for diagnosis.

In culture, *P. sulphurascens* grows fairly rapidly with an even advancing zone. The mat is white at first, becoming buff coloured then brown or yellow brown within 4 weeks

(Nobles, 1948). In comparison, cultures of *P. weirii* from cedar grow more slowly, are darker than those of *P. sulphurascens*, and have columnar tufts of hyphae. Setal hyphae of the cedar species are shorter than those of *P. sulphurascens* (Larsen and Lombard, 1989). In both cases, the hyphae are hyaline to yellow brown, with simple septa and no clamp connections. Setal hyphae are numerous, slender, pointed, thick walled and dark brown (Nobles, 1948).

The growth of *O. tomentosa* in culture is slow compared with *P. sulphurascens*; the advancing zone is even with occasional fans of faster growth. The hyphae are hyaline to dark brown and the hyphal mat ranges in colour from honey yellow to dark brown, and with age sometimes producing fruiting body primordia. In cultures more than 3 weeks old, the hyphae are branched with numerous chlamydospore-like swellings (Nobles, 1948). With the exception of the chlamydospore-like swellings, the cultural characteristics of *O. tomentosa* are similar to those of *P. pini*, and the decay caused by both fungi produces a very similar pattern, so they can be difficult to distinguish in wood samples or in culture. A number of methods have been developed to differentiate *P. pini* and various species of *Onnia*, including differential culture media (Hunt, 1997) and polyclonal antibodies (Hunt *et al.*, 1999); these require material derived from cultures. Germain *et al.* (2002) developed a PCR-based method to distinguish these two fungi directly from colonized wood. The method utilizes species-specific primers for the internal transcribed spacer (ITS) region of the ribosomal DNA (rDNA) gene of *O. tomentosa* and has been shown to distinguish this fungus from several other closely related species.

9.3 Infection Biology

Both *P. sulphurascens* and *O. tomentosa* enter their hosts through the root systems and spread from tree to tree across root contacts. Both fungi produce mycelium on the root surface, which initiates new infections. This ectotrophic mycelium is much more extensive in *P. sulphurascens* (Buckland *et al.*, 1954) than in *O. tomentosa*, which has less extensive intra-bark mycelium (Lewis *et al.*, 1992). This difference in surface mycelium explains the faster rate of spread that is found in *P. sulphurascens*. For both species, once a root is infected, the fungus also advances within the heartwood, and upon death of the root, spreads radially into the sapwood.

Lewis *et al.* (1992) found that for *O. tomentosa* in spruce, small roots (less than 4 cm in diameter) were most likely to become infected in a natural setting, although Whitney (1967) observed that larger roots (7–8.5 cm in diameter) become infected at a higher rate than small roots when they are artificially inoculated. Penetration of the bark by *O. tomentosa* occurs through small feeder roots (Lewis *et al.*, 1992) and in areas where the bark integrity is disrupted (Oullette *et al.*, 1971). Root contacts between infected and healthy roots only result in infection of the healthy root when ectotrophic or intra-bark mycelium are present. In spruce, the infection progresses into larger lateral roots and colonization of the heartwood progresses. Stain of the heartwood occurs up to 1 m in advance of the decay. The butt of the tree becomes colonized and butt rot can extend several metres up the tree. Disease development in pine is different in that the fungus is more commonly found killing cambium and phloem rather than causing rot in the heartwood. Pine roots from infected trees have mycelium in the bark at a higher rate than spruce roots, and discrete decay columns are often lacking in pine, which tends to have discontinuous bands of stain and advanced decay (Lewis and Hansen, 1991a).

The infection process for *P. sulphurascens* is similar, although this fungus appears better able to directly penetrate and kill the root bark; however, it also enters through feeder roots and injured bark. The ultrastructure of the host–pathogen interaction has been studied for *P. sulphurascens* but not for *O. tomentosa*. Within 2 days of inoculation, hyphae of *P. sulphurascens* are abundant and strongly adhered to the root surface. Several hyphal morphologies are present, and some of these form appressoria-like structures which directly penetrate the host bark. Hyphae penetrate cells and spaces between cells.

Within 7 days of inoculation, extensive colonization of the root cortex occurs, and a range of haustorial structures are present, as well as narrow hyphal tips that initiate disintegration of host cell walls and membranes during the infection stage. Necrosis of infected tissues is evident by 14 days post inoculation accompanied by significant cortical cell decay (Sturrock *et al.*, 2007; Islam *et al.*, 2009).

Trees have evolved a variety of defence mechanisms in response to continuous pressure from insects, pathogens and abiotic damage agents. These mechanisms can be pre-existing or induced, and can be physical barriers such as necrophylactic periderms or chemical defences such as antibiotic compounds or hydrolytic enzymes (Franceschi *et al.*, 2005). Some of the defence responses are general while others are specifically induced in response to some elicitors, such as cell wall compounds. The defence mechanisms that are employed by trees attacked by *P. sulphurascens* and *O. tomentosa* have not been well studied, especially for the latter fungus. The production of resin (a combination of terpenes and resin acids) is a defensive response to attack by pathogens such as *Armillaria* spp. and bark beetles. Resins can be preformed or induced upon attack, and can be observed streaming down trees that produce them as a defence response. Neither *P. sulphurascens* nor *O. tomentosa* appear to elicit a strong resin defence response in their main hosts, although resin soaking of pine roots infected with *O. tomentosa* is more common than in spruce (Schulting, 1987). Whitney and Denyer (1969) observed that while resin impregnation of wood wounded for artificial inoculation interfered with the development of some fungi, including *O. circinata*, it did not seem to affect the rate of colonization by *O. tomentosa* in spruce. Necrophylactic periderm formation is another common defence response in trees, and while it has been shown to be effective at limiting the spread of *Armillaria ostoyae* (Romagn.) Herink in larch (*Larix* spp.) and other resistant hosts (Robinson and Morrison, 2001), it does not appear to be a common defence response for *P. sulphurascens* and *O. tomentosa*. This may be due to the different mode of attack by these two fungi, which kill root cambium but also invade and decay the wood for nutrition, compared with *A. ostoyae*, which in its parasitic phase, attacks mainly the cambium, where necrophylactic periderms are able to form.

9.4 Epidemiology

As stated previously, both *P. sulphurascens* and *O. tomentosa* spread by root contacts, and the former is particularly effective at this mode of spread owing to the extensive network of ectotrophic mycelium produced on infected roots. In a classic study that involved the excavation of Douglas-fir and other conifer roots, Wallis and Reynolds (1965) observed ectotrophic mycelium on the root collar, on the large lateral and tap roots and on roots as small as 1 mm in diameter. The mycelium was extensive in mineral soil to as deep as 125 cm, but less common in the duff layer. The fungus can directly penetrate the bark of uninjured roots, and infection points along roots can be found well in advance of decay in the root. The rate of invasion is higher when roots are injured (Wallis and Reynolds, 1962).

P. sulphurascens is able to colonize dead roots and woody material in the soil although the penetration of wood is more limited than colonization of the bark surface, particularly as the length of time from mortality to inoculation increases. After 1 year, the colonization of dead roots by *P. sulphurascens* is sporadic (Wallis and Reynolds, 1965). The fungus is not able to spread far, if at all, in soil.

Once a tree is killed by *P. sulphurascens*, the fungus can live saprophytically in the root system for decades, up to 50 years for large root systems (Hansen, 1976, 1979). As the root systems age beyond several decades, the fungus becomes more limited to the interior of the roots and less extensive on the root surface where it can readily cause infection of healthy roots that come into contact with the colonized roots. The longevity of *P. sulphurascens* in killed roots is attributed, at least in part, to the production of zone lines by the fungus (Nelson, 1964). These lines separate *P. sulphurascens* from antagonistic organisms or adverse environmental conditions, and result from enzymatic activity that oxidizes phenolic substrates and ultimately yields dark melanin pigments, which resist microbial attack and serve as a physical barrier (Li, 1981). This ability to persist in dead stumps for multiple decades, and to invade dead roots and woody material from host species for a limited period of time, means that laminated root rot is a 'disease of the site' because it persists from one rotation or disturbance event to the next. Spread of the fungus by root contacts is by far the dominant means of spread for *P. sulphurascens*. Although fruiting bodies are produced, especially in coastal and wetter ecosystems, initiation of new infection centres by spores is thought to be a very rare event. Evidence for this is found in the lack of success of artificial inoculations using spores (Nelson, 1976) and the lack of genotypic diversity within infection centres (Bae *et al.*, 1994).

Root contacts are also an important means of spread by *O. tomentosa*, although the opportunities for successful infection of healthy roots in contact with infected roots is lower with this fungus than for *P. sulphurascens* owing to the limited amount of ectotrophic mycelium or intra-bark mycelium produced by *O. tomentosa*. Further, while *O. tomentosa* can directly penetrate bark, this usually occurs only on small roots with thin bark. The infection of larger roots requires access by the fungus through feeder roots, or other disruptions in the bark integrity (Lewis *et al.*, 1992). Like *P. sulphurascens*, *O. tomentosa* can persist in stumps for decades following death of the tree. Viable mycelium exists in stumps for at least 30 years, and is more common in spruce stumps than in pine stumps (Lewis and Hansen, 1991a). In spruce, once the infected tree is harvested or killed by the pathogen, the fungus in the root heartwood spreads out to the sapwood where it becomes more available to cause infection of healthy roots that may contact the infected root. Dead infected pine stumps do not appear to support the same radial growth of *O. tomentosa* in the root system. For this reason, and because of the greater length and horizontal nature of spruce root systems, spread of infection from stumps occurs at a greater rate and for greater distances from the inoculum source in spruce than in pine (Lewis and Hansen, 1991a).

Although vegetative spread by *O. tomentosa* across root contacts is less efficient and therefore slower than that by *P. sulphurascens*, unlike *P. sulphurascens*, *O. tomentosa* also spreads by spores. Direct evidence of the natural infection of host trees by basidiospores of *O. tomentosa* is lacking, and attempted inoculations of spruce roots by basidiospore suspensions were only successful in deep root wounds (Whitney, 1966). However, there is considerable vegetative compatibility and molecular evidence that supports the regular initiation of new infection centres by basidiospores, as described in the next paragraph.

Most wood-decaying basidiomycetes produce basidiospores that germinate to form haploid, homokaryotic hyphae. These hyphae grow until they meet other hyphae that have a different mating type allele. The two compatible hyphae fuse, cellular plasma combines and the nuclei become paired but not joined. The resultant dikaryon (two haploid nuclei) is what normally colonizes the host. When two unrelated dikaryons meet, a vegetative incompatibility response is triggered and a reaction line forms between the two genotypes, which prevents nuclear exchange (Todd and Rayner, 1980). This incompatibility system is thought to enhance the ability of individuals to compete for resources, and limit the spread of deleterious mycopathogens such as viruses. Lewis and Hansen (1991b) found unique vegetative compatibility groups among different disease centres, and these also showed different protein banding patterns. Larger and more diffuse disease centres comprised more than one vegetative compatibility group. Unique genotypes are normally produced through sexual reproduction, and in the basidiomycetes, the product of sexual reproduction is the basidiospore. Therefore, the most likely explanation for the common existence of unique genotypes of *O. tomentosa* within a single stand is the establishment of new infections through basidiospore dispersal and, ultimately, their germination and infection of host roots. DNA-based molecular approaches have supported that conclusion; these include identification of unique genotypes using random amplified polymorphic DNA (RAPD) (Gibson, 2005) and single strand conformation polymorphism (SSCP)

(Gibson, 2005; Germain *et al.*, 2009). Gibson (2005) found high genotypic variation in all six populations tested, moderate levels of gene diversity, significant linkage disequilibrium and significant deviation from Hardy–Weinberg equilibrium. Most of the genetic variation (70%) occurred within populations, indicating that vegetative reproduction was common, but accompanied by frequent sexual reproduction and the establishment of new genotypes. Both unmanaged, old-growth spruce forests and young spruce plantations showed the same population structure. In a study of *O. tomentosa* populations from a spruce plantation in Quebec, SSCP was carried out for two mitochondrial and three nuclear loci. Most of the spatially clustered samples were in the same genet, but multiple genets were found with an average size of 3.4 m based on the maximum distance between two samples in the same genet. The nuclear loci were in Hardy–Weinberg equilibrium in the larger genets, but the smaller genets were deficient in heterozygotes (Germain *et al.*, 2009). Collectively, these results support the initial development of infection centres by basidiospores, followed by vegetative expansion, although the mechanism of infection by the spores is unknown.

Among the root decay fungi that attack conifers, *P. sulphurascens* is one of the fastest at root contact spread due to its extensive network of ectotrophic mycelium, which facilitates root-to-root spread. While the fungus can grow approximately 40 cm along a root in 1 year (Wallis and Reynolds, 1962), the rate of disease advance is usually slower and quite variable. Disease expansion rates in a mountain hemlock stand in Oregon reached 34 cm year^{-1}, yet a nearby stand with greater species diversity showed only a 23 cm year^{-1} rate of advance (McCauley and Cook, 1980). Coastal Douglas-fir plantations in BC that were monitored for almost 30 years showed variable expansion rates, with some centres or edges of centres not advancing at all, and others advancing up over 60 cm year^{-1} (Bloomberg, 1990). Laminated root rot centres in high elevation, mixed conifer stands advanced at a rate of approximately 30 cm year^{-1} (Nelson and Hartman, 1975).

Less is known about the rate of spread of *O. tomentosa*, but due to its limited ectotrophic mycelium, the rate is most likely to be considerably slower, at least for vegetative spread, than it is for *P. sulphurascens*. Based on inoculation studies, Whitney (1962) determined that spread of the fungus along root systems of spruce trees averaged 4 cm year^{-1}, with a range of less than 1–15 cm year^{-1}. Trees planted within 4 m of a colonized spruce stump and within 3 m of a colonized pine stump had between a 10% and a 25% chance of becoming infected over a 30 year period (Lewis and Hansen, 1991a), suggesting that the maximum rate of primary vegetative spread in spruce and pine plantations in BC is about 10 cm year^{-1}. The relatively slow spread rate of *O. tomentosa*, and the fact that it also spreads by spores, results in a patchy distribution of disease centres. The disease centres themselves are often diffuse and difficult to identify as actual centres, and this pattern of distribution has earned the disease the common name of 'stand opening disease' (Whitney, 1962). Stand openings range in size from a few trees to coalesced patches of trees that can be a hectare or more in size. Dead, dying and fallen trees are found within the infested areas, with symptomatic trees at the margins, although symptoms are not definitively expressed in all infected trees.

Laminated root rot infection centres are sometimes very discrete, with well-defined margins and an advancing zone of infection (Childs, 1963; McCauley and Cook, 1980). In these forest types, in addition to the advance of the fungus from the infection centre into the surrounding healthy forest, there is often regeneration of trees within the infection centre, which, as they grow, come into contact with inoculum and are killed. This phenomenon is described as a wave pattern of mortality by Dickman and Cook (1989), and it can result in concentric bands of asymptomatic or uninfected trees, and bands where regeneration has been killed. In other cases, the disease is distributed throughout a stand, without any association with disease centres (Thies and Nelson, 1997).

The distribution and rate of spread of both diseases are affected by a number of site and stand factors. Stands that have a higher proportion of susceptible species are more likely to support the expansion of root disease than those with significant numbers of immune or resistant species owing to the lower probability of root contacts with susceptible species and the slower rate of spread of the hyphae in resistant hosts (McCauley and Cook, 1980; Bernier and Lewis, 1999). However, species composition is only one of several factors that determine root disease incidence, and even if susceptible species are present, other factors, such as moisture regime, aspect and competition from other biotic agents can limit disease development on the susceptible hosts (Byler *et al.*, 1990).

Host species density is not thought to be an important factor in disease incidence or spread, although van Groenewoud and Whitney (1969) reported a positive relationship between host density and disease incidence for *O. tomentosa*; conversely, Bernier and Lewis (1999) and Lewis *et al.* (2004) did not find a significant association with host species density. For both diseases, larger trees have been shown to have a higher probability of infection (Bloomberg and Reynolds, 1985; Lewis *et al.*, 2004), which may explain the lack of a relationship with host density. Larger trees have larger root systems, and therefore a higher likelihood of contacting inoculum.

Unlike the relationship between secondary insects, such as the Douglas-fir beetle, and the predisposition of weakened trees to attack by these bark beetles, tree vigour does not play a role in the susceptibility of trees to either *P. sulphurascens* or *O. tomentosa*. This idea was suggested for *P. sulphurascens* as a result of the observation that young trees regenerating in mortality centres do not become infected for many years, and that trees regenerating near the 'killing front' of the infection centre were more vigorous, as indicated by fast growth (Waring *et al.*, 1987). However, there are no direct observations of the predisposition of host trees to laminated root rot, and a direct test of the vigour hypothesis found no relationship between tree vigour and either inoculation success or colonization rate (Goheen and Hansen, 1993b). Likewise, there is no evidence to support an effect of tree vigour on the susceptibility of spruce or pine to *O. tomentosa*.

As the predominant means of spread for both root pathogens occurs at root contacts, the conditions of the site and soil are expected to affect disease incidence and severity. In a study of the incidence of tomentosus root disease in spruce forests in central BC, Bernier and Lewis (1999) determined that soil moisture regime, soil texture and slope position (which is related to soil moisture) all had significant influence on disease incidence. The incidence of *O. tomentosa* was low in wet sites with a level or depressed mesoslope (local) position, and highest in the more dry sites, although it should be noted that, overall, the study area was a relatively wet forest type. A comparable study in Alaska produced similar results; tomentosus root disease incidence was negatively related to percentage soil moisture content, and disease was found more commonly on upper slope positions (Lewis *et al.*, 2004). *O. tomentosa* appears to be most common on well-drained soils that are moist enough to support a high population of spruce, but not wet enough to limit fungal growth.

P. sulphurascens, like *O. tomentosa*, is more commonly found on fresh or dry soils in the Douglas-fir forests of coastal BC, rather than on very wet or very dry soils (Bloomberg and Beale, 1985). Further south, in the forests of the Oregon coast range, the incidence of *P. sulphurascens* is greatest on ridges and least at lower slope positions (Kastner *et al.*, 1994). This concurs with the findings of Byler *et al.* (1990), who observed *P. sulphurascens* in grand fir (*Abies grandis* (Douglas ex D. Don) Lindl.) and hemlock (*Tsuga* spp.) forest types, but not in the Douglas-fir forest types (despite the susceptibility of this species to the fungus) which were found primarily on southerly aspects and would, therefore, be drier. The availability of host species, and the effect of environmental factors such as soil moisture and aspect on host distribution, make it difficult to separate the influence of site factors from the influence of host availability on disease incidence. It is unlikely that a single factor, or even a group of factors, can reliably predict the incidence or severity of root disease, though given the lack of symptom expression in some infected trees, the cost of surveys, and a desire to enable proactive disease management rather than reactive management, predictive susceptibility rating systems are a desirable and useful tool. Various models have been developed that rely on a combination of site and species factors (e.g. Williams and Marsden, 1982) to predict the occurrence of root disease, although because of the problem of interactions among species composition and site factors, a more reliable approach may be to use comprehensive ecosystem classifications that integrate biophysical factors, rather than modelling them individually. For example, Byler *et al.* (1990) used habitat types and Bloomberg and Beale (1985) used the British Columbia Biogeoclimatic Ecosystem Classification system to rate laminated root rot hazard; Bernier and Lewis (1999) applied the same ecosystem classification system to tomentosus root disease.

9.5 Management Strategies and Tactics

Most of the management strategies for *P. sulphurascens* and *O. tomentosa* are based on knowing that the disease is present on a site in the first place, and then managing species composition and inoculum levels to limit spread and severity of the disease. It is important to note that both of these root diseases are persistent and can live off root systems for decades.

9.5.1 Avoidance

There are two main elements to avoidance-based management strategies for *P. sulphurascens* and *O. tomentosa*. One is access to and use of reliable survey and detection methods, and the second is the application of hazard or susceptibility rating systems to avoid sites that are most likely to carry existing inoculum for the disease. In most cases, avoidance management strategies must be used in conjunction with other management strategies because of the persistence of inoculum on infested sites.

As discussed earlier, both fungi can infect hosts and cause significant growth and butt cull losses with little external evidence of infection. This makes detection based on remote sensing or just external indicators during ground-based surveys unreliable, particularly at the smaller site-level scale where accurate mapping of disease distribution is important for the development of site-specific management prescriptions (Thies and Sturrock, 1995).

The normal distribution pattern for *O. tomentosa* consists of small (a few trees) disease centres that sometimes coalesce into larger stand openings. These disease centres themselves are often patchily distributed (Lewis and Hansen, 1991a) within a single stand as a result of changes in species composition or site conditions. The lack of large, discrete disease centres, and the lack of reliable crown symptoms, makes the quantification of disease incidence and severity difficult.

In the case of *P. sulphurascens*, obvious disease centres do occur in some forest types, such as dry Douglas-fir or grand fir sites (Williams and Marsden, 1982) and the mountain hemlock forests of the Oregon Cascades (Dickman and Cook, 1989), to the extent where mortality centres are easily recognizable from aerial photographs (Hansen and Goheen, 2000). However, in some forest types, laminated root rot also occurs without association to apparent disease centres, and scattered diseased trees can occur over large areas (Thies and Nelson, 1997). Similar to the situation with *O. tomentosa*, this latter distribution pattern is more difficult to identify as laminated root rot, and much more difficult to quantify.

Reliable line intercept survey methods that use above-ground symptoms have been developed for *P. sulphurascens* that are effective when the forest type is conducive to the development of discrete disease centres (Bloomberg *et al.*, 1980), but do not work well when the disease is scattered and not associated with disease centres.

No specific survey method has been developed for *O. tomentosa*, and at an operational level, surveys for this disease are rarely carried out before harvest. The lack of apparent infection centres and the lack of obvious crown symptoms produced by this fungus until just a few years before death result in a lack of detection during normal pre-harvest and silvicultural operations in North America. The impact of tomentosus root disease, while shown through research to be significant, occurs over such a long time frame, and has relatively subtle effects compared with those of other disturbance agents such as spruce beetle, that it often goes unnoticed during preparations for harvesting. In western Canada, in regions where tomentosus root disease is common, forest plantations are a relatively recent landscape feature (less than 40 years old), and all of the current forest harvest, except on the coast of BC, is still in primary (unmanaged) forests. While it has been shown that the fungus is able to attack planted or natural regeneration trees that contact inoculum in colonized roots of harvested trees, the rate of spread is much slower than that of *P. sulphurascens*. This, and the relative youth of plantations where the effects of *O. tomentosa* may not yet be evident, suggest that the full impact of tomentosus root disease in managed forests is not only unknown, but is also not adequately considered in the development of silvicultural prescriptions. Older spruce plantations in eastern Canada have shown significant losses to *O. tomentosa*, and genetic evidence suggests that most of the infestation of plantations occurs through inoculum harboured in stumps from the previous stand. There is also some evidence from spruce plantations on former agricultural land that infections can be initiated from outside the plantations through the introduction of basidiospores and successful establishment of new infections.

Although pre-harvest surveys for tomentosus root disease are not carried out on a regular basis, and are very expensive owing to the high labour cost involved in sampling roots, post-harvest stump-top surveys have been used to quantify disease incidence and determine the distribution across a harvested area. Stump top surveys in spruce will give a reliable estimate of disease distribution because approximately 80% of spruce stumps whose root systems were infected before harvest will have evidence of advanced decay at the stump surface, compared with only about 35% of pine stumps (Lewis and Hansen, 1991a; Whitney, 1997). The purpose of pre-harvest or post-harvest surveys is to determine whether some treatment is required to limit the spread of the pathogen from colonized stumps. Harvest sites can be stratified into areas that require the use of a resistance or eradication strategy, and areas where inoculum is lacking and therefore can be regenerated with the host species.

Avoidance strategies can also use predictive tools, particularly at landscape scales, to identify areas that are most likely to be infested with root disease. As discussed above, a number of models for estimating the incidence of disease have been developed. In the case of *O. tomentosa*, one model uses site series, i.e. a combination of soil moisture and nutrient regimes, to classify areas by the probability of root disease being present (Bernier and Lewis, 1999). Similarly, Bloomberg and Beale (1985) identified ecosystems that were more likely to have *P. sulphurascens* present. These models can be used to identify areas that should be avoided for harvesting, or at least for the re-establishment of a host species, or that may need more detailed root disease surveys before management.

9.5.2 Exclusion

Exclusion disease management methods require the prevention of the introduction of inoculum to a site. This management strategy does not apply well to laminated and tomentosus root rots because of the persistence of inoculum in stumps from one rotation to the next. On an international scale though, *P. sulphurascens* is listed (under the name of *P. weirii*) as an A1 pest recommended for quarantine in the area covered by the European and Mediterranean Plant Protection Organization (EPPO, 2011). The A1 designation is for pests that are not yet present in the area covered by the quarantine, and is intended to prevent spread of this pathogen into the area under quarantine.

9.5.3 Eradication

Eradication methods aimed at reducing inoculum are common strategies for laminated and tomentosus root diseases. The goal in this case is not total eradication of inoculum, but a significant reduction in inoculum volume. The most traditional method of inoculum reduction for root diseases has been stump removal. The removal of colonized stumps and large roots with the use of an excavator significantly reduces inoculum remaining in an infested site, and although small pieces of colonized root and wood may still be present, they normally cannot support the establishment of new infections for many years postharvest.

Stump removal has been shown to be effective in reducing mortality from *P. sulphurascens*, and the amount of inoculum is reduced sufficiently enough to enable successful establishment of primary commercial species such as Douglas-fir and grand fir (Morrison *et al.*, 1988; Thies *et al.*, 1994; Cleary *et al.*, 2013). Less information is available regarding the efficacy of stump removal for *O. tomentosa*, and the treatment is currently only being used in trials and not on an operational basis, although it does appear to be effective in reducing mortality from the pathogen. Stump removal is very expensive, which is one of the limitations on its use, particularly for the smaller diameter spruce common to sub-boreal spruce forests affected by tomentosus root disease.

As well as using an excavator to remove existing colonized stumps and large roots, stump removal is also carried out by push falling whole trees using a large excavator, which pushes the tree over then picks it up and knocks the soil from the root system. Both methods are expensive and cause more site disturbance than conventional felling and skidding. Soil compaction is the most important negative effect of stump removal and can have a negative effect on tree growth (Smith and Wass, 1991, 1994). Stump removal via conventional harvesting followed by destumping or push falling is cost-effective on high quality, heavily infested sites where susceptible host species are the desired species, and with coarse, well-drained soils that are less susceptible to compaction (Thies and Sturrock, 1995). This combination of attributes occurs more frequently for laminated root rot forest types than it does for tomentosus root disease types, and may explain why stump removal is not currently employed operationally for tomentosus root disease.

Two alternatives to physical stump removal that are aimed at inoculum reduction have been studied for laminated root rot. One involves the use of chemical fumigants, applied to holes drilled in the stumps, to kill the fungus. Several fumigants, including chloropicrin (trichloronitromethane), Vapam (sodium *N*-methyldithiocarbamate) and Vorlex (methylisothiocyanate) have been shown to eliminate viable *P. sulphurascens* mycelium from Douglas-fir stumps and major roots (Thies and Nelson, 1982, 1987). However a follow-up study to determine the effect of stump fumigation on infection rate and seedling growth found that there was no difference in infection rate and seedling growth between treated and control plots (Thies and Westlind, 2006). The authors suggest that this was due to small roots harbouring viable mycelium throughout the stand in sufficient numbers to initiate new infections on replacement trees. Despite the repeated observation that chloropicrin used to treat colonized stumps did not significantly affect the surrounding soil biota or vegetation (Ingham and Thies, 1996; Wilson *et al.*, 1999), public opinion on the use of pesticides, particularly biocides, and their demonstrated lack of efficacy in preventing infection of regeneration trees do not support the use of this method of inoculum reduction.

The second alternative to mechanical stump removal is the use of competing fungi that are inoculated into colonized stumps to compete with the pathogen. Field studies demonstrated that *Trichoderma viride* Pers., a well-known antagonist to other fungi, could establish in stumps colonized by *P. sulphurascens* when inoculated into holes drilled in the stumps, though the penetration depth by the antagonistic fungus was limited to the top portion of the stump and did not reduce inoculum sufficiently to affect the rate of infection of regeneration trees (Nelson and Thies, 1985). Although the replacement of *O. tomentosa* by brown rot fungi was observed in old (20 or more years) spruce stumps (Lewis and Hansen, 1991a), this has not been studied as a means of inoculum reduction for tomentosus root disease. While the use of antagonistic fungi remains an option for limiting inoculum volume, the lack, so far, of any demonstrated efficacy, and of the development of efficient and effective treatment procedures, precludes the implementation of this method.

9.5.4 Protection

Protection-based management involves methods that protect susceptible trees from infection despite the presence of viable inoculum. For laminated root rot and tomentosus root disease, there are no treatment methods available that directly protect seedlings or trees from infection. Further, it has been demonstrated that fertilization treatments aimed at enhancing the growth and vigour of trees are also not effective in reducing infection by *P. sulphurascens* (Goheen and Hansen, 1993b; Thies and Westlind, 2006), and it is expected that the same is true for seedling or tree fertilization and *O. tomentosa*, although this has not been studied.

In BC, in cutover areas infested by tomentosus root disease, avoidance planting is one of the recommended treatment options. This strategy involves the identification of the root disease in stump tops, then avoiding the planting of susceptible species within a 5 m radius of colonized stumps. This helps to prevent infection of regeneration trees and the further spread of the fungus into the plantation. The method is based on providing time for the fungus to die out of colonized stumps before the roots of regeneration trees grow sufficiently to contact inoculum. Due to the extensive network of ectotrophic mycelium produced by *P. sulphurascens*, and the size of root systems in some areas where this fungus is common, this approach is not normally recommended for the management of laminated root rot.

9.5.5 Resistance

The use of resistant, tolerant or immune species for the regeneration of cutover areas is a common, widely accepted management strategy for both laminated root rot and tomentosus root disease. This strategy will help to reduce inoculum for future rotations, and will limit the losses to disease. The choice of which species to plant depends on species suitability for the specific site conditions, and management objectives for the site. Immune species, such as alder (*Alnus* spp.), can be established for a short rotation, which combines resistance management methods with eradication methods as a result of the reduction in viable inoculum during the immune species crop rotation.

Species options tend to be quite limited for areas where tomentosus root disease is found, given its boreal and sub-boreal distribution. Tree species diversity is lower in these forests, which leaves fewer options for alternative species than in more temperate forests. Lodgepole pine (*P. contorta* Douglas ex Loudon), despite its susceptibility to infection by *O. tomentosa*, is the most commonly used alternative species in western Canadian forests. Although it appears to be as susceptible to infection by *O. tomentosa* as spruce, and it suffers a higher rate of mortality, it does not become infected as frequently, presumably because of its more vertical rooting habit, which is less conducive to root contacts (Lewis and Hansen, 1991a).

For laminated root rot, alternative species preferences west of the Cascade mountains and Coast Range mountains include red alder (*A. rubra* Bong.), western redcedar and western white pine (*P. monticola* Douglas ex D. Don). In eastern forests, lodgepole pine, ponderosa pine (*P. ponderosa* Lawson & C. Lawson) and white pines are preferred, as well as western larch (*L. occidentalis* Nutt.) and various hardwoods (Thies and Sturrock, 1995).

Limited evidence exists to suggest that some genotypes of Douglas-fir may be more resistant than others to infection by *P. sulphurascens*, and that these resistant types produce more defence compounds such as phenols (Entry *et al.*, 1994). However, the development of resistant varieties has not been further explored as a possible management approach for laminated root rot or for tomentosus root disease.

In areas with low-to-moderate levels of root disease in a recently cut block, where susceptible hosts are a preferred regeneration species due to site conditions, these species can be established in combination with other resistant, tolerant or immune species. The mixing of species reduces the chances of root contacts between healthy and infected roots from colonized stumps of the previous rotation, and tree-to-tree secondary spread within the plantation. The same approach can be applied during precommercial thinning operations, which can be focused on favouring the retention of immune and low susceptibility species. This will also reduce the likelihood of root contacts between susceptible species and act as a barrier to spread of the disease. This mixed species method is expected to be more successful with *O. tomentosa* than with *P. sulphurascens* because the latter is more efficient at spreading by root contacts, and mixtures of susceptible species with immune or resistant species were shown to not be effective in reducing losses to laminated root rot (Hansen, 1975; Wallis, 1976).

9.5.6 Therapy

Therapy involves the treatment of currently infected plants with the aim of curing the plant from infection. In general, therapy is a relatively uncommon management approach for forest tree diseases, and this is particularly true for root rots. Therapeutic treatments are expensive, especially in forests, as each infected tree has to be identified and treated individually. Given the extent to which root rot can invade the root system of a tree, successful treatment of the entire root system is unlikely, and few agents are available that would be specific to the fungus yet not harm the host. One study exists which demonstrates that there is a potential for the therapeutic treatment of laminated root rot using the fumigation of live colonized trees. Douglas-fir trees treated with fumigants such as chloropicrin and methylisothiocyanate showed a greater survival rate than untreated infected trees, although there was some damage to host tissues as well (Harrington and Thies, 2007).

9.5.7 Integrated disease management

P. sulphurascens and *O. tomentosa* are native pathogens and natural components of forest ecosystems. They have evolved with their hosts over millennia and are integral to natural ecosystem processes. They are viewed as pests because they interfere with human-imposed expectations from forests, and their relative impact on forested ecosystems is sometimes exacerbated by management practices that are aimed at particular resource objectives.

Root diseases are affected by fire, not directly through changes in inoculum volume, but indirectly through the species changes that occur following a stand-replacing fire. For forest types in which laminated root rot and tomentosus root disease are found, the most common pioneer species that are the first colonizers of burned areas are species such as pines and hardwoods, which are resistant, tolerant or immune to damage by the pathogens. Fire has therefore played an important historic role in limiting the development of inoculum in forests. The impact of root disease is enhanced, therefore, when fire is removed or reduced in the landscape as a result of fire suppression.

In the absence of fire, stands mature and biotic agents of disturbance become more dominant as modifiers of forest structure and composition. Successional pathways can lead to greater abundance of host species in some cases, such as in the replacement of pine by more shade-tolerant spruce in sub-boreal forests subject to tomentosus root disease, but it can also lead to a reduction in the abundance of susceptible species, such as the trend towards western hemlock in coastal forests where *P. sulphurascens* is found.

In addition to fire and its impact on inoculum reduction through succession, root disease in natural forests can also be regulated by successional processes that result from the disease itself. Openings in canopies caused by root disease may be colonized by broadleaves and other non-host vegetation, which provides some time for the inoculum in dead tree root systems to become less viable, such that by the time a susceptible host tree is established in the opening, infection is less likely to occur. The colonization by brushy species of canopy openings resulting from tomentosus root disease occurs frequently in sub-boreal/boreal spruce forests, and is thought to help reduce spread of the fungus. Silvicultural practices that promote the rapid regeneration of cutover areas eliminate this natural inoculum reduction process when host species are used for regeneration.

Partial cutting (felling) is another silvicultural practice that can significantly affect the severity and impacts of both root diseases. Partial cutting involves removing some proportion of the trees in a stand, either as a precommercial thinning of the stand to reduce density, or as a commercial harvest. In forests that are affected by root disease, partial cutting can result in the reduction of inoculum if immune or highly resistant species are retained, as discussed above, but it can also result in an increase in inoculum as the fungus can spread more quickly as a saprophyte in the root systems of infected cut trees than it can spread parasitically in live trees. This phenomenon is a serious concern in the management of root disease caused by some *Armillaria* species due to the significant increase in saprophytic activity by that fungus once the host has been killed, but it is less of a concern with *P. sulphurascens* because this fungus exhibits saprophytic behaviour at the initiation of infection and inoculum builds up as the root system becomes more extensively colonized. *O. tomentosa* also works as a saprophyte and a parasite at the same time while a tree's root system becomes progressively colonized. Newbery *et al.* (2007) demonstrated that forests with low-to-moderate levels of root disease caused by *O. tomentosa* did not respond differently to partial cut harvesting than similar forests without root disease. In fact, partial cutting appeared to reduce the impact of root disease, presumably due to a reduction of root contacts between susceptible hosts.

Harvesting activities can also result in greater exposure of the residual trees to windthrow, particularly along the edges of clear-cut (clear-felled) areas. Depending on the system used, partial cutting can also increase the likelihood of windthrow, and the presence of root disease and decay fungi can exacerbate the damage caused. Whitney *et al.* (2002) found that root disease caused by a number of fungi, including *O. tomentosa*, was an important contributor to black spruce mortality and windfall in leave strips after strip cutting, and that the incidence of uprooting was greater in leave strips than in uncut forests. However, single tree selection methods, which do not result in a discrete edge, did not affect the rate of mortality in *O. tomentosa*-infected forests, including mortality resulting from uprooting (Newbery *et al.*, 2007).

Good inventories are also essential for root disease management, for prediction of future growth and yield, and for making management decisions about treatment costs and benefits. The effects of laminated root rot have been incorporated into the Western Root Disease Model (WRDM) in the USA (Frankel, 1998), and in BC a root rot simulator has been developed for *P. sulphurascens*, which predicts spread and impact by this fungus. These models use information such as host abundance, stand structure, and site and disease conditions to determine the future rate of disease spread over time. The models can also account for management activities and can model the effects of various treatments on root disease severity. Models for tomentosus root disease are not yet at the same stage of development.

Root disease management requires an understanding of the interactions of root disease and fire or fire suppression, the natural roles that root disease fungi play in ecosystem function, and reliable inventory information. One-size-fits-all silviculture prescriptions or poor management decisions based on inadequate information may lead to an increase in damage caused by these root diseases.

References

Aho, P.E. (1971) *Decay of Engelmann Spruce in the Blue Mountains of Oregon and Washington.* Research Paper PNW-116, US Department of Agriculture, Forest Service, Pacific Northwest Research Station, Portland, Oregon.

Angwin, P.A. and Hansen, E.M. (1988) Population structure of *Phellinus weirii*. In: Morrison, D. (ed.) *Proceedings of the Seventh International Conference on Root and Butt Rots, Vernon and Victoria, British Columbia, Canada, August 9–16, 1988*. Forestry Canada, Pacific Forestry Centre, Victoria, British Columbia, Canada, pp. 371–380.

Bae, H., Hansen, E.M. and Strauss, H. (1994) Restriction fragment length polymorphisms demonstrate single origin of infection centers in *Phellinus weirii*. *Canadian Journal of Botany* 72, 440–447.

Bernier, D. and Lewis, K.J. (1999) Site and soil characteristics related to the incidence of *Inonotus tomentosus*. *Forest Ecology and Management* 120, 131–142.

Bier, J.E. and Buckland, D.C. (1947) Relation of research in forest pathology to the management of second growth forests. I: *Poria weirii* root rot, an important disease affecting immature stands of Douglas-fir. *British Columbia Lumberman* 31, 49–51.

Blanchette, R.A. (1980) Wood decay: a submicroscopic view. *Journal of Forestry* 78, 734–737.

Bloomberg, W.J. (1990) Effect of stand conditions on advance of *Phellinus weirii* in Douglas-fir plantations. *Phytopathology* 80, 553–559.

Bloomberg, W.J. and Beale, J.D. (1985) Relationship of ecosystem to *Phellinus weirii* root rot on southern Vancouver Island. In: Thies, W. (ed.) *Proceedings of the Western International Forest Disease Work Conference*. US Department of Agriculture, Forest Service, Pacific Northwest Forestry Range Experimental Station, Portland, Oregon, pp. 10–28.

Bloomberg, W.J. and Reynolds, G. (1985) Growth loss and mortality in laminated root rot infection centres in second growth Douglas-fir on Vancouver Island. *Forest Science* 31, 497–508.

Bloomberg, W.J., Cumberbirch, P.M. and Wallis, G.W. (1980) *A Ground Survey Method for Estimating Loss Caused by* Phellinus weirii *Root Rot. I. Development of Survey Design*. Information Report BC-R-3, Canadian Forest Service, Pacific Forest Research Centre, Victoria, British Columbia.

Boyce, J.S. (1963) Red root and butt rot in a Georgia slash pine plantation. *Plant Disease Reporter* 47, 572–573.

Buckland, D.C., Molnar, A.C. and Wallis, G.W. (1954) Yellow laminated root rot of Douglas-fir. *Canadian Journal of Botany* 32, 69–81.

Byler, J.W., Marsden, M.A. and Hagle, S.K. (1990) The probability of root disease on the Lolo National Forest, Montana. *Canadian Journal of Forest Research* 20, 987–994.

Childs, T.W. (1963) *Poria weirii* root rot. *Phytopathology* 53, 1124–1127.

Clark, J.W. (1958) The occurrence and investigations of physiological strains of *Poria weirii* Murr. PhD thesis, Yale University, New Haven, Connecticut.

Cleary, M.R., Arhipova, N., Morrison, D.J., Thomsen, I.M., Sturrock, R.N., Vasaitis, R., Gaitnieks, T. and Stenlid, J. (2013) Stump removal to control root disease in Canada and Scandinavia: a synthesis of results from long-term trials. *Forest Ecology and Management* 290, 5–14.

Davidson, A.G. and Redmond, D.R. (1957) Decay of spruce in the maritime provinces. *Forestry Chronicle* 33, 373–380.

Dickman, A. and Cook, S. (1989) Fire and fungus in a mountain hemlock forest. *Canadian Journal of Botany* 67, 2005–2016.

Doğan, H.H. and Karadelev, M. (2009) *Phellinus sulphurascens* (*Hymenochaetaceae, Basidiomycota*): a very rare wood-decay fungus in Europe collected in Turkey. *Turkish Journal of Botany* 33, 239–242.

Entry, J.A., Vance, N.C. and Copes, D.L. (1994) Resistance of graft-compatible and graft-incompatible *Pseudotsuga menziesii* rooted cuttings to *Phellinus weirii*. *Canadian Journal of Forest Research* 24, 878–881.

EPPO (2011) EPPO Standards: EPPO A1 and A2 Lists of Pests Recommended for Regulation as Quarantine Pests. PM 1/2(20) English. European and Mediterranean Plant Protection Organization, Paris. Available at: http://archives.eppo.org/EPPOStandards/PM1_GENERAL/pm1-02(20)_A1A2_2011.pdf (accessed 16 January 2011).

Franceschi, V.R., Krokene, P., Christiansen, E. and Krekling, T. (2005) Anatomical and chemical defenses of conifer bark against bark beetles and other pests. *New Phytologist* 167, 353–376.

Frankel, S. (1998) *User's Guide to the Western Root Disease Model, Version 3.0*. General Technical Report PSW-GTR-165, US Department of Agriculture, Forest Service, Pacific Southwest Station, Berkeley, California.

Germain, H., Laflamme, G., Bernier, L., Boulet, G. and Hamelin, R.C. (2002) DNA polymorphism and molecular diagnosis in *Inonotus* spp. *Canadian Journal of Plant Pathology* 24, 194–199.

Germain, H., Bergeron, M.-J., Bernier, L., Laflamme, G. and Hamelin, R.C. (2009) Patterns of colonization and spread in the fungal spruce pathogen *Onnia tomentosa*. *Molecular Ecology* 18, 4422–4433.

Gibson, S. (2005) An examination of genetic variation and disease spread in *Inonotus tomentosus* in the sub-boreal spruce zone of British Columbia. MSc thesis, University of Northern British Columbia, Prince George, British Columbia, Canada.

Gilbertson, R.L. (1974) *Fungi that Decay Ponderosa Pine*. University of Arizona Press, Tucson, Arizona.

Gilbertson, R.L. and Ryvarden, L. (1987) *North American Polypores*. Fungiflora, Oslo, Norway.

Goheen, D.J. and Hansen, E.M. (1993a) Effects of pathogens and bark beetles on forests. In: Schowalter, T. and Filip, G. (eds) *Beetle–Pathogen Interactions in Conifer Forests*. Academic Press, London, pp. 175–196.

Goheen, D.J. and Hansen, E.M. (1993b) Tree vigor and susceptibility to infection by *Phellinus weirii*: results of field inoculations. In: Johansson, M. and Stenlid, J. (eds) *Proceedings of the Eighth International Conference on Root and Butt Rots, Wik, Sweden and Haikko, Finland, August 9–16, 1993*. Swedish University of Agricultural Sciences, Uppsala, Sweden, pp. 45–51.

Hagle, S.K. (2009) Laminated root rot, ecology and management. In: *Forest Insect and Disease Management Guide for the Northern and Central Rocky Mountains*. US Department of Agriculture, Forest Service, Northern Region, State and Private Forestry, Missoula, Montana, Chapter 11.2.

Hansen, E.M. (1975) Phellinus (Poria) weirii *Root Rot in Douglas-fir-alder Stands 10-17 Years Old*. Research Note PNW-250, US Department of Agriculture, Forest Service, Pacific Northwest Forest and Range Experiment Station, Portland, Oregon.

Hansen, E.M. (1976) Twenty-year survival of *Phellinus* (*Poria*) *weirii* in Douglas-fir stumps. *Canadian Journal of Forest Research* 6, 123–128.

Hansen, E.M. (1979) Survival of *Phellinus weirii* in Douglas-fir stumps after logging. *Canadian Journal of Forest Research* 9, 484–488.

Hansen, E.M. and Goheen, E.M. (2000) *Phellinus weirii* and other native root pathogens as determinants of forest structure and process in western North America. *Annual Review of Phytopathology* 38, 515–539.

Harrington, C.A. and Thies, W.G. (2007) Laminated root rot and fumigant injection affect survival and growth of Douglas-fir. *Western Journal of Applied Forestry* 22, 220–227.

Holah, J.C., Wilson, M.V. and Hansen, E.M. (1993) Effects of a native forest pathogen, *Phellinus weirii*, on Douglas-fir forest composition in western Oregon. *Canadian Journal of Forest Research* 23, 2473–2480.

Hunt, R.S. (1997) Differential medium for *Phellinus pini* and *Inonotus tomentosus. Canadian Journal of Plant Pathology* 19, 307–309.

Hunt, R.S., Ekramoddoullah, A.K. and Zamani, A. (1999) Production of a polyclonal antibody to *Phellinus pini* and examination of its potential use in diagnostic assays. *European Journal of Forest Pathology* 29, 259–272.

Ingersoll, C.A., Wilson, M.V. and Thies, W.G. (1996) Effects of *Phellinus weirii* on early successional vegetation following timber harvest. *Canadian Journal of Forest Research* 26, 233–326.

Ingham, E.R. and Thies, W.G. (1996) Responses of soil foodweb organisms in the first year following clearcutting and application of chloropicrin to control laminated root rot. *Applied Soil Ecology* 3, 35–47.

Islam, M.A., Sturrock, R.N., Holmes, T.A. and Ekramoddoullah, A.K.M. (2009) Ultrastructural studies of *Phellinus sulphurascens* infection of Douglas-fir roots and immunolocalization of host pathogenesis-related proteins. *Mycological Research* 113, 700–712.

James, R.L., Stewart, C.A. and Williams, R.E. (1984) Estimating root disease losses in northern Rocky Mountain national forests. *Canadian Journal of Forest Research* 14, 652–655.

Jeong, J.W., Lim, Y.M., Lee, J.S. and Jung, H.S. (2005) Phylogeny of *Phellinus* and related genera inferred from combined data of ITS and mitochondrial SSU rDNA sequences. *Journal of Microbiology and Biotechnology* 15, 1028–1038.

Kastner, W.W., Goheen, D.J. and Edmonds, R.L. (1994) Relationship between occurrence of laminated root rot and site characteristics in Douglas-fir forests of the northern Oregon Coast Range. *Western Journal of Applied Forestry* 9, 14–17.

Lachance, D. (1978) The effect of decay on growth rate in a white spruce plantation. *Forestry Chronicle* 54, 20–23.

Larsen, M.J. and Lombard, F.F. (1989) Taxonomy and nomenclature of *Phellinus weirii* in North America. In: Morrison, D.J. (ed.) *Proceedings of the Seventh International Conference on Root and Butt Rots, Vernon and Victoria, British Columbia, Canada, August 9–16, 1988*. Forestry Canada, Pacific Forestry Centre, Victoria, British Columbia, Canada, pp. 573–578.

Larsen, M.J., Lombard, F.F. and Clark, J.W. (1994) *Phellinus sulphurascens* and the closely related *P. weirii* in North America. *Mycologia* 86, 121–130.

Lewis, K.J. (1997) Growth reduction in spruce infected by *Inonotus tomentosus* in central British Columbia. *Canadian Journal of Forest Research* 27, 1669–1674.

Lewis, K.J. and Hansen, E.M. (1991a) Survival of *Inonotus tomentosus* in stumps and subsequent infection of young stands in north central British Columbia. *Canadian Journal of Forest Research* 21, 1049–1057.

Lewis, K.J. and Hansen, E.M. (1991b) Vegetative compatibility groups and protein electrophoresis indicate a role for basidiospores in spread of *Inonotus tomentosus* in spruce forests of British Columbia. *Canadian Journal of Botany* 69, 1756–1763.

Lewis, K.J. and Lindgren, B.S. (2002) Relationship between spruce beetle and tomentosus root disease: two natural disturbance agents of spruce. *Canadian Journal of Forest Research* 32, 31–37.

Lewis, K.J., Morrison, D.J. and Hansen, E.M. (1992) Spread of *Inonotus tomentosus* from infection centres in spruce forests in British Columbia. *Canadian Journal of Forest Research* 22, 68–72.

Lewis, K.J., Trummer, L.M. and Thompson, R.D. (2004) Incidence of tomentosus root disease relative to spruce density and slope position in south-central Alaska. *Forest Ecology and Management* 194, 159–167.

Lewis, K.J., Thompson, R.D. and Trummer, L. (2005) Growth response of spruce infected by *Inonotus tomentosus* in Alaska and interactions with spruce beetle. *Canadian Journal of Forest Research* 35, 1455–1463.

Li, C.Y. (1981) Phenoloxidase and peroxidase activities in zone lines of *Phellinus weirii*. *Mycologia* 73, 811–821.

McCauley, K.J. and Cook, S.A. (1980) *Phellinus weirii* infestation of two mountain hemlock forests in the Oregon Cascades. *Forest Science* 26, 23–29.

Morrison, D.J., Wallis, G.W. and Weir, L.C. (1988) *Control of Armillaria and Phellinus Root Diseases: 20-Year Results from the Skimikin Stump Removal Experiment*. Information Report BC-X-302, Canadian Forest Service, Pacific and Yukon Region, Victoria, British Columbia, Canada.

Murrill, W.A. (1914) An enemy of the western red cedar. *Mycologia* 6, 93–94.

Nelson, E.E. (1964) Some probable relationships of soil fungi and zone lines to survival of *Poria weirii* in buried wood blocks. *Phytopathology* 54, 120–121.

Nelson, E.E. (1976) Colonization of wood disks initiated by basidiospores of *Phellinus weirii* (laminated root rot). *Forest Science* 22, 407–411.

Nelson, E.E. and Hartman, T. (1975) Estimating spread of *Poria weirii* in a high elevation mixed conifer stand. *Journal of Forestry* 73, 141–142.

Nelson, E.E. and Thies, W.G. (1985) Colonization of *Phellinus weirii*-infested stumps by *Trichoderma viride*. I. Effect of isolate and inoculum base. *European Journal of Forest Pathology* 15, 425–431.

Newbery, J.E., Lewis, K.J. and Walters, M.B. (2007) *Inonotus tomentosus* and the dynamics of unmanaged and partial-cut wet sub-boreal spruce-fir forests. *Canadian Journal of Forest Research* 37, 2663–2676.

Nobles, M.K. (1948) Studies in forest pathology VI. Identification of cultures of wood-rotting fungi. *Canadian Journal of Research* 26, 281–431.

Oullette, G.B., Bard, G. and Cauchon, R. (1971) Self-strangulation of roots: points of entry of root-rot fungi in the Grand'mere, white spruce plantations. *Phytoprotection* 52, 119–124.

Robinson, R.M. and Morrison, D.J. (2001) Lesion formation and host response to infection by *Armillaria ostoyae* in the roots of western larch and Douglas-fir. *Forest Pathology* 31, 371–385.

Robinson-Jeffrey, R.C. and Loman, A.A. (1963) Fungi isolated in culture from red heartwood stain and advanced decay of lodgepole pine in Alberta. *Canadian Journal of Botany* 41, 1371–1375.

Ross, E.W. (1966) Incidence of *Polyporus tomentosus* in slash pine plantations in the southeastern United States. *Plant Disease Reporter* 50, 527.

Schulting, P.J. (1987) Relative susceptibility of interior spruce (*Picea glauca* × *engelmannii* Engelm.) and lodgepole pine (*Pinus contorta* var. *latifolia* Engelm.) to *Inonotus tomentosus* (Fr.) Teng in central British Columbia. MSc thesis, University of British Columbia, Vancouver, British Columbia, Canada.

Smith, R.B. and Wass, E.F. (1991) *Impacts of Two Stumping Operations on Site Productivity in Interior British Columbia*. Information Report BC-X-327, Canadian Forest Service, Pacific and Yukon Region, Pacific Forest Research Centre, Victoria, British Columbia.

Smith, R.B. and Wass, E.F. (1994) *Impacts of a Stump Uprooting Operation on Properties of a Calcareous Loamy Soil and on Planted Seedling Performance*. Information Report BC-X-344, Canadian Forest Service, Pacific and Yukon Region, Pacific Forest Research Centre, Victoria, British Columbia.

Sturrock, R.N., Islam, M.A. and Ekramoddoullah, A.K.M. (2007) Host–pathogen interactions in Douglas-fir seedlings infected by *Phellinus sulphurascens*. *Phytopathology* 97, 1406–1414.

Thies, W.G. (1983) Determination of growth reduction in Douglas-fir infected by *Phellinus weirii*. *Forest Science* 29, 305–315.

Thies, W.G. and Nelson, E.E. (1982) Control of *Phellinus weirii* in Douglas-fir stumps by the fumigants chloropicrin, allyl alcohol, Vapam, or Vorlex. *Canadian Journal of Forest Research* 12, 528–532.

Thies, W.G. and Nelson, E.E. (1987) Reduction of *Phellinus weirii* inoculum in Douglas-fir stumps by the fumigants chloropicrin, Vorlex, or methylisothiocyanate. *Forest Science* 33, 316–329.

Thies, W.G. and Nelson, E.E. (1997) Laminated root rot: new considerations for surveys. *Western Journal of Applied Forestry* 12, 49–51.

Thies, W.G. and Sturrock, R.N. (1995) *Laminated Root Rot in Western North America*. General Technical Report PNW-GTR-349, US Department of Agriculture, Forest Service, Pacific Northwest Research Station, Portland, Oregon.

Thies, W.G. and Westlind, D.J. (2006) Application of chloropicrin to Douglas-fir stumps to control laminated root rot does not affect infection or growth of regeneration 16 growing seasons after treatment. *Forest Ecology and Management* 235, 212–218.

Thies, W.G., Nelson, E.E. and Zabowski, D. (1994) Removal of stumps from a *Phellinus weirii* infested site and fertilization affect mortality and growth of planted Douglas-fir. *Canadian Journal of Forest Research* 24, 234–239.

Todd, N.K. and Rayner, A.D. (1980) Fungal individualism. *Science Progress* 66, 331–354.

van Groenewoud, H. and Whitney, R.D. (1969) White spruce mortality in Saskatchewan and Manitoba. *Pulp and Paper Magazine of Canada*, April, pp. 1–4.

Wallis, G.W. (1976) Phellinus (Poria) weirii *Root Rot: Detection and Management Proposals in Douglas-fir Stands*. Forestry Technical Report 12, Canadian Forest Service, Pacific Forest Research Centre, Victoria, British Columbia, Canada.

Wallis, G.W. and Reynolds, G. (1962) Inoculation of Douglas-fir roots with *Poria weirii*. *Canadian Journal of Botany* 40, 637–645.

Wallis, G.W. and Reynolds, G. (1965) The initiation and spread of *Poria weirii* root rot of Douglas fir. *Canadian Journal of Botany* 43, 1–9.

Waring, R.H., Cromack, K. Jr, Matson, P.A., Boone, R.D. and Stafford, S.G. (1987) Responses to pathogen-induced disturbance: decomposition, nutrient availability, and tree vigour. *Forestry* 60, 219–227.

Whitney, R.D. (1962) Studies in forest pathology. XXIV. *Polyporus tomentosus* Fr. as a major factor in stand-opening disease of white spruce. *Canadian Journal of Botany* 40, 1631–1658.

Whitney, R.D. (1966) Germination and inoculation tests with basidiospores of *Polyporus tomentosus*. *Canadian Journal of Botany* 44, 1333–1343.

Whitney, R.D. (1967) Comparative susceptibility of large and small spruce roots to *Polyporus tomentosus*. *Canadian Journal of Botany* 45, 2227–2229.

Whitney, R.D. (1980) *Polyporus tomentosus* root and butt rot of trees in Canada. In: Dimitri, L. (ed.) *Proceedings from the Fifth International Conference on Problems of Root and Butt Rot in Conifers, Kassel, Federal Republic of Germany, August 7–12, 1978*. Hessian Forstliche Laboratory, Hann, Munden, pp. 283–297.

Whitney, R.D. (1993) Damage by tomentosus root rot in white spruce plantations in Ontario, and the effects of thinning on the disease. *Forestry Chronicle* 69, 445–449.

Whitney, R.D. (1995) Root-rotting fungi in white spruce, black spruce, and balsam fir in northern Ontario. *Canadian Journal of Forest Research* 25, 1209–1230.

Whitney, R.D. (1997) Relationship between decayed roots and aboveground decay in three conifers in Ontario. *Canadian Journal of Forest Research* 27, 1217–1221.

Whitney, R.D. and Denyer, W.B. (1969) Resin as a barrier to infection of white spruce by heart rotting fungi. *Forest Science* 15, 266–267.

Whitney, R.D. and van Groenewoud, H. (1964) The rate of advance of stand-opening disease over a ten-year period in white spruce at Candle Lake, Saskatchewan. *Forestry Chronicle* 40, 308–312.

Whitney, R.D., Fleming, R.L., Zhou, K. and Mossa, D.S. (2002) Relationship of root rot to black spruce windfall and mortality following strip clear-cutting. *Canadian Journal of Forest Research* 32, 283–294.

Williams, R.E. and Marsden, M.A. (1982) Modelling probability of root disease center occurrence in northern Idaho forests. *Canadian Journal of Forest Research* 12, 876–882.

Wilson, M.V., Ingersoll, C.A. and Thies, W.G. (1999) Testing for effects of stump fumigation with chloropicrin on vegetation in an early seral Douglas-fir stand. *Canadian Journal of Forest Research* 29, 1254–1258.

10 Heart Rots, Sap Rots and Canker Rots

Rimvydas Vasaitis*
Swedish University of Agricultural Sciences, Uppsala, Sweden

10.1 Pathogens, Significance and Distribution

This chapter is mainly devoted to stem rots in temperate and boreal forests of the northern hemisphere, primarily focusing on stands subjected to various management regimes but, to a certain extent, also to nature conservation. Decays of urban and amenity trees in recreational areas are therefore omitted, as well as tree care and 'hazard tree' management practices. For this, refer to Burdekin (1979), Schwarze *et al.* (2000), Sinclair and Lyon (2005), Terho (2009), and the Western Hazard Tree Workshop (2010).

Heart rot is decay in the central stem wood or heartwood. The term heartwood here applies to a cylinder of wood in the centre of tree stems that contains no functional (living) cells and serves merely to provide mechanical support for the crown of the tree (Tainter and Baker, 1996). In certain cases, heart rot-causing fungi in living stems do attack sapwood, which then results in the formation of sap rot. The attacked sapwood might be either dead (e.g. beneath disrupted or dead bark) or functional (e.g. encasing rotten heartwood). When functional sapwood is killed by heart rot fungus this results in development of stem canker, and such a pattern of the decay is then called a canker rot. Consequently, the same fungus in the living tree might simultaneously cause heart rot, sap rot and canker rot. A typical example of this type of fungus is the polypore *Fomitiporia robusta* (P. Karst.) Fiasson & Niemelä in living English oak (*Quercus robur* L.) (Sunhede and Vasiliauskas, 2002; Fig. 10.1).

10.1.1 Ecological groups of heart rot fungi

In a living tree, the heartwood is encased by functionally intact sapwood, which is not susceptible to decay owing to its high moisture content and physiological properties. Moreover, the sapwood is covered by bark, which represents an additional effective barrier to fungal infections. Manion (1991) overviewed four different concepts of the infection and colonization of a living tree by a decay fungus: (i) the 'Haddow–Etheridge' concept, which implies latent establishment in an intact tree (Haddow, 1938b; Etheridge and Craig, 1976); (ii) the 'Hartig' concept, which implies entry through heartwood-exposing injury (Hartig, 1874); (iii) the 'Shigo' concept

* E-mail: rimvys.vasaitis@slu.se

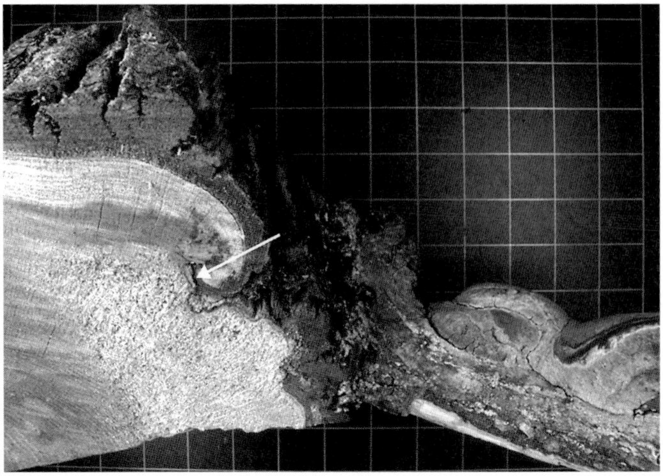

Fig. 10.1. Stem cross-section of *Quercus robur* L. showing heart rot, sap rot and canker rot caused by *Fomitiporia robusta* (P. Karst.) Fiasson & Niemelä. The arrow indicates the part of the stem where the cambium and functional sapwood were killed, leading to canker formation. Note the presence of fruiting bodies (conks) of the fungus on a branch. The black squares in the background are 3 × 3 cm. Modified from Vasiliauskas *et al.* (2003). Credits: Audrius Menkis.

of fungal succession on sapwood wounds (Shigo, 1966); and (iv) the 'Boddy–Rayner' concept, explaining fungal colonization of a tree in terms of wood moisture content (Boddy and Rayner, 1983). It was suggested that it is possible that any of the four concepts might be correct, at least with regard to certain decay phenomena. In fact, while the first three concepts deal in particular with infection courts of fungi on living trees, the fourth mainly reflects patterns of their spread inside a tree. Here, an attempt will be made to demonstrate that all three of the first-mentioned infection concepts are indeed correct and that none of them contradicts any other. Moreover, each of the first three concepts represents a characteristic attribute for respective ecologically distinct groups of stem decay fungi, while the fourth provides a general explanation for the different colonization patterns observed and strategies exhibited by representatives of those ecological groups.

Based on infection pathways, heart rot fungi are traditionally categorized as 'true heart rots' that infect trees through natural infection courts, and 'wound decay fungi' (or in other words, 'wound heart rots') that infect trees through mechanical wounds (Hansen and Lewis, 1997). Yet, such categorization could be modified and expanded while adapting it to the three infection concepts discussed above, by distinguishing: (i) true heart rots that gain entry to the heartwood of an intact tree using known or considered natural courts of infection and correspond to the Haddow–Etheridge concept; (ii) true heart rots that directly colonize heartwood exposed by natural injuries, such as broken large branches and tops, fire scars and stem cracks, and are thus are grouped under the classical Hartig concept; and (iii) wound heart rots entering heartwood through exposed dead sapwood, which comply with the Shigo infection concept. Table 10.1 presents the most relevant stem rot fungi in respect of those three categories, and includes also a fourth category, root rots and butt rots, which are not covered by other chapters in this volume.

The suggested categorization agrees well with the opinion on stem decay infection courts expressed in the classical work of Boyce (1961). Hyphae from germinating spores cannot penetrate bark or living sapwood. Consequently, a spore must alight on heartwood or exposed deadwood directly connected with heartwood for decay to start in a living tree. However, all the dead branches form bridges from the outside heartwood, as

Table 10.1. Main heart rot agents in forest stands of the northern hemisphere. Apart from the citations in the table, the fungi listed are referred to as heart rot and stem decay agents on living trees in classic compendia on forest pathology by Boyce (1961), Hepting (1971) and Sinclair and Lyon (2005).

Fungus	Geographic distribution	Principal host genera	Forest	Referred to as rot agents of living trees	Remarks
1. True heart rots of intact tree stems (infection according the 'Haddow–Etheridge' concept)					
Echinodontium tinctorium (Ellis & Everh.) Ellis & Everh.	Western N. America, Mexico to Alaska	*Abies*	Old natural[a]	Allen *et al.*, 1996; Hansen and Lewis, 1997; Holsten *et al.*, 2001	Inoculated on to living stems to promote wildlife habitats (Filip *et al.*, 2011)
Fomes fomentarius s.l., clade A (in Judova *et al.*, 2012)	Temperate and boreal north hemisphere	*Alnus*, *Betula*, *Fagus*	Old natural (when on living trees)	Butin, 1995; Allen *et al.*, 1996; Holsten *et al.*, 2001; Kuzmichev *et al.*, 2004; Dai *et al.*, 2007	In Europe, biodiversity component of old-growth beech (*Fagus sylvatica* L.) forest (Skov and Thomsen, 2003)
Fomitiporia robusta (P. Karst.) Fiasson & Niemelä	Temperate Eurasia	*Quercus*	Old natural	Černý, 1989; Kuzmichev *et al.*, 2004; Dai *et al.*, 2007	On over-mature oaks (Sunhede and Vasiliauskas, 2002)
Fomitopsis officinalis (Vill.) Bondartsev & Singer	Temperate and boreal Asia and western N. America	Wide range, coniferous	Old natural	Allen *et al.*, 1996; Holsten *et al.*, 2001; Kuzmichev *et al.*, 2004; Dai *et al.*, 2007	Inoculated on to living stems to promote wildlife habitats (Filip *et al.*, 2011)
Fomitopsis pinicola (Sw.) P. Karst.	Temperate and boreal north hemisphere	Wide range, coniferous and deciduous (when on dead wood)	Old natural (when on living trees)	Allen *et al.*, 1996; Holsten *et al.*, 2001; Kuzmichev *et al.*, 2004; Dai *et al.*, 2007	Inoculated on to living stems to promote wildlife habitats (Filip *et al.*, 2011); mortality agent in pristine forests of north Eurasia (Renvall *et al.*, 1991; Dai, 1996; Lännenpää *et al.*, 2008)
Inonotus obliquus (Ach. ex Pers.) Pilát	Temperate and boreal north hemisphere	*Betula*	Old natural	Allen *et al.*, 1996; Holsten *et al.*, 2001; Kuzmichev *et al.*, 2004; Dai *et al.*, 2007	
Mensularia radiata (Sowerby) Lázaro Ibiza	Temperate Eurasia	*Alnus*	Managed and natural	Phillips and Burdekin, 1982; Dai *et al.*, 2007; Arhipova *et al.*, 2011a, 2012	
Phellinus chrysoloma (Fr.) Donk	Temperate and boreal Europe	*Picea*	Old natural	Kuzmichev *et al.*, 2004; Dai *et al.*, 2007; Tomšovský *et al.*, 2010a	In Europe, indicator of forest value for nature conservation (Nitare, 2000); mortality agent in pristine forests (Lännenpää *et al.*, 2008)
Phellinus hartigii (Allesch. & Schnabl) Pat.	Temperate Eurasia, boreal N. America	*Abies*, *Tsuga*	Old natural	Černý, 1989; Allen *et al.*, 1996; Holsten *et al.*, 2001; Kuzmichev *et al.*, 2004; Dai *et al.*, 2007	On *Abies* in Eurasia, mainly on *Tsuga* in N. America

Continued

Table 10.1. Continued.

Fungus	Geographic distribution	Principal host genera	Forest	Referred to as rot agents of living trees	Remarks
Phellinus himalayensis Y.C.Dai	Temperate Asia	*Picea*	Natural	Dai *et al.*, 2007; Tomšovský *et al.*, 2010a	
Phellinus igniarius s.l.	Temperate and boreal Asia and N. America	Wide range, deciduous	Natural mature	Allen *et al.*, 1996; Holsten *et al.*, 2001; Kuzmichev *et al.*, 2004; Dai *et al.*, 2007	Limited material from USA indicates certain speciation within the group (Fischer and Binder, 2004)
Phellinus jezoënsis (Yamano) Parmasto	Temperate Asia	*Picea*	Old natural	Dai *et al.*, 2007; Tomšovský *et al.*, 2010a	
Phellinus nigricans (Fr.) P. Karst.	Boreal and temperate Europe	*Betula*	Old natural	Niemelä 1975; Černý,1989; Tomšovský *et al.*, 2010b	
Phellinus populicola Niemelä	Temperate and boreal north hemisphere	*Populus*	Old natural	Manion, 1991; Tainter and Baker, 1996; Fischer and Binder, 2004; Dai *et al.*, 2007	In Europe, indicator of forest value for nature conservation (Nitare, 2000); red listed (Larsson, 1997)
Phellinus tremulae (Bondartsev) Bondartsev & P.N. Borisov	Temperate and boreal Eurasia and N. America	*Populus*	Natural	Niemelä, 1974; Černý, 1989; Allen *et al.*, 1996; Kuzmichev *et al.*, 2004; Dai *et al.*, 2007	Destructive pathogen, limiting economic life of aspen (*Populus* spp.) to 40–50 years (Manion, 1991; Tainter and Baker, 1996)
Porodaedalea laricis (Jacz. ex Pilát) Niemelä	Temperate and boreal Eurasia	*Larix*	Old natural	Dai *et al.*, 2007; Tomšovský *et al.*, 2010a	
Porodaedalea pini (Brot.) Murrill	Temperate and boreal Eurasia	*Pinus*	Old natural	Kuzmichev *et al.*, 2004; Dai *et al.*, 2007; Tomšovský *et al.*, 2010a	In Europe, indicator of forest value for nature conservation (Nitare, 2000)
Porodaedalea pini s.l.	Temperate and boreal N. America	*Picea*, *Pinus*, *Pseudotsuga*, *Tsuga*		Allen *et al.*, 1996; Tainter and Baker, 1996; Holsten *et al.*, 2001	Consists of five host-specific groups (Fischer, 1996); in the USA, inoculated on to living stems to promote wildlife habitats (Filip *et al.*, 2011)
2. True heart rots associated with heartwood-exposing natural injury (broken branches and tops, stem cracks, fire scars; infection according the 'Hartig' concept)					
Fistulina hepatica (Schaeff.) With.	Temperate Europe and N. America	*Quercus*	Old natural	Phillips and Burdekin, 1982; Butin, 1995	In the past, referred to in Europe as important pathogen of mature oaks (*Quercus* spp.) (Aufsess, 1973; Černý, 1989); red listed in north Europe (Larsson, 1997) and indicator of forest value for nature conservation (Nitare, 2000)

Fomitopsis cajanderi (P. Karst.) Kotl. & Pouzar	Western N. America, Pacific Northwest	Coniferous	Natural	Scharpf, 1993; Hansen and Lewis, 1997	Inoculated on to living stems to promote wildlife habitats (Filip *et al.*, 2011)
Hapalopilus croceus (Pers.) Donk	Temperate Europe and N. America	*Quercus*	Old natural	Černý, 1989; Sunhede and Vasiliauskas, 2003b	In Europe, in nature conservation areas (Jahn, 1990); red listed (Larsson, 1997) and indicator of forest value for nature conservation (Nitare, 2000)
Inonotus dryophilus (Berk.) Murril	Temperate Europe and N. America	*Quercus*	Old natural	Berry and Lombard, 1978; Kuzmichev *et al.*, 2004	In the past, referred to in Europe as important pathogen of mature oaks (Aufsess, 1973; Granatov, 1973); red listed in north Europe (Larsson, 1997)
Laetiporus conifericola Burds. & Banik	Western N. America, California to Alaska	*Abies*, *Picea*, *Tsuga*	Old natural	Burdsall and Banik, 2001; Holsten *et al.*, 2001	On mature and over-mature trees
Laetiporus cremeiporus Y. Ota & T.Hatt.	Cool temperate and boreal Asia	*Quercus*	Natural	Dai *et al.*, 2007; Ota *et al.*, 2009	
Laetiporus gilbertsonii Burds.	Western N. America, Mexico to Washington	*Quercus*	Natural	Burdsall and Banik, 2001	
Laetiporus huroniensis Burds. & Banik	Eastern N. America, Great Lakes area	*Tsuga*	Old natural	Burdsall and Banik, 2001	On mature and over-mature trees
Laetiporus montanus Černý ex Tomšovský & Jankovský	Cool temperate and boreal Eurasia	*Abies*, *Larix*, *Picea*, *Pinus*	Old natural	Černý, 1989; Dai *et al.*, 2007; Ota *et al.*, 2009	In Europe, on over-mature trees
Laetiporus sulphureus s.l., clades C and E (in Vasaitis *et al.*, 2009)	Temperate Europe and N. America	*Quercus*	Old natural	Vasiliauskas *et al.*, 2003; Banik *et al.*, 2010	In Europe, on over-mature oaks
Laetiporus versisporus (Lloyd) Imazeki	Warm temperate Asia	*Quercus*	Natural	Dai *et al.*, 2007; Ota *et al.*, 2009	

Continued

Table 10.1. Continued.

Fungus	Geographic distribution	Principal host genera	Forest	Referred to as rot agents of living trees	Remarks
Pholiota adiposa (Batsch) P. Kumm., *P. squarrosa* (Vahl) P. Kumm.	Temperate and boreal northern hemisphere	*Abies*, *Betula*, *Tsuga*	Old natural	Scharpf, 1993; Tainter and Baker, 1996; Holsten *et al.*, 2001; Kuzmichev *et al.*, 2004; Dai *et al.*, 2007	Root/butt rot of deciduous trees (Philips and Burdekin, 1982; Butin, 1995); stem rot of alder/spruce (Arhipova *et al.*, 2011a; Cleary *et al.*, 2013); wound rot of birch (*Betula*) (Shigo, 1966)
Stereum gausapatum (Fr.) Fr.	Temperate Europe and N. America	*Quercus*	Managed and natural	Aufsess, 1973; Berry and Lombard, 1978; Phillips and Burdekin, 1982	Produces fruiting bodies on old standing trunks and still attached branches (Eriksson *et al.*, 1984)
Xylobolus frustulatus (Pers.) Boidin	Temperate Europe and N. America	*Quercus*	Natural	Aufsess, 1973; Berry and Lombard, 1978; Butin,1995	In Europe, on old-growth oaks (Sunhede and Vasiliauskas, 1996)
3. Wound heart rots associated with artificial mechanical damage (logging, bark stripping, pruning wounds; infection according the 'Shigo' concept)					
Amylostereum areolatum (Chaillet ex Fr.) Boidin	Temperate and boreal Eurasia	*Picea*, *Pinus*	Managed	Hansen and Lewis, 1997; Vasiliauskas, 1998a, 2001 and references therein	Fruiting bodies common on woody debris of spruce (*Picea* spp.) (Breitenbach and Kränzlin, 1986; Jahn, 1990)
Amylostereum chailletii (Pers.) Boidin	Temperate and boreal northern hemisphere	Wide range, coniferous	Managed	Hansen and Lewis, 1997; Vasiliauskas, 1998a, 2001 and references therein	Fruiting bodies common on woody debris of fir (*Abies* spp.) (Breitenbach and Kränzlin, 1986); often on left logs (Eriksson and Ryvarden, 1973)
Chondrostereum purpureum (Pers.) Pouzar	Temperate and boreal northern hemisphere	*Alnus*, *Betula*	Managed	Shigo, 1966; Phillips and Burdekin, 1982; Butin, 1995; Arhipova *et al.*, 2011a, 2012	Fruiting bodies common on stumps (Eriksson and Ryvarden, 1973); on dead wood of broadleaved trees (Breitenbach and Kränzlin, 1986)
Coniophora puteana (Schumach.) P. Karst., *C. arida* (Fr.) P. Karst.	Temperate and boreal western N. America and Europe	*Fraxinus*, *Quercus*, *Picea*, *Tsuga*	Natural and managed	Černý, 1989; Vasiliauskas, 1998b,c; Vasiliauskas and Stenlid, 1998b; Cleary *et al.*, 2013	Fruiting bodies on dead wood (Jahn, 1990; Ryman and Holmåsen, 1992); consistent latent infections to living trees/intact sapwood; mortality agents in pristine boreal conifer forests (Holsten *et al.*, 2001; Lännenpää *et al.*, 2008)

Pleurotus ostreatus (Jacq.) P. Kumm., *P. dryinus* (Pers.) P. Kumm.	Temperate and boreal northern hemisphere	Deciduous and coniferous	Managed and natural	Phillips and Burdekin, 1982; Černý, 1989; Vasiliauskas, 1989; Butin, 1995; Dai *et al.*, 2007	Fruiting bodies on stumps and lying stems; cultivated on dead wood as a culinary species (Butin, 1995)
Postia caesia (Schrad.) P. Karst.	Boreal and temperate Eurasia	*Picea*	Managed	Vasiliauskas, 1989; Butin, 1995	Fruiting bodies on rotten conifer and hardwood logs and slash (Ryvarden and Gilbertson, 1994); on stumps (Breitenbach and Kränzlin, 1986)
Postia stiptica (Pers.) Jülich	Temperate and boreal northern hemisphere	*Picea*	Managed	Phillips and Burdekin, 1982; Černý, 1989; Butin, 1995; Vasiliauskas, 1998b	Fruiting bodies on rotten conifers and hardwoods (Ryvarden and Gilbertson, 1994); on stumps (Breitenbach and Kränzlin, 1986)
Resinicium bicolor (Alb. & Schwein.) Parmasto	Temperate and boreal northern hemisphere	*Picea*	Managed	Vasiliauskas, 1998a,b; Cleary *et al.*, 2013	Fruiting bodies on rotten barkless stumps (Breitenbach and Kränzlin, 1986); important as decomposer of dead wood (Eriksson *et al.*, 1981)
Sistotrema brinkmannii (Bres.) J. Erikss.	Temperate and boreal northern hemisphere	*Picea*	Managed	Vasiliauskas, 1998a,b	Fruiting bodies mostly on much decayed wood (Eriksson *et al.*,1984); on dead wood of broadleaves and conifers (Breitenbach and Kränzlin, 1986)
Stereum hirsutum (Willd.) Pers.	Temperate and boreal northern hemisphere	*Quercus*	Managed	Shigo, 1966; Jahn, 1990; Vasiliauskas, 1998c	Fruiting bodies on dead deciduous wood (Eriksson *et al.*, 1984; Breitenbach and Kränzlin, 1986)
Stereum sanguinolentum (Alb. & Schwein.) Fr.	Boreal and temperate Eurasia	Wide range, coniferous	Managed	Phillips and Burdekin, 1982; Butin, 1995; Allen *et al.*, 1996; Hansen and Lewis, 1997; Dai *et al.*, 2007	Especially common on newly dead logs and stumps (Eriksson *et al.*, 1984); typical decomposer of logging residues (Vasiliauskas, 1989)
Trametes versicolor (L.) Lloyd	Temperate and boreal northern hemisphere	*Betula*, *Quercus*	Managed	Shigo, 1966; Phillips and Burdekin, 1982; Butin, 1995; Vasiliauskas, 1998c	Probably the most common wood rotting fungus on dead hardwoods (Ryvarden and Gilbertson, 1994)
4. Root and butt rots (not covered by other chapters in this volume)					
Climacocystis borealis (Fr.) Kotl. & Pouzar	Temperate and boreal northern hemisphere	*Picea*, *Pinus*, *Tsuga*	Old natural	Černý, 1989; Holsten *et al.*, 2001; Kuzmichev *et al.*, 2004	In Europe, indicator of forest value for nature conservation (Nitare, 2000)

Continued

Table 10.1. Continued.

Fungus	Geographic distribution	Principal host genera	Forest	Referred to as rot agents of living trees	Remarks
Ganoderma applanatum (Pers.) Pat.	Temperate and boreal northern hemisphere	Wide range, deciduous and coniferous	Natural and managed	Allen *et al.*, 1996; Holsten *et al.*, 2001; Kuzmichev *et al.*, 2004; Dai *et al.*, 2007	In North America, serious pathogen of deciduous forests (Ross, 1976; Manion, 1991); urban in Europe (Terho, 2009)
Grifola frondosa (Dicks.) Gray	Temperate Europe and N. America	*Quercus*	Old natural	Berry and Lombard, 1978; Phillips and Burdekin, 1982; Černý, 1989; Butin, 1995	Red listed in northern Europe (Larsson, 1997); indicator of forest value for nature conservation (Nitare, 2000)
Inonotus leporinus (Fr.) Gilb. & Ryvarden	Boreal and temperate Eurasia	*Picea*	Old natural	Černý, 1989; Kuzmichev *et al.*, 2004; Dai *et al.*, 2007	Mortality agent in pristine forests of north Europe (Lännenpää *et al.*, 2008); red listed (Larsson, 1997) and indicator of forest value for nature conservation (Nitare, 2000)
Perenniporia subacida (Peck) Donk	Boreal and temperate N. America and Asia	Wide range, coniferous	Old natural	Scharpf, 1993; Allen *et al.*, 1996; Hansen and Lewis, 1997; Dai *et al.*, 2007	Old-growth conifer forests
Phaeolus schweinitzii (Fr.) Pat.	Temperate and boreal northern hemisphere	Wide range, coniferous	Old natural	Scharpf, 1993; Allen *et al.*, 1996; Hansen and Lewis, 1997; Holsten *et al.*, 2001; Kuzmichev *et al.*, 2004; Dai *et al.*, 2007	In Europe, indicator of forest value for nature conservation (Nitare, 2000)
Pseudoinonotus dryadeus (Pers.) T. Wagner & M. Fisch.	Western N. America, Mexico to British Columbia, Europe, Asia	*Abies*, *Quercus*	Old natural	Hansen and Lewis, 1997; Kuzmichev *et al.*, 2004; Dai *et al.*, 2007	Red listed in northern Europe (Larsson, 1997); indicator of forest value for nature conservation (Nitare, 2000)
Sparassis crispa (Wulfen) Fr.	Boreal and temperate Eurasia	*Larix*, *Pinus*	Old natural	Phillips and Burdekin, 1982; Černý, 1989; Butin, 1995; Dai *et al.*, 2007	In Europe, indicator of forest value for nature conservation (Nitare, 2000)

[a]In geographic regions with intensive forestry, old natural forest is as a rule synonymous with protected forest.

does the exposed dead sapwood where the bark has been knocked off by injuries such as lightning wounds, breaking branches, falling trees and many others. Broken tops give direct access to the heartwood, as do fire scars when the fire has eaten deeply enough into the tree. Notably, the injuries considered in this fundamental work as the most important are those arising during natural ecosystem processes, and are typical attributes of old-growth forest. Consequently, the work of Boyce (1961), which has already been cited, reflects the infection pathways of the first two heart rot categories in Table 10.1.

Recent decades of forestry have been marked by a rapid increase in the area of forest plantations that is intensively managed for wood and/or wildlife. Characteristic of such stands is the presence on living stems of logging and bark stripping wounds, and in certain forest areas trees without such injuries have become uncommon (Steyrer, 1992). The wounds result in sapwood exposure, thus providing access for decay fungi to the heartwood. The exposed sapwood then becomes subject to fungal succession, and certain fungi, upon reaching the heartwood, are able to cause heart rot in living wounded trees (Shigo, 1965; Vasiliauskas, 1998a, 2001). Such heart rot will subsequently be referred to as a wound heart rot. Traits and impacts of all three categories of heart rots, and of certain butt rots and root rots, as well as of the saprotrophs (agents of sap rot in dead stems) will be discussed elsewhere in this chapter. It will be demonstrated that fungi from each group possess a number of rather distinct ecological traits and differ markedly in their infection biology, host colonization strategies, ecological roles and pathological impacts.

10.1.2 Pathogens

All the wound heart rot pathogens listed in the Table 10.1 were traditionally known as saprotrophic decomposers of various kinds of woody debris. Yet, seemingly, those species are adapted to penetrate the stem of a tree through exposed dead sapwood, become established and cause heart rot in a living stem. Still, due to the presence of a large number of wounded trees in managed forests, these fungi emerged as pathogens, threatening sustainable production of wood over large territories, e.g. in Europe (Vasiliauskas, 2001). In a way, these saprobes should now be considered to be more important heart rot pathogens than the classical true heart rot fungi that have been previously treated in texts of forest pathology.

Apart of those listed in Table 10.1, there are a few other heart rots that are deemed relevant for forest health in certain geographic areas. For example, *Inonotus andersonii* (Ellis & Everh.) Černý and *Perenniporia compacta* (Overh.) Ryvarden & Gilb. were reported as the most destructive heart rots of oaks (*Quercus* spp.) in the central USA (Berry and Lombard, 1978); *I. glomeratus* (Peck) Murrill is a common cause of canker rot in maples (*Acer* spp.) and beeches (*Fagus* spp.) in north-east USA (Manion, 1991); *Oxyporus populinus* (Schumach.) Donk cause hollows in the stems of maple and other hardwoods (Manion, 1991; Tainter and Baker, 1996); and *I. hispidus* (Bull.) P. Karst. is a heart rot agent on oaks, birches (*Betula* spp.), ashes (*Fraxinus* spp.) and other hardwoods in the USA and Europe (Butin, 1995; Tainter and Baker, 1996) – although in north-eastern Europe this fungus is rare and is red listed (Larsson, 1997). The only ascomycete to be mentioned in this chapter is *Kretzschmaria deusta* (Hoffm.) P.M.D. Martin, a cause of butt rots of many deciduous trees in the USA and Europe (Manion, 1991; Butin, 1995; Tainter and Baker, 1996). On conifers, conspicuous heart rot pathogens include *Tyromyces amarus* (Hedgc.) J. Lowe on cedar (*Chamaecyparis* spp.), and *Poria sequoiae* Bonar on redwood (*Sequoia sempervirens* (Lamb. ex D. Don) Endl.), both occurring in old-growth stands of the Pacific coast, and both strictly specific to their respective hosts (Scharpf, 1993; Tainter and Baker, 1996). A relevant example of butt heart rot of regional importance is *Hericium abietis* (Weir ex Hubert) K.A. Harrison, which infects old-growth true firs (*Abies* spp.) on the Pacific coast (Scharpf, 1993; Hansen and Lewis, 1997), and hemlock (*Tsuga* spp.) and spruce (*Picea* spp.) in Alaska (Holsten *et al.*, 2001).

10.1.3 Habitats

The infection courts of true and wound heart rot fungi fundamentally differ. As a rule, representatives of the first group are not capable of entering (younger) trees via artificial sapwood wounds, while the representatives of the second group are unable to infect (older) trees via natural pathways. As true heart rots are normally associated either with intact ageing stems or with stem injuries arising during natural ecosystem disturbance, one might argue that true heart rot fungi have evolved simultaneously with the evolution of forest ecosystems and are natural components of these ecosystems. Because with stand age numbers of both ageing weakened trees and of natural injuries in stands tend to increase, true heart rots become increasingly common in older forests, and currently represent the typical attributes of natural old-growth forest. In fact, the same also applies to root and butt heart rots. In comparison, wound heart rot fungi infect heartwood through artificially inflicted mechanical injuries, such as logging and bark stripping wounds (note that bark stripping does not normally occur in pristine forests) (Shigo, 1966; Vasiliauskas, 1998a, 2001), pruning damage (Risley and Silverborg, 1958; Metzler, 1997) or parent stumps (Berry and Lombard, 1978), so they are characteristic pathogens in forest stands managed for timber and big game. Stands are subjected to logging damage (and pruning) from the age of around 30 years, and to bark stripping damage mostly at the age of 30–50 years (Vasiliauskas, 1989, 1998a, 2001). Consequently, wound heart rots are typical attributes of younger forests. Therefore, different fungi infect tree stems in natural forests and managed stands, and the communities of heart rots in those habitats differ fundamentally.

Such a trend had already been noted in an early study by Maloy and Robinson (1968), in which the true heart rot *Echinodontium tinctorium* (Ellis & Everh.) Ellis & Everh. was not once isolated from the heartwood of 15–30 year old stems of grand fir (*Abies grandis* (Douglas ex D. Don) Lindl.) with open and healed scars, despite them showing decay similar to that caused by the fungus (a white rot); instead, only the wound heart rot fungus *Stereum sanguinolentum* (Alb. & Schwein.) Fr. was isolated from those trees, while *E. tinctorium* was associated with heart rot only in trees over 150 years of age. Although there were singular cases when true heart rot fungi (*Fomitopsis pinicola* (Sw.) P. Karst. and *Phellinus chrysoloma* (Fr.) Donk) were observed or isolated from logging and bark stripping damaged spruce (Vasiliauskas, 1989; Vasiliauskas and Stenlid, 1998a), it is likely that the trees were entered through natural infection courts; no other true (or root and butt) heart rots of spruce have been found associated with superficial sapwood wounds. Furthermore, the true heart rots of oak *F. robusta* and *Laetiporus sulphureus s.l.* were reported as dominant in overmature stands (Sunhede and Vasiliauskas, 1996), but they were never observed in middle-aged stands suffering from bark stripping (Vasiliauskas, 1998c). In birch stands in north Europe, none of the dominant decayers of mature forest, such as *Fomes fomentarius* (L.) J. Kickx f., *Inonotus obliquus* (Ach. ex Pers.) Pilát, *Phellinus igniarius s.l.* and *Piptoporus betulinus* (Bull.) P. Karst. (Niemelä, 1975; Roll-Hansen, 1981), were detected in discolorations resulting from broken branches and logging wounds in managed younger stands (Hallaksela and Niemistö, 1998; Vasaitis *et al.*, 2012a). Similarly, in beech forests in south Germany, true heart rot *F. fomentarius* was reported to be the most common fungus in virgin forests and strict forest reserves, but was almost missing in stands with a high management intensity (Müller *et al.*, 2007).

Thus, true heart rots are a characteristic feature of old forest ecosystems, which, as a rule, are now under a protection regime. Consequently, this group of fungi should nowadays be viewed from the perspective of nature conservation rather than forest pathology. In contrast, wound heart rots are already introduced to young stands by management activities, and are a characteristic feature of forests under anthropogenic pressure.

10.1.4 Impact

True heart rots and root and butt heart rots

In individual stems, the volume of decay and losses in wood production caused by these heart rots vary among fungal and tree species, but are usually significant (Berry and Lombard, 1978). Consequently, heart rots in forest stands were traditionally referred to as the most damaging of all types of tree disease, and much more important in mature and over-mature stands than in young stands (Tainter and Baker, 1996). It is assumed that losses due to decay increase with increasing tree age and are usually insignificant in trees less than 90 years old (Hansen and Lewis, 1997). According to graphs developed by Hepting and Hedgcock (1937), cull percentage in natural stands of oaks (northern red oak, *Q. rubra* L.; black oak, *Q. velutina* Lam.; white oak, *Q. alba* L.; and chestnut oak, *Q. prinus* L.) and tulip tree (*Liriodendron tulipifera* L.) would consistently increase with increasing percentage of trees with basal fire scars and stem diameter. Then, provided that 100% of trees possess fire scars: in a stand with an average tree diameter of 30 cm, the percentage cull would typically consist of 10–15%; in a stand of average tree diameter 50 cm, it would be 15–20%; and in a stand of average tree diameter of 75 cm, it would be 20–25%. In old-growth stands of Douglas-fir (*Pseudotsuga menziesii* (Mirb.) Franco) in the US Pacific Northwest, the volume of cull caused by *Porodaedalea pini* (Brot.) Murrill, *Fomitopsis officinalis* (Vill.) Bondartsev & Singer, *F. cajanderi* (P. Karst.) Kotl. & Pouzar and *Phaeolus schweinitzii* (Fr.) Pat. was reported to comprise only 4% in 80–120 year old stands, but 34–85% in 160–200 year old and older stands (Tainter and Baker, 1996). Notably, the first three species of the true heart rot fungi just listed (as well as others) are now being used (and by forest pathologists themselves) for artificial inoculations of Douglas-fir and other conifers in the region so as to promote stem decay in trees to be subsequently used as wildlife habitats (Filip *et al.*, 2011). Consequently, the impacts of these disease agents currently clearly fall outside the range of forest pathology.

Moreover, as old-growth forests disappear or are being excluded from harvesting, there is simultaneously a decrease in losses caused by true heart rot fungi. For example, in the early 1900s in the USA, heart rot in the southern pines averaged about 20% of the volume of the harvested timber for sawn timber production, but in the late 1900s losses were less than 1% (Tainter and Baker, 1996). In Europe, heart-rotting fungi of old natural (often montane) habitats have also previously been considered to be important forest pathogens. In a book by Černý (1989), for instance, *P. pini*, *P. schweinitzii*, *Onnia triquetra* (Pers.) Imazeki, *Phellinus vorax* (Harkn.) Černý and *Laetiporus montanus* Černý ex Tomšovský & Jankovský are the only heart rot fungi attributed to pine (*Pinus* spp.), and it is even implied that forests should be protected against these fungi. Conversely, current understanding would suggest that the occurrence of each of those fungi in a forest stand is a good indicator for the habitat to be considered for nature conservation (Nitare, 2000).

Apart from the impact of true heart rot fungi on and their beneficial roles in wildlife and habitat conservation, there is an increasing understanding of their relevance in natural stand development. Certain studies suggest that the role of these fungi in forest dynamics is more important than that of fire. Lewis and Lindgren (1999) investigated the importance of decay fungi as agents of gap formation and facilitators of uneven-aged stand structure in sub-boreal forests of British Columbia. The results indicated that decay fungi, in particular *E. tinctorium*, play an important role in removing fir from the overstorey and in enabling spruce recruitment. In pristine boreal forests of Europe, *Coniophora* spp., *P. chrysoloma*, *F. pinicola* and *Inonotus leporinus* (Fr.) Gilb. & Ryvarden were the main mortality agents of spruce, playing a key role in forest succession (Lännenpää *et al.*, 2008). *F. pinicola* has been reported as a mortality agent of old-growth spruce in pristine northern forests of Europe (Kotiranta and Niemelä, 1981; Renvall *et al.*, 1991), China (Dai, 1996) and Alaska (Holsten *et al.*, 2001). Also, fallen trunks provide habitats for many organisms (e.g. decay fungi inhabiting wood at late decomposition stages) and serve as nurse logs for subsequent forest regeneration (Vasiliauskas *et al.*, 2004).

Among the true heart rots of economic importance, the most notorious is *Phellinus tremulae* (Bondartsev) Bondartsev & P.N. Borisov, which often almost totally destroys the timber harvest in aspen (*Populus* spp.) stands that are already aged 40–50 years, which are distributed over large areas of the whole northern hemisphere (Niemelä, 1974; Tainter and Baker, 1996; Kuzmichev *et al.*, 2004; Dai *et al.*, 2007). In contrast, no heart rots were detected in 30 year old stands of birch, and the observed discolorations of heartwood, although common, were restricted and had no effect on wood quality (Hallaksela and Niemistö, 1998). For other tree species, a recent study in north-eastern Europe showed that a large proportion (75% on average) of trees in mature and over-mature stands of European (black) alder (*Alnus glutinosa* (L.) Gaertn.) is damaged by a heart rot, and that the principal agent is *Mensularia radiata* (Sowerby) Lázaro Ibiza. Decay columns of an average length of 7.7 m long develop in stems, and 4.2 m of this comprises the advanced (spongy) stages of rot (Arhipova *et al.*, 2012). This implies that the yield loss for a fully stocked 80 year old European alder stand would comprise 49.2% of the total stand volume, while the losses from advanced spongy rot would be 30.5%. According to a heart rot study in grey alder (*A. incana* (L.) Moench), the cull volume would be 32 m^3 ha^{-1} at stand age of 45–50 years and 60 m^3 ha^{-1} at age 60–65 years, corresponding to about 10% and 20% of the standing volume of a pure fully stocked stand (Arhipova *et al.*, 2011b). However, as the principal fungus isolated from decay columns was *Chondrostereum purpureum* (Pers.) Pouzar, it is unlikely that the observed heart rots were true heart rots.

Wound heart rots

Wound heart rot is a problem of intensively managed forest. Typical examples are European forests, where trees commonly suffer mechanical injuries inflicted during various types of selective harvesting, and bark stripping by deer and moose. In regions where shelterwood forestry is combined with game management (as is traditional in central and northern Europe), the proportion of wounded stems in the total growing stock could reach 20–50%, or even 60–90% (Vasiliauskas, 2001; Vasaitis *et al.*, 2012b). The pathological impacts of wound heart rots have been reviewed by Vasiliauskas (2001), and it has been demonstrated that yield losses can be significant, depending on the number of wounded trees, the time since the damage occurred, stand age and forest productivity. For example, in a productive fully stocked 80 year old stand of spruce with 100% of stems wounded 30 years previously, cull volume would comprise 25% of the total standing volume (Fig. 10.2), while in a 70 year old stand of oak and a 100 year old stand of ash, the proportions of cull volume would comprise 10% and 60%, respectively. Note that, as tree growth exceeds heart rot spread, the proportion of cull volume decreases in older stands, although because the most valuable butt log is affected, the financial losses have been reported to increase with time (Vasiliauskas, 2001). Also more recent studies have shown that bark stripping and subsequent rot has decreased the financial value of 30 year old stems of fir and ash by 22.5% and 25.7%, respectively, and that the losses tend to increase in older trees (Barszcz and Jamrozy, 2001). In this respect, birch seems to be an exception, as wound heart rot does not usually develop beyond its logging wounds, and discoloration is mainly restricted to the damaged portion of a stem (Shigo, 1966; Vasaitis *et al.*, 2012a). Consequently, the associated loss of wood production is reported to remain low during about 10 years after the damage has occurred.

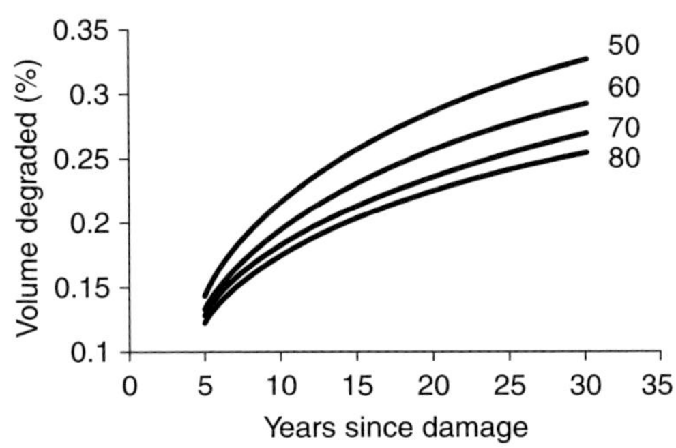

Fig. 10.2. Percentage of cull volume comprising wound heart rot in fully stocked *Picea abies* (L.) Karst. stands containing 1% of stems with logging and/or bark stripping wounds in relation to stand age (50, 60, 70 and 80 years) and years since the damage (reprinted from Vasiliauskas, 2001).

The overall large impact of wound heart rots on forestry has been known for a long time. Thus, in Austria in 1961–1970, production loss due to logging damage has been estimated to comprise 30 million m^3 (Schimitschek, 1975). The situation with bark stripping damage is more complex and the losses cannot be expressed solely by decayed wood volume. Because, as a rule, the injury is located at a stem height of around 1.5 m, decay development makes such stems prone to wind and snow damage at a premature age. This leads to the destruction of middle-aged stands by wind and snow, and young forest has to be clear-felled and replanted.

10.2 Diagnosis

The presence of fruiting bodies and injuries are the symptoms that a tree is infected by a heart rot fungus, and the fruiting bodies also provide information on the identity of the fungus. Correlations between the presence of fruiting bodies (even their number and size) with the extent of heart rot in a stem were analysed by Boyce (1961). The analysis reflects data from different tree species and pathogens, but mainly true heart rots in mature and over-mature natural forests, and generally indicates that fruiting bodies of true heart rot fungi are formed after the tree stem has been extensively colonized. The presence of fruiting bodies of true heart rot fungi therefore implies that the larger part or entire trunk of the tree has already been transformed to cull status (Allen *et al.*, 1996). For example, the stems of firs and aspens bearing several fruiting bodies of *E. tinctorium*, *P. tremulae* and *P. pini* from a wood yield point of view are referred to as a total loss (Tainter and Baker, 1996). Conversely, the presence of fruiting bodies of root and butt heart rot fungi in most cases indicates that the decay column is restricted to the butt region (Phillips and Burdekin, 1982; Allen *et al.*, 1996). However, one should keep in mind that intact trees infected by true heart rot usually show neither fruiting bodies nor other external symptoms (Hansen and Lewis, 1997).

Logging, bark stripping and other similar wounds are indicators of wound heart rot. Fruiting bodies of wound heart rot fungi are seldom produced on a surface of exposed sapwood, and the main value in wound heart rot diagnostics is not the presence of fruiting bodies, but the size (exposed sapwood area) and age of the injury. The pathological consequences of the mechanical wounding of a number of forest trees and decay agents have been reviewed by Vasiliauskas (2001). This review demonstrates that the majority of sapwood wounds have already been entered by stain or decay fungi by 2 years after injury, and that decay columns develop during subsequent years. For example, over 80% of logging and bark stripping wounds on spruce are colonized by decay fungi, which in 10–25 years spread over the heartwood and by 3–5 m longitudinally; the length of heart rot correlates positively with the wound dimensions. Bark stripping wounds on oak are typically associated with over 1 m long decay columns (Vasiliauskas, 1998c), and on fir with 2–3 m long decay columns (Barszcz and Jamrozy, 2001; Pach, 2005), while in ash they result in 8–9 m long heartwood discolorations (Vasiliauskas and Stenlid, 1998b). In comparison, in stems of birch, discoloration associated with logging damage is largely confined to the limits of the wound (Shigo, 1966; Vasaitis *et al.*, 2012a), and the same applies to beech (Schumann and Dimitri, 1993). In addition, the studies cited above indicate that following wound closure, decay development in spruce, fir and oak is confined. Moreover, a recent study by Vasaitis *et al.* (2012b) demonstrated that spruce wounds of initial width <4 cm will usually occlude without being infected by decay, while no wounds that are initially 15 cm wide will occlude during 20 years (Fig. 10.3). Apart from wound decay diagnostics, the probability of wound occlusion brings into any plan certain disease management implications that are discussed later.

For the identification of all types of heart rots in the absence of fruiting bodies, samples of decayed wood need to be taken from a tree, followed by fungal isolation into pure culture and identification either by culture morphological characteristics (e.g. Stalpers, 1978; Vasiliauskas *et al.*, 1998), or by culture analysis using molecular methods (e.g. Lygis *et al.*, 2005;

Fig. 10.3. Occlusion of *Picea abies* (L.) Karst. stem wounds. (A)–(F), 19 year old wounds: occluded of 1 × 15 (A), 2 × 15 (B), 3 × 15 (C) and 4 × 15 (E) cm initial size, and open of 3 × 15 (D) and 5 × 15 (F) cm initial size; G–J, 20 year old wounds of 15 × 20 cm initial size: nearly occluded (G), occluding (H), and non-occluding (I) and (J); size of paper sheet 5 × 5 cm. Note the exit holes of siricid woodwasps in (F) and (J), indicating that the trees are decayed by *Amylostereum* wound heart rot, and that the pathogens were introduced to trees by the insects. Credits: Vaidotas Lygis.

Nicolotti *et al.*, 2010). Recently, molecular methods have been developed for detection of the latent presence of decay fungi directly from environmental samples taken from visually healthy trees (Nicolotti *et al.*, 2010; Parfitt *et al.*, 2010), although these methods are laborious and too costly to be applied in practical forestry.

10.3 Infection Biology and Epidemiology

10.3.1 Courts of infection

True heart rots in intact trees

Traditionally, it has been assumed that the entry point of heart rot to an otherwise intact stem has to be via a branch stub large enough to connect the heartwood with the outside environment (Boyce, 1961). This assumption was based on the simple fact that the fruiting body of a heart rot fungus is typically formed on or in the vicinity of a branch stub. However, some early studies have already indicated different modes of the establishment of heart rot fungi in intact trees. For example, Basham (1966) reported that while *P. pini* was the most common fungus isolated from heart rot columns of pine, it was only occasionally detected in branch stubs, and only in those encased by decayed heart rot, so that it had evidently entered the stubs from established rot and not *vice versa*. Haddow (1938a) reported that *P. pini* infects pines at young age through dead twigs (of only a few millimetres in diameter) and weevil-killed leaders. Furthermore, it was shown that another heart rot fungus, *E. tinctorium*, establishes in hemlock branchlets 0.5–1.5 mm in diameter and remains dormant until it is embedded by heartwood at some point in the future. Significant heart rot develops later, after the infected trees become old and are stressed through wounding or other factors (Etheridge and Craig, 1976; Holsten *et al.*, 2001). Moreover, dwarf mistletoe (*Arceuthobium* spp.) and fir broom rust (*Melampsorella caryophyllacearum* (DC.) J. Schröt.) infections were reported to serve as entry points for *Phellinus hartigii* (Allesch. & Schnabl) Pat. to stems of fir (Gilbertson and Ryvarden, 1987; Černý, 1989), and small twigs were demonstrated to be infection foci of *P. tremulae* for stems of aspen (Holmer *et al.*, 1994). On old-growth oaks, fruiting bodies of *F. robusta* were commonly observed on intact living branches 5–30 cm thick (Sunhede and Vasiliauskas, 2002). Hence, the question of whether branch stubs serve both as an entry and exit point for a fungus, or only as an exit point following a substantial period of latent establishment in a living tree, has remained open (Rayner and Boddy, 1988).

Currently, the evidence is increasingly accumulating on the latent presence of decay fungi in the functional sapwood of intact trees, and such data is summarized in Table 10.2. Apart from the true heart rot agents that have already been discussed, *F. fomentarius*, *F. pinicola* and *M. radiata* were consistently detected in sound wood (and *F. pinicola* also in healthy phloem) of living trees, either by isolations or molecular methods. It should be noted that for those heart rot fungi the main infection court was traditionally thought to be stem injuries and large wounds on which fruiting bodies formed (Schwarze *et al.*, 2000). In the case of *F. fomentarius* though (for example), neither stem dissection nor rot column analysis could trace infections to stem wounds in beech (Skov and Thomsen, 2003). It is, therefore, likely that a period of latent establishment precedes the development of active decay in true heart rots, thus complying with the Haddow–Etheridge concept, although the pathways followed by most of these are unknown. A study by Oses *et al.* (2008) demonstrated the presence of fungal hyphae attached to the inner cell surface of healthy living tree xylem before the induction of wood degradation. Thin periderm, lenticels, scars on twigs, leaves and buds were all considered as infection courts for the latent establishment of propagules of decay fungi (Baum *et al.*, 2003; Parfitt *et al.*, 2010).

Notably, in order to be able to develop subsequent active decay, except for homothallic fungi, at least two spores of opposite mating types must penetrate, come in contact with each other and form a stable heterokaryon inside the xylem of a living tree. For example, heterokaryons of *P. betulinus* were regularly detected in intact birch, and there were no cases where the same genotype was obtained from more than one tree, suggesting that genetically diverse basidiospores act as infective agents (Adams, 1982). One possible scenario could then be the airborne arrival of basidiospores, which land on the tree surface, germinate and colonize certain areas of the surface by vegetative spread of homokaryotic mycelium; this is

Table 10.2. Decay fungi detected as latent in visually healthy intact trees.

Fungus	Tree species	Substrate	Reference	Remarks
1. True heart rots of intact tree stems				
Echinodontium tinctorium (Ellis & Everh.) Ellis & Everh.	*Tsuga heterophylla* (Raf.) Sarg.	Dead 1 mm diameter branchlets	Etheridge and Craig, 1976	The branchlets are subsequently encased by healthy stemwood
Fomes fomentarius (L.) J. Kickx f.	*Acer campestre* L., *Betula* spp., *Fraxinus excelsior* L., *Malus domestica* Borkh., *Quercus robur* L., *Salix* spp.	Healthy xylem/ functional sapwood of branches and stems	Parfitt *et al.*, 2010	Also rot of intact *Alnus incana* and *A. glutinosa* stems (Arhipova *et al.*, 2011a, 2012)
	Fagus sylvatica L.	Healthy xylem	Baum *et al.*, 2003; Parfitt *et al.*, 2010	
Fomitopsis pinicola (Sw.) P. Karst.	*Picea abies* (L.) Karst.	Sound wood of intact stems	Vasiliauskas *et al.*, 2001	Also rot of intact *A. glutinosa* and *A. incana* (Arhipova *et al.*, 2011a, 2012), *Betula pendula* (Lygis *et al.*, 2004a), *P. abies* (Cleary *et al.*, 2013) stems and wounded *P. abies* stems (Vasiliauskas and Stenlid, 1998a)
		Intact stem phloem	R. Vasaitis *et al.*, 2012, unpublished results	
	Alnus glutinosa (L.) Gaertn.	Sound wood of intact stems	Arhipova *et al.*, 2012	
Mensularia radiata (Sowerby) Lázaro Ibiza	*Alnus glutinosa*	Sound wood of intact stems	Arhipova *et al.*, 2012	Also rot of intact *A. glutinosa*, *A. incana* (Arhipova *et al.*, 2011a, 2012) and *B. pendula* (Lygis *et al.*, 2004a) stems
Phellinus tremulae (Bondartsev) Bondartsev & P.N. Borisov	*Populus tremula* L.	Healthy-looking twigs	Holmer *et al.*, 1994	Entering stem via twigs
Porodaedalea pini s.l. (Brot.) Murrill	*Pinus strobus* L.	Twig stubs, 0.2–1.3 cm diameter	Haddow, 1938a,b	The twig stubs are subsequently encased by healthy stemwood
2. True heart rots associated with heartwood-exposing natural injury (broken branches and tops, stem cracks, fire scars)				
Stereum gausapatum (Fr.) Fr.	*Acer campestre*, *Fagus sylvatica*, *Fraxinus excelsior*, *Quercus robur*, *Salix* spp.	Functional sapwood of branches and stems	Parfitt *et al.*, 2010	Associated with heartwood-exposing injuries (Berry and Lombard, 1978)
3. Wound heart rots associated with artificial mechanical damage (logging, bark stripping, pruning wounds)				
Coniophora puteana (Schumach.) P. Karst.	*Fagus sylvatica*	Xylem of intact twigs, branches and stems	Chapela and Boddy, 1988; Hendry *et al.*, 2002; Baum *et al.*, 2003	Also rot of intact (Cleary *et al.*, 2013) and wounded (Vasiliauskas, 1998b) *P. abies* stems, and wounded *F. excelsior* stems (Lygis *et al.*, 2005); fruiting bodies on stem wounds of *Q. robur* (Vasiliauskas, 1998c)
	Picea abies	Intact stem phloem	R. Vasaitis *et al.*, 2012, unpublished results	

Resinicium bicolor (Alb. & Schwein.) Parmasto	*Picea abies*	Intact stem phloem	R. Vasaitis *et al.*, 2012, unpublished results	Also rot of wounded *P. abies* stems (Vasiliauskas,1998b)
Sistotrema brinkmannii (Bres.) J. Erikss.	*Alnus incana* (L.) Moench	Sound wood of intact stems	Arhipova *et al.*, 2011a	Also rot of wounded *P. abies* stems (Vasiliauskas *et al.*, 1996; Vasiliauskas,1998b)
4. Root and butt rots				
Ganoderma applanatum (Pers.) Pat.	*Fraxinus excelsior*	Sound wood of intact stems	Lygis *et al.*, 2005	Living *F. excelsior* with dieback symptoms
5. Decay fungi producing fruiting bodies on dead decomposed wood and generally considered as saprotrophs (Eriksson and Ryvarden, 1976; Eriksson *et al.*, 1978, 1981, 1984; Breitenbach and Kränzlin, 1986; Jahn, 1990; Ryman and Holmåsen, 1992; Ryvarden and Gilbertson, 1993, 1994)				
Bjerkandera adusta (Willd.) P. Karst.	*Alnus glutinosa*	Sound wood of intact stems	Arhipova *et al.*, 2012	Rot of intact *A. glutinosa*, *A. incana*, *P. abies* (Arhipova *et al.*, 2011a,b, 2012) and *B. pendula* stems (Lygis *et al.*, 2004a); known decay agent of living trees (Phillips and Burdekin, 1982; Dai *et al.*, 2007)
	Alnus incana		Arhipova *et al.*, 2011a	
	Fraxinus excelsior		Lygis *et al.*, 2005	
	Picea abies	Sound wood of intact stems	N. Arhipova *et al.*, 2012, unpublished results	
		Intact stem phloem	R. Vasaitis *et al.*, 2012, unpublished results	
Coprinellus disseminatus (Pers.) J.E. Lange	*Alnus incana*	Sound wood of intact stems	Arhipova *et al.*, 2011a	Rot of intact *A. incana* and wounded *B. pendula* stems (Vasaitis *et al.*, 2012a)
	Fraxinus excelsior		Lygis *et al.*, 2005	Living *F. excelsior* with dieback symptoms
	Picea abies, *Pinus sylvestris* L.		N. Arhipova *et al.*, 2012, unpublished results	In *P. sylvestris* determined to genus level
Gloeophyllum sp.	*Picea abies*	Intact stem phloem	R. Vasaitis *et al.*, 2012, unpublished results	Rot of intact *P. abies* stems (Arhipova *et al.*, 2011b)
Gymnopilus penetrans (Fr.) Murrill	*Pinus sylvestris*	Sound wood of intact stems	N. Arhipova *et al.*, 2012, unpublished results	
Merulius tremellosus Schrad.	*Alnus incana*		Arhipova *et al.*, 2011a	Rot of intact *A. incana* stems (Arhipova *et al.*, 2011a)
	Pinus sylvestris		Lygis *et al.*, 2004b	
Mycoacia uda (Fr.) Donk	*Fraxinus excelsior*		Lygis *et al.*, 2005	*F. excelsior* with dieback symptoms, and stem wounds
Oxyporus latemarginatus (Durieu & Mont.) Donk	*Alnus incana*		Arhipova *et al.*, 2011a	
Peniophora cinerea (Pers.) Cooke	*Alnus glutinosa*		Arhipova *et al.*, 2012	Rot of intact *A. glutinosa* and *A. incana* stems (Arhipova *et al.*, 2011a, 2012), and wounded *F. excelsior* stems (Lygis *et al.*, 2005)
	Alnus incana		Arhipova *et al.*, 2011a	

Continued

Table 10.2. Continued.

Fungus	Tree species	Substrate	Reference	Remarks
	Fraxinus excelsior		Lygis *et al.*, 2005	Living *F. excelsior* with dieback symptoms
Peniophora limitata (Chaillet ex Fr.) Cooke	*Fraxinus excelsior*		Lygis *et al.*, 2005	*F. excelsior* with dieback symptoms; fruiting bodies on *F. excelsior* wounds (Vasiliauskas and Stenlid, 1998b)
Phanerochaete sordida (P. Karst.) J. Erikss. & Ryvarden	*Alnus glutinosa*		Arhipova *et al.*, 2012	
	Alnus incana		Arhipova *et al.*, 2011a	
	Picea abies, *Pinus sylvestris*		N. Arhipova *et al.*, 2012, unpublished results	
Phellinidium ferrugineofuscum (P. Karst.) Fiasson & Niemelä	*Betula pendula* Roth		Lygis *et al.*, 2004a	Fruiting bodies on *Picea* spp.; in Europe, indicator of forest value for nature conservation (Nitare, 2000)
Phlebia radiata Fr.	*Fagus sylvatica*	Healthy branch xylem	Baum *et al.*, 2003	Rot of intact *P. abies* stems (Arhipova *et al.*, 2011b)
	Pinus sylvestris	Sound wood of intact stems	Lygis *et al.*, 2004b	
Phlebia rufa (Pers.) M.P. Christ.	*Fraxinus excelsior*	Sound wood of intact stems	Lygis *et al.*, 2005	Living *F. excelsior* with dieback symptoms
Phlebiopsis gigantea (Fr.) Jülich	*Pinus mugo* Turra		V. Lygis *et al.*, 2012, unpublished results	
Pholiota highlandensis (Peck) Singer	*Pinus mugo*		V. Lygis *et al.*, 2012, unpublished results	Fruiting bodies on burned wood (Ryman and Holmåsen, 1992)
Piptoporus betulinus (Bull.) P. Karst.	*Betula* spp.	Latent in intact trunks	Adams, 1982	Restricted to *Betula* spp.; reported also as sap rot and heart rot agent (Roll-Hansen, 1981; Butin, 1995; Kuzmichev *et al.*, 2004; Dai *et al.*, 2007)
Stereum rugosum Pers.	*Acer campestre*, *Betula* spp., *Fagus sylvatica*, *Fraxinus excelsior*, *Malus domestica*, *Quercus robur*, *Salix* spp.	Functional sapwood of branches and stems	Parfitt *et al.*, 2010	
Trametes hirsuta (Wulfen) Lloyd	*Fraxinus excelsior*	Sound wood of intact stems	Lygis *et al.*, 2005	Vigorous trees showing external *Armillaria* spp. rhizomorphs

later is fertilized by newly landing spores, resulting in formation of heterokaryons. In fact, mass production and efficient spore dispersal is a general feature of wood decay basidiomycetes. For example, a 1 mm^2 pore layer of *P. tremulae* fruiting body could produce over 100,000 basidiospores a day (Manion, 1991). Seasonal spore production from a single *F. pinicola* fruiting body was estimated to exceed 10^{11} basidiospores a day (Nordén, 1997). Skov and Thomsen (2003) investigated spore spread of *F. fomentarius* in Denmark, and concluded that there are spores of *F. fomentarius* almost everywhere due to its immense spore production. Consequently, it is realistic to speculate that numerous genetically diverse basidiospores of true heart rot fungi would be normally present on trees in forest stands. However, the infection venues of true heart rot fungi in intact trees, the mechanisms of their symptomless persistence and the factors triggering decay development are mostly unknown to date, and much remains to be explored under the Haddow–Etheridge infection concept.

True heart rots associated with natural injury

From this category, only *Stereum gausapatum* (Fr.) Fr. was detected in functional sapwood of intact trees (Table 10.2). Apart from five different forest tree species (hedge maple, *Acer campestre* L.; European beech, *Fagus sylvatica* L.; European ash, *Fraxinus excelsior* L.; English oak; and willows, *Salix* spp.) the study, based on direct molecular analyses of intact xylem, consistently revealed the latent presence of the fungus in elder (*Sambucus nigra* L.) and hazel (*Corylus avellana* L., also known as common filbert) (Parfitt *et al.*, 2010). This, besides demonstrating *S. gausapatum* as one of the currently known most efficient colonizers of intact woody plants, also puts under serious doubt the categorization of the species as true heart rot associated with natural injury. The basis for this was that some other representatives of the genus are typical wound pathogens, and that the fungus commonly occurs in oaks with broken tops and fire scars (Berry and Lombard, 1978), although the suggested venues might in fact well be the exit points. The same study also noted the common association of *S. gausapatum* with dead and sound branch stubs, and in Europe the fungus was reported to be one of the most common pathogens of old-growth oaks (Aufsess, 1973), typically producing fruiting bodies on branches that are still attached (Eriksson *et al.*, 1984). Therefore, it seems likely that the fungus does comply with the Haddow–Etheridge infection concept, and could be put under true heart rots of intact trees. Notably, none of the rest of fungi from true heart rots associated with natural injury has been so far detected in intact trees, so seemingly are able to enter trees only through damage that exposes heartwood, so complying with the Hartig infection concept. These findings explicitly demonstrate that while infecting trees true heart rots principally exhibit different entrance modes and strategies.

Wound heart rots

Logging and bark stripping wounds represent injury in which bark is knocked or ripped off the stem, exposing the underlying dead sapwood. While in the living stem, functional sapwood is more resistant to decay than heartwood, following the death of a tree the situation reverses, and sapwood becomes the most susceptible (Rayner and Boddy, 1988). When bark is ripped off from a certain part of a living stem, as typically occurs during logging and bark stripping damage, this results in rapid death of the exposed portion of the underlying sapwood, which, due to desiccation and loss of physiological reactions, becomes accessible to growth over long distances by the hyphae of decay fungi (Metzler *et al.*, 2012). Certain fungal species (e.g. *Stereum*, *Amylostereum* and *Postia* spp.) by means of such dead sapwood acting as a bridge, grow into the heartwood and cause heart rot in mechanically wounded, but otherwise vigorous, trees (Vasiliauskas, 1998a, 2001 and references therein). As noted by Shigo (1966), such an infection pattern apparently involves fungal colonization, competition and succession in exposed dead sapwood, while newly formed functional sapwood remains decay free (Fig. 10.4; Plate 12). This pattern of infection presumes colonization of the exposed wound surface by airborne spores.

Fig. 10.4. Cross-section of *Picea abies* (L.) Karst. stem with logging wound. The arrow indicates the surface of the wound (exposed non-functional sapwood) through which *Stereum sanguinolentum* (Alb. & Schwein.) Fr. invades the heartwood and starts heart rot formation. The injured tree sustains growth and the layer of newly formed functional sapwood remains free of decay. Credits: Vytautas Bardauskas.

Among wound heart rot fungi, three species, i.e. *Coniophora puteana* (Schumach.) P. Karst., *Resinicium bicolor* (Alb. & Schwein.) Parmasto and *Sistotrema brinkmannii* (Bres.) J. Erikss., were also either isolated from intact stems or detected in healthy phloem using molecular methods (Table 10.2), thus apparently not complying with airborne arrival to the wound surface and participation in fungal succession. Although each of those fungi were demonstrated to develop distinct decay columns in wounded Norway spruce (*Picea abies* (L.) Karst.) (Vasiliauskas, 1998b), and *C. puteana* was reported to fruit occasionally on stem wounds of oak (Vasiliauskas, 1998c) and ash (Vasiliauskas and Stenlid, 1998b), once again the question is whether wounds for those fungi served as entry points. Moreover, the fungus was isolated from heart rot of pine as well (Basham, 1966). So, as *C. puteana* is also known as a heart rot and mortality agent in pristine old-growth boreal forests (Holsten *et al.*, 2001; Lännenpää *et al.*, 2008), it could probably be classified as a true heart rot of intact stems, while *R. bicolor* and *S. brinkmannii* could be classified as saprotrophs that are both able to establish latently in living trees and able to cause limited decay following wounding. None of the rest of the wound heart rots has been so far detected in intact trees, so presumably they are able to enter living trees only by hyphal growth through dead exposed sapwood (Metzler *et al.*, 2012).

Apparently, true heart rots on dead exposed sapwood are unable to participate in succession and compete with numerous saprotrophs, in particular with heartwood-accessing wound heart rot fungi, which otherwise are normally typical saprobic decomposers of various types of woody debris in forest ecosystems. Although some true heart rot fungi were occasionally observed or isolated from logging and bark stripping wounds, as for instance *F. pinicola* and *P. chrysoloma* on Norway spruce (Vasiliauskas, 1989; Vasiliauskas and Stenlid, 1998a), in those instances mechanical wounds were apparently used by the fungi as exit and dispersal points. Therefore, one might conclude that wound heart rots and true heart rots seemingly possess mutually exclusive routes of infection.

Root and butt heart rots

Of those fungi listed in Table 10.1, only *Ganoderma applanatum* (Pers.) Pat. has been detected as latent in the healthy wood of intact tree stems (Table 10.2). It is known that the fungus infects trees through injuries of the largest roots (<7.5 cm diameter) (Ross, 1976), and so in the cases that have been observed, it could have entered the apparently intact stems through root damage in the subsoil. Similarly, observed infections of the fungus in urban trees were in most cases associated with roots (Terho, 2009). Interestingly, the study by Parfitt *et al.* (2010) mentioned above detected the latent presence of *Hericium* spp. in intact sapwood of ash, beech, hazel and maple, and discussed the possible relationship of the detected fungus to *H. erinaceus* (Bull.) Pers., which is known to cause butt heart rot of oak and poplar (*Populus* spp.) in North America (Berry and Lombard, 1978; Tainter and Baker, 1996). Of this ecological group, *Grifola frondosa* (Dicks.) Gray typically causes butt heart rot in oaks with no sign of stem injuries, apparently penetrating into the stem via the heartwood of large dead or injured roots (Schwarze *et al.*, 2000). Moreover, soil infestations of basidiospores are reported as a potential source of infection to tree roots by *P. schweinitzii* (Barrett and Uscuplic, 1971), and a similar mode of infection has also been considered for *Pseudoinonotus dryadeus* (Pers.) T. Wagner & M. Fisch. (Schwarze *et al.*, 2000), *Climacocystis borealis* (Fr.) Kotl. & Pouzar and *I. leporinus*; for the latter species, tree-to-tree infections via root contacts have been discussed too (Černý, 1989). Despite rather limited knowledge on the entry courts of listed root rot and butt rot fungi to living trees, it is obvious that this ecological group of pathogens possess very specific infection pathways, in this respect clearly distinguishing them from the rest of the heart rot fungi.

Saprotrophs or sap rots

Traditionally, it was assumed that following the death of a tree (or a certain part of that tree, e.g. a branch), the substrate immediately becomes subject to colonization by abundant basidiospores representing high diversity of fungal species and genotypes. This assumption is generally confirmed by observations of numerous fungal individuals on the surface of woody substrates, and of their competition and succession (Cooke and Rayner, 1984; Rayner and Boddy, 1988). Such a mode of colonization would generally resemble the infection pattern occurring on dead sapwood exposed on living stems during the initial stages of its colonization by wound heart rot and other fungi. However, recent studies are increasingly revealing the latent presence in intact trees of wood decay fungi generally known as saprotrophs (Table 10.2). This situation provides them with a colonization advantage over those wood fungi that arrive at the substrate by means of airborne spores following the death of a tree. Hence, those saprotrophic fungi might have developed infection courts and strategies that are characteristic of true heart rots in intact trees. Although, in contrast to the latter, they are not able to decay the heartwood of a living tree, they quickly initiate sap rots in trees subjected to stress, e.g. attack by insects and/or pathogens. In addition, insects themselves are also being considered as potential direct vectors of sap rot. Five such common and widely distributed decomposers of dead wood as *F. pinicola*, *S. sanguinolentum*, *Bjerkandera adusta* (Willd.) P. Karst., *Phlebiopsis gigantea* (Fr.) Jülich and *Trichaptum abietinum* (Dicks.) Ryvarden have been occasionally detected on hibernating individuals of the bark beetle *Ips typographus* (Linnaeus) (Persson *et al.*, 2009). Two of these five fungi (*F. pinicola* and *S. sanguinolentum*) are listed in Table 10.1 as important rots, while three of them have been also reported to be latent inhabitants of intact trees and are listed in Table 10.2 (*F. pinicola*, *B. adusta* and *P. gigantea*), putting into question whether they have been beetle-introduced to trees, or acquired by the beetles while attacking the trees.

10.3.2 Access to host

True heart rots

In living intact trunks, the opportunities for infection by heart rot fungi could be limited, and allow extensive establishment of only a limited number of genotypes (Rayner and Boddy, 1988). Thus, in living stems of aspen and oak, each examined for heart rot columns of *P. tremulae*, *F. robusta*, *Inonotus dryophilus* (Berk.) Murrill and *Laetiporus sulphureus* (Bull.) Murrill, the heart rot fungus constituted a single fungal individual, or genet (Holmer *et al.*, 1994; Sunhede and Vasiliauskas, 2002, 2003a; Vasiliauskas *et al.*, 2003). This demonstrates that every observed infection to each entry point was accomplished by a single heterokaryotic mycelium, and that subsequently the same court was not reused by any other mycelium of the same or another pathogen. Moreover, due to the apparently limited amount of natural entry points per tree, a heart rot infected stem typically harbours one or a few genets of the pathogen. For example, the studies cited above report one to four genets per tree (mostly one) for *P. tremulae* and *F. robusta*, and exclusively one per tree for *I. dryophilus* and *L. sulphureus*.

In the case of *L. sulphureus*, this was rather surprising as the extensive decay columns in the oaks examined were associated with heartwood-exposing butt cracks over 1 m long, obviously accessible to multiple infections by genetically diverse airborne basidiospores. Yet, despite a presumably high infection pressure, only a single genotype of the fungus has established as an extensive decay column in each tree examined, so selection must be rigorous. An interesting parallel could be drawn here to the root-infecting heart rot fungus *Heterobasidion annosum* (Fr.) Bref. *s.l.* Infections by the basidiospores of this pathogen result in numerous genets on stump surfaces (Swedjemark and Stenlid, 2001), but very few of those genets possess the ability for further growth into the roots. As a result, root systems and groups of trees over a discrete territory around the centre of multiple infections (basidiospore-infected stumps) are typically colonized by a single fungal genet (Vasiliauskas and Stenlid, 1998d; Lygis *et al.*, 2004a; see Gonthier and Thor, Chapter 7, this volume). Therefore, while high diversity of the pathogen was observed on the court of infection exposed to the environment, only single genotypes were present in the heartwood of infected trees. This might also apply to true heart rot fungi entering the tree through heartwood-exposing injuries, as with *L. sulphureus*, although their mycelial interactions, early colonization patterns and mechanisms of genotypic selection during the initial phases of infection remain largely unknown. Even less is known about root and butt heart rot fungi. Nevertheless, distinct genotypes of *P. schweinitzii* were detected in different trees in the same plantation, and always as a single genet per tree, thereby implying individual infection by just one heterokaryotic mycelium from soil that is otherwise infested with genetically diverse basidiospores (Barrett and Uscuplic, 1971).

Wound heart rots

Artificial bark wounds exposing sapwood are readily colonized by numerous fungi and multiple infections occur. For example, wood samples just 6–8 cm long and of 0.5 cm diameter taken from decay columns above mechanical stem wounds normally showed the presence of three to five different fungal species (Vasiliauskas *et al.*, 1996; Vasiliauskas and Stenlid, 1998a,b). Moreover, some such samples were infected by two or even three different genets of the wound heart rot fungus *S. sanguinolentum* (Vasiliauskas and Stenlid, 1998c). These observations clearly demonstrate that the dead sapwood on a living stem is easily accessible to multiple colonization by wood-inhabiting fungi, including wound heart rot pathogens. In this respect, a parallel could be drawn with the sapwood of a dead trunk exposed to airborne infections by sap rot fungi. Apparently, large amounts of airborne spores land on the wound surface and germinate, and numerous individual mycelia grow into the functionally dead wood of an injured tree stem. As the mycelia of wound heart rot fungi are competitive with the mycelia of strictly saprotrophic fungi, they grow into the heartwood and initiate heart rot. However, more detailed studies on how many wound heart rot genets form decay columns in individual stems are lacking.

10.3.3 Stem colonization

True heart rots

According to their ecological traits, fungi are categorized as combative (establishing and persisting in a substrate), stress tolerant (growing under stress conditions) and ruderal (early arrival, reproduction and replacement) (Cooke and Rayner, 1984). During the colonization of a tree, wood decay fungi typically combine more than one of those traits, but in different sequences. Thus, true heart rots of intact trees exhibit stress-tolerant traits, while latently establishing in intact xylem and then, during subsequent growth under stressful conditions, establish rot inside intact stems. Yet, following the death of a tree, they switch to combative mode, and are able to persist on dead wood for decades. For example, field observations of *F. robusta* showed that its fruiting body continued development after 25 years of growth on a standing oak, producing annual growth layers on the fallen trunk for over 12 years, and was still living when collected (Sunhede and Vasiliauskas, 2002). In contrast, true heart rots associated with heartwood exposure must apparently also be combative during the initial stages of infection in order to replace earlier arrived ruderal fungi or other heart rot fungi. Then, like the fungi from previous group (true heart rots of intact trees), they switch to stress-tolerant growth while developing heart rot inside living stems, and, again to combative mode while continuing to decompose and produce fruiting bodies on dead trees for many years. Little is known about whether root and butt heart rots infect subsoil roots under stress-tolerant or combative conditions, but following infection they follow the strategies of true heart rots described above.

Wound heart rots

Following establishment in the heartwood, all wound heart rot fungi exhibit stress tolerance as heart rots in living stems, followed by exhibiting combative mode as sap rots on dead wood. Yet, they possess notably different strategies during the initial infection of exposed sapwood. Thus, *C. purpureum* and *S. sanguinolentum* during the infection phase combine ruderal and combative traits: rapid arrival, establishment and defending captured wood from competitors. *Pleurotus* spp. and *Postia* spp. behave differently; they are combative species, replacing less combative (ruderal) fungi in exposed sapwood. Such modes of infection can be also characterized, respectively, as primary and secondary resource capture (*sensu* Cooke and Rayner, 1984). *Amylostereum* spp. also exhibit a combative strategy during the initial colonization of wounded trees, yet those fungi possess a unique infection pathway, as they are introduced to wounds by siricid woodwasps, inserted as heterokaryons during oviposition (Plate 12). In northern Europe, the spread of these wound heart rots is largely dependent on insects, as genetic diversity in fungal populations is low and fruiting bodies rare (Vasiliauskas *et al.*, 1998). For over 100 years, *Amylostereum* spp. have been known as an economically important cause of sap rot in plantations of southern pines, causing consistently increasing problems, particularly in the southern hemisphere (Slippers *et al.*, 2012). Following tree death, all wound heart rot fungi are known to persist and regularly produce fruiting bodies on decomposing wood.

Saprotrophs or sap rots

Despite the general assumption that standing dead trunks become subject to massive colonization by airborne spores of ruderal and combative decay fungi, the evidence is increasingly accumulating on the latent presence of decay fungi in the sound xylem of living intact trees (Table 10.2). This implies a stress-tolerant condition during latent establishment in a living tree for these saprotrophic species too. Such a situation would provide the possibility for early-established combative fungi to rapidly colonize sapwood following death of a tree, thereby providing clear advantage over subsequently arriving competitors. Then, the combative saprotrophs would persist and produce fruiting bodies on decomposing substrate over the years, as is also the typical case for all heart rot fungi. A recent study on fungal succession in Norway spruce attacked by *I. typographus* (R. Vasaitis *et al.*, 2012, unpublished results) has demonstrated that a number of decay fungi invade the sapwood of dying trunks (with still green, but yellowing, crowns) within weeks of beetle attack, spreading in stems radially in succession after ophiostomatoid fungi (which are known to kill sapwood following beetle attack). Among these decay fungi, *F. pinicola*, *S. brinkmannii* and *Gloeophyllum sepiarium* (Wulfen) P. Karst. are species with known latent establishment in intact phloem and sapwood. Moreover, in the sapwood of dying spruce, the wound heart rot and sap rot agents *S. sanguinolentum*, *Amylostereum areolatum* (Chaillet ex Fr.) Boidin and *Schizophyllum commune* Fr. were also detected, and the first two of which could have been introduced to dying trees by insect attack as well as being present through latent establishment (Persson *et al.*, 2009; Slippers *et al.*, 2012).

Among the saprotrophs known for stress-tolerant latent establishment in intact visually healthy trees, *B. adusta*, *Coprinellus disseminatus* (Pers.) J.E. Lange, *Gloeophyllum* sp., *Merulius tremellosus* Schrad., *Peniophora cinerea* (Pers.) Cook, and *Phlebia radiata* Fr. have been previously isolated from heart rotten but otherwise intact stems, and *Mycoacia uda* (Fr.) Donk and *Peniophora limitata* (Chaillet ex Fr.) Cooke have been isolated from wounded stems (Table 10.2). However, the possible roles of these fungi in heart rot formation remain unclear, and are probably negligible. Fruiting bodies of *P. betulinus* have been regularly observed on dying birch, and in some texts the fungus is pointed out as a pathogen, yet the question remains open as to whether the death of a tree activates decay and fruiting of the fungus on trunks, or vice versa (Rayner and Boddy, 1988). For others of the fungi in this group, any association with decay in living trees so far has not been reported. Therefore, one might generally conclude that all the sap rot fungi listed are indeed true saprotrophs, using a stress-tolerant mode of initial colonization of living trunks, followed by extensive spread in wood after the death of a tree, and persistence and the regular production of fruiting bodies on dead decomposing wood.

Colonization strategies

There are four types of strategies exhibited by wood decay fungi during the colonization of a living tree stem. These are formulated as: (i) unspecialized opportunism, (ii) specialized opportunism, (iii) heart rot and (iv) active pathogenesis. As the living tree consists of a heterogeneous assemblage of microenvironments, each posing distinctive constraints on growth, in order to successfully colonize a tree, individual fungi are typically capable of combining several strategies (Rayner and Boddy, 1988). During the initial colonization, true heart rot fungi exhibit specialized opportunism, combined with either stress tolerance (when entering intact trees), or combative behaviour (when colonizing exposed heartwood of selective tree taxa). Apart from heartwood colonization, true heart rots that enter intact trees often invade functional sapwood by mechanisms of active pathogenesis, developing distinct canker rots, e.g. *Fomitiporia*, *Phellinus* and *Porodeadalea* spp., and *E. tinctorium* and *I. obliquus*. Initial colonization by wound heart rot fungi typically includes unspecialized opportunism during the combative (primary or secondary) capture of exposed dead sapwood, followed by sapwood colonization and spread inside the stem. Heart rot is a characteristic of all heart rot fungi. After entering the heartwood, all heart rots exhibit stress-tolerant growth, apparently overcoming the presence of fungitoxic compounds and the constraints of the gaseous regime, and developing a heart rot column, while functionally intact sapwood remains decay free (Rayner and Boddy, 1988; Plate 12).

10.3.4 Development of fruiting bodies

Following colonization of a tree, the life cycle of heart rot fungi is completed by fruiting body production and the abundant release to the environment of airborne spores that will subsequently cause tree infections. True heart rots have two distinct types of fruiting bodies: fleshy annual and hard perennial (conks). Fleshy annual fruiting bodies are characteristic of species that invade stems through natural heartwood-exposing injuries (except for *Fomitopsis cajanderi*, Table 10.1). Seemingly, those fungi in a colonized trunk have a permanent unrestricted access to the outer environment, which enables their fruiting (usually directly upon an injury) and the regular seasonal dispersal of basidiospores. Interestingly, true heart rot species that invade intact trees typically produce perennial conks. The situation for those fungi in tree stems is radically different, as the need to breach functionally intact sapwood becomes paramount if exit structures, notably fruiting bodies, are to be produced (Rayner and Boddy, 1988). Such venues are provided either by branch stubs originating from the heartwood, or by breaching the functional sapwood during active pathogenesis and canker formation. Consequently, perennial conks of the heart rots are typically produced on and around branch stubs and cankers. Thus, the formation of perennial fruiting bodies on a living tree could be regarded as the long-term investment of a fungus. For example, actively growing conks of the heart/sap rot fungus *F. robusta* on living oaks are known to exceed an age of 35 years (Sunhede and Vasiliauskas, 2002).

All wound heart rot fungi have access to the outer environment through dead colonized sapwood. The fruiting bodies of all wound heart rotting fungi are fleshy and annual, and are developed directly on the wound surface. Notably, this fruiting on the wound surface seldom occurs. In a study by Vasiliauskas (1989), more than 1000 Norway spruces with logging and bark stripping wounds were monitored over two vegetation seasons, and fruiting bodies were observed on only 79 of them (<10%). Only 6% of the bark stripping wounds examined on stems of ash and oak showed fruiting bodies of wood decay fungi (Vasiliauskas, 1998c; Vasiliauskas and Stenlid, 1998b). Apparently, wounds on a standing tree do not normally provide the appropriate microenvironment, in particular moisture, for the formation of fleshy annual fruiting bodies. One should bear in mind that in primeval forest ecosystems, these decay fungi evolved as pioneer combative wood decomposers of dead wood, and are adapted to fruit abundantly under increased moisture conditions, e.g. on downed decomposed woody debris. In contrast, wound heart rot fungi can be generally characterized by their relatively rapid reproduction rates, and produce fruiting bodies on wounds that are less than 10 years old. Such observations are available for *C. puteana*, *Stereum hirsutum* (Willd.) Pers. and *Trametes versicolor* (L.) Lloyd on birch, oak and ash (Shigo, 1966; Vasiliauskas, 1998c; Vasiliauskas and Stenlid, 1998b), and for *S. sanguinolentum* on Norway spruce (Vasiliauskas *et al.*, 1996).

10.3.5 Host range

True heart rots of intact trees

The ability to enter intact stems through natural courts of infection presumably requires adaptation to a certain host range. In order to be host specific, heart rot fungi must possess tolerance both to specific microenvironmental conditions and to variations in these conditions while entering and spreading inside a tree (Rayner and Boddy, 1988). Consequently, true heart rots that invade intact trees usually have one or few hosts on which they predominantly occur, while being able occasionally to occur also on other tree species. In forest stands in Europe, typical examples of such host selectivity are *P. hartigii* on silver fir (*Abies alba* Mill.), *F. robusta* on oak, *P. chrysoloma* on Norway spruce, *P. pini* on pine, *Phellinus nigricans* (Fr.) P. Karst. on birch, and *P. tremulae* and *Phellinus populicola* Niemelä on aspen (Niemelä, 1974, 1975; Butin, 1995). Although *Phellinus igniarius* (L.) Quél. was traditionally considered to be a harmful pathogen of numerous woody species (Niemelä, 1975; Butin, 1995), recent studies based on molecular methods have demonstrated that the *P. igniarius* group in Europe consists of different species that are mostly host specific (Tomšovský *et al.*, 2010b). In North America, a broader host range was reported for true heart rots in intact stems. Thus, it was noted that *E. tinctorium* typically occurs on true firs, hemlock and spruce, but *P. pini* was usually considered as having an even broader host range and capable of causing heart rot in many tree species (Manion, 1991; Tainter and Baker, 1996; Hansen and Lewis, 1997). While the potential for cryptic species and host specificity in *E. tinctorium* has been not investigated, recent research on *P. pini* has demonstrated that in North America the fungus is indeed an aggregate of over ten different species that are mainly host specific (Fischer, 1996; Tomšovský *et al.*, 2010a). In this respect, the exceptions among invaders of intact trees could be *F. fomentarius* and *F. pinicola*, the fruiting bodies of which are frequently observed on the wood of numerous hosts (Ryvarden and Gilbertson, 1993). However, a recent study based on molecular methods has revealed the presence of cryptic species and a certain host selectivity in *F. fomentarius* (Judova *et al.*, 2012). In contrast, early studies had already demonstrated that specimens of *F. pinicola* collected from conifers and deciduous trees were interfertile (Mounce and Macrae, 1938), indicating that individuals of the fungus colonizing broad spectrum of hosts represent a single although highly versatile biological entity.

True heart rots associated with natural injury

Among the heart rots associated with natural injury, there are species with both a narrow and a wide host range. *Fistulina hepatica* (Schaeff.) With., *Hapalopilus croceus* (Pers.) Donk, *I. dryophilus*, *S. gausapatum* and *Xylobolus frustulatus* (Pers.) Boidin are mostly found on oaks, while *F. cajanderi* causes rot in many coniferous species, although mainly in Douglas-fir. Traditionally, *L. sulphureus* was a classical example of a fungus that causes heart rot and decomposes the wood of many deciduous and conifers (Ryvarden and Gilbertson, 1993). Currently, at least seven distinct species are recognized in the *L. sulphureus* complex, some distributed over a distinct geographic range and some probably invasive, and most of them apparently possessing a certain host selectivity. Host specificity reported so far for root and butt heart rots is based on observations of fruiting bodies; some of them, such as *G. frondosa* or *I. leporinus*, are seemingly more host specific, while others, such as *Perenniporia subacida* (Peck) Donk and *P. schweinitzii*, are less host specific. For this group of fungi, the possible existence of cryptic species with certain host specificity has not been investigated and cannot be excluded.

Wound heart rots

As wound heart rot fungi produce fruiting bodies on the decomposed wood of numerous tree species, in the mycological literature they are traditionally referred to as common and widely spread saprotrophic wood decomposers with a broad host range. It seems that while gaining access to living trees they are more specialized. *Amylostereum* and *Postia* spp., and *S. sanguinolentum*, are primarily reported as wound pathogens of Norway spruce and silver fir, with *C. purpureum* found on alder (*Alnus* spp.) and birch (Table 10.1; Vasiliauskas, 2001 and references therein). A special case among the wound heart rot fungi is *C. puteana*, a globally distributed generalist and versatile saprotroph on a wide variety of substrates and environments, growing in woody buildings, but also on dead and living trees in managed forest stands and in old-growth pristine forests. Yet the fungus also possesses certain features characteristic of true heart rots of intact stems, as it is capable of establishing latently in living trees and is known to cause heart rot of over-mature trees. Interestingly, the presence of cryptic species with apparent hybridization has also been reported for *C. puteana* (Kauserud *et al.*, 2007).

10.4 Management Strategies and Tactics

10.4.1 True heart rots and root and butt rots

Certain silvicultural practices have been suggested to control true heart rots. Those include avoiding the planting of susceptible species on high hazard sites with high incidence of the pathogens, selective harvesting to promote a multiple aged stand and uneven canopy structure, and spacing to allow symmetrical crown development (Hansen and Lewis, 1997). Yet, as the losses significantly increase in mature and over-mature stands, the only realistic and economically feasible measure would be to shorten the rotation age or, in other words, to harvest trees before the losses become excessive (Boyce, 1961; Tainter and Baker, 1996; Hansen and Lewis, 1997; Holsten *et al.*, 2001). In particular, reducing the rotation age would be applicable to fast-growing forest plantations, e.g. of aspen and alder, where significant losses already occur at a relatively young age (Arhipova *et al.*, 2011a, 2012). Moreover, as aspen clones seem to differ in susceptibility to *P. tremulae*, breeding for resistance of this tree against the pathogen could be considered (Manion, 1991).

Some texts suggest the removal from forest stands of trees with fruiting bodies of heart rot fungi (Černý, 1989; Hansen and Lewis, 1997), which apparently contradicts currently ongoing programmes for their artificial inoculation (Filip *et al.*, 2011). Moreover, it has already been pointed out by Boyce (1961) that such an operation is unlikely to be economically feasible. Furthermore, the aim of reducing infection potential is unlikely to be achieved as the fruiting bodies typically hide up in the tree crowns and produce huge numbers of spores with a highly efficient airborne spread capacity (Sunhede and

Vasiliauskas, 2002). One should also keep in mind that nowadays the majority of true heart rots occur in over-mature old-growth forests, which become increasingly relevant for biodiversity and nature conservation, and to large extent are supposed to be under a protection regime. Indeed, surveys in such forests increasingly reveal the presence of rare and protected heart rot fungi, and point to the necessity for these sites to be put under protective status (Sunhede and Vasiliauskas, 2003b). In fact, this could currently be seen as an alternative management strategy for true heart rots and most butt heart rots. In fact, measures for the restoration of populations of certain fungal species (e.g. *H. croceus* in Europe) are presently being considered.

10.4.2 Wound heart rots

The absence of artificial wounding would be a prerequisite for complete wound heart rot control in forest stands, but this is unrealistic. The aim should, therefore, be directed at avoiding and minimizing stand damage. Studies on how logging damage could be reduced by the planning of forest operations, the training and adequate supervision of workers, use of different harvesting methods and treatment of wounds have been reviewed, and a number of silvicultural and operational options discussed (Dimitri, 1983; Vasiliauskas, 2001). In addition to this, for broadleaves it has been pointed out that seedlings of seed origin generally contain less rot than trees of sprout origin; additionally, sprouts from small stumps (<10 cm and not >15 cm in diameter) should be favoured as crop trees; single sprouts are preferable to fused sprouts (Boyce, 1961; Tainter and Baker, 1996). Another option for wound heart rot management strategy would be the planting of trees less susceptible to logging damage, and in this respect, several tree species have been investigated. For example, in the USA, a higher proportion of residual trees was injured during thinning in a birch stand than in similar treatments in a mixed beech and oak stand (Ostrofsky *et al.*, 1986). In Europe, silver fir has been shown to suffer fewer injuries during harvesting operations than Norway spruce (Kohnle and Kändler, 2007) and, what is more, wounds on silver fir were shown to be less susceptible to rot fungi (Metzler *et al.*, 2012). As a management implication, it was concluded that the economic risks in respect of wounding are expected to be lower in silver fir and, in particular, that timely removal of wounded silver fir is less critical than that of Norway spruce.

A recent study has provided criteria for the assignment of fresh mechanical stem wounds of Norway spruce either to decay risk or to non-risk categories, on which a decision could be based for retaining or removing spruce in damaged stands during selective felling (Vasaitis *et al.*, 2012b). The study demonstrated that in cases where 5 cm wide wounds were inflicted on stems of Norway spruce, the injuries were unlikely to occlude during two subsequent decades; neither did wounds of a 15 cm initial width occlude at all, and these were 100% invaded by decay (Figure 10.3). Consequently, in order to avoid yield losses, trees with such wounds should be considered for removal during any selective felling. The management of Norway spruce stands with old wounds in areas susceptible for bark stripping should primarily remove stems with old open wounds during any partial harvesting, while retaining trees with closed wounds (Vasaitis *et al.*, 2012b). For silver fir, it takes a much shorter time to occlude bark stripping wounds – 4–10 years for a 2 cm wide wound, and 6–12 years for a 5 cm wide wound (Pach, 2008), so on heavily damaged sites, trees with such injuries should be retained. In plantations of Sitka spruce (*Picea sitchensis* (Bong.) Carrière) all stem wounds with an initial size of <60 cm^2 were fully closed over a 15 year period, but none larger than 180 cm^2 were occluded (Welch *et al.*, 1997); this finding should also be considered during management decisions. Wounded oak, beech and birch, due to a low susceptibility to wound heart rot, should not subject to any particular strategy in relation to other stems in a stand during forestry operations, while all viable ash, despite extensive wound heartwood discoloration due to the catastrophic epidemics of dieback currently devastating this tree species on a continental scale (Bakys *et al.*, 2009), must be retained in stands despite any mechanical wounding.

Among other measures for wound heart rot management, the shaping of irregular mechanical wounds by cutting off loose pieces of bark that have been knocked has been suggested to enhance callus formation and wound closure (Manion, 1991; Schumann and Dimitri, 1993). Recommendations for pruning suggest that the pruning cuts should be close but not injure the bark of the main stem, that tearing of the bark below the pruned branch should be avoided, and that loose bark should be removed from around the wound (Manion, 1991). In north Europe, the pruning of birch resulted in only limited development of discoloration in stems (Schatz *et al.*, 2008), as little as that developing from natural branch stubs (Raulo *et al.*, 1978), thereby implying that this tree species is as resistant to pruning wounds as to logging wounds (Vasaitis *et al.*, 2012a). According to studies in the USA, the incidence of discoloration in pruning wounds of birch is also low, and pruning of small trees with small branches was recommended (Solomon and Shigo, 1976). In Norway spruce, the pruning of green branches proved to be excellent in this respect, as long as branch collar and bark injuries, and too long branch stubs were avoided (Metzler, 1997).

Finally, one should be aware that at least in central and north Europe, three of the most important forest diseases and pests are present, namely: (i) annosum root rot (*H. annosum s.l.*), (ii) the pine weevil (*Hylobius abietis* Linnaeus) and (iii) the big game (red deer and moose). Among other damages caused (e.g. the destruction of forest plantations and regeneration, travel accidents, etc.), the third pest category is one of the principal causes for wound heart rot and is incompatible with sustainable forest management (R. Vasaitis, 2012, unpublished results). However, in comparison with root rot and the pine weevil, this third category of the most destructive forest pests is the easiest one to physically eliminate or be sustained within game reserves. This would significantly improve the health condition of European forests.

References

Adams, T.J.H. (1982) *Piptoporus betulinus*, some aspects of population biology. PhD thesis, University of Exeter, Exeter, UK.

Allen, E., Morrison, D. and Wallis, G. (1996) *Common Tree Diseases of British Columbia*. Canadian Forest Service, Pacific Forestry Centre, Victoria, British Columbia, Canada.

Arhipova, N., Gaitnieks, T., Donis, J., Stenlid, J. and Vasaitis, R. (2011a) Decay, yield loss and associated fungi in stands of grey alder (*Alnus incana*) in Latvia. *Forestry* 84, 337–348.

Arhipova, N., Gaitnieks, T., Donis, J., Stenlid, J. and Vasaitis, R. (2011b) Butt rot incidence, causal fungi, and related yield loss in *Picea abies* stands of Latvia. *Canadian Journal of Forest Research* 41, 2337–2345.

Arhipova, N., Gaitnieks, T., Donis, J., Stenlid, J. and Vasaitis, R. (2012) Heart rot and associated fungi in *Alnus glutinosa* stands in Latvia. *Scandinavian Journal of Forest Research* 27, 327–336.

Aufsess von, H. (1973) Important fungal damages on old oaks. *Forstwissenschaftliches Centralblatt* 92, 153–169 [in German, with English summary].

Bakys, R., Vasaitis, R., Barklund, P., Thomsen, I.M. and Stenlid, J. (2009) Occurrence and pathogenicity of fungi in necrotic and non-symptomatic shoots of declining common ash (*Fraxinus excelsior*) in Sweden. *European Journal of Forest Research* 128, 51–60.

Banik, M.T., Lindner, D.L., Ota, Y. and Hattori, T. (2010) Relationships among North American and Japanese *Laetiporus* isolates inferred from molecular phylogenetics and single-spore incompatibility reactions. *Mycologia* 102, 911–917.

Barrett, D.K. and Uscuplic, M. (1971) The field distribution of interacting strains of *Polyporus schweinitzii* and their origin. *New Phytologist* 70, 581–598.

Barszcz, P. and Jamrozy, G. (2001) Depreciation of fir and ash wood tapped by red deer in Beskid Sadecki forests. *Sylwan* 12, 47–57 [in Polish, with English summary].

Basham, J.T. (1966) Heart rot of jack pine in Ontario. I. The occurrence of basidiomycetes and microfungi in defective and normal heartwood of living jack pine. *Canadian Journal of Botany* 44, 275–295.

Baum, S., Sieber, T.N., Schwarze, F.W.M.R. and Fink, S. (2003) Latent infections of *Fomes fomentarius* in the xylem of European beech (*Fagus sylvatica*). *Mycological Progress* 2, 141–148.

Berry, F.H. and Lombard, F.F. (1978) *Basidiomycetes Associated with Decay of Living Oak Trees*. Research Paper NE-413, US Department of Agriculture, Forest Service, Northeastern Forest Experiment Station, Broomall, Pennsylvania.

Boddy, L. and Rayner, A.D.M. (1983) Origins of decay in living deciduous trees: the role of moisture content and re-appraisal of the expanded concept of tree decay. *New Phytologist* 94, 623–641.

Boyce, J.S. (1961) *Forest Pathology*, 3rd edn. McGraw-Hill, New York.

Breitenbach, J. and Kränzlin, F. (1986) *Fungi of Switzerland, Volume 2. Non-gilled Fungi*: Heterobasidiomycetes, Aphyllophorales, Gastromycetes. Verlag Mykologia, Lucerne, Switzerland.

Burdekin, D.A. (1979) *Common Decay Fungi in Broadleaved Trees*. Arboricultural Leaflet 5, Forestry Commission, UK. Her Majesty's Stationery Office, London.

Burdsall, H.H. and Banik, M.T. (2001) The genus *Laetiporus* in North America. *Harvard Papers in Botany* 6, 43–55.

Butin, H. (1995) *Tree Diseases and Disorders. Causes, Biology, and Control in Forest and Amenity Trees*. Oxford University Press, Oxford, UK.

Černý, A. (1989) *Parasitic Wood-Decay Fungi*. Statne Zemedelske Nakladatelstvo, Prague [in Czech].

Chapela, I.H. and Boddy, L. (1988) Fungal colonization of attached beech branches. II. Spatial and temporal colonization of communities arising from latent invaders in bark and functional sapwood, under different moisture regimes. *New Phytologist* 110, 47–57.

Cleary, M.R., Arhipova, N., Morrison, D.J., Thomsen, I.M., Sturrock, R.N., Vasaitis, R., Gaitnieks, T. and Stenlid, J. (2013) Stump removal to control root disease in Canada and Scandinavia: a synthesis of results from long-term trials. *Forest Ecology and Management* 290, 5–14.

Cooke, R.C. and Rayner, A.D.M. (1984) *Ecology of Saprotrophic Fungi*. Longman, London.

Dai, Y.C. (1996) Changbai wood rotting fungi 7. A checklist of the polypores. *Fungal Science* 11, 79–105.

Dai, Y.C., Cui, B.K., Yuan, H.S. and Li, B.D. (2007) Pathogenic wood-decaying fungi in China. *Forest Pathology* 37, 105–120.

Dimitri, L. (1983) Wound decay following tree injury in forestry: establishment, significance and possibilities of its prevention. *Forstwissenschaftliches Centralblatt* 102, 68–79 [in German, with English summary].

Eriksson, J. and Ryvarden, L. (1973) *The Corticiaceae of North Europe, Vol. 2*. Fungiflora, Oslo, Norway.

Eriksson, J. and Ryvarden, L. (1976) *The Corticiaceae of North Europe, Vol. 4*. Fungiflora, Oslo, Norway.

Eriksson, J., Hjortstam, K. and Ryvarden, L. (1978) *The Corticiaceae of North Europe, Vol. 5*. Fungiflora, Oslo, Norway.

Eriksson, J., Hjortstam, K. and Ryvarden, L. (1981) *The Corticiaceae of North Europe, Vol. 6*. Fungiflora, Oslo, Norway.

Eriksson, J., Hjortstam, K. and Ryvarden, L. (1984) *The Corticiaceae of North Europe, Vol. 7*. Fungiflora, Oslo, Norway.

Etheridge, D.E. and Craig, H.M. (1976) Factors influencing infection and initiation of decay by the Indian paint fungus (*Echinodontium tinctorium*) in western hemlock. *Canadian Journal of Forest Research* 6, 299–318.

Filip, G., Chadwick, K., Zambino, P., Omdal, D., Ramsey-Kroll, A., Schmitt, C., Maffei, H., Saavedra, A., Rall, W. and Parks, C. (2011) *Seven- to 14-Year Effects of Artificially Inoculating Living Conifers to Promote Stem Decay and Subsequent Wildlife Use in Oregon and Washington Forests*. US Department of Agriculture, Forest Service, Forest Health Protection, Pacific Northwest Region, Portland, Oregon.

Fischer, M. (1996) Molecular and microscopical studies in the *Phellinus pini* group. *Mycologia* 88, 230–238.

Fischer, M. and Binder, M. (2004) Species recognition, geographic distribution and host–pathogen relationships: a case study in a group of lignicolous basidiomycetes, *Phellinus s.l. Mycologia* 96, 799–811.

Gilbertson, R.L. and Ryvarden, L. (1987) *North American Polypores, Part 2*. Fungiflora, Oslo, Norway.

Granatov, L.B. (1973) Most important heart rots of oak in Tula region, biology of the pathogens. PhD thesis, Moscow Forest Research Institute, Moscow, Russia [in Russian].

Haddow, W.R. (1938a) On the classification, nomenclature, hosts and geographical range of *Trametes pini* (Thore) Fries. *Transactions of the British Mycological Society* 22, 182–193.

Haddow, W.R. (1938b) The disease caused by *Trametes pini* (Thore) Fries in white pine (*Pinus strobus* L.). *Transactions of the Royal Canadian Institute* 22, 21–80.

Hallaksela, A.M. and Niemistö, P. (1998) Stem discoloration of planted silver birch. *Scandinavian Journal of Forest Research* 13, 169–176.

Hansen, E.M. and Lewis, K.J. (1997) *Compendium of Conifer Diseases*. APS Press, St Paul, Minnesota.

Hartig, R. (1874) *Wichtige Krankheiten der Waldbäume* [*Important Diseases of Forest Trees*]. Springer Verlag, Berlin.
Hendry, S.J., Boddy, L. and Lonsdale, D. (2002) Abiotic variables effect differential expression of latent infections in beech (*Fagus sylvatica*). *New Phytologist* 155, 449–460.
Hepting, G.H. (1971) *Diseases of Forest and Shade Trees of the United States*. Agriculture Handbook 386, US Department of Agriculture, Forest Service, Washington DC.
Hepting, G.H. and Hedgcock, G.G. (1937) *Decay in Merchantable Oak, Yellow-Poplar, and Basswood in the Appalachian Region*. Technical Bulletin 570, US Department of Agriculture, Forest Service, Washington, DC.
Holmer, L., Nitare, L. and Stenlid, J. (1994) Population structures and decay pattern in *Phellinus tremulae* as determined by somatic incompatibility. *Canadian Journal of Botany* 72, 1391–1396.
Holsten, E., Hennon, P., Trummer, L. and Schultz, M. (2001) *Insects and Diseases of Alaskan Forests*. Technical Publication, R10-TP-87, US Department of Agriculture, Forest Service, Juneau, Alaska.
Jahn, H. (1990) *Fungi on Trees*, 2nd edn. Patzer Verlag, Berlin [in German].
Judova, J., Dubikova, K., Gaperova, S., Gaper, J. and Pristas, P. (2012) The occurrence and rapid discrimination of *Fomes fomentarius* genotypes by ITS-RFLP analysis. *Fungal Biology* 116, 155–160.
Kauserud, H., Svegården, I.B., Decock, C. and Hallenberg, N. (2007) Hybridization among cryptic species of the cellar fungus *Coniophora puteana* (*Basidiomycota*). *Molecular Ecology* 16, 389–399.
Kohnle, U. and Kändler, G. (2007) Is silver fir (*Abies alba*) less vulnerable to extraction damage than Norway spruce (*Picea abies*)? *European Journal of Forest Research* 126, 121–129.
Kotiranta, H. and Niemelä, T. (1981) Composition of polypore communities of four forest areas in southern Central Finland. *Karstenia* 21, 31–48.
Kuzmichev, E.P., Sokolova, E.S. and Mozolevskaya, E.G. (2004) *Diseases of Woody Plants, Vol. 1. Diseases and Insect Pests in Forests of Russia*. VNIILM, Moscow [in Russian].
Lännenpää, A., Aakala, T., Kauhanen, H. and Kuuluvainen, T. (2008) Tree mortality agents in pristine Norway spruce forests in northern Fennoscandia. *Silva Fennica* 42, 151–163.
Larsson, K.-H. (1997) *Red-Listed Fungi in Sweden*. ArtDatabanken, Uppsala, Sweden [in Swedish].
Lewis, K.J. and Lindgren, B.S. (1999) Influence of decay fungi on species composition and size class structure in mature *Picea glauca* × *engelmannii* and *Abies lasiocarpa* in sub-boreal forests of central British Columbia. *Forest Ecology and Management* 123, 135–143.
Lygis, V., Vasiliauskas, R. and Stenlid, J. (2004a) Planting *Betula pendula* on pine sites infested by *Heterobasidion annosum*: disease transfer, silvicultural evaluation, and community of wood-inhabiting fungi. *Canadian Journal of Forest Research* 34, 120–130.
Lygis, V., Vasiliauskas, R. and Stenlid, J. (2004b) Silvicultural and pathological evaluation of Scots pine afforestations mixed with deciduous trees to reduce the infections by *Heterobasidion annosum s.s. Forest Ecology and Management* 201, 275–285.
Lygis, V., Vasiliauskas, R., Larsson, K.-H. and Stenlid, J. (2005) Wood-inhabiting fungi in stems of *Fraxinus excelsior* in declining ash stands of northern Lithuania, with particular reference to *Armillaria cepistipes*. *Scandinavian Journal of Forest Research* 20, 337–346.
Maloy, O.C. and Robinson, V.S. (1968) Microorganisms associated with heart rot in young grand fir. *Canadian Journal of Botany* 46, 306–309.
Manion, P.D. (1991) *Tree Disease Concepts*, 2nd edn. Prentice-Hall, Englewood Cliffs, New Jersey.
Metzler, B. (1997) Quantitative assessment of fungal colonization in Norway spruce after green pruning. *European Journal of Forest Pathology* 27, 1–11.
Metzler, B., Hecht, U., Nill, M., Brüchert, F., Fink, S. and Kohnle, U. (2012) Comparing Norway spruce and silver fir regarding impact of bark wounds. *Forest Ecology and Management* 274, 99–107.
Mounce, I. and Macrae, R. (1938) Interfertility phenomena in *Fomes pinicola* (Sw.) Cooke. *Canadian Journal of Botany* 16, 354–376.
Müller, J., Engel, H. and Blaschke, M. (2007) Assemblages of wood-inhabiting fungi related to silvicultural management intensity in beech forests in southern Germany. *European Journal of Forest Research* 126, 513–527.
Nicolotti, G., Gonthier, P. and Guglielmo, F. (2010) Advances in detection and identification of wood rotting fungi in timber and standing trees. In: Gherbawy, Y. and Voigt, K. (eds) *Molecular Identification of Fungi*. Springer Verlag, Berlin, pp. 251–276.
Niemelä, T. (1974) On Fennoscandian polypores. III. *Phellinus tremulae* (Bond.) Bond. & Borisov. *Annales Botanici Fennici* 11, 202–215.
Niemelä, T. (1975) On Fennoscandian polypores. IV. *Phellinus igniarius*, *P. nigricans* and *P. populicola* n.sp. *Annales Botanici Fennici* 12, 93–122.

Nitare, J. (2000) *Indicator Species for Assessing the Nature Conservation Value of Woodland Sites*. Skogsstyrelsen, Jönköping [in Swedish, with English summary].
Nordén, B. (1997) Genetic variation within and among populations of *Fomitopsis pinicola* (Basidiomycetes). *Nordic Journal of Botany* 17, 319–329.
Oses, R., Valenzuela, S., Freer, J., Sanfuentes, E. and Rodríguez, J. (2008) Fungal endophytes in xylem of healthy Chilean trees and their possible role in early wood decay. *Fungal Diversity* 33, 77–86.
Ostrofsky, W.D., Seymour, R.S. and Lemin, R.C. Jr (1986) Damage to northern hardwoods from thinning using whole-tree harvesting technology. *Canadian Journal of Forest Research* 16, 1238–1244.
Ota, Y., Hattori, T., Banik, M.T., Hagedorn, G., Sotome, K., Tokuda, S. and Abe, Y. (2009) The genus *Laetiporus* (*Basidiomycota, Polyporales*) in East Asia. *Mycological Research* 113, 1283–1300.
Pach, M. (2005) Extent and dynamics of wood decay spreading inward in fir stems as a result of bark stripping by ungulates. *Sylwan* 5, 23–35 [in Polish, with English summary].
Pach, M. (2008) The rate of bark-stripping wound closure in fir and some factors affecting it. *Sylwan* 4, 46–57 [in Polish, with English summary].
Parfitt, D., Hunt, J., Dockrell, D., Rogers, H.J. and Boddy, L. (2010) Do all trees carry the seeds of their own destruction? PCR reveals numerous wood decay fungi latently present in sapwood of a wide range of angiosperm trees. *Fungal Ecology* 3, 338–346.
Persson, Y., Vasaitis, R., Långström, B., Öhrn, P., Ihrmark, K. and Stenlid, J. (2009) Fungi vectored by the bark beetle *Ips typographus* following hibernation under the bark of standing trees and in the forest litter. *Microbial Ecology* 58, 651–659.
Phillips, D.H. and Burdekin, D.A. (1982) *Diseases of Forest and Ornamental Trees*. Mcmillan Press, London.
Raulo, J., Saarnio, R. and Ylitalo, T. (1978) Sealing of pruned branch stumps in curly birch and subsequent spread of discolouration into the stemwood. *Silva Fennica* 12, 257–263 [in Finnish, with English summary].
Rayner, A.D.M. and Boddy, L. (1988) *Fungal Decomposition of Wood: Its Biology and Ecology*. John Wiley, Chichester, UK.
Renvall, P., Renvall, T. and Niemelä, T. (1991) Basidiomycetes at the timberline in Lapland 2. An annotated checklist of the polypores of northeastern Finland. *Karstenia* 31, 13–28.
Risley, H. and Silverborg, S.B. (1958) *Stereum sanguinolentum* on Norway spruce following pruning. *Phytopathology* 48, 337–338.
Roll-Hansen, F. (1981) *Diseases of Forest Trees*. Landbruksforlaget, Oslo [in Norwegian].
Ross, W.D. (1976) Relation of aspen root size to infection by *Ganoderma applanatum*. *Canadian Journal of Botany* 54, 745–751.
Ryman, S. and Holmåsen, I. (1992) *Fungi, a Fieldbook*, 3rd edn. Interpublishing, Stockholm [in Swedish].
Ryvarden, L. and Gilbertson, R.L. (1993) *European Polypores, Part 1*. Fungiflora, Oslo.
Ryvarden, L. and Gilbertson, R.L. (1994) *European Polypores, Part 2*. Fungiflora, Oslo.
Scharpf, R.F. (1993) *Diseases of Pacific Coast Conifers*. Agriculture Handbook 521, US Department of Agriculture, Forest Service, Washington, DC.
Schatz, U., Heräjärvi, H., Kannisto, K. and Rantatalo, M. (2008) Influence of saw and secateur pruning on stem discolouration, wound cicatrisation and diameter growth of *Betula pendula*. *Silva Fennica* 42, 295–305.
Schimitschek, E. (1975) Report on damages in forest stands. *Allgemeine Forst- und Jagdzeitung* 146, 72–75 [in German].
Schumann, G. and Dimitri, L. (1993) Wounds and wound rot in beech. *Allgemeine Forst-Zeitschrift für Waldwirtschaft und Umweltvorsorge* 48, 456–460.
Schwarze, F.W.M.R., Engels, J. and Mattcheck, C. (2000) *Fungal Strategies of Wood Decay in Trees*. Springer, Berlin.
Shigo, A.L. (1965) The pattern of decays and discolorations in northern hardwoods. *Phytopathology* 55, 648–652.
Shigo, A.L. (1966) *Decay and Discoloration Following Logging Wounds on Northern Hardwoods*. Research Paper NE-47, US Department of Agriculture, Forest Service, Northeastern Forest Experiment Station, Upper Darby, Pennsylvania.
Sinclair, W.A. and Lyon, H.H. (2005) *Diseases of Trees and Shrubs*, 2nd edn. Cornell University Press, Ithaca, New York.
Skov, S. and Thomsen, I.M. (2003) The role of *Fomes fomentarius* in beech forests in Denmark. In: Thomsen, I.M. (ed.) *Forest Health Problems in Older Forest Stands. Proceedings of the Nordic/Baltic Forest Pathology Meeting, Denmark, September 2002*. Report No. 13, Danish Centre for Forest, Landscape and Planning, Copenhagen, pp. 73–76.

11 Ceratocystis Diseases

Thomas C. Harrington*
Iowa State University, Ames, Iowa, USA

11.1 Pathogens, Significance and Distribution

Most species of *Ceratocystis* are at least weakly pathogenic to woody plants, and some cause substantial losses. The genus *Ceratocystis* defines a closely related group of ecologically unique ascomycetes, and there are a number of epidemiological aspects and management considerations that are common to the diseases they cause. *Ceratocystis* spp. are generally associated with insects, mostly with beetles (Coleoptera). In their native ecosystems, these fungi are primarily wound colonizers and do not cause substantial damage to native hosts. However, some species have caused substantial tree mortality after being introduced to new forest ecosystems or to new cultivated hosts, and humans have fostered epidemics by plant wounding and mechanical transmission of the pathogens.

The focus of this chapter is the epidemiology and management of forest tree diseases caused by *Ceratocystis* spp., with emphasis on the three most important and studied diseases: oak wilt, canker stain on plane tree, and Ceratocystis wilt on eucalyptus. Kile (1993) wrote on diseases caused by *Ceratocystis* spp., and Harrington (2009) gave a brief overview of the biology of the genus. A recent symposium proceedings and a review on oak wilt cover a wealth of further information (Billings and Appel, 2009; Juzwik *et al.*, 2011). Panconesi *et al.* (2003) provide an account of canker stain on plane tree in Europe. Ceratocystis wilt on eucalyptus is an emerging disease in Brazil (Ferreira *et al.*, 2011).

Ceratocystis once included the much larger genus of *Ophiostoma*, which, along with the anamorph genera *Leptographium* and *Raffaelea*, are mostly saprophytic bark and ambrosia beetle associates, with a few plant pathogens: the cause of Dutch elm disease (*Ophiostoma ulmi* (Buisman) Nannf. and *O. novo-ulmi* Brasier), blackstain root disease on conifers (varieties of *Leptographium wageneri* (W.B. Kendr.) M.J. Wingf.), and laurel wilt (*Raffaelea lauricola* T.C. Harr., Fraedrich & Aghayeva) (Harrington, 2005; Harrington *et al.*, 2010). The biology of *Ceratocystis* spp. differs substantially from that of *Ophiostoma* spp., but the two genera have converged in respect of long-necked perithecia (sexual fruiting bodies) with sticky ascospore masses at their tip for insect dispersal. In contrast to *Ophiostoma* and its strong ties to bark beetles, most species of *Ceratocystis* produce fruity

* E-mail: tcharrin@iastate.edu

volatiles, presumably to attract their vectors, and their relationships with insects are varied (Harrington, 2009).

A number of asexual states have been noted for *Ceratocystis* spp., but these have been reduced to the genus *Thielaviopsis* (Paulin-Mahady *et al.*, 2002; Harrington, 2009). All *Ceratocystis* spp. have one or two endoconidial states, in which asexual spores (conidia) are produced within deep-seated phialides. The cylindrical to barrel-shaped endoconidial states were earlier placed in the anamorph genus *Chalara*, which is now restricted to anamorphs of discomycetes. In addition to the thin-walled and short-lived endoconidia, many species of *Ceratocystis* form aleurioconidia, which are pigmented, thick-walled, clamydospore-like spores borne on specialized conidiophores.

Phylogenetic analyses (Harrington, 2009) indicate that there are at least five species complexes within *Ceratocystis*, but *C. fagacearum* (Bretz) J. Hunt (the cause of oak wilt), *C. adiposa* (E.J. Butler) C. Moreau, and some ambrosia beetle symbionts (members of the anamorph genus *Ambrosiella*) do not appear to have close relatives. The species complexes include the *C. coerulescens* (Münch) B.K. Bakshi complex, the large *C. fimbriata* Ellis & Halst. complex, the *C. paradoxa* (Dade) C. Moreau complex (on monocotyledon hosts, including palms), the *C. moniliformis* (Hedgc.) C. Moreau complex (saprophytic sap-stain fungi or weakly pathogenic wound colonizers) and a group of apparently asexual species, the *Thielaviopsis* complex (soil-borne pathogens of minor importance to forestry).

11.1.1 The *Ceratocystis coerulescens* complex

Most of these species cause blue-stain in the sapwood of *Pinaceae*, and some are associated with tree-killing bark beetles (Harrington and Wingfield, 1998). Angiosperm pathogens include *C. virescens* (R.W. Davidson) C. Moreau, the cause of sap streak of sugar maple (*Acer saccharum* Marsh.), and the related asexual species *Thielaviopsis australis* (J. Walker & Kile) A.E. Paulin, T.C. Harr. & McNew, which causes myrtle wilt on myrtle beech (*Nothofagus cunninghamii* (Hook.) Oerst.). *C. virescens* is native to eastern North America (Harrington *et al.*, 1998) and has been locally important in reducing sugar yield and lumber quality in sugar maple ('sugar bush') stands managed for the collection of sap for maple syrup from Wisconsin to Maine (Houston, 1993). Myrtle wilt is important in the dynamics of myrtle beech-dominated temperate rainforests of Australia (Victoria and Tasmania), but is not of major commercial importance (Kile, 1993).

11.1.2 *Ceratocystis fagacearum*

The oak wilt pathogen is well placed in the genus *Ceratocystis*, but it is not closely related to any other species. It is unique in its capacity to cause a true vascular wilt disease and in its tight association with sap beetles (Coleoptera: Nitidulidae) (Harrington, 2009). Oak wilt is known only in the eastern USA, where it is a major cause of oak (*Quercus* spp.) mortality in many locations. The susceptibility of European and other exotic oaks has been demonstrated (MacDonald *et al.*, 2001), and Chinese chestnut (*Castanea mollissima* Blume) is also susceptible, suggesting that *C. fagacearum* could be a threat to other forest ecosystems if accidentally introduced.

It has been suggested that *C. fagacearum* evolved in Mexico or Central America and was naturally introduced to the eastern USA in recent times (Juzwik *et al.*, 2008b; Harrington, 2009). The fungus was first reported in the Upper Mississippi River Valley and may have spread from there to the Appalachian Mountains. It was probably killing oak trees in the Upper Midwest in the late 1800s (Gibbs and French, 1980), but was not confirmed in Texas until 1961, although it may have been present there since the 1930s (Appel, 1995a). Oak wilt is scattered across much of the eastern USA, but it is still 'filling in' within that range and is also expanding at the fringes (Juzwik *et al.*, 2008b, 2011; Appel, 2009). The disease was recently reported in New York at about 300 km northeast of its previously known range (Jensen-Tracy *et al.*, 2009). However, mortality due to oak wilt is generally not severe at the eastern edge of its distribution, where incidence may actually be decreasing (Juzwik *et al.*, 2011).

The disease is known in 52 of 55 counties in West Virginia, but is sporadic in that region and losses are generally not heavy (Juzwik, 2009). In the Upper Midwest, loss of timber value due to oak wilt can be heavy (Haugen *et al.*, 2009), but losses of amenity trees are of greater economic importance. In Minnesota, oak wilt is in only 25 of the 65 counties with suitable forest types, but the disease can be severe, especially on sandy soils north of Minneapolis and east of Rochester (Juzwik, 2009). Members of the red oak group (section *Lobatae*), such as northern red oak (*Q. rubra* L.) and northern pin oak (*Q. ellipsoidalis* E.J. Hill), are highly susceptible, but some members of the white oak group (section *Quercus*), such as bur oak (*Q. macrocarpa* Michx.) and to a lesser extent white oak (*Q. alba* L.), are also affected. To the south, in the Ozarks of Missouri, the disease is widespread, but the disease centres are relatively small and losses are not great (Juzwik, 2009).

Oak wilt is very important in some locations in Texas, where the pathogen appears to be expanding its range (Appel, 2009). At least 2500 ha are affected by the disease in central Texas, but the wilt has not yet expanded to eastern Texas or Louisiana, where there are many susceptible oaks. The live oak (*Q. virginiana* Mill.) and the Texas live oak (*Q. fusiformis* Small) are members of section *Quercus* (series *Virentes*) but are susceptible to oak wilt, as are the epidemiologically important red oaks, Buckley oak (*Q. buckleyi* Nixon & Dorr) and blackjack oak (*Q. marilandica* Münchh.). Loss of property values and of historically significant trees, and the ecological impacts, have been considerable in central Texas, where thousands of aesthetically important oak trees are killed each year – probably millions of trees in total (Appel, 1995a; Juzwik *et al.*, 2011). The loss of oak habitat is a further threat to the endangered golden-cheeked warbler (*Dendroica chrysoparia*) (Greene and Reemts, 2009).

11.1.3 The *Ceratocystis fimbriata* complex

This complex is further broken up into clades: the African, Asian, North American (NAC) and Latin American (LAC) clades (Johnson *et al.*, 2005; Harrington, 2009). The African species *C. albofundus* M.J. Wingf., De Beer & M.J. Morris has been reported from Uganda and South Africa, where it can cause substantial mortality in plantations of black wattle (*Acacia mearnsii* De Wild.) (Barnes *et al.*, 2005). A canker disease of tapping panels on the rubber tree (*Hevea brasiliensis* (Willd. ex A. Juss.) Müll. Arg.) in Asia is also caused by a member of the complex (South and Sharples, 1925), and *C. larium* M. van Wyk & M.J. Wingf. was recently described from styrax (*Styrax benzoin* Dryand.) used for incense production (van Wyk *et al.*, 2009).

Three species in the NAC are capable of causing tree mortality. *C. populicola* A. Johnson & T.C. Harr. (Ceratocystis canker on poplars, *Populus* spp.) forms conspicuous cankers and causes some mortality on quaking aspen (*P. tremuloides* Michx.) in the USA and Canada (Hinds, 1972a; Johnson *et al.*, 2005). It has also been introduced to Poland, where it caused mortality of various *Populus* spp. in a limited area (Przybyl, 1984), but the pathogen apparently has not spread. *C. variospora* (R.W. Davidson) C. Moreau causes an important disease on almond and other stone fruits (*Prunus* spp.) in California, but it is only a minor wound pathogen on native forest trees in the Midwest (Johnson *et al.*, 2005). *C. smalleyi* J.A. Johnson & T.C. Harr. causes a wilt of bitternut hickory (*Carya cordiformis* (Wangenh.) K. Koch) and other hickories in the Midwest in association with the hickory bark beetle, *Scolytus quadrispinosus* Say (Johnson *et al.*, 2005).

The LAC includes *C. fimbriata s.s.*, *C. platani* (J.M. Walter) Engelbr. & T.C. Harr., and also *C. cacaofunesta* Engelbr. & T.C. Harr., the cause of Ceratocystis wilt or mal de machete on cacao (*Theobroma cacao* L.) (Engelbrecht and Harrington, 2005). *C. cacaofunesta* is native to the upper Amazon but has caused substantial mortality in cacao plantations wherever the fungus has been introduced (Engelbrecht *et al.*, 2007).

Ceratocystis platani

Canker stain is a lethal disease on sycamores or plane trees (*Platanus* spp.) in the USA and Europe. Epidemics were initially recognized in the eastern USA on the commonly planted street tree, London plane tree (*P. hybrida* Brot.; syn. *P. acerifolia* (Aiton) Willd.), a hybrid of the American sycamore (*P. occidentalis* L.) and the oriental plane tree (*P. orientalis* L.), though *C. platani* has also been found on American sycamore in natural riparian forests (Walter, 1946). An estimated 10,000 London plane trees out of an initial population of 153,000 were killed in the Philadelphia area, and some communities lost more than 80% of these trees during the 1930s and 1940s (Walter *et al.*, 1952). However, the disease is no longer common in eastern US cities.

American sycamore trees are moderately resistant to the disease, and are generally scattered in natural stands, and rarely wounded, so losses due to canker stain in natural forests were considered minimal by Walter (1946) and Walter *et al.* (1952). Later, McCracken and Burkhardt (1979) reported losses of up to 30% in natural stands, and the potential losses in commercial plantations are high, especially in short-rotation stands managed from coppice or 'silage sycamore' for fibre production. However, there are several canker diseases on American sycamore, and some of the reported canker stain mortality in plantations may have been due to the presence of bacterial leaf scorch (*Xylella fastidiosa*) followed by secondary pathogens (*Botryosphaeria* spp.) (Britton *et al.*, 1998) that cause cankers and xylem discoloration similar to that caused by *C. platani*.

C. platani is native to the eastern USA, but there have been at least two introductions: one to Modesto (California) and the other to Europe (Engelbrecht *et al.*, 2004; Ocasio-Morales *et al.*, 2007). In California, the disease was reported on street trees in the city of Modesto (Perry and McCain, 1988), where it has killed most of the London plane trees, the American sycamores and the California sycamores (*P. racemosa* Nutt.). The pathogen has probably been in Modesto for 50 years or more, but it has not been reported outside the city, though there is native riparian California sycamore nearby. The threat to native California sycamore and street trees in other communities remains substantial.

In Europe, the fungus was first recognized at Forte dei Marmi (Italy), although mortality attributable to canker stain was reported near Naples in the 1960s (Panconesi *et al.*, 2003). It has been speculated that the fungus was brought into the Naples/Caserta area from the USA on wooden packing material used for military supplies during World War II (Panconesi, 1999). Genetic analyses (Engelbrecht *et al.*, 2004) suggest that a single genotype of the pathogen was introduced into Europe and apparently spread throughout Italy, Switzerland and southern France, where the disease spread from the port of Marseilles (Ferrari and Pinchenot, 1976). In France alone, more than 50,000 trees have died, and the removal and replacement of each tree costs thousands of euros (Mollet, 2007). *C. platani* is now in the Toulouse region, and since 2006 has been found along the historic Canal du Midi, which is lined with more than 50,000 mature London plane trees, thousands of which have already died (A. Vigouroux, France, 2011, personal communication). There were earlier unconfirmed reports of canker stain in Spain, but the disease was recently confirmed in a small plantation near Girona (Riba, 2011).

Canker stain was reported from the Peloponnese peninsula in Greece in 2003, the fungus apparently having originated from Italy (Ocasio-Morales *et al.*, 2007). This was the first report for canker stain in the native range of oriental plane tree, though the disease was severe on naturalized oriental plane trees in Sicily (Granata and Pennisi, 1989). In Greece, it has caused substantial mortality of oriental plane trees in natural stands along streams and rivers, as well as in ornamental plantings of both London and oriental plane trees. The disease could cause devastating losses to this important and historic riparian species, which extends naturally east into Turkey.

Ceratocystis fimbriata s.s.

The name *C. fimbriata* is herein restricted to populations that are native to South and Central America and the Caribbean (Harrington *et al.*, 2011). On forest trees, *C. fimbriata* has its greatest economic impact on eucalyptus (*Eucalyptus* spp. and hybrids) plantations in Brazil, but it also can cause cankers (mouldy rot) on tapping panels of the rubber tree (Martin, 1949) and causes a lethal canker stain in gumhar (*Gmelina arborea* Roxb.) plantations in the lower Amazon (Muchovej *et al.*, 1978). Some strains on mango (*Mangifera indica* L.) in Brazil are closely related to the strains on eucalyptus (Ferreira *et al.*, 2010). A Brazilian strain introduced to Pakistan attacks mango and a valuable forest tree known as Indian rosewood (*Dalbergia sissoo* Roxb. ex DC.) (Poussio *et al.*, 2010).

C. fimbriata was first observed on dying eucalyptus trees in the south of Bahia, Brazil, in 1997 and is now common in former cerrado forests in Minas Gerais (Ferreira *et al.*, 2011). It has been discovered in six other Brazilian states, appears to be increasing in importance, and may now be considered the most important disease on *Eucalyptus* spp.

Several *Ceratocystis* spp. have been isolated from eucalyptus around the world, but these species are mostly simple wound colonizers that do not cause tree mortality (Harrington *et al.*, 2011). New *Ceratocystis* spp. that are probably synonymous with *C. fimbriata* have been found on wounds of eucalyptus in north-western South America, but it is not clear whether they are important pathogens. Reported mortality of eucalyptus in Uruguay, Congo, Uganda and South Africa is likely to have been caused by Brazilian strains of *C. fimbriata* (Roux *et al.*, 2000; Ferreira *et al.*, 2011; Harrington *et al.*, 2011).

11.2 Diagnosis

The important *Ceratocystis* spp. on forest trees colonize the sapwood and eventually kill their hosts. Crown or branch dieback often precedes death. Leaves on affected branches may fall off, or may be poorly developed or show chlorosis or other colorations before dying, giving a characteristic 'wilt' appearance to the crown (McCracken and Burkhardt, 1979; Houston, 1993). On some species, oak wilt may cause diagnostic leaf symptoms, including a bronzing of the leaf tips on members of the red oak group, and live oaks may show a striking necrosis and chlorosis along the leaf veins (Appel, 2009; Juzwik *et al.*, 2011). In general, sapwood discoloration is more characteristic of these diseases than leaf symptoms.

Only *C. fagacearum* causes a true vascular wilt (Harrington, 2009), and the deep sapwood staining following along the ray parenchyma, typical of other Ceratocystis diseases, is not evident in oak wilt. Oak wilt may be distinguished based on a ring of brown vascular discoloration visible in cross-sections of wilted branches, but the staining may not always be conspicuous (Juzwik *et al.*, 2011). The discoloration is often concentrated around certain vessels, and more or less evenly spaced within the growth ring. Oak wilt can sometimes be diagnosed by mat production under the bark of recently killed trees of the red oak group (Engelhard, 1955), but a more typical diagnostic characteristic is a cluster of symptomatic and killed trees in a discrete pocket (an infection centre).

Apart from the findings in oak wilt, the xylem discoloration caused by *Ceratocystis* spp. is not restricted to a particular annual ring. If there is branch dieback, extensive xylem discoloration can generally be found below the dying branch, on the main stem or, especially, at the base of the tree if the tree was infected through the root system (Walter, 1946). The discoloration is a combination of host response to infection and the pigmented hyphae and spores of the pathogen. Microscopic examination will show dark brown aleurioconidia in the stained sapwood (Panconesi *et al.*, 2003). Other fungi can induce xylem discoloration, especially in stressed hosts, but *Ceratocystis* spp. tend to stain the ray parenchyma more intensely than *Botryosphaeria* spp. or *Lasiodiplodia theobromae* (Pat.) Griffon & Maubl., which cause symptoms that are frequently confused with those caused by *Ceratocystis* spp.

In some hosts, *Ceratocystis* spp. can move out from the sapwood and attack the cambium and inner bark tissue, causing a canker. In aspen (*Populus* spp.), cankers may have substantial callus production and be flared at the margins (Hinds, 1972a). Long, inconspicuous cankers or a spiral–vertical series of small cankers may form in *Platanus* spp. that have little callus growth at the canker margins (Walter, 1946; McCracken and

Burkhardt, 1979; Panconesi, 1999). Older cankers on plane trees may be slightly sunken, with the dead bark dried and cracked.

Another common feature of these diseases is attack by ambrosia beetles (Curculionidae: Scolytinae and Platypodinae), which bore through the bark and into the xylem to excavate their galleries (Kile, 1993; Ocasio-Morales *et al.*, 2007). Chopping into the tree where the ambrosia beetle frass is seen may reveal that the beetle bore holes are precisely placed in streaks of discoloured xylem. As adult ambrosia beetles tunnel in the sapwood of killed or dying trees, they push their frass outside the tree, and the fine powder accumulates in bark crevices or at the base of the tree, or is perhaps left as a 'toothpick' sticking out from the tree trunk. The pathogen may be isolated from the ambrosia beetle frass.

Because the above symptoms are not unique, isolation of the causal agent from discoloured sapwood or twigs is necessary for accurate diagnosis. No selective agar media are available, although some species, notably members of the *C. fimbriata* complex, can be isolated on fresh slices of carrot root (Moller and DeVay, 1968a). Many important *Ceratocystis* pathogens are homothallic and produce perithecia and ascospores rapidly on isolation media or on discoloured wood chips. Incubation of colonized plant tissue in a moist chamber may result in the production of perithecia with sticky ascospore droplets within a week. Even for heterothallic species like *C. fagacearum*, conidia will form within a few days on agar media. However, with any sample of discoloured wood, the ubiquitous *Fusarium solani* (Mart.) Sacc. may also grow on agar media or carrot slices and could be mistaken as a pathogen. Members of the LAC of the *C. fimbriata* complex may be isolated from soil using the carrot baiting technique (Laia *et al.*, 2000) or from water using sycamore twigs (Grosclaude *et al.*, 1988).

C. fagacearum does not form aleurioconidia and is more sensitive to heat and drying than other *Ceratocystis* spp., and it does not survive long in stem tissues (Merek and Fergus, 1954). Samples of living, discoloured wood are best be kept on ice or refrigerated before plating. Chips of the discoloured xylem or discs from small branches can be aseptically excised or surface sterilized and placed in acidified medium. Like the other *Ceratocystis* spp., the oak wilt fungus has a fruity smell in culture, though it may not be as strong as the banana-like smell of the *C. fimbriata* complex.

There are many cryptic species in *Ceratocystis* that differ little in morphology. Thus, accurate identification relies on DNA sequencing. The internal transcribed spacer (ITS) region of the nuclear ribosomal DNA region (rDNA) is useful for identifications (Harrington *et al.*, 2011). For some species (such as *C. smalleyi* and *C. variospora*), the ITS-1 region has repetitive bases that may make PCR and DNA sequencing difficult. In the LAC, there are many regions of insertions/deletions in the ITS region, and there may be heterogeneity among the rDNA repeats, thus making sequencing difficult; the ITS region may then provide little useful phylogenetic information for the delineation of species.

11.3 Infection Biology

Inoculum may come in the form of conidia or ascospores carried by insects, or on tools and equipment, in frass of ambrosia beetles or other wood-boring insects, or in sawdust. All *Ceratocystis* spp. on woody hosts infect through wounds and move through sapwood tissue, mostly in ray parenchyma, and some also cause cankers (Harrington, 2009). However, *C. fagacearum* moves systemically through the host's xylem without attacking parenchyma tissue in the early stages of colonization. In some diseases in large trees, the pathogen may move from an infected to a healthy tree through functional root grafts (anastomosed roots). Species that form aleurioconidia are potentially soil-borne or water-borne and can directly infect roots.

11.3.1 Inoculum

Conidia and ascospores are produced on short-lived mats of mycelium that form on cut surfaces or on cankers (Harrington, 2009). Spores from mats may be important in dispersal via rain or animal vectors, or by the contamination of tools and equipment. However, the mats may be more important as a site for fungal feeding and the acquisition of spores by insects.

Black perithecia with long necks position wet masses of ascospores above the sporulating mat. Ascospores are thought to survive longer than conidia and may be more effectively dispersed (Merek and Fergus, 1954). Unlike conidia, ascospores are held together in a sticky, hydrophobic matrix, so the spores are not readily separated by water but, instead, have an affinity for the hydrophobic exoskeleton of insects. Many *Ceratocystis* spp. are capable of 'selfing' (homothallism), perhaps because of the unreliability of insect dispersal of conidia for the cross-fertilization of mycelial mats (Harrington, 2009). All species in the *C. fimbriata* complex and *C. virescens* are homothallic through unidirectional mating type switching. *C. fagacearum* is heterothallic, meaning that it can only reproduce sexually if two strains of opposite mating type come in contact; nitidulids visiting various mats acquire conidia and cross-fertilize mats. The importance of ascospores in the epidemiology of oak wilt is supported by the fact that the two mating types occur in nature in roughly equal proportions (Yount, 1954; Appel *et al.*, 1985).

Members of the *C. fimbriata* complex produce long-lived aleurioconidia on mats and in wood, and these probably contribute to survival of the fungus in the frass of wood-boring insects and sawdust (Harrington, 2009). Fungal propagules expelled from the trees during tunnel cleaning by adult ambrosia beetles may be dispersed by wind or rain splash for relatively short distances. The frass is highly infectious when placed on wounds.

T. australis does not produce aleurioconidia, but viable conidia and conidiophores are expelled by ambrosia beetle tunnelling (Kile, 1993). *C. fagacearum* was not isolated from frass expelled by ambrosia beetles attacking trees with oak wilt (Peplinski and Merrill, 1974), although it was isolated from the mycangia of ambrosia beetles, but it is unlikely that ambrosia beetles introduce *C. fagacearum* into living oaks (Harrington, 2009).

Aleurioconidia in frass or sawdust are likely sources of soil-borne inoculum. Members of the LAC are mostly soil-borne pathogens, as exemplified in Ceratocsystis wilt on coffee (*Coffea*) and citrus (*Citrus*) (Marin *et al.*, 2003), mango (*Mangifera*) and eucalyptus (Laia *et al.*, 2000; Ferreira *et al.*, 2011), and canker stain on plane tree (Mutto Accordi, 1989). The infection of roots via soil-borne inoculum has not been well studied, but it occurs quickly and commonly in eucalyptus planted on former forested sites (Ferreira *et al.*, 2011), and wounding may not be necessary.

11.3.2 Insect vectors

Most *Ceratocystis* spp. produce fruity odours in the form of small chain aliphatic alcohols and their esters, or fusel oils (fusel alcohols), which are thought to attract fungal-feeding insects (Lin and Phelan, 1992; Kile, 1993). Ethyl acetate and ethyl alcohol are particularly common. These compounds are toxic and may reduce grazing by insects that are not regular fungal feeders, thus leaving the mats for vectors, such as nitidulids, that typically tolerate high concentrations of mycotoxins (Harrington, 2009).

Many different groups of insects can acquire spores of *Ceratocystis* spp., but most of them do not transmit the pathogen to new wounds and are not vectors. Drosophilid flies (Diptera) and nitidulids have frequently been associated with mycelial mats, but most *Ceratocystis* spp. do not have specific insect vectors (Kile, 1993; Harrington, 2009). Nitidulids are commonly believed to be the primary vectors, but it is not always clear that they transmit spores to suitable infection courts, i.e. fresh wounds. *C. variospora* and *C. populicola* have been associated with nitidulids, but also with other insects (Moller and DeVay, 1968b; Hinds, 1972b; Johnson *et al.*, 2005). Nitidulids are contaminated with *C. albofundus* (Heath *et al.*, 2009) and *C. platani* (Crone and Bachelder, 1961), but the importance of these potential vectors is not known.

The association of nitidulids with *C. fagacearum* is particularly strong (Harrington, 2009). The fungus has a unique capability of producing mats under the bark of freshly killed trees and exposing the mats by splitting the bark (Gibbs and French, 1980; Juzwik *et al.*, 2011). Pressure pads form in the middle of the mats on the xylem and phloem, and as the pads grow and push against each other, the bark is pushed away and cracks open. The sweet, melon-like odour emitted from the mat is believed to be attractive to nitidulids, which can enter through the cracked outer bark.

Many species of nitidulids inhabit *C. fagacearum* mats, but only a few species carry spores of the pathogen to fresh wounds (Juzwik *et al.*, 2004; Ambourn *et al.*, 2005; Hayslett *et al.*, 2008, 2009; Juzwik *et al.*, 2011). *Colopterus truncatus* (Randall) is an important vector in the Upper Midwest, Missouri and Texas; *Carpophilus sayi* (Parsons) also is a likely vector in the Upper Midwest; and *Colopterus niger* (Say) and *C. semitectus* (Say) also may be important in Missouri.

Only four species of *Ceratocystis* (*C. smalleyi*, *C. laricicola* Redfern & Minter, *C. polonica* (Siemaszko) C. Moreau and *C. rufipennis* M.J. Wingf., T.C. Harr. & H. Solheim) are known to be adapted to bark beetle vectors (Harrington, 2009). These species lack fruity odours, perhaps because their fusel oils are toxic, and bark beetles may not tolerate these volatiles. *C. smalleyi* is associated with the hickory bark beetle and mortality of bitternut hickory, but the aetiology of this mortality is not well understood, and the vector relationships need further study (Johnson *et al.*, 2005).

Bark beetles may be significant vectors of *C. fagacearum* in regions where nitidulids are less effective (Rexrode and Jones, 1970). However, there is a growing consensus that bark beetles are much less important than nitidulids as vectors of *C. fagacearum* (Gibbs and French 1980; Hayslett *et al.*, 2009; Juzwik *et al.*, 2011).

11.3.3 Wound colonization and xylem movement

Fresh wounds to the xylem are the most important infection court for most species of *Ceratocystis*. On natural hosts, wound colonization is typically limited to a small area around the wound site, and the pathogen quickly sporulates on the exposed xylem and inner bark tissue around the wound, often under flaps of the bark surrounding the wound (Johnson *et al.*, 2005). On highly susceptible hosts, a *Ceratocystis* sp. may extensively colonize sapwood tissue, causing a discoloration or stain and wilting and death of branches or whole trees as the host responds with tyloses and gums (Clérivet *et al.*, 2000). Some of these diseases may also involve necrosis of the inner bark (cankers), which may be subtle (e.g. on plane tree and eucalyptus) or conspicuous (e.g. on almond and aspen) due to host responses, such as gummosis or rings of flared callous tissues, leaving a perennial or target canker.

Infected trees may die over a period of months to years. Vertical colonization of London plane tree may be greater than 2 m a year, and a tree of 30–40 cm diameter may be girdled within 2–3 years (Panconesi *et al.*, 2003). Large American sycamores die within 6 years (McCracken and Burkhardt, 1979). *C. fagacearum* moves systemically through the non-living vessels as conidia, at least in the early stages of colonization, and longitudinal colonization may be particularly rapid. Red oaks and live oaks usually die within months, but bur oak survives longer, and white oak may survive many years with oak wilt (Juzwik *et al.*, 2011).

Phytotoxins have been implicated in a number of diseases caused by *Ceratocystis* spp. Limp leaves and bright coloration patterns on the leaves of affected branches are evident in many of these diseases, including canker stain on plane tree and oak wilt. Aromatic fusel oils produced by *Ceratocystis* spp. are thought to be important in the attraction of insects, but these same compounds are phytotoxic and may have an important role in host–pathogen interaction (Tabachnik and DeVay, 1980). There has been substantial work on the importance of cerato-platinin in canker stain on plane tree (Bernardi *et al.*, 2011).

Sapwood movement through functional root grafts can lead to extensive mortality in stands of mature trees. *T. australis* moves readily from tree to tree through functional root grafts (Kile, 1993), as does *C. platani* (Mutto Accordi, 1986). The oak wilt pathogen also moves through functional root grafts between trees (Juzwik *et al.*, 2011), and self-grafting (grafts among roots of the same tree) is important in the colonization of the root system of an individual oak tree (Blaedow and Juzwik, 2010).

11.4 Epidemiology

With many different potential sources of inoculum and dispersal mechanisms, Ceratocystis disease cycles are complicated and varied. Epidemics may occur where several favourable factors coincide, such as abundant sporulating mats, vector activity, wounding, mechanical transmission, root grafting or high levels of soil-borne inoculum. There is a history of localized outbreaks of short duration for some of these diseases.

11.4.1 Sporulation and vector activity

Sporulating mats are obvious sources of inoculum, but the dispersal of spores from mats to wounds has not been clearly demonstrated with many of these diseases. There is little doubt, however, that sporulating mats are important in oak wilt and canker stain on the plane tree.

Sporulating mats on firewood or logs may lead to the introduction of *C. fagacearum* to a new area (Appel, 1995b; Juzwik *et al.*, 2011). A recently discovered focus of oak wilt 300 km north-east of its previous known range was associated with the movement of firewood (Jensen-Tracy *et al.*, 2009). However, *C. fagacearum* only produces sporulating mats when the bark/wood interface is moist (Gibbs and French, 1980), so the intercontinental movement of the pathogen in debarked logs is not likely. In general, large sporulating mats of *C. fagacearum* are produced on members of the red oak group, suggesting that the fungus is naturally adapted to red oaks (Harrington, 2009; Juzwik *et al.*, 2011). No mats, or only a small number of small mats, are found on members of the white oak group (Engelhard, 1955).

Most oak wilt mats are produced in the spring and autumn in Minnesota. Spring mat production is dependent on minimum February temperatures and spring precipitation (Juzwik, 2009). Mats are also produced in the summer, but the most important nitidulid species are not frequently contaminated between late August and mid-October, and fresh wounds on oak are not as attractive to nitidulids at that time of year (Ambourn *et al.*, 2005; Juzwik, 2009). Mats produced in the spring are most important, and optimal mat production in the spring is usually on trees that became infected or wilted the summer before. The highest frequency of pathogen-contaminated nitidulids on oak wounds was in May in Minnesota and April in central Missouri (Juzwik *et al.*, 2004; Hayslett *et al.*, 2008, 2009).

Colopterus truncatus is an important vector because a high percentage of the population may be contaminated with *Ceratocystis*, and the beetle is strongly attracted to fresh wounds on oak. It has been found on wounds within 10 min of their creation (Juzwik *et al.*, 2011). Even so, the transmission of *C. fagcearum* is inefficient, sporadic and mostly local. Much higher numbers of contaminated *C. truncatus* were found in oak wilt stands than in stands free of oak wilt (Ambourn *et al.*, 2005). There is a higher efficiency of vector spread over shorter distances, and longer distance spread is spotty (Menges and Loucks, 1984; Shelstad *et al.*, 1991; Juzwik, 2009). Nitidulid transmission may be mostly limited to 1 km (Juzwik *et al.*, 1985), meaning that local sanitation programmes may be effective.

In Texas, live oaks are frequently killed, but they do not produce significant numbers of sporulating mats. Therefore, Buckley oak and blackjack oak provide most of the inoculum for insect dispersal, and these hosts are the target of sanitation programmes (Appel, 2009). Mats typically form in Texas from late winter through to the spring, and nitidulids are active and carry *C. fagacearum* from February to July (Hayslett *et al.*, 2009; Juzwik *et al.*, 2011).

Bark beetles probably do not play a major role in oak wilt epidemics, in part because the fungus is not adapted for sporulation in bark beetle galleries, and even if contaminated, bark beetles would not frequently feed in the xylem of healthy oak species and introduce spores to the vessels. In oak wilt infection centres in Minnesota, only four to 13 adults per thousand of the small oak beetle, *Pseudopityophthorus minutissimus* (Zimmermann), were contaminated with *C. fagacearum* in May and June (Ambourn *et al.*, 2006). In Europe, the bark beetle *Scolytus intricatus* (Ratzeburg) has been suggested as a potential vector of *C. fagacearum* because it has a life history and behaviour more suited to overland transmission than *Pseudopityophthorus* spp. (Webber and Gibbs, 1989). However, it is questionable whether *C. fagacearum* could become established in an ecosystem without suitable nitidulid vectors.

The thin bark of London plane tree allows for subtle cankers with inconspicuous sporulating mats, which produce inoculum that can be acquired by tools and equipment. *C. platani* sporulates within 24–48 h on canker surfaces, wounded parts of diseased trees and pruning cuts (Panconesi *et al.*, 2003). In the eastern USA, sporulation occurs during periods of high moisture content from May until October (Walter, 1946). These spores may be spread by insects or rain and enter soil or waterways (Walter, 1946; Crone and Bachelder, 1961; Panconesi *et al.*, 2003), but are probably most importantly spread by mechanical transmission on tools and equipment. Spores from mats do not normally spread far by rain or insects; only wounds on trees within 8 m of diseased London plane tree became infected in a field trial reported by Walter (1946).

11.4.2 Frass of wood-boring insects and sawdust

The activities of ambrosia beetles and other wood-boring insects appear to be particularly important in diseases caused by members of the LAC. The contaminated frass of the boring insects may be important for the wound colonization of nearby trees, for contributing to soil-borne inoculum and for contamination of waterways. Similarly, sawdust created in the removal of diseased trees is important in canker stain on plane trees.

The boring activities of the oak pinhole borer (*Platypus cylindricus* Burmeister; Curculionidae: Platypodinae) may be contributing to the rapid spread of canker stain in natural riparian forests in Greece (Ocasio-Morales *et al.*, 2007). Other wood-boring insects are thought to be similarly important in Italy (Panconesi *et al.*, 2003). More important in urban forests are the aleurioconidia that are released in sawdust. *C. platani* may remain viable in killed trees for 2 years or more (Grosclaude *et al.*, 1993; Panconesi, 1999), so the trees remain a continued source of inoculum as they are degraded by the activities of insects. However, removal of the trees produces abundant infectious sawdust that can blow in the wind to infect wounds or enter waterways.

The frass of ambrosia beetles attacking diseased eucalyptus trees is known to harbour viable inoculum of *C. fimbriata*, and frass from ambrosia beetles, termites and other wood-boring insects probably contributes to soil-borne inoculum (Ferreira *et al.*, 2011). Although soil-borne inoculum from cleared natural forests or cultivated woody crops may lead to high levels of mortality of planted eucalyptus, it is not clear whether rotations of susceptible eucalyptus clones increase the levels of soil-borne inoculum. This is of particular concern where *C. fimbriata* has been brought into new regions or to former pasture sites in eucalyptus cuttings because such introductions could lead to high disease levels in later rotations.

There has been speculation that *C. platani* was introduced into Italy from the eastern USA on solid wood packing material (Panconesi, 1999). Although it is possible that *C. platani* produced mycelial mats on the cut surfaces of such wood, it is unlikely that the fungus would produce a fresh mat with spores on dried wood after arrival in Europe. It is more likely that ambrosia beetle activity in the wood released *C. platani* in frass (Ocasio-Morales *et al.*, 2007). The sporulation of *C. fimbriata* on or in eucalyptus wood shipped from Brazil to South Africa may have led to the introduction of the pathogen (van Wyk *et al.*, 2009), but the fungus could also have been introduced in eucalyptus cuttings (Ferreira *et al.*, 2011; Harrington *et al.*, 2011).

11.4.3 Wounding and new infection centres

Many epidemics caused by *Ceratocystis* species have been associated with a high incidence of wounding (Kile, 1993). Wounding is important for sap streak of maple in managed sugar bush stands in which the bases of trees are often wounded by equipment during the collection of maple sap. Most of the infections are along roads or near work buildings, and wounds made in the spring when the soils are wet are particularly subject to infection (Houston, 1993). In Wisconsin, 5% of wounded maple trees had sap streak (Mielke and Charette, 1989).

Wounding during commercial harvesting and timber stand improvements can lead to substantial levels of oak wilt in Wisconsin and Michigan (Haugen *et al.*, 2009). However, the pruning of amenity trees, tree damage during building construction and other human activities are more commonly associated with new oak wilt infection centres in non-commercial forests. Wounds on oak are suitable for infection for only a few days (Zuckerman, 1954; Cobb *et al.*, 1965), and timing is critical. The highest frequency of successful nitidulid transmission is during the spring, when the mycelial mats are most abundant, there are large numbers of nitidulids carrying the fungus and wounds to the xylem are most susceptible to infection (Juzwik, 2009).

Because the vector–pathogen relationship is inefficient, new introductions to an oak stand are rare but important events in predicting losses (Menges and Loucks, 1984). The initiation rates of new oak wilt infection centres vary greatly among regions, and these rates generally relate to the severity of the disease. Juzwik (2009) summarized reported rates of new foci that ranged from 0.42 ha^{-1} yr^{-1} in Minnesota and Wisconsin to <0.07 in Missouri and <0.006 in Illinois. Rates of new infection centres were much lower in Pennsylvania and West Virginia and even lower still in North Carolina and Tennessee. In a survey of the Fort Hood Military Installation in Texas, there was only one new disease centre per 150 ha^{-1} $year^{-1}$ (0.007 foci ha^{-1} $year^{-1}$) (Appel, 2009), but the large and rapidly expanding infection centres resulted in substantial mortality of live oaks.

Although some stream-side American sycamores were infected through natural wounds made by driftwood, human-caused wounds were the only important infection court in the London plane tree (Walter, 1946). At an undisturbed (no wounds) study site of London plane trees with high levels of disease, no new infections occurred after 4 years. Only small wounds are needed for infection, and the thin outer bark of the London plane tree makes it highly susceptible to wounding. Wounds are only susceptible to infection for a few days (Walter, 1946; Vigouroux and Rouhani, 1987).

In the eastern USA, it was realized that pruning activities not only provided an infection court for the pathogen, but that the wound dressings commonly applied to pruning cuts carried inoculum in the form of sawdust that had fallen into the open cans of pruning paint that hung from the belts of tree trimmers (Walter, 1946; Walter *et al.*, 1952). Pruning followed by the application of tree paint to the wounds so caused led to higher infection levels than did pruning alone (Walter *et al.*, 1952). Inoculated wounds and unpainted wounds on London plane trees were not infected from 1 December to 15 February in New Jersey, though wounds sealed during this period with sterile tree paint resulted in 25–50% infection, apparently due to the insulation of the wounds from cold winter temperatures (Walter, 1946).

11.4.4 Mechanical transmission

Although *C. fagacearum* could possibly be moved tree to tree on pruning equipment (Bretz, 1951), it survives in sawdust for <1 day (Merek and Fergus, 1954) and only sporulates under the bark. *C. variospora* and other species that are known to be mechanically transmitted form aleurioconidia in sapwood, but sporulation on the outside of the tree may be more important (Teviotdale and Harper, 1991).

The mechanical transmission of *C. platani* is central to epidemics of canker stain on London plane trees. Contaminated pruning saws, tree paint and climbing ropes were important in the early epidemics in the eastern USA. The pathogen infected 40% of the wounds made by saws that had been previously used on diseased trees (Walter, 1946). Sawdust from diseased trees contaminated tree paint, in which *C. platani* was viable for months, and treating pruning wounds with this paint greatly increased infection probability (Walter *et al.*, 1952). Climbing ropes used on diseased trees and then used on healthy trees on the same day resulted in infection of 50% of the trees, and on the next day, 25% of trees were infected via the ropes (Walter, 1946). The fungus is also transmitted on saws in plantations of American sycamore managed from coppice.

The epidemiology of canker stain in Europe is complicated by modern machinery, especially terracing (earth-moving) equipment, which may acquire the fungus from sporulating cankers or bits of wood and mechanically transmit it to new locations as it wounds the thin-barked trees (Blankart and Vigouroux, 1982; Panconesi *et al.*, 2003). Such equipment is frequently moved within and among cities and is believed responsible for initiation of many new disease centres. The very rapid spread of *C. platani* from city to city in Italy (Panconesi, 1999) was most likely due to mechanical transmission on tools and equipment. The recent introduction to Gerona (Spain) was associated with terracing equipment (A. Vigouroux, France, 2011, personal communication).

Mechanical transmission is likely to be responsible for the dramatic rates of tree mortality within cities. In Gloucester (New Jersey), a canker stain epidemic began in 1926 in a 1915 planting of 279 London plane trees (Walter *et al.*, 1952). Some 14 years later, in 1940, the percentage of trees that were symptomatic or killed was 63%; after 17 years it was 73%, after 20 years it was 86% and after 23 years it was 91%. In the first 20 years after canker stain was discovered in Forte dei Marmi (Italy), 90% of the London plane trees had died (Panconesi, 1999). Well after the initial establishment of canker stain in Marseilles, 1850 mature London plane trees (about 13% of the initial population) were killed between 1960 and 1972 (Ferrari and Pichenot, 1976).

C. fimbriata will produce mycelial mats on wounded eucalyptus trees and stumps, so mechanical transmission may be important. Eucalyptus trees in plantations are commonly wounded during weed control operations and other activities. High incidence of Ceratocystis wilt has been noted in plantations that were cut and managed for sprout production for rooted cuttings (Ferreira *et al.*, 2011), and these stools may have been infected by contaminated saws during the harvest or by contaminated scissors and other tools used to collect the cuttings for rooting. It is likely that the fungus is spread throughout nurseries as workers harvest cuttings from hedge plants with contaminated tools.

11.4.5 Vegetatively propagated material

An often unappreciated fact is that many species in the *C. fimbriata* complex have been introduced to new areas in vegetatively propagated material. Corms, storage roots and slips of vegetable and ornamental crop plants have been known to harbour these fungi (Engelbrecht and Harrington, 2005; Thorpe *et al.*, 2005). Mechanical transmission may facilitate contamination of the propagated material in the nursery. Source plants, rooted cuttings and grafted scions may be asymptomatic when heavily watered, allowing for the unwitting distribution of thousands of infected plants.

C. fimbriata may have been moved from Brazil to Pakistan and Oman in infected mango stock (Ferreira *et al.*, 2010; Harrington *et al.*, 2011), *C. cacaofunesta* has been moved among various research stations in Latin America in cuttings from cacao trees (Englebrecht *et al.*, 2007), *C. populicola* was found in vegetatively propagated poplar in Canada and Poland (Przybyl, 1984; Vujanovic *et al.*, 1999; Johnson *et al.*, 2005), and the recent introduction of *C. platani* to Greece was thought to be on grafted cuttings of London plane tree from a nursery in Italy (Ocasio-Morales *et al.*, 2007). Infected eucalyptus cuttings have been shown to be important in bringing *C. fimbriata* to new regions and to former pastureland that had little or no soil-borne inoculum of the fungus (Ferreira *et al.*, 2010, 2011; Harrington *et al.*, 2011).

11.4.6 Root graft transmission

Movement through functional root grafts is potentially important for diseases affecting trees of sufficient age and size. The initiation of new disease centres is relatively rare in myrtle wilt, but many massive trees can be killed through the expansion of infection centres by root graft transmission (Kile *et al.*, 1989).

Many more oak trees are infected by *C. fagacearum* through root grafts than through wounds (Bruhn *et al.*, 1991; Appel, 1995b). Root grafting varies greatly among oak species, from 70% in northern pin oak to only 6% in bur oak (Parmeter *et al.*, 1956). A higher percentage of oak trees infected through root graft transmission and larger disease centres were found in stands with higher percentages of red oak (Menges and Loucks, 1984). The probability of transmission between two trees of the same species increases with tree size and decreases with distance between the two trees, but the speed of underground transmission is not related to these two factors (Menges and Kuntz, 1985). More root graft transmission is seen in lighter textured soils than in heavier soils (Prey and Kuntz, 1995). Some grafting also occurs between different species of oaks, but the importance of interspecific grafting in transmission is unknown (Juzwik, 2009).

The regional importance of oak wilt may correspond with the amount of root grafting of susceptible species. Expansion of disease foci averages 1.9–7.6 m yr^{-1} in Minnesota and up to 12 m yr^{-1} in Michigan, with the highest rates found on the sandiest soils where there may be mortality of 8–11 red oaks ha^{-1} yr^{-1} (Bruhn and Heyd, 1992; Juzwik, 2009). The mortality rate is lower in Pennsylvania (1–3 oaks $centre^{-1}$ yr^{-1}) and West Virginia (0.19–0.39 oaks $centre^{-1}$ yr^{-1}) (Jones, 1971; Mielke *et al.*, 1983). Expansion rates of up to 50 m yr^{-1} were estimated in infection centres in live oak stands in Texas (Appel, 2009).

As an affected tree dies, water flow is disrupted owing to the formation of tyloses, and the transpiration pull of the tree diminishes. This pulls the water and the fungus into the root system of a grafted healthy (transpiring) tree. Cutting a diseased tree also facilitates such undesirable pathogen movement into a root-grafted, healthy tree (Nair and Kuntz, 1975). *C. fagacearum* may survive for up to 5 years in the grafted roots of killed trees (Skelly and Wood, 1974).

Root graft transmission is also important in the movement of *C. platani* (Mutto Accordi, 1986). Clonal propagation of the London plane tree may facilitate root graft transmission because trees planted along a street are likely to be genetically identical and likely to form functional grafts. The pathogen can quickly move from tree to tree along avenues of mature London plane trees, and sanitation measures must include not only symptomatic trees but also their asymptomatic neighbours.

Root graft transmission of American sycamore in natural forest stands may not be important because the trees are generally scattered in floodplain forests. Plantations of American sycamore are typically harvested within 15 years, so root grafting may not be common. Eucalyptus plantations in Brazil are typically harvested after 7 years, and grafting is not likely to be common or important.

11.4.7 Soil-borne and waterborne inoculum

Aleurioconidia of *C. platani* are abundant in stained sapwood and, once liberated, can infest soil (Mutto Accordi, 1989) and waterways (Grosclaude *et al.*, 1988, 1991; Vigouroux and Stojadinovic, 1990). As *Platanus* spp. are riparian and frequently planted along waterways, the pathogen may be able to spread quickly in water. Canker stain is spreading rapidly through the London plane trees that line the Canal du Midi (A. Vigouroux, France, 2011, personal communication). The production of ambrosia beetle frass from diseased oriental plane trees in Greece is apparently leading to the rapid spread of *C. platani* on this species in natural watercourses (Ocasio-Morales *et al.*, 2007). The similarly rapid spread of canker stain on naturalized oriental plane trees in Sicily was also thought to be due to water movement (Granata and Pennisi, 1989).

C. fimbriata appears to be native to certain forest types of Brazil (Ferreira *et al.*, 2010; Harrington *et al.*, 2011). Eucalyptus planted to former cerrado forest sites in Minas Gerais can have very high levels of disease due to soil-borne inoculum (Ferreira *et al.*, 2011). Susceptible clones planted on former forest sites may begin to show symptoms and mortality at 15–20 months after planting, and the disease then progresses in a manner typical of a soil-borne disease; that is, a monomolecular model in which infections are due to an initial inoculum level, and the inoculum level does not increase during the rotation of the crop (Ferreira, 2009). Greater than 60% mortality has been noted 3 years after planting (Ferreira *et al.*, 2011). It is not known whether soil inoculum levels increase during eucalyptus rotations or during harvest of the plantation, but this could be a major consideration for future rotations of that species.

11.5 Management Strategies and Tactics

11.5.1 Avoidance

Wound avoidance is key to management of some diseases caused by *Ceratocystis* spp., including sap streak of sugar maple (Houston, 1993). There was no difference in sap streak incidence between stands logged during frozen versus non-frozen conditions in Wisconsin (Mielke and Charette, 1989), but Houston (1993) found that most infections occurred through wounds made during spring (the mud season) when soils were wet and roots were frequently injured by tractors and other equipment. Avoidance of such activity at that time of the year was recommended.

Most oak wounds infected by *C. fagacearum* are made in spring or early summer, so harvests should be curtailed during these periods (Haugen *et al.*, 2009; Cummings Carlson and Martin, 2010). In urban settings, pruning wounds made in the spring and early summer have a much higher probability of infection than pruning wounds made in late summer, autumn and winter. In a wounding study in Minnesota (Juzwik *et al.*, 1985), wounds made from May to mid June became infected, and the advertised slogan in Minnesota is 'Do not prune in May or June'. Based on the activity of vectors, April is the month of highest risk of wound infection in Missouri, but avoidance of wounds from April to June has been recommended (Hayslett *et al.*, 2008, 2009). In Texas, wounding should be avoided in late winter and spring (Hayslett *et al.*, 2009). It is not advisable, however, to delay treatment of storm-damaged trees. When pruning is absolutely necessary, a protective layer of latex paint can be applied immediately after pruning to protect the wound from infection (French and Juzwik, 1999).

The canker stain epidemics in eastern USA during the 1920s to the 1940s were associated with pruning wounds (Walter *et al.*, 1952), and avoidance of pruning was an important management strategy. Sporulation on cankers occurred during the moistest periods from May through to October, but Walter (1946) recommended that pruning should be done between 1 December and 15 February. Sanitation was the primary management tool, however, because total avoidance of wounding was not feasible.

In Europe, pruning was just part of the problem because terracing and other equipment had proven to be major wounding agents and a means of mechanical transmission (Blankart and Vigouroux, 1982; Panconesi *et al.*, 2003). None the less, it is recommended that pruning should be avoided or done in the coldest and driest months of the year, along with the application of fungicide to the pruning cut (Panconesi, 1999).

Ceratocystis wilt on eucalyptus on pastureland in Brazil can be avoided if disease-free cuttings are planted (Ferreira *et al.*, 2011). Disease incidence is low in such locations because soil inoculum levels are low due to the absence of woody hosts. Clean nursery stock can be produced if care is taken to avoid the collection of cuttings from diseased mother trees. Carefully monitored mini hedges in nurseries appear to be the best source of disease-free cuttings (Ferreira *et al.*, 2011).

11.5.2 Exclusion

Ceratocystis spp. have been frequently introduced to new areas by human activity, and international quarantines have been considered or are in place. Infected planting stock (especially rooted cuttings or grafted scions) is the most likely pathway of introduction, but restrictions on the movement of such stock are inadequate. The frass from wood-boring insects or spore-producing mats on wood taken from diseased trees is another potential pathway. Lastly, contaminated pruning tools and equipment can introduce pathogens such as *C. platani*.

Although the biology and history of oak wilt suggests that accidental introduction of *C. fagacearum* over long distances is unlikely (Harrington, 2009), *C. fagacearum* is categorized as an A1 quarantine pest by the European and Mediterranean Plant Protection Organization (EPPO) and is of quarantine significance for the Inter-African Phytosanitary Council (IAPSC) and the North American Plant Protection Organization (NAPPO). EPPO (2010) has requirements for the import of plants for planting and seed of *Quercus* spp. from the USA because of *C. fagacearum*, even though oaks are not normally vegetatively propagated, oak wilt does not commonly occur in nurseries, and seed transmission of the pathogen is highly unlikely. Options for importing oak wood (timber) from the USA include bark removal in conjunction with heat treatment, radiation or fumigation, though treatment of debarked logs may not be necessary. Mycelial mats form under the bark but only when the sapwood and inner bark tissue are moist (Gibbs and French, 1980), and the fungus does not survive long in stem wood, especially if it is debarked (Merek and Fergus, 1954). EPPO also includes in its regulations the quarantine of potential vectors – the timberworm *Arrhenodes minutus* (Drury) and the bark beetles *Pseudopityophthorus minutissimus* and *P. pruinosus* (Eichhoff) – though these are not considered to be efficient vectors. Nitidulid vectors are not listed.

In contrast to *C. fagacearum*, *C. platani* has been frequently introduced to new regions through human activity. Planting material of *Platanus* spp. should be obtained from regions where canker stain does not occur (EPPO/CABI, 1997). The recent introduction in Greece is likely to have been through infected planting stock from Italy, where canker stain is widespread (Ocasio-Morales *et al.*, 2007). It is further recommended that pruning tools are disinfected, even when used in uninfested regions (EPPO/CABI, 1997). All terracing machinery used in the vicinity of canker stain should be treated before moving it to another site. Blankart and Vigouroux (1982) recommended cleaning with a water jet and 8-hydroxyquinoline sulfate. Saws and climbing ropes are other important means of mechanical transmission and introductions to new areas. The single introduction of *C. platani* to Europe from the USA on solid wood packing material is unlikely to be repeated because there are no longer epidemics of canker stain in urban areas of the USA to supply cheap wood for packaging. Also, there are now better safeguards for treating solid wood packing material (FAO, 2009), which should help to control the activity of wood-boring insects and the release of *C. platani* in insect frass.

11.5.3 Eradication

Complete eradication of an exotic *Ceratocystis* sp. from even a limited area is a difficult task, but local introductions may be eliminated if recognized quickly, and sanitation practices have proven effective in managing local epidemics.

Because overland spread of *C. fagacearum* by nitidulids tends to be localized, the probability of infection can be reduced if symptomatic oak trees in the area are removed before mats form. Removal of all symptomatic and recently killed trees is generally recommended, as well as that of neighbouring asymptomatic trees that may already be infected. The goal is twofold: to remove potential mat-producing trees and to produce a barrier to root graft transmission. Trees that die in late summer are most likely to retain sufficient moisture and produce mats the next spring, so they are the focus of sanitation practices. Treatment of stumps of diseased trees with ammate, aresenite, or 2,4,5-trichlorophenoxyacetic acid (2,4,5-T) has been shown to somewhat reduce the survival time of *C. fagacearum* in roots (Merek and Fergus, 1954).

Portions of diseased trees greater than 5 cm diameter should be removed or destroyed before spring mat production and nitidulid activity (Haugen *et al.*, 2009). Cutting the diseased trees while still alive risks the rapid movement of conidia in the sapwood into adjacent, root-grafted trees, so it is recommended that roots are severed before felling. Mats can be produced on felled trees left on site (Morris, 1955), but the wood of diseased trees can be utilized for wood products, or chipped, burned or buried, so long as the bark is removed and the wood is dry before mats can form in the spring (French and Juzwik, 1999). Firewood can be used if it is debarked to promote drying or if stacked firewood is covered with plastic to prevent access by nitidulids (Haugen *et al.*, 2009).

An alternative to felling and removing symptomatic trees is girdling, which generally reduces the moisture content of the trees and lessens the probability of mat production (Gillespie *et al.*, 1957). Basal girdling of diseased red oaks in Texas reduced mat production, but was not effective in reducing disease levels, probably because most of the trees there are infected through root graft transmission (Greene *et al.*, 2008). Girdling trees that are part of a large, common root system probably does not slow underground spread of the pathogen.

Attempts have been made to eradicate *C. platani* from pockets of infection in southern Europe, but these attempts have mostly proven unsuccessful and are quite costly. The eradication of small, isolated infection centres that have not yet spread far, such as the recent effort in north-eastern Spain (Riba, 2011), can be successful. The widespread nature of the affected area in the Peloponnese peninsula in Greece may have led to the decision not to eradicate *C. platani*.

Sanitation was effective in managing canker stain in urban settings in the eastern USA (Walter, 1946). If a branch is infected, a London plane tree could be saved by severing the branch, if there was a clearance of 1 m between the cankered branch and the nearest trunk, but most symptomatic trees should be removed. Tools and equipment used on diseased trees should be disinfected with alcohol, sodium hypochlorite or other chemicals. These efforts began to quickly quell the epidemics in the eastern USA (Walter *et al.*, 1952), but a well-designed sanitation programme was not effective in stopping the epidemic of canker stain in Modesto, California (Perry and McCain, 1988), and canker stain remains a major issue in Europe in spite of sanitation efforts.

The prompt removal of diseased trees is recommended in Europe. In addition to the removal of symptomatic trees, at least one healthy plane tree should be removed on either side of the infected trees (Panconesi, 1999). Treatments with glyphosate or other herbicides are used to kill the root systems and prevent further spread via root graft transmission. When diseased trees are removed, a minimum of cutting is recommended, and all sawdust generated should be collected in tarps and properly disposed of (Plate 13). Traffic in the area should be minimized to prevent the wind movement of sawdust, and trees should not be sawn on a windy day. As much of the root system should be removed as possible, as well as removing the contaminated soil. The entire area should be treated with disinfectant or fungicide. Equipment can be treated with alcohol or sodium hypochlorite (Panconesi, 1999; Panconesi *et al.*, 2003).

Care should be taken to avoid mechanical transmission during sanitation practices in diseased commercial stands of American sycamore. Initially, McCracken and Burkhardt (1979) recommend felling diseased trees and suggested that the trees need not be removed because the fungus does not remain alive long. However, it is now known that the fungus survives for long periods as aleurioconidia in sapwood, and that the logs should be removed or otherwise disposed of before ambrosia beetles and other wood-boring insects enter and release contaminated frass.

11.5.4 Protection

Fungicide injections

Wilts of trees are difficult to manage with fungicides because of the large mass of xylem tissue to treat and the difficulty of delivering a sustained dose of protection throughout the tree at a reasonable cost. The high cost and the injury to the tree in the injection process must be balanced with the likelihood that the treatment will lower the risk of infection. Preventive treatment for oak wilt may only delay disease development rather than give complete protection. Therapeutic or protective fungicide treatments for London plane trees have not met with success (Panconesi, 1999).

The most effective treatments for oak wilt have been below-ground injection of large volumes of dilute systemic triazoles (namely, propiconazole) into root flares. Soil application or injections of lower volumes of more highly concentrated propiconazole may also be effective (Wilson, 2005), but even the high volume injections do not completely stop colonization of the host nor completely eradicate the fungus (Blaedow *et al.*, 2010). Beside the cost, the limitations of propiconazole are its somewhat restricted movement into the root system (where preventive doses are needed) and its short-lived nature in a living tree. The concentration of propiconazole in injected trees decreases substantially in the first year and is below detectable levels within 2 years.

Preventive treatment of red oaks within grafting distance of diseased red oaks only delays progression of the disease, so treatments may need to be applied every 2 years (Eggers *et al.*, 2005; Blaedow *et al.*, 2010). Preventive treatment of bur and white oak may be effective (Eggers *et al.*, 2005), but because the disease progresses slowly in these species, arborists often wait until the trees show symptoms and then apply a therapeutic treatment (Haugen *et al.*, 2009). Injection of propiconazole is widely practised in Texas, but treatments generally delay development of disease symptoms rather than prevent infection (Appel and Kurdyla, 1992; Billings *et al.*, 2001). The pathogen may pass through the root system of an injected live oak and move on to the next tree (Juzwik *et al.*, 2011). Spring injections may work better than injections later in the season (Billings *et al.*, 2001).

Wound dressings

Fungicidal wound dressings have been recommended for the protection of pruning or other wounds from infection by *C. variospora* (DeVay *et al.*, 1968) and other *Ceratocystis* spp. When wounding cannot be avoided, a variety of wound treatments are known to be effective in preventing infection by *C. fagacearum*, but latex paint is the recommended treatment because of its effectiveness and its lack of toxicity to the tree (French and Juzwik, 1999; Camilli *et al.*, 2007).

Wound dressings also can inhibit the growth of *C. platani* (Davis and Peterson, 1973). Walter (1946) recognized that wound dressings commonly had sawdust with viable *C. platani* inoculum that could survive for months. He recommended the addition of phenylmercury nitrate to kill the pathogen in tree paint, but such highly toxic compounds are no longer used. Panconesi (1999) recommended a fungicide dressing for the protection of wounds against canker stain.

Trenching and other barriers

Reduction in the root graft transmission of the oak wilt pathogen by the use of root-free zones has been practised for many years and can be highly effective in reducing losses (Bretz, 1951; Gehring, 1995; Cummings Carlson and Martin, 2010; Juzwik *et al.*, 2011). Generally, a trench is made to delimit infected from healthy trees. The placement of the trench can be calculated based on the probability of disease spread (Bruhn *et al.*, 1991), but generally involves a primary barrier of at least one ring of healthy trees beyond the symptomatic trees and a secondary barrier placed inside this to help protect that ring of healthy trees. For live oaks in Texas, the primary barrier should be a minimum of 30 m from a symptomatic tree because of rapid spread of the pathogen in the common root systems of these hosts (Juzwik *et al.*, 2011).

Trenches 1.5 m deep are recommended (Bruhn and Heyd, 1992), but shallower trenches still lower the risk of transmission. However, many sites are too steep or have other physical limitations that make trenching impracticable. Vibratory ploughs with a long blade (Plate 14), belt trenchers, or heavy rotating discs that can cut through rock have been recommended, but such equipment is not available in all regions (Juzwik *et al.*, 2011). Backhoes can also be used, but site disturbance is substantial with such equipment. Plastic barriers placed inside the trenches increase their effectiveness but add to the cost (Wilson, 2005). Killing trees by herbicide treatment has not been considered effective in protecting healthy trees at the edge of expanding oak wilt centres (Haugen *et al.*, 2009).

Soil trenching was effective in reducing oak wilt spread in live oak at Fort Hood Military Installation, where 11.2 km of trenches were put in place over 4 years (Greene and Reemts, 2009). The Texas Oak Wilt Suppression Project has resulted in over 1000 km of trenches, and a recent evaluation found a nearly 80% success rates, that is, slightly more than 20% of the trenches were breached by oak wilt within 3 years (Billings, 2009). Such 'breakouts' are usually attributed to insufficient depth of the trenches.

11.5.5 Resistance

With most Ceratocystis diseases, there is substantial variation in aggressiveness in the pathogen and also in resistance among host species and within host species or hybrids. For instance, variation in resistance to *C. populicola* was noted among poplar clones in Poland (Przybyl, 1984).

C. fagacearum has a broad host range among *Quercus* spp. and also can attack chestnuts (*Castanea* spp.) under experimental conditions, but host species vary widely in susceptibility. Such variation in resistance among species can be used to favour more resistant species in plantings and in stand management. A diversity of oak species lessens the risk of root graft transmission. Commercial forest stands should encourage diversity and avoid dense stands of highly susceptible species, such as red oak. Breeding oak for resistance to oak wilt may be feasible. There appears to be some inheritance of tolerance in Texas live oak, but it is not clear whether it will be possible to breed tolerant trees (Gray and Appel, 2009). The genetic uniformity of *C. fagacearum* should favour the durability of any resistance that is deployed (Harrington, 2009).

The high susceptibility of (the hybrid) London plane trees to canker stain comes from the oriental plane tree component. Genetic variation is limited in London plane tree (Panconesi *et al.*, 2003), but there is some variability in susceptibility to canker stain in existing cultivars of the species (Pilotti *et al.*, 2009). There is more resistance available in American sycamore, which is being utilized to develop resistant London plane trees (Vigouroux and Olivier, 2004). Resistant American sycamores were crossed with oriental planes and screened through repeated inoculations, leading to the London plane cultivar 'Vallis Clausa', which has been commercially available for 5 years. About 2000 trees have been planted in France without reported canker stain (A. Vigouroux, France, 2011, personal communication). Pure American sycamore is not a suitable street tree in Europe because of its high susceptibility to anthracnose caused by *Apiognomonia veneta* (Sacc. & Speg.) Höhn., but 'Vallis Clausa' has good resistance to anthracnose and some resistance to powdery mildew and to lacebugs (*Corythucha ciliata* (Say)). More resistant clones are to be developed. The limited availability and higher cost of London plane trees that are resistant to canker stain suggests that only limited plantings can be made in the near future. Because the European population of *C. platani* originated from a single genotype and the species reproduces asexually or through selfing (Engelbrecht *et al.*, 2004), the pathogen has limited potential to overcome the resistance developed in 'Vallis Clausa'.

In contrast, soil-borne populations of *C. fimbriata* in Brazil can be quite diverse, and individual isolates vary greatly in aggressiveness to eucalyptus and other hosts (Harrington *et al.*, 2011). Planting *Eucalyptus* clones on former forested sites naturally infested with soil-borne inoculum may select for strains of the fungus that are particularly aggressive to eucalyptus, and such strains have been distributed in infected rooted cuttings. However, in both *E. grandis* W. Hill ex Maid. and *E. urophylla* S.T. Blake, there are clones that range from highly resistant to highly susceptible, and in crosses of the two species, there is a high degree of heritability of resistance (Goncalves-Rosado *et al.*, 2010). There is great variability among commercially available hybrid clones, but there is some clone × isolate interaction, which may make selection for resistance more difficult (Zauza *et al.*, 2004). Some of the most susceptible clones have already been eliminated from commercial use. Efforts are under way to deploy resistance to particular local genotypes or populations of *C. fimbriata* (Alfenas and Ferreira, 2008), which may be the best way to manage the disease when there are high levels of aggressive strains in the soil (Ferreira *et al.*, 2011).

11.5.6 Therapy

The oak wilt pathogen moves too quickly to remove affected limbs in the hope of restricting the spread of *C. fagacearum* within the crown. Fungicide treatment of symptomatic red oaks is also not recommended. However, therapeutic treatments of white and bur oaks with propiconazole are effective in delaying symptom development and mortality (Osterbauer *et al.*, 1994; Eggers *et al.*, 2005), though follow-up treatments may be needed.

Pruning out infected limbs has been recommended for almond trees with mallet canker (DeVay *et al.*, 1968) and in canker stain of plane trees (Walter, 1946), but *C. platani* generally moves too quickly in London plane tree to use this therapy on a routine basis. Therapeutic fungicide treatments are not considered to be effective for canker stain (Panconesi *et al.*, 2003).

11.5.7 Integrated disease management

As with the management of most forest diseases, an integrated approach involving several strategies is necessary to control diseases caused by *Ceratocystis* spp., especially in urban settings. Both oak wilt and canker stain require integrated programmes of detection, sanitation and prevention, and these only succeed with educational efforts and coordination among public and private interests (Walter *et al.*, 1952; Panconesi, 1999; Haugen *et al.*, 2009; Koch *et al.*, 2010; Juzwik *et al.*, 2011).

Juzwik (2009) reviewed some of the models and risk maps for oak wilt, and the measures that can be utilized to manage the disease in high-risk areas. Barriers to prevent underground spread (e.g. trenching to sever root grafts) may not be effective without sanitation. In Wisconsin, where the disease risk is high, there are appropriate guidelines for harvesting and timber stand improvement (Juzwik *et al.*, 2008a). Integrated measures in central Texas include avoiding wounding in the spring, treating fresh wounds with latex paint, cautious removal of firewood, trenching and, occasionally, fungicide injections (Appel, 2009).

Various formal suppression programmes, largely supported by the US Department of Agriculture Forest Service and state governments, have been enacted for oak wilt (Starkey, 2009). Early projects in eastern states were abandoned as largely ineffective, but with a better understanding of epidemiology and management techniques, successful programmes have been developed in Minnesota and Texas. In both cases, most of the spread of the pathogen is by root contact, and most direct control involves trenching.

The Texas Oak Wilt Suppression Project is an integrated, multi-agency management effort that began in 1988 and is coordinated by the Texas Forest Service (Billings, 2009). The prevention and control of oak wilt are based upon public education on proper timing of pruning, treatment of wounds, destroying diseased red oaks, proper handling of firewood, propiconazole injections and the planting of resistant species. Direct control efforts are based on detection, field evaluations and the control of expanding infection centres, primarily through root trenching.

Although the integrated management strategies outlined by Walter (1946) eventually led to a dramatic reduction of canker stain of plane trees in urban areas in the eastern USA, similar efforts in California and Europe have not had the same desired effect. Some factors contributing to the difficulties in Europe may include the mobility of construction equipment, the complexities of international commerce under the European Union (EU), and the difficulties of forming and executing the large, multilevel programmes needed to manage a problem as large as canker stain. None the less, education has led to early detection, and coordinated efforts have led to more effective sanitation, and prevention measures have undoubtedly reduced losses. London plane trees that are resistant to canker stain are another promising tool.

The epidemiology of Ceratocystis wilt on eucalyptus is still being unravelled, but an integrated approach that includes the sanitization of tools and use of clean nursery stock has been proposed (Ferreira *et al.*, 2011). Highly susceptible clones of otherwise productive hybrids have been abandoned in regions with a high level of disease, and resistant clones are being developed (Zauza *et al.*, 2004; Goncalves-Rosado *et al.*, 2010). It is to be hoped that this disease will be another case where integrated strategies lessen the importance of a disease caused by *Ceratocystis* spp.

References

Alfenas, A.C. and Ferreira, E.M. (2008) Emerging diseases in eucalyptus plantations. *Tropical Plant Pathology* 33, 25–28.

Ambourn, A.K., Juzwik, J. and Moon, R.D. (2005) Seasonal dispersal of the oak wilt fungus by *Colopterus truncatus* and *Carpophilus sayi* in Minnesota. *Plant Disease* 89, 1067–1076.

Ambourn, A.K., Juzwik, J. and Eggers, J.E. (2006) Flight periodicities, phoresy rates, and levels of *Pseudopityophthorus minutissimus* branch colonization in oak wilt centers. *Forest Science* 52, 243–250.

Appel, D.N. (1995a) The oak wilt enigma: perspectives from the Texas epidemic. *Annual Review of Phytopathology* 33, 103–118.

Appel, D.N. (1995b) Epidemiology of oak wilt in Texas. In: Appel, D.N. and Billings, R.F. (eds) *Oak Wilt Perspectives: The Proceedings of the National Oak Wilt Symposium, June 22–25, 1992, Austin, Texas.* Information Development, Inc., Houston, Texas, pp. 21–28.

Appel, D.N. (2009) Oak wilt biology, impact, and host pathogen relationships: a Texas perspective. In: Billings, R.F. and Appel, D.N. (eds) *National Oak Wilt Symposium, June 4–7, 2007, Austin, Texas. The Proceedings of the Second National Oak Wilt Symposium.* Texas Forest Service Publication 166, College Station, Texas, pp. 37–48. Also available (dated 2008 and with different pagination) at: http://www.texasoakwilt.org/Professionals/NOWS/conference_assets/NOWS_Proceedings.pdf (pp. 43–54; accessed 3 December 2012).

Appel, D.N. and Kurdyla, T. (1992) Intravascular injection with propiconazole in live oak for oak wilt control. *Plant Disease* 76, 1120–1124.

Appel, D.N., Dress, C.F. and Johnson, J. (1985) An extended range for oak wilt and *Ceratocystis fagacearum* compatibility types in the United States. *Canadian Journal of Botany* 63, 1325–1328.

Barnes, I., Nakabonge, G., Roux, J., Wingfield, B.D. and Wingfield, M.J. (2005) Comparison of populations of the wilt pathogen *Ceratocystis albofundus* in South Africa and Uganda. *Plant Pathology* 54, 189–195.

Bernardi, R., Baccelli, I., Carresi, L., Comparini, C., Pazzagli, L. and Scala, A. (2011) Cerato-platanin elicits transcription of defence-related genes earlier than *Ceratocystis platani* on *Platanus acerifolia. Forest Pathology* 41, 255–261.

Billings, R.F. (2009) The Texas Cooperative Oak Wilt Suppression Project: lessons learned in the first twenty years. In: Billings, R.F. and Appel, D.N. (eds) *National Oak Wilt Symposium, June 4–7, 2007, Austin, Texas. The Proceedings of the Second National Oak Wilt Symposium.* Texas Forest Service Publication 166, College Station, Texas, pp. 219–233. Also available (dated 2008 and with different pagination) at: http://www.texasoakwilt.org/Professionals/NOWS/conference_assets/NOWS_Proceedings.pdf (pp. 225–239; accessed 3 December 2012).

Billings, R.F. and Appel, D.N. (eds) (2009) *National Oak Wilt Symposium, June 4–7, 2007, Austin, Texas. The Proceedings of the Second National Oak Wilt Symposium.* Texas Forest Service Publication 166, College Station, Texas, Texas. Also available (dated 2008 and with different pagination) at: http://www.texasoakwilt.org/Professionals/NOWS/conference_assets/NOWS_Proceedings.pdf (accessed 3 December 2012).

Billings, R.F., Gehring, E.H., Cameron, R.S. and Gunter, J.T. (2001) Current practices in managing oak wilt: federal cost share programs, trenching, chemical injection, and the Texas suppression program. In: Ash, C.L. (ed.) *Shade Tree Wilt Diseases*. APS Press, St Paul, Minnesota, pp. 117–129.
Blaedow, R.A. and Juzwik, J. (2010) Spatial and temporal distribution of *Ceratocystis fagacearum* in roots and root grafts of oak wilt affected red oaks. *Arboriculture and Urban Forestry* 36, 28–34.
Blaedow, R.A., Juzwik, J. and Barber, B. (2010) Propiconazole distribution and effects on *Ceratocystis fagacearum* survival in roots of treated red oaks. *Phytopathology* 100, 979–985.
Blankart, D. and Vigouroux, A. (1982) Lutte contre le chancre coloré du platane [The fight against plane canker stain]. *Phytoma – La Defense des Végétaux* 343, 51.
Bretz, T.W. (1951) Oak wilt. *Journal of Forestry* 49, 169–171.
Britton, K.O., Leininger, R., Chang, C.J. and Harrington, T.C. (1998) Association of *Xylella fastidiosa, Ceratocystis fimbriata* f. *platani* and *Botryosphaeria rhodina* with declining sycamore plantations in the south-eastern USA. In: *Seventh International Plant Pathology Congress, 9–16 August 1998, Edinburgh, UK*. British Society for Plant Pathology, Birmingham, UK, abstract 3.7.50.
Bruhn, J.N. and Heyd, R.L. (1992) Biology and control of oak wilt in Michigan red oak stands. *Northern Journal of Applied Forestry* 9, 47–51.
Bruhn, J.N., Pickens, J.B. and Stanfield, D.B. (1991) Probit analysis of oak wilt transmission through root grafts in red oak stands. *Forest Science* 37, 28–44.
Camilli, K., Appel, D.N. and Watson, W.T. (2007) Studies on pruning cuts and wound dressing for oak wilt control. *Aboriculture and Urban Forestry* 33, 132–139.
Clérivet, A., Déon, V., Alami, I., Lopez, F., Geiger, J.P. and Nicole, M. (2000) Tyloses and gels associated with cellulose accumulation in vessels are responses of plane tree seedlings (*Platanus × acerifolia*) to the vascular fungus *Ceratocystis fimbriata* f. sp. *platani*. *Trees* 15, 25–31.
Cobb, F.W. Jr, Fergus, C.L. and Stambaugh, W.J. (1965) Factors affecting infection of red and chestnut oaks by *Ceratocystis fagacearum*. *Phytopathology* 55, 1194–1199.
Crone, L.J. and Bachelder, S. (1961) Insect transmission of canker stain fungus, *Ceratocystis fimbriata* f. *platani*. *Phytopathology* 51, 576.
Cummings Carlson, J.C. and Martin, A.J. (2010) *Oak Wilt Management: What Are the Options?* University of Wisconsin Extension Publication G3590, University of Wisconsin, Madison, Wisconsin.
Davis, S.H. Jr and Peterson, J.L. (1973) A tree wound dressing to prevent spread of the *Ceratocystis* causing canker stain disease of the plane tree. *Plant Disease Reporter* 57, 28–30.
DeVay, J.E., English, H., Lukezic, F.L. and O'Reilly, H.J. (1968) Ceratocystis canker of almond trees. *Phytopathology* 58, 949–954.
Eggers, J., Juzwik, J., Bernick, S. and Mordaunt, L. (2005) *Evaluation of Propiconazole Operational Treatments of Oaks for Oak Wilt Control*. Research Note NC-390, US Department of Agriculture, Forest Service, North Central Research Station, St Paul, Minnesota.
Engelbrecht, C.J. and Harrington, T.C. (2005) Intersterility, morphology and taxonomy of *Ceratocystis fimbriata* on sweet potato, cacao and sycamore. *Mycologia* 97, 57–69.
Engelbrecht, C.J.B., Harrington, T.C., Steimel, J. and Capretti, P. (2004) Genetic variation in eastern North American and putatively introduced populations of *Ceratocystis fimbriata* f. *platani*. *Molecular Ecology* 13, 2995–3005.
Engelbrecht, C.J.B., Harrington, T.C., Alfenas, A.C. and Suarez, C. (2007) Genetic variation of populations of the cacao wilt pathogen, *Ceratocystis cacaofunesta*. *Plant Pathology* 56, 923–933.
Engelhard, A.W. (1955) Occurrence of oak wilt fungus mats and pads on members of the red and white oak groups in Iowa. *Plant Disease Reporter* 39, 254–255.
EPPO (2010) *Quercus* and *Castanea*. *EPPO Bulletin* 40, 376–386.
EPPO/CABI (1997) *Ceratocystis fimbriata* f. sp. *platani*. In: *Quarantine Pests for Europe*, 2nd edn. CAB International, Wallingford, UK, pp. 674–677.
FAO (2009) Regulation of wood packaging material in international trade. Revision of ISPM No. 15. In: *International Standards for Phytosanitary Measures 1 to 32*. Food and Agriculture Organization of the United Nations, Rome, pp. 200–215.
Ferrari, J.P. and Pichenot, M. (1976) The canker stain disease of plane tree in Marseille and in the south of France. *European Journal of Forest Pathology* 6, 18–25.
Ferreira, E.M., Harrington, T.C., Thorpe, D.J. and Alfenas, A.C. (2010) Genetic diversity and interfertility among highly differentiated populations of *Ceratocystis fimbriata* in Brazil. *Plant Pathology* 59, 721–735.
Ferreira, M.A. (2009) Estrutura genética de populações de *Ceratocystis fimbriata* e padrão espaço-temporal da murcha-de-Ceratocystis [Genetic structure of populations of *Ceratocystis fimbriata* and spatio-temporal pattern of Ceratocystis wilt]. PhD thesis, Universidade Federal de Viçosa, Minas Gerais, Brazil.

Ferreira, M.A., Harrington, T.C., Alfenas, A.C. and Mizubuti, E.S.G. (2011) Movement of genotypes of *Ceratocystis fimbriata* within and among *Eucalyptus* plantations in Brazil. *Phytopathology* 101, 1005–1012.

French, D.W. and Juzwik, J. (1999) *Oak wilt in Minnesota*. Report MI-03174, University of Minnesota Extension Service, University of Minnesota, St Paul, Minnesota.

Gehring, E. (1995) Evaluation of suppression project treatments. In: Appel, D.N. and Billings, R.F. (eds) *Oak Wilt Perspectives: The Proceedings of the National Oak Wilt Symposium, June 22–25, 1992, Austin, Texas*. Information Development, Inc., Houston, Texas, pp. 147–154.

Gibbs, J.N. and French, D.W. (1980) *The transmission of oak wilt*. Research Paper NC-185, US Department of Agriculture, Forest Service, North Central Research Station, St Paul, Minnesota.

Gillespie, W.H., Shigo, A.L. and Tryon, E.H. (1957) The degree of mat-production control obtained by girdling oak wilt trees in West Virginia and some factors influencing mat formation in girdled trees. *Plant Disease Reporter* 41, 362–367.

Goncalves-Rosado, C.C., da Silva Guimaraes, L.M., Titon, M., Lau, D., Rosse, L., de Resende, M.D.V. and Alfenas, A.C. (2010) Resistance to Ceratocystis wilt (*Ceratocystis fimbriata*) in parents and progenies of *Eucalyptus grandis* × *E. urophylla*. *Silvae Genetica* 59, 99–106.

Granata, G. and Pennisi, A.M. (1989) Estese morie di platani orientali in forestazioni naturali causate da *Ceratocystis fimbriata* (Ell. et Halst.) Davidson f. *platani* Walter [Extensive die-offs of oriental plane trees in natural forests caused by *Ceratocystis fimbriata* (Ell. et Halst.) Davidson f. *platani* Walter]. *Informatore Fitopatologico* 12, 59–61.

Gray, M.C. and Appel, D.N. (2009) Attempts to develop an oak wilt resistant live oak. In: Billings, R.F. and Appel, D.N. (eds) *National Oak Wilt Symposium, June 4–7, 2007, Austin, Texas. The Proceedings of the Second National Oak Wilt Symposium*. Texas Forest Service Publication 166, College Station, Texas, pp. 185–194. Also available (dated 2008 and with different pagination) at: http://www.texasoakwilt.org/Professionals/NOWS/conference_assets/NOWS_Proceedings.pdf (pp. 191–200; accessed 3 December 2012).

Greene, T.A. and Reemts, C.M. (2009) Oak wilt research at Fort Hood: inoculum sources at landscape scale. In: Billings, R.F. and Appel, D.N. (eds) *National Oak Wilt Symposium, June 4–7, 2007, Austin, Texas. The Proceedings of the Second National Oak Wilt Symposium*. Texas Forest Service Publication 166, College Station, Texas, pp. 79–88. Also available (dated 2008 and with different pagination) at: http://www.texasoakwilt.org/Professionals/NOWS/conference_assets/NOWS_Proceedings.pdf (pp. 85–94; accessed 3 December 2012).

Greene, T.A., Reemts, C.M. and Appel, D.N. (2008) Efficacy of basal girdling to control oak wilt fungal mat production in Texas red oak (*Quercus buckleyi*) in central Texas. *Southern Journal of Applied Forestry* 32, 168–172.

Grosclaude, C., Olivier, R., Pizzuto, J.C., Romiti, C. and Madec, S. (1988) Detection of *C. fimbriata* f.sp. *platani* by trapping. Application to the study of the persistence of the parasite in infected wood. *European Journal of Forest Pathology* 18, 385–390.

Grosclaude, C., Olivier, R., Pizzuto, J.C. and Romiti, C. (1991) Etude expérimentale du transport de l'inoculum de *Ceratocystis fimbriata* f. *platani* par l'eau d'une rivière [Experimental study on the transport of inoculum of *Ceratocystis fimbriata* f. *platani* by river water]. *European Journal of Forest Pathology* 21, 168–171.

Grosclaude, C., Olivier, R. and Romiti, C. (1993) Chancre coloré du platane. Transmission et survie hypogeés de l'agent responsible [Canker stain of plane. Transmission and hypogeal survival of the agent responsible]. *Phytoma – La Défense des Végétaux* 455, 57–59.

Harrington, T.C. (2005) Ecology and evolution of mycophagous bark beetles and their fungal partners. In: Vega, F.E. and Blackwell, M. (eds) *Ecological and Evolutionary Advances in Insect–Fungal Associations*. Oxford University Press, New York, pp. 257–291.

Harrington, T.C. (2009) The genus *Ceratocystis*. Where does the oak wilt fungus fit? In: Billings, R.F. and Appel, D.N. (eds) *National Oak Wilt Symposium, June 4–7, 2007, Austin, Texas. The Proceedings of the Second National Oak Wilt Symposium*. Texas Forest Service Publication 166, College Station, Texas, pp. 21–35. Also available (dated 2008 and with different pagination) at: http://www.texasoakwilt.org/Professionals/NOWS/conference_assets/NOWS_Proceedings.pdf (pp. 27–41; accessed 3 December 2012).

Harrington, T.C. and Wingfield, M.J. (1998) The *Ceratocystis* species on conifers. *Canadian Journal of Botany* 76, 1446–1457.

Harrington, T.C., Steimel, J. and Kile, G.A. (1998) Genetic variation in three *Ceratocystis* species with outcrossing, selfing and asexual reproductive strategies. *European Journal of Forest Pathology* 28, 217–226.

Harrington, T.C., Aghayeva, D.N. and Fraedrich, S.W. (2010) New combinations in *Raffaelea*, *Ambrosiella*, and *Hyalorhinocladiella*, and four new species from the redbay ambrosia beetle, *Xyleborus glabratus*. *Mycotaxon* 111, 337–361.

Harrington, T.C., Thorpe, D.J. and Alfenas, A.C. (2011) Genetic variation and variation in aggressiveness to native and exotic hosts among Brazilian populations of *Ceratocystis fimbriata*. *Phytopathology* 101, 555–566.

Haugen, L., O'Brien, J., Pokorny, J., Mielke, M. and Juzwik, J. (2009) Oak wilt in the North Central Region. In: Billings, R.F. and Appel, D.N. (eds) *National Oak Wilt Symposium, June 4–7, 2007, Austin, Texas. The Proceedings of the Second National Oak Wilt Symposium*. Texas Forest Service Publication 166, College Station, Texas, pp. 149–157. Also available (dated 2008 and with different pagination) at: http://www.texasoakwilt.org/Professionals/NOWS/conference_assets/NOWS_Proceedings.pdf (pp. 155–163; accessed 3 December 2012).

Hayslett, M.C., Juzwik, J. and Moltzan, B. (2008) Three *Coleopterus* beetle species carry the oak wilt fungus to fresh wounds on red oak in Missouri. *Plant Disease* 92, 270–275.

Hayslett, M.C., Juzwik, J., Moltzan, B., Appel, D. and Camilli, K. (2009) Insect vectors of the oak wilt fungus in Missouri and Texas. In: Billings, R.F. and Appel, D.N. (eds) *National Oak Wilt Symposium, June 4–7, 2007, Austin, Texas. The Proceedings of the Second National Oak Wilt Symposium*. Texas Forest Service Publication 166, College Station, Texas, 166, pp. 103–111. Also available (dated 2008 and with different pagination) at: http://www.texasoakwilt.org/Professionals/NOWS/conference_assets/NOWS_Proceedings.pdf (pp. 109–120; accessed 3 December 2012).

Heath, R.N., Wingfield, M.J., van Wyk, M. and Roux, J. (2009) Insect associates of *Ceratocystis albofundus* and patterns of association in a native savanna ecosystem in South Africa. *Environmental Entomology* 38, 356–364.

Hinds, T.E. (1972a) Ceratocystis canker of aspen. *Phytopathology* 62, 213–220.

Hinds, T.E. (1972b) Insect transmission of *Ceratocystis* species associated with aspen cankers. *Phytopathology* 62, 221–225.

Houston, D.R. (1993) *Recognizing and Managing Sapstreak Disease of Sugar Maple*. Research Paper NE-675, US Department of Agriculture, Forest Service, Northeastern Forest Experiment Station, Hamden, Connecticut.

Jensen-Tracy, S., Kenaley, S., Hudler, G., Harrington, T. and Logue, C. (2009) First report of the oak wilt fungus, *Ceratocystis fagacearum*, in New York state. *Plant Disease* 93, 428.

Johnson, J.A., Harrington, T.C. and Engelbrecht, C.J.B. (2005) Phylogeny and taxonomy of the North American clade of the *Ceratocystis fimbriata* complex. *Mycologia* 97, 1067–1092.

Jones, T.W. (1971) *An Appraisal of Oak Wilt Control Programs in Pennsylvania and West Virginia*. Research Paper NE-204, US Department of Agriculture, Forest Service, Northeastern Forest Experiment Station, Upper Darby, Pennsylvania.

Juzwik, J. (2009) Epidemiology and occurrence of oak wilt in Midwestern, Middle and South Atlantic states. In: Billings, R.F. and Appel, D.N. (eds) *National Oak Wilt Symposium, June 4–7, 2007, Austin, Texas. The Proceedings of the Second National Oak Wilt Symposium*. Texas Forest Service Publication 166, College Station, Texas, pp. 49–59. Also available (dated 2008 and with different pagination) at: http://www.texasoakwilt.org/Professionals/NOWS/conference_assets/NOWS_Proceedings.pdf (pp. 55–66; accessed 3 December 2012).

Juzwik, J., French, D.W. and Jeresek, J. (1985) Overland spread of the oak wilt fungus in Minnesota. *Journal of Arboriculture* 11, 323–327.

Juzwik, J., Skalbeck, T.C. and Neuman, M.F. (2004) Sap beetle species (*Coleoptera: Nitidulidae*) visiting fresh wounds on healthy oaks during spring in Minnesota. *Forest Science* 50, 757–764.

Juzwik, J., Cummings Carlson, J. and Scanlon, K. (2008a) Case study: risk analysis and guidelines for harvest activities in Wisconsin oak timberlands to minimize oak wilt threat. *Forest Encyclopedia Network: Environmental Threats: Case Studies*. US Department of Agriculture, Forest Service, Asheville, North Carolina. Online publication. Available at: http://www.forestryencyclopedia.net/p/p5/p3389/p3464 (accessed 3 December 2012).

Juzwik, J., Harrington, T.C., MacDonald, W.L. and Appel, D.N. (2008b) The origin of *Ceratocystis fagacearum*, the oak wilt fungus. *Annual Review of Phytopathology* 46, 13–26.

Juzwik, J., Appel, D.N., MacDonald, W.L. and Burks, S. (2011) Challenges and successes in managing oak wilt in the United States. *Plant Disease* 95, 888–900.

Kile, G.A. (1993) Plant diseases caused by species of *Ceratocystis sensu stricto* and *Chalara*. In: Wingfield, M.J., Seifert, K.A. and Webber, J.F. (eds) *Ceratocystis and Ophiostoma: Taxonomy, Ecology, and Pathogenicity*. APS Press, St Paul, Minnesota, pp. 173–183.

Kile, G.A., Packham, J.M. and Elliott, H.J. (1989) Myrtle wilt and its possible management in association with human disturbance of rainforest in Tasmania. *New Zealand Journal of Forest Science* 19, 256–264.

Koch, K.A., Quiram, G.L. and Venette, R.C. (2010) A review of oak wilt management: a summary of treatment options and their efficacy. *Urban Forestry and Urban Greening* 9, 1–8.

Laia, M.L., Alfenas, A.C. and Harrington, T.C. (2000) Isolation, detection in soil, and inoculation of *Ceratocystis fimbriata*, causal agent of wilting, die-back and canker in *Eucalyptus*. *Fitopatologia Brasileira* 25, 384.

Lin, H. and Phelan, P.L. (1992) Comparison of volatiles from beetle-transmitted *Ceratocystis fagacearum* and four non-insect dependent fungi. *Journal of Chemical Ecology* 18, 1623–1632.

MacDonald, W.L., Pinon, J., Tainter, F.H. and Double, M.L. (2001) European oaks – susceptible to oak wilt? In: Ash, C.L. (ed.) *Shade Tree Wilt Diseases*. APS Press, St Paul, Minnesota, pp. 131–137.

Marin, M., Castro, B., Gaitan, A., Preisig, O., Wingfield, B.D. and Wingfield, M.J. (2003) Relationships of *Ceratocystis fimbriata* isolates from Colombian coffee-growing regions based on molecular data and pathogenicity. *Journal of Phytopathology* 151, 395–405.

Martin, W.J. (1949) *Moldy Rot of Tapping Panels of* Hevea *Rubber-Trees*. Circular no. 798, US Department of Agriculture, Washington, DC.

McCracken, F.I. and Burkhardt, E.C. (1979) Limiting canker stain disease losses of sycamore in the midsouth. *Southern Journal of Applied Forestry* 3, 26–28.

Menges, E.S. and Kuntz, J.E. (1985) Predictive equations for local spread of oak wilt in southern Wisconsin. *Forest Science* 31, 43–51.

Menges, E.S. and Loucks, O.R. (1984) Modeling a disease-caused patch disturbance: oak wilt in the Midwestern United States. *Ecology* 65, 487–498.

Merek, E.L. and Fergus, C.L. (1954) Longevity of the oak wilt fungus in diseased trees. *Phytopathology* 44, 328.

Mielke, M.E. and Charette, D.A. (1989) The incidence of sapstreak disease of sugar maple in Menominee County, Wisconsin, and its relationship to wounds and season of logging. *Northern Journal of Applied Forestry* 6, 65–67.

Mielke, M.E., Hayes, C. and Rexrode, C.E. (1983) Local spread of oak wilt in northeastern West Virginia during 1970–1982. *Plant Disease* 67, 1222–1223.

Moller, W.J. and DeVay, J.E. (1968a) Carrot as a species-selective isolation medium for *Ceratocystis fimbriata*. *Phytopathology* 58, 123–124.

Moller, W.J. and DeVay, J.E. (1968b) Insect transmission of *Ceratocystis fimbriata* in deciduous fruit orchards. *Phytopathology* 58, 1499–1508.

Mollet, J.M. (2007) Les dégâts dus au chancre coloré du platane: coûts économiques, esthétiques et symboliques [Damage due to canker stain of plane: economic costs, aesthetics and symbolism]. In: *Colloque National 'Chancre Coloré du Platane', 11 octobre 2007, ENSAT, Toulouse*. Association Française de Protection des Plantes (AFPP), Alfortville, pp. 41–60.

Morris, C.L. (1955) Control of mat formation by the oak wilt fungus by early felling of diseased trees. *Plant Disease Reporter* 39, 258–260.

Muchovej, J.J., Albuquerque, F.C. and Ribeiro, G.T. (1978) *Gmelina arborea* a new host of *Ceratocystis fimbriata*. *Plant Disease Reporter* 62, 717–719.

Mutto Accordi, S. (1986) Spread of *Ceratocystis fimbriata* f.sp. *platani* through root anastomoses. *Informatore Fitopatologico* 36, 53–58.

Mutto Accordi, S. (1989) The survival of *Ceratocystis fimbriata* f.sp. *platani* in the soil. *Informatore Fitopatologico* 39, 57–62.

Nair, V.M.G. and Kuntz, J.E. (1975) Recent advances in oak wilt research. In: *Advances in Mycology and Plant Pathology*. Sager Printers, New Delhi, India, pp. 231–240.

Ocasio-Morales, R., Tsopelas, P. and Harrington, T.C. (2007) The origin of *Ceratocystis platani* on native *Platanus orientalis* in Greece and its impact on natural forests. *Plant Disease* 91, 901–907.

Osterbauer, N.K., Salisbury, T. and French, D.W. (1994) Propiconazole as a treatment for oak wilt in *Quercus alba* and *Q. macrocarpa*. *Journal of Arboriculture* 20, 202–203.

Panconesi, A. (1999) Canker stain of plane tree: a serious danger to urban plantings in Europe. *European Journal of Plant Pathology* 81, 3–15.

Panconesi, A., Moricca, S., Dellavalle, I. and Torraca, G. (2003) The epidemiology of canker stain of plane trees and its spread from urban plantings to spontaneous groves and natural forests. In: Balder, H., Strauch, K.H. and Backhaus, G.F. (eds) *Second International Symposium on Plant Health in Urban Horticulture, Berlin, Germany, August 27–29, 2003. Mitteilungen aus der Biologischen Bundesanstalt für Land- und Forstwirtschaft Berlin Dahlem* 394, 84–91.

Parmeter, J.R. Jr, Kuntz, J.E. and Riker, A.J. (1956) Oak wilt development in bur oaks. *Phytopathology* 46, 423–436.

Paulin-Mahady, A.E., Harrington, T.C. and McNew, D. (2002) Phylogenetic and taxonomic evaluation of *Chalara*, *Chalaropsis*, and *Thielaviopsis* anamorphs associated with *Ceratocystis*. *Mycologia* 94, 62–72.

Peplinski, J.D. and Merrill, W. (1974) Nonsurvival of *Ceratocystis fagacearum* in frass of oak bark beetles and ambrosia beetles. *Phytopathology* 64, 1528–1530.

Perry, E. and McCain, A.H. (1988) Incidence and management of canker stain in London plane trees in Modesto, California. *Journal of Arboriculture* 14, 18–19.

Pilotti, M., Brunetti, A., Tizzani, L. and Marani, O. (2009) *Platanus* × *acerifolia* genotypes surviving to inoculation with *Ceratocystis platani* (the agent of canker stain): first screening and molecular characterization. *Euphytica* 169, 1–17.

Poussio, G.B., Kazmi, M.R., Akem, C. and Fateh, F.S. (2010) First record of *Ceratocystis fimbriata* associated with shisham (*Dalbergia sissoo*) decline in Pakistan. *Australasian Plant Disease Notes* 5, 63–65.

Prey, A. and Kuntz, J.E. (1995) Epidemiology of oak wilt outside of Texas. In: Appel, D.N. and Billings, R.F. (eds) *Oak Wilt Perspectives: The Proceedings of the National Oak Wilt Symposium, June 22–25, 1992, Austin, Texas*. Information Development, Inc., Houston, Texas, pp. 15–20.

Przybyl, K. (1984) Development of the fungus *Ceratocystis fimbriata* in shoots of poplar clones with differing resistance. *European Journal of Forest Pathology* 14, 177–183.

Rexrode, C.O. and Jones, T.W. (1970) Oak bark beetles – important vectors of oak wilt. *Journal of Forestry* 68, 294–297.

Riba, J.M. (2011) El chancro colorado del plátano; *Ceratocytis fimbriata* f.sp. *platani*; afectaciones en Girona [Canker stain of plane; *Ceratocytis fimbriata* f.sp. *platani*; occurrences in Girona]. *PARJAP, Boletín de la Asociación Española de Parques y Jardines* 61, 6–10.

Roux, J., Wingfield, M.J., Bouillet, J.P., Wingfield, B.D. and Alfenas, A.C. (2000) A serious new wilt disease of *Eucalyptus* caused by *Ceratocystis fimbriata* in Central Africa. *Forest Pathology* 30, 175–184.

Shelstad, D., Queen, L., French, D. and Fitzpatrick, D. (1991) Describing the spread of oak wilt using a geographic information system. *Journal of Arboriculture* 17, 192–199.

Skelly, J.M. and Wood, F.A. (1974) Longevity of *Ceratocystis fagacearum* in ammate treated and nontreated root systems. *Phytopathology* 64, 1483–1485.

South, F.W. and Sharples, A. (1925) *The 'Mould Rot' Disease of* Hevea brasiliensis *in Malaya*. Bulletin No. 37, Department of Agriculture, Straits Settlements and Federated Malay States, Federated Malay States Government Press, Kuala Lumpur.

Starkey, D.A. (2009) USDA Forest Service perspective on oak wilt suppression. In: Billings, R.F. and Appel, D.N. (eds) *National Oak Wilt Symposium, June 4–7, 2007, Austin, Texas. The Proceedings of the Second National Oak Wilt Symposium*. Texas Forest Service Publication 166, College Station, Texas, pp. 235–243. Also available (dated 2008 and with different pagination) at: http://www.texasoakwilt.org/Professionals/NOWS/conference_assets/NOWS_Proceedings.pdf (pp. 241–249; accessed 3 December 2012).

Tabachnik, M. and DeVay, J.E. (1980) Black root rot development in cotton roots caused by *Thielaviopsis basicola* and the possible role of methyl acetate in pathogenesis. *Physiological Plant Pathology* 16, 109–117.

Teviotdale, B.L. and Harper, D.H. (1991) Infection of pruning and small bark wounds in almond by *Ceratocystis fimbriata*. *Plant Disease* 75, 1026–1030.

Thorpe, D.J., Harrington, T.C. and Uchida, J.Y. (2005) Pathogenicity, phylogenetics and human dispersal of *Ceratocystis fimbriata* on the family *Araceae*. *Phytopathology* 95, 316–323.

van Wyk, M., Wingfield, B.D., Clegg, P.A. and Wingfield, M.J. (2009) *Ceratocystis larium* sp. nov., a new species from *Styrax benzoin* wounds associated with incense harvesting in Indonesia. *Personia* 22, 75–82.

Vigouroux, A. and Olivier, R. (2004) First hybrid plane trees to show resistance against canker stain (*Ceratocystis fimbriata* f.sp. *platani*). *Forest Pathology* 34, 307–319.

Vigouroux, A. and Rouhani, H. (1987) Observation de sensibilités differentielles de quelques organes de *Platanus acerifolia* vis-à-vis de *Ceratocystis fimbriata* f. *platani* [Observation of the different susceptibilities of various organs of *Platanus acerifolia* to *Ceratocystis fimbriata* f. *platani*]. *European Journal of Forest Pathology* 17, 181–184.

Vigouroux, A. and Stojadinovic, B. (1990) Possibilités d'infection du platane pour *Ceratocystis fimbriata* f. *platani* après contamination de l'eau où se développent des racines blesses [Possibilities of the infection of plane by *Ceratocystis fimbriata* f. *platani* after contamination by water or the development of wounded roots]. *European Journal of Forest Pathology* 20, 118–121.

Vujanovic, V., St.-Arnaud, M., Charlebois, D. and Fortin, E. (1999) First report of *Ceratocystis fimbriata* infecting balsam poplar (*Populus balsamifera*). *Plant Disease* 83, 879.

Walter, J.M. (1946) *Canker stain of plane trees*. Circular 742, US Department of Agriculture, Washington, DC.
Walter, J.M., Rex, E.G. and Schreiber, R. (1952) The rate of progress and destructiveness of canker stain of plane-trees. *Phytopathology* 42, 236–239.
Webber, J.F. and Gibbs, J.N. (1989) Insect dissemination of fungal pathogens of trees. In: Wilding, N., Collins, N.M., Hammond, P.M. and Webber, J.F. (eds) *Insect–Fungus Interactions*. Academic Press, London, pp. 161–193.
Wilson, A.D. (2005) Recent advances in the control of oak wilt in the United States. *Plant Pathology Journal* 4, 177–191.
Yount, W.L. (1954) Identification of oak wilt isolates as related to kind of inoculum and pattern of disease spread. *Plant Disease Reporter* 38, 293–296.
Zauza, E.A.V., Alfenas, A.C., Harrington, T.C., Mizubuti, E.S. and Silvai, J.F. (2004) Resistance of *Eucalyptus* clones to *Ceratocystis fimbriata*. *Plant Disease* 88, 758–760.
Zuckerman, B.M. (1954) Relation of type and age of wound to infection by *Endoconidiophora fagacearum* Bretz. *Plant Disease Reporter* 38, 290–292.

12 Dutch Elm Disease and Other Ophiostoma Diseases

Thomas Kirisits*
University of Natural Resources and Life Sciences, Vienna (BOKU), Vienna, Austria

12.1 Pathogens, Significance and Distribution

Ophiostoma and related genera of ascomycetes and anamorphic fungi, subsequently referred to as *Ophiostoma s.l.*, include facultative parasites and saprotrophs that mainly live in the phloem and sapwood of angiosperms and gymnosperms. Most species are associated with phloem-feeding and wood-inhabiting beetles (Coleoptera), particularly bark and ambrosia beetles (Scolytinae and Platypodinae) (Kirisits, 2004; Linnakoski, 2011). Other species occur independently of obvious insect attacks, for example on wounded living trees and logs (Uzunovic *et al.*, 1999a; Kamgan Nkuekam, 2007).

Species of *Ophiostoma* cause two important types of diseases or damages, namely vascular wilt diseases and sap-stain or blue-stain, which can on living trees be referred to as vascular stain diseases (Webber and Gibbs, 1989; Kirisits, 2004). In the former category, the causal pathogens spread systemically in the vascular, water-conducting system of a tree, make it dysfunctional and rapidly lead to tree death. Blue-stain refers to dark discoloration of the sapwood of live or recently dead trees and logs, mainly conifers. Here, the causative fungi do not spread systemically but expand by hyphal growth in the xylem.

This chapter focuses on the epidemiology and management of both vascular wilt diseases and blue-stain caused by *Ophiostoma s.l.* The former include Dutch elm disease (DED) and the recently emerging laurel wilt (LW). The discourse on blue-stain will consider guilds of species that are similar in their ecology. Though not affecting live trees, blue-stain on logs during storage will also be considered because it results in substantial economic losses to forestry.

Although morphologically similar to *Ceratocystis*, *Ophiostoma s.l.* groups phylogenetically within the *Ophiostomatales*, distantly from *Ceratocystis*, which is placed in the *Microascales* (Paulin-Mahady *et al.*, 2002). The most obvious morphologically distinguishing features between *Ceratocystis* and *Ophiostoma* are the anamorph characteristics. *Ceratocystis* spp. have *Thielaviopsis* anamorphs (Paulin-Mahady *et al.*, 2002), and most have loose, unspecific and varied associations with insects (see Harrington, Chapter 11, this volume), in contrast to *Ophiostoma* spp., many of which have tight relationships with beetles (Kirisits, 2004; Linnakoski, 2011). Based on multi-gene

* E-mail: thomas.kirisits@boku.ac.at

phylogenies, Zipfel *et al.* (2006) divided *Ophiostoma s.l.* into the genera *Ophiostoma s.s.*, *Grosmannia* and *Ceratocystiopsis*. Most species in these three phylogenetic lineages differ morphologically from each other. *Ophiostoma* species have *Pesotum* and/or *Sporothrix* or *Hyalorhinocladiella* anamorphs, while most *Grosmannia* spp. are characterized by *Leptographium* anamorphic stages, and only a few have *Pesotum* states. *Ceratocystiopsis* spp. are well defined by small perithecia with short necks and falcate ascospores, as well as *Hyalorhinocladiella* or *Sporothrix* anamorphs (Zipfel *et al.*, 2006). Ecologically, *Ophiostoma*, *Grosmannia*, *Ceratocystiopsis* and their related anamorphic fungi are fairly homogenous (Kirisits, 2004; Linnakoski, 2011). Most nutritionally important fungi of ambrosia beetles, known as ambrosia fungi, belong to the anamorphic genera *Ambrosiella* and *Raffaelea*. Phylogenetically, *Raffaelea* forms a lineage within *Ophiostoma*, while *Ambrosiella* spp. are related to *Ceratocystis* (Harrington *et al.*, 2010).

A few species within *Ophiostoma s.l.* cause true vascular wilt diseases. On angiosperms, these include three DED fungi and the LW pathogen. The three varieties of *Leptographium wageneri* (W.B. Kendr.) M.J. Wingf., causal agents of blackstain root disease on conifers (see Eckhardt, Chapter 13, this volume), move slowly and systemically through the tracheids of their hosts and are thus pathologically similar to the vascular wilt pathogens on angiosperm trees that have just been mentioned.

12.1.1 Dutch elm disease

DED is caused by three closely related *Ophiostoma* species, *O. ulmi* (Buisman) Nannf., *O. novo-ulmi* Brasier and *O. himal-ulmi* Brasier & M.D. Mehrotra. The former two are introduced tree pathogens, which became associated with Eurasian and North American bark beetles and have been causing enormous damage on various elm species (*Ulmus* spp.) outside their natural ranges (Brasier, 2000). *O. ulmi* may originate from subtropical or tropical areas in Asia (Brasier, 2000), or from Japan, where it has recently been discovered (Masuya *et al.*, 2010). *O. novo-ulmi* has been suggested to be native in the Himalayas or other temperate mountain ranges in Asia (Brasier, 2000). *O. himal-ulmi* occurs endemically on Himalayan elm (*U. wallichiana* (Planch.)) in the western Himalayas, where it does not cause obvious disease (Brasier, 1996, 2000). The DED fungi are intersterile and differ slightly in their morphology, as well as in many of their molecular and phenotypic traits (Brasier, 1991, 2000; Harrington *et al.*, 2001). While *O. ulmi* is a weakly to moderately virulent pathogen, *O. novo-ulmi* and *O. himal-ulmi* display high levels of virulence to European and North American elms (Brasier, 1991, 2000).

O. ulmi and *O. novo-ulmi* have been responsible for two different pandemics of DED (Brasier, 2000). *O. ulmi* appeared in around 1910 in north-western Europe and subsequently spread within continental Europe and to western Asia (Gibbs, 1978; Brasier, 2000). It was first recorded in the UK in 1927 and in North America in 1930 (Gibbs, 1978). The epidemic associated with *O. novo-ulmi* was first recognized in the early 1970s in the UK (Gibbs, 1978; Brasier, 2000), but had probably already started in the 1940s at two separate locations (Brasier, 2000). Based on a partial reproductive barrier and other characteristics, *O. novo-ulmi* is separated into two forms, now recognized as subspecies (Brasier, 1991; Brasier *et al.*, 2004). *O. novo-ulmi* subsp. *novo-ulmi* Brasier is thought to have been introduced into the Moldova–Ukraine area in eastern Europe, from where it spread across Europe and into eastern and central Asia. At around the same time, *O. novo-ulmi* subsp. *americana* Brasier & S.A. Kirk appeared in the southern Great Lakes area, subsequently spread across North America and was later introduced into western Europe, where it expanded mainly eastwards and southwards through the continent. *O. novo-ulmi* subsp. *americana* was also introduced into New Zealand (Gadgil *et al.*, 2000) and was recently discovered in Japan (Masuya *et al.*, 2010). As *O. novo-ulmi* spread, it continuously replaced the less virulent and less competitive *O. ulmi*, which has become extinct in most areas where it once occurred (Brasier, 2000).

O. ulmi caused considerable losses in elm populations. However, at least in the UK, a remission of disease intensity from the mid

1930s onwards was observed, and by 1960 only an estimated 10–20% of the elm population had died (Peace, 1960; Gibbs, 1978). In Europe and parts of Asia, the second pandemic of DED caused by *O. novo-ulmi* was more severe and the majority of mature elm trees were killed. In the UK alone, it was estimated that 25 out of the estimated 30 million elms had died by the 1990s (Brasier, 1996). In North America, the epidemic caused by *O. novo-ulmi* was less recognized as American elm species displayed high susceptibility even to *O. ulmi* (Gibbs, 1978; Brasier, 2000). Although the impact of DED on the elm populations has been immense, their survival as species is not threatened. Abundant natural regeneration, suckering and sprouting of elm trees is often seen, and many trees form seeds before they become affected by the disease (Brasier, 1996; Röhrig, 1996; Collin *et al.*, 2000; Kirisits and Konrad, 2004).

12.1.2 Laurel wilt

Raffaelea lauricola T.C. Harr., Fraedrich & Aghayeva is a recently discovered vascular wilt pathogen of redbay (*Persea borbonia* (L.) Spreng.) and other species in the *Lauraceae* family in the USA (Fraedrich *et al.*, 2008; Harrington *et al.*, 2010). LW was first recognized in 2003, when it caused extensive mortality of redbay in South Carolina and Georgia. The disease is now widespread in the south-eastern Atlantic Coastal Plain region of the USA. *R. lauricola* came to North America together with its vector, the redbay ambrosia beetle (RAB) *Xyleborus glabratus* Eichhoff, which is native to South-east Asia (Fraedrich *et al.*, 2008; Mayfield *et al.*, 2009). The LW pathogen was recently confirmed to occur in association with its vector in Taiwan and Japan (Harrington *et al.*, 2011).

In areas where the RAB and *R. lauricola* have been established, high mortality rates (above 90%) of redbay were recorded within 2 years, suggesting a dramatic future decline of the species (Fraedrich *et al.*, 2008). An extinction of redbay is unlikely, however, as small seedlings and sprouts are rarely attacked by the RAB. The impact on other *Lauraceae* native to North America is not precisely known as yet, but it may also become high in the future (Fraedrich *et al.*, 2008). Moreover, the disease is a serious potential threat to commercial plantations of avocado (*Persea americana* Mill.) in the USA and other parts of the world (Mayfield *et al.*, 2009).

12.1.3 Blue-stain and vascular stain diseases

Numerous members of *Ophiostoma s.l.* other than the DED and LW fungi incite blue-stain or vascular stain on live trees or cause losses during storage of logs and timber (Harrington, 1993; Seifert, 1993; Kirisits, 2004). Compared with the DED and LW fungi, these species display relatively low levels of virulence to their hosts and some are even saprotrophs, which spread in tissues of dying and recently dead trees (Gibbs, 1993; Harrington, 1993; Kirisits, 2004). Although the phytopathogenicity of various blue-stain fungi has been proven (Harrington, 1993; Krokene, 1996; Kirisits, 2004; Six and Wingfield, 2011), tree mortality always occurs in association with mass attacks of their insect vectors.

Examples of insect-associated species within *Ophiostoma s.s.* that display low levels of virulence to their hosts include *O. bicolor* R.W. Davidson & D.E. Wells, *O. brunneociliatum* Math.-Käärik, *O. ips* (Rumbold) Nannf., *O. minus* (Hedgc.) Syd. & P. Syd., *O. montium* (Rumbold) Arx, *O. piceae* (Münch) Syd. & P. Syd. and *O. pseudotsugae* (Rumbold) Arx. Many species, particularly taxa with *Pesotum* and/or *Sporothrix* or *Hyalorhinocladiella* anamorphs, occur on the wounds of trees and on logs, independent of obvious attacks by insects. Examples are *O. piceae*, *O. floccosum* Math.-Käärik, *O. quercus* (Georgev.) Nannf. and other members of the *O. piceae* complex, and *O. minus*, *O. multiannulatum* (Hedgc. & R.W. Davidson) Hendr., *O. piliferum* (Fr.) Syd. & P. Syd. and *O. pluriannulatum* (Hedgc.) Syd. & P. Syd. Numerous *Grosmannia* and *Leptographium* spp. occur in the niches mentioned above for *Ophiostoma s.s.* (Krokene, 1996; Jacobs and Wingfield, 2001; Kirisits, 2004; Linnakoski, 2011).

Besides *R. lauricola*, *Raffaelea* includes at least two other pathogens, *R. quercivora* Kubono & Shin. Ito and *R. quercus-mongolicae* K.H. Kim, Y.J. Choi & H.D. Shin, which infect oaks (*Quercus* spp.) in Japan and Korea, respectively (Harrington *et al.*, 2010). Other *Raffaelea* spp. can cause wood discoloration around the breeding galleries of the ambrosia beetles they are associated with (Harrington *et al.*, 2010). Likewise, bark beetle-associated *Hyalorhinocladiella* spp., previously assigned to *Ambrosiella*, also cause sap-stain (Mathiesen-Käärik, 1953; Harrington *et al.*, 2010).

The habitats of species of *Ophiostoma s.l.* and *Ceratocystis* can overlap. Examples include the mycobiota of some aggressive conifer bark beetles, conifer logs and wounds on gymnosperms and angiosperms (Roll-Hansen and Roll-Hansen, 1980a,b; Harrington and Wingfield, 1998; Uzunovic *et al.*, 1999a; Kirisits, 2004; Kamgan Nkuekam, 2007). When members of *Ophiostoma s.l.* and *Ceratocystis* occur in the same niches, the *Ceratocystis* spp. usually display higher levels of virulence and are more important as tree pathogens (Krokene, 1996; Kirisits, 2004; Kamgan Nkuekam, 2007; see Harrington, Chapter 11, this volume).

12.2 Diagnosis

12.2.1 Dutch elm disease

The DED symptoms caused by *O. ulmi* and *O. novo-ulmi* were similar, but frequent recovery of trees from the disease was reported only after natural infection with *O. ulmi* (Peace, 1960). The symptom description given below follows Elgersma (1981), Butin (1995), Tainter and Baker (1996), Nierhaus-Wunderwald and Engesser (2003) and Sinclar and Lyon (2005).

When infection occurs in spring or early summer, foliage symptoms develop within a few weeks, but infections in late summer, when elms are generally less susceptible, are often recognized in the following spring only. This is because the pathogen spreads more slowly in the summer wood, and fall leaf discoloration can obscure foliage symptoms. On dry sites and in years with hot and dry weather, drought and heat stress can cause leaf yellowing, early leaf shedding and dieback of branches that resemble the external symptoms of DED. As pathogen spread is slower and disease development more chronic on drought-stressed trees, these conditions can also mask the external symptoms of DED.

Following pathogen transmission by elm bark beetles, external symptoms are initially restricted to one or a few individual branches, often in the middle to upper crown, which is referred to as branch flagging. Symptoms gradually evolve on other branches, linked to the systemic spread of the pathogen, until the entire crown and finally the main stem are affected. In the case of root transmission from infected neighbouring trees (Fig. 12.1), severe external symptoms develop rapidly in the entire crown. Leaves wilt, turn from light green over yellow to brown, shrivel, curl and become dry. Most of the leaves drop but some remain attached to the twigs for a long time. At later stages of infection, dead tips of withered elm shoots and twigs, still bearing leaves on their tops, bend downwards in a hook-like manner. Observation of sites of maturation feeding of *Scolytus* spp. in twig crotches (Fig. 12.1), or by *Hylurgopinus rufipes* (Eichhoff) on branches and small stems, provides evidence of the activity of the vectors of the DED pathogens.

When the outer bark of recently infected twigs and branches is pulled off, a brown-to-black streaking is seen on the wood surface, which follows the grain of the newly formed layer of spring wood. This streaking on the wood surface is best observed in spring and early summer. Later in the year, after the formation of the summer wood, the streaking on the wood surface is often no longer visible, although a brown-to-black discoloration of the vessels of one or several growth rings occurs in cross-sections of twigs and branches. The discoloration affects the spring wood vessels or the entire annual growth ring. Following annual wood growth, discoloration appears either as a continuous narrow ring or as small, discrete, intermittent spots or bands. Trees can die as soon as in the year of infection, or after several years. In trees that are in decline for several years before death,

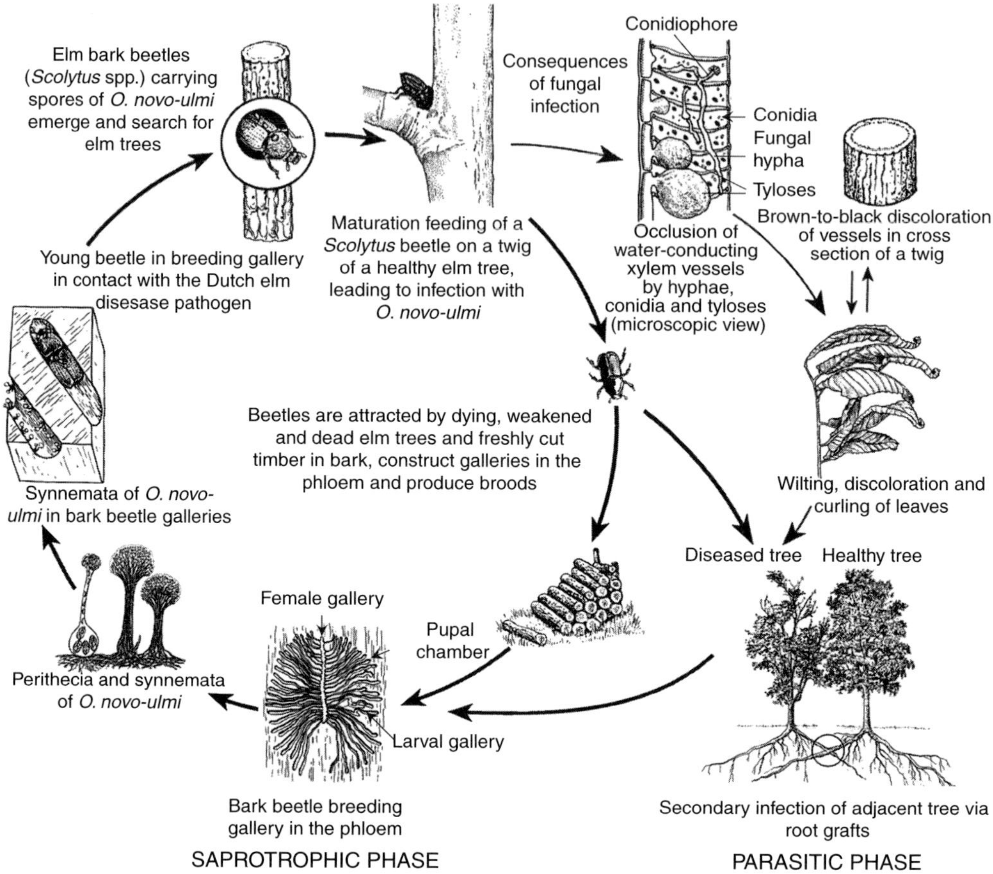

Fig. 12.1. Disease cycle of Dutch elm disease caused by *Ophiostoma novo-ulmi* Brasier vectored by elm bark beetles (*Scolytus* spp.) on susceptible elm species (*Ulmus* spp.). The North American elm bark beetle, *Hylurgopinus rufipes* (Eichhoff), not shown in the figure, visits larger branches and small stems for maturation feeding. Figure adapted from Nierhaus-Wunderwald and Engesser (2003), with permission of the Swiss Federal Institute for Forest, Snow and Landscape Research (WSL), Birmensdorf, Switzerland.

sparse sprouting and foliation as well as early leaf shedding are additional symptoms.

Dying and recently dead elms are colonized by elm bark beetles for breeding. Entrance holes and boring frass expelled as fine brown powder by adult beetles indicate early infestations. When the outer bark gradually becomes full of holes, juvenile beetles are leaving the trees. Upon exposure of the inner phloem, bark beetle breeding galleries, often specific for individual species, can be seen (Fig. 12.1; Postner, 1974). Signs of woodpecker attacks – the birds remove the outer bark to expose bark beetle galleries – help to locate infested trees, especially in the wintertime. Synnemata of the *Pesotum* stage and the less conspicuous conidiophores of the *Sporothrix* stage, as well as perithecia of *O. novo-ulmi*, can often be observed in bark beetle breeding systems using a strong lens or dissecting microscope.

Other diseases, especially Verticillium wilt caused by *Verticillium dahliae* Kleb. and *V. albo-atrum* Reinke & Berthold, can cause symptoms resembling those of DED. For unambiguous identification, the DED fungi are isolated by aseptically removing chips, slivers or discs of discoloured wood and placing them on standard laboratory media such as malt extract agar (Brasier, 1981). *Ophiostoma s.l.* unlike many other fungi can tolerate high concentrations of cycloheximide, and adding this antibiotic in low concentrations provides a selective medium for isolation (Brasier, 1981; Harrington, 1981). Alternatively, infected, streaked wood or phloem samples containing bark beetle galleries can be incubated in a moist chamber. *Pesotum* synnemata and *Sporothrix* conidiophores will develop within 1–2 weeks. *O. novo-ulmi* is heterothallic but, when both mating types are present, perithecia may also develop.

DED can be confused with elm yellows (see Griffiths, Chapter 4, this volume), but in the latter disease yellowing of the foliage usually occurs on all branches at once, while branch flagging is more typical in DED (Tainter and Baker, 1996). The phloem of trees affected by elm yellows turns brown, but wood discoloration and streaking does not occur; in elm yellows, the discoloured phloem also displays an odour of wintergreen or one resembling that of maple syrup (Tainter and Baker, 1996; Sinclair and Lyon, 2005).

12.2.2 Laurel wilt

Symptoms of LW differ slightly on various hosts, but those on redbay are described here (Fraedrich *et al.*, 2008; Mayfield *et al.*, 2009). In the early stages of the disease, external symptoms may occur on individual branches or appear on branch tips throughout the crown. Leaves wilt, show a reddish-brown to purplish-brown discoloration and frequently remain on the trees long after they have died. During spring and summer, the entire crown displays symptoms and trees die within a few weeks to several months. A dark black-to-brown discoloration develops in the sapwood of twigs, branches and stems, which follows the grain of the wood. Initially seen as localized streaks in the outermost sapwood, the discoloration extends more extensively through the entire sapwood area in the later stages of the disease. Small, inconspicuous entrance holes of the RAB occur in low numbers on branches and stems at early stages of disease. Dying and dead trees are intensively attacked by the RAB and native ambrosia beetles, and this is recognizable by the white boring dust, which emerges as loose piles or is compacted to dowel-like or toothpick-like structures emerging from the beetles' entrance holes (Mayfield *et al.*, 2009).

R. lauricola can be readily isolated from discoloured sapwood or from individuals of the RAB (Fraedrich *et al.*, 2008). Like *Ophiostoma s.l.*, *Raffaelea* spp. also tolerate high concentrations of cycloheximide, which provides the basis for a selective medium. The RAB carries five other probably non-pathogenic *Raffaelea* spp., which can be distinguished from *R. lauricola* based on comparisons of colony morphology, micromorphology and DNA sequences (Harrington and Fraedrich, 2010; Harrington *et al.*, 2010).

12.2.3 Blue-stain and vascular stain diseases

Live trees that are mass attacked by phloem-feeding and wood-boring beetles show foliage discoloration and fading as well as signs of insect attacks (brown or white sawdust, holes in the bark, breeding galleries in the phloem and/or wood). Staining of the phloem and wood surface develops around insect galleries. Discoloration is restricted to the sapwood and ranges in colour from grey to grey blue to blue to brown to black. On stem cross-sections, a deeply penetrating stain is observed, occurring as wedge-shaped, pie-shaped or linear zones (Jacobs and Wingfield, 2001; Sinclair and Lyon, 2005; Uzunovic *et al.*, 2008). Later, these areas frequently coalesce, resulting in discoloration of the entire sapwood within a few weeks (Sinclair and Lyon, 2005). Sap-stain associated with ambrosia beetles can look different, as fungi are inoculated into the deeper wood layers and discoloration may be restricted to the areas around the insect galleries (Mathiesen-Käärik, 1953; Uzunovic *et al.*, 2008).

Patterns of deep blue-stain are similar on logs (with or without insect attacks) and around wounds on live trees. Drier logs and sawn timber are affected by surface blue-stain. This type of discoloration is restricted to the wood surface, where dark spots and streaks, accompanied by intense sporulation, occur (Uzunovic *et al.*, 2008). In contrast to deep blue-stain, discoloured layers can be planed away (Uzunovic *et al.*, 2008). Blue-stain is caused by darkly pigmented, melanized fungal hyphae that are concentrated in ray parenchyma cells whose content is used by the fungi for nutrition (Butin, 1995; Uzunovic *et al.*, 2008). Tracheids are more rarely colonized by hyphae but serve as paths to reach new parenchyma cells. Blue-stain fungi do not cause wood decay, and their effects on structural wood properties are negligible (Butin, 1995; Uzunovic *et al.*, 2008).

Other fungi can cause wood discolorations similar to blue-stain (Butin, 1995; Sinclair and Lyon, 2005; Uzunovic *et al.*, 2008) and definite diagnosis therefore requires detection or isolation of the causal agents. Diagnosis can be confirmed by direct observation of long-necked perithecia, synnemata and conidiophores of *Ophiostoma s.l.* in bark beetle galleries and on blue-stained wood. Typically, several fungi occur in infected phloem and sapwood, belonging to various lineages within *Ophiostoma s.l.* and sometimes also to *Ceratocystis* (Solheim, 1993; Harrington and Wingfield, 1998; Kirisits, 2004).

Media with cycloheximide are selective for species of *Ophiostoma s.l.*, but *Ceratocystis* spp. are sensitive to this antibiotic and will not grow (Harrington, 1981; Jacobs and Wingfield, 2001; Zipfel *et al.*, 2006). Isolation can be done directly from insects, infected phloem and sapwood, and also from ascospores or asexual spores (conidia) occurring in insect galleries or on the wood surface. Isolations at the early stages of blue-stain development and from the progressing edge of stain into the sapwood will yield the most virulent species; these are able to colonize moist sapwood and are primarily responsible for the discoloration (Solheim, 1993). For species identification, DNA barcoding based on the internal transcribed spacer (ITS) region of the ribosomal DNA (rDNA) has become a standard method. *Ophiostoma s.l.* contains many cryptic, morphologically similar and closely related species, and contemporary descriptions of new taxa often rely on multi-gene phylogenies (Linnakoski, 2011).

12.3 Infection Biology and Epidemiology

12.3.1 Inoculum

Ascospores and conidia are abundantly produced by members of *Ophiostoma s.l.* (Malloch and Blackwell, 1993). For some lineages within or related to *Ophiostoma*, for example *Pesotum*, *Leptographium*, *Raffaelea* and *Hyalorhinocladiella*, only the anamorphic state is known. Some species of *Ophiostoma s.l.* produce conidia on more than one anamorphic stage. *O. ulmi*, *O. novo-ulmi* and members of the *O. piceae* species complex form conidia both on synnemata (*Pesotum* states) and on conidiophores (*Sporothrix* states) (Brasier, 1981; Harrington *et al.*, 2001). Many species also have yeast stages and produce blastospores by budding (Brasier, 1981; Fraedrich *et al.*, 2008). Numerous *Ophiostoma s.l.* species are homothallic, while other taxa, including *O. ulmi*, *O. novo-ulmi* and species of the *O. piceae* complex, are heterothallic (Brasier, 1981; Harrington *et al.*, 2001). In these species, perithecia are only formed when both mating types are present.

Members of *Ophiostoma s.l.* are adapted to dispersal by insects and other arthropods. Ascospores and conidia accumulate as mucilaginous, sticky masses on the apices of long-necked perithecia, synnemata and conidiophores, and adhere easily to the bodies of arthropods passing by (Malloch and Blackwell, 1993). *Sporothrix* states have more dry conidia, which may, at least to some extent, be dispersed by abiotic means (Gibbs, 1993; Kamgan Nkuekam, 2007).

Species of *Ophiostoma s.l.* do not form any long-lived resting spores. Overwintering and survival occur in or on vectors and hosts. Saprotrophic survival is relatively short, especially after dispersing insects have left the joint habitats. In the late stages of host colonization, other ecological groups dominate and the frequencies of *Ophiostoma* spp. decline over time (Solheim, 1993).

12.3.2 Dispersal

A few groups of beetles are involved in dispersing species of *Ophiostoma s.l.* to new habitats. Within these groups, their relationships with *Ophiostoma* spp. vary greatly and range from casual to intimate and from specific to unspecific (Jacobs and Wingfield, 2001; Kirisits, 2004; Viiri, 2004). Typically, one insect carries several fungi (Kirisits, 2004; Linnakoski, 2011). Fungal dispersal occurs externally (exozoically) by spores adhering to the surface of insect bodies, internally (endozoically) by spores passing the gut remaining viable, or in mycangia (Francke-Grosmann, 1967; Kirisits, 2004; Webber, 2004). The latter are specialized structures in the integument of the insects that serve for the preservation, transport and transmission of fungal associates (Francke-Grosmann, 1967; Kirisits, 2004).

The best known insect associates of *Ophiostoma s.l.* are true or phloem-feeding bark beetles (Curculionidae, Scolytinae). Many aggressive, tree-killing scolytines on conifers, including especially species in the genera *Dendroctonus* and *Ips*, but also secondary and even predominantly saprophagous species, consistently carry *Ophiostoma* spp. (Mathiesen-Käärik, 1953; Krokene, 1996; Kirisits, 2004; Linnakoski, 2011; Six and Wingfield, 2011). True bark beetles mainly transmit fungi exozoically or endozoically, but some species possess mycangia (Francke-Grosmann, 1967; Solheim, 1993; Kirisits, 2004).

All vectors of *O. ulmi* and *O. novo-ulmi* are true bark beetles, which disseminate pathogen propagules exozoically and endozoically (Webber, 2004). In North America, the native *H. rufipes*, *Scolytus multistriatus* (Marsham), introduced from Europe to North America, and probably also *S. schevyrewi* Semenov, introduced from Asia to the USA, act as vectors (Tainter and Baker, 1996; Jacobi *et al.*, 2007). In Eurasia, *S. scolytus* (Fabricius), *S. multistriatus*, *S. kirshi* Skalitzky, *S. laevis* Chapuis, *S. pygmaeus* (Fabricius) and other *Scolytus* spp. transmit the DED fungi (Postner, 1974; Webber, 2004). Other potential vectors are *Pteleobius kraatzii* (Eichhoff) and *P. vittatus* (Fabricius), but these are less important than *Scolytus* spp. (Postner, 1974; Webber, 2004). In Japan, *O. ulmi* and *O. novo-ulmi* have been found in association with *S. esuriens* Blandford (Masuya *et al.*, 2010). In the Himalayas, *O. himal-ulmi* is associated with *S. chelogaster* (Schedl), *S. kashmirensis* (Schedl) and *S. himal-ulmi* Schedl (Webber, 2004). All elm bark beetles are secondary insects, which breed in stressed and recently dead trees (Postner, 1974).

Xylomycetophagous bark beetles and platypodine beetles (Curculionidae, Scolytinae and Platypodinae) form the ecological group of 'ambrosia beetles'. They colonize the sapwood (platypodines also colonize the heartwood), where larvae feed on mutualistic, nutritionally important ambrosia fungi, including *Raffaelea* spp. (Mathiesen-Käärik, 1953; Francke-Grosmann, 1967; Postner, 1974; Schwerdtfeger, 1981; Kirisits, 2004; Harrington *et al.*, 2010; Edmonds *et al.*, 2011). Xylomycetophagous bark beetles are also associated with species of *Ophiostoma s.l.* (Mathiesen-Käärik, 1953; Kirisits, 2004; Linnakoski, 2011). Ambrosia fungi are primarily dispersed in mycangia of the beetles (Francke-Grosmann, 1967; Kirisits, 2004; Harrington *et al.*, 2010). For example, females of the RAB carry blastospores of *R. lauricola* in paired mandibular mycangia (Fig. 12.2; Fraedrich *et al.*, 2008).

Ambrosia beetles mainly attack recently dead trees, fresh logs and severely stressed trees (Postner 1974; Schwerdtfeger, 1981; Edmonds *et al.*, 2011), but in recent years infestations of live trees, especially angiosperms, have been increasing (Fraedrich *et al.* 2008; Hulcr and Dunn, 2011). Besides *R. lauricola* and the RAB, *R. quercivora*, associated with *Platypus quercivorus* (Murayama) in Japan, and *R. quercus-mongolicae*, carried by *P. koryoensis* (Murayama) in Korea, are the best known examples of such emerging insect–fungal pathogen complexes (Harrington *et al.*, 2010; Hulcr and Dunn, 2011).

Bark weevils (Curculionidae) and a few longhorn beetles (Cerambycidae) on conifers are also known to carry species of *Ophiostoma s.l.* (Mathiesen-Käärik, 1953; Jacobs and Wingfield, 2001; Jacobs *et al.*, 2003; Viiri, 2004; Jankowiak and Kolařík, 2010). No mycangia are known in these groups of insects, and fungal dissemination is thought to occur exozoically and endozoically (Jacobs and Wingfield, 2001; Viiri, 2004; Jankowiak and Kolařík, 2010).

Dispersal by abiotic agents has been suggested where *Ophiostoma* spp. occur independently of insect infestations (Dowding,

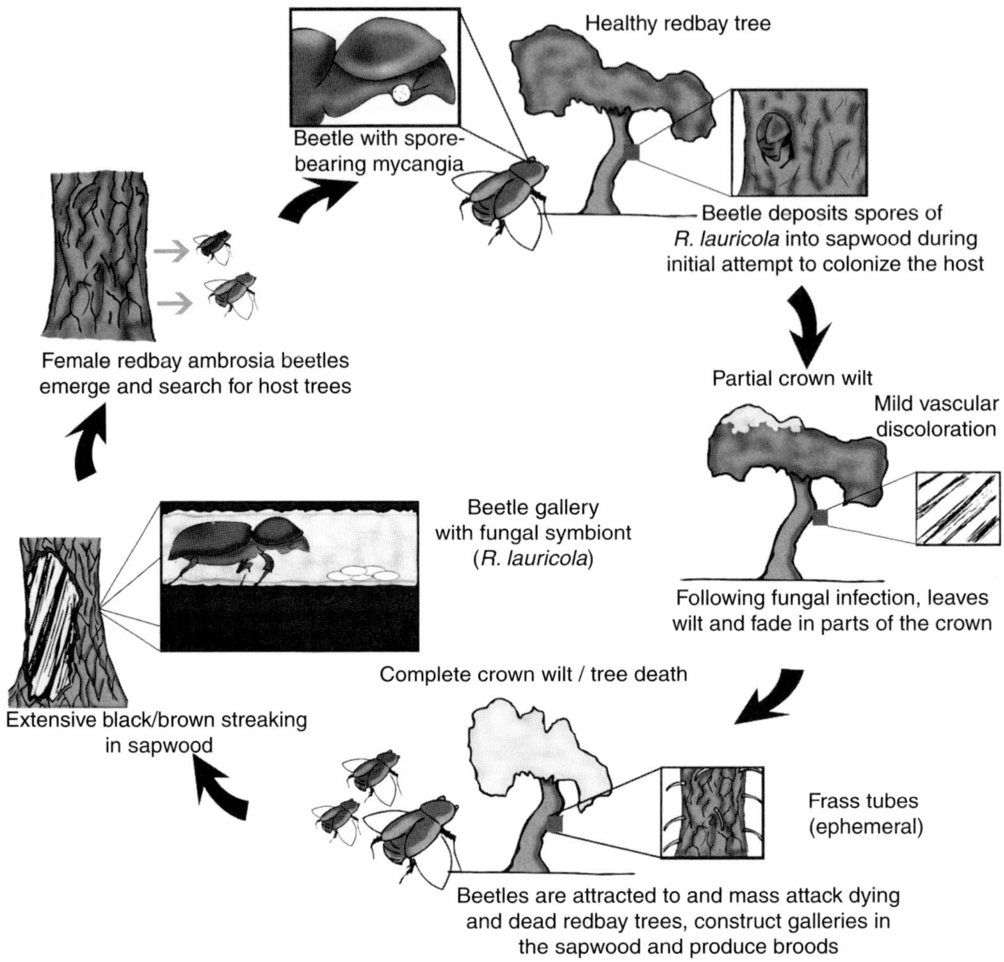

Fig. 12.2. Suggested disease cycle of laurel wilt in redbay (*Persea borbonia* (L.) Spreng.), caused by *Raffaelea lauricola* T.C. Harr., Fraedrich & Aghayeva, vectored by the redbay ambrosia beetle (*Xyleborus glabratus* Eichhoff). Reprinted with kind permission of Marc Hughes (University of Florida, Department of Plant Pathology, Institute of Food and Agricultural Sciences), Jeffrey Thomas and Albert E. Mayfield (Florida Department of Agriculture and Customer Services, Division of Forestry). University of Florida Dissertation by Marc Hughes (currently unpublished).

1969; Gibbs, 1993; Seifert, 1993; Uzunovic *et al.*, 2008). Spore dispersal of *Ophiostoma* spp. via dry air is insignificant, while the transport of spores, mainly conidia, in misty air, by water splash or water films, at least over short distances, is more likely to occur (Dowding, 1969; Gibbs, 1993; Uzunovic *et al.*, 2008). Despite this, casual spore dissemination by arthropods such as mites may be more important than dispersal by abiotic agents (Roll-Hansen and Roll-Hansen, 1980b; Gibbs, 1993; Powell *et al.*, 1995). *Ophiostoma* spp. are also disseminated by contaminated harvesting equipment (Uzunovic *et al.*, 2004, 2008).

12.3.3 Infection and spread in host tissues

Wounding is always required to enable *Ophiostoma* spp. to infect trees (Webber and Gibbs, 1989; Tainter and Baker, 1996). In species carried by beetles, wounding and inoculation are simultaneously done by the insects. For species dispersed by other means, harvesting, logging and bark peeling wounds as well as log ends are entrance courts for infections (Vasiliauskas and Stenlid, 1998; Kamgan Nkuekam, 2007; Uzunovic *et al.*, 2008).

Transmission of the DED fungi to healthy elm trees occurs during the maturation feeding of elm bark beetles. *Scolytus* spp. do their maturation feeding preferentially in the crotches of thin twigs or at the base of leaves on newly formed shoots in the tree crown (Fig. 12.1; Postner, 1974; Webber and Gibbs, 1989; Butin, 1995; Webber, 2004). *H. rufipes* feeds on the phloem of larger branches and small stems (Tainter and Baker, 1996; Sinclair and Lyon, 2005). The xylem is exposed by the beetles' feeding, and the trees are inoculated with the DED fungi. Infection can also occur when breeding attacks are initiated on stressed but previously uninfected trees (Tainter and Baker, 1996).

Elm bark beetle species vary in their efficiency to act as pathogen vectors, depending both on the percentage of insects carrying any inoculum and the quantities of spores attached to individual beetles. *S. scolytus* is considered to be the most efficient vector in Europe (Webber and Gibbs, 1989; Webber, 2004). However, vector efficiency varies in different environments, and in northern Italy, for example, *S. multistriatus* and *S. pygmaeus* are important vectors (Faccoli and Battisti, 1997). In North America, the introduced *S. multistriatus* and probably also *S. schevyrewi* are considered to be more important vectors than the native *H. rufipes* (Tainter and Baker, 1996; Jacobi *et al.*, 2007). Moser *et al.* (2010) suggested that two phoretic mites, *Proctolaelaps scolyti* Evans and *Tarsonemus crassus* (Schaarschmidt), which occur particularly on *S. scolytus*, may increase the chances of successful infections, as they carry additional inoculum of *O. novo-ulmi*.

Elm trees are most susceptible to infections during a period of a few weeks in spring, when the leaves are formed, maximum growth occurs and large spring wood vessels are formed (Tainter and Baker, 1996; Ghelardini and Santini, 2009). Following infection, *O. novo-ulmi* forms its yeast stage, and produces hyphae and conidia in the vessels; it spreads systemically in the vascular system, both passively via the sap flow and by hyphal growth, and passes through pits between vessels and penetrates cell walls (Tainter and Baker, 1996; Sinclair and Lyon, 2005). When the pathogen reaches the root system, spores can be carried upstream with the sap flow, disseminating the pathogen further within the tree (Sinclair and Lyon, 2005). Vessel occlusion and xylem dysfunction are caused by the physical presence of *O. novo-ulmi* hyphae and spores, by pectolytic enzymes degrading the cell walls, by toxins such as cerato-ulmin produced by the pathogen, and by tyloses, gels, gums and phenols formed by the host in response to infection (Fig. 12.1; Elgersma, 1981; Tainter and Baker, 1996; Sinclair and Lyon, 2005). Elm bark beetles are attracted to dying and recently dead trees, breed in the phloem, acquire inoculum upon pupation, emerge and subsequently infect new, healthy elm trees. Secondary infections occur via root grafts between adjacent trees (Tainter and Baker, 1996; Sinclair and Lyon, 2005).

Single attacks of RAB females on stems and branches of healthy redbay trees and other host species are an essential element of the disease cycle of LW (Fig. 12.2; Fraedrich *et al.*, 2008; Mayfield *et al.*, 2009). These single infestations most likely do not lead to breeding by the beetles, but they result in infection of host trees with the LW pathogen. From the infection sites, *R. lauricola* spreads systemically in the vascular system, and declining trees are in turn mass infested by the RAB. Secondary pathogen spread via roots has not yet been definitely documented in LW, but is likely (Mayfield *et al.*, 2009).

Lethal infections of other *Ophiostoma* pathogens always occur in association with mass infestations of bark-colonizing and wood-colonizing beetles, which are usually mediated by aggregation pheromones (Altenkirch *et al.*, 2002; Edmonds *et al.*, 2011). The mass attacks of the insects overcome the tree's resistance mechanisms, and their pathogenic fungal associates accelerate tree death, but whether the fungi play a decisive role in the establishment of the beetles is debated (Krokene 1996; Kirisits, 2004; Lieutier *et al.*, 2009; Six and

pruning of declining and dead branches from disease-free trees, and the preventive felling of trees with low vigour and of declining and damaged trees, are usually included in sanitation campaigns (Magasi *et al.*, 1993; Tainter and Baker, 1996). Disinfecting the felling and pruning equipment used for preventive, sanitation and curative treatments with 70% alcohol or sodium hypochlorite is recommended (Tainter and Baker, 1996; Nierhaus-Wunderwald and Engesser, 2003).

If elm material is already colonized by bark beetles, sanitation measures have to be completed before the juvenile insects emerge (Postner, 1974; Magasi *et al.*, 1993; Tainter and Baker, 1996). All breeding material that is already infested or is potential breeding material needs to be disposed of or treated (Postner, 1974; Magasi *et al.*, 1993; Tainter and Baker, 1996). Debarking makes stems and branches unsuitable for breeding. If they are already infested, and where the 'white stages' (larvae and pupae) of bark beetles are present, these are killed by debarking, but if breeding galleries already contain young adults, the peeled off bark needs to be burned, buried or chopped. The latter treatments are generally suitable for the disposal of infested and uninfested elm material. Where sanitation generates large quantities of material, its utilization for various purposes is desirable (Tainter and Baker, 1996).

In North America, chlorpyrifos sprayings on the lower boles of healthy elm trees have been applied to kill *H. rufipes* adults overwintering in the outer layers of thick bark (Tainter and Baker, 1996). Besides environmental concerns, this measure is expensive because large numbers of trees need to be treated for it to be effective. Traps equipped with aggregation pheromones are used to monitor the presence, relative abundance and flight periods of bark beetles, thus providing information to evaluate the effectiveness of sanitation efforts (Magasi *et al.*, 1993; Tainter and Baker, 1996; Gadgil *et al.*, 2000). Pheromone trapping is, however, not an effective control measure to reduce levels of DED (Magasi *et al.*, 1993; Tainter and Baker, 1996).

When LW was discovered, it was already too widespread to consider eradication of the RAB and *R. lauricola* (Mayfield *et al.*, 2009). Likewise, sanitation by the removal of symptomatic trees has not been an effective measure for reducing disease levels. Cleaning and sterilizing felling and pruning tools after they have been used on an infected tree and before using them again on an uninfected specimen is recommended (Mayfield *et al.*, 2009).

The focus of sanitation against insects carrying blue-stain fungi differs from that described for the vectors of the DED fungi, where disease prevention and reduction can be achieved. For the vectors of blue-stain fungi, the insects themselves are the primary target of the sanitation measures, usually with the most important aim of preventing a build-up of populations that would result in intensive outbreaks on living trees (Postner, 1974; Schwerdtfeger, 1981; Speight and Wainhouse, 1989; Altenkirch *et al.*, 2002; Wainhouse, 2005; Edmonds *et al.*, 2011). While measures can generally be directed towards a large number of insects, they are most important for the control of aggressive bark beetle species such as *Ips typographus* (Linnaeus) in Eurasia and *Dendroctonus* spp. in North America. These insects usually occur on dominant forest tree and major timber species, mostly conifers. The platypodid *Platypus quercivorus* that occurs on oaks in Japan is a notable exception. Sanitation is almost always combined with salvaging to recover as much timber value as possible (Nyland, 2002).

The sanitation practices for insects carrying blue-stain fungi are similar to those outlined above for the vectors of the DED fungi. If the removal of breeding material is not possible (e.g. in areas inaccessible to harvesting and logging machines) or not desirable (e.g. in national parks), trees can be entirely or partly debarked to make them unsuitable for breeding. The use of synthetic aggregation pheromones and various trapping techniques may in some situations be effective for lowering damage levels by bark and ambrosia beetles, but they are usually recommended only for monitoring the flight periods and relative abundance of the insects, and as measures complementing the removal of breeding material (Speight and Wainhouse, 1989; Altenkirch *et al.*, 2002; Wainhouse, 2005). The spraying of infested boles and logs with pyrethroids can, in exceptional cases, be used to kill juvenile bark beetles upon emergence from their breeding sites (Altenkirch *et al.*, 2002).

12.4.4 Protection

Fungicides

Injections or infusions of systemic fungicides into the vascular system of elms can prevent infections by *O. novo-ulmi*, but treatments are expensive and cause mechanical injury to trunks, and some fungicides are phytotoxic and damage trees (Tainter and Baker, 1996; Stennes, 2000; Stipes, 2000). Treatments therefore focus on large, particularly valuable shade and landscape trees. The prophylactic use of fungicides is much more promising than their therapeutic use (Tainter and Baker, 1996; Stennes, 2000; Stipes, 2000).

Thiabendazole and propiconazole have proved to be particularly effective in protecting elms from DED (Stennes, 2000; Stipes, 2000; Scheffer *et al.*, 2008). Thiabendazole provides protection for 2–3 years, whereas the persistence of propiconazole is probably shorter (Stennes, 2000; Stipes, 2000), but while thiabendazole damages cambium and wood around injection sites, the phytotoxicity of propiconazole is low (Stennes, 2000; Stipes, 2000). For a uniform distribution of fungicides in the vascular system, injections or infusions are done at root flares around the entire tree circumference, with treatment points 10–20 cm apart (Tainter and Baker, 1996; Stennes, 2000; Stipes, 2000). The infusion of redbay trees with propiconazole has provided protection against artificially inoculated *R. lauricola* (Mayfield *et al.*, 2009).

For environmental reasons, the application of fungicides in forests has declined in most countries and few, if any, products are registered for specific purposes (Altenkirch *et al.*, 2002; Uzunovic *et al.*, 2008). Because of this, the protection of fresh logs with anti-sap-stain products on forest landings is rarely done. The use of fungicides is restricted to special situations, which usually lie outside the key responsibilities of forest managers. For example, in New Zealand, logs of Monterey pine (*P. radiata* D. Don) assigned for export are chemically treated as soon as possible after felling (Uzunovic *et al.*, 2004, 2008).

Insecticides

As with fungicides, the general use of insecticides has been decreasing (Altenkirch *et al.*, 2002). Various insecticides (DDT, malathion, methoxychlor and chlorpyrifos) have been used in DED city management programmes in North America, either as entire tree sprayings to prevent beetles from maturation feeding or for lower trunk treatments to prevent adult *H. rufipes* from overwintering (Magasi *et al.*, 1993; Tainter and Baker, 1996). Insecticide application was effective for additionally reducing the number of new DED infections, but is now of minor importance (Magasi *et al.*, 1993; Tainter and Baker, 1996; Stipes, 2000). Solla *et al.* (2000) reported on the use of methoxychlor in clonal banks in Spain to protect elm genotypes from feeding by *Scolytus* spp. In North America, systemic insecticides injected into trees were tested, but results were unsatisfactory (Tainter and Baker, 1996).

Other than for elm bark beetles, the application of insecticides to protect living trees from attacks by insects carrying species of *Ophiostoma s.l.* is of no practical importance. Insecticides can, however, be useful for protecting logs from becoming infested by ambrosia and bark beetles (Wauer, 2000; Altenkirch *et al.*, 2002). In central Europe, synthetic pyrethroids such as alpha-cypermethrin and lambda-cyhalothrin, which operate as contact insecticides, are now most frequently used against bark and ambrosia beetles (Wauer, 2000; Altenkirch *et al.*, 2002).

A new method for protecting logs from insect attacks is based on polyester nets, in the fibres of which low concentrations of alpha-cypermethrin are embedded (Wimmer, 2012). Log piles are covered with such nets and landing insects are killed upon contact with the net during a period of up to 6 months. The nets can also be used for sanitation. Logs that are already infested are covered, and juvenile beetles are killed upon emergence (Wimmer, 2012). Because of undesirable non-target effects, insecticides should be seen as a last resort for the management of bark and ambrosia beetles, and be used just in specific cases and if other measures are not feasible (Wauer, 2000; Altenkirch *et al.*, 2002).

Biological protection

A biological product, Dutch Trig®, is used in practice to protect valuable elm trees from DED (Scheffer *et al.*, 2008). Dutch Trig® is based on isolate WCS850, representing a natural hyaline form of *Verticillium albo-atrum* originating from potato in the Netherlands (Scheffer *et al.*, 2008). When various elm species and clones were first challenged with isolate WCS850, and subsequently inoculated with *O. novo-ulmi*, disease severity was substantially and consistently lower than in trees that were pretreated only with water (Scheffer *et al.*, 2008). *V. albo-atrum* isolate WCS850 does not spread extensively in inoculated trees, its presence is transient and it does not display phytopathogenicity (Scheffer *et al.*, 2008). Although not fully understood as yet, a physiological response of the tree leading to induced resistance (Hubbes, 2004) is the likely mechanism involved in the protective effect against subsequent inoculation with *O. novo-ulmi* (Scheffer *et al.*, 2008).

In field trials in Dutch cities and in Denver, Colorado (USA), injections with Dutch Trig® reduced annual infections by *O. novo-ulmi* to below 1% among treated trees, while infection rates were significantly higher on untreated elm specimens (Scheffer *et al.*, 2008). Dutch Trig® is registered as a commercially available biological control product in the Netherlands, Germany, the USA and Canada. As part of integrated DED control programmes, the biological agent is injected once a year into the xylem of elm trees (Scheffer *et al.*, 2008).

In Canada, albino strains of *O. piliferum* and *Ceratocystis resinifera* T.C. Harr. & M.J. Wingf. were effective in protecting freshly felled conifer logs from blue-stain (Morin *et al.*, 2006; Uzunovic *et al.*, 2008). The mode of action of the albino strains is competition for resources and 'niche occupation'. The biological control agents, applied at high concentrations to the cut ends and total log surface area as soon as possible after felling, quickly colonize the sapwood, utilize the nutrients in the ray parenchyma cells and thereby prevent colonization by naturally occurring blue-stain fungi (Morin *et al.*, 2006; Uzunovic *et al.*, 2008). Wood brightness and quality are not negatively affected and wood permeability is increased, which is advantageous for some uses. In Canada, the albino strain of *O. piliferum* received temporary registration as the anti-sap-stain biological product Sylvanex for use on freshly felled pine logs (Uzunovic *et al.*, 2008).

Protection from root graft infections

Preventing root graft infections among neighbouring trees is an essential component for managing DED in urban areas (Tainter and Baker, 1996). Root graft transmissions are most easily avoided if infected trees are felled in the early stages of disease, before the pathogen has reached the root system.

Spread of the fungus can be prevented by the mechanical disruption of root systems between infected and healthy trees using trenchers or vibratory ploughs, or by killing roots with chemicals (Tainter and Baker, 1996). Sodium *N*-methyldithiocarbamate (Vapam) applied through a series of holes in the middle of the root region between diseased and healthy trees is effective and used for this purpose in the USA; it is not used in Europe owing to negative environmental side effects (Tainter and Baker, 1996; Scheffer *et al.*, 2008). Due to the phytotoxicity of Vapam, it is recommended to keep a distance of 2 to 3 m from healthy trees. Two weeks should pass after chemical treatment before the infected tree is removed, otherwise there is the increased danger that the healthy tree takes up inoculum through root connections (Tainter and Baker, 1996). An additional mechanical or chemical barrier can be initiated between the first and second group of apparently healthy trees, to safeguard against the risk that trees adjacent to the diseased tree are already infected but symptoms are not yet visible (Tainter and Baker, 1996).

Wound dressings and protection against bark peeling

It is recommended that pruning and surgery wounds created on elm trees during periods when elm bark beetles are active are covered with a wound dressing, to decrease their attraction to the insects (Tainter and Baker, 1996).

Wound dressings, applied as soon as possible after harvesting and logging wounds have been inflicted, can protect residual trees from a range of fungal infections (Vasiliauskas, 2001; Altenkirch *et al.*, 2002). Trees are typically treated to protect them from wound decay fungi (Vasiliauskas, 2001; Altenkirch *et al.*, 2002; see Vasaitis, Chapter 10, this volume), but not specifically from blue-stain fungal infections. It is, however, likely that effective prevention of decay using wound sealants (Vasiliauskas, 2001; see Vasaitis, Chapter 10, this volume) will also prevent blue-stain on live trees.

The protection of trees against bark peeling by ungulates is also primarily done to prevent wounds, which, in turn, are infected by decay fungi (see Vasaitis, Chapter 10, this volume), but at the same time entrance courts for blue-stain fungi into live trees are avoided. Altenkirch *et al.* (2002) provided an overview of mechanical, mechanical–biological and chemical means of protecting trees from bark peeling.

Log storage

If rapid processing is not possible, best storage practices can prevent attacks by insects as well as sap-stain and decay on logs by creating unfavourable conditions for these damaging agents (Wauer, 2000; Uzunovic *et al.*, 2008). Colonization by most insects and by blue-stain fungi can be prevented if wood moisture is either very low (below 19–20% of dry weight) or is kept high (Seifert, 1993; Wauer, 2000; Uzunovic *et al.*, 2008). Low temperatures also stop or restrict insect activities and fungal growth.

In the vegetation period, logs with bark can only be stored for several weeks to a few months before quality losses occur (Wauer, 2000; Uzunovic *et al.*, 2008). 'Sour-felling', referring to a harvesting method where delimbing is delayed until 1–2 months after felling, has the aim of preventing or reducing blue-stain and bark beetle attacks (Uzunovic *et al.*, 2008). This could be due to the lower levels of bark damage, lower wood moisture or lowered nutrient contents in sour-felled trees compared with delimbed control logs (Uzunovic *et al.*, 2008).

Immediate debarking and rapid drying of logs, facilitated by airy stacking in half-shaded, windy and dry storage places, excludes bark and ambrosia beetle attacks, can prevent blue-stain and allows log storage for at least a year (Wauer, 2000; Uzunovic *et al.*, 2008). However, substantial checking may occur and blue-stain fungi will infect logs if the drying process is too slow. Wood drying is particularly slowed down and the risk of fungal infections increased during warm and humid or rainy weather periods (Wauer, 2000; Uzunovic *et al.*, 2008). Long-term weather conditions cannot be predicted though, so the debarking and drying of logs are risky procedures.

Stacking can be a simple means to reduce attacks on logs in the inner parts of the pile by bark beetles (Wainhouse, 2005). Similarly, the debarking of only those logs that form the top layers of large stacks can greatly reduce overall bark beetle infestations of log piles (Speight and Wainhouse, 1989). Insect attacks are also largely prevented when log stacks are covered with a fleece (Wauer, 2000).

In areas where temperatures are below freezing for extensive periods, logs can be cold stored for a long time, with a low risk of quality losses resulting from insect and fungal colonization. In northern countries, cold storage can be extended into the vegetation period by protecting logs in a cover of snow, thereby maintaining low temperatures in the piles (Uzunovic *et al.*, 2008). In Canada, compacted snow on log piles is additionally covered with a sawdust layer to isolate the snow from solar radiation, in order to maintain protection long into the summer (Uzunovic *et al.*, 2008).

Wet storage is a very efficient method of storing fresh logs for up to several years (Wauer, 2000). If done thoroughly, insect attacks and fungal infections can be largely prevented as a result of the low oxygen content of water-saturated wood (Seifert, 1993; Wauer, 2000; Uzunovic *et al.*, 2008). It has become the standard method of timber storage following storm calamities and other natural disasters, which have led to an unexpected, excessive supply of timber and present considerable logistical challenges (Speight and Wainhouse, 1989; Wauer, 2000; Wainhouse, 2005). Besides maintaining the quality of the timber, wet storage overcomes difficulties in the timely processing of logs and buffers timber markets to some degree, thereby contributing to the avoidance of dramatic reductions of timber prices.

Logs can be stored in ponds and lakes, but storage on landings, in stacks that are constantly sprinkled with water (Plate 15), is now the preferred method (Speight and Wainhouse, 1989; Seifert, 1993; Wauer, 2000; Wainhouse, 2005; Uzunovic *et al.*, 2008). Sprinkling must be permanently done during periods when temperatures are over 0°C or 4°C, and also during mild periods in winter when temperatures are above freezing. To provide effective protection, log surfaces and, particularly, also the front ends must be permanently and thoroughly wetted. Both forms of water storage usually require legal permission by authorities due to their potential negative side effects on water quality and aquatic organisms (Wauer, 2000; Uzunovic *et al.*, 2008).

Wrapping and airtight sealing of fresh logs of high quality in a double layer of UV-resistant polyethylene foil (Plate 16) is a recently developed, alternative method of long-term storage (Wauer, 2000; Uzunovic *et al.*, 2008). Due to respiration and fermentation, CO_2 levels increase and oxygen content is reduced to less than 0.1% within a few days inside the foil, thus preventing the development of insects and fungi causing stain and decay for up to even 4 years of storage (Wauer, 2000; Uzunovic *et al.*, 2008). This method is expensive, but it does not require legal permission and is flexible as regards the quantities of stored wood, the location of the landing and the timber species (Wauer, 2000).

12.4.5 Resistance

In the 20th century, various programmes have been initiated in Europe and North America to breed elms resistant to DED. The first programme, in the Netherlands, aimed to select resistant individuals of field elm, and focused later on hybrids between the field elm and the wych elm, resulting in the release of clones that were resistant to *O. ulmi* but not to *O. novo-ulmi* (Mittempergher and Santini, 2004). Later, Asian elm species, especially Himalayan elm, were included as a source of DED resistance in the Dutch breeding programme, leading to a number of new releases of cultivars with moderately to high resistance levels to the disease (Mittempergher and Santini, 2004). Other European initiatives have been focusing on hybridizing European with Asian elm species, mainly Siberian elm, to obtain resistant cultivars adapted to the Mediterranean climate (Solla *et al.*, 2000; Mittempergher and Santini, 2004).

In North America, efforts have concentrated on species from Asia, mainly Siberian elm, Chinese elm (*U. parvifolia* Jacq.) and Japanese elm (*U. davidiana* Planch. var. *japonica* (Rehder) Nakai), selecting cultivars within these species and breeding hybrids within Asian species or between Asian and European genotypes (Townsend, 2000; Mittempergher and Santini, 2004). Identifying resistant individuals of the highly susceptible American elm, which can only with great difficulty be crossed with Eurasian species, was generally disappointing, but finally resulted in a few cultivars with partial resistance to DED (Townsend, 2000; Mittempergher and Santini, 2004; Sinclair and Lyon, 2005).

The various breeding programmes gave rise to about 20–25 elm cultivars with diverse genetic background and varying resistance to DED, some being highly resistant. Cultivars adapted to the climatic conditions in various areas are mainly suited as shade and landscape trees (Mittempergher and Santini, 2004). Realistically, their value for forestry and timber production is limited.

In Europe, some genotypes of the native elm species displaying relatively high resistance to DED, particularly within *U. minor*, have in recent years been proposed for use in hedges and in breeding programmes (Pinon *et al.*, 2005; Santini *et al.*, 2005; Solla *et al.*, 2005; Ghelardini and Santini, 2009). In the UK, mature individuals of some elm taxa that have survived the second epidemic of DED are propagated and used for replanting (Brasier, 1996). As European white elm is less preferred by bark beetles for maturation feeding than are other elm species (Webber, 2004), this species may be used as a hedgerow and shade tree. Finally, Siberian elm forms natural hybrids with field elm in Europe and with slippery elm (*U. rubra* Muhl.) in North America, and these hybridizations are likely to give rise to genotypes showing varying resistance levels to DED (Mittempergher and Santini, 2004).

Susceptibility to blue-stain or vascular stain is usually not decisive for determining the use of certain tree species in silviculture. However, the susceptibility of live trees to attack by bark and ambrosia beetles, in combination with other characteristics, can be essential for tree species selection.

12.4.6 Therapy

Cure from DED is possible, but only in the early stages of the disease and when infections result from bark beetle maturation feeding (Campana and Stipes, 1981b; Tainter and Baker, 1996). Where *O. novo-ulmi* has extensively spread in the crown or has reached the main stem, rescue of the tree is no longer possible. Likewise, root graft infections cannot be stopped by therapeutic treatments. Due to the high costs, therapy is limited to valuable individual trees, usually shade trees or elm genotypes maintained in conservation and breeding programmes (Campana and Stipes, 1981b; Tainter and Baker, 1996; Solla *et al.*, 2000).

The pruning of branches showing initial wilting symptoms can prevent the downward expansion of infections into the main stem (Campana and Stipes, 1981b; Tainter and Baker, 1996). If possible, the excised branch should include at least a 2–3 m long section without xylem discoloration (Campana and Stipes, 1981b; Tainter and Baker, 1996). In practice, branches are often cut at the main stem, which increases the chances of removal of all the infected xylem tissue (Tainter and Baker, 1996). Sometimes, surgery reveals that an infection has just extended from a branch to the main bole, but is still confined to a limited area below the pruning site (Campana and Stipes, 1981b). In this case, further spread can be stopped by appropriately placing 2.5 cm deep chainsaw cuts into the wood. This may lead to compartmentalization of the pathogen and cure of the tree (Campana and Stipes, 1981b).

When infection has not yet extended into the roots, the root system of elms can be saved by timely felling. The trees will then sprout again from the stump and/or form root suckers; this is desirable for coppice silviculture, for example in hedges. Likewise, the loss of elm genotypes propagated by rooted cuttings or by seeds and maintained in conservation orchards can be avoided by this measure (Solla *et al.*, 2000). Timely felling is likely also to be effective in rescuing root systems of redbay and other lauraceous species infected by *R. lauricola*.

The chemotherapy of infected trees with systemic fungicides is difficult and often fails to arrest infections (Campana and Stipes, 1981b; Tainter and Baker, 1996). In North America, thiabendazole and propiconazole have been shown to be at least partly effective for curative use (Stennes, 2000). Chemotherapy is always more effective when it is combined with surgery; pruning is best done after fungicide injection (Campana and Stipes, 1981b; Stennes, 2000). Trees wilting before the end of June or early July should not be selected for treatment because *O. novo-ulmi* is likely to be already widely established (Tainter and Baker, 1996; Stennes, 2000). Highly valuable mature elm specimens with early wilting symptoms on the top of branches in 5–15% of the crown, limited to a single crown part and with symptomatic branches relatively far away from the main bole, are suitable candidates for fungicide injections (Tainter and Baker, 1996; Stennes, 2000).

12.4.7 Integrated disease management

In natural environments, management of DED to minimize the level of infection is virtually impossible. In urban areas, however, integrated disease management (IDM) can be effective in reducing the mortality rates due to DED. Successful long-term and still continuing IDM programmes have been conducted in a number of cities both in Europe and North America (Gibbs, 1978; Speight and

Wainhouse, 1989; Magasi *et al.*, 1993; Brasier, 1996; Tainter and Baker, 1996; Stennes, 2000; Stipes, 2000; Scheffer *et al.*, 2008).

Intensive surveying for diseased trees, rigorous sanitation and the prevention of root graft transmissions are the most important measures for managing DED in urban areas (Campana and Stipes, 1981b; Magasi *et al.*, 1993; Tainter and Baker, 1996). In some cities, protective treatment of trees with Dutch Trig® is used widely (Scheffer *et al.*, 2008). The application of systemic fungicides and possibly also of insecticides can complement integrated management of DED. In addition, the use of a variety of tree species in new plantings to include only modest proportions of susceptible elm species and the planting of elm cultivars with high resistance levels to DED make urban tree populations more resilient both to DED and to other biotic and abiotic damaging factors (Campana and Stipes, 1981b; Manion, 1991). Keeping records of infected trees and possibly trapping elm bark beetles allow evaluation of the effectiveness of the measures taken and adjustment of management programmes (Magasi *et al.*, 1993; Scheffer *et al.*, 2008).

The development of IDM strategies, involving both researchers and practitioners, is envisaged for LW (Mayfield *et al.*, 2009). Due to the lack of effective measures, the disease is presently running its course in many forested areas. Activities focus on slowing the human-mediated long-range spread of the RAB and *R. lauricola*, monitoring their geographic spread, assessing their impact on various host species, considering and preparing germplasm conservation programmes for severely affected hosts, searching for putative resistant genotypes, testing possible disease management measures (e.g. pesticides, sanitation, silvicultural methods, insect trapping) in various situations, and the transfer of knowledge to practitioners and the general public (Mayfield *et al.*, 2009).

The management of blue-stain and vascular stain diseases relies on the combination of several strategies, which can mostly be implemented by routine forestry practices (Speight and Wainhouse, 1989; Altenkirch *et al.*, 2002; Wainhouse, 2005; Uzunovic *et al.*, 2008). For diseases and damage occurring in association with insect attacks, silviculture should aim to initiate and maintain forest stands that are resilient to attack by bark and ambrosia beetles as well as to other phloem-inhabiting and wood-inhabiting beetles. Contingency plans for catastrophic events, rapid and thorough salvaging and the sanitation of damaged trees as well as long-term storage of timber are key elements for lessening the impact of blue-stain and vascular stain resulting from beetle attacks. Best standards for silvicultural and wildlife management, hunting, harvesting, log processing and storage practices will minimize the occurrence of blue-stain in association with wounds, and on logs and in stored timber.

References

Altenkirch, W., Majunke, C. and Ohnesorge, B. (2002) *Waldschutz auf ökologischer Grundlage* [*Forest Protection on an Ecological Basis*]. Verlag Eugen Ulmer, Stuttgart, Germany.

Beal, E., Webber, J.F. and Eaton, R.A. (2010) Comparative susceptibility of pine, spruce and larch to sapstain. *Forest Pathology* 40, 116–128.

Brasier, C.M. (1981) Laboratory investigation of *Ceratocystis ulmi*. In: Stipes, R.J. and Campana, R.J. (eds) *Compendium of Elm Diseases*. APS Press, St Paul, Minnesota, pp. 76–79.

Brasier, C.M. (1991) *Ophiostoma novo-ulmi* sp. nov., causative agent of current Dutch elm disease pandemics. *Mycopathologia* 115, 151–161.

Brasier, C.M. (1996) New horizons in Dutch elm disease control. In: *Report on Forest Research 1996*. Forestry Commission, UK. Edinburgh, UK, pp. 20–28.

Brasier, C.M. (2000) Intercontinental spread and continuing evolution of the Dutch elm disease pathogens. In: Dunn, C.P. (ed.) *The Elms: Breeding, Conservation, and Disease Management*. Kluwer Academic Publishers, Dordrecht, The Netherlands, pp. 61–72.

Brasier, C.M. (2008) The biosecurity threat to the UK and global environment from international trade in plants. *Plant Pathology* 57, 792–808.

Brasier, C.M. and Gibbs, J.N. (1973) Origin of the Dutch elm disease epidemic in Britain. *Nature* 242, 607–609.

Brasier, C., Buck, K., Paoletti, M., Crawford, L. and Kirk, S. (2004) Molecular analysis of evolutionary changes in populations of *Ophiostoma novo-ulmi*. In: Gil, L., Solla, A. and Oullette, G. (eds) *New Approaches to Elm Conservation. Investigación Agraria: Sistemas y Recursos Forestales* 13, pp. 93–103.

Buchel, A. (2000) The species of the genus *Ulmus* L. In: Dunn, C.P. (ed.) *The Elms: Breeding, Conservation, and Disease Management*. Kluwer Academic Publishers, Dordrecht, The Netherlands, pp. 351–358.

Butin, H. (1995) *Tree Diseases and Disorders. Causes, Biology, and Control in Forest and Amenity Trees*. Oxford University Press, Oxford, UK.

Campana, R.J. and Stipes, R.J. (1981a) Dutch elm disease. Prospects for the future. In: Stipes, R.J. and Campana, R.J. (eds) *Compendium of Elm Diseases*. APS Press, St Paul, Minnesota, p. 9.

Campana, R.J. and Stipes, R.J. (1981b) Dutch elm disease. Control. In: Stipes, R.J. and Campana, R.J. (eds) *Compendium of Elm Diseases*. APS Press, St Paul, Minnesota, pp. 18–22.

Collin, E., Bilger, I., Eriksson, G. and Turok, J. (2000) The conservation of elm genetic resources in Europe. In: Dunn, C.P. (ed.) *The Elms: Breeding, Conservation, and Disease Management*. Kluwer Academic Publishers, Dordrecht, The Netherlands, pp. 281–293.

Dowding, P. (1969) The dispersal and survival of spores of fungi causing bluestain in pine. *Transactions of the British Mycological Society* 52, 125–137.

Edmonds, R.L., Agee, J.K. and Gara, R.I (2011) *Forest Health and Protection*, 2nd edn. Waveland Press, Long Grove, Illinois.

Elgersma, D.M. (1981) Dutch elm disease. Host susceptibility and resistance. In: Stipes, R.J. and Campana, R.J. (eds) *Compendium of Elm Diseases*. APS Press, St Paul, Minnesota, pp. 17–18.

Faccoli, M. and Battisti, A. (1997) Observations on the transmission of *Ophiostoma ulmi* by the smaller elm bark beetles (*Scolytus* spp.). In: Grégoire, J.-C., Liebhold, A.M., Stephen, F.M., Day, K.R. and Salmon, S.M. (eds) Proceedings: *Integrating Cultural Tactics into the Management of Bark Beetles and Reforestation Pests, Vallombrosa, Italy, September 1–3, 1996*. General Technical Report NE-236, US Department of Agriculture, Forest Service, Northeastern Forest Experiment Station, Radnor, Pennsylvania, pp. 172–176.

FAO (2011) *Guide to Implementation of Phytosanitary Standards in Forestry*. FAO Forestry Paper 164, Food and Agriculture Organization of the United Nations, Rome.

Fraedrich, S.W., Harrington, T.C., Rabaglia, R.J., Ulyshen, M.D., Mayfield, A.E. III, Hanula, J.L., Eickwort, J.M. and Miller, D.R. (2008) A fungal symbiont of the redbay ambrosia beetle causes a lethal wilt in redbay and other *Lauraceae* in the southeastern United States. *Plant Disease* 92, 215–224.

Francke-Grosmann, H. (1967) Ectosymbiosis in wood-inhabiting insects. In: Henry, S.M. (ed.) *Symbiosis, Volume 2*. Academic Press, New York, pp. 141–205.

Gadgil, P.D., Bulman, L.S., Dick, M.A. and Bain, J. (2000) Dutch elm disease in New Zealand. In: Dunn, C.P. (ed.) *The Elms: Breeding, Conservation, and Disease Management*. Kluwer Academic Publishers, Dordrecht, The Netherlands, pp. 189–199.

Ghelardini, L. and Santini, A. (2009) Avoidance by early flushing: a new perspective on Dutch elm disease research. *iForest* 2, 143–153.

Gibbs, J.N. (1978) Intercontinental epidemiology of Dutch elm disease. *Annual Review of Phytopathology* 16, 287–307.

Gibbs, J.N. (1993) The biology of ophiostomatoid fungi causing sapstain in trees and freshly cut logs. In: Wingfield, M.J., Seifert, K.A. and Webber, J.F. (eds) Ophiostoma *and* Ceratocystis*: Taxonomy, Ecology, and Pathogenicity*. APS Press, St Paul, Minnesota, pp. 153–160.

Harrington, T.C. (1981) Cycloheximide sensitivity as a taxonomic character in *Ceratocystis*. *Mycologia* 73, 1123–1129.

Harrington, T.C. (1993) Diseases of conifers caused by species of *Ophiostoma* and *Leptographium*. In: Wingfield, M.J., Seifert, K.A. and Webber, J.F. (eds) Ophiostoma *and* Ceratocystis*: Taxonomy, Ecology, and Pathogenicity*. APS Press, St Paul, Minnesota, pp. 161–172.

Harrington, T.C. and Fraedrich, S.W. (2010) Quantification of propagules of the laurel wilt fungus and other mycangial fungi from the redbay ambrosia beetle, *Xyleborus glabratus*. *Phytopathology* 100, 1118–1123.

Harrington, T.C. and Wingfield, M.J. (1998) The *Ceratocystis* species on conifers. *Canadian Journal of Botany* 76, 1446–1457.

Harrington, T.C., McNew, D., Steimel, J., Hofstra, D. and Farrell, R. (2001) Phylogeny and taxonomy of the *Ophiostoma piceae* complex and the Dutch elm disease fungi. *Mycologia* 93, 111–136.

Harrington, T.C., Aghayeva, D.N. and Fraedrich, S.W. (2010) New combinations in *Raffaelea*, *Ambrosiella*, and *Hyalorhinocladiella*, and four new species from the redbay ambrosia beetle, *Xyleborus glabratus*. *Mycotaxon* 111, 337–361.

Harrington, T.C., Yun, H.Y., Lu, S.S., Goto, H., Aghayeva, D.N. and Fraedrich, S.W. (2011) Isolations from the redbay ambrosia beetle, *Xyleborus glabratus*, confirm that the laurel wilt pathogen, *Raffaelea lauricola*, originated in Asia. *Mycologia* 103, 1028–1036.

Harwood, T.D., Tomlinson, I., Potter, C.A. and Knight, J.D. (2011) Dutch elm disease revisited: past, present and future management in Great Britain. *Plant Pathology* 60, 545–555.

Hintikka, V. (1974) *Ceratocystis ulmi* in Finland. *Karstenia* 14, 32.

Hubbes, M. (2004) Induced resistance for the control of Dutch elm disease. In: Gil, L., Solla, A. and Oullette, G. (eds) *New Approaches to Elm Conservation. Investigación Agraria: Sistemas y Recursos Forestales* 13, pp. 185–196.

Hulcr, J. and Dunn, R.R. (2011) The sudden emergence of pathogenicity in insect–fungus symbioses threatens naive forest ecosystems. *Proceedings of the Royal Society B, Biological Sciences* 278, 2866–2873.

Jacobi, W.R., Koski, R.D., Harrington, T.C. and Witcosky, J.J. (2007) Association of *Ophiostoma novo-ulmi* with *Scolytus schevyrewi* (Scolytidae) in Colorado. *Plant Disease* 91, 245–247.

Jacobs, K. and Wingfield, M.J. (2001) Leptographium *Species: Tree Pathogens, Insect Associates, and Agents of Blue-stain*. APS Press, St Paul, Minnesota.

Jacobs, K., Seifert, K.A., Harrison, K.J. and Kirisits, T. (2003) Identity and phylogenetic relationships of ophiostomatoid fungi associated with invasive and native *Tetropium* species (Coleoptera: Cerambycidae) in Atlantic Canada. *Canadian Journal of Botany* 81, 316–329.

Jankowiak, R. and Kolařik, M. (2010) Diversity and pathogenicity of ophiostomatoid fungi associated with *Tetropium* species colonizing *Picea abies* in Poland. *Folia Microbiologica* 55, 145–154.

Kamgan Nkuekam, G. (2007) *Ceratocystis* and *Ophiostoma* species infecting wounds on hardwood trees, with particular reference to South Africa. MSc thesis, University of Pretoria, Pretoria, South Africa.

Kausrud, K., Økland, B., Skarpaas, O., Grégoire, J.-C., Erbilgin, N. and Stenseth, N.C. (2012) Population dynamics in changing environments: the case of an eruptive forest species. *Biological Reviews* 87, 34–51.

Kirisits, T. (2004) Fungal associates of European bark beetles with special emphasis on the ophiostomatoid fungi. In: Lieutier, F., Day, K.R., Battisti, A., Grégoire, J.-C. and Evans, H.F. (eds) *Bark and Wood Boring Insects in Living Trees in Europe, a Synthesis*. Kluwer Academic Publishers, Dordrecht, The Netherlands, pp. 181–235.

Kirisits, T. and Konrad, H. (2004) Dutch elm disease in Austria. In: Gil, L., Solla, A. and Oullette, G. (eds) *New Approaches to Elm Conservation. Investigación Agraria: Sistemas y Recursos Forestales* 13, pp. 81–92.

Krokene, P. (1996) The role of bluestain fungi in tree-killing by bark beetles. PhD thesis, University of Oslo, Oslo.

Lee, K. and Gibbs, J.N. (1996) An investigation of the influence of harvesting practice on the development of bluestain in Corsican pine logs. *Forestry* 69, 137–141.

Lieutier, F., Yart, A. and Salle, A. (2009) Stimulation of tree defences by ophiostomatoid fungi can explain attack success of bark beetles on conifers. *Annals of Forest Science* 66, 801.

Linnakoski, R. (2011) Bark beetle-associated fungi in Fennoscandia with special emphasis on species of *Ophiostoma* and *Grosmannia*. PhD thesis, Dissertationes Forestales 119, University of Eastern Finland, Joensuu, Finland.

Magasi, L.P., Harrison, K.J., Urquhart, D.A. and Murray, D.M. (1993) *Three Decades of Dutch Elm Disease in Fredericton, N.B.: 1961–1990*. Information Report M-X-185E, Canadian Forest Service, Maritimes Region, Fredericton, New Brunswick, Canada.

Malloch, D. and Blackwell, M. (1993) Dispersal biology of the ophiostomatoid fungi. In: Wingfield, M.J., Seifert, K.A. and Webber, J.F. (eds) Ophiostoma *and* Ceratocystis: *Taxonomy, Ecology, and Pathogenicity*. APS Press, St Paul, Minnesota, pp. 195–206.

Manion, P.D. (1991) *Tree Disease Concepts*, 2nd edn. Prentice-Hall, Englewood Cliffs, New Jersey.

Masuya, H., Brasier, C., Ichihara, Y., Kubono, T. and Kanzaki, N. (2010) First report of the Dutch elm disease pathogens *Ophiostoma ulmi* and *O. novo-ulmi* in Japan. *Plant Pathology* 59, 805.

Mathiesen-Käärik, A. (1953) Eine Übersicht über die gewöhnlichsten mit Borkenkäfern assoziierten Bläuepilze in Schweden und einige für Schweden neue Bläuepilze [An overview of the most common bark beetle-associated blue-stain fungi in Sweden and some blue-stain fungi new for Sweden]. *Meddelanden från Statens Skogforskningsinstitutut* 43, 1–74.

Mayfield, A. *et al.* (2009) Recovery plan for laurel wilt on redbay and other forest species caused by *Raffaelea lauricola*, vector *Xyleborus glabratus*. National Plant Disease Recovery System, a cooperative project of The American Psychopathological Society and the United States Department of Agriculture. Available at: www.ars.usda.gov/SP2UserFiles/Place/00000000/opmp/ForestLaurelWilt100107.pdf (accessed 13 July 2012).

Mittempergher, L. and Santini, A. (2004) The history of elm breeding. In: Gil, L., Solla, A. and Oullette, G. (eds) *New Approaches to Elm Conservation. Investigación Agraria: Sistemas y Recursos Forestales* 13, pp. 161–177.

Morin, C., Tanguay, P., Breuil, C., Yang, D.-Q. and Bernier, L. (2006) Bioprotection of spruce logs against sapstain using an albino strain of *Ceratocystis resinifera*. *Phytopathology* 96, 526–533.

Moser, J.C., Konrad, H., Blomquist, S.R. and Kirisits, T. (2010) Do mites phoretic on elm bark beetles contribute to the transmission of Dutch elm disease? *Naturwissenschaften* 97, 219–227.

Nierhaus-Wunderwald, D. and Engesser, R. (2003) *Ulmenwelke – Biologie, Vorbeugung und Gegenmaßnahmen, 2. überarbeitete Auflage* [*Dutch Elm Disease – Biology, Prevention and Control Measures, 2nd, Updated Edition*]. Merkblatt für die Praxis 20, Eidgenössische Forschungsanstalt für Wald, Schnee und Landschaft WSL, Birmensdorf, Switzerland.

Nyland, R.D. (2002) *Silviculture: Concepts and Applications*, 2nd edn. McGraw-Hill, New York.

Paulin-Mahady, A.E., Harrington, T.C. and McNew, D. (2002) Phylogenetic and taxonomic evaluation of *Chalara*, *Chalaropsis*, and *Thielaviopsis* anamorphs associated with *Ceratocystis*. *Mycologia* 94, 62–72.

Peace, T.R. (1960) *The Status and Development of Elm Disease in Britain*. Forestry Commission Bulletin 33, Her Majesty's Stationery Office, London.

Pinon, J., Husson, C. and Collin, E. (2005) Susceptibility of native French elm clones to *Ophiostoma novo-ulmi*. *Annals of Forest Science* 62, 689–696.

Postner, M. (1974) Scolytidae (Ipidae), Borkenkäfer [Scolytidae (Ipidae), bark beetles]. In: Schwencke, W. (ed.) *Die Forstschädlinge Europas, Band 2* [*The Forest Pests of Europe, Volume 2*]. Paul Parey Verlag, Hamburg/Berlin, pp. 334–482.

Powell, M.A., Eaton, R.A. and Webber, J.F. (1995) The role of micro-arthropods in the defacement of sawn lumber by sapstain and mould fungi. *Canadian Journal of Forest Research* 25, 1148–1156.

Reay, S.D., Thwaites, J.M. and Farrell, R.L. (2005) A survey of *Ophiostoma* species vectored by *Hylastes ater* to pine seedlings in New Zealand. *Forest Pathology* 35, 105–113.

Röhrig, E. (1996) Die Ulmen in Europa. Ökologie und epidemische Erkrankung [The elms in Europe: ecology and epidemic disease]. *Forstarchiv* 67, 179–198.

Roll-Hansen, F. and Roll-Hansen, H. (1980a) Microorganisms which invade *Picea abies* in seasonal stem wounds. I. General aspects. *Hymenomycetes*. *European Journal of Forest Pathology* 10, 321–339.

Roll-Hansen, F. and Roll-Hansen, H. (1980b) Microorganisms which invade *Picea abies* in seasonal stem wounds. II. Ascomycetes, fungi imperfecti, and bacteria. General discussion, *Hymenomycetes* included. *European Journal of Forest Pathology* 10, 396–410.

Santini, A., Fagnani, A., Ferrini, F., Ghelardini, L. and Mittemphergher, L. (2005) Variation among Italian and French elm clones in their response to *Ophiostoma novo-ulmi* inoculation. *Forest Pathology* 35, 183–193.

Scheffer, R.J., Voeten, J.G.W.F. and Guries, R.P. (2008) Biological control of Dutch elm disease. *Plant Disease* 92, 192–200.

Schwerdtfeger, F. (1981) *Die Waldkrankheiten*, 4. Auflage. Paul Parey Verlag, Hamburg/Berlin.

Seifert, K.A. (1993) Sapstain of commercial lumber by species of *Ophiostoma* and *Ceratocystis*. In: Wingfield, M.J., Seifert, K.A. and Webber, J.F. (eds) Ophiostoma *and* Ceratocystis*: Taxonomy, Ecology, and Pathogenicity*. APS Press, St Paul, Minnesota, pp. 141–151.

Sinclair, W.A. and Lyon, H.H. (2005) *Diseases of Tree and Shrubs*, 2nd edn. Cornell University Press, Ithaca, New York.

Six, D.L. and Wingfield, M.J. (2011) The role of phytopathogenicity in bark beetle–fungus symbioses: a challenge to the classic paradigm. *Annual Reviews of Entomology* 56, 255–272.

Smalley, E.B. and Guries, R.P. (2000) Asian elms: sources of disease and insect resistance. In: Dunn, C.P. (ed.) *The Elms: Breeding, Conservation, and Disease Management*. Kluwer Academic Publishers, Dordrecht, The Netherlands, pp. 215–230.

Solheim, H. (1993) Ecological aspects of fungi associated with the spruce bark beetle *Ips typographus* in Norway. In: Wingfield, M.J., Seifert, K.A. and Webber, J.F. (eds) Ophiostoma *and* Ceratocystis*: Taxonomy, Ecology, and Pathogenicity*. APS Press, St Paul, Minnesota, pp. 235–242.

Solheim, H., Eriksen, R. and Hietala, A.M. (2011) Dutch elm disease has currently a low incidence on wych elm in Norway. *Forest Pathology* 41, 182–188.

Solla, A., Burón, M., Iglesias, S. and Gil, L. (2000) Spanish program for the conservation and breeding of elms against DED. In: Dunn, C.P. (ed.) *The Elms: Breeding, Conservation, and Disease Management*. Kluwer Academic Publishers, Dordrecht, The Netherlands, pp. 295–303.

Solla, A., Bohnens, J., Collin, E., Diamandis, S., Franke, A., Gil, L., Burón, M., Santini, A., Mittempergher, L., Pinon, J. and Vanden Broeck, A. (2005) Screening European elms for resistance to *Ophiostoma novo-ulmi*. *Forest Science* 51, 134–141.
Speight, M.R. and Wainhouse, D. (1989) *Ecology and Management of Forest Insects*. Clarendon Press, Oxford, UK.
Stennes, M.A. (2000) Dutch elm disease chemotherapy with Arbotect 20-S® and Alamo®. In: Dunn, C.P. (ed.) *The Elms: Breeding, Conservation, and Disease Management*. Kluwer Academic Publishers, Dordrecht, The Netherlands, pp. 173–188.
Stipes, R.J. (2000) The management of Dutch elm disease. In: Dunn, C.P. (ed.) *The Elms: Breeding, Conservation, and Disease Management*. Kluwer Academic Publishers, Dordrecht, The Netherlands, pp. 157–172.
Tainter, F.H. and Baker, F.A. (1996) *Principles of Forest Pathology*. John Wiley, Hoboken, New Jersey.
Tomlinson, I. and Potter, C. (2010) 'Too little, too late'? Science, policy and Dutch elm disease in the UK. *Journal of Historical Geography* 36, 121–131.
Townsend, A.M. (2000) USDA genetic research on elms. In: Dunn, C.P. (ed.) *The Elms: Breeding, Conservation, and Disease Management*. Kluwer Academic Publishers, Dordrecht, The Netherlands, pp. 271–278.
Uzunovic, A., Yang, D.-Q., Gagné, P., Breuil, C., Bernier, L., Byrne, A., Gignac, M. and Kim, S.H. (1999a) Fungi that cause sapstain in Canadian softwoods. *Canadian Journal of Microbiology* 45, 914–922.
Uzunovic, A., Webber, J.F., Peace, A.J. and Dickinson, D.J. (1999b) The role of mechanized harvesting in the development of bluestain in pine. *Canadian Journal of Forest Research* 29, 242–251.
Uzunovic, A., O'Callahan, D. and Kreber, B. (2004) Mechanical tree harvesters spread fungal inoculum onto freshly felled Canadian and New Zealand pine logs. *Forest Products Journal* 54(11), 34–40.
Uzunovic, A., Byrne, T., Gignac, M. and Yang, D.-Q. (2008) *Wood Discolorations and Their Prevention, with an Emphasis on Bluestain*. Special Publication SP-50, FPInnovations™, Vancouver, British Columbia, Canada.
Vasiliauskas, R. (2001) Damage to trees due to forestry operations and its pathological significance in temperate forests: a literature review. *Forestry* 74, 319–336.
Vasiliauskas, R. and Stenlid, J. (1998) Fungi inhabiting stems of *Picea abies* in a managed stand in Lithuania. *Forest Ecology and Management* 109, 119–126.
Vasiliauskas, R., Stenlid, J. and Johansson, M. (1996) Fungi in bark peeling wounds of *Picea abies* in central Sweden. *European Journal of Forest Pathology* 26, 285–296.
Viiri, H. (2004) Fungi associated with *Hylobius abietis* and other weevils. In: Lieutier, F., Day, K.R., Battisti, A., Grégoire, J.-C. and Evans, H.F. (eds) *Bark and Wood Boring Insects in Living Trees in Europe, a Synthesis*. Kluwer Academic Publishers, Dordrecht, The Netherlands, pp. 381–393.
Wainhouse, D. (2005) *Ecological Methods in Forest Pest Management*. Oxford University Press, Oxford, UK.
Wauer, A. (2000) Verfahren der Rundholzlagerung [Procedures of round timber storage]. *Berichte aus der Bayerischen Landesanstalt für Wald und Forstwirtschaft* 29, 1–91.
Webber, J.F. (2004) Experimental studies on factors influencing the transmission of Dutch elm disease. In: Gil, L., Solla, A. and Oullette, G. (eds) *New Approaches to Elm Conservation. Investigación Agraria: Sistemas y Recursos Forestales* 13, 197–205.
Webber, J.F. and Gibbs, J.N. (1989) Insect dissemination of fungal pathogens of trees. In: Wilding, N., Collins, N.M., Hammond, P.M. and Webber, J.F. (eds) *Insect–Fungus Interactions*. Academic Press, London, pp. 161–193.
Wimmer, V. (2012) Wirkung eines neu entwickelten Insektizidnetzes gegen ausgewählte Forstinsekten der Familien Curculionidae und Cleridae [Effect of a newly developed insecticide net against selected forest insects of the families Curculionidae and Cleridae]. MSc thesis, University of Natural Resources and Life Sciences, Vienna (BOKU), Vienna, Austria.
Zhou, X.D., de Beer, Z.W., Wingfield, B.D. and Wingfield, M.J. (2001) Ophiostomatoid fungi associated with three pine-infesting bark beetles in South Africa. *Sydowia* 53, 290–300.
Zipfel, R.D., de Beer, Z.W., Jacobs, K., Wingfield, B.D. and Wingfield, M.J. (2006) Multi-gene phylogenies define *Ceratocystiopsis* and *Grosmannia* distinct from *Ophiostoma*. *Studies in Mycology* 55, 75–97.

13 Blackstain Root Disease and Other Leptographium Diseases

Lori G. Eckhardt*
Auburn University, Auburn, Alabama, USA

13.1 Pathogens, Significance and Distribution

The distribution of *Leptographium* is worldwide where appropriate hosts are present (Jacobs and Wingfield, 2001), with many new discoveries in recent years (Masuya *et al.*, 2004; Zhou *et al.*, 2008). Most are known from the northern hemisphere where coniferous species are most common (Kendrick, 1962; Harrington and Cobb, 1986, 1987), while virtually all that have been recorded from the southern hemisphere have been introduced with pine-infesting bark beetles (Jacobs and Wingfield, 2001). Most *Leptographium* spp. are found inhabiting conifers, with only few known to colonize hardwood (broadleaved) hosts (Wiehe, 1949; Webber *et al.*, 1999; Jacobs *et al.*, 2006). Some *Leptographium* spp. are associated with serious diseases that can result in major economic losses (Harrington and Cobb, 1988; Solheim, 1992a,b; Wingfield *et al.*, 1993), while other species are considered to play an important role in disease (Wingfield and Marasas, 1980, 1981; Klepzig *et al.*, 1991; Eckhardt *et al.*, 2007). Most are best known for their associations with blue-stain of sapwood in conifers. *Leptographium* spp. are commonly associated with insects (Seifert, 1993; Jacobs and Wingfield, 2001).

Ophiostomatoid fungi (*Ascomycota*, *Ophiostomataceae*) include species of *Ophiostoma*, *Ceratocystiopsis* and *Grosmannia*, with anamorphs of the genera *Leptographium*, *Hyalorhinocladiella*, *Sporothrix* and *Pesotum* (Wingfield *et al.*, 1993; Zipfel *et al.*, 2006). Previously thought to be closely related to the genus *Ceratocystis*, fine morphological characteristics (de Hoog, 1974), as well as biochemical (Smith, 1967; Jewell, 1974; Harrington, 1981) and DNA sequence data (Hausner *et al.*, 1993; Spatafora and Blackwell, 1994) indicate that similarities between *Ceratocystis* and the *Ophiostomataceae s.s.* are the result of convergent evolution. Until recently, members of the *Ophiostomataceae* producing *Leptographium* anamorphs were considered to be species of *Ophiostoma* (Jacobs and Wingfield, 2001). The genus *Leptographium* has been shown to be a monophyletic lineage (Jacobs *et al.*, 2001; Zipfel *et al.*, 2006), and all *Ophiostoma* spp. with *Leptographium* anamorphs have been transferred to *Grosmannia*, with *Ophiostoma s.s.* and *Ceratocystiopsis* applied to species with *Pesotum*, *Sporothrix* or *Hyalorhinocladiella* anamorphs (Zipfel *et al.*, 2006).

The epidemiology and management of forest tree diseases associated with *Leptographium* spp. are the focus of this chapter,

* E-mail: eckhalg@auburn.edu

with emphasis on the few that are involved in root disease complexes. *L. wageneri* (W.B. Kendr.) M.J. Wingf. is responsible for blackstain root disease and is considered a virulent primary pathogen of several western conifer species (Cobb, 1988). White pine root disease associated with *L. procerum* (W.B. Kendr.) M.J. Wingf. on eastern white pine (*Pinus strobus* L.) causes major economic losses in Christmas tree plantations in the eastern USA (Lackner and Alexander, 1982). Other *Leptographium* spp., such as *L. terebrantis* S.J. Barras & T.J. Perry, *Grosmannia alacris* T.A. Duong, Z.W. de Beer & M.J. Wingf. (Duong *et al.*, 2012) and *G. huntii* (Rob.-Jeffr.) Zipfel, Z.W. de Beer & M.J. Wingf. are involved in complicated insect–fungal complexes, such as red pine (*P. resinosa* Aiton) decline (Klepzig *et al.*, 1991) and loblolly pine (*P. taeda* L.) decline (Eckhardt *et al.*, 2007). Other than *L. wageneri*, none of these fungi have been considered as a primary cause of disease, but a few have been found to be moderate to highly pathogenic (Harrington and Cobb, 1983; Wingfield, 1986; Bennet and Tattar, 1988; Ross *et al.*, 1992; Smalley *et al.*, 1993; Nevill *et al.*, 1995; Eckhardt *et al.*, 2004a; Matusick and Eckhardt, 2010; Matusick *et al.*, 2010, 2012).

13.1.1 *Leptographium wageneri*

L. wageneri was first reported in 1939 on pine species in California (Wagener and Mielke, 1961) and later described from other conifers (Harrington and Cobb, 1987). Kendrick (1962) described the causal agent of blackstain root disease as *Verticicladiella wageneri* W.B. Kendr., but the species was later transferred to *Leptographium* (Wingfield, 1985). Although other species of *Leptographium* had been isolated from symptomatic trees, Harrington and Cobb (1983) showed conclusively that blackstain root disease is caused by *L. wageneri*. The role of the fungus was overlooked for some time owing to the presence of bark beetles, and tree death was attributed to beetle infestation (Cobb, 1988). Blackstain root disease is restricted to the western USA (Walters and Walters, 1977; Harrington, 1982; Cobb, 1988) where the fungus spreads rapidly, and is capable of causing extensive losses in the forest (Byler *et al.*, 1979; Cobb *et al.*, 1982; Cobb, 1988).

Blackstain root disease occurs on trees of all ages and predisposes the host to further attacks by bark beetles (Helms *et al.*, 1971; Morrison and Hunt, 1988). While *L. wageneri* has been grouped with the major root pathogens, it also shows similar symptoms to vascular wilt pathogens in broadleaves (Leaphart, 1960; Smith, 1967; Harrington, 1982). It colonizes only the current year's xylem tracheids, which results in a black streaked pattern of the outer xylem wood and quickly girdles the infected tissue (Wagener and Mielke, 1961; Smith, 1967; Goheen and Cobb, 1978; Hessburg and Hansen, 1982; Cobb *et al.*, 1984; Bertagnole *et al.*, 1987). In contrast, most other *Leptographium* spp. colonize phloem and xylem tracheids, as well as ray parenchyma cells, causing a wedge-shaped appearance in the wood (Jacobs and Wingfield, 2001). Three varieties of *L. wageneri* have been described: *L. wageneri* var. *wageneri* (W.B. Kendr.) M.J. Wingf., which occurs on twoneedle pinyon pine (*P. edulis* Engelm.) and singleleaf pinyon pine (*P. monophylla* Torr. & Frém.) (Kendrick, 1962; Harrington, 1993); *L. wageneri* var. *pseudotsugae* T.C. Harr. & F.W. Cobb, which occurs on Douglas-fir (*Pseudotsuga menziesii* (Mirb.) Franco) (Cobb and Platt, 1967; Harrington and Cobb, 1987; Harrington, 1993); and *L. wageneri* var. *ponderosum* (T.C. Harr. & F.W. Cobb) T.C. Harr. & F.W. Cobb, which occurs on hard pines, i.e. ponderosa pine (*P. ponderosa* Lawson & C. Lawson), lodgepole pine (*P. contorta* Douglas ex Loudon), Jeffrey pine (*P. jeffreyi* Balf.) (Harrington and Cobb, 1987; Harrington, 1993). All three varieties are capable of infecting species other than those from which they were isolated (Cobb and Platt, 1967; Harrington and Cobb, 1984; Diamandis *et al.*, 1987).

Blackstain root disease, like all other diseases caused by ophiostomatoid fungi, is associated with insect transmission (Hansen *et al.*, 1988). Two root weevils (*Pissodes fasciatus* LeConte and *Steremnius carinatus* (Boheman)) and one root-feeding bark beetle (*Hylastes nigrinus* (Mannerheim)) carry and transmit the fungus between hosts (Witcosky *et al.*, 1986). Transmission occurs during maturation feeding and breeding activities, which are common actions associated with root and lower stem feeding insects. Attracted to host volatiles produced by injured or diseased roots, the insect vectors burrow through the soil and transport the fungus to host tissues (Witcosky and Hansen, 1985). *L. wageneri* may move short distances through root grafts (Jacobs and Wingfield, 2001); however, contact between hosts is not necessary for spread (Hessburg and Hansen, 1986).

13.1.2 *Leptographium procerum*

Originally described from jack pine (*Pinus banksiana* Lamb.) in Quebec (Kendrick, 1962), *Verticicladiella procera* W.B. Kendr. was later transferred to *Leptographium* by Wingfield (1985). *L. procerum* is commonly found inhabiting pine roots throughout much of the USA and other countries, including Canada, European countries, New Zealand and South Africa (Jacobs and Wingfield, 2001). The species has been isolated from approximately 20 pinaceous host species, such as red pine (Sinclair and Hudler, 1980), Scots pine (*P. sylvestris* L.) (Sinclair and Hudler, 1980), shortleaf pine (*P. echinata* Mill.) (Alexander *et al.*, 1988), slash pine (*P. elliottii* Engelm.) (Barnard *et al.*, 1991), longleaf pine (*P. palustris* Mill.) (Otrosina *et al.*, 1999), and loblolly pine (Alexander *et al.*, 1988). Many studies have been undertaken to examine the interactions between *L. procerum*, its vectors and eastern white pine in white pine root disease (Dochinger, 1967; Houston, 1969; Towers, 1977; Weaver *et al.*, 1981; Lackner and Alexander, 1983, 1984; Meyer *et al.*, 1983; Horner *et al.*, 1987), and questions raised as to the effectiveness of *L. procerum* as a pathogen (Wingfield, 1983; Harrington, 1988).

L. procerum also has been isolated from a number of curculionid vector species (Jacobs and Wingfield, 2001). In the south-eastern USA, it has been found associated with damaging above-ground bark beetles such as *Dendroctonus frontalis* Zimmermann (Otrosina *et al.*, 1997), *D. terebrans* (Olivier), and *Ips typographus* (Linnaeus) (Harrington, 1988), among others. It also has been isolated from stem-damaging and root-feeding weevils such as *Hylobius pales* (Herbst) (Lackner and Alexander, 1982), *Pachylobius picivorus* (Germar) (Wingfield, 1983) and *Pissodes nemorensis* Germar (Nevill and Alexander, 1992a). Finally, *L. procerum* is commonly found on root-feeding bark beetles around the world including *Hylastes* and *Hylurgus* spp. (Jacobs and Wingfield, 2001).

The pathogenicity of *L. procerum* is relatively well known. As stated previously, the fungus has been associated with many conifer species in several different systems. The wide occurrence of disease association with *L. procerum* has prompted the investigation of its pathogenicity on many different host species as well, most notably on eastern white pine (Wingfield, 1986; Nevill and Alexander, 1992b), lodgepole pine (Bertagnole *et al.*, 1983), loblolly pine (Eckhardt *et al.*, 2004a; Matusick *et al.*, 2010, 2012), Douglas-fir (Harrington and Cobb, 1983), longleaf pine (Otrosina *et al.*, 2002; Matusick *et al.*, 2010, 2012) and red pine (Klepzig *et al.*, 1996). Although some results illustrate *L. procerum* as being capable of severe infections (Lackner and Alexander, 1981; Eckhardt *et al.*, 2004a), most authors have suggested that it is a mild wound pathogen (Wingfield, 1986; Nevill *et al.*, 1995; Klepzig *et al.*, 1996; Matusick and Eckhardt, 2010; Matusick *et al.*, 2010, 2012). Some evidence suggests that pathogenicity differences exist among isolates (Lackner and Alexander, 1984); however, similar results were not supported in Wingfield (1986) when eight different isolates were tested. Previous discrepancies in findings on the pathogenicity of *L. procerum* can be attributed to the inoculation technique used (Wingfield, 1986). Several authors found *L. procerum* severely infected pines when a root-dip inoculation technique was used (Lackner and Alexander, 1982; Eckhardt *et al.*, 2004a). In contrast, *L. procerum* appeared to be mildly pathogenic when stem wound inoculations (Wingfield, 1986; Nevill *et al.*, 1995; Otrosina *et al.*, 2002; Matusick and Eckhardt, 2010) or root wound inoculations were used (Matusick *et al.*, 2010, 2012). Root inoculation seems to be an appropriate method based on the ability of *L. procerum* to persist in forest soil, although soil-borne propagules have been determined to be unimportant in its spread (Lewis *et al.*, 1987).

13.1.3 *Leptographium terebrantis*

Like *L. procerum*, *L. terebrantis* is commonly found associated with dying pine trees throughout the USA and Canada (Harrington, 1988). It has been isolated from dying Scots pine in Massachusetts (Highley and Tattar, 1985), from red pine in Wisconsin (Klepzig *et al.*, 1991) and also from eastern white pine (Harrington, 1988). In the western states of the USA and Canada it has been found in twoneedle pinyon pine, Douglas-fir (Harrington, 1988) and lodgepole pine (in British Columbia) (Morrison and Hunt, 1988). Three of the main south-eastern pine species, i.e. loblolly pine (Eckhardt *et al.*, 2007), longleaf pine (Otrosina *et al.*, 2002) and slash pine have been infected by *L. terebrantis* (L.G. Eckhardt, 2012, unpublished results).

L. terebrantis has been isolated from a wide variety of insect vectors within the USA. It has been associated with *D. terebrans* (Barras and Perry, 1971) and *D. frontalis* (Otrosina *et al.*, 1997) feeding above ground, as well as from *D. valens* LeConte, *I. pini* (Say), and *Hylurgops* and *Hylobius* spp. (Jacobs and Wingfield, 2001). Like *L. procerum*, it has been isolated from weevil species such as *H. pales* and *P. picivorus* as well as from the root-feeding bark beetles *Hylastes tenuis* Eichhoff and *H. salebrosus* Eichhoff in the south-eastern USA (Eckhardt *et al.*, 2004b).

The pathogen is found in several pine systems throughout the USA, including loblolly pine (Nevill *et al.*, 1995; Eckhardt *et al.*, 2004a; Matusick and Eckhardt, 2010; Matusick *et al.*, 2010, 2012), Douglas-fir (Harrington and Cobb, 1983), red pine (Klepzig *et al.*, 1996), eastern white pine (Wingfield, 1983, 1986), longleaf pine (Otrosina *et al.*, 2002; Matusick and Eckhardt, 2010; Matusick *et al.*, 2010), ponderosa pine (Owen *et al.*, 1987; Parmeter *et al.*, 1989, 1992), lodgepole pine, western white pine (*Pinus monticola* Douglas ex. D. Don), western hemlock (*Tsuga heterophylla* (Raf.) Sarg.), mountain hemlock (*T. mertensiana* (Bong.) Carrière), white spruce (*Picea glauca* (Moench) Voss), subalpine fir (*Abies lasiocarpa* (Hook.) Nutt.), grand fir (*A. grandis* (Douglas ex D. Don) Lindl.), western larch (*Larix occidentalis* Nutt.), western redcedar (*Thuja plicata* Donn ex D. Don) (Morrison and Hunt, 1988) and Japanese black pine (*Pinus thunbergii* Parl.) (Rane and Tattar, 1987). *L. terebrantis* has been shown to be moderately pathogenic to some hosts (Wingfield, 1983; Nevill *et al.*, 1995; Eckhardt *et al.*, 2004a; Matusick and Eckhardt, 2010; Matusick *et al.*, 2010, 2012), primarily in the eastern states, but was found to be weakly or non-pathogenic to many western conifer species (Harrington and Cobb, 1983; Morrison and Hunt, 1988). Evidence suggests that *L. terebrantis* has extremely variable pathogenicity, based primarily on the host (Morrison and Hunt, 1988 and others).

13.1.4 The *Grosmannia serpens* complex

Duong *et al.* (2012) found, using multiple gene sequence data, that there is strong phylogenetic evidence supporting five cryptic species in the *G. serpens* complex (i.e. *G. serpens* Goid., *G. alacris*, *L. gibbsii* T.A. Duong, Z.W. de Beer & M.J. Wingf., *L. yamaokae* T.A. Duong, Z.W. de Beer & M.J. Wingf. and *L. castellanum* T.A. Duong, Z.W. de Beer & M.J. Wingf.). In the literature, *G. serpens* has been associated with root disease in Italy (Goidanich, 1936; Lorenzini and Gambogi, 1976), South Africa (Wingfield and Knox-Davis, 1980; Wingfield and Marasas, 1980) and the USA (Eckhardt *et al.*, 2007). The species in Italy is *G. serpens*, while that in South Africa and the USA is *G. alacris* (Duong *et al.*, 2012).

G. serpens was first described associated with a root disease of pines in Italy (Goidanich, 1936; Lorenzini and Gambogi, 1976). In 1980, a disease similar to the one in Italy was reported on Monterey pine (*Pinus radiata* D. Don) and maritime pine (*P. pinaster* Aiton) in South Africa, and the causal agent was described as *Verticicladiella alacris* M.J. Wingf. & Marasas (Wingfield and Knox-Davis, 1980; Wingfield and Marasas, 1980). Based on morphological characters, *V. alacris* was reduced to synonymy with *L. serpens* (Goid.) Siemaszko, the anamorph of *G. serpens* (Wingfield and Marasas, 1981), and then changed to *G. alacris* (Duong *et al.*, 2012).

Grosmannia spp. have been found to be vectored by insect species throughout the world. Unlike *L. procerum* and *L. terebrantis*, *G. serpens* and *G. alacris* appear to only be vectored by root-feeding bark beetles (Jacobs and Wingfield, 2001). In the south-eastern USA, Eckhardt *et al.* (2004b) found that *H. tenuis* and *H. salebrosus* transported *G. alacris* consistently, while *H. pales* and *P. picivorus* did not. *Hylastes angustatus* (Herbst.) (Wingfield *et al.*, 1988), *H. ater* (Paykull) (Wingfield and Gibbs, 1991) and *H. linearis* Erichson (Wingfield and Knox-Davies, 1980) have been known to vector *G. alacris* in other parts of the world.

The pathogenicity of the *G. serpens* complex is largely unknown (Wingfield *et al.*, 1988). The lack of pathogenicity data reflects the lack of consistent isolation of the pathogen from diseased trees. Few studies have included *G. alacris* in inoculation experiments (Wingfield and Knox-Davies, 1980; Zhou *et al.*, 2002; Eckhardt *et al.*, 2004a; Matusick and Eckhardt, 2010; Matusick *et al.*, 2010, 2012). In limited root inoculation tests, *G. alacris* was found to produce dark lesions extending an average of 20 cm after 6 months (Wingfield and Knox-Davies, 1980). A similar technique was used to inoculate pine branches in South Africa (Zhou *et al.*, 2002) and loblolly pine stems in the USA (Eckhardt *et al.*, 2004a). In the South African studies, *G. alacris* produced lesions 1.5–3.7 cm long after 6 weeks, though the authors concluded that it was non-pathogenic to the pine species tested (Zhou *et al.*, 2002). Eckhardt *et al.* (2004a) obtained similar results with *G. alacris*, in which lesions averaging 3.0 cm were produced after 3 weeks. The three inoculation tests used similar techniques on different regions of the tree. Despite differences in experiment length, the cumulative results suggest that *G. alacris* grows most successfully in pine roots, from which it is most commonly isolated (Wingfield and Knox-Davies, 1980; Eckhardt *et al.*, 2004a). Because limited pathogenicity tests have been undertaken and the results have been conflicting, Wingfield *et al.* (1988) suggested that the pathogenicity of *G. alacris* has not been conclusively established and that the combination of insect feeding activity and the subsequent colonization of the fungi may result in the death of the tree. However, a larger root inoculation study by Matusick *et al.* (2010, 2012) concluded that *G. alacris* was highly pathogenic to southern pines in the USA and may suggest a more important role of the pathogen in tree mortality.

13.1.5 *Grosmannia huntii*

Originally described as *Ceratocystis huntii* Rob.-Jeffr. (Robinson-Jeffrey and Grinchenko, 1964), this species has a *Leptographium* anamorph (Wingfield, 1985) that is less frequently observed. *G. huntii* is much less well known than the other species mentioned. The sexual and asexual states were first described in lodgepole pine attacked by the mountain pine beetle *D. ponderosae* Hopkins in British Columbia (Robinson-Jeffery and Grinchenko, 1964). In the USA, it has also been collected in Arizona, Colorado, New York, Oregon and Washington (Davidson and Robinson-Jeffrey, 1965). It has been isolated from black spruce (*Picea mariana* (Mill.) Britton, Sterns & Poggenb.) (Olchowecki and Reid, 1974), jack pine (*P. banksiana* Lamb.) (Olchowecki and Reid, 1974), western white pine, eastern white pine and ponderosa pine (Robinson-Jeffrey and Grinchenko, 1964). *G. huntii* was also isolated from diseased red pine roots in Wisconsin (Klepzig *et al.*, 1991), and from pine roots in the south-eastern USA, including loblolly pine and longleaf pine (Menard, 2007; Zanzot, 2009).

G. huntii has been isolated from a wide variety of insect vectors in the USA; it has been transported by *D. ponderosae* (Robinson-Jeffrey and Grinchenko, 1964), *I. pini* (Davidson and Robinson-Jeffrey, 1965; Harrington, 1988), *Tomicus piniperda* (Linnaeus) (Gibbs and Inman, 1991) and several *Hylastes* spp. (Harrington, 1988; Jacobs *et al.*, 1998; Eckhardt *et al.*, 2007; Zanzot *et al.*, 2010).

13.1.6 *Leptographium calophylli*

L. calophylli (Wiehe) J.F. Webber, K. Jacobs & M.J. Wingf. is responsible for a disease known as Takamaka disease, which occurs on Alexandrian laurel (*Calophyllum inophyllum* L.) trees, which are indigenous to Seychelles and Mauritius in Africa (Wiehe, 1949; Wainhouse *et al.*, 1998; Webber *et al.*, 1999). The fungus was initially identified as a species of *Haplographium* (Wiehe, 1949), but Gams (1971) transferred it to *Verticillium*. Webber *et al.* (1999) found that it is unlike other *Verticillium* spp. and subsequently described the fungus as *L. calophylli*.

Takamaka disease is characterized by wilting of the crowns, in which the leaves lose their shine, curl inwards and dry out. The leaf drying is sudden and the dried leaves remain attached to the trees for up to 2 weeks (Wiehe, 1949). There are no apparent lesions on the trunks, branches or roots associated with this disease. However, brownish streaks are visible in the tracheids of symptomatic trees (Wiehe, 1949).

Fungal infection occurs through wounds on the branches and twigs as a result of wounding by strong winds or bark beetle activity. Bark beetle tunnels are frequently associated with the disease, and beetle feeding and breeding habits cause additional wounding (Wiehe, 1949). The bark beetle *Cryphalus trypanus* Sampson has been identified as the principal vector of *L. calophylli*. Because the disease is very localized in character it will not be discussed any further.

13.2 Diagnosis

Symptoms associated with blackstain root disease include reduced leader and branch growth, chlorosis, reduced needle size, needle retention, needle cast, resinous lesions on the lower stems of trees (Plate 17), and a pronounced reduction in height growth (Leaphart, 1960; Cobb and Platt, 1967; Hunt and Morrison, 1980; Witcosky, 1989; Lawson and Cobb, 1987a). It is common for young seedlings and saplings to die only a few months after infection. Large trees can die in 2 years or may survive for up to 10 years. More commonly, bark beetles kill the trees before pronounced foliar symptoms are present. Infected trees appear to form more heartwood than uninfected trees, which reduces their water conduction capacity (Lawson and Cobb, 1987b). *L. wageneri* causes severe reduction in photosynthesis and transpiration as a result of water stress and stomatal closure (Helms *et al.*, 1971), which is most probably the result of the action of phytotoxins (Cobb, 1988). Root diseases associated with other ophiostomatoid fungi have very similar characteristics, although blackstain root disease is the most damaging (Wingfield *et al.*, 1988).

Symptoms associated with white pine root disease include extended periods of bud break, the retardation of shoot elongation, crooking of growing shoots, retention of needles, needle wilt, browning of needles, resin-soaked black-streaked wood at the bases of stems and basal cankers (Towers, 1977; Anderson and Alexander, 1979; Mackenzie and Dick, 1984; Alexander *et al.*, 1988).

Symptoms associated with other Leptographium diseases include the colonization of the root system and, over time, the trees experience growth loss and are predisposed to other pests; they decline and die prematurely. Affected seedlings and saplings show no apparent symptoms but die rapidly as a result of insect feeding activity, the formation of subsequent wound courts, fungal colonization and complete vascular occlusion of the root collar and roots. Maturing and older trees have a different complex of symptomology. Crown conditions change dramatically from a full green needle flush to becoming sparse with only the current year's needles present, and these become short and chlorotic. In some pine species, the needles wilt and remain on the trees after death. Maturing trees may present no crown symptoms but have limited height and stem growth. Trees often linger for several years after crown symptoms appear and before premature death. The primary roots have evidence of excessive insect feeding and breeding activity, which creates wound courts. Roots exhibit extensive areas of vascular staining and resinosis near wound courts, and insect feeding activity and galleries. Root death occurs past the point of occlusion caused by fungal growth and a resin response, leading to the loss of fine roots and mycorrhizal associations (Eckhardt *et al.*, 2007; Menard, 2007; Zanzot, 2009). The symptoms can be scattered in a random pattern across the landscape or be characterized by distinct infections centres (Klepzig *et al.*, 1991; Eckhardt *et al.*, 2007).

13.3 Infection Biology

Inoculum comes in the form of asexual spores (conidia) or ascospores carried by insects. These fungi infect host tissue as the insects feed and tunnel, forming galleries. The more insect activity that occurs the higher the level of inoculum that can be introduced.

White pine root disease begins with a dark brown discoloration of the cambium at the base of the trees and in severe infection there is marked resin exudation (Alexander *et al.*, 1988). Colonized roots are resin soaked and cross-sections of the stems reveal prominent wedges of blue-stained wood. Discoloration of the sapwood is consistent with the patterns and physiology of blue-stain fungi (Alexander *et al.*, 1988). Electron microscopical examination has shown that *L. procerum* erodes the cell walls and spreads from cell to cell via pits (Kilbertus *et al.*, 1980). Reduced water potential in symptomatic trees supports the idea that the root disease is associated with xylem dysfunction (Horner *et al.*, 1987). Tree death occurs when the xylem is blocked by resin (Alexander *et al.*, 1988).

Other species of ophiostomatoid fungi colonize phloem, xylem tracheids and ray parenchyma cells, causing a wedge-shaped appearance in the wood (Jacobs and Wingfield, 2001). The tree reacts to the fungus to produce resin, which forms an occluded lesion, and the fungal colonization appears as sapwood staining in the phloem. In this way, the fungus may aid its insect vectors in root colonization and brood fecundity (Eckhardt *et al.*, 2004b), but this has not been clearly established. Successful infection is influenced by many factors, both biotic and abiotic. Host vigour may determine its selection by the insects, along with the ability of the fungi to successfully infect the host's tissue. Abiotic stresses from weather, soil and topographic position are contributing factors to reducing host vigour. Once established in the host roots, the fungus may grow, parasitize cells and use their nutrients, or disrupt vascular flow and kill tree roots.

13.3.1 Inoculum

Fruiting bodies in members of the *Ophiostomataceae* are typically darkly pigmented perithecia with spherical bases and long necks (Upadhyay, 1981; Jacobs and Wingfield, 2001), but they are not produced in stromata (Kirk *et al.*, 2001). The asci are evanescent and ascospores are released in a gelatinous matrix through the ostiole at the perithecial apex, thus facilitating vectoring by insects. All anamorphic states are hyphomycetous, varying from micronemata (*Sporothrix* spp.), to synnemata (*Pesotum* spp.) to mononemata (*Leptographium* spp.).

Leptographium spp. typically have darkly pigmented mycelia and mononematous conidiophores that produce either hyaline or light-coloured single-celled conidia from one to several ranks of metulae (branches) (Kendrick, 1962; Wingfield, 1985). The conidia in *Leptographium* spp. are borne in slimy masses which are also well suited to insect vectors, particularly bark beetles (Molnar, 1965; Malloch and Blackwell, 1993; Wingfield, 1993).

13.3.2 Insect vectors

Insects are commonly associated with *Leptographium* species (Kendrick, 1962; Harrington, 1988; Klepzig *et al.*, 1991; Wingfield and Gibbs, 1991; Eckhardt *et al.*, 2007). There are three hypotheses to explain the relationship between the fungi and the insects. The first is that the fungi are transported with little primary benefit to the beetle (Mathre, 1964; Hinds, 1972; Goheen and Cobb, 1978; Witcosky and Hansen, 1985; Lewis and Alexander, 1986), although they may serve as food for the insects and play some role in brood development (Leach *et al.*, 1934; Nelson, 1934; Eckhardt *et al.*, 2004b). The second is that the association is just coincidental (Harrington, 1993). The third is that phytopathogenicity may mediate competitive interactions among fungi and support survival and efficient resource capture in living, defensive trees (Six and Wingfield, 2011).

The conidia of *Leptographium* spp. are sticky and adhere to the body of insects

(Harrington, 1993; Malloch and Blackwell, 1993). However, several insects carry these fungi in mycangia (Francke-Grosmann, 1965; Whitney and Farris, 1970; Barras and Perry, 1971; Six and Paine, 1996; Ross and Solheim, 1997), and mycangial fungi have been shown to be important to the beetles nutritionally.

Bark beetles associated with *Leptographium* spp. occur mostly on conifers (Grosmann, 1931; Harrington, 1988; Paine *et al.*, 1990). These beetles can be either primary or secondary pests. Primary bark beetles attack and kill healthy trees, while secondary beetles attack already weakened trees, rarely killing them (Berryman, 1972; Paine *et al.*, 1990). While most beetles have quite specific associations with *Leptographium* fungi, other associations are diverse and appear to be coincidental (Harrington, 1988). Most root-feeding bark beetles and weevils are considered to be secondary pests and have a casual relationship with *Leptographium* spp.

Several studies show that infection by root disease and blue-stain fungi predispose trees to further attack by bark beetles (Francke-Grosmann, 1965; Livingston *et al.*, 1983; Kulhavy *et al.*, 1984; Krokene and Solheim, 1996). Otrosina *et al.* (1997) showed that roots infected with *L. procerum* and *L. terebrantis* might predispose trees to further bark beetle attack by diminishing tree defences. Cobb *et al.* (1974) indicated that there is a high degree of association between root disease and *Dendroctonus* spp. that infest trees. Krokene (1996) and Kronkene and Solheim (1996) showed that aggressive beetles vector pathogenic fungi, whereas non-aggressive beetles tend to carry less pathogenic fungi.

13.4 Epidemiology

The occurrence of Leptographium decline and disease presents similar aetiology regardless of the fungal and host species involved. These factors contribute variably to the symptoms and presence of disease. Contributions by the multifarious and complicated factors associated with pine decline complexes, such as insect vector response, host selectivity and virulence of the fungi, are the focus of current studies.

The abiotic factors that predispose pine to susceptibility to vector insects and fungal infection may result in changes in pine physiology that lead to further biological consequences (Hodges *et al.*, 1979; Eckhardt and Menard, 2008; Eckhardt *et al.*, 2009). Decline symptoms appear to be more pronounced on areas with steeper slopes and a south-facing aspect. This observation led to the hypothesis that topographic factors may influence the severity of decline. The topographic features of slope and aspect have proved to be the most relevant abiotic factors in predisposing pines to stresses that reduce tree vigour to a level where tree defences against subsequent biotic stresses are overcome (Eckhardt and Menard, 2008).

Stress within these sites can be increased by the addition of a disturbance, e.g. drought, storm damage, fire or anthropogenic activity, that acts as an inciting factor. The additional stress can be related to increased insect activity and abundance, root infections and decline symptomology. The level to which these disturbances change the severity of pine decline vary depending upon: (i) the type of disturbance; (ii) the simultaneous occurrence of multiple disturbances; (iii) the severity of static abiotic stress (slope and aspect); (iv) epizootic insect populations; (v) the timing of disturbance relative to insect activity; and (vi) host age and vigour. Sites with static abiotic topography stress that also have anthropogenic disturbances, e.g. stand thinning or prescribed fire, have sharply spiking pest insect populations that remain high for up to 3 years after the event (Eckhardt, 2003). These sites also have higher levels of isolation rates for ophiostomatoid species from the insects and roots, more evidence of infected roots and more crown symptomology associated with decline (Menard *et al.*, 2006; Eckhardt *et al.*, 2007).

The pattern of stress related to pine decline and the complex factor associations of site, insect, pathogen and host can be described, and it occurs in multiple variations; further, it can and does lead to mortality levels that are both economically and environmentally problematic (Menard *et al*., 2006; Eckhardt *et al*., 2007, 2009).

13.5 Management Strategies and Tactics

Disease strategies for blackstain root disease may include the replacement of old trees by more vigorous trees that are less prone to attack by bark beetles, the spacing of trees to prevent disease spread through root contact, the planting of mixed stands (Goheen *et al*., 1978), the minimization of stand and site disturbance and the selection of disease-resistant trees (Cobb, 1988; Hansen *et al*., 1988). Sanitation through the removal of diseased trees or chemical treatment has also been suggested (Witcosky, 1989). For Douglas-fir, thinning after insect flight will reduce the activity of vectors. If species other than Douglas-fir are planted, and site disturbances and tree injury are minimized, this will also reduce the incidence of blackstain root disease. Management to produce a vigorous crop in a shorter rotation will probably reduce the incidence of *L. wageneri* and may also reduce the presence of *D. ponderosae*, which seems to prefer blackstain root diseased trees. Diseased stands should be harvested at the pathological rotation age, i.e. when volume loss to blackstain equals growth. The rapid loss of viability of the fungus in stumps means that infested sites may be regenerated without delay to lodgepole pine or to another tree species (Hunt and Morrison, 1986). An integrated pest management plan, making use of sanitation, resistant species and desirable cultural practices provides an ideal strategy for reducing the impact of blackstain root disease.

The control and management of white pine root disease includes the planting of trees on sites that are suitable for the species, the control of weevils and bark beetles, the removal of slash in and around the plantation and weed control (Alexander *et al*., 1988). It is also advisable to allow sites to lie fallow for a year, or to consider planting trees that are not susceptible to the disease (Lewis, 1985). White pine root disease affects trees more seriously when they are planted on wet sites, and poor site drainage will promote disease development (Anderson and Alexander, 1979; Alexander *et al*., 1988; Smith, 1991).

Control and management procedures for other Leptographium diseases are similar to the measures outlined above. Management to produce a vigorous crop and minimize site disturbance and tree injury are likely to reduce vector populations and fungal infection. Planting trees on sites that are suitable for that species will help, as well as 'thinning when planting', i.e. planting at a lower density so as to reduce the need for a precommercial thinning, thus eliminating a stand disturbance that attracts vector insects. Thinning the stand when insect populations are low or after spring and/or autumn insect flights decreases the vector population spike (Zeng, 2012). The removal of slash in and around stands reduces vector populations (L.G. Eckhardt, 2012, unpublished results) and thereby the available inoculum in the stand. Planting less susceptible tree species or resistant trees may also reduce the impact of root infection (Singh, 2012). An integrated pest management plan provides an ideal strategy for reducing the impact of *Hylastes* spp. and the fungi that they carry.

References

Alexander, S.A., Homer, W.E. and Lewis, K.J. (1988) *Leptographium procerum* as a pathogen of pines. In: Harrington, T.C. and Cobb, F.W. Jr (eds) *Leptographium Root Diseases on Conifers*. APS Press, St Paul, Minnesota, pp. 97–112.

Anderson, R.L. and Alexander, S.A. (1979) *How to Identify and Control White Pine Root Decline*. Forestry Bulletin SA-FR/P6, US Department of Agriculture, Forest Service, Southeastern Area, State and Private Forestry, Atlanta, Georgia.

Barnard, E.L., Gilly, S.P. and Dixon, W.N. (1991) Incidence of *Heterobasidion annosum* and other root-infecting fungi in residual stumps and roots in thinned slash pine plantations in Florida. *Plant Disease* 75, 823–828.

Barras, S.J. and Perry, T. (1971) *Leptographium terebrantis* sp. nov. associated with *Dendroctonus terebrans* in loblolly pine. *Mycopathologia et Mycologia Applicata* 43, 1–10.

Bennet, E.M. and Tattar, T.A. (1988) Bluestain fungi and insect vector interaction in Japanese black and Scots pine mortality. *Arboricultural Journal* 12, 237–247.

Berryman, A.A. (1972) Resistance of conifers to invasion by bark beetle–fungi associations. *BioScience* 22, 598–602.

Bertagnole, C.L., Woo, J.Y. and Partridge, A.D. (1983) Pathogenicity of five *Verticicladiella* species to lodgepole pine. *Canadian Journal of Botany* 61, 1861–1867.

Bertagnole, C.L., Partridge, A.D. and LeTourneau, D.J. (1987) Interacting root pathogens associated with blackstain disease of *Pinus ponderosa* Laws. *Phytopathology* 77, 1237.

Byler, J.W., Cobb, F.W. Jr and Rowney, D.L. (1979) *An Evaluation of Blackstain Root Disease on the Georgetown Divide, Eldorado County, CA*. Report No. 79, US Department of Agriculture, Forest Service, Region 5, San Francisco, California.

Cobb, F.W. Jr (1988) *Leptographium wageneri*, cause of blackstain root disease: a review of its discovery, occurrence and biology with emphasis on pinyon and ponderosa pine. In: Harrington, T.C. and Cobb, F.W. Jr (eds) *Leptographium Root Diseases on Conifers*. APS Press, St Paul, Minnesota, pp. 41–62.

Cobb, F.W. Jr and Platt, W.D. (1967) Pathogenicity of *Verticicladiella wageneri* to Douglas-fir. *Phytopathology* 57, 998–999.

Cobb, F.W. Jr, Parmeter, J.R. Jr, Wood, D.L. and Stark, R.W. (1974) Root pathogens as agents predisposing ponderosa pine and white fir to bark beetles. In: Kuhlman, E.G. (ed.) *Proceedings of the Fourth International Conference on* Fomes annosus, *September 17–22, 1973, Athens, Georgia*. US Department of Agriculture, Forestry Service, Washington, DC, pp. 8–15.

Cobb, F.W. Jr, Slaughter, G.W., Rowney, D.L. and DeMars, C.J. (1982) Rate of spread of *Ceratocystis wageneri* in ponderosa pine stands in central Sierra Nevada. *Phytopathology* 72, 1359–1362.

Cobb, F.W. Jr, Goheen, D.J. and Harrington, T.C. (1984) Blackstain root disease caused by *Verticicladiella wageneri*. In: *Proceedings of the Fourth International Congress for Plant Pathology, University of Melbourne, Melbourne, Victoria, Australia, August 17–24, 1983*. Rowprint Services, North Melbourne, Victoria, Australia, p. 84 (abstract).

Davidson, R.W. and Robinson-Jeffrey, R.C. (1965) New records of *Ceratocystis europhioides* and C. *huntii* with *Verticicladiella* imperfect stages from conifers. *Mycologia* 57, 488–490.

de Hoog, G.S. (1974) *The Genera* Blastobotrys, Sporothrix, Calcarisporium *and* Calcarisporiella *gen. nov.* Studies in Mycology 7, Centraalbureau Voor Schimmelcultures, Baarn, The Netherlands.

Diamandis, S., Epstein, L. and Cobb, F.W. Jr (1987) Germination of conidia of *Verticicladiella wageneri* on root surfaces. *Phytopathology* 77, 1238 [abstract].

Dochinger, L.S. (1967) *Leptographium* root decline of eastern white pine. *Phytopathology* 57, 809.

Duong, T.A., de Beer, Z.W., Wingfield, B.D. and Wingfield, M.J. (2012) Phylogeny and taxonomy of species in the *Grosmannia serpens* complex. *Mycologia* 104, 715–732.

Eckhardt, L.G. (2003) Biology and ecology of *Leptographium* species and their vectors as components of loblolly pine decline. PhD thesis, Louisiana State University, Baton Rouge, Louisiana.

Eckhardt, L.G. and Menard, R.D. (2008) Topographic features associated with loblolly pine decline in central Alabama. *Forest Ecology and Management* 255, 1735–1739.

Eckhardt, L.G., Jones, J.P. and Klepzig, K.D. (2004a) Pathogenicity of *Leptographium* species associated with loblolly pine decline. *Plant Disease* 88, 1174–1178.

Eckhardt, L.G., Goyer, R.A., Klepzig, K.D. and Jones, J.P. (2004b) Interactions of *Hylastes* species (Coleoptera: Scolytidae) with *Leptographium* species associated with loblolly pine decline. *Journal of Economic Entomology* 97, 468–474.

Eckhardt, L.G., Weber, A.M., Menard, R.D., Jones, J.P. and Hess, N.J. (2007) Insect–fungal complex associated with loblolly pine decline in central Alabama. *Forest Science* 53, 84–92.

Eckhardt, L.G., Menard, R.D. and Gray, E.D. (2009) Effects of oleoresins and monoterpenes on *in vitro* growth of fungi associated with pine decline in the southern United States. *Forest Pathology* 39, 157–167.

Francke-Grosmann, H. (1965) Über symbiosen von xylo-mycetophagen und ploeophagen Scolytoidea mit holzbewohnenden pilzen [About symbioses of xylomycetophagous and phloeophagous Scolytidae with wood-dwelling fungi]. International Symposium Berlin-Dahlem, *Material und Organismen*, Supplement 1, pp. 503–522.

Gams, W. (1971) *Cephalosporium*-artige Schimmelpilze (*Hyphomycetes*) [*Cephalosporium*-like fungi (*Hyphomycetes*)]. G. Fisher, Stuttgart, Germany.

Gibbs, J.N. and Inman, A. (1991) The pine shoot beetle *Tomicus piniperda* as a vector of bluestain fungi to windblown pine. *Forestry* 64, 239–249.
Goheen, D.J. and Cobb, F.W. Jr (1978) Occurrence of *Verticicladiella wageneri* and its perfect state, *Ceratocystis wageneri* sp. nov., in insect galleries. *Phytopathology* 68, 1192–1195.
Goheen, D.J., Cobb, F.W. Jr and McKibbin, G.N. (1978) Influence of soil moisture on infection of ponderosa pine by *Verticicladiella wageneri. Phytopathology* 68, 913–916.
Goidanich, G. (1936) Il genere di Ascomiceti *'Grosmannia'* G. Goid [The ascomycete genus *'Grosmannia'* G. Goid]. *Bollettino della Stazione di Patologia Vegetale* 16, 26–60.
Grosmann, H. (1931) Contributions to the knowledge concerning the life partnership between bark beetles and fungi. *Zeitschrift fur Parasitenkunde* 3, 56–102.
Hansen, E.M., Goheen, D.J., Hessburg, P.F., Witcosky, J.J. and Schowalter, T.D. (1988) Biology and management of blackstain root disease in Douglas-fir. In: Harrington, T.C. and Cobb, F.W. Jr (eds) *Leptographium Root Diseases on Conifers*. APS Press, St Paul, Minnesota, pp. 63–80.
Harrington, T.C. (1981) Cycloheximide sensitivity as a taxonomic character in *Ceratocystis. Mycologia* 73, 1123–1129.
Harrington, T.C. (1982) *Verticicladiella wageneri*: taxonomy and vector relations. PhD thesis, University of California, Berkeley, California.
Harrington, T.C. (1988) *Leptographium* species, their distribution hosts and insect vectors. In: Harrington, T.C. and Cobb, F.W. Jr (eds) *Leptographium Root Diseases on Conifers*. APS Press, St Paul, Minnesota, pp. 1–39.
Harrington, T.C. (1993) Diseases of conifers caused by species of *Ophiostoma* and *Leptographium.* In: Wingfield, M.J., Seifert, K.A. and Webber, J.F. (eds) Ceratocystis *and* Ophiostoma*: Taxonomy, Ecology and Pathogenicity*. APS Press, St Paul, Minnesota, pp. 161–172.
Harrington, T.C. and Cobb, F.W. Jr (1983) Pathogenicity of *Leptographium* and *Verticicladiella* spp. isolated from roots of western North American conifers. *Phytopathology* 73, 596–599.
Harrington, T.C. and Cobb, F.W. Jr (1984) Host specialization of three morphological variants of *Verticicladiella wageneri. Phytopathology* 74, 286–290.
Harrington, T.C. and Cobb, F.W. Jr (1986) Varieties of *Verticicladiella wageneri. Mycologia* 78, 562–567.
Harrington, T.C. and Cobb, F.W. Jr (1987) *Leptographium wageneri* var. *pseudotsugae,* var. nov., cause of blackstain root disease on Douglas-fir. *Mycotaxon* 30, 501–507.
Harrington, T.C. and Cobb, F.W. Jr (eds) (1988) *Leptographium Root Diseases on Conifers*. APS Press, St Paul, Minnesota.
Hausner, G., Reid, J. and Klassen, G.R. (1993) On the subdivision of *Ceratocystis s.l.*, based on partial ribosomal DNA sequences. *Canadian Journal of Botany* 71, 52–63.
Helms, J.A., Cobb, F.W. Jr and Whitney, H.S. (1971) Effect of infection *by Verticicladiella wageneri* on the physiology of *Pinus ponderosa. Phytopathology* 61, 920–925.
Hessburg, P.F. and Hansen, E.M. (1982) Histopathology of *Verticicladiella wageneri* in Douglas-fir. *Phytopathology* 72, 930.
Hessburg, P.F. and Hansen, E.M. (1986) Soil temperature and rate of colonization of *Ceratocystis wageneri* in Douglas-fir. *Phytopathology* 76, 627–631.
Highley, L. and Tattar, T.A. (1985) *Leptographium terebrantis* and black turpentine beetles associated with bluestain and mortality of black and Scots pine on Cape-Cod, Massachusetts. *Plant Disease* 69, 528–530.
Hinds, T.E. (1972) Insect transmission of *Ceratocystis* species associated with aspen canker. *Phytopathology* 62, 221–225.
Hodges, J.D., Elam, W.W., Watson, W.F. and Nebeker, T.E. (1979) Oleoresin characteristics and susceptibility of four southern pines to southern pine beetle (Coleoptera: Scolytidae) attacks. *Canadian Entomologist* 111, 889–896.
Horner, W.E., Alexander, S.A. and Lewis, K.J. (1987) Colonization patterns of *Verticicladiella procera* in Scots and eastern white pine and associated resin-soaking, reduced sapwood moisture content, and reduced needle water potential. *Phytopathology* 77, 557–560.
Houston, D.R. (1969) Basal canker of white pine. *Forest Science* 15, 66–83.
Hunt, R.S. and Morrison, D.J. (1980) *Blackstain Root Disease in British Columbia.* Forest Pest Leaflet 67, Canadian Forest Service, Pacific Forestry Centre, Victoria, British Columbia, Canada.
Hunt, R.S. and Morrison, D.J. (1986) Blackstain root disease on lodgepole pine in British Columbia. *Canadian Journal of Forest Research* 16, 996–999.
Jacobs, K. and Wingfield, M.J. (2001) Leptographium *Species: Tree Pathogens, Insect Associates and Agents of Bluestain.* APS Press, St Paul, Minnesota.

Jacobs, K., Wingfield, M.J., Crous, P.W. and Harrington, T.C. (1998) *Leptographium engelmannii*, a synonym of *Leptographium abietinum*, and description of *Leptographium hughesii* sp. nov. *Canadian Journal of Botany* 76, 1660–1667.
Jacobs, K., Wingfield, M.J. and Wingfield, B.D. (2001) Phylogenetic relationships in *Leptographium* based on morphological and molecular characters. *Canadian Journal of Botany* 79, 719–732.
Jacobs, K., Eckhardt, L.G. and Wingfield, M.J. (2006) *Leptographium profanum* sp. nov., a new species from hardwood roots in North America. *Canadian Journal of Botany* 84, 759–766.
Jewell, T.R. (1974) A qualitative study of cellulose distribution in *Ceratocystis* and *Europhium*. *Mycologia* 66, 139–146.
Kendrick, B.W. (1962) The *Leptographium* complex. *Verticicladiella* Hughes. *Canadian Journal of Botany* 40, 771–797.
Kilbertus, G., Mangenot, F. and Radtke, D. (1980) L'alteration des bois d'oeuvre de hêtre (*Fagus silvatica* L.) par *Fusarium solani* (Mart.) Sacc. et *Verticicladiella procera* Kendrick [The alteration of lumber beech (*Fagus sylvatica* L.) by *Fusarium solani* (Mart.) Sacc. and *Verticicladiella procera* Kendrick]. *Cryptogamie Mycologie* 1, 97–104.
Kirk, P.M., Cannon, P.F., David, J.C. and Stalpers, J.A. (2001) *Ainsworth and Bisby's Dictionary of the Fungi*, 9th edn. CAB International, Wallingford, UK.
Klepzig, K.D., Raffa, K.F. and Smalley, E.B. (1991) Association of an insect–fungal complex with red pine decline in Wisconsin. *Forest Science* 37, 1119–1139.
Klepzig, K.D., Smalley, E.B. and Raffa, K.F. (1996) Interactions of ecologically similar saprogenic fungi with healthy and abiotically stressed conifers. *Forest Ecology and Management* 86, 163–169.
Krokene, P. (1996) The role of bluestain fungi in tree-killing by bark beetles. PhD thesis, University of Oslo, Oslo.
Krokene, P. and Solheim, H. (1996) Fungal associates of five bark beetle species colonizing Norway spruce. *Canadian Journal of Forest Research* 26, 2115–2122.
Kulhavy, D.L., Partridge, A.D. and Stark, R.W. (1984) Root diseases and blister rust associated with bark beetles (Coleoptera: Scolytidae) in western white pine in Idaho. *Environmental Entomology* 13, 813–817.
Lackner, A.L. and Alexander, S.A. (1981) *Verticicladiella procera* pathogenic on *Pinus strobus* and *P. taeda*. *Phytopathology* 71, 233.
Lackner, A.L. and Alexander, S.A. (1982) Occurrence and pathogenicity of *Verticicladiella procera* in Christmas tree plantations in Virginia. *Plant Disease* 66, 211–212.
Lackner, A.L. and Alexander, S.A. (1983) Root disease and insect infestations on air pollution sensitive *Pinus strobus* and studies of pathogenicity of *Verticicladiella procera*. *Plant Disease* 67, 679–681.
Lackner, A.L. and Alexander, S.A. (1984) Incidence and development of *Verticicladiella procera* in Virginia Christmas tree plantations. *Plant Disease* 68, 210–212.
Lawson, T.T. and Cobb, F.W. Jr (1987a) Growth reductions in young Douglas-fir infected with *Verticicladiella wageneri*. In: *Abstracts of the Annual Meeting of the APS, and of the North Central Division, August 2–6, 1987*. *Phytopathology* 77, 1727, abstract 322.
Lawson, T.T. and Cobb, F.W. Jr (1987b) Pathological heartwood formation in pole-sized Douglas-fir infected with *Verticicladiella wageneri*. *Phytopathology* 77, 1727 [Abstract].
Leach, J.G., Orr, L.W. and Christensen, C. (1934) The interrelationships of bark beetles and blue-staining fungi in felled Norway pine timber. *Journal of Agricultural Research* 49, 315–341.
Leaphart, C.D. (1960) A root stain disease of eastern white pine. *Plant Disease Reporter* 44, 704–706.
Lewis, K.J. (1985) Studies on the spread of *Verticicladiella procera* by soilborne and insectborne propagules. MSc thesis, Virginia Polytechnic Institute and State University, Blacksburg, Virginia.
Lewis, K.J. and Alexander, S.A. (1986) Insects associated with the transmission of *Verticicladiella procera*. *Canadian Journal of Forest Research* 16, 1330–1333.
Lewis, K.J., Alexander, S.A. and Horner, W.E. (1987) Distribution and efficacy of propagules of *Verticicladiella procera* in soil. *Phytopathology* 77, 552–556.
Livingston, W.H., Mangini, A.C., Kinzer, H.G. and Mielke, M.E. (1983) Association of root diseases and bark beetles (*Coleoptera*: *Scolytidae*) with *Pinus ponderosa* in New Mexico. *Plant Disease* 67, 674–676.
Lorenzini, G. and Gambogi, P. (1976) Un caso di 'moria' di *Pinus pinea* associata alla presenza di *Verticicladiella* sp. (nota preliminare) [A case of 'blight' of *Pinus pinea* associated with the presence of *Verticicladiella* sp. (preliminary note)]. *Informatore Fitopatologico* 5, 5–8.
Mackenzie, M. and Dick, M. (1984) Verticicladiella root disease. *Forest Pathology in New Zealand* 8, 1–4.

Malloch, D. and Blackwell, M. (1993) Dispersal biology of the ophiostomatoid fungi. In: Wingfield, M.J., Seifert, K.A. and Webber, J.F. (eds) Ceratocystis *and* Ophiostoma*: Taxonomy, Ecology and Pathogenicity.* APS Press, St Paul, Minnesota, pp. 195–206.

Masuya, H., Wingfield, M.J., Kubono, T. and Ichihara, Y. (2004) *Leptographium pruni*, sp. nov. from bark beetle-infested *Prunus jamasakura* in Japan. *Mycologia* 96, 548–557.

Mathre, D.E. (1964) Survey of *Ceratocystis* spp. associated with bark beetles in California. *Contributions of the Boyce Thompson Institute* 22, 353–362.

Matusick, G. and Eckhardt, L.G. (2010) Variation in virulence among four root-inhabiting ophiostomatoid fungi on *Pinus taeda* L., *P. palustris* Mill., and *P. elliottii* Englem. seedlings. *Canadian Journal of Plant Pathology* 32, 361–367.

Matusick, G., Eckhardt, L.G. and Somers, G. (2010) The pathogenicity of four *Leptographium* species to mature longleaf pine roots. *Forest Ecology and Management* 260, 2189–2195.

Matusick, G., Somers, G. and Eckhardt, L.G. (2012) Root lesions in large loblolly pine (*Pinus taeda* L.) following inoculation with four root-inhabiting ophiostomatoid fungi. *Forest Pathology* 42, 37–43.

Menard, R.D. (2007) An assessment of the risk mapping system for the use of managing loblolly pine decline sites within red-cockaded woodpecker habitat. MSc thesis, Louisiana State University, Baton Rouge, Louisiana.

Menard, R., Eckhardt, L.G. and Hess, N. (2006) *Assessment of Loblolly Pine on Fort Benning Military Reservation, Fort Benning, Georgia.* Report 2006-02-01, US Department of Agriculture, Forest Service, Forest Health Protection, Pineville, Louisiana.

Meyer, G.J., Hindal, D.F. and Quinn, D.O. (1983) Occurrence of white pine root decline in western Virginia. *Phytopathology* 73, 967.

Molnar, A.C. (1965) Pathogenic fungi associated with a bark beetle on alpine fir. *Canadian Journal of Botany* 43, 563–570.

Morrison, D.J. and Hunt, R.S. (1988) *Leptographium* species associated with root disease of conifers in British Columbia. In: Harrington, T.C. and Cobb, F.W. Jr (eds) *Leptographium Root Diseases on Conifers.* APS Press, St Paul, Minnesota, pp. 97–112.

Nelson, R.M. (1934) Effect of bluestain fungi on southern pines attacked by bark beetles. *Phytopathologische Zeitschrift* 7, 327–353.

Nevill, R.J. and Alexander, S.A. (1992a) Transmission of *Leptographium procerum* to eastern white pine by *Hylobius pales* and *Pissodes nemorensis* (Coleoptera: Curculionidae). *Plant Disease* 76, 307–310.

Nevill, R.J. and Alexander, S.A. (1992b) Distribution of *Hylobius pales* and *Pissodes nemorensis* (Coleoptera: Curculionidae) within Christmas tree plantations with procerum root disease. *Environmental Entomology* 2, 1077–1085.

Nevill, R.J., Kelley, W.D., Hess, N.J. and Perry, T.J. (1995) Pathogenicity to loblolly pines of fungi recovered from trees attacked by southern pine beetles. *Southern Journal of Applied Forestry* 19, 78–83.

Olchowecki, A. and Reid, J. (1974) The genus *Ceratocystis* in Manitoba. *Canadian Journal of Botany* 52, 1675–1711.

Otrosina, W.J., Hess, N.J., Zarnoch, S.I., Perry, T.J. and Jones, J.P. (1997) Bluestain fungi associated with roots of southern pine trees attacked by the southern pine beetle, *Dendroctonus frontalis. Plant Disease* 81, 942–945.

Otrosina, W.J., Bannwart, D. and Roncadori, R.W. (1999) Root-infecting fungi associated with a decline of longleaf pine in the southeastern United States. *Plant and Soil* 217, 145–150.

Otrosina, W.J., Walkinshaw, C.H., Zarnoch, S.J., Sung, S. and Sullivan, B.T. (2002) Root disease, longleaf pine mortality, and prescribed burning. In: Outcalt, K.W. (ed.) *Proceedings of the Eleventh Biennial Southern Silvicultural Research Conference.* General Technical Report SRS-48, US Department of Agriculture, Forest Service, Southern Research Station, Asheville, North Carolina, pp. 551–557.

Owen, D.R., Lindahl, K.Q., Wood, D.L. and Parmeter, J.R. Jr (1987) Pathogenicity of fungi isolated from *Dendroctonus valens, D. brevicomis,* and *D. ponderosae* to ponderosa pine seedlings. *Phytopathology* 77, 631–636.

Paine, T.D., Raffa, K.F. and Harrington, T.C. (1990) Interactions among scolytid bark beetles, their associated fungi, and live host conifers. *Annual Review of Entomology* 42, 179–206.

Parmeter, J.R. Jr, Slaughter, G.W., Chen, M.-M., Wood, D.L. and Stubbs, H.A. (1989) Single and mixed inoculations of ponderosa pine with fungal associates of *Dendroctonus* spp. *Phytopathology* 79, 768–772.

Parmeter, J.R. Jr, Slaughter, G.W., Chen, M.-M. and Wood, D.L. (1992) Rate and depth of sapwood occlusion following inoculation of pine with bluestain fungi. *Forest Science* 38, 34–44.

Rane, K.K. and Tattar, T.A. (1987) Pathogenicity of bluestain fungi associated with *Dendroctonus terebrans*. *Plant Disease* 71, 879–883.

Robinson-Jeffrey, R.C. and Grinchenko, A.H.H. (1964) A new fungus in the genus *Ceratocystis* occurring on bluestained lodgepole pine attacked by bark beetles. *Canadian Journal of Botany* 42, 527–532.

Ross, D.W. and Solheim, H. (1997) Pathogenicity to Douglas-fir of *Ophiostoma pseudotsugae* and *Leptographium abietinum*, fungi associated with the Douglas-fir beetle. *Canadian Journal of Forest Research* 27, 39–43.

Ross, D.W., Fenn, P. and Stephen, F.M. (1992) Growth of southern pine beetle associated fungi in relation to induced wound response in loblolly pine. *Canadian Journal of Forest Research* 22, 1851–1859.

Seifert, K.A. (1993) Sapstain of commercial lumber by species of *Ophiostoma* and *Ceratocystis*. In: Wingfield, M.J., Seifert, K.A. and Webber, J.F. (eds) Ceratocystis *and* Ophiostoma*: Taxonomy, Ecology and Pathogenicity*. APS Press, St Paul, Minnesota, pp. 141–151.

Sinclair, W.A. and Hudler, G.W. (1980) Tree and shrub pathogens new or noteworthy in New York State. *Plant Disease* 64, 590–592.

Singh, A. (2012) Variation in resistance of *Pinus taeda* and *Pinus elliotii* families against *Leptographium* root infection fungi and the assessment of morphological traits linked to *Leptographium* infection. MSc thesis, Auburn University, Auburn, Alabama.

Six, D.L. and Paine, T.D. (1996) *Leptographium pyrinum* is a mycangial fungus of *Dendroctonus adjunctus*. *Mycologia* 88, 739–744.

Six, D.L. and Wingfield, M.J. (2011) The role of phytopathogenicity in bark beetle–fungus symbioses: a challenge to the classic paradigm. *Annual Review of Entomology* 56, 255–272.

Smalley, E.B., Raffa, K.F., Proctor, R.H. and Klepzig, K.D. (1993) Tree responses to infection by species of *Ophiostoma* and *Ceratocystis*. In: Wingfield, M.J., Seifert, K.A. and Webber, J.F. (eds) Ceratocystis *and* Ophiostoma*: Taxonomy, Ecology and Pathogenicity*. APS Press, St Paul, Minnesota, pp. 207–217.

Smith, R.S. Jr (1967) Verticicladiella root disease of pines. *Phytopathology* 57, 935–938.

Smith, V.L. (1991) Occurrence of procerum root disease caused by *Leptographium procerum* on white pine in Connecticut. *Plant Disease* 75, 431.

Solheim, H. (1992a) Fungal succession in sapwood of Norway spruce infested by the bark beetle *Ips typographus*. *European Journal of Forest Pathology* 22, 136–148.

Solheim, H. (1992b) The early stages of fungal invasion in Norway spruce infested by the bark beetle *Ips typographus*. *Canadian Journal of Botany* 70, 1–5.

Spatafora, J.W. and Blackwell, M. (1994) The polyphyletic origins of ophiostomatoid fungi. *Mycological Research* 98, 1–9.

Towers, B. (1977) The occurrence of *Verticicladiella procera* in Pennsylvania: 1976. *Plant Disease Reporter* 61, 477.

Upadhyay, H.P. (1981) *A Monograph of* Ceratocystis *and* Ceratocystiopsis. University of Georgia Press, Athens, Georgia.

Wagener, W.W. and Mielke, J.L. (1961) A staining fungus root disease of ponderosa, Jeffrey and pinyon pines. *Plant Disease Reporter* 45, 831–835.

Wainhouse, D., Murphy, S., Grieg, B., Webber, J.F. and Vielle, M. (1998) The role of the bark beetle *Cryphalus trypanus* in the transmission of the vascular wilt pathogen of takamaka (*Calophyllum inophyllum*) in the Seychelles. *Forest Ecology and Management* 108, 193–199.

Walters, J.W. and Walters, N.R. (1977) *Verticicladiella wageneri* in the southwest. *Plant Disease Reporter* 61, 419.

Weaver, M.J., Stipes, R.J., Amos, D.F. and Heikkenen, H.J. (1981) Factors associated with declining eastern white pine (*Pinus strobus*) on Virginia landscapes. *Phytopathology* 71, 772.

Webber, J.F., Jacobs, K. and Wingfield M.J. (1999) A re-examination of the vascular wilt pathogen of takamaka (*Calophyllum inophyllum*). *Mycological Research* 103, 1588–1592.

Whitney, H.S. and Farris, S.H. (1970) Maxillary mycangium in the mountain pine beetle. *Science* 167, 54–55.

Wiehe, P.O. (1949) Wilt of *Calophyllum inophyllum* L. var. *tacamaca* (Willd.) R.E.V. caused by *Haplographium calophylli* sp. nov. in Mauritius. *Mycological Papers* 29, 1–12.

Wingfield, M.J. (1983) Association of *Verticicladiella procera* and *Leptographium terebrantis* with insects in the Lake States. *Canadian Journal of Forest Research* 13, 1238–1245.

Wingfield, M.J. (1985) Re-classification of *Verticicladiella* based on conidial development. *Transactions of the British Mycological Society* 85, 81–93.

Wingfield, M.J. (1986) Pathogenicity of *Leptographium procerum* and *Leptographium terebrantis* on *Pinus strobus* seedlings and established trees. *European Journal of Forest Pathology* 16, 299–308.

1

Plate 1. Virus-induced leaf symptoms: (**A**) chlorotic ringspots in *Quercus robur*, (**B**) chlorotic line pattern in CLRV-infected *Fagus sylvatica*, (**C**) extensive intercostal chlorosis of CLRV-infected *Betula pubescens*, (**D**) chloroses of ArMV-infected *Betula pendula*, (**E**) chlorotic ringspots in *Ulmus laevis*, (**F**) chlorotic leaf pattern of CLRV-infected *Sambucus nigra*, (**G**) chlorotic and necrotic spots in CLRV-infected *Sorbus aucuparia*, (**H**) chlorotic ringspots in EMARaV-infected *S. aucuparia*, (**I**) little leaf and staghead induced by PopMV in *Populus nigra*, and (**J**) virus-like symptoms in *S. aucuparia* associated with ApMV.

6

7

Plate 6. Manual stump treatment with urea solution against *Heterobasidion annosum s.l.* on a *Picea abies* stump. Credits: Paolo Gonthier.

Plate 7. Single-grip harvester performing stump treatment against *Heterobasidion annosum s.l.* in thinning. Credits: Stefan Örtenblad.

8

9

Plate 8. *Armillaria ostoyae* mycelial fans. Credits: Philippe Legrand.
Plate 9. *Armillaria gallica* sub-cortical rhizomorphs. Credits: Jean-Jacques Guillaumin.

10

11

Plate 10. Typical appearance of upturned trees whose root systems have been infected with (**A**) *Phellinus sulphurascens* and (**B**) *Onnia tomentosa*. Note that *P. sulphurascens* causes more extensive root decay resulting in the 'root ball' appearance. Credits: (**A**) Canadian Forestry Service/Natural Resources Canada, (**B**) Kathy J. Lewis.

Plate 11. (**A**) Dark brown setal hyphae in wood colonized by *Phellinus sulphurascens*, (**B**) yellow white mycelium of *Onnia tomentosa* in the bark of colonized spruce root, and (**C**) laminated decay pattern caused by *P. sulphurascens*. Credits: (**A, C**) Canadian Forestry Service/Natural Resources Canada, (**B**) Kathy J. Lewis.

12

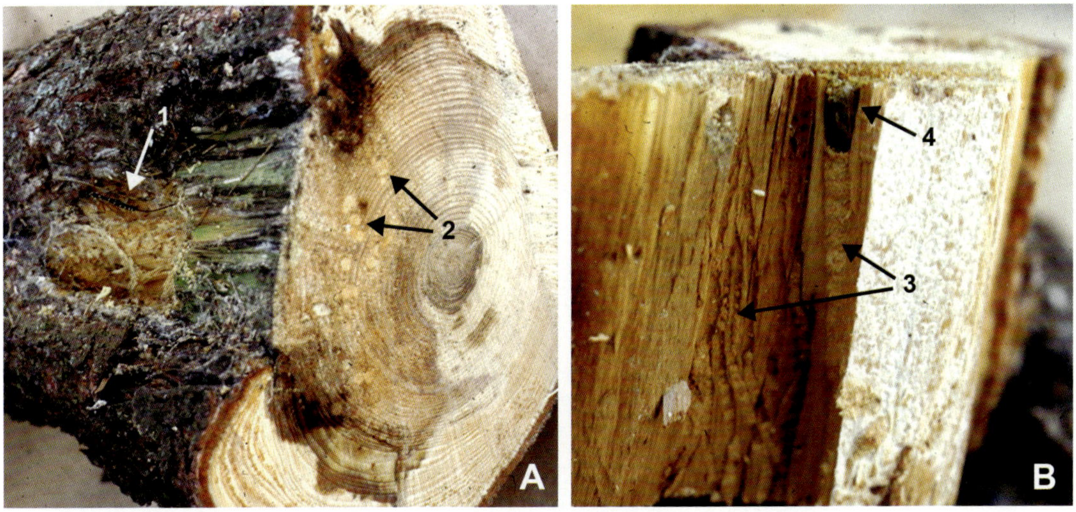

13

Plate 12. Cross-section (**A**) and longitudinal section (**B**) of wound heart rot columns caused by *Amylostereum areolatum* in stems of *Picea abies*. Arrows indicate: (**1**) *Rhyssa* sp., a parasitic wasp (*Ichneumonidae*), looking for larvae of *Sirex juvencus* inside the wound to put the egg in (note woodpecker activity on a surface of the wound, also in a hunt for the larvae); (**2**) and (**3**) larval tunnels of *S. juvencus* filled with a fine frass; (**4**) dead adult of *S. juvencus* in the pupal chamber. Note that larval tunnels are always confined within a heart rot column. Credits: Vytautas Bardauskas.

Plate 13. *Platanus hybrida* infected by *Ceratocystis platani*. Sawdust generated during felling is collected in tarps. Credits: Giovanni Nicolotti.

14

15

Plate 14. Vibratory plow used in Minnesota to sever root grafts between healthy oaks and trees with oak wilt. The blade is inserted into the ground and vibrates as it is pulled through the forest floor with minimal soil disturbance. Credits: Charles Evenson.

Plate 15. Storage of *Picea abies* logs in stacks constantly sprinkled with water to protect them against insect infestations and infections by blue-stain and other fungi.

16

17

Plate 16. Wrapping and airtight sealing of fresh logs of high quality in a double layer of UV-resistant polyethylene foil is a relatively new long-term storage method to protect logs against insect infestations and infections by blue-stain and other fungi.

Plate 17. Resinous lesion associated with blackstain root disease on *Pinus Ponderosa*. Credits: Paolo Gonthier.

18

19

Plate 18. Examples of canker (**A**, **B**) and dieback (**C**) of eucalyptus trees caused by species of the *Botryosphaeriaceae*.

Plate 19. Bark necrosis (cankers) on chestnut trees caused by virulent strains (**A**) and hypovirulent strains (**B**) of *Cryphonectria parasitica*. Credits: Phytopathology WSL.

20

21

Plate 20. Biological control of chestnut blight by inoculation of virulent cankers with hypovirus-infected *Cryphonectria parasitica* strains. Credits: Phytopathology WSL.

Plate 21. Development of the necrotic process on young *Cupressus sempervirens* stems 30 days (**A**), 60 days (**B**) and 90 days (**C**) after the artificial inoculation with *Seiridium cardinale*.

22

23

Plate 22. Crown symptoms caused by *Seiridium cardinale* on *Cupressus sempervirens*.
Plate 23. Canopy dieback in *Pinus radiata* due to pitch canker. Each dead branch corresponds to at least one infection by *Gibberella circinata*.

24

25

Plate 24. Discolored tissue corresponds to a lesion produced by *Gibberella circinata*, which was revealed by removing the bark at an infection site on a branch of *Pinus radiata*.
Plate 25. Stand of *Larix decidua* declining as a consequence of stems girdled by *Lachnellula willkommii*. Murau, Styria, Austria. Credits: Thomas L. Cech, BFW.

26

28

27

Plate 26. Stand of *Larix decidua* sanitary thinned resulting in efficient reduction of larch canker. Neckenmarkt, Burgenland, Austria. Credits: Hannes Schönauer, BFW.
Plate 27. Stem breakage of a *Populus tremuloides* tree at a Hypoxylon canker.
Plate 28. Young Hypoxylon stem canker on *Populus tremuloides* that originated in a branch gall resulting from the oviposition of a poplar gall saperda (*Saperda inornata*).

29

30

Plate 29. An aerial spray trial with copper fungicides against Dothistroma needle blight on *Pinus radiata* in Kaingaroa Forest, central North Island, New Zealand. Compartment 904B. (**A**) Unsprayed treatment and (**B**) single spray treatment (see van der Pas *et al.*, 1984).

Plate 30. *Lithocarpus densiflorus* mortality caused by *Phytophthora ramorum* at Shell Beach, Tomales Bay State Park, Marin County, California. Credits: Doug Schmidt, Garbelotto Laboratory, UC Berkeley.

31

32

Plate 31. An oak injected with phosphonates to control *Phytophthora ramorum*. Credits: Doug Schmidt, Garbelotto Laboratory, UC Berkeley.

Plate 32. Fungicide treatment against birch rust in a nursery. Credits: Erkki Oksanen, METLA.

Wingfield, M.J. (1993) *Leptographium* species as anamorphs of *Ophiostoma*: progress in establishing acceptable generic and species concepts. In: Wingfield, M.J., Seifert, K.A. and Webber, J.F. (eds) Ceratocystis *and* Ophiostoma*: Taxonomy, Ecology and Pathogenicity*. APS Press, St Paul, Minnesota, pp. 43–51.

Wingfield, M.J. and Gibbs, J.N. (1991) *Leptographium* and *Graphium* species associated with pine-infesting bark beetles in England. *Mycological Research* 95, 1257–1260.

Wingfield, M.J and Knox-Davies, P.S. (1980) Root disease, associated with *Verticicladiella alacris*, of pines in South Africa. *Plant Disease* 64, 569–571.

Wingfield, M.J. and Marasas, W.F.O. (1980) *Verticicladiella alacris* sp. nov., associated with a root disease of pines in South Africa. *Transactions of the British Mycological Society* 75, 21–28.

Wingfield, M.J. and Marasas, W.F.O. (1981) *Verticicladiella alacris*, a synonym of *Verticicladiella serpens*. *Transactions of the British Mycological Society* 76, 508–510.

Wingfield, M.J., Capretti, P. and MacKenzie, M. (1988) *Leptographium* spp. as root pathogens on conifers. An international perspective. In: Harrington, T.C. and Cobb, F.W. Jr (eds) *Leptographium Root Diseases on Conifers*. APS Press, St Paul, Minnesota, pp. 113–128.

Wingfield, M.J., Seifert, K.A. and Webber, J.F. (eds) (1993) Ceratocystis *and* Ophiostoma. *Taxonomy, Ecology and Pathogenicity*. APS Press, St Paul, Minnesota.

Witcosky, J.J. (1989) Root beetles, stand disturbance, and management of blackstain root disease in plantations of Douglas-fir. In: Alfaro, R.I. and Glover, S.G. (eds) *Insects Affecting Reforestation: Biology and Damage. Proceedings of a Meeting of the IUFRO Working Group on Insects Affecting Reforestation (S2.07–03), held under the auspices of the eighteenth International Congress of Entomology, July 3–9, 1988, Vancouver, BC*. Forestry Canada, Pacific and Yukon Region, Pacific Forestry Centre, Victoria, British Columbia, Canada, pp. 58–70.

Witcosky, J.J. and Hansen, E.M. (1985) Root-colonizing insects recovered from Douglas-fir in various stages of decline due to blackstain root disease. *Phytopathology* 75, 399–402.

Witcosky, J.J., Schowalter, T.D. and Hansen, E.M. (1986) *Hylastes nigrinus* (Coleoptera: Scolytidae), *Pissodes fasciatus*, and *Steremnius carinatus* (Coleoptera: Curculionidae) as vectors of blackstain root disease of Douglas-fir. *Environmental Entomology* 15, 1090–1095.

Zanzot, J.W. (2009) Biology and ecology of root-feeding beetles and ophiostomatoid fungi in sandhills longleaf pine stands. PhD thesis, Auburn University, Auburn, Alabama.

Zanzot, J.W., Matusick, G. and Eckhardt, L.G. (2010) Ecology of root-feeding beetles and their associated fungi on longleaf pine in Georgia. *Environmental Entomology* 39, 415–423.

Zeng, Y. (2012) Silvicultural disturbances effect on root-feeding bark beetle populations and the incidence of ophiostomatoid fungal species contributing to southern pine decline in *Pinus taeda* stands. MSc thesis, Auburn University, Auburn, Alabama.

Zhou, X.D., De Beer, W., Wingfield, B.D. and Wingfield, M.J. (2002) Infection sequence and pathogenicity of *Ophiostoma ips*, *Leptographium serpens* and *L. lundbergii* to pines in South Africa. *Fungal Diversity* 10, 229–240.

Zhou, X.D., Jacobs, K., Kirisits, T., Chhetri, D.B. and Wingfield, M.J. (2008) *Leptographium bhutanense* sp. nov., associated with the root collar weevil *Hylobitelus chenkupdorjii* on *Pinus wallichiana* in Bhutan. *Persoonia* 21, 1–8.

Zipfel, R.D., de Beer, W., Jacobs, K., Wingfield, B.D. and Wingfield, M.J. (2006) Multi-gene phylogenies define *Ceratocystiopsis* and *Grosmannia* distinct from *Ophiostoma*. *Studies in Mycology* 55, 75–97.

14 Cankers and Other Diseases Caused by the *Botryosphaeriaceae*

James W.M. Mehl,* Bernard Slippers, Jolanda Roux and Michael J. Wingfield
University of Pretoria, Pretoria, South Africa

14.1 Pathogens, Significance and Distribution

Species of the ascomycete family *Botryosphaeriaceae* are important pathogens of trees and other woody plants. They have been reported from more than 1000 plant hosts. These include commercially important trees utilized in agriculture and forestry as well as those in native forests. Various disease symptoms have been linked to infection by species in the family, including cankers, dieback and blight of individual plant parts or entire trees (Slippers and Wingfield, 2007) (Plate 18). Several publications have predicted the increasing importance of the *Botryosphaeriaceae* in the light of global climate change (Desprez-Loustau *et al.*, 2007; Slippers and Wingfield, 2007; Thompson *et al.*, 2010). Furthermore, recent research has linked species in the family to outbreaks of new diseases (Alvarez-Loayza *et al.*, 2008; Dakin *et al.*, 2010; Piškur *et al.*, 2011) or to the re-emergence of known diseases (Giraud, 2009).

A cryptic and more recently recognized aspect of the life cycle of the *Botryosphaeriaceae* is their ability to infect host plants and to persist endophytically as latent pathogens. Before this discovery, it was thought that these fungi were systemic pathogens (Webb, 1983; Shearer *et al.*, 1987). Since they were first recognized as tree pathogens, the *Botryosphaeriaceae* have been treated as wound-infecting opportunists, although it was also known that they were capable of infecting via natural openings (stomata, lenticels and others). The first evidence of endophytism in these fungi emerged when Petrini and Fisher (1988) isolated *Diplodia pinea* (Desm.) J. Kickx f. (syn. *Sphaeropsis sapinea* (Fr.) Dyko & B. Sutton) from healthy stems of Scots pine (*Pinus sylvestris* L.). Later, Fisher *et al.* (1993) isolated *Botryosphaeria dothidea* (Moug.) Ces. & De Not. from leaves, twig bark and asymptomatic xylem of shining gum (*Eucalyptus nitens* (H. Deane & Maiden) Maiden). Subsequent studies have confirmed that virtually every species in the family occupies a plant host endophytically for at least part of its life cycle (Slippers and Wingfield, 2007).

Several reviews addressing various aspects of the *Botryosphaeriaceae* have appeared in the past decade (Crous *et al.*, 2006; Slippers and Wingfield, 2007; Phillips *et al.*, 2008; Slippers *et al.*, 2009). These have underscored the importance of the *Botryosphaeriaceae* as plant pathogens. Some of them have addressed the taxonomy of the family (Crous *et al.*, 2006; Phillips *et al.*, 2008) by clarifying the identity or resolving the taxonomy of individual species

* E-mail: james.mehl@fabi.up.ac.za

(de Wet *et al.*, 2003; Slippers *et al.*, 2004a; Phillips *et al.*, 2006), while others have clarified host associations (Slippers *et al.*, 2004b, 2007, 2009). However, the ecological role of most of these fungi remains poorly understood (Slippers and Wingfield, 2007).

At least 20 genera are classified in the family *Botryosphaeriaceae*. Molecular tools have enabled the systematic grouping of these genera within the last decade, and clarified many complex taxonomic questions. In many cases, the sexual state linked to an asexual genus remains undiscovered (e.g. *Endomelanconiopsis, Lasiodiplodia, Pseudofusicoccum*), while in other cases asexual states have been linked as synanamorphs or to sexual morphs (e.g. *Botryosphaeria* with *Fusicoccum*, *Phaeobotryosphaeria* with *Sphaeropsis* and *B. dothidea*, *Neofusicoccum* and *Neoscytalidium* with *Dichomera*) (Phillips, 2002; Slippers *et al.*, 2004a; Phillips *et al.*, 2005, 2008; Crous *et al.*, 2006; Pavlic *et al.*, 2008). Regardless of the taxonomic issues, most of these fungi are known pathogens or have been isolated from diseased plants.

14.2 Diagnosis

Numerous symptoms accompany infection by the *Botryosphaeriaceae*, and these typically become obvious once plants are subjected to stress. Disease symptoms have been comprehensively listed by Slippers and Wingfield (2007) and include: cankers and dieback of twigs, branches and the stem (Plate 18); root rot and cankers of the roots; blue-stain; blossom blights, fruit rots and seed capsule abortion; seedling damping-off and blight; and tree death. The diversity of *Botryosphaeriaceae* associated with most of these disease symptoms necessitates laboratory isolation and subsequent DNA-based diagnostics.

The association of one or more species of the *Botryosphaeriaceae* with disease symptoms typically commences with the collection of diseased material. For comparative purposes, asymptomatic or seemingly healthy material should also be collected. Material is then stored at 4°C and can be processed immediately after collection or left for a week or more. During processing, material can be sectioned, surface disinfected to eliminate epiphytic fungi and placed on to any generalist media (e.g. 2% malt extract agar, potato dextrose agar) for isolation. Media can be supplemented with antibiotics to inhibit the growth of potentially competitive bacteria. As a general guide, isolates with a 'fluffy' mycelium, either white in colour to creamy or pigmented 'greenish brown' or grey to grey black potentially represent the *Botryosphaeriaceae* and can be grouped according to culture morphology (Slippers and Wingfield, 2007).

Isolates can be induced to sporulate by transferring a mycelial plug to water agar overlaid with sterile pine needles (Pavlic *et al.*, 2004) or sterile host material, either branch sections (Mehl *et al.*, 2011) or leaves (Inderbitzin *et al.*, 2010). Plates can then be placed under continual near-ultraviolet light (Pavlic *et al.*, 2004), in the dark, or under artificial laboratory bench light (Inderbitzin *et al.*, 2010). The resulting pycnidia are useful for studying conidial morphology and making single spore cultures for DNA-based studies.

The sequencing of species of the *Botryosphaeriaceae* was initially confined to the internal transcribed spacer (ITS) ribosomal DNA (rDNA) locus (consisting of ITS1, 5.8S and ITS2) (Jacobs and Rehner, 1998; Zhou and Stanosz, 2001). Virtually every subsequent study has used several loci, including ITS, because combined sequence data from multiple loci results in improved resolution among taxa (de Wet *et al.*, 2003; Inderbitzin *et al.*, 2010; Mehl *et al.*, 2011). Additional loci used in phylogenetic species recognition include the mitochondrial small subunit (mtSSU) (Zhou and Stanosz, 2001), protein coding genes (for translation elongation factor 1α (TEF-1α), β-tubulin, actin, calmodulin, chitin synthase, glyceraldehyde-6-phosphate dehydrogenase, histone-3 and heat shock protein) (de Wet *et al.*, 2003; Inderbitzin *et al.*, 2010), various unknown microsatellite-containing loci (de Wet *et al.*, 2003; Pavlic *et al.*, 2009) and data from the rDNA locus (small subunit (SSU), ITS and large subunit (LSU)) (Zhou and Stanosz, 2001; Crous *et al.*, 2006; Phillips *et al.*, 2008). Concordance among multiple loci has been used to convincingly argue for the existence of cryptic species, an approach known as genealogical concordance phylogenetic species recognition (GCPSR), or applied as a genealogical sorting index (de Wet *et al.*, 2003; Slippers *et al.*, 2004b; Pavlic *et al.*, 2009; Sakalidis *et al.*, 2011). Regardless of the loci selected for phylogenetic analyses, most sequence data pertaining to species of the *Botryosphaeriaceae* on GenBank (the US National Institutes of Health (NIH) genetic sequence database provided by the US National Center for Biotechnology Information (NCBI)) are limited to the ITS and TEF-1α loci. Much work is therefore required to fill the gaps and to facilitate analyses that incorporate sequence data from the other loci listed.

PCR-RFLPs (restriction fragment length polymorphisms) were first developed by Jacobs (2002) to distinguish among species infecting mangoes (*Mangifera indica* L.). Typically, the ITS region is amplified, sometimes in combination with another gene region (Pavlic *et al.*, 2007), and digested using the restriction endonuclease *Cfo*I or various combinations of other enzymes (Slippers *et al.*, 2004b, 2007; Alves *et al.*, 2005). This method, however, does not distinguish among all species using the banding pattern from a particular locus (Slippers *et al.*, 2004b), thereby necessitating the amplification of other loci and their subsequent restriction. Furthermore, sequence data are initially required to test whether restriction sites for the selected endonucleases occur within the locus to be amplified. Consequently, the method has little relevance in screening plant material for which the full complement of the associated *Botryosphaeriaceae* has not been determined.

Species-specific primers, as the name implies, enable the detection of an individual species of the *Botryosphaeriaceae*. These primers have been developed based on variations in sequence data of the ITS locus and tested using nested PCR (Ma and Michailides, 2002; Flowers *et al.*, 2003). One study developed primers based on data from the mtSSU locus (Smith and Stanosz, 2006), while another exploited sequence variation within the same locus to delineate among several species using melting analysis of the PCR products (Luchi *et al.*, 2011). Other than the study of Ma and Michailides (2002), these studies have tested the ability of the primers developed to detect species *in planta* using wood, bark or bud samples. Primers tested are usually extremely sensitive. Real-time quantitative PCR has enabled detection at as little as 0.1 pg fungal DNA, or roughly two to three asexual spores (conidia) (Ridgway *et al.*, 2011; Spagnolo *et al.*, 2011). These studies have been confined to the *Botryosphaeriaceae* on pines (*Pinus* spp.) (Flowers *et al.*, 2003; Luchi *et al.*, 2005, 2011; Smith and Stanosz, 2006) and grapevines (Ridgway *et al.*, 2011; Spagnolo *et al.*, 2011), and have only concentrated on a few species: *D. pinea*, *D. seriata* De Not., *D. scrobiculata* de Wet, Slippers & M.J. Wingf. (Flowers *et al.*, 2003; Luchi *et al.*, 2005, 2011; Smith and Stanosz, 2006), *Neofusicoccum parvum* (Pennycook & Samuels) Crous, Slippers & A.J.L. Phillips/ *N. ribis* (Slippers, Crous & M.J. Wingf.) Crous, Slippers & A.J.L. Phillips and *B. dothidea* (Ridgway *et al.*, 2011; Spagnolo *et al.*, 2011). There is a need to consider species from many other genera within the family.

The molecular tools outlined above have enabled rapid identification and delineation amongst the *Botryosphaeriaceae*. They can potentially be used for quarantine purposes to screen plant material for these fungi. However, there is a danger of obtaining false positive or negative results and the limitations of these techniques should thus be carefully considered.

14.3 Infection Biology

Based on the classification of Rodriguez *et al.* (2009), the *Botryosphaeriaceae* are best grouped with the class 3 endophytes. These are defined as non-clavicipitaceous endophytes that have a broad host range, occupy the shoots of plants, have limited *in planta* colonization, high *in planta* biodiversity, grow intercellularly and are transmitted horizontally. Like other endophytes, the *Botryosphaeriaceae* germinate epiphytically, penetrate plants and grow intercellularly (Arnold and Herre, 2003). The ability of these fungi to invade plant tissue and persist therein is competitively advantageous, especially when they later switch to a saprophytic life cycle.

The *Botryosphaeriaceae* have two main strategies for infecting plants. The first involves infection via natural apertures and endophytic persistence within asymptomatic tissue (Flowers *et al.*, 2003; Bihon *et al.*, 2011a). Infection generally requires the presence of free water (Rayachhetry *et al.*, 1996a). Conidia can be trapped among bud scales resulting in subsequent infection of the buds (Beisel *et al.*, 1984; Michailides, 1991; Michailides and Morgan, 1993). Other natural openings, such as stomata on leaves and shoots, and lenticels either on fruit or stems, as well as the inflorescences, pedicels or peduncles of fruit trees, provide avenues for infection (Weaver, 1979; Michailides, 1991; Johnson and Kotzé, 1994; Kim *et al.*, 1999; Ploetz, 2003). The process begins with conidial germination: germ tubes emerge from either one or both ends of the conidia (Milholland, 1970; Kim *et al.*, 1999). In the case of stomata that may be closed, germ tubes encircle stomates or grow over them (Brookhouser and Peterson, 1970; Milholland, 1970). Utilizing the peduncle, the *Botryosphaeriaceae* gain entrance to fruits to initiate stem end rots (Johnson and Kotzé, 1994) once fruit begin ripening or after harvest (Menge and Ploetz, 2003; Timmer *et al.*, 2003). Direct appressorial penetration of fruit has been reported when high levels of inoculum are present, but appears to be rare (Johnson and Kotzé, 1994; Menge and Ploetz, 2003).

The second strategy for plant infection by the *Botryosphaeriaceae* is via wounds (Epstein *et al.*, 2008). Wounds associated with diseases caused by these fungi include those resulting from human activities such as pruning, hand-picking of fruit, grafting, or via scars when scales, buds, petioles and fruit abscise (Michailides, 1991; Johnson, 1994; Pusey *et al.*, 1995; Menge and Ploetz, 2003). Large numbers of isolates have been obtained from graft unions, suggesting transmission by grafting tools (Aroca *et al.*, 2006). These high numbers could, however, result from stress arising from graft incompatibility and the presence of *Botryosphaeriaceae* occupying the tissue endophytically. Other sources of wound infections include those attributed to branch splitting resulting from wind, frost and hail damage (Swart and Wingfield, 1991; Menge and Ploetz, 2003) although even these wounds could have enabled endophytic *Botryosphaeriaceae* to emerge. Finally, wounds naturally formed during root growth can also enable infection from inoculum present on residual diseased plant material on the ground (Whitelaw-Weckert *et al.*, 2006).

14.3.1 Histopathology

The infection process of the *Botryosphaeriaceae* argues against a strict classification of these fungi as class 3 endophytes. Class 3 endophytes have been reported to only grow intercellularly (Rodriguez *et al.*, 2009). In contrast, the *Botryosphaeriaceae* can grow both intercellularly and intracellularly. Additionally, tissue specificity occurs among different species of class 3 endophytes, promoting the absence of niche overlap and cooperation, rather than competition, between these fungi. However, different species of the *Botryosphaeriaceae* can occupy the same tissue sample simultaneously (Pusey, 1993; Smith *et al.*, 2001).

After penetration via the stomata, species of the *Botryosphaeriaceae* produce an appressorium-like hyphal mass in the stomatal pit (Brookhouser and Peterson, 1970; Milholland, 1970). Mycelium originating from this mass then grows into the mesophyll (Brookhouser and Peterson, 1970). In leaves, the mycelium continues to grow, advancing through the intercellular spaces of the mesophyll and intracellular spaces of the midrib to colonize the oil cells. Pycnidial stromata are initiated in the oil cells that develop and these stromata rupture the leaf epidermis (Rayachhetry *et al.*, 1996a). In fruits, fungal progress is rapid as the fruits are used as a source of nutrition. In avocadoes (*Persea americana* Mill.), the *Botryosphaeriaceae* move via the vascular bundles to the base and stem ends to encompass the fruit (Johnson, 1994), while in citrus (*Citrus*) the centre columnella enables progression through the fruit (Timmer *et al.*, 2003).

Host response to infection by the *Botryosphaeriaceae* involves the formation of barriers. In fruits, cell walls produce protuberances that block fungal growth and confine the invading mycelium. These protuberances are marked by the formation of red halos surrounding the circular lesions caused by infection (Kim *et al.*, 2001). In other plant organs, hyperplasia occurs that involves the differentiation of parenchyma cells to form new periderm and the production of gum ducts proximal or distal to a site of infection (Biggs and Britton, 1988). The new periderm comes into contact, and becomes continuous, with the old periderm. It is constituted by three to five layers of suberized phellem cells (Biggs and Britton, 1988; Rayachhetry *et al.*, 1996a). The new periderm might delimit these fungi and prevent their entry into the deeper plant tissues, but is not always effective in doing so.

Formation of the new periderm produced by plants in response to infection by the *Botryosphaeriaceae* can be ineffective and allow these fungi to invade the cortex. This is as a result of defects or cracks in the periderm due to the rapid proliferation of callus (Milholland, 1972). Hyphae invade the primary phloem and vascular cambium. In the case of wounds, the cell walls of xylem vessels and tracheids adjacent to the wounds are extensively lignified, and the vascular cambium beneath the new periderm re-differentiates to xylem and phloem. These tissues are subsequently invaded via the pits, becoming necrotic and collapsing to create gaps (Milholland, 1972; Rayachhetry *et al.*, 1996a). In addition, the accumulation of parenchyma cells results in disorganization of the cortex and phloem. The parenchyma cells are eventually crushed, producing large cavities that are occupied by mycelium (Milholland, 1970; Atia *et al.*, 2003). The cortex collapses and is filled with a watery substance, while the periderm expands until it ruptures (Brown and Hendrix, 1981). The end result is a canker.

Entry of the *Botryosphaeriaceae* into the cortex allows invasion of the systemic tissues. The mycelium invades the xylem via the xylem rays (Milholland, 1970; Ramos *et al.*, 1991; Rayachhetry *et al.*, 1996a). The xylem vessels are then occupied by an abundance of tyloses (Ramos *et al.*, 1991; Rayachhetry *et al.*, 1996a; Atia *et al.*, 2003) that either partially or completely block the vessels. Vessels can be occupied by several mycelial strands, which then branch to invade the adjacent tracheids (Milholland, 1972; Ramos *et al.*, 1991; Rayachhetry *et al.*, 1996a). Rapid mycelial growth occurs in the xylem as a result of the availability of nutrients (Cedeño *et al.*, 1996).

The formation of the new periderm produced by plants in response to infection by the *Botryosphaeriaceae* can be successful in restricting these fungi to the epidermis. As a result, hyphae are not observed in the cortex (Milholland, 1972). However, enzymatic activity by the *Botryosphaeriaceae* results in degradation of the epidermal cell wall, thus enabling entry of the mycelium into the epidermal cells. This allows for their persistence in the bark, as latent pathogens. Swelling subsequently occurs, due to separation of the cuticular membrane from the underlying epidermal cell wall, and the cuticular membrane becomes folded and creased (Chou, 1978). This gives old cankers a characteristically swollen, crusty appearance (Brown and Britton, 1986).

14.3.2 The role of stress

Various stresses enable endophytes, including the *Botryosphaeriaceae,* to cause disease (Schulz *et al.*, 1998). Some species in the family can cause disease in the absence of stress (Mullen *et al.*, 1991), but disease typically only manifests once plants are stressed. Stress affects both cuticle penetrability and the persistence of endophytes in plant tissue (Arnold and Herre, 2003).

Levitt (1972) and Schoeneweiss (1975) noted that stress results in a strain on plants that is either plastic or elastic. Plastic strains result in irreversible physical or chemical change(s) in physiology as opposed to elastic strains, where physical or chemical changes are reversible when stress is removed. Knowledge of when an elastic strain transitions to a plastic strain and the implicit physiological limits differ among plants (Paoletti *et al.*, 2001).

Moisture stress caused by drought is most commonly associated with disease reports resulting from infection by the *Botryosphaeriaceae*. Moisture stress contributes to nutrient stress by causing deficiencies, as plants require water to translocate nutrients. Moisture stress also impedes plant defence; wound periderm formation is halted or inhibited (Parker, 1961; Puritch and Mullick, 1975; Paoletti *et al.*, 2001), along with inhibited production of defensive compounds (Madar *et al.*, 1995). Upon exposure to moisture stress, plants seemingly cope with the initial loss of turgidity and reduction of xylem water potentials. Initially, the strain remains elastic, with increased colonization of the wood near the cambium and bark by endophytic *Botryosphaeriaceae* as water potentials further decrease to a particular threshold level. Maintenance of these low water potentials for more than 3 days results in a plastic strain and predisposition to disease (Crist and Schoeneweiss, 1975; Schoeneweiss, 1975, 1981a,b; Paoletti *et al.*, 2001). The water potentials differ depending on the plant pathosystem concerned (Paoletti *et al.*, 2001).

Early studies by Schoeneweiss (Crist and Schoeneweiss, 1975; Schoeneweiss, 1975, 1981a,b) focused on the effects of low temperatures in plant pathosystems involving the *Botryosphaeriaceae*. Recent studies have, however, considered higher temperatures in the context of global climate change (Desprez-Loustau *et al.*, 2007; Botella *et al.*, 2010). For low temperatures, Rayachhetry *et al.* (1996b) found that short-term exposure (3 days) to low temperatures (6°C) in melaleuca (*Melaleuca* spp.) resulted in increased canker length and sapwood discoloration. In other treatments involving longer periods of exposure, symptoms did not manifest. Similarly, Schoeneweiss (Crist and Schoeneweiss, 1975; Schoeneweiss, 1975, 1981a,b) found increased wood colonization in mountain ash (*Sorbus* spp.), European white birch (*Betula pendula* Roth), sweetgum (*Liquidambar* spp.) and red osier dogwood (*Cornus sericea* L.) at low temperatures. He postulated that freezing results in dehydration of plant cells (Schoeneweiss, 1981a,b). Studies that considered elevated temperatures (Desprez-Loustau *et al.*, 2007; Botella *et al.*, 2010) have predicted that the *Botryosphaeriaceae* will increasingly become more prominent and significant. Apart from the increased likelihood of moisture stress occurring, higher temperatures favour the mycelial growth of these fungi. Typically, the temperature optima for growth are 25–30°C. Future studies should, therefore, consider what effect short-term periods of exposure to heat stress have on plant pathosystems involving the *Botryosphaeriaceae*.

A number of diseases caused by the *Botryosphaeriaceae* result in defoliation (Gerlach *et al.*, 1974; Shahin and Claflin, 1980; Chakraborty *et al.*, 2004). Defoliation produces stress, and results in an avenue for additional infections and in decreased starch content, in terms of both soluble and stored carbohydrates (Schoeneweiss, 1975; Old *et al.*, 1990). The effect of defoliation, as with other stresses, varies depending on the pathosystem. Downy birch (*Betula pubescens* Ehrh.) stems inoculated with *B. dothidea* and defoliated for more than 4 weeks were girdled by the pathogen and died, while those defoliated for 4 weeks and allowed to recover survived (Crist and Schoeneweiss, 1975). Similarly, punktrees (*Melaleuca quinquenervia* (Cav.) S.F. Blake) inoculated with *N. ribis* and exposed to complete defoliation suffered 90% mortality within 4 weeks (Rayachhetry *et al.*, 1996b). Defoliation and infection by the *Botryosphaeriaceae* initially produces an elastic strain. Sustained defoliation, however, results in a plastic strain, and potentially high levels of mortality among infected plants (Crist and Schoeneweiss, 1975; Schoeneweiss, 1975, 1981b).

There are a number of other stresses that can allow disease expression by the *Botryosphaeriaceae*. The transplanting of plants frequently necessitates extensive pruning of roots, which results in a loss of stored root food reserves and reduced root surface for nutrient uptake (Schoeneweiss, 1981b). Wounding resulting from hail, snow, insects, or combinations thereof, both produces stress and provides an avenue for further infections by the *Botryosphaeriaceae* (Nicholls and Ostry, 1990; Menge and Ploetz, 2003). Improper planting soils that are too shallow, too hard or too sandy to retain water also place stress on trees (Wright and Marks, 1970). Poor training and thinning of trees results in shading, weakening of branches, reduced vigour and overstocking (Wright and Marks, 1970; Crist and Schoeneweiss, 1975). Competition for resources, including water, can arise from neighbouring vegetation (Blodgett *et al.*, 1997). High levels of nitrogen and phosphorus (Van Dijk *et al.*, 1992), resulting from fertilization, can reduce the concentrations of lignin and total soluble phenolics, rendering plants susceptible to infection by various fungi, including the *Botryosphaeriaceae* (Blodgett *et al.*, 2005). Nutrient deficiencies also cause stress (Menge and Ploetz, 2003). Other diseases can weaken and stress plants that are already infected endophytically by the *Botryosphaeriaceae* (Menge and Ploetz, 2003). Damage can also arise from the application of chemicals (Menge and Ploetz, 2003). Finally, in the case of juvenile tropical palms (*Iriartea deltoidea* Ruiz & Pav.), high light intensities stress the plants, rendering them susceptible to disease caused by *Diplodia mutila* (Fr.) Mont. (syn. *Botryosphaeria stevensii* Shoemaker) (Alvarez-Loayza *et al.*, 2011).

14.3.3 Insect associations

The *Botryosphaeriaceae* are principally dispersed by water and wind. They are not specifically adapted for insect dispersal and it is unlikely that insects play any significant role in their dissemination. Nevertheless, there are several reports of *Botryosphaeriaceae* associated with insect damage (Feci *et al.*, 2003; Romero *et al.*, 2005; Moral *et al.*, 2010). In each of these cases, damage caused by these insects enabled either entry of the *Botryosphaeriaceae*, or disease expression when the pathogen was already present as an endophyte. Disease expression is also likely to facilitate the abscission of insect-damaged plant parts, as noted by Slippers and Wingfield (2007).

Several recent publications (Adair *et al.*, 2009; Janson *et al.*, 2009, 2010; Heath and Stireman, 2010) have reported the transmission of *B. dothidea* by gall midges. These insects possess specialized structures (mycangia) on the ovipositor that house fungal conidia of unknown origin. Evidently, oviposition includes the inoculation of a gall with conidia of the fungus. However, this is seemingly casual, as *B. dothidea* was not isolated from every gall tested and other fungi were also isolated from the same galls (Adair *et al.*, 2009; Heath and Stireman, 2010). For one of the midges, *B. dothidea* is a source of sterol for developing larvae, as shown by the sterol profile consisting of ergosterol and its precursors/metabolites (Janson *et al.*, 2009). This is not surprising: insects are unable to produce their own sterols so phytophagous insects typically ingest sterols derived from plants (phytosterols) and then enzymatically convert them to cholesterol (Clayton, 1964).

14.3.4 Melanization

Several genera of the *Botryosphaeriaceae* produce pigmented spores or hyaline spores that become pigmented at maturity, e.g. *Barriopsis*, *Diplodia*, *Lasiodiplodia*, *Neodeightonia* (Crous *et al.*, 2006; Phillips *et al.*, 2008). A number of species of the *Botryosphaeriaceae* also produce pigmented cultures. Pigmentation in fungi results from the production of melanin (Bell and Wheeler, 1986).

Melanin enables appressorial penetration of the plant host epidermis or cellular membranes (Bell and Wheeler, 1986; Butler *et al.*, 2001). This is accomplished by the accumulation of glycerol in the melanized appressorium, which generates intense pressure on the epidermis or membrane under attack, owing to melanin forming a barrier that is impermeable to glycerol (Nosanchuk and Casadevall, 2003). Consequently, melanins can play a prominent role in the virulence and success of the *Botryosphaeriaceae* as pathogens.

14.3.5 Phytotoxins

Several species of the *Botryosphaeriaceae* are known to produce various non-specific phytotoxins (Evidente *et al.*, 1996, 1999, 2003a, 2010; Veiga *et al.*, 2007; Djoukeng *et al.*, 2009) (Table 14.1). There is a strong likelihood that many other phytotoxins are also produced by the *Botryosphaeriaceae*, and are awaiting discovery and characterization.

14.4 Epidemiology

Species of the *Botryosphaeriaceae* can overwinter or survive adverse environmental conditions on a variety of substrates, often endophytically. These include infected and/or pruned wood (Epstein *et al.*, 2008), bark

Table 14.1. Known toxins produced by species of the *Botryosphaeriaceae*. Listed, along with toxins, are their associated phytopathogenic effects, host species and any recorded antimycotic activity.

Species and toxin	Disease symptoms	Inoculated host(s)	Antimicrobial against phytopathogenic fungi	References
***Diplodia cupressi* A.J.L. Phillips & A. Alves**				
Sphaeropsidin A	Non-specific; produces a variety of disease symptoms, including necrosis and dieback	Various *Cupressus* spp.	Yes	Evidente *et al.*, 1996
Sphaeropsidin B Sphaeropsidin C	Yellowing, browning, necrosis and dieback	*Cupressus* and *Quercus* spp.	Yes	Evidente *et al.*, 1997
Sphaeropsidin D	Chlorosis and subsequent necrosis	*Hesperocyparis macrocarpa* (Hartw.) Bartel	NT[a]	Evidente *et al.*, 2002
Sphaeropsidin F	Chlorosis	*Cupressus sempervirens* L.	NT	Evidente *et al.*, 2003b
Sapinopyridione	Chlorosis Chlorosis and dieback	*Hesperocyparis arizonica* (Greene) Bartel *H. macrocarpa* and *C. sempervirens*	NT	Evidente *et al.*, 2006
Sphaeropsidone	Browning and necrosis Chlorosis	*H. macrocarpa* *H. arizonica*	Yes	Evidente *et al.*, 1998
Episphaeropsidone	Necrosis Browning and necrosis	*H. arizonica* and *H. macrocarpa* *C. sempervirens*	Yes	Evidente *et al.*, 1998
***Diplodia pinea* (Desm.) J. Kickx f. (syn. *Sphaeropsis sapinea* (Fr.) Dyko & B. Sutton)[b]**				
Sapinofuranone A	Browning on excised twigs, similar to sapwood staining	*Cupressus* spp.	NT	Evidente *et al.*, 1999
Sapinofuranone B	Yellowing of twigs, withering of needles	*Pinus* spp.		
Mellein (3*R*,4*R*)-4-hydroxymellein (3*R*,4*S*)-4-hydroxymellein	Chlorosis and subsequent necrosis	*Pinus radiata* D. Don (cuttings)	NT	Cabras *et al.*, 2006
***Diplodia seriata* De Not. (syn. *Botryosphaeria obtusa* (Schwein.) Shoemaker)**				
Mellein (3*R*,4*R*)-4-hydroxymellein 5-hydroxymellein 7-hydroxymellein 4,7-hydroxymellein	Necrotic lesions, characteristic of infection[c]	*Malus pumila* Mill.	NT	Venkatasubbaiah *et al.*, 1991; Djoukeng *et al.*, 2009

Continued

Table 14.1. Continued.

Species and toxin	Disease symptoms	Inoculated host(s)	Antimicrobial against phytopathogenic fungi	References
4-hydroxybenzaldehyde				
Tyrosol				
***Diplodia mutila* (Fr.) Mont. (syn. *Botryosphaeria stevensii* Shoemaker)[d]**				
Diplopyrone	Necrotic lesions on leaves within 4 days, wilting of cuttings within 8 days	*Quercus suber* L.	NT	Evidente *et al.*, 2003a
***Lasiodiplodia theobromae* (Pat.) Griffon & Maubl.**				
(3*S*,4*R*)-3-carboxy-2-methylene-heptan-4-olide[e]	Lasiodiplodin inhibits ATP synthesis by interfering with electron flow and phosphorylation	NT	NT	Aldridge *et al.*, 1971; Barbosa *et al.*, 2003; He *et al.*, 2004; Veiga *et al.*, 2007
Mellein				
Lasiodiplodin				
De-*O*-methyl-lasiodiplodin				
(3*R*,4*R*)-hydroxymellein				
Botryosphaeran				
***Neofusicoccum parvum* (Pennycook & Samuels) Crous, Slippers & A.J.L. Phillips**				
(3*R*,4*R*)-4-hydroxymellein		Tomato (*Solanum lycopersicon* L.) (cuttings)	NT	Evidente *et al.*, 2010
(3*R*,4*S*)-4-hydroxymellein				
Isosclerone				
Tyrosol				

[a]NT, not tested.
[b]It is unclear which morphotype(s) of the fungus were tested: whether A, C or both.
[c]Toxin production has not been correlated with virulence.
[d]The fungus also produces sphaeropsidins A and C (Evidente *et al.*, 1997).
[e]Toxicity remains doubtful as it was evaluated using bananas (*Musa*) (He *et al.*, 2004).

(Srivastava, 1972), infected rotting/mummified fruit, twigs, buds, petioles, flowers, inflorescences and rachises (Michailides, 1991). Inoculum on these substrates can be sustained for up to 6 years, while infected plant parts typically persist for longer than healthy plant parts (Michailides, 1991; Ahimera *et al.*, 2003) and are not abscised. Inoculum on infected fruits can be more abundant than that on infected leaves (Ahimera *et al.*, 2003). Principally, inoculum consists of conidia housed in pycnidia. Latent propagules can also be introduced into an orchard through infected seedlings, rootstocks or scions (Cysne *et al.*, 2010). Inoculum may also be present on residual diseased plant material present on the ground, thereby enabling root infection (Hodges, 1983; Whitelaw-Weckert *et al.*, 2006). Root-to-root transmission might also occur (Wingfield and Knox-Davies, 1980) although this has not been shown experimentally. Primary infections occur in early spring and summer, and secondary infections in late summer and autumn (Michailides and Morgan, 1993).

During spore release, conidia of the *Botryosphaeriaceae* are typically exuded from pycnidia in a gelatinous matrix as a ribbon-like mass or cirrhus (Michailides and Morgan, 1993; Úrbez-Torres *et al.*, 2010). This gelatinous matrix or mucilage encompasses conidia within the pycnidia, and contains glucides and proteins. The matrix is believed to prevent conidia from desiccating, losing their viability and germinating (Fournet, 1969; Fitt and McCartney, 1986). During periods of high relative humidity, the mucilage swells to produce the ribbon-like cirrhi; when wet, it dissolves, thereby freeing conidia to germinate (Fitt and McCartney, 1986).

Dispersal of the conidia of *Botryosphaeriaceae*, including the initiation of spore release, has been correlated with rainfall, or occurs as a result of overhead irrigation events, as water dissolves the mucilage to free the conidia. Spore release usually begins within 1 hour of water beginning to fall and lasts for up to 2 hours after water has stopped falling (Amponsah *et al.*, 2009; Úrbez-Torres *et al.*, 2010). The release can be elevated during night-time rainfall events as the humidity is higher while temperatures are lower and wind is reduced (Holmes and Rich, 1970). It has been hypothesized that conidia are dispersed over relatively short distances because of their dependence on rain splash (Úrbez-Torres *et al.*, 2010). Consequently, these conidia are involved in further disease cycles and infections of the plant concerned (Michailides *et al.*, 2001), in which they can persist endophytically. Methods of dissemination other than rain splash include wind blowing during rain, infected pruning shears and insects (Swart *et al.*, 1987; Epstein *et al.*, 2008).

The sexual spores (ascospores) of the *Botryosphaeriaceae* are produced in perithecia (pseudothecia). Ascospore release can be stimulated by fog (Úrbez-Torres *et al.*, 2010), rainfall, dew or mist (Pusey, 1989). Ascospores are generally found in much smaller numbers than conidia (Parker and Ramsdell, 1977; Michailides *et al.*, 2001), but appear to be released during periods of rapid air movement or wind and optimal temperatures, in the spring (Pusey, 1989). These climatic factors can facilitate long-distance dispersal of the pathogen as the ascospores are airborne (Pusey, 1989; Michailides *et al.*, 2001; Ma and Michailides, 2002).

14.5 Management Strategies and Tactics

14.5.1 Avoidance

Strategies that minimize or avoid infection by the *Botryosphaeriaceae* are the most important means of disease management and control as the fungi persist endophytically once plants are infected. Minimizing exposure to stress is an important first step to dealing with diseases of the *Botryosphaeriaceae* because stress is clearly the dominant factor inciting disease expression. In commercial forestry, site–species matching is used to anticipate the site and climatic variation requirements of different species and provenances, and thus minimize stress in clonal plantings (Little *et al.*, 2003). A second strategy is to utilize appropriate planting densities and spacings to avoid potential competition between plants for nutrients and/or water. Thinning can be undertaken in situations where this is unavoidable (Dawson and Were, 1998). Stands should also not be left to over mature (Slippers *et al.*, 2009).

Some control strategies used to manage diseases of *Botryosphaeriaceae* in commercial agricultural orchards could be applied in seed orchards and nurseries. For example, grapevine rootstocks are treated with hot water or a chemical or biological drench to kill inoculum and thereby minimize infection (Fourie and Halleen, 2004). Appropriate hygiene practices, such as disinfecting grafting shears, aim to avoid potential scion-to-scion transmission of these fungi (Fourie and Halleen, 2004; Cysne *et al.*, 2010). Soil treatment involving fumigation, soil solarization or sanitation can be applied to reduce inoculum on infected plant debris at the soil surface, as these sources would otherwise infect rootstocks or seedlings via planting wounds (Whitelaw-Weckert *et al.*, 2006). Soil preparation may also entail the addition of any fertilizer that is required to avoid nutrient stress, and irrigation soon after planting to avoid moisture stress. Strategies for irrigation that reduce the humidity available to the *Botryosphaeriaceae* for spore release, mycelial growth and infection should, therefore, limit or eliminate a wet tree canopy and/or orchard floor. These include the use of drip or micro-jet irrigation (preferred) (Michailides and Morgan, 2004), or else sprinklers with a low trajectory angle (Michailides and Morgan, 1993, 2004), and irrigation done for short time periods, preferably early in the day, to minimize the duration of leaf wetness (Michailides and Morgan, 1992). Pressure applied in irrigation systems can also be lowered to reduce any fogging and humidity that would otherwise enable infection (Michailides and Morgan, 2004). In pistachio (*Pistacia vera* L.) orchards, it has been recommended that dead plant material be burned (Michailides and Morgan, 2004), and sanitation should be practised throughout the growing period to control and manage infection by these fungi.

Pruning in seed orchards should be done in winter as low temperatures minimize the dispersal of inoculum and infection by the *Botryosphaeriaceae* (Swart and Wingfield, 1991), and should be done so that only a moderate amount (25%) of the crown is removed. Regular (39%) or heavy (49%) pruning can produce physiological stress and result in disease expression by latent *Botryosphaeriaceae* (Chou and MacKenzie, 1988). Pruning wounds provide an additional avenue for infection by the *Botryosphaeriaceae*, so coating wounds with a fungicidal paint may reduce infection by these and other pathogens (Epstein *et al.*, 2008). In commercial forestry, these strategies for pruning are impractical, both economically and in terms of labour.

Horizontal transmission of the *Botryosphaeriaceae* between trees inevitably results in newly established plantation or orchard trees being infected by these fungi. The infections originate from neighbouring flora, including windbreaks, that contribute a mix of isolates over the lifetime of a tree (Stanosz *et al.*, 2005, 2007; Burgess *et al.*, 2006; Pavlic *et al.*, 2007; Bihon *et al.*, 2011a). Some studies have shown that species of *Botryosphaeriaceae* infect related hosts (within the same plant family), while other studies show that species are not host specific and originate from unrelated neighbouring plants (Damm *et al.*, 2007; Inderbitzin *et al.*, 2010). Clearly, neighbouring trees and plants will need to be considered as an extension in any management strategy against the *Botryosphaeriaceae*. This is especially important in nursery settings where young plants can be heavily infected by inoculum originating from surrounding plants (Stanosz *et al.*, 2005, 2007).

14.5.2 Exclusion

The global movement of people and products continues to facilitate the movement of fungal pathogens, including the *Botryosphaeriaceae*, across geographic or ecological borders (Anderson *et al.*, 2004). Global trade has increased the frequency of introductions of exotic species and is expected to continue rising (Cunningham *et al.*, 2003; Anderson *et al.*, 2004). Pathogens, including the *Botryosphaeriaceae*, are typically introduced via infected plant material, including imported planting or nursery material. In some cases, such pathogens can then shift or jump hosts to cause devastating diseases on native vegetation (Slippers *et al.*, 2005). There are already several examples of such diseases caused by the *Botryosphaeriaceae* in which non-native trees are planted in new areas in close proximity to native trees that are broadly related to the non-natives, albeit not at the scale of epidemics (Burgess *et al.*, 2006; Pavlic *et al.*, 2007; Pérez *et al.*, 2010).

Seed infection by the *Botryosphaeriaceae* and the possibility of disseminating more virulent and pathogenic genotypes of these fungi through international seed trade is a concern (Burgess *et al.*, 2001; Burgess and Wingfield, 2002). A species that is suspected to have been spread via seed and associated material is *D. pinea*, which infects and causes diseases of pine trees. This fungus infects both immature cones (Smith *et al.*, 1996), from which it has been isolated from the outer scales of second-year cones, and the pith tissue of third-year seed cones, ovuliferous scales, seeds and seed wings (Smith *et al.*, 2002). While the fungus sporulates prolifically on cones (Smith *et al.*, 2002), conidia are passed on at low frequencies (2.3%) to the seed (Bihon *et al.*, 2011b). Conidia may also be present in coarse woody debris collected with cones and seeds (Burgess and Wingfield, 2002; Bihon *et al.*, 2011b). Seed infection is also caused by early harvesting of the cones (Fraedrich, 1996). The emerging seedlings are not systematically infected, although overall germination is decreased (Bihon *et al.*, 2011b). In seeds infected by either *D. pinea* or *Lasiodiplodia theobromae* (Pat.) Griffon & Maubl., both the embryo and gametophyte tissues are destroyed, effectively killing the seeds (Miller and Bramlett, 1979; Rees, 1988; Rees and Webber, 1988; Fraedrich, 1996). Consequently, quarantine for the *Botryosphaeriaceae* should consider the global spread of seeds, along with any plant debris accompanying those seeds.

Quarantine will prove challenging in limiting the global movement of the *Botryosphaeriaceae*, because of the extensive dissemination of propagation material for nurseries – such as rootstocks, scions and other nursery material. The *Botryosphaeriaceae* are noticeably conspicuous in their absence among lists of introduced pathogens (Brasier, 2008), probably as a result both of their ability to remain latent within visibly symptomless material and of a poor understanding of their native ranges. Molecular tools such as those outlined in this chapter will enable quarantine authorities to detect these fungi within apparently healthy plant material.

14.5.3 Eradication

Eradication for the *Botryosphaeriaceae* is most probably impossible. No eradication programme has been attempted for these fungi. The fact that they persist endophytically means that complete detection and eradication within either a plantation or orchard is impossible. Also, inoculum frequently originates from neighbouring flora (including plantations, orchards, windbreaks, individual plants and trees) (Burgess *et al.*, 2006; Pavlic *et al.*, 2007), and the removal of these sources, including native flora, is obviously not possible.

14.5.4 Protection

Fungicidal applications have been evaluated in the control of the *Botryosphaeriaceae* in forestry, but have been shown to be ineffective. These trials have demonstrated that fungicides, while effective *in vitro*, are ineffective *in vivo* (Peterson, 1977; Palmer *et al.*, 1986; Hartman *et al.*, 2009). Their application is also impractical, when the labour, risks to the people involved and costs are considered. This is because trees host a multitude of isolates that infect over the lifetime of a tree (Bihon *et al.*, 2011a) and fungicide application over the entire lifetime of a tree is simply impossible. Nevertheless, fungicides can be applied to manage these fungi in young plants in nurseries (Swart and Wingfield, 1991). In agriculture, in contrast, several fungicides have been registered for application to various crops to control diseases of the *Botryosphaeriaceae* (Michailides and Morgan, 2004; Kim *et al.*, 2005; van Niekerk *et al.*, 2006).

14.5.5 Resistance

Plant breeding provides a basis for evaluating the resistance and tolerance to species of the *Botryosphaeriaceae* in clonal forestry and agriculture. These are easily evaluated using pathogenicity trials involving stem inoculations with agar plugs of species of *Botryosphaeriaceae*, which provide an opportunity to screen for resistance before commercialization and field planting (Slippers *et al.*, 2009). In two studies (Mohali *et al.*, 2009; Rodas *et al.*, 2009), variation was evident both among species that were inoculated and in clonal tolerance to infection. Clonal tolerance, however, should be consistent regardless of the site where a stand is established (Rodas *et al.*, 2009).

There can be considerable variation in pathogenicity among isolates of species of the *Botryosphaeriaceae*, so much so that this can influence virulence on the trees that are inoculated. This has been evident from pathogenicity trials undertaken on a variety of plant hosts (Pavlic *et al.*, 2007; Stanosz *et al.*, 2007; Mohali *et al.*, 2009; Mehl *et al.*, 2011). The variation can be confounded by both site factors and other environmental factors when a trial is undertaken (Rodas *et al.*, 2009). Consequently, a large number of isolates should be used in trials in both commercial forestry and agricultural planting materials. Challenging a clone, clonal rootstock or scion with multiple isolates, some potentially highly pathogenic, will provide a better indication of resistance and tolerance mechanisms than simply using one or two isolates, even if their pathogenicity is well established.

14.5.6 Integrated disease management

An integrated approach is recommended to ensure that diseases of the *Botryosphaeriaceae* are controlled to an acceptable level. Strategies for management in plantations include site–species matching (Little *et al.*, 2003), the use of appropriate planting densities and spacing (Dawson and Were, 1998) and consideration of neighbouring plants that are sources of inoculum (Burgess *et al.*, 2006; Pavlic *et al.*, 2007). In seed orchards and nurseries, strategies include sanitation, hygiene practices such as the disinfection of grafting shears and pruning tools (Fourie and Halleen, 2004; Cysne *et al.*, 2010), wound treatment with a fungicidal paint (Epstein *et al.*, 2008), pruning at an appropriate time and at a moderate level (Swart and Wingfield, 1991), and irrigation strategies that minimize exposure to the water or humidity that are available for spore release and mycelial growth (Michailides and Morgan, 2004). Chemical control through fungicide use is restricted to nurseries, owing to the practical impossibilities of application to plantations and seed orchards (Swart and Wingfield, 1991). Trials on nursery stock and planting material and field trials on forestry stock should be undertaken at multiple locations to avoid the confounding of results (Rodas *et al.*, 2009) and should utilize multiple isolates.

References

Adair, R.J., Burgess, T., Serdani, M. and Barber, P. (2009) Fungal associations in Asphondylia (Diptera: Cecidomyiidae) galls from Australia and South Africa: implications for biological control of invasive acacias. *Fungal Ecology* 2, 121–134.

Ahimera, N., Driever, G.F. and Michailides, T.J. (2003) Relationships among propagule numbers of *Botryosphaeria dothidea*, latent infections, and severity of panicle and shoot blight in pistachio orchards. *Plant Disease* 87, 846–853.

Aldridge, D.C., Galt, S., Giles, D. and Turner, W.B. (1971) Metabolites of *Lasiodiplodia theobromae*. *Journal of the Chemical Society C: Organic*, 1623–1627.

Alvarez-Loayza, P., White, J.F. Jr, Bergen, M. and Cadenas, C. (2008) *Diplodia mutila* causing seedling mortality of the palm *Iriartea deltoidea*. *Plant Pathology* 57, 382.

Alvarez-Loayza, P., White, J.F. Jr, Torres, M.S., Balslev, H., Kristiansen, T., Svenning, J.-C. and Gil, N. (2011) Light converts endosymbiotic fungus to pathogen, influencing seedling survival and niche-space filling of a common tropical tree, *Iriartea deltoidea*. *PLoS One* 6(1): e16386.

Alves, A., Phillips, A.J.L., Henriques, I. and Correia, A. (2005) Evaluation of amplified ribosomal DNA restriction analysis as a method for the identification of *Botryosphaeria* species. *FEMS Microbiology Letters* 245, 221–229.

Amponsah, N.T., Jones, E.E., Ridgway, H.J. and Jaspers, M.V. (2009) Rainwater dispersal of *Botryosphaeria* conidia from infected grapevines. *New Zealand Plant Protection* 62, 228–233.

Anderson, P.K., Cunningham, A.A., Patel, N.G., Morales, F.J., Epstein, P.R. and Daszak, P. (2004) Emerging infectious diseases of plants: pathogen pollution, climate change and agrotechnology drivers. *Trends in Ecology and Evolution* 19, 535–544.

Arnold, A.E. and Herre, E.A. (2003) Canopy cover and leaf age affect colonization by tropical fungal endophytes: ecological pattern and process in *Theobroma cacao* (*Malvaceae*). *Mycologia* 95, 388–398.

Aroca, A., García-Figueres, F., Bracamonte, L., Luque, J. and Raposo, R. (2006) A survey of trunk disease pathogens within rootstocks of grapevines in Spain. *European Journal of Plant Pathology* 115, 195–202.

Atia, M.M.M., Aly, A.Z., Tohamy, M.R.A., El-Shimy, H. and Kamhawy, M.A. (2003) Histopathological studies on grapevine die-back. *Zeitschrift fur Pflanzenkrankheiten und Pflanzenschutz* 110, 131–142.

Barbosa, A.M., Steluti, R.M., Dekker, R.F.H., Cardoso, M.S. and Corradi da Silva, M.L. (2003) Structural characterization of Botryosphaeran: α (1→3;1→6)-β-D-glucan produced by the ascomyceteous fungus, *Botryosphaeria* sp. *Carbohydrate Research* 338, 1691–1698.

Beisel, M., Hendrix, F.F. Jr and Starkey, T.E. (1984) Natural inoculation of apple buds by *Botryosphaeria obtusa*. *Phytopathology* 74, 335–338.

Bell, A.A. and Wheeler, M.H. (1986) Biosynthesis and functions of fungal melanins. *Annual Review of Phytopathology* 24, 411–451.

Biggs, A.R. and Britton, K.O. (1988) Presymptom histopathology of peach trees inoculated with *Botryosphaeria obtusa* and *B. dothidea*. *Phytopathology* 78, 1109–1118.

Bihon, W., Burgess, T.I., Slippers, B., Wingfield, M.J. and Wingfield, B.D. (2011a) Distribution of *Diplodia pinea* and its genotypic diversity within asymptomatic *Pinus patula* trees. *Australasian Plant Pathology* 40, 540–548.

Bihon, W., Slippers, B., Burgess, T.I., Wingfield, M.J. and Wingfield, B.D. (2011b) Sources of *Diplodia pinea* endophytic infections in *Pinus patula* and *P. radiata* seedlings in South Africa. *Forest Pathology* 41, 370–375.

Blodgett, J.T., Kruger, E.L. and Stanosz, G.R. (1997) *Sphaeropsis sapinea* and water stress in a red pine plantation in central Wisconsin. *Phytopathology* 87, 429–434.

Blodgett, J.T., Herms, D.A. and Bonello, P. (2005) Effects of fertilization on red pine defense chemistry and resistance to *Sphaeropsis sapinea*. *Forest Ecology and Management* 208, 373–382.

Botella, L., Santamaría, O. and Diez, J.J. (2010) Fungi associated with the decline of *Pinus halepensis* in Spain. *Fungal Diversity* 40, 1–11.

Brasier, C.M. (2008) The biosecurity threat to the UK and global environment from international trade in plants. *Plant Pathology* 57, 792–808.

Brookhouser, L.W. and Peterson, G.W. (1970) Infection of Austrian, Scots, and ponderosa pines by *Diplodia pinea*. *Phytopathology* 61, 409–414.

Brown, E.A. and Britton, K.O. (1986) *Botryosphaeria* diseases of apple and peach in the southeastern United States. *Plant Disease* 70, 480–484.

Brown, E.A. and Hendrix, F.F. (1981) Pathogenicity and histopathology of *Botryosphaeria dothidea* on apple stems. *Phytopathology* 71, 375–379.

Burgess, T.I. and Wingfield, M.J. (2002) Quarantine is important in restricting the spread of exotic seed-borne tree pathogens in the southern hemisphere. *International Forestry Review* 4, 56–65.

Burgess, T.I., Wingfield, B.D. and Wingfield, M.J. (2001) Comparison of genotypic diversity in native and introduced populations of *Sphaeropsis sapinea* isolated from *Pinus radiata*. *Mycological Research* 105, 1331–1339.

Burgess, T.I., Sakalidis, M.L. and Hardy, G.E.St.J. (2006) Gene flow of the canker pathogen *Botryosphaeria australis* between *Eucalyptus globulus* plantations and native eucalypt forests in Western Australia. *Austral Ecology* 31, 559–566.

Butler, M.J., Day, A.W., Henson, J.M. and Money, N.P. (2001) Pathogenic properties of fungal melanins. *Mycologia* 93, 1–8.

Cabras, A., Mannoni, M.A., Serra, S., Andolfi, A., Fiore, M. and Evidente, A. (2006) Occurrence, isolation and biological activity of phytotoxic metabolites produced *in vitro* by *Sphaeropsis sapinea*, pathogenic fungus of *Pinus radiata*. *European Journal of Plant Pathology* 115, 187–193.

Cedeño, L., Mohali, S.R. and Palacios-Prü, E. (1996) Ultrastructure of *Lasiodiplodia theobromae* causal agent of Caribbean pine blue stain in Venezuela. *Interciencia* 21, 264–271.

Chakraborty, M.R., Dutta, S. and Chatterjee, N.C. (2004) Dieback disease of bottle brush – a new record from India. *Journal of Mycology and Plant Pathology* 34, 146.

Chou, C.K.S. (1978) Penetration of young stems of *Pinus radiata* by *Diplodia pinea*. *Physiological Plant Pathology* 13, 189–192.
Chou, C.K.S. and MacKenzie, M. (1988) Effect of pruning intensity and season on *Diplodia pinea* infection of *Pinus radiata* stem through pruning wounds. *European Journal of Forest Pathology* 18, 437–444.
Clayton, R.B. (1964) The utilization of sterols by insects. *Journal of Lipid Research* 5, 3–19.
Crist, C.R. and Schoeneweiss, D.F. (1975) The influence of controlled stresses on susceptibility of European white birch stems to attack by *Botryosphaeria dothidea*. *Phytopathology* 65, 369–373.
Crous, P.W., Slippers, B., Wingfield, M.J., Rheeder, J., Marasas, W.F.O., Phillips, A.J.L., Alves, A., Burgess, T.I., Barber, P. and Groenewald, J.Z. (2006) Phylogenetic lineages in the *Botryosphaeriaceae*. *Studies in Mycology* 55, 235–253.
Cunningham, A.A., Daszak, P. and Rodriguez, J.P. (2003) Pathogen pollution: defining a parasitological threat to biodiversity conservation. *Journal of Parasitology* 89, S78–S83.
Cysne, A.Q., Cardoso, J.E., Aline de Holanda, N.M. and Farias, F.C. (2010) Spatial-temporal analysis of gummosis in three cashew clones at northeastern Brazil. *Journal of Phytopathology* 158, 676–682.
Dakin, N., White, D., Hardy, G.E.St.J. and Burgess, T.I. (2010) The opportunistic pathogen, *Neofusicoccum australe*, is responsible for crown dieback of peppermint (*Agonis flexuosa*) in Western Australia. *Australasian Plant Pathology* 39, 202–206.
Damm, U., Crous, P.W. and Fourie, P.H. (2007) *Botryosphaeriaceae* as potential pathogens of *Prunus* species in South Africa, with descriptions of *Diplodia africana* and *Lasiodiplodia plurivora* sp. nov. *Mycologia* 99, 664–680.
Dawson, I. and Were, J. (1998) Multiplication, that's the name of the game: guidelines for seed production of agroforestry trees. *Agroforestry Today* 10, 19–22.
Desprez-Loustau, M.-L., Robin, C., Reynaud, G., Déqué, M., Badeau, V., Piou, D., Husson, C. and Marçais, B. (2007) Simulating the effects of a climate-change scenario on the geographical range and activity of forest-pathogenic fungi. *Canadian Journal of Plant Pathology* 29, 101–120.
de Wet, J., Burgess, T.I., Slippers, B., Preisig, O., Wingfield, B.D. and Wingfield, M.J. (2003) Multiple gene genealogies and microsatellite markers reflect relationships between morphotypes of *Sphaeropsis sapinea* and distinguish a new species of *Diplodia*. *Mycological Research* 107, 557–566.
Djoukeng, J.D., Polli, S., Larignon, P. and Abou-Mansour, E. (2009) Identification of phytotoxins from *Botryosphaeria obtusa*, a pathogen of black dead arm disease of grapevine. *European Journal of Plant Pathology* 124, 303–308.
Epstein, L., Sukhwinder, K. and VanderGheynst, J.S. (2008) *Botryosphaeria*-related dieback and control investigated in noncoastal California grapevines. *California Agriculture* 62, 161–166.
Evidente, A., Sparapano, L., Motta, A., Giordano, F., Fierro, O. and Frisullo, S. (1996) A phytotoxic pimarane diterpene of *Sphaeropsis sapinea* f. sp. *cupressi*, the pathogen of a canker disease of cypress. *Phytochemistry* 42, 1541–1546.
Evidente, A., Sparapano, L., Fierro, O., Bruno, G., Giordano, F. and Motta, A. (1997) Sphaeropsidins B and C, phytotoxic pimarane diterpenes from *Sphaeropsis sapinea* f. sp. *cupressi* and *Diplodia mutila*. *Phytochemistry* 45, 705–713.
Evidente, A., Sparapano, L., Fierro, O., Bruno, G., Giordano, F. and Motta, A. (1998) Sphaeropsidone and episphaeropsidone, phytotoxic dimedone methylethers produced by *Sphaeropsis sapinea* f. sp. *cupressi* grown in liquid culture. *Phytochemistry* 48, 1139–1143.
Evidente, A., Sparapano, L., Fierro, O., Bruno, G. and Motta, A. (1999) Sapinofuranones A and B, two new 2(3H)-dihydrofuranones produced by *Sphaeropsis sapinea*, a common pathogen of conifers. *Journal of Natural Products* 62, 253–256.
Evidente, A., Sparapano, L., Bruno, G. and Motta, A. (2002) Sphaeropsidins D and E, two other pimarane diterpenes, produced *in vitro* by the plant pathogenic fungus *Sphaeropsis sapinea* f. sp. *cupressi*. *Phytochemistry* 59, 817–823.
Evidente, A., Maddau, L., Spanu, E., Franceschini, A., Lazzaroni, S. and Motta, A. (2003a) Diplopyrone, a new phytotoxic tetrahydropyranpyran-2-one produced by *Diplodia mutila*, a fungus pathogen of cork oak. *Journal of Natural Products* 66, 313–315.
Evidente, A., Sparapano, L., Andolfi, A., Bruno, G. and Motta, A. (2003b) Sphaeropsidin F, a new pimarane diterpene produced *in vitro* by the cypress pathogen *Sphaeropsis sapinea* f. sp. *cupressi*. *Australian Journal of Chemistry* 56, 615–619.
Evidente, A., Fiore, M., Bruno, G., Sparapano, L. and Motta, A. (2006) Chemical and biological characterisation of sapinopyridione, a phytotoxic 3,3,6-trisubstituted-2,4-pyridione produced by *Sphaeropsis sapinea*, a toxigenic pathogen of native and exotic conifers, and its derivatives. *Phytochemistry* 67, 1019–1028.

Evidente, A., Punzo, B., Andolfi, A., Cimmino, A., Melck, D. and Luque, J. (2010) Lipophilic phytotoxins produced by *Neofusicoccum parvum*, a grapevine canker agent. *Phytopathologia Mediterranea* 49, 74–79.

Feci, E., Smith, D. and Stanosz, G.R. (2003) Association of *Sphaeropsis sapinea* with insect-damaged red pine shoots and cones. *Forest Pathology* 33, 7–13.

Fisher, P.J., Petrini, O. and Sutton, B.C. (1993) A comparative study of fungal endophytes in leaves, xylem and bark of *Eucalyptus* in Australia and England. *Sydowia* 45, 338–345.

Fitt, B.D.L. and McCartney, H.A. (1986) Spore dispersal in splash droplets. In: Ayres, P.G. and Boddy, L. (eds) *Water, Fungi and Plants*. Cambridge University Press, Cambridge, UK, pp. 87–104.

Flowers, J.L., Hartman, J.R. and Vaillancourt, L.J. (2003) Detection of latent *Sphaeropsis sapinea* infections in Austrian pine tissues using nested-polymerase chain reaction. *Phytopathology* 93, 1471–1523.

Fourie, P.H. and Halleen, F. (2004) Occurrence of grapevine trunk disease pathogens in rootstock mother plants in South Africa. *Australasian Plant Pathology* 33, 313–315.

Fournet, J. (1969) Propriétés et rôle du cirrhe du *Septoria nodorum* Berk. [Properties and role of cirrhus of *Septoria nodorum* Berk.]. *Annales de Phytopathologie* 1, 87–94.

Fraedrich, S. (1996) Seedborne diseases of southern pines and developing strategies for their control. In: Landis, T.D. and South, D.B. (eds) *National Proceedings: Forest and Conservation Nursery Associations, 1996*. General Technical Report PNW 389, US Department of Agriculture, Forest Service, Pacific Northwest Research Station, Portland, Oregon, pp. 75–81.

Gerlach, W.W.P., Hoitink, H.A.J. and Ellett, C.W. (1974) Shoot blight and stem dieback of *Pieris japonica* caused by *Phytophthora citricola*, *P. citrophthora* and *Botryosphaeria dothidea*. *Phytopathology* 64, 1368–1370.

Giraud, M. (2009) Le black rot du pommier [Apple black rot]. *Infos-Ctifl* 257, 36–41.

Hartman, J.R., Vaillancourt, L.J., Flowers, J.L. and Bateman, A.M. (2009) Managing Diplodia tip blight of landscape Austrian pines. *Arboriculture and Urban Forestry* 35, 27–32.

He, G., Matsuura, H. and Yoshihara, T. (2004) Isolation of an α-methylene-γ-butyrolactone derivative, a toxin from the plant pathogen *Lasiodiplodia theobromae*. *Phytochemistry* 65, 2803–2807.

Heath, J.J. and Stireman, J.O. III (2010) Dissecting the association between a gall midge, *Asteromyia carbonifera*, and its symbiotic fungus, *Botryosphaeria dothidea*. *Entomologia Experimentalis et Applicata* 137, 36–49.

Hodges, C.S. (1983) Pine mortality in Hawaii associated with *Botryosphaeria dothidea*. *Plant Disease* 67, 555–556.

Holmes, J. and Rich, A.E. (1970) Factors affecting release and dissemination of *Physalospora obtusa* spores in a New Hampshire apple orchard. *Phytopathology* 60, 1052–1054.

Inderbitzin, P., Bostock, R.M., Trouillas, F.P. and Michailides, T.J. (2010) A six locus phylogeny reveals high species diversity in *Botryosphaeriaceae* from California almond. *Mycologia* 102, 1350–1368.

Jacobs, K.A. and Rehner, S.A. (1998) Comparison of cultural and morphological characters and ITS sequences in anamorphs of *Botryosphaeria* and related taxa. *Mycologia* 90, 601–610.

Jacobs, R. (2002) Characterisation of *Botryosphaeria* species from mango in South Africa. MSc thesis, University of Pretoria, Pretoria, South Africa.

Janson, E.M., Grebenok, R.J., Behmer, S.T. and Abbot, P. (2009) Same host–plant, different sterols: variation in sterol metabolism in an insect herbivore community. *Journal of Chemical Ecology* 35, 1309–1319.

Janson, E.M., Peeden, E.R., Stireman, J.O. III and Abbot, P. (2010) Symbiont-mediated phenotypic variation without co-evolution in an insect–fungus association. *Journal of Evolutionary Biology* 23, 2212–2228.

Johnson, G.I. (1994) *Dothiorella* stem canker and fruit rot. In: Ploetz, R.C., Zentmyer, G.A., Nishijima, W.T., Rohrbach, K.G. and Ohr, H.D. (eds) *Compendium of Tropical Fruit Diseases*. APS Press, St Paul, Minnesota, p. 76.

Johnson, G.I. and Kotzé, J.M. (1994) Stem-end rot. In: Ploetz, R.C., Zentmyer, G.A., Nishijima, W.T., Rohrbach, K.G. and Ohr, H.D. (eds) *Compendium of Tropical Fruit Diseases*. APS Press, St Paul, Minnesota, pp. 81–83.

Kim, K.W., Park, E.W. and Ahn, K.K. (1999) Pre-penetration behavior of *Botryosphaeria dothidea* on apple fruits. *Plant Pathology Journal* 15, 223–227.

Kim, K.W., Park, E.W., Kim, Y.H., Ahn, K.K., Kim, P.G. and Kim, K.S. (2001) Latency- and defense-related ultrastructural characteristics of apple fruit tissues infected with *Botryosphaeria dothidea*. *Phytopathology* 91, 165–172.

Kim, K.W., Kim, K.R. and Park, E.W. (2005) An infection model of apple white rot based on conidial germination and appressorium formation of *Botryosphaeria dothidea*. *Plant Pathology Journal* 21, 322–327.

Levitt, J. (1972) *Responses of Plants to Environmental Stresses*. Academic Press, New York.

Little, K.M., van Staden, J. and Clarke, G.P.Y. (2003) *Eucalyptus grandis* × *E. camaldulensis* variability and intra-genotypic competition as a function of different vegetation management treatments. *New Forests* 25, 227–242.
Luchi, N., Capretti, P., Surico, G., Orlando, C., Pazzagli, M. and Pinzani, P. (2005) A real-time quantitative PCR assay for the detection of *Sphaeropsis sapinea* from inoculated *Pinus nigra* shoots. *Journal of Phytopathology* 153, 37–42.
Luchi, N., Pratesi, N., Simi, L., Pazzagli, M., Capretti, P., Scala, A., Slippers, B. and Pinzani, P. (2011) High-resolution melting analysis: a new molecular approach for the early detection of *Diplodia pinea* in Austrian pine. *Fungal Biology* 115, 715–723.
Ma, Z. and Michailides, T.J. (2002) Characterization of *Botryosphaeria dothidea* isolates collected from pistachio and other plant hosts in California. *Phytopathology* 92, 519–526.
Madar, Z., Solel, Z., Riov, J. and Sztejnberg, A. (1995) Phytoalexin production by cypress in response to infection by *Diplodia pinea* f. sp. *cupressi* and its relation to water stress. *Physiological and Molecular Plant Pathology* 47, 29–38.
Mehl, J.W.M., Slippers, B., Roux, J. and Wingfield, M.J. (2011) *Botryosphaeriaceae* associated with *Pterocarpus angolensis* (kiaat) in South Africa. *Mycologia* 103, 534–553.
Menge, J.A. and Ploetz, R.C. (2003) Diseases of avocado. In: Ploetz, R.C. (ed.) *Diseases of Tropical Fruit Crops*. CAB International, Wallingford, UK, pp. 35–71.
Michailides, T.J. (1991) Pathogenicity, distribution, sources of inoculum, and infection courts of *Botryosphaeria dothidea* on pistachio. *Phytopathology* 81, 566–573.
Michailides, T.J. and Morgan, D.P. (1992) Effects of temperature and wetness duration on infection of pistachio by *Botryosphaeria dothidea* and management of disease by reducing duration of irrigation. *Phytopathology* 82, 1399–1406.
Michailides, T.J. and Morgan, D.P. (1993) Spore release by *Botryosphaeria dothidea* in pistachio orchards and disease control by altering the trajectory angle of sprinklers. *Phytopathology* 83, 145–152.
Michailides, T.J. and Morgan, D.P. (2004) Panicle and shoot blight of pistachio: a major threat to the California pistachio industry. Available at: http://www.apsnet.org/publications/apsnetfeatures/Pages/Pistachio.aspx (accessed 6 December 2012).
Michailides, T.J., Morgan, D.P. and Felts, D. (2001) Collection and characterization of *Botryosphaeria dothidea* from various hosts and pathogenicity studies on pistachio. *KAC Plant Protection Quarterly* 11, 3–8.
Milholland, R.D. (1970) Histology of *Botryosphaeria* canker of susceptible and resistant highbush blueberries. *Phytopathology* 60, 70–74.
Milholland, R.D. (1972) Histopathology and pathogenicity of *Botryosphaeria dothidea* on blueberry stems. *Phytopathology* 62, 654–660.
Miller, T. and Bramlett, D.L. (1979) Damage to reproductive structures of slash pine by two seed-borne pathogens: *Diplodia gossypina* and *Fusarium moniliforme* var. *subglutinans*. In: Bonner, F. (ed.) *Proceedings, IUFRO/USFS/Mississippi State University Symposium on Flowering and Seed Development in Trees, May 15–18, 1978*. Mississippi State University, Starkville, Mississippi, pp. 347–355.
Mohali, S.R., Slippers, B. and Wingfield, M.J. (2009) Pathogenicity of seven species of the *Botryosphaeriaceae* on *Eucalyptus* clones in Venezuela. *Australasian Plant Pathology* 38, 135–140.
Moral, J., Muñoz-Díez, C., González, N., Trapero, A. and Michailides, T.J. (2010) Characterization and pathogenicity of *Botryosphaeriaceae* species collected from olive and other hosts in Spain and California. *Phytopathology* 100, 1340–1351.
Mullen, J.M., Gilliam, C.H., Hagan, A.K. and Morgan-Jones, G. (1991) Canker of dogwood caused by *Lasiodiplodia theobromae*, a disease influenced by drought stress or cultivar selection. *Plant Disease* 75, 886–889.
Nicholls, T.H. and Ostry, M.E. (1990) *Sphaeropsis sapinea* cankers on stressed red and jack pines in Minnesota and Wisconsin. *Plant Disease* 74, 54–56.
Nosanchuk, J.D. and Casadevall, A. (2003) The contribution of melanin to microbial pathogenesis. *Cellular Microbiology* 5, 203–223.
Old, K.M., Gibbs, R., Craig, I., Myers, B.J. and Yuan, Z.Q. (1990) Effect of drought and defoliation on the susceptibility of eucalypts to cankers caused by *Endothia gyrosa* and *Botryosphaeria ribis*. *Australian Journal of Botany* 38, 571–581.
Palmer, M.A., Nicholls, T.H. and Croghan, C.F. (1986) Fungicidal control of shoot blight caused by *Sphaeropsis sapinea* on red pine nursery seedlings. *Plant Disease* 70, 194–196.
Paoletti, E., Danti, R. and Strati, S. (2001) Pre- and post-inoculation water stress affects *Sphaeropsis sapinea* canker length in *Pinus halepensis* seedlings. *Forest Pathology* 31, 209–218.

Parker, A.K. (1961) Bark moisture relations in disease development: present status and future needs. *Recent Advances in Botany* 2, 1535–1537.

Parker, P.E. and Ramsdell, D.C. (1977) Epidemiology and chemical control of *Godronia* (*Fusicoccum*) canker of highbush blueberry. *Phytopathology* 67, 1475–1480.

Pavlic, D., Slippers, B. Coutinho, T.A., Gryzenhout, M. and Wingfield, M.J. (2004) *Lasiodiplodia gonubiensis* sp. nov., a new *Botryosphaeria* anamorph from native *Syzygium cordatum* in South Africa. *Studies in Mycology* 50, 313–322.

Pavlic, D., Slippers, B., Coutinho, T.A. and Wingfield, M.J. (2007) *Botryosphaeriaceae* occurring on native *Syzygium cordatum* in South Africa and their potential threat to *Eucalyptus*. *Plant Pathology* 56, 624–636.

Pavlic, D., Wingfield, M.J., Barber, P., Slippers, B., Hardy, G.E.St.J. and Burgess, T.I. (2008) Seven new species of the *Botryosphaeriaceae* from baobab and other native trees in Western Australia. *Mycologia* 100, 851–866.

Pavlic, D., Slippers, B., Coutinho, T.A. and Wingfield, M.J. (2009) Multiple gene genealogies and phenotypic data reveal cryptic species of the *Botryosphaeriaceae*: a case study on the *Neofusicoccum parvum/N. ribis* complex. *Molecular Phylogenetics and Evolution* 51, 259–268.

Pérez, C.A., Wingfield, M.J., Slippers, B., Altier, N.A. and Blanchette, R.A. (2010) Endophytic and canker-associated *Botryosphaeriaceae* occurring on non-native *Eucalyptus* and native *Myrtaceae* trees in Uruguay. *Fungal Diversity* 41, 53–69.

Peterson, G.W. (1977) Infection, epidemiology, and control of Diplodia blight of Austrian, ponderosa and Scots pines. *Phytopathology* 67, 511–514.

Petrini, O. and Fisher, P.J. (1988) A comparative study of fungal endophytes in xylem and whole stem of *Pinus sylvestris* and *Fagus sylvatica*. *Transactions of the British Mycological Society* 91, 233–238.

Phillips, A.J.L. (2002) *Botryosphaeria* species associated with diseases of grapevines in Portugal. *Phytopathologia Mediterranea* 41, 3–18.

Phillips, A.J.L., Alves, A., Correia, A. and Luque, J. (2005) Two new species of *Botryosphaeria* with brown, 1-septate ascospores and *Dothiorella* anamorphs. *Mycologia* 97, 513–529.

Phillips, A.J.L., Oudemans, P.V., Correia, A. and Alves, A. (2006) Characterisation and epitypification of *Botryosphaeria corticis*, the cause of blueberry cane canker. *Fungal Diversity* 21, 141–155.

Phillips, A.J.L., Alves, A., Pennycook, S.R., Johnston, P.R., Ramaley, A., Akulov, A. and Crous, P.W. (2008) Resolving the phylogenetic and taxonomic status of dark-spored teleomorph genera in the *Botryosphaeriaceae*. *Persoonia* 21, 29–55.

Piškur, B., Pavlic, D., Slippers, B., Ogris, N., Maresi, G., Wingfield, M.J. and Jurc, D. (2011) Diversity and pathogenicity of *Botryosphaeriaceae* on declining *Ostrya carpinifolia* in Slovenia and Italy following extreme weather conditions. *European Journal of Forest Research* 130, 235–249.

Ploetz, R.C. (2003) Diseases of mango. In: Ploetz, R.C. (ed.) *Diseases of Tropical Fruit Crops*. CAB International, Wallingford, UK, pp. 327–363.

Puritch, G.S. and Mullick, D.B. (1975) Effect of water stress on the rate of non-suberized impervious tissue formation following wounding in *Abies grandis*. *Journal of Experimental Botany* 26, 903–910.

Pusey, P.L. (1989) Availability and dispersal of ascospores and conidia of *Botryosphaeria* in peach orchards. *Phytopathology* 79, 635–639.

Pusey, P.L. (1993) Role of *Botryosphaeria* species in peach tree gummosis on the basis of differential isolation from outer and inner bark. *Plant Disease* 77, 170–174.

Pusey, P.L., Kitajima, H. and Wu, Y. (1995) Fungal gummosis. In: Ogawa, J.M., Zehr, E.I., Bird, G.W., Ritchie, D.F., Uriu, K. and Uyemoto, J.K. (eds) *Compendium of Stone Fruit Diseases*. APS Press, St Paul, Minnesota, pp. 33–34.

Ramos, L.J., Lara, S.P., McMillan, R.T. Jr and Narayanan, K.R. (1991) Tip dieback of mango (*Mangifera indica*) caused by *Botryosphaeria ribis*. *Plant Disease* 75, 315–318.

Rayachhetry, M.B., Blakeslee, G.M. and Miller, T. (1996a) Histopathology of *Botryosphaeria ribis* in *Melaleuca quinquenervia*: pathogen invasion and host response. *International Journal of Plant Sciences* 157, 219–227.

Rayachhetry, M.B., Blakeslee, G.M. and Center, T.D. (1996b) Predisposition of melaleuca (*Melaleuca quinquenervia*) to invasion by the potential biological control agent *Botryosphaeria ribis*. *Weed Science* 44, 603–608.

Rees, A.A. (1988) Infection of *Pinus caribaea* seed by *Lasiodiplodia theobromae*. *Transactions of the British Mycological Society* 90, 321–324.

Rees, A.A. and Webber, J.F. (1988) Pathogenicity of *Sphaeropsis sapinea* to seed, seedlings and saplings of some central American pines. *Transactions of the British Mycological Society* 91, 273–277.

Ridgway, H.J., Amponsah, N.T., Brown, D.S., Baskarathevan, J., Jones, E.E. and Jaspers, M.V. (2011) Detection of botryosphaeriaceous species in environmental samples using a multi-species primer pair. *Plant Pathology* 60, 1118–1127.
Rodas, C.A., Slippers, B., Gryzenhout, M. and Wingfield, M.J. (2009) *Botryosphaeriaceae* associated with *Eucalyptus* canker diseases in Colombia. *Forest Pathology* 39, 110–123.
Rodriguez, R.J., White J.F. Jr, Arnold, A.E. and Redman, R.S. (2009) Fungal endophytes: diversity and functional roles. *New Phytologist* 182, 314–330.
Romero, M.A., Sánchez, M.E. and Trapero, A. (2005) First report of *Botryosphaeria ribis* as a branch dieback pathogen of olive trees in Spain. *Plant Disease* 89, 208.
Sakalidis, M.L., Hardy, G.E.St.J. and Burgess, T.I. (2011) Use of the genealogical sorting index (GSI) to delineate species boundaries in the *Neofusicoccum parvum–Neofusicoccum ribis* species complex. *Molecular Phylogenetics and Evolution* 60, 333–344.
Schoeneweiss, D.F. (1975) Predisposition, stress, and plant disease. *Annual Review of Phytopathology* 13, 193–211.
Schoeneweiss, D.F. (1981a) Infectious diseases of trees associated with water and freezing stress. *Journal of Arboriculture* 7, 13–18.
Schoeneweiss, D.F. (1981b) The role of environmental stress in diseases of woody plants. *Plant Disease* 65, 308–314.
Schulz, B., Guske, S., Dammann, U. and Boyle, C. (1998) Endophyte–host interactions II: defining symbiosis of the endophyte–host interaction. *Symbiosis* 25, 213–227.
Shahin, E.A. and Claflin, L.E. (1980) Twig blight of Douglas-fir: a new disease caused by *Dothiorella dothidea*. *Plant Disease* 64, 47–50.
Shearer, B.L., Tippett, J.T. and Bartle, J.R. (1987) *Botryosphaeria ribis* infection associated with death of *Eucalyptus radiata* in species selection trials. *Plant Disease* 71, 140–145.
Slippers, B. and Wingfield, M.J. (2007) *Botryosphaeriaceae* as endophytes and latent pathogens of woody plants: diversity, ecology and impact. *Fungal Biology Reviews* 21, 90–106.
Slippers, B., Crous, P.W., Denman, S., Coutinho, T.A., Wingfield, B.D. and Wingfield, M.J. (2004a) Combined multiple gene genealogies and phenotypic characters differentiate several species previously identified as *Botryosphaeria dothidea*. *Mycologia* 96, 83–101.
Slippers, B., Fourie, G., Crous, P.W., Coutinho, T.A., Wingfield, B.D., Carnegie, A.J. and Wingfield, M.J. (2004b) Speciation and distribution of *Botryosphaeria* spp. on native and introduced *Eucalyptus* trees in Australia and South Africa. *Studies in Mycology* 50, 343–358.
Slippers, B., Stenlid, J. and Wingfield, M.J. (2005) Emerging pathogens: fungal host jumps following anthropogenic introduction. *Trends in Ecology and Evolution* 20, 420–421.
Slippers, B., Smit, W.A., Crous, P.W., Coutinho, T.A., Wingfield, B.D. and Wingfield, M.J. (2007) Taxonomy, phylogeny and identification of *Botryosphaeriaceae* associated with pome and stone fruit trees in South Africa and other regions of the world. *Plant Pathology* 56, 128–139.
Slippers, B., Burgess, T.I., Pavlic, D., Ahumada, R., Maleme, H., Mohali, S., Rodas, C.A. and Wingfield, M.J. (2009) A diverse assemblage of *Botryosphaeriaceae* infect *Eucalyptus* in native and non-native environments. *Southern Forests* 71, 101–110.
Smith, D.R. and Stanosz, G.R. (2006) A species-specific PCR assay for detection of *Diplodia pinea* and *D. scrobiculata* in dead red and jack pines with collar rot symptoms. *Plant Disease* 90, 307–313.
Smith, H., Wingfield, M.J., Crous, P.W. and Coutinho, T.A. (1996) *Sphaeropsis sapinea* and *Botryosphaeria dothidea* endophytic in *Pinus* spp. and *Eucalyptus* spp. in South Africa. *South African Journal of Botany* 62, 86–88.
Smith, H., Crous, P.W., Wingfield, M.J., Coutinho, T.A. and Wingfield, B.D. (2001) *Botryosphaeria eucalyptorum* sp. nov., a new species in the *B. dothidea*-complex on *Eucalyptus* in South Africa. *Mycologia* 93, 277–285.
Smith, H., Wingfield, M.J. and Coutinho, T.A. (2002) The role of latent *Sphaeropsis sapinea* infections in post-hail associated dieback of *Pinus patula*. *Forest Ecology and Management* 164, 177–184.
Spagnolo, A., Marchi, G., Peduto, F., Phillips, A.J.L. and Surico, G. (2011) Detection of *Botryosphaeriaceae* species within grapevine woody tissues by nested PCR, with particular emphasis on the *Neofusicoccum parvum/N. ribis* complex. *European Journal of Plant Pathology* 129, 485–500.
Srivastava, D.N. (1972) Epidemiology and prevention of *Diplodia* stem-end rot of ripe mango fruits. *Acta Horticulturae* 24, 235–236.
Stanosz, G.R., Smith, D.R. and Albers, J.S. (2005) Surveys for asymptomatic persistence of *Sphaeropsis sapinea* on or in stems of red pine seedlings from seven Great Lakes region nurseries. *Forest Pathology* 35, 233–244.

Stanosz, G.R., Smith, D.R. and Leisso, R. (2007) Diplodia shoot blight and asymptomatic persistence of *Diplodia pinea* on or in stems of jack pine nursery seedlings. *Forest Pathology* 37, 145–154.

Swart, W.J. and Wingfield, M.J. (1991) Biology and control of *Sphaeropsis sapinea* on *Pinus* species in South Africa. *Plant Disease* 75, 761–766.

Swart, W.J., Wingfield, M.J. and Knox-Davies, P.S. (1987) Conidial dispersal of *Sphaeropsis sapinea* in three climatic regions of South Africa. *Plant Disease* 71, 1038–1040.

Thompson, S., Alvarez-Loayza, P., Terborgh, J. and Katul, G. (2010) The effects of plant pathogens on tree recruitment in the Western Amazon under a projected future climate: a dynamical systems analysis. *Journal of Ecology* 98, 1434–1446.

Timmer, L.W., Garnsey, S.M. and Broadbent, P. (2003) Diseases of citrus. In: Ploetz, R.C. (ed.) *Diseases of Tropical Fruit Crops*. CAB International, Wallingford, UK, pp. 163–195.

Úrbez-Torres, J.R., Battany, M., Bettiga, L.J., Gispert, C., McGourty, G., Roncoroni, J., Smith, R.J., Verdegaal, P. and Gubler, W.D. (2010) *Botryosphaeriaceae* species spore-trapping studies in California vineyards. *Plant Disease* 94, 717–724.

Van Dijk, H.F.G., Van der Gaag, M., Perik, P.J.M. and Roelofs, J.G.M. (1992) Nutrient availability in Corsican pine stands in the Netherlands and the occurrence of *Sphaeropsis sapinea*: a field study. *Canadian Journal of Botany* 70, 870–875.

van Niekerk, J.M., Fourie, P.H., Halleen, F. and Crous, P.W. (2006) *Botryosphaeria* spp. as grapevine trunk disease pathogens. *Phytopathologia Mediterranea* 45, S43–S54.

Veiga, T.A.M., Silva, S.C., Francisco, A.-C., Filho, E.R., Vieira, P.C., Fernandes, J.B., Silva, M.F.G.F., Müller, M.W. and Lotina-Hennsen, B. (2007) Inhibition of photophosphorylation and electron transport chain in thylakoids by lasiodiplodin, a natural product from *Botryosphaeria rhodina*. *Journal of Agricultural and Food Chemistry* 55, 4217–4221.

Venkatasubbaiah, P., Sutton, T.B. and Chilton, W.S. (1991) Effect of phytotoxins produced by *Botryosphaeria obtusa*, the cause of black rot of apple fruit and frogeye leaf spot. *Phytopathology* 81, 243–247.

Weaver, D.J. (1979) Role of conidia of *Botryosphaeria dothidea* in the natural spread of peach tree gummosis. *Phytopathology* 69, 330–334.

Webb, R.S. (1983) Seed capsule abortion and twig dieback of *Eucalyptus camaldulensis* in south Florida induced by *Botryosphaeria ribis*. *Plant Disease* 67, 108–109.

Whitelaw-Weckert, M.A., Sergeeva, V. and Priest, M.J. (2006) *Botryosphaeria stevensii* infection of Pinot Noir grapevines by soil–root transmission. *Australasian Plant Pathology* 35, 369–371.

Wingfield, M.J. and Knox-Davies, P.S. (1980) Association of *Diplodia pinea* with a root disease of pines in South Africa. *Plant Disease* 64, 221–223.

Wright, J.P. and Marks, G.C. (1970) Loss of merchantable wood in radiata pine associated with infection by *Diplodia pinea*. *Australian Forestry* 34, 107–119.

Zhou, S. and Stanosz, G.R. (2001) Relationships among *Botryosphaeria* species and associated anamorphic fungi inferred from the analyses of ITS and 5.8S rDNA sequences. *Mycologia* 93, 516–527.

15 Chestnut Blight

Simone Prospero and Daniel Rigling*
Swiss Federal Institute for Forest, Snow and Landscape Research (WSL), Birmensdorf, Switzerland

15.1 Pathogen, Significance and Distribution

The causal agent of chestnut blight, *Cryphonectria parasitica* (Murrill) M.E. Barr, is an ascomycete fungus (*Diaporthales*) that is native to eastern Asia (China and Japan). During the 20th century, this pathogen was accidentally introduced into North America and Europe through infected Asian chestnut plants (Griffin, 1986; Roane *et al.*, 1986; Heiniger and Rigling, 1994). Within 50 years of its first recovery in 1904 in the New York Botanical Garden, the fungus spread epidemically at a rate of more than 30 km $year^{-1}$ (Evans and Finkral, 2010) throughout the native distribution range of the American chestnut (*Castanea dentata* (Marsh.) Borkh.) in north-eastern USA.

In Europe, *C. parasitica* was officially first discovered in 1938 in Italy near Genoa, one of the main Italian commercial ports. Multiple introductions probably occurred both via North America and directly from Asia, and they did not only target Italy but also south-western France and Spain (Dutech *et al.*, 2010). As in North America, chestnut blight spread rapidly in Europe, and by the end of the 1960s the disease had already been recorded in many countries such as Albania, Bosnia-Herzegovina, Croatia, France, Greece, Hungary, Italy, Slovenia, Spain, Switzerland and Turkey (Heiniger and Rigling, 1994; Robin and Heiniger, 2001). The pathogen successively colonized the more distant or isolated chestnut stands in north-western Spain and Portugal, in northern Switzerland, in Germany, in northern France, in Austria, in eastern Europe (the Czech Republic, Macedonia, Romania and the Slovak Republic) and in the Aegean and Black Sea regions of Turkey (Heiniger and Rigling, 1994; Haltofová *et al.*, 2005; Akilli *et al.*, 2009; Erincik *et al.*, 2011). *C. parasitica* is also present in the most eastern distribution range of the European chestnut (*Castanea sativa* Mill.), and has recently been officially reported in Iran (Kazempour *et al.*, 2006) and Azerbaijan (Aghayeva and Harrington, 2008).

In Asia, the impact of *C. parasitica* on the native chestnut species is minor. Thanks to the co-evolution with the pathogen, the two main local host species, i.e. the Chinese chestnut (*C. mollissima* Blume) and the Japanese chestnut (*C. crenata* Siebold & Zucc.), are highly resistant to *C. parasitica* (Anagnostakis, 1992).

In North America, chestnut blight has had devastating consequences on the native American chestnut forests. The American

* E-mail: daniel.rigling@wsl.ch

chestnut, a tree that could easily grow to be 35 m high and more than 1 m in diameter, has been almost completely removed from the overstorey by the blight because of its high susceptibility to the disease (Anagnostakis, 1992). The dramatic dieback of American chestnut has resulted in irreversible changes in the forest ecosystem structure and functioning (Griffin, 1992). In addition to such drastic ecological impacts, chestnut blight has also had major economic consequences, including the loss of a high-quality food resource for people who collected chestnuts (Diamond *et al.*, 2000) and the loss of valuable timber (Holmes *et al.*, 2009). Chestnut blight in North America is considered a textbook example of the impact of invasive pathogens on native ecosystems.

When the *C. parasitica* epidemics started in Europe, a high chestnut mortality was observed and there was serious concern about the survival of European chestnut. Fortunately, later on, many chestnut stands in Europe recovered owing to the spontaneous appearance of hypovirulence (Heiniger and Rigling, 1994). The incidence of the disease still remained high, but healing cankers became dominant and the survival rate of the infected trees increased considerably. Although at present some tree mortality is still visible in chestnut stands, the survival of European chestnut as a major tree species does not seem to be endangered any more by chestnut blight in areas with widespread hypovirulence. In contrast, in most of the regions at the epidemic front, e.g. north-east Spain, Portugal and northern Switzerland, where the incidence of hypovirulence is still very low, chestnut blight still causes severe chestnut decline (Colinas and Uscuplic, 1999; Bragança *et al.*, 2007; Heiniger and Rigling, 2009).

The main hosts of *C. parasitica* are species in the genus *Castanea*, particularly the American and the European chestnut. In addition, *C. parasitica* has been found on several oak species (*Quercus* spp.) and in a few cases also on other host plants (Gryzenhout *et al.*, 2009). Generally, hosts other than *Castanea* spp. may be infected in areas where the disease pressure from chestnut trees is high (Turchetti and Maresi, 2008).

15.2 Diagnosis

C. parasitica infects the stem and branches of its host plants (Heiniger and Rigling, 1994; EPPO, 2005). On young grafted trees, infections are most frequently located in the region of the graft. In coppices, old orchard trees and high forests, infections can be observed at the base of the trunk, along the trunk or on branches in the crown. The symptoms induced by *C. parasitica* on European chestnut vary depending on the age of the infected stem/branch and on the virulence of the fungal strain involved. In its typical form, chestnut blight is always associated with extensive necrosis (cankers) of the bark. On young trees/branches, the smooth bark at the site of the infection turns from olive-green to bright brown, orange brown, yellow brown, or red brown. When older trees/branches become infected, the discoloration caused by the cankers is usually less pronounced. Virulent fungal strains grow in the cambium and in the bark tissue, and develop characteristic pale brown, flat mycelial fans that are visible after removing the outer bark. The cambium under the infected bark is killed and the bark sinks inwards. The cankers may enlarge rapidly and girdle the stem or branch, which induces the death of the plant parts distal to the infection. The leaves wilt and typically remain hanging on the infected branch, sometimes even until winter. Below the cankers, trees typically produce numerous epicormic (adventitious) shoots (Plate 19A).

When infections are caused by hypovirulent fungal strains, cankers may initially show similar characteristics to those induced by virulent fungal strains, but their expansion may successively slow down and new layers of bark may form under the affected areas, causing a swelling and cracking of the outer bark. Healed, inactive cankers are typically superficial, callused or swollen (Plate 19B) and mycelial fans are less frequent than in active cankers. Because the cambium is not colonized and killed by the fungus, the plant parts distal to the infection survive and no epicormic shoots are formed below the canker.

Masses of yellow orange to reddish brown pustules (stromata) harbouring the sexual (perithecia) and asexual (pycnidia) fruiting bodies of the fungus may develop on the infected bark, especially if the infection is caused by a virulent *C. parasitica* strain. On older trees/branches with a rough bark, stromata often form in the longitudinal cracks of the bark, even in the absence of a clearly visible lesion. The bark of recently dead or cut chestnut wood is also a favourable substrate for the development of stromata (Prospero *et al.*, 2006). Perithecia can be recognized by their ostioles, which appear, when using a hand-magnifying lens, as black dots on papillate protuberances on the surface of the stroma. As the ostioles of perithecia and pycnidia can look similar, the presence of perithecia should be confirmed by dissecting samples of stromata under a dissecting microscope. Asexual spores (conidia) are extruded as up to 1 cm long yellow tendrils during moist weather, but sexual ascospores are actively discharged into the air from the flask-shaped perithecia. On oak species, symptoms are localized and consist of slow-developing cankers that only rarely kill infected trees or branches.

The fruiting structures of *C. parasitica* can be microscopically characterized in the laboratory. The yellow to dark orange stromata (0.5–4 mm in diameter, up to 2.5 mm high) are embedded in the bark with only the upper part protruding from the surface. Pycnidia (100–300 μm in diameter) are formed as irregular cavities within the stromatic tissue and, in moist conditions, extrude the yellow tendrils of hyaline conidia through the pycnidial ostioles. Perithecia (300–400 μm in diameter) are globose and deeply immersed in the stromatic tissue. Each has a long cylindrical neck. For details on these macromorphological and micromorphological features refer to EPPO (2005).

In the absence of fruiting bodies, or for confirmation, *C. parasitica* may be identified after isolation in pure culture (EPPO, 2005). On potato dextrose agar (PDA), virulent isolates produce an initially white mycelium, which becomes subsequently yellow orange and, after 1–2 weeks, red orange. In hypovirulent isolates, the mycelium typically remains white. After about 5 days of incubation in the light at room temperature (20°C), pycnidia begin to form in culture and appear as a dense, orange-pigmented globose mat. Virulent isolates show an abundant production of pycnidia, which are typically formed on concentric rings if cultures are incubated in a light–dark cycle. In hypovirulent isolates, pycnidial production is largely reduced under low light conditions, but can be considerably stimulated with more intense light conditions (Hillman *et al.*, 1990). Perithecia are not produced in culture either in the virulent or the hypovirulent strains. DNA-based identification of pure cultures mainly rely on the sequence analysis of conserved DNA regions, for example, of the internal transcribed spacer (ITS) region or the β-tubulin gene.

Beside *C. parasitica*, the genus *Cryphonectria* includes other species (see Gryzenhout, Chapter 16, this volume). Of these, three saprotrophic species are frequently found when attempting to isolate *C. parasitica* from stromata on the bark of dead chestnut or oak wood, namely *C. radicalis* (Schwein.) M.E. Barr in Europe and North America, *C. naterciae* M.H. Bragança, E. Diogo & A.J.L. Phillips (probably a synonym of *C. decipiens* Gryzenh. & M.J. Wingf.) in Europe, and *C. japonica* (Tak. Kobay. & Kaz. Itô) Gryzenh. & M.J. Wingf. in Japan. For details on how to distinguish these other *Cryphonectria* spp. from *C. parasitica* refer to Hoegger *et al.* (2002), Gryzenhout *et al.* (2009) and Bragança *et al.* (2011).

15.3 Infection Biology

C. parasitica is a necrotrophic pathogen that requires dead host tissue to penetrate into the stems and branches of healthy chestnut trees (Fig. 15.1). Although the early stage of infection has been poorly documented, it is generally assumed that conidia or ascospores germinate in fresh wounds or growth cracks where an initial lesion is formed (Roane *et al.*, 1986).

Besides wounds in the bark, moribund chestnut tissue might also serve as an entry point for *C. parasitica* (Prospero *et al.*, 2006). Such tissue can be generated by the natural death of a tree's lower twigs or branches due to shading from the upper canopy or by sprout mortality resulting from competition.

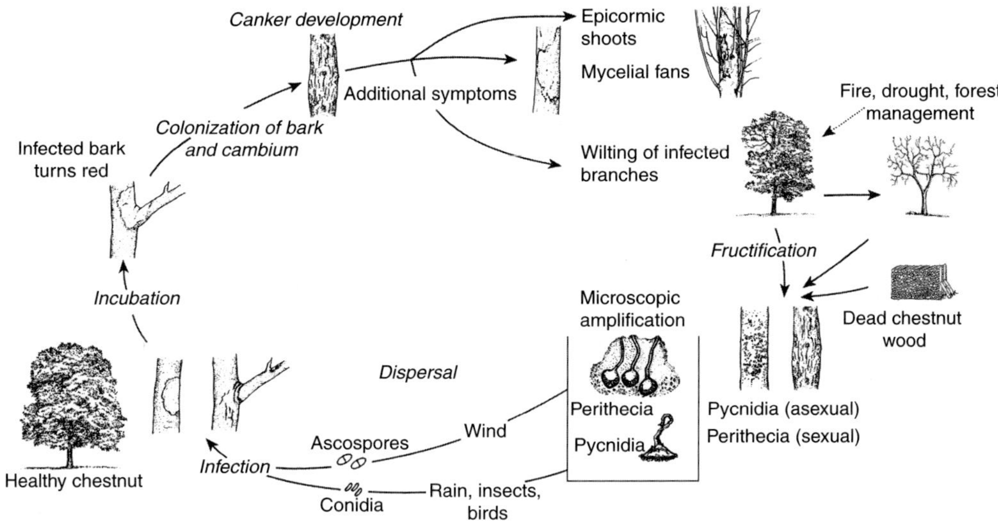

Fig. 15.1. Life cycle of *Cryphonectria parasitica* (Murrill) M.E. Barr, the causal agent of chestnut blight (adapted after the life cycle proposed by Ursula Heiniger).

C. parasitica infection has also been observed to develop on surviving chestnut trees after they were exposed to forest fires or drought conditions which apparently caused some stem sections to die (D. Rigling and M. Conedera, 2011, unpublished results).

Canker development has been studied in detail using histological methods (Hebard *et al.*, 1984). Once initially established in host tissue, *C. parasitica* produces its characteristic mycelial fans, which advance intercellularly in the bark and cambium of susceptible chestnut trees. Host defence reactions against *C. parasitica* infections include the lignification of cell walls and subsequent wound periderm formation. Histological examinations indicate that mycelial fans are able to penetrate through zones of lignified host cells and developing wound periderm. Once wound periderm is fully formed, no penetration of *C. parasitica* fans seems to occur. In susceptible chestnut species, wound periderm formation is continuously inhibited because the advancing mycelial fans kill the host tissue. The mycelial fans are thought to build up physical pressure and this physically splits the host cell. Growth of the mycelial fans in susceptible hosts is promoted by the ability of *C. parasitica* to utilize nutrients from dead host cells, especially tannins, which normally play a role in the host defence reaction (Roane *et al.*, 1986). Hebard *et al.* (1984) concluded that the rate and extent of mycelial fan formation was the most important factor determining canker enlargement.

Host tissues in front of the mycelial fans are presumably killed by the action of toxins and cell wall degrading enzymes, and are subsequently invaded by the fans (Roane *et al.*, 1986). One metabolite that is likely to be involved in the infection process is oxalic acid, which is secreted by the fungus at the advancing edge of the canker, and is thought to have a toxic effect on host tissue and to trigger cell wall degradation (Havir and Anagnostakis, 1983). Evidence for the role of oxalic acid in pathogenesis has recently been found through target mutation of the gene encoding oxaloacetate acetylhydrolase, which is the enzyme that produces oxalic acid (Chen *et al.*, 2010). *C. parasitica* strains carrying a knockout mutation in this gene displayed dramatically reduced virulence, which strongly suggests that oxalic acid plays a key role in the infection process. Other virulence factors identified by specific mutations include G-protein signalling (Gao and Nuss, 1996), a kexin-like protease involved in protein secretion (Jacob-Wilk *et al.*, 2009), an ST12 transcription factor homologue (Deng *et al.*, 2007), a tannic acid inducible laccase (Chung *et al.*, 2008) and a cylophilin (Chen *et al.*, 2011).

15.4 Epidemiology

All biotic and abiotic factors that may affect chestnut blight epidemics must be considered when developing disease management strategies (Milgroom and Cortesi, 2004). In particular, tackling those factors that reduce the rate of epidemic progress of *C. parasitica* should help to make chestnut blight control more effective.

15.4.1 Sexual and asexual reproduction

Both asexual conidia and sexual ascospores contribute to the spread of the chestnut blight fungus (Heiniger and Rigling, 1994). Conidia, which are released from the pycnidia during moist conditions, are mainly splash dispersed by rain over short distances, typically only a few metres, or washed down the branches and stems (Griffin, 1986; Anagnostakis, 1987). Thus, the same *C. parasitica* genotype can cause multiple infections in a single tree or sprout cluster (Milgroom and Lipari, 1995). If conidia are transported down to the ground, they may remain viable in the soil for a long time (Heald and Gardner, 1914). They can be transported over long distances via insects, birds or windborne dust (Heald and Studhalter, 1914; Russin *et al.*, 1984). Ascospores are actively discharged from the perithecia and may be carried by the wind over long distances, typically up to a few hundred metres. Heald *et al.* (1915) showed that, if bark dries slowly after rainfall, ascospore discharge could continue for up to 14 hours. Moreover, after a dry period, 4.5 mm of rain was sufficient to induce abundant expulsion of ascospores. In North America, ascospores were found to discharge during warm rains throughout spring, summer and autumn, with a peak in late summer or autumn (Heald and Gardner, 1913). In France, ascospores were also trapped between March and October, but the peak was already observed in May (Guérin *et al.*, 2001). This period of ascospore release corresponds to the period of highest receptivity of European chestnut for *C. parasitica* in Europe (Bazzigher, 1981; Guérin and Robin, 2003).

Crossing experiments in the laboratory have shown that *C. parasitica* has a sexual system resembling that of self-sterile (i.e. heterothallic) ascomycetes (McGuire *et al.*, 2001). Mating is controlled by a single locus (the *MAT* locus) with two alleles (*MAT-1* and *MAT-2*), and only matings between strains of the opposite mating type are successful (Marra and Milgroom, 2001). This mating system is what has mostly been found in the laboratory, but investigations in natural populations have shown that *C. parasitica* has rather a mixed mating system, with self-fertilization occurring at variable frequencies (Marra *et al.*, 2004). In *C. parasitica*, most self-fertile isolates are heterokaryotic for mating type (McGuire *et al.*, 2004). Heterokaryon formation may occur through recurrent biparental inbreeding or parasexual recombination involving mitotic crossing over (Milgroom *et al.*, 2009).

15.4.2 Invasion dynamics

Studies using vegetative compatibility (vc) types and neutral genetic markers have shown that new *C. parasitica* populations are mainly established by just one, or only a few, genotypes (haplotypes) (Hoegger *et al.*, 2000; Sotirovski *et al.*, 2004; Perlerou and Diamandis, 2006; Dutech *et al.*, 2008; Milgroom *et al.*, 2008; Prospero and Rigling, 2012). The low genetic diversity may reflect a founder effect, i.e. a loss of genetic variation that occurs when a new population is established by a very small number of individuals from a larger population. Alternatively, it has been hypothesized that specific genotypes might be better adapted to colonizing a new environment than others (Milgroom *et al.*, 2008). The mating opportunities of *C. parasitica* are limited because of the presence of a single/predominant mating type. Thus, at the beginning of a new epidemic, asexual reproduction may ensure reproduction and allow a rapid population growth. Subsequently, the introduction of the opposite mating type may increase the incidence of sexual reproduction. However, sexual reproduction might already occur in a founder population from the beginning, provided that both mating types have been concurrently introduced. This seems to be the case in Portugal where a sexually reproducing dominant vc type was found (Bragança *et al.*, 2007). Genetic and vc type diversity may also increase through mutations and/or additional immigration of genotypes. In northern Switzerland and Croatia, the highest diversity was observed in areas where genotypes from different genetic pools had come together (Ježić *et al.*, 2012; Prospero and Rigling, 2012).

15.4.3 Environmental factors

According to Anagnostakis and Aylor (1984), air temperature was the environmental factor that most influenced the spread of chestnut blight in North America. At an average daily temperature of 20°C, cankers on American chestnuts expanded at a rate of 1 mm day^{-1}. In a growth chamber experiment, canker expansion on European chestnut was found to be fastest at 27°C. However, the conditions for canker development were considerably less favourable below 20°C (Bazzigher, 1981). Frost and, in particular, frost injury have been associated with a higher and more severe blight incidence on Chinese and Japanese chestnuts, as well as with more frequent basal cankers on American chestnut sprouts (Roane *et al.*, 1986).

Another environmental factor that plays a central role in the development of most pathogenic fungi is the availability of water from humid air, rain or dew. In *C. parasitica*, moist conditions induce the release of the tendrils of conidia and the expulsion of the ascospores (Griffin, 1986). A reduced availability of water seems to affect the susceptibility of chestnut trees to *C. parasitica*. Field observations in Japan suggested that a summer drought considerably increases the incidence of blight on Japanese chestnut (Roane *et al.*, 1986). According to a growth chamber experiment, canker expansion on the excised dormant stems of American chestnut appears to be promoted by water stress in the bark (Gao and Shain, 1995). Waldboth and Oberhuber (2009) showed that severe droughts and chestnut blight synergistically promoted growth decline and mortality of European chestnut. In southern Switzerland, following the exceptional summer drought in 2003, a heavy *C. parasitica* sporulation on chestnut sprouts stressed by drought was observed. *C. parasitica* was noticed to respond in a similar way to fire disturbance (D. Rigling and S. Prospero, 2011, unpublished results).

Various bioclimatic and epidemiological models are currently being used to predict the future geographic distribution and impact of forest pathogens under a scenario of climate change. In the specific case of chestnut blight, Desprez-Loustau *et al.* (2007) showed that climate change in France will not necessarily lead to a range expansion of blight or to an increased incidence of the disease. In fact, the blight-promoting effects of warming might be counterbalanced by reduced water availability.

15.4.4 Host factors

In the absence of hypovirulence, host susceptibility is the main factor determining whether a chestnut blight epidemic will occur. Guérin and Robin (2003) performed artificial inoculations at regular intervals for 30 months both on living European chestnut sprouts in the field and on excised stem segments in growth chambers under controlled conditions. Their results revealed a seasonal variation in the susceptibility of European chestnut to *C. parasitica*. They found that there were more conidial infections and lesion growth on both the living sprouts and excised stems in spring and summer than there were in autumn and winter. This seasonal pattern in susceptibility, they concluded, is probably due to the intrinsic susceptibility of cortical tissues to *C. parasitica*, which may vary according to the nutritional value and water content of the bark. In susceptible chestnut species, the age of a wound seems to considerably influence the chance of *C. parasitica* becoming established (Bazzigher and Schmid, 1962). On European chestnut trees, the largest lesions developed after 1 day old wounds were inoculated.

In wounds older than 4 days, it was no longer possible for *C. parasitica* to establish.

15.5 Management Strategies and Tactics

15.5.1 Avoidance

The options for preventing *C. parasitica* infections are limited in areas where chestnut blight is present. Because the fungus infects the tree through wounds, chestnut grafts and pruning cuts are particularly vulnerable to the disease. Infections can be reduced as a first step by using disinfected tools for grafting and pruning. Of the different grafting techniques available, bud grafting might be less vulnerable to blight infection than bark grafting because it produces smaller wounds. It is also recommended that any wounding of chestnut trees should be avoided during the periods of dispersal of airborne ascospores, i.e. between March and October (Northern hemisphere).

15.5.2 Exclusion

Quarantine regulations have been established across the world to prevent the introduction and spread of *C. parasitica*. These regulations, which affect the movement and trading of chestnut wood and bark, seeds and living plants, vary slightly among countries and continents. In continents where chestnut blight is already present, e.g. North America or Europe, the regulations aim to prevent the spread of the pathogen into areas that are still disease free and to ensure that no additional genotypes enter. In Europe, *C. parasitica* is listed as an A2 quarantine organism by the European and Mediterranean Plant Protection Organization (EPPO, 2005). Quarantine regulations against *C. parasitica* within Europe target the trading of chestnut wood and bark, as well as of chestnut and oak plants, but not chestnut seeds and fruits, according to Council Directive 2000/29/EC (Council of the European Union, 2000; Seemann, 2001). Chestnut or oak plants intended for planting can only be moved within Europe if they are accompanied by a plant passport. This certifies that either 'the plants originate in areas known to be free from *C. parasitica*' or 'no observations of *C. parasitica* have been made at the place of production or its immediate vicinity since the beginning of the last complete cycle of vegetation'.

One major problem for quarantine regulations arises from the fact that chestnut plants can carry the pathogen without showing any symptoms (Bissegger and Sieber, 1994) and contaminated plants cannot be detected through visual inspection. Disease symptoms may subsequently develop only after several months. For example, in Australia, *C. parasitica* symptoms were detected on imported chestnut plants 16 months after the plants were placed into a post-entry quarantine facility (Cunnington and Pascoe, 2003).

15.5.3 Eradication

The history of eradication measures against *C. parasitica* mostly provides a picture of failures rather than successes. Experience from North America and Europe suggests that once chestnut blight has been established in a forest area, eradication is not realistic any more (Anagnostakis, 1987; Heiniger and Rigling, 2009). In contrast to forest areas, it might be feasible to eradicate chestnut blight in orchards, in particular if they are well separated from infected chestnut forests. Successful examples have been reported from western North America, which is far outside the natural range of American chestnut (Griffin, 1986). In this area, European and Asian chestnut cultivars were planted for nut production. After the first detection of chestnut blight in these orchards in 1934, eradication efforts (uprooting and burning of infected trees) were undertaken over many years. Newly infected trees were found until 1959, i.e. 25 years after the first eradication measures, but no further cankered trees were detected during annual inspections between 1960 and 1964, and chestnut blight was declared eradicated from these orchards.

15.5.4 Protection

Before *C. parasitica* can penetrate into a tree, the tree must have a wound. Thus, any measure that promotes a rapid sealing of the wounds can be expected to help in preventing chestnut blight infections. For this reason, wound seals are frequently used to protect grafts, pruning cuts or any sort of wound. Typically, these seals are quick drying, impervious to water and may or may not contain fungicides, which are, if used, usually copper-related compounds. However, despite the widespread application of wound seals, no sound data exist on the real efficacy of these methods to prevent *C. parasitica* infections.

In Japan, soil fertilization with nitrogen, phosphorus and potassium seemed to influence the incidence of blight and the dimensions of the lesions on artificially inoculated Japanese chestnut trees in orchards (Roane *et al.*, 1986). Specifically, increasing the levels of nitrogen in the soil negatively affected the incidence of *C. parasitica* and lesion length, which suggests that nitrogen may be important in promoting the expression of resistance to blight. In north-eastern Portugal though, no correlation was detected between soil type and the occurrence of chestnut blight in orchard-like groves (Bragança *et al.*, 2009).

15.5.5 Resistance

Resistant chestnut species occur in Asia, where *C. parasitica* is thought to have co-evolved with Chinese and Japanese chestnut (Anagnostakis, 1992). Chinese chestnut is considered more blight resistant than Japanese chestnut, although the blight resistance of both chestnut species has been reported to vary considerably (Graves, 1950; Huang *et al.*, 1996). Mapping the resistance in an interspecific cross between American and Chinese chestnuts using 241 polymorphic markers revealed three putative resistance loci (Kubisiak *et al.*, 1997). Several candidate genes possibly involved in blight resistance have recently been identified through a comparative transcriptome analysis of American and Chinese chestnuts (Barakat *et al.*, 2009).

The sources of blight resistance detected in Asian chestnuts have been used in breeding programmes in both the USA and Europe (Jaynes, 1978; Bazzigher and Miller, 1991). In France, interspecific crosses between Asian and European chestnut trees were originally conducted to create hybrid rootstocks resistant to ink disease (*Phytophthora* spp.) (Guedes-Lafargue *et al.*, 2005). These hybrids also exhibit some degree of resistance to chestnut blight (C. Robin, France, 2010, personal communication) and are increasingly being planted in chestnut orchards in Europe, particularly in France (Hennion, 2010).

In contrast to Europe, where breeding tends to be mainly for nut production in orchards, the goal of a major ongoing breeding programme in the USA is to restore American chestnut as a dominant forest tree. The American Chestnut Foundation initiated this programme in the 1980s to backcross the blight resistance of the Chinese chestnut into the American chestnut genome (Burnham, 1981; Hebard, 2005). At each backcross, the progeny were selected for resistance to chestnut blight by inoculating seedlings with virulent *C. parasitica* strains. Because blight resistance is only partially dominant though, the progeny from the backcrosses reached only moderate levels of resistance. Therefore, third backcross trees were intercrossed among themselves, which allowed the recovery of trees homozygous for blight resistance. First backcross trees derived from two earlier breeding programmes were initially used as the main sources of blight resistance. To avoid an overly narrow basis of blight resistance, which might be overcome by *C. parasitica*, additional sources of resistance from other Chinese and Japanese chestnut trees are being incorporated into the breeding programme. Large surviving American chestnut trees have been used by the American Chestnut Cooperators' Foundation as a possible alternative source for blight resistance (Griffin, 2000). The challenge of these breeding programmes is not only to produce blight resistant American chestnut, but also to capture much of the natural genetic diversity into the selected chestnut trees (Huang, 1998). The breeding process is likely to be accelerated by applying genomic tools, e.g. marker-aided selection, and genetic engineering for blight resistance, which is assumed to be feasible (Wheeler and Sederoff, 2009).

15.5.6 Therapy

Hypovirulence

To date, it is known that mycoviruses are frequently associated with plant pathogenic fungi (Ghabrial and Suzuki, 2009). In *C. parasitica*, several viruses have been described so far, with the most important for biological control in the family *Hypoviridae* (Hillman and Suzuki, 2004). Hypoviruses are unencapsidated positive-strand RNA viruses, which mostly occur as double-stranded RNA (dsRNA) replicative forms (Shapira *et al.*, 1991). Infection by a hypovirus can not only make the fungal strain less virulent, but also cause numerous phenotypic changes, including reduced orange pigmentation, reduced asexual sporulation and loss of female fertility (Nuss, 2005). However, the vegetative growth of *C. parasitica* on culture media is often not significantly influenced by the dsRNA (Robin *et al.*, 2010).

DIVERSITY AND DISTRIBUTION OF THE HYPOVIRUSES. Four distinct hypovirus species have been well characterized, namely *Cryphonectria hypovirus 1* (CHV-1), CHV-2, CHV-3 and CHV-4 (Hillman and Suzuki, 2004; Turina and Rostagno, 2007). These are taxonomically related but have different genome organizations. The hypovirus that has been best investigated is CHV-1, which is so far the only hypovirus detected in Europe, where it plays a central role in the biological control of chestnut blight (Dawe and Nuss, 2001; Milgroom and Cortesi, 2004). The origin of CHV-1 is probably Asia, where it has been found in *C. parasitica* and other *Cryphonectria* spp. (Peever *et al.*, 1998; Liu *et al.*, 2003).

Several subtypes of CHV-1 have been identified using a reverse transcription polymerase chain reaction (RT-PCR)/restriction fragment length polymorphism (RFLP) method and partial sequencing of the viral genome (Allemann *et al.*, 1999; Gobbin *et al.*, 2003). The Italian subtype (subtype I) is widespread in south and south-eastern Europe, e.g. Italy, Switzerland, south-eastern France, Bosnia-Herzegovina, Croatia, Slovenia, Macedonia and Greece (Allemann *et al.*, 1999; Krstin *et al.*, 2008, 2011; Robin *et al.*, 2010). Besides CHV-1 subtype I, additional subtypes have been found in France (subtypes F1 and F2) and in Spain and Germany (subtype E/D). The different subtypes vary in their virulence towards the fungal host (Bryner and Rigling, 2011). In particular, viral isolates in the French subtypes F1 and F2 seem to strongly inhibit the ability of *C. parasitica* to grow and sporulate on chestnut bark (Chen and Nuss, 1999; Robin *et al.*, 2010). Hypoviruses CHV-2 and CHV-3, which also cause hypovirulence, have been found almost exclusively in North America (Peever *et al.*, 1997; Smart *et al.*, 1999). The fourth hypovirus species (CHV-4) is frequent in the Appalachians in eastern North America, but it has no significant effect on virulence or on other phenotypic traits of *C. parasitica* (Linder-Basso *et al.*, 2005).

TRANSMISSION OF THE HYPOVIRUSES. Hypoviruses can be transmitted both vertically and horizontally. Vertical transmission may occur into asexually produced conidia, which probably play a major role in hypovirulence dissemination (Anagnostakis, 1988). Several studies have shown that the transmission rate is highly variable and it is rare for all conidia to be infected (Peever *et al.*, 2000; Prospero *et al.*, 2006). Hypovirus-infected strains are female sterile but can act as male partners (spermatia) in sexual crosses (Anagnostakis, 1988; Anagnostakis *et al.*, 1998). No hypovirus-infected ascospores have yet been observed (Anagnostakis, 1988; Carbone *et al.*, 2004; Prospero *et al.*, 2006).

Horizontal hypovirus transmission between a hypovirus-infected and a hypovirus-free fungal strain may occur through cytoplasmic exchange following hyphal contact between two mycelia (Van Alfen *et al.*, 1975). The formation of a stable anastomosis (hyphal fusion) and, consequently, the success of hypovirus transmission, is controlled by a vegetative incompatibility (vic) system involving at least six unlinked biallelic vic loci (Cortesi and Milgroom, 1998). Two *C. parasitica* individuals are compatible, i.e. they belong to the same vegetative compatibility (vc) type, if they exhibit the same combination of alleles at all vic loci. The success of hypovirus transmission is close to 100% when the two individuals are vegetatively compatible. If strains differ at one or more vic loci, transmission is still possible, but success rates gradually decrease (Liu and Milgroom, 1996). It should be noted that not all vic loci have the same effect on hypovirus transmission, and reciprocal transmission for most vic loci is not equal (Cortesi *et al.*, 2001). Hypovirus transmission between different vc types in the field has been demonstrated using genealogy-based methods (Carbone *et al.*, 2004).

DIVERSITY AND DISTRIBUTION OF VC TYPES. The incidence and geographic distribution of *C. parasitica* vc types have been extensively investigated in Europe. To allow a Europe-wide comparison, genetically defined testers for 64 different vc types (EU-1 to EU-64) were used for these studies (Cortesi and Milgroom, 1998). The distribution of vc types in Europe was found to have a clear geographic structure. On the one hand, the vc types EU-1, EU-2 and EU-5 are particularly frequent in central Europe, i.e. in northern Italy, Switzerland, Germany and south-eastern France (Cortesi *et al.*, 1998; Robin and Heiniger, 2001). In southern and eastern Europe, i.e. in Greece, southern Italy, the Slovak Republic, Macedonia, Bulgaria, Romania and parts of Turkey, on the other hand, vc type EU-12 is clearly predominant (Sotirovski *et al.*, 2004; Perlerou and Diamandis, 2006; Milgroom *et al.*, 2008; Erincik *et al.*, 2011). Populations located between these two main areas, e.g. in Bosnia-Herzegovina, Croatia or Slovenia, often comprise vc types from both areas and recombinant vc types have also been observed (Krstin *et al.*, 2008, 2011).

In western Europe, i.e. in Portugal, northern Spain and south-western France, the situation is markedly different. Here, the *C. parasitica* populations are dominated by vc types that do not or only rarely occur in the rest of Europe, namely EU-11, EU-33, EU-66 and EU-72. Numerous new vc types incompatible with all of the 64 European genetically defined testers have been detected (Robin *et al.*, 2000; Montenegro *et al.*, 2008; Robin *et al.*, 2009). A special situation regarding vc type diversity seems also to have unfolded in the Black Sea region. While in Turkey the vc type EU-1 is strongly predominant (Akilli *et al.*, 2009), in neighbouring Georgia a high vc type diversity is observed and new, still unknown vc types have been found (D. Rigling and S. Prospero, 2011, unpublished results). The finding of new vc types indicates that the genetics of vegetative compatibility cannot be completely explained by the six biallelic vic loci so far described. The geographic distribution of vc types supports the hypothesis that multiple *C. parasitica* introductions occurred in Europe.

In Europe, the number of vc types occurring varies considerably from country to country (Robin and Heiniger, 2001). Newly founded populations generally have a lower vc type diversity than old, established populations (Hoegger *et al.*, 2000; Milgroom *et al.*, 2008; Dutech *et al.*, 2010). North American *C. parasitica* populations are generally characterized by a higher vc diversity than European populations (Anagnostakis *et al.*, 1986; Milgroom and Cortesi, 1999). In China and Japan, vc type diversity in native *C. parasitica* populations is significantly higher than in Europe, and most vc types that occur are not compatible with any of the 64 European vc type testers (Liu and Milgroom, 2007).

OCCURRENCE OF NATURAL HYPOVIRULENCE. To date, natural hypovirulence has been widespread in many chestnut-growing areas of Europe. Its appearance in a chestnut stand usually happened 10–25 years after the first detection of chestnut blight (Robin and Heiniger, 2001). Where natural hypovirulence is present, the incidence of chestnut blight is characteristically still high, but disease severity is usually low. For example, in southern Switzerland where chestnut blight appeared in 1948 and natural hypovirulence was first noticed in 1975, healed cankers are currently widespread and mortality appears to affect only small trees and sprouts (Bissegger *et al.*, 1997). An inoculation experiment in a local chestnut coppice stand showed that, within 2 years of inoculation, 30 out of 35 artificially initiated virulent cankers had already become infected by naturally occurring hypoviruses (Fig. 15.2).

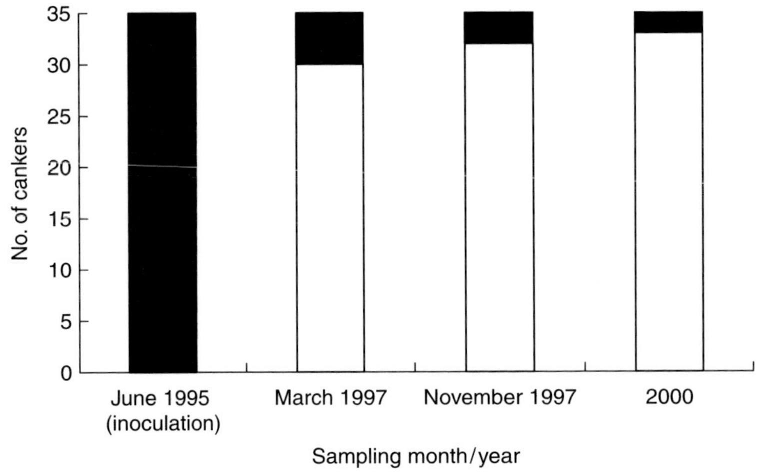

Fig. 15.2. Natural hypovirus infection of artificially initiated virulent cankers in southern Switzerland (D. Rigling, 2011, unpublished results). Chestnut stems in a coppice forest were inoculated in June 1995 with hypovirus-free strains of *Cryphonectria parasitica* (Murrill) M.E. Barr. The developing cankers were periodically sampled to determine whether they had become infected with a hypovirus. Key: □, hypovirus infected; ■, hypovirus free.

ARTIFICIAL INTRODUCTION OF HYPOVIRULENCE. In regions without natural hypovirulence, chestnut blight still causes mortality in chestnut stands. Currently, the hypovirus is mainly absent, or only present at low incidence, in *C. parasitica* populations at the epidemic front of chestnut blight, e.g. in northern France, northern Switzerland, Portugal, north-western Spain, Greece and parts of Turkey (Robin and Heiniger, 2001; Perlerou and Diamandis, 2006; Bragança *et al.*, 2007; Montenegro *et al.*, 2008). Efforts have been made to artificially introduce the hypovirus into blighted chestnut stands in several European countries.

The rationale of the biological control of chestnut blight is to inoculate (treat) virulent cankers with hypovirus-infected *C. parasitica* strains (Plate 20). Upon hyphal anastomosis, the hypovirus is horizontally transmitted to the virulent strain, which may result in the canker healing (Grente, 1965). For treatment, holes are usually created with a cork borer around the virulent canker and filled with slurry of the hypovirulent mycelium. As the hypovirus is best transmitted between fungal individuals belonging to the same vc type, the vc type(s) occurring in an infected chestnut stand need(s) to be determined before the treatment. Local virulent fungal strains of the dominant vc types are then converted into hypovirulent strains in the laboratory and used for the biological control treatments. If there are several vc types present in a specific chestnut growing region, mixtures of hypovirulent strains belonging to different vc types may be used for treatments (Heiniger and Rigling, 1994; Robin *et al.*, 2000; Turchetti and Maresi, 2008).

Beside the therapeutic effect of the hypovirus on cankers treated in this way, untreated cankers can also be expected to heal thanks to spontaneous hypovirus dissemination within treated chestnut stands. In fact, the success of a treatment is assessed by determining the incidence of healed cankers among the treated and the untreated cankers (Robin *et al.*, 2010).

SUCCESS OF BIOCONTROL THROUGH ARTIFICIAL HYPOVIRULENCE. The biological control of chestnut blight through the artificial application of hypovirulent *C. parasitica* strains is critically discussed in Milgroom and Cortesi (2004). In particular, these authors concluded that: (i) the success or failure of hypovirulence is determined by the characteristics of the three interacting trophic levels of the tree, fungus and hypovirus, and the environment; and (ii) the recovery of chestnut stands in Europe may be correlated with the presence of hypovirulence, but sound data unequivocally supporting this hypothesis is lacking. In recent years in Europe, intensive treatments have been conducted in both plantations for chestnut production (mostly in France) and in old orchards or coppice stands, e.g. in northern Switzerland, Slovakia, Greece, Turkey and Albania (Juhasova *et al.*, 2005; Heiniger and Rigling, 2009; Çeliker and Onoğur, 2010; Lushaj and Lushaj, 2010; Robin *et al.*, 2010).

In general, a high therapeutic efficacy of treatments is reported in coppice stands and on young trees with a smooth bark and clearly visible canker margins (Robin *et al.*, 2000; Hoegger *et al.*, 2003; Heiniger and Rigling, 2009). In contrast, on old trees, cankers are either hard to identify, e.g. because sporulation in the longitudinal cracks of the bark is diffuse, or they are situated on branches in the crown, which makes individual canker treatments particularly difficult. Treatment failures may also occur when vc type diversity in the target chestnut stand is high and/or when the cankers to be treated are caused by *C. parasitica* vc types for which no hypovirulent strains are available (Robin *et al.*, 2000). Field results suggest that the chances that artificially introduced hypoviruses will become established and spread can be enhanced by treating as many cankers as possible (Heiniger and Rigling, 2009; Perlerou and Diamandis, 2010). In northern Switzerland, after repeated systematic treatments of cankers over a 5 year period (2003–2008), hypovirus-infected *C. parasitica* cultures could be recovered from 60–90% of the treated cankers and from 20% to nearly 60% of the untreated cankers (Fig. 15.3). In southern Switzerland, where natural hypovirulence occurs, the incidence of untreated cankers infected with the hypovirus reached 50% to just over 70%.

The success of biological control seems to be affected not only by the intensity of the treatments, but also by other characteristics of the three trophic levels involved. Which hypovirus is used for biological control appears to be an important factor. MacDonald and Fulbright (1991) emphasized that the degree of viral suppression of fungal growth and sporulation will probably affect the ability of the hypovirus to persist and disseminate in a population of *C. parasitica*. Natural hypovirulence in Europe originates largely from the spread of the Italian subtype of CHV-1, which has a mild effect on *C. parasitica* (Allemann *et al.*, 1999; Chen and Nuss, 1999; Robin *et al.*, 2010). Several different hypovirus isolates in this subtype were successfully used in the biological control treatments in northern Switzerland described above and illustrated in Fig. 15.3. The actual disease incidence could be another critical factor for the success of hypovirulence. According to Griffin (1986), before hypovirulence is initiated and can spread, a critical mass of virulent *C. parasitica* thallus must develop on a tree or in a stand (the 'critical-canker-mass' hypothesis). Other, still poorly investigated, factors that might influence the outcome of biological control are the presence/absence of adequate vectors for the hypovirus (Scharf and Depalma, 1981; Russin *et al.*, 1984; Nannelli *et al.*, 1998), or specific site factors.

The apparent success of hypovirulence in Europe encouraged the artificial release of hypoviruses for the biological control of chestnut blight in North America (Connecticut, Michigan, Virginia, West Virginia, Wisconsin and Ontario) as early as the mid 1970s (Jaynes and Elliston, 1980; MacDonald and Fulbright, 1991). The European CHV-1 and CHV-3 from Michigan were the hypoviruses most commonly used in the trials (Liu *et al.*, 2002). Although conversion of virulent cankers took place, biological control using the hypovirulent *C. parasitica* strains introduced was, with a few exceptions (e.g. Michigan), unfortunately not sustainable because the released hypoviruses did not persist for long in the *C. parasitica* population (Nuss, 1992; Liu *et al.*, 2002). Various reasons have been proposed for explaining the differences between North America and Europe in the efficacy of the biological control of chestnut blight (Milgroom and Cortesi, 2004). These include the following: (i) the vc type diversity of the American *C. parasitica* population is too high; (ii) the hypovirus type used as a biocontrol agent may not be appropriate; (iii) the blight susceptibility of the American and the European chestnut differs; and (iv) the forest structure and chestnut management practices in Europe

Havir, E.A. and Anagnostakis, S.L. (1983) Oxalate production by virulent but not by hypovirulent strains of *Endothia parasitica*. *Physiological Plant Pathology* 23, 369–376.
Heald, F.D. and Gardner, M.W. (1913) The relative prevalence of pycnospores and ascospores of the chestnut blight fungus during the winter. *Phytopathology* 3, 296–305.
Heald, F.D. and Gardner, M.W. (1914) Longevity of pycnospores of the chestnut blight fungus in soil. *Journal of Agricultural Research* 2, 67–75.
Heald, F.D. and Studhalter, R.A. (1914) Birds as carriers of the chestnut blight fungus. *Journal of Agricultural Research* 2, 405–422.
Heald, F.D., Gardner, M.W. and Studhalter, R.A. (1915) Air and wind dissemination of ascospores of the chestnut blight fungus. *Journal of Agricultural Research* 3, 493–526.
Hebard, F.V. (2005) The backcross breeding program of the American Chestnut Foundation. *Journal of the American Chestnut Foundation* 19, 55–78.
Hebard, F.V., Griffin, G.J. and Elkins, J.R. (1984) Developmental histopathology of cankers incited by hypovirulent and virulent isolates of *Endothia parasitica* on susceptible and resistant chestnut trees. *Phytopathology* 74, 140–149.
Heiniger, U. and Rigling, D. (1994) Biological control of chestnut blight in Europe. *Annual Review of Phytopathology* 32, 581–599.
Heiniger, U. and Rigling, D. (2009) Application of the *Cryphonectria hypovirus* (CHV-1) to control the chestnut blight, experience from Switzerland. *Acta Horticulturae* 815, 233–245.
Hennion, B. (2010) Chestnut production in France: review, perspectives. *Acta Horticulturae* 866, 493–497.
Hillman, B.I. and Suzuki, N. (2004) Viruses of the chestnut blight fungus, *Cryphonectria parasitica*. *Advances in Virus Research* 63, 423–472.
Hillman, B.I., Shapira, R. and Nuss, D.L. (1990) Hypovirulence-associated suppression of host functions in *Cryphonectria parasitica* can be partially relieved by high light intensity. *Phytopathology* 80, 950–956.
Hoegger, P.J., Rigling, D., Holdenrieder, O. and Heiniger, U. (2000) Genetic structure of newly established populations of *Cryphonectria parasitica*. *Mycological Research* 104, 1108–1116.
Hoegger, P.J., Rigling, D., Holdenrieder, O. and Heiniger, U. (2002) *Cryphonectria radicalis*: rediscovery of a lost fungus. *Mycologia* 94, 105–115.
Hoegger, P.J., Heiniger, U., Holdenrieder, O. and Rigling, D. (2003) Differential transfer and dissemination of hypovirus and nuclear and mitochondrial genomes of a hypovirus-infected *Cryphonectria parasitica* strain after introduction into a natural population. *Applied and Environmental Microbiology* 69, 3767–3771.
Holmes, T.P., Aukema, J.E., Von Holle, B., Liebhold, A. and Sills, E. (2009) Economic impacts of invasive species in forests. Past, present, and future. *Annals of the New York Academy of Sciences* 1162, 18–38.
Huang, H.W. (1998) Restoring the American chestnut to its native range: genetic variation in the American chestnut and selection strategies for recurrent parents. *Journal of the American Chestnut Foundation* 12, 27–39.
Huang, H.W., Carey, W.A., Dane, F. and Norton, J.D. (1996) Evaluation of Chinese chestnut cultivars for resistance to *Cryphonectria parasitica*. *Plant Disease* 80, 45–47.
Jacob-Wilk, D., Turina, M., Kazmierczak, P. and Van Alfen, N.K. (2009) Silencing of *Kex2* significantly diminishes the virulence of *Cryphonectria parasitica*. *Molecular Plant–Microbe Interactions* 22, 211–221.
Jaynes, R.A. (1978) Selection and breeding blight resistance chestnut trees. In: MacDonald, W.L., Cech, F.C., Luchok, J. and Smith, C. (eds) *Proceedings of the American Chestnut Symposium, Morgantown, WV, January 4–5, 1978*. West Virginia University Books, Morgantown, West Virginia, pp. 4–6.
Jaynes, R.A. and Elliston, J.E. (1980) Pathogenicity and canker control by mixtures of hypovirulent strains of *Endothia parasitica* in American chestnut. *Phytopathology* 70, 453–456.
Jaynes, R.A. and Van Alfen, N.K. (1974) Control of American chestnut blight by trunk injection with methyl-2 benzimidazole carbamate (MBC). *Phytopathology* 64, 1479–1480.
Ježić, M., Krstin, L., Rigling, D. and Ćurković-Perica, M. (2012) High diversity in populations of the introduced plant pathogen, *Cryphonectria parasitica*, due to encounters of genetically divergent genotypes. *Molecular Ecology* 21, 87–99.
Juhasova, G., Adamcikova, K. and Robin, C. (2005) Results of biological control of chestnut blight in Slovakia. *Phytoprotection* 86, 19–23.
Kazempour, M.N., Khodaparast, S.A., Arefipour, M., Salehi, M., Amanzadeh, B., Ramazanie, M. and Shiraz, B.K. (2006) Occurrence of *Cryphonectria parasitica* the causal agent of chestnut blight in Iran. *Plant Pathology* 55, 815.
Krstin, L., Novak-Agbaba, S., Rigling, D., Krajačić, M. and Ćurković Perica, M. (2008) Chestnut blight fungus in Croatia: diversity of vegetative compatibility types, mating types and genetic variability of associated *Cryphonectria hypovirus 1*. *Plant Pathology* 57, 1086–1096.

Krstin, L., Novak-Agbaba, S., Rigling, D. and Ćurković Perica, M. (2011) Diversity of vegetative compatibility types and mating types of *Cryphonectria parasitica* in Slovenia and occurrence of associated *Cryphonectria hypovirus 1*. *Plant Pathology* 60, 752–761.

Kubisiak, T.L., Hebard, F.V., Nelson, C.D., Zhang, J.S., Bernatzky, R., Huang, H., Anagnostakis, S.L. and Doudrick, R.L. (1997) Molecular mapping of resistance to blight in an interspecific cross in the genus *Castanea*. *Phytopathology* 87, 751–759.

Linder-Basso, D., Dynek, J.N. and Hillman, B.I. (2005) Genome analysis of *Cryphonectria hypovirus 4*, the most common hypovirus species in North America. *Virology* 337, 192–203.

Liu, Y.-C. and Milgroom, M.G. (1996) Correlation between hypovirus transmission and the number of vegetative incompatibility (vic) genes different among isolates from a natural population of *Cryphonectria parasitica*. *Phytopathology* 86, 79–86.

Liu, Y.-C. and Milgroom, M.G. (2007) High diversity of vegetative compatibility types in *Cryphonectria parasitica* in Japan and China. *Mycologia* 99, 279–284.

Liu, Y.-C., Durrett, R. and Milgroom, M.G. (2000) A spatially-structured stochastic model to simulate heterogenous transmission of viruses in fungal populations. *Ecological Modelling* 127, 291–308.

Liu, Y.-C., Double, M.L., MacDonald, W.L. and Milgroom, M.G. (2002) Persistence of *Cryphonectria* hypoviruses after their release for biological control of chestnut blight in West Virginia forests. *Forest Pathology* 32, 345–356.

Liu, Y.-C., Linder-Basso, D., Hillman, B.I., Kaneko, S. and Milgroom, M.G. (2003) Evidence for interspecies transmission of viruses in natural populations of filamentous fungi in the genus *Cryphonectria*. *Molecular Ecology* 12, 1619–1628.

Lushaj, B.M. and Lushaj, A.B. (2010) Biological control of the chestnut blight canker, experience from Albania. *Acta Horticulturae* 866, 467–476.

MacDonald, W.L. and Fulbright, D.W. (1991) Biological control of chestnut blight: use and limitations of transmissible hypovirulence. *Plant Disease* 75, 656–661.

Marra, R.E. and Milgroom, M.G. (2001) The mating system of the fungus *Cryphonectria parasitica*: selfing and self-incompatibility. *Heredity* 86, 134–143.

Marra, R.E., Cortesi, P., Bissegger, M. and Milgroom, M.G. (2004) Mixed mating in natural populations of the chestnut blight fungus, *Cryphonectria parasitica*. *Heredity* 93, 189–195.

McGuire, I.C., Marra, R.E., Turgeon, B.G. and Milgroom, M.G. (2001) Analysis of mating-type genes in the chestnut blight fungus, *Cryphonectria parasitica*. *Fungal Genetics and Biology* 34, 131–144.

McGuire, I.C., Marra, R.E. and Milgroom, M.G. (2004) Mating-type heterokaryosis and selfing in *Cryphonectria parasitica*. *Fungal Genetics and Biology* 41, 521–533.

Milgroom, M.G. and Cortesi, P. (1999) Analysis of population structure of the chestnut blight fungus based on vegetative incompatibility genotypes. *Proceedings of the National Academy of Sciences of the United States of America* 96, 10518–10523.

Milgroom, M.G. and Cortesi, P. (2004) Biological control of chestnut blight with hypovirulence: a critical analysis. *Annual Review of Phytopathology* 42, 311–338.

Milgroom, M.G. and Lipari, S.E. (1995) Spatial analysis of nuclear and mitochondrial RFLP genotypes in populations of the chestnut blight fungus, *Cryphonectria parasitica*. *Molecular Ecology* 4, 633–642.

Milgroom, M.G., Sotirovski, K., Spica, D., Davis, J.E., Brewer, M.T., Milev, M. and Cortesi, P. (2008) Clonal population structure of the chestnut blight fungus in expanding ranges in southeastern Europe. *Molecular Ecology* 17, 4446–4458.

Milgroom, M.G., Sotirovski, K., Risteski, M. and Brewer, M.T. (2009) Heterokaryons and parasexual recombinants of *Cryphonectria parasitica* in two clonal populations in southeastern Europe. *Fungal Genetics and Biology* 46, 849–854.

Montenegro, D., Aguín, O., Sainz, M.J., Hermida, M. and Mansilla, J.P. (2008) Diversity of vegetative compatibility types, distribution of mating types and occurrence of hypovirulence of *Cryphonectria parasitica* in chestnut stands in NW Spain. *Forest Ecology and Management* 256, 973–980.

Nannelli, R., Turchetti, T. and Maresi, G. (1998) Corticolous mites (Acari) as potential vectors of *Cryphonectria parasitica* (Murr.) Barr hypovirulent strains. *International Journal of Acarology* 24, 237–244.

Nuss, D.L. (1992) Biological control of chestnut blight: an example of virus-mediated attenuation of fungal pathogenesis. *Microbiological Reviews* 56, 561–576.

Nuss, D.L. (2005) Hypovirulence: mycoviruses at the fungal–plant interface. *Nature Reviews Microbiology* 3, 632–642.

Payne, J.A., Jaynes, R.A. and Kays, S.J. (1983) Chinese chestnut production in the United States: practice, problems, and possible solutions. *Economic Botany* 37, 187–200.

Peever, T.L., Liu, Y.-C. and Milgroom, M.G. (1997) Diversity of hypoviruses and other double-stranded RNAs in *Cryphonectria parasitica* in North America. *Phytopathology* 87, 1026–1033.

Peever, T.L., Liu, Y.-C., Wang, K.R., Hillman, B.I., Foglia, R. and Milgroom, M.G. (1998) Incidence and diversity of double-stranded RNAs occurring in the chestnut blight fungus, *Cryphonectria parasitica*, in China and Japan. *Phytopathology* 88, 811–817.

Peever, T.L., Liu, Y.-C., Cortesi, P. and Milgroom, M.G. (2000) Variation in tolerance and virulence in the chestnut blight fungus–hypovirus interaction. *Applied and Environmental Microbiology* 66, 4863–4869.

Perlerou, C. and Diamandis, S. (2006) Identification and geographic distribution of vegetative compatibility types of *Cryphonectria parasitica* and occurrence of hypovirulence in Greece. *Forest Pathology* 36, 413–421.

Perlerou, C. and Diamandis, S. (2010) Spread of introduced hypovirulence vs natural hypovirulence in chestnut blight. *Acta Horticulturae* 866, 393–398.

Prospero, S. and Forster, B. (2011) Chestnut gall wasp (*Dryocosmus kuriphilus*) infestations: new opportunities for the chestnut blight fungus *Cryphonectria parasitica*? *New Disease Reports* 23, 35.

Prospero, S. and Rigling, D. (2012) Invasion genetics of the chestnut blight fungus *Cryphonectria parasitica* in Switzerland. *Phytopathology* 102, 73–82.

Prospero, S., Conedera, M., Heiniger, U. and Rigling, D. (2006) Saprophytic activity and sporulation of *Cryphonectria parasitica* on dead chestnut wood in forests with naturally established hypovirulence. *Phytopathology* 96, 1337–1344.

Roane, M.K., Griffin, G.J. and Elkins, J.R. (1986) *Chestnut Blight, Other Endothia Diseases, and the Genus* Endothia. APS Monograph Series. APS Press, St Paul, Minnesota.

Robin, C. and Heiniger, U. (2001) Chestnut blight in Europe: diversity of *Cryphonectria parasitica*, hypovirulence and biocontrol. *Forest Snow and Landscape Research* 76, 361–367.

Robin, C., Anziani, C. and Cortesi, P. (2000) Relationship between biological control, incidence of hypovirulence, and diversity of vegetative compatibility types of *Cryphonectria parasitica* in France. *Phytopathology* 90, 730–737.

Robin, C., Capdevielle, X., Martin, M., Traver, C. and Colinas, C. (2009) *Cryphonectria parasitica* vegetative compatibility type analysis of populations in south-western France and northern Spain. *Plant Pathology* 58, 527–535.

Robin, C., Lanz, S., Soutrenon, A. and Rigling, D. (2010) Dominance of natural over released biological control agents of the chestnut blight fungus *Cryphonectria parasitica* in south-eastern France is associated with fitness-related traits. *Biological Control* 53, 55–61.

Root, C., Balbalian, C., Bierman, R., Geletka, L.M., Anagnostakis, S., Double, M., MacDonald, W. and Nuss, D.L. (2005) Multi-seasonal field release and spermatization trials of transgenic hypovirulent strains of *Cryphonectria parasitica* containing cDNA copies of hypovirus CHV1-EP713. *Forest Pathology* 35, 277–297.

Russin, J.S., Shain, L. and Nordin, G.L. (1984) Insects as carriers of virulent and cytoplasmic hypovirulent isolates of the chestnut blight fungus. *Journal of Economic Entomology* 77, 838–846.

Scharf, C.S. and Depalma, N.K. (1981) Birds and mammals as vectors of the chestnut blight fungus (*Endothia parasitica*). *Canadian Journal of Zoology* 59, 1647–1650.

Scibilia, K.L. and Shain, L. (1989) Protection of American chestnut with hypovirulent conidia of *Cryphonectria* (*Endothia*) *parasitica*. *Plant Disease* 73, 840–843.

Seemann, D. (2001) Plant health and quarantine regulations of the European Union for *Cryphonectria parasitica*. *Forest Snow and Landscape Research* 76, 402–404.

Shapira, R., Choi, G.H. and Nuss, D.L. (1991) Virus-like genetic organization and expression strategy for a double-stranded-RNA genetic element associated with biological control of chestnut blight. *EMBO Journal* 10, 731–739.

Smart, C.D., Yuan, W., Foglia, R., Nuss, D.L., Fulbright, D.W. and Hillman, B.I. (1999) *Cryphonectria hypovirus 3*, a virus species in the family *Hypoviridae* with a single open reading frame. *Virology* 265, 66–73.

Sotirovski, K., Papazova-Anakieva, I., Grunwald, N.J. and Milgroom, M.G. (2004) Low diversity of vegetative compatibility types and mating type of *Cryphonectria parasitica* in the southern Balkans. *Plant Pathology* 53, 325–333.

Turchetti, T. and Maresi, G. (2008) Biological control and management of chestnut diseases. In: Ciancio, A. and Mukerji, K.J. (eds) *Integrated Management of Diseases Caused by Fungi, Phytoplasma and Bacteria (Integrated Management of Plant Pests and Diseases, Volume 3)*. Springer Science, Dordrecht, The Netherlands, pp. 85–118.

Turina, M. and Rostagno, L. (2007) Virus-induced hypovirulence in *Cryphonectria parasitica*: still an unresolved conundrum. *Journal of Plant Pathology* 89, 165–178.

Van Alfen, N.K., Jaynes, R.A., Anagnostakis, S.L. and Day, P.R. (1975) Chestnut blight – biological control by transmissible hypovirulence in *Endothia parasitica*. *Science* 189, 890–891.
Waldboth, M. and Oberhuber, W. (2009) Synergistic effect of drought and chestnut blight (*Cryphonectria parasitica*) on growth decline of European chestnut (*Castanea sativa*). *Forest Pathology* 39, 43–55.
Weidlich, W.H. (1978) A preliminary report on a method of biological control of chestnut blight not involving the use of a hypovirulent strain of *Endothia parasitica*. In: MacDonald, W.L., Cech, F.C., Luchok, J. and Smith, C. (eds) *Proceedings of the American Chestnut Symposium, Morgantown, WV, January 4–5, 1978*. West Virginia University Books, Morgantown, West Virginia, pp. 79–83.
Wheeler, N. and Sederoff, R. (2009) Role of genomics in the potential restoration of the American chestnut. *Tree Genetics and Genomes* 5, 181–187.

only been found rarely in this country, and then in parts where the climate is not strictly tropical to subtropical (Davison and Coates, 1991; Gryzenhout *et al.*, 2006d; Pegg *et al.*, 2010). Another likely origin was South America on native *Melastomataceae*, which is in the same order (*Myrtales*) as *Eucalyptus* (Wingfield *et al.*, 2001a). In Africa, the range of *C. cubensis* is thought to be the result of introductions (Gryzenhout *et al.*, 2006d; Nakabonge *et al.*, 2007; Roux and Apetorgbor, 2009). However, the description of *C. deuterocubensis* as a distinct species indicated that the originally described *C. cubensis* represented two species with separate life histories, geography and hosts, and refuted the hypotheses of an Asian or American origin. The separate life histories of these two species were also confirmed when limited gene flow was shown between *C. cubensis* and *C. deuterocubensis* (Gryzenhout *et al.*, 2006d; van der Merwe *et al.*, 2010).

C. cubensis was first reported from South Africa by Wingfield *et al.* (1989), where the pathogen was particularly devastating to promising clones of *Eucalyptus* spp. developed by South African forestry companies. Similar to the situation in Brazil, the occurrence of the pathogen necessitated an intensive yet successful disease screening programme to make the successful growing of *Eucalyptus* spp. possible in the subtropical areas of the country (van Heerden and Wingfield 2002; Wingfield, 2003; van Heerden *et al.*, 2005). Subsequent phylogenetic and taxonomic studies revealed that the South African fungus represented another distinct species, which was named *Chrysoporthe austroafricana* Gryzenh. & M.J. Wingf. (Gryzenhout *et al.*, 2004). Whereas previously this pathogen was thought to have been introduced into South Africa (van Heerden and Wingfield, 2001), and then had an impact on eucalyptus plantations, it was now clear that in South Africa a distinct and particularly virulent species affected eucalyptus. It is therefore important that this destructive species should not spread to other eucalyptus growing areas of the world, and especially not to Australasia, where *Eucalyptus* spp. are native (Gryzenhout *et al.*, 2004).

Surveys in South Africa and other parts of Africa found that *C. austroafricana* had a wide distribution in southern to central Africa on plantation-grown *Eucalyptus* spp. (Nakabonge *et al.*, 2006a; Vermeulen *et al.*, 2011). During the South African surveys, the fungus was discovered on native and widespread *Syzygium* spp., such as water pear (*S. guineense* Wall.) and water berry (*S. cordatum* Hochst.) (Heath *et al.*, 2006), and it was also later found on these hosts in other southern African countries (Nakabonge *et al.*, 2006a; Vermeulen *et al.*, 2011). Fortunately, the pathogen forms reduced cankers on these hosts, and so has a limited impact. All of these data indicate that this fungus most likely has an African origin (Heath *et al.*, 2006).

An additional two *Chrysoporthe* spp. were also discovered in Africa. *C. syzygiicola* Chungu, Gryzenh. & Jol. Roux and *C. zambiensis* Chungu, Gryzenh. & Jol. Roux were described from water berry and eucalyptus, respectively, in two different regions of Zambia (Chungu *et al.*, 2010). *C. austroafricana* co-occurs with these species in Zambia (Nakabonge *et al.*, 2006a; Chungu *et al.*, 2010; Vermeulen *et al.*, 2011). Both of the new species were shown to cause lesions on eucalyptus. *C. syzygiicola* was particularly virulent, with the selected isolates being more pathogenic than established virulent isolates of *C. austroafricana*. The distributions of these two species in Zambia are still thought to be restricted, but this is based on relatively limited sampling; it is thus possible that the two species that are currently known from eucalyptus and native water berry can cross-infect on these hosts, and may occur more widely in Zambia and possibly in neighbouring African countries. In the absence of further knowledge, *C. syzygiicola* and *C. zambiensis* should be treated as pathogens of significant quarantine importance, because they could pose a threat to clonal and hybrid eucalyptus programmes developed for *C. austroafricana* in South Africa, to fledging *Eucalyptus* spp. plantations in other African countries and to global eucalyptus forestry programmes.

C. cubensis and *C. deuterocubensis*, as well as *C. austroafricana*, *C. syzygiicola* and *C. zambiensis*, are known to occur in different parts of Africa. *C. cubensis* is known from the DRC (Myburg *et al.*, 2003; Roux *et al.*, 2003) and Ghana (Roux and Apetorgbor, 2009), while *C. deuterocubensis* occurs in Kenya,

Malawi, Zanzibar and the northern part of Mozambique (Myburg *et al.*, 2003; Nakabonge *et al.*, 2006a). Despite surveys in these countries, *C. cubensis* and *C. deuterocubensis* have not been found on native trees such as *Syzygium* spp., which are known to be susceptible to *Chrysoporthe* spp. (Heath *et al.*, 2006; Chungu *et al.*, 2010). A population diversity study based on isolate collections from these surveys indicated that the population of *C. deuterocubensis* is largely clonal (Nakabonge *et al.*, 2007). It would appear then that *C. cubensis* and *C. deuterocubensis* have been introduced into Africa, which implies that they could possibly pose a threat to susceptible native vegetation, although they have not yet been reported from these hosts. This would be similar to the situation with introduced chestnut blight in Europe and the USA (see Prospero and Rigling, Chapter 15, this volume). As in the case of *C. syzygiicola* and *C. zambiensis* described in the previous paragraph, *C. cubensis* and *C. deuterocubensis* could also threaten the forestry programme of South Africa, where they have not yet been detected.

In South America, *C. cubensis* co-occurs with three other *Chrysoporthe* spp. Like the situation in Africa, these additional species have limited geographical distributions, most likely as a result of under-sampling. *C. hodgesiana* (Gryzenh. & M.J. Wingf.) Chungu, Gryzenh. & M.J. Wingf. and *C. inopina* Gryzenh. & M.J. Wingf. are known only from Colombia and have not yet been reported from eucalyptus (Gryzenhout *et al.*, 2004, 2006d). In pathogenicity trials, *C. hodgesiana* (as *C. cubensis*) was shown to be able to infect the grand eucalyptus (*E. grandis* W. Hill ex Maid.) used in the forestry programmes of Colombia (Wingfield *et al.*, 2001a), but *C. inopina* has not yet been tested (Gryzenhout *et al.*, 2006d). *C. cubensis*, *C. hodgesiana* and *C. inopina* share similar native hosts, such as *Tibouchina lepidota* Baill. (Gryzenhout *et al.*, 2004, 2006d; Rodas *et al.*, 2005), and it is likely that *C. hodgesiana* and *C. inopina* occur more widely. All of these *Chrysoporthe* spp. were associated with dieback symptoms on native woody hosts, indicating that these fungi have the ability to cause disease and hence are of quarantine importance.

Chrysoporthe doradensis sp. nov. was reported by Gryzenhout *et al.* (2005a) associated with girdling canker symptoms and occasional dead trees of planted grand eucalyptus and Indonesian gum (*E. deglupta* Blume) in Ecuador; the species is so far only known from Ecuador. Pathogenicity trials showed that this pathogen forms girdling lesions on Indonesian gum seedlings, and longer lesions on princess-flower trees (*T. urvilleana* (DC.) Cogn.) (Gryzenhout *et al.*, 2005a). However, the current distribution and impacts that *C. doradensis* could have on eucalyptus-based forestry are unclear. The occurrence of other *Chrysoporthe* spp. on native *Melastomataceae* in neighbouring Colombia (Rodas *et al.*, 2005; Gryzenhout *et al.*, 2006d; van der Merwe *et al.*, 2012) raised the possibility that this species also occurs naturally in Ecuador on, for instance, native species of *Tibouchina* that grow in cooler areas of the country (Gryzenhout *et al.*, 2005a), or that this pathogen could possibly be found in other South American countries.

Holocryphia eucalypti (M. Venter & M.J. Wingf.) Gryzenh. & M.J. Wingf. (reported as *Endothia gyrosa*, *E. havanensis* Bruner and *Cryphonectria eucalypti* M. Venter & M.J. Wingf.) causes bark cracking, cankers and dieback on *Eucalyptus* spp. (Walker *et al.*, 1985; Gryzenhout *et al.*, 2003). So far, the pathogen is known only on *Eucalyptus* spp. and on the closely related *Corymbia* spp. (Gryzenhout *et al.*, 2009), but it also has the ability to naturally infect *Tibouchina* spp. (Heath *et al.*, 2007; Gryzenhout *et al.*, 2010b). It is the only member of the *Cryphonectriaceae* so far described that occurs naturally in native eucalyptus forests (Walker *et al.*, 1985; Old *et al.*, 1986; Yuan and Mohammed, 1999). *H. eucalypti* also causes disease problems elsewhere where *Eucalyptus* spp. are planted as exotics (Van der Westhuizen *et al.*, 1993; Gryzenhout *et al.*, 2003; Roux and Nakabonge, 2010), such as in planted eucalyptus in South Africa (Van der Westhuizen *et al.*, 1993), New Zealand (Gryzenhout *et al.*, 2010b) and Uganda (Roux and Nakabonge, 2010). In South Africa, the pathogen is thought to be introduced, as shown in a population study based on microsatellites (Nakabonge *et al.*, 2008).

The impact of *H. eucalypti* in natural forests is not significant (Walker *et al.*, 1985;

Old *et al.*, 1986; Yuan and Mohammed, 1997, 1999, 2000). In plantations, it is also not considered to be serious (Van der Westhuizen *et al.*, 1993; Yuan and Mohammed, 2000; Gryzenhout *et al.*, 2003; Carnegie, 2007), though it does have the capacity to cause tree death and to be problematic on marginal sites (Old *et al.*, 1990; Wardlaw, 1999; Wingfield and Roux, 2002; Gryzenhout *et al.*, 2003). This is because the pathogen is mostly opportunistic and associated with stress and wounding (Old *et al.*, 1990; White and Kile, 1993).

Species in *Celoporthe*, a genus morphologically similar to *Chrysoporthe*, are also known to affect eucalyptus. Recently, two new species, *Celoporthe eucalypti* S.F. Chen, Gryzenh., M.J. Wingf. & X.D. Zhou and *C. guangdongensis* S.F. Chen, Gryzenh., M.J. Wingf. & X.D. Zhou, were described from eucalyptus in China (Chen *et al.*, 2011), and were proven to be pathogenic in pathogenicity trials. Another species of *Celoporthe*, *C. syzygii* S.F. Chen, Gryzenh., M.J. Wingf. & X.D. Zhou, was described on the non-native ornamental black plum (*Syzygium cuminii* (L.) Skeels) in the same area and this species was shown to be as pathogenic to eucalyptus clones as the two *Celoporthe* spp. described from eucalyptus. Moreover, these *Celoporthe* spp. were shown to be similar in pathogenicity to the known eucalyptus pathogen *C. deuterocubensis* when results were compared with those from a previous trial in the same location involving *C. deuterocubensis* isolates from China (Chen *et al.*, 2010). Like *C. syzygii*, South African isolates of *C. dispersa* Nakab., Gryzenh., Jol. Roux & M.J. Wingf. were also shown to be pathogenic to a eucalyptus clone in artificial trials, although this species has not yet been found on eucalyptus (Nakabonge *et al.*, 2006b). Similarly, two other newly described species from Africa, *C. fontana* M. Verm., Gryzenh. & Jol. Roux and *C. woodiana* M. Verm., Gryzenh. & Jol. Roux, were also pathogenic to eucalyptus in pathogenicity trials (Vermeulen *et al.*, 2012a).

Recently, a new pathogen in the *Cryphonectriaceae*, *Cryptometrion aestuescens* Gryzenh. & M.J. Wingf., was discovered in Indonesia. *C. aestuescens* caused girdling cankers and resulted in the death of clonal hedges of eucalyptus. It was also detected in field plots and was proven to be pathogenic in field trials (Gryzenhout *et al.*, 2010a). An interesting aspect of this pathogen is that, unlike most other species in the *Cryphonectriaceae*, it was found in cooler areas of this tropical country and has an optimal growth temperature of 20°C. It is unclear whether *C. aestuescens* is native to Indonesia, but its limited impact on eucalyptus and absence from other areas of the world would suggest that it is. The ability of the species to grow optimally at lower temperatures also suggests that it could pose a threat to cold tolerant *Eucalyptus* spp.

Microthia havanensis (Bruner) Gryzenh. & M.J. Wingf. has been reported from diseased eucalyptus in Florida (Barnard *et al.*, 1987). It has subsequently also been found on eucalyptus in Hawaii and Mexico (Gryzenhout *et al.*, 2006e), and was originally described from eucalyptus in Cuba (Bruner, 1916). This pathogen, however, is usually associated with cankers covered with fruiting bodies of *C. cubensis* and *C. deuterocubensis* in Florida and Hawaii, respectively. It is thought – but has not been tested – that *M. havanensis* is opportunistic or saprophytic on these cankers, which are most probably caused by *Chrysoporthe* spp. A similar role is suspected for the association of *M. havanensis* with cankers on the naturally occurring firetree (*Morella faya* (Aiton) Wilbur) in the Azores and Madeira (Gardner and Hodges, 1990; Hodges and Gardner, 1992; Gryzenhout *et al.*, 2009).

Whereas all previously discovered species of *Cryphonectriaceae* were associated with woody tissue, a new species of *Cryphonectriaceae* has recently been described from leaves of *Eucalyptus* spp. *Foliocryphia eucalypti* Cheewangkoon & Crous was isolated from leaves of Tasmanian snow gum (*E. coccifera* Hook. F.) in Australia (Cheewangkoon *et al.*, 2009). *Aurantiosacculus eucalypti* (Cooke & Massee) Dyko & B. Sutton, *A. acutatus* Crous & Summerell and *A. eucalyptorum* Crous and C. Mohammed were found associated with leaf spots of *Eucalyptus* spp. in Australia (Crous *et al.*, 2012b). Similarly, *Chrysocrypta corymbiae* Crous & Summerell was isolated from leaf spots of a *Corymbia* sp. (a close relative of *Eucalyptus*) in Australia (Crous *et al.*, 2012c). Leaves were a previously unknown and unexplored substrate for members of the *Cryphonectriaceae*.

16.1.3 Impact of pathogens on other planted trees

Members of the *Cryphonectriaceae* do not commonly occur on other plantation tree species such as pines (*Pinus* spp.), wattles (*Acacia* spp.) or other planted *Acacia* spp., although one commercially grown tree genus, *Terminalia* (which, like *Eucalyptus*, resides in the *Myrtales*) is affected. *Terminalia* spp. are native to central African countries, where they have medicinal, sociological and economic importance, as summarized by Begoude *et al.* (2010). The trees are also grown in the Americas and in Africa for their excellent wood qualities, and disease resistance (Lamb and Ntima, 1971; Groulez and Wood, 1985). *Aurifilum marmelostoma* Begoude, Gryzenh. & Jol. Roux has recently been described causing bark cracking and cankers on Ivory Coast almond (*Terminalia ivorensis* A. Chev.) planted in Cameroon, and it was also associated with dead tissues of ornamental and non-native *Terminalia mantaly* H. Perier (Begoude *et al.*, 2010). *A. marmelostoma* was linked to other disease reports of Ivory Coast almond from Ghana (Ofosu-Asiedu and Cannon, 1976) and Kenya (Begoude *et al.*, 2010) as well. The presence of the pathogen in other African countries could imply that it is native to Africa; it could also have a wider host range than has yet been reported.

Ivory Coast almond is also planted in Ecuador, where a different member of the *Cryphonectriaceae*, which is morphologically quite distinct from *A. marmelostoma*, was associated with decline and diffuse basal stem cankers of the tree. This pathogen also fruited abundantly on stumps of superb terminalia (*T. superba* Engl. & Diels) (Gryzenhout *et al.*, 2005b). The pathogen, *Rostraureum tropicale* Gryzenh. & M.J. Wingf., was proven pathogenic in pathogenicity trials on superb terminalia. As already mentioned, *R. tropicale* is distinct from *A. marmelostoma* (which is currently known only from Africa), but is closely related to *R. longirostre* (Earle) Gryzenh. & M.J. Wingf., a saprotrophic fungus from the Caribbean and South America (Roane, 1986; Gryzenhout *et al.*, 2005b). It is thus likely that *R. tropicale* is a fungus native to Ecuador with the ability to infect Ivory Coast almond, though it was not clear from inoculation studies and disease symptoms whether this pathogen really caused the decline of the trees (Gryzenhout *et al.*, 2005b).

The impact of *A. marmelostoma* and *R. tropicale* is currently not that great due to the limited planting of *Terminalia* spp. in the world. It is unknown, however, what the effect would be should these pathogens be introduced into other countries that grow *Terminalia* spp., into the native ranges of these trees, or into continents where the individual pathogen species do not occur naturally. This is especially true in Africa, where these trees are often used and planted together with crops in rural areas in a 'taungya' system, and are crucial for the income and survival of subsistence farmer families (Lamb and Ntima, 1971).

16.1.4 Importance of native pathogens

Some *Cryphonectriaceae* have been discovered only on native woody plants. On such plants they are usually associated with mild dieback symptoms, or with cankers that do not necessarily kill the tree. Some of these species of *Cryphonectriaceae* are rarely found, mostly because only limited sampling could be done in the respective countries, or detection is difficult because they are not very conspicuous or do not fruit abundantly (we are here excluding *Cryphonectriaceae* found as saprophytes on dead woody material).

Examples in Africa of native pathogens include *Latruncellus arorae* M. Vermeulen, Gryzenh. & Jol. Roux on *Galpinia transvaalica* N.E. Br. in an isolated area in Swaziland. *L. arorae* was shown to be pathogenic to *G. transvaalica* and a eucalyptus clone in pathogenicity trials (Vermeulen *et al.*, 2011). *Chrysoporthe syzygiicola* occurs on water berry in one part of Zambia (Chungu *et al.*, 2010), as does *Celoporthe fontana* (Vermeulen *et al.*, 2012a). *Celoporthe dispersa* infected native *Syzygium* spp. and the rare tree *Heteropyxis canescens* Oliver (*Myrtales*) in different and isolated natural areas of South Africa (Nakabonge *et al.*, 2006b; Vermeulen *et al.*, 2011). This pathogen was also found on non-native *Tibouchina granulosa* (Desr.) Cogn. (*Myrtales*), and it was more pathogenic in greenhouse trials on a *Eucalyptus* clone than on *T. granulosa* (Nakabonge *et al.*, 2006b; Vermeulen *et al.*, 2011). Although *Celoporthe woodiana* is only known thus far on non-native *Tibouchina granulosa* (Desr.) Cogn. (*Myrtales*), and has been shown to be pathogenic in greenhouse trials on a eucalyptus clone (Nakabonge *et al.*, 2006b; Vermeulen *et al.*, 2011, 2012a), its grouping within the African clade of *Celoporthe* suggests that this fungus may also occur on a native host.

A new and damaging pathogen has recently been found on native Cape beech trees (*Rapanea melanophloeos* L., *Primulales*) in South Africa (Chen *et al.*, 2012). *Immersiporthe knoxdaviesiana* S.F. Chen, M.J. Wingf. & Jol. Roux was only found in a single location of the geographical range of the tree, and was particularly aggressive in field and artificial pathogenicity trials. This restricted distribution, and the fact that the disease has not been noted despite previous and geographically extensive surveys, could suggest that the pathogen has been introduced. It now threatens native Cape beech trees; a situation that is similar to the effect of *C. parasitica* on native American chestnuts in North America.

In the Americas, *Ursicollum fallax* Gryzenh. & M.J. Wingf. and *Microthia coccolobae* (Vizioli) Gryzenh. & M.J. Wingf. occur on native seagrape (*Coccoloba uvifera* (L.) L., *Polygonales*) in Florida and Bermuda, respectively (Barnard *et al.*, 1993; Gryzenhout *et al.*, 2006e). *E. gyrosa* occurs widely in North America (Roane, 1986; Gryzenhout *et al.*, 2009). In Colombia, *C. inopina* has a very restricted known occurrence on *T. urvilleana* (Gryzenhout *et al.*, 2006d). The weak pathogen *Aurapex penicillata* Gryzenh. & M.J. Wingf. fruits conspicuously and abundantly on various native melastomes (*Myrtales*), though it has rarely been found on *Eucalyptus* spp. (Gryzenhout *et al.*, 2006e). In pathogenicity trials, *A. penicillata* was shown to be weakly pathogenic, but it was able to induce lesions in grand eucalyptus (Gryzenhout *et al.*, 2006a).

Several *Cryphonectriaceae* species occur naturally in Asia (including South-east Asia) and Oceania. In Japan and China, members of the genus *Cryphonectria*, including *C. parasitica*, occur naturally on Japanese and Chinese oaks (Liu and Milgroom, 2007). *H. eucalypti* is a native pathogen of eucalyptus in Australia (Walker *et al.*, 1985); *Foliocryphia eucalypti*, *Chrysocrypta corymbiae* and *Aurantiosacculus* spp. from eucalyptus leaves most likely are also native (Cheewangkoon *et al.*, 2009; Crous *et al.*, 2012b,c). In neighbouring New Zealand, *Amphilogia gyrosa* (Berk. & Broome) Gryzenh., Glen & M.J. Wingf. and *A. major* Gryzenh., H.F. Glen & M.J. Wingf. are reported to be associated with mild root cankers on native *Elaeocarpus* spp. (Gilmour, 1966; Dingley, 1969; Pennycook, 1989; Gryzenhout *et al.*, 2005c).

Members of the genus *Chrysoporthe* are examples of fungi that occur as native pathogens, but that are also able to infect non-native *Eucalyptus* timber species. *C. austroafricana*, *C. cubensis* and *C. deuterocubensis* are all best known as serious canker pathogens of eucalyptus, but these pathogens all have native hosts in the same countries that are planted with eucalyptus. In the case of *C. austroafricana*, evidence points strongly to the fact that the pathogen is native to Africa on *Syzygium* spp. (Heath *et al.*, 2006; Nakabonge *et al.*, 2006a; Vermeulen *et al.*, 2012b). Similarly, *C. cubensis* and *C. deuterocubensis* have a high phenotypic diversity in countries of South America and Asia, respectively (van Heerden *et al.*, 1997). Native hosts of *C. cubensis* in South and Central America include members of the *Melastomataceae*, namely *Miconia* spp., *Rhynchanthera mexicana* DC., *Clidemia sericea* D. Don and *Tibouchina* spp. (Rodas *et al.*, 2005; Gryzenhout *et al.*, 2006d; van der Merwe *et al.*, 2012), while *C. deuterocubensis* has been found on native *Melastoma melabathricum* L. (which also belongs to the *Melastomataceae*) in Indonesia (Gryzenhout *et al.*, 2006d). Reports of the occurrence of *C. cubensis* on other *Myrtales*, such as *Psidium cattleianum* Sabine (Hodges, 1988), *Marlierea edulis* Nied. (Barreto *et al.*, 2006), and *T. granulosa* (Seixas *et al.*, 2004) could also represent host ranges of this fungus.

The unprecedented discoveries of Chrysoporthe canker pathogens of eucalyptus on native hosts demonstrates the danger of introducing these pathogens to such native hosts in other areas of the world (Gryzenhout *et al.*, 2009). For example, hosts of *Chrysoporthe* spp. such as *Tibouchina* spp. and *Lagerstroemia indica* L. (*Myrtales*), are beautiful ornamentals planted in various parts of the world (Gryzenhout *et al.*, 2006c). *Tibouchina* spp. have been shown to be specially susceptible to *Chrysoporthe* spp. outside South America, including *C. deuterocubensis* and *C. austroafricana* (Myburg *et al.*, 2002a; Gryzenhout *et al.*, 2006d), and also to *C. dispersa* and *H. eucalypti* in Australia and New Zealand (Nakabonge *et al.*, 2006b; Gryzenhout *et al.*, 2010b). In China, non-native ornamental black plum was shown to be infected with *C. deuterocubensis* (Chen *et al.*, 2010) and *C. syzygii* (Chen *et al.*, 2011). Other new species of *Celoporthe* (*C. eucalypti* and *C. guangdongensis*) described from commercial eucalyptus trees in China were also shown to be able to infect black plum, which is native to Bangladesh, India, Indonesia, Nepal, and Pakistan (Chen *et al.*, 2011).

Although native pathogens appear seemingly harmless and inconspicuous, the introduction of *C. parasitica* into North America and Europe, and the serious diseases caused by *Chrysoporthe* spp., show the potential devastating impact that native pathogens can have in foreign territory. They could infect the same or related hosts, or hosts previously not thought to be susceptible. Knowledge and monitoring of these species, and continued surveys for members of the *Cryphonectriaceae* in various countries of the world, are thus worthwhile. This is especially true because the indications are that the *Cryphonectriaceae* are vastly under-sampled (Gryzenhout *et al.*, 2006b, 2009), particularly in areas of the world where it is relatively poorly known for the fungi to occur on native woody plants, such as South-east Asia, Africa and South America.

16.2 Diagnosis

Pathogens in the *Cryphonectriaceae* are associated with bark and woody tissues. Typical symptoms include diffuse, sunken or deforming cankers on roots, trunks and branches, which extend both longitudinally and laterally. On eucalyptus, these pathogens are in some cases associated with cankers bleeding kino, such as in the case of *H. eucalypti* (White and Kile, 1993). Lesions often girdle the stem or branches, resulting in dieback and tree death (Sinclair and Lyon, 2005; Gryzenhout *et al.*, 2009). Symptoms are usually restricted to the cambium. Exceptions are *M. coccolobae* that can also infect fruits of seagrape trees (Vizioli, 1923), and recent species described from leaves of *Eucalyptus* and *Corymbia* spp. in Australia (Cheewangkoon *et al.*, 2009; Crous *et al.*, 2012b,c).

Usually members of the *Cryphonectriaceae* have complex, conspicuous orange fruiting bodies (Gryzenhout *et al.*, 2009). These bodies are usually stromatic, with dark perithecia embedded in stromatic tissue and perithecial necks protruding from the stromata as blackened ostiolar dots covered with orange or dark layers of tissue (Gryzenhout *et al.*, 2009). They can be distinguished from those of canker pathogens such as *Nectria* spp. that have orange to red, globose and simple perithecia (Sinclair and Lyon, 2005). Anamorphic structures are stromatal, often resembling pycnidia, and have narrowing necks or no necks (Gryzenhout *et al.*, 2009).

Cryphonectriaceae with blackened fruiting bodies can be easily confused with other fungi, specifically with other bark-inhabiting diaporthalean fungi such as *Diaporthe* spp. or *Valsa* spp. that also cause canker diseases on trees (Sinclair and Lyon, 2005). Fruiting bodies of the *Cryphonectriaceae* differ from those of the genera *Diaporthe* or *Valsa* in having orange, yellow or often red spore drops or tendrils, whereas those of *Diaporthe* spp. and *Valsa* spp. are usually pale. The sexual state of the *Cryphonectriaceae* may contain orange, erumpent tissue, whereas in the case of *Diaporthe* spp., only black perithecial necks can be seen protruding from the bark, and in *Valsa* spp., the stromata are usually blackened. *Cryphonectriaceae* with orange fruiting bodies may be confused with species in the *Harknessiaceae* (Crous *et al.*, 2012a) or with a newly described genus *Auratiopycnidiella* (Crous *et al.*, 2012b) of uncertain family in the *Diaporthales*, which has orange stromatic tissue. The structures of the *Cryphonectriaceae* will also discolour in KOH and lactic acid (Gryzenhout *et al.*, 2006b).

The fruiting bodies of pathogenic and saprotrophic species can usually be readily observed on the surface of cankers or dead woody material and in leaf spots (Cheewangkoon *et al.*, 2009; Gryzenhout *et al.*, 2009; Crous *et al.*, 2012b,c). Most of these fungi have conspicuous orange to red fruiting bodies, and exude spores in red to orange to yellow spore droplets or tendrils. Even those species with blackened fruiting bodies can usually be seen owing to the orange discoloration of the surrounding bark tissue after sporulation. However, species that are rare and found on native trees sometimes do not fruit abundantly and can thus be especially difficult to find when they have black fruiting bodies, for instance those of *Celoporthe* spp. It might also be possible that fruiting bodies of more than one species occur together, for instance *Celoporthe* spp. with *Chrysoporthe* spp. (Myburg *et al.*, 2003; Nakabonge *et al.*, 2006b; Chen *et al.*, 2011), *M. havanensis* with *C. cubensis* (Barnard *et al.*, 1987; Gryzenhout *et al.*, 2006e), *A. penicillata* with *Chrysoporthe* spp. (Gryzenhout *et al.*, 2006a) and *C. radicalis* with *C. parasitica* (Hoegger *et al.*, 2002; Sotirovski *et al.*, 2004; Ghezi *et al.*, 2010).

Genera in the *Cryphonectriaceae* are distinguished by structural differences seen in vertical sections, and the shapes and colours of the anamorph and/or teleomorph structures. Examples have been treated in detail by Gryzenhout *et al.* (2009). Once a generic placement has been made, species are usually distinguished through differences in spore and ascus shape, septation or size, and cultural characteristics (Gryzenhout *et al.*, 2009). These characteristics can sometimes be difficult to determine without detailed morphological studies; sectioning of these complex structures by hand or with equipment such as a cryostat, will then be necessary. The anamorph structures are usually the most important for morphological distinctions, particularly for generic identifications, but they can be absent from samples, in which case many of the genera appear identical, as is the case for *Cryphonectria*, *Microthia* and *Cryptometrion*, which all have similar stromatic structure and ascospores. The absence of the sexual state with the distinguishing characteristic, such as asci that differ in size in the case of *Chrysoporthe* spp., or red to orange discolorations of young conidiomata in the case of *Celoporthe* spp., may further hinder morphological identifications (Gryzenhout *et al.*, 2009).

DNA sequence comparisons, if available, offer a more certain means of identification for the *Cryphonectriaceae* (Gryzenhout *et al.*, 2009). Vouchered phylogenies have been published, in which genera, and often species, can be placed based only on data from the internal transcribed spacer (ITS) region of the ribosomal operon (Gryzenhout *et al.*, 2009; Begoude *et al.*, 2010; Chungu *et al.*, 2010; van der Merwe *et al.*, 2010; Chen *et al.*, 2011, 2012; Vermeulen *et al.*, 2011, 2012a). Data from ITS DNA sequences must be complemented by data from the regions of the household genes β-tubulin, histone H3 or translation elongation factors in cases where ITS data alone are not sufficient for identifications. This is especially the case for some species of *Chrysoporthe* and *Celoporthe*, which are often only distinguishable by single base pair differences fixed between species, and where such differences are not shown in the ITS sequence data for some species (Myburg *et al.*, 2002b; van der Merwe *et al.*, 2010; Chen *et al.*, 2011; Vermeulen *et al.*, 2012a).

Certain molecular detection techniques have been developed for some species in the *Cryphonectriaceae*, but do not exist for all genera or species in the family. A PCR-restriction fragment length polymorphism (RFLP) using the restriction enzymes *Alu*I and *Cfo*I has been used to distinguish *C. parasitica*, *E. gyrosa*, *C. cubensis* and *C. deuterocubensis* (Myburg *et al.*, 1999), although this study was done before numerous new species had been added to the *Cryphonectriaceae* (Gryzenhout *et al.*, 2009). Bragança *et al.* (2011), however, showed that this technique does not distinguish between *C. parasitica*, *C. radicalis* and *C. naterciae* M.H. Bragança, E. Diogo & A.J.L. Phillips. A PCR-RFLP with *Eco*RI and *Cfo*I has also been used to distinguish between *E. gyrosa* and *H. eucalypti* isolates when these were still considered to be one species (Venter *et al.*, 2001). The morphologically similar species *C. cubensis*, *C. deuterocubensis* and *C. austroafricana* can be distinguished with a PCR-RFLP (with *Hind*III and *Alu*I) developed by van der Merwe *et al.* (2010), which is useful to confirm identities of these three species in areas where they could be co-occurring. Care should be taken though because this technique has not been tested on other species of *Chrysoporthe*.

Other techniques than PCR-RFLP include the following. Popov *et al.* (2010) developed species-specific primers to distinguish *C. parasitica* from other fungi, in particular from other canker pathogens occurring in the European chestnut (*Castanea sativa* Mill.) in northern Turkey, but these have not been tested on other members of the *Cryphonectriaceae*. Bragança *et al.* (2011) could distinguish between *C. parasitica*, *C. radicalis* and *C. naterciae* with microsatellite-primed PCR fingerprinting. DNA sequencing is thus the only method that will accurately distinguish between all the species in the *Cryphonectriaceae* – and possible new taxa – when used in the phylogenies that are already published. Such an approach is more accurate because BLAST (Basic Local Alignment Search Tool) searches on the GenBank database (the US National Institutes of Health (NIH) genetic sequence database provided by the US National Center for Biotechnology Information (NCBI)) can often be misleading.

16.3 Infection Biology

Owing to the similar biology of the other species in the *Cryphonectriaceae* to that of *C. parasitica*, it can be assumed that the same principles of dispersal and infection (see Prospero and Rigling, Chapter 15, this volume) can be applied to other pathogens in the *Cryphonectriaceae* (Bruner, 1917; Gryzenhout *et al.*, 2009). Like *C. parasitica*, all species of the *Cryphonectriaceae* form slimy conidial tendrils or droplets, which suggest that insects are important in movement of spores, with flying insects more likely to be involved in long- and short-distance spore movement. Silvicultural practices such as pruning can also lead to new infections through spore movement and the creation of new infection courts (Nutman and Roberts, 1952). Spore tendrils or droplets harden after time, and can easily break off after mechanical disturbance and be distributed by wind, dust or soil. Rain or moisture dissolves these droplets further, resulting in additional spore distribution, usually over short distances. As with *C. parasitica*, ascospores of other pathogens in the *Cryphonectriaceae* do not typically form slimy droplets, but are forcefully expelled. Ascospores carried on wind are, therefore, the primary mode of long-distance dispersal.

Anthropogenic factors are important for long-distance movement of the *Cryphonectriaceae* (Wingfield *et al.*, 2001b,c). Seedlings infected with *C. parasitica* are believed to have been responsible for the introduction of the fungus into the USA (Anagnostakis, 1987), and infected plant material such as seedlings, nuts and chestnut products is suspected to have caused the introduction of *C. parasitica* into Europe (Jaynes and DePalma, 1984; Heiniger and Rigling, 1994). These factors can be compounded by inconspicuous symptoms. Although not studied for the *Cryphonectriaceae* in general, *C. parasitica* has been shown to be able to exist as an endophyte in healthy tissue (Bissegger and Sieber, 1994), a fact that could have made its transport easier.

Species in the *Cryphonectriaceae* are not known to infect woody material spontaneously. Natural or mechanical wounds are usually required for entry (Stevens, 1917; Hunter *et al.*, 1976; White and Kile, 1993). Infections and symptom development are then worsened by drying conditions in the wood and warmer weather conditions, as has been demonstrated for *E. gyrosa* (Hunter *et al.*, 1976; Hunter and Stipes, 1978; Appel and Stipes, 1984).

16.4 Epidemiology

The different hosts affected by the *Cryphonectriaceae* vary from native woody shrubs and trees, to important timber species, to trees planted as ornamentals or in orchards, for instance chestnut trees for chestnut production. Silvicultural practices are applied in forests and plantations to maintain tree vigour, health and timber yield, while orchard practices require that trees are productively grown over long periods so they are constantly pruned and the fruits harvested. Epidemiological approaches to maintaining the health of these trees thus vary according to the host affected. However, few epidemiological studies have been done on the pathogens in the *Cryphonectriaceae*, and the newly described *Cryphonectriaceae* from leaves further illustrate how little is known. Studies on chestnut blight, therefore, provide insight into the epidemiology of other members of the *Cryphonectriaceae* that affect woody tissue and bark, of which most occur in natural forests or tree plantations.

Some members of the *Cryphonectriaceae*, such as *C. austroafricana* on *T. granulosa*, and *E. gyrosa* on oaks, chestnuts (*Castanea* spp.) and formosan sweetgum (*Liquidambar formosana* Hance.), infect street or garden ornamentals (Myburg *et al.*, 2002a; Gryzenhout *et al.*, 2009), or have the potential to do so, as has been shown by other species of *Chrysoporthe*. The only species in the *Cryphonectriaceae* that infects orchards besides *C. parasitica* are *C. cubensis* and *C. deuterocubensis*, which damage clove trees. Most tree species affected by *Cryphonectriaceae* are plantation trees, which are routinely and completely felled and harvested for timber or pulp, or coppiced to stimulate regeneration of the shoots. Plantations are essentially agricultural systems established over vast areas and where chemical disease control is unfeasible, unlike disease control in orchard or ornamental trees. Planting disease-resistant material from selected seed lots, hybrids and clonal propagation is the only viable option and has been very successful in managing plantation diseases caused by the *Cryphonecriaceae*, especially *C. cubensis* and *C. deuterocubensis* on eucalyptus. In large plantations, material from different genetic sources is planted in mosaic blocks, thereby protecting the various monocultures from changes in the genomics of the local pathogens, or protecting a particular genotype of the host against the introduction of new genotypes of the pathogen (Chou, 1981; Leakey, 1987; Wingfield, 1990; Denison and Kietzka, 1993).

In most natural ecosystems where *Cryphonectriaceae* occur, careful monitoring of individual trees is not possible because there is no management or monitoring as in the case of the well-managed forest systems of the northern hemisphere. For instance, attempts are made to save even individual trees by means of silvicultural practices or applications of hypovirulent strains against chestnut blight in natural forest scenarios, orchards and on ornamentals (Griffin, 1986). This needs to be done continuously because trees are infected repeatedly, and branches usually have multiple cankers. Such an approach cannot always be taken elsewhere though. In fact, in Africa and South America forested areas are often threatened by slash and burn practices and medicinal plant and firewood harvesting. This puts these systems under more pressure, which may lead to higher disease incidence or to exposure to cross-infection from other pathogens, or may enable the native pathogens from these systems to cross-infect new trees. Other factors such as timber theft and trading may also represent uncontrolled avenues for native pathogens to be transported to other areas where they could cause infections.

The optimum environmental conditions for the *Cryphonectriaceae* are very diverse. As an example, tropical to subtropical pathogens include species in *Chrysoporthe* and *A. penicillata*, *A. marmelostoma* and *R. tropicale*, while *H. eucalypti* and *Celoporthe* spp. also occur in less humid but warm areas. *Cryphonectria* and *Endothia* spp. are known from temperate areas. Interestingly, *C. aestuescens*, despite being found in tropical Indonesia, was shown to have a lower optimum temperature, indicating a preference for cooler conditions.

Based on the conditions where *C. austroafricana* has been known to occur, bioclimatic modelling was used to predict the occurrence of this pathogen in South Africa (van Staden *et al.*, 2004). *C. austroafricana* is predicted to occur only in relatively limited, subtropical areas of South Africa and not in colder areas. However, the modelling showed that climate change would shift the occurrence of this pathogen to other areas where susceptible species occur that are currently not affected.

Environmental conditions and the general health of the host play an important role in diseases caused by *Cryphonectriaceae*. *C. parasitica* and *Chrysoporthe* spp. are primary pathogens on the hosts with which they did not evolve, such as American chestnut (*Castanea dentata* (Marsh.) Borkh.) and eucalyptus, respectively. On their natural hosts these pathogens are associated with mild disease symptoms. Other pathogens, such as *H. eucalypti* and *E. gyrosa*, are considered to be opportunistic (Appel and Stipes, 1986; Old *et al.*, 1990; Yuan and Mohammed, 2000). Environmental stress conditions such as wounding, drought, pathogen or insect damage and defoliation, and marginal planting sites will increase disease severity (Appel and Stipes, 1986; Old *et al.*, 1990; White and Kile, 1993; Wingfield and Roux, 2002; Gryzenhout *et al.*, 2003).

Significant host–environment interactions have been demonstrated in field inoculations for isolates of *H. eucalypti* and *C. austroafricana* (van Heerden and Wingfield, 2002; Gryzenhout *et al.*, 2003; van Heerden *et al.*, 2005). These interactions could influence the performance of specific species, hybrids or clones, even though they have been selected for disease resistance. The possibility of such interactions also indicates that climate change could have an influence on the severity of disease when environmental conditions become stressful, as has been shown to be likely for *C. parasitica* (Desprez-Loustau *et al.*, 2007) and *C. austroafricana* (van Staden *et al.*, 2004). This could be equally likely for diseases caused by other opportunistic canker pathogens in the *Cryphonectriaceae*, which could become more serious under more stressful conditions (Sturrock *et al.*, 2011).

It is important in the epidemiology of the *Cryphonectriaceae* to consider the occurrence of these pathogens on other hosts or their ability to infect such hosts. In many cases, these hosts represent native or naturally occurring trees that could serve as alternative hosts where the fungus could survive in the absence of its better known commercial hosts. Conversely, such hosts could harbour pathogens with an unprecedented ability to infect commercial trees. These pathogens could, therefore, continue to infect planted introduced hosts, as has been shown for *Chrysoporthe* spp. Native hosts could also represent caches of genetic diversity, especially in the case of native pathogens that infect non-native trees. In some cases, tree species could also be susceptible even if the fungus is not known to infect them, as has been demonstrated in a series of cross-inoculations on related and unrelated woody host species of *C. cubensis* in Brazil (Hodges *et al.*, 1986; Seixas *et al.*, 2004). In the case of alternative ornamental hosts, these represent unexpected avenues of introductions when humans move infected planting stock.

The presence of species in countries or continents, distribution within a country, extended host ranges of pathogens and population diversity studies are all crucial for devising effective policies against possible introductions of pathogens of the *Cryphonectriaceae* and for restricting further movement (Wingfield *et al.*, 2001b,c). In the case of most species of the *Cryphonectriaceae*, such data simply do not exist, while for other species such as *Chrysoporthe* spp., *C. parasitica* and *H. eucalypti*, numerous surveys over the past few decades have led to better distribution and host data, and enabled population studies that could test hypotheses on movement and centres of diversity (Anagnostakis *et al.*, 1986; Milgroom, 1995; Liu *et al.*, 1996; van Heerden *et al.*, 1997, 2001; van Zyl *et al.*, 1998; Liu and Milgroom, 2007; Nakabonge *et al.*, 2007, 2008). The approach in which data are continuously generated as a means of combating future epidemics is proactive. Continued monitoring for the presence of *Cryphonectriaceae* and the possible discovery of novel hosts would thus be beneficial, and would build on a growing collection of published data, as summarized by Gryzenhout *et al.* (2009).

16.5 Management Strategies and Tactics

The following management strategies (other than those followed for forest pathogens in general) are relevant to diseases caused by *Cryphonectriaceae* in planted areas. They are mostly applied to plantations of eucalyptus – as none of the pathogens in the *Cryphonectriaceae* causes devastation in natural areas except for *C. parasitica*. The *Cryphonectriaceae* are also not known as nursery diseases of plantation species or of newly planted trees in the field, although some, such as *H. eucalypti* and *C. aestuescens*, have been shown to infect clonal banks or stressed seedlings.

Very little has been researched for members of the *Cryphonectriaceae* on the various aspects of disease management, such as avoidance, exclusion, eradication and protection of trees. The vast majority of research in management is more relevant to chestnut blight in natural forests and orchards. Quarantine regulations among countries and continents, and within a country, and management practices for *C. parasitica* – such as chemical control, site selection, the thinning of trees and branches, selection of the correct site, soil and soil moisture conditions, and sanitation practices and wound treatments – are all universal and basic for any successful management programme (Griffin 1986, 2000; see Prospero and Rigling, Chapter 15, this volume). However, some specific management practices and tactics have been developed for members of the *Cryphonectriaceae* other than *C. parasitica*, and these are discussed below.

16.5.1 Resistance

For pathogens in the *Cryphonectriaceae* that affect plantation species, hybridization and breeding for resistance both between and within species, and clonal propagation of resistant or tolerant lines, are the most effective strategies for disease control (Wingfield and Roux, 2002; Wingfield, 2003). This has been successfully shown and developed for *Chrysoporthe* spp. by means of targeted inoculations (Alfenas *et al.*, 1983; van Heerden and Wingfield, 2002; Wingfield, 2003; van Heerden *et al.*, 2005; Chen *et al.*, 2010; Guimarães *et al.*, 2010; Pegg *et al.*, 2010). Differences in resistance to *Celoporthe* spp. (Chen *et al.*, 2011), *H. eucalypti* (Gryzenhout *et al.*, 2003) and *C. aestuescens* (Gryzenhout *et al.*, 2010a) have also been found in the particular clones tested, and breeding programmes against these pathogens are thus possible.

Breeding programmes against *C. austroafricana* have been so successful that the disease is now rarely found in South Africa on eucalyptus (Wingfield and Roux, 2002), although it is still a threat in neighbouring countries such as Mozambique and Zambia (Roux *et al.*, 2005; Nakabonge *et al.*, 2006a), where intensive breeding and selection have not yet been done. Neither have the resistant lines in South Africa been tested against other *Chrysoporthe* spp. that also occur in Africa, such as *C. cubensis*, *C. deuterocubensis*, *C. syzygiicola* and *C. zambiensis* (Vermeulen *et al.*, 2011). Likewise, eucalyptus programmes in Brazil, where most resistance breeding has taken place, are similarly threatened by the presence of other *Chrysoporthe* spp. in South America, as well as *Chrysoporthe* spp. from Africa and Asia.

16.5.2 Therapy

Therapy methods, including the applications of fungicides or biocontrol using hypovirulent isolates, are used only for *C. parasitica*. The close relatedness of *Chrysoporthe* spp. to *C. parasitica*, however, has prompted searches for similar hypoviruses. Viruses were found in isolates of *C. cubensis* in Brazil that caused altered cultural morphology such as slow growth, different cultural margins and a change of colour from orange to white; these isolates were also shown to be less virulent than uninfected isolates (van Zyl *et al.*, 1999). Two mitoviruses were isolated from a collection of *C. austroafricana* isolates from South Africa, but they did not cause a reduction in virulence (van Heerden, 2004). Isolates of *C. austroafricana* could though be transfected with a single-stranded transcript of *Cryphonectria hypovirus 1* (CHV-1-EP713) (van Heerden *et al.*, 2001) and successfully converted to hypovirulence. None of these viruses or transfected isolates has been applied in the field as a biocontrol method against *C. austroafricana* or *C. cubensis*.

16.5.3 Integrated disease management

In most cases, resistance is the best option for managing pathogens in the *Cryphonectriaceae* that affect plantations. This should be done continuously, and new hybrids and clones should be screened for resistance by means of artificial inoculation tests (Wingfield, 1990, 2003). Careful site matching of tree species is also necessary to lower any plant stress that could predispose them to these canker pathogens (Wingfield and Roux, 2002). The reduction of planting and environmental stress, and of pathogen and pest damage, by means of appropriate silvicultural practices will further promote tree health. Continued monitoring is crucial to detect new disease outbreaks and to widen knowledge of the geographical and host ranges of these pathogens, which are of such great quarantine importance. Those pathogens occurring in surrounding native vegetation, or with the potential to infect surrounding vegetation, should be treated with extra care because native hosts could represent a cache of inoculum.

References

Alfenas, A.C., Jeng, R. and Hubbes, M. (1983) Virulence of *Cryphonectria cubensis* on *Eucalyptus* species differing in resistance. *European Journal of Forest Pathology* 13, 197–205.

Anagnostakis, S.L. (1987) Chestnut blight: the classical problem of an introduced pathogen. *Mycologia* 79, 23–37.

Anagnostakis, S.L., Hau, B. and Krantz, J. (1986) Diversity of vegetative compatibility groups of *Cryphonectria parasitica* in Connecticut and Europe. *Plant Disease* 70, 536–538.

Appel, D.N. and Stipes, R.J. (1984) Canker expansion on water-stressed pin oaks colonized by *Endothia gyrosa*. *Plant Disease* 68, 851–853.

Appel, D.N. and Stipes, R.J. (1986) A description of declining and blighted pin oaks in eastern Virginia. *Journal of Arboriculture* 12, 155–158.

Barnard, E.L., Geary, T., English, J.T. and Gilly, S.P. (1987) Basal cankers and coppice failure of *Eucalyptus grandis* in Florida. *Plant Disease* 71, 358–361.

Barnard, E.L., Barnard, M.R., El-Gholl, N.E. and Ash, E.C. (1993) Preliminary investigations of some Florida isolates of *Cryphonectria* and/or *Endothia* spp. In: Kelley, W., Miller, T., Smith, T., Hess, N. and Shain, L. (eds) *Castanea, Coccoloba, Eucalyptus and Quercus spp. Southwide Forest Disease Workshop, 13–15 January 1993, Auburn, Alabama. Program and Abstracts.* School of Forestry, Auburn University, Alabama, abstract from Paper Session V. Miscellaneous Topics, p. 38. Available at: https://fp.auburn.edu/sfws/enebak/swfdw/programs/1993_Jan_13-15.pdf (accessed 10 December 2012).

Barreto, R.W., Rocha, F.B. and Ferreira, F.A. (2006) First record of natural infection of *Marlierea edulis* by the eucalyptus canker fungus *Chrysoporthe cubensis*. *Plant Pathology* 55, 577.

Begoude, A.D., Gryzenhout, M., Wingfield, M.J. and Roux, J. (2010) *Aurifilum*, a new fungal genus in the *Cryphonectriaceae* from *Terminalia* species in Cameroon. *Antonie Van Leeuwenhoek Journal of Microbiology* 98, 263–278.

Bissegger, M. and Sieber, T.N. (1994) Assemblages of endophytic fungi in coppice shoots of *Castanea sativa*. *Mycologia* 86, 648–655.

Bragança, H., Rigling, D., Diogo, E., Capelo, J., Phillips, A. and Tenreiro, R. (2011) *Cryphonectria naterciae*: a new species in the *Cryphonectria–Endothia* complex and diagnostic molecular markers based on microsatellite-primed PCR. *Fungal Biology* 115, 852–861.

Bruner, S.C. (1916) A new species of *Endothia*. *Mycologia* 8, 239–242.

Bruner, S.C. (1917) *Una Enfermedad Gangrenosa de los Eucaliptos* [*A Gangrenous Disease of Eucalypts*]. Bulletin 37, Estación Experimental Agronómica, Santiago de las Vegas, Cuba, pp. 1–33.

Carnegie, A.J. (2007) Forest health condition in New South Wales, Australia, 1996–2005. II. Fungal damage recorded in eucalyptus plantations during forest health surveys and their management. *Australasian Plant Pathology* 36, 225–239.

Cheewangkoon, R., Groenewald, J.Z., Summerell, B.A., Hyde, K.D., To-anun, C. and Crous, P.W. (2009) *Myrtaceae*, a cache of fungal biodiversity. *Persoonia* 23, 55–85.

Chen, S.F., Gryzenhout, M., Roux, J., Xie, Y.J., Wingfield, M.J. and Zhou, X.D. (2010) Identification and pathogenicity of *Chrysoporthe cubensis* on *Eucalyptus* and *Syzygium* spp. in South China. *Plant Disease* 94, 1143–1150.

Chen, S.F., Gryzenhout, M., Roux, J., Xie, Y.J., Wingfield, M.J. and Zhou, X.D. (2011) Novel species of *Celoporthe* from *Eucalyptus* and *Syzygium* trees in China and Indonesia. *Mycologia* 103, 1384–1410.

Chen, S.F., Wingfield, M.J., Roets, F. and Roux, J. (2012) A serious canker disease caused by *Immersiporthe knoxdaviesiana* gen. et sp. nov. (*Cryphonectriaceae*) on native *Rapanea melanophloeos* in South Africa. *Plant Pathology*. In press: doi: 10.1111/j.1365–3059.2012.02671.x.

Chou, C.K.S. (1981) Monoculture, species diversification, and disease hazards in forestry. *New Zealand Journal of Forestry* 26, 20–36.

Chungu, D., Gryzenhout, M., Muimba-Kankolongo, A., Wingfield, M.J. and Roux, J. (2010) Taxonomy and pathogenicity of two novel *Chrysoporthe* species from *Eucalyptus grandis* and *Syzygium guineense* in Zambia. *Mycological Progress* 9, 379–393.

Crous, P.W., Summerell, B.A., Shivas, R.G., Carnegie, A.J. and Groenewald, J.Z. (2012a) A re-appraisal of *Harknessia* (*Diaporthales*), and the introduction of *Harknessiaceae* fam. nov. *Persoonia* 28, 49–65.

Crous, P.W., Summerell, B.A., Alfenas, A.C., Edwards, J., Pascoe, I.G., Porter, I.J. and Groenewald, J.Z. (2012b) Genera of diaporthalean coelomycetes associated with leaf spots of tree hosts. *Persoonia* 28, 66–75.

Crous, P.W. *et al.* (2012c) Fungal planet description sheets: 107–127. *Persoonia* 28, 138–182.

Davison, E.M. and Coates, D.J. (1991) Identification of *Cryphonectria cubensis* and *Endothia gyrosa* from eucalyptus in Western Australia using isozyme analysis. *Australasian Plant Pathology* 20, 157–160.

Denison, N.P. and Kietzka, J.E. (1993) The development and utilisation of vegetative propagation in Mondi for commercial afforestation programmes. *South African Forestry Journal* 165, 47–54.

Desprez-Loustau, M.-L., Robin, C., Reynaud, G., Déqué, M., Badeau, V., Piou, D., Husson, C. and Marçais, B. (2007) Simulating the effects of a climate-change scenario on the geographical range and activity of forest-pathogenic fungi. *Canadian Journal of Plant Pathology* 29, 101–120.

Dingley, J.M. (1969) *Records of Plant Diseases in New Zealand.* Bulletin 192, New Zealand Department of Scientific and Industrial Research, Wellington, New Zealand.

Florence, E.J.M., Sharma, J.K. and Mohanan, C. (1986) A stem canker disease of *Eucalyptus* caused by *Cryphonectria cubensis* in Kerala. *Kerala Forestry Research Institute Scientific Paper* 66, 384–387.

Gardner, D.E. and Hodges, C.S. (1990) Diseases of *Myrica faya* (firetree, *Myricaceae*) in the Azores, Madeira and the Canary Islands. *Plant Pathology* 39, 326–330.

Ghezi, E., Khodaparast, S.A. and Zare, R. (2010) Distribution and severity of damage by *Cryphonectria parasitica* in the chestnut stands in Guilan Province, Iran. *Forest Pathology* 40, 450–457.

Gilmour, J.W. (1966) *The Pathology of Forest Trees in New Zealand. The Fungal, Bacterial, and Algal Pathogens.* Technical Paper 48, New Zealand Forest Research Institute, Wellington, New Zealand.

Griffin, G.J. (1986) Chestnut blight and its control. *Horticultural Reviews* 8, 291–336.

Griffin, G.J. (2000) Blight control and restoration of the American chestnut. *Journal of Forestry* 98, 22–27.

Groulez, J. and Wood, P.J. (1985) Terminalia superba, *a Monograph*. Commonwealth Forestry Institute, Oxford, UK.

Gryzenhout, M., Eisenberg, B.E., Coutinho, T.A., Wingfield, B.D. and Wingfield, M.J. (2003) Pathogenicity of *Cryphonectria eucalypti* to *Eucalyptus* clones in South Africa. *Forest Ecology and Management* 176, 427–437.

Gryzenhout, M., Myburg, H., van der Merwe, N.A., Wingfield, B.D. and Wingfield, M.J. (2004) *Chrysoporthe*, a new genus to accommodate *Cryphonectria cubensis. Studies in Mycology* 50, 119–142.

Gryzenhout, M., Myburg, H., Wingfield, B.D., Montenegro, F. and Wingfield, M.J. (2005a) *Chrysoporthe doradensis* sp. nov. pathogenic to *Eucalyptus* in Ecuador. *Fungal Diversity* 20, 39–57.

Gryzenhout, M., Myburg, H., Wingfield, B.D., Montenegro, F. and Wingfield, M.J. (2005b) *Rostraureum tropicale* gen. sp. nov. (*Diaporthales*) associated with dying *Terminalia ivorensis* in Ecuador. *Mycological Research* 109, 1029–1044.

Gryzenhout, M., Glen, H.F., Wingfield, B.D. and Wingfield, M.J. (2005c) *Amphilogia* gen. nov. for *Cryphonectria*-like fungi from *Elaeocarpus* spp. in New Zealand and Sri Lanka. *Taxon* 54, 1009–1021.

Gryzenhout, M., Myburg, H., Rodas, C.A., Wingfield, B.D. and Wingfield, M.J. (2006a) *Aurapex penicillata* gen. sp. nov. from native *Miconia theaezans* and *Tibouchina* spp. in Colombia. *Mycologia* 98, 105–115.

Gryzenhout, M., Myburg, H., Wingfield, B.D. and Wingfield, M.J. (2006b) *Cryphonectriaceae* (*Diaporthales*), a new family including *Cryphonectria, Chrysoporthe, Endothia* and allied genera. *Mycologia* 98, 239–349.

Gryzenhout, M., Wingfield, B.D. and Wingfield, M.J. (2006c) New taxonomic concepts for the important forest pathogen *Cryphonectria parasitica* and related fungi. *FEMS Microbiology Letters* 258, 161–172.

Gryzenhout, M., Rodas, C.A., Portales, J.M., Clegg, P., Wingfield, B.D. and Wingfield, M.J. (2006d) Novel hosts of the *Eucalyptus* canker pathogen *Chrysoporthe cubensis* and a new *Chrysoporthe* species from Colombia. *Mycological Research* 110, 833–845.

Gryzenhout, M., Myburg, H., Hodges, C.S., Wingfield, B.D. and Wingfield, M.J. (2006e) *Microthia, Holocryphia* and *Ursicollum*, three new genera on *Eucalyptus* and *Coccoloba* for fungi previously known as *Cryphonectria. Studies in Mycology* 55, 35–52.

Gryzenhout, M., Wingfield, B.D. and Wingfield, M.J. (2009) *Taxonomy, Phylogeny, and Ecology of Bark-infecting and Tree Killing Fungi in the* Cryphonectriaceae. APS Press, St Paul, Minnesota.

Gryzenhout, M., Tarigan, M., Clegg, P.A. and Wingfield, M.J. (2010a) *Cryptometrion aestuescens* gen. sp. nov. (*Cryphonectriaceae*) pathogenic to *Eucalyptus* in Indonesia. *Australasian Plant Pathology* 39, 161–169.

Gryzenhout, M., Vermeulen, M., Dick, M. and Wingfield, M.J. (2010b) The *Eucalyptus* canker pathogen *Holocryphia eucalypti* on *Eucalyptus* in New Zealand. *Australasian Plant Disease Notes* 5, 5–8.

Guimarães, L.M. da Silva, de Resende, M.D.V., Lau, D., Rosse, L.N., Alves, A.A. and Alfenas, A.C. (2010) Genetic control of *Eucalyptus urophylla* and *E. grandis* resistance to canker caused by *Chrysoporthe cubensis*. *Genetics and Molecular Biology* 33, 525–531.

Heath, R.N., Gryzenhout, M., Roux, J. and Wingfield, M.J. (2006) Discovery of the Cryphonectria canker pathogen on native *Syzygium* spp. in South Africa. *Plant Disease* 90, 433–438.

Heath, R.N., Roux, J., Gryzenhout, M., Carnegie, A.J., Smith, I.W. and Wingfield, M.J. (2007) *Holocryphia eucalypti* on *Tibouchina urvilleana* in Australia. *Australasian Plant Pathology* 36, 560–564.

Heiniger, U. and Rigling, D. (1994) Biological control of chestnut blight in Europe. *Annual Review of Phytopathology* 32, 581–599.

Hodges, C.S. (1980) The taxonomy of *Diaporthe cubensis*. *Mycologia* 72, 542–548.

Hodges, C.S. (1988) *Preliminary Exploration for Potential Biological Control Agents for* Psidium cattleianum. Technical Report 66, Cooperative National Park Resources Studies Unit, University of Hawaii at Manoa, Honolulu, Hawaii.

Hodges, C.S. and Gardner, D.E. (1992) Survey for potential biological control agents for *Myrica faya* in the Azores and Madeira islands. Report submitted to the Cooperative National Park Resources Unit, University of Hawaii at Manoa, Department of Botany, Honolulu, Hawaii.

Hodges, C.S., Reis, M.S., Ferreira, F.A. and Henfling, J.D.M. (1976) O cancro do eucalipto causado por *Diaporthe cubensis* [Canker of eucalyptus caused by *Diaporthe cubensis*]. *Fitopatalogia Brasileira* 1, 129–167.

Hodges, C.S., Alfenas, A.C. and Ferreira, F.A. (1986) The conspecificity of *Cryphonectria cubensis* and *Endothia eugeniae*. *Mycologia* 78, 334–350.

Hoegger, P.J., Rigling, D., Holdenrieder, O. and Heiniger, U. (2002) *Cryphonectria radicalis*: rediscovery of a lost fungus. *Mycologia* 94, 105–115.

Hunter, P.P. and Stipes, R.J. (1978) The effect of month of inoculation with *Endothia gyrosa* on development of pruned branch canker of pin oak (*Quercus palustris*). *Plant Disease Reporter* 62, 940–944.

Hunter, P.P., Griffin, G.J. and Stipes, R.J. (1976) The influence of osmotic water potential on the linear growth of *Endothia* species. *Phytopathology* 66, 1418–1421.

Jaynes, R.A. and DePalma, N.K. (1984) Natural infection of nuts of *Castanea dentata* by *Endothia parasitica*. *Phytopathology* 74, 296–299.

Lamb, A.F.A. and Ntima, O.O. (1971) *Fast Growing Timber Trees of the Lowland Tropics, No. 5*: Terminalia ivorensis. Commonwealth Forestry Institute, Oxford, UK.

Leakey, R.R.B. (1987) Clonal forestry in the tropics: a review of developments, strategies and opportunities. *Commonwealth Forestry Review* 66, 61–75.

Liu, Y.-C. and Milgroom, M.G. (2007) High diversity of vegetative compatibility types in *Cryphonectria parasitica* in Japan and China. *Mycologia* 99, 279–284.

Liu, Y.-C., Cortesi, P., Double, M.L., MacDonald, W.L. and Milgroom, M.G. (1996) Diversity and multilocus genetic structure in populations of *Cryphonectria parasitica*. *Phytopathology* 86, 1344–1351.

Micales, J.A., Stipes, R.J. and Bonde, M.R. (1987) On the conspecificity of *Endothia eugeniae* and *Cryphonectria cubensis*. *Mycologia* 79, 707–720.

Milgroom, M.G. (1995) Population biology of the chestnut blight fungus, *Cryphonectria parasitica*. *Canadian Journal of Botany* 73, S311–S319.

Myburg, H., Wingfield, B.D. and Wingfield, M.J. (1999) Phylogeny of *Cryphonectria cubensis* and allied species inferred from DNA analysis. *Mycologia* 91, 243–250.

Myburg, H., Gryzenhout, M., Heath, R.N., Roux, J., Wingfield, B.D. and Wingfield, M.J. (2002a) Cryphonectria canker on *Tibouchina* in South Africa. *Mycological Research* 106, 1299–1306.

Myburg, H., Gryzenhout, M., Wingfield, B.D. and Wingfield, M.J. (2002b) β-tubulin and histone H3 gene sequences distinguish *Cryphonectria cubensis* from South Africa, Asia and South America. *Canadian Journal of Botany* 80, 590–596.

Myburg, H., Gryzenhout, M., Wingfield, B.D. and Wingfield, M.J. (2003) Conspecificity of *Endothia eugeniae* and *Cryphonectria cubensis*: a re-evaluation based on morphology and DNA sequence data. *Mycoscience* 44, 187–196.

Nakabonge, G., Roux, J., Gryzenhout, M. and Wingfield, M.J. (2006a) Distribution of Chrysoporthe canker pathogens on *Eucalyptus* and *Syzygium* spp. in eastern and southern Africa. *Plant Disease* 90, 734–740.

Nakabonge, G., Gryzenhout, M., Roux, J., Wingfield, B.D. and Wingfield, M.J. (2006b) *Celoporthe dispersa* gen. et sp. nov. from native *Myrtales* in South Africa. *Studies in Mycology* 55, 255–267.

Nakabonge, G., Gryzenhout, M., Wingfield, B.D., Wingfield, M.J. and Roux, J. (2007) Genetic diversity of *Chrysoporthe cubensis* in eastern and southern Africa. *South African Journal of Science* 103, 261–264.

Nakabonge, G., Burgess, T., Gryzenhout, M., Wingfield, B.D., Wingfield, M.J. and Roux, J. (2008) Population structure of the fungal pathogen *Holocryphia eucalypti* in Australia and South Africa. *Australasian Plant Pathology* 37, 154–161.

Nutman, F.J. and Roberts, F.M. (1952) Acute die-back of clove trees in the Zanzibar Protectorate. *Annals of Applied Biology* 39, 599–607.

Ofosu-Asiedu, A. and Cannon, P. (1976) *Terminalia ivorensis* decline in Ghana. *PANS: Pest Articles and News Summaries* 22, 239–242. Centre for Overseas Pest Research, Ministry of Overseas Development, London.

Old, K.M. and Davison, E.M. (2000) Canker diseases of eucalyptus. In: Keane, P.J., Kile, G.A., Podger, F.D. and Brown, B.N. (eds) *Diseases of Eucalypts*. CSIRO Publishing, Collingwood, Victoria, Australia, pp. 241–258.

Old, K.M., Murray, D.I.L., Kile, G.A., Simpson, J. and Malafant, K.W.J. (1986) The pathology of fungi isolated from eucalyptus cankers in south-eastern Australia. *Australian Forest Research* 16, 21–36.

Old, K.M., Gibbs, R., Craig, I., Myers, B.J. and Yuan, Z.Q. (1990) Effect of drought and defoliation on the susceptibility of eucalyptus to cankers caused by *Endothia gyrosa* and *Botryosphaeria ribis*. *Australian Journal of Botany* 38, 571–581.

Old, K.M., Wingfield, M.J. and Yuan, Z.Q. (2003) *A Manual of Diseases of Eucalypts in South-East Asia.* Centre for International Forestry Research (CIFOR), Bogor, Indonesia.

Pegg, G.S., Gryzenhout, M., O'Dwyer, C., Drenth, A. and Wingfield, M.J. (2010) The *Eucalyptus* canker pathogen *Chrysoporthe cubensis* discovered in eastern Australia. *Australasian Plant Pathology* 39, 343–349.

Pennycook, S.R. (1989) *Plant Diseases Recorded in New Zealand, Volume 2, Part II. Fungal Plant Diseases.* Plant Diseases Division, Department of Scientific and Industrial Research, Auckland, New Zealand.

Popov, A.P., Tsvetkov, I.L., Belov, A.A., Konichev, A.S., Ivanushkina, N.E., Kochkina, G.A. and Ozerskaya, S.M. (2010) Molecular genetic identification of the phytopathogenic fungus *Cryphonectria parasitica. Microbiology* 79, 223–228.

Reid, J. and Booth, C. (1969) Some species segregated from the genera *Cryptospora*, *Cryptosporella*, and *Sillia*. *Canadian Journal of Botany* 47, 1055–1060.

Roane, M.K. (1986) Taxonomy of the genus *Endothia*. In: Roane, M.K., Griffin, G.J. and Elkins, J.R. (eds) *Chestnut Blight, Other* Endothia *Diseases, and the Genus* Endothia. APS Press, St Paul, Minnesota, pp. 28–39.

Roane, M.K., Stipes, R.J., Phipps, P.M. and Miller, O.K. Jr (1974) *Endothia gyrosa*, causal pathogen of pin oak blight. *Mycologia* 66, 1042–1047.

Rodas, C.A., Gryzenhout, M., Myburg, H., Wingfield, B.D. and Wingfield, M.J. (2005) Discovery of the *Eucalyptus* canker pathogen *Chrysoporthe cubensis* on native *Miconia* (*Melastomataceae*) in Colombia. *Plant Pathology* 54, 460–470.

Roux, J. and Apetorgbor, M. (2009) First report of *Chrysoporthe cubensis* from *Eucalyptus* in Ghana. *Plant Pathology* 59, 806.

Roux, J. and Nakabonge, G. (2010) First report of *Holocryphia eucalypti* on *Eucalyptus grandis* in Uganda. *Plant Pathology* 59, 409.

Roux, J., Myburg, H., Wingfield, B.D. and Wingfield, M.J. (2003) Biological and phylogenetic analyses suggest that two *Cryphonectria* spp. cause cankers of *Eucalyptus* in Africa. *Plant Disease* 87, 1329–1332.

Roux, J., Meke, G., Kanyi, B., Mwangi, L., Mbaga, A., Hunter, G.C., Nakabonge, G., Heath, R.N. and Wingfield, M.J. (2005) Diseases of plantation forestry tree species in eastern and southern Africa. *South African Journal of Science* 101, 409–413.

Seixas, C.D.S, Barreto, R.W., Alfenas, A.C. and Ferreira, F.A. (2004) *Cryphonectria cubensis* on an indigenous host in Brazil: a possible origin for *Eucalyptus* canker disease? *Mycologist* 18, 39–45.

Sharma, J.K., Mohanan, C. and Florence, E.J.M. (1985) *Disease Survey in Nurseries and Plantations of Forest Tree Species Grown in Kerala*. Research Report 36, Kerala Forest Research Institute, Kerala, India.

Sinclair, W.A. and Lyon, H.H. (2005) *Diseases of Trees and Shrubs*, 2nd edn. Cornell University Press, Ithaca, New York.

Sotirovski, K., Papazova-Anakieva, I., Grünwald, N.J. and Milgroom, M.G. (2004) Low diversity of vegetative compatibility types and mating type of *Cryphonectria parasitica* in the southern Balkans. *Plant Pathology* 53, 325–333.

Stevens, N.E. (1917) Some factors influencing the prevalence of *Endothia gyrosa. Bulletin of the Torrey Botanical Club* 44, 127–144.

Stipes, R.J. and Phipps, P.M. (1971) A species of *Endothia* associated with a canker disease of pin oak (*Quercus palustris*) in Virginia. *Plant Disease Reporter* 55, 467–469.

Sturrock, R.N., Frankel, S.J., Brown, A.V., Hennon, P.E., Kliejunas, J.T., Lewis, K.J., Worrall, J.J. and Woods, A.J. (2011) Climate change and forest diseases. *Plant Pathology* 60, 133–149.

van der Merwe, N.A., Gryzenhout, M., Steenkamp, E.T., Wingfield, B.D. and Wingfield, M.J. (2010) Multigene phylogenetic and population differentiation data confirm the existence of a cryptic species within *Chrysoporthe cubensis. Fungal Biology* 114, 966–979.

van der Merwe, N.A., Steenkamp, E.T., Rodas, C., Wingfield, B.D. and Wingfield, M.J. (2012) Host switching between the native and non-native trees in a population of the canker pathogen *Chrysoporthe cubensis* from Colombia. *Plant Pathology*. In press: doi: 10.1111/j.1365-3059.2012.02657.x.

Van der Westhuizen, I.P., Wingfield, M.J., Kemp, G.H.J. and Swart, W.J. (1993) First report of the canker pathogen *Endothia gyrosa* on *Eucalyptus* in South Africa. *Plant Pathology* 42, 661–663.

van Heerden, S.W. (2004) Studies on *Cryphonectria cubensis* in South Africa with special reference to mycovirus infection: Chapter 2, Molecular characterisation of mitoviruses co-infecting South African isolates of the *Eucalyptus* canker pathogen *Cryphonectria cubensis*. PhD thesis, Faculty of Natural and Agricultural Science, University of Pretoria, Pretoria, South Africa, pp. 33–76. Available at: http://upetd.up.ac.za/thesis/submitted/etd-08132008-115331/unrestricted/thesis.pdf (accessed 10 December 2012).

van Heerden, S.W. and Wingfield, M.J. (2001) Genetic diversity of *Cryphonectria cubensis* isolates in South Africa. *Mycological Research* 105, 94–99.

van Heerden, S.W. and Wingfield, M.J. (2002) Effect of environment on the response of *Eucalyptus* clones to inoculation with *Cryphonectria cubensis*. *Forest Pathology* 32, 395–402.

van Heerden, S.W., Wingfield, M.J., Coutinho, T., van Zyl, L.M. and Wright, J.A. (1997) Diversity of *Cryphonectria cubensis* isolates in Venezuela and Indonesia. In: *Proceedings of IUFRO Conference on Silviculture and Improvement of Eucalypts, Salvador Bahia, Brazil, 24–29 August 1997, Volume 2*. Embrapa, Colombo, Brazil, pp. 142–146.

van Heerden, S.W., Geletka, L.M., Preisig, O., Nuss, D.L., Wingfield, B.D. and Wingfield, M.J. (2001) Characterization of South African *Cryphonectria cubensis* isolates infected with a *C. parasitica* hypovirus. *Phytopathology* 91, 628–632.

van Heerden, S.W., Amerson, H.V., Preisig, O., Wingfield, B.D. and Wingfield, M.J. (2005) Relative pathogenicity of *Cryphonectria cubensis* on *Eucalyptus* clones differing in their resistance to *C. cubensis. Plant Disease* 89, 659–662.

van Staden, V., Erasmus, B.F.N., Roux, J., Wingfield, M.J. and van Jaarsveld, A.S. (2004) Modelling the spatial distribution of two important South African plantation forestry pathogens. *Forest Ecology and Management* 187, 61–73.

van Zyl, L.M., Wingfield, M.J., Alfenas, A.C. and Crous, P.W. (1998) Population diversity among Brazilian isolates of *Cryphonectria cubensis*. *Forest Ecology and Management* 112, 41–47.

van Zyl, L.M., Wingfield, M.J., Alfenas, A.C. and Crous, P.W. (1999) Hypovirulence detected in Brazilian isolates of *Cryphonectria cubensis*. *Plant Pathology* 48, 267–272.

Venter, M., Wingfield, M.J., Coutinho, T.A. and Wingfield, B.D. (2001) Molecular characterization of *Endothia gyrosa* isolates from *Eucalyptus* in South Africa and Australia. *Plant Pathology* 50, 211–217.

Vermeulen, M., Gryzenhout, M., Wingfield, M.J. and Roux, J. (2011) New records of *Cryphonectriaceae* from southern Africa including *Latruncellus aurorae* gen. sp. nov. *Mycologia* 103, 554–569.

Vermeulen, M., Gryzenhout, M., Wingfield, M.J. and Roux, J. (2012a) Species delineation in the tree pathogen genus *Celoporthe* (*Cryphonectriaceae*) in southern Africa. *Mycologia*. In press: doi: 10.3852/12-206.

Vermeulen, M., Gryzenhout, M., Wingfield, M.J. and Roux, J. (2012b) Population structure of *Chrysoporthe austroafricana* in southern Africa determined using vegetative compatibility groups (VCGs). *Forest Pathology*. In press: doi: 10.1111/efp.12006.

Vizioli, J. (1923) Some *Pyrenomycetes* of Bermuda. *Mycologia* 15, 107–119.

Walker, J., Old, K.M. and Murray, D.I.L. (1985) *Endothia gyrosa* on *Eucalyptus* in Australia with notes on some other species of *Endothia* and *Cryphonectria*. *Mycotaxon* 23, 353–370.

Wardlaw, T.J. (1999) *Endothia gyrosa* associated with severe stem cankers on plantation grown *Eucalyptus nitens* in Tasmania, Australia. *European Journal of Forest Pathology* 29, 199–208.

Weir, J.R. (1925) Notes on the parasitism of *Endothia gyrosa* (Schw.) Fr. *Phytopathology* 15, 489–491.
White, D.A. and Kile, G.A. (1993) Discolouration and decay from artificial wounds in 20-year-old *Eucalyptus regnans*. *European Journal of Forest Pathology* 23, 431–440.
Wingfield, M.J. (1990) Current status and future prospects for forest pathology in South Africa. *South African Journal of Science* 86, 60–62.
Wingfield, M.J. (2003) Daniel McAlpine Memorial Lecture. Increasing threat of diseases to exotic plantation forests in the southern hemisphere: lessons from Cryphonectria canker. *Australasian Plant Pathology* 32, 133–139.
Wingfield, M.J. and Roux, J. (2002) Plantation diseases of South African forest plantations. In: Own, D.L. (ed.) *South African Forestry Handbook 1*. The South African Forestry Institute, Pretoria, South Africa, pp. 241–252.
Wingfield, M.J., Swart, W.J. and Abear, B. (1989) First record of Cryphonectria canker of *Eucalyptus* in South Africa. *Phytophylactica* 21, 311–313.
Wingfield, M.J., Rodas, C., Myburg, H., Venter, M., Wright, J. and Wingfield, B.D. (2001a) First report of Cryphonectria canker on *Tibouchina* in Colombia. *Forest Pathology* 31, 297–306.
Wingfield, M.J., Slippers, B., Roux, J., and Wingfield, B.D. (2001b) Worldwide movement of exotic forest fungi, especially in the tropics and the southern Hemisphere. *BioScience* 51, 134–140.
Wingfield, M.J., Roux, J., Coutinho, T., Govender, P. and Wingfield, B.D. (2001c) Plantation disease and pest management in the next century. *South African Forestry Journal* 190, 67–71.
Yuan, Z.Q. and Mohammed, C. (1997) Investigation of fungi associated with stem cankers of eucalypts in Tasmania, Australia. *Australasian Plant Pathology* 26, 78–84.
Yuan, Z.Q. and Mohammed, C. (1999) Pathogenicity of fungi associated with stem cankers of eucalypts in Tasmania, Australia. *Plant Disease* 83, 1063–1069.
Yuan, Z.Q. and Mohammed, C. (2000) The pathogenicity of isolates of *Endothia gyrosa* to *Eucalyptus nitens* and *E. globulus*. *Australasian Plant Pathology* 29, 29–35.

17 Cypress Canker

Roberto Danti,* Gianni Della Rocca and Alberto Panconesi
Plant Protection Institute, National Research Council (CNR), Florence, Italy

17.1 Pathogens, Significance and Distribution

Bark canker of cypress (*Cupressus* and related genera) is a pandemic disease due to the worldwide spread of the fungal pathogen *Seiridium cardinale* (W.W. Wagener) B. Sutton & I.A.S. Gibson and of two other similar species, *S. cupressi* (Guba) Boesew. (teleomorph: *Lepteutypa cupressi* (Nattrass, C. Booth & B. Sutton) H.J. Swart) and *S. unicorne* (Cooke & Ellis) B. Sutton. *S. cardinale* is the most widespread and aggressive of the three species and has caused epidemics in many regions into which it was introduced, especially during the second half of the 19th century (Panconesi, 1990; Graniti, 1998).

Several other pathogenic fungi are known to be responsible for cankers on cypress as well. Among these, *Diplodia cupressi* A.J.L. Phillips & A. Alves, *Kabatina thujae* R. Schneid. & Arx, *Pestalotiopsis funerea* (Desm.) Steyaert, *Phomopsis occulta* (Sacc.) Traverso and *P. juniperivora* G. Hahn are the most common. This chapter, however, will take into consideration only Seiridium canker of cypress because of its spread at a global level and the serious damage and economic losses that it causes to many *Cupressaceae* in forests, windbreaks and ornamental plantings.

Cypress has suffered attacks of bark canker for more than 80 years. The first epidemic of *S. cardinale* was reported in California in 1928 on Monterey cypress (*Hesperocyparis macrocarpa* (Hartw.) Bartel), which over a period of only a few years was completely destroyed in the plantations located in inland districts (Wagener, 1939). Monterey cypress, which is extremely susceptible to canker and widely traded for ornamental purposes, had a major role in spreading the disease to other host species worldwide. In the course of the following decades, the disease spread progressively over the five continents: to New Zealand, France and Chile; to Italy, Argentina and Greece (Birch, 1933; Barthelet and Vinot, 1944; Grasso, 1951; Saravi Cisneros, 1953; Anastassiadis, 1963; Mujica *et al.*, 1980) and, subsequently, to the entire Mediterranean basin and to other countries in central and northern Europe; and to Canada, north and South Africa and Australia (Torres, 1969; Strouts, 1970; Sutton and Gibson, 1972; Funk, 1974; Caetano, 1980; Solel *et al.*, 1983; Wingfield and Swart, 1988; Danti *et al.*, 2009). *S. cardinale* canker has been reported on species of *Cupressus*, *Hesperocyparis*, *Chamaecyparis*, *Cryptomeria*, *Juniperus*, *Thuja*, *Callitropsis* ×*leylandii* (A.B. Jacks. & Dallim.) D.P. Little

* E-mail: danti@ipp.cnr.it

(*nootkatensis* × *macrocarpa*) (also known as Leyland cypress and ×*Cupressocyparis*) and *Callitropsis*.

The impact caused by epidemics of *S. cardinale* has been particularly serious in the Mediterranean area, where disease incidence in the 1980s reached 70% and 90% in some wooded and ornamental plantations of central Italy and Greece, respectively (Xenopoulos and Diamandis, 1985; Panconesi and Raddi, 1991). In this area, the Italian cypress (*Cupressus sempervirens* L.) has been grown as a domestic and multi-purpose tree since antiquity, mainly for its ornamental and symbolic role, but additionally for the production of timber, for reforestation and for barriers and hedges. In several regions, e.g. in Tuscany, where Italian cypress is a typical feature of the landscape and an important element of gardens and formations of artistic, cultural and historic interest, *S. cardinale* has represented a serious threat. Even today, the cypress canker represents a factor in the deterioration of cypress, thus limiting its planting and cultivation both in the Mediterranean basin and many other regions of the world, especially in areas where the climatic conditions are particularly favourable to the pathogen.

S. cupressi was reported for the first time in Kenya in 1942 (Wimbush, 1944) in badly damaged plantations of Monterey cypress and cedar-of-Goa (*Hesperocyparis lusitanica* (Mill.) Bartel). The current distribution area includes east and South Africa, Australia (Cunnington, 2007) and New Zealand (Chou, 1989). The host genera, *Cupressus*, *Hesperocyparis*, *Callitropsis* ×*leylandii*, *Chaemacyparis* and *Juniperus*, belong to the *Cupressaceae*. In 1984, the fungus was also reported on a few Italian cypress trees on the island of Kos in Greece (Graniti, 1986).

The geographical distribution of *S. unicorne* seems to include North America, New Zealand, Australia, Japan and Europe (Fuller and Newhook, 1954; Sutton, 1980; Van der Werff, 1984; Chou, 1989; Caetano *et al.*, 1991; Tabata, 1991; Tisserat *et al.*, 1991; Della Rocca *et al.*, 2011b), and its reported host range includes eight different families. According to Chou (1990), *S. unicorne* is unable to cause infection on species that do not belong to the *Cupressaceae* and many reports of the fungus on hosts of other families that have been made in the past may have been misidentifications of other similar fungi. New Zealand is the only country where the three *Seiridium* spp. are known to occur simultaneously (Boesewinkel, 1983).

17.2 Diagnosis

The diagnosis of Seiridium canker of cypress is based on observation of symptoms on affected trees, on the morphological and cultural characteristics of fungi, and on the use of molecular techniques based on analysis of specific genomic sequences. The symptoms on trees produced by the three *Seiridium* spp. are very similar, though *S. unicorne* generally gives rise to less virulent cankers and to less severe damage than do the other two species.

The first evidence of infection produced by *S. cardinale* is the appearance on the bark of a reddish brown necrotic spot around the entry point of the pathogen, often associated with resin emission. The necrosis spreads out progressively and becomes fusiform and slightly depressed. Longitudinal cracks subsequently appear, from which resin flows more or less abundantly, depending on the genotype and on the reactivity of the host (Plate 21). In young plants, the canker may even girdle the stem or branches in a relatively short time, whereas in adult trees canker development requires more time and, in addition, depends on the individual reaction of the host. Large cankered organs often appear to be greatly deformed and depressed, with bark cracks and resin emission as a result of the action of the pathogen and the host's reaction. When a canker has girdled a branch or a stem, the portion above the canker dies; its foliage turns first pale green, then yellow, brownish red and, lastly, ash grey. Within a few months, the foliage falls off and the affected branch (or stem) becomes completely bare. Both early symptoms (yellow crowns) and advanced development of symptoms (bare branches) can frequently be observed on the same tree (Plate 22).

Infections can occur in any part of the crown and the trunk, but are generally more frequent on the outer branches and twigs, which are more exposed to the fungal inoculum and also have a less thickened periderm. A useful diagnostic trait is represented by the acervuli, the asexual fruiting bodies of the fungus that develop on the surface of the necrotic bark and on cones during the spring and autumn. Acervuli appear as black sub-epidermal or sub-peridermal pustules (0.2–3 mm in size) that open, when mature, with the breakage of the upper tissues. However, similar acervuli are produced by other fungal pathogens of cypress as well, e.g. *P. funerea*.

In the genus *Seiridium*, the conidia (the spores produced by the acervuli) are five euseptate, fusiform, and with four brown median cells and conic, hyaline terminal cells. The main distinctive trait among the three species of *Seiridium* causing canker of cypress trees is the morphology of the terminal (apical and basal) cells of the conidia. *S. cardinale* has very short appendages (1 μm long), sometimes absent, whereas *S. cupressi* and *S. unicorne* have appendages up to 13 μm in length (Fig. 17.1), although in *S. cupressi* they generally follow the curve of the conidia, and in *S. unicorne* the apical appendage frequently forms a right angle with the axis of the conidium (Guba, 1961; Boesewinkel, 1983). Sutton (1980) and Chou (1989) considered the orientation of the conidial appendages a too variable trait, and treated *S. cupressi* and *S. unicorne* as a single species.

Some cultural differences have been pointed out among the three species in colony colour and morphology (off-white with grey-green shades and with a salmon orange reverse in *S. cardinale*; salmon or apricot with grey and off-white areas in *S. cupressi*; olive green or dark green in *S. unicorne*) and in radial growth, which, measured after 6 days at 26°C, was higher in *S. cardinale* (31 mm diameter) than in *S. cupressi* (13 mm diameter) and *S. unicorne* (18 mm diameter) (Boesewinkel, 1983). Graniti (1986) also observed that in *S. cupressi* the acervuli are formed close to the margin of the cankered area and are slightly larger (up to 2 mm) than in *S. cardinale* (up to 1.5 mm), which forms acervuli over the entire surface of the canker.

Molecular studies based on the sequencing of the genes coding for structural proteins (histones and β-tubulins) have supported the separation into three distinct species of the fungi responsible for cypress canker (Barnes *et al.*, 2001). Analysis of sequence data showed that *Seiridium* isolates resided in two major clades: one accommodated *S. unicorne* and the other consisted in two sub-clades, which included the other two species separately. Existence of three species was confirmed by the restriction fragment length polymorphism (RFLP) technique, based on the digestion of β-tubulin amplicons with a single restrictive

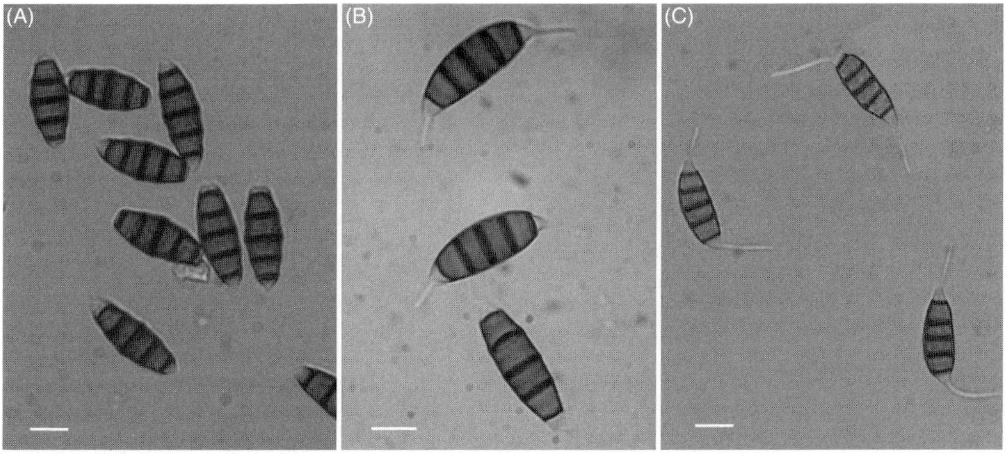

Fig. 17.1. Conidia of *Seiridium* spp. observed under a light microscope. (A) *Seiridium cardinale* (W.W. Wagener) B. Sutton & I.A.S. Gibson, size 21–30 × 8–10 μm; (B) *S. cupressi* (Guba) Boesew., size 27–32 × 7.5–10 μm; and (C) *S. unicorne* (Cooke & Ellis) B. Sutton, size 25–27 × 7.5–10 μm. Bars, 10 μm.

endonuclease (Krokene *et al.*, 2004). Studies on the *in vitro* production of toxins seem to support the separation of the fungi affecting cypress into three distinct species too. The differentiated production of toxins exhibited by the three fungi suggests the possible use of this chemotaxonomic means for their identification. It should be noted here that as *S. cupressi* has been often misidentified as *S. unicorne* on the basis of morphological traits (Graniti, 1986; Cunnington, 2007), the real nature of epidemics attributed to *S. unicorne* in some countries should be verified by the use of suitable diagnostic techniques.

17.3 Infection Biology

The biology, epidemiology and pathogenicity of *S. cardinale* have been widely studied, whereas relatively few studies have involved *S. cupressi* and *S. unicorne*.

The occurrence of *S. cardinale* infections is favoured by the presence of small wounds in the periderm of stems and branches through which the conidia or mycelium enter the inner bark. Although it acts mainly as a wound pathogen, the fungus can also reach the living bark through natural openings, such as stomata and lenticels, when the temperature and relative humidity (RH) are optimal for the germination of conidia and mycelial growth (25°C and 100% RH, respectively) (Panconesi, 1990). Young trunks and branches are particularly exposed to infections, because they are more greatly subjected to bark injuries due to cold, hail, or to forced growth by fertilizers, insects, rodents, etc.

During milder and more humid periods, such as the spring and autumn, the acervuli develop on the surface of the necrotic bark and on maturing cones, which are easily colonized by the fungus (Fig. 17.2). The conidia are disseminated by rainwater that flows downwards on to the surface of stems and branches, by windborne raindrops, by insects, and presumably by birds and small mammals, and can give rise to new infections on the same trees or on nearby trees. The progressive occurrence and development of a series of infections on a tree can cause its death in a relatively short time, depending on the age, susceptibility and environmental conditions.

In addition it has been observed that the fungus can remain viable for several years after the death of the infected tissues, and can give rise to subsequent sporulations in previously formed acervular craters, thereby contributing to the maintenance of a high amount of inoculum. Forms in which the fungus has been conserved have been observed in dead tissues, and include resting mycelium, chlamydospores and sclerotia (Panconesi, 1990).

Well-known vectors of the pathogen are bark beetles of the genus *Phloeosinus* (*P. aubei* (Perris), *P. thujae* (Perris) and *P. armatus* Reitter). During their life cycles, these insects can carry the inoculum of the fungus from the cankered trees, where the insects reproduce, to healthy trees, where they feed, thus contributing to the spread of the pathogen. Other cone-spermophagous insects, e.g. *Orsillus maculatus* (Fieber) and *Cydia cupressana* Kearfott, are known to carry the conidia of the fungus even over long distances. A potential role in the spread of *S. cardinale* is likely for some other cone insects too (Roques and Battisti, 1999).

The conidia of *S. cardinale* germinate at temperatures ranging from 5°C to 35°C, with an optimum at 25°C. Germination of the conidia is greatly influenced by the RH: when the RH drops from 100% to 80%, the percentage of germinated conidia decreases drastically (from 98% to 60%), and germination is completely prevented when the RH drops below 50% (Panconesi, 1990). *S. cardinale* is more thermophilic than either *S. cupressi* or *S. unicorne*, and can cause infections at a temperature up to 30°C. In contrast, *S. cupressi* shows a maximum pathogenicity between 20°C and 25°C, while infections develop more slowly at 30°C, whereas for *S. unicorne in vitro* germination and growth are higher at 20°C, and infective capacity is maximum between 25°C and 30°C (Graniti, 1998).

Based on the results of artificial inoculations on different host species, *S. cardinale* is more aggressive than the other two species under Mediterranean climatic conditions. However, the bark necrosis caused by *S. cupressi* grows constantly during the summer, whereas

Fig. 17.2. (A) Acervuli of *Seiridium cardinale* (W.W. Wagener) B. Sutton & I.A.S. Gibson developed on the surface of a cankered stem; (B) section of a sub-epidermal erumpent acervulum of *S. cardinale* observed by light microscopy. Bar, 30 μm.

the development of the *S. cardinale* necrosis slows down during the warmest months. The necrotrophic action of *S. cupressi* thus takes advantage of the host summer dormancy. In the same Mediterranean climate, *S. unicorne* has shown a reduced virulence compared with the other two *Seiridium* spp. (Ponchet *et al.*, 1990; Xenopoulos, 1991).

S. cardinale shows its necrotrophic action in the bark tissues located between the outer xylem and the outer bark, which become brown-purplish in colour (hence the name *S. cardinale*) and also resin soaked. Based on ultrastructural observations, the hyphae of the fungus are concentrated particularly in the phloem, in the parenchyma, in the cambium and in the medullary rays, including the fibres (Mutto Accordi and Panconesi, 1987). The hyphae develop mostly in the intercellular spaces, colonizing and degrading the cell walls through the production of cell wall degrading enzymes (CWDEs), such as cellulase, xylanase and polygalacturonase (Magro *et al.*, 1982).

The longitudinal growth of the fungus is favoured by the structure of the sieve plates of the phloem through which the hyphae are able to pass easily. Wagener (1939), Moriondo (1972), Grasso *et al.* (1979) and Solel *et al.* (1983) described resin exudation and the formation of resin ducts as a result of the infection process. In the genus *Cupressus*, and in related genera, no constitutive resin ducts are present in the xylem and the phloem but the local formation of resin ducts is induced as a defence response in the phloem of the Italian cypress as a result of *S. cardinale* infections or mechanical wounds (Moriondo, 1972).

The death of the distal portion of a cankered twig, branch or stem generally occurs when the necrotic process has girdled the bark of an affected organ. It has been observed that this is due to the occlusion of the pit membranes of the tracheids by metabolites produced by the fungus and by the response of the tree to the pathogen, which blocks the sap flow in the xylem vessels. The wilting cannot be attributed to the sparse hyphal colonization of the sapwood, or to the occasional accumulation of tannins, or to tylosis or resin production, which have not been observed in the xylem (Madar *et al.*, 1990). Even in dead trees or seriously affected trees, the wood is not decayed by the fungus and the timber is marketable.

The necrosis of the host cells that occurs before colonization by the fungus, and the development of foliar symptoms on branches and twigs that are not yet girdled, suggest that toxins produced by the fungus take part in the pathogenesis, and that these toxins diffuse to the adjacent tissues and possibly move to the leaves with transpiration. Nine non-host-specific toxins produced *in vitro* have been purified and characterized (Graniti, 1998). Each of the three *Seiridium* spp. produces, *in vitro*, at least one major toxin plus other secondary toxins. *S. cardinale* mainly produces seiridin and isoseiridin. *S. cupressi* mainly produces cyclopaldic acid and, secondarily, sericuprolide, in addition to small amounts of seiridin, isoseiridin and seiricardin. *S. unicorne* generally shows a lower production of toxins, among which are seiridin, isoseiridin and seiricardin-c. Among the three species, *S. cupressi* is able to produce a wider range of toxins (Graniti, 1998). In several bioassays, the application of seiridin and cyclopaldic acid induced symptoms that were similar to those observed in nature, and were more severe than symptoms induced by other toxins (Sparapano and Evidente, 1995).

Inoculations with toxin-deficient mutants of *S. cupressi* produced a milder symptomatology in leaves and branches than did inoculations of wild isolates, even if the development of cankers remained unchanged. The virulence of *S. cardinale* isolates was correlated with the amount of toxins (seiridin, isoseiridin) produced *in vitro*. Most isolates of *S. cardinale* produced both seiridin and isoseiridin, but in variable proportions, and 5% of the isolates did not produce any toxin, whereas other isolates produced either only seiridin or only isoseiridin (Sparapano and Evidente, 1995; Graniti, 1998). Toxins may, therefore, be virulence factors for the three species of *Seiridium* that are pathogenic on cypresses. This hypothesis finds confirmation in the fact that *S. unicorne*, the least aggressive species of the three, which is unable to produce cyclopaldic acid or any significant amount of seiridin, causes cankers in trunks and branches, but does not induce systemic symptoms on the crown as do the other two species (Graniti, 1998).

17.4 Epidemiology

For almost 80 years, cypress has suffered from destructive bark canker epidemics that have arisen as a result of the introduction of the various causal pathogens to different regions of the world. Epidemics were favoured by climatic conditions suited to the pathogens and by the density and continuity of a susceptible host over a vast area.

The Mediterranean climate, which is characterized by a marked thermal range between summer and winter, and by mild, rainy and humid seasons (usually the spring and autumn), is an essential requirement for epidemic development of *S. cardinale* canker. The humid and rainy seasons favour the sporulation and germination of the conidia, which can retain their germinability and pathogenicity for several months, even under unfavourable conditions (Panconesi *et al.*, 1993). In the northern part of the Mediterranean, cypress trees are predisposed to infections by the winter cold and by frequent late frost, which favour the occurrence of bark lesions. Generally, a clear increase of new infections can be observed during the course of particularly humid and rainy springs that have been preceded by very cold winters. As both the sporulation and germination of *S. cardinale* conidia take place within a temperature range of 5°C to 30°C, infections in the Mediterranean environment may occur for most of the year, if RH conditions are suitable.

The lack of one or more favourable factors can prevent the spread of the fungus. Microclimatic conditions may play a major role in the development of the disease and explain the presence of nuclei of the host that have escaped the disease not far from areas that have been severely affected.

In Greece, the severity of the *S. cardinale* attacks that have occurred over recent decades seems to be closely connected with the climatic conditions in the different districts of the country. Higher incidences, from 25% to 50%, have been observed along the western coast, which is exposed to more humid winds, whereas along the drier eastern coast, cypress canker has occurred only sporadically. The most serious attacks were reported in several isolated sites with a particularly favourable microclimatic conditions, such as the small valleys of Megalopolis (in the Peloponnese) and Karystos (on the island of Eubea), where incidences of 80% and 95% were recorded, respectively, at the end of the 1980s. At Karystos, 80% of the infected trees died (Xenopolous and Diamandis, 1985).

In Tuscany (Italy), the first epidemics of *S. cardinale* canker developed in areas characterized by greater humidity, such as the plain of Florence. Cypress canker, which was reported here in 1951, spread rapidly as a typical polyetic disease (i.e. the disease occurred over several growing seasons, with the inoculum produced in one season carried over to the next, leading to a build-up of an inoculum over the years) during the following decades; by the end of the 1980s, it was in many forests and plantations, with more than 50% of the trees affected, and peaks of disease occurrence of 70–80% (Panconesi and Raddi, 1991). In the other provinces, the disease incidence increased subsequently. According to the regional monitoring network of forest health (META Service), the mean incidence in the region for the 2006–2009 period was 22–23%, with marked differences between the various provincial districts due to the variation in the microclimatic conditions in a very heterogeneous territory, and also to the effect of the scattered sanitation interventions that have been carried out so far.

In Italy as a whole, cypress canker is reported throughout the territory (Panconesi, 1990), even if in most regions the sporadic presence of cypresses has contributed to limiting its spread. The incidence of the disease rose only in districts where the host population is more dense and climatic conditions are suitable for the pathogen, such as around the Garda Lake (from 14% to 17%), in the province of Trento (from 9% to 53%) (Santini *et al.*, 1997a) and in Sicily, where scattered peaks exceeding 50% were reported in plantations of Monterey cypress (Sidoti *et al.*, 2007). Currently, serious damage is frequently recorded in north and central Italy on Leyland cypress, which was widely planted in barriers during the last two decades because of its rapid growth, but is very susceptible to *S. cardinale* canker.

The spread and severity of attacks of canker has been maintained rather low in the warmer and drier regions of north Africa, where the disease has also been reported (Panconesi, 1990; Danti *et al.*, 2009).

The absence of epidemics involving native populations of Italian cypress can be attributed to geographic and climatic barriers that limit the spread of the pathogen. The native populations of Italian cypress are mostly residual, often located in areas with limiting pedo-climatic conditions, e.g. in Crete. In these environments, *S. cardinale* presumably has less adaptive capacity than its host, which slows down its development, thus causing only sporadic and limited infections. In addition, it has been observed that in natural areas of Italian cypress, the number of *S. cardinale* conidia vectored by insects has markedly decreased (Santini and Di Lonardo, 2000).

Similarly, in California, *S. cardinale* canker heavily damaged the inland plantations of Monterey cypress, whereas the coastal native populations remained mostly unaffected (Wagener, 1939). In Monterey cypress, the absence of the disease in native populations does not seem to be due to a form of genetic resistance, given the marked susceptibility shown by this species to artificial inoculations with *S. cardinale* (Grasso *et al.*, 1979; Raddi, 1979). In its restricted native range along the coast of the Monterey Bay, Monterey cypress is subjected to the influence of the cold ocean winds that even during the summer prevent the temperatures from rising to values that are more favourable to fungal growth and sporulation. A few kilometres inland though, the climatic conditions change drastically and become similar to those typical of the Mediterranean, with marked differences between summer and winter, and with humid and rainy seasons, so that the epidemic spread of the pathogen is favoured.

Climatic conditions favourable to the pathogen have also been found in the state of Victoria in south-eastern Australia (Purnell, 1960; Swart, 1973), where Monterey cypress has been severely damaged, so that its use in new plantations has been discouraged. Here, cypress canker seems to be caused especially by *S. cupressi*, and to a lesser extent by *S. cardinale* and *S. unicorne* (Cunnington, 2007). In New Zealand, where the disease was reported in 1933 (Birch, 1933), Seiridium canker is now widespread throughout the territory on various *Cupressaceae*, with the greatest incidence of the disease observed in the northern part of the country, around Auckland, where the temperatures are warmer (Van der Werff, 1984).

17.5 Management Strategies and Tactics

17.5.1 Avoidance

S. cardinale is a primary pathogen that has exhibited greater aggressiveness on plants growing fast in non-limiting edaphic conditions, e.g. on fertile and deep soils (Panconesi, 1990). Cultural measures aimed at modifying the environmental conditions favourable to the occurrence of new infections and to the development of the disease may include: (i) locating nurseries or new plantations in very sunny and airy sites, far from diseased adult cypresses and other inoculum sources of the pathogen; (ii) avoiding valley-bottom zones that are subject to stagnation of humidity and thermal inversion; (iii) not making overly dense sowings/plantings; and (iv) maintaining sufficiently large spaces between the plants, both between the rows and along the rows, in order to facilitate air flow among the crowns.

17.5.2 Exclusion

Today, globalization (transport, trade, tourism, etc.) and the opening up of new borders (as in the European Union) increase the risk of introduction of new plant diseases. The introduction to new countries and regions of cypress plants or other propagation materials may be a cause of the spread of bark canker if they are not subjected to sanitary inspections.

S. cardinale has now attained a pandemic distribution and has been reported in many temperate regions that are suitable for the cultivation of *Cupressaceae*. Phytosanitary measures may be useful for preventing further spread of the pathogen in territories that are potentially favourable to its development, e.g. Japan, where until now only the presence of *S. unicorne* (*sensu* Sutton) has been reported (Tabata, 1991).

Furthermore, the introduction of new isolates characterized by greater virulence into a given region may favour the outbreak of new epidemics, to the detriment of plants that have escaped the previous waves of the disease. The introduction of isolates of a complementary mating type can allow sexual reproduction to occur in the pathogen population, thereby promoting a greater variability in the fungus and conferring on it a greater evolutionary potential in terms of virulence. The *S. cardinale* teleomorph has been reported only once in California by Hansen (1956), as *Leptosphaeria* sp., although it was not officially recognized because it was incompletely described. Therefore, studying the variability, reproduction and distribution of the pathogen may help to prevent both its introduction into new regions and the introduction of new genotypes into regions where the pathogen is already present.

S. cupressi is widespread in different continents, but at present it has not been reported in Europe. The pathogen was found in the 1980s on Kos (Greece) on several cankered Italian cypresses (Graniti, 1986), which were promptly felled in order to prevent the disease from spreading further. The possible introduction of *S. cupressi* into the Mediterranean area, which has been already devastated by the epidemic of *S. cardinale*, would represent a serious danger for cypress trees, and would also complicate the genetic improvement of cypress for resistance to *S. cardinale* canker that has been developed over the past three decades. As has already been suggested by Graniti (1986), *S. cupressi* should be included in the European and Mediterranean Plant Protection Organisation (EPPO) A1 list in order to adopt all the sanitary measures necessary for preventing its introduction into the European countries and its possible spread. The threat represented by the possible introduction of *S. cupressi* should be seriously considered both because of the virulence exhibited by the fungus in other regions, and because of the theoretical possibility that the onset of new epidemics may be favoured by the spread of ascospores produced by the teleomorph and their distribution by the wind (Moricca *et al.*, 2000).

S. unicorne is less aggressive than the other two species of *Seiridium*, and has never been associated with epidemics. Compared with the other two species, the possible introduction of *S. unicorne* into new regions is less worrisome.

17.5.3 Eradication

Sanitation is the main method for controlling cypress canker, and this is aimed at eliminating or reducing the inoculum sources of the pathogen in a territory, or in more or less restricted areas – in forest stands, in ornamental plantations, or in single trees. This is an extinction control strategy that involves the mechanical removal, by means of the felling and/or pruning of the infected trees or parts of the crown, and providing for all resulting infected material to be destroyed by burning.

In the forest stands, all cypress trees affected by the disease should be felled, because cutting off the infected organs would involve obvious operational difficulties in entering such locations with the proper vehicles, e.g. elevated platforms. The procedure is carried out in the same way in the nursery owing to the need to eliminate definitively every source of infection. In adult plantations, where it is possible to use an elevated platform, and particularly in plantations that have an ornamental or historic role, or are monuments (rows, groups, hedges, single trees), the procedure is different, with all cypresses compromised by canker being felled, while crowns that are affected to a limited extent are subjected to pruning. This is done in an attempt to recover some of the affected trees.

By reducing the inoculum of the pathogen in the environment, sanitation makes it possible to attain a series of objectives: (i) reduce pathogen spread and control the epidemic in a given area; (ii) safeguard the cypresses from deterioration due to the uncontrolled action of bark canker; (iii) recover diseased trees through a well-performed pruning, which currently is the only therapeutic method available; (iv) increase the genetic value of seed stands through the elimination of susceptible trees; and (v) reduce the populations of bark beetles, which are important vectors of the pathogen.

The effectiveness of sanitation is linked to the prompt, careful and swift execution of cuttings/fellings, as well as to the extent of the area that it is essential to treat, with the overall aim of reducing the amount of inoculum in the environment as much as possible, and thereby limiting the destructive potential of the disease. In most cases, i.e. when confined to the peripheral portions of the crown, infections are easy to control with a prompt pruning of the affected organs. Interventions will become gradually more difficult and will have more uncertain results when the disease has involved larger branches and the stem.

In California, as a result of the epidemic of *S. cardinale* in the extensive plantations of Monterey cypress, it was possible to adopt drastic measures for eradicating the inoculum through the elimination of all infected trees and by discouraging the use of susceptible hosts in sensitive areas of the inland districts (Wagener, 1939). In Italy, in contrast, the use of such drastic measures would have radically changed the features of the landscape by removing one of its main characterizing elements.

17.5.4 Protection

Chemical control

Trials of chemical treatments for protecting cypresses from *S. cardinale* canker have been promoted basically by the need to support a considerable nursery production of *Cupressaceae*, as this has significant economic weight in the Mediterranean regions (Della Rocca *et al.*, 2011a).

A range of studies was undertaken in the past to evaluate the preventive and curative effectiveness of various fungicides for controlling *S. cardinale* canker. These studies revealed the effectiveness of systemic benzimidazoles and thiophanate-methyl for the preventive control of infections. Good effectiveness was also exhibited by several contact fungicides (i.e. chlorothalonil, dichochlofuanide, mancozeb), whereas limited results were obtained with copper products (Panconesi and Parrini, 1979; Govi and Deserti, 1980; McCain, 1984; Panconesi and Raddi, 1986).

In recent years, the European Directive 2009/128/EC (European Parliament and Council of the European Union, 2009) required a drastic reduction in chemicals used in agriculture, including benzimidazole and many contact fungicides that have been shown to be effective against *S. cardinale* canker. A recent experiment was carried out in order to test several compounds never evaluated before for the protection of cypress against canker (Della Rocca *et al.*, 2011a). This study provided evidence of good results for azoxystrobin, which was as effective as thiophanate-methyl in reducing canker development when sprayed on to plants before test inoculations. Boscalid was found to be able to prevent *in vitro* conidial germination and mycelial growth of *S. cardinale*, and appeared to be a promising contact fungicide for the prevention of cypress canker infections. Azoxystrobin and boscalid are both included in a less hazardous class than thiophanate-methyl, and have fewer risks for the environment and for users.

Chemical control constitutes an irreplaceable method for controlling bark canker of cypress in the nursery, where fertilization, watering and a high plant density may increase the predisposition of plants to *S. cardinale* infections, but it is characterized by important limitations. While it is particularly suitable for young plants in nurseries or in plantations, it can only seldom be proposed for adult trees that have a particular landscape value, or value as a monument or of historic interest; even then, spraying the upper and inner part of the crowns properly is technically difficult and time-consuming. Chemical treatments cannot really be advised for the protection of adult plantations either, especially in urban sites, or in forest stands, owing to the unavoidable dispersal of the sprayed products into the environment and to the difficulties of executing such an exercise.

Biological control

In one of the few studies conducted on this topic to date (Magro *et al.*, 1984), volatile and non-volatile metabolites of *Trichoderma viride* Pers. were capable *in vitro* of inhibiting germination and mycelial growth of *S. cardinale. T. viride*, which is commonly found as an epiphyte in the phylloplane of cypresses, has revealed a potential usefulness for the control of *S. cardinale*. However, no experiment has ever been carried out aimed at verifying the antagonist action of *T. viride in planta*, or at studying an effective form of application.

A recent study showed that the bacterium *Pseudomonas chlororaphis* subsp. *aureofaciens* prevented both the conidial germination and mycelial growth of *S. cardinale in vitro* through the production of phenazine-1-carboxylic acid (Raio *et al.*, 2011). Furthermore, when applied *in planta* before test inoculations, it also prevented the development of cankers on Italian cypress stems. Further, the bacterium was able to colonize the cypress phylloplane of Italian cypress as an epiphyte and appears to be a promising means for the control of cypress canker, to be eventually used within an integrated disease control strategy.

Fertilization

Cultural measures aimed at promoting the vigour of plants may promote the development of cankers. In order to reduce the predisposition of plants to infections, it is advisable not to apply excessive watering and fertilization, especially in the nursery. In particular, the excessive use of nitrogen promotes plant growth, slowing down the lignification of the tissues. The abundance of parenchymal tissues that results predisposes the bark to lesions due to forced growth and to cold, thus favouring the occurrence of infections and their subsequent development.

17.5.5 Resistance

Among the cypress species, a high interspecific variability in response to *S. cardinale* inoculations has been demonstrated. North American species phylogenetically close to Monterey cypress were found to be extremely susceptible, whereas the American species belonging to the *Hesperocyparis arizonica* (Greene) Bartel group were found to be more or less susceptible, as well as Italian cypress. Instead, the Asiatic species (*Cupressus chengiana* S.Y. Hu, *C. duclouxiana* Hickel, *C. funebris* Endl., *C. torulosa* D. Don) and North African species (*Cupressus sempervirens* L. var. *dupreziana* (A. Camus) Silba, *C. sempervirens* L. var. *atlantica* (Gaussen) Silba), in addition to a couple of North American species (*H. glabra* (Sudw.) Bartel and *H. bakeri* (Jeps.) Bartel) showed good resistance to *S. cardinale* (Allemand, 1979; Andréoli and Ponchet, 1991).

Variability of the pathogen

Several trials, based on artificial inoculations of Italian cypress clones with some Mediterranean *S. cardinale* isolates, revealed significant differences among isolates (in terms of canker size) only on tolerant and susceptible clones, but not on resistant clones. No host specificity at species or clone level have ever been revealed for any *S. cardinale* isolate (Raddi and Panconesi, 1984; Ponchet *et al.*, 1990). The absence of the sexual reproduction has contributed to lowering the variability of virulence within the Mediterranean *S. cardinale* population. The substantial evenness of the fungus has favoured the execution of the genetic improvement programme for cypress, and the criteria for selecting resistant genotypes that have been followed for years are currently still effective.

A low variability among *S. cardinale* isolates was also revealed by a molecular approach, but to date most studies have been based on an analysis of conserved DNA regions, which are more suited to the discrimination of taxa at species or subspecies level (the genes for the internal transcribed spacer regions ITS1 and ITS2, 5.8S, histone and β-tubulin) (Viljoen *et al.*, 1993; Moricca *et al.*, 2000; Barnes *et al.*, 2001; Krokene *et al.*, 2004). The use of random amplification of polymorphic DNA (RAPD) did not reveal any informative polymorphism among 77 isolates of *S. cardinale* from northern Italy (Pedron *et al.*, 2007). Recently, eight specific microsatellites have been developed for *S. cardinale* (Della Rocca *et al.*, 2009), and these have been employed for studies on the population genetics of 98 isolates from Europe and California. This has revealed the presence of a number of multi-locus genotypes and a higher genotypic diversity in the California population (Della Rocca *et al.*, 2011c). The absence of linkage disequilibrium calculated for the *S. cardinale* population in California also supported the potential occurrence of sexual reproduction of the fungus in this region. A better understanding of the dynamics of the pathogen might be derived from a comparison of the populations that have settled on the different continents.

The mechanisms of resistance

Inoculation tests have demonstrated that no immunity to *S. cardinale* infections exists within the host populations, and that in resistant species most plants are able to recover completely from the canker by healing the necrotic lesion, whereas in susceptible species this happens only in a few plants. A detailed histological study of the host reaction processes induced as a result of infection with *S. cardinale* was carried out by Ponchet and Andréoli (1990). The tree defence response, based on the formation of necrophylactic periderm (also called neoperiderm), is activated by the host to compartmentalize the damaged tissues, as in the case of any sort of wound, and is not a specific response to *S. cardinale* infection. The vigour of this reaction determines the resistance of a cypress tree to bark canker, because only plants that are able to set up effective barriers and then compartmentalize the infected tissues can then succeed in repairing the lesion. In inoculated plants, the development of necrophylactic periderm is influenced in timing (a delay) and intensity by the pathogen (Madar and Liphschitz, 1989).

In susceptible plants, necrophylactic periderm is not formed, or it is discontinuous and not very thick. The reaction is weak, so compartmentalization is ineffective and does not prevent the mycelium of the fungus from continuing its development in the bark tissues. In resistant plants, necrophylactic periderm is multilayered, thickened and well integrated with the constitutive periderm. The necrosis is thus confined by a solid barrier and the healthy tissues can start a repairing process through the annual production of new layers of phloem and xylem, until the wound is completely healed. After several (usually 4–5) years, the stem has recovered, and only a vertical scar can be seen on the bark.

Between susceptibility and resistance, a range of intermediate responses reveals that the defence reaction of cypress to *S. cardinale* infections is a gradual process. An intermediate response can tend towards susceptibility or resistance, depending also on the environmental conditions; it is defined as tolerance when plants are able to maintain their normal growth rate even when affected by several active cankers. This suggests that the mechanism that governs the reaction of cypress to *S. cardinale* infections is a quantitative trait under polygenic control and that its phenotypic expression is influenced by the genotype and by the environment. So most of the clones showed a change to their response to canker when inoculated in different environments, and the host–pathogen–environment interaction was significant for most of the tested genotypes. Often, a resistant clone may need a longer time to heal the canker completely in more humid and cooler sites than it needs in sites with mild winters and droughty summers (Santini *et al.*, 1997b). It is, therefore, necessary that candidates are evaluated on different sites to identify the clones with the highest level of resistance for each environment.

Patented varieties and other selections

Starting in 1986, a series of canker-resistant clones of Italian cypress were patented by the Plant Protection Institute of the Italian National Research Council (CNR) and put on sale. Several of these clones, such as 'Bolgheri', 'Italico', 'Mediterraneo', 'Le Crete 1' and 'Le Crete 2' (Panconesi and Raddi 1990a,b; Danti *et al.*, 2006), are characterized by a fastigiate and columnar habit and are particularly suitable for ornamental plantings. In particular, 'Le Crete 2' was selected for its tolerance to clayey substrates, and is propagated by rooted cuttings. 'Agrimed n. 1' is particularly suitable for windbreaks, by virtue of its typical crown architecture, whereas 'Florentia' and 'Etruria' are characterized by a wider crown habit, and are suitable for soil protective plantings as well as for hedges. In France, a multi-clonal variety, called 'Mistral', consisting of five canker-resistant clones, has been patented for windbreak use, whereas the 'Sancorey' clone with its fastigiate crown shape is appreciated as an ornamental.

The constant effort of the breeder is to enlarge the genetic base of resistance as much as possible. Today, the wide availability of cypress germplasm selected by IPP-CNR for resistance to *S. cardinale* canker makes it possible to successfully use cypress to cope with a series of current and emerging problems in the Mediterranean countries. In the past decade, a number of experimental field plantings have been set up in various Mediterranean countries for the evaluation of 100 Italian cypress canker-resistant clones for the protection of crops from strong winds, the production of quality timber, the mitigation of fire risks through cypress barriers, and also for combating desertification, protecting soils from erosion and screening buildings and infrastructures.

Through a complex programme of controlled crosses on 1500 mother plants, the (narrow sense) heritability of *S. cardinale* canker resistance estimated for Italian cypress was found to be a little lower than 0.2. Thus, the general combining ability (GCA) of clones was evaluated with the aim of selecting the best parents to be assigned to the large-scale production of improved seed in clonal orchards. The first cypress clonal orchard has recently been realized in Italy, using genotypes selected from the local seed stand.

17.5.6 Therapy

The only possibility of assisting a cankered tree to recover consists of the mechanical removal of the parts affected by the infection. In the crown, this can be done through proper pruning, as previously described. When a canker is located in the lower portion of a large-sized trunk and its width does not exceed 30–40% of the trunk's circumference, it is possible to attempt therapy through a delicate conservation intervention, aimed at removing only the cankered tissues. The intervention consists first of delimiting the borders of the necrotic area by means of appropriate debarking of the outer bark. Subsequently, at a distance of very few centimetres from the border of the necrosis, the healthy bark is cut down to the wood below and the canker is removed with a scalpel. Lastly, a protective mastic is applied to the wound, e.g. vinyl glue, to which a fungicide may be added. This operation, which requires the operators to have a certain technical background, is not easy to perform, especially on trunks with an irregular cross-section or inside dense crowns. It is justified, therefore, only for trees of particular aesthetic value, or with value as monuments.

None of the fungicides evaluated to date has ever shown a therapeutic effect on active growing cankers. Several systemic products, such as benzimidazoles, thiophanate-methyl and recently, azoxystrobin (see under Section 17.5.4, Protection, *Chemical control*), have been shown only to slow down the growth of the canker during to the first colonization phases of the bark by the fungus (Panconesi and Parrini, 1979; Della Rocca *et al.*, 2011a).

17.5.7 Integrated disease management

The control of cypress canker has been based on variable integration of the different methods of disease control, depending on the context in which the trees to be protected are located, e.g. plantations with a landscape or historic value, or a value as monuments, seed stands and other types of wooded formations, plants in gardens, nurseries, etc.

The protection of plantations that have an ornamental and landscape value, or that are a part of historic villas and gardens, is based on the eradication of the inoculum sources of the disease through sanitation. This will also have a therapeutic aim due to the need to recover affected cypresses that are not yet compromised by canker. As part of the proper measures, sanitation should be extended to the adjacent areas as well, so as to create a (sort of) protective belt, and canker-resistant varieties should be used to replace all the trees that have been removed. In addition, for particularly valuable plantations, it should be possible to carry out a selection for resistance in order to use the original selected genotypes for replacements, through the creation of experimental fields assigned to the collection of germplasm. This strategy is currently used for the preservation of cypress trees that are monuments at historic villas, and in the boulevards and gardens of Tuscany. In the case of cypresses in private or public gardens, in addition to sanitation and to the use of resistant varieties, chemical prevention can be adopted to protect young plants during the seasons that are most favourable to the occurrence of infections. The execution of chemical treatments is even easier on hedges that are regularly pruned.

The protection of cypresses in forest stands (either seed stands or protective stands) is based on the removal of all infected trees (sanitation), irrespective of how diseased they are. In nurseries, the control of cypress canker is based on the adoption of cultural practices aimed at reducing the risk of new infections and on chemical prevention by means of spring and autumn treatments, whose frequency will depends on weather conditions. The prompt elimination of symptomatic trees and the sanitation of cypresses situated in the surrounding area will help to eradicate inoculum of the pathogen. For new plantings, the use of varieties selected for resistance is advisable. For ornamental purposes and windbreaks, it is possible to use the patented varieties available on the market that come from Italy and France.

References

Allemand, P. (1979) Relations phylogéniques dans le genre *Cupressus* (*Cupressaceae*) [Phylogenetic relationships in the genus *Cupressus* (*Cupressaceae*)]. In: Grasso, V. and Raddi, P. (eds) *Seminario: Il Cipresso: Malattie e Difesa, 23–24 Settembre. CEE-AGRIMED, Firenze, Italy*. Tipografia l'Artigiano, Firenze, Italy, pp. 51–67.

Anastassiadis, B. (1963) A new for Greece disease of the cypress. *Annales de l'Institut Phytopathologique Benaki (N.S.)* 5, 164–166.

Andréoli, C. and Ponchet, J. (1991) Potential use of exotic cypress species resistant to canker disease. In: Panconesi, A. (ed.) *Il Cipresso: Proposte di Valorizzazione Ambientale e Produttiva nei Paesi Mediterranei della Comunità Economica Europea*. CNR, Regione Toscana, CEE, Firenze, Italy, pp. 150–157.

Barnes, I., Roux, J. and Wingfield, M.J. (2001) Characterization of *Seiridium* spp. associated with cypress canker based on β-tubulin and histone sequences. *Plant Disease* 85, 317–321.

Barthelet, J. and Vinot, M. (1944) Notes sur les maladies des cultures méridionales [Note on the diseases of southern cultures]. *Annales des Epiphyties* 10, 18–20.

Birch, T.T.C. (1933) Gummosis diseases of *Cupressus macrocarpa*. *New Zealand Journal of Forestry* 3, 108–113.

Boesewinkel, H.J. (1983) New records of the tree fungi causing cypress canker in New Zealand, *Seiridium cupressi* (Guba) comb. nov. and *S. cardinale* on *Cupressocyparis* and *S. unicorne* on *Cryptomeria* and *Cupressus*. *Transactions of the British Mycological Society* 80, 544–547.

Caetano, M.F.F. (1980) Uma grave doença das Cupressáceas em Portugal [A serious disease of *Cupressaceae* in Portugal]. *Agros* 63, 5–10.

Caetano, M.F., Ramos, P. and Pinto-Ganhão, J. (1991) The phytosanitary situation of cypress in Portugal and the new prospects. *Il Cipresso: Proposte di Valorizzazione Ambientale e Produttiva nei Paesi Mediterranei della Comunità Economica Europea*. CNR, Regione Toscana, CEE, Firenze, Italy, pp. 81–88.

Chou, C.K.S. (1989) Morphological and cultural variation of *Seiridium* spp. from cankered *Cupressaceae* hosts in New Zealand. *European Journal of Forest Pathology* 19, 435–445.

Chou, C.K.S. (1990) Pathogenic variation of *Seiridium* spp. isolated from cankered *Cupressaceae* hosts in New Zealand. *European Journal of Forest Pathology* 20, 32–43.

Cunnington, J.H. (2007) *Seiridium cupressi* is the common cause of cypress canker in south-eastern Australia. *Australasian Plant Disease Notes* 2, 53–55.

Danti, R., Panconesi, A., Di Lonardo, V., Della Rocca, G. and Raddi, P. (2006) 'Italico' and 'Mediterraneo': two *Seiridium cardinale* canker-resistant cypress cultivars of *Cupressus sempervirens*. *HortScience* 41, 1357–1359.

Danti, R., Della Rocca, G. and El Wahidi, F. (2009) *Seiridium cardinale* newly reported on *Cupressus sempervirens* in Morocco. *Plant Pathology* 58, 1174.

Della Rocca, G., Buonamici, A., Cossu, C., Vendramin, G.G. and Danti, R. (2009) Development and characterization of eight microsatellite markers for the fungus *Seiridium cardinale*. *Molecular Ecology Resources* 9, 1375–1429.

Della Rocca, G., Di Lonardo, V. and Danti, R. (2011a) Newly-assessed fungicides for the control of *Seiridium cardinale* cypress canker. *Phytopathologia Mediterranea* 50, 65–73.

Della Rocca, G., Danti, R. and Garbelotto, M. (2011b) First report of *Seiridium unicorne* causing bark cankers on a Monterey cypress in California. *Plant Disease* 95, 619.
Della Rocca, G., Eyre, C.A., Danti, R. and Garbelotto, M. (2011c) SSR and sequence analyses of the fungal pathogen *Seiridium cardinale* indicate California is the source of the cypress canker epidemic for the Mediterranean region. *Phytopathology* 101, 1408–1417.
European Parliament and Council of the European Union (2009) Directive 2009/128/EC of the European Parliament and of the Council of 21 October 2009 establishing a framework for Community action to achieve the sustainable use of pesticides. *Official Journal of the European Communities* 24.11.2009, L 309/71–86.
Fuller, C.E.K. and Newhook, F.J. (1954) A report on cypress canker in New Zealand. *New Zealand Journal of Agriculture* 88, 211–220.
Funk, A. (1974) Microfungi associated with dieback of native *Cupressaceae* in British Columbia. *Canadian Plant Disease Survey* 54, 166–168.
Govi, G. and Deserti, F. (1980) Interventi chimici contro il *Coryneum cardinale* del cipresso [Chemical control interventions for *Coryneum cardinale* on cypress]. *Informatore Fitopatologico* 10, 19–21.
Graniti, A. (1986) *Seiridium cardinale* and other cypress cankers. *EPPO Bulletin* 16, 479–486.
Graniti, A. (1998) Cypress canker: a pandemic in progress. *Annual Review of Phytopathology* 36, 91–114.
Grasso, V. (1951) Un nuovo agente patogeno del *Cupressus macrocarpa* Hartw. in Italia [A new pathogenic agent of *Cupressus macrocarpa* Hartw. in Italy]. *Italia Forestale e Montana* 6, 63–65.
Grasso, V., Panconesi, A. and Raddi, P. (1979) Testing for resistance to cypress canker disease in Italy. *Phytopathologia Mediterranea* 18, 166–171.
Guba, E.F. (1961) *Monograph of Monochaetia and Pestalotia*. Harvard University Press, Cambridge, Massachusetts.
Hansen, H. (1956) The perfect stage of *Coryneum cardinale. Phytopathology* 46, 636–637.
Krokene, P., Barnes, I., Wingfield, B.D. and Wingfield, M.J. (2004) A PCR-RFLP based diagnostic technique to rapidly identify *Seiridium* species causing cypress canker. *Mycologia* 96, 1352–1354.
Madar, Z. and Liphschitz, N. (1989) Histological studies of *Cupressus sempervirens* L. affected by *Diplodia pinea* f.sp. *cupressi* and *Seiridium cardinale. IAWA Bulletin* 10, 183–192.
Madar, Z., Solel, Z. and Sztejnberg, A. (1990) The effect of *Diplodia pinea* f.sp. *cupressi* and *Seiridium cardinale* on water flow in cypress branches. *Physiological and Molecular Plant Pathology* 37, 389–398.
Magro, P., Di Lenna, P. and Marciano, P. (1982) Cell wall-degrading enzymes produced by *Seiridium cardinale*, agent of the cypress canker. *European Journal of Forest Pathology* 12, 150–156.
Magro, P., Di Lenna, P. and Marciano, P. (1984) *Trichoderma viride* on cypress shoots and antagonistic action against *Seiridium cardinale. European Journal of Forest Pathology* 14, 165–170.
McCain, A.H. (1984) Cypress canker control with fungicides. *Journal of Arboriculture* 10, 212–214.
Moricca, S., Børja, I., Vendramin, G.G. and Raddi, P. (2000) Differentiation of *Seiridium* species associated with virulent canker on cypress in the Mediterranean region by PCR-SSCP. *Plant Pathology* 49, 774–781.
Moriondo, F. (1972) Il cancro del cipresso da *Coryneum cardinale* Wag. I° contributo: la progressione del processo infettivo nei tessuti caulinari [Cypress canker caused by *Coryneum cardinale.* Part 1: Development of infection in the stem tissues]. *Annali Accademia Italiana Scienze Forestali* 21, 399–426.
Mujica, F., Vergara, C. and Oehrens, E. (1980) *Flora Fungosa Chilena*, 2nd edn. Publicación Ciencias Agricolas 5, Universidad de Chile, Santiago.
Mutto Accordi, S. and Panconesi, A. (1987) Ultrastructural modifications in *Cupressus sempervirens* tissues invaded by *Seiridium cardinale. European Journal of Forest Pathology* 17, 193–204.
Panconesi, A. (1990) Pathological disorders in the Mediterranean basin. In: Ponchet, J. (ed.) *Agriculture – AGRIMED Research Programme: Progress in EEC Research on Cypress Diseases.* Report EUR 12493 EN, Commission of the European Communities, Brussels/Luxembourg, pp. 54–81.
Panconesi, A. and Parrini, C. (1979) Nuove esperienze di lotta chimica contro *Seiridium* (*Coryneum*) *cardinale* [New experiences in the chemical control of *Seiridium* (*Coryneum*) *cardinale*]. *Informatore Fitopatologico* 29, 13–17.
Panconesi, A. and Raddi, P. (1986) The influence of some chemical treatments on cypress canker disease development. *European Journal of Forest Pathology* 16, 83–86.
Panconesi, A. and Raddi, P. (1990a) Una realtà presente per il futuro del cipresso. Selezionati cloni resistenti al cancro del cipresso [A reality for the future of cypress. Selected clones resistant to cypress canker]. *Cellulosa e Carta* 1, 3–5.
Panconesi, A. and Raddi, P. (1990b) Agrimed n. 1 e Bolgheri: due nuove selezioni resistenti al cancro del cipresso [Agrimed n. 1 and Bolgheri: two new selections resistant to cypress canker]. *Cellulosa e Carta* 1, 7–12.

Panconesi, A. and Raddi, P. (1991) Cypress canker disease: biological and epidemiological aspects. In: Panconesi, A. (ed.). *Il Cipresso: Proposte di Valorizzazione Ambientale e Produttiva nei Paesi Mediterranei della Comunità Economica Europea*. CNR, Regione Toscana, CEE, Firenze, Italy, pp. 49–60.

Panconesi, A., Santini, A. and Casini, N. (1993) Conservazione della germinabilità e della patogenicità dei conidi di *Seiridium cardinale* (Wag.) Sutton & Gibson, in relazione alla sua diffusione epidemiologica [Conservation of the germinability and pathogenicity of conidia of *Seiridium cardinale* (Wag.) Sutton & Gibson, in relation to its epidemiological distribution]. *Informatore Fitopatologico* 43, 45–49.

Pedron, L., Piva, G., and La Porta, N. (2007) The genetic structure of cypress canker fungus in Italy using RAPD and minisatellite markers. *Acta Silvatica Lignaria Hungarica*, Special Edition, 159–168.

Ponchet, J. and Andréoli, C. (1990) Compartmentalization and reactions in the host. In: Ponchet, J. (ed.) *Agriculture – AGRIMED Research Programme: Progress in EEC Research on Cypress Diseases*. Report EUR 12493 EN, Commission of the European Communities, Brussels/Luxembourg, pp. 96–111.

Ponchet, J., Andréoli, C., Xenopoulos, S., Caetano, M.F., Raddi, P. and Panconesi, A. (1990) Pathogenic variability in *Seiridium* sp. In: Ponchet, J. (ed.) *Agriculture – AGRIMED Research Programme: Progress in EEC Research on Cypress Diseases*. Report EUR 12493 EN, Commission of the European Communities, Brussels/Luxembourg, pp. 112–126.

Purnell, H.M. (1960) Cypress canker. *Journal of the Department of Agriculture, Victoria* 58, 287–288.

Raddi, P. (1979) Variabilità della resistenza al cancro nell'ambito del cipresso comune (*C. sempervirens*) e di altre specie [Variability of resistance to canker within the common cypress (*C. sempervirens*) and other species]. In: Grasso, V. and Raddi, P. (eds) *Seminario: Il Cipresso: Malattie e Difesa, 23–24 Settembre. CEE-AGRIMED, Firenze, Italy*. Tipografia l'Artigiano, Firenze, Italy, pp. 185–193.

Raddi, P. and Panconesi, A. (1984) Pathogenicity of some isolates of *Seiridium* (*Coryneum*) *cardinale*, agent of cypress canker disease. *European Journal of Forest Pathology* 14, 348–354.

Raio, A., Puopolo, G., Cimmino, A., Danti, R., Della Rocca, G. and Evidente, A. (2011) Biocontrol of cypress canker by the phenazine producer *Pseudomonas chlororaphis* subsp. *aureofaciens* strain M71. *Biological Control* 58, 133–138.

Roques, A. and Battisti, A. (1999) Cypress pests. In: Teissier Du Cros, E. (ed.) *Cypress: A Practical Handbook*. EU FAIR 3 Concerted Action. Studio Leonardo, Firenze, Italy, pp. 75–95.

Santini, A. and Di Lonardo, V. (2000) Genetic variability of the 'bark canker resistance' character in several natural provenances of *Cupressus sempervirens*. *Forest Pathology* 30, 87–96.

Santini, A., Bertolini, F. and Ambrosi, P. (1997a) Il cancro del cipresso in provincia di Trento [Cypress canker in the province of Trento]. *Sherwood* 23, 19–22.

Santini, A., Casini, N., Di Lonardo, V. and Raddi, P. (1997b) Canker resistance stability of some *Cupressus sempervirens* clones to *Seiridium cardinale*. *Journal of Genetics and Breeding* 51, 269–277.

Saravi Cisneros, R. (1953) Cancrosis de los cipreses provocada por *Coryneum cardinale* Wag. en la provincia de Buenos Aires (Argentina) [Cypress canker provoked by *Coryneum cardinale* Wag. in the province of Buenos Aires (Argentina)]. *Revista de la Facultad de Agronomia de La Plata (Tercera Epoca)* 29, 107–119.

Sidoti, A., Alfieri, G., Marranca, S. and Scalici, G. (2007) Diffusione del cancro corticale del cipresso in popolamenti demaniali della Sicilia [Spread of cypress canker in state plantations in Sicily]. *Sicilia Foreste* 47, 59–62.

Solel, Z., Messinger, R., Golan, Y. and Madar, Z. (1983) *Coryneum* canker of cypress in Israel. *Plant Disease* 67, 550–551.

Sparapano, A. and Evidente, A. (1995) Studies on structure–activity relationship of seiridins, phytotoxins produced by three species of *Seiridium*. *Natural Toxins* 3, 166–173.

Strouts, R.G. (1970) *Coryneum* canker of *Cupressus*. *Plant Pathology* 19, 149–150.

Sutton, B.C. (1980) *The Coelomycetes. Fungi Imperfecti with Pycnidia, Acervuli and Stromata*. Commonwealth Mycological Institute, Kew, UK [now CAB International, Wallingford, UK].

Sutton, B.C. and Gibson, I.A.S. (1972) *Seiridium cardinale*. *CMI Description of Pathogenic Fungi and Bacteria* No. 326. Commonwealth Mycological Institute, Kew, UK [now CAB International, Wallingford, UK].

Swart, H.J. (1973) The fungus causing cypress canker. *Transactions of the British Mycological Society* 61, 71–82.

Tabata, M. (1991) Distribution and host range of *Seiridium unicorne* in Japan. *Transactions of the Mycological Society of Japan* 32, 259–264.

Tisserat, N.A., Nus, A. and Barnes, L.W. (1991) A canker disease of the *Cupressaceae* in Kansas and Texas caused by *Seiridium unicorne*. *Plant Disease* 75, 138–140.

Torres, J. (1969) Grave enfermedad de los cipreses en España [Severe disease of cypresses in Spain]. *Boletín del Servicio de Plagas Forestales* 12, 97–99.

Van der Werff, H.S. (1984) Cypress canker in New Zealand plantations. *New Zealand Journal of Science* 18, 101–108.

Viljoen, C.D., Wingfield, B.D. and Wingfield, M.J. (1993) Comparison of *Seiridium* isolates associated with cypress canker using sequence data. *Experimental Mycology* 17, 323–328.

Wagener, W.W. (1939) The canker of *Cupressus* induced by *Coryneum cardinale* n. sp. *Journal of Agricultural Research* 58, 1–46.

Wimbush, S.H. (1944) Canker of Monterey cypress in Kenya. *Empire Forestry Journal* 23, 74.

Wingfield, M.J. and Swart, W.J. (1988) Cypress canker in South Africa. *Abstracts of 5th International Congress of Plant Pathology, Kyoto, Japan, August 20–27, 1988*, pp. 361.

Xenopoulos, S. (1991) Pathogenic variability of various isolates of *Seiridium cardinale, S. cupressi* and *S. unicorne* inoculated on selected *Cupressus* clones and seedlings. *European Journal of Forest Pathology* 21, 129–135.

Xenopoulos, S. and Diamandis, S. (1985) A distribution map for *Seiridium cardinale* causing the cypress canker in Greece. *European Journal of Forest Pathology* 15, 223–226.

18 Pitch Canker

Thomas R. Gordon*
University of California, Davis, California, USA

18.1 Pathogen, Significance and Distribution

Pitch canker is a disease of forest trees caused by *Gibberella circinata* Nirenberg & O'Donnell ex Britz, T.A. Cout., M.J. Wingf. & Marasas, a fungus commonly known by the name of its anamorph, *Fusarium circinatum* Nirenberg & O'Donnell.

The host range of *G. circinata* extends to at least 60 species of *Pinus* (Hepting and Roth, 1953; Muramoto *et al.*, 1993; Guerra-Santos, 1999; Hodge and Dvorak, 2000; Gordon *et al.*, 2001; Enebak and Carey, 2003; Enebak and Stanosz, 2003) and also includes Douglas-fir (*Pseudotsuga menziesii* (Mirb.) Franco), the only known non-pine host (Gordon *et al.*, 2006). None of the other plant species tested have been found to be susceptible to disease caused by *G. circinata* (McCain *et al.*, 1987; Porter *et al.*, 2009). Under experimental conditions, *G. circinata* is capable of colonizing maize (*Zea mays* L.) asymptomatically (Swett and Gordon, 2009), but the extent to which this occurs in nature is unknown. Thus, the only documented hosts of *G. circinata* under natural conditions are coniferous trees.

Pitch canker was first recognized as a disease in the state of North Carolina in the USA (Hepting and Roth, 1946) and was later shown also to occur in Haiti (Hepting and Roth, 1953). Pitch canker thereafter became more widespread in the south-eastern USA (SE USA) and was the cause of a devastating epidemic affecting slash pine (*Pinus elliottii* Engelm.) in Florida in the 1970s. Pitch canker was discovered in California in 1986 (McCain *et al.*, 1987) and since that time the pathogen has gained a worldwide distribution, with confirmed reports from Japan (Kobayashi and Muramoto, 1989), South Africa (Viljoen *et al.*, 1994), Mexico (Guerra-Santos, 1999), Spain (Landeras *et al.*, 2005), South Korea (Lee *et al.*, 2000), Chile (Wingfield *et al.*, 2002), Italy (Carlucci *et al.*, 2007) and Portugal (Bragança *et al.*, 2009). The origin of the pitch canker pathosystem is not known, but several lines of evidence suggest that *G. circinata* may have emerged as a pathogen of pines in Mexico or Central America (Gordon *et al.*, 2001).

In Mexico, pitch canker is found in native pine forests, but is not typically seen to be highly damaging, as might be expected where host and pathogen have co-evolved over an extended period of time. Furthermore, the Mexican population of *G. circinata* is genetically diverse (Britz *et al.*, 2001), consistent with Mexico being the centre of origin of the pathogen. If *G. circinata* first manifested virulence to

* E-mail: trgordon@ucdavis.edu

pine in this area, it may have followed a host jump from a grass such as Mexican teosinte (*Zea mexicana* (Schrad.) Kuntze), which has a geographic range overlapping that of many pine species. *G. circinata* has not been isolated from Mexican teosinte, but *G. subglutinans* (E.T. Edwards) P.E. Nelson, Toussoun and Marasas is a commensal associate of teosinte and is interfertile with *G. circinata* (Desjardins *et al.*, 2000). Most progeny in the F_1 generation resulting from a hybrid cross between *G. subglutinans* and *G. circinata* are avirulent on pine, but virulence can be restored at a high frequency in progeny of a backcross to the *G. circinata* parent (Friel *et al.*, 2007). The close relationship between *G. circinata* and *G. subglutinans* suggests that these taxa may share a recent common ancestor. Finally, *G. circinata* can colonize the roots, stems and ears of maize (a horticultural derivate of teosinte), so may have ancestral traits shared with the grass-colonizing strains from which it was derived (Swett and Gordon, 2009).

Pitch canker continues to be an important problem in the SE USA, where it affects many species of pine, including slash pine and loblolly pine (*P. taeda* L.) (Dwinell *et al.*, 1985). The canker can be a cause of mortality in seedlings in nurseries at emergence and may also be associated with transplant failure when infected seedlings are used in reforestation. Mature trees are affected in plantations and seed orchards. Pitch canker is seen in native forests in the SE USA, but is not regarded as a serious problem in that environment. In California, the disease affects susceptible species in landscapes and in native populations, with Monterey pine (*P. radiata* D. Don) being the most widely and severely affected in both situations (Gordon *et al.*, 2001). The native range of Monterey pine is quite limited, with only three mainland populations, all of them in California, and all now affected by pitch canker. The entirety of the native range includes two small populations on islands off the coast of Mexico, from which pitch canker has not been reported. Many pines native to California are at least somewhat susceptible to pitch canker, based on greenhouse tests, but most do not occur in areas where the disease is present and so have not yet been challenged under natural conditions (Gordon *et al.*, 2001).

In South Africa, the pitch canker problem began in nurseries to which the pathogen may have been introduced with seed of Gregg's pine (*P. greggii* Engelm. ex Parl.) from Mexico (Wingfield *et al.*, 1999). *G. circinata* was the cause of extensive seedling mortality in South Africa in the early 1990s (Viljoen *et al.*, 1994) and the pathogen eventually became established in plantations of Monterey pine (Coutinho *et al.*, 2007), perhaps as a consequence of the movement of infected but symptomless seedlings from nurseries. It remains unclear how important a problem pitch canker may become for plantation-grown pines in South Africa.

As in South Africa, the pitch canker pathogen in Chile was found first in nurseries (Wingfield *et al.*, 2002). It has not yet been reported to occur in plantations, but this possibility represents a serious concern, given the very extensive plantings of the highly susceptible Monterey pine. New Zealand and Australia also have large commitments to Monterey pine as a plantation species, but pitch canker has yet to become established in either country. Pitch canker is having a significant impact on pine production in Spain, where Monterey pine and pines that are native to Spain are affected by the disease (Pérez-Sierra *et al.*, 2007). Recent reports from Italy (Carlucci *et al.*, 2007) and Portugal (Bragança *et al.*, 2009) suggest that *G. circinata* is becoming more widespread in Europe. In Japan, pitch canker is reported to be a cause of dieback in luchu pine (*P. luchuensis* Mayr) (Kobayashi and Muramoto, 1989), which is native to that country. In South Korea, the disease is damaging to pitch pine (*P. rigida* Mill.) (Lee *et al.*, 2000) in plantations and seed orchards.

Although the pitch canker pathogen has recently attained a much wider geographic distribution, many countries where susceptible hosts are abundant are not yet affected. On a local scale, it seems likely that the canker disease has not yet met its ecological limits in many of the areas where it is found.

Pitch canker can be responsible for significant economic losses (Dwinell *et al.*, 1985). In nurseries, *G. circinata* can cause extensive seedling mortality, especially in highly susceptible species such as Monterey pine and Mexican weeping pine (*P. patula* Schiede ex

Schltdl. & Cham.). In plantations, pitch canker can reduce volume growth; this is attributable to a loss of photosynthetic capacity due to canopy dieback. Mortality is typically associated with cankers on the main stem, but secondary factors such as drought and/or bark beetles may also contribute to the demise of a tree (Storer *et al.*, 2002).

18.2 Diagnosis

G. circinata can infect susceptible pines at all stages of growth. Seedlings may be infected by propagules residing on or in seeds, or in the soil. Seedling infection can result in pre-emergence and post-emergence mortality, but it is also possible for infected seedlings to survive and remain symptomless for an extended period of time. Symptoms on seedlings killed by *G. circinata* at or shortly after emergence are not particularly distinctive and laboratory examination is required to confirm a diagnosis. Culturing on a *Fusarium* selective medium (Aegerter and Gordon, 2006), which contains pentachloronitrobenzene and is inhibitory to most fungi other than *Fusarium* spp., allows for reliable recovery of *G. circinata* from infected tissue. Most often the pathogen is found to be colonizing the root collar. Depending on the soil in which seedlings were growing and the condition of the sample, the inclusion of additional antibiotics (Correll *et al.*, 1991) to suppress bacteria may be advisable.

The facility with which *G. circinata* can be identified based on morphological criteria depends on the prevalence of other *Fusarium* spp. The formation of microconidia in false heads borne on polyphialides is sufficient to distinguish the anamorph of *G. circinata* from *F. proliferatum* (Matsush.) Nirenberg ex Gerlach & Nirenberg, which can also be associated with pine seedlings and may be a cause of mortality. Morphological differences between these two species are readily visualized on carnation leaf agar amended with 1.5% KCl (Nelson *et al.*, 1983). Some species, however, are more difficult to separate from *G. circinata* based solely on morphological criteria, and molecular diagnostic methods can help to achieve a definitive diagnosis (Steenkamp *et al.*, 1999; Schweigkofler *et al.*, 2004; Ramsfield *et al.*, 2008; Ioos *et al.*, 2009).

Under experimental conditions, trees infected at the seedling stage have been shown to survive for up to 2 years without manifesting symptoms (Storer *et al.*, 1998a; C.L. Swett and T.R. Gordon, 2011, unpublished results). When latent infections become active, above-ground symptoms develop uniformly throughout the canopy. Needles become chlorotic, wilt and eventually die. This results from growth of the pathogen at the root collar, which girdles the tree and prevents transport of water to the shoot. An accumulation of host-produced resin is usually evident on the main stem of the tree near the soil line. Removing bark will reveal resin-soaked, discoloured tissue, from which the pathogen can be recovered.

Whereas *G. circinata* is extremely important as a soil-borne and seed-borne pathogen in nurseries, it is more familiar as a cause of canopy dieback. Dieback results from a lesion, usually within one or two whorls of a branch tip, which girdles and kills the affected branch distal to the point of infection (Plate 23). Axial growth of the pathogen usually does not progress past the nodes distal and proximal to the infection site (Barrows-Broaddus and Dwinell, 1983), probably because of the accumulation of host-produced resin. Although resin appears not to directly impede growth of *G. circinata* (Barrows-Broaddus and Dwinell, 1983), it will restrict the flow of water, which ultimately leads to desiccation of the infected tissue and the pathogen. Pitch canker intensifies through multiple infections on the same tree. Infections can occur at any point along a branch, though the pathogen appears to grow more rapidly in succulent, current year growth than in older, more lignified tissue. The coalescence of multiple infections can girdle large-diameter branches. Severely diseased trees may suffer top kill resulting from the girdling of the main stem and/or attack by engraver beetles (*Ips* spp.); cankers lower on the trunk can result in death of the entire tree.

A diagnostic feature of branches killed by pitch canker is a mass of resin at the junction of living and symptomatic tissue, which corresponds to the site of infection (Plate 24). Other causes of tip dieback, such as western gall rust (caused by *Endocronartium harknessii* (J.P. Moore) Y. Hiratsuka), and shoot blight caused by *Diplodia pinea* (Desm.) J. Kickx f., typically lack this characteristic symptom. Paring away the bark at the infection site reveals a discoloured, resin-soaked lesion, from which *G. circinata* can be cultured. Selective media can be used for this purpose, but water agar is often sufficient. In either case, the pathogen will grow from lesion margins and produce diagnostic microconidiophores. Where tissue was recently infected, growth of the pathogen may be apparent after only 24 h of incubation. The older the infection, the longer it may take for the pathogen to emerge.

In addition to affecting vegetative tissues, the pitch canker fungus can infect and kill both female strobili and mature cones (Dwinell *et al.*, 1985; Barrows-Broaddus, 1990). Infected cones may be misshapen and stunted, and often abort before reaching maturity (Correll *et al.*, 1991). *G. circinata* can also infect cones without doing visible damage, and seeds from such cones can carry the pathogen internally (Storer *et al.*, 1998a). Infected seeds may or may not show visible evidence of deterioration.

18.3 Infection Biology

G. circinata is considered to be a wound pathogen, with little or no capacity to infect intact host surfaces. Wounds may be caused by weather events such as hail, by silvicultural practices such as cone removal or by the activity of biotic agents, especially insects. In seed orchards, infections of loblolly pine and slash pine have been associated with wounds caused by mechanical shakers and by the removal of cones from branches (Dwinell *et al.*, 1985). The deodar weevil (*Pissodes nemorensis* Germar) (Blakeslee and Foltz, 1981) and *Rhyacionia* spp. (Runion *et al.*, 1993) are reported to be potential wounding agents and vectors of pitch canker in the SE USA. In South Africa, the deodar weevil was found to be infesting trees in plantations that were suffering from pitch canker (Coutinho *et al.*, 2007). The twig beetle (*Pityophthorus* spp.) and engraver beetle have been shown to vector pitch canker in California (Gordon *et al.*, 2001), as have the cone beetle (*Conophthorus radiatae* Hopkins) and the dry twig and cone beetle (*Ernobius punctulatus* Fall) (Hoover *et al.*, 1995, 1996). Also in California, the spittlebug (*Aphrophora canadensis* Walley) is an effective wounding agent (Storer *et al.*, 1998b), but has not been shown to act as a disease vector. Numerous other insects carry the pathogen and might play a role in its dissemination. In Spain, *Ips sexdentatus* Borner and *Pityophthorus pubescens* Marsham, among several other pine-associated insect species, are reported to carry spores of *G. circinata* (Romón *et al.*, 2007).

Under experimental conditions, the risk of infection is influenced by the nature, size and age of a wound, and by the ambient conditions that pertain after wounding. For example, pruning wounds on Monterey pine became less susceptible over time but still sustained a 30% infection rate 28 days after cuts were made (Sakamoto and Gordon, 2006). Wound size had a significant effect on the frequency of infection in Monterey pine, with 1.6 mm diameter wounds sustaining a higher infection rate than wounds 0.5 mm in diameter (Sakamoto and Gordon, 2006). In loblolly pine, pinhole wounds (3–5 cm deep) were readily infected, whereas wounds made by removing needle fascicles or lateral branches became infected at a much lower rate (Kuhlman, 1987). Pinhole wounds on branches of loblolly pine and slash pine remained susceptible for 21 days, though at a much reduced rate (20%) compared with fresh wounds (70%) (Kuhlman, 1987).

In Monterey pine, wounds 0.3 mm in diameter and 1 mm deep became infected at a high rate (82%) when inoculated immediately after wounding, but at a much lower rate (12.5%) if inoculations were delayed by 48 h (Inman *et al.*, 2008). The interval of wound susceptibility defines the period within which spores must germinate, and subsequent growth must be sufficient to establish an infection. Spore germination and growth both proceed very slowly at 10°C and more rapidly as the temperature increases up to 20°C (Inman *et al.*, 2008). For this reason, infection rates tend to be lower in winter than during warmer periods (Inman *et al.*, 2008). However, higher temperatures will favour infections only if wounds are deep enough to reach moisture within the plant or if ambient humidity is high and/or free moisture is present. Thus, infections mediated by *Pityophthorus* spp., which create only very shallow wounds on healthy branches, occur at a higher frequency when relative humidity is at or close to 100% (Sakamoto *et al.*, 2007). In contrast, where inoculum was applied to deeper wounds created mechanically, the effect of ambient humidity on infection frequency was found not to be significant (Sakamoto and Gordon, 2006).

Temperature and moisture requirements for infection are consistent with the widespread occurrence of pitch canker in the SE USA, where rainfall is common during warm periods. Conversely, in California, precipitation occurs primarily during the coolest months of the year and pitch canker is restricted to the central coast, where moderate temperatures coincide with high humidity and/or condensation provided by moist marine air. Even though the present distribution of pitch canker implies a climatic limitation on the geographic range of the disease, whether or not such limitations remain effective may be contingent upon the activity of insect vectors and wounding agents. Whereas twig beetles create shallow wounds on healthy branches, which they find unsuitable for colonization, the cone beetle will move the pathogen deeply into host tissue wherein spore germination will not be dependent on ambient moisture. Therefore, if the range of the cone beetle expands, or other wounding agents with similar feeding/breeding habitats develop an association with *G. circinata*, pitch canker may become problematic for susceptible trees over a much wider area. In California, this could include coastal areas north of 39°N latitude, where the absence of pitch canker in stands of susceptible species (planted Monterey pine and native bishop pine (*P. muricata* D. Don)) presumably reflects the fact that temperatures are relatively cool during periods when moisture is available, which limits the opportunities for infection of shallow wounds.

18.4 Epidemiology

18.4.1 Pathogen dissemination and the initiation of epidemics

Central to the epidemiology of any disease are the means by which the pathogen can move between susceptible individuals. *G. circinata* produces asexual spores (conidia), both microconidia and macroconidia, and it is these propagules that allow for local dispersal from infected to non-infected trees. Although the literature does not provide a detailed assessment of the relative importance of the two spore types, microconidia are produced more abundantly in culture and have been observed more frequently than macroconidia on most pine hosts (Barrows-Broaddus and Dwinell, 1983). However, macroconidia formed in sporodochia are reported to be common on infected branches of slash pine (Blakeslee *et al.*, 1978), which may reflect an influence of host species and/or environmental conditions on sporulation patterns of the pathogen. Both microconidia and macroconidia are well adapted for splash dispersal and attachment to motile organisms. Both also can become airborne and thus contribute to the propagules that are recovered on plates left exposed to the air in infested areas (Correll *et al.*, 1991). Spores deposited from the air are assumed to be responsible for the inoculum that can be recovered from host surfaces by rinsing bark or needles with water and culturing the 'rinsate' on a selective medium. Likewise, wind dispersal of spores can explain the common occurrence of superficially infested seeds. In a stand where pitch canker is present, there is a high probability that seeds collected from trees in that stand will carry inoculum of *G. circinata*, even when the tree from which cones are collected appears to be disease free (Storer *et al.*, 1998a).

It is not known how far spores may be dispersed, but being relatively thin walled and hyaline, it seems likely that their viability will decline rather quickly over time in a dry atmosphere. Therefore, wind-driven dispersal of spores is probably most important for movement over short distances within an infested stand and much less important for movement between widely separated stands. The attachment of spores to insects may extend the range of dispersal somewhat, but movement of inoculum across continents or oceans would remain only a remote possibility. Such long-distance dispersal has most likely been the result of the movement of infected or infested host material, with seedborne inoculum posing the greatest risk for the introduction of the pathogen into new areas. Furthermore, because the pathogen can cryptically infect seedlings, seedling nurseries can serve as highly effective staging areas for local and regional dissemination of the pathogen following its introduction on seed. Such a scenario appears to have facilitated the establishment of pitch canker in California.

G. circinata was most likely introduced to California on seed imported from the SE USA for the purpose of producing seedlings to be returned to the SE USA for use in reforestation. That such seed lots can carry *G. circinata* was implied by recovery of the pathogen from killed seedlings of longleaf pine (*P. palustris* Mill.) grown in a California nursery without a history of pitch canker (T.R. Gordon, 2011, unpublished results). Where infested seed is sown, growth of the pathogen on infected seedlings could create a reservoir of inoculum in soil. Susceptible trees subsequently grown at that location would consequently be at risk of infection. One species likely to be included in this rotation would be Monterey pine, which is commonly sold in California as Christmas trees. Infected but symptomless Monterey pine seedlings shipped to 'choose and cut' Christmas tree farms would serve as vehicles for wider dissemination of the pathogen.

Infected transplants established at Christmas tree farms that became diseased would not be marketable, but those trees that remain pre-symptomatic could be sold to customers who would then transport them and the pathogen to other locations. This mode of dissemination fits very well with the historical distribution of vegetative compatibility groups (VCGs) of *G. circinata* in California. Based on a collection of isolates obtained between 1987 and 1989, the California population comprised five VCGs (Correll *et al.*, 1992). One of these, VCG C3, was found on a Christmas tree farm in Los Angeles County (LAC) and was also the only VCG recovered from an isolated stand of Monterey pine in Santa Barbara County (SBC), approximately 150 km north of the LAC site. It later became apparent that Monterey pine seedlings from an infested site in LAC had been shipped to a Christmas tree farm in SBC, where VCG C3 was also found, and this may explain how the isolated infestation in SBC became established. A later survey revealed a much larger infestation north of SBC, in which the population was composed almost exclusively of VCG C3 (Gordon *et al.*, 1996). Taken together, these facts are consistent with the establishment of pitch canker in a pine seedling nursery and its subsequent dispersal on infected but symptomless planting stock. After its arrival in a new area, the pathogen could have been further disseminated by windborne spores and also by insects. Once pitch canker had killed large trees, their removal and disposal, often as logs for firewood, offered additional opportunities for the pathogen to spread.

Because many pine-associated insects are attracted to dead and declining trees, they may acquire and distribute inoculum produced on a diseased host. For example, trees suffering from pitch canker are commonly attacked by engraver beetles, which may accelerate the decline and death of a tree. Colonization by engraver beetles may also occur after diseased trees are felled. In either case, logs cut from infected trees provide a breeding substrate for engraver beetles, and their progeny may emerge carrying the pathogen. Although species of *Ips* are well documented as vectors of pitch canker (Fox *et al.*, 1991), they mostly attack trees under stress. Consequently, engraver beetles may be more important in intensifying pitch canker in trees that are already diseased (or weakened by other factors such as drought) rather than as initiators of an epidemic in a stand of healthy trees. This concept is also supported by the fact that engraver beetles favour large-diameter branches and the main stem of trees, whereas the first infections on a tree are usually found on young branches. Twig beetles (*Pityophthorus* spp.) are better candidates for vectoring the pathogen to narrow diameter branches.

Considerable evidence supports an important role for twig beetles as vectors of pitch canker both in the initial and intensifying phases of an epidemic. Early studies of pitch canker in California established an association between species of *Pityophthorus* and symptomatic branches (McCain *et al.*, 1987). The placement of infected branch tips in rearing chambers documented the emergence of twig beetles, most of which were carrying propagules of *G. circinata* (McNee *et al.*, 2002). However, because twig beetles immediately seek declining host material in which to breed, it was unclear how they could deliver the pathogen to vigorously growing young branches, which are the type commonly observed to sustain infections.

The capacity for *Pityophthorus* spp. to function as vectors of pitch canker appears to be a consequence of a limitation on their ability to locate a suitable host. This limitation was revealed by experiments showing that twig beetles were recovered at a comparable frequency from traps with and without host material (Bonello *et al.*, 2001a), which suggests that the insects were not guided by volatiles generated by dying branches. This being the case, it is reasonable to suppose that twig beetles visit branches more or less randomly and determine the suitability of the substrate only after landing. If this determination involves 'tasting', it could result in a wound that would serve as an infection court. To test this hypothesis, spores were deposited on the surface of Monterey pine trees maintained in a growth chamber, and inoculated trees were either exposed to *P. setosus* Blackman, or not. Nearly all of the trees (89%) visited by *P. setosus* developed infections, whereas 94% of those that were not exposed did not (Sakamoto *et al.*, 2007). The few trees not visited by *P. setosus* that did have infections were found to be infested by mealy bugs (Pseudoccidae), which presumably were responsible for creating infection courts. Experiments employing *P. carmeli* Swaine indicated that this species was as efficient as *P. setosus* in mediating infections by *G. circinata* (Sakamoto *et al.*, 2007).

The studies described above document that at least two species of *Pityophthorus* have the ability to act as vectors or wounding agents. Support for the operation of this mode of disease transmission in nature derives from experiments conducted in a naturally infested stand of Monterey pine. In these experiments, a pheromone (pityol) was used to increase the frequency with which selected trees were visited by twig beetles. After 3 years, trees baited with pityol sustained a significantly greater number of infections caused by *G. circinata* than trees that were not so baited (Storer *et al.*, 2004).

18.4.2 Disease progress

To better understand the dynamics of disease development in planted stands of Monterey pine and to gauge the ultimate impact of pitch canker on this susceptible species in California, monitoring plots were established, within which individual trees were evaluated periodically over a 4 year period. The trend observed in most plots was for many trees to become infected before the disease intensified in already diseased trees (Storer *et al.*, 2002). This pattern was attributed to random branch selection by insects vectoring the pathogen, which, when disease incidence was low, mostly resulted in infections on previously uninfected trees. As the incidence of pitch canker increased, randomly selected branches were more likely to be on trees that were already diseased, rather than on one of the few disease-free trees. As a consequence of this pattern, disease incidence increased more rapidly than disease severity. Over the 4 year period (1992–1996), 97% of the trees sustained at least one infection, indicating they were susceptible to pitch canker. This observation, coupled with a high infection rate in all plots, which was independent of the initial level of disease (Storer *et al.*, 2002), suggested an ultimately very severe impact on populations of Monterey pine. However, in the years that followed, increases in disease severity did not remain on the same steep trajectory. Limitations on the development of pitch canker appear to have resulted both from intraspecific differences in susceptibility (Storer *et al.*, 1999), with some trees being relatively resistant, and from the fact that many severely affected trees eventually recovered from the disease.

Recovery from pitch canker in Monterey pine was first noted when the plots described above were revisited in 1999. At that time, many trees that had been severely diseased 3 years earlier were observed to be entirely free of symptoms (Gordon *et al.*, 2001). This remission of disease could be attributed to the fact that earlier infections were no longer visible and no new infections had occurred. On close inspection, formerly symptomatic branch tips corresponding to infections that were no longer active could be identified from beneath the canopy. The dead tip was otherwise obscured by extensive growth of subtending branches. Recovery from pitch canker had been previously reported in the SE USA, particularly in loblolly pine (Dwinell *et al.*, 1985), but was not expected to occur in a highly susceptible species such as Monterey pine. The recovery of once severely diseased trees implied they had become more resistant to pitch canker, which could be explained by the operation of systemic induced resistance (SIR).

Previous work had shown that SIR occurs in Monterey pine in response to inoculations with *G. circinata* under controlled conditions (Bonello *et al.*, 2001b). Trees that were previously inoculated sustained shorter lesions on subsequent challenge with the pathogen than trees that had not been previously exposed. The observations of disease remission described above suggested that SIR was occurring under natural conditions as well. As a test of this hypothesis, trees observed to have recovered from pitch canker were inoculated, and lesion lengths were measured to provide an indication of susceptibility. Nearly all of the trees tested developed only very short lesions, none of which grew sufficiently to girdle the inoculated branch. By this measure, trees were judged to be resistant to pitch canker, supporting the hypothesis that SIR was operative in nature and was facilitating the recovery of trees that were previously highly susceptible to and severely affected by pitch canker (Gordon *et al.*, 2011). This report appears to have been the first documentation that SIR actually occurs in nature.

If exposure to *G. circinata* renders trees more resistant to pitch canker, then trees in areas where the disease is well established should be more resistant on average than trees in areas where the disease has arrived only recently. A comparison of Monterey pine stands differing in the residence time of pitch canker, confirmed that the median lesion length was significantly shorter on trees in 'old' than in 'new' areas (Gordon *et al.*, 2011). Whereas SIR is likely to be an essential element in recovery from pitch canker, studies of disease development in both native and planted stands of Monterey pine indicate that, to the extent that SIR was operative, it did not prevent initially rapid increases in disease incidence and severity (Storer *et al.*, 2002; Wikler *et al.*, 2003). Furthermore, these studies provided evidence that variable rates of disease development could be due to factors other than differences in genetic resistance or the history of exposure to the pathogen.

Where native forests interface with urban areas, trees are exposed to a range of conditions owing to varying degrees of landscape modification. To assess the relative importance of these modifications on the development of pitch canker, plots were established in stands of Monterey pine that corresponded to the following landscape types: (i) wildland, wherein trees were largely unaffected by human intervention; (ii) light urban, corresponding to isolated stands of native trees within residential areas; (iii) heavy urban, wherein native trees were retained on residential properties and integrated into managed landscapes; and (iv) golf, corresponding to native trees bordering a golf course. Plots were also categorized as coastal (within 1.5 km of the water) or inland (≥1.5 km from the water). Data gathered over the course of 3 years indicated that the severity of pitch canker was influenced both by landscape type and location (coastal versus inland). Taken as a group, managed plots (golf, urban and heavy urban) had higher levels of disease than wildland plots, independently of geographical location, and coastal plots had higher levels of disease than inland plots, regardless of landscape type (Wikler *et al.*, 2003). All the managed stands were fragmented, whereas the trees in wildland plots were mostly within relatively large, contiguous stands. Thus, many more trees in managed stands occupied edges which may render them more prone to stress and may also enhance their exposure to inoculum.

It is possible that greater disease severity in coastal than inland locations is an artefact of differences in initial inoculum levels. In other words, disease may have been more severe in coastal stands only because that is where the pathogen initially became established. However, if that was true, the geographic effect, which was first documented over a 3-year period ending in 1999 (Wikler *et al.*, 2003), would be expected to erode over time, but in fact the differential between coastal and inland plots was little changed when a survey was conducted in 2010 (T.R. Gordon, 2011, unpublished results). This argues for consideration of physical and/or biological factors associated with proximity to the coast that might influence the development of pitch canker. One possibility would be differences in the activity of insect vectors or wounding agents. For example, the spittlebug *Aphrophora canadensis*, which creates wounds that can serve as infection courts (Storer *et al.*, 1998b), appears to be more common in coastal areas than further inland and therefore may contribute to a geographic effect on disease severity. Variation in microclimate may also be important. In particular, longer periods of free moisture on host surfaces owing to frequent intervals of fog in the coastal zone could increase the efficiency of the infection process relative to drier inland locations. This concept is supported by studies under controlled conditions that document a significant effect of ambient humidity on the frequency of infections mediated by twig beetles (Sakamoto *et al.*, 2007).

18.4.3 Global risk assessment

Based on the distribution of pitch canker in California on both local and regional scales (Gordon *et al.*, 2001), and the known requirements for spore germination and growth of *G. circinata* (Inman *et al.*, 2008), the optimal conditions for infection are relatively warm temperatures and free moisture on the surface of a susceptible host. The more frequently such conditions prevail, the greater the risk of infection and the more likely it is that pitch canker will cause significant damage to susceptible species. Formulated on the basis of such considerations, the model CLIMEX was used to characterize the risk of pitch canker on a global scale (Ganley *et al.*, 2009). Predictions of the model were consistent with the distribution of pitch canker in North America, where the disease has been established long enough to have approached the ecological limits on its geographic range. The model further predicts that pitch canker constitutes a limited risk to continental Europe, where the disease has recently become established, although pockets of optimal climate were found in Spain, Portugal, France, Italy and Georgia. In Chile, where pitch canker is a problem in nurseries but not as yet in plantations, the climate was found to be suboptimal. In contrast, regions within Australia and New Zealand, where the highly susceptible Monterey pine is widely planted and pitch canker does not yet occur, appear to have a climate that would be optimal for the disease. CLIMEX has also been used to predict how climate change may influence the risk of pitch canker in Australasia in the future (Ganley *et al.*, 2011).

There are two important caveats with respect to conclusions drawn from models based on climate limitations. Firstly, as noted above, insects that colonize healthy tissue may negate environmental limitations on infection. Insects creating galleries, and for that matter any manner of wounding that allows the pathogen to access moisture within a tree, can effectively shield the fungus from the limiting effects of dry conditions. Second, weather will have little or no effect on the activity of *G. circinata* as a soil-borne pathogen. Thus, countries, or regions within countries, with a climate deemed unsuitable for pitch canker may nevertheless sustain serious damage from seedling mortality in nurseries and a higher rate of transplant failure as a consequence of latent infections.

18.5 Management Strategies and Tactics

18.5.1 Exclusion

In countries such as New Zealand and Australia where pitch canker does not occur, and in many other countries where the pathogen is not yet widely distributed, exclusion of the disease is still possible. At the international level, quarantine measures can help to reduce the risk of movement of the pathogen between countries by enforcing appropriate restrictions on and monitoring of imported plant material. Of particular concern are seeds, seedlings and scion wood.

Notwithstanding the efficacy of some seed treatments (see below), it is advisable to avoid moving pine seed from infested to non-infested areas. This restriction should be applied to pines regardless of their susceptibility to pitch canker because resistance to the disease would not preclude the deposition of airborne spores on seed. If pine seed is to be imported into non-infested countries, even seed originating from areas presumed to be free of pitch canker should be subjected to appropriate fungicidal treatments. Furthermore, imported seed should be sown and maintained under quarantine conditions so any seedling diseases can be investigated to determine whether *G. circinata* was the cause. Recovery of the pathogen from a seedling indicates that nearby symptomless seedlings may also be infected. Thus, further sampling should be undertaken to determine the extent of the infestation. If limited in size, the affected area(s) could be treated as described below, but for more extensive infestations, destruction of the crop may be justified. If any seedlings are to be shipped from a site where pitch canker has been confirmed to occur, a representative sample of seedlings should be tested for the presence of latent infections. The size of the sample should be scaled according to the detection threshold deemed to provide the desired level of safety.

Both seedlings and scion wood constitute a risk for the introduction of *G. circinata*. Seedlings can sustain latent infections, so a visual assessment cannot ensure that they are pathogen free. Relatively resistant species are of particular concern because infections will frequently fail to induce symptoms. For example, Douglas-fir is only marginally susceptible to pitch canker, and trees in the field rarely show symptoms, even where diseased Monterey pine occurs in the same stand. Yet in provenance tests, lesions developed on every inoculated individual (Gordon *et al.*, 2006). The potential for Douglas-fir to serve as a carrier of *G. circinata* was realized when infected scion wood was imported into New Zealand (Vogler *et al.*, 2004). The infection was detected in quarantine and destroyed so introduction of the pathogen was prevented.

The orchard in California from which the scion wood originated was identified, and the pathogen was confirmed to occur there, but it could be recovered from only a single branch (Vogler *et al.*, 2004). Although that branch appeared unhealthy, it did not display symptoms typical of pitch canker. However, washing needles on symptomless Douglas-fir branches did recover spores of *G. circinata* in some cases (T.R. Gordon, 2011, unpublished results). These findings suggest that *G. circinata* was growing and sporulating on trees that lacked detectable symptoms of pitch canker, which serves to underscore the fact that no practical form of inspection can ensure that host material is free of the pathogen.

18.5.2 Eradication

Once *G. circinata* is recognized as a cause of disease in a new location, it is probably too well established for eradication to be a realistic option, at least on a national or regional scale. In a nursery, establishment is most likely to follow from the sowing of contaminated seed, leading to infestation of the soil. Experiments in which soil was inoculated with spore suspensions and maintained under greenhouse conditions showed that the pathogen remained viable at low levels for up to a year (T.R. Gordon, 2011, unpublished results). Survival intervals could be longer where the pathogen resides within infected host material. Consequently, if one relies only on rotations away from pine to reduce soil inoculum densities, a 1 year interval may not be sufficient.

Elimination of the pathogen in soil can be achieved by fumigation with a fungicidal product such as chloropicrin, but collateral suppression of the resident microflora may render fumigated soil more conducive to the establishment of subsequent introductions of the pathogen. For this reason, management practices that maintain a robust soil microflora may better serve to reduce the risk of establishment of *G. circinata*. However, where a recently discovered infestation affects only a small area, aggressive eradicative measures may be justified. To this end, all infected trees should be removed, including as much of the root system as is practicable, and either burned or placed in sealed containers for disposal off-site. Care should be taken not to expand the infested area by any movement of pathogen-contaminated soil. The affected site should be treated with a material that will kill the pathogen. Metam sodium can be effective for this purpose, and may be preferable to more potent fumigants because it has a less suppressive effect on the bacterial component of the microflora.

18.5.3 Protection

As noted above, *G. circinata* can be found on seeds in areas where pitch canker occurs, regardless of the disease status of the tree from which seed is obtained. When trees are healthy, the infestation tends to be superficial, but seed obtained from cones on infected branches may harbour deeper infestations, which will survive exposure to surface-acting treatments such as sodium hypochlorite (Storer *et al.*, 1998a). However, two seed treatments were found to be effective in eliminating both internal and external infestations of Monterey pine seed. Benomyl, and a combination of benzimidazole, carboxin and thiram, increased seedling emergence to 53% and 57%, respectively, compared with only 43% for untreated seed (T.R. Gordon and M.A. Dick, 2011, unpublished results). *G. circinata* was not isolated from any seedlings emerging from the treated seed. Thiabendazole suspended in 10% dimethyl sulfoxide was reported to significantly reduce but not eliminate infestations of *G. circinata* in seeds of longleaf pine (Runion and Bruck, 1988).

Good sanitation practices in seedling nurseries can help to prevent the establishment of the pitch canker pathogen and should also be part of a programme for the management of an existing infestation. Obviously, this should include the use of pathogen-free seed. It is also important to ensure that irrigation water is not a source of inoculum. This may require the implementation of a sterilization procedure, as can be accomplished by exposing water to ozone or ultraviolet light or treatment with hydrogen peroxide. Where growth media and containers are used, these should be pathogen free, using sterilization if necessary. The regular removal of diseased plants is also advisable, to minimize build-up of pathogen inoculum.

The incidence of infection in plantations and seed orchards may be reduced by practices that limit opportunities for insects to create infection courts. Accordingly, it would be appropriate to perform silvicultural operations when conditions are not conducive to the activity of insects or to infection by *G. circinata*. Insecticides may be useful in some situations. For example, soil applications of carbofuran significantly reduced damage by the pine tip moth (*Rhyacionia* spp.) and the incidence of infection by *G. circinata* in 1 year old loblolly pine (Runion *et al.*, 1993). Fungicides can also be used to reduce the frequency of infection by *G. circinata*, although this has not been shown to be cost-effective. Applications of the fungicide thiabendazole reduced the incidence of pitch canker, but the magnitude of the benefit was judged insufficient to justify the cost of application (Runion *et al.*, 1993). Comprehensive control of pitch canker using fungicides would be difficult to achieve because nearly all aerial surfaces would require protection for an extended period of time.

Where canopy dieback is limited to small-diameter branches, it should be possible to rid a tree of disease by pruning. Of course pre-symptomatic infections will be missed and new infections can occur. Consequently, multiple rounds of pruning might be required to eliminate all infections and maintain a tree in a symptom-free condition. To assess the efficacy of this approach, all infected branches were removed from trees in planted stands of Monterey pine. At one location, where most trees were disease free, 57 symptomatic branches were removed from four trees in the initial round of pruning. Thereafter, 101, 69, 370 and 588 diseased branches were removed 7, 14, 22, and 43 months later, respectively (Gordon *et al.*, 2001). Whereas only four trees in the stand had symptoms of pitch canker on the first branch removal date, 43 months later, 48 of the 51 trees were diseased. Similar results were obtained at three other locations, showing that eradicative pruning was not effective in eliminating the disease or even in reducing its incidence or severity. It is possible that disease would have progressed more rapidly in the absence of pruning, but it is doubtful the benefit gained would be great enough to justify the costs in most situations. A valuable landscape tree might be an exception to this generality. If pruning restores the aesthetic value of a tree, costs might be repaid if the tree recovers and the expense of removal and replacement is thereby avoided.

Rapidly growing trees appear to be more susceptible to pitch canker, and high soil fertility has been associated with more severe disease (Fraedrich and Witcher, 1982). Consequently, the growth benefit accruing from inputs such as irrigation, fertilization and weed control could be offset by enhanced susceptibility to pitch canker. Nitrogen can render trees more susceptible to pitch canker when applied as fertilizer but also, as a consequence of exposure to volatile forms, as may occur in proximity to ventilated poultry farms (Lopez-Zamora *et al.*, 2007). Trees subjected to drought stress may be at greater risk of damage from pitch canker (Dwinell and Phelps, 1977; Owen and Adams, 2001), so stocking densities should be matched to the available water supply. Stress may help to explain the observation that pitch canker is more severe in plantings started with seeds from non-local provenances. This observation suggests that genotypes that are not well adapted to a planting site may be more prone to damage from pitch canker (Dwinell *et al.*, 1985).

18.5.4 Resistance

In plantations and seed orchards, genetic resistance offers an attractive approach to disease management because it can be highly effective, requires no inputs during the production cycle and engenders no adverse environmental impacts. Pine species show a wide range of variation in susceptibility to pitch canker so it may be possible to reduce the risk of damage by growing species (or hybrids) that are relatively resistant. To this end, greenhouse tests of young trees can be used to gauge the susceptibility of trees under field conditions. For example, a comparison of Monterey pine with three other pine species (Canary Island pine, *P. canariensis* C. Sm.; Aleppo pine, *P. halepensis* Mill.; and Italian stone pine, *P. pinea* L.) showed a strong correlation between lesion lengths resulting from branch inoculations on trees maintained in a greenhouse and the incidence of pitch canker on the same species in a landscape, where they were exposed to the pathogen under natural conditions (Gordon *et al.*, 1998a). Among species grown in plantations in the SE USA, inoculations of 1 year old seedlings revealed loblolly pine to be the most resistant, with shortleaf pine (*P. echinata* Mill.) and Virginia pine (*P. virginiana* Mill.) the most susceptible, and slash pine ranked as intermediate (Dwinell, 1978). Even for susceptible species, such as slash pine and Monterey pine, intraspecific variation in susceptibility has been demonstrated (Rockwood *et al.*, 1988; Gordon *et al.*, 1998b; Matheson *et al.*, 2006; Roux *et al.*, 2007), and can be exploited to develop planting stock with higher levels of resistance than is currently available. In Monterey pine, families that manifested rapid growth were, in some cases, also among the most resistant to pitch canker (Matheson *et al.*, 2006). So it may be possible to develop families and genotypes that are less prone to damage caused by pitch canker without compromising commercially important characteristics.

18.5.5 Integrated disease management

In addition to inherent genetic resistance, SIR might be exploited to render trees less disposed to damage from pitch canker. Methyl jasmonate can elevate resistance in Monterey pine to infection by *D. pinea* (Gould *et al.*, 2008) and salicylic acid enhances the resistance of Monterey pine to *G. circinata* (T.R. Gordon, 2011, unpublished results). These findings confirm the efficacy of chemical inducers, at least under controlled conditions. However, Gould *et al.* (2008) recorded a transient decline in photosynthetic rates and growth of seedling pines following the application of methyl jasmonate, indicating that elevated resistance may carry a penalty in the form of reduced productivity. This being the case, one possible strategy could be to apply an inducer of resistance only after infections have been observed within a plantation, with the goal of moderating further development of disease. In this way, the cost associated with SIR would only be incurred when there was a reasonable probability that it would be repaid. Knowing more about the magnitude of any reduction in growth rate and the duration and generality/specificity of the elevated resistance will facilitate a more informed assessment of the feasibility of deploying chemical induction of resistance as part of a disease management programme.

References

Aegerter, B.J. and Gordon, T.R. (2006) Rates of pitch canker induced seedling mortality among *Pinus radiata* families varying in levels of genetic resistance to *Gibberella circinata* (anamorph *Fusarium circinatum*). *Forest Ecology and Management* 235, 14–17.

Barrows-Broaddus, J. (1990) Colonization of cones and seed of loblolly pine following inoculation with *Fusarium subglutinans*. *Plant Disease* 74, 1002–1005.

Barrows-Broaddus, J. and Dwinell, L.D. (1983) Histopathology of *Fusarium moniliforme* var. *subglutinans* in four species of southern pines. *Phytopathology* 73, 882–889.

Blakeslee, G.M. and Foltz, J.L. (1981) The deodar weevil, a vector and wounding agent associated with pitch canker of slash pine. *Phytopathology* 71, 861.

Blakeslee, G.M., Kratka, S.H., Schmidt, R.A. and Moses, C.S. (1978) Sporodochia of the pitch canker fungus (*Fusarium moniliforme* var. *subglutinans*) as found in diseased slash pines in Florida. *Plant Disease Reporter* 62, 656–657.

Bonello, P., McNee, W.R., Storer, A.J., Wood, D.L. and Gordon, T.R. (2001a) The role of olfactory stimuli in the location of weakened hosts by twig-infesting *Pityophthorus* spp. *Ecological Entomology* 26, 8–15.

Bonello, P., Gordon, T.R. and Storer, A.J. (2001b) Systemic induced resistance in Monterey pine. *Forest Pathology* 31, 99–106.

Bragança, H., Diogo, E., Moniz, F. and Amaro, P. (2009) First report of pitch canker on pines caused by *Fusarium circinatum* in Portugal. *Plant Disease* 93, 1079.

Britz, H., Coutinho, T.A., Gordon, T.R. and Wingfield, M.J. (2001) Characterisation of the pitch canker fungus, *Fusarium circinatum*, from Mexico. *South African Journal of Botany* 67, 609–614.

Carlucci, A., Colatruglio, L. and Frisullo, S. (2007) First report of pitch canker caused by *Fusarium circinatum* on *Pinus halepensis* and *P. pinea* in Apulia (southern Italy). *Plant Disease* 91, 1683.

Correll, J.C., Gordon, T.R., McCain, A.H., Fox, J.W., Koehler, C.S., Wood, D.L. and Schultz, M.E. (1991) Pitch canker in California: pathogenicity, distribution, and canker development on Monterey pine (*Pinus radiata*). *Plant Disease* 75, 676–682.

Correll, J.C., Gordon, T.R. and McCain, A.H. (1992) Genetic diversity in California and Florida populations of the pitch canker fungus, *Fusarium subglutinans* f.sp. *pini*. *Phytopathology* 82, 415–420.

Coutinho, T.A., Steenkamp, E.T., Mongwaketsi, K., Wilmot, M. and Wingfield, M.J. (2007) First outbreak of pitch canker in a South African pine plantation. *Australasian Plant Pathology* 36, 256–261.

Desjardins, A.E., Plattner, R.D. and Gordon, T.R. (2000) *Gibberella fujikuroi* mating population A and *Fusarium subglutinans* from teosinte species and maize from Mexico and Central America. *Mycological Research* 104, 865–872.

Dwinell, L.D. (1978) Susceptibility of southern pines to infection by *Fusarium moniliforme* var. *subglutinans*. *Plant Disease Reporter* 62, 108–111.

Dwinell, L.D. and Phelps, W.R. (1977) Pitch canker of slash pine in Florida. *Journal of Forestry* 75, 488–489.

Dwinell, L.D., Barrows-Broaddus, J. and Kuhlman, E.G. (1985) Pitch canker: a disease complex of southern pines. *Plant Disease* 69, 270–276.

Enebak, S.A. and Carey, W.A. (2003) Pitch canker caused by *Fusarium circinatum* identified on spruce pine in Alabama. *Plant Disease* 87, 449.

Enebak, S.A. and Stanosz, G.R. (2003) Responses of conifer species of the Great Lakes region of North America to inoculation with the pitch canker pathogen *Fusarium circinatum*. *Forest Pathology* 33, 333–338.

Fox, J.W., Wood, D.L., Koehler, C.S. and O'Keefe, S.T. (1991) Engraver beetles (Scolytidae: *Ips* species) as vectors of the pitch canker fungus, *Fusarium subglutinans*. *Canadian Entomologist* 123, 1355–1367.

Fraedrich, B.R. and Witcher, W. (1982) Influence of fertilization on pitch canker development on three southern pine species. *Plant Disease* 66, 938–940.

Friel, C.J., Desjardins, A.E., Kirkpatrick, S.C. and Gordon, T.R. (2007) Evidence for recombination and segregation of virulence to pine in a hybrid cross between *Gibberella circinata* and *G. subglutinans*. *Mycological Research* 111, 827–831.

Ganley, R.J., Watt, M.S., Manning, L. and Iturritxa, E. (2009) A global climatic risk assessment of pitch canker disease. *Canadian Journal of Forest Research* 39, 2246–2256.

Ganley, R.J., Watt, M.S., Kriticos, D.J., Hopkins, A.J.M. and Manning, L. (2011) Increased risk of pitch canker to Australasia under climate change. *Australasian Plant Pathology* 40, 228–237.

Gordon, T.R., Storer, A.J. and Okamoto, D. (1996) Population structure of the pitch canker pathogen, *Fusarium subglutinans* f. sp. *pini*, in California. *Mycological Research* 100, 850–854.

Gordon, T.R., Okamoto, D., Storer, A.J. and Wood, D.L. (1998a) Susceptibility of five landscape pines to pitch canker disease, caused by *Fusarium subglutinans* f.sp. *pini*. *HortScience* 33, 868–871.

Gordon, T.R., Wikler, K.R., Clark, S.L., Okamoto, D., Storer, A.J. and Bonello, P. (1998b) Resistance to pitch canker disease, caused by *Fusarium subglutinans* f.sp. *pini*, in Monterey pine (*Pinus radiata*). *Plant Pathology* 47, 706–711.

Gordon, T.R., Storer, A.J. and Wood, D.L. (2001) The pitch canker epidemic in California. *Plant Disease* 85, 1128–1139.

Gordon, T.R., Kirkpatrick, S.C., Aegerter, B.J., Wood, D.L. and Storer, A.J. (2006) Susceptibility of Douglas fir (*Pseudotsuga menziesii*) to pitch canker, caused by *Gibberella circinata* (anamorph = *Fusarium circinatum*). *Plant Pathology* 55, 231–237.

Gordon, T.R., Kirkpatrick, S.C., Aegerter, B.J., Fisher, A.J., Storer, A.J. and Wood, D.L. (2011) Evidence for the occurrence of induced resistance to pitch canker, caused by *Gibberella circinata* (anamorph *Fusarium circinatum*), in populations of *Pinus radiata*. *Forest Pathology* 41, 227–232.

Gould, N., Reglinski, T., Spiers, M. and Taylor, J.T. (2008) Physiological trade-offs associated with methyl jasmonate-induced resistance in *Pinus radiata*. *Canadian Journal of Forest Research* 38, 677–684.

Guerra-Santos, J.J. (1999) Pitch canker on Monterey pine in Mexico. In: Devey, M.E., Matheson, A.C. and Gordon, T.R. (eds) *Current and Potential Impacts of Pitch Canker in Radiata Pine, Proceedings of the IMPACT Monterey Workshop, California, USA, 30 November to 3 December 1998*. Forestry and Forest Products Technical Report 112, CSIRO, Canberra, Australia, pp. 58–61.

Hepting, G.H. and Roth, E.R. (1946) Pitch canker, a new disease of some southern pines. *Journal of Forestry* 44, 724–744.

Hepting, G.H. and Roth, E.R. (1953) Host relations and spread of the pine pitch canker disease. *Phytopathology* 43, 475.

Hodge, G.R. and Dvorak, W.S. (2000) Differential responses of Central American and Mexican pine species and *Pinus radiata* to infection by the pitch canker fungus. *New Forests* 19, 241–258.

Hoover, K., Wood, D.L., Fox, J.W. and Bros, W.E. (1995) Quantitative and seasonal association of the pitch canker fungus, *Fusarium subglutinans* f.sp. *pini* with *Conophthorus radiatae* (Coleoptera: Scolytidae) and *Ernobius punctulatus* (Coleoptera: Anobiidae) which infest *Pinus radiata*. *Canadian Entomologist* 127, 79–91.

Hoover, K., Wood, D.L., Storer, A.J., Fox, J.W. and Bros, W.E. (1996) Transmission of pitch canker fungus, *Fusarium subglutinans* f.sp. *pini* to Monterey pine, *Pinus radiata*, by cone and twig infesting beetles. *Canadian Entomologist* 128, 981–994.

Inman, A.R., Kirkpatrick, S.C., Gordon, T.R. and Shaw, D.V. (2008) Limiting effects of low temperature on growth and spore germination in *Gibberella circinata*, the cause of pitch canker in pine species. *Plant Disease* 92, 542–545.

Ioos, R., Fourrier, C., Iancu, G. and Gordon, T.R. (2009) Sensitive detection of *Fusarium circinatum* in pine seed by combining an enrichment procedure with a real-time polymerase chain reaction using dual-labeled probe chemistry. *Phytopathology* 99, 582–590.

Kobayashi, T. and Muramoto, M. (1989) Pitch canker of *Pinus luchuensis*, a new disease of Japanese forests. *Forest Pests* 38, 169–173.

Kuhlman, E.G. (1987) Effect of inoculation treatment with *Fusarium moniliforme* var. *subglutinans* on dieback of loblolly and slash pine seedlings. *Plant Disease* 71, 161–162.
Landeras, E. *et al.* (2005) Outbreak of pitch canker caused by *Fusarium circinatum* on *Pinus* spp. in northern Spain. *Plant Disease* 89, 1015.
Lee, J.K., Lee, S.H., Yang, S.I. and Lee, Y.W. (2000) First report of pitch canker disease on *Pinus rigida* in Korea. *Plant Pathology Journal* 16, 52–54.
Lopez-Zamora, I., Bliss, C., Jokela, E.J., Comerford, N.B., Grunwald, S., Barnard, E. and Vasquez, G.M. (2007) Spatial relationships between nitrogen status and pitch canker disease in slash pine planted adjacent to a poultry operation. *Environmental Pollution* 147, 101–111.
Matheson, A.C., Devey, M.E., Gordon, T.R., Werner, W., Vogler, D.R., Balocchi, C. and Carson, M.J. (2006) Heritability of response to inoculation by pine pitch canker of seedlings of radiata pine. *Australian Forestry Journal* 69, 101–106.
McCain, A.H., Koehler, C.S. and Tjosvold, S.A. (1987) Pitch canker threatens California pines. *California Agriculture* 41, 22–23.
McNee, W.R., Wood, D.L., Storer, A.J. and Gordon, T.R. (2002) Incidence of the pitch canker pathogen and associated insects in intact and chipped Monterey pine branches. *Canadian Entomologist* 134, 47–58.
Muramoto, M., Tashiro, T. and Minamihashi, H. (1993) Distribution of *Fusarium moniliforme* var. *subglutinans* in Kagoshima prefecture and its pathogenicity to pines. *Journal of the Japanese Forestry Society* 75, 1–9.
Nelson, P.E., Toussoun, T.A. and Marasas, W.F.O. (1983) Fusarium *Species: An Illustrated Manual for Identification.* Pennsylvania State University Press, University Park, Pennsylvania.
Owen, D. and Adams, D. (2001) Impact of pitch canker on ornamental Monterey pines in Santa Cruz County, California, U.S., 1987–2000. *Journal of Arboriculture* 27, 298–305.
Pérez-Sierra, A., Landeras, E., León, M., Berbegal, M., Garcia-Jiménez, J. and Armengol, J. (2007) Characterization of *Fusarium circinatum* from *Pinus* spp. in northern Spain. *Mycological Research* 111, 832–839.
Porter, B., Wingfield, M.J. and Coutinho, T.A. (2009) Susceptibility of South African native conifers to the pitch canker pathogen, *Fusarium circinatum*. *South African Journal of Botany* 75, 380–382.
Ramsfield, T.D., Dobbie, K., Dick, M.A. and Ball, R.D. (2008) Polymerase chain reaction-based detection of *Fusarium circinatum*, the causal agent of pitch canker disease. *Molecular Ecology Resources* 8, 1270–1273.
Rockwood, D.L., Blakeslee, G.M., Lowerts, G.A., Underhill, E.M. and Oak, S.W. (1988) Genetic strategies for reducing pitch canker incidence in slash pine. *Southern Journal of Applied Forestry* 12, 28–32.
Romón, P., Iturrondobeitia, J.C., Gibson, K., Lindgren, B.S. and Goldarazena, A. (2007) Quantitative association of bark beetles with pitch canker fungus and effects of verbenone on their semiochemical communication in Monterey pine forests in northern Spain. *Environmental Entomology* 36, 743–750.
Roux, J., Eisenberg, B., Kanzler, A., Nel, A., Coetzee, V., Kietzka, E. and Wingfield, M.J. (2007) Testing of selected South African *Pinus* hybrids and families for tolerance to the pitch canker pathogen, *Fusarium circinatum*. *New Forests* 33, 109–123.
Runion, G.B. and Bruck, R.I. (1988) Effects of thiabendazole-DMSO treatment of longleaf pine seed contaminated with *Fusarium subglutinans* on germination and seedling survival. *Plant Disease* 72, 872–874.
Runion, G.B., Cade, S.C. and Bruck, R.I. (1993) Effects of carbofuran and thiabendazole on incidence of pitch canker in loblolly pine. *Plant Disease* 77, 166–169.
Sakamoto, J.M. and Gordon, T.R. (2006) Factors influencing infection of mechanical wounds by *Fusarium circinatum* on Monterey pines (*Pinus radiata*). *Plant Pathology* 55, 130–136.
Sakamoto, J.M., Gordon, T.R., Storer, A.J. and Wood, D.L. (2007) The role of *Pityophthorus* spp. as vectors of pitch canker affecting *Pinus radiata*. *Canadian Entomologist* 139, 864–871.
Schweigkofler, W., O'Donnell, K. and Garbelotto, M. (2004) Detection and quantification of *Fusarium circinatum*, the causal agent of pine pitch canker, from two California sites by using a real-time PCR approach combined with a simple spore trapping method. *Applied and Environmental Microbiology* 70, 3512–3520.
Steenkamp, E.T., Wingfield, B.D., Coutinho, T.A., Wingfield, M.J. and Marasas, W.F.O. (1999) Differentiation of *Fusarium subglutinans* f.sp. *pini* by histone gene sequence data. *Applied and Environmental Microbiology* 65, 3401–3406.
Storer, A.J., Gordon, T.R. and Clark, S.L. (1998a) Association of the pitch canker fungus, *Fusarium subglutinans* f.sp. *pini* with Monterey pine seeds and seedlings in California. *Plant Pathology* 47, 649–656.

Storer, A.J., Wood, D.L., Wikler, K.R. and Gordon, T.R. (1998b) Association between a native spittlebug (Homoptera: Cercopidae) on Monterey pine and an introduced tree pathogen which causes pitch canker disease. *Canadian Entomologist* 130, 783–792.

Storer, A.J., Bonello, P., Gordon, T.R. and Wood, D.L. (1999) Evidence of resistance to the pitch canker pathogen (*Fusarium circinatum*) in native stands of Monterey pine (*Pinus radiata*). *Forest Science* 45, 500–505.

Storer, A.J., Wood, D.L. and Gordon, T.R. (2002) The epidemiology of pitch canker of Monterey pine in California. *Forest Science* 48, 694–700.

Storer, A.J., Wood, D.L. and Gordon, T.R. (2004) Twig beetles, *Pityophthorus* spp. (Coleoptera: Scolytidae), as vectors of the pitch canker pathogen in California. *Canadian Entomologist* 136, 685–693.

Swett, C.L. and Gordon, T.R. (2009) Colonization of corn (*Zea mays*) by the pitch canker pathogen, *Fusarium circinatum*: insights into the evolutionary history of a pine pathogen. *Phytopathology*, 99, S126–S127 [abstract].

Viljoen, A., Wingfield, M.J. and Marasas, W.F.O. (1994) First report of *Fusarium subglutinans* f.sp. *pini* on pine seedlings in South Africa. *Plant Disease* 78, 309–312.

Vogler, D.R., Gordon, T.R., Aegerter, B.J., Kirkpatrick, S.C., Lunak, G.A., Stover, P. and Violett, P. (2004) First report of the pitch canker fungus (*Fusarium circinatum*) in the Sierra Nevada of California. *Plant Disease* 88, 772.

Wikler, K.R., Storer, A.J., Newman, W., Gordon, T.R. and Wood, D.L. (2003) The dynamics of an introduced pathogen in a native Monterey pine (*Pinus radiata*) forest. *Forest Ecology and Management* 179, 209–221.

Wingfield, M.J., Wingfield, B.D., Coutinho, T.A., Viljoen, A., Britz, H. and Steenkamp, E. (1999) Pitch canker: a South African perspective. In: Devey, M.E., Matheson, A.C. and Gordon, T.R. (eds) *Current and Potential Impacts of Pitch Canker in Radiata Pine, Proceedings of the IMPACT Monterey Workshop, California, USA, 30 November to 3 December 1998*. Forestry and Forest Products Technical Report 112, CSIRO, Canberra, Australia, pp. 62–69.

Wingfield, M.J., Jacobs, A., Coutinho, T.A., Ahumada, R. and Wingfield, B.D. (2002) First report of the pitch canker fungus, *Fusarium circinatum*, on pines in Chile. *Plant Pathology* 51, 397.

19 Larch Canker

Thomas L. Cech*
Institute of Forest Protection, Federal Research and Training Centre for Forests, Natural Hazards and Landscape, Vienna, Austria

19.1 Pathogen, Significance and Distribution

The term 'larch canker' or 'European larch canker' is used for a canker disease of larches (*Larix* spp.) and golden larches (*Pseudolarix* spp.) caused by the fungal species *Lachnellula willkommii* (Hartig) Dennis (*Ascomycota*) in Europe, Asia (the European part, central Siberia, the Far East and Japan) and North America. The pathogen is probably native to Asia (Kobayashi, 1970; Kurkela, 1983), and has been introduced to the north-east and north-west of the USA and west Canada. In Europe, it probably became established from the 18th century onwards. At present, it is widespread over the whole distribution area of European larch (*Larix decidua* Mill.) in continental Europe.

The increasing problems with larch canker, especially in north European countries (south and south-west of Scandinavia) during the 19th century were due to the fact that European larch and exotic larch species were widely used for afforestation in that area (Yde-Andersen, 1979a). The disease probably reached Denmark by the 1800s (Oppermann, 1923), and was certainly found there around 1830 (Rostrup, 1880). In Norway, the disease was reported as widespread in the second half of the 19th century (Brunchorst, 1888). In Sweden, it was almost certainly present by about the mid 1800s (Schotte, 1917).

In the UK, the disease had already been mentioned before the 1800s (Oppermann, 1923), and in 1904 and 1907 it was spread from there to Massachusetts (USA), in batches of infected European larch and Japanese larch (*L. kaempferi* (Lam.) Carrière) plants. The presence of the fungus in the USA was discovered in 1927 (Spaulding and Siggers, 1927). The disease was effectively eradicated from Massachusetts by 1965 (Tegethoff, 1965), but a new infestation was found in Canada in 1980 (Magasi and Pond, 1982). Subsequently, infestations were observed in the coastal areas of eastern Maine (USA), as well as in southern New Brunswick and central Nova Scotia (Canada) in 1980 (Miller-Weeks and Stark, 1983). In 1992, larch canker was discovered on Prince Edward Island (Canada). In North America, *L. willkommii* is now established in New Brunswick, Prince Edward Island, and Nova Scotia harming especially tamarack (*L. laricina* (Du Roi) K. Koch), and in five counties of Maine (Chaloux, 2010; Canadian Forest Service, 2011).

In China, *L. willkommii* is listed among 19 invasive alien species originating from

* E-mail: thomas.cech@bfw.gv.at

North and South America (Xu *et al.*, 2006). In Japan, the pathogen was first collected on the branches of Japanese larch in Nagano prefecture in 1957 (Ito and Zinno, 1957). Thereafter, serious outbreaks of the canker disease caused by this fungus were found in natural forests and plantations of Japanese larch at altitudes of 1400 to 1800 m a.s.l. around Mount Yatsugatake, Mount Fuji and Mount Asama, in the central parts of the main island of Japan (Ito, 1961; Ito *et al.*, 1963).

Reports of larch canker on Douglas-fir (*Pseudotsuga menziesii* (Mirb.) Franco) by Massee and Crossland (1905), Hoogh (1924) and Jörstad (1925) are most likely influenced by the fact that at the time it was not realized that, in addition to *L. willkommii*, other *Lachnellula* spp. occur on larch, of which some also occur on other conifer species (Yde-Andersen, 1979a). *L. willkommii* has also been reported from pitch pine (*Pinus rigida* Mill.) and even Scots pine (*P. sylvestris* L.), but this has not been proven (Ostaff, 1985).

The identity of the fungus was unclear for a long time, especially because a number of related and similar species occur on susceptible trees, although most of them are saprophytic or only weakly pathogenic (Hahn and Ayers, 1943). According to Baral (2000), in addition to *L. willkommii*, the following *Lachnellula* spp. are reported from larch and golden larch species: *L. agassizii* (Berk. & M.A. Curtis) Dennis (host *Larix* not confirmed), *L. angustispora* Raitv., *L. arida* (W. Phillips), *L. calyciformis* (Willd.) Dharne, *L. ellisiana* (Rehm) Baral comb. nov., *L. flavovirens* (Bres.) Dennis, *L. gallica* (P. Karst. & Har.) Dennis, *L. laricis* (Cooke) Dharne, *L. minuta* Dharne, *L. occidentalis* (G.G. Hahn & Ayers) Dharne, *L. resinaria* var. *resinaria* (Cooke & W. Phillips) Rehm, *L. subtilissima* (Cooke) Dennis, *L. suecica* (de Bary ex Fuckel) Nannf. and *L. tuberculata* Dharne. Of these species, the following show a tendency towards facultative pathogenicity on larches, while the remainder are saprophytic: *L. agassizii*, *L. arida*, *L. ellisiana*, *L. flavovirens*, *L. occidentalis* and *L. subtilissima*. *L. willkommii*, however, is the only species that induces cankers on larches (Dharne, 1965).

19.2 Diagnosis

As a first step in diagnosis of the disease, living trees and bark-bearing wood must be inspected for the presence of cankers and of the fruiting bodies of the pathogen on dead or dying branches and stems; in addition, trees must be inspected for withered, dead and/or discoloured needles. Larch cankers can been observed on young trees in nurseries, in plantations, in natural stands and even in the crowns of very old trees.

L. willkommii causes large numbers of cankers to form on the branches and trunks of infected trees. Cankered areas appear as a first symptom because the fungus kills the growing bark. Swellings develop on twigs and branches, while on larger stems sunken cankers appear. The first circular or elliptical depressions often form around dwarfed shoots. Intense resin flow goes along with the cankers, which gives them a shiny, slightly bluish appearance. In the later stages of the disease the resin-covered surface turns blackish, and the bark cracks and is detached. A ridge of new wood develops around enlarging cankers on stems and trunks as the tree grows. Needles above the canker shrivel up and die or turn yellow early. Cankers on the stem lead to a distortion and to the production of epicormic shoots. Some cankers heal, appearing as rough areas on the bark surface (Peace, 1962). If the stem or trunk is girdled, branches, seedlings and young trees will die (Plate 25) (Sinclair and Lyon, 2005).

After the establishment of *Lachnellula* mycelium in living bark, the production of abnormal wood starts immediately (Yde-Andersen, 1979a). The wood surrounding the cankers differs in several ways from normal wood. Where the cambium has died, of course no wood is produced, but at the edges of wounds abnormal wood occurs. The first wood cells formed here in the spring are thick walled with simple pits rather like ray cells. Between these cells there may be intercellular spaces, which are often filled with resin, but normal resin ducts also occur. Afterwards, tracheids are formed, and these are normal apart from spiral callosities of the walls (Hiley, 1919).

As a second step in disease diagnosis, the fruiting bodies and ascospores that have been found have to be investigated for identification of the species. Apothecia are produced one to several years after infection (Sylvestre-Guinot and Delatour, 1983). They are white and hairy and resemble small cups with yellow orange centres. Descriptions based on features of the apothecia and ascospores of *Lachnellula* spp. are available (Buczacki, 1975; Oguchi, 1981; Minter, 2005), and a dichotomous key to *Lachnellula* spp. worldwide is also available on the Internet (Baral, 2000).

The diagnostic features of *L. willkommii* are as follows. The apothecia (which are associated with the canker lesions) are scattered or often radial grouped, erumpent, at first globular but, on enlarging, opening in moist conditions to form a convex to flat very pale yellowish to straw coloured disc with a short stalk, 1–6 mm in diameter with a moderately thick, irregular chalky-white rim. The exciple (the rim or outer covering of the apothecium) is white, densely hairy, the excipular hairs are cylindrical, hyaline, thin walled, with slightly swollen tips, persisting, 3–4 μm wide, and white, cream, beige, pale brown to grey in dried specimens. The asci are clavate, inamyloid, obtusely rounded at the apices, eight spored and 126–180 × 9–12 μm. Paraphyses exceed the asci and are flexuous–filiform, septate, obtuse, scarcely swollen at the tips, 160–215 μm long, 1–3 μm wide and 4 μm wide at tip (Buczacki, 1975; Chalkley, 2010). Ascospores are obliquely uniseriate, smooth, hyaline, at first continuous, elongate–elliptic, fusiform, obtuse or pointed at the extremities and 15–28 × 6–9 μm.

Small white to yellow stromata bearing the microconidial form of the fungus emerge in cankers (Hahn and Ayers, 1943; Sylvestre-Guinot and Delatour, 1983), followed by production of the apothecia (Yde-Andersen, 1979b). Spermatia are 2–8 × 1–2 μm, continuous, hyaline, allantoid to elliptic–oblong on short, subulate, simple or branching sporophores in irregular loculi contained in whitish erumpent, waxy or fleshy stromata. The function of such spermatia is unknown (Buczacki, 1975).

On the basis of fruiting body morphology, *L. willkommii* can be easily distinguished from *L. arida* and *L. flavovirens* as it has white instead of brown external hairs and broader ascospores. Among the other white-haired species, it can be distinguished from most species, except for *L. occidentalis*, by its broad, cylindrical ascospores. The morphological differences between *L. willkommii* and *L. occidentalis* are slight and still deserve clarification (Baral, 2000). Failures in correct determination are probably frequent, and reports in the literature might often represent *L. occidentalis*. Dharne's clear distinction by the inamyloid asci of *L. willkommii* becomes relative owing to the variation in *L. occidentalis*. Thus, the main distinction is physiological, i.e. the pathogenicity of *L. willkommii* and the saprophytic nature of *L. occidentalis* (Hahn and Ayers, 1934; Kujala, 1950; Seaver, 1951; Manners, 1953; Dharne, 1965; Baral, 1984, 2000; Breitenbach and Kränzlin, 1984; Schmid-Heckel, 1985).

Fresh ascospores of *L. willkommii* germinate readily in artificial media, for instance in 3% malt extract agar with an admixture of beef extract and citric acid (Hiley, 1919), in 3% malt agar (Hahn and Ayers, 1934) and in dextrose agar (Ito *et al.*, 1963). Hiley (1919) found that germination could take place throughout a temperature range of 13–27°C; according to Ito *et al.* (1963) the range was from 0°C to 30°C, with an optimum at 20°C; and according to Sylvestre-Guinot (1981) the range was from 4°C to 28°C, with an optimum at 16°C. Germination is possible both in the light and in darkness (Ito *et al.*, 1963), but ceases if the pH value of the medium exceeds 7.7.

Before the germination of *L. willkommii*, often one, more rarely two, transverse septa are formed in the ascospores, and the germination takes place by one, two or, rarely, several vegetative hyphae being put forth. These vegetative hyphae may be either terminal or lateral (Hiley, 1919; Ito *et al.*, 1963). After *in vitro* germination, a richly ramified submersed mycelium is produced and, in addition, a white velvety aerial mycelium, in which fleecy mycelium tufts sometimes occur (Hahn and Ayers, 1934; Manners, 1953; Ito *et al.*, 1963; Dharne, 1965). On 2% malt extract agar, dense chalky white colonies with a velvety aerial mycelial growth are formed. *L. occidentalis* differs from *L. willkommii* in that a yellow stain of the aerial mycelium develops with age, and the tendency to produce microconidia is lower (Dharne, 1965).

The optimum temperature for the mycelial growth of *L. willkommii* varies: according to Hiley (1919) it is between 14°C and 26°C, while Ito *et al.* (1963) mentioned 20°C and Dharne (1965) 12–15°C. The minimum temperature required for mycelial growth can be 0°C, according to Ito *et al.* (1963), and the maximum is above 30°C.

The anamorphic stage of *L. willkommii* (*Naemospora* sp.) frequently develops after about 2 months in culture (Hiley, 1919; Hahn and Ayers, 1934; Dharne, 1965). Microconidia are produced in irregular brownish labyrinthiform cavities, either solitary or in small groups. On certain media (e.g. malt agar with larch twigs added), apothecia are produced in Petri dishes (Dharne, 1965).

As the morphology and location of the fruiting bodies usually enable a diagnosis of *L. willkommii*, the available molecular methods are sparse. Harrison and DeVerno (1999) reported a method to identify *L. willkommii* using random amplified polymorphic DNA (RAPD) fingerprinting. With this method, a distinction between *L. willkommii* and *L. occidentalis* is possible, even if there are only either immature or overripe fruiting bodies or vegetative mycelium available.

19.3 Infection Biology

The infection biology of *L. willkommii* is not totally clear, especially on the question of whether infection is possible without any wounding (e.g. through needle scars subsequent to needle fall) or not (Sylvestre-Guinot and Delatour, 2002). The dominant infective propagules are the ascospores of the fungus.

Airborne spores infect the living bark predominantly through short shoots (Hopp, 1956; Zycha, 1960; Ito *et al.*, 1963; Robak, 1964; Buczacki, 1973d; Yde-Andersen, 1980; Sylvestre-Guinot and Delatour, 2002). Massee (1902) took ascospores of *L. willkommii*, and placed them on fresh cuts in the bark; cankers were regularly produced using this method, but only on shoots and twigs younger than 5 years old. Several observations on the development of cankers show that their initiation on larch species primarily takes place on parts of the trees (branches or stems) younger than 6–8 years (Hopp, 1956; Zycha, 1960; Buczacki, 1973c).

Sylvestre-Guinot (1986) found infections of *L. willkommii* to be predominantly associated with micro damage by weevils (e.g. *Hylobius abietis* Linnaeus), less common with short shoots and more rare with galls of the larch bark moth *Cydia zebeana* Ratzeburg. Wounded trees are susceptible to infection at all seasons of the year (Hahn and Ayers, 1943). Hartig (1880) inserted blocks of bark tissue from cankers into fresh artificial wounds on larch branches, which caused cankers in the following season. Sylvestre-Guinot and Delatour (1990) studied the susceptibility of larch to *L. willkommii* by wounding experiments. Three types of artificial wounds (cut to bark, puncturing of bark, removal of green needles) were inoculated with ascospores of *L. willkommii*. Infection occurred all year round, in all types of wounds, but symptoms could become visible from 1 to 4 years following inoculation. However, the infection rate after 4 years was related to the month of inoculation, with more infections when inoculation took place at the beginning of the growing season (April, May) or in October (Sylvestre-Guinot and Delatour, 1990). The findings above are consistent with numerous observations of outbreaks of larch canker that have been made following abnormal frost events leading to wounds, or at least cambial necroses, at the bases of shoots.

The pathogen can also be transmitted by mycelium growing from one branch into another, if the branches of two neighbouring trees touch each other (Surini, 2004). The pathogen is able to survive as a saprophyte for some (unknown) period of time (US Department of Agriculture, 1991), but it is quickly inhibited by competing microorganisms (Yde-Andersen, 1979c; Sinclair and Lyon, 2005). It is capable of infecting the xylem (Blanchette, 2001) and so could be present in logs without bark. However, it is not easily isolated from any wood or bark tissues other than those of active canker margins. The development of new fruiting bodies has not been observed to occur on dead wood or bark (Ostrofsky, 2009).

According to Langner (1936), the destruction of even a small amount of bark tissue by the fungal mycelium leads to the production of wood and, similarly, of cork tissue around the infection spot. Toxins produced by the mycelium diffuse through the cork layers into the surrounding living bark leading to a yellowish discoloration and reduction in the frost hardiness of the tissues present without killing them directly. The following frost kills the tissues, which are invaded by the mycelium after the frost event. In the next growing season, the host will again react by forming a layer of cork around the wound between the infected and the non-infected tissue, and the sequence of events described will be ready to start all over again the next winter.

As already mentioned, the cankers may continue their growth year after year, but growth takes place especially in the spring before flushing and not at all in the summer (Hartig, 1880). The rate of extension may vary from canker to canker, but seems to be independent of the growth of the attacked tree (Hopp, 1956; Buczacki, 1973a). Hartig (1880) found that the rate of longitudinal extension varied with the age of the cankers; during the first year after initiation they grew about 10 mm year^{-1}, and after 10 years the extension was about 25 mm year^{-1}. Hopp (1956) found that the average annual extension was 11 mm, but the variation from canker to canker was between 3 and 33 mm. Sometimes the extension ceases, either along the whole of the edge of the canker or along part of it, resulting in the start of healing.

19.4 Epidemiology

In larch canker, the development of epidemics is influenced by: (i) host susceptibility based on resistance; (ii) climatic factors; (iii) site conditions; and (iv) silvicultural management.

All of these factors are relevant for successful disease management strategies, but, as concluded by most researchers, host susceptibility is regarded as the most essential aspect. The age of the host has no influence on whether infection may occur or not (Yde-Andersen, 1980). In terms of the risk of larch canker epidemics, host susceptibility, both at the interspecific and the intraspecific level, is a crucial aspect in artificial stands, but is much less important in natural stands. Numerous observations and experimental trials have shown that the selection of larch species or provenances for afforestation is the primary factor for the health of the growing stand in relation to infection with larch canker. Although there are no totally resistant larch species or provenances known, there are considerable differences in canker resistance that are worth taking into consideration when larch stands are to be established by planting.

Climate is mainly related to the success of infection, which is strongly dependent on sufficient air humidity on the one hand and the frequency of wounding on the other, with the probability of frost damage to the fore. These aspects are relevant at the macroclimatic as well as the microclimatic level, as demonstrated by regional outbreaks of larch canker epidemics following abnormal humid years ('canker years') and the spectrum of factors creating local humid or frost-endangered conditions. The latter are narrowly linked to site properties.

Because *L. willkommii* is principally propagated by wind, and the spores are disseminated each year to infect new trees, larch canker is mainly a problem in mild oceanic climates with frequent precipitation throughout the vegetation period, or in areas or stands with high humidity due to mist or fog. Robak (1964) demonstrated the diverse climatic situation in Norway, which has an oceanic climate along the west coast and a continental climate in the east, as a crucial factor for frost damage favouring *L. willkommii* epidemics. The region west of Norway's north–south running mountains is characterized by mild winter temperatures leading commonly to incomplete dormancy of the larches; buds may already be swelling in November or December, and flushing in late March, with green needle tips sometimes visible as early as in February. Spring frost is rare except in local frosty sites. In continental areas, in contrast, full dormancy is regular and spring frosts are quite common, but of no consequence to the larches as they flush late (Robak, 1964).

Abnormal frost events may trigger larch canker outbreaks regardless of plant provenance or site. As described by Bibelriether and Sperber (1962), larches in Spessart (Germany) had been subject to extreme frost in February and March 1956 after a very mild January. In spring 1956, they flushed only partially, with only some regeneration following during the summer. In the subsequent years there was a high increase in frequency of larch cankers in the stands (Bibelriether and Sperber, 1962). The period of host susceptibility is not known exactly, nor whether the susceptibility is influenced by climatic conditions or not. It appears probable, however, that mild winters predispose the host to infection.

Ascospore release can occur at any time of the year when temperature and rainfall conditions are suitable (Sylvestre-Guinot, 1981; Sylvestre-Guinot and Delatour, 2002), but there are seasonal aspects which are relevant for hygienic measures. The range of temperature when ascospores are released from the apothecia is wide, with reported minima of 0°C (Ito *et al.*, 1963) or 4°C (Sylvestre-Guinot, 1981). As Sylvestre-Guinot found in studies on ascospore release in relation to temperature and humidity, temperatures within the range of 0°C to 30°C play a negligible role compared with humidity. Spore release is stopped only if the mean daily temperature falls below 0°C (Sylvestre-Guinot, 1981). Precipitation, in contrast, directly triggers ascospore production. Rain following drought periods of at least 5–13 days induces spore production within 5–8 h. At the end of rainfall, spore production usually continues for 24–48 h (Sylvestre-Guinot, 1981). There is not much limiting effect of drought periods on ascospore viability; spores of *L. willkommii* can withstand desiccation of 8 months (von Gaisberg and Dieterich, 1928). So the relation between canker growth and climatic conditions is likely to be a consequence of the influence of climate on wound healing. Therefore, dry and warm summers favouring length growth of larch also favour callusing and consequently the chances for recovery (Wachter, 1964).

Sites providing optimal conditions for infection are usually windward exposed, whereas downwind sites are often canker free, because they are less subject to rain (Wachter, 1964). Following Schober (1977), a humid and cool climate, especially the accumulation of cold air, sites in narrow valleys, common fog, wet soils, high vegetation and high frost risk are the main factors that favour outbreaks of larch canker. Thus, a high stand density can also improve the spread of canker by providing damp conditions that favour spore dispersal and germination. Soil quality per se seems to be of minor influence to the severity of larch canker: numerous observations show that stands on nutrient rich soils are not more or less damaged than stands on poor soils (Wachter, 1964).

19.5 Management Strategies and Tactics

When larch stands are being artificially established, choice of site and selection of larch species or provenances are the most essential aspects of disease management. Silviculture has, instead, to be focused on either hygienic measures (sanitation of existing infective material) or measures to enable a quick and strong growth of the young larches.

19.5.1 Avoidance

Measures to avoid outbreaks of larch canker, apart from the selection of suitable species, hybrids or provenances, may be considered from two aspects: selection of the planting site and silvicultural treatment (Young, 1969). Sites likely to be most free from damaging frosts are the middle portions of long convex slopes, or similar situations where free drainage of air always occurs. Sites to be avoided are those where the pooling of cold air leads to frosty conditions on clear cold nights during the growing season. These include: (i) hollows, wherever they may occur; (ii) the foot of long slopes; (iii) valley bottoms, particularly if the slope of the valley is slight, or if the mouth of the valley is constricted; (iv) plains and plateaux lacking air drainage; (v) shelves on hillsides; and (vi) narrow steep-sided glens and gullies. Furthermore, sites close to rivers or lakes very often are susceptible to larch canker outbreaks (Schober, 1977).

Planting larches on sites where larch canker is present on adult trees close by is likely to result in infection of the crop. If larches are established near to a diseased larch stand, it should be done on the downwind side of the stand to avoid infection by windblown spores. Larches should be planted only on warm and sunny sites with low humidity. Sites tending to frost damage should be avoided most of all (Wachter, 1964). Good nutrition may reduce canker development, for instance by enabling the formation of strong and thick bark and rapid growth (Wachter, 1964). High stand density should be avoided, first by planting at not less than 3 m spacing within the row and with 2–3 m between the rows (500–2000 plants ha^{-1}). There should be no mixing with spruce (*Picea* spp.), and early weeding, pruning of dead branches and continuous thinning should be done (Schober, 1977).

Great attention should be paid to weeding and cleaning so that the stand will attain and pass through the thicket stage in the shortest possible time. To this end, the young trees must be kept free for the first few years from interference by grass, bracken, brambles and other low vegetation; it is important in the early stages that the tree should retain its complete branch system, thus ensuring maximum growth. Cleanings should be carried out when necessary to avoid weakening the larches by undue competition from hardwood (broadleaved) coppice. Early brashing and systematic heavy thinnings are also important in maintaining healthy conditions (Plate 26). Individual trees, free from cankers and of fine quality, are often found standing alone in a matrix of other species, which suggests that the risk of infection by canker may be reduced by planting larch in a mixture (Young, 1969).

19.5.2 Exclusion

Quarantine measures against the introduction of larch canker are required by law especially in countries that have either already faced the introduction of *L. willkommii* (e.g. the USA, Canada), or harbour larch species with a high risk of getting infected. Quarantine measures appear to have been successful in limiting the spread of the pathogen to coastal parts of mainland Nova Scotia and New Brunswick (Canadian Forest Service, 1998).

Introduction of the pathogen can result from trade in nursery stock, but also from trade in larch wood and wood products bearing bark (Chalkley, 2010) and, theoretically, from wood free from bark as well (Blanchette, 2001). Canada restricts the movement of such materials from the provinces of New Brunswick and Nova Scotia (Canadian Food Inspection Agency, 2011). Nevertheless, the pathogen is able to spread naturally from tree to tree on land; airborne ascospores may also have been the means by which the pathogen crossed water to reach Prince Edward Island in Canada. Living trees and wood bearing bark must be inspected for the presence of cankers and/or the fruiting bodies of the pathogen. The Canadian Food Inspection Agency regulations provide guidelines, noting that young cankers appear as swellings on twigs and branches, or as depressions on larger stems accompanied by exuding resin (Chalkley, 2010). The fruiting bodies in or around the canker during most of the year are also observed. Inspectors must also watch for withered, dead and/or discoloured needles, and check dead or dying branches and stems carefully. Quarantine regulations in both Canada and the USA are established to prevent further introductions of *L. willkommii* by importation, as well as its spread from the areas in which it already occurs (Canadian Food Inspection Agency, 2011). In Canada, these controls relate to all species, hybrids and horticultural varieties of *Larix* spp. and *Pseudolarix* spp., including plants, plant parts (branches, twigs, scions, logs with bark, pulpwood, isolated bark), plant propagation material and seed with debris, as well as firewood, wood chips, bark chips, logs, telephone poles, railway ties and lumber (Canadian Food Inspection Agency, 2011).

Rhatigan *et al*. (1998) tested the efficiency of methyl bromide in treating of wood against several pathogens by growing them in 1 cm media-amended Dahurian larch (*Larix gmelinii* (Rupr.) Rupr.) wood cubes and exposing them to initial concentrations of 0, 79, or 237 mg l^{-1}

of methyl bromide for 8, 24, or 72 h. All fumigations were performed at 18°C in sealed glass jars. Fumigant concentrations in the jars and sorption into the wood cubes were closely monitored. After fumigation, the cubes were quartered and plated on selected media. The percentage of cube quarter sections with no visible growth after 2 weeks was used as a bioassay of fumigation efficacy. Survival was noted at concentration by time (C × T) values as high as 1230 mg h l^{-1} for *L. willkommii*. As previous studies have shown methyl bromide to diffuse slowly through wet wood, these values may be difficult to achieve in the centre of logs by using conventional tent fumigation techniques (Rhatigan *et al.*, 1998). Nevertheless, methyl bromide fumigation is used in the USA to control pests and pathogens in timber, including *L. willkommii*.

19.5.3 Eradication

As the successful eradication of larch canker in the USA before 1965 showed, intense and repeated removal and destruction of infected branches is a way to wipe out or at least reduce the pathogen inoculum. Cankered plantations that have reached the pole stage invariably improve after a heavy thinning in which as many as possible of the badly infected trees are removed. In some cases, in order to leave only healthy trees, this may so reduce the number of trees that underplanting with shade-bearing conifers is practicable and necessary to make full use of the land (Young, 1969). Following Sylvestre-Guinot (1981), for any measures targeted on a reduction of inoculum, either winter periods with temperatures below 0°C, or drought periods in summer, remain appropriate.

Before eradication, a careful monitoring of the infested stands is required. Infected trees should be felled, and branches cut off. The branch and top area of the stems should be burned on-site, while the larger stem material can be milled into dimension planks or landscape timbers. Slab material generated from the milling should also be burned on-site in winter or early spring and the milling equipment should be disinfected (Ostrofsky, 2009). Pruning tools may also provide a mode of spread, as spores on the tools can be easily transferred to another tree (Surini, 2004). Tools should therefore be disinfected before use. The cleaning of pruning tools is best done after each cutting of an infested branch, by using 70% ethanol. Any wounds cut on the remaining trees should be closed by fungicide mastic (Surini, 2004).

19.5.4 Protection

There is no recommendation for any chemical protective treatment of larch canker except for the closure of wounds by fungicide-containing seals and possible preventive control at the time of sporulation (Surini, 2004).

19.5.5 Resistance

Resistance is the main strategy for managing larch canker, but only if stands are established artificially. This concerns the selection of species of larch as well as the selection of provenances or hybrids.

Larch species susceptibility

Species of larch show high differences in canker susceptibility. Generally, *L. willkommii* is reported from the following hosts: *Larix decidua*, *L. gmelinii*, *L. kaempferi*, *L. kurilensis* Mayr, *L. laricina*, *L.* × *marschlinsii* Coaz, *L. occidentalis* Nutt., *L. sibirica* Lebed., *L. sukaczewii* Dyl. and *Pseudolarix amabilis* (J. Nelson) Rehder. The fungus has been observed in the USA on Chinese golden larch (*P. amabilis*) (Miller and Aldrich, 1936). Hence, no *Larix* or *Pseudolarix* spp. is known to be completely resistant.

Even for Japanese larch, which was regarded as totally resistant in Europe (Chalkley, 2010), some outbreaks of larch canker are known from Germany (Schober, 1953) and from the USA (Hahn and Ayers, 1934). Notwithstanding, the main conclusion with Japanese larch in Europe is that the attacks of the fungus are of no importance (Schober, 1953; Phillips and Burdekin, 1982). In Japan, in contrast, the disease occurs naturally on Japanese larch, where it is found in both natural and artificial stands, and all the trees might be attacked (Ito *et al.*, 1963; Kurkela, 1983).

Hybrids between European and Japanese larches are also considered more or less resistant to the canker (Larsen, 1927, 1956), though, as experiments in Denmark showed, stands of the hybrid may sometimes be almost as heavily attacked as stands of superior provenances of European larch (Göhrn, 1957; Yde-Andersen, 1979a). Nevertheless, hybrid larch is much less susceptible to the canker than other larches. Sylvestre-Guinot *et al.* (1999) confirmed this in experiments in which they inoculated different hybrid larch provenances. After 18 months, the infection rate was 27.9% and most (80–90%) of the experimental clones were found to be receptive. After 39 months, nearly all the infected sites had recovered and the final infection rate was 2.2%. Sylvestre-Guinot and Delatour (2002) also confirmed that there are interspecific differences in susceptibility, with European larch being the most susceptible, then Japanese larch and hybrid larch the least susceptible.

Provenances of European larch and susceptibility

Considerable differences in disease severity of larch canker had already been observed in stands of European larch in the early 19th century. At that time, large-scale decline was common in northern European countries and in parts of Germany. Likewise, the pathogen was already widespread in central and east European countries (especially in the Alps), but here it had a low impact on the stands (Yde-Andersen, 1979a). Host provenance related to a variation in susceptibility of the host to the disease was discussed only in the early 20th century, with the initial research to clarify this being conducted by Münch (1933, 1935a,b).

Münch examined diseased European larch stands in eastern and north-eastern Germany. He found that the majority of the stands younger than 80 years were badly affected by canker, whereas older ones were practically healthy. Furthermore, he observed that canker severity was much higher in stands that had originated by planting than in stands that had been established by sowing. Realizing that the more healthy stands derived from larches from the Sudeten mountains and the diseased ones from alpine regions in Tyrol, he concluded that provenance played the major role in disease disposition. As to the most likely cause of the much higher incidence of larch canker in the younger stands, Münch pointed to the fact that during the past 80 years seeds from Tyrol had been used.

Before that time, seeds from the Sudeten mountains had been used for afforestation. Münch's opinion was that the reason for the higher canker susceptibility of Tyrolean larch was to be found in the climatic conditions of its origin. Tyrolean larch was regarded as a provenance adapted to a continental climate, whereas Sudeten larch was regarded as a provenance adapted to more oceanic conditions. Thus, Tyrolean larch growing in an oceanic climate was damaged by both late and early frosts with consequent larch canker. The results of eight provenances trials with European larch in Württemberg (Germany), aged from 7 to 14 years, confirmed the high susceptibility of Tyrolean larch to canker (Zimmerle, 1941). Schober (1949) examined 111 stands of European larch aged 1 to 180 years in the north-western part of Germany. He found, like Münch, a heavy incidence of canker in stands younger than about 100 years, whereas the older stands were almost without cankers.

Schober (1977) mentioned the results of experiments established in the 1960s in Germany and in other countries. These experiments comprised provenances of European larch from the Alps, the Sudeten mountains, the Carpathian mountains and Poland. The susceptibility of the provenances grown under oceanic conditions decreased in the following order: progeny of larch from the Alps, progeny of larch from the Carpathian and the Sudeten mountains, and progeny of larch from Poland. In the progenies of European larch from the Alps, there were about 41 times as many trees with stem canker as in progenies from the other places of origin (Schober, 1977). So in stands of European larch in its natural range, the disease is of minor importance (Yde-Andersen, 1979a). Where European larch is grown in regions with an oceanic climate, there is a different situation, because under these conditions there is a prevailing risk of epidemic outbreaks of the disease. The most extensive attacks are likely on provenances from the Alps, and less severe attacks are likely to occur on provenances from Moravia and the Sudeten mountains or from selected native stands. The least extensive attacks are to be expected on Polish provenances. Trials in both Denmark and the UK showed that the progeny of selected native stands are attacked only to a moderate degree. This is the case even when they originate from regions now known to be less suitable. It is thought that this is the result of positive selection during one or more generations in the present area of cultivation through disease and thinning (Yde-Andersen, 1979a).

Other studies have investigated the generally low susceptibility of Polish (European) larch provenances from the Sudeten mountains to larch canker (Schober, 1977; Sylvestre-Guinot and Delatour, 1983; Pâques *et al.*, 1999). According to Kulej (2006), who analysed data on the degree of infection of 24 Polish larch provenances in respect of resistance to larch canker, there is no relationship between resistance and the regions in Poland from which the provenances were derived, as even in the Holy mountains or the Sudeten mountains there are some provenances with low resistance and others with higher resistance. However, Kulej's conclusion was that the resistance of larch to *L. willkommii* is determined by its genotype. Thus, provenance plays the first role with respect to susceptibility, followed by site conditions (climate), and cultivation methods (Kulej, 2006).

Expression of resistance

Kulej found a strong correlation between the seed weight of the parental (European larch) trees and resistance of the progeny to larch canker (Kulej, 2001, 2006). Buczacki (1973b), studying European larch, found that the most vigorous trees were the most affected. Contrary to that, Weisgerber and Sindelar (1992) stated that larches with a high growth potential were less attacked by *L. willkommii* than slower growing trees. Kulej (2001) did not find a significant correlation between growth potential of the larches and resistance. However, there is a generally positive correlation between growth vigour and incidence of canker healing (Buczacki, 1973b). Day and Peace (1934) and Schober (1949) observed that cankers on the stems of dominant trees were more inclined to heal than cankers on the stems of suppressed trees. However, the healing seemed primarily to take place when the trees had only a slight diameter increment (Yde-Andersen, 1979a).

Another important factor seems to be the thickness of the bark (Orzel and Kulej, 1999). Kulej (2006) found that certain larch provenances showing high resistance had the thickest bark of all provenances tested. It seems probable that this factor is related to infection biology, as infections are limited to not or less suberized bark areas, which are less protected from frost damage (Wachter, 1964). Therefore, Wachter (1964) recommends the use of larch provenances with a fast juvenile growth, which develop a thick bark in a short time. Furthermore, their winter dormancy should be resistant to breakage.

Breeding for resistance

A reduction of the risk of larch canker by hybridization between species or provenances of low susceptibility and high susceptibility is one aim in larch breeding programmes, which are especially conducted in Europe.

In north-western European countries, where larch is non-native or where it is native but also used as an exotic, breeding has been or is still active: in these countries, either European larch or Japanese larch or both are used together with their hybrids. Breeding has been strongly focused on low elevation reforestation sites (0–1000 m a.s.l.). As a consequence, the choice of the right species (including hybrids) and populations was strategic: adaptability for lowland milder site conditions was and is a priority with a special focus on spring phenology and on canker resistance (Pâques *et al.*, 2013).

Interspecific hybridization is one of the aims for reducing canker susceptibility. Natural hybridization between European and Japanese larches had already been observed in Scotland in the 1900s (Wright, 1976). Artificial crossing between the two species (European larch female × Japanese larch male) was successfully conducted for the first time in Poland in the early 20th century. Interspecific hybrids were mostly developed in western Europe where European larch is native but where local populations are not suitable for lowland ecological conditions, as in France or Germany, and in oceanic countries such as Belgium, Denmark and the UK, where larch is not native and where European larch mostly fails as a result of canker (Pâques *et al.*, 2013). The fast juvenile growth of hybrid larch, together with its low susceptibility to canker, makes it a good candidate for reforestation in more oceanic conditions, but also under more continental conditions where the choice of species is more open (Pâques *et al.*, 2013).

In eastern European countries on the contrary, conservation programmes for native resources are the priority. There, European larch is the favourite species, with little interest in either interspecific or intraspecific hybrids. In some countries, such as Slovakia, the spread of non-native species is prohibited in national parks and nature reserves, and their use is limited or discouraged elsewhere. The establishment of Japanese larch or hybrid populations may create risks: according to Pâques *et al.* (2013), the place of hybrid larch in European forests is put in question from the viewpoint of forest sustainability, because local genetic resources of European larches might be threatened by a bastardization process with the hybrids. The issue is serious in central Europe (Poland, the Czech Republic), and it might become a threat in some regions of the Alps too, where Sudeten (and hybrid) larch is introduced as a response to the susceptibility of local populations to larch canker. Furthermore, in countries that more and more replace artificial regeneration of forests by natural regeneration, hybrid larches formerly introduced by afforestation are not sufficiently known in respect of their successful reproduction in the F_1 and especially the F_2 generations (Pâques *et al.*, 2013). In some other countries, such as Austria, Italy and Scandinavia, the planting of larch is so marginal that here is no need for breeding (Pâques *et al.*, 2013).

19.5.6 Therapy

Individual trees can be freed from *L. willkommii* by the removal of cankers, unless the infection affects the basal parts of the stem.

19.5.7 Integrated disease management

As a consequence of the fact that a variety of factors favouring larch canker outbreaks exist, which may be equally decisive in determining an outbreak, a combination of strategies should preferably be used depending on the specific situation. In North America, avoiding the introduction and spread of the pathogen dominates all other targets. Therefore, strict regulations on the import and transport of infectious material, combined with great and expensive efforts to eradicate or limit the spread of the pathogen, are essential aspects of disease management. In western Europe, where larch stands are often established by planting for economic use, the selection of suitable species or provenances should be combined with the avoidance of sites favouring the disease, i.e. sites subjected to frost or high humidity. In eastern Europe, where the protection of natural larch stands prevails, the management of larch canker should be aimed at avoiding the introduction of foreign larch genotypes, in combination with silvicultural measures to maintain a stand climate that keeps resistance to larch canker outbreaks at a stable and high level. In the area of the Alps, similar strategies are appropriate. In addition, the role of larch as a stand-forming tree species in alpine protection forests should be considered more than at present, so the incidence of larch canker should be kept as low as possible.

References

Baral, H.O. (1984) Taxonomische und ökologische Studien über die Koniferen bewohnenden europäischen Arten der Gattung *Lachnellula* Karsten [Taxonomic and ecological studies on the conifer inhabiting European species of the genus *Lachnellula* Karsten]. *Beiträge zur Kenntnis der Pilze Mitteleuropas* 1, 143–156.

Baral, H.O. (2000) Dichotomous key to *Lachnellula* (worldwide). Available at: http://www.ascofrance.com/uploads/forum_file/7637.doc (accessed 5 March 2012).

Bibelriether, H. and Sperber, G. (eds) (1962) *Lärche und Strobe im Spessart* [*Larch and Weymouth Pine in the Spessart Region*]. *Forstwissenschaftliche Forschungen (Beihefte zum Forstwissenschaftlichen Centralblatt)* 16. Parey, Hamburg, Germany.

Blanchette, J.Y. (2001) Microscopic observations of the early stages of the infection process by *Lachnellula willkommii* (Hartig) Dennis artificially inoculated on larch (*Larix laricina* (Du Roi) K. Koch) seedlings. PhD thesis, University of New Brunswick, Fredericton, New Brunswick, Canada.

Breitenbach, J. and Kränzlin, F. (1984) Pilze der Schweiz. Beitrag zur Kenntnis der Pilzflora der Schweiz. Band 1. Ascomyceten (Schlauchpilze) [Fungi of Switzerland. Contribution to the Knowledge of the Fungal Flora of Switzerland. Volume 1. Ascomycetes (Ascomycota)]. Verlag Mykologia, Lucerne, Switzerland, pp. 937–938.

Brunchorst, J. (1888) Oversigt over de i Norge optraedende, ökonomisk vigtige plantesygdomme [Overview of Norway performers, economically important plant diseases]. *Bergens Museums Årsberetning* 1887, 6.

Buczacki, S.T. (1973a) Some factors governing mycelial establishment and lesion extension in the larch canker disease. *European Journal of Forest Pathology* 3, 39–49.

Buczacki, S.T. (1973b) Some aspects of the relationship between growth vigour, canker, and dieback of European larch. *Forestry* 46, 71–79.

Buczacki, S.T. (1973c) A microecological approach to larch canker biology. *Transactions of the British Mycological Society* 61, 315–329.

Buczacki, S.T. (1973d) Observation on the infection biology of larch canker. *European Journal of Forest Pathology* 3, 228–232.

Buczacki, S.T. (1975) *Trichoscyphella willkommii*. In: CMI Descriptions of Pathogenic Fungi and Bacteria No. 450, Commonwealth Mycological Institute, Kew, UK [now CAB International, Wallingford, UK], pp. 1–2.

Canadian Food Inspection Agency (2011) D-02-12: Import requirements for non-processed wood and other non-propagative wood products, except solid wood packaging material, from all areas other than the continental United States. Effective Date: July 18, 2011 (6th Revision) Available at: http://www.inspection.gc.ca/plants/plant-protection/directives/forestry/d-02-12/eng/1312383927173/1312384324372 (accessed 13 December 2012).

Canadian Forest Service (1998) *Forest Health Highlights in Canada*. Forest Health Network, Atlantic Forestry Centre, Fredericton, New Brunswick, Canada.

Canadian Forest Service (2011) Forest Invasive Alien Species: European larch canker. Available at: http://www.exoticpests.gc.ca/es-details/disease/1000096 (accessed 12 December 2012).

Chalkley, D. (2010) Invasive fungi. European larch canker-*Lachnellula willkommii*. US National Fungus Collections Fungal Databases, USDA ARS Systematic Mycology and Microbiology Laboratory, Agricultural Research Service, US Department of Agriculture, Beltsville, Maryland. Available at: /sbmlweb/fungi/index.cfm (accessed 5 March 2012).

Chaloux, P. (2010) European Larch Canker (ELC) (*Lachnellula willkommii*) – Five townships in Maine added to the quarantine area. National Plant Protection Organization's Phytosanitary Alert System. Available at: http://www.pestalert.org/oprDetail.cfm?oprID=432&keyword=Lachnellula (accessed 5 March 2012).

Day, W.R. and Peace, T.R. (1934) *The Experimental Production and the Diagnosis of Frost Injuries on Forest Trees*. Clarendon Press, Oxford, UK.

Dharne, C.G. (1965) Taxonomic investigations on the discomycetous genus *Lachnellula* Karst. *Journal of Phytopathology* 53, 101–144.

Göhrn, V. (1957) Proveniensforsög med laerk [Provenance trials with larch]. *Forstlige Forsögsvesen Danmark* 23, 1–124.

Hahn, G.G. and Ayers, T.T. (1934) *Dasyscyphae* on conifers in North America. I. The large-spored, white-excipled species. *Mycologia* 26, 73–101.

Hahn, G.G. and Ayers, T.T. (1943) Role of *Dasyscyphus willkommii* and related fungi in the production of canker and dieback of larches. *Journal of Forestry* 41, 483–495.

Harrison, K.J. and DeVerno, L.L. (1999) RAPD fingerprinting of European larch canker, *Lachnellula willkommii* (Hartig) Dennis: a quarantine-related forest pathogen. In: N.E. Carter (ed.) *Proceedings, Northeastern Forest Pest Council Annual Meeting, March 9–11, 1998, Fredericton, NB*. New Brunswick Department of Natural Resources and Energy, Fredericton, New Brunswick, Canada.

Hartig, R. (1880) Die Lärchenkrankheiten, insbesondere der Lärchenkrebspilz *Peziza willkommii* R.Hartig [Larch diseases, especially the canker fungus of larches *Peziza willkommii* R.Hartig]. *Untersuchungen aus dem forstbotanischen Institut zu München* 1, 63–87.

Hiley, W.E. (1919) *The Fungal Diseases of Common Larch*. Clarendon Press, Oxford, UK.

Hoogh, J.De (1924) De groene Douglas (*Pseudotsuga taxifolia* Britt.) in Nederland [The Douglas fir (*Pseudotsuga taxifolia* Britt.) in the Netherlands]. *Mededelingen Rijksboshbouwproefstation* 1, 113–114.

Hopp, P.J. (1956) *Beiträge zur Kenntnis des durch* Dasyscypha willkommii *(Hartig) verursachten Lärchenkrebses* [*Contribution to the Knowledge of the Damage Caused by Larch Canker* Dasyscypha willkommii (*Hartig*)]. Instititut für Waldbau-Technik, Göttingen, Germany.

Ito, K. (1961) Larch canker. *Forest Protection News* 10, 220–223 [in Japanese].

Ito, K. and Zinno, Y. (1957) *Dasyscypha* on *Larix leptolepis* in Japan. *Journal of the Japanese Forestry Society* 39, 452–455 [in Japanese].

Ito, K., Zinno, Y. and Kobayashi, T. (1963) Larch canker in Japan. *Bulletin of the Government Forest Experiment Station, Tokyo* 155, 23–47.

Jörstad, I. (1925) Norske Skogsykdommer. I. Nålesykdommer bevirket av rustsopper, ascomyceter og fungi imperfecti [Norwegian forest diseases. I. Diseases of conifers caused by rusts, ascomycetes, and fungi imperfecti.]. *Meddelelser fra Det Norske Skogforsøksvesen* 2, 19–186.

Kobayashi, T. (1970) Evidence that the larch canker fungus is native in Japan. *Phytopathologische Zeitschrift* 69, 366–368.

Kujala, V. (1950) *Über die Kleinpilze der Koniferen in Finnland,* Ascomycetes, Fungi, Imperfecti, Uredinales [*On the Microfungi of Conifers in Finland,* Ascomycetes, Fungi, Imperfecti, Uredinales]. Metsätieteellinen Tutkimuslaitos, Helsinki, Finland.

Kulej, M. (2001) Zmienność oraz wartość hodowlana modrzewi różnych pochodzeñ z terenu Polski w warunkach siedliskowych Beskidu Sądeckiego. [Variability and breeding quality of the diverse provenances of larch in the habitats of the Beskid Sądecki Mts.]. Thesis, Agricultural Academy in Cracow, Cracow, Poland. Also published as *Zeszyty Naukowe Akademii Rolniczej im. H. Kołłątaja w Krakowie: Rozprawy* No. 273.

Kulej, M. (2006) Resistance of larches of Polish provenances to larch canker *Lachnellula willkommii* (Hartig) Dennis under mountain conditions of the Sacz Beskid. *Electronic Journal of Polish Agricultural Universities* 9(2), article 29. Available at: http://www.ejpau.media.pl/volume9/issue2/art-29.html (accessed 5 March 2012).

Kurkela, T. (1983) European larch canker – its biology, hosts, site relationships, implications to forestry and the microbiology of cankers. European larch canker workshop. Canadian Forest Service. *MFRC Workshop Proceedings* 3, 28–34.

Langner, W. (1936) Untersuchungen über Lärchen-, Apfel- und Buchenkrebs [Investigations on larch, apple and beech canker]. *Phytopathologische Zeitschrift* 9, 111–146.

Larsen, C.S. (1927) Fra en rejse i Skotland i efterårct 1926 [From a trip to Scotland in autumn 1926]. *Dansk Skovforeningens Tidsskrift* 12, 255–272.

Larsen, C.S. (1956) *Genetics in Silviculture*. Oliver and Boyd, London, UK.

Magasi, L.P. and Pond, S.E. (1982) European larch canker: a new disease in Canada and a new North American host record. *Plant Disease* 66, 339–340.

Manners, J.G. (1953) Studies on Larch canker. I. The taxonomy and biology of *Trichoscyphella willkommii* (Hart.) Nannf. and related species. *Transactions of the British Mycological Society* 36, 362–374.

Massee, G. (1902) Larch and spruce and fir canker. *Journal of Board of Agriculture* (London) 9, 176–188.

Massee, G. and Crossland, C. (1905) *Dasyscypha calycina* Fckl. (*P. willkommii* Hartig). Fungus flora of Yorkshire. *Botanical Transactions of the Yorkshire Naturalists' Union* 4, 270–271.

Miller, J.A. and Aldrich, K.F. (1936) *Pseudolarix amabilis*, a new host for *Dasyscypha willkommii*. *Science* 83, 499.

Miller-Weeks, M. and Stark, D. (1983) European larch canker in Maine. *Plant Disease* 67, 448.

Minter, D.W. (2005) *Descriptions of Pathogenic Fungi and Bacteria* 165, Sheets 1641–1650. CAB International, Wallingford, UK.

Münch, E. (1933) Das Lärchenrätsel als Rassenfrage [The larch riddle as a provenance question]. *Tharandter Forstliches Jahrbuch* 84, 438–531.

Münch, E. (1935a) Das Lärchenrätsel als Rassenfrage [The larch riddle as a provenance question]. *Zeitschrift für Forst- und Jagdwesen* 67, 421–442.
Münch, E. (1935b) Das Lärchenrätsel als Rassenfrage [The larch riddle as a provenance question]. *Zeitschrift für Forst- und Jagdwesen* 67, 483–500.
Oguchi, T. (1981) Studies on the species of *Lachnellula* in Hokkaido: their morphology, physiology and pathogenicity. *Bulletin of the Hokkaido Forest Experimental Station* 19, 189–238.
Oppermann, A. (1923) Dyrkning af laerki Danmark [The cultivation of larch in Denmark]. *Forstlige Forsögsvesen i Danmark* 7, 1–324.
Orzel, S. and Kulej, M. (1999) Bark thickness of larches of various Polish provenances in an experimental area of the Forest Experimental Station in Krynica. *Folia Forestalia Polonica*, 41, 25–35.
Ostaff, D.P. (1985) Age distribution of European larch canker in New Brunswick. *Plant Disease* 69, 796–798.
Ostrofsky, W.D. (2009) An eradication plan for larch infected with the larch canker pathogen (*Lachnellula willkommii*) in Brunswick, Cumberland County, Maine. (Appendix C, excerpted from: Forest and Shade Tree Insect & Disease Conditions for Maine: A Summary of the 2008 Situation). Available at: http://www.maine.gov/doc/mfs/BrunswickELC.htm (accessed 5 March 2012).
Pâques, L.E., Sylvestre-Guinot, G. and Delatour, C. (1999) Variabilité clonale de la race polonica du mélèze d'Europe pour la résistance à *Lachnellula willkommii* [Clonal variability of a Polish race of European larch for resistance to *Lachnellula willkommii*]. *Annales des Sciences Forestières* 56, 155–166.
Pâques, L.E., Foffová, E., Heinze, B., Lelu-Walter, M.A., Liesebach, M. and Philippe, G. (2013) Larches (*Larix* sp.) breeding monograph. In: Pâques, L.E. (ed.) *Forest Tree Breeding in Europe. Current State-of-the-Art and Perspectives Series: Managing Forest Ecosystems, Vol. 25.* Springer, Dordrecht, The Netherlands.
Peace, T.R. (1962) *Pathology of Trees and Shrubs.* Clarendon Press, Oxford, UK.
Phillips, D.H. and Burdekin, D.A. (1982) *Diseases of Forest and Ornamental Trees.* Palgrave Macmillan, London.
Rhatigan, R.G., Morrell, J.J. and Filip, G.M. (1998) Toxicity of methyl bromide to four pathogenic fungi in larch heartwood. *Forest Products Journal* 48, 63–67.
Robak, H. (1964) Some observations on larch canker and climate. In: *Proceedings of the FAO/IUFRO Symposium on Internationally Dangerous Forest Diseases and Insects, Vol. 1.* Report No: FO-FAO/FORPEST/64, FAO, Rome, pp. 1–8.
Rostrup, E. (1880) Sygdomme hos Skovtraeerne foraarsagede af ikke rustagtige Snyltesvampe [Diseases of forest trees caused by not-rust-like parasitic fungi]. *Tidsskrift for Skovbrug* 4, 1–86.
Schmid-Heckel, H. (1985) Zur Kenntnis der Pilze in den Nördlichen Kalkalpen. Nationalpark Berchtesgaden [On knowledge of fungi in the northern limestone Alps. Berchtesgaden National Park]. *Forschungsberichte* 8, 1–201.
Schober, R. (1949) *Die Lärche, eine ertragskundlich-biologische Untersuchung* [*Larch: An Investigation of Growth, Yield and Biology*]. Verlag M. & H. Schaper, Hanover, Germany.
Schober, R. (1953) *Die japanische Lärche, eine biologisch-ertragskundliche Untersuchung* [*Japanese Larch: An Investigation of Growth, Yield and Biology*]. *Schriftenreihe der Forstlichen Fakultät der Universität Göttingen* No. 7/8. J.D. Sauerländer's Verlag, Frankfurt, Germany.
Schober, R. (1977) *Vom II. Internationalen Lärchenprovenienzversuch. Ein Beitrag zur Lärchenherkunftsfrage* [*The Second International Larch Provenance Trial. A Contribution to the Larch Origin Question*]. J.D. Sauerländer's Verlag, Frankfurt, Germany.
Schotte, G. (1917) Lärken och dess betydelse for svensk skogshushållning [Larch and its importance for Swedish forest management]. *Meddelanden från Statens Skogsforskningsinstitut* 13, 1–841.
Seaver, F.J. (1951) *The North American Cup Fungi (Inoperculates).* Hafner Publishing Company, New York.
Sinclair, W.A. and Lyon, H. (2005) *Diseases of Trees and Shrubs.* Cornell University Press, Ithaca, New York.
Spaulding, P. and Siggers, P.V. (1927) The European larch canker in America. *Science* 66, 480–481.
Surini, T. (2004) Le Chancre du Mélèze – Larch Canker. *Lachnellula willkommii.* Available at: http://inforets.free.fr/article.php3?id_article=192 (accessed 5 March 2012).
Sylvestre-Guinot, G. (1981) Étude de l'émission des ascospores du *Lachnellula willkommii* (Hartig) Dennis dans l'est de la France [Study on the emission of ascospores of *Lachnellula willkommii* (Hartig) Dennis in the east of France]. *European Journal of Forest Pathology* 11, 275–283.
Sylvestre-Guinot, G. (1986) Étude de sites d'infection du *Lachnellula willkommii* (Hartig) Dennis chez le *Larix decidua* Miller [Study on the sites of infection of *Lachnellula willkommii* (Hartig) Dennis on *Larix decidua* Miller]. *Annales de Science Forestière* 43, 199–206.

Sylvestre-Guinot, G. and Delatour, C. (1983) Possibilités d'appréciation de la sensibilité du genre *Larix* au *Lachnellula willkommii* (Hartig) Dennis par inoculations artificielles [Feasibility of evaluating the susceptibility of *Larix* to *Lachnellula willkommii* by artificial inoculation]. *Annales des Sciences Forestières* 40, 337–354.

Sylvestre-Guinot, G. and Delatour, C. (1990) Réceptivité des blessures aux ascospores de *Lachnellula willkommii* (Hartig) Dennis chez *Larix decidua* Mill. [Receptivity of wounds to ascospores of *Lachnellula willkommii* (Hartig) Dennis on *Larix decidua* Mill.]. *Annales des Sciences Forestières* 47, 57–66.

Sylvestre-Guinot, G. and Delatour, C. (2002) Twenty years of research on larch canker in France. Improvement of larch (*Larix* sp.) for better growth, stem form and wood quality. In: *Proceedings of International Symposium, Gap (Hautes-Alpes), Auvergne & Limousin, France, 16–21 September, 2002*, pp. 194–203.

Sylvestre-Guinot, G., Pâques, L.E. and Delatour, C. (1999) Résistance du mélèze hybride inoculé par *Lachnellula willkommii* [Resistance of hybrid larch inoculated with *Lachnellula willkommii*]. *Annales des Sciences Forestières* 56, 485–492.

Tegethoff, A.C. (1965) Resurvey for European larch canker in Essex County, Massachusetts, 1965. *Plant Disease Reporter* 49, 834–835.

US Department of Agriculture (1991) *Pest Risk Assessment of the Importation of Larch from Siberia and the Soviet Far East*. Miscellanous Publication No. 1495, US Department of Agriculture, Forest Service, Washington. DC.

von Gaisberg, E. and Dieterich, V. (1928) *Beiträge zur Biologie des Lärchenkrebspilzes*, Dasyscypha willkommii *Hrtg*. [*Contribution to the Biology of the Larch Canker Fungus* Dasyscypha willkommii *Hrtg*.]. Mitteilungen der Württembergischen Forstlichen Versuchsanstalt, Tübingen, Germany.

Wachter, H. (1964) Über den Stammkrebsbefall an der europäischen Lärche [On the stem canker infestation of the European larch]. *Forstwissenschaftliches Centralblatt* 83, 257–281.

Weisgerber, H. and Sindelar, J. (1992) IUFRO's role in coniferous tree improvement: history, results, and future trends of research and international cooperation with European larch (*Larix decidua* Mill.). *Silvae Genetica* 41, 150–161.

Wright, J.W. (1976) *Introduction to Forest Genetics*. Academic Press, London.

Xu, H., Qiang, S., Han, Z., Guo, J., Huang, Z., Sun, H., He, S., Ding, H., Wu, H. and Wan, F. (2006) The status and cause of species invasion in China. *Biodiversity and Conservation* 15, 2893–2904.

Yde-Andersen, A. (1979a) Host spectrum, host morphology and geographic distribution of larch canker, *Lachnellula willkommii*: a literature review. *European Journal of Forest Pathology* 9, 211–219.

Yde-Andersen, A. (1979b) Disease symptoms, taxonomy and morphology of *Lachnellula willkommii*. *European Journal of Forest Pathology* 9, 220–228.

Yde-Andersen, A. (1979c) *Lachnellula willkommii* canker formation and the role of microflora: a literature review. *European Journal of Forest Pathology* 9, 347–355.

Yde-Andersen, A. (1980) Infection process and the influence of frost damage in *Lachnellula willkommii*: a literature review. *European Journal of Forest Pathology* 10, 28–36.

Young, C.W.T. (1969) *Larch Canker and Dieback*. Forestry Commission Leaflet 16, Stationery Office, London.

Zimmerle, H. (1941) *Beiträge zur Biologie der Europaeischen Lärche in Württemberg* [*Contribution to the Biology of European Larch in Württemberg*]. Mitteilungen der Wuerttembergschen Forstlichen Versuchsanstalt, Tübingen, Germany.

Zycha, H. (1960) Zur Frage der Infektion beim Lärchenkrebs [On the question of infection in larch cancer]. *Phytopathologische Zeitschrift* 37, 61–74.

20 Hypoxylon Canker

Michael E. Ostry*
USDA Forest Service, Northern Research Station, St Paul, Minnesota, USA

20.1 Pathogen, Significance and Distribution

Entoleuca mammata (Wahlenb.) J.D. Rogers & Y.M. Ju (syn. *Hypoxylon mammatum* (Wahlenb.) P. Karst.) causes the most damaging canker disease of quaking aspen (*Populus tremuloides* Michx.) – Hypoxylon canker – in many areas of North America. A study by Anderson (1964) in the Lake States (Michigan, Minnesota and Wisconsin) estimated that Hypoxylon canker killed 1–2% of the aspen volume each year, which is equivalent to 31% of the net annual growth. The estimated yearly volume loss in Ontario (Canada) was 2 million m^3 (Pitt *et al.*, 2001). Bigtooth aspen (*P. grandidentata* Michx.) is occasionally infected but the disease is not nearly as damaging to this species as it is to quaking aspen.

Quaking and bigtooth aspens are the most abundant commercially and ecologically important species in the region. Aspens are critical to numerous wildlife species for food and habitat, and contribute substantially to many local and regional economies. It was estimated in 1971 that tree mortality attributed to Hypoxylon canker in the Lake States was equal to a loss of 4.4 million US$ a year at harvest (Marty, 1972). Since that time, the value of aspen for pulp and paper and solid wood products has significantly increased, with a corresponding increase in its utilization by the forest industry.

Hypoxylon canker is widely distributed in the north-eastern and Lake States regions of the USA but less common in the western USA (Hinds, 1985) and, for reasons unknown, it is absent from Alaska (Hinds and Laurent, 1978). The disease is widely distributed in Canada (Bier, 1940). In Europe, it has been reported affecting European aspen (*P. tremula* L.) from Finland, Germany and Sweden (Miller, 1961), and from France (Pinon, 1979). Other hosts in Europe include white poplar (*P. alba* L.), black cottonwood (*P. balsamifera* L. subsp. *trichocarpa* (Torr. & A. Gray ex Hook.) Brayshaw) and hybrid aspen (*P. tremula* × *P. tremuloides*) (Kasanen *et al.*, 2004). The fungus has also been collected in Italy and Switzerland (Kasanen *et al.*, 2004).

Various hybrid *Populus* spp. are occasionally damaged by *E. mammata*. Hypoxylon cankers were found on plantation-grown trees of *P. nigra* L. var. *betulifolia* × *P. nigra* 'Volga', *P. nigra* var. *betulifolia* × *P. balsamifera* subsp. *trichocarpa*, *P. maximowiczii* A. Henry × (*P.* ×*berolinensis* C. Koch), *P. balsamifera* subsp. *balsamifera* × (*P.* ×*berolinensis*), and *P. deltoides* Bartram ex Marsh. × *P. nigra* 'Incrassata' (Ostry and McNabb, 1986). Experimental evidence

* E-mail: mostry@fs.fed.us

for confirming reports of the saprophytic or pathogenic relationships of *E. mammata* on several hardwood (broadleaved) species other than *Populus* have largely been inconclusive (Ostry and Anderson, 2009). Its pathogenicity on aspen, along with several key morphological features, were evidence that led Rogers and Ju (1996) to remove the aspen fungus from the genus *Hypoxylon* and place it in the genus *Entoleuca*.

The incidence and impact of Hypoxylon canker is greatest in the first 20 years of a developing aspen stand. Stem cankers on trees of this age are generally low on the stem, resulting in the death of affected trees; this contrasts with the upper stem cankers that develop in older trees and generally are not lethal if the trees develop new leaders (Anderson and Martin, 1981; Ostry and Anderson, 2009). In addition to trees in native forests, aspens in plantations and ornamental landscape plantings are vulnerable to damage by Hypoxylon canker.

20.2 Diagnosis

Cankers can develop anywhere on the branches and stems of aspens of all ages. Symptoms of infection by *E. mammata* are quite variable depending on the stage of disease development. Young cankers first become visible as slightly sunken, yellowish orange areas with irregular margins. Later, the outermost bark (periderm) within the canker becomes blistered, eventually cracking open, and exposing a powdery grey mat of fungal tissue, conidial pillars and conidia. Conidia are single celled, hyaline, fusiform to ellipsoid and range from 5.5–8.0 × 1.5–4.0 μm in size.

As patches of bark flake off, cankers become rough and black in the centre while the bark at the advancing margins of the enlarging cankers becomes yellowish orange. In the oldest areas of canker tissue, perithecial stromata develop in the crumbling cortex. The patches of hard, cushion-like stromata are white when young and turn grey to black as they age. Ascospores are single celled, dark brown, elongate ellipsoid and range from 9.0–12.0 × 20.0–33.0 μm in size.

Callus occasionally develops at the margins of cankers but usually the fungus invades new tissues and the canker expands too rapidly for callus to develop, resulting in characteristic diffuse cankers that can eventually girdle and kill affected trees. Variation in callus production and wound closure may explain differences in canker resistance and susceptibility among aspen clones. Cutting into cankers exposes white or grey mycelial fans that usually extend beyond the visible canker margin, and the diagnostic mottled black and yellowish cream colour of the sapwood.

'Flags', i.e. dead leaves remaining on branches girdled by a canker, are a common disease symptom in the crowns of affected trees, and are especially visible during the dormant season. Wood decay by *E. mammata* (Merrill *et al.*, 1964) often results in branch and stem breakage at cankers and is also a damage characteristic of the disease (Plate 27).

20.3 Infection Biology

Even after many years of research, questions remain concerning some aspects of the infection biology of *E. mammata*, but the consensus of investigators is that ascospores infect wounded xylem (Ostry and Anderson, 2009). Hubbes (1964) described *E. mammata* as a sapwood parasite that required a wound into the wood or dead bark to avoid toxic compounds in the living bark that inhibit growth of the fungus (Hubbes, 1962, 1969).

Toxins produced by *E. mammata* are thought to be involved in pathogenesis but their specific role still remains unclear (Griffin and Manion, 1985; Bélanger *et al.*, 1989, 1990; Mottet *et al.*, 1991; Pinon and Manion, 1991; Kruger and Manion, 1993a,b). Hubbes (1964) provided the first evidence that diffusible substances produced by *E. mammata* inhibited the toxic bark effects and prevented callus formation at wound sites on aspen. It was postulated that the toxin was important in the initial process of infection and disease development (Schipper, 1975). Cell-free culture extracts of *E. mammata* caused aspen bark necrosis and collapse, and inhibited wound callus formation (Schipper, 1978).

Although most commonly encountered as a stem canker, the fungus can infect twigs and branches throughout the crowns of trees. Advanced stem cankers are often centred on a branch stub. Cankers commonly originate on branches and, if they are close by, eventually grow into the main stems of affected trees. Dead wood is not infected by *E. mammata* and cankers do not expand far into dead wood. Cankers have been found originating at the base of young, unwounded branches (Manion, 1975). The absence of the green layer above branch axils, which, if present, consists of cortex and part of the secondary phloem containing host defence compounds, may also provide an entry court for the fungus (French and Oshima, 1959).

The ideal wound for infection of aspen by *E. mammata* was described by Bagga and Smalley (1969). Their hypothesis was that bark-boring insects could cause deep wounds and tunnels with high relative humidity into which the fungus could penetrate and colonize trees. Because *E. mammata* utilizes cellulose, cellobiose and glucose in the xylem, but not sucrose in the phloem (Bagga, 1968; Schipper and Anderson, 1971), wounds deep into the xylem not only provide access to the sugars it requires for growth, but the pathogen also avoids the toxins in the bark. The fungus forms sheets of hyphae that decay the woody xylem, obtaining nutrients that allow it to grow into the phloem and eventually producing enzymes that detoxify the fungistatic compounds in the green bark layer (Ostry and Anderson, 1998).

Several species of insects make oviposition wounds on aspen that have been demonstrated to be frequent sites for infection (Anderson *et al.*, 1979; Ostry and Anderson, 1983, 1986). The habits of these insects and of downy woodpeckers (*Picoides pubescens* (Linnaeus)) foraging on the insect larvae partially explain the distribution of Hypoxylon canker within affected trees and aspen stands (Ostry *et al.*, 1982; Ostry and Anderson, 1998). On average, there is a latent period of 26 months from infection by ascospores through galls resulting from oviposition by the poplar gall saperda (*Saperda inornata* Say) until symptom expression (Ostry and Anderson, 1995), which has contributed to the past difficulty in determining the origin of cankers (Plate 28).

Approximately a year after symptoms develop, hyphal pegs form and break through the bark periderm revealing the grey mat of fungal tissue and exposing the conidia. Conidia are not infectious, nor are they involved in the spread of the fungus in the field. Rather, they are thought to function as spermatia for the production of ascospores that develop in the perithecial stromata that replace the conidial mats 1–2 years later (Griffin *et al.*, 1992).

20.4 Epidemiology

There have been many studies of Hypoxylon canker conducted in natural stands and many of the results have been contradictory. The wide geographical range of aspen, with many site and environmental variables, its clonal habit and variation in disease susceptibility, and differences in the ages of trees under study have confounded interpretations of the research results. In addition, care must be exercised in reading these reports because it is not always clear whether the authors were including branch cankers as well as stem cankers in their analyses. Furthermore, variation in disease resistance among aspen clones expressed as rapid stem death, in contrast to tolerance in which diseased stems can support cankers for longer periods of time, has complicated studies of the disease. More importantly, it is not always evident whether the results are expressed in terms of the prevalence (ratio of living infected trees to total living trees at a given time) or incidence (ratio of new infected trees to the initial total of trees during a given time interval).

Factors contributing to moisture stress were found to increase canker susceptibility (Bagga and Smalley, 1974). Moisture stress, through its effect on increasing host amino acids favourable for the growth of *E. mammata* (Griffin *et al.*, 1986; Bélanger *et al.*, 1990), and in reducing levels of bark compounds inhibitory to the growth of *E. mammata* (Kruger and Manion, 1994), has been suggested as a major factor in increasing tree susceptibility and canker elongation. Moisture stress has also been shown to influence infection by ascospores (Bier and Rowat, 1962b).

Ascospores are liberated from perithecia during wet weather throughout the year, including during the winter months in Minnesota at temperatures as low as –4°C (Wood and French, 1965a), although infection of trees during those months has not been demonstrated. In the field, branch galls resulting from oviposition by the poplar gall saperda on aspens inoculated with ascospores resulted in canker development, thus providing evidence of the critical role of insect wounds and ascospores in the biology of the disease (Ostry and Anderson, 1995).

Detailed study of the increase in canker incidence in an aspen plantation (Manion and Blume, 1975) and in naturally occurring clones (Falk *et al.*, 1989), and genetic studies of the population structure of *E. mammata* within plantations (Griffin *et al.*, 1992; Ostry and Anderson, 2009) have revealed that the spread of the fungus and increase in disease incidence were random and that each canker in the populations studied was genetically different. It was concluded that cankers result from infection by single ascospores and that somatic incompatibility keeps individual canker isolates genetically isolated, thereby preventing pathogenic races from developing. Evidence for outcrossing in populations of *E. mammata* was obtained in experimental pairings of single ascospore-derived strains from the same perithecium (Bier, 1940; Sharland and Rayner, 1989).

Yearly fluctuations in the prevalence of Hypoxylon canker (i.e. 'wave years') reported by Schmiege and Anderson (1960) were suggested to be the result of cankers that developed under favourable conditions for infection in the galleries made by various insects (Nord and Knight, 1972).

A conceptual model incorporating various pre-infection and post-infection factors in the Hypoxylon canker disease cycle was developed to illustrate how the host, pathogen and environment interact within individual trees and within stands. Pre-infection factors, such as early wound responses, defence compounds and resistance to wood-boring insects are difficult to select for but may correlate with natural canker prevalence better than post-infection factors, such as rate of canker elongation, callus formation or branch death (Ostry and Anderson, 2009).

20.5 Management Strategies and Tactics

A large body of literature on Hypoxylon canker produced by numerous researchers has added to our understanding of this disease. However, it is difficult to provide examples of direct, operational control efforts or changes in management directed at minimizing its impact as a result of this research. More often, the loss in volume and quality resulting from a high incidence of white trunk rot caused by *Phellinus tremulae* (Bondartsev) Bondartsev & P.N. Borisov is the major factor influencing harvest age over much of the range of aspen, imposing what has been referred to as the 'pathological rotation' age of aspen.

Aspen has several unique traits that complicate silviculture and favour single-entry, even-aged management (Mowrer, 1988). Among these traits are the clonal growth of aspen, its root sucker regeneration, its extreme shade intolerance, the thin bark that makes it vulnerable to damage by many biotic and abiotic factors and its ability to highly self-regulate its stand density over time.

In the past, the large areas of aspen in the Lake States that established and developed on disturbed sites after the harvest of pine (*Pinus* spp.) and following the many ensuing slash fires have been managed extensively, applying no specific silvicultural practices other than the stands being predominantly regenerated using clear-cutting (clear-felling). For the most part, dependable, natural regeneration was obtained by this method, and stands were allowed to develop until the next felling cycle.

More recently, there has been an interest in applying alternative silvicultural systems for aspen management (Ruark, 1990; Zasada *et al.*, 2001; Gradowski *et al.*, 2010). These systems of intensive management may make it necessary to develop and apply strategies such as genetic improvement and clone manipulation to minimize the impact of Hypoxylon canker and other damaging agents.

20.5.1 Avoidance

The prevalence of Hypoxylon canker and mortality of infected trees have been subjects of study by many investigators who have attempted to correlate site and stand factors with disease incidence and severity. A greater prevalence of Hypoxylon canker on poorer sites has been reported (Lorenz and Christensen, 1937; Gruenhaugen, 1945). Study of the cause of aspen clone deterioration in the Lake States revealed that under similar environmental conditions, genetic differences among clones determined clone longevity, and that stocking differences among clones were attributed to variation in insect damage and to thinning by Hypoxylon canker (Shields and Bockheim, 1981).

Perala (1984) reported significantly less injury by both wood-boring insects and Hypoxylon canker on an excellent aspen site (50 years old, site index 27 m) compared with a good site (50 years old, site index 23 m). However, in an earlier study, a large number of aspen field plots established across the range of aspen in the Lake States to examine the prevalence of Hypoxylon canker as affected by stand and site characteristics yielded ambiguous results (Anderson, 1964). There was no correlation between canker prevalence and site index. The only consistent relationships were that canker prevalence fluctuated widely from year to year and prevalence was less in older stands. The author suggested that climate, microclimate and weather were most likely more important in the occurrence of canker than soil and other characteristics affecting aspen growth and vigour. Looking back on this study though, it was done before the research that demonstrated the importance of wounds made by various insects, which, in turn, influence the incidence of the disease in aspen stands (Anderson *et al.*, 1979; Ostry and Anderson, 1998).

Inoculations of greenhouse-grown aspen plants revealed that factors related to host moisture stress increased susceptibility to infection and canker development (Bagga and Smalley, 1974). A survey of natural aspen stands in central New York for the prevalence of canker revealed that several physical and chemical factors contributing to the availability and retention of moisture were negatively correlated with canker prevalence (Bruck and Manion, 1980). Several interacting soil and host variables were used to develop a prediction model for evaluating sites for the risk of Hypoxylon canker.

One of the most frequent observations on the incidence of Hypoxylon canker cited is its relationship with stand density. Greater disease has been reported among trees along stand edges and in low-density stands by many investigators (Ostry and Anderson, 2009). Based on these reports, aspen management recommendations often include statements of regenerating and maintaining high stocking densities and a closed canopy to avoid excessive losses to Hypoxylon canker (Perala, 1977).

20.5.2 Exclusion

Hypoxylon canker can be found throughout most of the range of aspen in North America except Alaska. Because historically it was assumed that the disease occurred widely, long-distance dispersal and geographic spread of the pathogen was not studied in detail and no explanation for its absence in Alaska has been provided.

Molecular analysis and historical evidence suggests that *E. mammata* was introduced to Europe several centuries ago (Kasanen *et al.*, 2004). The latent period in this disease that was previously mentioned (Ostry and Anderson, 2009) means that the pathogen could inadvertently be moved to unaffected areas on infected, asymptomatic plants. Therefore, reasonable care should be taken not to establish plantations or landscape plantings with infected stock.

20.5.3 Eradication

The removal of infected trees during thinning to reduce inoculum within aspen stands has been suggested (Bier, 1940). However, the removal of infected trees from a stand isolated from other infected aspens failed to eliminate ascospores that were thought to have originated on overlooked cankers and carried into the stand by wind (Froyd and French, 1967). Felling infected trees did not significantly reduce spore inoculum because ascospores continued to be ejected from cankers on these trees and could be collected on spore traps for 2 years after felling (Froyd and French, 1967). The authors concluded that eradication was not an efficient measure of control.

As previously noted, population studies of *E. mammata* in plantations (Griffin *et al.*, 1992; Ostry and Anderson, 2009) demonstrated that the presence of a fungal isolate from an infected tree does not increase the risk of infection of adjacent trees. The availability of suitable wounds, rather than the presence of inoculum, plays a major role in the random occurrences of the disease within stands. Thus, eradication of *E. mammata* inoculum is not a strategy for managing healthy aspen.

20.5.4 Protection

There have been only a few research studies published on the use of chemical protectants against Hypoxylon canker. Cycloheximide was found to inhibit mycelial growth of *E. mammata in vitro*, and cankers did not develop on inoculated, sterilized black cottonwood cuttings dipped in actdione unless the cuttings were water stressed (Bier, 1962). In addition, it was found that the antibiotic treatment was injurious to the developing shoots and roots of the treated cuttings.

Saprophytic fungi on well-watered greenwood cuttings of black cottonwood prevented canker development when inoculated with *E. mammata*, but were not inhibitory when the cuttings were water stressed (Bier and Rowat, 1962a, 1963a). The authors proposed that the antagonistic activity of the facultative parasites would provide an opportunity for biological control using natural bark extracts. Some authors (Bier and Rowat, 1963b; Bier, 1965) have suggested that the susceptibility or resistance of living bark of quaking aspen to *E. mammata* may be related to the presence of microorganisms, which may either be directly antagonistic to *E. mammata* (as just noted) or act via the inactivation of bark pyrocatechol, which is fungistatic towards *E. mammata* (Hubbes, 1962).

Naturally occurring bacteria and fungi on or in *E. mammata* stromata and aspen bark were found to be inhibitory to ascospore germination (Wood and French, 1965b). The authors suggested that at certain times conditions within infection courts might differentially favour the germination of ascospores over development of these naturally occurring microorganisms.

Fertilization by treatments with NPK (nitrogen + phosphorus + potassium) did not result in significant reduction in the incidence of Hypoxylon canker on inoculated aspen in a field test in Michigan (Teachman *et al.*, 1980). The authors postulated that lack of host moisture stress and clonal variation in host susceptibility to infection may have been among the factors confounding the effects of fertilization on canker development.

20.5.5 Resistance

Recognition of the wide variation in many traits among natural aspen clones, including damage by insect pests and pathogens, has provided optimism for the selection and breeding for Hypoxylon canker resistance. In a study in Michigan, Copony and Barnes (1974) found that the number of aspen stems within various clones affected by Hypoxylon canker ranged from 9% to 90%. They reported that 55% to 80% of the variation in the disease among clones was genetic. Discussing the relationship between stand density and canker incidence, the authors suggested that genetic control of variation in suckering ability may be one mechanism involved, with clones producing dense suckers experiencing less damage by Hypoxylon canker than clones producing fewer suckers.

Other investigators question whether canker incidence in the field is an accurate measure of resistance to natural infection (Manion and Griffin, 1992). Results from the inoculation of branches on 100 aspen clones were not correlated with the natural prevalence of canker (French and Hart, 1978). In another study, canker expansion on inoculated stems of three of five clones was not correlated with the natural prevalence of cankers on those clones, although results from the inoculation of grafted plants of the five clones in the greenhouse were (Enebak *et al.*, 1999).

Aspen breeding in the Lake States was initiated in 1955, supported in part by the forest industry (Li *et al.*, 1993a). Large family differences in Hypoxylon canker incidence were found in a 30 year old aspen field trial of 25 full-sib families (Li *et al.*, 1993b). Families that self-pruned well had a lower incidence of Hypoxylon canker than other families. The authors suggested that trees that self-pruned had fewer branches on which cankers could develop and grow into the main stem. Interspecies crosses, however, have not been as successful in aspen breeding because of the high susceptibility of aspens to damage by several diseases (as well as to Hypoxylon canker) (Ostry and Anderson, 2009).

Screening aspen for resistance to Hypoxylon canker has been conducted by inoculating seedlings (Enebak *et al.*, 1996), branches (French and Manion, 1975; Valentine *et al.*, 1976; French and Hart, 1978; Griffin *et al.*, 1984) and main stems (Enebak *et al.*, 1999). In these tests, mycelium of *E. mammata* was introduced through mechanical wounds and canker expansion or the production of canker-inhibiting callus were measured. The results have been variable and open to multiple interpretations. Results from the inoculation of seedlings of two full-sib families with a 30 year canker history in the field were correlated, but highly dependent on seedling age, with the more lignified 9 month old seedlings being better predictors than the 3 month old seedlings (Enebak *et al.*, 1996). Chemical and morphological barriers to *E. mammata* in two aspen genotypes with different levels of field resistance varied by age of tissue and correlated with canker history (Enebak *et al.*, 1997).

A study of the early wound responses of greenhouse-grown aspen genotypes resistant and susceptible to *E. mammata* demonstrated differences in the development of an intact and localized lignified barrier zone and wound callus with phenolic substances that correlated with canker resistance (Bucciarelli *et al.*, 1998, 1999). The levels of defence-related genes were found to be associated with aspen genotype, wounding and time after wounding, and were roughly associated with Hypoxylon canker resistance (Thamarus and Furnier, 1998).

No differences in natural canker incidence were found among trees in a plantation of nearly 600 aspen progeny from controlled crosses of parents that were either both diseased, or only one diseased or both healthy after 20 years (Ostry and Anderson, 2009). Canker incidence among parent trees propagated via root cuttings from diseased and disease-free trees was also similar, illustrating the difficulty of selecting parent trees based on canker prevalence in the field (Manion and Griffin, 1992).

The establishment of stands of improved aspen will require the recognition of not only the genetic component of the disease cycle but also the spatial and environmental factors influencing the disease. One approach that may minimize the impact of Hypoxylon canker in the first 15–20 years of a plantation of widely spaced aspen would be to establish the plantation with resistant progeny selected from long-term tests of clones. After coppicing, spatial resistance (McNabb *et al.*, 1982) resulting from dense sucker reproduction will then further reduce disease incidence.

20.5.6 Therapy

With the exception of pruning infected branches before the canker reaches the main stem (Ostry and Anderson, 1979), or excising the fungus in the early stages of the disease on high-value trees (Hinds and Krebill, 1975), treatment for the direct control of Hypoxylon canker is not available. There have been a few published reports on the investigation of the use of antibiotics to control the canker, but this line of research produced little success.

Treatment of cankers on aspens in Colorado with cycloheximide was not effective in inhibiting the disease (Hinds and Peterson, 1966). The application of cycloheximide in fuel oil directly on to stem cankers arrested canker enlargement, and treated trees survived longer than non-treated trees during a small-scale field test in New York (Brown and Silverborg, 1970). Although the authors stated that their results were limited and inconclusive, they did suggest that direct control of Hypoxylon canker may be useful on high-value trees.

In Minnesota, three formulations of phytoactin and two of semicarbazone in an oil emulsion were aerially applied by helicopter to trees with cankers during the summer (Anderson, 1971). The treatments were not effective in controlling the cankers or increasing the survival of infected trees.

20.5.7 Integrated disease management

As no direct control measures are available, the concept of integrated pest management and silvicultural strategies to manage aspen have been suggested to minimize the impact of Hypoxylon canker (Ostry, 1986). Results from numerous studies and surveys (Schreiner, 1925; Day and Strong, 1959; Copony and Barnes, 1974; Bruck and Manion, 1980; Brandt *et al.*, 2003) have indicated that the disease is less severe in well-stocked stands, so managers have been encouraged to obtain and retain optimum stand stocking. It should be pointed out, however, that there is evidence that infected stems die more rapidly when shaded (Anderson and Anderson, 1968; Anderson, 1972), and therefore infected stems in denser stands may be removed (i.e. die) more quickly than in more open stands. Surveys would thus give the impression that these stands had a lower prevalence of canker than the less stocked stands in which infected stems survive for longer periods of time.

Spatial resistance (McNabb *et al.*, 1982), that is, the interactions of environmental and biotic factors with tree density, can influence disease incidence and this relationship has been proposed as influencing the incidence and severity of Hypoxylon canker in aspen stands. For example, persistent branches on the lower stems of aspen in low-density stands may provide entry courts such as insect oviposition wounds (Ostry and Anderson, 1998) that account for a greater incidence of canker than among trees in dense stands that promote self-pruning (Ostry and Anderson, 1979).

The practice of thinning aspen in general (Perala, 1991; Penner *et al.*, 2001) and, particularly, its effect on the occurrence of Hypoxylon canker (Pitt *et al.*, 2001; Ostry *et al.*, 2004) has been controversial owing to the economic aspects, variation in site, clone genetics, an incomplete understanding of the biology of the disease and contradictory results from research plots. Anderson (1964) reported on two thinning studies: one with significant effects and the other with less certain results. An increase in Hypoxylon canker in thinned plots was reported early in a study in Minnesota (Anderson and Anderson, 1968), but further analysis after 47 years revealed that canker incidence was affected to greater extent by clonal differences in canker susceptibility than to the thinning treatment (Ostry *et al.*, 2004). It has been suggested that forest managers consider factors that affect aspen regeneration densities and rotation length when thinning is planned (Bates *et al.*, 1989). Cleland *et al.* (2001) suggested that the thinning of aspen stands should be delayed until the lower branches have naturally self-pruned.

In Ontario, precommercial thinning resulted in an elevated incidence (based on the initial trees in the stand) of Hypoxylon canker on living trees but no significant differences in the prevalence (based on the number of surviving trees) of canker among the living trees (Pitt *et al.*, 2001). The authors concluded that even though thinning as a management tool could be used without concern for occurrence of the canker, thinning and tree mortality attributed to Hypoxylon canker and other causes may increase the risk of achieving less than optimum stocking for maximum fibre yield at the end of the rotation, and these losses should be factored into determining the desired final densities. Partial felling in aspen resulted in damage by sun scald, logging wounds invaded by canker fungi and infestations by wood-boring insects, all of which weakened the residual trees (Walters *et al.*, 1982).

Clonal differences in the prevalence of Hypoxylon canker, as well as in growth and yield, have been the basis for the suggestion of clone management in aspen stands in which superior clones are favoured over disease-susceptible clones (Perala, 1977). By not harvesting inferior, disease-susceptible clones, the root suckers from the harvested superior clones are allowed to expand and predominate on the site. It has been recommended that if 15–25% of the aspens in a stand is affected by Hypoxylon canker, the stand should be harvested early (Schipper and Anderson, 1976). If more than 25% of the aspens are affected the site should be converted to other species, assuming that clone genetics is responsible for the high disease incidence.

Whether there is a site relationship to canker incidence is uncertain. However, managing aspen on the best sites will increase yields and avoid losses caused by insect pests and other pathogens that are more severe on lower quality sites. Studies have shown that aspen that had a history of defoliation by the forest tent caterpillar (*Malacosoma disstria* Hübner) had an increased susceptibility to wood-boring insects and Hypoxylon canker (Churchill *et al.*, 1964; Anderson and Martin, 1981).

Based on the importance of insect wounds as infection sites, aspen stands should be monitored for insect outbreaks that can increase the risk of Hypoxylon canker. Depending on the current age and stocking of affected stands, it may be desirable to harvest early with the objective of obtaining uniform, full stocking in the regenerating stand. It has been suggested that shearing (i.e. clear-cutting using specialized equipment) young understocked (non-commercial) aspen stands, so that they resprout from the remaining root systems, can be used to regenerate these stands to full stocking and restore their productivity (Perala, 1983). Delaying the harvesting of an aspen stand may be warranted to avoid exposing a young, regenerating stand to known current or expected outbreaks of defoliating insects, such as the forest tent caterpillar, and to subsequent increases of stem-boring insects (Churchill *et al.*, 1964) or to emergences of the periodic cicada (*Magicicada septendecim* Linnaeus); these can all increase the risk of aspen stands to damage by Hypoxylon canker (Ostry and Anderson, 1983).

Hypoxylon canker is just one of the many potentially damaging agents of aspen. Managers need to consider the many values of both 'healthy' and diseased or infested aspen stems for fibre production, wildlife habitat, soil stabilization, aesthetics, etc., throughout its range on various sites. How associated insect pests and pathogens interact with aspen on different sites and under different environmental conditions will determine the vulnerability and ultimate impact to individual trees, clones and stands. Managers will need to determine in many cases whether practical strategies to minimize potential damage are available, warranted or even desirable.

References

Anderson, G.W. (1971) Aerially applied antibiotics fail to control Hypoxylon canker on aspen. *Plant Disease Reporter* 55, 650.

Anderson, G.W. (1972) Development of *Hypoxylon mammatum* cankers on artificially shaded aspen stems. *Forest Science* 18, 316–318.

Anderson, G.W. and Anderson, R.L. (1968) Relationship between density of quaking aspen and incidence of Hypoxylon canker. *Forest Science* 14, 107–112.

Anderson, G.W. and Martin, M.P. (1981) Factors related to incidence of Hypoxylon cankers in aspen and survival of cankered trees. *Forest Science* 27, 461–476.

Anderson, N.A., Ostry, M.E. and Anderson, G.W. (1979) Insect wounds as infection sites for *Hypoxylon mammatum* on trembling aspen. *Phytopathology* 69, 476–479.

Anderson, R.L. (1964) Hypoxylon canker impact on aspen. *Phytopathology* 54, 253–257.

Bagga, D.K. (1968) Physiology of *Hypoxylon pruinatum* and its pathogenesis on quaking aspen. PhD thesis, University of Wisconsin, Madison, Wisconsin.

Bagga, D.K. and Smalley, E.B. (1969) Factors affecting canker development on *Populus tremuloides* artificially inoculated with *Hypoxylon pruinatum*. *Canadian Journal of Botany* 47, 907–914.

Bagga, D.K. and Smalley, E.B. (1974) The development of Hypoxylon canker of *Populus tremuloides*: role of interacting environmental factors. *Phytopathology* 64, 658–662.

Bates, P.C., Blinn, C.R., Alm, A.A. and Perala, D.A. (1989) Aspen stand development following harvest in the Lake States region. *Northern Journal of Applied Forestry* 6, 178–183.

Bélanger, R.R., Falk, S.P., Manion, P.D. and Griffin, D.H. (1989) Tissue culture and leaf spot bioassays as variables in regression models explaining *Hypoxylon mammatum* incidence on *Populus tremuloides* clones in the field. *Phytopathology* 79, 318–321.

Bélanger, R.R., Manion, P.D. and Griffin, D.H. (1990) Amino acid content of water-stressed plantlets of *Populus tremuloides* clones in relation to clonal susceptibility to *Hypoxylon mammatum in vitro*. *Canadian Journal of Botany* 68, 26–29.

Bier, J.E. (1940) *Studies in Forest Pathology. III. Hypoxylon Canker of Poplar*. Publication 691/Technical Bulletin 27, Canadian Department of Agriculture, Ottawa, Ontario, Canada.

Bier, J.E. (1962) Acti-dione and natural bark extracts in the control of Hypoxylon canker of poplar. *Forestry Chronicle* 38, 363–365.

Bier, J.E. (1965) Some inoculum factors in pathogenicity studies of *Hypoxylon pruinatum* (Klotzsche) Cke. on *Populus tremuloides* Michx. *Canadian Journal of Botany* 43, 877–883.

Bier, J.E. and Rowat, M.H. (1962a) The relation of bark moisture to the development of canker diseases caused by native, facultative parasites. VII. Some effects of the saprophytes on the bark of poplar and willow on the incidence of Hypoxylon canker. *Canadian Journal of Botany* 40, 61–69.

Bier, J.E. and Rowat, M.H. (1962b) The relation of bark moisture to the development of canker diseases caused by native, facultative parasites. VIII. Ascospore infection of *Hypoxylon pruinatum* (Klotzsch) Cke. *Canadian Journal of Botany* 40, 897–901.

Bier, J.E. and Rowat, M.H. (1963a) Further effects of bark saprophytes on Hypoxylon canker. *Forest Science* 9, 263–269.

Bier, J.E. and Rowat, M.H. (1963b) Some inoculum and substrate factors in the cultural inhibition of *Hypoxylon pruinatum* (Klotzsche) Cke. by pyrocatechol. *Canadian Journal of Botany* 41, 1585–1596.

Brandt, J.P., Cerezke, H.F., Mallett, K.I., Volney, W.J.A. and Weber, J.D. (2003) Factors affecting trembling aspen (*Populus tremuloides* Michx.) health in the boreal forest of Alberta, Saskatchewan, and Manitoba, Canada. *Forest Ecology and Management* 178, 287–300.

Brown, D.H. and Silverborg, S.B. (1970) Effect of cycloheximide on Hypoxylon cankers on aspen. *Plant Disease Reporter* 54, 462–465.

Bruck, R.I. and Manion, P.D. (1980) Interacting environmental factors associated with the incidence of Hypoxylon canker on trembling aspen. *Canadian Journal of Forest Research* 10, 17–24.

Bucciarelli, B., Jung, H.G., Ostry, M.E., Anderson, N.A. and Vance, C.P. (1998) Wound response characteristics as related to phenylpropanoid enzyme activity and lignin deposition in resistant and susceptible *Populus tremuloides* inoculated with *Entoleuca mammata* (*Hypoxylon mammatum*). *Canadian Journal of Botany* 76, 1282–1289.

Bucciarelli, B., Ostry, M.E., Fulcher, R.G., Anderson, N.A. and Vance, C.P. (1999) Histochemical and microspectrophotometric analyses of early wound responses of resistant and susceptible *Populus tremuloides* inoculated with *Entoleuca mammata* (=*Hypoxylon mammatum*). *Canadian Journal of Botany* 77, 548–555.

Churchill, G.B., John, H.H., Duncan, D.P. and Hodson, A.C. (1964) Long-term effects of defoliation of aspen by the forest tent caterpillar. *Ecology* 45, 630–633.

Cleland, D.T., Leefers, L.A. and Dickmann, D.I. (2001) Ecology and management of aspen: a Lake States perspective. In: Shepperd, W.D., Binkley, D., Bartos, D.L., Stohlgren, T.J. and Eskew, L.G. (eds) *Sustaining Aspen in Western Landscapes: Symposium Proceedings*. Proceedings RMRS-P-18, US Department of Agriculture, Forest Service, Rocky Mountain Research Station, Grand Junction, Colorado, pp. 81–99.

Copony, J.A. and Barnes, B.V. (1974) Clonal variation in the incidence of Hypoxylon canker on trembling aspen. *Canadian Journal of Botany* 52, 1475–1481.

Day, M.W. and Strong, F.C. (1959) A study of Hypoxylon canker on aspen. *Quarterly Bulletin of the Michigan Agricultural Experiment Station* 41, 870–877.

Enebak, S.A., Li, B. and Ostry, M.E. (1996) Seedling response of two trembling aspen (*Populus tremuloides*) families to infection by *Hypoxylon mammatum*. *European Journal of Forest Pathology* 26, 245–252.

Enebak, S.A., Bucciarelli, B., Ostry, M.E. and Li, B. (1997) Histological analyses of the host response of two aspen genotypes to wounding and inoculation with *Hypoxylon mammatum*. *European Journal of Forest Pathology* 27, 337–345.

Enebak, S.A., Ostry, M.E. and Anderson, N.A. (1999) Inoculation methods for selecting *Populus tremuloides* resistant to Hypoxylon canker. *Canadian Journal of Forest Research* 29, 1192–1196.
Falk, S.P., Griffin, D.H. and Manion, P.D. (1989) Spatial pattern analysis and sampling of Hypoxylon canker in naturally occurring clones of *Populus tremuloides*. *Plant Disease* 73, 31–33.
French, D.W. and Oshima, N. (1959) Host bark characteristics and infection by *Hypoxylon pruinatum* (Klot) Cke. *Forest Science* 5, 255–258.
French, J.R. and Hart, J.H. (1978) Variation in resistance of trembling aspen to *Hypoxylon mammatum* identified by inoculating naturally occurring clones. *Phytopathology* 68, 485–489.
French, J.R. and Manion, P.D. (1975) Variability of host and pathogen in Hypoxylon canker of aspen. *Canadian Journal of Botany* 53, 2740–2744.
Froyd, J.D. and French, D.W. (1967) Ejection and dissemination of ascospores of *Hypoxylon pruinatum*. *Canadian Journal of Botany* 45, 1507–1517.
Gradowski, T., Lieffers, V.J., Landhäusser, M., Sidders, D., Volney, J. and Spence, J.R. (2010) Regeneration of *Populus* nine years after variable retention harvest in boreal mixedwood forests. *Forest Ecology and Management* 259, 383–389.
Griffin, D.H. and Manion, P.D. (1985) Host–pathogen interactions as measured by bioassay of metabolites as produced by *Hypoxylon mammatum* with its host *Populus tremuloides*. *Phytopathology* 75, 674–678.
Griffin, D.H., Manion, P.D., Valentine, F.A. and Gustavson, L. (1984) Canker elongation, branch death, and callus formation as resistance or susceptibility responses in *Populus tremuloides* and virulence or avirulence characteristics of *Hypoxylon mammatum*. *Phytopathology* 74, 683–687.
Griffin, D.H., Quinn, K. and McMillen, B. (1986) Regulation of hyphal growth rate of *Hypoxylon mammatum* by amino acids: stimulation by proline. *Experimental Mycology* 10, 307–314.
Griffin, D.H., Quinn, K.E., Gilbert, G.S., Wang, C.J.K. and Rosemarin, S. (1992) The role of ascospores and conidia as propagules in the disease cycle of *Hypoxylon mammatum*. *Phytopathology* 82, 114–119.
Gruenhagen, R.H. (1945) *Hypoxylon pruinatum* and its pathogenesis on poplar. *Phytopathology* 35, 72–89.
Hinds, T.E. (1985) Diseases. In: DeByle, N.V. and Winokur, R.P. (eds) *Aspen: Ecology and Management in the Western United States*. General Technical Report RM-119, US Department of Agriculture, Forest Service, Rocky Mountain Forest and Range Experiment Station, Fort Collins, Colorado, pp. 87–106.
Hinds, T.E. and Krebill, R.G. (1975) *Wounds and Canker Diseases on Western Aspen*. Forest Pest Leaflet 152, US Department of Agriculture, Forest Service, Washington, DC.
Hinds, T.E. and Laurent, T.H. (1978) Common aspen diseases found in Alaska. *Plant Disease Reporter* 62, 972–975.
Hinds, T.E. and Peterson, R.S. (1966) Antibiotic tests on western gall rust and aspen cankers. *Plant Disease Reporter* 50, 741–744.
Hubbes, M. (1962) Inhibition of *Hypoxylon pruinatum* by pyrocatechol isolated from bark of aspen. *Science* 136, 156.
Hubbes, M. (1964) New facts on host–parasite relationships in the Hypoxylon canker of aspen. *Canadian Journal of Botany* 42, 1489–1494.
Hubbes, M. (1969) Benzoic and salicylic acids isolated from a glycoside of aspen bark and their effect on *Hypoxylon pruinatum*. *Canadian Journal of Botany* 47, 1295–1301.
Kasanen, R., Hantula, J., Ostry, M.E., Pinon, J. and Kurkela, T. (2004) North American populations of *Entoluca mammata* are genetically more variable than populations in Europe. *Mycological Research* 108, 766–774.
Kruger, B.M. and Manion, P.D. (1993a) Genetic control of *Populus tremuloides* sensitivity to metabolites of *Hypoxylon mammatum*. *Canadian Journal of Botany* 71, 1276–1279.
Kruger, B.M. and Manion, P.D. (1993b) Sensitivity of *Populus tremuloides* to toxic metabolites of *Hypoxylon mammatum* and susceptibility to *H. mammatum* infection. *Canadian Journal of Botany* 71, 1298–1303.
Kruger, B.M. and Manion, P.D. (1994) Antifungal compounds in aspen: effect of water stress. *Canadian Journal of Botany* 72, 454–460.
Li, B., Wyckoff, G.W. and Einspahr, D.W. (1993a) Hybrid aspen performance and genetic gains. *Northern Journal of Applied Forestry* 10, 117–122.
Li, B., Wyckoff, G.W. and Humenberger, E. (1993b) Early selection for quaking aspen. In: Mohn, C.A. (ed.) *Proceedings of the Second Northern Forest Genetics Association Conference, Roseville, Minnesota. July 29–30, 1993*. University of Minnesota, St Paul, Minnesota, pp. 32–46.
Lorenz, R.C. and Christensen, C.M. (1937) *A Survey of Forest Free Diseases and Their Relation to Stand Improvement in the Lake and Central States*. Bureau of Plant Industry Publication, US Department of Agriculture, Washington, DC.

Manion, P.D. (1975) Two infection sites of *Hypoxylon mammatum* in trembling aspen (*Populus tremuloides*). *Canadian Journal of Botany* 53, 2621–2624.
Manion, P.D. and Blume, M. (1975) Epidemiology of Hypoxylon canker of aspen. *Proceedings of the American Phytopathological Society* 2, 101.
Manion, P.D. and Griffin, D.H. (1992) Resistance in aspen to Hypoxylon canker. In: Blanchette, R.A. and Biggs, A.R. (eds) *Defense Mechanisms of Woody Plants Against Fungi*. Springer Verlag, New York, pp. 308–320.
Marty, R. (1972) The economic impact of Hypoxylon canker on the Lake States resource. In: *Aspen: Symposium Proceedings*. General Technical Report NC-1, US Department of Agriculture, Forest Service, North Central Forest Experiment Station, St Paul, Minnesota, pp. 21–26.
McNabb, H.S. Jr, Hall, R.B. and Ostry, M.E. (1982) Biological and physical modifications of the environment and resulting effect upon the host–parasite interactions in short-rotation tree crops. In: Heybroek, H.M., Stephan, B.R. and von Weissenberg, K. (eds) *Proceedings of the Third International Workshop on the Genetics of Host–Parasite Interactions in Forestry, Wageningen, The Netherlands, 14–21 September 1980*. Centre for Agricultural Publishing and Documentation (Pudoc), Wageningen, The Netherlands, pp. 60–71.
Merrill, W., French, D.W. and Wood, F.A. (1964) Decay of wood by species of the *Xylariaceae*. *Phytopathology* 54, 56–58.
Miller, J.H. (1961) *A Monograph of the World Species of* Hypoxylon. University of Georgia Press, Athens, Georgia.
Mottet, M.J., Dessureault, M. and Bélanger, R. (1991) Infection of *Populus* seedlings with ascospores of *Hypoxylon mammatum*: the role of toxins. *Canadian Journal of Plant Pathology* 13, 281.
Mowrer, H.T. (1988) Is managing aspen density worthwhile? In: Schmidt, W.C. (ed.) *Proceedings – Future Forests of the Mountain West: a Stand Culture Symposium, 1986 September 29–October 3, Missoula, MT*. General Technical Report INT-243, US Department of Agriculture, Forest Service, Intermountain Research Station, Ogden, Utah, pp. 201–209.
Nord, J.C. and Knight, F.B. (1972) The importance of *Saperda inornata* and *Oberea schaumii* (Coleoptera: Cerambycidae) galleries as infection courts of *Hypoxylon pruinatum* in trembling aspen, *Populus tremuloides*. *Great Lakes Entomologist* 5, 87–92.
Ostry, M.E. (1986) Aspen management minimizes disease impact. In: *Proceedings, Integrated Pest Management Symposium for Northern Forests, 1986 March 24–27, Madison, Wisconsin*. Cooperative Extension Service, University of Wisconsin, Madison, Wisconsin, pp. 191–199.
Ostry, M.E. and Anderson, G.W. (1979) Hypoxylon canker incidence on pruned and unpruned aspen. *Canadian Journal of Forest Research* 9, 290–291.
Ostry, M.E. and Anderson, N.A. (1983) Infection of trembling aspen by *Hypoxylon mammatum* through cicada oviposition wounds. *Phytopathology* 73, 1092–1096.
Ostry, M.E. and Anderson, N.A. (1986) Infection of aspen by *Hypoxylon mammatum* through treehopper wounds. *Phytopathology* 76, 957.
Ostry, M.E. and Anderson, N.A. (1995) Infection of *Populus tremuloides* by *Hypoxylon mammatum* ascospores through *Saperda inornata* galls. *Canadian Journal of Forest Research* 25, 813–816.
Ostry, M.E. and Anderson, N.A. (1998) *Interactions of Insects, Woodpeckers, and Hypoxylon Canker on Aspen*. Research Paper NC-331, US Department of Agriculture, Forest Service, North Central Research Station, St Paul, Minnesota.
Ostry, M.E. and Anderson, N.A. (2009) Genetics and ecology of the *Entoleuca mammata–Populus* pathosystem: implications for aspen improvement and management. *Forest Ecology and Management* 257, 390–400.
Ostry, M.E. and McNabb, H.S. Jr (1986) Populus *Species and Hybrid Clones Resistant to* Melampsora, Marssonina, *and* Septoria. Research Paper NC-272, US Department of Agriculture, Forest Service, North Central Research Station, St Paul, Minnesota.
Ostry, M.E., Daniels, K. and Anderson, N.A. (1982) Downy woodpeckers – a missing link in a forest disease life cycle? *Loon* 54, 170–175.
Ostry, M.E., Anderson, N.A., Rugg, D.J. and Ward, K.T. (2004) *Long-term Response of Precommercially Thinned Aspen Clones to Hypoxylon Canker*. Research Paper NC-341, US Department of Agriculture, Forest Service, North Central Research Station, St Paul, Minnesota.
Penner, M., Robinson, C. and Woods, M. (2001) The response of good and poor aspen clones to thinning. *Forestry Chronicle* 77, 874–884.
Perala, D.A. (1977) *Manager's Handbook for Aspen in The North-Central States*. General Technical Report NC-36, US Department of Agriculture, Forest Service, North Central Forest Experiment Station, St Paul, Minnesota.

Perala, D.A. (1983) *Shearing Restores Full Productivity to Sparse Aspen Stands.* Research Note NC-296, US Department of Agriculture, Forest Service, North Central Forest Experiment Station, St Paul, Minnesota.
Perala, D.A. (1984) How endemic injuries affect early growth of aspen suckers. *Canadian Journal of Forest Research* 14, 755–762.
Perala, D.A. (1991) Déjà vu: 'does it pay to thin young aspen?' In: Haavisto, V.F., Smith, C.R. and Mason, C. (eds) *Space to Grow: Spacing and Thinning in Northern Ontario: Proceedings of a Symposium, June 18–20, 1990, Sault Ste. Marie, Ontario.* Joint Report 15, Forestry Canada/Ontario Ministry of Natural Resources, Ontario, Canada, pp. 139–144.
Pinon, J. (1979) The origin and the main characters of French isolates of *Hypoxylon mammatum* (Wahl) Miller. *European Journal of Forest Pathology* 9, 129–142.
Pinon, J. and Manion, P.D. (1991) *Hypoxylon mammatum* and its toxins – recent advances in understanding their relationships with canker disease of poplar. *European Journal of Forest Pathology* 21, 202–209.
Pitt, D., Weingartner, D. and Greifenhagen, S. (2001) Precommercial thinning of trembling aspen in northern Ontario, Part 2. Interactions with Hypoxylon canker. *Forestry Chronicle* 77, 902–910.
Rogers, J.D. and Ju, Y.-M. (1996) *Entoleuca mammata* comb. nov. for *Hypoxylon mammatum* and the genus *Entoleuca*. *Mycotaxon* 59, 441–448.
Ruark, G.A. (1990) Evidence for the reserve shelterwood system for managing quaking aspen. *Northern Journal of Applied Forestry* 7, 58–62.
Schipper, A.L. Jr (1975) Changes in dehydrogenase and peroxidase activities of aspen infected with *Hypoxylon mammatum*. *Phytopathology* 65, 440–445.
Schipper, A.L. Jr (1978) A *Hypoxylon mammatum* pathotoxin responsible for canker formation in quaking aspen. *Phytopathology* 68, 866–872.
Schipper, A.L. Jr and Anderson, G.W. (1971) Alteration of sugar translocation in aspen by *Hypoxylon mammatum*. *Phytopathology* 61, 366–368.
Schipper, A.L. Jr and Anderson, R.L. (1976) *How to Identify Hypoxylon Canker of Aspen.* US Department of Agriculture, Forest Service, North Central Forest Experiment Station, St Paul, Minnesota.
Schmiege, D.C. and Anderson, G.W. (1960) *The Forest Insect and Disease Situation, Lake States 1959.* Station Paper 79, US Department of Agriculture, Forest Service, Lake States Forest Experiment Station, St Paul, Minnesota.
Schreiner, E.J. (1925) Preliminary survey of Hypoxylon poplar canker in Oxford County, Maine. *Mycologia* 17, 218–220.
Sharland, P.R. and Rayner, A.D.M. (1989) Mycelial interactions in outcrossing populations of *Hypoxylon*. *Mycological Research* 93, 187–198.
Shields, W.J. and Bockheim, J.G. (1981) Deterioration of trembling aspen clones in the Great Lakes region. *Canadian Journal of Forest Research* 11, 530–537.
Teachman, G.E., Frederick, D.J., Perkis, W.E. and Jurgensen, M.F. (1980) Effects of fertilization on Hypoxylon canker of trembling aspen. *Plant Disease* 64, 284–286.
Thamarus, K.A. and Furnier, G.R. (1998) Temporal and genotypic variation of wound-induced gene expression in bark of *Populus tremuloides* and *P. grandidentata*. *Canadian Journal of Forest Research* 28, 1611–1620.
Valentine, F.A., Manion, P.D. and Moore, K.E. (1976) Genetic control of resistance to Hypoxylon infection and canker development in *Populus tremuloides*. In: *Proceedings of the Twelfth Lake States Forest Tree Improvement Conference, August 1975.* General Technical Report NC-26, US Department of Agriculture, Forest Service, North Central Forest Experiment Station, St Paul, Minnesota, pp. 132–146.
Walters, J.W., Hinds, T.E., Johnson, D.W. and Beatty, J. (1982) *Effects of Partial Cutting on Diseases, Mortality, and Regeneration of Rocky Mountain Aspen Stands.* Research Paper RM-240, US Department of Agriculture, Forest Service, Rocky Mountain Forest and Range Experiment Station, Fort Collins, Colorado.
Wood, F.A. and French, D.W. (1965a) Hypoxylon canker of aspen: seasonal development of cankers and ascospore ejection in winter. *Phytopathology* 55, 771–774.
Wood, F.A. and French, D.W. (1965b) Microorganisms associated with *Hypoxylon pruinatum*. *Mycologia* 57, 766–775.
Zasada, J.C., David, A.J., Gilmore, D.W. and Landhäusser, S.M. (2001) Ecology and silviculture of natural stands of *Populus* species. In: Dickmann, D.I., Isebrands, J.G., Eckenwalder, J.E. and Richardson, J. (eds) *Poplar Culture in North America.* NRC Press, Ottawa, Canada, pp. 119–151.

21 Branch and Tip Blights

Paolo Capretti,[1]* Alberto Santini[2] and Halvor Solheim[3]
[1]*University of Firenze, Florence, Italy;* [2]*Plant Protection Institute, National Research Council (CNR), Florence, Italy;* [3]*The Norwegian Forest and Landscape Institute, Ås, Norway*

21.1 Pathogens, Significance and Distribution

The green portions of tree crowns have many vital functions. The crown is the place where photosynthesis, a large part of respiration and physiological activities occur, enabling tree growth. The young shoots consist of very delicate tissues that are protected by thin layers of rhytidome, and at the tips by cuticle and waxes. This modest protection enables the exchange of gases between the inner tissues that are rich in water and the organic products of photosynthesis, including sugars, and the environment, where a wide range of organisms is constantly looking for sources of energy. As a consequence, these organisms have modified their lifestyles in the course of evolution, acquiring the capacity to parasitize green tissue.

In the temperate zone, the main agents causing damage to forest tree crowns, with the exception of abiotic agents and insects, are the *Ascomycota*, followed by the rust-causing *Basidiomycota*. This chapter will mainly focus on *Gremmeniella abietina* (Lagerb.) M. Morelet, *Diplodia pinea* (Desm.) J. Kickx f. (syn. *Sphaeropsis sapinea* (Fr.) Dyko & B. Sutton), *Sirococcus conigenus* (Pers.) P.F. Cannon & Minter, and *Venturia populina* (Vuill.) Fabric., which are among the best known agents of serious epidemics on conifers and on poplars (*Populus* spp.).

Crown diseases are often more serious in conifers than in broadleaves. Conifers, with few exceptions, do not renew the green, photosynthezing part of their crowns every year, but retain their needles for 2 or more years. Damage to the crown, especially to the buds and the young shoots, may halt growth and indirectly weaken the tree, rendering it more liable to other diseases. In stands, infections lasting more than a few seasons may be severe enough to weaken the tree, leaving it prone to secondary soil-borne parasites or bark beetles.

The fungi responsible for branch and tip blights are necrotrophic pathogens that in natural forests are generally related to modest and scattered damage as a consequence of episodic stress events (Donaubauer, 1972; Moriondo, 1976). Species such as *G. abietina* and *D. pinea* are considered extremely dangerous and have a wide distribution in temperate forests, particularly in uniform monoculture plantations located on sites characterized by high humidity during the flushing period and mild temperatures during the summer

* E-mail: paolo.capretti@unifi.it

(Gibbs, 1984). To understand the reasons for their present distribution it is necessary to reconsider the increased human disturbance to forest ecosystems over the past two centuries as well as the increased planting of tree monocultures and of exotic tree species (Ostergern and Le Bossé, 2011). Also, realistically, the intensification of international trade in plant material has facilitated the introduction of pathogenic species to new areas (Stenlid *et al.*, 2011; Santini *et al.*, 2013).

These pathogens have changed from their original behaviour as weakly damaging to harmful organisms. The process probably started after the 1850s when, owing to an increasing demand for pulpwood, firewood and saw timber, large parts of the temperate forests of central and northern Europe, as well as of North America, were reforested with introduced, i.e. non-native, species after several years of intensive forestry (Thomsen, 2009; Ostergern and Le Bossé, 2011). In most cases, exotic conifers were used in plantations and, together with the seeds or planting material of exotics, pathogens were accidentally introduced into the new forest stands through nurseries. In Europe, the high demand for timber, including in the metal smelting industries, reduced forest land. During the course of the 18th and 19th centuries, afforestation programmes gradually started over large parts of central and north Europe (Ostergern and Le Bossé, 2011). Since that period, various tree species were imported from neighbouring countries, mostly from the southern regions, thus favouring the spread of harmful microorganisms across the continent and introducing tree provenances that were not always suitable for their new environments (Donaubauer, 1972; Gibbs, 1984). Furthermore, exotic tree species introduced into a new environment also encountered organisms that were previously unable to cause severe damage to the native species with which they co-evolved, but were able to become particularly aggressive pathogens for the new hosts.

Austrian pine (*Pinus nigra* Arnold) and Scots pine (*P. sylvestris* L.) are among the most common species in reforestation programmes in northern and central Europe and North America. Mycological reports from the late 1800s and early 1900s provide evidence that these species have been used by fungal pathogens as a bridge to spread and colonize new environments across the world (Gibbs, 1984; O'Brien, 1984). *G. abietina* and *D. pinea* can be cited in this context as representative examples of pathogens introduced into new environments.

The anamorphic stage of *G. abietina*, *Brunchorstia pinea* (P. Karst.) Höhn, was first reported on Austrian pine in Norway (Brunchorst, 1888). All Fennoscandian countries experienced major outbreaks of *G. abietina* from the 1880s to the beginning of the 20th century and, particularly in Denmark, *G. abietina* disease episodes were initiated on Austrian pine (Thomsen, 2009), which was widely planted outside its natural range (Gibbs, 1984). In North America, the disease, known as Scleroderris canker, was noticed in Michigan in 1951 (O'Brien, 1984). The original source of the infection is unknown (Laflamme, 2009), though there are reports of extensive conifer plantations having been established with young trees (including Austrian pine and Scots pine) imported from Germany since the 1870s (O'Brien, 1984; Tainter and Baker, 1996).

Up to now, three species of the genus *Gremmeniella* have been identified: *G. laricina* (Ettl.) Petrini, L.E. Petrini, Lafl. & Ouell. with host genus *Larix*, *G. juniperina* K. Holm & L. Holm with host genus *Juniperus* and *G. abietina* with *Abies*, *Larix*, *Picea* and *Pinus* as host genera (Petrini *et al.*, 1989). *G. abietina* comprises two varieties: *G. abietina* var. *abietina* (Lagerb.) M. Morelet, found mainly on pines (*Pinus* spp.), but also on spruce (*Picea* spp.) and fir (*Abies* spp.), and *G. abietina* var. *balsamea* Petrini, L.E. Petrini, Lafl. & Ouell., which is found on spruces and firs (Petrini *et al.*, 1989). Three races are distinguished within *G. abietina* var. *abietina* that belong to different continents: there are Asian, North American and European races. The Asian race is established in Japan on Todo fir (*Abies sachalinensis* (Fr. Schmidt) Mast.) (Yokota *et al.*, 1974), while the North American race is found in various pine stands (Laflamme *et al.*, 1998). During the 1980s and 1990s it became clear that there were differences within the European race, and three biotypes were distinguished: Alpine, European (type A, Large Tree Type-LTT) and Northern (type B, Small Tree Type-STT)

(Morelet, 1980a; Uotila, 1983; Hellgren and Högberg, 1995; Hamelin *et al.*, 1996). A fourth biotype has been newly found based on isolates from Aleppo pine (*Pinus halepensis* Mill.) in Spain (Botella *et al.*, 2010).

According to Laflamme (2010), the differences in race within *G. abietina* just described should be raised to species level, so he suggested four species: one on Todo fir in Japan, and three indigenous species in North America with different host species/genera, on balsam fir (*Abies balsamea* (L.) Mill.), spruce and pine. The name *G. abietina* is reserved as a European species because it was described from Sweden (Lagerberg, 1914). In Europe we may have at least two different species (G. Laflamme, Canada, 2012, personal communication), the European biotype as *G. abietina* and the Alpine and Northern biotypes as one species, but more work has to be done on the European biotypes. The ecological differences between the European and Northern biotypes suggest that these types might be separate species. They cross poorly and natural hybrids have not been observed, although their distributions overlap in many areas in Scandinavia (Kaitera *et al.*, 2000; Uotila *et al.*, 2000).

G. abietina has a wide distribution in the northern hemisphere, but many of the taxonomic units (variety, race, biotype) have a limited distribution. The Asian race on Todo fir is found only on Sakhalin Island in Japan (Yokota *et al.*, 1974). The European biotype has a wide distribution in Europe and has also been introduced to North America, where there is a report from New York state, in 1975 (Skilling, 1977); this biotype is now spread over large areas in north-eastern North America. The Northern biotype is restricted to northern Europe, mainly at moderate altitudes, while the Alpine biotype is found at high altitudes in the central European Alps.

The anamorphic fungus *D. pinea* was cited by Desmazieres (1842) in France on Scots pine, and later in Belgium and Italy. The disease, presently known also as Sphaeropsis tip blight, was described in the UK on Scots pine. Later on, starting from the 1920s, the fungus, after accidental introduction, probably via pine seeds or planting material, was reported in several Commonwealth countries, with particularly serious damage in South Africa (1909), New Zealand (1919) and Australia (1936), mostly on Monterey pine (*P. radiata* D. Don). On the same host, the fungus was reported from Argentina in 1910 (Waterman, 1943; Capretti, 1956; Punithalingam and Waterston, 1970). In the USA and Canada, the disease was related to the presence of European pine species (Austrian pine and Scots pine) that are quite often reported as hosts of the pathogen in North America, starting from the 1930s (Waterman, 1943; Punithalingam and Waterston, 1970; Stanosz *et al.*, 2001).

The taxonomy and host range of the *Botryosphaeriaceae* is still under revision and therefore it is quite difficult to depict a complete map of their distribution (see Mehl *et al.*, Chapter 14, this volume). Generally, *D. pinea*, one of the most representative species of this family, is considered to be cosmopolitan and is distributed in natural and native pine forests over the world (Punithalingam and Waterston, 1970). However, in South Africa *D. pinea* has been found on *Prunus* spp. (Damm *et al.*, 2007) and, potentially, might also have a different host range in the northern hemisphere.

A closely related species of *Diplodia*, *D. scrobiculata* J. de Wet, Slippers & M.J. Wingf. is relatively common in the eastern USA and has occasionally been found in France and other Mediterranean countries (Morelet and Chandelier, 1993). In Italy, the fungus has been found not only on coniferous genera but also on angiosperm genera such as *Olea* and *Arbutus* (Lazzizera *et al.*, 2008; Linaldeddu *et al.*, 2010). Recently, *D. scrobiculata* has been detected in the southern hemisphere, most probably introduced with germplasm of Monterey pine from California (de Wet *et al.*, 2003; Bihon *et al.*, 2011).

S. conigenus is an anamorphic fungus responsible of shoot blight on seedlings, saplings and mature trees of several species of conifers in the temperate and boreal forests of the northern hemisphere. Presently, Sirococcus tip blight has been reported on different species, both in nurseries and plantations, in North America and Europe. Reports are mostly from temperate and alpine forests, but also from the Mediterranean basin on Aleppo pine in Spain, Italy and even northern Africa (Morelet,

1972). Possibly, following its main host genera (*Picea* and *Pinus* spp.), the natural distribution of this fungus has been influenced by anthropogenic activities as well (Danti and Capretti, 1998; Halmschlager *et al.*, 2000; Konrad *et al.*, 2007; Rossman *et al.*, 2008; Lilja *et al.*, 2010).

Territorial overlapping in the distribution of fungal pathogens causing similar damage occurs also on poplars affected by Venturia shoot blight. Owing to their high dissemination potential, populations of *Venturia* spp. often overlap with each other at least in Europe and North America. Recently, molecular analysis showed that there are several species within this genus (Tenzer *et al.*, 1999; Kasanen *et al.*, 2001; Blenis, 2007; Lilja *et al.*, 2010).

21.2 Diagnosis

Traditionally, diagnostics in plant pathology is based on observation of symptoms and the identification of organisms that eventually are present on the affected organs. The procedure can be completed by pathogenicity tests following Koch's postulates.

Diagnosis at the crown level might appear difficult as most of the branch and tip blight diseases produce aspecific symptoms. Generally shoots, needles and foliage show symptoms of disease a few weeks after flushing. They consist of discoloration, deformation or reduction of growth, and necrosis of tissues, buds and brachiblasts. In some cases, on conifers, it is possible to observe resin flow as well. Often, signs consisting of fungal fruiting bodies also occur around the infected tissue.

Sometimes symptoms occur in the lower part or the less illuminated inner part of the crown. These symptoms are mostly caused by weak parasites that cause the premature defoliation of old needles and can be mistaken as being associated with environmental factors such as water stress (Butin, 1995). In the most serious cases, there is acute damage, with the withering of young needles that make up the external portion of the crown, and also the death of the shoots and buds. Such severe symptoms are usually caused by virulent fungi, or strong environmental stresses that may kill the tree.

On broadleaves, fungal infections do not usually occur on the terminal shoot, but mostly on new shoots, and as a result the secondary branches do not develop. The buds and the tips of the twigs die, and adventitious buds and sometimes witches' brooms form. Diseases of the leaves and shoots, besides reducing tree growth, also generally cause necrosis and severe premature leaf-fall. The defoliation and necrosis of the shoots that typically occur in spring, such as those caused by *Venturia* spp. on the *Salicaceae* or on ash (*Fraxinus* spp.), or by *Gnomonia* spp. on plane tree (*Platanus* spp.), oak (*Quercus* spp.) and beech (*Fagus* spp.) are injurious not only for the stunting that they invariably cause, which may be very severe, but especially because they reduce the lignification of the new shoots, which make them more prone to damage by cold, and by mildew affecting the new growth in summer.

Of the various fungi causing branch and tip blight, many produce characteristic colonies *in vitro*, or assume conidial shapes that enable them to be identified morphologically. However, the taxonomy of some of these species is so complex that they can only be identified to the genus level, as with some *Diplodia* spp. – *D. pinea* and *D. scrobiculata*, for example – which can only be distinguished by molecular markers. During the last few years, several diagnostic methods based on nucleic acid have been developed (Lundquist and Hamelin, 2005). These can be used to make pathogen-specific probes and identify DNA sequences of the pathogens. In some cases, these sequences are diagnostic at the species level, or sequence analysis might allow also the differentiation of biotypes or races. In more than a few cases, molecular tools allow the detection of fungal pathogens in asymptomatic tissues (Luchi *et al.*, 2005, 2011), and are extremely useful in uncovering the mechanisms behind invasions by exotic forest pathogens, and for understanding their biology, ecology, and epidemiology (Lundquist and Hamelin, 2005). Molecular methods are particularly useful in nursery seedling inspection, quarantine control and forest disease management programmes that require sensitive detection protocols to limit the spread of pathogens. They could also have broad applications as laboratory support for large-scale monitoring programmes. A few examples are reported below for the most common pathogens.

Some molecular tests have been developed for the identification of *G. abietina* on infected plant material or pure cultures. Different races/biotypes of *G. abietina* can also be distinguished that allow for epidemiological and aetiological studies (Hellgren and Högberg, 1995; Hamelin *et al.*, 1996, 1998, 2000; Hantula *et al.*, 1996; Hantula and Müller, 1997; Zeng *et al.*, 2005; Børja *et al.*, 2006).

Probably due to the number of environments and the hosts associated, the identification of the causal organisms responsible for Sphaeropsis tip blight has been controversial for long time, and some taxonomic aspects have only just recently been clarified. *D. pinea* and *D. scrobiculata* can now be distinguished using a PCR-based assay developed for the specific detection of the pathogens from pine host tissues (Smith and Stanosz, 2006). Recently a high resolution melting (HRM) analysis was also set up that allows the distinction of *D. pinea* from closely related species; this shows specific melting profiles for each pathogen, thus allowing a rapid screening in the host (Luchi *et al.*, 2011).

Based on cultural characteristics, conidial morphology and DNA sequences, *S. conigenus* was thought until recently to have two groups: a P group found in the genera *Larix*, *Picea*, *Pinus* and in deodar cedar (*Cedrus deodara* (Roxb.) G. Don f.), and a T group found in western hemlock (*Tsuga heterophylla* (Raf.) Sarg.) and atlas cedar (*Cedrus atlantica* (Endl.) Manetti ex Carrière). Now, using internal transcribed spacer (ITS) markers, a third group, the S group, was found on *Picea* spp. in Canada and Switzerland (Smith *et al.*, 2003; Konrad *et al.*, 2007). Further, these markers allowed Konrad *et al.* (2007) to distinguish different haplotypes in the P group, some of which corresponded to geographical boundaries or to particular host species, e.g. Aleppo pine in Spain and different host species in Canada.

PCR primers were also designed for the detection and identification of *S. conigenus* (formerly P group) and the closely related fungus *S. tsugae* Rossman, Castl., D.F. Farr & Stanosz (formerly T group) in tissues of an apparently healthy host (Smith and Stanosz, 2008). Rossman *et al.* (2008) surveyed the morphological and nucleotide sequence data of these groups and, based on their findings, re-described *S. conigenus* and formally named the S and T groups as *S. piceicola* Rossman, Castl., D.F. Farr & Stanosz and *S. tsugae*, respectively (Lilja *et al.*, 2010).

The taxonomy of *Venturia* spp. is still relatively undocumented and consequently the diagnosis of this species is only relatively correct. In northern Europe, a majority of the shoot blight symptoms on hybrid aspen and European aspen (*Populus tremula* L.) are caused by *V. tremulae* Aderh. (Kasanen *et al.*, 2001, 2004). Molecular analysis showed variation within the ITS sequence suggesting that there are more species within the genus *Venturia*. At least two species occur on hybrid aspen and European aspen: *V. tremulae* (anamorph: *Pollaccia radiosa* (Lib.) E. Bald. & Cif.) and *V. macularis* (Fr.) E. Mül. & Arx, which has been considered to be a saprophyte or non-aggressive pathogen (Morelet, 1985). On *Populus* section *Tacamahaca*, the disease is caused by *V. populina* (anamorph: *Pollaccia elegans* Servazzi) and *V. inopina* G. Newc. (Newcombe, 2003).

21.3 Infection Biology

The agents causing damage to forest trees at crown level reach the succulent tissues of the host shoots either directly through the stomata, leaf scars and buds, or by enzyme action through the cuticle, or indirectly by means of animal vectors. The fungi then act by forming intercellular or superficial mycelium that absorbs nutrients through its haustoria, penetrating the parenchyma and the mesophyllum. This absorption of nutrients soon leads to the death of the meristem tissue (Butin, 1995; Tainter and Baker, 1996).

The fungi causing crown blight sometimes have a sexual form in their life cycle, and in both autumn and spring they form fruiting bodies on the shoots, buds and needles.

After the fungi become established they mostly produce asexual reproductive structures such as acervuli or pycnidia, by which they spread on the trees, and which produce asexual spores (conidia) that give rise to secondary infections. After the growing season, the fungus overwinters in a saprophytic stage – either as mycelium in the tree tissues, or as asexual reproductive structures or fruiting bodies on dead buds and infected twigs, on which it sometimes forms small lesions; it can sometimes be found on the cones as well (Butin, 1995; Tainter and Baker, 1996). The ascospore stage, though rare in some species, enables the fungus to recombine genetically and to develop new forms of the disease it causes. The conidial form, in contrast, enables the fungus to grow and spread around a locus of infection and to do so more than once during the growing season, if the environmental conditions are suitable for mycelial growth.

The conidia and/or ascospores of *G. abietina* are dispersed during most of the growing season, depending on the race/biotype and geography (Uotila, 1992; Petäistö and Heinonen, 2003), and spore dispersal is dependent on weather conditions, especially on moisture (Laflamme and Archambault, 1990; Petäistö and Heinonen, 2003). The fungus enters the apical buds and developing shoots via germinating conidia or ascospores, especially during cool, wet springs and summers. Pine seedlings are most susceptible to infection when buds form or during active shoot growth (Petäistö, 1999; Petäistö and Laine, 1999). After spore germination, the fungus penetrates the periderm and subsequently invades the shoot tissues and proceeds downwards into the stem and needle fascicles. Shoots start dying from the tips in the following spring (Von Lang and Schütt, 1974; Von Siepmann, 1976; Patton *et al.*, 1984).

The first symptoms of infection by *G. abietina* can be seen on needles when their bases turn orange to brown, while the tip may be still green for a while. During the summer, the entire needle becomes brown and finally falls off. Small, black pycnidia appear at the base of dead needles or on dead shoot tips throughout the year, but more commonly in spring and early autumn. After the shoots die, the entire branch may be infected, causing a significant loss of foliage; further weakening of the trees could be due to secondary attacks by other fungi and insects. The fungus overwinters as mycelium in the conifer host or as immature fruiting bodies. *G. abietina* is capable of infecting the host while it is actively growing, but rapid development of disease symptoms can take place while the host plant is dormant. Mortality incurred during the development of an epidemic depends on the size of the host plant at the time of infection (Donaubauer, 1972; Moriondo, 1976; Barklund and Rowe, 1981). The Northern biotype and the North American race of *G. abietina* infect seedlings, small trees and lower branches, and develop under snow cover. Trees weakened by various factors are affected, and infection may also take place in wounds (Roll-Hansen *et al.*, 1992).

G. abietina was first described after studying the cause of extensive top dying in Norway spruce (*Picea abies* (L.) Karst.) in southern Sweden (Lagerberg, 1914). This was a huge infection over large areas, and the typical symptom was a dead top after the infection had encircled the main stem on the previous year's shoot. Solheim (1986) observed that the attacked trees had longer current shoots than unaffected trees. Norway spruce may also be infected as seedlings, both 1 year old and 2 years old. Even though the main disease symptoms are different in Norway spruce from those in pine, the earliest symptoms are quite similar: needles turn brown, starting from the base, in the midsection of the shoot (Børja *et al.*, 2006; Petäistö, 2008). While pine seedlings are affected from the top, Norway spruce seedlings are attacked around the middle of the current shoot (Børja *et al.*, 2006; Petäistö, 2008).

D. pinea is a weak parasite that preferably invades tissues with high nutrient content, such as the mesophyllum, vascular cambium, medullary rays and phloem of plants weakened by abiotic factors. Fungal penetration occurs through stomata or small wounds (Capretti, 1956). The pathogen produces pycnidia in the chlorophyll-containing parenchymatous tissue of needles and green shoots. Germinating conidia infect young tissue and mycelium develops in the stomatal chambers of needles and young shoots, colonizes the resin ducts, pervades secretory cells and grows intercellularly and intracellularly in the mesophyllum and chlorophyllous parenchyma tissues (Punithalingam and Waterston, 1970).

In spring, the shoots show diffuse dieback and browning, and produce resinous gums. The infection usually starts from the younger shoots of the lower branches, but then spreads acropetally through the entire crown. The tree takes on a brownish red necrotic aspect; the dry twigs and needles may remain attached until the following growing season, but they become twisted. The fungus also infects the cones. In late summer, small black pustules form on infected tissue; these represent the pycnidia of the fungus. On infected twigs, small cankers form, and these may ascend to infect larger branches. Infected wood is dark in colour, and when the infection is severe, the tree dies. A typical symptom on Italian stone pine (*P. pinea* L.) is the abortion of the cones before they are mature. On young seedlings, *D. pinea* girdles the stems, killing the plants. In nursery seedlings, *D. pinea* infects the stems, collars and roots directly, killing the seedlings.

S. conigenus infects seeds and the current-year shoots with conidia borne in small, black pycnidia on cone scales, needles or shoots when the humidity is high. Part of the new infections in the nursery originate from contaminated seeds. After seed germination, conidia formed in pycnidia on diseased seedlings spread the disease to new seedlings (Sutherland *et al.*, 1981; Sutherland, 1987). *S. conigenus* affects various stages and organs of its hosts. A close relationship between calcium and magnesium deficiency in the needles and disease severity is suggested (Anglberger *et al.*, 2002, 2003). Apart from infection of the current-year shoots, the fungus can spread into branches, where it may cause cankers. It also occurs on cones and in seeds, which can lead to serious seedling disease, particularly in container nurseries (Sutherland *et al.*, 1981). However, *S. conigenus* can also be isolated from apparently healthy needles and twigs of Norway spruce, which suggests endophytic behaviour (Halmschlager *et al.*, 2000; Anglberger *et al.*, 2003).

Venturia spp. overwinter in the sexual stage and the fruiting bodies are embedded in the host bark or leaf tissue which the pathogen has killed. As the fruiting bodies develop further, the plant surface breaks open and ascospores are released from the ostiole of the fruiting body. Sporulation occurs in rainy weather throughout summer (R. Kasanen and T. Kurkela, Finland, 2012, personal communication). Primary infections by ascospores are followed by conidial production.

21.4 Epidemiology

The likelihood of a crown infection in a given environment depends not only on the occurrence of the inoculum of the pathogen and its amount, but also to a large extent on the weather (Desprez-Loustau *et al.*, 2006; Sturrock *et al.*, 2011). When the climate is temperate and cool with rainfall occurring evenly over the growing season, the trees grow well but many pathogens also complete their life cycle. In contrast, the Mediterranean climate, with its hot and dry periods, cuts short the life cycle of many pathogens, lowering the risk of infection in the aerial organs of trees during these periods. An exception to this rule is found in places where water is a constant presence, such as nurseries, or stands beside rivers or lakes.

Some of the most important environmental factors affecting crown diseases are:

- Humidity of the air, or relative humidity (RH): this allows the growth of the shoots and young leaves, but also favours the growth of fungal fruiting bodies, and the release and germination of spores and conidia, all of which increase the possibility of trees becoming infected.
- Temperature: this factor affects the activity of both the pathogen and the host tree. Temperature extremes, both in summer or in winter, slow down the growth of trees and of pathogenic fungi.
- The amount of light in the crown: this factor favours the growth of the coating of needles and leaves (i.e. cuticle, wax), and at the same time enhances evaporation from the crown and better ventilation. Sunlight, especially its UV

component, also inhibits the germination of fungal and bacterial propagules.

Some forest stands characteristics also affect the course of crown diseases:

- Monospecific versus polyspecific cultures: monospecific stands are at greater risk of infection, and the risk becomes very much greater in monoclonal stands (e.g. poplar plantations).
- Tree age: young trees are more susceptible to disease, not only because their organs are younger, but also because, being lower, they are closer to the ground, which is more humid.
- Stand density: this factor affects the humidity of the stand at crown level and also the amount of illumination in the green portion of the crown.
- Host vigour: this determines the ease with which infections occur. Except for the biotrophic agents (rusts and mildews), weakened trees are more easily infected with crown disease fungi.
- Exposure, elevation, and the depth and duration of snow cover: these factors determine the temperature, the humidity and hence the duration of the growing season for the trees, and the possibility of pathogens completing their life cycles.

Some anthropogenic activities also affect the incidence of crown diseases:

- Fertilization of the soil: nitrogen fertilizers anticipate or prolong the life cycle of trees and increase the water content of the tissues, but at the same time reduce the thickness of the coating that protects the green parts of the trees.
- Air pollution: this factor reduces host vigour but it also reduces the spread and viability of pathogen propagules.
- Global warming: depending on the warming scenario (warmer and drier versus warmer and wetter climate), different classes of pathogens will be favoured (Sturrock *et al.*, 2011).

Climatic conditions such as wet spring and cool summer months, high precipitation, high RH and fog are reported to favour serious *G. abietina* outbreaks (Butin and Hackelberg, 1978; Uotila, 1988). If free water is available in the form of rain for 4–8 h at approximately 17°C, major spore release will take place (Skilling, 1972). Winter storage temperatures influence the infection rate in seedlings; those overwintering at –7°C and –3°C are more diseased than those kept at 0°C (Petäistö and Laine, 1999). In addition, factors that retard shoot maturation increase host plant susceptibility to *G. abietina* infection (Donaubauer, 1972; Uotila, 1988). Examples of these include increased nitrogen levels (Ylimartimo, 1991; Barklund, 1993), incomplete lignification due to rapid growth, low temperatures and light intensity during the summer (Petäistö and Repo, 1988; Gezelius and Näsholm, 1993).

In southern Europe, climatic conditions are not usually conducive for the disease. However, epidemic events have occurred on Italian stone pine and Aleppo pine in both Italy and Spain (Capretti, 1984; Botella *et al.*, 2010). It is thought that these epidemics were favoured by unusually low air temperatures from late winter to the spring, which affected these thermophilic pine species. The source of the inoculum possibly originated from the near mountain chains of the Alps and the Pyrenees, where the disease is common on pines (Morelet, 1980b; Capretti, 1984; Botella *et al.*, 2010). One of the worst epidemics in northern Europe occurred in spring 2001 on large Scots pines in south-eastern Norway (Solheim, 2001) and in adjacent parts of Sweden (Wulff *et al.*, 2006). The autumn of 2000 had been mild and wet with a sudden onset of frost periods, and the fluctuation between mild and cold periods continued throughout the whole winter. These climatic conditions are known to be conducive to infection by *G. abietina* (Donaubauer, 1972; Marosy *et al.*, 1989).

The European biotype of *G. abietina* was introduced to North America in 1975 in New York state (Skilling, 1977), and in 1978 it was found in Quebec (Canada) (Lachance, 1979). Even though it has been treated as a quarantine pathogen, it has spread over large areas in north-eastern North America. In Newfoundland, only the European biotype of the fungus is found (Warren *et al.*, 2011). Scleroderris canker was first recorded there in 1979 and by 1980 the province has already established a quarantine zone (Singh *et al.*, 1980; Warren *et al.*, 2011). Over successive years, however, the disease has intensified and become more widespread, and in 1998 a new quarantine zone had to be established. This held until 2007 when there was an outbreak of *G. abietina* in a pine plantation located 400 km outside the quarantine zone. How the disease arrived is uncertain, but campsites for hunting groups from areas with the disease could be the source; hunters arrived with a lot of firewood from areas with potentially diseased pine trees (Warren *et al.*, 2011).

D. pinea spreads through conidia, which are windborne and released in the spring and summer. Rainy hot weather favours the onset of infection. Infection is also favoured by wounds or decline. Declining trees, especially when they receive too much shade or are water stressed, are particularly susceptible (Maresi *et al.*, 2007). The fungus can also survive as a saprophyte, and the inoculum survives for many years on infected cones lying on the ground (Santini *et al.*, 2013).

A low light intensity, either alone or in conjunction with low temperature before inoculation predisposed seedlings to *Sirococcus* infection (Wall and Magasi, 1976). In Austria, imbalanced tree nutrition was associated with an increased incidence of Sirococcus shoot blight on Norway spruce (Anglberger *et al.*, 2002, 2003).

21.5 Management Strategies and Tactics

Whether crown diseases will occur in a forest stand depends to a fairly large extent on environmental factors. Consequently, there are a number of silvicultural approaches that can deal with these diseases (Tainter and Baker, 1996).

In stands where the pathogen is well established, one of the first things to do is to sanitize the trees, and to reduce the amount of inoculum by thinning the stand and pruning infected tree parts. To prevent a recurrence of the disease, infected material (e.g. the cones) has to be removed from the ground, and fallen leaves and needles worked under the soil. To prevent diseases in the future, or when replacing dead trees, nursery material should be chosen whose species and provenance are suitable for the stand, and also resistant to the pathogen.

Currently, the various crown diseases must be countered both by legal regulations and silvicultural practices. These have to prevent further spread of the pathogen and to ensure the effectiveness of control measures, while respecting the environment and safeguarding the equilibrium of the ecosystem. In the future, abiotic diseases associated with environmental extremes are expected to increase (Boland *et al.*, 2004; Desprez-Loustau *et al.*, 2007). Climate change might also improve the establishment of introduced species in new geographical areas. With time, we expect more non-native pathogens to be introduced into the forests of temperate regions (Desprez-Loustau *et al.*, 2007; La Porta *et al.*, 2008; Santini *et al.*, 2008b).

Control measures will vary depending on whether the forest is extensive or is intended for exploitation. In extensive forests, parasites are a natural occurrence and are often compatible with good ecological equilibrium. Control measures will not be designed to eliminate forest diseases altogether, but rather to keep them within sustainable limits. Conversely, in productive stands, control measures should safeguard production, keeping it to economically sustainable levels, while always respecting the environment.

As part of a balanced protection against disease, the forest environment should be constantly monitored, and the incidence of harmful parasites determined from time to time. Monitoring is now facilitated by innovative techniques to identify tree pathogens using molecular techniques (Lundquist and Hamelin, 2005), to detect and locate diseased trees (e.g. the global positioning system, GPS; geographic information systems, GIS; aerial photography; light detection and ranging, LIDAR), and to devise simulation models (Nevalainen, 2002). Depending on the diagnosis, control measures should be taken as part of other silvicultural practices, and can sometimes be supplemented with other protective measures.

Chemical control measures are becoming ever more circumscribed by laws aimed at protecting biological equilibrium. In Europe, under Council Directive 1991/414/EEC chemical control is only permitted in nurseries and in forest stands for commercial wood production (Council of the European Communities, 1991).

Research in biological control is very active, though it has still not borne fruit. In the last few years it has received a significant boost from studies on endophytic fungi in forest trees. Many of these fungi are strong antagonists of tree pathogens and represent an inexhaustible source of secondary metabolites that can be used as biological drugs.

Concurrently, various studies have examined the phenomenon of systemic induced resistance (SIR), which results from the elicitation of plant defence reactions to any stresses of biotic or abiotic origin. This classically occurs when an inducing agent, pathogenic or not, is applied before a challenge inoculation with a pathogen (Bonello *et al.*, 2008). Because SIR is a general phenomenon that can protect the plant against several pathogens, it sustains more attention today that any other mode of action of biocontrol agents. However, it must be emphasized that SIR and other modes of biological control are not mutually exclusive; for example, SIR might only exert a complementary effect to microbial antagonism (Alabouvette *et al.*, 2006).

The genetic improvement of plants by a process of selection has raised many hopes, but at the moment exists only for poplar and some forest trees in new stands (Santini *et al.*, 2005, 2008a; Ghelardini *et al.*, 2006; Laflamme *et al.*, 2006). This project can in any case only be expected to produce results over the long term.

In the last few years, an increasing body of legislation has defined the rules of quarantine more precisely, but such laws can only be effective if they are accompanied by advances in the diagnosis of the diseases that are subject to quarantine. Since the 1960s, the European Union (EU) has issued a series of directives (the latest Council Directive 2000/29/EC) to regulate the import, export and transit of quarantine products, in certain cases imposing specific rules of quarantine (Council of the European Union, 2000).

In existing stands, many pathogens and their harmful effects can be countered by monitoring the trees, and identifying diseases as soon as they appear, as their causal agents may spread very rapidly.

The cultural practices that are recommended are generally those that will ensure tree vigour, so that the trees react robustly to any diseases, especially those caused by weak parasites. This can be done by avoiding environmental stresses, for instance by regular fertilizing, hoeing, control of insects and, where necessary, supplementary watering. At present, control measures are limited to reducing the sources of inoculum and removing plant material fallen to the ground (leaves, twigs, etc.), as well as all branches with fungal stromata, cankers, lesions or witches' brooms, and to eradicating and eliminating dead or declining trees.

Because diseased seedlings may appear green and healthy immediately after winter, their detection in the short time between lifting and outplanting in the springtime remains a problem. Symptoms are dependent on weather conditions, and it may be several weeks before diseases such as Scleroderris canker become clearly visible (Lilja *et al.*, 2010). Modern tools, based on molecular biology, can help to identify diseases in seedlings, although their current cost may prohibit their routine application. Only healthy seedlings survive after outplanting and can guarantee the future of nurseries and artificial reforestation techniques. Changes and adaptations in forest nursery technology have occurred since the 2000s, but these may need to occur at an accelerated rate because of rapid changes in climate.

G. abietina may be controlled in the nursery using the fungicide chlorothalonil applied about seven times from May to mid August (Skilling and Waddell, 1970, 1974). At the forest scale, however, once *G. abietina* is established in a plantation it is almost impossible to control. The use of chemicals is not practicable in plantation crops where careful selection of disease-free planting material, as well as the selection of planting sites at some distance from affected plantations, are important considerations.

In nature, *D. pinea* only infects trees under stress or already declining trees, so it can be controlled by maintaining tree health, avoiding excess stagnant water and removing any sources of inoculum, such as infected cones on the ground on which the fungus remains viable for long periods of time (Santini *et al.*, 2008b). In nurseries, where damage caused by this fungus is greater, chemical control is feasible. Copper-based products can be applied from April or May; alternatively, fungicides can be sprayed at fortnightly intervals, applying not more than four sprayings (Stanosz and Smith, 1996; Moret and Muñoz, 2007).

The post-germination spread of *S. conigenus* occurs from disease loci originating from seed-borne inocula (Sutherland *et al.*, 1981; Sutherland, 1987). Thus, the mishandling of cones or seeds is one source of the pathogen. The contamination of seeds often occurs via the soil during cone collection or in storage (Sutherland *et al.*, 2002). The inclusion of old, infested cones in current year collections can also be a source of *S. conigenus* (Sutherland *et al.*, 2002). Strict hygiene during collection and storage, and culling of damaged seeds, will decrease the risk of seed-borne inocula. Sometimes, the disease can be traced to infected cones on nearby forest trees. In such cases it may be worthwhile removing such trees (Hamm *et al.*, 1990).

V. populina still causes serious defoliation every spring in stands where the cultivation of 'Canadian' clones of *Populus* is more intense. Therefore, the only feasible management strategy is to use mixtures of resistant or moderately susceptible clones (Lilja *et al.*, 2010).

References

Alabouvette, C., Olivain, C. and Steinberg, C. (2006) Biological control of plant diseases: the European situation. *European Journal of Plant Pathology* 114, 329–341.

Anglberger, H., Halmschlager, E., Hietz, P. and Mattanovich, J. (2002) Effect of fertilisation on the resistance of spruce seedlings to shoot dieback caused by *Sirococcus conigenus*. In: Uotila, A. and Ahola, V. (eds) *Proceedings of the IUFRO Working Party 7.02.02 Shoot and Foliage Diseases, Meeting at Hyytiälä, Finland, 17–22 June, 2001*. The Finnish Forest Research Institute Research Papers 829, Metsäntutkimuslaitoksen Tiedonantoja, Vantaaa, Finland, pp. 152–157.

Anglberger, H., Sieghardt, M., Katzensteiner, K. and Halmschlager, E. (2003) Needle nutrient status of *Sirococcus* shoot blight-diseased and healthy Norway spruces. *Forest Pathology* 33, 21–29.

Barklund, P. (1993) Nutrient status in Scots pine stands attacked by *Gremmeniella abietina*. In: Barklund, B., Livsey, S., Karlman, M. and Stephan, R. (eds) *Proceedings of an International Symposium on Shoot Diseases of Conifers. IUFRO Working Party S2.06.02, Canker and Shoot Blight of Conifers, Garpenberg (Sweden), 10–15 June 1991*. Sveriges Lantbruksuniversitet, Uppsala, Sweden, pp. 101–106.

Barklund, P. and Rowe, J. (1981) *Gremmeniella abietina* (*Scleroderris lagerbergii*), a primary parasite in a Norway spruce dieback. *European Journal of Forest Pathology* 11, 97–108.

Bihon, W., Slippers, B., Burgess, T., Wingfield, M.J. and Wingfield, B.D. (2011) *Diplodia scrobiculata* found in the southern hemisphere. *Forest Pathology* 41, 175–181.

Blenis, P.V. (2007) Impact of simulated aspen shoot blight on trembling aspen. *Canadian Journal of Forest Research* 37, 719–735.

Boland, G.J., Melzer, M.S., Hopkin, A., Higgins, V. and Nassuth, A. (2004) Climate change and plant disease in Ontario. *Canadian Journal of Plant Pathology* 26, 335–350.

Bonello, P., Capretti, P., Luchi, N., Martini, V. and Michelozzi, M. (2008) Systemic effects of *Heterobasidion annosum s.s.* infection on severity of *Diplodia pinea* tip blight and terpenoid metabolism in Italian stone pine (*Pinus pinea*). *Tree Physiology* 28, 1653–1660.

Børja, I., Solheim, H., Hietala, A. and Fossdal, C.G. (2006) Etiology and real-time polymerase chain reaction-based detection of *Gremmeniella*- and *Phomopsis*-associated disease in Norway spruce seedlings. *Phytopathology* 96, 1305–1314.

Botella, L., Tuomivirta, T.T., Kaitera, J., Carrasco Navarro, V., Diez, J.J. and Hantula, J. (2010) Spanish population of *Gremmeniella abietina* is genetically unique but related to type A in Europe. *Fungal Biology* 114, 778–789.

Brunchorst, J. (1888) Über eine neue verheerende Kranheit der Schwarzföhre (*Pinus austriaca* Höss) [On a new devastating disease of black pine (*Pinus austriaca* Höss)]. *Bergens Museum Aarsberetning* 6, 1–16.

Butin, H. (1995) *Tree Diseases and Disorders: Causes, Biology and Control in Forest and Amenity Trees*. Oxford University Press, Oxford, UK.

Butin, H. and Hackelberg, L. (1978) Über den Verlauf einer *Scleroderris-lagerbergii*-Epidemie in einem Schwarzkiefernbestand [On the progress of an outbreak of *Scleroderris-lagerbergii* in a stand of black pine]. *European Journal of Forest Pathology* 8, 369–379.

Capretti, C. (1956) *Diplodia pinea* (Desm.) Kickx agente del disseccamento di varie specie del gen. *Pinus* e di altre conifere [*Diplodia pinea* (Desm.) Kickx the cause of a dieback of different species of the genus *Pinus* and other conifers]. *Annali Accademia di Scienze Forestali* 5, 171–202.

Capretti, P. (1984) The movement of *Gremmeniella abietina* from the Alps to the Apennines. In: Manion, D.P. (ed.) *Proceedings of an International Symposium on Scleroderris Canker of Conifers, held in Syracuse, USA, June 21–24, 1983*. Martinus Nijhoff/Dr. W. Junk Publishers, The Hague, The Netherlands, pp. 153–157.

Council of the European Communities (1991) Council Directive 91/414/EEC of 15 July 1991 concerning the placing of plant protection products on the market. *Official Journal of the European Communities* 34, 19.8.1991, L 230/1–32.

Council of the European Union (2000) Council Directive 2000/29/EC of 8 May 2000 on protective measures against the introduction into the Community of organisms harmful to plants or plant products and against their spread within the Community. *Official Journal of the European Communities* 10.7.2000, L 169/1–112.

Damm, U., Crous, P.W. and Fourie, P.H. (2007) *Botryosphaeriaceae* as potential pathogens of *Prunus* species in South Africa, with descriptions of *Diplodia africana* and *Lasiodiplodia plurivora* sp. nov. *Mycologia* 99, 664–680.

Danti, R. and Capretti, P. (1998) Shoot blight of *Pinus halepensis* Mill. in the Italian peninsula. In: Laflamme, G., Bérubé, J.A. and Hamelin, R.C, (eds) *Foliage, Shoot and Stem Diseases of Trees: Proceedings of the IUFRO Working Party 7.02.02 Meeting, May 25–31, 1997, Quebec City, Quebec*. Information Report LAU-X-122, Canadian Forest Service, Laurentian Forestry Centre, Sainte-Foy, Quebec, Canada, pp. 103–107.

de Wet, J., Burges, T., Slippers, B., Preisig, O., Wingfield, B.D. and Wingfield, M.J. (2003) Multiple gene genealogies and microsatellite markers reflect relationship between morphotypes of *Sphaeropsis sapinea* and distinguish a new species of *Diplodia*. *Mycological Research* 107, 557–566.

Desmazières, J.B. (1842) Neuvième notice sur quelques plantes cryptogames la plupart inédies, récemment découvertes en France, et qui vont paraître en nature, dans la collection publiée par l'auteur [Ninth notice on some cryptogamic plants most unpublished, recently discovered in France, and which occur naturally, in the collection published by the author]. *Annales des Sciences Naturelles* 17, 91–118.

Desprez-Loustau, M.-L., Marçais, B., Nageleisen, L.-M., Piou, D. and Vannini, A. (2006) Interactive effects of drought and pathogens in forest trees. *Annals of Forest Science* 63, 597–612.

Desprez-Loustau, M.-L., Robin, C., Reynaud, G., Déqué, M., Badeau, V., Piou, D., Husson, C. and Marçais, B. (2007) Simulating the effects of a climate-change scenario on geographical range and activity of forest-pathogenic fungi. *Canadian Journal of Plant Pathology* 29, 101–120.

Donaubauer, E. (1972) Environmental factors influencing outbreak of *Scleroderris lagerbergii* Gremmen. *European Journal of Forest Pathology* 2, 21–25.

Gezelius, K. and Näsholm, T. (1993) Free amino acids and protein in Scots pine seedlings cultivated at different nutrient availabilities. *Tree Physiology* 13, 71–86.

Ghelardini, L., Falusi, M. and Santini, A. (2006) Variation in timing of bud-burst of *Ulmus minor* clones from different geographical origins. *Canadian Journal of Forest Research* 36, 1982–1991.

Gibbs, J. (1984) *Brunchorstia* dieback in Europe. In: Manion, D.P. (ed.) *Proceedings of an International Symposium on Scleroderris Canker of Conifers, held in Syracuse, USA, June 21–24, 1983*. Martinus Nijhoff/ Dr.W. Junk Publishers, The Hague, The Netherlands, pp. 32–41.

Halmschlager, E., Gabler, A. and Andrae, F. (2000) The impact of *Sirococcus* shoot blight on radial and height growth of Norway spruce (*Picea abies*) in young plantations. *Forest Pathology* 30, 127–133.

Hamelin, R.C., Lecours, N., Hansson, P., Hellgren, M. and Laflamme, G. (1996) Genetic differentiation within the European race of *Gremmeniella abietina*. *Mycological Research* 100, 49–56.

Hamelin, R.C., Lecours, N. and Laflamme, G. (1998) Molecular evidence of distinct introductions of the European race of *Gremmeniella abietina* into North America. *Phytopathology* 88, 582–588.

Hamelin, R.C., Bourassa, M., Railm, M., Dusabenyagasani, M., Jacobi, V. and Laflamme, G. (2000) PCR detection of *Gremmeniella abietina*, the causal agent of Scleroderris canker of pine. *Mycological Research* 104, 527–532.

Hamm, P.B., Campbell, S.J. and Hansen, E.M. (1990) *Growing Healthy Seedlings. Identification and Management of Pests in Northwest Forest Nurseries*. US Department of Agriculture, Forest Service, Pacific Northwest Region and Forest Research Laboratory, College of Forestry, Oregon State University, Corvallis, Oregon.

Hantula, J. and Müller, M.M. (1997) Variation within *Gremmeniella abietina* in Finland and other countries as determined by random amplified microsatellites (RAMS). *Mycological Research* 101, 169–175.

Hantula, J., Dusabenyagasani, M. and Hamelin, R.C. (1996) Random amplified microsatellites (RAMS) – a novel method for characterizing genetic variation within fungi. *European Journal of Forest Pathology* 26, 159–166.

Hellgren, M. and Högberg, N. (1995) Ecotypic variation of *Gremmeniella abietina* in northern Europe: disease patterns reflected by DNA variation. *Canadian Journal of Botany* 73, 1531–1539.

Kaitera, J., Seitamäki, L. and Jalkanen, R. (2000) Morphological and ecological variation of *Gremmeniella abietina* var. *abietina* in *Pinus sylvestris*, *Pinus contorta* and *Picea abies* sapling stands in northern Finland and the Kola Peninsula. *Scandinavian Journal of Forest Research* 15, 13–19.

Kasanen, R., Hantula, J. and Kurkela, T. (2001) The occurrence of an undescribed species of *Venturia* in blighted shoots of *Populus tremula*. *Mycological Research* 105, 338–343.

Kasanen, R., Hantula, J., Vourinen, M., Stenlid, J., Solheim, H. and Kurkela, T. (2004) Migrational capacity of Fennoscandian populations of *Venturia tremulae*. *Mycological Research* 108, 64–70.

Konrad, H., Stauffer, C., Kirisits, T. and Halmschlager, E. (2007) Phylogeographic variation among isolates of the *Sirococcus conigenus* P group. *Forest Pathology* 37, 22–39.

La Porta, N., Capretti, P., Thomsen, I.M., Kasanen, R., Hietala, A.M. and Von Weissenberg, K. (2008) Forest pathogens with higher damage potential due to climate change in Europe. *Canadian Journal of Plant Pathology* 30, 177–195.

Lachance, D. (1979) Découverte de la souche européenne de *Gremmeniella abietina* au Québec [Discovery of the European strain of *Gremmeniella abietina* in Quebec]. *Phytoprotection* 60, 168.

Laflamme, G. (2009) Earliest documented report of Scleroderris canker in North America: damage believed until now to be caused by summer frost. *Phytoprotection* 90, 89–95.

Laflamme, G. (2010) Solutions to the imbroglio over the nomenclature of *Gremmeniella abietina*. *Phytopathology* 100, S194–S195.

Laflamme, G. and Archambault, L. (1990) Evaluation of microclimatic factors affecting ascospore release of *Gremmeniella abietina* var. *balsamea*. *Canadian Journal of Plant Pathology* 12, 190–194.

Laflamme, G., Hopkin, A.A. and Harrison, K.J. (1998) Status of the European race of *Scleroderris* in Canada. *Forestry Chronicle* 74, 561–566.

Laflamme, G., Rioux, D., Simard, M., Bussières, G. and Mallett, K. (2006) Resistance of *Pinus contorta* to the European race of *Gremmeniella abietina*. *Forest Pathology* 36, 83–96.

Lagerberg, T. (1914) Granens topptorka [On top dry-out of spruce in Sweden]. *Meddelanden från Statens Skogsförsöksanstalt* 10, 9–44 [in Swedish, with German summary].

Lazzizera, C., Frisullo, S., Alves, A., Lopes, J. and Phillips, A.J.L. (2008) Phylogeny and morphology of *Diplodia* species on olives in southern Italy and description of *Diplodia olivarum* sp. nov. *Fungal Diversity* 31, 63–71.

Lilja, A., Poteri, M., Petäistö, R.-L., Rikala, R., Kurkela, T. and Kasanen, R. (2010) Fungal diseases in forest nurseries in Finland. *Silva Fennica* 44, 525–545.

Linaldeddu, B.T., Scanu, B. and Franceschini, A. (2010) First report of *Diplodia scrobiculata* causing canker and branch dieback on strawberry tree (*Arbutus unedo*) in Italy. *Plant Disease* 94, 919.

Luchi, N., Capretti, P., Surico, G., Orlando, C., Pazzagli, M. and Pinzani, P. (2005) A real-time quantitative PCR assay for the detection of *Sphaeropsis sapinea* from inoculated *Pinus nigra* shoots. *Journal of Phytopathology* 153, 37–42.

Luchi, N., Pratesi, N., Simi, L., Pazzagli, M., Capretti, P., Scala, A., Slippers, B. and Pinzani, P. (2011) High-resolution melting analysis: a new molecular approach for the early detection of *Diplodia pinea* in Austrian pine. *Fungal Biology* 115, 715–723.

Lundquist, J.E. and Hamelin, R.C. (2005) *Forest Pathology: From Genes to Landscapes*. APS Press, St Paul, Minnesota.

Maresi, G., Luchi, N., Pinzani, P., Pazzagli, M. and Capretti, P. (2007) Detection of *Diplodia pinea* in asymptomatic pine shoots and its relation to the Normalized Insolation Index. *Forest Pathology* 37, 272–280.

Marosy, M., Patton, R.F. and Upper, C.D. (1989) A conducive day concept to explain the effect of low temperature on the development of Scleroderris shoot blight. *Phytopathology* 79, 1293–1301.

Morelet, P.M. (1972) *Ascochyta piniperda* sur pin d'Alep en Provence et au Maroc [*Ascochyta piniperda* on *Pinus halepensis* in Provence and Morocco]. *Bulletin de la Société des Sciences Naturelles et d'Archeologie de Toulon et du Var* 198, 8–9.

Morelet, P.M. (1980a) La maladie à *Brunchorstia*. I. Position systématique et nomenclature du pathogen [*Brunchorstia* infection. I. Systematic position and nomenclature of the pathogen]. *European Journal of Forest Pathology* 10, 268–277.

Morelet, P.M. (1980b) La maladie à *Brunchorstia*. II. Répartition en France [*Brunchorstia* infection. II. Distribution in France]. *European Journal of Forest Pathology* 10, 354–359.

Morelet, P.M. (1985) Les *Venturia* des peupliers de la section *Leuce*. I. Taxonomie [*Venturia* species of poplars of the section *Leuce*. I. Taxonomy]. *Cryptogamie Mycologie* 6, 101–117.

Morelet, P.M. and Chandelier, P. (1993) A case of variability in *Sphaeropsis sapinea*. *European Journal of Forest Pathology* 23, 317–320.

Moret, A. and Muñoz, Z. (2007) Control of *Diplodia pinea* and *D. scrobiculata* in *Pinus halepensis* by 5-chloro-salicylic acid. *Phytopathologia Mediterranea* 46, 150–156.

Moriondo, F. (1976) Le malattie infettive nei rimboschimenti [Infectious diseases in reforestation]. *Annali Accademia Italiana di Scienze Forestali* 12, 181–196.

Nevalainen, S. (2002) The incidence of *Gremmeniella abietina* in relation to topography in southern Finland. *Silva Fennica* 36, 459–473.

Newcombe, G. (2003) Native *Venturia inopina* sp. nov., specific to *Populus trichocarpa* and its hybrids. *Mycological Research* 107, 108–116.

O'Brien, T.J. (1984) Historical and current *Scleroderris* situations in the United States. In: Manion, D.P. (ed.) *Proceedings of an International Symposium on Scleroderris Canker of Conifers, held in Syracuse, USA, June 21–24, 1983*. Martinus Nijhoff/Dr. W. Junk Publishers, The Hague, The Netherlands, pp. 26–31.

Ostergern, R.C. and Le Bossé, M. (2011) *The Europeans: A Geography of People, Culture and Environment*. Guilford Press, New York.

Patton, R.F., Spear, R.N. and Blenis, P.V. (1984) The mode of infection and early stages of colonization of pines by *Gremmeniella abietina*. *European Journal of Forest Pathology* 14, 193–202.

Petäistö, R.-L. (1999) Growth phase of bare-root Scots pine seedlings and their susceptibility to *Gremmeniella abietina*. *Silva Fennica* 33, 179–185.

Petäistö, R.-L. (2008) Infection of Norway spruce container seedlings by *Gremmeniella abietina*. *Forest Pathology* 38, 1–15.

Petäistö, R.-L. and Heinonen, J. (2003) Conidial dispersal of *Gremmeniella abietina*: climatic and microclimatic factors. *Forest Pathology* 33, 353–373.

Petäistö, R.-L. and Laine, A. (1999) Effects of winter storage temperature and age of *Pinus sylvestris* seedlings on the occurrence of disease induced by *Gremmeniella abietina*. *Scandinavian Journal of Forest Research* 14, 227–233.

Petäistö, R.-L. and Repo, T. (1988) Stress combinations and the susceptibility of Scots pine to *Ascocalyx abietina*. *Mitteilungen der Forstlichen Bundesversuchsanstalt Wien* 162, 103–118.

Petrini, O., Petrini, L.E., Laflamme, G. and Ouellette, G.B. (1989) Taxonomic position of *Gremmeniella abietina* and related species: a reappraisal. *Canadian Journal of Botany* 67, 2805–2814.

Punithalingam, E. and Waterston, J.M. (1970) *Diplodia pinea*. *Descriptions of Pathogenic Fungi and Bacteria* 273. Commonwealth Mycological Institute, Kew, UK [now CAB International, Wallingford, UK].

Roll-Hansen, F., Roll-Hansen, H. and Skrøppa, T. (1992) *Gremmeniella abietina*, *Phacidium infestans* and other causes of damage in alpine, young pine plantations in Norway. *European Journal of Forest Pathology* 22, 77–94.

Rossman, A.Y., Castlebury, L.A., Farr, D.F. and Stanosz, G.R. (2008) *Sirococcus conigenus*, *Sirococcus piceicola* sp. nov. and *Sirococcus tsugae* sp. nov. on conifers: anamorphic fungi in the *Gnomoniaceae*, *Diaporthales*. *Forest Pathology* 38, 47–60.

Santini, A., Fagnani, A., Ferrini, F., Ghelardini, L. and Mittempergher, L. (2005) Variation among Italian and French elm clones in their response to *Ophiostoma novo-ulmi* inoculation. *Forest Pathology* 35, 183–193.

Santini, A., La Porta, N., Ghelardini, L. and Mittempergher, L. (2008a) Breeding against Dutch elm disease adapted to the Mediterranean climate. *Euphytica* 163, 45–56.

Santini, A., Pepori, A., Ghelardini, L. and Capretti, P. (2008b) Persistence of some pine pathogens in coarse woody debris in a *Pinus pinea* forest. *Forest Ecology and Management* 256, 502–506.

Santini, A. *et al.* (2013) Biogeographical patterns and determinants of invasion by forest pathogens in Europe. *New Phytologist* 197, 238–250.

Singh, P., Dorworth, C.E. and Skilling, D.D. (1980) *Gremmeniella abietina* in Newfoundland. *Plant Disease* 64, 1117–1118.

Skilling, D.D. (1972) Epidemiology of *Scleroderris lagerbergii*. *European Journal of Forest Pathology* 2, 16–21.

22.3 Infection Biology

Dothistroma spp. produce conidia that are dispersed by rain splash (Gadgil, 1967; Ivory, 1972). In general, dispersal in the field is very localized. Gibson *et al.* (1964) found that most conidia spread within 50–100 m of the source. Long-distance dispersal of conidia is thought to occur through conidia being carried in clouds, fog or mist (Gibson *et al.*, 1964; Gilmour, 1967a), or by human transportation of infected seedlings (Gilmour, 1967a; Ivory, 1967). Dale *et al.* (2011) suggested that long-distance dispersal in Canada is through ascospores of the teleomorph, with local spread primarily via conidia.

After the germination of propagules, germ tubes grow over the needle surface and enter through stomatal pores (Fig. 22.2). This takes approximately 3 days, depending on the temperature and humidity levels. There are conflicting reports over whether stomata influence the direction of growth (Gadgil, 1967; Peterson, 1969, 1978; Karadzić, 1989; Muir and Cobb, 2005), and it is possible that differences in pine species and in study conditions (e.g. field versus laboratory) affected these findings. Hyphae generally grow over the needle surface for up to 10 days and may produce secondary conidia (Gadgil, 1967; Peterson, 1969). After entering the stomata, the hyphae branch and grow mainly in the mesophyll tissue (Gadgil, 1967; Ivory, 1972; Muir and Cobb, 2005). The mesophyll tissue is killed in advance of the mycelium and the overall spread of the fungus in the needle is limited to this necrotic mesophyll tissue, though recent results suggest the early post-penetration phase is biotrophic rather than necrotrophic (R.E. Bradshaw and M.S. Kabir, 2011, unpublished results). Franich *et al.* (1986) hypothesized that a rapid plant response by an accumulation of benzoic acid prevents colonization of those cells by the pathogen. The stromata develop underneath the epidermis then erupt through, tearing the plant tissue (Fig. 22.3). Conidia are released during rain splash on to fresh needles where they germinate and cause new infections. Needles attached to the tree, or suspended in the tree, are the principal source of inoculum. Needles on the forest floor are quickly colonized by saprophytes and spore production stops after about 2 months (Gadgil, 1970).

The role of dothistromin, the polyketide toxin produced by DNB pathogens and responsible for the red coloration in DNB-infected pine needles is still unknown. Although the injection of purified dothistromin into pine needles induced disease symptoms (Shain and Franich, 1981), dothistromin-deficient mutants of *D. septosporum* were able to cause DNB in Monterey pine in a glasshouse study (Schwelm *et al.*, 2009). Furthermore, in some inoculation experiments where conditions that were highly conducive to infection and rapid disease development on Monterey pine seedlings were employed, red banding did not occur (P.D. Gadgil, New Zealand, 2011, personal communication). These observations suggest that dothistromin is not a pathogenicity factor required for manifestation of the disease. However, recent results suggest that dothistromin is a virulence factor that affects lesion formation and sporulation (M.S. Kabir and R.E. Bradshaw, 2013, unpublished results). It is also possible that dothistromin has a role in stromatal development. Another hypothesis for the role of dothistromin is that it allows *D. septosporum* to protect its niche from competing microorganisms.

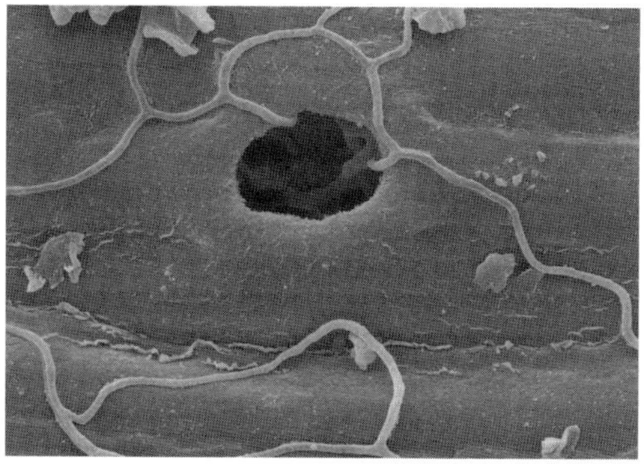

Fig. 22.2. Scanning electron micrograph of *Dothistroma septosporum* (Dorog.) M. Morelet hyphae on the surface of a *Pinus radiata* D. Don needle, and hyphal penetration into a stomatal pore. Hyphal growth within the substomatal cavity is also evident. Magnification 1500×. Credits: Shahjahan Kabir and Doug Hopcroft, Massey University, New Zealand.

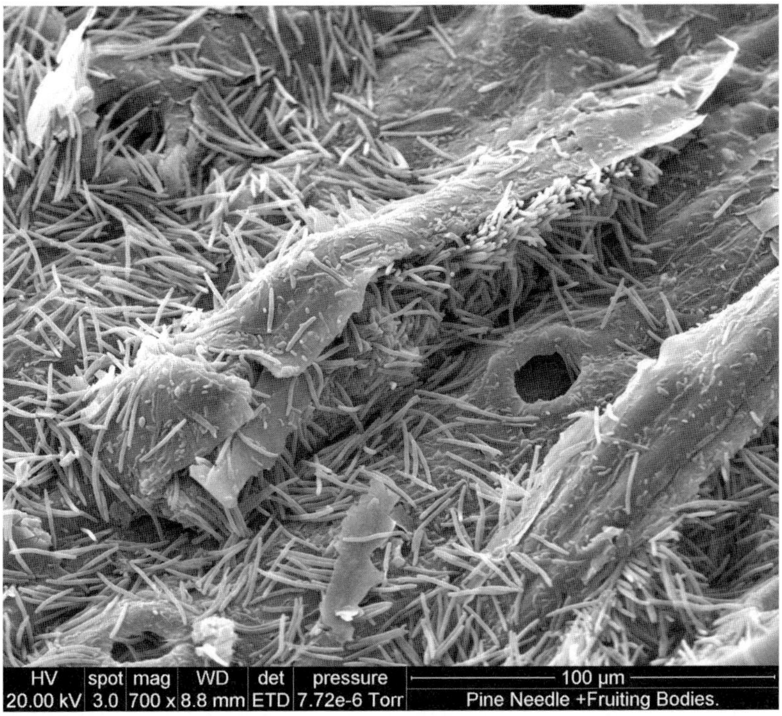

Fig. 22.3. Scanning electron micrograph of erumpent *Dothistroma septosporum* (Dorog.) M. Morelet stromata and released conidia. Magnification 700×. Credits: Shahjahan Kabir and Doug Hopcroft, Massey University, New Zealand.

Dothistromin is a broad-spectrum toxin that can inhibit the growth of other fungi, including known pine needle endophytes, in culture (Stoessl *et al.*, 1990; Schwelm *et al.*, 2009).

22.4 Epidemiology

While most pine species can be naturally infected by *Dothistroma* spp., susceptibility varies widely between host species (Table 22.1). Watt *et al.* (2009) provided a comprehensive host list grouped into three susceptibility categories. Between countries, susceptibility within the same host species appears to vary. For instance, Peterson (1967) reported that Scots pine (*Pinus sylvestris* L.) growing in severely infected Corsican pine stands in Nebraska (USA) were almost free of disease, as did Lang and Karadzic (1987) in Germany. In contrast, Skelly (1972) reported *Dothistroma* causing moderate damage and the death of about 30 Scots pines in a Christmas tree plantation in Virginia (USA); some Scots pine plantations in the UK have also suffered moderate damage (A. Brown, UK, 2011, personal communication). Watt *et al.* (2009) listed eastern white pine (*P. strobus* L.) as either susceptible, moderately susceptible or mildly susceptible. It is evident, therefore, that susceptibility varies considerably between provenances or genotypes within the same species. Such differences offer an opportunity for the forester to reduce disease by selecting less susceptible hosts, or less susceptible genotypes within a species, for planting.

Climate has a major effect on forest diseases (Hepting, 1963), but foliar fungi such as the DNB pathogens may be more responsive to climate than pathogens that affect other parts of their hosts. The ability of *Dothistroma* spp. to infect, and the time it takes to complete an infection cycle, are both strongly tied

Table 22.1. Susceptibility of common pines (*Pinus* spp.) to *Dothistroma* spp.[a]

Degree of susceptibility	Host species
Very high–high	*P. attenuata* Lemmon, *P. contorta* Douglas ex Loudon var. *latifolia* Engelm. ex S. Watson, *P. halepensis* Mill., *P. nigra* Arnold subsp. *laricio* Maire, *P. ponderosa* Lawson & C. Lawson, *P. jeffreyi* Balf., *P. radiata* D. Don (exhibits resistance after about 15 years), *P. thunbergii* Parl.
Medium	*P. canariensis* C.Sm., *P. caribaea* Morelet, *P. densiflora* Siebold & Zucc., *P. mugo* Turra, *P. muricata* D. Don, *P. palustris* Mill., *P. pinaster* Aiton, *P. pinea* L.
Low–very low	*P. arizonica* Engelm., *P. contorta*, *P. coulteri* D. Don, *P. elliottii* Engelm., *P. hartwegii* Lindl., *P. koraiensis* Siebold & Zucc., *P. lambertiana* Douglas, *P. michoacana* Martínez, *P. montezumae* Lamb., *P. monticola* Douglas ex D. Don, *P. nigra*, *P. patula* Schiede ex Schltdl. & Cham., *P. pseudostrobus* Lindl., *P. strobiformis* Engelm., *P. strobus* L., *P. sylvestris* L., *P. taeda* L.

[a]Sources: data compiled primarily from Watt *et al.* (2009), with additional material sourced from Gilmour (1967c).

to temperature and precipitation (Peterson, 1973; Gadgil, 1974, 1977). Rogerson (1954) noted that disease in Kansas increased markedly during the wet year of 1951, but during the following dry year of 1952 there was no new disease development. Murray and Batko (1962) recorded severe infection in years where rainfall from June to September ranged from 315 to 450 mm, but no infection in years where rainfall over those months was only 160 to 175 mm.

In the 1960s to the 1980s, considerable research on the environmental conditions required for *Dothistroma* spp. germination and infection was carried out using laboratory-based pathogen assays (Gadgil, 1967, 1974; Ivory, 1972; Gadgil and Holden, 1976; Peterson, 1976; Franich *et al.*, 1986). Results varied with respect to the optimal conditions that enabled germination of conidia and subsequent infection of needle tissue. Such variation may have been because of differences in technique (Peterson, 1976) or differences between strains of the pathogens or hosts. However, it is clear that germination and successful infection can occur across a wide range of temperatures, with a minimum of 4.4°C reported in controlled conditions (Gadgil, 1968) and a range of 7–28°C in field conditions (Peterson, 1967; Gilmour, 1981). The optimal temperature for germination is around 18–20°C. Light is not needed for germination (Ivory, 1972; Gadgil and Holden, 1976), but moisture is required (Gibson *et al.*, 1964).

The time between conidial germination and the production of conidia-bearing stromata is governed by temperature, needle wetness and host. In ideal controlled conditions, inoculated seedlings can develop stromata in less than 3 weeks (Gadgil, 1974, 1977; Gadgil and Holden, 1976). Parker (1972), experimenting on 4 month old Monterey pine seedlings, found that varying amounts of infection were achieved in 32 days, depending on the temperature and humidity regime tested. Gadgil (1974, 1977) found that infection was favoured at around 16–20°C when needles were kept continually moist for 12 weeks. The same author showed that long needle wetness periods are needed for rapid stromatal formation: stromata were first seen 2–3 weeks after inoculation in plants kept continuously wet over that period, compared with 7 weeks or longer for plants that were wetted infrequently and then kept dry. In the field situation in New Zealand, Gilmour (1981) showed that the length of the pre-reproduction period on Monterey pine was negatively correlated with mean air temperature, with the longest period being 15–16 weeks over winter, and the shortest, about 5 weeks, in the early summer. Likewise, in Kenya, Hocking and Etheridge (1967) found fruiting bodies on healthy Monterey pine seedlings exposed to high inoculum in the field in summer after 5 weeks, while Ivory (1972) found that it took 10 weeks for symptoms to develop during the colder months. In Finland, symptoms appeared

1 month after inoculation, with stromata appearing after another 2–4 weeks (Müller *et al.*, 2009). In eastern Nebraska, infection on Austrian pine and ponderosa pine was first seen after 11–16 weeks (Peterson, 1973). In New Zealand, a relationship between DNB and rainfall during the warmer months has been demonstrated (Fig. 22.4), and this has also been shown in Chile (L.S. Bulman, P.D. Gadgil and R. Ahumada, 2011, unpublished results), where disease levels increased soon after the onset of high rainfall in the autumn. In Australia, Marks and Hepworth (1986) reported an outbreak of DNB in many Monterey pine plantations in Victoria in early 1984, after rainfall of over 100 mm was recorded in every month except one from March 1983 to January 1984.

Light is needed for infection. While germination and growth on the needle surface can occur in complete darkness, adequate light with a suitable photoperiod is needed for infection to be successful. Gadgil and Holden (1976) used a photoperiod of 16 h with a photosynthesizing period of 12 h at an intensity of about 180 W m^{-2} at top plant height. At a light intensity of 58 W m^{-2} only 20% of inoculated needles were infected, compared with 80–90% in treatments exposed to 133 and 181 W m^{-2}. Inhibition of disease development has been seen on inoculated plants growing in greenhouses in New Zealand; when some of these plants were placed outside in full sunlight symptoms developed rapidly, while those left in the greenhouse remained healthy (Ivory, 1972). In field studies, Gibson *et al.* (1964) was the first to suggest shade-induced resistance, and Ivory (1972) subsequently found that shaded branches remained disease free 100 days after inoculation. It was thought the shading effect might provide an opportunity to control the disease, but this suggestion was rejected by Gibson (1966), who found that height growth was so seriously affected by shading that benefits from reduced disease were easily overridden.

Even without climate change, which in many parts of the world is predicted to bring warmer and wetter conditions, DNB has the potential to develop in regions where it has not been a problem before. The potential geographic areas with suitable climatic conditions for DNB have been evaluated (Watt *et al.*, 2009). From reports of DNB worldwide and using a CLIMEX distribution

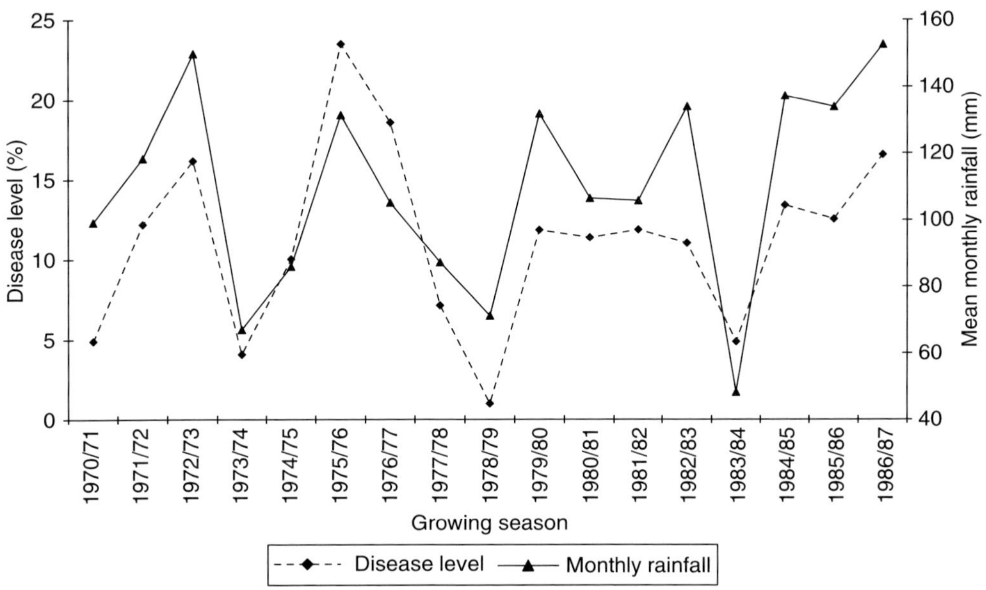

Fig. 22.4. Relationship between rainfall from November to February (summer rainfall) and disease severity of Dothistroma needle blight in New Zealand.

Ganley, R.J., Brunsfield, S.J. and Newcombe, G. (2004) A community of unknown, endophytic fungi in western white pine. *Proceedings of the National Academy of Sciences of the United States of America* 101, 10107–10112.

Ganley, R.J., Sniezko, R.A. and Newcombe, G. (2008) Endophyte-mediated resistance against white pine blister rust in *Pinus monticola*. *Forest Ecology and Management* 255, 2751–2760.

Garcia, J. and Kummerow, J. (1970) Infection of Monterey pine graftings with *Dothistroma pini*. *Plant Disease Reporter* 54, 403–404.

Gibson, I.A.S. (1966) *The Effect of Shade on the Incidence of Dothistroma Blight of Pines*. Mycological Note No. 45(2), East African Agricultural Forestry Research Organisation, Muguga Kikuyu, Kenya.

Gibson, I.A.S. (1967) The distribution, impact and control of a foliage disease of pines in Africa caused by *Dothistroma pini* Hulbary. In: *Proceedings of the Fourteenth IUFRO Congress, Munich, September 4–9 1967*. Part 5, Section 24, pp. 209–220. IUFRO, Munich, Germany.

Gibson, I.A.S. (1971) Field control of Dothistroma blight of *Pinus radiata* using copper fungicide sprays. *East African Agricultural Forestry Journal* 36, 247–274.

Gibson, I.A.S. (1974) Impact and control of Dothistroma blight of pines. *European Journal of Forest Pathology* 4, 89–100.

Gibson, I.A.S., Christensen, P.S. and Munga, F.M. (1964) First observations in Kenya on a foliage disease of pines caused by *Dothistroma pini* Hulbary. *Commonwealth Forestry Review* 43, 31–48.

Gilmour, J.W. (1967a) Distribution and significance of the needle blight of pines caused by *Dothistroma pini* in New Zealand. *Plant Disease Reporter* 51, 727–730.

Gilmour, J.W. (1967b) Distribution, impact and control of *Dothistroma pini* in New Zealand. *Proceedings of the Fourteenth IUFRO Congress, September 4–9 1967*. Part 5, Section 24, pp. 221–248. IUFRO, Munich, Germany.

Gilmour, J.W. (1967c) Host list for *Dothistroma pini* in New Zealand. Research Leaflet No. 16, New Zealand Forest Service, Forest Research Institute, Rotorua, New Zealand.

Gilmour, J.W. (1981) The effect of season on infection of *Pinus radiata* by *Dothistroma pini*. *European Journal of Forest Pathology* 11, 265–269.

Groenewald, M., Barnes, I., Bradshaw, R.E., Brown, A.V., Dale, A., Groenewald, J.Z., Lewis, K.J., Wingfield, B.D., Wingfield, M.J. and Crous, P.W. (2007) Characterization and distribution of mating type genes in the Dothistroma needle blight pathogens. *Phytopathology* 97, 825–834.

Hanso, M. and Drenkhan, R. (2008) First observations of *Mycosphaerella pini* in Estonia. *Plant Pathology* 57, 1177.

Harrington, T.C. and Wingfield, M.J. (1998) Diseases and the ecology of indigenous and exotic pines. In: Richardson, D.M. (ed.) *Ecology and Biogeography of* Pinus. Cambridge University Press, Cambridge, UK, pp. 381–404.

Hepting, G.H. (1963) Climate and forest diseases. *Annual Review of Phytopathology* 1, 31–50.

Hirst, P., Richardson, T.E., Carson, S.D. and Bradshaw, R.E. (1999) *Dothistroma pini* genetic diversity is low in New Zealand. *New Zealand Journal of Forestry Science* 29, 459–472.

Hocking, D. and Etheridge, D.E. (1967) Dothistroma needle blight of pines. I. Effect and etiology. *Annals of Applied Biology* 59, 133–141.

Hood, I.A. and Ramsden, M. (1996) *Dothistroma Needle Blight on* Pinus radiata *at Gambubal Forest*. Interim Report, QFRI Disease Management Research Trials, Queensland Forest Research Institute, Indooroopilly, Australia.

Ioos, R., Fabre, B., Saurat, C., Fourrier, C., Frey, P. and Marçais, B. (2010) Development, comparison, and validation of real-time and conventional PCR tools for the detection of the fungal pathogens causing brown spot and red band needle blights of pine. *Phytopathology* 100, 105–114.

Ito, K., Zinno, Y. and Suto, Y. (1975) Dothistroma needle blight of pines in Japan. *Bulletin of the Government Forest Experiment Station* 272, 123–140.

Ivkovic, M., Baltunis, B., Gapare, W.J., Sasse, J., Dutkowski, G., Elms, S. and Wu, H. (2010) Breeding against Dothistroma needle blight of radiata pine in Australia. *Canadian Journal of Forest Research* 40, 1653–1660.

Ivory, M.H. (1967) A new variety of *Dothistroma pini* in Kenya. *Transactions of the British Mycological Society* 50, 289–297.

Ivory, M.H. (1968) Reaction of pines in Kenya to attack by *Dothistroma pini* var. *keniensis*. *East African Agriculture and Forestry Journal* 33, 236–244.

Ivory, M.H. (1972) Resistance to Dothistroma needle blight induced in *Pinus radiata* by maturity and shade. *Transactions of the British Mycological Society* 59, 205–212.

Jayawickrama, K.J.S. (2001) Estimated among-family and within-family variances and heritabilities from three radiata pine clonal trials. *Forest Genetics* 8, 247–257.

Karadzić, D. (1987) Effectiveness of some fungicides in the control of *Dothistroma pini* Hulbary in *Pinus nigra* plantations. *Zastita Bilja* 38, 15–31.

Karadzić, D. (1989) *Scirrhia pini* Funk et Parker. Life cycle of the fungus in plantations of *Pinus nigra* Arn. in Serbia. *European Journal of Forest Pathology* 19, 231–236.

Kirisits, T. and Cech, T.L. (2006) Is Dothistroma needle blight going to become an important forest disease in Austria? *Forstschutz Aktuell* 36, 20–26 [in German].

Lang, K.J. and Karadzić, D. (1987) Is *Dothistroma pini* a danger to *Pinus sylvestris*? *Forstwissenschaftliches Centralblatt* 106, 45–50.

Libby, W.J., Bannister, M.H. and Linhart, Y.B. (1968) The pines of Cedros and Guadalupe Islands. *Journal of Forestry* 66, 846–853.

Liu, J.J. and Ekramoddoullah, A.K.M. (2011) Genomic organization, induced expression and promoter activity of a resistance gene analog (*PmTNL1*) in western white pine (*Pinus monticola*). *Planta* 233, 1041–1053.

Lorenz, R.W. (1970) The growth of ponderosa pine in Illinois limited by the needle blight. *Transactions of the Illinois State Academy of Science* 63, 401–408.

Markovskaja, S. and Treigiene, A. (2009) New data on invasive pathogenic fungus *Dothistroma septosporum* in Lithuania. *Botanica Lithuanica* 15, 41–45.

Marks, G.C. (1966) Report on visit to New Zealand December 1965 to study Dothistroma needle blight. Forest Pathology Report, New Zealand Forest Service, Forest Research Institute, Rotorua, New Zealand.

Marks, G.C. and Hepworth, G. (1986) Effect of certain features of a stand of *Pinus radiata* on *Dothistroma septospora* needle blight. *Australian Forest Research* 16, 223–229.

Marks, G.C. and Smith, I.W. (1987) Effect of canopy closure and pruning on *Dothistroma septospora* needle blight of *Pinus radiata* D. Don. *Australian Forest Research* 17, 145–150.

Marks, G.C., Smith, I.W. and Cook, I.O. (1989) Spread of *Dothistroma septospora* in plantations of *Pinus radiata* in Victoria between 1979 and 1988. *Australian Forestry* 52, 10–19.

McDougal, R.L., Schwelm, A. and Bradshaw, R.E. (2011a) Dothistromin biosynthesis genes allow inter- and intraspecific differentiation between Dothistroma pine needle blight fungi. *Forest Pathology* 41, 407–416.

McDougal, R.L., Yang, S., Schwelm, A., Stewart, A. and Bradshaw, R.E. (2011b) A novel GFP-based approach for screening biocontrol microorganisms *in vitro* against *Dothistroma septosporum*. *Journal of Microbiological Methods* 87, 32–37.

Muir, J.A. and Cobb, J.F.W. (2005) Infection of radiata and bishop pine by *Mycosphaerella pini* in California. *Canadian Journal of Forest Research* 35, 2529–2538.

Müller, M.M., Hantula, J. and Vuorinen, M. (2009) First observations of *Mycosphaerella pini* on Scots pine in Finland. *Plant Disease* 93, 322.

Munk, A. (1957) Danish *Pyrenomyctes*. A preliminary flora. *Dansk Botanisk Arkiv* 17, 312.

Murray, J.S. (1967) *Dothistroma pini* Hulbary – its occurrence in Europe. *Proceedings of the Fourteenth IUFRO Congress, September 4–9 1967*. Part 5, Section 24, pp. 265–268. IUFRO, Munich, Germany.

Murray, J.S. and Batko, S. (1962) *Dothistroma pini* Hulbary: a new disease on pine in Britain. *Forestry* 3, 57–65.

Neumann, F.G., Collett, N.G. and Smith, I.W. (1993) The *Sirex* wasp and its biological control in plantations of radiata pine variably defoliated by *Dothistroma septospora* in north-eastern Victoria. *Australian Forestry* 56, 129–139.

Old, K.M. and Dudzinski, M.J. (1999) Forest pathogen introductions to Australia: experiences, threats and counter measures. In: Devey, M.E., Matheson, A.C. and Gordon, T.R. (eds) *Current and Potential Impacts of Pitch Canker in Radiata Pine. Proceedings of the IMPACT Monterey Workshop, California, USA, 30 November to 3 December 1998*. Forestry and Forest Products No. 112, CSIRO, Canberra, Australia, pp. 11–20.

Olsen, P.F. (1971) *Dothistroma* spraying experience – 1966 to 1970. *New Zealand Journal of Forestry* 16, 101–104.

Parker, A.K. (1972) Artificial inoculation of *Pinus radiata* with *Scirrhia* (*Dothistroma*) *pini*: effect of relative humidity and temperature on incubation. *Phytopathology* 62, 1160–1164.

Parker, A.K. and Collis, D.G. (1966) Dothistroma needle blight of pines in British Columbia. *Forestry Chronicle* 42, 160–161.

Paterson, D.N. (1966) Crude estimates of genetic gains from plus trees selected in the East African tree breeding programme. Technical Note 18, East African Agricultural Forestry Research Organisation, Muguga Kikuyu, Kenya.

Pehl, L., Burgermeister, W. and Wulf, A. (2004) *Mycosphaerella* needle fungi – identification by means of ITS-RFLP patterns. *Nachrichtenblatt des Deutschen Pflanzenschutzdienstes* 56, 239–244.

Peterson, G.W. (1967) Dothistroma needle blight of pines in North America. *Proceedings of the Fourteenth IUFRO Congress, September 4–9 1967*. Part 5, Section 24, pp. 269–278. IUFRO, Munich, Germany.

Peterson, G.W. (1969) Growth of germ tubes of *Dothistroma pini* conidia positively directed towards stomata of Austrian and ponderosa pine needles. *Phytopathology* 59, 1044.

Peterson, G.W. (1973) Infection of Austrian and ponderosa pines by *Dothistroma pini* in eastern Nebraska. *Phytopathology* 63, 1060–1063.

Peterson, G.W. (1976) *Some Reflections on Biological Studies of Needle Diseases*. Research Note RM-323, US Department of Agriculture, Forest Service, Rocky Mountain Forest and Range Experiment Station, Fort Collins, Colorado.

Peterson, G.W. (1978) Development of *Dothistroma pini* upon and within needles of Austrian and ponderosa pines in eastern Nebraska. *Phytopathology* 68, 1422–1430.

Poole, A.L. (1966) Needle blight of pines, a new record. *FAO Plant Protection Bulletin* 14, 123.

Power, A.B. and Dodd, R.S. (1984) Early differential susceptibility of juvenile seedlings and more mature stecklings of *Pinus radiata* to *Dothistroma pini*. *New Zealand Journal of Forestry Science* 14, 223–228.

Ray, J.W. and Vanner, A.L. (1987) Improvements in spray technology used to control Dothistroma needle blight in New Zealand pine plantations. In: Green, G.W. (ed.) *Proceedings – Symposium on the Aerial Application of Pesticides in Forestry, Ottawa, Canada, 20–22 October 1987*. Associate Committee on Agriculture and Forestry Aviation, AFA-TN-18, National Research Council of Canada, NRC 29197, Ottawa, Ontario, Canada, pp. 325–328.

Rogerson, C.T. (1954) Kansas mycological notes, 1952. *Transactions of the Kansas Academy of Science* 57, 280–284.

Schischkina, A.K. and Tzanava, N.I. (1966) Study of the red needle spot of pine in Georgia. *Trudy Instytuta Zashchity rastenii: Tbilisii* 18, 285–295.

Schwelm, A., Barron, N.J., Baker, J., Dick, M., Long, P.G., Zhang, S. and Bradshaw, R.E. (2009) Dothistromin toxin is not required for Dothistroma needle blight in *Pinus radiata*. *Plant Pathology* 58, 293–304.

Shain, L. and Franich, R.A. (1981) Induction of Dothistroma blight symptoms with dothistromin. *Physiological Plant Pathology* 19, 49–55.

Shaw, C.G. and Toes, E.H.A. (1977) Impact of Dothistroma needle blight and Armillaria root rot on diameter growth of *Pinus radiata*. *Phytopathology* 67, 1319–1323.

Skelly, J.M. (1972) *Dothistroma pini* found infecting Scotch pine, *Pinus sylvestris*, a first report. *Phytopathology* 62, 671–672.

Solheim, H. and Vuorinen, M. (2011) Rødbandsjuke er påvist i Norge [Red band needle blight is present in Norway]. *SKOGeiernen* 1, 26–28 [in Norwegian].

Stergiopoulos, I. and de Wit, P.J.G.M. (2009) Fungal effector proteins. *Annual Review of Phytopathology* 47, 233–263.

Stoessl, A., Abramowski, Z., Lester, H.H., Rock, G.L. and Towers, G.H.N. (1990) Further toxic properties of the fungal metabolite dothistromin. *Mycopathologia* 112, 179–186.

Thomas, J.E. and Lindberg, G.D. (1954) A needle disease of pines caused by *Dothistroma pini*. *Phytopathology* 44, 333.

Thompson, J.H. (1966) *Forest Disease Control*. Memorandum, File 5200, US Department of Agriculture, Forest Service, Washington, DC.

Thyr, B.D. and Shaw, C.G. (1964) Identity of the fungus causing red band disease on pines. *Mycologia* 56, 103–109.

van der Pas, J.B. (1981) Reduced early growth rates of *Pinus radiata* caused by *Dothistroma pini*. *New Zealand Journal of Forestry Science* 11, 210–220.

van der Pas, J.B., Bulman, L. and Horgan, G.P. (1984) Estimation and cost benefits of spraying *Dothistroma pini* in tended stands of *Pinus radiata* in New Zealand. *New Zealand Journal of Forestry Science* 14, 23–40.

Van Haverbeke, D.F. and Boldt, C.E. (1968) Vigor and density of shelterbelt conifers can be improved. *Journal of Forestry* 66, 187–192.

Vidakovic, M., Krstinic, A., Halambek, M. and Borzan, Z. (1986) Uspijevanje nekih vrsta i hibrida dvoiglicavih borova na Durdevackim pijescima [Growth of some two-needled pines and their hybrids on the Durdevac sands]. *Anali za Sumarstvo* 12, 71–86.

Villebonne, D. and Maugard, F. (1999) Rapid development of Dothistroma needle blight (*Scirrhia pini*) on Corsican pine (*Pinus nigra* subsp. *laricio*) in France. *La Santé des Forêts, Annual Report 1998*. Les Cahiers du DSF 1, DERF, Paris, pp. 30–32.

Wagener, W.W. (1967) *Red Band Needle Blight of Pines...a Tentative Appraisal for California*. Research Note PSW-153, US Department of Agriculture, Forest Service, Pacific Southwest Forest and Range Experiment Station, Berkeley, California.

Wallis, C.M., Reich, R.W., Lewis, K.J. and Huber, D.P.W. (2010) Lodgepole pine provenances differ in chemical defense capacities against foliage and stem diseases. *Canadian Journal of Forest Research* 40, 2333–2344.

Watt, M.S., Kriticos, D.J., Alcaraz, S., Brown, A.V. and Leriche, A. (2009) The hosts and potential geographic range of Dothistroma needle blight. *Forest Ecology and Management* 257, 1505–1519.

Watt, M.S., Bulman, L.S. and Palmer, D.J. (2011) The economic cost of Dothistroma needle blight to the New Zealand forest industry. *New Zealand Journal of Forestry* 56, 20–22.

Welsh, C., Lewis, K.J. and Woods, A. (2009) The outbreak history of Dothistroma needle blight; an emerging forest disease in northwestern British Columbia, Canada. *Canadian Journal of Forest Research* 39, 2505–2519.

Whyte, A.D.G. (1976) Spraying pine plantations with fungicides – the manager's dilemma. *Forest Ecology and Management* 1, 7–19.

Wilcox, M.D. (1982) Genetic variation and inheritance of resistance to Dothistroma needle blight in *Pinus radiata*. *New Zealand Journal of Forestry Science* 12, 14–35.

Wilcox, P.L., Amerson, H.V., Kuhlman, E.G., Liu, B.H., O'Malley, D.M. and Sederoff, R.R. (1996) Detection of a major gene for resistance to fusiform rust disease in loblolly pine by genomic mapping. *Proceedings of the National Academy of Sciences of the United States of America* 93, 3859–3864.

Woods, A.J. (2003) Species diversity and forest health in northwest British Columbia. *Forestry Chronicle* 79, 892–897.

Woods, A.J., Coates, K.D. and Hamann, A. (2005) Is an unprecedented Dothistroma needle blight epidemic related to climate change? *BioScience* 55, 761–769.

Woollons, R.C. and Hayward, W.J. (1984) Growth losses in *Pinus radiata* stands unsprayed for *Dothistroma pini*. *New Zealand Journal of Forestry Science* 14, 14–22.

23 Other Foliar Diseases of Coniferous Trees

Miroslava Bednářová, Miloň Dvořák, Joseph Janoušek and Libor Jankovský*
Mendel University, Brno, Czech Republic

23.1 Pathogens, Significance and Distribution

23.1.1 Main needle pathogens

Foliar diseases of conifers are common in nurseries, plantations, forest stands and urban/amenity plantings. Among the negative effects of these diseases on trees, defoliation, reduced vigour and predisposition to biotic and abiotic stresses are the most important. Damage of the needles inflicts significant aesthetic loss for amenity trees and Christmas trees. Most conifers form perennial foliage, and therefore foliar diseases are an essential factor influencing tree health. Any defoliation is easy to see and has an effect on the increment and vitality of the trees. The youngest foliage is most effective in assimilation, and at the same time the most sensitive to diseases (Wang *et al.*, 1995; Jach and Ceulemans, 2000; Zha *et al.*, 2002). However, in coniferous trees, as much as 70% of the canopy may consist of old (>1 year old) needles (Schulze *et al.*, 1977).

Changes in needle longevity or some degree of defoliation affect the carbon balance, rate of nutrient cycling and, ultimately, the net primary production of a forest ecosystem, and they also influence physiological processes within the canopy (Chabot and Hicks, 1982; Reich *et al.*, 1995). Complete defoliation of a coniferous tree is usually fatal. Fungal pathogens may play an important role in the regeneration of natural stands too. Based on studies in Russia, Ranta and Saloniemi (2005) suggested that pathogenic fungi that affect foliage may affect the spatial distribution and regeneration dynamics of natural Scots pine (*Pinus sylvestris* L.) stands.

Air pollutants such as sulfur dioxide, nitrogen oxides and ozone can cause spots on needles similar to those seen in the initial stages of fungal infection. Some biotic agents, such as fungi, bacteria, phytoplasmas, algae, viruses, sucking insects (e.g. aphids), scale insects and spider mites, especially the spruce spider mite *Oligonychus ununguis* (Jacobi), may kill needles and produce severe defoliation, discoloration, needle spots, or casts (Agrios, 1997).

Most needle casts are caused by fungi of the *Ascomycota*, such as Dothistroma needle blight (see Bulman *et al.*, Chapter 22, this volume), Lophodermium needle cast, Cyclaneusma needle cast, Elytroderma and other needle casts on pines (*Pinus* spp.), Swiss and Rhabdocline needle casts on Douglas-fir (*Pseudotsuga menziesii* (Mirb.) Franco), etc. No foliage diseases

* E-mail: jankov@mendelv.cz

on conifers are caused by *Basidiomycota*, except those caused by rust fungi. Rusts infect needles of conifers frequently, e.g. *Coleosporium* spp. on pines, *Chrysomyxa* spp. on spruces (*Picea* spp.), *Pucciniastrum* spp. on firs (*Abies* spp.), *Melampsora* spp. on pines, firs, spruces and larches (*Larix* spp.), etc. (see Hamelin, Chapter 26, this volume).

The majority of foliage pathogens belong to the *Leotiomycetes* – order *Rhytismatales* (genera *Bifusella, Davisomycella, Elytroderma, Hypoderma, Hypodermella, Isthmiella, Lirula, Lophodermella, Lophodermium, Lophomerum, Meloderma, Ploioderma, Virgella*) and order *Helotiales* (*Didymascella, Meria, Phacidium, Rhabdocline, Tiarosporella*), and to the *Dothideomycetes* – order *Capnodiales* (*Mycosphaerella* and its anamorphs *Cercoseptoria, Dothistroma* and *Lecanosticta*), order *Dothideales* (*Phaeocryptopus, Sclerophoma*) and order *Pleosporales* (*Hendersonia, Rhizosphaera*). The *Sordariomycetes* are represented by order *Diaporthales* (genus *Valsa*).

Needle casts are transmitted by wind or water aerosol. Specific climatic and ecological conditions are necessary for the emergence and development of the infection. Heat and excessive moisture with dew formation promote the growth and spreading of foliage diseases. The severity of the infection is closely related to the climatic conditions; the rate of infection varies annually depending on the weather. Some factors, such as drought stress, hail storms, sucking insects and spider mite outbreaks might predispose needles to fungal infection as well. Most fungi are opportunistic pathogens, which may enter the tree through the open wounds. The biology and life cycle, such as the sporulation of asexual and sexual stages of fungal pathogens on the needles of conifers, are restricted by the phenology of the trees and also to specific seasons. Successful infection depends on climatic factors; thus, the severity of infection primarily varies from year to year and symptoms also differ within the year.

There is a broad spectrum of fungi that colonize living green tissues as endophytes, and are later involved in the decomposition of cellulose and also lignin in dead needles; these can be sometimes confused with saprophytic fungi owing to their production of fruiting bodies in dead tissues. Several foliage fungi have endophytic phases, and under specific conditions they might develop symptoms in green needles as weak parasites. Endophytes continue with the decomposition of dead needles as the principal needle decomposers (Sieber, 1988; Livsey and Barklund, 1992; Jurc *et al.*, 1996; Sieber *et al.*, 1999; Müller and Hallaksela, 2000; Müller *et al.*, 2001; Kowalski and Zych, 2002; Osono and Hirose, 2009).

Most fungal species causing needle cast are associated with one host species or host genus. *Cyclaneusma minus* (Butin) DiCosmo, Peredo & Minter and *Lophodermium seditiosum* Minter, Staley & Millar usually infect two-needle pines, *Phaeocryptopus gaeumannii* (T. Rohde) Petr. and *Rhabdocline pseudotsugae* Syd. infect Douglas-firs exclusively; some diseases of fir needles also have strong host specificity. However, Dothistroma needle blight caused by *Mycosphaerella pini* Rostr. ex Munk (the sexual stage of *D. septosporum* (Dorog.) M. Morelet) is reported from more than 85 species of pines, spruces and Douglas-firs (Watt *et al.*, 2009; see Bulman *et al.*, Chapter 22, this volume). Also, *Sydowia polyspora* (Bref. & Tavel) E. Müll. is widespread in a broad range of host species, such as pines, firs, etc.

Various conifers have different susceptibilities to needle diseases. The most susceptible trees among economically important conifers are pines. It is not possible to produce pines in nurseries without regular control of these diseases. Larches seem to be very susceptible to needle diseases, although loss of needles is not as fatal as in the case of evergreen conifers, even though premature defoliation reduces larch growth. Norway spruce (*Picea abies* (L.) Karst.) is quite resistant to needle diseases.

Climate has been recognized as an important environmental determinant in the distribution of pathogens and the development of diseases (Desprez-Loustau *et al.*, 2007; De Wolf and Isard, 2007). Some climatic models predict an abundance of various pathogens in certain areas as a result of forecast climatic change, and this prediction can be applied for some needle diseases, such as Swiss needle cast of Douglas-firs. Attention is being paid to pathogens that have increased in incidence or in geographical or host range. Climatic change might enable the spread of the disease by surpassing a natural environmental threshold that has previously restricted a pathogen's development in northern temperate regions (Woods *et al.*, 2005; Watt *et al.*, 2010, 2011a,b).

Needle diseases of conifers are of increasing economic importance as a result of the establishment of intensive plantations for timber and pulp production throughout the world. The introduction of new conifer species, including hybrids and improved trees, to plantations in geographically new areas also brings the risk of spreading of new invasive diseases and pests.

23.1.2 Distribution of needle pathogens

Africa

The genera *Cyclaneusma*, *Dothistroma* and *Lophodermium* are reported from most African countries that have pine plantations. *Mycosphaerella gibsonii* H.C. Evans is reported as an introduced pathogen from East and South Africa (Evans, 1984; Crous *et al.*, 1990; Roux *et al.*, 2005).

America, North

There is large diversity of needle pathogens in North America due to a broad spectrum of native host species in the USA and Canada. Needle casts with economic impact on pines in forest nurseries, plantations of Christmas trees and also in forest stands are caused predominantly by *Lophodermium* spp., *Mycosphaerella dearnessii* M.E. Barr and *M. pini*. A broad spectrum of needle pathogens with minor impact includes native needle casts on firs, such as *Lirula abietis-concoloris* (Mayr ex Dearn.) Darker, *L. nervisequa* (DC.) Darker, *Phacidium infestans* P. Karst. and *Virgella robusta* (Tubeuf) Darker. On spruces, less significant needle casts, such as *Lirula macrospora* (R. Hartig) Darker, *Lophodermium crassum* Darker, *L. piceae* (Fuckel) Höhn. and *Rhizosphaera kalkhoffii* Bubák were observed. In recent years, some new fungi have been reported on spruce, such as *Stigmina lautii* B. Sutton on blue spruce (*Picea pungens* Engelm.) (Hodges, 2002; Walla and Kinzer, 2008). Brown felt blight caused by *Herpotrichia juniperi* (Duby) Petr. is regularly reported on several genera of conifers, while *Neopeckia coulteri* (Peck) Sacc. is reported on the foliage of pines only. Cedar leaf blight caused by *Didymascella thujina* (E.J. Durand) Maire is a needle disease that occurs on several members of the *Cupressaceae* and has little or no economic importance, though it sometimes causes damage in urban and amenity plantings.

America, South and Central

Cyclaneusma needle cast and Dothistroma needle blight are reported from plantations of pine in South and Central America. *M. dearnessii* is probably native in the mountain areas of Honduras and Guatemala, and *M. gibsonii* is reported from Jamaica and Nicaragua (EPPO, 1980).

Asia

The spectrum of needle diseases such as Dothistroma needle blight and Cyclaneusma needle cast that occur in South-east Asia in plantations of introduced pines is similar to that which occurs in other regions of the southern hemisphere where pines have been introduced. Native Khasi pine (*Pinus kesiya* Royle ex Gordon) is susceptible to Dothistroma needle blight. *M. gibsonii* is reported from many countries in Asia (Evans, 1984; Suto and Ougi, 1998; EPPO, 2008). *M. laricis-leptolepidis* Kaz. Itô, K. Sato & M. Ota is actually reported in native larches from China, the Korean peninsula and Japan (EPPO, 1978) as a serious disease and threat for larches in other geographic areas. Several new species of fungi on needles, which are of uncertain risk to conifers in other areas, have recently been described from China and Japan (Kaneko, 2003; Cheng *et al.*, 2005; Hou *et al.*, 2005, 2006; Hou and Piepenbring, 2006, 2009; Luo *et al.*, 2010; Fan *et al.*, 2011).

Europe

Lophodermium needle cast, caused by *L. seditiosum*, is the most important pathogen of Scots pines in nurseries. In the last 20 years, Dothistroma needle blight has also spread on Scots pine. In some areas, outbreaks of *Meloderma desmazieri* (Duby) Darker have been noted on plantations of eastern white pine (*Pinus strobus* L.). The anamorph of *M. dearnessii* can be found in several countries, but without important economic impact. *C. minus* and *C. niveum* (Pers.) DiCosmo, Peredo & Minter, *Lophodermella sulcigena* (Link) Tubeuf, *L. conjuncta* (Darker) Darker, *P. infestans* and rusts, such as *Coleosporium* spp., have only local significance (Butin, 1995). In some years, local outbreaks of *Mycosphaerella laricina* (R. Hartig) Mig., and *Meria laricis* Vuill. occur on larches. In native stands of larch in mountain areas, *Hypodermella laricis* Tubeuf causes larch needle cast. A broad spectrum of fir needle casts, such as *L. nervisequa*, *Rhizosphaera macrospora* Gourb. & M. Morelet and *Valsa abietis* Fr., rarely *Phaeocryptopus nudus* (Peck) Petr., and the rust *Pucciniastrum epilobii* G.H. Otth, causes no serious economic damage. *S. polyspora* causes severe damage of needles on grand firs (*Abies grandis* (Douglas ex D. Don) Lindl.) in plantations of Christmas trees (Cech, 1992; Perny *et al.*, 2002; Talgø *et al.*, 2010).

Oceania, New Zealand and Australia

Needle casts are a serious threat on plantations of introduced pines and Douglas-firs in Oceania, New Zealand and Australia. *P. gaeumannii* was found for the first time on Douglas-fir in New Zealand in 1959, and now is distributed throughout most of the country and causes severe economic loss (Ridley and Dick, 2001; Gadgil, 2005; Watt *et al.*, 2010). Some other fungal pathogens of Douglas-fir, such as *R. kalkhoffii* and *R. pini* (Corda) Maubl., are of little economic impact, as they usually infect dead tissue and sometimes colonize green needles damaged by climatic factors (Gilmour, 1966; Gadgil, 2005). *C. minus* and *Lophodermium pinastri* (Schrad.) Chevall. are reported from Monterey pine (*Pinus radiata* D. Don) plantations in New South Wales, Australia (Choi and Simpson, 1991).

23.2 Foliar Diseases of Douglas-fir

Swiss needle cast (SNC), caused by *Phaeocryptopus gaeumannii*, was first recorded in Switzerland in 1925, but its first occurrence was retrospectively noticed in the Coast Range in Oregon in 1916 (Boyce, 1940). The disease was then noted in other parts of Europe and in New Zealand (Hood and Kershaw, 1975). Later, it was confirmed also in Asia, i.e. Turkey (Temel *et al.*, 2003), and in South America, i.e. Chile (Osorio, 2007). From 1985, the incidence of the disease has been increasing and its manifestation has been more severe (Hansen *et al.*, 2000; Manter *et al.*, 2003b). Population studies of *P. gaeumannii* discovered the sympatric existence of two isolated lineages, one causing SNC in all areas of occurrence, while the other is limited to only the area of the most severe outbreak in the Oregon Coast Range (Winton *et al.*, 2006).

Rhabdocline needle cast (RNC) is caused by *R. pseudotsugae* and was first noted in Scotland in 1914, then in Germany in 1926, and in 1930s also in the USA. Nowadays, it is widespread all over North America and Europe (Parker and Reid, 1969; Tainter and Baker, 1996). The causal agents of RNC have a somewhat difficult taxonomy. Parker and Reid (1969) described two species, *R. pseudotsugae* and *R. weirii* A.K. Parker & J. Reid, with five subspecies: *R. pseudotsugae* subsp. *pseudotsugae*, subsp. *epiphylla*, *R. weirii* subsp. *weirii* (which is the only one associated with an anamorph *Rhabdogloeum*), subsp. *oblonga* and subsp. *obovata*. Stone and Gernandt (2005) considered this organization of species and subspecies to be unreliable, because it relies too much on such a variable marker, as is ascus iodine reaction. Based on a deeper morphological analysis, which also used molecular genetic methods, the elevation of these subspecies to the rank of species was proposed, i.e. *R. pseudotsugae*, *R. epiphylla* (A.K. Parker & J. Reid) J.K. Stone & Gernandt, *R. weirii*, *R. oblonga* (A.K. Parker & J. Reid) J.K. Stone & Gernandt and *R. obovata* (A.K. Parker & J. Reid) J.K. Stone & Gernandt. All of these species except for *R. pseudotsugae* are known only from North America.

Dothistroma rhabdoclinis Butin is a hyperparasite of *R. pseudotsugae*, forming spermogonia of the synamorph *Asteromella* on the upper side of RNC-affected needles and conidiomata on the underside. *D. rhabdoclinis* penetrates through the intercellular space, there parasitizing on *R. pseudotsugae* hyphae and preventing the formation of ascomata. Hyphae of *D. rhabdoclinis* penetrate through and subsequently damage the parenchymal cells, endodermis and vascular tissues, causing overall browning of the needle and its premature cast. The biological role of this organism is not clear, because on the one hand it accelerates the development of RNC at the needle level, while on the other hand it reduces the occurrence of *R. pseudotsugae* inoculum (Butin *et al.*, 2000).

Some other needle-affecting fungi, such as *Rhizosphaera pseudotsugae* Butin & Kehr, *Phoma eupyrena* Sacc. and *P. infestans* have lesser impact on Douglas-fir. *Rhizosphaera pseudotsugae* manifests similarly to SNC on the stomata on the underside of the needle, forming globular, dark conidiomata of irregular size (Butin and Kehr, 2000). *P. eupyrena* causes blights on the seedlings of Douglas-fir and other woody plants. The fungus causes yellowing and browning of the needles, which are subsequently cast. Dark, conidia-producing pycnidia grow on the dead needles. Effective control measures against *P. eupyrena* are mulching or just leaving the mosses on the ground, thus preventing the formation of soil cones around the bases of seedling stems after the rain, where the infection survives (Kliejunas *et al.*, 1985).

23.2.1 Diagnosis

SNC and RNC on Douglas-fir differ from each other in the development of chlorosis on needles and in fruiting bodies.

At the end of winter and in spring, yellow flecks on needles infected by RNC spread out, becoming sharply delineated and getting purplish brown. Yellow, oblong apothecia, 2–5 mm long, form in late April and early May. They open in wet conditions and release ascospores, which are at first hyaline and one celled, and later two celled, light brown and 18–20 × 6.5–7.5 μm in size.

In the case of SNC, chlorotic spots are diffuse across the whole needle width, without sharp delineation and the chlorosis affects nearly the whole needle at the same time. Needles are gradually cast 2–3 years after the infection. Stomata on the undersides of the needle are mostly clogged by tiny, spherical, dark pseudothecia 50–100 μm in size. Ascospores are at first hyaline, later brownish and 9–10 × 3.5–6 μm in size. The needles turn first yellow, then brown and are then cast (Butin, 1995; Tainter and Baker, 1996; Stone *et al.*, 2008a).

Real-time PCR with appropriately selected specific primers can be used quite effectively to assess the proportion of infection by *P. gaeumannii* in the host. The detection of the pathogen is possible as early as 1 month after the infection (Winton *et al.*, 2002). Real-time PCR has also been used also for the detection of *R. pseudotsugae* and *Rhabdocline oblonga*. Three copies of DNA are enough for detection with real-time PCR (Catal *et al.*, 2010).

The amount of the pathogenic fungus in the host needle tissues is possible to assess by analysis of the ergosterol contained in the cellular walls of the fungi. However, this method is not specific, as ergosterol is found not only in pathogenic fungi but also in endophytic fungi (Manter *et al.*, 2001).

23.2.2 Infection biology

Ascospores of *P. gaeumannii* are released from pseudothecia from April to September, but most intensively in June and July (Michaels and Chastagner, 1984). Germinating ascospores of *P. gaeumannii* settle most frequently on new needles and produce suprastomatal appressoria, from which penetration pegs enter the needles. Initial infection occurs from late May until early July and coincides with sprouting and shoot elongation. Colonization inside the needles is exclusively intercellular, in both the palisade parenchyma layer and the spongy mesophyll, and increases continuously. Pseudothecia are formed in peristomatal chambers and are visible 9 months after the initial colonization, i.e. in early spring. Other intact stomata are invaded by epiphytic hyphae (Stone *et al.*, 2008a).

The primary pathological effect of *P. gaeumannii* appears to be clogging of the stomata by fruiting bodies, subsequently leading to carbon starvation and the reduction of photosynthesis, resulting in premature shedding of foliage (Hansen *et al.*, 2000). Net respiration exceeds assimilation when at least 25% of stomata are occluded (Manter *et al.*, 2003a). This point of carbon equilibrium is crucial for foliage retention. At this point, abscission of the needles starts (Cannell and Morgan, 1990). Reduced photosynthesis as photo-oxidative damage may also contribute to premature needle loss in sun-exposed foliage (Manter, 2002). Manter *et al.* (2000) found no effect of fungal presence on gas exchange until the emergence of the fungal pseudothecia, which physically block the stomata and reduce carbon dioxide assimilation.

Ascospores of *Rhabdocline* spp. are windblown from late May until early July. In moist conditions, they germinate and penetrate through the mesophyll and intercellular space of the needles, causing autumn yellowing of the needles. Hyphae grow into parenchymal cells of the needle mesophyll. In spring, apothecia form on the underside of the needles, releasing new ascospores during the period of sprouting of Douglas-firs.

23.2.3 Epidemiology

In North America, SNC is spread through the Pacific coast range in the foggy zone. The (unusual) windblown transfer of spores to new stands might occur as a result of damp airflow through infected stands. High moisture enables the attachment and survival of spores on the needle surface, their germination and penetration through the stomata. A moderate winter enables pathogen development in the intercellular spaces of the needles and faster fructification in the spring (Shaw *et al.*, 2011).

Douglas-firs are susceptible to needle blights from the age of 8 years, and severely affected stands of 80 years old are no exception. Late-shooting trees are more resistant because the shooting season then misses the season of the release of *Rhabdocline pseudotsugae* ascospores. Similarly, when weather fluctuations delay or accelerate shooting, infection might be avoided (Shaw *et al.*, 2011).

23.2.4 Management strategies and tactics

Rocky Mountain Douglas-fir (*P. menziesii* var. *glauca*), which is highly susceptible to RNC, might be substituted by the more resistant coast Douglas-fir variety (*P. menziesii* var. *viridis*) (Butin, 1995), although this is susceptible to SNC. Many provenance tests have been performed that have focused on Douglas-fir resistance to needle casts. Hoff (1986) stated that in Idaho the most resistant Douglas-firs are from the lowlands as compared with those from middle and high elevations.

Chlorothalonil-based fungicides are recommended against SNC and RNC; benomyl and mancozeb can also be used. Azoxystrobin-based preparations are less effective. Repeated spraying is necessary, three or four times from the shooting season until shoot ripening in the case of RNC. For SNC, the first spraying must be applied when the shoots are 1–5 cm long, and spraying must be repeated in 2–3 weeks, and again after another 2–3 weeks. Low stand density has a positive effect on the severity of the infection; however, thinning does not always suppress propagation of the pathogen (Hood and Kershaw, 1975; Hood, 1997; Bates and Despot, 2005; Stone *et al.*, 2007).

A disease occurrence model for SNC has been formulated until 2050 (Stone *et al.*, 2008b), and has been successfully tested in Oregon and New Zealand (Manter *et al.*, 2003b, 2005). Moderate winter temperature stimulates the virulence of *P. gaeumannii* and, according to the current climate change scenarios, the model predicted further dissemination and a more severe impact of this pathogen on Douglas-fir stands. Manter *et al.* (2005) pursued creation of a model based on temperature and location. This geographic information system (GIS)-like model is able to predict the occurrence of SNC infection for a short period of time.

23.3 Foliar Diseases of Pines

Pines are the most susceptible to needle cast of all coniferous species. There are several species with economic impact, such as *L. seditiosum*, *C. minus*, *M. gibsonii*, *M. dearnessii* and *M. pini*, and these require the use of the control measures or phytosanitary precautions in nurseries and plantations of Christmas trees. A broad spectrum of endophytes or weak parasites produces fruiting bodies abundantly in dying and dead needles.

Lophodermium needle cast serves as an 'umbrella' disease, and can be treated as covering control against pathogens infecting needles in nurseries from July to September; Dothistroma needle blight is similarly an 'umbrella' disease that can be treated as covering control against other species infecting needles from May and June.

Some pathogens of pine needles are mentioned as quarantine species, e.g. *M. dearnessii* and *M. gibsonii* are quarantine species for Europe, and *L. seditiosum* for New Zealand.

23.3.1 Lophodermium needle cast

Lophodermium is a large and complex genus in the family *Rhytismataceae*, comprising several pathogenic species and endophytes of pines. *L. seditiosum* is the only pine-inhabiting species of the genus known to be pathogenic, especially in Europe on two- and three-needle pines, and causes needle cast in natural stands of pines. The pathogen is of economic significance in nurseries and Christmas tree plantations, sometimes causing losses in already established cultures, and in some years also in natural stands. In the past, it was confused with *L. pinastri*, which is now considered to be an endophyte, fruiting on ageing needles and following *L. seditiosum*. *L. conigenum* (Brunaud) Hilitzer is reported as a saprophyte from dying needles and twigs. Some other species, e.g. *L. australe* Dearn., *L. baculiferum* Mayr, *L. durilabrum* Darker, *L. molitoris* Minter, *L. nitens* Darker, *L. pini-excelsae* S. Ahmad, *L. pini-pumilae* Sawada and *L. staleyi* Minter are found on pines, but cause no serious disease (Minter and Millar, 1978, 1980a,b, 1984, 1993; Minter *et al.*, 1978; Minter, 1981; Butin, 1995). *L. seditiosum* produces severe defoliation of young needles on seedlings and young pines in nurseries and also in forest stands. Control is necessary in nurseries, especially in Europe for native two-needle pines, such as Scots pine, mugo pine (*Pinus mugo* Turra) and Austrian pine (*Pinus nigra* Arnold).

L. seditiosum is widespread in Europe, North America and Asia (China, Japan, Korea), and while it is not present in New Zealand and Australia, it is mentioned as a quarantine pathogen there. Also, *L. pinastri*, *L. conigenum* and *L. molitoris* are reported from that region (Johnston *et al.*, 2003; Simpson and Grgurinovic, 2004).

Diagnosis

Lophodermium needle cast is connected with 'spring needle cast'. The first symptoms of infection occur on needles during the summer, manifesting as tiny yellow spots around the points of infection in September on current year green needles; however, these symptoms are easily confused with spots caused by ozone, winter burn, scale insects (*Leucaspis* spp.), pine weevils (*Brachonyx pineti* (Paykull)), spruce spider mites, conifer aphids (*Schizolachnus pineti* (Fabricius)), pine needle beetles (*Cryptocephalus pini* (Linnaeus)), and pine leaf beetles (*Luperus pinicola* (Duftschmid)). In seedlings, nearly all of the needles are infected, but in older trees (up to 10 years), only needles in the lower third of the crown are infected. Needles that are entirely infected turn brown rapidly in the early spring. Only oblong, black pycnidia of the anamorph *Leptostroma austriacum* Oudem., which has (micro)conidia 5–8 × 1 μm in size, form at this time on dead needles; in the summer, these are followed by black, oval to navicular hysterothecia, 1–1.5 mm long, which start to form on second year needles in litter or on dead needles dried on twigs. The hysterothecia open under most conditions by shiny longitudinal straight slits. The asci release one-celled, acicular, gently undulating ascospores, which are (90) 100–130 (150) μm long and coated with slime. *L. pinastri* produces black diagonal lines between individual hysterothecia and in this differs from *L. seditiosum*. However, exact identification is based on PCR or PCR–restriction fragment length polymorphism (RFLP) (Stenström and Ihrmark, 1997; Johnston *et al.*, 2003). Similar hysterothecia are produced by *Meloderma desmazieri* on five-needle pines, although in that case the hysterothecia are surrounded by black rings.

Infection biology

L. seditiosum produces conidia released from pycnidia on needles from the early spring. However, the conidia are not able to infect green needles and they have no epidemiological significance; infected needles are cast mostly in May. Infectious ascospores are released from hysterothecia on second year needles, usually lying on the ground, from (June) July to September (Minter, 1981; Butin, 1995). During this period, young, current year needles are infected by released airborne ascospores in wet weather/under humid conditions. On a needle, ascospores form a short germ tube ending in an appressorium-like structure and penetrating needles through stomata in the cuticle and epidermis (Diwani and Millar, 1990).

Epidemiology

Heavy rains and high humidity during the summer and autumn create favourable conditions for infection and support the release and spread of ascospores (Diwani and Millar, 1990; Van Maanen and Gourbière, 2000). Outbreaks occur especially in secondary stands and plantations. Damage in natural stands established from natural regeneration is not so common, in spite of the fungus being commonly present in these stands as well. Weakened seedlings, e.g. by an after-planting reaction, together with stress factors and extreme climatic conditions, such as drought and late frost in clear-cuts, predispose trees to infection. Good nutrition is mentioned as one of the factors influencing the level of infection; pine stands on pure sandy soils are usually more susceptible to Lophodermium needle cast. Infected plants can be revitalized with minimum loss if the year of severe infection is succeeded by a year with a lower level of infection. Severity of the damage decreases with increasing age of pines: trees older than 10 years are not susceptible to the infection any more.

Management strategies and tactics

Avoidance is based on the elimination of infected pine stands and individual trees in the vicinity of forest nurseries and the rotation of woody plants on beds in nurseries. Good soil and appropriate fertilization in nurseries improve the health of seedlings and their resistance to infection (Podger and Wardlaw, 1990). It is necessary to adequately space trees when planting, and to encourage drying and minimize disease transmission by removing weeds around trees to allow good air circulation and drying of foliage.

Windbreaks in nurseries may be reservoirs of inoculum. In young pine stands, mingled broadleaves reduce infection. A layer of fallen leaves covering infected needles captures ascospores and also accelerates the decomposition of needles and spreading of antagonistic saprophytes on needles, thus not allowing the full development of hysterothecia. If the infection occurs in the first year after planting, the source is from infected stock in the nursery. The ideal conditions for development of hysterothecia are in pure soils (sandy soils, peat bog, etc.), covered by mosses. In stands such as these, conditions are ideal for pathogenic fungi to fully develop and carry out their life cycle on needles lying in moss beds, which maintain ideal humidity and concurrently reduce the activity of saprophytic fungi by the production of antimicrobial compounds.

Prevention of the introduction of the disease is based on using uninfected stock from disease-free nurseries. To avoid introduction into new areas, it is necessary to strictly observe phytosanitary precautions, such as prohibiting the introduction of any livestock and also the importation of organic substrates, soil and mulch based on pines. The main source of Lophodermium needle cast inoculum for infections in plantations and nearby nurseries are young infected fallen needles (Diwani and Millar, 1990).

Control measures must be applied in nurseries and also in plantations of Christmas trees in order to keep stock healthy. Spraying must be timed to occur at the time of ascospore release, usually in the summer from (June) July to September, every 2–3 weeks. Control in stands and plantations should be applied under specific conditions and when there are severe outbreaks. Greater control efficiency can be achieved by monitoring meteorological data so that pathogen occurrence (i.e. ascospore release) can be predicted. Copper-based fungicides, such as copper oxychloride, are effective for control, even when there is some risk of injury from tissue burns, if they are applied to the young needles. The application of organic fungicides based on propiconazole, fluazinam, azoxystrobin, mancozeb, benomyl, carbendazim + triadimefon, chlorothalonil and flutriafol showed high efficacy in needle cast control, both preventive and curative (Boudier, 1987; Stenström and Arvidsson, 2001).

Different susceptibility of provenances of Scots pine to the disease was observed. However, no resistant clone has been selected up to now (Petkov, 1993; Agúndez *et al.*, 1994), and some varieties of pines are known to be highly susceptible, e.g. some varieties of Scots pines originating from southern Europe, if they are grown under wet conditions.

23.3.2 Cyclaneusma needle cast

C. minus is a fungus that causes Cyclaneusma needle casts of pines. It can be found in all continents in natural forest stands, and also in plantations of introduced pine species.

Although a large number of pine species has been recorded as hosts of *C. minus*, including mountain pine (*Pinus uncinata* Mill. ex Mirb.) in subalpine stands (Giordano and Gonthier, 2011), under experimental conditions, the fungus has been shown to be pathogenic only to Monterey pine (Gadgil, 1984) and Scots pine (Karadžić, 1981).

Diagnosis

In Monterey pine, symptoms first appear on 1 year old or older needles in the central and lower parts of the crown. Needles turn mottled yellow green at first, and then mottled yellow brown a few weeks later. In highly susceptible trees, almost the whole crown can be affected. Transverse reddish bands are also commonly seen. In some years and in some localities, needles finally become a uniform reddish brown rather than the more usual mottled yellow brown. Needles showing symptoms are readily detached from the tree and most are shed prematurely, generally in the spring. By early summer, infected trees hold only the newly flushed foliage. The susceptibility of the trees to the disease varies. Resistant trees, which do not develop symptoms of infection, are not necessarily immune to infection by *C. minus*, although *C. minus* populations in needles of healthy trees are less than those in diseased trees (van der Pas *et al.*, 1984). The severity of the infection varies considerably from year to year.

In Scots pine, symptoms of infection are found on needles of all ages in Europe and on 1 year old or older needles in North America. The first symptoms appear as small, light green spots, which coalesce and turn the needle dusty yellow with transverse brown bands. Finally, the needles become brown. Infected needles are usually cast within a few months of the appearance of the symptoms.

Sub-epidermal apothecia are scattered in needles when they are fully open, and are waxy, concolorous with (the same colour as) the needle surface and 0.2–1.7 mm long × 0.2–0.25 mm wide. Ascomata push through the needle epidermis, which is torn but remains hinged, and opens by a single longitudinal split lifting two parallel lobes. Mature ascomata swell when moist and push back the hinged epidermal flaps. Ascospores are cylindrical to filiform, curved, 0–2 septate, smooth, 75–95 × 2.5–3 μm, thin walled, hyaline and with guttules. Pycnidia are scattered, deeply immersed, globose to sub-globose, 0.1–0.2 mm in diameter, with walls composed of hyaline, pseudoparenchymatous cells, 2–3 μm in diameter. *C. minus* could be confused with the similar *C. niveum*, which has larger apothecia, 0.3–1.0 mm long, and ascospores 75–120 × 3–4 μm in size and sickle shaped, rather than bacilliform conidia 9–21 × 1 μm in size (Millar and Minter, 1980; Minter and Millar, 1980c; DiCosmo *et al.*, 1983; Butin, 1995).

Infection biology and epidemiology

In New Zealand, *C. minus* infects susceptible trees of Monterey pine at temperatures ranging from 10°C to 25°C. Current year needles are resistant to the infection until they are 8 or 9 months old. They become infected from autumn to early winter, begin to develop symptoms of infection by midwinter, and are usually cast in the early spring, when they are about 1 year old (Gadgil, 1984; Bulman, 1988). Symptoms appear on needles in the summer of the second growing season, even though the needles become susceptible to the infection in the midsummer of the first growing season. These needles are cast at the end of the summer or in the autumn, when they are about 2 years old. In Europe, Karadžić (1981) reported that symptoms on Scots pine needles appear 2–3 months after the initial infection. Rack (1981) found that *C. minus* could be first isolated from current year needles when they were about 6 months old. However, *C. minus* is commonly reported as an endophyte in needles (Jurc *et al.*, 1996; Sieber *et al.*, 1999; Kowalski and Zych, 2002).

Management strategies and tactics

To achieve an almost disease-free stand, it is necessary to delay first thinning until the age of 7 or 8 years, when symptoms of infection can be easily detected. The ideal silvicultural regime for the control of the disease includes greatly delayed first thinning until the seventh year of age, followed by second thinning at the age of 9 or 10 years, to the final stocking of 250 stems ha^{-1} to remove the remaining disease-susceptible trees (Bulman, 2001).

23.3.3 Cercospora needle blight

M. gibsonii (anamorph: *Cercoseptoria pini-densiflorae* (Hori & Nambu) Deighton) is a pathogen that causes Cercospora needle blight or brown needle blight of pine, particularly at the later nursery stage, and has become a major obstacle to the production of pine seedlings, especially maritime pine (*P. pinaster* Aiton), Japanese black pine (*P. thunbergii* Parl.) and Japanese red pine (*P. densiflora* Siebold & Zucc.) in Japan and Taiwan. It also attacks maritime pine, Monterey pine, Canary Island pine (*P. canariensis* C. Sm.), Okinawa pine (*P. luchuensis* Mayr), Chinese red pine (*P. massoniana* Lamb.), Merkus pine (*Pinus merkusii* Jungh. & Vriese ex Vriese), red pine (*P. resinosa* Aiton), eastern white pine and Scots pine. Resistance has been reported in sand pine (*P. clausa* (Chapm. ex Engelm.) Vasey ex Sarg.), slash pine (*P. elliottii* Engelm.), Khasi pine, Mexican weeping pine (*P. patula* Schiede ex Schltdl. & Cham.) and pitch pine (*P. rigida* Mill.). The incidence of mortality in some nurseries can reach 80%. Severe defoliation in young plantations of Monterey pine has occurred in Tanzania, resulting in reduced growth and sometimes even death of the trees. The disease is important on Merkus pine and Caribbean pine (*P. caribaea* Morelet) in nurseries in west Malaysia (EPPO, 1980). The fungus is widely distributed in tropical and subtropical areas like Central and South America, but it is also described from Asia, Africa and Oceania (Ivory, 1975; EPPO, 1980; Evans, 1984; Crous *et al.*, 1990; Roux *et al.*, 2005).

Diagnosis

Lesions occur especially on seedlings 1–2 years old; they are initially pale green and later yellowish brown to grey, and appear towards the distal part of the needles. They coalesce, which results in complete needle necrosis and subsequent needle cast. Dark brown stromata fill the stomatal cavities, and numerous fruiting bodies appear as sooty spots on the lesions. There is never any reddish tint to the necrotic needle tissues, such as may occur in other infections. The disease may be confused with Dothistroma needle blight or brown spot needle blights, but conidia are pale yellowish olivaceous, long obclavate, straight or slightly curved with three to seven septae and truncated or rounded, and with an unthickened base and obtuse tip; they are 20–68 (mostly 40–50) × 2.5–4.5 μm in size.

Infection biology and epidemiology

M. gibsonii overwinters as mycelial masses or immature stromata in the tissue of infected needles. The primary infection source consists of airborne conidia produced in the spring from infected needles and spread by wind or rain splash. Transplanted seedlings with diseased needles also serve as an infection source. Conidia germinate between 10°C and 35°C (25°C being optimal). The incubation period varies with the environment, but is thought to be around 6 weeks. Following infection, stromata form in the stomatal cavities and bear dense conidiophores.

Management strategies and tactics

It is unlikely that the fungus could spread from Africa and Asia in the form of the windborne spores. It could be imported on infected seedlings and on cut branches of pines, and because of the long incubation period, latent infection presents a hazard.

In nurseries, control can be obtained by using mancozeb or copper-based fungicides (Reddy and Pandey, 1973). Once the infection occurs, it is important to remove and destroy all diseased seedlings.

M. gibsonii is listed as an A1 quarantine pest by the European and Mediterranean Plant Protection Organization (EPPO), but it is also a quarantine disease to the Comunidad Andina (CA) and Comité Regional de Sanidad Vegetal del Cono Sur (COSAVE). It is potentially dangerous to planted pines in these areas.

23.3.4 Brown spot needle blight

M. dearnessii (anamorph: *Lecanosticta acicola* (Thüm.) Syd.) is a fungal pathogen that causes brown spot needle blight on pines. Typical symptoms on green needles are brown spots, and later strips surrounded by a yellow halo. As the infection progresses, tips or whole needles die and are cast.

Diagnosis

M. dearnessii can be easily confused with *Dothistroma* spp. because the symptoms vary enormously depending on the host species. However, acervuli are green inside and conidia are sub-hyaline to dark brown, echinulate to verrucose or tuberculate, thick walled, straight to curved, with one to five septa, fusiform to cylindrical, (10) 12–45 (55) × 2–4.5 μm in size and with a rounded apex and truncate base; ascospores are hyaline, smooth, single septate, usually four guttulate, oblong to cuneate and 7.5–14 × 2–3.5 μm in size (EPPO, 2008; Ioos *et al.*, 2010).

Infection biology and epidemiology

M. dearnessii is a pathogen of a wide range of pines, such as longleaf pine (*P. palustris* Mill.), ponderosa pine (*P. ponderosa* Lawson & C. Lawson), Monterey pine, eastern white pine, Scots pine, loblolly pine (*P. taeda* L.), Japanese black pine, and many other Central and North American pine species and their hybrids (Sinclair and Lyon, 2005). It is supposed that *M. dearnessii* originates from Central America and southern North America (Evans, 1984). The pathogen has also been introduced to various regions of the world, e.g. to China, Europe and Japan (Huang *et al.*, 1995; Suto and Ougi, 1998; Drenkhan and Hanso, 2009). It spreads mostly through conidia or ascospores by rain splash or by wind for long distances (Siggers, 1944). High numbers of infections occur during hot and wet periods (Kais, 1975).

Management strategies and tactics

Brown spot needle blight causes damage in natural stands, plantations and nurseries. It can seriously limit the growth of pines during their early years because of greater humidity near the ground, which favours the occurrence of the disease. The disease can also cause death of the affected tree in the case of a severe infection that has been going on for several years.

The disease can be treated with application of fungicides, e.g. Bordeaux mixture, or with fire in young plantations (Siggers, 1944). In many countries where the disease is not widespread, infected trees have to be eradicated according to the phytosanitary rules that prevail. The resistance varies between and within species of pines, and thus selection or breeding can present effective protection against brown spot needle blight (Boyer, 1972). *M. dearnessii* is listed as an A2 quarantine pest by EPPO.

23.3.5 Other pathogens of pines

Bifusella linearis (Peck) Höhn. causes tar spot needle cast or Bifusella blight on five-needle pines in the western USA and Canada. The disease appears under moist conditions, and because its effects are usually limited to the defoliation of 2–3 year old needles, it is not devastating. The fungus forms narrow, shiny black, elongate hysterothecia of variable length, produced on the lower surface of dead needles, frequently connected with crust-like mycelial formations; these are the most distinctive characteristic of this disease (Bega, 1978; Minter and Millar, 1984). There are currently no fungicide recommendations for this fungus. Other species from the genus *Bifusella* are *B. pini* (Dearn.) Darker and *B. saccata* (Darker) Darker. These species are unlikely to be serious pathogens. Both can be found on pines in North America (Bega, 1978; Minter, 1988, 1993).

Elytroderma deformans (Weir) Darker causes Elytroderma needle blight. Damage varies greatly in severity. The disease is severe in western North America (Canada, the USA and Mexico). The disease reduces the crown growth of two- and three-needle pines. It also reduces the vigour of the trees and causes increased susceptibility to other pathogens and bark beetles. Typical symptoms of Elytroderma disease are red brown 'flags', which appear in the spring. Infected needles are reddish brown during the spring and early summer of their second year. Infection occurs in late summer, but symptoms do not appear until early spring of the following year. Groups of needles turn red and die, forming distinct flags. The fungus is perennial in the twigs, often stimulating them to form relatively small, open and tufted witches' brooms. The fungus overwinters in three ways: (i) in the cast needles; (ii) as a new infection in green current year needles; and (iii) as a perennial infection in buds and twigs (Hunt, 1978).

In spring, the 1 year old needles that were infected in the previous year, acquire a bright red discoloration, while the new needles that begin to form remain green. Pycnidial blisters are concolorous and about 1 mm long. Under wet conditions, pycnidia exude spermatia, which do not cause infection. It has been suggested that if the spermatia do not find environmental conditions satisfactory for dissemination and fertilization, new infections will not survive and ascospores will not develop (Lightle, 1954; Roth, 1959; Waters, 1962). During the summer, hysterothecia 1–8 × 1 mm in size begin to form and later open by a longitudinal fissure. Asci are clavate and 140–240 × 30–45 μm in size; ascospores are 90–120 × 6–9 μm in size, elongated and cylindrical with one central septum. Ascospores are released in late spring and early summer and in this time period cause infection in recently formed needles. A strong infection results in such a severe cast that only current year needles remain; this is called the 'lion's tail' symptom. Needles and shoots are shorter and twigs may be thickened and curved upwards. On some trees, loss of apical dominance occurs, so that lateral buds grow and the whole tree resembles a broom. Practical control depends on silvicultural precautions, such as clear-cuttings (clear-fellings) in mature stands and thinning to promote vigorous growth in young stands.

Herpotrichia juniperi and *Neopeckia coulteri* cause the pine needle disease known as brown felt blight. *H. juniperi* attacks many coniferous species, including cypresses (*Cupressus* spp.), firs, hemlocks (*Tsuga* spp.), junipers (*Juniperus* spp.), pines and spruces. *N. coulteri* infests only pines. Fungal growth is prolific on branches that are buried under the snow. As the snow melts and the felts are exposed, fungal development ceases. The fungi produce thick, grey, felt-like mats of mycelium that smother needles, twigs and branches. *N. coulteri* is able to penetrate living needles, while *H. juniperi* attacks needles and twigs that are already dead. Freshly exposed felts are dark brown or black, weathering to grey brown. Spherical, black fruiting bodies 0.25–0.5 mm in size develop on and in mycelial mats in the second year after mat formation. These become conspicuous as the felt weathers away, and produce ascospores during the summer and early autumn. Infectious inoculum appears as ascospores and fragments of mycelium that remain as the felts disintegrate. Long-lasting snow cover is a favourable environment for brown felt development. Branch death may occur on small seedlings and on the lower branches of older trees covered in snow. This disease may cause great damage in forest nurseries, where bare-root seedlings are produced (Butin, 1995).

Lophodermella arcuata (Darker) Darker causes needle cast of five-needle pine, such as eastern white pine in North America. Fruiting bodies are reported to develop rapidly on first year needles, which turn brown in the spring. Ascospores are released in July and August and infect 2–5 month old needles. Most needles are cast by September of their second year, so that the life cycle is completed in 15 or 16 months. The effect of the disease on radial growth may be confused with drought effects. The disease has been considered unimportant in natural stands (Shaw and Leaphart, 1960; Burleigh *et al.*, 1982; Minter and Millar, 1993).

Lophodermella cerina (Darker) Darker causes needle cast of lodgepole pine (*Pinus contorta* Douglas ex Loudon), slash pine, ponderosa pine and loblolly pine on the western coast of the USA. Symptoms on first year needles have been reported to appear in November, with short, oval, light brown fruiting bodies visible by late February and prominent by the end of March (Minter and Millar, 1993).

Lophodermella concolor (Dearn.) Darker can cause severe defoliation in Canada and the USA. Growth reduction and mortality may come after repeated epidemic outbreaks, particularly in young trees. *L. concolor* is restricted to the two-needle pine species. Trees of all ages can be infected, but the pathogen is especially associated with young stands. Needles infected in the previous year turn reddish brown in May or June. In July, needles become straw-coloured. Although new needles are infected during their first year of growing, there is little visual evidence of the infection at this time. Damage occurs in the needle tissue during the winter, and the needles turn red in the spring. Affected needles on the previous year's growth are cast during the summer and the remaining new needles look like a lion's tail. Fruiting bodies are formed on dead or dying needles, appearing as shallow oval depressions, which are concolorous with the needle surface. Infected needles are shed at about the time that the apothecia mature, and spores are released during periods of wet weather and infect new growth (Minter and Millar, 1993; Hunt, 1995).

Elytroderma needle cast symptoms are similar to those of *Lophodermella*. However, the long, black fruiting structures of *Elytroderma* are very different from the small, inconspicuous *Lophodermella* fruiting structures. Other fungi often invade needles infected with *Lophodermella*, which may cause confusion over the fruiting structures. *Hendersonia pinicola* (Dearn.) Trotter, itself not an important pathogen, may be found on infected needles, where fruiting structures appear as small black spots.

Lophodermella conjuncta (Darker) Darker causes pine needle blight leading to premature needle cast. It is reported from Scandinavia, the UK, Estonia and Switzerland. Host species are Scots pine, mugo pine and Austrian pine. This disease becomes evident in the summer, when resinous red bands develop on older needles. This happens in the year after the infection. Yellowing and then browning of needles occur, connected to the bands, but the discoloration spreads rapidly, until whole needles are dead. The banding symptom may cause confusion with Dothistroma needle blight. Fruiting bodies are produced on the second year needles from mid-August. Infected needles are cast prematurely during the third season (Mitchell *et al.*, 1978; Millar and Minter, 1980).

Lophodermella montivaga Petr. can be distinguished from *L. concolor* by its elongate, brown fruiting bodies and tendency to have a more northern distribution. In the case of *L. montivaga*, needles change colour from orange to brown quickly, about a month after the infection. Needles infected by *L. concolor* discolour in the early spring of the following year (Minter and Millar, 1993).

Lophodermella sulcigena (Link) Höhn. causes needle blight on various pine species, such as Scots pine, mugo pine and lodgepole pine. The disease leads to premature needle cast; sometimes it is called 'Swedish pine cast'. *L. sulcigena* is a European species and occurs in central Europe and Scandinavia (Mitchell *et al.*, 1976; Millar and Minter, 1978).

Meloderma desmazieri causes a needle cast on pines that is widely distributed throughout Europe, North and South America, Africa and Australia. Symptoms of the disease first develop from the previous year's infection like yellow spots on the needles in the early spring. These spots turn brown and enlarge, giving severely infected trees a reddish tinge. Hysterothecia appear on the dead parts of needles in February and ripen from May to July. Spores are spread during the summer, when needles become infected. The yellow spots do not develop until the following spring (Minter and Gibson, 1978a). The infection also penetrates to the current year shoots and causes shoot dieback. The Chinese pine-nut (*Pinus koraiensis* Siebold & Zucc.) and eastern white pine are highly susceptible.

Phacidium infestans infects needles of a broad spectrum of conifers. In winter, the infection develops only on needles under the snow cover. After the spring thaw, infected needles remain on the trees, the needle tissues die and the needles turn bleached brown. Apothecia form in the late summer as dark brown, round, flatted fructifications disrupting the epidermis, and arranged around the middle rib on the underside of the needle. Hyaline, elongate, slightly flattened ascospores 15–26 × 5–9 μm in size are released from August to October and infect current year needles, thus completing the cycle in a year. *P. infestans* var. *abietis* includes two formae speciales (f.spp.): North American *P. infestans* var. *abietis* f.sp. *abietis* on firs and Douglas-firs, and Japanese *P. infestans* var. *abietis* f.sp. *piceae* on firs, spruces and pines (Minter and Millar, 1980d; Roll-Hansen, 1987, 1989; Butin, 1995). Disease is common in natural Swiss stone pines (*Pinus cembra* L.) in the Alps.

Ploioderma lethale (Dearn.) Darker causes needle browning on pines in the spring. Current year needles are infected during the summer. Later, in winter and early spring, before new needles begin to grow, infected needles turn brown from their tips. The dying needle or needle tip has a mottled appearance. Needles in various stages of dieback are often shed prematurely. The fruiting bodies of *P. lethale* are shiny black, long and narrow (Minter and Gibson, 1978b). The ecology of a similar fungus, *Ploioderma hedgcockii* (Dearn.) Darker, is different: hysterothecia occur on green needles, which turn yellowish all over and are shed prematurely. Fruiting bodies are shiny black and elliptical (Boyce, 1969; Cannon and Minter, 1984). Both fungi are described from North America.

Some other species, such as *Sydowia polyspora* (reported as *Sclerophoma pithyophila* (Corda) Höhn.), occur on damaged needles. Apothecia of the probably endophytic fungus *Cenangium acuum* Cooke & Peck may be found on dead needles. Apothecia are individual, substantially protruding above the surface of the needles and 0.5–3 mm in diameter. Pycnidia of *Diplodia pinea* (Desm.) J. Kickx f. might be abundant on dead needles that have dried on twigs after thinning.

23.4 Foliar Diseases of Spruces

A healthy spruce retains its needles for 5–7 years; however, inappropriate site conditions, such as drought and air pollution, may cause dramatic defoliation, starting usually from the older needles. Spruces, particularly Norway spruce, are relatively resistant to needle diseases, especially in comparison with pines.

Spruces are host species of the aecial stage of several rust fungi that cause local outbreaks, such as *Chrysomyxa* spp. (see Hamelin, Chapter 26, this volume). Some spruces, such as blue spruce, are also sensitive to Dothistroma needle blight in plantations of Christmas trees, if neighbouring pines are infected; Schrenk's spruce (*Picea schrenkiana* Fisch. & C.A. Mey.) is highly susceptible to Dothistroma needle blight.

More than 50 species have been reported as endophytes in green spruce needles. The most common are *Darkera parca* H.S. Whitney, J. Reid & Piroz., *L. macrospora*, *Lophodermium piceae*, *Penicillium glaucoalbidum*

(Desm.) Houbraken & Samson, *R. kalkhoffii* and *Sydowia polyspora* (Kowalski and Lang, 1984; Sieber, 1988; Müller and Hallaksela, 1998; Müller *et al.*, 2001).

Seven species of *Hypodermataceae*, i.e. *Isthmiella crepidiformis* (Darker) Darker, *Lirula brevispora* Ziller, *L. macrospora*, *Lophodermium crassum* Darker, *L. piceae*, *Lophomerum darkeri* Ouell. and *L. septatum* Tehon ex Ouell., are reported from spruce needles as potentially pathogenic (Ziller, 1969a).

23.4.1 Diagnosis, infection biology and epidemiology

Cucurbitaria piceae Borthw., which occurs on spruce species in Europe, causes the abortion of buds. In spring, infected buds swell, twist and warp spirally; they are covered with dark black basal stromata. On black ascostroma, sessile, discontinuous, in clusters, glabrous, dark brown to grey black, more or less spherical pycnidia appear, about 0.5 mm in diameter, with a circular aperture 40–60 μm in diameter. Conidia are filamentous, (150) 220–320 × 5–8 μm, hyaline and slightly tapered at the top. Perithecia form after pycnidia, asci are eight spored, 180–250 × 25–30 μm in size; ascospores are elongated, conical to fusiform, multi-celled, brown or dark brown and 35–50 × 12–15 μm in size (Shoemaker, 1967).

D. parca (anamorph: *Tiarosporella parca* (Berk. & Broome) H.S. Whitney, J. Reid & Piroz.) is reported on spruce needles from Canada and Europe. Its significance as a parasite is unknown; it is mostly reported as an endophyte or weak parasite, producing fruiting bodies on dead needles. Sub-epidermal blackish, round-shaped pycnidia 0.5 mm in diameter form abundantly on dead needles. Conidia are hyaline, cylindrical (20) 23–40 × 4–6 (7) μm in size, truncate at the apex, with a sub-denticulate scar at the apex, at first granular, and enveloped in a mucilaginous sheath. Apothecia are scattered, shiny black, elliptical to elongate, 2 mm in size; ascospores are hyaline, but may become pale brown, single celled, elliptical (17) 20–35 (41) × 8–15 μm in size (Whitney *et al.*, 1975; Cech and Tomiczek, 1988; Karadžić, 1998; Müller and Hantula, 1998).

L. macrospora is widespread in spruce needles and occasionally strikes blue spruce and white spruce (*Picea glauca* (Moench) Voss) in the USA and Canada. In Europe, the disease causes occasional defoliation of blue spruce in plantations; strong infection causes browning and cast of all second year needles in Norway spruces grown under wet conditions as well, usually on young trees growing under the canopy and suffering from lack of light and high humidity, but without any serious outbreaks. Sprouting needles are infected by ascospores and in early spring of the year following the infection, needles begin to turn from yellow to brown or bronze in Norway spruce; in blue spruce they become a typical yellow and later a pinkish brown colour, similar to the symptoms of *R. kalkhoffii* infection, and infected needles turn brown from May of the next year. Pycnidia of the anamorphic stage *Hypodermina hartigii* Hilitzer occur on brown needles in July and produce ovoid, hyaline 3 × 2 μm sized spermatia. Hysterothecia start to form subsequently from July on the underside of the needles remaining on twigs, as 2–15 mm long, shiny black, elongated fruiting bodies, opening by longitudinal slits under moist conditions. Fruiting bodies ripen in April and May of the third year following the infection; ascospores are acicular and 75 × 1.5 μm, in 100 × 15–21 μm sized asci (Cannon and Minter, 1984).

L. piceae is present latently on the needles as an endophyte or weak pathogen, and after death of the needles plays the role of the pioneer decomposer. If the parasitic role prevails, first symptoms occur as pink reddish or red brown strips on green or yellowing needles. Usually, it is individual needles in shadow or needles suffering from abiotic factors that are infected. This latent stage lasts for a relatively long time, while mycelia penetrate through the needle. The fungus produces blackish spermogonia on needles from the autumn to spring, while infected needles remain on the twigs. Spermatia are slightly curved, 3–5 × 1.4 μm in size. Subsequently, infected needles fall off, with dead needles rarely staying on twigs, and black transverse zones occur on needles together with the formation of black elliptic, navicular hysterothecia up to 1.5 mm in size, which open in wet conditions by a longitudinal slit; needle-shaped ascospores 80–100 × 1.8–2.5 μm are released from May to June and infect new needles.

R. kalkhoffii is an opportunistic fungal pathogen with worldwide distribution, causing needle cast in blue spruce and Engelmann spruce (*Picea engelmannii* Parry ex Engelm.). Norway spruce is relatively resistant. The coincidence and consequences of extreme climatic factors or any other stress factors play important roles in development of the disease (Diamandis, 1978). Typical of mass outbreaks of *R. kalkhoffii* on blue spruce is the reddening of second year needles in the lower part of the crown, which proceeds up the crown. In the case of severe infection, only current year needles are present. Fruiting bodies often occur on needles dying as a result of some other factor and quickly spread through the dead tissue. The fungus overwinters in infected needles both on the tree and on those fallen onto the ground. Black, roundish pycnidia 80–150 μm in diameter grow through the lines of stomata, especially in spring, although they are to be found during the whole year. The conidia are roundish to ovoid, hyaline and 7–10 × 3–5 μm in size. Spores infect current year needles or stressed or senescent needles of any age, usually in the spring. Symptoms arise in late summer as yellow mottling; on blue spruce, the infected needles typically turn purple brown in the whole lower part of crown. Infected needles turn brown during late winter or early spring and they are cast the next summer, 1 year after the infection. *R. kalkhoffii* might be confused with *R. pini* (Corda) Maubl., which occurs on spruces, pines and firs, while *R. macrospora* occurs mainly on firs, but occasionally also on spruces (Diamandis and Minter, 1980; Butin, 1995).

L. macrospora and *R. kalkhoffii* infect a broad spectrum of spruces only occasionally; none the less they cause considerable and economically significant damage on the needles of blue spruce and Engelmann spruce. In young spruces, infection by *L. macrospora* often leads to complete defoliation of the second year part of the shoots. *R. kalkhoffii* causes severe defoliation of the lower part of the crown in blue spruce.

23.4.2 Management strategies and tactics

Control measures should be applied to blue spruces in Christmas trees plantations, if severe symptoms of needle cast occur, by spraying in the end of May, when needles are half elongated, and again 4–6 weeks later. Other control measures include supporting good ventilation inside plantations by spacing trees properly, and retaining moisture by mulching and irrigation. Control includes infections by *Lirula* and *Rhizosphaera*, and also that of some other infections that break out at the same time of the year. For *Lirula*, it is necessary to repeat the spraying for 3 subsequent years after the year when the infection occurs owing to its 3 year cycle.

23.5 Foliar Diseases of Firs

There is a great diversity of fungi living on fir needles. However, they are not the cause of any serious outbreaks in forest stands and forest nurseries. There are some diseases that cause severe losses in plantations of Christmas trees and in urban/amenity plantings (Cech, 1992; Perny *et al.*, 2002; Talgø *et al.*, 2010). Several fungal pathogens on fir needles are reported from firs in the USA and Canada, such as *Lirula abietis-concoloris*, *L. mirabilis* (Darker) Darker, *L. nervata* (Darker) Darker, *L. nervisequa*, *Lophodermium decorum* Darker and *Virgella robusta* (Darker, 1967; Bega, 1978; Funk, 1985; Scharpf, 1988). From Europe, *Delphinella abietis* (O. Rostr.) E. Müll, *L. nervisequa*, *Nematostoma parasiticum* (R. Hartig) M.E. Barr., *Phacidium infestans*, *P. nudus*, *Penicillium glaucoalbidum*, *Rhizosphaera abietis* L. Mangin & Har., *R. kalkhoffii*, *R. macrospora* and *Valsa abietis* are reported (Cech, 1992; Butin, 1995; Perny *et al.*, 2002; Talgø *et al.*, 2010). Several species of heteroecious rust fungi, e.g. *Pucciniastrum epilobii* G.H. Otth. and *P. goeppertianum* (Kuehn) Kleb., occur on fir needles in the aecial stage. More than 120 fungal species are recorded with a connection to fir needles as endophytes living in needles or colonizers of green needles and needles in litter (Sieber-Canavesi and Sieber, 1993).

23.5.1 Diagnosis, infection biology, epidemiology and management

D. abietis is reported from many host species of young firs up to 15 years of age from Europe (Solheim and Skage, 2002) and North America (Funk, 1985). The fungus usually causes no serious damage, but it may bring about severe needle cast in plantations of Christmas trees. Infection is connected with (and similar to) that by late frost: current year needles in the lower part of the crown are damaged. The fungus might penetrate from needles to shoots and cause a dieback. Round, black pycnidia 150–200 μm in size produce round to spindle hyaline conidia 10–16 × 4–6.5 μm in size. Numerous round-shaped black pseudothecia cover especially the upper surface of dead needles on current year shoots. Pseudothecia are 0.3–0.7 mm in diameter, ascospores are hyaline, uniseptate and 10–20 × 4.5–5 μm in size. The removal of infected shoots and control by the application of fungicides is recommended in the spring, at the time of ascospore release (usually from May to June) (Butin, 1995; Ellis and Ellis, 1997).

Isthmiella faullii (Darker) Darker is reported from the USA and Canada. Symptoms are similar to those caused by *L. nervisequa* (Cannon and Minter, 1984; Albers *et al.*, 1996), which are described below.

Lirula nervata is a typical species on balsam fir (*Abies balsamea* (L.) Mill.). A 3 year life cycle is characteristic of this species. During the summer, ascospores infect current year needles, which remain symptomless until the following spring, when needles turn yellow and in the summer produce characteristic single pycnidia on the upper surface of the needles with single-celled, hyaline, ovate conidia 4.5–6.0 × 1.0–1.2 μm in size. From the first summer, typical oblong, black fruiting bodies start to develop in the underside of the needle. Asci are cylindrical to clavate, eight spored and 130–208 × 17–27 μm. Following the third year, fruiting bodies release filiform, clavate, hyaline ascospores sized 70–90 × 2.5–3.5 μm, surrounded by a conspicuous gelatinous sheath 3–6 μm thick, and infect new needles (Albers *et al.*, 1996).

L. nervisequa produces local defoliation of young and also mature trees of silver fir (*A. alba* Mill.) in Europe. Browning of needles commences from June to July. The fungal species has 3 year life cycle similar to that of *L. nervata*. Dry dark to blackish fruiting bodies form in the second year on the underside of browned needles, up to about 10 mm long, opening by a single longitudinal split, with lips absent. Ascospores are hyaline, aseptate, smooth, filiform, 75–90 × 3–4 μm, enveloped in a mucous sheath 3–4.5 μm thick (Minter and Millar, 1984).

N. parasiticum causes needle browning in several species of firs, such as silver fir, Nordmann fir (*A. nordmanniana* (Steven) Spach), Christmas tree (*A. veitchii* Lindl.) and noble fir (*A. procera* Rehder); it can infect also spruce (Butin, 1995).

P. nudus causes complete red browning of the last year's needles. There are blackish, sub-globose, depressed pseudothecia, 50–125 μm in diameter and growing abundantly through the rows of stomata in the underside part of the dead, brown needles on twigs. Ascospores are hyaline to sub-hyaline, 9–18 × 4–7 μm in size and slightly constricted at the middle septum. An anamorphic stage is unknown. The fungus has a wide distribution. It is reported from the USA, Japan, Taiwan and Europe (Petrak, 1925, 1938; Barr, 1968; Chen, 1974).

Several species of *Rhizosphaera*, such as *R. abietis*, *R. macrospora* and *R. oudemansii* Maubl. (Gourbière and Morelet, 1979; Diamandis and Minter, 1980; Martinez and Ramirez, 1984; Butin, 1995), occur as weak parasites of needles in various fir species and usually do not cause problems, in spite of being common pathogens. Round, black, tiny pycnidia form abundantly in tens to hundreds from stomatal lines on the underside of the needles. Damage includes discoloration of the needles, which eventually turn brown.

Some species of the genus *Tiarosporella* occur on conifers. *T. abietis* H.S. Whitney, J. Reid & Piroz. (teleomorph: *Darkera abietis* H.S. Whitney, J. Reid & Piroz.) is reported from needles of balsam fir, white fir (*A. concolor*

(Gord. & Glend.) Lindl. ex Hildebr.), grand fir and subalpine fir (*Abies lasiocarpa* (Hook.) Nutt.) from Canada (Whitney *et al.*, 1975); there are no records from other continents. *T. durmitorensis* Karadžić was described as a very aggressive pathogen colonizing needles of all ages of silver fir and Macedonia fir (*A. borisii-regis* Mattf.) in high-elevation alpine stands in the Durmitor Mountains in Montenegro. This species is very similar to *T. abietis* from Canada, but forms larger conidia (Karadžić, 1998).

Valsa abietis occurs commonly on needles, where the fungus produces the anamorph *Cytospora abietis* Sacc. The fungus infects needles and also shoots, whereas the teleomorphic stage develops in dead twigs only. There are three to ten pycnidia per needle only, which are 0.5–1 mm in size; conidia in tendrils are sausage shaped and 4–4.5 × 1–1.5 μm in size. Infections are caused by ascospores only.

Virgella robusta is characterized by dark hysterothecia on the underside of the needle and by brown pycnidia in two rows, also on the lower surface of the needles. The disease is known from the USA and Canada from several species of fir, such as Pacific silver fir (*Abies amabilis* (Douglas ex Loudon) Douglas ex Forbes), white fir, grand fir, subalpine fir and noble fir (Bega, 1978; Minter and Cannon, 2001).

23.5.2 Other foliar diseases of firs

Current season needle necrosis (CSNN) causes severe damage of fir needles in plantations of Christmas trees in Europe and Canada (Cech, 1992; Perny *et al.*, 2002; Cech *et al.*, 2010; Talgø *et al.*, 2010). The first symptoms appear in early summer as yellowish bands on needles, quickly spreading to reddish brown necrotic lesions killing usually the upper third to the upper half of the needle, when some needles are cast prematurely. Sensitivity of the trees to the infection is individual. Defoliation and brown needle tips dramatically reduce the quality of Christmas trees in the market. CSNN is reported mainly from grand fir, Nordmann fir, noble fir, Bornmueller's fir (*A. bornmuelleriana* Mattf.) and also Douglas-fir. Symptoms of CSNN on silver fir are rare. It is possible to isolate a broad spectrum of fungi, e.g. *Cladosporium* spp., *Epicoccum* spp., *Fusarium* spp., *Penicillium glaucoalbidum*, *Pestalotiopsis funerea* (Desm.) Steyaert, *Phomopsis* spp., *Sclerophoma* spp. (Talgø *et al.*, 2010), from infected needles as secondary invaders. In Europe, the disease has been connected with *Kabatina abietis* Butin & Pehl (Butin and Pehl, 1993; Perny *et al.*, 2002). From the USA, *Hormonema merioides* A. Funk, T.A.D. Woods & Hopk. (Funk, 1985) is reported as the disease agent. Sequencing of the internal transcribed spacer (ITS) ribosomal DNA of *K. abietis* resulted in the finding of *Hormonema dematioides* Lagerb. & Melin, an anamorph of *Sydowia polyspora* that includes anamorphs such as *K. abietis* and *H. merioides*, and also *Sclerophoma pithyophila* (Talgø *et al.*, 2010).

Control is applied in nurseries only, and includes removing infected parts and spraying from May to July.

23.6 Foliar Diseases of Larches

The spectrum of fungal needle pathogens on larches is broad and includes needle blights and rusts; also aphids, such as *Adelges geniculatus* (Ratzeburg), moth larvae such as *Zeiraphera diniana* (Hubner) and *Coleophora laricella* (Hubner), the sawfly *Pristiphora laricis* (Hartig) (Perny, 2004) and some spider mites are common on young needles and cause similar symptoms to fungi, e.g. local spots and chlorosis together with premature defoliation. Besides these pests, several species of needle cast, such as *Hypodermella laricis*, *Lophodermium laricinum* Duby, *Meria laricis*, *Mycosphaerella laricina* (Butin, 1995; Cech, 2004), and some other minor species are reported. Larch needle casts frequently occur together, because they require similar conditions for infection. The needles of larches are hosts for the aecial stages of rusts, such as *Melampsora caprearum* Thüm., *M. epitea* Thüm., *M. medusae* Thüm., *M. minutissima* (Opiz) Bubák, *M. populnea* (Pers.) P. Karst. and

Melampsoridium betulinum (Pers.) Kleb. (Butin 1995; see Hamelin, Chapter 26, this volume).

23.6.1 Diagnosis, infection biology and epidemiology

Hypodermella laricis infects sprouting needles early in the spring. Infection causes yellowing of the needles on branches. Infected needles turn first yellow, and later are blushing and browning; they drop within 2 weeks of infection. All infected needles are usually found in the lower part of the crown; however, the entire crown may be affected. Fruiting bodies form on the needles that are still on the tree as small, elliptical, black, subcuticular hysterothecia, more or less in a continuous row along the main venation and sized 0.5–0.8 × 0.2–0.3 mm. In comparison with *Lophodermium laricinum*, the fruiting bodies are smaller; asci are 80–112 × 20–24 µm in size, with four hyaline, single celled, tear-shaped ascospores sized 60–100 × 6 µm and coated with a gelatinous sheath 5 µm thick. Pycnidia are black, 120–300 × 80–120 µm in size; conidia are hyaline, elongated, pyriform and 4–5 × 1 µm in size. Sometimes, *H. laricis* occurs together with *L. laricinum* in the same needle. Based on the symptoms only, *H. laricis* can be confused with other needle casts and also with frost damage. Hypodermella needle cast of larch is reported from mountain areas in Europe and also from the USA and Canada (Minter, 1988; Garbutt, 1996).

Lophodermium laricinum is a weak pathogen reported from mountain areas in Europe, and also from North America (Ziller, 1969b). The fungus causes yellow discoloration on needles and produces fruiting bodies and thin, black transversal dark lines in the dead needles. Fruiting bodies, formed on dropped needles, and sometimes also on needles dried on the twigs, are oblong, rhomboidal, rounded at the top, 0.5–1.5 × 0.25–0.75 mm in size and nested under the cuticle up to the epidermis. Asci are long, club shaped, 95–115 × 9–12 µm in size; ascospores are acicular and 70–105 × 2–2.5 µm in size.

Meria laricis is an anamorphic fungus, the teleomorphic stage being unknown. Sequence analysis of the ITS region indicates that *M. laricis* form a monophyletic clade within the ascomycete genus *Rhabdocline* (Gernandt *et al.*, 1997). This pathogen affects seedlings up to 2 years old in nurseries, and later is located on the lower part of the crown only (Cech, 2004; Lilja *et al.*, 2010). Under favourable climatic conditions, *M. laricis* might cause local outbreaks in natural stands as well (Maresi *et al.*, 2004). The yellowish brown discoloration of needles might be easily confused with frost damage, but no fruiting bodies are formed in frost-damaged needles. Infection starts in the spring, usually by conidia which are formed in the last year's needles dropped on to the ground.

Symptoms of *M. laricis* infection are manifest first by yellowing of needles, often from the middle; sometimes, needles blush and then drop off. Most needles are shed in the same year that they die, but needle casting may continue throughout the growing season. The fungus penetrates through the sub-epidermis of the infected needles, creating small whitish or colourless bodies on the needle surface, which are stored in colourless hyphae with four septae that penetrate through stomata. Conidia are narrowed in the middle and 8–10 × 2.5–3 µm in size (Butin, 1995). Outbreaks are supported by rainfall and excess humidity in the spring, and the severity of the infection differs in individual years according to the weather. *M. laricis* also invades needles at aphid feeding sites, causing premature needle fall. Disease is widespread in Europe on the European larch (*Larix decidua* Mill.) and on introduced Japanese larch (*Larix kaempferi* (Lam.) Carrière); in the USA and Canada it occurs on western larch (*L. occidentalis* Nutt.) and European larch; in New Zealand, Australia and Oceania, it occurs on Japanese larch (CABI, 1970).

Mycosphaerella laricina is the causal agent of larch needle blight. The fungus is one of the most important pathogen of larch needles in some areas, causing premature needle cast, especially in wet years or at damp locations. First symptoms appear on green needles in June and July as red brown spots or bands, in which very small black pustules sized 100–300 µm are forming; these contain cylindrical conidia with one to four septae, up to 30 µm in length. Together with the conidia, small, single-celled, round-shaped microconidia of the synanamorphic stage may occur. Usually, the lower part of the crown is affected; infected needles are prematurely cast in July and August. Round-shaped pseudothecia, 100–150 μm in size and nested in the needle tissue, are formed on dropped needles the following year from March to May. They ripen until the summer, and new asci, sized 50–60 μm, are produced from June, bearing eight hyaline ascospores per ascus. Ascospores are two celled and 15–17 × 5–6 μm in size. They are windblown and germinate in needles exposed to moisture; mycelium penetrates through the stomata into the needles and brown spots are soon apparent. Symptoms might be confused with frost damage or some other biotic factors and needle blights. Laboratory examination is necessary, as is the case with other needle diseases. *M. laricina* commonly colonizes needles sucked by aphids (Cech *et al.*, 2010). Several local outbreaks have been reported that were dependent on the weather (Cech, 2004; Krehan and Cech, 2004; Maresi *et al.*, 2004). *M. laricina* is reported from the USA and Canada (Patton and Spear, 1983; Bergdahl *et al.*, 1985; Ostry *et al.*, 1985; Palmer *et al.*, 1986).

Mycosphaerella laricis-leptolepidis (anamorphs: *Phoma yano-kubotae* Kitajima, *Phyllosticta laricis* Sawada) causes needle cast of Japanese larch and is native in some areas in China, Korea and Japan. Larches 10–20 years old are most endangered. The disease develops in July, mostly in the upper part of the crown. Symptoms are brown spots surrounded by a yellow halo on infected green needles. Spots successively blend and needles turn brown; spermogonia develop later in the upper part of dying needles. Spermogonia are thick walled and sized 83–165 × 75–143 μm. Spermatia are hyaline, rod shaped and 3–5 × 0.5–1 μm in size. Pseudothecia occur individually or in groups, are partially erumpent, globose and sized 88–156 × 84–142 μm. Asci are clavate and 49–99 × 7–12 μm in size. Ascospores are hyaline, unequally two celled, constricted at the septum and sized 11–18 × 3–5 μm. Three to four sprays with copper-based fungicides during June and July have proven effective in preventing development of the disease. There are also selected clones of Japanese larch that are more resistant to the disease in Korea and Japan (EPPO, 1978; CABI, 1998). Introduction of the disease to another region will bring significant economic losses, so avoidance based on quarantine precaution and excluding any import of sensitive plant material is necessary.

23.6.2 Management strategies and tactics

Most infections are concentrated from late May to July, the time of sprouting. Therefore, it is necessary to apply control measures at this time, when the infection cycles of all pathogens begin. The severity of infections varies greatly from year to year according to the weather conditions.

Control for *Meria laricis* is necessary in nurseries only; the application of copper-based fungicides brings good results, if they are applied from early June to August. Sulfur has also been successfully applied. Systemic fungicides based on benomyl, carbendazim, chlorothalonil, cyproconazole, propiconazole and their cocktail combinations, and triadimefon were successfully applied, with four sprayings during the growing season. The best results were obtained by using fungicides with a base of cyproconazole. *M. laricis* has been reported to have developed resistance to benomyl, so combined preparations, and alternating applications (e.g. bavistin and zineb) were particularly effective in preventing this disease (Ramson and Burth, 1970; Boudier, 1981).

Reduction of infection can be achieved by growing larches in combination with some broadleaved trees; a recommended mix is with beech (*Fagus* spp.), fallen leaves of which cover the infected needles and accelerate their decomposition, so that the conidia of pathogens cannot penetrate through the layer of the leaves. Also recommended is the burning of infected needles, thus preventing spreading conidia as a main source of the infection in the next spring (Butin, 1995).

Susceptibility differs by species and provenance of larches as a host species, e.g. European larch is more severely affected in the USA by *Mycosphaerella laricina* than are tamarack (*Larix laricina* (Du Roi) K. Koch) and Japanese larch. European larches from Austrian, French and Italian seed sources were more severely affected than those originating from the former Czechoslovakia (Czech Republic and Slovakia now) and Poland (Palmer *et al.*, 1986).

Mycosphaerella laricis-leptolepidis is considered to be a potentially invasive species in non-Asian regions; in the EPPO region it is listed as an A1 pest (EPPO, 1978), and as potentially dangerous to the native European larch. Precautions are taken to prevent the importation of infected plants or substrates, including needles, as per current phytosanitary regulations.

23.7 Foliar Diseases of Yews

There are no serious foliage diseases that have been reported from yews (*Taxus* spp.). *Dothiora taxicola* (Sacc.) M.E. Barr is regarded as the most dangerous pathogen affecting needles and shoots. *Cryptocline taxicola* (Allesch.) Petr. is known to occur on needles of the English yew (*Taxus baccata* L.) in Europe and North America. There are reports of foliage diseases caused by *Diplodia taxi* (Sowerby) De Not., *D. taxicola* and *Phyllosticta concentrica* Sacc. on Pacific yew (*Taxus brevifolia* Nutt.) from the USA (Callen, 1938; Brandenburger, 1985; Farr *et al.*, 1989; Pehl and Wulf, 2002).

The white snow mould *Phacidium taxicola* Dearn. & House, and the brown felt blights *Herpotrichia juniperi* and *Neopeckia coulteri*, result in localized damage to foliage buried under the snow (Farr *et al.*, 1989).

On dead or dying needles some secondary fungi occur, e.g. *Anthostomella taxi* Grove, *Guignardia philoprina* (Berk. & M.A. Curtis) Aa, *Leptosphaeria coniothyrium* (Fuckel) Sacc., *Penicillium taxi* R. Schneid. and *Pestalotiopsis funerea* (Callen, 1938; Thomas and Polwart, 2003).

Dothiora taxicola causes damage of individual needles and twigs on young yews and individual adult trees. It is possible to record the differences in severity of the infection between years. The previous year's infected needles are separately scattered on a twig, and turn yellow, then reddish, and then dry up and fall off. Perithecia, forming in the palisade parenchyma on the upper side of the needles, are at first evident as spherical nipples. Perithecia are gregarious, epiphyllous, subconical to oval, 120–180 × 30–40 µm in size and not protruding above the epidermis. The wall consists of at least six outer layers of dark brown or black, laterally compressed, thick-walled cells, 5–15 × 5–9 µm in size, and two to four inner rows of hyaline, thin-walled cells, 12–17 × 7.5–13 µm in size. Asci are arranged in clumps, club shaped, often curved, thin walled, thickened at the apex, eight spored and sized 45–65 × 10–15 µm. Paraphyses are absent. Spores are at first uniseptate, later three septate (in rare cases five septate after germination), often slightly constricted in the middle and 15–20 × 5–6.5 µm in size. Black pycnidia of the anamorph *Cytospora taxifolia* Cooke & Massee are solitary, scattered, completely sunk in the needle tissue and occupying almost the whole thickness of the needle. The ostiolum is open, usually on the lower surface, irregularly spherical to elliptical and sized 350–400 × 180–200 µm. Spores are hyaline, rod shaped and 3–4 × 0.7–1 µm in size with three to five incomplete septae (Callen, 1938).

Cryptocline taxicola is a weak parasite, occurring during the humid years, generally in the lower part of the crown. Acervuli form on the upper and also on the lower surfaces of necrotic and chlorotic needles, initially pale green, later reddish brown. Fruiting bodies are blackish, circular to subcircular, 150–350 µm in diameter and surrounded by brown circles. After cultivation in a wet chamber, a cream-coloured spore mass emerges from the circular to subcircular acervuli, and smooth-walled and hyaline conidia are formed on phialidic, cylindrical conidiophores. Conidia are 15.6 × 6.85 µm, truncate at the base and obtuse at the apex. Neither precautions nor control measures against this disease are taken, although during some years, the needle cast caused by this fungus has an epidemic character (Callen, 1938; Pehl and Wulf, 2002; Bukvayová, 2007). The parasite is also reported on Pacific yew from North America (Vujanovic and St-Arnaud, 2001).

23.8 Foliar Diseases of *Cupressaceae*

Members of the *Cupressaceae* family, e.g. juniper (*Juniperus* spp.), arborvitae (*Thuja* spp.) and cedar (*Chamaecyparis* spp.), are not so susceptible to needle diseases. For example, the occurrence of infections by some ubiquitous circum-boreal pathogens, such as *Herpotrichia juniperi*, *Neopeckia coulteri* and *Phacidium infestans*, is not very significant and happens only locally at places with prolonged snow cover. On arborvitae (*Thuja occidentalis* L.) and western redcedar (*Thuja plicata* Donn ex D. Don), blights caused by *Stigmina thujina* (Dearn.) B. Sutton and *Lepteutypa cupressi* (Nattrass, C. Booth & B. Sutton) H.J. Swart, are also not important. *Pestalotiopsis funerea* colonizes single scales and also twigs as a weak parasite, and is associated with redcedar dieback; it commonly occurs on freshly dead scales and parts of the tree (Boyce, 1961; Hepting, 1971). In contrast, *Pestalotia peregrina* Ellis & G. Martin is reported as a severe dieback agent of western redcedar in India (Minore, 1983).

Several species of *Lophodermium* are reported from the *Cupressaceae*. *L. juniperinum* (Fr.) De Not. is a widespread colonizer of suppressed needles of junipers, cedars and incense cedars (*Calocedrus* spp.) in Europe, the USA, and Canada. Hysterothecia mature from May to July. There are just a few indications that this fungus is parasitic (Bega, 1978; Cannon and Minter, 1984; Butin, 1995). Eight species of needle cast are reported from junipers in China, such as *Hypoderma junipericola* C.L. Hou, Y.R. Lin. & M. Piepenbr., *Lophodermium harbinense* Y.R. Lin., *L. jianchuanense* C.L. Hou & M. Piepenbr., *L. juniperinum* and *Soleella junipericola* C.L. Hou & M. Piepenbr. (Hou *et al.*, 2005). *Lophodermium juniperinum* has been found on redcedar in Australia (Stahl, 1966).

Didymascella thujina is the cause of Didymascella leaf blight or Keithia disease on arbovitae. The fungus is native to Canada and the USA and has been introduced into central and northern Europe. Early symptoms are found in the spring in the form of white spots forming on the upper side of individual scales of the last year's shoots. Infected needles eventually die. Apothecia on a single scale are roundish to oval, elongated, 0.5–2 mm in diameter and dark olive brown. Ascospores are at first hyaline, later sub-hyaline, brownish and with a septum on the apical tip. Didymascella leaf blight is an airborne disease that propagates and is spread in humid conditions, and can affect large areas; of greater concern is its occurrence in nurseries, where conditions are ideal for fungal growth and propagation. The disease is rarely lethal for a host older than 4–5 years, as susceptibility decreases with age. Didymascella leaf blight is the most important foliage disease of seedlings of arbovitae. The disease can be controlled by fungicides, as has been effectively done, particularly in nurseries. However, the timing of control measures must respect the long period of spore production (Minore, 1983; Butin, 1995; Minter, 1997).

Didymascella tetraspora (W. Phillips & Keith) Maire is similar to *D. thujina*, but with four-spored asci, and it is known from junipers in Europe (Butin, 1995).

Pseudocercospora juniperi (Ellis & Everh.) B. Sutton and Hodges is the causal agent of a Cercospora needle blight. The disease is common on species of junipers, arborvitae, cedar, cypress, bald cypress (*Taxodium* spp.), Japanese cedar (*Cryptomeria* spp.), redwood (*Sequoia* spp.) and other genera. This needle disease occurs throughout the USA and Canada, but has not been reported outside these countries. The fungus overwinters on the infected needles of living trees. The first symptoms appear in late summer on the inner branch needles and towards the lower part of the tree as a browning of the needles, which are eventually cast. The infection slowly spreads upwards and outwards from the lower part of the crown until, in severe cases, only the needles at the tips of the upper branches remain green. Stromata appear as tiny, greenish, dark pustules 100 μm in diameter on the upper surface of the needles or on small twigs. The yellow brown conidia are cylindrical to obclavate, echinulate and 40×6 μm in size. Conidia are present throughout the spring and summer, and are spread by the wind. Infection usually occurs during periods of wet weather (Hodges, 1962). Cercospora needle blight can be controlled by spraying with copper fungicides in two applications in mid June, before fresh symptoms appear, and in early July up to development of the new shoots. Fungicides based on organic compounds, such as azoxystrobin, myclobutanil, propiconazole and thiophanate methyl have shown good results.

References

Agrios, G.N. (1997) *Plant Pathology*, 4th edn. Academic Press, New York.

Agúndez, D., Alía, R., Stephan, R., Gil, L. and Pardos, J.A. (1994) Ensayo de procedencias españolas y alemanas de *Pinus sylvestris* L.: comportamiento en vivero y supervivencia en montes [Study of Spanish and German provenances of *Pinus sylvestris* L.: performance in the nursery and survival in stands]. *Ecología* 8, 245–257.

Albers, M., Albers, J., Cummings Carlson, J., Haugen, L. and Wenner, N. (1996) *How to Identify and Manage Needlecast Diseases on Balsam Fir.* [Report] NA-FR-02-96, US Department of Agriculture, Forest Service, Northeastern Area, Newtown Square, Pennsylvania.

Barr, M.E. (1968) The *Venturiaceae* in North America. *Canadian Journal of Botany* 46, 799–864.

Bates, R.M. and Despot, D.A. (2005) Effects of azoxystrobin application rate and treatment interval on the control of *Rhabdocline pseudotsugae* on Douglas-fir Christmas trees. *Plant Health Progress* 2005, 1–9.

Bega, R.V. (1978) *Diseases of Pacific Coast Conifers*. Agriculture Handbook 521, US Department of Agriculture, Forest Service, Washington, DC.

Bergdahl, D.R., Halik, S.S. and Smeltzer, D.L.K. (1985) *Mycosphaerella laricina* on European larch in Vermont. *Plant Disease* 69, 83.

Boudier, B. (1981) Essai de lutte contre le dessèchement des aiguilles dû à *Meria laricis* sur mélèzes d'Europe (*Larix decidua*) en pépinières [Study on the control of needle drying caused by *Meria laricis* in European larch (*Larix decidua*) in nurseries]. *Revue Forestière Française* 5, 394–399.

Boudier, B. (1987) Lutte contre 'le rouge cryptogamique du pin' [Fight against the red pine fungus]. *Revue Forestière Française* 39, 107–112.

Boyce, J.S. (1940) A needle-cast of Douglas-fir associated with *Adelopus gaeumannii*. *Phytopathology* 30, 649–659.

Boyce, J.S. (1961) *Forest Pathology*, 3rd edn. McGraw-Hill, New York.

Boyce, J. S. (1969) *Needle Cast of Southern Pines*. Forest Pest Leaflet 28, US Department of Agriculture, Forest Service, Washington, DC.

Boyer, W.D. (1972) *Brown-Spot Resistance in Natural Stands of Longleaf Pine Seedlings*. Research Note SO-142, US Department of Agriculture, Forest Service, Southern Forest Experiment Station, New Orleans, Louisiana.

Brandenburger, W. (1985) *Parasitische Pilze an Gefäßpflanzen in Europa* [*Parasitic Fungi of Vascular Plants in Europe*]. Gustav Fischer Verlag, Stuttgart, Germany, pp. 32–33.

Bukvayová, N. (2007) *Cryptocline taxicola* (All.) Petr. – a new plant pathogen reported in Slovak Republic. *Plant Protection Science* 43, 122–124.

Bulman, L.S. (1988) Incidence and severity of Cyclaneusma needle-cast in fifteen *Pinus radiata* plantations in New Zealand. *New Zealand Journal of Forestry Science* 18, 92–100.

Bulman, L.S. (2001) Silvicultural control. In: Bulman, L.S. and Gadgil, P.D. (eds) *Cyclaneusma Needle-cast in New Zealand*. Forest Research Bulletin 222, New Zealand Forest Service, Forest Research Institute, pp. 55–62.

Burleigh, J.R., Wood, W. and Sebesta, P. (1982) Growth retardation of plantation-grown sugar pine by *Lophodermella arcuata*. *Plant Disease* 66, 680–682.

Butin, H. (1995) *Tree Diseases and Disorders: Causes, Biology and Control in Forest and Amenity Trees*. Oxford University Press, Oxford, UK.

Butin, H. and Kehr, R. (2000) *Rhizosphaera pseudotsugae* sp. nov. and related species. *Mycological Research* 104, 1012–1016.

Butin, H. and Pehl, L. (1993) *Kabatina abietis* sp. nov., associated with browning of fir needles. *Mycological Research* 97, 1340–1342.

Butin, H., Kehr, R. and Pehl, L. (2000) *Dothistroma rhabdoclinis* sp. nov. associated with *Rhabdocline pseudotsugae* on Douglas-fir. *Forest Pathology* 30, 195–203.

CABI (1970) Meria laricis *Vuill. Distribution Maps of Plant Diseases* 379. CAB International, Wallingford, UK.

CABI (1998) Mycosphaerella laricis-leptolepidis *Kaz. Ito, K. Sato & M. Ota. Distribution Maps of Plant Diseases* 767. CAB International, Wallingford, UK.

Callen, E.O. (1938) Some fungi on the yew. *Transactions of the British Mycological Society* 22, 94–106.

Cannell, M.G.R. and Morgan, J. (1990) Theoretical study of variables affecting the export of assimilates from branches of *Picea*. *Tree Physiology* 6, 257–266.

Cannon, P.F. and Minter, D.W. (1984) *Descriptions of Pathogenic Fungi and Bacteria* 80, Sheets 793–799. CAB International, Wallingford, UK.

Catal, M., Adams, G.C. and Fulbright, D.W. (2010) Evaluation of resistance to Rhabdocline needlecast in Douglas-fir variety Shuswap, with quantitative polymerase chain reaction. *Phytopathology* 100, 337–344.

Cech, T.L. (1992) Nadelschäden an *Abies grandis* [Needle damage in *Abies grandis*]. *Forstschutz Aktuell* 9/10, 7–8.

Cech, T.L. (2004) Nadelschütten der Lärche [Needle damage in larch]. *Forstschutz Aktuell* 32, 11–13.

Cech, T.L. and Tomiczek, C. (1988) *Tiarosporella parca* (Berk. and Br.) Whitneyerster Nachweis in Österreich [*Tiarosporella parca* (Berk. and Br.) Whitneyerster Nachweis in Austria]. *European Journal of Forest Pathology* 18, 382–384.

Cech, T.L., Krehan, H., Perny, B. and Steyrer, G. (2010) Kronenschaden bei Larchen in Osterreich weit verbreitet [Widespread crown damage of larch in Austria]. *Forstschutz Aktuell* 50, 2–6.

Chabot, B.F. and Hicks, D.J. (1982) The ecology of leaf life spans. *Annual Review of Ecology and Systematics* 13, 229–259.

Chen, Z.-C. (1974) Notes on new Formosan forest fungi. I. Black mildews on *Abies kawakamii*. *Taiwania* 19, 115–122.

Cheng, L.H., Ying, R.L. and Piepenbring, M. (2005) Species of *Rhytismataceae* on needles of *Juniperus* spp. from China. *Canadian Journal of Botany* 83, 37–46.

Choi, D. and Simpson, J.A. (1991) Needle cast of *Pinus radiata* in New South Wales. *Australian Journal of Botany* 39, 137–152.

Crous, P.W., Wingfield, M.J. and Swart, W.J. (1990) Shoot and needle diseases of *Pinus* spp. in South Africa. *South African Forestry Journal* 154, 60–66.

Darker, G.D. (1967) A revision of the genera of the *Hypodermataceae*. *Canadian Journal of Botany* 45, 1399–1444.

Desprez-Loustau, M.-L., Robin, C., Reynaud, G., Déqué, M., Badeau, V., Piou, D., Husson, C. and Marçais, B. (2007) Simulating the effects of a climate-change scenario on the geographical range and activity of forest-pathogenic fungi. *Canadian Journal of Plant Pathology* 29, 101–120.

De Wolf, E.D. and Isard, S.A. (2007) Disease cycle approach to plant disease prediction. *Annual Review of Phytopathology* 45, 203–220.

Diamandis, S. (1978) 'Top-dying' of Norway spruce, *Picea abies* (L.) Karst., with special reference to *Rhizosphaera kalkhoffii* Bubák. II. Status of *R. kalkhoffii* in 'top-dying' of Norway spruce. *European Journal of Forest Pathology* 8, 345–356.

Diamandis, S. and Minter, D.W. (1980) *Descriptions of Fungi and Bacteria* 66, Sheets 656–657. CAB International, Wallingford, UK.

DiCosmo, F., Peredo, H. and Minter, D.W. (1983) *Cyclaneusma* gen. nov., *Naemacyclus* and *Lasiostictis*, a nomenclatural problem resolved. *European Journal of Forest Pathology* 13, 206–212.

Diwani, S.A. and Millar, C.S. (1990) Sources of inoculum of *Lophodermium seditiosum* on *Pinus sylvestris*. *European Journal of Forest Pathology* 20, 1–7.

Drenkhan, R. and Hanso, M. (2009) Recent invasion of foliage fungi of pines (*Pinus* spp.) to the Northern Baltics. *Forestry Studies/Metsanduslikud Uurimused* 51, 49–64.

Ellis, M.B. and Ellis, J.P. (1997) *Microfungi on Land Plants*. Richmond Publishing Co. Ltd., Slough, UK.

EPPO (1978) Data sheets on quarantine organisms No. 16, *Mycosphaerella larici-leptolepis*. *EPPO Bulletin* 8(2).

EPPO (1980) Data sheets on quarantine organisms No. 7, *Cercospora pini-densiflorae*. *EPPO Bulletin* 10(1).

EPPO (2008) *Mycosphaerella dearnessii* and *Mycosphaerella pini*. *EPPO Bulletin* 38, 349–362.

Evans, H.C. (1984) *The Genus* Mycosphaerella *and its Anamorphs* Cercoseptoria, Dothistroma *and* Lecanosticta *on Pines*. Mycological Papers 153, Commonwealth Mycological Institute, Kew, UK [now CAB International, Wallingford, UK].

Fan, L., Lin, Y.-R., Wang, S. and Hou, C.-L. (2011) Two species of *Lirula* on *Abies* from Yunnan, Southwest China. *Mycological Progress* 11, 279–286.

Farr, D.F., Bills, G.F., Chamuris, G.P. and Rossman, A.Y. (1989) *Fungi on Plants and Plant Products in the United States*. APS Press, St Paul, Minnesota.

Funk, A. (1985) *Foliar Fungi of Western Trees*. Information Report BC-X-265, Canadian Forest Service, Pacific Forest Research Centre, Victoria, British Columbia, Canada.

Gadgil, P.D. (1984) Cyclaneusma (Naemacyclus) needle-cast of *Pinus radiata* in New Zealand. 1: Biology of *Cyclaneusma minus*. *New Zealand Journal of Forestry Science* 14, 179–196.

Gadgil, P.D. (2005) *Fungi of New Zealand 4: Fungi on Trees and Shrubs in New Zealand*. Fungal Diversity Press, Hong Kong.

Garbutt, R. (1996) *Foliage Diseases in Western Larch in British Columbia*. Forest Pest Leaflet 71, Canadian Forest Service, Pacific Forestry Centre, Victoria, British Columbia, Canada.

Gernandt, D.S., Camacho, F.J. and Stone, J.K. (1997) *Meria laricis*, an anamorph of *Rhabdocline*. *Mycologia* 89, 735–744.

Gilmour, J.W. (1966) *The Pathology of Forest Trees in New Zealand. The Fungal, Bacterial and Algal Pathogens*. Technical Paper 48, New Zealand Forest Service, Forest Research Institute, Rotorua, New Zealand.

Giordano, L. and Gonthier, P. (2011) An outbreak of *Cyclaneusma minus* needle cast on Swiss mountain pine (*Pinus uncinata*) in Italy. *Journal of Plant Pathology* 93, 74.

Gourbière, F. and Morelet, M. (1979) Le genre *Rhizosphaera* Mangin et Hariot. 1. *R. oudemansii* et *R. macrospora* nov. sp. [The genus *Rhizosphaera* Mangin & Hariot. 1. *R. oudemansii* and *R. macrospora* nov. sp.]. *Revue de Mycologie* 43, 81–95.

Hansen, E.M., Stone, J.K., Capitano, B.R., Rosso, P., Sutton, W., Winton, L., Kanaskie, A. and McWilliams, M.G. (2000) Incidence and impact of Swiss needle cast in forest plantations of Douglas-fir in coastal Oregon. *Plant Disease* 84, 773–778.

Hepting, G.H. (1971) *Diseases of Forest and Shade Trees of the United States*. Agricultural Handbook 386, US Department of Agriculture, Forest Service, Washington, DC.

Hodges, C.S. (1962) Comparison of four similar fungi from *Juniperus* and related conifers. *Mycologia* 54, 62–69.

Hodges, C.S. (2002) First Report of *Stigmina lautii* in the United States. *Plant Disease* 86, 699.

Hoff, R.J. (1986) *Susceptibility of Inland Douglas-fir to Rhabdocline Needle Cast*. Research Note INT-375, US Department of Agriculture, Forest Service, Intermountain Research Station, Ogden, Utah.

Hood, I.A. (1997) Swiss needle cast. In: Hansen, E.M. and Lewis, K.J. (eds) *Compendium of Conifer Diseases*. APS Press, St Paul, Minnesota, pp. 55–56.

Hood, I.A. and Kershaw, D.J. (1975) Distribution and infection period of *Phaeocryptopus gaeumannii* in New Zealand. *New Zealand Journal of Forest Science* 5, 201–208.

Hou, C.-L. and Piepenbring, M. (2006) Five new species of *Hypoderma* (*Rhytismatales*, *Ascomycota*) with a key to *Hypoderma* species known from China. *Nova Hedwigia* 82, 91–104.

Hou, C.-L. and Piepenbring, M. (2009) *Lophodermium puerense* sp. nov. on needles of *Pinus yunnanensis* var. *tenuifolia* from southwest Yunnan. *Mycotaxon* 107, 259–262.

Hou, C.-L., Lin, Y.R. and Piepenbring, M. (2005) Species of *Rhytismataceae* on needles of *Juniperus* spp. from China. *Canadian Journal of Botany* 83, 37–46.

Hou, C.-L., Gao, J. and Piepenbring, M. (2006) Four rhytismataceous ascomycetes on needles of pines from China. *Nova Hedwigia* 83, 512–522.

Huang, Z.-Y., Smalley, E.B. and Gurries, R.P. (1995) Differentiation of *Mycosphaerella dearnessii* by cultural characters and RAPD analysis. *Phytopathology* 85, 522–527.

Hunt, R.S. (1978) *Elytroderma Disease of Pines*. Forest Pest Leaflet 27, Forestry Canada, Forest Insect and Disease Survey, Victoria, British Columbia, Canada.

Hunt, R.S. (1995) *Common Pine Needle Casts and Blights in the Pacific Region*. Forest Pest Leaflet 43, Canadian Forest Service, Pacific Forestry Centre, Victoria, British Columbia, Canada.

Ioos, R., Fabre, B., Saurat, C., Fourrier, C., Frey, P. and Marçais, B. (2010) Development, comparison, and validation of real-time and conventional PCR tools for the detection of the fungal pathogens causing brown spot and red band needle blights of pine. *Phytopathology* 100, 105–114.

Ivory, M.H. (1975) The pathology of *Pinus* spp. in West Malaysia. *Commonwealth Forestry Review* 54, 154–165.

Jach, M.E. and Ceulemans, R. (2000) Effects of season, needle age and elevated atmospheric CO_2 on photosynthesis in Scots pine (*Pinus sylvestris*). *Tree Physiology* 20, 145–157.

Johnston, P.R., Park, D., Dick, M.A., Ortiz-García, S. and Gernandt, D.S. (2003) Identifying pine-inhabiting *Lophodermium* species using PCR-RFLP. *New Zealand Journal of Forestry Science* 33, 10–24.

Jurc, M., Jurc, D., Gogala, N. and Simončič, P. (1996) Air pollution and fungal endophytes in needles of Austrian pine. *Phyton* (Horn – Austria) 36, 111–114.

Kais, A.G. (1975) Environmental factors affecting brown-spot infection on longleaf pine. *Phytopathology* 65, 1389–1392.

Kaneko, S. (2003) Two new species of *Lirula* on *Abies* from Japan. *Mycoscience* 44, 335–338.

Karadžić, D.M. (1981) Infection of *Pinus sylvestris* by *Naemacyclus minor*. In: Millar, C.S. (ed.) *Current Research on Conifer Needle Diseases. Proceedings of the IUFRO Working Party on Needle Diseases, Sarajevo, 1980*. Aberdeen University, Aberdeen, UK, pp. 99–101.

Karadžić, D.M. (1998) *Tiarosporella durmitorensis* sp. nov. – a new pathogenic fungus on needles of *Abies alba*. *European Journal of Forest Pathology* 28, 145–152.

Kliejunas, J.T., Allison, J.R., McCain, A.H. and Smith, R.S. (1985) Phoma blight of fir and Douglas-fir seedlings in a California nursery. *Plant Disease* 69, 773–775.

Kowalski, T. and Lang, K.J. (1984) Die Pilzflora von Nadeln, Trieben und Ästen unterschiedlich alter Fichten (*Picea abies* (L.) Karst.) mit besonderer Berücksichtigung vom Fichtensterben betroffener Altbäume [The fungal flora of needles, shoots and branches of different aged spruce (*Picea abies* (L.) Karst.) with special emphasis on the old dying spruce trees affected]. *Forstwissenschaftliches Centralblatt* 103, 349–360.

Kowalski, T. and Zych, P. (2002) Endophytic fungi in needles of *Pinus nigra* growing under different site conditions. *Polish Botanical Journal* 47, 251–257.

Krehan, H. and Cech, T.L. (2004) Larchenschaden in der Obersteiermark. Ein Fallbeispiel für komplexe Einwirkungen von Schadursachen [Larch damage in Upper Styria. A case study of the complex effects of harmful causes]. *Forstschutz Aktuell* 32, 4–8.

Lightle, P.C. (1954) The pathology of *Elytroderma deformans* on ponderosa pine. *Phytopathology* 44, 557–569.

Lilja, A., Poteri, M., Petäistö, R.-L., Rikala, R., Kurkela, T. and Kasanen, R. (2010) Fungal diseases in forest nurseries in Finland. *Silva Fennica* 44, 525–545.

Livsey, S. and Barklund, P. (1992) *Lophodermium piceae* and *Rhizosphaera kalkhoffii* in fallen needles of Norway spruce (*Picea abies*). *European Journal of Forest Pathology* 22, 204–216.

Luo, J.T., Lin, Y.R., Shi, G.K. and Hou, C.L. (2010) *Lophodermium* on needles of conifers from Yunnan Province, China. *Mycological Progress* 9, 235–244.

Manter, D.K. (2002) Energy dissipation and photoinhibition in Douglas-fir needles with a fungal-mediated reduction in photosynthetic rates. *Journal of Phytopathology* 150, 674–679.

Manter, D.K., Bond, B.J., Kavanagh, K.L., Rosso, P.H. and Filip, G.M. (2000) Pseudothecia of Swiss needle cast fungus, *Phaeocryptopus gaeumannii*, physically block stomata of Douglas-fir, reducing CO_2 assimilation. *New Phytologist* 148, 481–491.

Manter, D.K., Kelsey, R.G. and Stone, J.K. (2001) Quantification of *Phaeocryptopus gaeumannii* colonization in Douglas-fir needles by ergosterol analysis. *Forest Pathology* 31, 229–240.

Manter, D.K., Bond, B.J., Kavanagh, K.L., Stone, J.K. and Filip, G.M. (2003a) Modeling the impacts of the foliar pathogen, *Phaeocryptopus gaeumannii*, on Douglas-fir physiology: net canopy carbon assimilation, needle abscission and growth. *Ecological Modelling* 164, 211–226.

Manter, D.K., Winton, L.M., Filip, G.M. and Stone, J.K. (2003b) Assessment of Swiss needle cast disease: temporal and spatial investigations of fungal colonization and symptom severity. *Journal of Phytopathology* 151, 344–351.

Manter, D.K., Reeser, P.W. and Stone, J.K. (2005) A climate-based model for predicting geographic variation in Swiss needle cast severity in the Oregon Coast Range. *Phytopathology* 95, 1256–1265.

Maresi, G., Capretti, P., Ambrosi, P. and Minerbi, S. (2004) Larch crown diseases in Trentino and South Tyrol. *Journal of Forest Science* 50, 313–318.

Martinez, A.T. and Ramirez, C. (1984) *Rhizosphaera oudemansii* (*Sphaeropsidales*) associated with a needle cast of Spanish *Abies pinsapo*. *Mycopathologia* 83, 175–182.

Michaels, E. and Chastagner, G.A. (1984) Seasonal availability of *Phaeocryptopus gaeumannii* ascospores and conditions that influence their release. *Plant Disease* 68, 942–944.

Millar, C.S. and Minter, D.W. (1978) Lophodermella sulcigena. *Descriptions of Fungi and Bacteria* 57, Sheet 562. CAB International, Wallingford, UK.

Millar, C.S. and Minter, D.W. (1980). Naemacyclus minor. *Descriptions of Fungi and Bacteria* 66, Sheet 659. CAB International, Wallingford, UK.

Minore, D. (1983) *Western Redcedar: a Literature Review*. General Technical Report PNW-150, US Department of Agriculture, Forest Service, Pacific Northwest, Forest and Range Experiment Station, Portland, Oregon.

Minter, D.W. (1981) *Lophodermium* on pines. *Mycological Papers* 147, Commonwealth Mycological Institute, Kew, UK [now CAB International, Wallingford, UK], pp. 1–54.

Minter, D.W. (1988) *Descriptions of Fungi and Bacteria* 95, Sheets 941–950. CAB International, Wallingford, UK.
Minter, D.W. (1993) Bifusella pini. *Descriptions of Fungi and Bacteria* 115, Sheet 1141. CAB International, Wallingford, UK.
Minter, D.W. (1997) Rhytisma salicinum. *Descriptions of Fungi and Bacteria* 134, Sheet 1339. CAB International, Wallingford, UK.
Minter, D.W. and Cannon, P.F. (2001) Virgella robusta. *Descriptions of Fungi and Bacteria* 148, Sheet 1480. CAB International, Wallingford, UK.
Minter, D.W. and Gibson, I.A.S. (1978a) Meloderma desmazieresii. *Descriptions of Fungi and Bacteria* 57, Sheet 569. CAB International, Wallingford, UK.
Minter, D.W. and Gibson, I.A.S. (1978b) Ploioderma lethale. *Descriptions of Fungi and Bacteria* 57, Sheet 570. CAB International, Wallingford, UK.
Minter, D.W. and Millar, C.S. (1978) *Descriptions of Fungi and Bacteria* 57, Sheets 563–568. CAB International, Wallingford, UK.
Minter, D.W. and Millar, C.S (1980a) Ecology and biology of three *Lophodermium* species on secondary needles of *Pinus sylvestris. European Journal of Forest Pathology* 10, 169–181.
Minter, D.W. and Millar, C.S. (1980b) A study of three pine inhabiting *Lophodermium* species in culture. *Nova Hedwigia* 32, 361–368.
Minter, D.W. and Millar, C.S. (1980c) Naemacyclus niveus. *Descriptions of Fungi and Bacteria* 66, Sheet 660. CAB International, Wallingford, UK.
Minter, D.W. and Millar, C.S. (1980d) Phacidium infestans. *Descriptions of Pathogenic Fungi and Bacteria* 66, Sheet 652. CAB International, Wallingford, UK.
Minter, D.W. and Millar, C.S. (1984) *Descriptions of Fungi and Bacteria* 79, Sheets 782–786. CAB International, Wallingford, UK.
Minter, D.W. and Millar, C.S. (1993) *Descriptions of Fungi and Bacteria* 115, Sheets 1144–1147. CAB International, Wallingford, UK.
Minter, D.W., Staley, J.M. and Millar, C.S. (1978) Four species of *Lophodermium* on *Pinus sylvestris. Transactions of the British Mycological Society* 71, 295–301.
Mitchell, C.P., Millar, C.S. and Haworth, M.N. (1976) Effect of the needle-cast fungus *Lophodermella sulcigena* on growth of Corsican pine. *Forestry* 49, 153–158.
Mitchell, C.P., Millar, C.S. and Williamson, B. (1978) The biology of *Lophodermella conjuncta* Darker on Corsican pine needles. *European Journal of Forest Pathology* 8, 108–118.
Müller, M.M. and Hallaksela, A.-M. (1998) Diversity of Norway spruce needle endophytes in various mixed and pure Norway spruce stands. *Mycological Research* 102, 1183–1189.
Müller, M.M. and Hallaksela, A.-M. (2000) Fungal diversity in Norway spruce: a case study. *Mycological Research* 104, 1139–1145.
Müller, M.M. and Hantula, J. (1998) Diversity of *Tiarosporella parca* in Finland, Norway and Switzerland. *Mycological Research* 102, 1165–1168.
Müller, M.M., Valjakka, R., Suokko, A. and Hantula, J. (2001) Diversity of endophytic fungi of single Norway spruce needles and their role as pioneer decomposers. *Molecular Ecology* 10, 1801–1810.
Osono, T. and Hirose, D. (2009) Ecology of endophytic fungi associated with leaf litter decomposition. In: Rai, M. (ed.) *Applied Mycology*. CAB International, Wallingford, UK, pp. 92–109.
Osorio, O.M. (2007) Detection of the needle cast fungus *Phaeocryptopus gaeumannii* in plantations of *Pseudotsuga menziesii* in Valdivia, Chile. *Bosque* 28, 69–74.
Ostry, M.E., Nicholls, T.H. and Palmer, M.A. (1985) Needlecast of western larch seedlings caused by *Mycosphaerella laricina. Plant Disease* 69, 542.
Palmer, M.A., Ostry, M.E., Robbins, K. and Nicholls, T.H. (1986) Occurrence and development of *Mycosphaerella laricina* on larch in the north central United States. *Plant Disease* 70, 921–923.
Parker, A.K. and Reid, J. (1969) The genus *Rhabdocline* Syd. *Canadian Journal of Botany* 47, 1533–1545.
Patton, R.F. and Spear, R.N. (1983) Needle cast of European larch caused by *Mycosphaerella laricina* in Wisconsin and Iowa. *Plant Disease* 67, 1149–1153.
Pehl, L. and Wulf, A. (2002) Nadelschaden an *Taxus baccata* L. durch *Cryptocline taxicola* (All.) Petr. [Needle damage to *Taxus baccata* L. by *Cryptocline taxicola* (All.) Petr.]. *Nachrichtenblatt des Deutschen Pflanzenschutzdienstes* 54, 266–268.
Perny, B. (2004) Schädlinge an Lärchennadeln [Pests of larch needles]. *Forstschutz Aktuell* 32, 14–18.
Perny, B., Cech, T., Donaubauer, E. and Tomiczek, C. (2002) *Krankheiten und Schädlinge in Christbaumkulturen*, 2nd edn [*Diseases and Pests in Christmas Tree Plantations*, 2nd edn]. Bundesforschungs- und Ausbildungszentrum für Wald, Naturgefahren und Landschaft, Vienna, Austria.

Petkov, P. (1993) Comparative investigations on the resistance of Scots pine (*Pinus sylvestris*) provenances to the *Lophodermium pinastri* complex. *Nauka za Gorata* 30, 63–68.
Petrak, F. (1925) Mykologische Notizen 429. Über die Nebenfruchtformen von *Adelopus nudus* (Peck) Theiss. [Mycological notes 429. On the fruiting structures of *Adelopus nudus* (Peck) Theiss.]. *Annales Mycologici* 23, 50–51.
Petrak, F. (1938) Beiträge zur Systematik und Phylogenie der Gattung *Phaeocryptopus* Naumov [Contributions to the systematics and phylogeny of the genus *Phaeocryptopus* Naumov]. *Annales Mycologici* 36, 9–26.
Podger, F.D. and Wardlaw, T.J. (1990) Spring needle-cast of *Pinus radiata* in Tasmania: II. Effects of fertilisers and thinning on disease severity, and the impact of disease on growth. *New Zealand Journal of Forestry Science* 20, 206–219.
Rack, K. (1981) Interactions between *Naemacyclus minor* and *Lophodermium pinastri* during 'pine needle cast'. In: Millar, C.S. (ed.) *Current Research on Conifer Needle Diseases. Proceedings of the IUFRO Working Party on Needle Diseases, Sarajevo, 1980*. Aberdeen University, Aberdeen, UK, pp. 103–111.
Ramson, A. and Burth, U. (1970) New fungicides for crops, and extension of the range of application of preparations already approved. Test results 1969. *Nachrichtenblatt des Deutschen Pflanzenschutzdienstes* 24, 209–216.
Ranta, H. and Saloniemi, I. (2005) Distribution of fungal foliage and shoot pathogens in a natural Scots pine population in relation to environmental variables. *Canadian Journal of Forest Research* 35, 503–510.
Reddy, M.A.R. and Pandey, P.C. (1973) Cercospora needle blight of radiata pine in India. *Indian Forester* 99, 308–309.
Reich, P.B., Koike, T., Gower, S.T. and Schoettle, A.W. (1995) Cause and consequences of variation in conifer leaf life span. In: Smith, W.K. and Hinckley, T.M. (eds) *Ecophysiology of Coniferous Forests*. Academic Press, New York, pp. 225–254.
Ridley, G.S. and Dick, M.A. (2001) *An Introduction to the Diseases of Forest and Amenity Trees in New Zealand*. Forest Research Bulletin 220, New Zealand Forest Service, Forest Research Institute, Rotorua, New Zealand.
Roll-Hansen, F. (1987) *Phacidium infestans* and *Phacidium abietis*. Hosts, especially *Abies* species in Norwegian nurseries. *European Journal of Forest Pathology* 17, 311–315.
Roll-Hansen, F. (1989) *Phacidium infestans*. A literature review. *European Journal of Forest Pathology* 19, 237–250.
Roth, L.F (1959) Perennial infection of ponderosa pine by *Elytroderma deformans*. *Forest Science* 5, 182–191.
Roux, J., Meke, G., Kanyi, B., Mwangi, L., Mbaga, A., Hunter, G.C., Nakabonge, G., Heath, R.N. and Wingfield, M.J. (2005) Diseases of plantation forestry trees species in eastern and southern Africa. *South African Journal of Science* 101, 409–413.
Scharpf, R.F. (1988) Epidemiology of *Lirula abietis-concoloris* on white fir in California. *Plant Disease* 72, 855–858.
Schulze, E.-D., Fuchs, M. and Fuchs, M.I. (1977) Spatial distribution of photosynthetic capacity and performance in a mountain spruce forest of northern Germany. III. The significance of the evergreen habit. *Oecologia* 30, 239–248.
Shaw, C.G. and Leaphart, C.D. (1960) Two serious foliage diseases of western white pine in the Inland Empire. *Plant Disease Reporter* 44, 655–659.
Shaw, D.C., Filip, G.M., Kanaskie, A., Maguire, D.A. and Littke, W.A. (2011) Managing an epidemic of Swiss needle cast in the Douglas-fir region of Oregon: the role of the Swiss needle cast cooperative. *Journal of Forestry* 109, 109–119.
Shoemaker, R.A. (1967) *Cucurbitaria piceae* and associated *Sphaeropsidales* parasitic on spruce buds. *Canadian Journal of Botany* 45, 1243–1248.
Sieber, T. (1988) Endophytische Pilze in Nadeln von gesunden und geschädige Fichten (*Picea abies* (L.) Karsten) [Endophytic fungi in the needles of healthy and damaged spruce (*Picea abies* (L.) Karsten)]. *European Journal of Forest Pathology* 18, 321–342.
Sieber, T.N., Rys, J. and Holdenrieder, O. (1999) Mycobiota in symptomless needles of *Pinus mugo* ssp. *uncinata*. *Mycological Research* 103, 306–310.
Sieber-Canavesi, F. and Sieber, T.N. (1993) Successional patterns of fungal communities in needles of European silver fir (*Abies alba* Mill.). *New Phytologist* 125, 149–161.
Siggers, P.V. (1944) *The Brown Spot Needle Blight of Pine Seedlings*. Technical Bulletin 870, US Department of Agriculture, Forest Service, Washington, DC.
Simpson, J.A. and Grgurinovic, C.A. (2004) First record of *Lophodermium conigenum* on *Pinus* in Australia. *Australasian Plant Pathology* 33, 447–448.

Sinclair, W.A. and Lyon, H.H. (2005) *Diseases of Trees and Shrubs*, 2nd edn. Comstock Publishing Associates, Cornell University Press, Ithaca, New York.

Solheim, H. and Skage, J.O. (2002) Losses caused by the needle blight fungus *Delphinella abietis* in a greenery trial of Nobel fir in western Norway. *Forest Pathology* 32, 373–377.

Stahl, W. (1966) Needle-cast fungi on conifers in the Australian Capital Territory. *Australian Forestry* 30, 20–32.

Stenström, E. and Arvidsson, B. (2001) Fungicidal control of *Lophodermium seditiosum* on *Pinus sylvestris* seedlings in Swedish forest nurseries. *Scandinavian Journal of Forest Research* 16, 147–154.

Stenström, E. and Ihrmark, K. (1997) Detection of *Lophodermium seditiosum* from pine needles using PCR techniques. In: Duhne, H.-W., Adam, G., Diekmann, M., Frahm, J., Mauler-Machnik, A. and van Halteren, P. (eds) *Diagnosis and Identification of Plant Pathogens. Proceedings of the Fourth International Symposium of the European Foundation for Plant Pathology, September 9–12, 1996, Bonn, Germany*. Kluwer, Dordrecht, The Netherlands, pp. 223–226.

Stone, J.K. and Gernandt, D.S. (2005) A reassessment of *Hemiphacidium, Rhabdocline*, and *Sarcotrichila* (*Hemiphacidiaceae*). *Mycotaxon* 91, 115–126.

Stone, J.K., Reeser, P.W. and Kanaskie, A. (2007) Fungicidal suppression of Swiss needle cast and pathogen reinvasion in a 20-year-old Douglas-fir stand in Oregon. *Western Journal of Applied Forestry* 22, 248–252.

Stone, J.K., Capitano, B.R. and Kerrigan, J.L. (2008a) The histopathology of *Phaeocryptopus gaeumannii* on Douglas-fir needles. *Mycologia* 100, 431–444.

Stone, J.K., Coop, L.B. and Manter, D.K. (2008b) Predicting effects of climate change on Swiss needle cast disease severity in Pacific Northwest forests. *Canadian Journal of Plant Pathology* 30, 169–176.

Suto, Y. and Ougi, D. (1998) *Lecanosticta acicola*, causal fungus of brown spot needle blight in *Pinus thunbergii*, new to Japan. *Mycoscience* 39, 319–325.

Tainter, F.H. and Baker, F.A. (1996) *Principles of Forest Pathology*. John Wiley, Hoboken, New Jersey.

Talgø, V., Chastagner, G.A., Thomsen, I.B., Cech, T.L., Riley, K., Lange, K., Klemsdal, S.S. and Stensvand, A. (2010) *Sydowia polyspora* associated with current season needle necrosis (CSNN) on true fir (*Abies* spp.). *Fungal Biology* 114, 545–554.

Temel, F., Stone, J.K. and Johnson, G.R. (2003) First report of Swiss needle cast caused by *Phaeocryptopus gaeumannii* on Douglas-fir in Turkey. *Plant Disease* 87, 12.

Thomas, P.A. and Polwart, A. (2003) *Taxus baccata* L. *Journal of Ecology* 91, 489–524.

van der Pas, J.B., Bulman, L. and Slater-Hayes, J.D. (1984) Cyclaneusma (Naemacyclus) needle-cast of *Pinus radiata* in New Zealand. 3. Incidence and severity of the needle-cast. *New Zealand Journal of Forestry Science* 14, 10–214.

Van Maanen, A. and Gourbière, F. (2000) Balance between colonization and fructification in fungal dynamics control: a case study of *Lophodermium pinastri* on *Pinus sylvestris* needles. *Mycological Research* 104, 587–594.

Vujanovic, V. and St-Arnaud, M. (2001) First report of *Cryptocline taxicola* infecting Pacific yew (*Taxus brevifolia*) in eastern North America. *Plant Disease* 85, 922.

Walla, J.A. and Kinzer, K.M. (2008) Symptoms and signs of *Stigmina lautii* on spruce needles in North Dakota. *Phytopathology* 98, 164.

Wang, K., Kellomäki, S. and Laitinen, K. (1995) Effects of needle age, long-term temperature and CO_2 treatments on the photosynthesis of Scots pine. *Tree Physiology* 15, 211–218.

Waters, C.W. (1962) Significance of life history studies of *Elytroderma deformans*. *Forest Science* 8, 250–254.

Watt, M.S., Kriticos, D.J., Alcaraz, S., Brown, A.V. and Leriche, A. (2009) The host and potential geographic range of Dothistroma needle blight. *Forest Ecology and Management* 257, 1505–1519.

Watt, M.S., Stone, J.K., Hood, I.A. and Palmer, D.J. (2010) Predicting the severity of Swiss needle cast on Douglas-fir under current and future climate in New Zealand. *Forest Ecology and Management* 260, 2232–2240.

Watt, M.S., Palmer, D.J. and Bulman, L.S. (2011a) Predicting the severity of Dothistroma needle blight on *Pinus radiata* under future climate in New Zealand. *Forest Ecology and Management* 261, 1792–1798.

Watt, M.S., Stone, J.K., Hood, I.A. and Manning, L.K. (2011b) Using a climatic niche model to predict the direct and indirect impacts of climate change on the distribution of Douglas-fir in New Zealand. *Global Change Biology* 17, 3608–3619.

Whitney, H.S., Reid, J. and Pirozynski, K.A. (1975) Some new fungi associated with needle blight of conifers. *Canadian Journal of Botany* 53, 3051–3063.

Winton, L.M., Stone, J.K., Watrud, L.S. and Hansen, E.M. (2002) Simultaneous one-tube quantification of host and pathogen DNA with real-time polymerase chain reaction. *Phytopathology* 92, 112–116.

Winton, L.M., Hansen, E.M. and Stone, J.K. (2006) Population structure suggests reproductively isolated lineages of *Phaeocryptopus gaeumannii*. *Mycologia* 98, 781–791.

Woods, A., Coates, K.D. and Hamann, A. (2005) Is an unprecedented Dothistroma needle blight epidemic related to climate change? *BioScience* 55, 761–769.

Zha, T., Wang, K.-Y., Ryyppö, A. and Kellomäki, S. (2002) Impact of needle age on the response of respiration in Scots pine to long-term elevation of carbon dioxide concentration and temperature. *Tree Physiology* 22, 1241–1248.

Ziller, W.G. (1969a) Studies of hypodermataceous needle diseases. II. *Lirula brevispora* sp. nov., causing needle blight of spruce. *Canadian Journal of Botany* 47, 261–262.

Ziller, W.G. (1969b) *Sarcotrochila alpina* and *Lophodermium laricinum* causing larch needle blight in North America. *Plant Disease Reporter* 53, 237–239.

24 Foliar Diseases of Broadleaved Trees

Tadeusz Kowalski*
University of Agriculture in Krakow, Poland

24.1 Pathogens, Significance and Distribution

Fungi that cause foliar diseases of broadleaved trees comprise a differentiated group of organisms. This is shown by their morphological characteristics, which are indicative of their taxonomy, as well as by their life cycles, epidemiology, host range and disease symptoms on the host plants. Foliar diseases of broadleaved trees are caused mostly by members of the *Ascomycota*. On leaves, these fungi reproduce either sexually (the teleomorph) or asexually (the anamorph), or produce both states (teleomorph and anamorph). In the last situation, both states can be produced simultaneously or in succession, at a specific point in time or under specific conditions. In many cases, the two reproductive states have different names, which may cause confusion of nomenclature.

Among the *Basidiomycota*, fungi that cause foliar diseases are mostly the rust fungi (see Hamelin, Chapter 26, this volume). Other members of the *Basidiomycota*, including *Microstroma juglandis* (Bérenger) Sacc., which occurs on leaves of walnut (*Juglans* spp.), and *Exobasidium* spp., which occurs on the leaves of rhododendron (*Rhododendron* spp.) and of other *Ericaceae*, are of minor significance (Mielke and Ostry, 2004; Sinclair and Lyon, 2005).

From the point of view of nutritional preferences, two broad categories of foliar disease pathogens can be distinguished, i.e. biotrophs, which obtain nutrients from living cells, and necrotrophs, which typically start to colonize living plants, and then destroy and kill the host tissue and obtain nutrients from the dead host cells. The foliar pathogens discussed in this chapter are almost all necrotrophs. Only species of *Taphrina* and ascomycetous species of *Pachysacca*, occurring on eucalyptus (*Eucalyptus* spp.), are biotrophs (Keane, 1997). Among other biotrophs, the rust fungi (*Pucciniales*) are described separately (see Hamelin, Chapter 26, this volume), and powdery mildews are discussed within nursery diseases (see Lilja and Poteri, Chapter 27, this volume), as from a practical point of view their importance in nursery environments is greater than it is in forest ecosystems.

Pathogens that cause foliar diseases have different host ranges. A broad range of hosts is known for some species of *Colletotrichum*, e.g. *C. gloeosporioides* (Penz.) Penz. & Sacc. (teleomorph: *Glomerella cingulata* (Stoneman) Spauld. & H. Schrenk) and *C. acutatum* J.H. Simmonds, which have been recorded

* E-mail: rltkowal@cyf-kr.edu.pl

on 115 species of woody plants (Sinclair and Lyon, 2005). In contrast, a narrow host range is characteristic for many species of *Taphrina*, which are mostly restricted to a single host species. For example, among the pathogens of alder (*Alnus* spp.), *T. tosquinetii* (Westend.) Tul. occurs on European alder (*A. glutinosa* (L.) Gaertn.), *T. sadebeckii* Johanson and *T. epiphylla* (Sadeb.) Sacc. occur on grey alder (*A. incana* (L.) Moench) and *T. viridis* (Sadeb.) Maire occurs on green alder (*A. viridis* (Chaix) DC.) (Bacigálová *et al.*, 2003).

Frequently, within a pathogen species, formae speciales (f.spp.) can be distinguished. This classification is applied to pathogens that are adapted to specific hosts. They are often distinguished by host preference, usually at species level, but scarcely or not at all distinguished by morphological criteria. A few formae speciales can be differentiated within *Marssonina brunnea* (Ellis & Everh.) Magnus (teleomorph: *Drepanopeziza punctiformis* Gremmen). *M. brunnea* f.sp. *brunnea* Spiers infects mostly balsam poplar (*Populus balsamifera* L.), eastern cottonwood (*P. deltoides* Bartram ex Marsh.) and Lombardy poplar (*P. nigra* L.); *M. brunnea* f.sp. *trepidae* Spiers infects European aspen (*P. tremula* L.) and quaking aspen (*P. tremuloides* Michx.). Many formae speciales identified within species of *Colletotrichum* infect both annual and perennial plants (Sutton, 1992). Some strains within *Rhytisma acerinum* (Pers.) Fr. have been differentiated on the basis of host preferences. However, no formae speciales have so far been distinguished in *R. acerinum* (Wulf, 1994).

An increasing range of hosts has been observed for some foliar pathogens. The host range of *Cristulariella depraedans* (Cooke) Höhn., which had been known for more than 100 years and included mainly various maple species (*Acer* spp.), but also *Bucida buceras* L. and *Aruncus dioicus* (Walter) Fernald (Redhead, 1975), is now being amended. Studies in Germany in 1996–1999 showed that *C. depraedans* has a much broader potential host range than known previously. Lang (2000) reported the occurrence of *C. depraedans* on 21 woody and herbaceous plants not belonging to the maples. Studies in Poland also confirmed that the host range of *C. depraedans* is not restricted to plants within the genus *Acer*. The fungus was observed on ten species of plants and four of them, common filbert (or hazel) (*Corylus avellana* L.), European beech (*Fagus sylvatica* L.), dwarf honeysuckle (*Lonicera xylosteum* L.) and European bird cherry (*Prunus padus* L.), were new additions to its host range (Kowalski and Bartnik, 2008). In such cases it is hard to know whether the increasing host range results from more detailed studies made recently or from changes of environment that allow the pathogen to infect more species of plants.

Our understanding of the host range of some pathogens may be confused by the fact that they cause diseases on some tree species but occur on a broader spectrum of plants as endophytes (Sieber, 2007). Moreover, our knowledge of the actual interspecific and intraspecific diversity and taxonomy of many fungal pathogens is incomplete. This refers particularly to genera that include many species, e.g. *Colletotrichum* and *Mycosphaerella* (Sutton 1992; Crous, 2009). Incorrect classification of a fungus often prevents the identification of an accurate host range.

The significance of foliar diseases of broadleaved trees depends mostly on the size of the epidemic area, the severity of leaf infection, tree age, the time of infection in the growing season, and the longevity of the disease – which may occur in one or a few successive seasons. A low severity of disease, appearing as a few spots on leaves or the local infection of single leaves, causes minor damage to a tree. A higher severity of disease resulting from the advanced colonization of leaves causes reduction in photosynthesis and plant growth. Occasionally, a severe disease outbreak causes premature defoliation or dieback of various parts of the tree. Infection during extreme conditions may result in death of trees, regardless of their age. Weakening of an affected tree makes it more susceptible to attack by insects and other fungi. Damage is greater if the disease occurs over two or more successive years (Butin, 1995; Spiers, 1998; Cellerino, 1999; Holdenrieder *et al.*, 2004; Sinclair and Lyon, 2005).

Leaf spot infections can lead to premature defoliation if they start early in the growing season. If a tree loses its leaves early in the season, it may produce new ones and its health is not seriously affected. If it loses them in midsummer though, the production of new leaves may prevent the tree from fully hardening off before the onset of cold weather or may cause a reduction in the amount of reserves stored, which often leads to an increased risk of frost damage.

The scale of growth reduction resulting from a foliar disease depends on the species of tree or conditions of growth. Some fungi infect only leaves. Many foliar pathogens, however, particularly those belonging to the genera *Venturia*, *Drepanopeziza*, *Apiognomonia*, *Gnomonia*, *Mycosphaerella* and *Glomerella*, also infect buds, stems and branches, causing cankers and stem-tip dieback. The economic importance of pathogens in this latter group is usually more significant than that of pathogens infecting only leaves.

Species such as *Dicarpella dryina* Belisario & M.E. Barr, *Didymosphaeria petrakiana* Sacc., *Linospora tetraspora* G.E. Thomps., *L. ceuthocarpa* (Fr.) Lind and *Pyrenopeziza betulicola* Fuckel usually occur locally and are of minor economic importance. The lesser importance of such species as *M. juglandis* is because of their ineffectiveness in killing leaf tissues and causing crown defoliation, resulting in minimal effect on the tree, probably through only minor reduction in photosynthesis (Butin and Kehr, 1995; Newcombe, 1998b; Pappinen *et al.*, 2002; Mielke and Ostry, 2004; Kowalski, 2006b).

The virulence of a particular fungus can be increased after transfer to and dispersal through another continent, where susceptible plants previously grew with no contact with the pathogen. A classic example is *Drepanopeziza punctiformis*, which is native to the USA where it has caused no noticeable damage but, after its spread to Europe in 1958, it started to cause dramatic damage in plantations of poplar (van der Meiden, 1964; Butin, 1995; Cellerino, 1999). Similarly, the introduction of *Venturia saliciperda* Nüesch from Europe to North America at the beginning of the 20th century caused epidemics in native willow (*Salix* spp.) populations (Butin, 1960; Kenaley *et al.*, 2010).

Currently, foliar diseases of broadleaved trees are of increasing economic importance because of the establishment of forest tree plantations throughout the world, primarily to provide structural timber and fibre for pulp and paper production and wood biomass for energy production. Trees such as poplar, willow, birch (*Betula* spp.) and maple are usually grown and tested in Europe (Christersson, 2006; Helby *et al.*, 2006; Walle *et al.*, 2007). Apart from the genera *Populus*, *Salix* and *Eucalyptus*, plantations of sycamore (*Platanus* spp.) and sweetgum (*Liquidambar* spp.) are currently being established in some regions of the USA (Davis and Trettin, 2006; Rockwood *et al.*, 2006). In the tropics and the southern hemisphere, the most widely established tree species in plantations, apart from the genus *Pinus*, are *Eucalyptus* spp. and *Acacia* spp. Increasingly important foliar pathogens in such plantations include *Mycosphaerella* spp., *Phaeoseptoria eucalypti* Hansf. and *Aulographina eucalypti* (Cooke & Massee) Arx & E. Müll. (Lundquist and Purnell, 1987; Wingfield *et al.*, 2001; Carnegie and Ades, 2003; Carnegie and Keane, 2003).

Additional aspects of the destructiveness of foliar pathogens that should be emphasized include: (i) limitations to the cultivation of tree species that are susceptible to the local foliar pathogens (Butin, 1995); (ii) reductions in the ornamental value of trees resulting from foliar diseases; (iii) reductions in crop yield of fruit trees as disease severity increases, e.g. in walnuts after the occurrence of walnut anthracnose caused by *Gnomonia leptostyla* (Fr.) Ces. & De Not. (Mielke and Ostry, 2004); (iv) difficulties in the breeding of trees highly susceptible to foliar diseases for resistance to other diseases, e.g. the case of several hybrids of aspen susceptible to *Apioplagiostoma populi* (E.K. Cash & Waterman) M.E. Barr causing bronze leaf disease – here, susceptibility to *A. populi* was a genetic barrier to hybridization for resistance to *Entoleuca mammata* (Wahlenb.) J.D. Rogers & Y.M. Ju, which causes canker (Smith *et al.*, 2002); (v) inoculum production on leaves by pathogens causing severe symptoms on other plant parts, e.g. *Hymenoscyphus pseudoalbidus* V. Queloz, C.R. Grünig, R. Berndt, T. Kowalski, T.N. Sieber & O. Holdenrieder (anamorph: *Chalara fraxinea* T. Kowalski), which is considered to be the main causal agent of the ash (*Fraxinus* spp.) decline that spread in Europe from 1992 (Kowalski, 2006a; Kowalski and Holdenrieder, 2009; Queloz *et al.*, 2011).

The geographical distributions of foliar pathogens throughout the world vary. Distribution is determined mostly by the occurrence of suitable plants. Most foliar pathogens show a high degree of host specialization. They are highly virulent on one or on only a limited number of host plants (Knogge, 1996). Species with a very broad host range have a better chance of a broader spread and distribution. Thus, some of them, including *Botrytis cinerea* Pers. and *Glomerella cingulata*, occur in several continents. Nevertheless, some fungi with a narrow host range may also spread worldwide, e.g. *Drepanopeziza punctiformis* and *D. populi-albae* (Kleb.) Nannf., which infect poplars. Spread into and across new areas sometimes takes a long time. In New Zealand, *D. punctiformis* and *D. populi-albae* were recorded for the first time in 1975 and 1985, respectively, despite poplars having been introduced there between 1840 and 1850 (Spiers, 1998).

The establishment of intensive forest plantations throughout the world, including those based either on chosen native or non-native tree species, especially *Eucalyptus*, *Acacia* and *Salix* spp., will enforce changes in the distribution of the pathogens associated with these plants. New species of foliar pathogens may appear where harmless species were previously present (Wingfield *et al.*, 2001). In the past, the establishment of poplar plantations was largely responsible for affecting the distribution of many poplar-specific pathogens (Dance, 1961). The introduction and spread of certain fungal species not only resulted from the cultivation of particular species of poplar but also of their hybrids (Cellerino, 1999).

Global climatic conditions, including temperature and moisture in particular, form the next most important determinant of foliar pathogen distribution. They affect the virulence of a pathogen, which can play different roles even on susceptible tree species, depending on host genotype and environment. Under unfavourable climatic conditions, a pathogen may be harmless or absent. Thus, some fungal diseases occur only locally, in certain countries or certain regions of a continent (Butin, 1995; Newcombe, 1998b; Spiers, 1998; Mottet *et al.*, 2007; Feau *et al.*, 2010).

Quarantine regulations applied by particular countries also affect a pathogen's distribution if they prevent its introduction from one region or continent to another; this usually restricts the area and borders of its occurrence. It seems that quarantine regulations prevented the transfer of *Davidiella populorum* (G.E. Thomps.) Aptroot (anamorph: *Septoria musiva* Peck) from North America, where it is common on poplars, to Europe (Cellerino, 1999; EPPO, 2005). Another pathogen, *Stegophora ulmea* (Fr.) Syd. & P. Syd., occurs only in North America and China, even though susceptible species of elm (*Ulmus* spp.) also grow in other parts of the world (EPPO, 2005).

The distribution and frequency of occurrence of foliar pathogens are also affected by other factors connected with the specific requirements necessary for a pathogen's growth and expansion, with ecological interactions occurring in specific locations at specific times and with the conditioning necessary for the disease process. The specific factors may include vectors necessary for dispersal of the inoculum or the occurrence of conditions that predispose the plant to infection. In the case of sooty moulds, for example, the fungi use plant exudates and the sugary honeydew secreted by plants or sap-sucking insects such as aphids and scales.

The area over which many pathogens occur has increased over the past few decades, and a tendency for steady further increase in the expansion of the range of pathogens should be expected. This may result partly from global climatic changes, which may increase the predisposition of plants to infection and may create the conditions necessary for pathogen growth and distribution (La Porta *et al.*, 2008; Desprez-Loustau *et al.*, 2010; Pautasso *et al.*, 2010). Increasing attention is being paid recently to the emerging infectious diseases caused by pathogens that have increased in incidence or in geographical or host range, have changed in their pathogenesis, have newly evolved or have been newly discovered or recognized (Brown, 1984; Holdenrieder *et al.*, 2004; Moslonka-Lefebvre *et al.*, 2011).

24.2 Diagnosis

Where fungal foliar pathogens are concerned, successful infection and an incubation period, which usually lasts from a few days to a few weeks, are followed by the appearance of disease symptoms resulting from physiological or structural changes. Aetiological signs then appear as fungal vegetative and generative structures on the colonized leaves. The type of disease symptom allows us to distinguish various categories of foliar disease (Sinclair and Lyon, 2005). Only some of them are clearly defined. Occasionally the disease symptoms are so characteristic that they allow identification of the causal agent, e.g. tar spot caused by *Rhytisma* spp. on maples, leaf blister caused by *Taphrina* spp. on poplar, white mould caused by *Microstroma* spp. on walnut and *Carya* spp., ink spot caused by *Ciborinia* spp. if sclerotia are formed on poplar leaves, and leaf blotch caused by *Linospora* spp. if pseudostromatic structures are formed on aspen leaves (Kojwang and Kurkela, 1984; Butin, 1995; Smith *et al.*, 2002; Sinclair and Lyon, 2005). Usually, however, identification of the causal agent on the basis of the macroscopic symptoms of disease is not possible. Laboratory observations are then necessary. It must be emphasized though that a single fungal pathogen can cause different disease symptoms on leaves.

Many foliar pathogens produce only anamorphic or spermatial states on the surface of necrotic areas (Sutton, 1992; Butin and Kehr, 1998). If fructification is absent on leaves, the induction of fructification on lesions under moist conditions (in moist chambers) or isolation of fungi on artificial media is necessary. As well as classical methods, molecular procedures are currently used for the detection and identification of fungi in plant tissues (Innis *et al.*, 1999).

The most common symptoms of foliar diseases occurring on forest, ornamental and shade trees are leaf spots. These are round blemishes restricted to areas of necrosis, usually on the leaf blade. In most leaf spot diseases, they are small, scattered, circular to oval, tan, dark brown, yellow, grey, purple or coal black, often with yellow, purple, red or reddish brown to black margins. Some spots are raised or may drop out leaving ragged holes. Some are shiny or marked with light and dark concentric zones. Spots may increase in number or size, or merge into large, angular to irregular necrotic areas, usually in the late summer and early autumn. Heavily infected leaves may turn yellow to brown, wither and fall prematurely (Butin, 1995; Sinclair and Lyon, 2005; Kowalski, 2006b).

An overview of the symptoms of the most important categories of foliar diseases is reported in Table 24.1.

It must be mentioned that the disease symptoms described in Table 24.1 result from direct infection of leaves by pathogens. There are, however, other fungi that, despite occurring on/in other parts of trees, cause systemic symptoms on the foliage resulting from primary infection elsewhere on the plant. An example is *Chondrostereum purpureum* (Pers.) Pouzar, i.e. the cause of silver leaf disease on various tree species. It infects the sapwood of tree trunks by wounds and causes wood decay, and its fungal toxins are transported to the leaves where they cause a silvery sheen. Other examples include *Verticillium dahliae* Kleb. and *V. albo-atrum* Reinke & Berthold on various species, *Ophiostoma novo-ulmi* Brasier on elm, and *Ceratocystis fagacearum* (Bretz) J. Hunt on oak (*Quercus* spp.), which cause changes in the tree crown and leaf wilt resulting from plugging the xylem tissue with gum, tyloses and extensions of the xylem cell wall (Butin, 1995; Sinclair and Lyon, 2005).

24.3 Infection Biology

24.3.1 Primary infection

The development of primary infections on new leaves in spring is dependent on the location and conditions in which the pathogen overwintered, and on its inoculum potential, which is related to the number of spores

Table 24.1. Main features of the most important foliar diseases of broadleaved trees.

Foliar disease	Fungal pathogen	Host(s)	Main symptoms	Sign(s), characters of fruiting bodies	References
Cristulariella leaf spot	*Cristulariella depraedans* (Cooke) Höhn.	*Acer* spp. and other plant species	Roundish, greyish-white, necrotic spots	Stalked heads, 0.1–0.2 mm in diameter, black sclerotia	Butin, 1995; Kowalski and Bartnik, 2008
Elm black spot	*Stegophora ulmea* (Fr.) Syd. & P. Syd.	*Ulmus* spp.	Small, irregular, greyish to black spots, often surrounded by a yellow halo	Acervuli, perithecia	McGranahan and Smalley, 1984; Sinclair and Lyon, 2005
Ink spot	*Ciborinia whetzelii* (Seaver) Seaver	*Populus tremuloides* Michx., *P. grandidentata* Michx., *P. deltoides* Bartram ex Marsh., *P. balsamifera* L., *P. nigra* L.	Tan to brown necrotic areas	Saucer-shaped, nearly black sclerotia, apothecia	Sinclair and Lyon, 2005
Leaf blotch of horse chestnut	*Guignardia aesculi* (Peck) V.B. Stewart	*Aesculus* spp.	Irregular, reddish-brown areas, distortion and partial shrivelling of leaflets	Round, black pycnidia, macroconidia with long appendages, rod-shaped microconidia, pseudothecia	Butin, 1995; Sinclair and Lyon, 2005
Marssonina leaf spot	*Drepanopeziza punctiformis* Gremmen	*Populus nigra*, *P. deltoides*, *P. balsamifera*, Euramerican hybrids of poplar	Punctiform, blackish-brown, angular spots	Acervuli with unequally two-celled macroconidia, immersed in the leaf tissue, apothecia	Butin, 1995; Feau *et al.*, 2010
Poplar scab	*Venturia macularis* (Fr.) E. Müll. & Arx	*Populus tremuloides*, *P. grandidentata*, *P. alba* L., *P. deltoides*, hybrid poplars	Variably sized spots, distortion of leaves, curling and dieback of new terminal shoots	Irregular, olive-green acervuli, pseudothecia	Anderson and Anderson, 1980
Septoria leaf spot	*Davidiella populorum* (G.E. Thomps.) Aptroot	*Populus* spp.	Circular to angular, reddish-brown to black spots, with yellow margin	Pycnidia, pseudothecia	Sinclair and Lyon, 2005; Feau *et al.*, 2010
Sooty mould	*Capnodium*, *Antennariella*, *Aureobasidium*, *Cladosporium*, *Alternaria* and other *Ascomycota* and anamorphic fungi	Various species of trees and shrubs	Greyish-black, powdery or crust-like coating on leaves and young, non-woody shoots	Hyphae, spores, fruiting bodies	Reynolds, 1976

Continued

Table 24.1. Continued.

Foliar disease	Fungal pathogen	Host(s)	Main symptoms	Sign(s), characters of fruiting bodies	References
Sycamore anthracnose	*Apiognomonia veneta* (Sacc. & Speg.) Höhn.	*Platanus* spp.	Death of veins and surrounding tissue of leaves, blight of buds and shoots, sunken cankers on twigs	Pycnidia, perithecia	Sinclair and Lyon, 2005
Tar spot	*Rhytisma acerinum* (Pers.) Fr.	*Acer* spp.	Roundish necrotic areas	Circular, black tar-like stromata, pycnidia, saucer-shaped apothecia	Wulf, 1994; Butin, 1995; Sinclair and Lyon, 2005
Tubakia leaf spot	*Dicarpella dryina* Belisario & M.E. Barr	*Quercus* spp., *Castanea crenata* Siebold & Zucc.	Roughly circular, dark to reddish-brown spots, or irregular blotches	Black pycnothyria, perithecia	Holdenrieder and Kowalski, 1989; Sinclair and Lyon, 2005; Kowalski 2006b
White mould	*Microstroma juglandis* (Bérenger) Sacc.	*Juglans* spp., *Carya* spp.	Yellowish discoloration on upper surface, whitish growth on lower surface of leaves	A naked layer of basidia, basidiospores in the snow-white coating	Mielke and Ostry, 2004; Sinclair and Lyon, 2005
Yellow blister of poplar	*Taphrina populina* (Fr.) Fr.	*Populus* spp.	Circular, blister-like spots, yellowish depressions on lower surface of leaves	A naked single layer of asci in depressions of blister, the yeast-like phase	Butin, 1995; Sinclair and Lyon, 2005

able to germinate and infect a plant. For the majority of necrotrophic foliar pathogens, their spores are produced in the spring, on fallen leaves, after overwintering on the ground or in the litter (Anderson and Anderson, 1980; Ostry, 1987). In species of *Grovesinia* the spores are produced on sclerotia formed on leaves, and in species of *Ciborinia* they are produced on sclerotia fallen from leaves to the ground (Butin, 1995; Sinclair and Lyon, 2005).

Spores are spread primarily by contact, air currents, rain, splashing water, and human and other non-human vectors, including insects. The distance of dispersal ranges from a few centimetres to hundreds of kilometres (Nagarajan and Singh, 1990). For primary infections, the distance between the source of the inoculum and the location of the infected plant is important. If the inoculum source occurs on fallen leaves in the litter, the leaves in the lower part of the crown are often infected more severely than those in the upper part of the crown (Wulf, 1994; Butin, 1995; Cellerino, 1999; Smith *et al*., 2002; Sinclair and Lyon, 2005; Feau *et al*., 2010).

Some pathogens overwinter on the above-ground parts of trees. In *Taphrina* spp., the fungus overwinters on twigs or among bud scales in the form of blastospores produced by budding ascospores. After contact of a blastospore with the young leaf, the fungus begins its mycelial parasitic phase. In *Apioplagiostoma populi*, the fungus overwinters on dead infected leaves attached to the tree and produces primary inoculum on the leaf surface in the spring. The group of fungi that overwinter on the above-ground parts of trees also includes all those that infect shoots, twigs and small branches, as well as leaves, e.g. the genera *Apiognomonia*, *Gnomonia*, *Dicarpella*, *Drepanopeziza*, *Glomerella* and *Venturia*, or the species *Davidiella populorum*, *Dicarpella dryina*, *Septotinia populiperda* Waterman & E.K. Cash, *Septotis podophyllina* (Ellis & Everh.) Arx and *A. populi* (Ostry, 1987; Holdenrieder and Kowalski, 1989; Von Sury and Flückiger, 1991; Butin, 1995; Smith *et al*., 2002). These fungi are active not only during the growing season but also in the winter during periods of mild weather, forming cankers and causing twig dieback. Their fruiting bodies, which are filled with spores, are produced on dead twigs and necrotic areas in the spring. Spores are released and disseminated to the young expanding shoots and leaves.

Some potential pathogens can overwinter in plant tissue in an endophytic form and may change their behaviour to parasitic in the spring. Examples are: *D. dryina*, which on beech can move from the overwintered terminal buds to the current year's twigs (Kaneko and Kaneko, 2004); *Glomerella cingulata* and *Diplodina acerina* (Pass.) B. Sutton, which overwinter in buds of sycamore maple (*A. pseudoplatanus* L.) (Wulf, 1994); and *Stegophora ulmea*, which overwinters in buds of elms (McGranahan and Smalley, 1984). In New Zealand, *Drepanopeziza populi-albae*, after surviving the winter in buds, may infect the emerging foliage within a month of bud burst (Spiers, 1998).

In *Microstroma juglandis*, basidiospores are responsible for spread of the disease. In species of *Rhytisma*, *Ciborinia* and *A. populi* only ascospores are able to infect. Some foliar pathogens, including some species of *Septoria*, *Discula*, *Monostichella* and *Aureobasidium*, produce only asexual spores (conidia). The majority of *Ascomycota* can produce both conidia and ascospores on dead leaves in the spring, and both kinds of spores can cause primary infection of leaves. Teleomorph states are formed by different fungi with different frequencies and abundance, mostly depending on the environmental conditions (Wulf, 1994; Butin, 1995; Sinclair and Lyon, 2005).

The time of fruiting body maturation and production of spores is usually coincident with the time of emergence of new leaves (Butin, 1995; Smith *et al*., 2002). The amount of spore discharge usually increases temporarily during intensive rainfall and decreases during droughts (Wulf, 1994). The period in which spores are liberated depends on the species of fungus and weather conditions. In pathogens such as *Venturia macularis* (Fr.) E. Müll. & Arx and *Rhytisma punctatum* (Pers.) Fr., the period is relatively short, lasting 40 h and 5 days, respectively (Woo and Partridge, 1969; Anderson and Anderson, 1980). This short period of spore release seems to result from a need for the fungus to adapt to the limited period in which the plant is susceptible to infection.

24.3.2 Secondary infection

New generations of spores are produced on necrotic areas on leaves and on shoots by many polycyclic foliar pathogens during the growing season. They are usually conidia produced on single conidiophores or layers of conidiophores grouped in sporodochia, acervuli, pycnothyria or pycnidia. Sporadically, sexual spores can be produced in ascomata, e.g. in *Glomerella* and *Mycosphaerella* spp. Spores are released continuously and disseminated to other leaves and young shoots, usually by wind and rain. Dispersal in *Venturia macularis* occurs in airborne water droplets carrying clumps of conidia (Dance, 1961) and in *Davidiella populorum* by rain splashes carrying conidia released from pycnidia (Feau *et al.*, 2010). Thus, secondary infections occur and there may be many repeated cycles of disease throughout the growing season. This may intensify the disease level, particularly later in the growing season. Only monocyclic fungi, such as species of *Rhytisma*, *Ciborinia* and *Taphrina*, are incapable of regenerating spores and reinfecting plants during the growing season.

Spore deposition and retention on leaves are affected by the properties of the leaf surface. In *Discula quercina* (Cooke) Sacc., which colonizes the leaves of white oak (*Q. alba* L.), infection is greater along the veins than on the lamina because a greater density of hairs and glands along the veins helps spore-laden rainwater to accumulate (Wilson and Carroll, 1994).

24.3.3 Attachment and penetration

The structure and topography of the leaf surface and leaf anatomy play an important role in infection. The most important characters are: the presence of trichomes and hydathodes, the density and structure of stomata, the thickness, morphology and composition of cuticle, the presence, thickness and composition of epicuticular waxes, and abrasion resistance (Andrews, 1992; Wilson and Carroll, 1994; Lindow and Brandl, 2003; Lindow, 2006). Waxy cuticle not only prevents water loss from the plant, but also prevents direct contact of the pathogen with epidermal cells and thereby limits infection. The hydrophobic nature of the cuticle prevents water collecting on the leaf surface, which is an important defence against many pathogens that require standing water on the leaf surface for spore germination. The presence of trichomes on the leaf surface may generally facilitate the attachment of spores and further development of the hyphae. In black walnut (*J. nigra* L.), however, trichomes on juvenile leaves are so dense that they hamper conidial attachment and decrease the possibility of infection (Neely, 1986). Additionally, trichomes apparently help to retain water on the plant surface and so provide nutrients for microbial growth (Lindow and Brandl, 2003).

Four stages crucial to the establishment of a fungal pathogen in the plant can be identified: attachment of the fungus to the surface of leaf, germination (often associated with formation of fungal infection structures), penetration of the host and colonization of the host plant tissues (Schäfer, 1994). Adhesion of spores to the leaf surface is often helped by changes in the properties or development of the fungus stimulated by the host plant. This can be demonstrated in *Apiognomonia errabunda* (Roberge ex Desm.) Höhn., the conidia of which attach differently to host and non-host plant surfaces; adhesion to the host surface was greatest at pH 5 and decreased with increasing pH, and a non-protein molecule present on the conidial surface was actively recognized by the host (Toti *et al.*, 1992). Several compounds have already been proven to be essential or helpful in the adhesion and penetration phases. These include hydrophobins, enzymes (e.g. cutinase), polysaccharides, melanins, glycerol and components of signal transduction pathways, which is the process by which an extracellular signalling molecule activates a membrane receptor that, in turn, alters intracellular molecules and creates a response.

Hydrophobins, a class of small cysteine-rich proteins expressed only in filamentous fungi, are found on the outer surface of conidia and of the hyphal wall and seem to be involved in the attachment of spores to hydrophobic surfaces of plants (Wösten, 2001). Fungal extracellular degradative, hydrolytic enzymes, e.g. cutinase, also play an important role in pathogenesis. Cutin monomers released from the cuticle by small amounts of cutinase on the surfaces of fungal spores can greatly increase the amount of cutinase secreted by the spore and can advance disintegration of the plant's structural barriers, i.e. the cuticle and epidermal cell wall (Knogge, 1996). The conidial matrix contains other enzymes, including cellulase, non-specific esterase, invertase, pectinesterase and pectic lyase, which are also known to be involved in fungal adhesion and penetration. Spores and hyphae of a large number of plant pathogens have mucilaginous sheaths. Mucilage is composed primarily of polysaccharides with neutral or carboxyl functional groups, as found for example on the conidia of *Apiognomonia errabunda* (Toti *et al.*, 1992).

Fungal pathogens have evolved various strategies and different organs for invading plants and optimizing the colonization of their tissues (Knogge, 1996). Some fungi, e.g. *D. populorum*, enter the host through stomata (Feau *et al.*, 2010). Other pathogens invade leaves through wounds. Stanosz (1993) reported severe infection of sugar maple (*A. saccharum* Marsh.) leaves by *Discula campestris* (Pass.) Arx through wounds made after the leaves had been infested with the pear thrips, *Taeniothrips inconsequens* (Uzel). The production of hydrolytic enzymes may be combined with the formation of hyphal 'pressing' organs, or appressoria, which produce minute infection pegs that grow and enter the host using turgor pressure. An appressorium is a prerequisite for invasion by many pathogens (Emmett and Parbery, 1975). Hyaline appressoria may be formed by the *Erysiphales*, the cause of powdery mildews, and the *Pucciniales*, the cause of rusts. Melanized appressoria may be formed by many other pathogens, and are particularly prominent in *Colletotrichum*. Melanin in the appressorium seems to be an essential active compound in mechanical penetration by the fungus. It helps to achieve a high turgor pressure, which may reach more than 8 MPa. Mechanical piercing through the cuticle and cell wall is possibly assisted by the activity of hydrolytic enzymes (a topic reviewed in Mendgen and Deising, 1993).

The infection process has been studied in detail for many foliar pathogens (Black and Neely, 1978; Seifers and Ammon, 1980; Kojwang and Kurkela, 1984; Wulf, 1994). It often begins with the deposition and germination of ascospores. In *Rhytisma acerinum* and *Pleuroceras pseudoplatani* (Tubeuf) M. Monod, the ascospores are deposited almost exclusively on the upper surface of leaves, particularly along veins in the case of *P. pseudoplatani*. Ascospore densities decrease with the height of trees (Wulf, 1994). Primary infections by *P. pseudoplatani* initiate primary spots at the base of the leaf, close to the vein, and colonization of the vein collenchyma is initiated from these first spots. The infection process on sycamore leaves by *Apiognomonia veneta* (Sacc. & Speg.) Höhn., as described by Seifers and Ammon (1980), starts with the germination of spores on the leaf surface; spore germination was 10% after 3 h and 100% after 6 h at 22–23°C. Growth of the germ tube was random and not associated with any particular morphological feature of the leaf surface. Penetration through stomatal openings was not observed.

24.3.4 Colonization

Differences between obligate biotrophic and necrotrophic fungi occur in the colonization of host tissues. Although both types of pathogens produce a dense mycelial net of intercellular and intracellular hyphae that grow in the mesophyll, biotrophs are known additionally to produce haustoria (Perfect and Green, 2001). Haustoria of *Pachysacca* spp. penetrate eucalyptus leaf cells located immediately below the fungal stromata (Keane, 1997). The biotrophic *Taphrina* spp. discussed in this chapter do not form haustoria.

Several molecules, e.g. phytotoxins, hormone-like compounds and enzymes, have already been proven to be essential for colonization, particularly by necrotrophic pathogens. They generally inactivate mechanisms of plant defence, changing the plant's morphology or physiology to the benefit of the pathogen (Knogge, 1996; Hanson, 2008; Grad *et al.*, 2009). A phytotoxin as referred to here is a substance produced by a fungus that is toxic to the plant. There are, however, also phytotoxins that induce plant resistance to pathogens. Some phytotoxins are host selective, while others are active in a wide range of plants, on both host and non-host species (Walton, 1996). The change from a biotrophic to necrotrophic lifestyle can be accomplished within a few days (Horbach *et al.*, 2011). In pathogens that have intercellular growth, the particular kinds of leaf tissues may be colonized in succession. *Pleuroceras pseudoplatani*, for example, first colonizes the spongy mesophyll and then the palisade mesophyll (Wulf, 1994).

Forest trees may form symbiotic associations with endophytic fungi that live in healthy tissues in the form of quiescent mycelia (Sieber, 2007). Endophyte communities in angiosperms are frequently dominated by members of the *Ascomycota*, especially species of the *Diaporthales* (Sieber, 2007). Leaves of forest trees are often colonized initially by endophytic species of *Phloeospora*, *Gnomonia*, *Venturia*, *Apiognomonia*, *Dicarpella*, *Phyllosticta* and *Glomerella* (Pehl and Butin, 1994; Wulf, 1994; Kowalski, 2003, 2004; Sieber, 2007). Under favourable conditions, or after wounding of leaves, they may change their lifestyle to necrotrophic, with disease symptoms appearing. Factors that activate such endophytes include damage by gall-forming insects or mites (Table 24.2). Leaf spots or symptoms characteristic of anthracnose may appear where damage has occurred (Butin, 1992; Pehl and Butin, 1994; Wulf, 1994; Kowalski, 2003, 2004).

24.4 Epidemiology

The amount of disease that develops in a plant community is determined by the host, the pathogen and the environment. These three equally important components are referred to as the disease triangle, a central principle of plant pathology (Parker and Gilbert, 2004; Eastburn *et al.*, 2011). The results of interactions among these three elements determine the risk of epidemics. Additionally, the effects of human activity on the risk of epidemics should be emphasized. These may result from management and cultural practices, which can change the conditions of plant growth.

Table 24.2. Examples of gall-making insects/mites and associated fungal endophytes on the foliage of broadleaved trees.

Tree species	Gall-making insect/mite	Fungal endophytes	References
Acer pseudoplatanus L.	*Harrisomyia vitrina* (Kieffer)	*Diplodina acerina*, *Discula campestris*, *Aureobasidium apocryptum*	Pehl and Butin, 1994; Wulf, 1994
	Drisina glutinosa Giard	*Glomerella cingulata*, *Diplodina acerina*	Kowalski, 2003
Fagus sylvatica L.	*Mikiola fagi* (Hartig), *Hartigiola annulipes* (Hartig)	*Apiognomonia errabunda*	Pehl and Butin, 1994
Quercus robur L.	*Cynips divisa* Hartig, *Neuroterus numismalis* (Fourcroy)	*Discula quercina*	Pehl and Butin, 1994
	Trioza remota Foerster	*Discula quercina*, *Aureobasidium apocryptum*	Butin, 1992; Pehl and Butin, 1994
Tilia cordata Mill.	*Eriophyes lateannulatus* Schulze	*Apiognomonia errabunda*	Pehl and Butin, 1994
Ulmus glabra Huds.	*Byrsocrypta ulmi* (Linnaeus), *Physemococcis ulmi* (Kieffer), *Schizoneura ulmi* (Linnaeus)	*Glomerella cingulata*, *Asteromella* spp., *Ascochyta ulmella*	Kowalski, 2004

Advances in transportation technology and expanding global trade have greatly accelerated the rate of biological invasions, which can lead to the spread of disease epidemics (Parker and Gilbert, 2004). The most important aspects concerning the three main components of the disease triangle in foliar diseases of broadleaved trees are discussed below, with emphasis on how they relate to foliar disease management in cultivated trees.

24.4.1 Host

Susceptibility to certain pathogens is the most important character of the plant in determining infection. Epidemic development is likely only if a large number of individuals in a plant community are susceptible. However, plants have developed diverse defence strategies to resist pathogens or other extraneous agents. They include multiple constitutive, preformed structural and biochemical barriers, and inducible defences produced in response to infection (Andrews, 1992; Lindow and Brandl, 2003; Eyles *et al.*, 2010).

Generally, the host response involves different mechanisms developed to combat either biotrophic or necrotrophic pathogens. An effective defence mechanism against biotrophic pathogens is often in the form of a hypersensitive response of the host plant. This involves the programmed death of invaded cells followed by inactivation of the pathogen. In contrast, necrotrophic pathogens benefit from the death of the host's plant cells, and their development is limited by defence responses activated by plant hormones such as jasmonic acid, and ethylene signalling (Glazebrook, 2005; De Wit *et al.*, 2009).

Induced defence mechanisms in plants are much differentiated. They are classified and categorized in various ways (Eyles *et al.*, 2010). The common response to infection is the production of secondary metabolites, including terpenoids, phenolics and alkaloids, or proteins and peptides (Eyles *et al.*, 2010). Defence in the inducible protein-based category is expressed in Canadian poplar (*P.* × *canadensis* Moench; syn. *Populus euramericana* Guinier), which produces protein-related substances in response to *Drepanopeziza punctiformis* (Yuan *et al.*, 2008). One type of biochemical response that is strongly associated with defence is the accumulation of phytoalexins (Grayer and Kokubun, 2001). Inducible anatomical defences include the formation of mechanical and structural barriers. A well-recognized form of anatomical modification is the lignification and suberization of the cell wall and formation of a wound periderm (Eyles *et al.*, 2010). This category of induced defence occurs in juvenile leaves of southern Australian shining gum (*Eucalyptus nitens* (H. Deane & Maiden) Maiden), in which lignification, suberization and the formation of necrophylactic periderm are induced in response to *Mycosphaerella* spp. (Smith *et al.*, 2007).

Many broadleaved trees exhibit varying levels of susceptibility to foliar pathogens because of genetic variation, preformed constitutive barriers and inducible defences (Ostry and Ward, 2003; Feau *et al.*, 2010). In pathogens with a broad host range, variation in susceptibility can be seen at the genus level, e.g. in *Cristulariella depraedans*, which occurs on more than 20 host plants but is most pathogenic on species of *Acer*, which are the most susceptible (Butin,1995; Lang, 2000; Kowalski and Bartnik, 2008).

The way in which a plant responds to a particular pathogen often depends on the plant's age. Some pathogens are particularly dangerous in forest nurseries, or in plantations or groups of young trees (Alfieri, 1978; Ostry, 1987; Butin, 1995; Sinclair and Lyon, 2005). The level of infection in the growing season also depends on leaf age. Generally, juvenile plant tissue is more susceptible to pathogen invasion because of poorly developed natural defence mechanisms (Griffiths, 1981). Greater susceptibility of juvenile plant tissue can be seen in the epidemiology of many foliar diseases of trees (Anderson and Anderson, 1980; Carnegie and Ades, 2002; Smith *et al.*, 2007). This does not apply, however, to walnut anthracnose caused by *Gnomonia leptostyla*, in which the juvenile foliage of black walnut has been observed to be less severely infected than more mature foliage. Older leaves are more susceptible partly because of a decreased concentration of juglone (5-hydroxy-1,4-naphthoquinone), which

Different foliar fungal pathogens require different temperatures for successful development at each stage of their life cycles. The examples below illustrate some requirements. The optimal temperature for growth of the mycelium of *Pleuroceras pseudoplatani* is 15–20°C (Wulf, 1994); it is 8°C for the germination of ascospores of *Stegophora ulmea* (McGranahan and Smalley, 1984) and 21°C for ascospores of *Septotis populiperda* (Moesz & Smarods) B. Sutton (Schwenke, 1959). Liberation of the ascospores of *Apioplagiostoma populi* decreases as the mean weekly temperature approaches 20°C (Smith *et al.*, 2002). The infection of black walnut by *Gnomonia leptostyla* was most successful after incubation of the pathogen at 21°C *in vitro* (Black and Neely, 1978). The temperature for the reproduction of *Taphrina populina* (Fr.) Fr. cannot be more than 15–20°C (Cellerino, 1999).

Moisture also has a profound effect on the potential for epidemic development. The involvement of moisture in fungal disease epidemiology usually relates to rainfall and its amount and frequency, relative humidity, dew, fog and leaf wetness. These factors affect many fungal biological processes, including the production of spores, liberation and dispersal of inoculum, longevity of spore dispersal, spore germination and the process of infection (Aylor, 1990; Huber and Gillespie, 1992). For example, *G. leptostyla* is able to infect susceptible tissue only within the narrow range of relative humidity, 98–100% (Black and Neely, 1978), and the minimal and optimal humidity for infection from ascospores of *S. populiperda* were 90–92% and 98–99%, respectively (Schwenke, 1959).

Disease development by many fungal foliar pathogens is dependent on an appropriate set of weather conditions, and the occurrence of these conditions at the most suitable season and for an adequate period of time. Many reports describe the coincidence between severe epidemics and occurrence of a complete set of suitable weather conditions. Changing weather conditions cause variations in disease severity both through the year and between years (Melchior *et al.*, 1983; Lunquist and Purnell, 1987; Butin, 1995).

Sturrock *et al.* (2011) differentiated two groups of pathogens: those indirectly influenced by climate and those directly influenced. Those in the first group tend to infect hosts that are stressed by environmental factors or other biotic agents affected by climate, as illustrated by *Davidiella populorum*, which causes significantly larger cankers on water-stressed trees than on non-stressed trees (Maxwell *et al.*, 1997; Sturrock *et al.*, 2011). The second group is illustrated by *Calonectria quinqueseptata* Figueiredo & Namek, which is the causal agent of severe leaf blight of many tree species in tropical and subtropical forests, primarily in Australia and South-east Asia, and causes severe epidemics associated with warm temperatures and long periods of leaf wetness. Modelling has suggested that small increases in temperature and precipitation can substantially increase the area in which there is a high probability of severe epidemics (Booth *et al.*, 2000; Sturrock *et al.*, 2011).

There is much evidence for rapid climate change. As the temperature rises globally, it is expected that mean precipitation will decrease, with more rain falling in winter while the summers become drier; carbon dioxide (CO_2) and ozone (O_3) may increase as well (La Porta *et al.*, 2008; Pautasso *et al.*, 2010; Eastburn *et al.*, 2011). Increasingly more attention is being paid to the influence of global warming on the development of plant disease epidemics and the mechanisms that alter plant–pathogen relationships in natural plant communities (Manning and Tiedemann, 1995; Pautasso *et al.*, 2010; Eastburn *et al.*, 2011). Studies suggest that if global warming promotes factors that cause or accelerate tissue necrosis, we may expect a greater risk of epidemics caused by necrotrophic pathogens (Eastburn *et al.*, 2011).

Increased O_3 in the atmosphere weakens plants and so may increase the risk of epidemics caused by necrotrophs. In the case of leaf spot caused by *Drepanopeziza punctiformis*, the exposure of poplars to O_3 applied by fumigation increased the susceptibility of old leaves but not young leaves. Elevated O_3 also enhanced the germination of conidia, increasing the number of germ tubes, but had no effect on pathogen growth *in vitro* (Beare *et al.*, 1999). The incidence of canker caused by *D. populorum* increased on shoots of hybrid

Populus deltoides × *P. nigra* after chronic exposure of rooted cuttings to O_3 (Woodbury *et al.*, 1994).

Increased CO_2 in the atmosphere stimulates plant growth and so will increase leaf number and area, canopy size and density. This will lead to a lower temperature and higher humidity within the canopy, which may promote some fungal diseases (Manning and Tiedemann, 1995; Eastburn *et al.*, 2011). Some exceptions to the general rule have, however, already been observed. McElrone *et al.* (2005) reported that an increased level of CO_2 significantly reduced the incidence and severity of leaf spot caused by *Phyllosticta minima* (Berk. & M.A. Curtis) Underw. & Earle on red maple (*Acer rubrum* L.) through changes in host physiology and leaf chemistry. This pathogen infects red maple leaves by conidial germ tubes that enter through the stomata. Increased CO_2 reduced stomatal conductance by 21–36%, and provided smaller openings for infecting germ tubes; it also reduced leaf nitrogen (N), and increased the C:N ratio and total phenolic and tannin content. McElrone *et al.* (2005) concluded that reduced stomatal opening and altered leaf chemistry that occurs with elevated CO_2 may also result in reduced incidence and severity of diseases caused by other pathogens that target the stomata.

Water stress in plants can increase predisposition to disease (Boyer, 1995; Desprez-Loustau *et al.*, 2006). This may be essential for opportunistic pathogens that can infect only weakened plants. Von Sury and Flückiger (1991) reported that water stress can substantially affect the interaction between London plane trees (*Platanus hybrida* Brot.) and *Apiognomonia veneta*; the water stress suppressed leaf blight but promoted the formation of cankers on shoots and sporulation of the pathogen.

Air pollution (apart from climate change) is also an important environmental factor that significantly affects the incidence and severity of foliar diseases. Each interaction between air pollutant and disease is unique. Pollutants may reduce or intensify disease incidence. Air pollutants directly affect the fungus and its host, modifying their physiology and biochemistry (Heagle, 1973; Bytnerowicz, 1996; Kosiba, 2007). Many reports show that SO_2 and O_3 affect disease incidence in the field, mainly in rusts, mildews and foliar disease caused by *Venturia inaequalis* (Cooke) G. Winter, *Rhytisma acerinum* and *Diplocarpon rosae* F.A. Wolf (Heagle, 1973; Coleman *et al.*, 1988; Kosiba, 2007). Some reports show that a plant weakened by these pollutants tends to be more susceptible to weak pathogens that are able only to infect weakened plants (Heagle, 1973; Manning and Tiedemann, 1995).

Edaphic (soil) factors, particularly site fertility, affect the growth of trees and their susceptibility to pathogens. An imbalance in the nutrient status of the soil may result from incorrect fertilization or deposition from the emission of pollutants. This may predispose trees to infection by fungal pathogens (Baule and Fricker, 1967; Huber and Watson, 1974; Dimitri, 1977; Poteri and Rousi, 1996; Hagan *et al.*, 2008).

There is no general rule, however, about relationships between fertilization and disease predisposition. Observations on foliar diseases of forest and shade trees often refer to rusts and mildews (Baule and Fricker, 1967; Poteri and Rousi, 1996), and it is generally claimed that vigorous trees are more susceptible to obligate parasites but less susceptible to facultative parasites (Hesterberg and Jurgensen, 1972). Many examples illustrate these relationships. One concerns the relationship between the potassium (K) content of poplar leaves and the number of infections by the leaf spot pathogen *D. punctiformis*; leaves with low K content were infected more often than leaves with higher K (van der Meiden, 1964). In quickly growing poplar, a balanced and sufficient nutritional supply protected trees from severe attack by *D. punctiformis* (Garbaye and Pinon, 1973). Higher nitrogen fertilization enhanced the resistance of black walnut to anthracnose caused by *G. leptostyla* (Neely, 1981, 1986) and also that of flowering dogwood (*Cornus florida* L.) to Cercospora leaf spot caused by *Pseudocercospora cornicola* (Tracy & Earle) Y.L. Guo & X.J. Liu (Hagan *et al.*, 2008). The form of nitrogen fertilizer used may affect disease incidence too (Huber and Watson, 1974). Higher phosphorus fertilization and the application of trace elements decreased the susceptibility of *Eucalyptus* spp. to the leaf spot pathogens *Mycosphaerella cryptica* and *Phaeoseptoria eucalypti* (Nichol *et al.*, 1992; Carnegie and Ades, 2001). Studies of Carnegie and Ades (2001) suggest that a low level of phosphorus may predispose Tasmanian bluegum (*Eucalyptus globulus* Labill.) to infection by *M. cryptica*.

Biotic factors

The most important biotic factors include insects, fungi and bacteria. The significance of insects as vectors of inoculum is slight for foliar diseases. Some foliar pathogens may use wounds caused by insects feeding on leaves as routes for infection, or benefit from the weakened condition of plants damaged by insects, even if the damage occurs in a plant part other than the leaves (Stanosz, 1993; Butin, 1995; Sinclair and Lyon, 2005).

Insects and mites that cause the formation of galls are a significant group but are involved only indirectly in the epidemiology of foliar diseases. Gall-forming invertebrates (see Table 24.2), may damage leaves and stimulate the activity of some endophytic fungi, which may then cause necrosis, produce secondary inoculum and increase the potential risk for other pathogens (Butin, 1992; Pehl and Butin, 1994; Wulf, 1994; Kowalski, 2003, 2004; Sieber, 2007).

The infection of leaves by fungal pathogens is affected by fungi and bacteria inhabiting the phylloplane. Their survival in the phylloplane is influenced by climate and environmental factors such as moisture, pH, light, temperature and nutrient availability (Lindow and Brandl, 2003; Lindow, 2006).

24.5 Management Strategies and Tactics

24.5.1 Avoidance

Increasing knowledge of interactions among host, pathogen and environment makes it possible to apply certain principles to reduce losses from epidemic diseases (Berger, 1977).

If weather conditions or other factors in a particular region tend to increase a plant's susceptibility to a pathogen, it is advisable to avoid the cultivation of that plant in the region. In Europe, poplar trees should no longer be grown for wood production in the Alps and surrouding regions at elevations of 200–800 m a.s.l., or in the eastern Po valley, where *Venturia populina* is especially aggressive (Cellerino, 1999). In New Zealand, the cultivation of poplar and willow clones susceptible to species of *Marssonina* should be restricted to districts with dry summers where infection is reduced by host plant escape (Spiers, 1998). There are many examples of disease avoidance in the cultivation of forest trees, including for species of *Populus, Platanus, Ulmus* and *Juglans* (Santamour, 1976, 1984; Strobl and Fraser, 1989; Smith *et al.*, 2002; Sinclair and Lyon, 2005).

Locations such as cold pockets or other situations that favour dew formation should be avoided when establishing new plantations (Lundquist and Purnell, 1987; Smith *et al.*, 2007).

24.5.2 Exclusion

Historical analysis of the appearance of pathogens causing serious epidemics and losses in different continents makes it possible to identify pathogens as native or non-native, i.e. already present, or introduced from other regions or continents (Sturrock *et al.*, 2011). Many recent reports emphasize the impact of invasions of alien pathogen species (Desprez-Loustau *et al.*, 2007; Jones and Baker, 2007; Kirisits, 2010; Pautasso *et al.*, 2010). In Europe, the number of non-native pathogen species differs among countries and ranges from seven (in Latvia) to 198 (in the UK) (Desprez-Loustau *et al.*, 2007). In the USA, 49% of all at-risk tree species are threatened in part by non-native pathogens (Wilcove *et al.*, 1998). Several dangerous, non-native foliar pathogens have also been introduced to Australia (Brown, 1984; Spiers, 1998).

Exclusion can be achieved by quarantine regulations imposed at both national and international levels. The list of foliar pathogens of broadleaved trees subject to quarantine includes *Davidiella populorum* and *Stegophora ulmea*, which are absent from Europe. According to Cellerino (1999), countries at risk from the accidental introduction of *D. populorum* must apply severe quarantine measures and set strict limits on the exchange of biological material. Countries in north and central Europe, which grow many of the poplar hybrids from sections *Aigeiros* × *Tacamahaca*, which have been shown to be susceptible to *D. populorum* in North America, are at particular risk. Europe also grows the European white elm (*Ulmus laevis* Pall.) and the Wych elm (*U. glabra* Huds.), which are highly susceptible to *S. ulmea*, and so is at risk from this pathogen (EPPO, 2005). More details about phytosanitary regulations are given by MacLeod *et al.* (2010).

It is claimed that a substantial reduction in the international transfer of plants is the best way of reducing risk from the introduction of pathogens to new geographic regions (Waggoner and Aylor, 2000; Brasier, 2008). Monitoring is particularly important in the case of endophytes present on imported symptomless plants and which may present a threat in new regions (Sieber, 2007). Local certification programmes should ensure the use of pathogen-free seeds and other planting material (Cook, 2000; Mielke and Ostry, 2004) after short or long distance continental transfer of foliar pathogens present in seeds, cuttings or plants (Spiers, 1998; Mottet *et al.*, 2007). Other measures that can be applied as part of an exclusion strategy include elements of biological control. Some of the aspects mentioned here are shared by other strategies, e.g. an antagonist that eliminates or controls a pathogen may be included as part of either exclusion or protection strategies, respectively.

24.5.3 Eradication

The removal and destruction of fallen leaves is recommended to reduce the source of inoculum for most foliar diseases (Alfieri, 1978; Butin, 1995; Mottet *et al.*, 2007). Where new plantations are being established, the inoculum source should be removed before planting. In established plantations, inoculum sources should be removed in the spring, before the development of new leaves. Leaves that fall early in the growing season because of high disease severity should also be removed. The leaves can be raked and burnt, buried or ploughed (Sivanesan, 1990). Many foliar pathogens, particularly those that cause anthracnose, scab and shoot blight, can overwinter in cankers and on the dead twigs and branches that make up an important source of inoculum. In these cases, symptomatic parts and tissues should be pruned and removed completely by the spring. This practice reduces the number of fungal spores available and prevents new infections on the emerging shoots and leaves in that season.

Infected and dead parts of plants can also be removed in the growing season. This not only helps to eliminate primary inoculum but also reduces the substrate available for pathogen reproduction. Pruning is particularly important in the control of diseases on nursery stock. It is economically and aesthetically important also in plantations and in groups of ornamental and shade trees.

Eradication can be achieved only rarely in forest stands; an example is the early stage of invasion of a non-native pathogen that would cause a new disease in a restricted area. The elimination of a host plant in such a case is possible and can be justified both ecologically and economically. This is illustrated by the elimination of *Ulmus* bonsais infected by *Stegophora ulmea*, the causal agent of elm black spot. The pathogen was introduced to Europe on elm bonsais in 2000. Eradication has prevented the spread of *S. ulmea* to cultivated *Ulmus* spp., which are highly susceptible to the disease (EPPO, 2005).

The eradication of eastern cottonwood trees in the region around Normandin, in Quebec (Canada), has helped to control *Davidiella populorum*, which causes Septoria leaf spot. Until recently, the distribution of *D. populorum* was sympatric with the northern limit of the range of eastern cottonwood (Feau *et al.*, 2010), but since 2000 the pathogen has spread beyond its northern range. The eradication of eastern cottonwood outside the usual range of *D. populorum* consisted of cutting down the trees, burying the litter layer and destroying the stumps, sprouts and suckers. These eradication practices almost completely eliminated the pathogen within a year (Mottet *et al.*, 2007).

24.5.4 Protection

Protection can be achieved in many ways. The application of protectant fungicides before infection may be used where foliar pathogens are a threat. Successful crop protection depends on the appropriate choice of chemical product, and the amount used per hectare and per year, which is determined by the dosage and number of applications, and the application technique.

Bordeaux mixture was the first foliar fungicide (De Waard *et al.*, 1993). New fungicides, which are more effective, less harmful to the environment, easier to apply and cheaper have been continually developed. New approaches to chemical control often result from better understanding of the pathogen's biology, especially its life cycle. Some examples of the chemical control of foliar pathogens of broadleaved trees are discussed below. However, for environmental and economic reasons, the possibilities for the application of fungicides in forestry is very limited. Fungicides are applied mostly in nurseries and valuable young stands, but also in short-rotation plantations and to valuable ornamentals, and where disease occurs in recently transplanted trees.

Poplar plantations, which are susceptible to *Davidiella populorum*, *Drepanopeziza punctiformis* or *D. populi-albae*, have been treated with benomyl, chlorothalonil, captafol, basic copper chloride, thiophanate methyl, and dithiocarbamates such as maneb or mancozeb (Carlson, 1974; Spiers and Wenilam, 1983; Ostry, 1987; Cellerino, 1999; Feau *et al.*, 2010). Despite the positive effects of their application, it is claimed that chemical control can be less effective in conditions of high pathogen density and high crown density (Ostry, 1987). Effective chemical control of *Stegophora ulmea* on *Ulmus* spp. has been achieved with copper, as Bordeaux mixture, and with various forms of sulfur, benomyl or chlorothalonil (Alfieri, 1978). Effective chemical control of most anthracnose pathogens can be achieved by spraying with fungicides containing mancozeb, chlorothalonil, copper sulfate, copper hydroxide or Bordeaux mixture, or lime sulfur. The first spray should be applied at bud break and then twice or three times more during the growing season at 12–14 day intervals. Specific recommendations on the use of fungicides, depending on species of host and causal agents, and the locality, should be followed (Jacobi, 2003). One to two sprays applied in the autumn at the end of leaf fall are recommended to prevent infection through leaf scars (Paetzholdt and Sell, 1977).

Chlorothalonil has been used effectively for control of a broad range of fungal pathogens, including *Mycosphaerella cryptica* and *Aulographina eucalypti* in young eucalyptus plantations and nurseries in Australia and New Zealand (Carnegie and Ades, 2003). The non-systemic fungicide thiram gave excellent control of *Calonectria morganii* Crous, Alfenas & M.J. Wingf., also a cause of disease on eucalyptus trees (Peerally, 1991). In India, fungicides were evaluated *in vitro* for the control of *Cylindrocladium* spp. (Sharma and Mohanan, 1991). Mancozeb, followed by propineb, prochloraz, dichlofluanid and captafol were the most effective inhibitors of *Alternaria tenuissima* (Kunze) Wiltshire, the cause of a leaf spot disorder of *Pseudopanax* 'Gold Splash' (a native shrub of New Zealand with export potential) in greenhouse tests (Everett and Neilson, 1996). The control of *Taphrina* spp. requires the application of fungicide during the dormant period, from January to early April. A spray in the autumn can also be justified. Later applications, after bud opening and leaf emergence, were ineffective (Tainter and Baker, 1996). Entirely different treatment timing would be needed for the protection of ash leaves and stems against *Hymenoscyphus pseudoalbidus*. This is because of the time of ascospore liberation and dispersal, which occur mainly between early July and late September (Kowalski and Holdenrieder, 2009; Queloz *et al.*, 2011).

There is a recent tendency to develop and produce a diverse range of chemical products with novel modes of action. These new classes of fungicides, which include anilinopyrimidines, phenoxyquinolines, oxazolidinediones, spiroketalamines, scytalone dehydratase inhibitors, phenylpyrroles, strobilurins and activators of systemic acquired resistance are expected to have significant impact on disease control in the future (Gullino *et al.*, 2000).

Cultural practices are among the most effective approaches to minimizing the impact of foliar pathogens on plants. They are important in nurseries, forest stands, plantations and groups of shade and ornamental trees. Most cultural practices should tend to reduce free moisture on leaves. Increased humidity, caused by reduced wind flow (airflow) in the canopy, may increase the susceptibility of trees in dense stands to foliar pathogens. Pruning tends to thin dense tree canopies. It reduces leaf shading between trees, increases air circulation, reduces moisture within the canopy, improves the drying of leaf surfaces and reduces the number of dead leaves collecting in the canopy. Weeds, which can increase the relative humidity around seedlings, should be removed mechanically or chemically (Mielke and Ostry, 2004). To help the crowns develop, increased row spacing or increased separation of small groups of trees should be applied when introducing new trees to existing groups of trees. In most parts of Europe, planting trees in rows in an east–west direction will enable the wind to blow through.

Overall tree vigour should be maintained by fertilizing, mulching and watering. Irrigation systems that increase the wetness of leaves should be avoided (Paetzholdt and Sell, 1977). Improved plant growth and mineral nutrition, tolerance of stresses and diseases, including those caused by foliar pathogens, and decreased predisposition to diseases, can all result partly from mycorrhizal associations. Some cultural practices may affect various processes contributing to mycorrhizal formation. The benefits of habitat modifications aimed at mycorrhizal formation, or direct inoculation with mycorrhizal fungi, have been documented for a wide range of tree species in different countries (Smith and Read, 2007).

Decreased disease severity may be achieved by appropriate fertilization, which should, however, be adjusted to the local habitat and forestry practices (Hagan *et al.*, 2008). Fertilization has been an accepted agronomic practice for many decades, and numerous investigations have shown its effects on disease incidence (Hesterberg and Jurgensen, 1972). In forestry, most studies on the effects of fertilization have been centred on intensively managed nurseries, although some of them also include forest stands and shade trees (Baule and Fricker, 1967).

No general rule can be made about the relationship between fertilization and disease resistance. Each case must be assessed individually, and depends on soil type, soil fertility and pH. The main effects of fertilizers on diseases are indirect and include the modification of pathogen metabolism (Hesterberg and Jurgensen, 1972). Biotrophic and necrotrophic fungi react differently to the physiological state of the host plant, and detailed knowledge of the threat (biotroph or necrotroph) is essential for further management. Nitrogen fertilization requires experience and knowledge of the effects of different forms of nitrogen on the severity of particular diseases (Huber and Watson, 1974). The effects of macro-elements and trace elements, particularly if they can decrease disease severity, should be included in decisions on further management (van der Meiden, 1964; Garbaye and Pinon, 1973; Neely, 1981, 1986; Nichol *et al.*, 1992; Carnegie and Ades, 2001; Hagan *et al.*, 2008).

Legislation that restricts fungicide use leads to increased interest in the biological control of plant diseases (Boland, 1990; Butt *et al.*, 2001). There are various naturally occurring phenomena, including hyperparasitism, antibiosis, competition for nutrients, induction of plant resistance and improvement of plant vigour which can be applied directly or indirectly in biological control of foliar pathogens (Fravel, 1988; Campbell, 1989). Hyperparasitism has been reported mostly in rusts and mildews (Kranz, 1981). Antibiosis and antagonists operating in competition for nutrients are observed in the phyllosphere, rhizosphere, and in living symptomless and symptomatic (necrotic) leaf tissues (Andrews, 1992; Köhl and Fokkema, 1994; Van Loon *et al.*, 1998; Sieber, 2007).

Biocontrol in the phyllosphere is aimed at the prevention of infection by the pathogen or the suppression of pathogen inoculum (Jacobsen, 2006). The shortness of time available in which a pathogen and antagonistic microbial population can interact on or in a healthy leaf before successful penetration (usually a few hours) is a factor limiting successful biocontrol of foliar pathogens (Köhl and Fokkema, 1994). Successful colonization of the phyllosphere by bacteria and fungi is affected by weather-related environmental factors, e.g. moisture and temperature, and by light and nutrient availability (Köhl and Fokkema, 1994; Lindow and Brandl, 2003; Lindow, 2006). The common phyllosphere microbiota may include *Ascomycota* and *Basidiomycota* yeast genera, e.g. *Aureobasidium*, *Cryptococcus*, *Rhodotorula* and *Sporobolomyces*, and filamentous fungi, particularly in the genera *Cladosporium*, *Epicoccum* and *Trichothecium*, with different densities and diversity depending on the plant species (Köhl and Fokkema, 1994; Yang *et al.*, 1994). Gram-positive rod-shaped bacteria in the genus *Bacillus* are among the important antagonists of foliar pathogens (Omoifo and Ikotun, 1987; Zhang *et al.*, 2008). Several commercial antagonist-based products are registered and manufactured (Demoz and Korsten, 2006). *B. subtilis* QST strain (Serenade) has been used as a biocontrol agent against anthracnose diseases in the USA (Douglas, 2008). *Trichoderma polysporum* (Link) Rifai, *T. harzianum* Rifai, *Clonostachys rosea* f. *catenulata* (J.C. Gilman & E.V. Abbott) Schroers and *Streptomyces griseoviridis* are active ingredients in the commercial products BinabT, Gliomix and Mycostop, which are used successfully against *Botrytis cinerea* in Sweden (Capieau, 2004).

Although the number of biocontrol products is increasing, they still represent only about 1% of sales of all crop protection products (Fravel, 2005). The scale of biological control in forestry practice is currently also small (Cook, 1993). However, considerable advances are being made to improve biological control through matching pathogens, their antagonists and climates, and increasing the number of antagonists or modifying their habitat. New strategies and technologies, including the genetic engineering of antagonists, give new perspectives and promise even greater advances in the near future (Lorito *et al.*, 2010).

24.5.5 Resistance

Genetic resistance to infection resulting from the co-evolution of forest trees and their pathogens has been chiefly responsible for disease control in natural ecosystems. The problem of resistance of trees to foliar pathogens is very extensive because of the broad range of mechanisms that contribute to this phenomenon (Van Loon *et al.*, 1998; Punja, 2001; Eyles *et al.*, 2010). Aspects that should be considered concern adaptive resistance, resistance by escape (Newcombe, 1998a; Feau *et al.*, 2010), and spatial resistance, the latter described by Ostry and Anderson (2009) as the interactions of environmental and biotic factors with tree density that influence disease incidence.

It is well known that monocultures put plants at risk from various diseases (El-Ghany *et al.*, 2009). Infection can be reduced by planting mixtures of several cultivars or species (Spiers, 1998). An example is the experience that has been gained from eucalyptus plantations, which provides a basis for practical control of leaf diseases. The Tasmanian bluegum and the shining gum are the main *Eucalyptus* species chosen for cultivation in cool temperate zones in Australia (Smith *et al.*, 2007). These species have different susceptibilities to species of *Mycosphaerella*. Tasmanian bluegum exhibits a range of symptoms, while shining gum, in the same planting areas, appears to be tolerant. Although the cultivation of Tasmanian bluegum in high-risk areas would be preferable from a silvicultural point of view, shining gum is grown more often, in the expectation of avoiding potential problems caused by *Mycosphaerella* spp. (Smith *et al.*, 2007).

In Tasmania, particularly in the north-western region, serious defoliation of native eucalyptus trees infected by *Mycosphaerella* spp. has been associated with wet and humid conditions. Intensive production of timber there requires the development of management strategies to avoid or reduce the impact of these pathogens. A detailed taxonomic survey of *Mycosphaerella* spp. from young eucalyptus plantations was the first step in recognizing the potential risk in Tasmania (Milgate *et al.*, 2001). Also in Australia – in New South Wales, blackbutt (*Eucalyptus pilularis* Sm.) is an important commercial plantation species. It tends to suffer less from leaf damage by insects and foliar diseases than do species in the subgenus *Symphyomyrtus*.

Its greater resistance, together with the inherent quality of its wood, makes it an ideal species for growing in its natural range, where insect damage and fungal diseases are most likely to occur (Henson and Smith, 2007).

Differences in susceptibility to *Mycosphaerella* spp. among and between provenances of shining gum have been recorded in South Africa. Selection of the more resistant provenances seems to be the solution to avoid severe damage by *Mycosphaerella* spp. (Lundquist and Purnell, 1987). As well as this, cultivation of Tasmanian bluegum has been largely abandoned on high-altitude sites in this region, where it can be damaged by frost (Lundquist and Purnell, 1987). It should be emphasized that cultivars of forest and shade trees susceptible to foliar pathogens have sometimes been obtained as a consequence of successful breeding for resistance to other pathogens.

Although a range of different strategies for protecting trees from pathogens might be available, resistant varieties or clones are often considered the best approach to preventing disease (Ostry, 1987; Royle and Ostry, 1995; Punja, 2001; Carnegie and Ades, 2002; Noël *et al.*, 2005; Mottet *et al.*, 2007). For this reason, there are numerous breeding programmes for improved resistance to foliar pathogens in trees worldwide. Poplar has been the object of the most intensive tree improvement programmes because of the worldwide importance of the genus and its hybrids, in forests and plantations, and their susceptibility to many fungal pathogens (Royle and Ostry, 1995; Feau *et al.*, 2010). The number of poplar cultivars registered for commercial use in all European Union (EU) countries has increased to 130. Observations on their behaviour towards major pathogens, including *Drepanopeziza punctiformis*, have been recorded and reported (Pinon and Valadon, 1997). In Quebec, a poplar breeding programme aimed at the development and selection of poplar clones adapted to local climatic conditions was instituted in 1969. Some 5000 clones have been evaluated since then, and 40 of them have been recommended in several regions where 1500 ha of poplars are being planted each year. There has also been intensive tree improvement work on poplar in other provinces of Canada (Feau *et al.*, 2010).

Selection and breeding programmes usually deal with resistance to the most destructive foliar pathogens, but there are also examples of classical plant breeding and of the application of genetic engineering, which are mentioned now. Laborious cross-breeding between the American sycamore (*Platanus occidentalis* L.) and the oriental plane tree (*P. orientalis* L.) has resulted in two cultivars (Columbia and Liberty) that are highly resistant to *Apiognomonia veneta* (Santamour, 1976, 1984). An investigation into genetic resistance of black walnut to *Gnomonia leptostyla*, and the selection of resistant clones for use as parental stock in seed orchards began after the failure to register benomyl-based fungicide, which, although very effective, could not be applied at the fruit-bearing stage. Significant differences in susceptibility of black walnut clones to *G. leptostyla* were observed (Funk *et al.*, 1981).

Inoculation experiments with 15 poplar species in China showed that some of them are very susceptible while others are slightly, moderately or highly resistant to *D. punctiformis*. Resistance was, remarkably, negatively correlated with iron (Fe) and calcium (Ca) contents (Yu *et al.*, 1997). Ostry *et al.* (1988), using an *in vitro* leaf disk bioassay, was able to distinguish poplar clones that were highly, moderately or only slightly resistant to *Davidiella populorum*. Resistant clones can be used in further breeding and selection.

The resistance of forest trees to fungal pathogens is a genetically determined trait and, as such, it can be modified by genetic transformation. While plant–pathogen molecular interactions, resulting from their long- term co-evolution, are complex, various plant resistance mechanisms, as well as ways of overcoming fungal resistance exist. Numerous plant, fungal or bacterial genes may, therefore, be used in order to enhance the resistance of forest trees to fungal disease. Examples are: the wheat-derived oxalate oxidase (OxO) (Liang *et al.*, 2001) gene; the amaranth (*Amaranthus*)-derived Ac-AMP1.2 and synthetic ESF12 antimicrobial peptide genes (Liang *et al.*, 2002); the motherwort (*Leonurus japonicus* Houttuyn)-derived LJAMP2 antimicrobial peptide gene (Jia *et al.*, 2010) for poplar; the sugarbeet

(*Beta vulgaris* L.)-derived chitinase 4 cDNA for birch (Pappinen *et al.*, 2002); and the endochitinase (ech42) gene derived from the biocontrol fungus *T. harzianum* for spruce and poplar (Noël *et al.* 2005). Once transferred into the host genome, all of these genes have been proven to increase resistance against various pathogens as compared with control plants. Moreover, a non-specific resistance mechanism may be introduced by transformation with a single gene only that improves resistance against two or more unrelated pathogens at once, as demonstrated by Noël *et al.* (2005). That makes the resistance enhancement of trees via genetic transformation a potentially very effective tool – a tool that can provide an alternative to fungicide-based chemical control and can help to reduce losses caused by phytopathogenic fungi. The advances made by the application of a broad range of cloned genes to enhance the resistance of transgenic plants against fungal pathogens, as well as the challenges of genetic engineering and its prospects, have been reviewed by Punja (2001).

24.5.6 Therapy

Therapy for foliar diseases can be applied only on a limited scale in forests. It can be justified in nurseries, plantations or valuable shade and ornamental trees. Many positive effects can be achieved by therapy that uses the application of systemic fungicides, which can be applied to the soil, seeds or foliage. Systemic chemicals may be eradicants or protectants. If eradicants are applied early (e.g. at the beginning of leaf development) or, alternatively, protectants are applied, it is difficult to distinguish between the level of preventive therapy and remedial therapy achieved. Nevertheless it is important to alternate fungicides with different modes of action to delay the development of pathogen strains resistant to novel fungicides.

Systemic fungicides have both advantages and disadvantages. Several remain, however, as useful components of crop protection programmes. The efficacy of the systemic fungicide benomyl in controlling epidemics of *Venturia macularis* on quaking aspen in northern Minnesota (USA) was tested in the late 1970s (Anderson and Anderson, 1980). Trees were sprayed with the fungicide at 2.4 g l^{-1} at 2–3 week intervals during the growing season. Increased growth in height, decreased shoot blight and superior form of the sprayed trees, compared with the formation of multiple leaders in unsprayed trees, were observed. It was noted that the final spray in most years could be applied about 1 July.

The efficacy of the protectant fungicides chlorothalonil and benomyl in reducing infection by *Mycosphaerella* spp. in young eucalyptus plantations in southern Australia has been studied (Carnegie and Ades, 2003). Trees were treated with a mixture of chlorothalonil (2.5 kg ha^{-1}) and benomyl (0.5 kg ha^{-1}) every 3 weeks, from October 1992 to May 1993. The mixture of the two fungicides was effective in controlling *Mycosphaerella* spp. in Tasmanian bluegum plantations. Poplar plantations infected by *Drepanopeziza punctiformis* or *D. populi-albae* can be sprayed with systemic fungicides, including triforine, dodine and triadimenol (Cellerino, 1999).

Systemic fungicides can be applied for the effective elimination of leaf endophytes. Treatment can be justified particularly against endophytes that tend to change their lifestyle to become pathogenic after trees are weakened; for example, these fungicides can be used against the causal agents of anthracnoses (Sieber, 2007). The therapeutic effects of systemic fungicides can be achieved not only by spraying but also by injection. This technique can be applied to protect large, highly valuable shade trees. The effectiveness of injections depends not only on the chemical product but also on the timing of its application.

The removal of diseased branches may be an effective therapeutic option against those pathogens that infect stems and branches as well as leaves. A therapeutic use of physical methods, e.g. thermotherapy, should also be considered for destroying foliar pathogens present in seeds or cuttings to prevent their transportation over long distances (Spiers, 1998).

24.5.7 Integrated disease management

The first and most important step in managing a plant disease is the correct identification of the causal agent. This can be difficult, particularly with fungal genera that have many species or with a disease that can have different causal agents. A method for quick and simple identification should be developed (Cai *et al.*, 2009). Understanding the pathogen's life cycle is the next essential step. This will allow the occurrence of the pathogen to be monitored. Increasing global trade may increase the chance of encountering some non-native and introduced pathogens. The effective monitoring of populations may help the discovery of introduced pathogens in their initial stage of spread. More accurate forecasting systems for plant disease should be developed to help forest workers to determine when to apply control measures to suppress tree diseases. Systems for early disease forecasting are based mainly on correlations between disease occurrence and specific environmental conditions (Berger, 1977). Current disease forecasting systems should be part of any chemical control programme and include inputs intended to improve fungicide efficiency. Inputs would result from evaluation of inoculum potential, disease assessments, the construction of temperature-driven models for the prediction of pathogen development or activity, monitoring weather for the prediction of pathogen behaviour, the identification of conditions for infection and economic injury, and action thresholds to aid in decision making for applying fungicides.

Action thresholds for controlling foliar diseases in broadleaved trees depend on two considerations: economic and aesthetic. Both are determined by the point at which the amount of pathogen infestation makes it necessary – from an economic or aesthetic point of view – to apply control measures. In either case, a combination of management procedures appropriate for the particular situation and conditions should be used. Options include mechanical, physical, cultural and biological controls. The component procedures of integrated disease management (IDM) can be applied either simultaneously or with spatial or temporal differences. Chemical control should be an additional option, when others tactics have failed. Before applying costly treatments, which may also have side effects for the environment, one should keep in mind that the severity of leaf diseases frequently differs between years and that many trees can compensate for leaf losses very well.

The chosen procedure may be a component of one or more of the strategies mentioned below. Mechanical procedures for the control of foliar diseases in broadleaved trees include the removal of infected leaves and branches that are sources of pathogen inoculum, or removal of the pathogen itself from a site (Alfieri, 1978; Sivanesan, 1990; Butin, 1995; Mottet *et al.*, 2007). This would be part of an eradication strategy. Physical methods, e.g. thermotherapy of seeds or cuttings, applied for the elimination of endophytes that are a potential threat to leaves, are part of a therapeutic strategy. Cultural practices usually include planting on sites with no history of disease (avoidance strategy), thinning by pruning the lower branches for better air circulation or fertilizing (protection strategy), and breeding for resistance (resistance strategy). Chemical control forms part of a protection strategy or a therapeutic strategy. Biological control, which can be considered as resulting from an effect on either the pathogen or plant, can be, respectively, part of an exclusion strategy or a protection strategy.

The concept of integrated control, as IDM, extends to many classes of pests and pathogens, and therefore also includes tactics additional to those discussed above (Kogan, 1998; Gray *et al.*, 2009).

References

Alfieri, S.A. (1978) *Black Spot of Elm*. Plant Pathology Circular 189, Florida Department of Agriculture and Consumer Services, Gainesville, Florida.

Anderson, N.A. and Anderson, R.L. (1980) Leaf and shoot blight of aspen caused by *Venturia macularis* in northern Minnesota. *Plant Disease* 64, 558–559.

Andrews, J.H. (1992) Biological control in the phyllosphere. *Annual Review of Phytopathology* 30, 603–635.

Aylor, D.E. (1990) The role of intermittent wind in the dispersal of fungal pathogens. *Annual Review of Phytopathology* 28, 73–92.
Bacigálová, K., Lopandic, K., Rodrigues, M.G., Fonseca, A., Herzberg, M., Pinsker, W. and Prillinger, H. (2003) Phenotypic and genotypic identification and phylogenetic characterisation of *Taphrina* fungi on alder. *Mycological Progress* 2, 179–196.
Baule, H. and Fricker, C. (1967) *Die Düngung von Waldbäumen* [*The Fertilization of Forest Trees*]. BLV Bayerischer Landwirtschaftsverlag GmbH, Munich, Germany.
Beare, J.A., Archer, S.A. and Bell, J.N.B. (1999) *Marssonina* leafspot disease of poplar under elevated ozone: pre-fumigated host and *in vitro* studies. *Environmental Pollution* 105, 409–417.
Berger, R.D. (1977) Application of epidemiological principles to achieve plant disease control. *Annual Review of Phytopathology* 15, 165–183.
Black, W.M. and Neely, D. (1978) Effects of temperature, free moisture, and relative humidity on the occurrence of walnut anthracnose. *Phytopathology* 68, 1054–1056.
Boland, G.J. (1990) Biological control of plant diseases with fungal antagonists: challenges and opportunities. *Canadian Journal of Plant Pathology* 12, 295–299.
Booth, T.H., Jovanovic, T., Old, K.M. and Dubzinski, M.J. (2000) Climatic mapping to identify high-risk areas for *Cylindrocladium quinqueseptatum* leaf blight on eucalypts in mainland South East Asia and around the world. *Environmental Pollution* 108, 365–372.
Boyer, J.S. (1995) Biochemical and biophysical aspects of water deficits and the predisposition to disease. *Annual Review of Phytopathology* 33, 251–274.
Brasier, C.M. (2001) Rapid evolution of introduced plant pathogens via interspecific hybridization. *BioScience* 51, 123–133.
Brasier, C.M. (2008) The biosecurity threat to the UK and global environment from international trade in plants. *Plant Pathology* 57, 792–808.
Brown, J.F. (1997) Airborne inoculum. In: Brown, J.F. and Ogle, H.J. (eds) *Plant Pathogens and Plant Diseases*. Rockvale Publications, Armidale, New South Wales, Australia, pp. 207–218.
Brown, J.S. (1984) Recent invasions of Australia and New Zealand by pathogenic fungi and counter measures. *EPPO Bulletin* 14, 417–428.
Butin, H. (1960) Die Krankheiten der Weide und deren Erreger [The diseases of willow and their pathogens]. *Mitteilungen aus der Biologischen Bundesanstalt für Land- und Forstwirtschaft Berlin-Dahlem* 98, 1–46.
Butin, H. (1992) Effect of endophytic fungi from oak (*Quercus robur* L.) on mortality of leaf inhabiting gall insects. *European Journal of Forest Pathology* 22, 237–246.
Butin, H. (1995) *Tree Diseases and Disorders: Causes, Biology and Control in Forest and Amenity Trees*. Oxford University Press, Oxford, UK.
Butin, H. and Kehr, R. (1995) Leaf blotch of lime associated with *Asteromella tiliae* comb. nov. and the latter's connection to *Didymosphaeria petrakiana*. *Mycological Research* 99, 1191–1194.
Butin, H. and Kehr, R. (1998) *Gloeosporidina platani* sp. nov., the spermatial state of the anthracnose fungus *Apiognomonia veneta* (Sacc. & Speg.) Höhn. *European Journal of Forest Pathology* 28, 297–305.
Butt, T.M., Jackson, C. and Magan, N. (2001) Introduction – Fungal biological control agents: progress, problems and potential. In: Butt, T.M., Jackson, C. and Magan, N. (eds) *Fungi as Biocontrol Agents*. CAB International, Wallingford, UK, pp. 1–8.
Bytnerowicz, A. (1996) Physiological aspects of air pollution stress in forests. *Phyton* 36, 15–22.
Cai, L. *et al.* (2009) A polyphasic approach for studying *Colletotrichum*. *Fungal Diversity* 39, 183–204.
Campbell, R. (1989) *Biological Control of Microbial Plant Pathogens*. Cambridge University Press, Cambridge, UK.
Capieau, K. (2004) Biological control of grey mould in Swedish forest nurseries. PhD thesis, Swedish University of Agricultural Sciences, Uppsala, Sweden.
Carlson, L.W. (1974) Fungicidal control of poplar leaf spots. *Canadian Plant Disease Survey* 54, 81–85.
Carnegie, A.J. and Ades, P.K. (2001) Added phosphorus is associated with reduced severity of *Mycosphaerella cryptica* in *Eucalyptus globulus*. *Australian Forestry* 64, 203–208.
Carnegie, A.J. and Ades, P.K. (2002) The proportion of leaf spots caused by *Mycosphaerella cryptica* and *M. nubilosa* on *Eucalyptus globulus*, *E. nitens* and their F1 hybrids in a family trial in Tasmania, Australia. *Australasian Mycologist* 21, 53–63.
Carnegie, A.J. and Ades, P.K. (2003) *Mycosphaerella* leaf disease reduces growth of plantation-grown *Eucalyptus globulus*. *Australian Forestry* 66, 113–119.
Carnegie, A.J. and Keane, P.J. (2003) Variation in severity of target spot, caused by *Aulographina eucalypti*, in a eucalypt species and provenance trial in Victoria. *Australasian Plant Pathology* 32, 393–402.

Cellerino, G.P. (1999) *Review of Fungal Diseases in Poplar*. Food and Agriculture Organization of the United Nations, Rome, Italy.

Christersson, L. (2006) Biomass production of intensively grown poplars in the southernmost part of Sweden: observations of characters, traits and growth potential. *Biomass and Bioenergy* 30, 497–508.

Cline, S. and Neely, D. (1984) Relationship between juvenile-leaf resistance to anthracnose and the presence of juglone and hydrojuglone glucoside in black walnut. *Phytopathology* 74, 185–188.

Coleman, J.S., Jones, C.G. and Smith, W.H. (1988) Interactions between an acute ozone dose, eastern cottonwood, and *Marssonina* leaf spot: implications for pathogen community dynamics. *Canadian Journal of Botany* 66, 863–868.

Colhoun, J. (1973) Effects of environmental factors on plant disease. *Annual Review of Phytopathology* 11, 343–364.

Cook, R.J. (1993) Making greater use of introduced microorganisms for biological control of plant pathogens. *Annual Review of Phytopathology* 31, 53–80.

Cook, R.J. (2000) Advances in plant health management in the twentieth century. *Annual Review of Phytopathology* 38, 95–116.

Crous, P.W. (2009) Taxonomy and phylogeny of the genus *Mycosphaerella* and its anamorphs. *Fungal Diversity* 38, 1–24.

Dance, B.W. (1961) Spore dispersal in *Pollaccia radiosa* (Lib.) Bald. and Cif. *Canadian Journal of Botany* 39, 1429–1435.

Davis, A.A. and Trettin, C.C. (2006) Sycamore and sweetgum plantation productivity on former agricultural land in South Carolina. *Biomass and Bioenergy* 30, 769–777.

Demoz, B.T. and Korsten, L. (2006) *Bacillus subtilis* attachment, colonization, and survival on avocado flowers and its mode of action on stem-end rot pathogens. *Biological Control* 37, 68–74.

Desprez-Loustau, M.-L., Marçais, B., Nageleisen, L.M., Piou, D. and Vannini, A. (2006) Interactive effects of drought and pathogens in forest trees. *Annals of Forest Science* 63, 597–612.

Desprez-Loustau, M.-L., Robin, C., Buée, M., Courtecuisse, R., Garbaye, R., Suffert, F., Sache, I. and Rizzo, D.M. (2007) The fungal dimension of biological invasions. *Trends in Ecology and Evolution* 22, 472–480.

Desprez-Loustau, M.-L., Courtecuisse, R., Robin, C., Husson, C., Moreau, P.A., Blancard, D., Selosse, M.A., Lung-Escarmant, B., Piou, D. and Sache, I. (2010) Species diversity and drivers of spread of alien fungi (*sensu lato*) in Europe with a particular focus on France. *Biological Invasions* 12, 157–172.

De Waard, M.A., Georgopoulos, S.G., Hollomon, D.W., Ishii, H., Leroux, P., Ragsdale, N.N. and Schwinn, F.J. (1993) Chemical control of plant diseases: problems and prospects. *Annual Review of Phytopathology* 31, 403–421.

De Wit, P.J., Mehrabi, R., van den Burg, H.A. and Stergiopoulos, I. (2009) Fungal effector proteins: past, present and future. *Molecular Plant Pathology* 10, 735–747.

Dimitri, L. (1977) Influence of nutrition and application of fertilizers on the resistance of forest plants to fungal diseases. *European Journal of Forest Pathology* 7, 177–186.

Dinoor, A. and Eshed, N. (1984) The role and importance of pathogens in natural plant communities. *Annual Review of Phytopathology* 22, 443–466.

Douglas, S.M. (2008) *Anthracnose Diseases of Trees*. The Connecticut Agricultural Experiment Station, New Haven, Connecticut. Available at: http://www.ct.gov/caes/lib/caes/documents/publications/fact_sheets/plant_pathology_and_ecology/anthracnose_diseases_of_trees_06-03-11r.pdf (accessed 19 December 2012).

Eastburn, D.M., McElrone, A.J. and Bilgin, D.D. (2011) Influence of atmospheric and climatic change on plant–pathogen interactions. *Plant Pathology* 60, 54–69.

El-Ghany, A., Eman, T.M., El-Taher, M. and El-Sheikh, H.H. (2009) Efficacy of fungal rust disease on willow plant in Egypt. *Australian Journal of Basic and Applied Sciences* 3, 1527–1539.

Emmett, R.W. and Parbery, D.G. (1975) Appressoria. *Annual Review of Phytopathology* 13, 147–165.

EPPO (2005) Data sheets on quarantine pests. *Stegophora ulmea*. *EPPO Bulletin* 35, 416–418.

Everett, K.R. and Neilson, H.F. (1996) Evaluation of fungicides for control of *Alternaria* leaf spot of *Pseudopanax*. *New Zealand Journal of Crop and Horticultural Science* 24, 267–272.

Eyles, A., Bonello, P., Ganley, R. and Mohammed, C. (2010) Induced resistance to pests and pathogens in trees. *New Phytologist* 185, 893–908.

Feau, N., Hamelin, R.C., Vandecasteele, C., Stanosz, G.R. and Bernier, L. (2005) Genetic structure of *Mycosphaerella populorum* (anamorph *Septoria musiva*) populations in north central and northeastern North America. *Phytopathology* 95, 608–616.

Feau, N., Mottet, M-J., Périnet, P., Hamelin, R.C. and Bernier, L. (2010) Recent advances related to poplar leaf spot and canker caused by *Septoria musiva*. *Canadian Journal of Plant Pathology* 32, 122–134.

Fravel, D.R. (1988) Role of antibiosis in the biocontrol of plant diseases. *Annual Review of Phytopathology* 26, 75–91.

Fravel, D.R. (2005) Commercialization and implementation of biocontrol. *Annual Review of Phytopathology* 43, 337–359.

Funk, D.T., Neely, D. and Bey, C.F. (1981) Genetic resistance to anthracnose of black walnut. *Silvae Genetica* 30, 115–117.

Garbaye, J. and Pinon, J. (1973) Nutrition minérale et sensibilité à *Marssonina brunnea* (Ell. et Ev.) P. Magn. de *Populus* × *euramericana* (Dode) Guinier, CV. 'I-214': étude préliminaire sur jeunes plants en milieu contrôlé [Mineral nutrition and susceptibility to *Marssonina brunnea* (Ell. et Ev.) P. Magn. of *Populus* × *euramericana* (Dode) Guinier, CV. 'I-214': preliminary study on young plants in a controlled environment]. *Annales des Sciences Forestières* 30, 423–431.

Glazebrook, J. (2005) Contrasting mechanisms of defense against biotrophic and necrotrophic pathogens. *Annual Review of Phytopathology* 43, 205–227.

Grad, B., Kowalski, T. and Kraj, W. (2009) Studies on secondary metabolite produced by *Chalara fraxinea* and its phytotoxic influence on *Fraxinus excelsior*. *Phytopathologia* 54, 61–69.

Gray, M.E., Ratcliffe, T. and Rice, M.E. (2009) The IPM paradigm: concepts, strategies and tactics. In: Radcliffe, E.B., Hutchison, W.D. and Cancelado, R.E. (eds) *Integrated Pest Management: Concepts, Tactics, Strategies and Case Studies*. Cambridge University Press, Cambridge, UK, pp. 1–13.

Grayer, R.J. and Kokubun, T. (2001) Plant–fungal interactions: the search for phytoalexins and other antifungal compounds from higher plants. *Phytochemistry* 56, 253–263.

Griffiths, E. (1981) Iatrogenic plant diseases. *Annual Review of Phytopathology* 19, 69–82.

Gullino, M.L., Leroux, P. and Smith, C.M. (2000) Uses and challenges of novel compounds for plant disease control. *Crop Protection* 19, 1–11.

Hagan, A.K., Akridge, J.R. and Bowen, K.L. (2008) *Influence of Nitrogen Fertilization Rate on Disease Incidence, Flower Bud Set, and Growth of Flowering Dogwood in Alabama*. Bulletin 671, Alabama Agricultural Experiment Station, Auburn, Alabama.

Hanson, R.J. (2008) *The Chemistry of Fungi*. RSC Publishing, Cambridge, UK.

Heagle, A.S. (1973) Interactions between air pollutants and plant parasites. *Annual Review of Phytopathology* 11, 365–388.

Helby, P., Rosenqvist, H. and Roos, A. (2006) Retreat from *Salix*. Swedish experience with energy crops in the 1990s. *Biomass and Bioenergy* 30, 422–427.

Henson, M. and Smith, H.J. (2007) Achievements in forest tree genetic improvement in Australia and New Zealand 1: *Eucalyptus pilularis* Smith tree improvement in Australia. *Australian Forestry* 70, 4–10.

Hesterberg, G.A. and Jurgensen, M.F. (1972) The relation of forest fertilization to disease incidence. *Forestry Chronicle* 48, 92–96.

Holdenrieder, O. and Kowalski, T. (1989) Pycnidial formation and pathogenicity in *Tubakia dryina*. *Mycological Research* 92, 166–169.

Holdenrieder, O., Pautasso, M., Weisberg, P.J. and Lonsdale, D. (2004) Tree diseases and landscape processes: the challenge of landscape pathology. *Trends in Ecology and Evolution* 19, 446–452.

Horbach, R., Navarro-Quesada, A.R., Knogge, W. and Deising, H.B. (2011) When and how to kill a plant cell: infection strategies of plant pathogenic fungi. *Journal of Plant Physiology* 168, 51–62.

Huber, D.W. and Watson, R.D. (1974) Nitrogen form and plant disease. *Annual Review of Phytopathology* 12, 139–165.

Huber, L. and Gillespie, T.J. (1992) Modeling leaf wetness in relation to plant disease epidemiology. *Annual Review of Phytopathology* 30, 553–577.

Innis, M.A., Gelfand, D.H. and Sninsky, J.J. (1999) *PCR Applications: Protocols for Functional Genomics*. Academic Press, San Diego, California.

Jacobi, W.R. (2003) *Aspen and Poplar Leaf Spots*. Fact Sheet 2.920, Colorado State University Cooperative Extension, Fort Collins, Colorado. Available at: http://www.ext.colostate.edu/pubs/garden/02920.html (accessed 19 December 2012).

Jacobsen, B.J. (2006) Biological control of plant diseases by phyllosphere applied biological control agents. In: Bailey, M.J., Lilley, A.K., Timm-Wilson, T.M. and Spencer-Phillips, P.T.N. (eds) *Microbial Ecology of Aerial Plant Surfaces*. CAB International, Wallingford, UK, pp. 133–147.

Jia, Z., Gou, J., Sun, Y., Yuan, L., Tang, Q., Yang, X. and Pei, Y. (2010) Enhanced resistance to fungal pathogens in transgenic *Populus tomentosa* Carr. by overexpression of an nsLTP-like antimicrobial protein gene from motherwort (*Leonurus japonicus*). *Tree Physiology* 30, 1599–1605.

Jones, D.R. and Baker, R.H.A. (2007) Introductions of non-native plant pathogens into Great Britain, 1970–2004. *Plant Pathology* 56, 891–910.

Kaneko, R. and Kaneko, S. (2004) The effect of bagging branches on level of endophytic fungal infection in Japanese beech leaves. *Forest Pathology* 34, 65–78.

Keane, P.J. (1997) Diseases in natural plant communities. In: Brown, J.F. and Ogle, H.J. (eds) *Plant Pathogens and Plant Diseases*. Rockvale Publications, Armidale, New South Wales, Australia, pp. 518–532.

Keane, P.J. and Kerr, A. (1997) Factors affecting disease development. In: Brown, J.F. and Ogle, H.J. (eds) *Plant Pathogens and Plant Diseases*. Rockvale Publications, Armidale, New South Wales, Australia, pp. 287–298.

Kenaley, S.C., Hudler, G.W., O'Brien, D.D., Cameron, K.D. and Smart, L.B. (2010) *Willow Scab*. Department of Cornell University Plant Disease Diagnostic Clinic, Cornell University, Ithaca, New York. Available at: http://plantclinic.cornell.edu/factsheets/willowscab.pdf (accessed 12 December 2012).

Kirisits, T. (2010) Eingeschleppte Krankheitserreger an Waldbäumen und Klimawandel [Introduced pathogens in forest trees and climate change]. In: Rabitsch, W. and Essl, F. (eds) *Aliens: Neobiota und Klimawandel-Eine Verhängnisvolle Affäre?* Bibliothek der Provinz, Weitra, Austria, pp. 59–69.

Knogge, W. (1996) Fungal infection of plants. *Plant Cell* 8, 1711–1722.

Kogan, M. (1998) Integrated pest management: historical perspectives and contemporary developments. *Annual Review of Phytopathology* 43, 243–270.

Köhl, J. and Fokkema, N.J. (1994) Fungal interaction on living and necrotic leaves. In: Blakeman, J.P. and Williamson, B. (eds) *Ecology of Plant Pathogens*. CAB International, Wallingford, UK, pp. 321–334.

Kojwang, H.O. and Kurkela, T. (1984) *Linospora ceuthospora* on aspen (*Populus tremula*) in Finland. *Karstenia* 24, 33–40.

Kosiba, P. (2007) Impact of air pollution on the occurrence of *Rhytisma acerinum* 'tar-spot' on maple leaves. *Acta Societatis Botanicorum Poloniae* 76, 333–343.

Kowalski, T. (2003) Endophytic fungi: V. Mycobiota in living leaves of *Acer pseudoplatanus* and in necrotic tissues around galls of *Drisina glutinosa*. *Phytopathologia Polonica* 29, 23–35.

Kowalski, T. (2004) Endophytic fungi: VI. Mycobiota in living leaves of *Ulmus glabra* and in necrotic tissues associated with gall-making insects. *Phytopathologia Polonica* 32, 61–73.

Kowalski, T. (2006a) *Chalara fraxinea* sp. nov. associated with dieback of ash (*Fraxinus excelsior*) in Poland. *Forest Pathology* 36, 264–270.

Kowalski, T. (2006b) *Tubakia dryina*, symptoms and pathogenicity to *Quercus robur*. *Acta Mycologica* 41, 299–304.

Kowalski, T. and Bartnik, C. (2008) *Cristulariella depraedans* as causal agent of leaf spots of a maple and other trees and shrubs. *Acta Mycologica* 43, 5–12.

Kowalski, T. and Holdenrieder, O. (2009) The teleomorph of *Chalara fraxinea*. *Forest Pathology* 39, 304–308.

Kranz, J. (1981) Hyperparasitism of biotrofic fungi. In: Blakeman, J.P. (ed.) *Microbial Ecology of the Phylloplane*. Academic Press, London, UK, pp. 327–352.

Krupinsky, J.M. (1989) Variability in *Septoria musiva* in aggressiveness. *Phytopathology* 79, 413–416.

Lang, K.J. (2000) New hosts of *Cristulariella depraedans*. *Forest Pathology* 30, 117–120.

La Porta, N., Capretti, P., Thomsen, I.M., Kasanen, R., Hietala, A.M. and von Weissenberg, K. (2008) Forest pathogens with higher damage potential due to climate change in Europe. *Canadian Journal of Plant Pathology* 30, 177–195.

Liang, H., Maynard, C.A., Allen, R.D. and Powell, W.A. (2001) Increased *Septoria musiva* resistance in transgenic hybrid poplar leaves expressing a wheat oxalate oxidase gene. *Plant Molecular Biology* 45, 619–629.

Liang, H., Catranis, C.M., Maynard, C.A. and Powell, W.A. (2002) Enhanced resistance to the poplar fungal pathogen, *Septoria musiva*, in hybrid poplar clones transformed with genes encoding antimicrobial peptides. *Biotechnology Letters* 24, 383–389.

Lindow, S. (2006) Phyllosphere microbiology: a perspective. In: Bailey, M.J., Lilley, A.K., Timm-Wilson, T.M. and Spencer-Phillips, P.T.N. (eds) *Microbial Ecology of Aerial Plant Surfaces*. CAB International, Wallingford, UK, pp. 1–20.

Lindow, S.E. and Brandl, M.T. (2003) Microbiology of the phyllosphere. *Applied and Environmental Microbiology* 69, 1875–1883.

Lorito, M., Woo, S.L., Harman, G.E. and Monte, E. (2010) Translational research on *Trichoderma*: from 'omics to the field. *Annual Review of Phytopathology* 48, 395–417.

Lundquist, J.E. and Purnell, R.C. (1987) Effects of *Mycosphaerella* leaf spot on growth of *Eucalyptus nitens. Plant Disease* 71, 1025–1029.

MacLeod, A., Pautasso, M., Jeger, M.J. and Haines-Young, R. (2010) Evolution of the international regulation of plant pests and challenges for future plant health. *Food Security* 2, 49–70.

Manning, W.J. and Tiedemann, A.V. (1995) Climate change: potential effects of increased atmospheric carbon dioxide (CO_2), ozone (O_3), and ultraviolet-B (UV-B) radiation on plant diseases. *Environmental Pollution* 88, 219–245.

Maxwell, D.L., Kruger, E.L. and Stanosz, G.R. (1997) Effects of water stress on colonization of poplar stems and excised leaf disks by *Septoria musiva. Phytopathology* 87, 381–388.

McElrone, A.J., Reid, C.D., Hoye, K.A., Hart, E. and Jackson, R.B. (2005) Elevated CO_2 reduces disease incidence and severity of a red maple fungal pathogen via changes in host physiology and leaf chemistry. *Global Change Biology* 11, 1828–1836.

McGranahan, G.H. and Smalley, E.B. (1984) Conidial morphology, axenic growth, and sporulation of *Stegophora ulmea. Phytopathology* 74, 1300–1303.

Melchior, G.H., Stephan, B.R. and Mohrdiek, O. (1983) Growth performance and *Marssonina* attack of *Populus deltoides* Bartr. grown in northern Germany. *Silvae Genetica* 32, 65–71.

Mendgen, K. and Deising, H. (1993) Infection structures of fungal plant pathogens: a cytological and physiological evaluation. *New Phytologist* 124, 193–213.

Mielke, M.E. and Ostry, M.E. (2004) Diseases of intensively managed eastern black walnut. In: Michler, C.H., Pijut, P.M., Van Sambeek, J.W., Coggeshall, M.V., Seifert, J., Woeste, K., Overton, R. and Ponder, F. Jr (eds) *Proceedings of the Sixth Walnut Council Research Symposium*. General Technical Report NC-243, US Department of Agriculture, Forest Service, North Central Research Station, St Paul, Minnesota, pp. 110–113.

Milgate, A.W., Yuan, Z.Q., Vaillancourt, R.E. and Mohammed, C. (2001) *Mycosphaerella* species occurring on *Eucalyptus globulus* and *Eucalyptus nitens* plantations of Tasmania, Australia. *Forest Pathology* 31, 53–64.

Möbius, N. and Hertweck, C. (2009) Fungal phytotoxins as mediators of virulence. *Current Opinion in Plant Biology* 12, 390–398.

Moslonka-Lefebvre, M., Finley, A., Dorigatti, I., Dehnen-Schmutz, K., Harwood, T., Jeger, M.J., Xu, X., Holdenrieder, O. and Pautasso, M. (2011) Networks in plant epidemiology: from genes to landscapes, countries, and continent. *Phytopathology* 101, 392–403.

Mottet, M.J., Lamontagne, D., Caron, F., Fauchon, A. and Périnet, P. (2007) Risk management of *Septoria* canker in Québec hybrid poplar plantations. In: *Conference Handbook – 2007 Annual Meeting of the Poplar Council of Canada – Poplar Culture: A Collaborative Effort from Clone to Mill, Rivière-du-Loup et Québec, Canada, September 16–21, 2007*. Poplar Council of Canada / Ministère des Ressources Naturelles et de la Faune, Direction de la Recherche Forestière, Quebec, Canada, p. 59.

Nagarajan, S. and Singh, D.V. (1990) Long-distance dispersion of rust pathogens. *Annual Review of Phytopathology* 28, 139–153.

Neely, D. (1981) Application of nitrogen fertilizer to control anthracnose of black walnut. *Plant Disease* 65, 580–581.

Neely, D. (1986) Total leaf nitrogen correlated with walnut anthracnose resistance. *Journal of Arboriculture* 12, 312–315.

Newcombe, G. (1998a) A review of exapted resistance to diseases of *Populus*. *European Journal of Forest Pathology* 28, 209–216.

Newcombe, G. (1998b) Southerly extension of poplar leaf blight (*Linospora tetraspora*) in the Pacific Northwest. *Plant Disease* 82, 590.

Nichol, N.S., Wingfield, M.J. and Swart, W.J. (1992) The effect of site preparation and fertilisation on the severity of *Phaeoseptoria eucalypti* on *Eucalyptus* species. *European Journal of Forest Pathology* 22, 424–431.

Noël, A., Levasseur, C., Le, V.Q. and Séguin, A. (2005) Enhanced resistance to fungal pathogens in forest trees by genetic transformation of black spruce and hybrid poplar with a *Trichoderma harzianum* endochitinase gene. *Physiological and Molecular Plant Pathology* 67, 92–99.

Omoifo, C. and Ikotun, T. (1987) Inhibition of growth of some plant pathogens by antagonistic microorganisms. *Journal of Basic Microbiology* 27, 515–519.

Ostry, M.E. (1987) Biology of *Septoria musiva* and *Marssonina brunnea* in hybrid *Populus* plantations and control of *Septoria* canker in nurseries. *European Journal of Forest Pathology* 17, 158–165.

Ostry, M.E. and Anderson, N.A. (2009) Genetics and ecology of the *Entoleuca mammata–Populus* pathosystem: implications for aspen improvement and management. *Forest Ecology and Management* 257, 390–400.

Ostry, M.E. and Ward, K.T. (2003) Field performance of *Populus* expressing somaclonal variation in resistance to *Septoria musiva*. *Plant Science* 164, 1–8.

Ostry, M.E., McRoberts, R.E., Ward, K.T. and Resendez, R. (1988) Screening hybrid poplars *in vitro* for resistance to leaf spot caused by *Septoria musiva*. *Plant Disease* 72, 497–499.
Paetzholdt, M. and Sell, P. (1977) Platanenkrebs überwunden [Sycamores overcome canker]. *Deutsche Baumschule* 9/77, 282–283.
Pappinen, A., Degefu, K., Keinonen, Y., Syrjälä, L. and von Weissenberg, K. (2002) Transgenic silver birch (*Betula pendula*) expressing sugarbeet chitinase 4 shows enhanced resistance to *Pyrenopeziza betulicola*. *Plant Cell Reports* 20, 1046–1051.
Parker, I.M. and Gilbert, G.S. (2004) The evolutionary ecology of novel plant–pathogen interactions. *Annual Review of Ecology, Evolution, and Systematics* 35, 675–700.
Pautasso, M., Dehnen-Schmutz, K., Holdenrieder, O., Pietravalle, S., Salama, N., Jeger, M.J., Lange, E. and Hehl-Lange, S. (2010) Plant health and global change – some implications for landscape management. *Biological Reviews* 85, 729–755.
Peerally, A. (1991) The classification and phytopathology of *Cylindrocladium* species. *Mycotaxon* 40, 323–366.
Pehl, L. and Butin, H. (1994) Endophytische Pilze in Blätern von Laubbäumen und ihre Beziehungen zu Blattgallen (Zoocecidien) [Endophytic fungi in leaves of deciduous trees and their relationships with leaf galls (zoocecids)]. *Mitteilungen aus der Biologischen Bundesanstalt für Land- und Forstwirtschaft Berlin-Dahlem* 297, 1–56.
Perfect, S.E. and Green, J.R. (2001) Infection structures of biotrophic and hemibiotrophic fungal plant pathogens. *Molecular Plant Pathology* 2, 101–108.
Pinon, J. and Valadon, A. (1997) Comportement des cultivars de peupliers commercialisables dans l'Union européenne vis-à-vis de quelques parasites majeurs [Performance of commercial cultivars of poplars in the European Union in relation to various major parasites]. *Annals of Forest Science* 54, 19–38.
Poteri, M. and Rousi, M. (1996) Variation in *Melampsoridium* resistance among European white-birch clones grown in different fertilization treatments. *European Journal of Forest Pathology* 26, 171–181.
Punja, Z.K. (2001) Genetic engineering of plants to enhance resistance to fungal pathogens – a review of progress and future prospects. *Canadian Journal of Plant Pathology* 23, 216–235.
Queloz, V., Gruenig, C.R., Berndt, R., Kowalski, T., Sieber, T.N. and Holdenrieder, O. (2011) Cryptic speciation in *Hymenoscyphus albidus*. *Forest Pathology* 41, 133–142.
Redhead, S.A. (1975) The genus *Cristulariella*. *Canadian Journal of Botany* 53, 700–707.
Reynolds, D.R. (1976) Observations on growth forms of sooty mold fungi. *Nova Hedwigia* 26, 179–193.
Rockwood, D.L., Carter, D.R., Langholtz, M.H. and Stricker, J.A. (2006) *Eucalyptus* and *Populus* short rotation woody crops for phosphate mined lands in Florida USA. *Biomass and Bioenergy* 30, 728–734.
Royle, D.J. and Ostry, M.E. (1995) Disease and pest control in the bioenergy crops poplar and willow. *Biomass and Bioenergy* 9, 69–79.
Santamour, F.S. Jr (1976) Resistance to sycamore anthracnose disease in hybrid *Platanus*. *Plant Disease Reporter* 60, 161–162.
Santamour, F.S. Jr (1984) 'Columbia' and 'Liberty' plane trees. *HortScience* 19, 901–902.
Schäfer, W. (1994) Molecular mechanisms of fungal pathogenicity to plants. *Annual Review of Phytopathology* 32, 461–477.
Schwenke, H.J. (1959) Untersuchungen über Sporenkeimung und Myzelwachstum von *Septotis populiperda* Waterman et Cash [Studies on spore germination and mycelial growth of *Septotis populiperda* Waterman et Cash]. *Phytopathologische Zeitschrift* 35, 389–400.
Seifers, D. and Ammon, V. (1980) Mode of penetration of sycamore leaves by *Gloeosporium platani*. *Phytopathology* 70, 1050–1055.
Sharma, J.K. and Mohanan, C. (1991) *In vitro* evaluation of fungicides against *Cylindrocladium* spp. causing diseases of *Eucalyptus* in Kerala, India. *European Journal of Forest Pathology* 21, 17–26.
Sieber, T.N. (2007) Endophytic fungi in forest trees: are they mutualists? *Fungal Biology Reviews* 30, 1–15.
Sinclair, W.A. and Lyon, H.H. (2005) *Diseases of Trees and Shrubs*, 2nd edn. Cornell University Press, Ithaca, New York.
Sivanesan, A. (1990) *Mycosphaerella populorum*. CMI Descriptions of Pathogenic Fungi and Bacteria No. 988. *Mycopathologia* 109, 57–58.
Smith, A.H., Gill, W.M., Pinkard, E.A. and Mohammed, C.L. (2007) Anatomical and histochemical defence responses induced in juvenile leaves of *Eucalyptus globulus* and *Eucalyptus nitens* by *Mycosphaerella* infection. *Forest Pathology* 37, 361–373.
Smith, J.A., Blanchette, R.A., Ostry, M.E. and Anderson, N.A. (2002) Etiology of bronze leaf disease of *Populus*. *Plant Disease* 86, 462–469.
Smith, S.E. and Read, D.J. (2007) *Mycorrhizal Symbiosis*. Elsevier, Oxford, UK.

Spiers, A.G. (1998) *Melampsora* and *Marssonina* pathogens of poplars and willows in New Zealand. *European Journal of Forest Pathology* 28, 233–240.
Spiers, A.G. and Wenilam, H.T. (1983) Fungicidal control of *Marssonina brunnea* on poplar seed. *European Journal of Forest Pathology* 13, 344–348.
Stanosz, G.R. (1993) Symptoms, association, and pathogenicity of *Discula campestris*, a cause of sugar maple seedling anthracnose. *Plant Disease* 77, 1022–1026.
Strobl, S. and Fraser, K. (1989) Incidence of *Septoria* canker of hybrid poplars in eastern Ontario. *Canadian Plant Disease Survey* 69, 109–112.
Sturrock, R.N., Frankel, S.J., Brown, A.V., Hennon, P.E., Kliejunas, J.T., Lewis, K.J, Worrall, J.J. and Woods, A.J. (2011) Climate change and forest diseases. *Plant Pathology* 60, 133–149.
Sutton, B.C. (1992) The genus *Glomerella* and its anamorph *Colletotrichum*. In: Bailey, J.A. and Jeger, M.J. (eds) *Colletotrichum: Biology, Pathology and Control*. CAB International, Wallingford, UK, pp. 1–26.
Tainter, F.H. and Baker, F.A. (1996) *Principles of Forest Pathology*. John Wiley, Hoboken, New Jersey.
Timmermann, V., Børja, I., Hietala, A.M., Kirisits, T. and Solheim, H. (2011) Ash dieback: pathogen spread and diurnal patterns of ascospore dispersal, with special emphasis on Norway. *EPPO Bulletin* 41, 14–20.
Toti, L., Viret, O., Chapela, I.H. and Petrini, O. (1992) Differential attachment by conidia of the endophyte, *Discula umbrinella* (Berk. & Br.) Morelet, to host and non-host surfaces. *New Phytologist* 121, 469–475.
van der Meiden, H.A. (1964) Die Bedeutung von einigen Blattkrankheiten für die Pappelwirtschaft [The meaning of some leaf diseases for the economics of poplar]. *Der Forst- und Holzwirt* 19, 257–260.
Van der Plank, J.E. (1963) *Plant Diseases: Epidemics and Control*. Academic Press, New York.
Van Loon, L.C., Bakker, P.A.H.M. and Pieterse, C.M.J. (1998) Systemic resistance induced by rhizosphere bacteria. *Annual Review of Phytopathology* 36, 453–483.
Von Sury, R. and Flückiger, W. (1991) Effects of air pollution and water stress on leaf blight and twig cankers of London planes (*Platanus acerifolia* (Ait.) Willd.) caused by *Apiognomonia veneta* (Sacc. & Speg.) Höhn. *New Phytologist* 118, 397–405.
Waggoner, P.E. and Aylor, D.E. (2000) Epidemiology: a science of patterns. *Annual Review of Phytopathology* 38, 71–94.
Walle, I.V., Van Camp, N., Van de Casteele, L., Verheyen, K. and Lemeur, R. (2007) Short-rotation forestry of birch, maple, poplar and willow in Flanders (Belgium). I – Biomass production after 4 years of tree growth. *Biomass and Bioenergy* 31, 267–275.
Walton, J.D. (1996) Host-selective toxins: agents of compatibility. *Plant Cell* 8, 1723–1733.
Ward, K.T. and Ostry, M.E. (2005) Variation in *Septoria musiva* and implications for disease resistance screening of poplars. *Plant Disease* 89, 1077–1082.
Wilcove, D.S., Rothstem, D., Dubow, J., Phillips, A. and Losos, E. (1998) Quantifying threats to imperiled species in the United States. *BioScience* 48, 607–615.
Wilson, D. and Carroll, G.C. (1994) Infection studies of *Discula quercina*, an endophyte of *Quercus garryana*. *Mycologia* 86, 635–647.
Wingfield, M.J., Slippers, B., Roux, J. and Wingfield, B.D. (2001) Worldwide movement of exotic forest fungi especially in the tropics and southern hemisphere. *BioScience* 51, 134–140.
Woo, J.V. and Partridge, A.D. (1969) The life history and cytology of *Rhytisma punctatum* on bigleaf maple. *Mycologia* 61, 1085–1095.
Woodbury, P.B., Laurence, J.A. and Hudler, G.W. (1994) Chronic ozone exposure increases the susceptibility of hybrid *Populus* to disease caused by *Septoria musiva*. *Environmental Pollution* 86, 109–114.
Wösten, H.A.B. (2001) Hydrophobins: multipurpose proteins. *Annual Review of Microbiology* 55, 625–646.
Wulf, A. (1994) *Pilzbedingte Blattkrankheiten an Ahorn unter Besonderer Berücksichtigung des Bergahorns* (Acer pseudoplatanus *L.*) [Fungal leaf diseases of maple with special consideration of the sycamore maple (Acer pseudoplatanus *L.*)]. Schriften aus der Forstlichen Fakultät der Universität Göttingen und der Niedersächsischen Forstlichen Versuchsanstalt. J.D. Sauerländer's Verlag, Frankfurt, Germany.
Yang, D., Bernier, L. and Dessureault, M. (1994) Biological control of *Septoria* leaf spot of poplar by *Phaeotheca dimorphospora*. *Plant Disease* 78, 821–825.
Yu, X., Cunti, X., Haifeng, L. and Chuanwei, W. (1997) Study on resistance of poplars to *Marssonina brunnea*. *Journal of Forestry Research* 8, 231–234.
Yuan, K.U.N., Zhang, B.O., Zhang, Y., Cheng, Q., Wang, M. and Huang, M. (2008) Identification of differentially expressed proteins in poplar leaves induced by *Marssonina brunnea* f.sp. *multigermtubi*. *Journal of Genetics and Genomics* 35, 49–60.
Zhang, C.X., Zhao, X., Jing, Y.X., Chida, T., Chen, H. and Shen, S.H. (2008) Phenotypic and biological properties of two antagonist *Bacillus subtilis* strains. *World Journal of Microbiology and Biotechnology* 24, 2179–2181.

25 Oomycete Diseases

Katherine J. Hayden,[1]* Giles E.St.J. Hardy[2] and Matteo Garbelotto[1]
[1]University of California, Berkeley, California, USA; [2]Murdoch University, Murdoch, Western Australia

25.1 Pathogens, Significance and Distribution

The most important oomycete forest pathogens comprise two genera: *Pythium* and the formidable genus *Phytophthora*, whose name appropriately means 'plant destroyer'. *Pythium* spp. cause seed and root rots and damping-off diseases that thwart seedling establishment, and have been implicated in helping to drive forest diversity patterns through increased disease pressures on seedlings closest to their mother tree (Janzen, 1970; Connell, 1971). In contrast, *Phytophthora* spp. can cause disease at every life stage of forest trees, from root to crown, and from trunk cankers to foliar blights (Erwin and Ribeiro, 1996). They are remarkably flexible and effective pathogens with an unusual genetic architecture that may favour the rapid evolution of pathogenicity (Jiang *et al.*, 2008; Raffaele *et al.*, 2010; Seidl *et al.*, 2011). Outbreaks of disease caused by *Phytophthora* spp. (especially when they have been introduced to new systems) have been documented with dramatic, and sometimes disastrous, effects since the mid 1800s, when European and North American chestnuts (*Castanea* spp.) began dramatic declines from chestnut ink disease, a root rot caused in part by the extreme generalist *Phytophthora cinnamomi* Rands (Crandall *et al.*, 1945; Anagnostakis, 1995). *P. cinnamomi* is notorious for the massive mortality it has caused in jarrah (*Eucalyptus marginata* Donn ex Sm.) forests in Western Australia, where it was first observed in the 1920s (Podger, 1972). *P. cinnamomi* causes root disease in agricultural and forest systems worldwide with varying degrees of virulence, but as Phytophthora dieback it has been seen to kill 50–75% of the species in sites in Western Australia, in some cases leaving every tree and much of the understorey dead (Weste, 2003). Shearer *et al.* (2004) estimate that of the 5710 described plant species in the South West Botanical Province of Western Australia, approximately 2300 species are susceptible, 800 of which are highly susceptible.

More recently, the trunk canker caused by *Phytophthora ramorum* Werres, De Cock & Man in't Veld has caused a devastating die-off of oaks (*Quercus* spp.) and tanoak (*Lithocarpus densiflorus* (Hook. & Arn.) Rehder) in western North America (Rizzo and Garbelotto, 2003), spreading from a relatively minor foliar blight of ornamentals in nurseries to a fatal scourge in US wildlands and UK gardens. In contrast to *P. cinnamomi*, which is a root pathogen

* E-mail: khayden@berkeley.edu

transmitted by soil and water, infection by and transmission of *P. ramorum* occur primarily above ground. While the two species have much in common, e.g. extremely broad host ranges and disastrous consequences of introduction on native ecosystems, their contrasting modes of transmission span the range for oomycete pathogens. Here, we treat these two species and the extensive literature surrounding them as case studies for oomycete diseases and their management.

A great many more species of *Phytophthora* have an impact on forest systems, and new species are discovered almost yearly. Most of these are likely to have been established for some time (even if they are relatively recent introductions), but have only just been noticed because of increased sampling and new molecular tools (Jung *et al.*, 2002, 2011; Hansen *et al.*, 2003; Balci *et al.*, 2008; Burgess *et al.*, 2009; Scott *et al.*, 2009). Others are entirely new, however, including a novel *Phytophthora* species complex that has begun to decimate alders (*Alnus* spp.) in Europe (Brasier *et al.*, 1995, 1999). In addition to the more detailed descriptions of *P. cinnamomi* and *P. ramorum*, we briefly describe a selection of those species currently known to have the greatest impacts on forest systems. For a more complete description of the biology and disease control of older *Phytophthora* spp., refer to Erwin and Ribeiro (1996).

P. cinnamomi has a worldwide distribution, and is the cause of Phytophthora root and collar rot (synonyms: ink disease of hardwoods, Phytophthora dieback, littleleaf disease of pines, stripe canker of cinnamon). *P. cinnamomi* was confirmed as a cause of ink disease, along with *P. cambivora* (Petri) Buisman in Europe, in the early 1900s (Day, 1938), and as the cause of 'jarrah dieback' in 1965 (Podger *et al.*, 1965; Podger, 1972). The pathogen causes root rots on a huge number of species (Hardham, 2005). Disease severity ranges from asymptomatic to fatal, depending on the host and environmental conditions, with among the worst observed in the jarrah 'graveyards' of Western Australia. It has been documented to live saprophytically, and to persist in moist soil for as long as 6 years (Zentmyer and Mircetich, 1966). However, the work of McCarren *et al.* (2005) provides strong evidence that *P. cinnamomi* is unable to survive saprophytically in the absence of host plants. In dry conditions, it persists in the hardpan layer, where deep lateral roots may be in contact with free water even when the surface soil is dry (Shea *et al.*, 1983; Kinal *et al.*, 1993; Shearer *et al.*, 2010). It favours warmer temperatures than most *Phytophthora* spp., but has a wide range of conducive temperatures, facilitating its worldwide distribution.

P. ramorum is the cause of sudden oak death trunk canker, ramorum blight and dieback, and is distributed in nurseries in the western USA, western Canada and throughout Europe. It is present in gardens in the UK and wildlands in the USA. Diseases caused by *P. ramorum* emerged in the mid 1990s as foliar and twig blight of nursery ornamentals in Europe (Werres *et al.*, 2001), and as fatal canker disease of oaks and tanoaks in a 300 km stretch of the California coast (Rizzo *et al.*, 2002). The twig and foliar blight infects hosts in nearly every plant family, and is the primary source of inoculum even in the US epidemic (Garbelotto *et al.*, 2002). The canker form in true oaks has not been documented to contribute to pathogen spread, but tanoaks are distinctly susceptible to both the sporulating foliar and twig disease and the fatal canker (Davidson *et al.*, 2008). The disease has been responsible for the deaths of hundreds of thousands, if not millions of oaks and tanoaks (Plate 30), the near extirpation of tanoak in some parts of its range, and millions of US$ incurred in costs due to quarantine and the monitoring of nurseries.

Pythium spp. (e.g. *P. debaryanum* R. Hesse, *P. irregulare* Buisman, *P. ultimum* Trow) are world-wide distributed root pathogens that are a major cause of seed rots and damping-off diseases. These diseases are a major cause of seedling mortality, and have long been thought to help drive diversity in tropical forests through density-dependent mortality (Augspurger, 1984; Bell *et al.*, 2006). They have also been documented in temperate systems (Packer and Clay, 2000, 2003). See Gilbert (2002) for a review of the role and limitations of pathogens as natural enemies in driving forest diversity, as proposed by Janzen (1970) and Connell (1971). Put simply, the Janzen–Connell hypothesis posits density-dependent feedback, wherein the highest density of seeds and their pathogens or other natural enemies will both occur close to a parent tree. Seeds may have to travel some distance to escape their enemies, thus helping to drive forest diversity.

For forest managers, seed rots and damping-off diseases are a hindrance to re-seeding efforts, and may inspire the use of nursery-grown seedlings rather than starting new trees from seed, bringing with them the potential to spread nursery diseases into wildlands. *Pythium* spp. have been implicated in the decline of spruce (*Picea* spp.) and beech (*Fagus* spp.) in Bavaria (Nechwatal and Osswald, 2001), the decline of Spanish oaks (Romero *et al.*, 2007), and as contributing to growth reductions of Scots pine (*Pinus sylvestris* L.) in Scotland, especially when mixed with *P. cinnamomi* or *Fusarium* spp. (Chavarriaga *et al.*, 2007). Features of the *P. ultimum* genome sequence are consistent with its status as a necrotroph and generalist pathogen of seedlings; it lacks cutinases to break down plant cuticles, lacks xylanase to digest complex polysaccharides, and does not have the RXLR effectors (effector proteins that can be delivered to the inside of host cells to manipulate innate immunity) that characterize *Phytophthora* pathogenicity (Lévesque *et al.*, 2010).

A number of well-known *Phytophthora* spp., other than *P. cinnamomi* and *P. ramorum*, are described below. *Phytophthora alni* Brasier & S.A. Kirk is distributed in Europe (Brasier *et al.*, 2004b) and Alaska (Adams *et al.*, 2008). This oomycete is a self-fertile complex of hybrids causing fatal root and collar rots in alders. First isolated in 1993 and 1994 in the UK (Brasier *et al.*, 1995), the new species was eventually found to be a recent hybrid of *P. cambivora* and a species similar to *P. fragariae* Hickman (Brasier *et al.*, 1999), probably with several instances of hybridization and backcrossing (Ioos *et al.*, 2006). Diseases caused by the *P. alni* hybrid complex are characterized by root rot, dark and sometimes bleeding cankers on the trunk and/or crown decline, although asymptomatic infections are common, and are an important source of inoculum in infested watersheds (Lonsdale, 2003; Elegbede *et al.*, 2010).

Phytophthora cactorum (Lebert & Cohn) J. Schröt. is globally distributed and causes disease on a wide range of species, from forest trees to ornamentals. It has been associated with canker diseases on maple (*Acer* spp.) (Caroselli and Howard, 1939; Erwin and Ribeiro, 1996), beech (Jung *et al.*, 2006), horse chestnut (*Aesculus hippocastanum* L.) (Brasier and Strouts, 1976) and white poplar (*Populus alba* L.) (Cerny *et al.*, 2008), and with declining oaks in Europe (Jung *et al.*, 1996); it has long been associated with root rots of conifers.

Phytophthora cambivora has a worldwide distribution and, as already noted, is co-implicated with *P. cinnamomi* as the cause of ink disease of chestnut in Europe, both in its historic form, dating to before 1800 (Day, 1938; Crandall *et al.*, 1945), and in its recent resurgence (Vettraino *et al.*, 2001, 2005). The infection spreads from roots into the collar of the tree, and frequently causes death within 2 years. The disease may also move more slowly, and cause the slow decline of the crown, followed by death within several years (Day, 1938). Ink disease is characterized by a black exudate from a lesion that spreads upwards from the soil. The pathogen is soilborne and heterothallic, and is favoured by wet soil and by cooler temperatures than is *P. cinnamomi* (Erwin and Ribeiro, 1996).

Alongside *P. cambivora* and *P. cinnamomi*, the more weakly pathogenic *P. citricola* Sawada is among the species most frequently isolated on symptomatic chestnuts in Europe (Vettraino *et al.*, 2001, 2005). It is homothallic, and produces oospores that may contribute to its relatively stronger ability to persist in soil during the dry season (Vettraino *et al.*, 2001). It is an important pathogen worldwide, and can infect a wide range of hosts. However, with the advent of molecular tools, isolates previously described as *P. citricola* based on morphological characteristics are now shown to be part of a species complex (Burgess *et al.*, 2009; Jung and Burgess, 2009) to include *P. plurivora* T. Jung & T.I. Burgess and *P. multivora* P.M. Scott & T. Jung (Scott *et al.*, 2009). For example, in Western Australia, the sequencing of the internal transcribed spacer (ITS) region of 73 isolates previously thought to be *P. citricola* revealed that no single isolate corresponded to *P. citricola*

(Burgess *et al.*, 2009), and it appears that *P. citricola* is not present in wildlands in Western Australia. Therefore, it is likely that many diseases previously linked to *P. citricola* may well not be, and more work is required on the *P. citricola* species complex to fully understand the diseases they cause.

Phytophthora kernoviae Brasier, Beales & S.A. Kirk is the cause of a trunk canker of beech trees, and of leaf and twig blight and dieback in the UK and New Zealand. Although homothallic, *P. kernoviae* shares many characteristics with *P. ramorum*, and was in fact first discovered in 2003 during a survey of symptomatic trees at sites in Cornwall that were first found to be infested with *P. ramorum* (Brasier *et al.*, 2004a). It is an aggressive pathogen of beech trees, causing large and fatal cankers. Like *P. ramorum*, however, its transmission is from foliar lesions on ornamentals, including *Rhododendron* spp. (Brasier *et al.*, 2005). In addition to foliar symptoms, root infections have been observed on common rhododendron (*R. ponticum* L.), the primary agent of spread to canker hosts in woodlands, and these potentially complicate removal and eradication efforts (Fichtner *et al.*, 2011).

Phytophthora lateralis Tucker & Milbrath causes a root rot of Port Orford cedar (*Chamaecyparis lawsoniana* (A. Murray bis) Parl.) in the Pacific Northwest of the USA and in British Columbia (Canada). It has recently been discovered in forest soils in Taiwan (Brasier *et al.*, 2010) and in a new outbreak in Brittany (France) (Robin *et al.*, 2011); otherwise, it is a nursery pathogen in Europe. *P. lateralis* is highly specific, infecting primarily *Chamaecyparis* spp., although Pacific yew (*Taxus brevifolia* Nutt.) is a minor host in North America (DeNitto and Kliejunas, 1991). Port Orford cedar is highly susceptible, and in Oregon forests it has caused major losses. There has been only one report of the aerial transmission of *P. lateralis* in the USA (Trione and Roth, 1957). Notably, a recent outbreak of disease caused by *P. lateralis* in France has been primarily foliar, and shows signs of aerial transmission (Robin *et al.*, 2011).

Phytophthora palmivora (E.J. Butler) E.J. Butler is the cause of rots, blights and cankers, from root to fruits. It infects plant hosts worldwide, but is most important as a forest pathogen in the tropics, where it is a major pathogen of rubber (*Hevea brasiliensis* (Willd. ex A. Juss.) Müll. Arg.) and cacao (*Theobroma cacao* L.) trees. It is heterothallic, and produces caducous (shedding) sporangia and chlamydospores. Importantly, Arnold *et al.* (2003) and Herre *et al.* (2007) demonstrated the role of endophytes in host resistance by showing that their presence protected cacao trees from disease caused by *P. palmivora*.

Phytophthora pinifolia Alv. Durán, Gryzenh. & M.J. Wingf. is remarkable because it is the first known oomycete foliar pathogen of pines (*Pinus* spp.). It was discovered in Chile in 2003 in plantations of Monterey pine (*P. radiata* D. Don), where it is beginning to have a major impact on forestry. All size classes of trees are susceptible, and disease is characterized by the rapid death and drop of needles in mature trees and the death of seedlings (Durán *et al.*, 2008).

25.2 Diagnosis

Infections caused by oomycetes are not always readily diagnosed, especially in novel outbreaks. The intrinsic difficulty lies in the different types of disease that may emerge when different pathogen species are involved, when outbreaks occur in sites that may be less or more conducive to disease because of soil and/or environmental parameters, when different hosts are involved, and even when different lineages or genotypes of the pathogen and/or the host are involved. The combination of two or more of these factors above may lead to symptoms and to epidemiological patterns that are different from those expected, and may be hard to identify.

The detection and identification of *Phytophthora* spp. have been recently reviewed by Martin *et al.* (2012). Soil-borne and waterborne species such as *P. cinnamomi*, *P. cambivora* and *P. lateralis* in general cause at first an underground infection of fine roots that may or may not, depending on host species and on site, proceed to an infection of the main roots, the root crown, and even of the lower part of the bole. In the absence of above-ground signs of infection, diagnosis must be based on retrieval of the pathogen directly from the soil, and on symptoms such as reduction in growth rate, early foliar senescence and, eventually, drying of the canopy as indicators of infection (Campbell and Copeland, 1954; Zentmyer, 1980; Brasier *et al.*, 1993). Some host species may be relatively tolerant of root infection by soil-borne *Phytophthora* spp., making diagnosis extremely difficult; in these cases symptoms often accelerate and lead to tree mortality only when other debilitating factors such as drought are superimposed on the root infection. Often, the only way to confirm root infections without extensive excavation is to bait the pathogen from the soil around trees. Baiting techniques to isolate *Phytophthora* spp. from soil are varied, and are described in detail by Tsao (1983) and Erwin and Ribeiro (1996), while modifications are described by Jung *et al.* (2000) and Scott *et al.* (2009).

In other pathosystems, including white oaks (*Q. alba* L.) in Mexico (Tainter *et al.*, 2000), chestnuts in North America (Crandall *et al.*, 1945), and manzanitas (*Arctostaphylos* spp.) in California (Swiecki *et al.*, 2003) infected by *P. cinnamomi*, Port Orford cedar infected by *P. lateralis* in western North America (Tucker and Milbrath, 1942; Zobel *et al.*, 1982; Hansen *et al.*, 2000; Oh *et al.*, 2006), alders infected by *P. alni* in Europe (Brasier *et al.*, 1995; Streito, 2003) and chestnuts infected by *P. cambivora* (Day, 1938; Robin *et al.*, 2006), host tolerance, while sometimes present and epidemiologically important, is quite low and infection of the roots underground leads to obvious symptoms and signs above ground in a broader range of environmental conditions. Outbreaks can be identified by thinning canopies and mortality of trees in groups, often associated with girdling lesions advancing from the root collar into the lower parts of the bole. Lesions are normally brown to cinnamon and black in colour, extensive in the phloem, while limited in the bark and the xylem, and are characterized by irregular margins at times marked by a dark line. Lesions can sometimes be identified as areas of sunken bark but, in general, lesion identification requires an excision of the outer layers of bark to expose the phloem. To confirm infection, it is often necessary to isolate the pathogen by plating a small chip from the edges of a lesion. Plating techniques and media selective for *Phytophthora* spp. are described in detail by Erwin and Ribeiro (1996).

One of the best known symptoms associated with above-ground stem colonization by *Phytophthora* spp. is the so-called 'bleeding', 'seeping', or 'gummosis' in some tree species, which can be described as the production of a dark and viscous substance on the outside of the bark. This 'bleeding' corresponds to the lesion in the phloem but in the absence of any obvious mechanical damage or insect-generated wound. Bleeding symptoms on the trunk, if correctly diagnosed, can help to determine the presence of an infection by a *Phytophthora* species; however, not all hosts will develop such symptoms, and there is considerable variation even within some species.

Similar girdling lesions and associated bleeding are also characteristically associated with infections by aerial Phytophthoras, or with aerial infection caused by soil-borne *Phytophthora* spp. In these cases, lesions can occur on the stems, branches and even twigs without any clear association with root crowns (collars). In fact, in the case of *P. ramorum*, large aerial trunk lesions disappear just below the soil level (Davidson *et al.*, 2003). Lesions caused by these aerial species can lead to the development of different symptoms, depending on the tolerance of the host and, especially, on the location of the lesion. As a result, highly susceptible hosts infected in the lower part of the main trunk may be rapidly girdled, with a simultaneous death of the entire canopy. Conversely, a lesion on a branch may lead to death of the branch alone, without immediate consequences for the whole tree. In most cases, aerial *Phytophthora* spp. girdle and kill the infected portion or the whole individual; thus the slow decline associated with root infection by some soil-borne *Phytophthora* spp. seems to be less frequently observed. In many cases, epicormic adventitious branches are produced right below the lesion; although these sprouts rarely survive long, they may help to identify a *Phytophthora* lesion. Drought-adapted species such as California coast live oak (*Q. agrifolia* Née) may be effectively girdled, but maintain a canopy that remains green and even breaks bud and grows for up to several years after the stem is girdled, thanks to the presence of stored carbohydrates. These trees undergo a progressive drying of the foliage, are often affected by accelerated decay processes, and become attractive to insect infestations even if the canopy has not turned brown. The presence of entry holes by bark beetles and of fruiting bodies of wood decaying fungi (both basidiomycetes and ascomycetes) in several portions of the main stem is a clear indication that the tree is functionally dead.

Another clear distinction between soil-borne or waterborne and aerial *Phytophthora* spp. is the pattern of spread of the disease. Infection by soil-borne or waterborne species appears to be clearly associated with bodies of water, areas where water accumulates – such as poorly drained draws (shallow gullies) and flood areas, underground water tables, and finally unpaved roads or even dirt tracks. This association is lacking for aerial species and is replaced by the spatial aggregation of infections in areas adjacent to sporulating hosts.

In complex forest ecosystems such as the coastal forests of California, or the jarrah forest and banksia (*Banksia* spp.) woodlands of Western Australia, an infestation by a *Phytophthora* spp. can also be diagnosed via susceptible indicator plants that develop symptoms more rapidly than the tree hosts of interest. Leaf lesions on California laurels (*Umbellularia californica* (Hook. & Arn.) Nutt.) and twig infections of tanoaks, Pacific madrones (*Arbutus menziesii* Pursh), and huckleberries (*Gaylussacia* spp.), just to name a few species, often precede lethal girdling stem lesions of oaks and tanoaks and can strengthen the diagnosis of an infestation by *P. ramorum*, even in the absence of lethal infections on tree hosts. Similarly, the mortality of shrubs and herbaceous hosts such as bull banksia (*B. grandis* Willd.) and balga (*Xanthorrhoea preissii* Endl.) in the jarrah forest of Western Australia can be rapid and precede any symptoms on trees, thus providing a valuable indicator of an expanding infestation (Shearer and Tippett, 1989).

Whether from soil, from an indicator plant, or from a tree, the identification of the *Phytophthora* species involved can be of paramount importance in predicting risk and in assessing the need for management. Many *Phytophthora* spp., both native and introduced, can coexist in forest ecosystems. Morphological identification of the pathogen in culture remains the best approach, but often the pathogen is extremely hard to isolate from some host species or at certain times of year (e.g. during dry spells, or at unfavourable temperatures). Additionally, new species are continually being described, while traditional taxonomic expertise is becoming more limited. For these reasons, diagnosis can be significantly enhanced by DNA-based techniques and immunoassays. Sequencing one or two diagnostic loci and comparing the sequences with those in public databases may suffice to identify a species that is not familiar to the researcher. When no cultures can be obtained, immunoassays and PCR-based protocols can confirm the presence of a *Phytophthora* species directly from infected plant tissue or from soil and water (Table 25.1). It should be highlighted that the pitfalls of diagnosis based solely on molecular techniques are many, and include both false negative and false positives diagnoses. Currently, the advantages and disadvantages of molecular identification approaches have been somewhat adequately investigated only for *P. ramorum* (Hayden *et al.*, 2006; Kox *et al.*, 2007; Lane *et al.*, 2007; Vettraino *et al.*, 2010). Hüberli *et al.* (2000) discuss the problems of misdiagnosis based on false negative isolation by culture for *P. cinnamomi*.

25.3 Infection Biology

Oomycetes are not true fungi; they are instead classified with the brown algae, or water moulds. Their vegetative hyphae resemble fungi, but unlike true fungi, *Pythium* and *Phytophthora* cell walls do not contain chitin; they require an outside source for sterols, and infection occurs through either direct germination of asexually produced sporangia, or through encystment and germination of the biflagellate, swimming zoospores that emerge from sporangia (Alexopoulos *et al.*, 1996; Erwin and Ribeiro, 1996).

Phytophthora spp. are generally hemibiotrophic, first infecting living tissue, and then

Table 25.1. Taxon-specific PCR-based methods for the detection of *Phytophthora* spp.

Target	References
Multiple taxa:	
P. cambivora, *P. cinnamomi*, *P. citricola*, *P. citrophthora*, *P. infestans*, *P. kernoviae*, *P. nemorosa*, *P. nicotianae*, *P. pseudosyringae*, *P. quercina*, *P. ramorum*	Érsek *et al.*, 1994; Böhm *et al.*, 1999; Schubert *et al.*, 1999; Nechwatal *et al.*, 2001; Ippolito *et al.*, 2004; Martin *et al.*, 2004; Schena *et al.*, 2006, 2008; Langrell *et al.*, 2011
P. alni complex	Ioos *et al.*, 2005; Bakonyi *et al.*, 2006
P. cinnamomi	Schubert *et al.*, 1999; Kong *et al.*, 2003; O'Brien, 2008; Williams *et al.*, 2009; Langrell *et al.*, 2011
P. lateralis	Winton and Hansen, 2001
P. ramorum	Kroon *et al.*, 2004; Martin *et al.*, 2004; Tomlinson *et al.*, 2005, 2007; Hayden *et al.*, 2006; Hughes *et al.*, 2006; Schena *et al.*, 2006; Belbahri *et al.*, 2007; Bilodeau *et al.*, 2007a,b

growing into and between plant cells while leaving a necrotic lesion behind, whereas *Pythium* spp. may be either hemibiotrophs or necrotrophs (Latijnhouwers *et al.*, 2003). Hyphae originating from direct germination of sporangia enter the plant through stomata, lenticels or other openings. Zoospores germinate by first encysting on the plant surface (roots, stems or leaves), after which either an appressorium forms to break the host cell wall, or a germ tube may enter through stomata. In good conditions for disease development, rapid cycling of infection followed by asexual sporulation allow for huge increases in inoculum loads over short time periods.

Zoospore encystment is triggered by pressure, electrochemical charge and chemical signalling. Zoospore movements have been documented to respond to chemical exudates from plants, to light, to gravity and to charge (Hardham *et al.*, 1994; Hardham, 2001, 2007). Zoospores can travel short distances (millimetres to centimetres) under their own power, or spores can passively travel long distances in flowing water. Propagules of *Phytophthora cinnamomi* have been documented in flowing groundwater up to 5 m below the soil surface, where free water may persist even when the surface dries (Shea *et al.*, 1983; Kinal *et al.*, 1993; Shearer *et al.*, 2010). Some species are also able to shed sporangia, which can then disperse by wind or water. On landing, new infections may occur either through direct germination or indirectly by the release of zoospores. Even after the wind dispersal of sporangia, neither direct germination nor zoospore release may occur in the absence of free water (Judelson and Blanco, 2005), and zoospores and sporangia are relatively fragile and short lived.

Oomycetes are persistent in soil, especially if it remains damp. *Pythium* spp. can persist as a saprobe or as resting oospores (Martin and Loper, 1999), and *Phytophthora* spp. may persist in soil or gravel for years as oospores, chlamydospores or, sometimes, saprophytically (Zentmyer and Mircetich, 1966; Hwang and Ko, 1978; Weste and Vithanage, 1979; Linderman and Davis, 2006; Shishkoff, 2007). Chlamydospores are asexual spores with a surprising degree of intraspecific variation in structure, wall thickness and tolerance of extreme environments, while oospores are produced by sexual recombination, and tend to be thick walled and quite persistent (McCarren *et al.*, 2005).

The frequency of chlamydospore and oospore production is governed by environmental conditions as well as by the biology of the species in question. The production of both structures tends to be favoured at temperatures somewhat below optimum growth temperatures, requires adequate moisture and nutrients, and is influenced by the chemical environment (Mircetich *et al.*, 1968; Tsao, 1969; Cother and Griffin, 1974; Ribeiro, 1983; Erwin and Ribeiro, 1996). Homothallic species can produce oospores by self-fertilization, but heterothallic species require contact between opposite mating types.

Both *Phytophthora ramorum* and *P. cinnamomi* are heterothallic, albeit that in their introduced ranges they appear to reproduce mostly asexually, even when both mating types are present (Dobrowolski *et al.*, 2003; Ivors *et al.*, 2006; Prospero *et al.*, 2007; Mascheretti *et al.*, 2008; Goss *et al.*, 2009). None the less, Brasier (1971, 1975) showed that A2 mating types of several heterothallic species, including *P. cinnamomi*, could be stimulated to produce self-fertilized oospores by antibiotic compounds from *Trichoderma* spp. Some plant root exudates have a similar effect; A2 *P. cinnamomi* cease the production of sporangia and generate self-fertilized oospores in response to exudates from avocado (*Persea americana* Mill.) (Zentmyer, 1979) and acacia roots (Jayasekera *et al.*, 2007).

25.4 Epidemiology

25.4.1 Mode of dispersal

A primary distinction in the epidemiology of oomycete pathogens is whether they are dispersed aerially, or solely by water and soil. Species with caducous sporangia, including most of the foliar oomycete pathogens (Erwin and Ribeiro, 1996), may travel by wind as well as by water splash, flowing water and in soil. For example, Davidson *et al.* (2005) documented *Phytophthora ramorum* inoculum in rainwater, streams and even in soil on hikers' shoes. A more recent genetic analysis of gene flow in *P. ramorum* revealed a bimodal peak, whereby most genotypic similarity was detected at 15 m, the range for water splash, with a second peak at 1000 m, the likely range for wind dispersal (Mascheretti *et al.*, 2008). Analysis of individual populations could not definitively distinguish between new infestations from wind- or human-mediated dispersal, although at least some wind transmission was probable in the study populations. New infestations of *P. ramorum* have been detected at 4000 m from the nearest known population in Oregon (Hansen *et al.*, 2008), where there is an active programme to eradicate the pathogen wherever it is detected (Goheen *et al.*, 2002; Kanaskie *et al.*, 2006, 2008). Hansen *et al.* (2008) observed that new Oregon infestations often appeared as the infection of several trees with the same pathogen genotype at once, suggesting that wind dispersal might occur as relatively rare, mass movement of spores by turbulence events. At distances greater than 1–4 km, human-mediated movement through nursery plants or soil is the most important mechanism for spread.

Soil-borne pathogens are more reliant on humans for both medium- and long-range dispersal, although waterways may carry propagules for longer distances, and spread from soil moisture within an infested site can be extremely efficient, reaching up to 100% infection within a short time (Colquhoun and Hardy, 2000; Hansen *et al.*, 2000). Roadways are especially important. An analysis of the spread of *Phytophthora lateralis* on Port Orford cedar in the USA determined that 26 of 36 new infestations were due to vehicle traffic on roads, and the rest were from foot traffic (Jules *et al.*, 2002). Roads have been similarly implicated in the spread of *Phytophthora cinnamomi* into the jarrah forest in Western Australia, through vehicle traffic and through the earth movement required for their construction (Shearer and Tippett, 1989). Bauxite mining in Western Australia is a particular problem for the spread of *P. cinnamomi* in the jarrah forest. The mining process moves huge quantities of soil, redirects drainage patterns, and necessitates the construction and heavy use of new roads, all of which have the potential to spread the pathogen and have been the subject of a containment effort (Colquhoun and Hardy, 2000; Colquhoun and Kerp, 2007).

Human activities may lead to unexpected or unusual disease forms. In nurseries with artificial irrigation and abundant water splash, needle infections of firs (*Abies* spp.) by *P. cinnamomi* have been reported (McCain and Scharph, 1986). Recently, Robin *et al.* (2011) reported foliar disease caused by *P. lateralis* on Port Orford cedar in landscapes in France, where it previously had been seen only in nurseries. Trione and Roth (1957) reported an infrequent aerial habit of *P. lateralis* in USA forests, but aerial disease has not since been reported, nor has it been reported to play any role in disease dynamics.

P. lateralis was previously confined to nurseries in Europe, so the new report is of both a new range and a new form. Evidently, some component of the new environment has triggered a shift in the pathogen's mode of transmission, with potentially important implications for epidemiology in the new range (Robin *et al.*, 2011).

The significance of the movement of nursery plants as a source of spread of *Phytophthora* spp. should not be underestimated. Nursery plants have played a major role in the evolution and spread of, among others, the *Phytophthora alni* complex (Brasier *et al.*, 1999; Brasier, 2000; Jung and Blaschke, 2004), have had a demonstrable role in the spread of *P. cinnamomi* (Hardy and Sivasithamparam, 1988) and are responsible for the rapid spread of *P. ramorum* throughout much of the USA and Europe (Ivors *et al.*, 2006; Mascheretti *et al.*, 2008, 2009). Brasier (2008) documents the many introductions of destructive pathogens from trade in plant products, and the threat that they pose to ecosystems and economies.

25.4.2 Host range

Pathogens with large host ranges will have different ecological and epidemiological dynamics from specialists. It is no coincidence that the extreme generalists *P. cinnamomi* and *P. ramorum* have both led to spreading epidemics of forest disease. *P. cinnamomi* is estimated to have more than 3000 plant hosts (Hardham, 2005), while the much more recently emerged *P. ramorum* has a smaller but ever-growing list of known hosts that encompasses nearly every family of plants (Garbelotto *et al.*, 2002; Grünwald *et al.*, 2008). A wide host range opens the range of niches available to the pathogen, while the probability of an individual spore encountering a susceptible host is increased, thereby increasing the probability of transmission. Generalists are more likely than specialists to become established in new environments because of the greater likelihood of finding a susceptible host. They may even be more likely to jump to new hosts, through inoculum rain from a nearby susceptible species that could eventually include a variant able to overcome the previously resistant host's defences (Parker and Gilbert, 2004).

None the less, all hosts may not be equal. Of special concern are tolerant species, which support pathogen infection with minimal or no disease symptoms or fitness effects, but nevertheless support pathogen sporulation and transmission. Roy and Kirchner (2000) modelled the potentially insidious effect of host tolerance on host–pathogen co-evolution; in contrast to host resistance, in which some intermediate level of virulence is theoretically optimal, tolerance interactions can be expected to lead to ever-increasing pathogen virulence and aggressiveness. Real interactions are most likely to involve some mixture of resistance and tolerance.

California laurel trees are a case in point: they restrict infection by *P. ramorum* to their leaves, where they experience only minor physiological damage from ramorum blight (DiLeo *et al.*, 2009). The pathogen sporulates prolifically from California laurel, however, and the trees have played a major role in the epidemiology of sudden oak death in California. As always, the interactions of environment, hosts and pathogens all play a role in disease dynamics. Davidson *et al.* (2011) attribute the quicker disease progress of sudden oak death in California redwood (*Sequoia sempervirens* (Lamb. ex D. Don) Endl.) forests than occurs in similar mixed evergreen forests to the greater propensity of California laurel leaves in evergreen forests to drop their leaves during the dry summer months. Because California laurel trees in redwood forests are more likely to keep their diseased leaves, there is a greater supply of inoculum at the ready when moist conditions in the spring once again favour disease. The sporulation of *P. ramorum* and *P. kernoviae* has been documented from entirely asymptomatic fruits and rhododendron leaves used as baits (Denman *et al.*, 2008); the role of asymptomatic infection in disease dynamics hence merits further study. A genetic analysis of a newly observed infestation of *P. ramorum* in northern California revealed that the pathogen population had in reality been in residence and reproducing for some time (S. Mascheretti and M. Garbelotto, 2011, unpublished results); it is entirely possible that this was in latent, asymptomatic infections.

25.5 Management Strategies and Tactics

25.5.1 Avoidance

A seemingly obvious technique for avoiding oomycete diseases is to choose forestry sites outside the pathogen's environmental limits (Table 25.2). Climate may affect both pathogen and host. Every pathogen has its own temperature requirements for growth and sporulation; temperature and phenology effects on host susceptibility can be just as important but are often overlooked. California coast live oaks in the western USA are most susceptible to sudden oak death in the springtime during a flush of active cambial growth (Dodd *et al.*, 2008). This period of increased susceptibility has an unfortunate synchronism with both the time for maximum susceptibility of the vector, California laurel, and the time of maximum pathogen sporulation (Hüberli *et al.*, 2012). Locating planting sites in regions that are unfavourable or marginal for pathogen growth and transmission may avoid or minimize disease. Pythium diseases of seedlings may be avoided to some extent by simply sowing seed when the weather shifts to slow pathogen growth and speed plant growth, depending on the particular host–pathogen combination (Tainter and Baker, 1996). In Australia, avoidance of *Phytophthora cinnamomi* involves extensive mapping to determine where the pathogen is present and where it is not present in wildlands. Strict hygiene and quarantine protocols are then implemented in those plant communities that are disease free, and considered to be susceptible or at risk to the pathogen (Dell *et al.*, 2005). In Oregon, consideration of the proximity to known infestations is recommended for avoidance of *P. lateralis*, as is the placement of sites upslope from roads and on convex slopes to minimize exposure to new inoculum (Hansen *et al.*, 2000).

Table 25.2. Avoidance strategies available to control oomycete diseases.

Avoidance strategy	References
Canopy architecture	
Open canopies to avoid soil-borne species; however, may aid transmission of aerially-dispersed species	Tainter and Baker, 1996; Kelly and Meentemeyer, 2002; Swiecki and Bernhardt, 2002, 2006; Rizzo *et al.*, 2005
Inoculum sources	
Monitoring/mapping pathogen locations; proximity to infectious hosts; thinning or removal of infectious hosts	Shearer and Tippett, 1989; Kelly and Meentemeyer, 2002; Goheen *et al.*, 2004; Meentemeyer *et al.*, 2004; Davidson *et al.*, 2005, 2008; Maloney *et al.*, 2005; Rizzo *et al.*, 2005; Valachovic *et al.*, 2008, 2010; Cobb *et al.*, 2010; Davis *et al.*, 2010
Site selection	
Climate unfavourable to pathogen; water dynamics to carry potentially infested water away	Tainter and Baker, 1996; Hansen *et al.*, 2000; Tooley and Browning, 2008; Fichtner *et al.*, 2009; Hüberli *et al.*, 2012
Soil characteristics	
Permeable, well drained, loamy, organic content	Campbell and Copeland, 1954; Hoitink *et al.*, 1977, 1997; Weste and Vithanage, 1979; Shea *et al.*, 1983; Hardy and Sivasithamparam, 1991; Kinal *et al.*, 1993; Tainter and Baker, 1996; Fichtner *et al.*, 2009; Shearer *et al.*, 2010; Shearer and Crane, 2011

Avoidance of aerially dispersed oomycetes

One of the most important tools for the avoidance of aerially dispersed Phytophthora diseases is the minimization of spore sources. Proximity to the infectious host California laurel is a major risk factor for sudden oak death in California (Kelly and Meentemeyer, 2002; Meentemeyer *et al.*, 2004; Davidson *et al.*, 2005, 2008; Maloney *et al.*, 2005; Cobb *et al.*, 2010; Davis *et al.*, 2010). Selectively thinning California laurel has been suggested as an inoculum reduction strategy (Rizzo *et al.*, 2005; Valachovic *et al.*, 2010). Thinning treatments can range from pruning the lower branches of California laurels, as recommended by Rizzo *et al.* (2005), to the complete removal of tanoak and California laurel hosts (Goheen *et al.*, 2004; Valachovic *et al.*, 2008, 2010). Thinning treatments that span this continuum have been initiated in California (Valachovic *et al.*, 2010; EFSA, 2011). Preliminary results have shown that removal of California laurel from a 10–15 m boundary surrounding oak trees reduces spore loads to levels that usually avoid infection of the trees (EFSA, 2011).

Aerially-dispersed oomycetes make use of the same resting soil structures as their soil-borne kin. Shishkoff (2007) found that chlamydospores of *Phytophthora ramorum* persisted for over a year in potting mix and sand, and germinated near plant roots. Chlamydospore survival has been shown to be decreased by soil drying (Fichtner *et al.*, 2007), and chlamydospores have been inactivated *in vitro* at temperatures at and below −10°C and at 40°C (Tooley and Browning, 2008). A study of the persistence of chlamydospores under forest conditions found that survival of *P. ramorum* chlamydospores was minimized by redwood soil (steaming the soil did not change suppression), and that the infection of plant material by splash from the soil was decreased but not eliminated by the presence of leaf litter (Fichtner *et al.*, 2009). None the less, spread of *P. ramorum* was found to be greater in redwood forests than in mixed evergreen forest, perhaps in part owing to the greater leaf abscission by foliar hosts in those forest types (Davidson *et al.*, 2011).

Closed canopies and dense stands promote Phytophthora and Pythium diseases; sunlight and open stands can help avoid root diseases, particularly damping-off (Tainter and Baker, 1996). However, increased airflow in open stands may in fact aid transmission of aerially-dispersed *Phythophthora* spp. (Kelly and Meentemeyer, 2002; Swiecki and Bernhardt, 2002, 2006; Rizzo *et al.*, 2005).

Avoidance of soil- and water-dispersed oomycetes

Soil characteristics play a large role in the development of soil-borne Phytophthora and Pythium diseases, and much can be accomplished by choosing appropriate sites for plantations or reforestation. Well-drained, loamy soils with a high organic content are widely recommended for avoiding root diseases. Good drainage minimizes the time in contact with free water that is required for oomycete infection; high organic content reduces oospore and chlamydospore survival times (Weste and Vithanage, 1979), and it also harbours microbes that may outcompete poorly saprophytic *Pythium* spp. Organic soils in the form of composts have long been found to suppress a number of *Phytophthora* and *Pythium* spp. (Hoitink *et al.*, 1977, 1997; Hardy and Sivasithamparam, 1991). The microbial community no doubt plays a large role in the protective effect; although in some cases the composts offer disease protection even after sterilization (Fichtner *et al.*, 2009).

A study of different disease progress rates for *Banksia* spp. inoculated with *Phytophthora cinnamomi* in different soil types from across a threatened park system supported the overall rating of soil types, with the greatest mortality in sands and the least mortality in loams (Shearer and Crane, 2011). An impermeable concretized duricrust has been linked to Phytophthora root and collar rot in eucalyptus and banksia forests (Shea *et al.*, 1983; Kinal *et al.*, 1993; Shearer *et al.*, 2010); free-flowing water collects above this layer, remaining in contact with deep taproots even as the surface soil dries, thereby facilitating disease. Likewise, the relationship between soil quality and risk for littleleaf disease caused by *P. cinnamomi* on pines in the south-east USA is well established, and soils are rated before planting on a scale of 1–100. Impermeable, highly eroded soils with strongly mottled, firm subsoil are most at risk (Campbell and Copeland, 1954; Tainter and Baker, 1996).

The reduction of spore loads by host removal has also been tried for soil-borne species. In Western Australia, attempts were made to remove the highly susceptible bull banksia using silvicultural techniques (stump poisoning or by fire) to reduce inoculum loads of *P. cinnamomi*, especially where jarrah grew on soils considered to be highly conducive to the pathogen (Shearer and Tippett, 1989).

25.5.2 Exclusion

Exclusion of nursery pathogens from forests

The exclusion of nursery pathogens from forested areas is a critical issue for forest health, as exemplified by the epidemics of both *Phytophthora cinnamomi* and *P. ramorum*. Long-distance movement of plants and soil should be avoided; planting projects using seedlings should use locally grown sources and scrupulously disease-free plants (Table 25.3). Seedlings should be sourced from nurseries that have a strong emphasis on hygiene; measures include the use of new seedling containers and of container media that have been adequately composted or steam pasteurized, irrigating with water known to be *Phytophthora* free, and keeping containers on benches off the ground (Hardy and Sivasithamparam, 2002). Even local sourcing for plants does not guarantee against the emergence of new pathogens, as nurseries can select for the evolution of new hybrids such as *P. alni*.

Exclusion of aerial Phytophthoras

The current policies for exclusion of *P. ramorum* have relied on inspection protocols for nurseries and quarantines at the level of political boundaries. Within the USA, there has been a concerted eradication effort in Oregon, and the regulation of movement of host plants among infested counties within states or across state borders. A coordinated effort to slow the spread of *P. ramorum* to uninfested regions has been lacking in the California, the largest part of the pathogen's naturalized range, although a number of small projects have been initiated at the local level (Rizzo *et al.*, 2005; Alexander and Lee, 2010; Valachovic *et al.*, 2010).

Exclusion of soil-borne oomycetes

Western Australia has been the site of a more successful coordinated effort to exclude soil-borne species from new locations. Bauxite mining in jarrah forest clears about 550 ha of

Table 25.3. Exclusion strategies available to control oomycete diseases.

Exclusion strategy	References
Forest sanitation	
Monitoring/mapping pathogen locations; restricting vehicle movement from infested to uninfested areas; cleaning vehicles before entering uninfested areas; preventing infested and uninfested soils from mixing; preventing water draining from infested to uninfested areas; education of public and forestry workers	Colquhoun and Hardy, 2000; Hansen *et al.*, 2000; Colquhoun and Kerp, 2007
Nursery sanitation	
Use of new seedling containers, container media pasteurized; irrigation water *Phytophthora* free; water splash kept off leaves and wetness time minimized; containers kept off the ground; suppressive composts or fungicides avoided; sustained heat treatment to kill resting structures in plant or soil material via composting, solarization, oven treatment or autoclaving	Le Bihan *et al.*, 1997; Hardy and Sivasithamparam, 2002; Harnik *et al.*, 2004; Swain *et al.*, 2006; Tooley *et al.*, 2008

forest and moves approximately 6 million m^3 of soil annually, so the potential for spreading *Phytophthora cinnamomi* is high (Colquhoun and Hardy, 2000). *P. cinnamomi* occurs in a mosaic throughout the jarrah forest and a range of procedures have been developed for every stage of the mining and restoration process to ensure that the pathogen is not spread. These procedures include: (i) knowing where the pathogen is through detailed mapping for the presence of *P. cinnamomi* using susceptible indicator species and soil baiting; (ii) restricting vehicle movement from infested to uninfested areas; (iii) cleaning vehicles before entering uninfested areas; (iv) preventing infested and uninfested soils from mixing; (v) preventing water draining from infested to uninfested areas; (vi) training all field staff and planners; and (vii) monitoring the spread of the disease attributable to mining and investigating the causes (Colquhoun and Kerp, 2007). As a result of these procedures, and despite the logistics of moving huge volumes of soil, it is estimated that the rates of spread of *P. cinnamomi* attributable to mining activities equate to 6×10^{-4} ha for every hectare mined (Colquhoun and Kerp, 2007). Therefore, it is possible to manage the soil-borne species of *Phytophthora* given adequate resources and infrastructure.

Port Orford cedar disease from *P. lateralis* has been the target of a similar effort to contain spread from new areas. The disease spreads primarily from mud on vehicles along roads and then through waterways (Jules *et al.*, 2002). Techniques in place for controlling spread have been outlined in Hansen *et al.* (2000), and include: (i) road closures – permanent and wet season; (ii) re-engineering roads, elevating and paving surfaces, and redirecting runoff away from Port Orford cedar stands; (iii) controlling the use of water for dust control and firefighting to prevent infested water sources from being sprayed on to stands; (iv) public education about risk of spread; and (v) restricting harvest operations.

Sanitation

For all oomycete diseases, the sanitation of infested soil, water, and plant materials is critical to prevent pathogen spread. Heat to 40°C effectively kills even resting chlamydospores of *Phytophthora ramorum* (Harnik *et al.*, 2004; Swain *et al.*, 2006; Tooley *et al.*, 2008). This temperature can be achieved via direct exposure to the sun, by a solarization process or by composting. Simply piling dead wood in the sun rather than in the shade is a wise choice for helping to minimize pathogen survival, or a more formal composting protocol can be used to assure the destruction of propagules in plant material (Swain *et al.*, 2006). Solarization has been shown to be as effective as chemical fumigants in eliminating *Pythium* spp. from nursery soil (Le Bihan *et al.*, 1997) and eradicating *Phytophthora cryptogea* Pethybr. & Laff. from infested plots of a loamy soil up to 30 cm depth (Kaewruang *et al.*, 1989).

In addition to the sanitation of water used for dust control and firefighting that was previously mentioned, landscape and park settings should avoid irrigating with stream or pond water during times of maximum sporulation, although there is some evidence that passing through sprinklers reduces inoculum levels of *P. ramorum*, and lowers disease incidence in nursery plants (Tjosvold *et al.*, 2008). Standard irrigation practices are recommended, including avoiding water splash on leaves, avoiding watering drought-adapted species during the dry season, or avoiding watering in the evenings, when free water and leaf wetness times will be longest and so allow the maximum opportunity for oomycete infection. Tjosvold *et al.* (2008) found that direct application of infested irrigation water to leaves of nursery stock was required for ramorum blight to occur. However, infection of the aerial parts of plants by *Phytophthora* via the roots should not be completely discounted, as *P. ramorum* infection from infested soil has been documented on various ornamental species (Parke and Lewis, 2007; Shishkoff, 2007). In production nurseries, it is critical to avoid the use of fungicides including phosphonates, which can suppress disease symptoms, and thus allow the sale and movement of apparently disease-free nursery stock into the wider environment.

25.5.3 Eradication

In contrast to its wide establishment at the time of its discovery in California, sudden oak death was first detected as nine infestations in a small area in Oregon (Goheen *et al.*, 2002; Hansen *et al.*, 2008). Consequently, an eradication programme for *P. ramorum* was initiated in Oregon, involving cutting and burning all host plants within 15–30 m (2001 and 2002) or 100 m (2003 onwards) of the detection site, followed by aerial detection surveys and stream leaf baiting to detect new infestations (Table 25.4). This programme has successfully eradicated the pathogen within microsites, and has doubtlessly slowed the spread of the epidemic in Oregon. Yet eradication has not been successful on a landscape scale: new infestations have been detected each year following the initiation of the eradication programme, at up to 4 km from the nearest known infestation (Kanaskie *et al.*, 2006, 2008; Hansen *et al.*, 2008).

A recent eradication programme for *P. cinnamomi* has also had local success. Working on the idea that denying the pathogen living roots for nutrition would ruin its competitive ability in soil, the programme involved: (i) physical and herbicidal removal of all plants from the site, to a distance of 4–10 m beyond the disease front; (ii) root barriers to prevent the intrusion of roots from adjacent sites; (iii) treatments of the soil surface with triadiazole and metalaxyl; (iv) fumigation of the deep soil and soil surface with metham-sodium. The pathogen was completely eradicated from one site, and largely controlled at the others (Dunstan *et al.*, 2010).

25.5.4 Protection

Systemic chemical protection: phosphonates

Systemically translocated chemicals containing phosphonates (including fosetyl-aluminium and potassium salts of phosphorous acid) are among the most important agents for controlling oomycete diseases (Cohen and Coffey, 1986; Smillie *et al.*, 1989; Guest and Grant, 1991) (Table 25.5). Phosphonates have two modes of action, depending on dose (Smillie *et al.*, 1989). At high concentrations, they decrease pathogen growth and sporulation directly (Wilkinson *et al.*, 2001a,b; Garbelotto *et al.*, 2009). At low concentrations, they act indirectly by stimulating host defences. There is some evidence that host defences may be stimulated by changes in pathogen metabolism, rather than by an effect of phosphonate on the host itself. Low concentrations of phosphonates have been found to alter the metabolism of *Phytophthora* spp. without slowing hyphal growth (Grant *et al.*, 1990).

Phosphonates have long been observed to have varying actions among individual hosts; one hypothesis is that the chemicals act primarily to stimulate extant host defences (Grant *et al.*, 1990; Guest and Bompeix, 1990; Daniel and Guest, 2005). A study by Pilbeam *et al.* (2011) compared the histological responses to phosphonates of resistant and susceptible lines of infected eucalyptus and revealed that while phosphonate application

Table 25.4. Eradication strategies available to control oomycete diseases caused by *Phytophthora* spp.

Eradication strategy	References
P. cinnamomi	
Physical and herbicidal removal of all plants to 4–10 m beyond disease front; root barriers; triadiazole and metalaxyl treatments of soil surface; fumigation of deep soil	Dunstan *et al.*, 2010
Locally successful; landscape effects unknown	
P. ramorum	
Physical, herbicide, and burn removal of hosts within 100 m of infestation; aerial monitoring for new sites	Goheen *et al.*, 2002; Kanaskie *et al.*, 2006, 2008; Hansen *et al.*, 2008
Locally successful, but spread continues at landscape level	

Table 25.5. Strategies for protection against and therapy for Phytophthora and Pythium diseases.

Protective or therapeutic strategy	References
Biological control	
Composts and organic soils suppress pathogen growth and survival; amending plant community to include hosts, which act to reduce environmental spore loads	Hoitink *et al.*, 1977, 1997; Shea *et al.*, 1978; Hardy and Sivasithamparam, 1991; D'Souza *et al.*, 2004; Fichtner *et al.*, 2009
Copper compounds	
Give surface protection and have a long use history; no therapeutic benefit; ecotoxic	McCallan, 1949; Hirst *et al.*, 1961; Frank *et al.*, 1976; Heitefuss, 1989; Erwin and Ribeiro, 1996; Howard *et al.*, 1998; Pietrzak and McPhail, 2004; Bünemann *et al.*, 2006; Garbelotto *et al.*, 2009
Phenylamide compounds (e.g. metalaxyl)	
Active only on oomycetes by inhibiting RNA synthesis; inhibit spore production and chlamydospore germination; protective and therapeutic; resistance in pathogens is common	Cohen and Coffey, 1986; Erwin and Ribeiro, 1996
Phosphonate compounds	
Act directly to suppress pathogen and indirectly to stimulate host defences; protective and therapeutic	Cohen and Coffey, 1986; Smillie *et al.*, 1989; Grant *et al.*, 1990; Guest and Grant, 1991; Tainter and Baker 1996; Jackson *et al.*, 2000; Wilkinson *et al.*, 2001a,b; Dobrowolski *et al.*, 2008; Suddaby *et al.*, 2008; Garbelotto and Schmidt, 2009; Garbelotto *et al.*, 2009; King *et al.*, 2010; Pilbeam *et al.*, 2011

equalized responses in resistant and susceptible lines after several days, their early histological responses were quite different. Responses were stronger in the susceptible line, and included increases in lignin and suberin production. The resistant line increased suberin alone, and the stimulation of mitosis and callus formation was observed in both lines.

In the field, 75% of tanoak trees in 32 plots that were treated once annually with a bark application of phosphonate (plus a surfactant) as described by Garbelotto and Schmidt (2009) were healthy after 5 years, compared with only 55% of untreated trees (M. Garbelotto and D.J. Schmidt, 2011, unpublished results).

Pathogen resistance to phosphonates is believed to be unlikely or slow to develop because of its complex mode of action, yet reduced sensitivity to the chemical has been documented. In particular, Dobrowolski *et al.* (2008) were able to experimentally select for phosphonate resistance in *P. cinnamomi* isolates, and documented reduced sensitivity to phosphonate in isolates originating in sites with long-term histories of phosphonate usage.

Systemic chemical protection: phenylamides

Phenylamides, of which metalaxyl is the most commonly used, are water soluble and are rapidly translocated into plant tissues from roots, shoots and leaves. They act by inhibiting ribosomal RNA synthesis in oomycetes, and only in oomycetes (Erwin and Ribeiro, 1996). They can be very effective at inhibiting sporangial production and chlamydospore germination at concentrations as low as 1 μg ml^{-1} (Cohen and Coffey, 1986). Resistance to these compounds has become a major problem, however, and necessitates the combination with or alternation of metalaxyl with other fungicides. Metalaxyl was found to be effective in controlling *P. ramorum in vitro*, but drenches were not effective for controlling sudden oak death on California coast live oak *in planta* (Garbelotto *et al.*, 2009).

Surface chemical protection: copper compounds

Copper sulfate, as the famous Bordeaux mixture, has been used to control plant disease since the very beginning of plant pathology. Copper compounds are used extensively as foliar sprays and trunk paints to control aerial *Phythophthora* spp., but they offer a strictly protective function, having no action once infection has occurred (McCallan, 1949; Heitefuss, 1989; Erwin and Ribeiro, 1996). Copper hydroxide is effective in controlling ramorum blight on California laurel leaves, resulting in almost complete reduction of lesions on treated leaves for at least 4 weeks after applying a foliar spray at 460 μg ml^{-1} (Garbelotto *et al.*, 2009). Soil treatments of copper sulfate have been shown to have some protective effect against *P. cinnamomi*: the pathogen was not fully killed by copper sulfate treatments *in vitro* or in gravel and topsoil, but reductions in plant infection rates were observed in treated soil (Howard *et al.*, 1998).

Copper is an environmental pollutant, and copper compounds have negative long-term impacts through their accumulation in soil and accompanying heavy metal toxicity (Hirst *et al.*, 1961; Frank *et al.*, 1976; Bünemann *et al.*, 2006). In soil layers that are slow to turn over, copper may be available to leaching for very long periods of time (Pietrzak and McPhail, 2004).

Biological protection

The role of biological compounds in protection from oomycete diseases is highlighted by the suppressive action of composts on *Phytophthora* spp. and *Pythium* spp. Unsterilized composts may suppress a number of *Phytophthora* and *Pythium* species (Hoitink *et al.*, 1977, 1997; Hardy and Sivasithamparam, 1991); documentation of suppression by sterilized soils is rarer (Fichtner *et al.*, 2009). Consequently, non-sterilized composts should not be incorporated into container media, as *Phytophthora* spp. could be spread in stock that is infested but made symptomless by suppression (Hardy and Sivasithamparam, 1991).

Plant species may also be used for biological control. *Acacia* spp. have been documented to suppress *P. cinnamomi* sporulation (Shea *et al.*, 1978) and to protect bull banksia from infection by *P. cinnamomi* when they were planted together in a field experiment (D'Souza *et al.*, 2004). Before being deployed to either reduce the number of susceptible hosts or to actively suppress the pathogen, a plant species must be determined to be truly resistant. Hosts with asymptomatic infections will not have the desired effect; these species will rather allow inoculum to continue to build (D'Souza *et al.*, 2005).

25.5.5 Resistance

A breeding programme for host resistance to manage *P. lateralis* disease in Port Orford cedars that relies on results from multiple inoculation techniques has experienced recent success, with resistant genotypes now commercially available (Sniezko *et al.*, 2012; Table 25.6). The success is a result from a concerted effort to identify resistant trees from survivors in diseased areas that begun in 1996, followed by propagation, crosses and multiple inoculation assays, and importantly, by a combination of greenhouse testing with field validation (Oh *et al.*, 2006; Sniezko, 2006). The programme has focused on quantitative (partial) resistance, which in this and other systems has been less likely to be overcome by the evolution of virulence in the pathogen (Carson and Carson, 1989; Kinloch *et al.*, 2008).

Heritable variation in resistance to oomycete pathogens has been observed in other systems. Resistance in Monterey pine to littleleaf disease has been a consideration for breeding since its documentation in 1984 (Butcher *et al.*, 1984). Selection and propagation efforts are likewise underway for chestnut trees resistant to ink disease caused by *P. cinnamomi* and *P. cambivora* (Robin *et al.*, 2006; Miranda-Fontaina *et al.*, 2007), while a genomics screening effort is in progress to find molecular markers for resistance to *P. cinnamomi* in chestnut that may be used for early screening (Olukolu *et al.*, 2012). A project

Table 25.6. Use of resistance to manage oomycete diseases caused by *Phytophthora* spp.

Resistance programme	References
Chestnut (*Castanea* spp.) resistance to *P. cinnamomi* and *P. cambivora* ink disease Jarrah (*Eucalyptus marginata* Donn ex Sm.) lines resistant to *P. cinnamomi* Monterey pine (*Pinus radiata* Donn ex Sm.) resistance to *P. cinnamomi* Port Orford cedar (*Chamaecyparis lawsoniana* (A. Murray bis) Parl.) resistant to *P. lateralis* based on multiple inoculation techniques; lines commercially available Tanoak (*Lithocarpus densiflorus* (Hook. & Arn.) Rehder) resistance to *P. ramorum*	Butcher *et al.*, 1984; Stukely and Crane, 1994; Oh *et al.*, 2006; Robin *et al.*, 2006; Sniezko, 2006; Miranda-Fontaina *et al.*, 2007; Stukely *et al.*, 2007; Hayden *et al.*, 2010; Olukolu *et al.*, 2012; Sniezko *et al.*, 2012

to survey variation in quantitative resistance to sudden oak death in tanoaks throughout their range had identified families that perform better than average in laboratory and field trials; these families have potential to form the basis of a resistance breeding programme (Hayden *et al.*, 2010, 2011).

25.5.6 Therapy

Phosphonate and phenylamide compounds may be used for therapy as well as for protection (Table 25.5). In New Zealand, stands of Monterey pine were rehabilitated by the aerial application of phosphonates applied at 568 kg ha^{-1} (Tainter and Baker, 1996). The application resulted in a suite of improvements that together limited the pathogen: crown improvement, root rejuvenation and mycorrhizal gains; and improved soil aeration and reduced waterlogging because of improved crown condition and increased transpiration. Phosphonate injections of oaks and tanoaks (Plate 31) can provide therapy if applied in the early stages of the disease (Garbelotto and Schmidt, 2009). Resistant and susceptible lines of eucalyptus seedlings show histological changes in response to phosphonate treatment after infection by *P. cinnamomi*; these promote callus formation by stimulating mitosis and the production of lignin (resistant lines), and lignin and suberin (susceptible lines) (Pilbeam *et al.*, 2011). In grass-tree (*Xanthorrhoea australis* R. Br.) seedlings, potassium phosphonate limited the growth of *P. cinnamomi* to the root cortex and the cell membrane retracted from the cell wall. Additionally, phenolic compounds and electron dense substances were deposited around the wall of infected and neighbouring cells. In contrast, in seedlings not treated with potassium phosphonate, hyphal growth became intracellular and progressed into the vascular tissue (Daniel *et al.*, 2005). At a broader scale, phosphonate applied as a foliar spray has been shown to control *P. cinnamomi* in a range of Australian native plant species in two natural plant communities (Tynan *et al.*, 2001).

Metalaxyl is especially useful for therapy because of its inhibitory effects on spore production and chlamydospore germination (Cohen and Coffey, 1986), and its rapid translocation to new tissues (Erwin and Ribeiro, 1996). Widespread problems with resistance do limit its usefulness though.

The amendment of soils may have some therapeutic effect. Tainter and Baker (1996) recommend 5–10–5 commercial (NPK) fertilizer at 2273 kg ha^{-1} applied with 454 kg ammonium sulfate and repeated every 4 years for the therapy of pine trees in the early stages of development of littleleaf disease. Increased nitrogen offsets the declining crown, and may be sufficient to maintain the trees for some time. An especially conducive year for disease may offset any therapeutic effect, however.

25.5.7 Integrated disease management

Oomycete diseases are so widespread and so versatile that no one management tactic can be completely successful. Any site-level management plan must take into account the environment, the host, the pathogen and human activities, and work to reduce inoculum load, likelihood of infection and severity of the disease altogether.

From a broad perspective, nurseries are the most important venues for disease management because of their role in incubating and spreading pathogens worldwide (Hardy and Sivasithamparam, 2002; Brasier, 2008; Frankel, 2008). Hardy and Sivasithamparam (2002) provide recommendations for nursery sanitation that should be followed for all commercial operations, whether for forest trees or ornamentals. While composts and chemical treatments may be useful to suppress disease in natural settings, they should be avoided in nursery settings because of their potential to disguise infected but asymptomatic plants that are harbouring resting spores (Hardy and Sivasithamparam, 1991). If they are to be effective, enacting limitations on the movement of pathogens from nurseries should not rest solely on foresters, but must include the landscape industry and components of public education and public policy.

Management programmes for outbreaks of soil-borne Phytophthora disease in wildlands have depended on regulatory and local support, as well as on extensive monitoring, sanitation and quarantine programmes (Hansen *et al.*, 2000; Colquhoun and Kerp, 2007; Frankel, 2008; Alexander and Lee, 2010). Large-scale management of aerial Phytophthoras provides a special problem because of the potential for long-distance dispersal and complex host relationships. Ndeffo Mbah and Gilligan (2010a,b), using the *P. ramorum* epidemic in the USA as a model system, provide decision-making models for the allocation of detection, eradication and treatment resources, recommending equal effort towards detection and eradication, with treatment directed at infectious foliar hosts.

Successes in Western Australia demonstrate that the management of Phytophthora disease can be accomplished on a landscape scale, given enough scientific knowledge and political will. While the number of known *Phytophthora* spp. is increasing steadily, an understanding of the ecology of soil-borne and aerial Phytophthoras is also growing, allowing further avenues for management (Rizzo *et al.*, 2005). Future management successes will depend on the combination of an increasing body of research with the political will to mount a coordinated response.

References

Adams, G.C., Catal, M., Trummer, L., Hansen, E.M., Reeser, P. and Worrall, J.J. (2008) *Phytophthora alni* subsp. *uniformis* found in Alaska beneath thinleaf alders. *Plant Health Progress*, doi: 10.1094/PHP-2008-1212-02-BR.

Alexander, J. and Lee, C.A. (2010) Lessons learned from a decade of sudden oak death in California: evaluating local management. *Environmental Management* 46, 315–328.

Alexopoulos, C.J., Mims, C.W. and Blackwell, M. (1996) *Introductory Mycology*. John Wiley, New York.

Anagnostakis, S. (1995) The pathogens and pests of chestnuts. *Advances in Botanical Research* 21, 125–145.

Arnold, A.E., Mejía, L.C., Kyllo, D., Rojas, E.I., Maynard, Z., Robbins, N. and Herre, E.A. (2003) Fungal endophytes limit pathogen damage in a tropical tree. *Proceedings of the National Academy of Sciences of the United States of America* 100, 15649–15654.

Augspurger, C.K. (1984) Seedling survival of tropical tree species: interactions of dispersal distance, light-gaps, and pathogens. *Ecology* 65, 1705–1712.

Bakonyi, J., Nagy, Z.Á. and Érsek, T. (2006) PCR-based DNA markers for identifying hybrids within *Phytophthora alni*. *Journal of Phytopathology* 154, 168–177.

Balci, Y., Balci, S., Blair, J.E., Park, S.-Y., Kang, S. and Macdonald, W.L. (2008) *Phytophthora quercetorum* sp. nov., a novel species isolated from eastern and north-central USA oak forest soils. *Mycological Research* 112, 906–916.

Belbahri, L., Calmin, G., Wagner, S., Moralejo, E., Woodward, S. and Lefort, F. (2007) Specific hybridization real-time PCR probes for *Phytophthora ramorum* detection and diagnosis. *Forest Pathology* 37, 403–408.

Bell, T., Freckleton, R.P. and Lewis, O.T. (2006) Plant pathogens drive density-dependent seedling mortality in a tropical tree. *Ecology Letters* 9, 569–574.

Bilodeau, G.J., Lévesque, C.A., de Cock, A.W.A.M., Brière, S.C. and Hamelin, R.C. (2007a) Differentiation of European and North American genotypes of *Phytophthora ramorum* by real-time polymerase chain reaction primer extension. *Canadian Journal of Plant Pathology* 29, 408–420.

Bilodeau, G.J., Lévesque, C.A., de Cock, A.W.A.M., Duchaine, C., Brière, S.C., Uribe, P., Martin, F.N. and Hamelin, R.C. (2007b) Molecular detection of *Phytophthora ramorum* by real-time polymerase chain reaction using TaqMan, SYBR Green, and molecular beacons. *Phytopathology* 97, 632–642.

Böhm, J., Hahn, A., Schubert, R., Bahnweg, G., Adler, N., Nechwatal, J., Oehlmann, R. and Obszwald, W. (1999) Real-time quantitative PCR: DNA determination in isolated spores of the mycorrhizal fungus *Glomus mosseae* and monitoring of *Phytophthora infestans* and *Phytophthora citricola* in their respective host plants. *Journal of Phytopathology* 147, 409–416.

Brasier, C.M. (1971) Induction of sexual reproduction in single A2 isolates of *Phytophthora* species by *Trichoderma viride*. *Nature New Biology* 231, 283.

Brasier, C.M. (1975) Stimulation of sex organ formation in *Phytophthora* by antagonistic species of *Trichoderma*. I. The effect *in vitro*. *New Phytologist* 74, 183–194.

Brasier, C.M. (2000) The rise of the hybrid fungi. *Nature* 405, 134–135.

Brasier, C.M. (2008) The biosecurity threat to the UK and global environment from international trade in plants. *Plant Pathology* 57, 792–808.

Brasier, C.M. and Strouts, R.G. (1976) New records of *Phytophthora* on trees in Britain. I. Phytophthora root rot and bleeding canker of horse chestnut (*Aesculus hippocastanum* L.). *Journal of Forest Pathology* 6, 129–136.

Brasier, C.M., Robredo, F. and Ferraz, J.F.P. (1993) Evidence for *Phytophthora cinnamomi* involvement in Iberian oak decline. *Plant Pathology* 42, 140–145.

Brasier, C.M., Rose, J. and Gibbs, J.N. (1995) An unusual *Phytophthora* associated with widespread alder mortality in Britain. *Plant Pathology* 44, 999–1007.

Brasier, C.M., Cooke, D.E.L. and Duncan, J.M. (1999) Origin of a new *Phytophthora* pathogen through interspecific hybridization. *Proceedings of the National Academy of Sciences of the United States of America* 96, 5878–5883.

Brasier, C.M., Denman, S., Brown, A. and Webber, J. (2004a) Sudden oak death (*Phytophthora ramorum*) discovered on trees in Europe. *Mycological Research* 108, 1108–1110.

Brasier, C.M., Kirk, S.A., Delcan, J., Cooke, D.E.L., Jung, T. and Man in't Veld, W.A. (2004b) *Phytophthora alni* sp. nov. and its variants: designation of emerging heteroploid hybrid pathogens spreading on *Alnus* trees. *Mycological Research* 108, 1172–1184.

Brasier, C.M., Beales, P.A., Kirk, S.A., Denman, S. and Rose, J. (2005) *Phytophthora kernoviae* sp. nov., an invasive pathogen causing bleeding stem lesions on forest trees and foliar necrosis of ornamentals in the UK. *Mycological Research* 109, 853–859.

Brasier, C.M., Vettraino, A.M., Chang, T.T. and Vannini, A. (2010) *Phytophthora lateralis* discovered in an old growth *Chamaecyparis* forest in Taiwan. *Plant Pathology* 59, 595–603.

Bünemann, E.K., Schwenke, G.D. and Van Zwieten, L. (2006) Impact of agricultural inputs on soil organisms – a review. *Soil Research* 44, 379–406.

Burgess, T.I., Webster, J.L., Ciampini, J.A., White, D., Hardy, G.E.St.J. and Stukely, M.J.C. (2009) Re-evaluation of *Phytophthora* species isolated during 30 years of vegetation health surveys in western Australia using molecular techniques. *Plant Disease* 93, 215–223.

Butcher, T., Stukely, M. and Chester, G. (1984) Genetic variation in resistance of *Pinus radiata* to *Phytophthora cinnamomi*. *Forest Ecology and Management* 8, 197–220.

Campbell, W.A. and Copeland, O.L. Jr (1954) *Littleleaf Disease of Shortleaf and Loblolly Pines*. Circular 970, US Department of Agriculture, Washington, DC.

Caroselli, N.E. and Howard, F.L. (1939) Bleeding canker of maples. *Bulletin Bartlett Tree Research Laboratory* 3, 44–48.

Carson, S.D. and Carson, M.J. (1989) Breeding for resistance in forest trees: a quantitative genetics approach. *Annual Review of Phytopathology* 27, 373–395.

Cerny, K., Strnadova, V., Gregorova, B., Holub, V., Tomsovsky, M., Mrazkova, M. and Gabrielova, S. (2008) *Phytophthora cactorum* causing bleeding canker of common beech, horse chestnut, and white poplar in the Czech Republic. *Plant Pathology* 58, 394.

Chavarriaga, D., Bodles, W.J., Leifert, C., Belbahri, L. and Woodward, S. (2007) *Phytophthora cinnamomi* and other fine root pathogens in north temperate pine forests. *FEMS Microbiology Letters* 276, 67–74.

Cobb, R.C., Meentemeyer, R.K. and Rizzo, D.M. (2010) Apparent competition in canopy trees determined by pathogen transmission rather than susceptibility. *Ecology* 91, 327–333.

Cohen, Y. and Coffey, M.D. (1986) Systemic fungicides and the control of oomycetes. *Annual Review of Phytopathology* 24, 311–338.

Colquhoun, I.J. and Hardy, G.E.St.J. (2000) Managing the risks of Phytophthora root and collar rot during bauxite mining in the *Eucalyptus marginata* (jarrah) forest of western Australia. *Plant Disease* 84, 116–127.

Colquhoun, I.J. and Kerp, N.L. (2007) Minimizing the spread of a soilborne plant pathogen during a large-scale mining operation. *Restoration Ecology* 15, S85–S93.

Connell, J.H. (1971) On the role of natural enemies in preventing competitive exclusion in some marine animals and in rain forest trees. In: Den Boer, P.J. and Gradwell, G.R. (eds) *Dynamics of Populations: Proceedings of the Advanced Study Institute on Dynamics of Numbers in Populations, Oosterbeek, the Netherlands, 7–18 September 1970*. Centre for Agricultural Publishing and Documentation (Pudoc), Wageningen, The Netherlands, pp. 298–312.

Cother, E.J. and Griffin, D.M. (1974) Chlamydospore germination in *Phytophthora drechsleri*. *Transactions of the British Mycological Society* 63, 273–279.

Crandall, B.S., Gravatt, G.F. and Ryan, M.M. (1945) Root disease of *Castanea* species and some coniferous and broadleaf nursery stocks, caused by *Phytophthora cinnamomi*. *Phytopathology* 35, 162–180.

Daniel, R. and Guest, D. (2005) Defence responses induced by potassium phosphonate in *Phytophthora palmivora*-challenged *Arabidopsis thaliana*. *Physiological and Molecular Plant Pathology* 67, 194–201.

Daniel, R., Wilson, B.A. and Cahill, D.M. (2005) Potassium phosphonate alters the defence response of *Xanthorrhoea australis* following infection by *Phytophthora cinnamomi*. *Australasian Plant Pathology* 34, 541–548.

Davidson, J.M., Werres, S., Garbelotto, M., Hansen, E.M. and Rizzo, D.M. (2003) Sudden oak death and associated diseases caused by *Phytophthora ramorum*. *Plant Health Progress* 23, doi: 10.1094/PHP-2003-0707-01-DG.

Davidson, J.M., Wickland, A.C., Patterson, H.A., Falk, K.R. and Rizzo, D.M. (2005) Transmission of *Phytophthora ramorum* in mixed-evergreen forest in California. *Phytopathology* 95, 587–596.

Davidson, J.M., Patterson, H.A. and Rizzo, D.M. (2008) Sources of inoculum for *Phytophthora ramorum* in a redwood forest. *Phytopathology* 98, 860–866.

Davidson, J.M., Patterson, H.A., Wickland, A.C., Fichtner, E.J. and Rizzo, D.M. (2011) Forest type influences transmission of *Phytophthora ramorum* in California oak woodlands. *Phytopathology* 101, 492–501.

Davis, F.W., Borchert, M., Meentemeyer, R.K., Flint, A. and Rizzo, D.M. (2010) Pre-impact forest composition and ongoing tree mortality associated with sudden oak death in the Big Sur region, California. *Forest Ecology and Management* 259, 2342–2354.

Day, W.R. (1938) Root-rot of sweet chestnut and beech caused by species of *Phytophthora*. I. Cause and symptoms of disease: its relation to soil conditions. *Forestry* 12, 101–116.

Dell, B., Hardy, G.E.St.J. and Vear, K. (2005) History of *Phytophthora cinnamomi* management in Western Australia. In: Calver, M.C., Bigler-Cole, H., Bolton, G., Dargavel, J., Gaynor, A., Horwitz, P., Mills, J. and Wardell-Johnston, G. (eds) *A Forest Conscienseness: Proceedings Sixth National Conference of the Australian Forest History Society Inc.* Millpress Science Publishers, Rotterdam, The Netherlands, pp. 391–406.

DeNitto, G.A. and Kliejunas, J.T. (1991) First report of *Phytophthora lateralis* on Pacific yew. *Plant Disease* 75, 968.

Denman, S., Moralejo, E., Kirk, S.A., Orton, E. and Whybrow, A. (2008) Sporulation of *Phytophthora ramorum* and *P. kernoviae* on asymptomatic foliage and fruit. In: Frankel, S.J., Kliejunas, J.T. and Palmieri, K.M. (eds) *Proceedings of the Sudden Oak Death Third Science Symposium, March 5–9, 2007, Santa Rosa, California*. General Technical Report PSW-GTR-214, US Department of Agriculture, Forest Service, Pacific Southwest Research Station, Albany, California, pp. 201–207.

DiLeo, M.V., Bostock, R.M. and Rizzo, D.M. (2009) *Phytophthora ramorum* does not cause physiologically significant systemic injury to California bay laurel, its primary reservoir host. *Phytopathology* 99, 1307–1311.

Dobrowolski, M.P., Tommerup, I.C., Shearer, B.L. and O'Brien, P.A. (2003) Three clonal lineages of *Phytophthora cinnamomi* in Australia revealed by microsatellites. *Phytopathology* 93, 695–704.

Dobrowolski, M.P., Shearer, B.L., Colquhoun, I.J., O'Brien, P.A. and Hardy, G.E.St.J. (2008) Selection for decreased sensitivity to phosphite in *Phytophthora cinnamomi* with prolonged use of fungicide. *Plant Pathology* 57, 928–936.

Dodd, R.S., Hüberli, D., Mayer, W., Harnik, T.Y., Afzal-Rafii, Z. and Garbelotto, M. (2008) Evidence for the role of synchronicity between host phenology and pathogen activity in the distribution of sudden oak death canker disease. *New Phytologist* 179, 505–514.

D'Souza, N.K., Colquhoun, I.J., Shearer, B.L. and Hardy, G.E.St.J. (2004) The potential of five western Australian native *Acacia* species for biological control of *Phytophthora cinnamomi*. *Australian Journal of Botany* 52, 267–279.

D'Souza, N.K., Colquhoun, I.J., Shearer, B.L. and Hardy, G.E.St.J. (2005) Assessing the potential for biological control of *Phytophthora cinnamomi* by fifteen native western Australian jarrah-forest legume species. *Australasian Plant Pathology* 34, 533–540.

Dunstan, W.A., Rudman, T., Shearer, B.L., Moore, N.A., Paap, T., Calver, M.C., Dell, B. and Hardy, G.E.St.J. (2010) Containment and spot eradication of a highly destructive, invasive plant pathogen (*Phytophthora cinnamomi*) in natural ecosystems. *Biological Invasions* 12, 913–925.

Durán, A., Gryzenhout, M., Slippers, B., Ahumada, R., Rotella, A., Flores, F., Wingfield, B.D. and Wingfield, M.J. (2008) *Phytophthora pinifolia* sp. nov. associated with a serious needle disease of *Pinus radiata* in Chile. *Plant Pathology* 57, 715–727.

EFSA (2011) Does the removal of bay laurel reduce the inoculum pressure of *Phytophthora ramorum* in oak woodlands? *European Food Safety Authority Journal* 9, 2186–2293.

Elegbede, C.F., Pierrat, J.-C., Aguayo, J., Husson, C., Halkett, F. and Marçais, B. (2010) A statistical model to detect asymptomatic infectious individuals with an application in the *Phytophthora alni*-induced alder decline. *Phytopathology* 100, 1262–1269.

Érsek, T., Schoelz, J.E. and English, J.T. (1994) PCR amplification of species-specific DNA sequences can distinguish among *Phytophthora* species. *Applied and Environmental Microbiology* 60, 2616–2621.

Erwin, D.C. and Ribeiro, O.K. (1996) *Phytophthora Diseases Worldwide*. APS Press, St Paul, Minnesota.

Fichtner, E.J., Lynch, S.C. and Rizzo, D.M. (2007) Detection, distribution, sporulation, and survival of *Phytophthora ramorum* in a California redwood–tanoak forest soil. *Phytopathology* 97, 1366–1375.

Fichtner, E.J., Lynch, S.C. and Rizzo, D.M. (2009) Survival, dispersal, and potential soil-mediated suppression of *Phytophthora ramorum* in a California redwood–tanoak forest. *Phytopathology* 99, 608–619.

Fichtner, E.J., Rizzo, D.M., Kirk, S.A. and Webber, J.F. (2011) Root infections may challenge management of invasive *Phytophthora* spp. in UK woodlands. *Plant Disease* 95, 13–18.

Frank, R., Braun, H.E., Ishida, K. and Suda, P. (1976) Persistent organic and inorganic pesticide residues in orchard soils and vineyards of southern Ontario. *Canadian Journal of Soil Science* 56, 463–484.

Frankel, S.J. (2008) Sudden oak death and *Phytophthora ramorum* in the USA: a management challenge. *Australasian Plant Pathology* 37, 19–25.

Garbelotto, M. and Schmidt, D. (2009) Phosphonate controls sudden oak death pathogen for up to 2 years. *California Agriculture* 63, 10–17.

Garbelotto, M., Davidson, J.M., Ivors, K., Maloney, P.E., Hüberli, D., Koike, S.T. and Rizzo, D.M. (2002) Non-oak native plants are main hosts for sudden oak death pathogen in California. *California Agriculture* 57, 18–23.

Garbelotto, M., Harnik, T.Y. and Schmidt, D.J. (2009) Efficacy of phosphonic acid, metalaxyl-M and copper hydroxide against *Phytophthora ramorum in vitro* and *in planta*. *Plant Pathology* 58, 111–119.

Gilbert, G.S. (2002) Evolutionary ecology of plant diseases in natural ecosystems. *Annual Review of Phytopathology* 40, 13–43.

Goheen, E.M., Hansen, E.M., Kanaskie, A., McWilliams, M.G., Osterbauer, N. and Sutton, W. (2002) Sudden oak death caused by *Phytophthora ramorum* in Oregon. *Plant Disease* 86, 441.

Goheen, E.M., Hansen, E.M., Kanaskie, A., McWilliams, M.G., Osterbauer, N., Sutton, W. and Rehms, L. (2004) An eradication strategy for *Phytophthora ramorum* in Oregon coastal forests. *Phytopathology* 94, S35.

Goss, E.M., Carbone, I. and Grünwald, N.J. (2009) Ancient isolation and independent evolution of the three clonal lineages of the exotic sudden oak death pathogen *Phytophthora ramorum*. *Molecular Ecology* 18, 1161–1174.

Grant, B.R., Dunstan, R.H., Griffith, J.M., Niere, J.O. and Smillie, R.H. (1990) The mechanism of phosphonic (phosphorous) acid action in *Phytophthora*. *Australasian Plant Pathology* 19, 115.

Tucker, C.M. and Milbrath, J.A. (1942) Root rot of *Chamaecyparis* caused by a species of *Phytophthora*. *Mycologia* 34, 94–103.

Tynan, K.M., Wilkinson, C.J., Holmes, J.M., Dell, B., Colquhoun, I.J., McComb, J.A. and Hardy, G.E.St.J. (2001) The long-term ability of phosphite to control *Phytophthora cinnamomi* in two native plant communities of Western Australia. *Australian Journal of Botany* 49, 761–770.

Valachovic, Y., Lee, C., Marshall, J. and Scanlon, H. (2008) Wildland management of *Phytophthora ramorum* in northern California forests. In: Frankel, S.J., Kliejunas, J.T. and Palmieri, K.M. (eds) *Proceedings of the Sudden Oak Death Third Science Symposium, March 5-9, 2007, Santa Rosa, California*. General Technical Report PSW-GTR-214, US Department of Agriculture, Forest Service, Pacific Southwest Research Station, Albany, California, pp. 305–312.

Valachovic, Y., Lee, C., Marshall, J. and Scanlon, H. (2010) Forest treatment strategies for *Phytophthora ramorum*. In: Frankel, S.J., Kliejunas, J.T. and Palmieri, K.M. (eds) *Proceedings of the Sudden Oak Death Fourth Science Symposium, June 15–18, 2009, Santa Cruz, California*. General Technical Report PSW-GTR-229, US Department of Agriculture, Forest Service, Pacific Southwest Research Station, Albany, California, pp. 239–248.

Vettraino, A.M., Natili, G., Anselmi, N. and Vannini, A. (2001) Recovery and pathogenicity of *Phytophthora* species associated with a resurgence of ink disease in *Castanea sativa* in Italy. *Plant Pathology* 50, 90–96.

Vettraino, A.M., Morel, O., Perlerou, C., Robin, C., Diamandis, S. and Vannini, A. (2005) Occurrence and distribution of *Phytophthora* species in European chestnut stands, and their association with ink disease and crown decline. *European Journal of Plant Pathology* 111, 169–180.

Vettraino, A.M., Sukno, S., Vannini, A. and Garbelotto, M. (2010) Diagnostic sensitivity and specificity of different methods used by two laboratories for the detection of *Phytophthora ramorum* on multiple natural hosts. *Plant Pathology* 59, 289–300.

Werres, S., Marwitz, R., Man in't Veld, W.A., De Cock, A.W.A.M., Bonants, P.J.M., de Weerdt, M., Themann, K., Ileva, E. and Baayen, R.P. (2001) *Phytophthora ramorum* sp. nov., a new pathogen on *Rhododendron* and *Viburnum*. *Mycological Research* 105, 1155–1165.

Weste, G. (2003) The dieback cycle in Victorian forests: a 30-year study of changes caused by *Phytophthora cinnamomi* in Victorian open forests, woodlands and heathlands. *Australasian Plant Pathology* 32, 247–256.

Weste, G. and Vithanage, K. (1979) Survival of chlamydospores of *Phytophthora cinnamomi* in several non-sterile, host-free forest soils and gravels at different soil water potentials. *Australian Journal of Botany* 27, 1–9.

Wilkinson, C.J., Holmes, J.M., Dell, B., Tynan, K.M., McComb, J.A., Shearer, B.L., Colquhoun, I.J. and Hardy, G.E.St.J. (2001a) Effect of phosphite on *in planta* zoospore production of *Phytophthora cinnamomi*. *Plant Pathology* 50, 587–593.

Wilkinson, C.J., Holmes, J.M., Tynan, K.M., Colquhoun, I.J., McComb, J.A., Hardy, G.E.St.J. and Dell, B. (2001b) Ability of phosphite applied in a glasshouse trial to control *Phytophthora cinnamomi* in five plant species native to western Australia. *Australasian Plant Pathology* 30, 343–351.

Williams, N., Hardy, G.E.St.J. and O'Brien, P.A. (2009) Analysis of the distribution of *Phytophthora cinnamomi* in soil at a disease site in western Australia using nested PCR. *Forest Pathology* 39, 95–109.

Winton, L.M. and Hansen, E.M. (2001) Molecular diagnosis of *Phytophthora lateralis* in trees, water, and foliage baits using multiplex polymerase chain reaction. *Forest Pathology* 31, 275–283.

Zentmyer, G.A. (1979) Stimulation of sexual reproduction in the A2 mating type of *Phytophthora cinnamomi* by a substance in avocado roots. *Phytopathology* 69, 1129–1131.

Zentmyer, G.A. (1980) Phytophthora cinnamomi *and the Diseases it Causes*. APS Press, St Paul, Minnesota.

Zentmyer, G.A. and Mircetich, S.A. (1966) Saprophytism and persistence in soil by *Phytophthora cinnamomi*. *Phytopathology* 56, 710–712.

Zobel, D.B., Roth, L.F. and Hawk, G.M. (1982) *Ecology, Pathology, and Management of Port Orford-Cedar* (Chamaecyparis lawsoniana*)*. General Technical Report PNW-184, US Department of Agriculture, Forest Service, Pacific Northwest Research Station, Portland, Oregon.

26 Tree Rusts

Richard C. Hamelin*
Laurentian Forestry Centre, Quebec and The University of British Columbia, Vancouver, Canada

26.1 Pathogens, Significance and Distribution

Rust fungi (order *Pucciniales*) are an extremely important and diverse group of plant pathogens that can affect a surprisingly diverse group of host trees, comprising gymnosperms, angiosperms and ferns, and cause a variety of different symptoms (Cummins and Hiratsuka, 2003). Forest pathologists have studied rusts for over a century and some of the earliest papers were published in the late 1800s and early 1900s. This attests to the economic and environmental importance of tree rusts. Not surprisingly, one of the first tree pathogen genomes to be sequenced was that of the poplar leaf rust fungus *Melampsora laricis-populina* Kleb. (Duplessis *et al.*, 2011a). The unique features of tree rusts, their impact and wide distribution make them an important group of pathogens.

26.1.1 Distribution

Tree rusts are widely distributed in Europe, North and South America, Africa, Asia and Australia. Their distribution often follows that of their hosts and many species have tracked their hosts in non-native environments. This is the case, for example, for *M. laricis-populina*, which is found on poplars (*Populus* spp.) in Europe (its centre of origin), North America, Australia and Africa, and also for the white pine blister rust fungus *Cronartium ribicola* J.C. Fisch., which occurs in Asia, Europe and North America.

The most important tree rusts, from both an economic and ecosystem perspective, belong to just a few genera – but primarily to *Cronartium* and *Melampsora*. *Cronartium*, and its anamorphic genus *Peridermium*, have aecial hosts in the *Pinaceae* and a variety of telial hosts, and cause some of the most severe diseases of pines (*Pinus* spp.), such as white pine blister rust, fusiform rust caused by *C. quercuum* f.sp. *fusiforme* Burds. & G.A. Snow and resin-top disease caused by *C. flaccidum* (Alb. & Schwein.) G. Winter. The genus *Melampsora* is estimated to comprise up to 32 species that can infect *Salicaceae* as telial hosts and the *Pinaceae* and flowering plants as aecial hosts (Vialle *et al.*, 2011). Some of the most important losses in poplar and willow (*Salix* spp.) plantations around the world are caused by Melampsora rusts, including *M. laricis-populina*, *M. medusae* Thüm., *M. occidentalis* H.S. Jacks. and the *M. epitea* Thüm. complex. *M. populnea* f.sp. *pinitorqua* Boerema & Verh., the pine twisting

* E-mail: richard.hamelin@ubc.ca

rust, unlike other members of this genus, causes its main economic losses on pines but only minor damage on poplar. There are several Melampsora poplar rusts that also have aecial hosts in other plants than the conifers. *M. allii-populina* Kleb., for example, has its aecial stage on members of onion family, i.e. *Allium* spp., and *M. populnea* f.sp. *rostrupii* Boerema & Verh. has its aecial stage on Dog's mercury, *Mercurialis perennis* L. (Vialle *et al.*, 2011).

There are several tree rust species in other genera, such as *Chrysomyxa*, *Coleosporium*, *Melampsorella*, *Melampsoridium* that have aecial conifer hosts and a variety of different telial hosts (Ziller, 1974; Cummins and Hiratsuka, 2003). Most species in these genera do not cause widespread damage but they can cause severe problems locally under certain circumstances. Some *Chrysomyxa* spp. have a circum-boreal distribution; *Chrysomyxa pyrolae* (DC.) Rostr. causes a serious disease of cones that affects seed production (Savile, 1950; Crane, 2001).

Recently, *Puccinia psidii* G. Winter has emerged as a serious threat to eucalyptus (*Eucalyptus* spp.) and species of *Myrtaceae*, which represent major plantation species for pulp and timber production (Booth *et al.*, 2000; Glen *et al.*, 2007; Wingfield *et al.*, 2008). This rust causes mostly a foliar and shoot disease but it can also induce mortality in heavily infected trees; in addition, it can also reduce fruit production. The pathogen has recently expanded its geographic range and is now present in South and Central America, in Florida (USA), and in several Hawaiian islands (Booth *et al.*, 2000; Uchida *et al.*, 2006; Glen *et al.*, 2007). One concern is that it could threaten the centre of biodiversity of *Eucalyptus* spp. and *Melaleuca* spp. in Australia. *P. psidii* certainly has the potential to cause extensive damage on both planted and native hosts in the southern hemisphere. It is already a significant quarantine organism and could rapidly become one of the most significant tree rust diseases worldwide if it becomes widely established.

26.1.2 Types of damage

The damage caused by tree rusts can range from barely noticeable to severe, and depends on the host infected. Rusts can inflict various types of damage, from cankers to brooms and defoliation (Ziller, 1974).

Cankers

The most severe forms of damage caused by tree rusts are cankers and galls, which are typically caused by members of the *Cronartium* genus on their aecial hosts (Peterson and Jewell, 1968). Cankers are caused following infection of needles and growth of the pathogen through the foliage into the branches or stems. Cankers can develop within a few months following inoculation (Hunt *et al.*, 2007), but they can take several years to develop as well; trees can also survive for a number of years with stem cankers. The rate of canker development is dependent on host morphology, vigour and genotype, and on the pathogen's aggressiveness. Very little is known about the correlations among canker growth, the interaction of host vigour and pathogen aggressiveness.

Stem cankers can result in the mortality of susceptible hosts, as well as growth reduction and stem deformation. White pine blister rust can cause nearly 100% mortality and result in massive failure of plantations, but some trees can survive for several years with stem cankers. Resistance mechanisms have been discovered based on the rate of canker development, and canker shape and morphology (Hunt, 1997).

Losses to fusiform rust occur mostly as mortality caused when young pines develop stem galls and either die before harvest age or become deformed and unmerchantable. However, trees with severe stem galls may die throughout the rotation and can become susceptible to windthrow when exposed to high-velocity winds (Schmidt, 2003).

Branch cankers are usually less damaging than stem cankers. Branch cankers and galls can be eliminated by self-pruning when the canopy closes, in particular if the rust is not aggressive.

Foliar infection and defoliation

Foliar infection and defoliation by rusts can result in growth losses and make hosts more susceptible to other pathogens. Losses are caused by the exploitation of the host resources by the pathogen, the reduction of the host's photosynthetic area and early defoliation. Growth reduction can be important: in poplars infected by poplar rusts it can be as high as 65% (Widin and Schipper, 1981). Heavy defoliation can also be associated with secondary infection, tree mortality or shoot tip mortality, causing loss of leaders and a bushy habit (Glen *et al.*, 2007). Defoliation can be particularly important when aecial and telial hosts are in close proximity, causing infections to begin early in the season. Heavy defoliation in nurseries can also result in mortality. Eucalyptus trees infected by *P. psidii* can suffer heavy mortality when infections occur in nurseries or in young plantations.

Most of the economic damage resulting from defoliation is on either the telial or the aecial host, but rarely on both, although there are some exceptions. Poplar rusts can cause economically important defoliation on both their poplar and their conifer hosts. This has been particularly important in Christmas tree plantations, where heavy infection by *M. occidentalis* in Douglas-fir (*Pseudotsuga menziesii* (Mirb.) Franco) can result in severe needle losses (Ziller, 1965). Several *Chrysomyxa* spp. produce their uredia and/or telia on cultivated rhododendrons (*Rhododendron* spp.) and cause economically important diseases on these hosts (Crane, 2001).

Brooming and systemic infections

Species of the genera *Chrysomyxa* and *Melampsorella* can cause systemic infections and infect cones or cause brooming. Chrysomyxa rusts are responsible for significant economic losses, especially in exotic spruce (*Picea* spp.) plantations (Takahashi and Saho, 1985). The mode of action is not well known but probably involves a hormonal modification in the host that results in hyperplasia. The resulting brooms can grow very large and cause breakages, which can provide a point of entry for other pathogens (Ziller, 1974).

26.1.3 Impact of tree rusts

Tree rusts have a severe impact on plantations that are exploited for commercial purposes. Forest managers have to balance the desired outcome of increasing growth versus the risk of increasing rust incidence and damage (Schmidt, 2003). This has long been recognized and is reflected in the extensive work conducted to control these important pathogens. More recently, however, it has been recognized that tree rusts can also affect keystone species and modify the fine balance in some fragile ecosystems, in particular when exotic rusts are introduced into native pathosystems (Keane *et al.*, 1994).

Economic impact

The economic impact of the major pine rusts is important and can be measured in millions of US$. The impact of white pine blister rust is so severe and widespread that several white pine species of western North America are no longer commercially significant (Zeglen *et al.*, 2010). It is difficult to evaluate the true economic losses because western white pine (*P. monticola* Douglas ex D. Don) trees now represent such a small proportion of the regeneration. Nevertheless, annual losses to white pine blister rust, including losses of potential revenues, have been evaluated at US$ 450 million (Colautti *et al.*, 2006).

Fusiform rust affects several hard pines, including the economically important loblolly pine (*P. taeda* L.) and slash pine (*P. elliottii* Engelm.) in the southern USA. Annual losses to fusiform rust have been estimated at over US$ 100 million (Cubbage *et al.*, 2000; Schmidt, 2003). About 3.5 million ha of the 9.6 million ha of slash and loblolly pines were found to have at least 10% of the trees infected (Anderson *et al.*, 1986). Fusiform rust has affected nearly all forest management practices, including nursery management, site preparation, species selection, fertilization, thinning, pruning, controlled burning and rotation length (Schmidt, 2003). This level of impact has greatly affected the research and production of pines. Management schemes can be modified to reduce impact; for example, infection by fusiform rust will affect the growing of pine intended for pulpwood production or sawlog production differently (Bridgwater *et al.*, 2005).

Grand eucalyptus (*E. grandis* W. Hill ex Maid.) is one of the species that is the most susceptible to *P. psidii*. This rust is regarded as potentially the most damaging agent to eucalyptus plantations worldwide and could greatly affect the production thereof. Although for the moment it is largely a potential threat, *P. psidii* has already been a limiting factor in the establishment of commercial eucalyptus plantations in Brazil (Glen *et al.*, 2007).

Impact on the ecosystem

In addition to their economic impact, tree rusts can have severe and permanent impacts on entire ecosystems. These impacts may be difficult to translate into economic losses, but could be even more important than those reported above.

White pine blister rust has recently been recognized as an important threat to fragile ecosystems. Whitebark pine (*P. albicaulis* Engelm.) is a keystone species in alpine ecosystems. It provides food for wildlife and cover for snow retention and watershed protection in high elevation ecosystems in North America. The population levels of this species are decreasing rapidly as a result of the exotic white pine blister rust, combined with an unprecedented epidemic of a mountain pine beetle (*Dendroctonus ponderosae* Hopkins) (Keane *et al.*, 1994; Tomback and Resler, 2007; McKinney *et al.*, 2009). As a result, these ecosystems are being destabilized by the perturbation of a food chain that is highly dependent on pine nuts, which provide food for nutcrackers, squirrels and bears. In addition, the death of high elevation pines often results in increased soil erosion and destabilization. Limber pine (*P. flexilis* James), bristlecone pine (*P. aristata* Engelm.) and foxtail pine (*P. balfouriana* Balf.) ecosystems are also currently threatened by white pine blister rust (Schoettle and Sniezko, 2007).

In a somewhat ironic twist, *P. psidii* had been initially viewed as a possible biocontrol agent in Florida against an invasive Australian weed, *Melaleuca quinquenervia* (Cav.) Blake (Rayachhetry *et al.*, 2001). However, it now attacks a native threatened species, i.e. twinberry (*Myrcianthes fragrans* (Sw.) McVaugh) (Glen *et al.*, 2007).

26.2 Diagnosis

It is tempting to say that the diagnosis of tree rusts is straightforward, as when the aeciospores, uredospores or teliospores are produced, it is relatively easy to identify the rusts, at least at the genus level. A stem canker on a pine can be identified as caused by *Cronartium* or *Peridermium*, and based on the host, shape of canker or gall, and sometimes the spore morphology, it can be identified to the species. Rusts on poplar can similarly be identified by knowing which host or hybrid is affected and by uredospore characteristics. The detection of *P. psidii* is also straightforward as the signs and symptoms are obvious and there are few rust diseases of myrtaceous plants.

There are several complications though. Low levels of disease in asymptomatic plant tissue or germplasm can be difficult to detect and this has been the reason behind the worldwide spread of white pine blister rust (Geils *et al.*, 2010). Also, early detection of a new introduction can be complicated, especially if a new host is infected. For heteroecious rusts, linking the telial and aecial hosts can be challenging (McDonald *et al.*, 2006). Finally, the paucity of morphological features in rust spores makes species recognition sometimes challenging (Vialle *et al.*, 2011).

26.2.1 Linking heteroecious rusts on different hosts

Because of the heteroecious nature of most tree rusts, it can be difficult to identify and link rusts on different hosts. This is an extremely important aspect for heteroecious rusts because the presence of the alternate host often dictates control measures. The recent discovery, after over a century of research, of new telial hosts for *C. ribicola* and *C. flaccidum* attests to the challenge of identifying and linking rusts on their hosts (Kaitera and Hantula, 1998; McDonald *et al.*, 2006). The finding that some species in the family *Orobanchaceae* (*Pedicularis* and *Castilleja* spp.) are functioning as alternate hosts in natural epidemics in North America was important and surprising, given the intensity of the studies on this pathogen and the amounts of money invested in programmes on alternate host eradication (McDonald *et al.*, 2006). Similarly, *C. flaccidum* was observed for the first time on small cow-wheat, *Melampyrum sylvaticum* L., and also on other telial hosts in this flowering herbaceous genus, and some of these hosts seem to be epidemiologically relevant (Kaitera and Hantula, 1998; Kaitera and Nuorteva, 2003; Kaitera *et al.*, 2005). These findings have a direct impact on risk assessment and site hazard rating, and on disease management strategies.

26.2.2 DNA barcoding to identify rust species

DNA barcoding, or the use of a short standard DNA region to identify taxa, is proving extremely powerful in the better characterization of fungi (Seifert, 2009; Vialle *et al.*, 2009). For rust species, this approach is particularly useful because of the problems mentioned previously. The overlap in morphological characters of the uredial and telial stages within the *Melampsora* has led to confusion in both species recognition and identification (Vialle *et al.*, 2011). DNA barcodes obtained from mitochondrial and nuclear genome regions of the *Melampsora* spp. on poplar provide a highly accurate means of identifying and resolving these taxa. Approximately 25% of the herbarium specimens tested with this approach had been previously misidentified, including some species targeted by quarantines (Feau *et al.*, 2009).

The concept of formae speciales (f. spp.) has been important in rust taxonomy to indicate host specialization. DNA barcoding in the genus *Chrysomyxa* revealed diversification within the species complex and speciation with regard to hosts (Feau *et al.*, 2011). Specialization was found on both the telial and, to a lesser extent, the aecial hosts. Unexpectedly, the spruce cone rusts *C. pyrolae* and *C. monesis* Ziller coalesced with the pine needle rusts belonging to the genus *Coleosporium*, whereas the microcyclic species *Chrysomyxa weirii* H.S. Jacks. was embedded within the genus *Melampsora* (Feau *et al.*, 2011). Such results were unexpected based on morphology and host range, but could provide important information on infection biology and epidemiology. For example, the *C. weirii* telial host, which is not known, could possibly be found within the *Salicaceae*.

Some rusts have overlapping aecial as well as telial hosts, which can complicate identification. This is the case for *M. laricis-populina* and *M. medusae* f.sp. *deltoidis* Shain. *M. laricis-populina* was reported for the first time in eastern North America on the telial host poplar (Innes *et al.*, 2004), but to determine whether the rust was completing its life cycle in Canada, it was important to establish whether or not it was present on the aecial host larch (*Larix* spp.). However, the native *M. medusae* f.sp. *deltoidis* also infects larch as its aecial host. The ability of *M. laricis-populina* to complete its life cycle on larch was confirmed by DNA barcoding (Grondin *et al.*, 2005). This proved that the rust is established, can overwinter and can reproduce sexually. These are important factors in understanding the epidemiology of the pathogen and its potential to spread and sexually reproduce in North America. The ability to complete its life cycle may allow *M. laricis-populina* to generate pathogenic variation that will challenge poplar breeders in this region.

26.2.3 PCR assays to detect rusts

PCR-based detection assays can greatly help in detecting and monitoring rusts in host tissues. Assays have been developed, mostly for tree rusts that require monitoring or are targeted by quarantines. A PCR assay was developed to detect *P. psidii* in host and non-host material (Langrell *et al.*, 2008). Real-time PCR assays were developed to detect and monitor the infection of poplars by poplar rusts. This allowed monitoring of the infection process and highlighted the differences in fungal biomass between resistant and susceptible hosts by generating growth curves even before symptoms become visible (Boyle *et al.*, 2005).

26.3 Infection Biology

Much knowledge has accumulated in recent years on the infection biology of rusts. This is due to the combination of histological and molecular techniques that make it possible to follow the infection process *in vivo* with unprecedented insights. Although it is difficult to generalize the infection process for different rusts on different hosts, recent developments in the best studied pathosystems will be highlighted in this section.

26.3.1 Life cycle

Rusts have some of the most complex life cycles of the fungal kingdom. The necessity of heteroecious rusts to infect completely unrelated telial and aecial hosts can appear as an evolutionary oddity, potentially requiring the pathogen to develop a different pathogenicity toolbox for infecting different hosts. Yet rusts are extremely successful and adaptable.

Details of the life cycles for individual rusts can be found in reference manuals (Ziller, 1974; Cummins and Hiratsuka, 2003). There are a variety of types of life cycle within the rusts. Heteroecious macrocyclic rusts have five spore types on two different hosts; within this group are several *Chrysomyxa* spp., *Cronartium* spp., *Melampsora* spp. and *Melampsorella* spp. Demicyclic rusts, which have four spore types on two different hosts, are exemplified by the cedar-apple rusts caused by *Gymnosporangium* spp. Some rusts are autoecious (possibly *P. psidii*), being able to complete their entire life cycle on a single host, or have a so-called endocyclic life cycle that produces one or two spore types on a single host (e.g. *Peridermium* spp.). As just noted, it is believed that *P. psidii* might be an autoecious rust, but for now its complete life cycle has not been observed; all spore stages have been observed on eucalyptus except for spermogonia (Glen *et al.*, 2007), but the latter are often cryptic and can be difficult to observe. Most genera comprise species with unresolved life cycles. There are no known aecial hosts for *C. weirii* and *Melampsora aecidioides* (DC.) J. Schröt., for example (Crane, 2001; Vialle *et al.*, 2011). Phylogenetic and DNA barcoding studies hold great promise for resolving such questions and give clues for the discovery of these aecial hosts (Feau *et al.*, 2009, 2011).

In heteroecious rusts, spermogonia and aecia develop on galls, cankers or needles/leaves of aecial hosts. Aeciospores generally infect young leaves of the telial hosts early in the spring, and within 5–10 days uredia begin to appear. Uredia and telia are produced on the telial host leaves, usually on the lower surfaces, but sometimes on both sides of the leaves. Uredospores can reinfect leaves of the same telial host repeatedly during the growing season. Teliospores produced either in a telial column or on sub-epidermal crusts germinate by forming basidiospores on basidia (Ziller, 1974). Basidiospores are discharged into the air and dispersed by wind to infect the aecial host. Rusts can either overwinter as telia, usually on the telial host leaves (e.g. *Melampsora* spp.) and produce basidiospores in the spring, or as mycelium within the aecial host (e.g. *Cronartium* spp.) (Ziller, 1974). In the latter case, basidiospore release and infection of the aecial host takes place in the autumn. Spermogonia are formed on aecial hosts one or more years following infection (*Cronartium* spp.) or shortly after aeciospore infection (*Melampsora* spp.). Spermatia are produced in an orange sweet exudate that attracts insects, which carry the spermatia from one spermogonium to another, effecting spermatization (Hunt, 1985). These spores are not infectious and their role is the dikaryotization of the fungal tissues. Aeciospores are produced the following year (for *Cronartium* spp.) or shortly after spermatization (*Melampsora* spp.). In demicyclic rusts, the repeating uredial stage is absent and in endocyclic rusts all spore stages are absent except for the aecial and sometimes the spermogonial stages.

Most rust genera comprise heteroecious and autoecious rusts, and it appears that these life cycles have appeared multiple times along the evolutionary history of the rusts. *Peridermium* endocyclic stem rusts on pines are present in different clades and share terminal nodes with their sister *Cronartium* spp. (Vogler and Bruns, 1998).

26.3.2 Infection process

In the pathosystems studied, infection of the telial and aecial hosts can take place either through stomata following spore germination and germ tube formation, or by direct penetration between epidermal cells following formation of an appressorium and an infection peg (Patton and Johnson, 1970; Miller *et al.*, 1980; Ragazzi and Fedi, 1992; Rinaldi *et al.*, 2007; Wingfield *et al.*, 2008). Penetration occurs early after spore germination and has been found to take place as early as 2 to 6 h post inoculation in *M. laricis-populina* (Rinaldi *et al.*, 2007).

As infection proceeds, hyphae grow in the intercellular space and extensive colonization and accumulation of fungal biomass can be measured or observed (Hudgins *et al.*, 2005; Rinaldi *et al.*, 2007; Jacobs *et al.*, 2009). In stem rusts, hyphal colonization can be observed in the vascular bundle and adjacent mesophyll cells, parenchyma cells and xylem rays, as well as in and between tracheids (Jackson and Parker, 1958; Hudgins *et al.*, 2005; Jacobs *et al.*, 2009). Normal wood development is affected through changes in cellular differentiation and development. There is evidence that cankers form as a result of increased tree hormone activity, including cytokinins and gibberellins, following fungal invasion (Seifers *et al.*, 1985; Myburg *et al.*, 2006). This abnormal division and swelling of cortical and phloem parenchyma cells and the accumulation of phenolic compounds is what ultimately cause canker or gall formation (Seifers *et al.*, 1985; Hudgins *et al.*, 2005). The gall tissues remain organized and contain typical xylem and phloem cells, but the proportion and size of cell types is abnormal (Jackson and Parker, 1958).

Haustoria, i.e. specialized structures that allow rusts to obtain sugars and nutrients located in the haustorial plasma membrane, are produced within the host cells (Staples, 2001). Exciting work shows that haustoria are not only involved in nutrition but can also be engaged in the suppression of host defence responses and in redirecting or reprogramming the host's metabolic flow (Voegele and Mendgen, 2003). Haustoria have been shown to deliver secreted proteins that act as effectors of pathogenicity into host cells during the infection process (Kemen *et al.*, 2005).

26.3.3 Genomic and molecular dissection of infection

A better understanding of rust biology and epidemiology would greatly enhance our ability to manage them in an integrated fashion. Genomic approaches have generated exciting developments in rust research. The complete genome of the poplar rust *M. laricis-populina* has been sequenced and has revealing several interesting features, including a genome architecture that could explain the rapid evolution of rusts and their ability to overcome resistance in their hosts (Duplessis *et al.*, 2011a). This genomic resource provides powerful tools that can be exploited for several purposes, including providing a better understanding of the infection process, as well as for the development of epidemiological knowledge and monitoring tools (Feau *et al.*, 2007; Duplessis *et al.*, 2009; Xhaard *et al.*, 2009; Boyle *et al.*, 2010; Hacquard *et al.*, 2010; Duplessis *et al.*, 2011a,b).

Many of the most innovative approaches in rust research have been conducted in the poplar rust pathosystem. Genomic studies are allowing a fine dissection of the infection and colonization processes and the identification of important genes and gene families that could help understand this rust pathosystem. Among the prominent features of the genome, almost half of the *M. laricis-populina* genome was composed of repeated elements, possibly accounting for its capacity to adapt and overcome host resistance (Pinon *et al.*, 1994; Gérard *et al.*, 2006). In addition, nitrogen and sulfur assimilation pathways were deficient, whereas expanded families of membrane transporters were observed. These features are consistent with the biotrophic lifestyle of the rusts and their dependence on the host for nutrient assimilation (Duplessis *et al.*, 2011a).

Among the most promising results, a large group of small-secreted proteins (SSPs) were discovered; these appear to be closely involved in host–pathogen interactions and are likely to comprise rust effectors of pathogenicity

(Duplessis *et al.*, 2011a). SSP transcripts were expressed throughout the infection process of *M. laricis-populina* on its poplar host in time-course experiments (Duplessis *et al.*, 2011b). The expression of SSP transcripts appeared to be coordinated during leaf infection and candidate SSP effectors accumulated in rust infection structures (Hacquard *et al.*, 2012). Transcripts encoding SSP were also observed in the mesophyll even at the time of sporulation and could indicate that they have a role in the suppression of host defence late in the infection stage (Hacquard *et al.*, 2010).

One of the hallmarks of effectors is that they are under strong positive selection because they constantly have to evolve to avoid host recognition. This was confirmed in several studies that showed significant positive selection for SSPs in data obtained from genome analysis and comparisons (Feau *et al.*, 2007; Joly *et al.*, 2010; Hacquard *et al.*, 2012). These SSPs appeared to be under different evolutionary constraints from the rest of the genome and displayed much lower homology to other fungi (Joly *et al.*, 2010). The SSP genes are likely to be involved in co-evolutionary arms races with host factors and could be useful in monitoring or for host screening.

26.3.4 Host specificity

Most rust fungi are very specialized and attack only a single host family, genus or a few species. Host specificity is an important aspect of rust biology, although there is still much to learn about this in rusts; recent studies have shown that some rusts previously thought to have narrow host ranges can be found on new and unsuspected hosts. The process of how this happens is yet unknown and could include host jumps or undetected or cryptic infections.

Pine stem rusts are generally limited to a few aecial hosts: five-needle pines for *C. ribicola*; and hard pines for *C. quercuum* f.sp. *fusiforme*, *C. comandrae* Peck and *C. comptoniae* Arthur (Ziller, 1974). However, some pine stem rusts can have surprisingly broad host ranges on their telial hosts. *C. ribicola* and *C. flaccidum* were both recently found to have new alternate hosts. The broader host range of *C. ribicola* includes native species of *Castilleja* and *Pedicularis* (McDonald *et al.*, 2006), and these infections appear to be epidemiologically important (Richardson *et al.*, 2007). This could be especially important in high elevation whitebark pine stands, where sufficient time was observed for *C. ribicola* to complete its life cycle, and natural *C. ribicola* infection was detected on *Pedicularis racemosa* Douglas ex Benth. and five species of *Castilleja* (Mulvey and Hansen, 2011).

Puccinia psidii is unusual in that it has an extremely wide host range. Its ability to infect commercial *Eucalyptus* spp. could be the result of bringing together the eucalyptus trees and the rust in a non-native environment where co-evolution and host–pathogen equilibrium have not developed (Wingfield *et al.*, 2008). This is very similar to the situation that developed for interamerican hybrid poplars, which are severely affected by non-native poplar rusts in Australia, New Zealand and South America.

26.4 Epidemiology

Most rusts cause few problems in their native natural environment, but can cause important epidemics when the balance between the pathogen, its hosts and the environment is disturbed (Castello *et al.*, 1995). The rapid increase and spread of several tree rusts have been paralleled by Powers *et al.* (1981) and MacDonald (2003). It is important to recognize these disturbances to better understand how these factors can be modulated or modified. The introduction of non-native rusts, as in the case of *C. ribicola*, is probably the most important factor affecting the balance. The introduced pathogen has not co-evolved with the host in its new environment and there are few other population regulators to prevent large-scale epidemics.

The reverse situation, where non-native hosts are deployed in plantations, can generate severe epidemics with native pathogens, and also with introduced pathogens that have followed their hosts. Furthermore, there are examples of host-specific pathogens native to areas where trees have been planted as non-natives that have undergone host jumps. This is the case with native and introduced poplar leaf rusts affecting hybrid poplars in Europe and North America, and with eucalyptus trees planted outside their native range in the southern hemisphere (Wingfield *et al.*, 2008).

With heteroecious rusts, another important cause of epidemics is the impact of forest management practices that affect the abundance or distribution of the alternate hosts. Fire control in the southern USA has resulted in increased in density of oak (*Quercus* spp.), the telial hosts of *C. quercuum* f.sp. *fusiforme*, in the underbrush of pine stands. This, combined with the extensive planting of loblolly and slash pines outside their natural range, has resulted in severe fusiform rust epidemics on the pines (MacDonald, 2003).

26.4.1 Conditions for infection

Rust fungi generally share similar requirements for the infection of their telial and aecial hosts. The necessity for a moist period is the most important factor for spore germination and infection. This is often the factor that is found to limit infection (Van Arsdel *et al.*, 1956; Peterson and Jewell, 1968). Most species studied also show some limitation with higher temperatures, and optimal temperatures for germination of teliospores and basidiospores are generally 15°C to 25°C.

The synchronicity of the host development for infection between the aecial and telial hosts is an additional important consideration that probably does not receive enough attention. The year-to-year variation, termed 'wave years', in stem rust incidence has been attributed to such phenomena (Peterson and Jewell, 1968). Severe rust outbreaks are often followed by years of low infections.

26.4.2 Landscape features

Some of the most interesting studies in pine rust epidemiology have been conducted in the USA Lake States. Three levels of climatic determinants were defined at the macro-scale (generally latitude and elevation), mesoscale (topographic features such as hills and slopes) and micro-scale (local features such as gaps and openings in the canopy) (Van Arsdel, 1965). Forest managers can exploit the local features to define site hazard and risks and adjust their planting schemes accordingly.

26.4.3 Dispersal

The different spore types produced by rusts allow for both local and long-distance dispersal (Nagarajan and Singh, 1990). For heteroecious rusts with an aecial economic host, such as pine rusts, dispersal of the basidiospore is an important aspect. Basidiospores of *C. ribicola* are believed to be dispersed over relatively short distances, but infections have been observed as far as 14 km from a source (Van Arsdel, 1965). The uredospores and aeciospores are considered to possess long-distance dispersal potential. Well-documented cases, such as the migration of Melampsora poplar rust from Australia to New Zealand suggest that they can spread over 500 km in single jumps (Wilkinson and Spiers, 1976; Viljanen-Rollinson and Cromey, 2002; Geils *et al.*, 2010).

26.5 Management Strategies and Tactics

26.5.1 Avoidance

Hazard and risk ratings provide a means for assessing the potential severity of rust infestations and their probable impacts on management. Determining the rust hazard of sites, and adjusting silvicultural and planting practices in light of this, has been particularly important for management of pine stem rusts. Hazard assessments rate the environmental potential for infection of susceptible pines and include rust incidence, climatic conditions, site characteristics and other factors, such as forest density. One of the most important factors though is probably the presence of inoculum sources nearby; for heteroecious rusts, this means presence of the telial host. Additional predictors can include, for example for white pine blister rust, elevation, slope, summer precipitation, stand density, average tree diameter, height and age (Jacobi *et al.*, 1993; Smith and Hoffman, 2001).

When landscape or other features result in more conducive conditions for rust, these can be used when selecting sites for planting, or to develop simulation models for describing the likely outcomes (Hunt, 1983; Zeglen *et al.*, 2010). For eastern white pine (*P. strobus* L.), moisture and temperature conditions were used to delineate hazard zones and provide guidelines for planting (Van Arsdel *et al.*, 1956). In British Columbia (Canada), the range of distribution of western white pine falls within high hazard zones, but rust tends to be more severe on slopes (Hunt, 1983; Zeglen *et al.*, 2010). Definition of site hazards for fusiform rust is also strongly governed by factors affecting presence of the telial host; the proximity, the orientation and the density of oaks were the best predictors of rust incidence (Hollis and Schmidt, 1977). This allowed the classification of sites into low, medium or high hazard, and the focusing of disease management on sites with a high hazard (Hollis and Schmidt, 1977; Froelich and Snow, 1986).

Silvicultural practices can also have an impact on rust incidence and severity, in particular via practices that influence the presence of the alternate host. Harvesting practices can affect the incidence of the alternate host (*Ribes* spp.) of white pine blister rust and, therefore, the rust incidence. For example, to reduce white pine blister rust incidence logging can be conducted in the winter to reduce ground disturbance, thereby decreasing the ability of *Ribes* spp. to colonize the site (Zeglen *et al.*, 2009). In addition, partial harvesting results in canopy retention, which reduces the seed germination of *Ribes* spp. because of the lower light level in the stands (Zeglen *et al.*, 2009). The incidence of fir broom rust caused by *Melampsorella caryophyllacearum* (DC.) J. Schröt. was also affected by site factors that influenced light index, which, in turn, was correlated with the presence of the alternate hosts (Oliva and Colinas, 2007). The correlation between rust severity and light index values of stands suggest that opening the canopy could reduce the presence of the alternate hosts in the understorey.

Infections that take place on seedlings can take over a year to develop into cankers. This can result in latent infection on seedlings and the introduction of infected seedlings into plantations. A possible means of preventing the establishment of rust in the plantations is to grow seedlings in indoor nurseries, where conditions can be controlled to prevent infection (Zeglen *et al.*, 2010).

26.5.2 Exclusion

Exclusion represents a very important means of tree rust management. In particular, quarantines can ensure that rusts are not accidentally introduced into non-native environments where they can become invasive. Several tree rust species are on the quarantine lists of the European and Mediterranean Plant Protection Organization (EPPO) and the North American Plant Protection Organization (NAPPO), e.g. many species of pine stem rusts (*Cronartium* spp.), cedar-apple rusts (*Gymnosporangium* spp.) of *Malus* spp. and poplar rusts (*Melampsora* spp.). Zones under quarantine must enforce the inspection and certification of imported and exported material to ensure that they remain pathogen free.

The physical exclusion of rust spores can be challenging because of the ability of rust spores to disseminate over long distances. However, there are some interesting examples of physical exclusion. The infection of spruce cones by spruce cone rust caused by *Pucciniastrum areolatum* (Fr.) G.H. Otth and *Chrysomyxa pyrolae* were hindered by bagging the tree flowers (Kaitera *et al.*, 2009). Tests are ongoing using seedling enclosures to physically block *C. ribicola* spores from white pine foliage and prevent lethal stem infections (Zeglen *et al.*, 2010). This would obviously only take place early in tree development to protect the seedlings and saplings when they are most susceptible. It has previously been shown that enclosing infected white pines in cages during the production of spermatia prevented the production of aecia (Hunt, 1985). Although this was not proposed as a control method and instead was used to demonstrate heterothallism in *C. ribicola*, it is one example of how exclusion can prevent the completion of the rust life cycle.

26.5.3 Eradication

One of the most massive eradication programmes to control a plant pathogen was the attempt to eradicate the alternate hosts of *C. ribicola*. In the USA, this programme started in the 1930s and continued until the 1970s, and millions of hectares were treated. A quarantine was established that authorized the destruction of *Ribes* spp., including the cultivated European black currant (*Ribes nigrum* L.) (Maloy, 1997). Eradication of alternate hosts has not been attempted, at least not on such a large scale, for any other tree rust.

Overall, large-scale eradication of the alternate host has proved to be a failure in reducing the spread of white pine blister rust. There are many reasons for this failure, but the most important ones are the difficulty of finding and reaching all the alternate hosts, and the difficulty in killing the alternate host in environmentally acceptable ways and in a manner that would have an impact on the rust epidemic (Maloy, 1997). Eradication has been especially problematic in western North America because of the difficult terrain that workers had to contend with, although local reduction or elimination of the alternate host (as opposed to eradication) has been effective in some instances. For example, the incidence of white pine blister rust was lower in Maine (USA) in areas where the eradication of *Ribes* spp. had been performed than in non-treated areas (Ostrofsky *et al.*, 1988).

Even if eradication of the *Ribes* spp. had been successful, it is possible that the impact on the epidemic could have been negated. The recent discovery of other alternate hosts of *C. ribicola* certainly shows that there is potential for host jumps; as previously mentioned in Section 26.3.4, telial hosts in the genera *Pedicularis* and *Castilleja* have been reported in North America (McDonald *et al.*, 2006; Zambino *et al.*, 2007). Although the importance of these telial hosts in the epidemiology of white pine blister rust has not been established, they could be important in high elevation ecosystems where *Ribes* spp. are absent. As also noted in Section 26.3.4, evidence was found for the natural infection of *C. ribicola* on *Pedicularis* spp. and *Castilleja* spp. in whitebark pine stands (in Washington and Oregon, USA), where there was sufficient time for the rust to complete its life cycle (Mulvey and Hansen, 2009). A compounding problem for the eradication *Ribes* spp. is the currant industry, which has grown large in the last few decades based on the commercial production of currants.

26.5.4 Protection

Chemical control

Tree rusts can be controlled under some circumstances with fungicides. In general, fungicide treatment is only desirable in nurseries, in stool beds, in intensive plantations, or on ornamentals. Some of the earliest attempts at controlling rusts with chemicals used antibiotics, in particular cycloheximide and phytoactin (Maloy, 1997). However, the efficiency of these treatments has never been demonstrated and control programmes using these chemicals were abandoned.

A more widespread and efficient class of compounds for rust control are the sterol-inhibiting fungicides. Triadimefon is one of the active ingredients that have been reported to be active against several rusts. Triadimefon has been applied as foliar spray, as soil treatment and as seed treatment, and has been found to protect southern pines against fusiform rust, white pine trees against white pine blister rust, maritime pine (*Pinus pinaster* Aiton) and Scots pine (*P. sylvestris* L.) against pine twisting rust (Kelley, 1980; Desprez-Loustau *et al.*, 1992; Bérubé, 1996). Interestingly, triadimefon has shown little impact on beneficial mycorrhizal fungi (Kelley, 1980; Desprez-Loustau *et al.*, 1992). Several systemic fungicides were also found to be translocated in young tissues of eucalyptus trees and reported to prevent infection by *Puccinia psidii* (Wingfield *et al.*, 2008).

Pellets impregnated with triadimefon and fertilizers have been developed and tested. These pellets provide an interesting approach as they are placed in the planting cavity at transplantation to provide long-term protection against pathogens and promote growth. Triadimefon-impregnated pellets have been tested on eastern white pine seedlings and found to lower infection and mortality rates, with effects that were reported to last for several years (Pitt *et al.*, 2006). In tests conducted with western white pine using similar pellets, good disease control was obtained but growth reduction was observed. Triadimefon has also been used against rusts on Christmas trees.

Thinning and pruning

Thinning and pruning have to be thoroughly conducted to be efficient (Hunt, 1991). Incomplete removal can negate the effect of the silvicultural management activities. Pruning branches that are infected is an effective way to reduce infections by removing the pathogen. If this is conducted early enough it can prevent the pathogen from reaching the stem. Pruning can protect the affected tree and prevent its death; it can also reduce the inoculum load by removing sporulating cankers and reduce disease spread.

Pruning has been a recommended control method against most stem rusts, including *Cronartium comandrae*, *C. flaccidum*, *C. quercuum* f.sp. *fusiforme* and *C. ribicola* (Zeglen *et al.*, 2010). This approach is particularly effective when the pruning is conducted on the lower branches of young trees (less than 30 years old) where infections can result in stem cankers (Hunt, 1991). The complete removal of lower branches can effectively increase survival and reduce disease incidence (Hunt, 1991; O'Hara *et al.*, 2010). In British Columbia, it is recommended that nearly two-thirds of the live crown of trees that are only 1–2 m high be removed, and that another 50% of the crown be removed when the trees are 4–5 m high (Zeglen *et al.*, 2010). Pruning also has the added benefit that it enhances wood quality: wounds caused by the pruning are occluded and wood formed subsequently is free of knots and has a higher value (O'Hara, 2007). Thinning plantations and removing infected trees can also reduce inoculum load and protect uninfected trees (Zeglen *et al.*, 2010), although it could result in increased infections because as the amount of thinning increases, airflow and spore circulation increases.

26.5.5 Resistance

Breeding trees for resistance to rust has been an important activity for several decades. Genetic resistance is considered one of the most feasible methods to manage tree rust and this trait is often the most important consideration in tree breeding programmes. Classical breeding can produce genetically diverse and resistant populations that can be used for reforestation or restoration of natural forests, although the cost of setting up and operating tree breeding programmes and the long generation times involved has hampered progress in several jurisdictions. Also, resistance must be durable for trees because they are to be confronted by evolving pathogens for decades or centuries. Unlike agricultural crops, which can be ploughed and replanted the following season, the breakdown of resistance in tree plantations can result in long-term damage.

Well-organized and funded breeding programmes have resulted in significant gains in resistance against poplar rusts, white pine blister rust and fusiform rust. These programmes have developed screening and selection experiments that exploit intraspecific and interspecific variation in resistance, and have identified several mechanisms of quantitative and qualitative resistance. Some of the resistance genes have been successfully incorporated and deployed into operational use.

Qualitative resistance to infection is expressed as hypersensitive response, a type of programmed cell death. Usually, qualitative resistance is conditioned by a major gene (so-called major gene resistance) and can be easily selected and used in crosses. Quantitative resistance can be expressed as variation in latent period, sporulation rate, and the number and size of uredia. Quantitative resistance is generally polygenic and requires quantitative breeding strategies; it is generally assumed to be non-race specific and presumed to exert lower selection pressures on the pathogen populations, and hence could be more durable. However, this simple explanation of qualitative and quantitative resistance has been challenged by molecular genetic and genomic analyses.

Poplar resistance to poplar leaf rust

Selection and breeding for resistance to poplar rusts has benefited from the ease of conducting screening because the economically important host is infected by the telial rust stage, and also because of the ease of clonally propagating resistant material.

Some of the most important progress in poplar breeding has occurred following interspecific crosses. The concept of resistance by exaptation (as opposed to co-evolved resistance) was invoked to explain the resistance afforded following transfer of resistance genes from non-co-evolved pathosystems (Newcombe, 1998). For example, the black cottonwood (*Populus balsamifera* L. subsp. *trichocarpa* (Torr. & A. Gray ex Hook.) Brayshaw), a poplar that is native to the Pacific Northwest of North America is resistant to *Melampsora medusae* f.sp. *deltoidis*, the eastern North American poplar leaf rust; the eastern cottonwood (*P. deltoides* Bartram ex Marsh.) is resistant to the Eurasian poplar leaf rust *M. laricis-populina*; and the Japanese poplar (*P. maximowiczii* A. Henry) is resistant to *M. occidentalis*, the western North American poplar leaf rust. Black cottonwood × eastern cottonwood hybrids exhibited resistance to pathogens native to the Pacific Northwest, including *M. occidentalis*, and the resistance gene (*MMD1*) was mapped to the black cottonwood parent (Newcombe *et al.*, 1996). This pattern changed when *M. medusae* f.sp. *deltoidis*, which co-evolved with eastern cottonwood, was introduced in the Pacific Northwest and started hybridizing with the native *M. occidentalis* (Newcombe *et al.*, 2000).

The genetic and genomic characterization of resistance has made tremendous progress in the poplar leaf rust pathosystem and will probably help us to understand resistance mechanisms and possibly serve as a model system for other rust pathosystems. Molecular markers have been used to map resistance loci in poplars (Cervera *et al.*, 1996, 2001; Tabor *et al.*, 2000). One exciting finding is the discovery that resistance genes could be clustered in the genome. Sequence analysis of DNA markers linked to the resistance locus revealed the presence of resistance genes of the nucleotide-binding site/leucine-rich repeat (NBS-LRR) class and transposons (Zhang *et al.*, 2001; Lescot *et al.*, 2004). Genes controlling quantitative and qualitative resistance, inherited from black and eastern cottonwood, respectively, were mapped to the same chromosome in a genomic region rich in genes for NBS-LRR proteins and serine threonine kinase (STK) genes (Bresson *et al.*, 2011). The genes for thaumatin-like proteins (TLPs), which are part of a group of defence pathogenesis-related proteins, with presumed antifungal properties, were clustered and were up-regulated during poplar leaf rust infection (Petre *et al.*, 2011).

White pine resistance to white pine blister rust

Overall, North American pines seem to have very low levels of resistance to *C. ribicola*, yet resistance has been detected in all of those species following extensive screening (Sniezko *et al.*, 2008). By contrast, Asian pine species exhibit high levels of survival in inoculation tests, with less than 10% of seedlings developing symptoms and over 70% of infected seedlings surviving the inoculations (Sniezko *et al.*, 2008). Resistance has been best described and exploited in sugar pine (*P. lambertiana* Douglas) and western white pines.

Both qualitative and quantitative resistances were found in the North American pine species. Different mechanisms of resistance to white pine blister rust were discovered in sugar pines. These included major gene resistance with a hypersensitive reaction and slow-rusting polygenic resistance (Kinloch and Byler, 1981). *Cr2*, a dominant resistance gene that produces a hypersensitive reaction in pine needles, was present at a low level in natural population of western white pine (Kinloch *et al.*, 1999). Interestingly, it was unevenly distributed: it was present in less than 0.1% in some southern populations, but it was undetectable further north in the Cascades, as well as in the Rocky Mountains and Coast Mountains (Kinloch *et al.*, 2003).

Given the high frequency of resistance to *C. ribicola* in the Asian pines, an interspecific hybridization programme was developed to introduce resistance genes from related white pine species to eastern white pine (Lu and Derbowka, 2009). Families resulting from crosses between eastern white pine and Bhutan pine (*P. wallichiana* A.B. Jacks.) that were backcrossed to eastern white pine exhibited better survival than the eastern white pine parents (Lu *et al.*, 2005). However, insufficient cold hardiness could limit the survival and growth of the interspecific hybrids, especially at northern sites (Lu and Sinclair, 2006).

A strong programme has been developed to breed currants that are resistant to *C. ribicola*. Generally, black currants are very susceptible to white pine blister rust, while red currants (*Ribes triste* Pall.) are less so. Rust-resistant, commercial black currant cultivars frequently show poor yields and a susceptibility to powdery mildew. Several resistant, ornamental currants are available (Maloy, 1997).

Southern pine resistance to fusiform rust

Resistance has been the best means of limiting damage by *C. quercuum* f.sp. *fusiforme* in pine plantations in the southern USA (Schmidt *et al.*, 1981). High levels of resistance to fusiform rust were found in loblolly and slash pines (Powers and Kraus, 1983). The absence of stem symptoms, an important resistance trait, was the most common resistance response in many resistant families (Kuhlman and Powers, 1988).

The genetics of resistance to *C. quercuum* f.sp. *fusiforme* has been thoroughly studied. A resistance locus, behaving as a single dominant gene, was mapped by association with genetic markers. This resistance gene is believed to be stable as the result of co-evolution between the pathogen and its host. Therefore, single qualitative resistance can be a stable gene, despite the presence of virulence in the pathogen population (Wilcox *et al.*, 1996). This could be the result of endemism of the rust on the native pines, but it is important because it points to the importance of avoiding the intensive selection pressure on the pathogen population.

Defeated rust resistance

One of the problems encountered with qualitative resistance is the appearance of races that can defeat the resistance. This has resulted in a 'boom and bust' cycle similar to that which occurs in agricultural crops. Poplar breeding programmes can barely stay ahead in the poplar rust race. Unfortunately, it appears that rusts have the capacity to evolve rapidly to overcome the deployed resistance. *Melampsora laricis-populina* has evolved races that can overcome the deployed resistance in poplar (Pinon, 1992); multiple races can overcome all the resistance genes deployed.

Races of *M. medusae* f.sp. *deltoidis* were identified that differed in their aggressiveness (latent period and uredia number per leaf disc), and a significant isolate × cultivar interaction was observed for both traits, indicating a potential for selection. Isolates from the northernmost sampling location appeared more aggressive than those from southern latitudes (Prakash and Thielges, 1987). Pinon (1992) studied race populations of *M. laricis-populina* in Europe over 4 years. He found that race E1 is ubiquitous in France (and probably in western Europe), E2 exists in at least the northern half of France, in Belgium and in the Netherlands, and E3 is present in the east of France and very probably in the west and south-west as well. Races E2 and E3 occurred irregularly among years on larch, the alternate host. There was some link between race populations on larch and on poplar: when races E2 and E3 were infrequent on poplar at the end of the growing season, they were practically undetectable on larch in the following spring. On universal poplar clones (clones susceptible to all known races), races E2 and E3 were in the minority but their counter-selection was not evident. On differential clones (susceptible only to E2 or E3) the compatible race was in the majority, but at the end of the growing season infections by incompatible races (that do not infect poplar growing actively in the greenhouse) were detected. Pinon (1992) discusses these results and proposes several hypotheses. The first is that specific resistance delays the epidemics in differential poplar clones. The second is that, up to now, no race of *M. laricis-populina* combining all the virulences has been found, which is in agreement with the low theoretical frequency of such a race. Third, Pinon suggests that as races E2 and E3 seem to remain relatively stable, they reflect stability within host populations.

Although quantitative resistance is believed to be more durable than qualitative resistance, large variance in the former was attributed to specific interactions in the quantitative resistance of eastern cottonwood to *M. medusae* f.sp. *deltoidis*, indicating the potential for pathogen adaptation (Hamelin *et al.*, 1992). Also, genes for quantitative rust resistance to *Melampsora* spp. may be clustered with qualitative resistance genes. A segregating defeated qualitative resistance gene inherited from eastern cottonwood had major beneficial effects on quantitative resistance, and when it was defeated by *M. laricis-populina*, resulted in lower levels of quantitative resistance (Dowkiw and Bastien, 2007).

Plant growth promoters

Plant growth-promoting rhizobacteria have been developed for their capacity to induce systemic protection against pathogens. In lobolly pine, bacterial isolates reduced the number of *C. quercuum* f.sp. *fusiforme* galls following inoculations. The mode of action appeared to have induced systemic resistance to fusiform rust in loblolly pine, resulting in less infection in treated than in untreated control seedlings (Enebak and Carey, 2000).

26.5.6 Therapy

Excising or scribing stem cankers or large branch cankers can be used to save trees with potentially lethal cankers. This involves removing the bark from the side edges of the canker and removing the inner bark down to the sapwood (Zeglen *et al.*, 2010). Because canker excision is time-consuming, it should be restricted to high value trees in parks or highly visible locations (Zeglen *et al.*, 2009).

26.5.7 Integrated disease management

As an example, management options have been proposed for white pine blister rust that integrate multiple approaches to mitigating the impacts of the rust in threatened remote high elevation five-needle pine ecosystems (Schoettle and Sniezko, 2007). These have included reducing the pathogen populations, managing forest composition, improving host vigour, introducing resistant stock by artificial regeneration, and diversifying age class structure to affect the natural selection process for resistance (Hunt *et al.*, 2010).

References

Anderson, R.L., McClure, J.P., Cost, N. and Uhler, R.J. (1986) Estimating fusiform rust losses in five southeast States. *Southern Journal of Applied Forestry* 10, 237–240.

Bérubé, J.A. (1996) Use of triadimefon to control white pine blister rust. *Forestry Chronicle* 72, 637–638.

Booth, T.H., Old, K.M. and Jovanovic, T. (2000) A preliminary assessment of high risk areas for *Puccinia psidii* (Eucalyptus rust) in the neotropics and Australia. *Agriculture, Ecosystems and Environment* 82, 295–301.

Boyle, B., Hamelin, R.C. and Seguin, A. (2005) *In vivo* monitoring of obligate biotrophic pathogen growth by kinetic PCR. *Applied and Environmental Microbiology* 71, 1546–1552.

Boyle, B., Levée, V., Hamel, L., Nicole, M. and Séguin, A. (2010) Molecular and histochemical characterisation of two distinct poplar Melampsora leaf rust pathosystems. *Plant Biology* 12, 364–376.

Bresson, A., Jorge, V., Dowkiw, A., Guerin, V., Bourgait, I., Tuskan, G.A., Schmutz, J., Chalhoub, B., Bastien, C. and Rampant, P.F. (2011) Qualitative and quantitative resistances to leaf rust finely mapped within two nucleotide-binding site leucine-rich repeat (NBS-LRR)-rich genomic regions of chromosome 19 in poplar. *New Phytologist* 192, 151–163.

Bridgwater, F., Kubisiak, T., Byram, T. and Mckeand, S. (2005) Risk assessment with current deployment strategies for fusiform rust-resistant loblolly and slash pines. *Southern Journal of Applied Forestry* 29, 80–87.

Castello, J.D., Leopold, D.J. and Smallidge, P.J. (1995) Pathogens, patterns, and processes in forest ecosystems. *BioScience* 45, 16–21.

Cervera, M.T., Gusmão, J., Steenackers, M., Peleman, J., Storme, V., Vanden Broeck, A., Van Montagu, M. and Boerjan, W. (1996) Identification of AFLP molecular markers for resistance against *Melampsora larici-populina* in *Populus*. *Theoretical and Applied Genetics* 93, 733–737.

Cervera, M.T., Storme, V., Ivens, B., Gusmão, J., Liu, B.H., Hostyn, V., Van Slycken, J., Van Montagu, M. and Boerjan, W. (2001) Dense genetic linkage maps of three *Populus* species (*Populus deltoides*, *P. nigra* and *P. trichocarpa*) based on AFLP and microsatellite markers. *Genetics* 158, 787–809.

Colautti, R.I., Bailey, S.A., van Overdijk, C.D.A., Amundsen, K. and Macisaac, H.J. (2006) Characterised and projected costs of nonindigenous species in Canada. *Biological Invasions* 8, 45–59.

Crane, P.E. (2001) Morphology, taxonomy, and nomenclature of the *Chrysomyxa ledi* complex and related rust fungi on spruce and *Ericaceae* in North America and Europe. *Canadian Journal of Botany* 79, 957–982.

Cubbage, F.W., Pye, J.M., Holmes, T.P. and Wagner, J.E. (2000) An economic evaluation of fusiform rust protection research. *Southern Journal of Applied Forestry* 24, 77–85.

Cummins, G. and Hiratsuka, Y. (2003) *Illustrated Genera of Rust Fungi*, 3rd edn. APS Press, St Paul, Minnesota.

Desprez-Loustau, M., Dupuis, F. and Viguie, A. (1992) Evaluation of single annual applications of sterol-inhibiting fungicides for control of pine twisting rust. *Plant Disease* 76, 376–382.

Dowkiw, A. and Bastien, C. (2007) Presence of defeated qualitative resistance genes frequently has major impact on quantitative resistance to *Melampsora larici-populina* leaf rust in *P.* × *interamericana* hybrid poplars. *Tree Genetics and Genomes* 3, 261–274.

Duplessis, S., Major, I., Martin, F. and Séguin, A. (2009) Poplar and pathogen interactions: insights from *Populus* genome-wide analyses of resistance and defense gene families and gene expression profiling. *Critical Reviews in Plant Sciences* 28, 309–334.

Duplessis, S. *et al.* (2011a) Obligate biotrophy features unraveled by the genomic analysis of rust fungi. *Proceedings of the National Academy of Sciences of the United States of America* 108, 9166–9171.

Duplessis, S., Hacquard, S., Delaruelle, C., Tisserant, E., Frey, P., Martin, F. and Kohler, A. (2011b) *Melampsora larici-populina* transcript profiling during germination and time course infection of poplar leaves reveals dynamic expression patterns associated with virulence and biotrophy. *Molecular Plant–Microbe Interactions* 24, 808–818.

Enebak, S.A. and Carey, W.A. (2000) Evidence for induced systemic protection to fusiform rust in loblolly pine by plant growth-promoting rhizobacteria. *Plant Disease* 84, 306–308.

Feau, N., Joly, D.L. and Hamelin, R.C. (2007) Poplar leaf rusts: model pathogens for a model tree. *Canadian Journal of Botany* 85, 1127–1135.

Feau, N., Vialle, A., Allaire, M., Tanguay, P., Joly, D.L., Frey, P., Calland, B.E. and Hamelin, R.C. (2009) Fungal pathogen (mis-) identifications: a case study with DNA barcodes on *Melampsora* rusts of aspen and white poplar. *Mycological Research* 113, 713–724.

Feau, N., Vialle, A., Allaire, M., Maier, W. and Hamelin, R.C. (2011) DNA barcoding in the rust genus *Chrysomyxa* and its implications for the phylogeny of the genus. *Mycologia* 103, 1250–1266.

Froelich, R.C. and Snow, G.A. (1986) Predicting site hazard to fusiform rust. *Forest Science* 32, 21–35.

Geils, B.W., Hummer, K.E. and Hunt, R.S. (2010) White pines, *Ribes*, and blister rust: a review and synthesis. *Forest Pathology* 40, 147–185.

Gérard, P.R., Husson, C., Pinon, J. and Frey, P. (2006) Comparison of genetic and virulence diversity of *Melampsora larici-populina* populations on wild and cultivated poplar and influence of the alternate host. *Phytopathology* 96, 1027–1036.

Glen, M., Alfenas, A.C., Zauza, E.A.V., Wingfield, M.J. and Mohammed, C. (2007) *Puccinia psidii*: a threat to the Australian environment and economy, a review. *Australasian Plant Pathology* 36, 1–16.

Grondin, J., Bourassa, M. and Hamelin, R.C. (2005) First report of the aecial state of *Melampsora larici-populina* on *Larix* spp. in North America. *Plant Disease* 89, 1242.

Hacquard, S.P., Delaruelle, C., Legué, V., Tisserant, E., Kohler, A., Frey, P., Martin, F. and Duplessis, S. (2010) Laser capture microdissection of uredinia formed by *Melampsora larici-populina* revealed a transcriptional switch between biotrophy and sporulation. *Molecular Plant–Microbe Interactions* 23, 1275–1286.

Hacquard, S.P. *et al.* (2012) A comprehensive analysis of genes encoding small secreted proteins identifies candidate effectors in *Melampsora larici-populina* (poplar leaf rust). *Molecular Plant–Microbe Interactions* 25, 279–293.

Hamelin, R.C., Shain, L. and Thielges, B.A. (1992) Adaptation of poplar leaf rust to eastern cottonwood. *Euphytica* 62, 69–75.
Hollis, C.A. and Schmidt, R.A. (1977) Site factors related to fusiform rust incidence in north Florida slash pine plantations. *Forest Science* 23, 69–77.
Hudgins, J.W., Mcdonald, G.I., Zambino, P.J., Klopfenstein, N.B. and Franceschi, V.R. (2005) Anatomical and cellular responses of *Pinus monticola* stem tissues to invasion by *Cronartium ribicola*. *Forest Pathology* 35, 423–443.
Hunt, R.S. (1983) White pine blister rust in British Columbia II. Can stands be hazard rated? *Forestry Chronicle* 59, 30–33.
Hunt, R.S. (1985) Experimental evidence of heterothallism in *Cronartium ribicola*. *Canadian Journal of Botany* 63, 1086–1088.
Hunt, R.S. (1991) Operational control of white pine blister rust by removal of lower branches. *Forestry Chronicle* 67, 284–287.
Hunt, R.S. (1997) Relative value of slow-canker growth and bark reactions as resistance responses to white pine blister rust. *Canadian Journal of Plant Pathology* 19, 352–357.
Hunt, R.S., Roke, G. and Jensen, G.D. (2007) Initiation of stem infection in western white pine by blister rust (*Cronartium ribicola*). *Canadian Journal of Plant Pathology* 29, 18–24.
Hunt, R.S., Geils, B.W. and Hummer, K.E. (2010) White pines, *Ribes*, and blister rust: integration and action. *Forest Pathology* 40, 402–417.
Innes, L., Marchand, L., Frey, P., Bourassa, M. and Hamelin, R.C. (2004) First report of *Melampsora larici-populina* on *Populus* spp. in eastern North America. *Plant Disease* 88, 85.
Jackson, L.W.R. and Parker, J.N. (1958) Anatomy of fusiform rust galls on loblolly pine. *Phytopathology* 48, 637–640.
Jacobi, W.R., Geils, B.W., Taylor, J.E. and Zentz, W.R. (1993) Predicting the incidence of comandra blister rust on lodgepole pine: site, stand, and alternate-host influences. *Phytopathology* 83, 630–637.
Jacobs, J.J., Burnes, T.A., David, A.J. and Blanchette, R.A. (2009) Histopathology of primary needles and mortality associated with white pine blister rust in resistant and susceptible *Pinus strobus*. *Forest Pathology* 39, 361–376.
Joly, D.L., Feau, N., Tanguay, P. and Hamelin, R. (2010) Comparative analysis of secreted protein evolution using expressed sequence tags from four poplar leaf rusts (*Melampsora* spp.). *BMC Genomics* 11, 422–437.
Kaitera, J. and Hantula, J. (1998) *Melampyrum sylvaticum*, a new alternate host for pine stem rust *Cronartium flaccidum*. *Mycologia* 90, 1028–1030.
Kaitera, J. and Nuorteva, H. (2003) *Cronartium flaccidum* produces uredinia and telia on *Melampyrum nemorosum* and on Finnish *Vincetoxicum hirundinaria*. *Forest Pathology* 33, 205–213.
Kaitera, J., Nuorteva, H. and Hantula, J. (2005) Distribution and frequency of *Cronartium flaccidum* on *Melampyrum* spp. in Finland. *Canadian Journal of Forest Research* 35, 229–234.
Kaitera, J., Tillman-Sutela, E. and Kauppi, A. (2009) Cone bagging hinders cone and rust development of *Picea abies*. *Baltic Forestry* 15, 28–32.
Keane, R.E., Morgan, P. and Menakis, J.P. (1994) Landscape assessment of the decline of whitebark-pine (*Pinus albicaulis*) in the Bob Marshall Wilderness Complex, Montana, USA. *Northwest Science* 68, 213–229.
Kelley, W.D. (1980) Evaluation of systemic fungicides for control of *Cronartium quercuum* f.sp. *fusiforme* on loblolly pine seedlings. *Plant Disease* 64, 773–775.
Kemen, E., Kemen, A.C., Rafiqi, M., Hempel, U., Mendgen, K., Hahn, M. and Voegele, R.T. (2005) Identification of a protein from rust fungi transferred from haustoria into infected plant cells. *Molecular Plant–Microbe Interactions* 18, 1130–1139.
Kinloch, B.B. and Byler, J.W. (1981) Relative effectiveness and stability of different resistance mechanisms to white pine blister rust in sugar pine. *Phytopathology* 71, 386–391.
Kinloch, B.B., Sniezko, R.A., Barnes, G.D. and Greathouse, T.E. (1999) A major gene for resistance to white pine blister rust in western white pine from the Western Cascade Range. *Phytopathology* 89, 861–867.
Kinloch, B.B., Sniezko, R.A. and Dupper, G.E. (2003) Origin and distribution of *Cr2*, a gene for resistance to white pine blister rust in natural populations of western white pine. *Phytopathology* 93, 691–694.
Kuhlman, E.G. and Powers, H.R. Jr (1988) Resistance responses in half-sib loblolly pine progenies after inoculation with *Cronartium quercuum* f.sp. *fusiforme*. *Phytopathology* 78, 484–487.
Langrell, S.R.H., Glen, M. and Alfenas, A.C. (2008) Molecular diagnosis of *Puccinia psidii* (guava rust) – a quarantine threat to Australian eucalypt and *Myrtaceae* biodiversity. *Plant Pathology* 57, 687–701.

Lescot, M., Rombauts, S., Zhang, J., Aubourg, S., Mathe, C., Jansson, S., Rouze, P. and Boerjan, W. (2004) Annotation of a 95-kb *Populus deltoides* genomic sequence reveals a disease resistance gene cluster and novel class I and class II transposable elements. *Theoretical and Applied Genetics* 109, 10–22.

Lu, P. and Derbowka, D. (2009) Breeding eastern white pine for blister rust resistance: a review of progress in Ontario. *Forestry Chronicle* 85, 745–755.

Lu, P. and Sinclair, R.W. (2006) Survival, growth and wood specific gravity of interspecific hybrids of *Pinus strobus* and *P. wallichiana* grown in Ontario. *Forest Ecology and Management* 234, 97–106.

Lu, P.X., Sinclair, R.W., Boult, T.J. and Blake, S.G. (2005) Seedling survival of *Pinus strobus* and its interspecific hybrids after artificial inoculation of *Cronartium ribicola. Forest Ecology and Management* 214, 344–357.

MacDonald, W.L. (2003) Dominating North American forest pathology issues of the 20th century. *Phytopathology* 93, 1039–1040.

Maloy, O.C. (1997) White pine blister rust control in North America: a case history. *Annual Review of Phytopathology* 35, 87–109.

McDonald, G.I., Richardson, B.A., Zambino, P.J., Klopfenstein, N.B. and Kim, M.S. (2006) *Pedicularis* and *Castilleja* are natural hosts of *Cronartium ribicola* in North America: a first report. *Forest Pathology* 36, 73–82.

McKinney, S.T., Fiedler, C.E. and Tomback, D.F. (2009) Invasive pathogen threatens bird–pine mutualism: implications for sustaining a high-elevation ecosystem. *Ecological Applications* 19, 597–607.

Miller, T., Patton, R.F. and Powers, H.R. (1980) Mode of infection and early colonization of slash pine-seedlings by *Cronartium quercuum* f.sp. *fusiforme*. *Phytopathology* 70, 1206–1208.

Mulvey, R. and Hansen, E. (2009) White pine blister rust on new telial hosts (*Castilleja* and *Pedicularis*) in whitebark pine ecosystems at Mt. Rainier and Crater Lake National Parks. *Phytopathology* 99, S90–S91.

Mulvey, R.L. and Hansen, E.M. (2011) *Castilleja* and *Pedicularis* confirmed as telial hosts for *Cronartium ribicola* in whitebark pine ecosystems of Oregon and Washington. *Forest Pathology* 41, 453–463.

Myburg, H., Morse, A.M., Amerson, H.V., Kubisiak, T.L., Huber, D., Osborne, J.A., Garcia, S.A., Nelson, C.D., Davis, J.M., Covert, S.F. and Van Zyl, L.M. (2006) Differential gene expression in loblolly pine (*Pinus taeda* L.) challenged with the fusiform rust fungus, *Cronartium quercuum* f.sp. *fusiforme*. *Physiological and Molecular Plant Pathology* 68, 79–91.

Nagarajan, S. and Singh, D.V. (1990) Long-distance dispersion of rust pathogens. *Annual Review of Phytopathology* 28, 139–153.

Newcombe, G. (1998) A review of exapted resistance to diseases of *Populus*. *European Journal of Forest Pathology* 28, 209–216.

Newcombe, G., Bradshaw, H.D., Chastagner, G.A. and Stettler, R.F. (1996) A major gene for resistance to *Melampsora medusae* f.sp. *deltoidae* in a hybrid poplar pedigree. *Phytopathology* 86, 87–94.

Newcombe, G., Stirling, B., Mcdonald, S. and Bradshaw, H.D. (2000) *Melampsora* × *columbiana*, a natural hybrid of *M. medusae* and *M. occidentalis*. *Mycological Research* 104, 261–274.

O'Hara, K.L. (2007) Pruning wounds and occlusion: a long-standing conundrum in forestry. *Journal of Forestry* 105, 131–138.

O'Hara, K.L., Grand, L.A. and Whitcomb, A.A. (2010) Pruning reduces blister rust in sugar pine with minimal effects on tree growth. *California Agriculture* 64, 31–36.

Oliva, J. and Colinas, C. (2007) Canopy openings may prevent fir broom rust (*Melampsorella caryophyllacearum*) infections. *European Journal of Forest Research* 126, 507–511.

Ostrofsky, W.D., Rumpf, T., Struble, D. and Bradbury, R. (1988) Incidence of white pine blister rust in Maine after 70 years of a *Ribes* eradication program. *Plant Disease* 72, 967–970.

Patton, R.F. and Johnson, D.W. (1970) Mode of penetration of needles of eastern white pine by *Cronartium ribicola*. *Phytopathology* 60, 977–982.

Peterson, R.S. and Jewell, F.F. (1968) Status of American stem rusts of pine. *Annual Review of Phytopathology* 6, 23–40.

Petre, B., Major, I., Rouhier, N. and Duplessis, S. (2011) Genome-wide analysis of eukaryote thaumatin-like proteins (TLPs) with an emphasis on poplar. *BMC Plant Biology* 11: 33. doi: 10.1186/1471-2229-11-33.

Pinon, J. (1992) Frequency and evolution of *Melampsora larici-populina* Klebahn races in north-western France. *Annals of Forest Science* 49, 1–15.

Pinon, J., Newcombe, G. and Chastagner, G.A. (1994) Identification of races of *Melampsora larici-populina*, the Eurasian poplar leaf rust fungus, on *Populus* species in California and Washington. *Plant Disease* 78, 101.

Pitt, D.G., Meyer, T., Park, M., MacDonald, L., Buscarini, T. and Thompson, D.G. (2006) Application of slow-release tablets to enhance white pine regeneration: growth response and efficacy against white pine blister rust. *Canadian Journal of Forest Research* 36, 684–698.

Powers, H.R. Jr and Kraus, J.F. (1983) Developing fusiform rust-resistant loblolly and slash pines. *Plant Disease* 67, 187–189.

Powers, H.R. Jr, Schmidt, R.A. and Snow, G.A. (1981) Current status and management of fusiform rust on southern pines. *Annual Review of Phytopathology* 19, 353–371.

Prakash, C.S. and Thielges, B.A. (1987) Pathogenic variation in *Melampsora medusae* leaf rust of poplars. *Euphytica* 36, 563–570.

Ragazzi, A. and Fedi, I.D. (1992) Penetration of *Cronartium flaccidum* into pine needles. *European Journal of Forest Pathology* 22, 278–283.

Rayachhetry, M.B, Van, T.K., Center, T.D. and Elliott, M.L. (2001) Host range of *Puccinia psidii*, a potential biological control agent of *Melaleuca quinquenervia* in Florida. *Biological Control* 22, 38–45.

Richardson, B.A., Zambino, P.J., Klopfenstein, N.B., Mcdonald, G.I. and Carris, L.M. (2007) Assessing host specialization among aecial and telial hosts of the white pine blister rust fungus, *Cronartium ribicola*. *Canadian Journal of Botany* 85, 299–306.

Rinaldi, C., Kohler, A., Frey, P., Duchaussoy, F., Ningre, N., Couloux, A., Wincker, P., Le Thiec, D., Fluch, S., Martin, F. and Duplessis, S. (2007) Transcript profiling of poplar leaves upon infection with compatible and incompatible strains of the foliar rust *Melampsora larici-populina*. *Plant Physiology* 144, 347–366.

Savile, D.B.O. (1950) North American species of *Chrysomyxa*. *Canadian Journal of Forest Research* 28, 318–330.

Schmidt, R.A. (2003) Fusiform rust of southern pines: a major success for forest disease management. *Phytopathology* 93, 1048–1051.

Schmidt, R.A., Powers, H.R. Jr and Snow, G.A. (1981) Application of genetic-disease resistance for the control of fusiform rust in intensively managed southern pine. *Phytopathology* 71, 993–997.

Schoettle, A.W. and Sniezko, R.A. (2007) Proactive intervention to sustain high-elevation pine ecosystems threatened by white pine blister rust. *Journal of Forest Research* 12, 327–336.

Seifers, D., Ammon, V., Spencer, J. and Walkinshaw, C. (1985) Alteration in auxin levels in selected southern pines inoculated with the fusiform rust fungus. *Agriculture, Ecosystems and Environment* 14, 131–138.

Seifert, K.A. (2009) Progress towards DNA barcoding of fungi. *Molecular Ecology Resources* 9, 83–89.

Smith, J.P. and Hoffman, J.T. (2001) Site and stand characteristics related to white pine blister rust in high-elevation forests of southern Idaho and western Wyoming. *Western North American Naturalist* 61, 409–416.

Sniezko, R., Kegley, A. and Danchok, R. (2008) White pine blister rust resistance in North American, Asian, and European species – results from artificial inoculation trials in Oregon. *Annals of Forest Research* 51, 53–66.

Staples, R.C. (2001) Nutrients for a rust fungus: the role of haustoria. *Trends in Plant Science* 6, 496–498.

Tabor, G.M., Kubisiak, T.L., Klopfenstein, N.B., Hall, R.B. and McNabb, H.S. Jr (2000) Bulked segregant analysis identifies molecular markers linked to *Melampsora medusae* resistance in *Populus deltoides*. *Phytopathology* 90, 1039–1042.

Takahashi, I. and Saho, H. (1985) Notes on the Japanese rust fungi (on *Picea* sp.), 9: dissemination of *Chrysomyxa abietis* (Wallroth) Unger. *Transactions of the Mycological Society of Japan* 26, 433–439.

Tomback, D.F. and Resler, L.M. (2007) Invasive pathogens at alpine treeline: consequences for treeline dynamics. *Physical Geography* 28, 397–418.

Uchida, J.Y., Zhong, S. and Killgore, E. (2006) First report of a rust disease on ohia caused by *Puccinia psidii* in Hawaii. *Plant Disease* 90, 524.

Van Arsdel, E.P. (1965) Micrometeorology and plant disease epidemiology. *Phytopathology* 55, 945–950.

Van Arsdel, E.P., Riker, A.J. and Patton, R.F. (1956) The effects of temperature and moisture on the spread of white pine blister rust. *Phytopathology* 46, 307–318.

Vialle, A., Feau, N., Allaire, M., Didukh, M., Martin, F., Moncalvo, J.M. and Hamelin, R.C. (2009) Evaluation of mitochondrial genes as DNA barcode for *Basidiomycota*. *Molecular Ecology Resources* 9, 99–113.

Vialle, A., Frey, P., Hambleton, S., Bernier, L. and Hamelin, R.C. (2011) Poplar rust systematics and refinement of *Melampsora* species delineation. *Fungal Diversity* 50, 227–248.

Viljanen-Rollinson, S.L.H. and Cromey, M.G. (2002) Pathways of entry and spread of rust pathogens: implications for New Zealand's biosecurity. *New Zealand Plant Protection* 55, 42–48.

Voegele, R.T. and Mendgen, K. (2003) Rust haustoria: nutrient uptake and beyond. *New Phytologist* 159, 93–100.

Vogler, D.R. and Bruns, T.D. (1998) Phylogenetic relationships among the pine stem rust fungi (*Cronartium* and *Peridermium* spp.). *Mycologia* 90, 244–257.

Widin, K.D. and Schipper, A.L. Jr (1981) Effect of *Melampsora medusae* leaf rust infection on yield of hybrid poplars in the north-central United States. *European Journal of Forest Pathology* 11, 438–448.

Wilcox, P.L., Amerson, H.V., Kuhlman, E.G., Liu, B.H., Omalley, D.M. and Sederoff, R.R. (1996) Detection of a major gene for resistance to fusiform rust disease in loblolly pine by genomic mapping. *Proceedings of the National Academy of Sciences of the United States of America* 93, 3859–3864.

Wilkinson, A.G. and Spiers, A.G. (1976) Introduction of poplar rusts *Melampsora larici-populina* and *Melampsora medusae* to New Zealand and their subsequent distribution. *New Zealand Journal of Science* 19, 195–198.

Wingfield, M.J., Slippers, B., Hurley, B.P., Coutinho, T.A., Wingfield, B.D. and Roux, J. (2008) Eucalypt pests and diseases: growing threats to plantation productivity. *Southern Forests* 70, 139–144.

Xhaard, C., Andrieux, A., Halkett, F. and Frey, P. (2009) Characterization of 41 microsatellite loci developed from the genome sequence of the poplar rust fungus, *Melampsora larici-populina*. *Conservation Genetics Resources* 1, 21–25.

Zambino, P.J., Richardson, B.A. and Mcdonald, G.I. (2007) First report of the white pine blister rust fungus, *Cronartium ribicola*, on *Pedicularis bracteosa*. *Plant Disease* 91, 467.

Zeglen, S., Hunt, R. and Cleary, M. (2009) White pine blister rust forest health stand establishment decision aid. *BC Journal of Ecosystems and Management* 10, 97–100.

Zeglen, S., Pronos, J. and Merler, H. (2010) Silvicultural management of white pines in western North America. *Forest Pathology* 40, 347–368.

Zhang, J., Steenackers, M., Storme, V., Neyrinck, S., Van Montagu, M., Gerats, T. and Boerjan, W. (2001) Fine mapping and identification of nucleotide binding site/leucine-rich repeat sequences at the *MER* locus in *Populus deltoides* 'S9-2'. *Phytopathology* 91, 1069–1073.

Ziller, W.G. (1965) Studies of western tree rusts. 6. The aecial host ranges of *Melampsora albertensis*, *M. medusae* and *M. occidentalis*. *Canadian Journal of Botany* 43, 217–230.

Ziller, W.G. (1974) *The Tree Rusts of Western Canada*. Canadian Forest Service, Ottawa, Ontario, Canada.

27 Seed, Seedling and Nursery Diseases

Arja Lilja[1]* and Marja Poteri[2]

[1]*Vantaa Research Unit, Finnish Forest Research Institute, Vantaa, Finland;*
[2]*Suonenjoki Research Unit, Finnish Forest Research Institute, Suonenjoki, Finland*

27.1 Introduction

In most cases of poor plant health, environmental factors such as temperature and moisture play an important role in the probability of infection and transmission of pathogens. Modern nurseries must optimize culture conditions to maximize seedling production. The use of greenhouses with temperature regulation, and the use of selected seeds, pathogen-free growth media and irrigation and fertilization result in good seedling growth but may also provide ideal conditions for many biotic diseases. In addition to these cultural conditions, abiotic stress caused by frost, heat, darkness, drought, wetness or injury can further make rapidly growing seedlings more susceptible to fungal attack.

One of the effects of climate change will be an increase in environmental extremes, most likely accompanied by an increase in abiotic diseases and, in turn, increased interactions between abiotic and biotic diseases (Boland *et al.*, 2004; Deprez-Loustau *et al.*, 2007). With global warming, some diseases and insects may be able to occur farther north or at higher elevations than under the present climatic conditions (Sturrock *et al.*, 2011). For example, an increase in the disease intensity of Dothistroma needle blight (see Bulman *et al.*, Chapter 22, this volume) has coincided with increased spring and summer rainfall as well as an increase in the daily minimum temperature during late summer in British Columbia (Canada) and the UK (Woods *et al.*, 2005; Brown and Webber, 2008). There are also increases in diseases such as rusts and powdery mildews, as well as in diseases caused by other fungi that infect the leaves and needles of trees under conditions of higher precipitation (Desprez-Loustau *et al.*, 2007). Climatic change might also improve the establishment of introduced species in new geographical areas. Most diseases known to cause problems in forestry are presently caused by fungi, but in the future the risk of viral diseases vectored by invasive insects might increase (Roos *et al.*, 2011). Furthermore, climate warming and instability, along with heavy rains, might favour *Phytophthora* spp., most of which are alien species (Sturrock *et al.*, 2011).

In many cases, it is easier to prevent nursery diseases than to cure them at the stage when the first symptoms are visible. Control and prevention requires an effective programme that integrates good hygiene, cultural practices, biological and chemical control and, possibly, the genetic resistance of host plants to the disease concerned. Strict hygiene, including

* E-mail: arja.lilja@metla.fi

the use of good quality seeds, pest-free growth media, clean containers and growing area surfaces, together with cultural practices such as proper irrigation, ventilation, fertilization, the use of appropriate seedling densities, and the removal of weeds, infected seedlings and host trees inside or around a nursery discourage or prevent the entry and survival of many pathogens. In the case of rust diseases, it is wise, in addition, to remove the primary or alternate host of the rust to an appropriate distance around the nursery. Inside nurseries, trees and plants used as windbreaks or ornamentals might be a source of inoculum causing diseases such as Sphaeropsis tip blight, Phytophthora rot and cankers or Scleroderris canker. However, many diseases – such as Scleroderris canker, grey mould, rusts, blights and needle casts – might require the annual application of fungicides to prevent infections. For environmental reasons though, the use of fungicides should be minimized; they can also be used in combination with biocides or can be partly replaced by biocontrol agents. A knowledge of pathogen life cycles and the conditions that predispose seedlings to disease guarantees effective and economic management.

27.2 Seed Diseases

27.2.1 Pathogens

Most seeds acquire on their surfaces spores or other inocula of various microorganisms, which become significant only if they become active (Urošević, 1961; Lilja *et al.*, 1995; Cram and Fraedrich, 2009). While the majority of pathogens associated with forest tree seeds are fungi, seed-borne viruses and numerous species of bacteria have also been reported (Bremer *et al.*, 1991; Lilja *et al.*, 1995). Although the amount of fungal contaminants might vary considerably, Urošević (1961) reported that the average number of spores g^{-1} seeds ranges in Norway spruce (*Picea abies* (L.) Karst.) and Scots pine (*Pinus sylvestris* L.) from 50 to 150,000. The main effects of seed-borne fungi are the damping-off and root rot diseases that they may cause; to some extent they can reduce seed viability during storage as well (Sutherland and Woods, 1978; Sutherland, 1979, 1987; Lilja *et al.*, 1995). A number of fungi attack cones, catkins, nuts or acorns already on the tree, and invade embryo and gametophyte tissue, resulting in destruction, low germination or mummification (Sutherland *et al.*, 2002; Tillman-Sutela *et al.*, 2004).

27.2.2 Infections on seed-producing trees

Examples of fungi attacking seeds on the tree are the anamorphic fungus *Sirococcus conigenus* (Pers.) P.F. Cannon & Minter, the ascomycetous *Ciboria* spp. and the basidiomycetous rust fungi. The cone rust fungi of Norway spruce – *Chrysomyxa pyrolae* (DC.) Rostr. and *Pucciniastrum areolatum* (Fr.) G.H. Otth – destroy seed production totally or cause problems in seed germination. According to a study by Tillman-Sutela *et al.* (2004), spores of *C. pyrolae* were abundant inside Norway spruce seeds that failed to germinate. In nature, *S. conigenus* is often encountered on cone scales; in nurseries, infections by the pathogen are most commonly seed-borne and these form the inoculum source and the initial disease centres for top blight and cankers on conifers (Sutherland *et al.*, 1981; Lilja *et al.*, 2005).

Infections by *Ciboria alni* (O. Rostr.) N.F. Buchw. and *C. betulae* (Woronin ex Navashin) W.L. White on alders (*Alnus* spp.) and birches (*Betula* spp.), respectively, damage seeds so that they do not germinate. Black rot caused by *C. batschiana* (Zopf) N.F. Buchw. is a serious disease on chestnut (*Castanea* spp.) nuts and oak (*Quercus* spp.) acorns (Knudsen *et al.*, 2004; Migliorini *et al.*, 2010). Infections are caused by ascospores and occur just before or after the nuts or acorns fall from trees. All *Ciboria* spp. form a pseudosclerotium on an infected seed, on which apothecia will form in the year after infection; the infection is accompanied by seed mummification.

27.2.3 Soil-borne infections

Several ascomycetous anamorphs, such as *Alternaria* spp., *Aspergillus* spp., *Fusarium* spp.,

Penicillium spp., *Trichoderma* spp. and *Trichothecium roseum* (Pers.) Link, as well as oomycetous microbes in the genera *Pythium* and *Phytophthora*, and also the basidiomycetous *Rhizoctonia* spp., are common in soil. Thus, seeds that originated from fruits or cones that have had contact with soil have a higher risk of acquiring an infection, especially if the seeds are not mature or are damaged by insects (Procházková and Jancarek, 1991; Fraedrich *et al.*, 1994; Sutherland *et al.*, 2002; Cram and Fraedrich, 2009).

27.2.4 Other seed-borne diseases

Pitch canker is another disease of conifers that can spread via infested seed (see Gordon, Chapter 18, this volume). This is a devastating disease of mature pines (*Pinus* spp.) and Douglas-fir (*Pseudotsuga menziesii* (Mirb.) Franco) and is caused by *Gibberella circinata* Nirenberg & O'Donnell ex Britz, T.A. Cout., M.J. Wingf. & Marasas (anamorph: *Fusarium circinatum* Nirenberg & O'Donnell). Pitch canker occurs on different pine species throughout the south-eastern USA, Chile, Japan, Mexico and South Africa (EPPO, 2005). In South Africa, the fungus has caused severe losses mainly in nurseries as a root rot pathogen (Viljoen *et al.*, 1994). Although *G. circinata* is a quarantine pathogen in many countries, including within the European Union (EU), it is now present in nursery seedlings of pines in Spain and Portugal (Pérez-Sierra *et al.*, 2007; Bragança *et al.*, 2009). It has also been found in nurseries – but not in the field – in Chile (Jacobs *et al.*, 2007). Observed symptoms on Monterey pine (*Pinus radiata* D. Don) and maritime pine (*P. pinaster* Aiton) were dieback and wilting of the basal needles and terminal shoots (Bragança *et al.*, 2009).

27.2.5 Damage in storage

Examples of fungi that spread among seeds in storage are *Aspergillus* spp., *Penicillium* spp., *Trichoderma* spp. and *Caloscypha fulgens* (Pers.) Boud. (Sutherland *et al.*, 2002). *C. fulgens* is a psychrophilic ascomycete that causes germination failure of conifer seeds mainly in North America; it originates as seed-borne infections from cones either when they have been left on the ground for extended periods of time or when they have become contaminated with soil by some other means (Sutherland and Woods, 1978; Sutherland, 1979). In Europe, *C. fulgens* has been found to be present only on imported seed lots (Schröder *et al.*, 2002; Talgø *et al.*, 2010). Further, during storage, *C. batschiana* can grow from infected seed lots to healthy ones (Sutherland *et al.*, 2002).

27.2.6 Control of seed-borne diseases

Conifer seeds acquire pathogens mainly during cone collection, processing and storage (Sutherland, 1979; Fraedrich and Miller, 1995). The inclusion of cones from the previous year infected with rusts or other pathogens might be a source of new infections (Sutherland *et al.*, 1981; Kaitera *et al.*, 2009). Acorns of oaks collected from nets above the forest floor may also be infected by *C. batschiana*, but the infection has been shown to be lower than for acorns collected from the forest floor (Procházková *et al.*, 2005).

Many seed-borne infections can be avoided by increasing the level of hygiene during seed and cone collection and processing (Fraedrich *et al.*, 1994; Sutherland *et al.*, 2002; Cram and Fraedrich, 2009). Treating seeds with fungicides might also kill beneficial microbes. Several biocontrol agents and water treatments have also been found effective against seed-borne infections (Knudsen *et al.*, 2004; Migliorini *et al.*, 2010).

In seed health testing, molecular methods based on PCR will in future replace and be complementary to traditional cultural and serological methods (Sutherland *et al.*, 2002; Glynn and Edwards, 2009; Munkvold, 2009). DNA-specific probes are available for many pathogens and real-time PCR has high sensitivity and the ability to quantify pathogen DNA and test for multiple pathogens simultaneously (Littke, 1996; Munkvold, 2009).

27.3 Seedling Diseases

27.3.1 Damping-off

Damping-off disease is caused by many soil-borne or seed-borne non-host-specific fungi and fungal-like microorganisms that affect germinants before or just after emergence (Mittal, 1983; Mittal and Wang, 1987, 1993; Lilja *et al.*, 1995; Salerno and Lori, 2007). Pre-emergence damping-off kills the seeds or germinants before they emerge, while the classic symptom of post-emergence damping-off is decay of the hypocotyl at ground level (Sutherland *et al.*, 2002). Sometimes there is no clear separation between post-emergence damping-off and root rot on young seedlings. Many pathogens that infect the succulent tissues of young seedlings, such as *Fusarium*, *Pythium*, *Rhizoctonia*, *Neonectria* and *Alternaria* spp., can cause both damping-off and root rot (Sutherland and Davis, 1991; Lilja *et al.*, 1992; Salerno and Lori, 2007; Kwaśna and Bateman, 2009).

27.3.2 Root rot

Several species of fungi and fungal-like microorganisms can cause root rot. Seedlings suffering from root rot occur commonly in small patches that are randomly distributed. The first sign of root disease is the loss of vigour. The shoot of infected seedlings might be wilting, stunted or chlorotic, and eventually the seedlings might die (Hamm and Hansen, 1982; Lilja *et al.*, 1992). In most cases, infection starts from the root tips, which become pigmented and cannot divide and grow, and seedlings have a root system with few laterals (Hietala, 1995, 1997). Sometimes, the whole root system is water soaked and dark, and the rot may reach the stem base (Hamm and Hansen, 1982; Lilja *et al.*, 1992).

Oomycetous pathogens such as *Pythium* spp. and *Phytophthora* spp. are common in decayed roots of conifers and hardwoods in nurseries all over the world; they thrive under wet conditions because infections are caused primarily by zoospores swimming in free water (Hamm and Hansen, 1982; Lilja, 1994; Jung *et al.*, 2005). Many uninucleate, binucleate and multinucleate *Rhizoctonia* spp. (teleomorph: *Ceratobasidium* spp. or *Thanatephorus* spp.) can infect roots too and cause losses in seedling production (Lilja *et al.*, 1992; Lilja, 1994; Hietala, 1995; Hietala *et al.*, 2001). The recently described ascomycetes, *Ilyonectria macrodidyma* (Halleen, Schroers & Crous) P. Chaverri & Salgado and *I. radicicola* (Gerlach & L. Nilsson) Chaverri & Salgado have been found among the most common species on both decayed and healthy looking roots of conifer seedlings (Menkis *et al.*, 2006; Menkis and Vasaitis, 2010; Chaverri *et al.*, 2011; Menkis and Burokiene, 2012). Other common fungi on seedlings with root rot are *Fusarium* spp. (Lilja *et al.*, 1992; Menkis *et al.*, 2006). Both of the genera *Ilyonectria* and *Fusarium* include many species that can be highly pathogenic in nurseries under certain circumstances (Unestam *et al.*, 1989; James *et al.*, 1991).

In addition to causing mortality and cull in the nursery, *Phytophthora* and *Rhizoctonia* root infections can greatly reduce the rate of survival of outplanted seedlings (Hansen *et al.*, 1980; Lilja and Rikala, 2000). Other pathogens that have been shown to carry over from nursery to field planting sites are *Ilyonectria* spp., *Calonectria kyotensis* Terash. and *Macrophomina phaseolina* (Tassi) Goid. (Saunders *et al.*, 1992; Barnard, 1994; Axelrood *et al.*, 1998; Menkis *et al.*, 2006; Menkis and Burokiene, 2012). The charcoal root rot pathogen *M. phaseolina* was present even on asymptomatic loblolly pine (*Pinus taeda* L.) seedlings and caused death in the field over two growing seasons (Barnard, 1994).

27.3.3 Control of damping-off and root rot

The control and prevention of damping-off and root rot requires an effective programme that integrates good hygiene and cultural practices, and biological and chemical control. One source of pathogens that cause damping-off and root rot is the seed (Fraedrich *et al.*, 1994; Sutherland *et al.*, 2002; Cram and Fraedrich, 2009). Thus, equipment and tools used during seed collection, storage and later on for cultivation should be kept clean and pathogen free to avoid contamination (Fraedrich *et al.*, 1994; Sutherland *et al.*, 2002). The risk of damping-off can also be reduced by sowing when the temperature is high enough to promote rapid and even germination (Perrin and Sampangi, 1986).

For controlling fungal infections after germination and seedling development, there is a need to assess the methods used for manipulating the moisture, aeration conditions and pH of the growth medium, as waterlogging and alkaline conditions favour the growth of many fungi and *Oomycota* (Hamm and Hansen, 1982; Unestam *et al.*, 1989; Lilja *et al.*, 1998, 2007). In greenhouse experiments with severe root dieback caused by uninucleate *Rhizoctonia* spp., seedlings grown in a mixture of peat and mineral wool have been shown to have less root damage than seedlings grown in pure peat (Venn *et al.*, 1986). The reason for the diminished root dieback in a peat/mineral wool medium on racks was assumed to be the presence of an air layer between the trays and the sand; the air layer reduced water retention and the risk of pathogen contamination from the soil surface underneath (Venn *et al.*, 1986). The use of compost to suppress soil-borne diseases such as damping-off and root rots caused by *Rhizoctonia* spp. and oomycetous pathogens has also been successfully demonstrated in peat-based growth media (Scheuerell *et al.*, 2005; Termorshuizen *et al.*, 2006).

As well as soil or other surfaces beneath the containers, reused containers and irrigation water serve as a carrier of inoculum (Venn *et al.*, 1986; Kohmann and Børja, 2002; Hong and Moorman, 2005; Rytkönen *et al.*, 2008). Containers and surfaces in the growing area should be kept free of organic matter and cleaned regularly by steam, hot water or disinfectants to be sure that all propagules of weeds, fungi and other microbes will die. Many technologies, including heat, chlorination, ozonization and different types of UV radiation have been tested for the elimination of microbes in irrigation water (Cohn and Hong, 2003; Hong *et al.*, 2003; Banihashemi *et al.*, 2010).

27.4 Other Nursery Diseases

27.4.1 Grey mould

Grey mould disease is caused by the anamorphic ascomycete *Botrytis cinerea* Pers.

Symptoms and epidemiology

The disease usually occurs on young seedlings of most tree species, or on the lower parts of older seedlings (Venn, 1979; Zhang, 1992; Zhang and Sutton, 1994; Zhang *et al.*, 1995). Early symptoms of the disease on Norway spruce include yellow spots on the needles (Petäistö *et al.*, 2004). Later, the seedlings might exhibit top dieback, with the youngest shoots of conifers wilting (Petäistö, 2006). On broadleaved trees, necrotic patches appear on the leaves or stem. Advanced infections can easily be identified by the presence of grey mycelium over the affected parts of plants and conidia borne on tree-like conidiophores.

Low light intensity either alone or coupled with stress caused by high temperature or prolonged drought are important predisposing factors for *B. cinerea* infection, as well as abiotic damage such as that caused by frost, fertilizers, herbicides or mechanical injuries (Zhang, 1992; Zhang and Sutton, 1994; Zhang *et al.*, 1995). In dense seedling stands, dead lower needles in a position where they lack light represent a source of inoculum of grey mould (Mittal *et al.*, 1987). *B. cinerea* can also infect plants in combination with other fungi, or colonize necroses caused by other diseases (Domański and Kowalski, 1988; Hansen and Hamm, 1988). The fungus is very commonly present on dead or dying vegetable matter so that conidia are always available when conditions are suitable for infection. The conidia germinate between 0°C and 25°C, and the rate is temperature dependent, being optimal at 7–20°C (Petäistö, 2006). Russell (1990) found that infection can take place after just 3 h at 15–20°C and 98% relative humidity (RH) if there is free water on plant surfaces.

Grey mould primarily affects the young developing parts of plants, and the risk of infection decreases when seedling tissues mature. It has been shown that the disease fails to develop in black spruce (*Picea mariana* (Mill.) Britton, Sterns & Poggenb.) seedlings with completely green foliage (Zhang, 1992; Zhang and Sutton, 1994), and Norway spruce seedlings were found to be tolerant at the end of the growing season (Petäistö *et al.*, 2004). *B. cinerea* might also be a very harmful pathogen during storage as it has an affinity for cold, moist and dark conditions and the infection commonly originates from the nursery period (Venn, 1979; Mittal *et al.*, 1987). After freezer storage, grey mould infections were also found to spread from diseased seedlings to healthy ones inside cardboard boxes during prolonged thawing and before outplanting if the air circulation inside the boxes does not keep the temperature and moisture under the critical values for the development of mould and the welfare of seedlings (Petäistö, 2006).

Control

Cultural disease management techniques, together with chemical or biological control, are needed to limit losses caused by *B. cinerea*. The removal of weeds and dead or dying plant material is important. Most crucial is to keep the microclimate within the canopy as dry and well aerated as possible in order to reduce infections (Mittal *et al.*, 1987). This is achieved by regulating both the seedling density and the irrigation. Proper ventilation is important during greenhouse periods. It has been recommended that the seedlings are irrigated in the morning instead of the evening to ensure the rapid drying of shoots. It is also possible to brush the tops of the seedling with plastic pipe or a wooden dowel after irrigation to dislodge any water droplets and to accelerate drying. Avoidance of predisposing factors such as low light intensity, high temperature or prolonged drought, as well as abiotic damages, such as those caused by frost, fertilizers, herbicides or mechanical injuries, helps to keep the seedlings less vulnerable to grey mould attack (Zhang, 1992; Zhang and Sutton, 1994; Zhang *et al.*, 1995). Several fungicides representing different chemical groups and modes of action are registered for the control of the disease, e.g. thiophanate methyl, fenhexamide, fludioxonil, boscalid, cyprodinil, mepanipyrim and pyrimethanil. To reduce the risk of development of grey mould resistance to fungicides, it is important to alternate active ingredients representing different chemical groups and avoid active ingredients that are known to have a poor effectiveness against *B. cinerea*. The use of only site-specific products should be restricted and instead these should be rotated with fungicides having a different target site or multi-site activity in disease control.

A mycoparasite, *Clonostachys rosea* (Link) Schroers, Samuels, Seifert & W. Gams f. *rosea* (Schroers *et al.*, 1999), has shown to reduce grey mould infections as well as fungicides do (Zhang *et al.*, 1994). Similarly, two microbiological products, i.e. GlioMix® (*Gliocladium* spp.) and Binab® TF WP (*Trichoderma harzianum* Rifai and *T. polysporum* (Link) Rifai), reduced the amount of diseased needles by over 90% compared with controls, and were as effective as the fungicide tolylfluanid (Capieau *et al.*, 2004). Control treatments are needed especially when humidity is high and temperature is below 18°C. Seedlings may benefit also if sprayings have been done before storage or before they are packed into tight boxes for delivery.

27.4.2 Phytophthora diseases

Several species of *Phytophthora* cause damages to tree seedlings in forest and ornamental nurseries around the world.

Symptoms and epidemiology

Phytophthora spp. can infect both the roots and the aerial parts of plants, causing root rot, necrotic lesions on the foliage and stems, cankers on the stems and the wilting of new growth. Typical for part of the infections is that diseased tissue has a reddish discoloration (Hamm and Hansen, 1982; Orlikowski *et al.*, 2004; Jung *et al.*, 2005; Moralejo *et al.*, 2009; Yakabe *et al.*, 2009). *Phytophthora* spp. spread mainly through the movement of infected plants, infested soil and water, although there are species that are transmitted aerially too (see Hayden *et al.*, Chapter 25, this volume).

The most common species causing root rot in Douglas-fir, western hemlock (*Tsuga heterophylla* (Raf.) Sarg.), sugar pine (*P. lambertiana* Douglas), western white pine (*P. monticola* Douglas ex D. Don), ponderosa pine (*P. ponderosa* Lawson & C. Lawson), grand fir (*Abies grandis* (Douglas ex D. Don) Lindl.), white fir (*A. concolor* (Gord. & Glend.) Lindl. ex Hildebr.), noble fir (*A. procera* Rehder), Shasta red fir (*A. magnifica* A. Murray bis var. *shastensis* Lemmon) and Pacific silver fir (*A. amabilis* (Douglas ex Loudon) Douglas ex Forbes), are *Phytophthora megasperma* Drechsler, *P. pseudotsugae* Hamm & E.M. Hansen, *P. cactorum* (Lebeert & Cohn) J. Schröt., *P. cryptogea* Pethybr. & Laff., *P. drechsleri* Tucker and *P. cinnamomi* Rands (Hamm and Hansen, 1982, 1986). With the exception of *P. cinnamomi*, these species have adapted to the environmental conditions found in nurseries in the Pacific Northwest USA (Hamm and Hansen, 1986). *P. cactorum*, *P. cinnamomi*, *P. cryptogea* and *P. megasperma* are found also in European nurseries, as well as *P. cambivora* (Petri) Buisman and *P. citrophthora* (R.E. Sm. & E.H. Sm.) Leonian (Jung, 2009; T. Jung, Germany, 2012, personal communication). Nursery seedlings that are infected but have survived in the nursery have a higher risk of dying after outplanting (Hansen *et al.*, 1980; Doggett *et al.*, 1997).

Recent molecular technology has confirmed that many species of *Phytophthora* previously described as *P. megasperma* and *P. citricola* Sawada represent complexes that have now been resolved and found to include more than one species (Hansen *et al.*, 1986, 2009; Föster and Coffey, 1993; Jung and Burgess, 2009; Hong *et al.*, 2010). Six subgroups were defined within the *P. megasperma* complex, characterized by distinctive morphologies, pathogenicities, protein compositions and molecular taxonomy based on mitochondrial and nuclear DNA polymorphisms (Hansen *et al.*, 1986; Föster and Coffey, 1993). Later, the subgroup DF occurring on Douglas-fir seedlings and herbaceous weeds was described as the new species *P. sansomeana* E.M. Hansen & Reeser (Hansen *et al.*, 2009). Similarly, both isoenzyme analysis and single-strand-conformation polymorphism (SSCP) of ribosomal DNA (rRNA) fingerprinting techniques found several distinct subgroups within the *P. citricola* complex, although in most cases no differences in morphology or host range were detected (Oudemans *et al.*, 1994; Kong *et al.*, 2003). Phylogenetic studies of the internal transcribed spacer (ITS 1 and ITS 2) regions and 5.8S gene of the rRNA operon, the mitochondrial *cox1* and β-tubulin genes revealed two new species in the *P. citricola* complex, namely *P. plurivora* T. Jung & T.I. Burgess and *P. multivora* P.M. Scott & T. Jung (Jung and Burgess, 2009; Scott *et al.*, 2009). Also, one *P. citricola* group was showed to be identical to *P. pini* Leonian, a species that has been recently resurrected (Hong *et al.*, 2010). *P. multivora* is widespread in natural ecosystems of western Australia, but also present in nurseries in Europe, as well as *P. pini* (Scott *et al.*, 2009; Hong *et al.*, 2010; T. Jung, Germany, 2012, personal communication). *P. plurivora* is abundant in semi-natural ecosystems and nurseries across Europe where it causes bark necrosis, fine root loss and dieback in numerous tree species (Nechwatal and Osswald, 2001; Jung and Blaschke, 2004; Jung *et al.*, 2005; Jung and Nechwatal, 2008). In Norway, it is evident that *P. plurivora* has escaped from nurseries to nature (Talgø *et al.*, 2009). Many isolates representing these new species have in previous studies been called *P. citricola* or, occasionally, *P. inflata* Caros. & Tucker (Jung and Burgess, 2009).

Many new *Phytophthora* spp. have also been found in studies on lethal disease of alders in Europe, and on declining oaks and European beech (*Fagus sylvatica* L.) in Europe and the USA (Jung *et al.*, 1999, 2002; Werres *et al.*, 2001; Brasier *et al.*, 2004; Belbahri *et al.*, 2006). Most of the *Phytophthora* spp. found during these studies have also been present in nurseries, and infected nursery stock has shown to be responsible for the dispersal of the pathogens into nature (Jung *et al.*, 2000, 2005; Balci and Halmschlager, 2003; Jung and Blaschke, 2004; Jung and Nechwatal, 2008; Jung, 2009; Yakabe *et al.*, 2009; Goss *et al.*, 2011). The most harmful have been invasive exotic species such as *P. ramorum* Werres, De Cock & Man in't Veld, the cause of 'sudden oak death' and other diseases on numerous hosts, as well as *P. alni* Brasier & S.A. Kirk, *P. pseudosyringae* T. Jung & Delatour and its relative *P. quercina* T. Jung (Davidson *et al.*,

2002; Jung *et al.*, 2003; Jung and Blaschke, 2004; Wickland *et al.*, 2008; Jung, 2009). The relatively low genetic variation found among the isolates of these invasive species proves that they have been dispersed by human activities (Prospero *et al.*, 1999; Cooke *et al.*, 2005; Wickland *et al.*, 2008; Jung, 2009; Linzer *et al.*, 2009).

Control

In general, the control of these pathogens integrates good hygiene and cultural practices, and chemical control. However, once established, *Phytophthora* spp. are difficult to eradicate (Vercauteren *et al.*, 2010). Thus, the best method to avoid Phytophthora diseases is exclusion, which aims to prevent the introduction and establishment of the pathogen. Preventing the establishment, as well as the effective eradication, of diseased plants requires robust, specific and sensitive detection techniques. Traditional methods are based on isolation into selective media directly from plants, soil, water or baits (Erwin and Ribero, 1996).

Commercially available immunological ELISA tests are useful and quick, if the need is to identify the possible pathogen at the genus level (Ali-Shtayeh *et al.*, 1991; Themann *et al.*, 2002; Kox *et al.*, 2007). They can also be used as a first screening before further testing (Kox *et al.*, 2007). The lateral flow device (LFD) can also be used for the detection of *Phytophthora* spp. in plant tissues in the field (Lane *et al.*, 2007; Tomlinson *et al.*, 2010a). Since the introduction of PCR, a variety of sensitive molecular techniques have been developed for pathogen screening in the laboratory, including the screening of *Phytophthora* spp. The topic has been reviewed extensively by Cooke *et al.* (2007) and O'Brien *et al.* (2009). Tomlinson *et al.* (2005, 2007, 2010a,b) developed a number of assays that can be used for infected plant material in the field. A simple and rapid method for DNA extraction is followed by a real-time PCR with a portable platform (Tomlinson *et al.*, 2005) or loop-mediated isothermal amplification (LAMP) of target DNA using labelled primers, and detection of the generically labelled amplification products by a sandwich immunoassay in a format suitable for LFD (Tomlinson *et al.*, 2010b). Riedel *et al.* (2010) are developing a miniature microarray system suitable for field work.

Hygiene includes the use of pathogen-free growth media, containers, irrigation water, tools, machinery and people. The detection and removal of diseased plants is also a part of hygiene practices. One important part of the management is to avoid saturated growth media and water flowing over the ground. Most crucial is to keep the microclimate within the canopy as dry and well aerated as possible in order to reduce moist conditions suitable for infection. It has been shown that the introduction and spread of oomycetes have occurred through the use of contaminated irrigation water (Ali-Shtayeh *et al.*, 1991; Themann *et al.*, 2002).

The chemical control of Phytophthora diseases should be planned carefully to minimize the risk of resistance development among pathogen populations. However, fungicide applications are an important part of the management of these oomycetous pathogens. The systemic fungicide metalaxyl has been used extensively for the control of Phytophthora root rots on conifers, but in time many species have been shown to have become tolerant to this active ingredient (Humphrey, 1980; Hunger *et al.*, 1982; Hamm *et al.*, 1984). Today, more promising fungicides are those based on phosphoric acid. These are *Phytophthora* specific, and at appropriate dosages they stimulate host resistance to the pathogen, and are decomposed to the phosphorus needed for plant growth. They have been successfully used to control root rot of pine seedlings as well as other Phytophthora-related diseases (Hardy *et al.*, 2001; Reglinski *et al.*, 2009). The mode of action of these fungicides is still under investigation (King *et al.*, 2010), and there is also some evidence of an insensitivity of oomycetes to phosphonates occurring with prolonged use (Dobrowolski *et al.*, 2008). According to the Fungicide Resistance Action Committee list, both phosphoric acid-based fungicides and ethyl phosphonates, like fosetyl-aluminium, are considered to have a low resistance risk. A novel fungicide belonging to the carboxylic acid amides has been promising in controlling oomycetes in the field (Sakai *et al.*, 2010).

27.4.3 Scleroderris canker

Scleroderris canker is caused by the ascomycete *Gremmeniella abietina* (Lagerb.) M. Morelet (anamorph: *Brunchorstia pinea* (P. Karst.) Höhn.).

Symptoms and epidemiology

Conifers are potential hosts of *G. abietina*. Although the disease symptoms that occur in the spring on nursery seedlings vary depending on tree species, generally infected needles become brown from the base and eventually drop off. In inoculation experiments, the browning of needles of Norway spruce first occurred in the midsection of the shoot, whereas on Scots pine it was at the top of the shoot (Petäistö, 2008). Buds also fail to burst and necrotic lesions might be present on Norway spruce shoots (Børja *et al.*, 2006). Typical for pine seedlings is a drooping of needles before they fall off.

Based on biochemical and DNA profiles, *G. abietina* has been divided into North American, Asian and European races (Hamelin *et al.*, 1993; Lecours *et al.*, 1994; Müller and Uotila, 1997). Within the European race, three separate biotypes have been recognized: one affecting large trees (LTT), one affecting small trees (STT) and an Alpine type (Hamelin *et al.*, 1996; see Capretti *et al.*, Chapter 21, this volume). The European race is more virulent than the others and in nurseries where both the LTT and STT types are present (Uotila, 1983; Terho and Uotila, 1999). However, epidemics following cool, cloudy and wet growing seasons and mild winters are mainly caused by the LTT type, which is shown to be very pathogenic to nursery seedlings (Kurkela, 1981; Terho and Uotila, 1999; Børja *et al.*, 2006).

G. abietina has a life cycle of 3–4 years. Apothecia and ascospores are not common on nursery seedlings because they form 3 years after infection, or later. Conidia have been shown to be formed within 1 year on pine and 2 years on spruce seedlings (Petäistö, 2008). Conidia are present in the air through the whole summer, and during humid conditions a peak in release has been found in June and July (Petäistö and Heinonen, 2003). However, there are no symptoms visible on seedlings during the first summer. The infection proceeds for several weeks and, in general, the fungus does not invade the host tissue in the apical buds and developing shoots before the end of growing season (Patton *et al.*, 1984). During their first growing season, Scots pine seedlings are most susceptible to infection when buds are formed (Petäistö, 1999). Second year seedlings, however, have a higher infection risk during active shoot growth (Petäistö and Laine, 1999). Similarly, on spruce the susceptible period begins later on the first year seedlings than on the second year seedlings (Petäistö, 2008). Cool and rainy summers favour the disease, especially if they occur successively. Low light intensity, high fertilization and all factors that delay shoot maturation increase host susceptibility to *G. abietina*. Winter storage temperatures may also influence infection rate; those overwintering at –7°C and –3°C were found to be more diseased than those kept at 0°C (Petäistö and Laine, 1999). Seedlings stored at lower temperatures broke bud later than those that were kept warmer, suggesting that a physiological factor related to growth initiation may increase vulnerability (Petäistö and Laine, 1999).

Infected seedlings usually appear green and healthy immediately after the snow has melted or they have been taken out of freezer storage. The development of symptoms depends on weather conditions and the disease is not clearly seen before seedling growth has started. A PCR-based detection of infection helps to cull symptomless but diseased seedling lots before outplanting (Hamelin *et al.*, 2000).

Control

The destruction of infected seedlings and trees in and around the nursery lowers the number of spores and conidia. However, control of Scleroderris canker requires repeated spray applications. Contact fungicides such as chlorothalonil, captan, mancozeb and inorganic copper have multi-site activity in preventing spore germination and should be applied before the pathogen has penetrated the plant tissues. There are several systemic fungicides that also have curative activity, including propiconazole, prochloraz and azoxystrobin. In controlling Scleroderris canker it is important to choose an appropriate fungicide according to the stage of infection.

27.4.4 Sphaeropsis tip blight

Sphaeropsis tip blight and canker is caused by the anamorphic fungus *Diplodia pinea* (Desm.) J. Kickx f. (syn. *Sphaeropsis sapinea* (Fr.) Dyko & B. Sutton).

Symptoms and epidemiology

The fungus infects the bark, shoots, needles and cones of many species of *Abies*, *Cedrus*, *Juniperus*, *Picea*, *Pseudotsuga* and of over 30 species of *Pinus* (Sutton, 1980; Stanosz *et al.*, 1999). The pathogen occurs worldwide, and in nurseries it is associated with shoot blight and collar rot. In shoot blight, the terminal buds and upper needles die, leading to death if the seedlings are small. In collar rot, the shoots and needles initially appear chlorotic, and later desiccated (Palmer and Nicholls, 1985).

The pathogen penetrates the host directly or through wounds (Smith *et al.*, 1996). Disease symptoms can be seen soon after infection, or *D. pinea* may live in seedlings, trees, cones and seeds as an endophyte for long periods of time (Smith *et al.*, 1996; Stanosz *et al.*, 2005). These latent infections have been implicated in the movement of the pathogen around the world (Burgess *et al.*, 2004). Disease expression is most commonly associated with stress or non-optimal growth conditions before or after infection (Smith *et al.*, 1996; Johnson *et al.*, 1997; Paoletti *et al.*, 2001). Host and environmental factors influence the frequency of *D. pinea* in asymptomatic shoots of mature trees as well. It was possible to isolate the pathogen from healthy-looking shoots in an urban area as well as from mature trees growing in sites that received a high amount of solar radiation (Flowers *et al.*, 2001; Maresi *et al.*, 2007).

Three different *D. pinea* morphotypes, i.e. A, B and C, have been described based on cultural and molecular characteristics and pathogenicity (Blodgett and Stanosz, 1997; Stanosz *et al.*, 1999; de Wet *et al.*, 2002). However, de Wet *et al.* (2003) have shown that only two morphotypes, i.e. A and C, are important pathogens and that the C type originating from Indonesia and Mexico is more virulent than the more common A type (Blodgett and Stanosz, 1997; de Wet *et al.*, 2002). According to multiple gene genealogies and microsatellite markers, the B morphotype isolates are distantly related to *D. pinea* and actually represent the new species *D. scrobiculata* J. de Wet, Slippers & M.J. Wingf. (de Wet *et al.*, 2003). In inoculation tests, *D. scrobiculata* was less pathogenic and showed antagonism against *D. pinea* (Santamaría *et al.*, 2011). When both pathogens occurred in a single seedling, symptom severity was less than when *D. pinea* alone was present (Santamaría *et al.*, 2011).

The disease is important in warm climates. The conidia are shown to germinate in water at temperatures ranging from 15°C to 35°C and the percentage of germination was highest at 25°C. Conidia started to germinate within 4 h at 25°C when the air humidity was over 90%, but there was no germination if the humidity was 75% or lower (Jiangyan *et al.*, 1999).

Control

One important factor in control of the disease is to keep the microclimate within the canopy as dry and well aerated as possible in order to reduce the germination of conidia. All infected tissues support the development of pycnidia and spore production. Cones of several ages, needles in the litter and diseased, previous year's seedlings can all serve as the source of the disease in forests (Palmer *et al.*, 1988; Munck *et al.*, 2009; Munck and Stanosz, 2010). In nurseries, the disease is transmitted mainly from mature trees, and infected cones serve as a primary infection source (Bihon *et al.*, 2011). One solution is to remove all diseased trees within and around the nursery to lower the amount of conidia and possible infection. If chemical control is needed, sprayings with contact fungicides, e.g. chlorothalonil and copper-based products, are started before bud break and repeated at intervals as recommended by the authorized product. If needed, sprayings can be supplemented with systemic fungicides such as triazole group products and thiophanate methyl.

27.4.5 Dothistroma needle blight

Dothistroma needle blight, which is caused by the two ascomycetes *Dothistroma septosporum* (Dorog.) M. Morelet and *D. pini* Hulbary, is an economically important disease causing premature defoliation (see Bulman *et al.*, Chapter 22, this volume).

Symptoms and epidemiology

Early symptoms are yellow bands with tan spots. Later the bands will turn red and fruiting bodies will be borne on red zones, clearly different from the rest of the needles. The teleomorph of *D. septosporum* is *Mycosphaerella pini* Rostr. ex Munk (Barnes *et al.*, 2004), whereas that of *D. pini* is unknown (Barnes *et al.*, 2008). These two species show only slight differences in conidial morphology and it is almost impossible to distinguish them from each other without sequence analysis (Barnes *et al.*, 2004). Both species have two mating types. The screening of isolates from collections showed that only one mating type of *D. septosporum* was present in Australia and New Zealand, but in collections from Europe and Canada where the sexual state is known, both mating types were found (Grünewald *et al.*, 2007). Intriguingly, collections from South Africa and the UK, where the sexual stage is unknown, included both mating types (Grünewald *et al.*, 2007). In *D. pini*, for which no teleomorph is known, both mating types were present in collections from the USA (Grünewald *et al.*, 2007).

Many coniferous hosts are susceptible to *D. septosporum* but over 60 species of two-, three- and five-needle pines are particularly vulnerable to the pathogen worldwide (Gibson, 1979; Ivory, 1994; Barnes *et al.*, 2008). *D. pini* has been reported mainly from exotic Austrian pine (*P. nigra* Arnold) in the north and central USA, and later it was also reported from pallas pine (*P. pallasiana* D. Don) in Ukraine and south-western Russia (Barnes *et al.*, 2004, 2008). More recent reports are from Austrian pine in France and Hungary, and from Austrian pine and mugo pine (*P. mugo* Turra) in Russia (Ioos *et al.*, 2010; Barnes *et al.*, 2011). Scots pine generally has a low susceptibility and the most serious *D. septosporum* epidemics in the field occur on pines that are cultivated outside their natural distribution area (Watt *et al.*, 2009). However, in inoculation trials, nursery seedlings of Scots pine with Finnish isolates from mature trees have been found vulnerable. Although the tree flora in Finland is dominated by Scots pine, the disease is rare in forests (Müller *et al.*, 2009).

In the UK, the disease was first observed on nursery stock (Murray and Batko, 1962). Since the late 1990s, the incidence of the disease has increased, particularly on Corsican pine (*P. nigra* Arnold subsp. *laricio* Maire) and lodgepole pine (*P. contorta* Douglas ex Loudon) in the forest and also on nursery seedlings around Europe (Brown and Webber, 2008). As Dothistroma needle blight is listed in the EU Plant Directive as a quarantine pest, a good practice is that susceptible pine species may not be moved out of the nursery until they have been demonstrated to be free of infection for one full growing cycle, which may be up to 2 years, depending on when the infection is first discovered.

In future, climate change might increase the risk of Dothistroma needle blight in Europe, although there will be a global reduction in the potentially suitable area (Watt *et al.*, 2011). Infection depends on several factors, including periods of needle wetness, temperature and availability of conidia (Bulman, 1993). In Finland, symptoms and conidial stromata are found on needles of forest trees after a mild and rainy winter (Müller *et al.*, 2009).

Control

Cultural practices aimed at decreasing needle wetness and maintaining a low nitrogen level are important in prevention of the disease. The use of genetic resistance has been promising for preventing or reducing damage by *D. septosporum* in areas where genetic variation of the pathogen is low, as in New Zealand

(Hirst *et al.*, 1999). Resistant tree clones have been identified in Austrian pine, ponderosa pine and Monterey pine in the USA (Peterson, 1982). It is now possible to obtain seeds of Austrian and ponderosa pines that have been identified as having high resistance to the disease (Peterson and Read, 1971; Peterson, 1982). Copper-based fungicides have been used for control of the disease (Peterson, 1967; Brown and Webber, 2008).

27.4.6 Snow blights

Snow blights are mainly seriously damaging where snow cover is deep and prolonged. The two ascomycetes *Phacidium infestans* P. Karst. and *Herpotrichia juniperi* (Duby) Petr. are the main causes of the diseases in nurseries.

Symptoms and epidemiology

Both *P. infestans* and *H. juniperi* spread from seedling to seedling under snow cover by a mycelium, which is white in colour in the case of *P. infestans* and light brown or colourless for *H. juniperi*, the latter causing a disease called brown felt. Seedlings suffering from snow blights are usually detected following snowmelt in the spring, and within days the mycelium of both fungi turns darker. Withered needles below the snow line are matted together with mycelium and infected seedlings occur in patches.

P. infestans can infect several conifers, although it has been considered to be an important pathogen especially of pines in the northern parts of Europe and Asia (Björkman, 1948; Roll-Hansen, 1987; Hanso, 2000). *H. juniperi* can also attack a wide range of conifers, including species in the genera *Abies*, *Juniperus*, *Picea*, *Pinus* and *Tsuga*. Both fungi can grow at low temperatures and, according to Vuorinen and Kurkela (1993), the CO_2 concentration (production) and activity of *P. infestans* under snow reached their maxima at –3°C to 0°C in late winter.

Infection is also possible without snow and the risk of damage is high during overwintering in freezer storage if affected material is present and the humidity is high enough. In Estonia, *H. juniperi* infections have been recorded up to a height of 3 m in Norway spruce in dense forest stands during moist winter conditions, where only a sparse amount of snow covered the ground (Hanso and Tõrva, 1975).

Control

Chemical control is needed in the autumn during spore dispersal, which takes place from the end of August or the beginning of September. Sprayings should be repeated every 2 weeks or until the seedlings are covered by the permanent snow. Active ingredients with both a preventive and curative activity can be rotated, e.g. chlorothalonil, triazole and imidazole group fungicides or azoxystrobin.

27.4.7 Needle casts

Needle cast disease is caused by a group of fungi that in the end cause infected conifers to shed their needles. At least three ascomycetes occur sporadically in nurseries and can cause severe damage. *Lophodermium seditiosum* Minter, Staley & Millar and *Cyclaneusma minus* (Butin) DiCosmo, Peredo & Minter attack pines, and *Meria laricis* Vuill. attacks larch (*Larix* spp.) seedlings (Minter, 1981a,b; Gernandt *et al.*, 1997).

Although needle cast infections do not always kill seedlings, e.g. the buds remain alive, seedlings do not necessarily survive planting stress and so infected seedlings should be culled before outplanting. Stenström and Ihrmark (2005) have developed species-specific PCR primers to detect latent infections of *L. seditiosum* to help culling, because seedlings can appear healthy in the early spring or in material stored below 0°C.

Lophodermium needle cast

L. seditiosum is an aggressive pathogen on two-, three- and some five-needle pines, while other *Lophodermium* spp. are mainly host-specific endophytes of conifers (Sieber, 1988; Müller and Hallaksela, 1998). In Lophodermium needle cast, infections occur preferentially on current year needles via ascospores, and yellow lesions, which turn brown, develop on the needles (Minter, 1981a,b). The yellow margin of the lesions is caused by starch-free cells in the mesophyll (Diwani and Millar, 1981).

Like most needle cast pathogens, *L. seditiosum* fruits mainly after abscission and litter fall in the summer and autumn (Minter and Millar, 1980; Sinclair *et al.*, 1987; Gourbière and Debouzie, 2003). Ascocarps of *L. seditiosum* are produced rapidly in the summer of the same year on dead portions of infected needles that fell in the early summer (Hanso, 1970; Diwani and Millar, 1990). High precipitation during late summer and autumn creates favourable conditions for infection (Diwani and Millar, 1981). The fruiting bodies beneath the epidermis are black and elliptical, and visible to naked eyes (Sinclair *et al.*, 1987).

Cyclaneusma needle cast

C. minus is commonly a weak pathogen that colonizes healthy-looking needles as an endophyte (Helander *et al.*, 1994; Sieber *et al.*, 1999). However, it has been shown to cause needle cast of many pine species in nurseries throughout the world. It infects all but the current year needles. Symptoms are seen in the early autumn as a yellowing of the older, interior needles. The infection has happened during the previous year or earlier (Sinclair *et al.*, 1987). Transverse sections appear on Scots pine needles, but this symptom is uncommon on other hosts. The apothecia of *C. minus* appear first as elliptical swellings, blisters on brown needles without any special colour. Later, at maturity, they open during wet conditions and release ascospores into the air, and these can infect new needles. Spore production and infection can take place whenever temperatures are above freezing and needles are wet.

Larch needle cast

M. laricis is clearly pathogenic and affects many larch species, including western larch (*Larix occidentalis* Nutt.), European larch (*L. decidua* Mill.) and Siberian larch (*L. sibirica* Ledeb.). Infection is also possible on Douglas-fir (Sinclair *et al.*, 1987). Symptoms appear first as necrotic lesions on the needles, which later turn brown, wither and soon fall from the tree. Heavily infected trees may be killed, and infection in seedlings always reduces their growth.

M. laricis does not reproduce sexually. Conidia are produced in the spring and these infect only immature needles (Gernandt *et al.*, 1997). Seedlings can escape infection if dry conditions prevail during the needle growth period because needles become resistant with maturity. The fungus penetrates through the stomata, and later white conidiophores emerge from the needles. In Finland, needle cast caused by *M. laricis* has mainly been a problem in Siberian larches during their second growing season.

Control of needle cast

The removal of fallen needles and avoiding having nurseries near conifer forest lower the number of spores or conidia of all needle cast pathogens. *M. laricis* mainly infects seedlings during their second growing season and larch needle cast can be avoided by producing only 1 year old seedlings. Needle cast diseases can be controlled in nurseries by repeated application of fungicides to the foliage at 2 weeks intervals if the humidity is high. Spraying should be started just before spore liberation, which depends on the pathogen species involved. Wide-spectrum fungicides that affect spore germination, e.g. chlorothalonil, mancozeb and copper-based products, may be used against needle cast. Systemic fungicides, e.g. azoxystrobin and triazole group fungicides, can also be used.

27.4.8 Powdery mildews

Powdery mildews are caused by a group of ascomycetous fungi and are common on many broadleaved tree species.

Symptoms and epidemiology

Infected leaves and young shoots display on their surface a whitish grey powdery-looking coating, which is formed of vegetative and sporulating hyphae. Besides leaf discoloration, heavy infections cause growth distortion, premature leaf fall and shoot drying. Leaf defoliation may prevent shoots becoming winter hardy and lost shoot tips may lead to multiple leaders and poor seedling quality.

During the growing season, the disease is spread mainly via single-celled oidium-type conidia, which are formed on infected grey coloured leaf surfaces. Powdery mildews can overwinter as mycelium in twigs and buds (Kerling, 1966; Nef and Perrin, 1999) but under favourable conditions they also produce cleistothecia later in the growing season and autumn. Cleistothecia are regarded as the main inoculum at the beginning of growing season (Marçais *et al.*, 2009).

Oaks are mainly infected by *Erysiphe alphitoides* (Griffon & Maubl.) U. Braun & S. Takam., which is most common in temperate regions (Takamatsu *et al.*, 2007). Due to the recent taxonomic studies based on DNA sequence data, the former *Microsphaera* and *Uncinula* species have been shown to belong to the genus *Erysiphe* (Braun, 2011; Desprez-Loustau *et al.*, 2011). *Phyllactinia* spp. have a broad host range (e.g. *Fagus*, *Aesculus* and *Fraxinus* spp.), while *Sawadaea tulasnei* (Fuckel) Homma infects almost solely *Aceraceae* (Hirose *et al.*, 2005).

Powdery mildews are spread easily by wind and infection does not need stressed or injured plant tissue. Newly flushed leaves are more susceptible than mature leaves and there is also interspecific and intraspecific variation in disease resistance between tree species. Climatic factors have an important role in disease expression. Dry periods or semi-arid environments may predispose to the disease as conidia can germinate without extra moisture.

Control

For management, early detection is important. Under nursery conditions, cultural practices such as reduced seedling density and low nitrogen fertilization cannot always be obtained and thus chemical control is needed, especially with species or genotypes that have shown high susceptibility. Sprayings with fungicides such as quinoxyfen, myclobutanil, kresoxim-methyl and sulfur compounds should be started as soon as the first symptoms appear, and it may be necessary to repeat treatments until the end of the growing season.

27.4.9 Melampsora rusts

Rusts on poplars

A number of *Melampsora* spp. around the world have poplars or aspens (*Populus* spp.) as their alternate hosts and, in general, they are more harmful to broadleaved trees than to conifers serving as their aecial hosts. The *M. populnea* (Pers.) P. Karst. complex is a group of morphologically similar rusts that produce their uredial–telial stage on European aspen (*P. tremula* L.) and white poplar (*P. alba* L.) and their hybrids (Liro, 1908; Longo *et al.*, 1975). The species in this group are *M. laricis-tremulae* Kleb., *M. magnusiana* G.H. Wagner, *M. populnea* f.sp. *pinitorqua* Boerema & Verh. and *M. populnea* f.sp. *rostrupii* Boerema & Verh. In a phylogenetic analysis, *M. laricis-tremulae* and *M. populnea* f.sp. *pinitorqua* were unresolved within a monophyletic clade (bootstrap support 100%), although their secondary hosts are different, i.e. larch and pine, respectively (Feau *et al.*, 2009). Other species that complete their life cycle on poplars are *M. laricis-populina* Kleb., *M. medusae* Thüm. and *M. occidentalis* H.S. Jacks. (see Hamelin, Chapter 26, this volume).

The Melampsora rust infections cause the leaves of susceptible poplar species and clones to shrivel and fall prematurely, thus causing substantial yield losses, and they might also cause cankers on bark (Sinclair *et al.*, 1987; Wang and van der Kamp, 1992). Several generations might be borne on leaves before leaf fall, because the uredospores can reinfect the alternate host several times during the summer (Walker *et al.*, 1974). Nursery seedlings have been shown to die after a series of infections (Walker *et al.*, 1974). The amount of disease, however, varies from year to year depending on weather conditions. Although optimal temperatures vary from one *Melampsora* species to another, all temperatures from 10°C to 25°C support the infection if the relative humidity is 80% or more (McCracken *et al.*, 1984). However, the resistance of cultivars of poplars has been shown in some cases to be very temperature sensitive, so that inoculations at temperatures over 20°C were less successful than inoculations at lower temperatures (Chandrashekar and Heather, 1981; Singh and Heather, 1982).

Rusts on conifers

Pine twisting rust caused by the aecial stage of *M. populnea* f.sp. *pinitorqua* occurs sporadically in nurseries and might then cause losses to many pine species, including Aleppo pine (*P. halepensis* Mill.), Corsican pine, maritime pine, Scots pine and Italian stone pine (*P. pinea* L.), which all are natural hosts of *M. populnea* f.sp. *pinitorqua* (Longo *et al.*, 1975). Similarly, *M. occidentalis* infects Douglas-fir and western larch in nurseries (Hamm *et al.*, 1990). *M. populnea* f.sp. *pinitorqua* attacks elongated pine shoots while *M. occidentalis* infects current year needles of conifer hosts. Infections are seen first as yellow spots. Later, aecidia with masses of yellow orange aeciospores appear. Shoots infected by *M. populnea* f.sp. *pinitorqua* may become girdled, and when the lesion is confined to one side of the shoot, normal growth on the opposite side results in bending or twisting of the shoot. Young 8–10 week old nursery seedlings die if infected by the fungus. Needles infected with *M. occidentalis* will drop and this might later lead to the culling of seedlings (Hamm *et al.*, 1990). The aeciospores infect nearby hardwood hosts, but not conifers. Both *M. populnea* f.sp. *pinitorqua* and *M. occidentalis* produce their uredial and telial stages on aspens or poplars. Aeciospores and uredospores are shown to disperse mainly during dry weather, although sudden rain can briefly increase the spore density of aeciospores (Kurkela, 1973). Many other *Melampsora* species have also their spermogonial–aecial stage on larch needles, e.g. *M. laricis-populina*, *M. medusae*, *M. laricis-tremulae* and *M. populnea* f.sp. *rostrupii*, but their role as nursery pathogens on larch is not clear.

Control

Infections of cultivated hosts can be reduced by eliminating alternate hosts in and around the nursery. In high-risk areas chemical control is needed. Both contact and curative fungicides such as chlorothalonil, triazoles, imidazoles and azoxystrobin can be used. It is important to avoid many successive sprayings using active ingredients that have the same single-site activity and instead to rotate fungicides with different site activities.

27.4.10 Birch rust

The birch rust fungus *Melampsoridium betulinum* (Pers.) Kleb. causes premature defoliation on different birch species. This pathogen is a reason for premature leaf fall, and it disturbs winter hardening of shoots and buds which may dry out during winter. Severely diseased seedlings have also reduced height growth after outplanting (Lilja, 1973).

Symptoms and epidemiology

Birch leaves are infected by uredospores, which are produced on the lower leaf surfaces throughout the late growing season. The first symptoms are yellow spots on leaves, which usually appear in the lower part of seedlings where excess moisture favours uredospore germination. Continuous uredospore production spreads leaf rust gradually to the uppermost leaves so that under favourable conditions all leaves may become infected. Leaf rust is more pronounced in the late summer when water condensation on leaf surfaces is increased. *M. betulinum* is reported to hibernate as mycelium in birch bud scales and twigs (Dooley, 1984) and as uredospores in the leaves of small birch germlings, which remain green overwinter (Liro, 1906). Telia are also formed on infected birch leaves and larch is described as an aecial stage host for *M. betulinum* (Klebahn, 1904).

Birch seedlings with low nitrogen content have been less infected than more fertilized seedlings (Poteri and Rousi, 1996), and there is also variation in disease resistance among different birch species and clones (Poteri, 1998).

Control

In nurseries, it is difficult to avoid primary infections and to control the spread of birch rust by cultural procedures such as reduced seedling density and reduced nitrogen fertilization. Rust epidemics, however, vary largely between different years and control measures are not necessary every year. Regular monitoring of the development of rust infection is important to start chemical control (Plate 32), if needed, in time. There are several active ingredients used for leaf rusts, including site-specific systemic azoxystrobin, trifloxystrobin, kresoxim-methyl, myclobutanil and propiconazole. The number of repeated fungicide sprayings depends on the timing of the first rust outbreak and the prevailing weather conditions. To lower the risk for fungicide resistance it is important to rotate fungicides between different chemical groups.

References

Ali-Shtayeh, M.S., MacDonald, J.D. and Kabashima, J. (1991) A method for using commercial ELISA test to detect zoospores of *Phytophthora* and *Pythium* species in irrigation water. *Plant Disease* 75, 305–311.

Axelrood, P.E., Chapman, W.K., Seifert, K.A., Trotter, D.B. and Shrimpton, G. (1998) *Cylindrocarpon* and *Fusarium* root colonization of Douglas-fir seedlings from British Columbia reforestation sites. *Canadian Journal of Forest Research* 28, 1198–1206.

Balci, Y. and Halmschlager, E. (2003) First report of *Phytophthora quercina* from oak forests in Austria. *Plant Pathology* 52, 403.

Banihashemi, Z., MacDonald, J.D. and Lagunas-Solar, M.C. (2010) Effect of high-power monochromatic (pulse UV laser) and low-power broadband UV radiation on *Phytophthora* spp. in irrigation water. *European Journal of Plant Pathology* 127, 229–238.

Barnard, E.L. (1994) Nursery-to-field carryover and post-outplanting impact of *Macrophomina phaseolina* on loblolly pine on a cutover forest site in north central Florida. *Tree Planters' Notes* 45, 68–71.

Barnes, I., Crous, P.W., Wingfield, B.D. and Wingfield, M.J. (2004) Multigene phylogenesis reveals that red band needle blight of *Pinus* is caused by two distinct species of *Dothistroma, D. septosporum* and *D. pini*. *Studies in Mycology* 50, 551–565.

Barnes, I., Kirisits, T., Akulov, A., Chhetri, D.B., Wingfield, B.D., Bulgakov, T.S. and Wingfield, M.J. (2008) New hosts and country records of the Dothistroma needle blight pathogens from Europe and Asia. *Forest Pathology* 38, 178–195.

Barnes, I., Kirisits, T., Wingfield, M.J. and Wingfield, B.D. (2011) Needle blight of pine caused by two species of *Dothistroma* in Hungary. *Forest Pathology* 41, 361–369.

Belbahri, L., Moralejo, E., Calmin, G., Oszako, T., García, J.A., Descals, E. and Lefort, F. (2006) *Phytophthora polonica*, a new species isolated from declining *Alnus glutinosa* stands in Poland. *FEMS Microbiology Letters* 261, 165–174.

Bihon, W., Slippers, B., Burgess, T., Wingfield, M.J. and Wingfield, B.D. (2011) Sources of *Diplodia pinea* endophytic infections in *Pinus patula* and *P. radiata* seedlings in South Africa. *Forest Pathology* 41, 370–375.

Björkman, E. (1948) Studier över snöskyttesvampens (*Phacidium infestans* Karst.) biologi samt metoder för snöskyttets bekämpande [Studies on the biology of the Phacidium blight (*Phacidium infestans* Karst.) and its prevention]. *Meddelanden från Statens Skogsforskningsinstitut* 37, 1–136 [in Swedish, with English summary].

Blodgett, J.T. and Stanosz, G.R. (1997) *Sphaeropsis sapinea* morphotypes differ in aggressiveness, but both infect nonwounded red or jack pines. *Plant Disease* 81, 143–147.

Boland, G.J., Melzer, M.S., Hopkin, A., Higgins, V. and Nassuth, A. (2004) Climate change and plant diseases in Ontario. *Canadian Journal of Plant Pathology* 26, 335–350.

Børja, I., Solheim, H., Hietala, A.M. and Fossdal, C.G. (2006) Etiology and real-time polymerase chain reaction-based detection of *Gremmeniella*- and *Phomopsis*-associated disease in Norway spruce seedlings. *Phytopathology* 96, 1305–1314.

Bragança, H., Diogo, E., Moniz, F. and Amaro, P. (2009) First report of pitch canker on pines caused by *Fusarium circinatum* in Portugal. *Plant Disease* 93, 1079.

Brasier, C.M., Kirk, S.A., Delcan, J., Cooke, D.E.L., Jung, T. and Man in't Veld, W.A. (2004) *Phytophthora alni* sp. nov. and its variants: designation of emerging heteroploid hybrid pathogens spreading on *Alnus* trees. *Mycological Research* 108, 1172–1184.

Braun, U. (2011) The current systematics and taxonomy of the powdery mildews (*Erysiphales*): an overview. *Mycoscience* 52, 210–212.
Bremer, K., Lehto, K. and Kurkela, T. (1991) Metsäpuiden virus- ja mykoplasmatauteja [Diseases caused by viruses and mycoplasmas in forest trees]. *Metsäntutkimuslaitoksen Tiedonantoja* 382, 1–15 [in Finnish].
Brown, A. and Webber, J. (2008) *Red Band Needle Blight of Conifers in Britain*. Forestry Commission Research Note FRCN002, Forest Research, Edinburgh, UK.
Bulman, L.S. (1993) Cyclaneusma needle-cast and Dothistroma needle blight in NZ pine plantations. *New Zealand Forestry* 38, 21–24.
Burgess, T.I., Wingfield, M.J. and Wingfield, B.D. (2004) Global distribution of *Diplodia pinea* genotypes revealed using simple sequence repeat (SSR) markers. *Australasian Plant Pathology* 33, 513–519.
Capieau, K., Stenlid, J. and Stenström, E. (2004) Potential for biocontrol of *Botrytis cinerea* in *Pinus sylvestris* seedlings. *Scandinavian Journal of Forest Research* 19, 312–319.
Chandrashekar, M. and Heather, W.A. (1981) Temperature sensitivity of reactions of *Populus* spp. to races of *Melampsora larici-populina*. *Phytopathology* 71, 421–424.
Chaverri, P., Salgado, C., Hirooka, Y., Rossman, A.Y. and Samuels, G.J. (2011) Delimitation of *Neonectria* and *Cylindrocarpon* (*Nectriacea, Hypocreales, Ascomycota*) and related genera with *Cylindrocarpon*-like anamorphs. *Studies in Mycology* 68, 57–78.
Cohn, R.D. and Hong, C.X. (2003) Efficacy of ultraviolet irradiance for disinfesting recycled irrigation water. *Phytopathology* 93, S123.
Cooke, D.E.I., Jung, T., Williams, N.A., Schubert, R., Oβwald, W. and Dunchan, J. (2005) Genetic diversity of European population of the oak fine-root pathogen *Phytophthora quercina*. *Forest Pathology* 35, 57–70.
Cooke, D.E.I., Schena, L. and Cacciola, S.O. (2007) Tools to detect, identify and monitor *Phytophthora* species in natural ecosystems. *Journal of Plant Pathology* 89, 13–28.
Cram, M.M. and Fraedrich, S.W. (2009) Seed diseases and seedborne pathogens of North America. *Tree Planters' Notes* 53, 35–44.
Davidson, J.M., Garbelotto, M., Hansen, E.M., Reeser, P. and Rizzo, D.M. (2002) Another canker-causing *Phytophthora* from California and Oregon forest trees. *Phytopathology* 92, S17–S18.
Desprez-Loustau, M.-L., Robin, C., Reynaud, G., Déqué, M., Badeau, V., Piou, D., Husson, C. and Marçais, B. (2007) Simulating the effects of a climate-change scenario on the geographical range and activity of forest-pathogenic fungi. *Canadian Journal of Plant Pathology* 29, 101–120.
Desprez-Loustau, M.-L., Feau, N., Mougou-Hamdane, A. and Dutech, C. (2011) Interspecific and intraspecific diversity in oak powdery mildews in Europe: coevolution history and adaptation to their hosts. *Mycoscience* 52, 165–173.
de Wet, J., Wingfield, M.J., Coutinho, T.A. and Wingfield, B.D. (2002) Characterisation of the 'C' morphotype of the pine pathogen *Sphaeropsis sapinea*. *Forest Ecology and Management* 161, 181–188.
de Wet, J., Burgess, T., Slippers, B., Preisig, O., Wingfield, B.D. and Wingfield, M.J. (2003) Multiple gene genealogies and microsatellite markers reflect relationships between morphotypes of *Sphaeraopsis sapinea* and distinguish a new species of *Diplodia*. *Mycological Research* 107, 557–566.
Diwani, S.A. and Millar, C.S. (1981) Biology of *Lophodermium seditiosum* in nurseries in N.E. Scotland. In: Millar, C.S. (ed.) *Current Research on Conifer Needle Diseases. Proceedings of the IUFRO Working Party on Needle Diseases, Sarajevo, 1980*. Aberdeen University, Aberdeen, UK, pp. 67–74.
Diwani, S.A. and Millar, C.S. (1990) Sources of inoculum of *Lophodermium seditiosum* on *Pinus sylvestris*. *European Journal of Forest Pathology* 20, 1–7.
Dobrowolski, M.P., Shearer, B.L., Colquhoun, I.J., O'Brien, P.A. and Hardy, G.E.St.J. (2008) Selection for decreased sensitivity to phosphite in *Phytophthora cinnamomi* with prolonged use of fungicide. *Plant Pathology* 57, 928–936.
Doggett, C.A., Cordell, C.E. and Rogers, D. (1997) Effects of Phytophthora root rot on survival and growth of Fraser fir Christmas trees. *Tree Planters' Notes* 48, 72–75.
Domański, S. and Kowalski, T. (1988) Untypical dieback of the current season's shoots of *Pinus sylvestris* in Poland. *European Journal of Forest Pathology* 18, 157–160.
Dooley, H.L. (1984) Temperature effects on germination of uredospores of *Melampsoridium betulinum* and on rust development. *Plant Disease* 68, 686–688.
EPPO (2005) Data sheets on quarantine pests. *Gibberella circinata*. *EPPO Bulletin* 35, 383–386.
Erwin, D.C. and Ribeiro, O.K. (1996) *Phytophthora Diseases Worldwide*. APS Press, St Paul, Minnesota.
Feau, N., Vialle, A., Allaire, M., Tanguay, P., Joly, D.L., Frey, P., Callan, B.E. and Hamelin, R.C. (2009) Fungal pathogen (mis-) identifications: a case study with DNA barcodes on *Melampsora* rusts of aspen and white poplar. *Mycological Research* 113, 713–724.

Flowers, J., Nuckles, E., Hartman, J. and Vaillancourt, L. (2001) Latent infection of Austrian and Scots pine tissue by *Sphaeropsis sapinea*. *Plant Disease* 85, 1107–1112.
Föster, H. and Coffey, M.D. (1993) Molecular taxonomy of *Phytophthora megasperma* based on mitochondrial and nuclear DNA polymorphisms. *Mycological Research* 97, 1101–1112.
Fraedrich, S.W. and Miller, T. (1995) Mycoflora associated with slash-pine seeds from cones collected at seed orchards and cone-processing facilities in the south-eastern USA. *European Journal of Forest Pathology* 25, 73–82.
Fraedrich, S.W., Miller, T. and Zarnoch, S.J. (1994) Factors affecting the incidence of black seed rot in slash pine. *Canadian Journal of Forest Research* 24, 1717–1725.
Gernandt, D.S., Camacho, F.J. and Stone, J.K. (1997) *Meria laricis*, an anamorph of *Rhabdocline*. *Mycologia* 89, 735–744.
Gibson, I.A.S. (1979) *Diseases of Forest Trees Widely Planted as Exotics in the Tropics and Southern Hemisphere. Part II. The Genus* Pinus. Commonwealth Mycological Institute/Commonwealth Forestry Institute, Oxford, UK.
Glynn, N.C. and Edwards, S.G. (2009) Evaluation of PCR assays for quantifying seed-borne infection by *Fusarium* and *Microdochium* seedling blight pathogens. *Journal of Applied Microbiology* 108, 81–87.
Goss, E.M., Larsen, M., Vercauteren, A., Werres, S., Heungens, K. and Grünwald, N.J. (2011) *Phytophthora ramorum* detections in Canada: evidence for migration within North America and from Europe. *Phytopathology* 101, 166–171.
Gourbière, F. and Debouzie, D. (2003) Local variations in microfungal populations on *Pinus sylvestris* needles. *Mycological Research* 107, 1221–1230.
Grünewald, M., Barnes, I., Bradshaw, R.E., Brown, A.V., Dale, A., Grünewald, J.Z., Lewis, K.J., Wingfield, B.D., Wingfield, M.J. and Crous, P.W. (2007) Characterization and distribution of mating type genes in the Dothistroma needle blight pathogens. *Phytopathology* 97, 825–834.
Hamelin, R.C., Quellette, G.B. and Bernier, L. (1993) Identification of *Gremmeniella abietina* races with random amplified polymorphic DNA markers. *Applied and Environmental Microbiology* 59, 1752–1755.
Hamelin, R.C., Lecours, N., Hansson, P., Hellgren, M. and Laflamme, G. (1996) Genetic differentiation within the European race of *Gremmeniella abietina*. *Mycological Research* 100, 49–56.
Hamelin, R.C., Bourassa, M., Dusabenyagasani, M., Jacobi, V. and Laflamme, G. (2000) PCR detection of *Gremmeniella abietina*, the causal agent of Scleroderris cancer of pine. *Mycological Research* 104, 527–532.
Hamm, P.B. and Hansen, E.M. (1982) Pathogenicity of *Phytophthora* species to Pacific Northwest conifers. *European Journal of Forest Pathology* 12, 167–174.
Hamm, P.B. and Hansen, E.M. (1986) *Phytophthora* root rot in forest nurseries of the Pacific Northwest. In: Landis, T.D. (ed.) *Proceedings: Combined Western Forest Nursery Council and Intermountain Nursery Association Meeting, August 12–15, 1986, Tumwater, Washington*, General Technical Report RM-137, US Department of Agriculture, Forest Service, Rocky Mountain Research Station, Fort Collins, Colorado, pp. 122–124.
Hamm, P.B., Cooley, S.J. and Hansen, E.M. (1984) Response of *Phytophthora* spp. to metalaxyl in forest tree nurseries in the Pacific Northwest. *Plant Disease* 68, 671–673.
Hamm, P.B., Campbell, S.J. and Hansen, E.M. (eds) (1990) *Growing Healthy Seedlings. Identification and Management of Pests in Northwest Forest Nurseries*. US Department of Agriculture, Forest Service, Pacific Northwest Region and Forest Research Laboratory, College of Forestry, Oregon State University.
Hansen, E.M. and Hamm, P.B. (1988) Canker disease of Douglas-fir seedlings in Oregon and Washington bareroot nurseries. *Canadian Journal of Forest Research* 18, 1053–1058.
Hansen, E.M., Roth, L.F., Hamm, P.B. and Julis, A.J. (1980) Survival, spread, and pathogenicity of *Phytophthora* spp. on Douglas-fir seedlings planted on forest sites. *Phytopathology* 70, 422–425.
Hansen, E.M., Brasier, C.M., Shaw, D.S. and Hamm, P.B. (1986) The taxonomic structure of *Phytophthora megasperma*: evidence for emerging biological species groups. *Transactions of the British Mycological Society* 97, 557–573.
Hansen, E.M., Wilcox, W.F., Reeser, P.W. and Sutton, W. (2009) *Phytophthora rosacearum* and *P. sansomeana*, new species segregated from the *Phytophthora megasperma* 'complex'. *Mycologia* 101, 129–135.
Hanso, M. (1970) Die ganzjährige Kiefernadelschütte in phytopathologischer Sicht [Year-round pine needle shed from the phytopathological viewpoint]. *Metsanduslikud Uurimused* 8, 260–274.
Hanso, M. (2000) *Phacidium* snow blight in the Baltic countries. *Metsanduslikud Uurimused* 34, 64–74.
Hanso, M. and Tõrva, A. (1975) Okaspuu – nogihallitus Eestis I [Summary – black snow mould in Estonia I]. *Metsanduslikud Uurimused* 12, 262–279.

Hardy, G.E.St.J., Barrett, S. and Shearer, B.L. (2001) The future of phosphite as a fungicide to control the soil borne plant pathogen *Phytophthora cinnamomi* in natural ecosystems. *Australasian Plant Pathology* 30, 133–139.

Helander, M.L., Sieber, T.N., Petrini, O. and Neuvonen, S. (1994) Endophytic fungi in Scots pine needles: spatial variation and consequences of simulated acid rain. *Canadian Journal of Botany* 72, 1108–1113.

Hietala, A.M. (1995) Uni- and binucleate *Rhizoctonia* spp. co-existing on the roots of Norway spruce seedlings suffering from root dieback. *European Journal of Forest Pathology* 25, 136–144.

Hietala, A.M. (1997) The mode of infection of a pathogenic uninucleate *Rhizoctonia* sp. in conifer seedling roots. *Canadian Journal of Forest Research* 27, 471–480.

Hietala, A.M., Vahala, J. and Hantula, J. (2001) Molecular evidence suggests that *Ceratobasidium bicorne* has an anamorph known as a conifer pathogen. *Mycological Research* 105, 555–562.

Hirose, S., Tanda, S., Kiss, L., Grigaliunaite, B., Havrylenko, M. and Takamatsu, S. (2005) Molecular phylogeny and evolution of the maple powdery mildew (*Sawadaea, Erysiphaceae*) inferred from nuclear rDNA sequences. *Mycological Research* 109, 912–922.

Hirst, P., Richardson, T.E., Carson, S.D. and Bradshaw, R.E. (1999) *Dothistroma pini* genetic diversity is low in New Zealand. *New Zealand Journal of Forestry Science* 29, 459–472.

Hong, C.X. and Moorman, G.W. (2005) Plant pathogens in irrigation water: challenges and opportunities. *Critical Reviews in Plant Sciences* 24, 189–208.

Hong, C.X., Richardson, P.A., Kong, P. and Bush, E.A. (2003) Efficacy of chlorine on multiple species of *Phytophthora* in recycled nursery irrigation water. *Plant Disease* 87, 1183–1189.

Hong, C., Gallegly, M.E., Richardson, P.A. and Kong, P. (2010) *Phytophthora pini* Leonian resurrected to distinct species status. *Mycologia* 103, 351–360.

Humphrey, W.A. (1980) Controlling *Phycomycetes* which attract conifers. *HortScience* 15, 397.

Hunger, R.M., Hamm, P.B., Horner, C.E. and Hansen, E.M. (1982) Tolerance of *Phytophthora megasperma* isolates to metalaxyl. *Plant Disease* 66, 645–649.

Ioos, R., Fabre, B., Saurat, C., Fourrier, C., Frey, P. and Marçais, B. (2010) Development, comparison, and validation of real-time and conventional PCR tools for the detection of the fungal pathogens causing brown spot and red band needle blights of pine. *Phytopathology* 100, 105–114.

Ivory, M.H. (1994) Records of foliage pathogens of *Pinus* species in tropical countries. *Plant Pathology* 43, 511–518.

Jacobs, A., Coutinho, T.A., Winfield, M.J., Ahumada, R. and Wingfield, B.D. (2007) Characterization of the pitch canker fungus, *Fusarium circinatum*, from Chile. *South African Journal of Science* 102, 253–257.

James, R.L., Dumroese, R.K. and Wenny, D.L. (1991) *Fusarium* diseases of conifer seedlings. In: Sutherland, J.R. and Glover, S.G. (eds) *Proceedings of the First Meeting of IUFRO Working Party S2.07-09. Diseases and Insects in Forest Nurseries, August 23–30, 1990, Victoria, BC.* Information Report BC-X-331, Forestry Canada, Pacific and Yukon Region, Pacific Forestry Centre, Victoria, British Columbia, pp. 181–190.

Jiangyan, C., Xiaodong, S., Shuhua, L., Guirong, L. and Guijun, X. (1999) Studies on the biology of *Sphaeropsis sapinea* on *Pinus sylvestris* var. *mongolica*. *Journal of Liaoning Forestry and Technology* 1999, 712.

Johnson, J.W., Gleason, M.L., Parker, S.K., Provin, E.B. and Iles, J.K. (1997) Duration of water stress affects development of Sphaeropsis canker on Scots pine. *Journal of Arboriculture* 23, 73–76.

Jung, T. (2009) Beech decline in central Europe driven by the interaction between *Phytophthora* infections and climatic extremes. *Forest Pathology* 39, 73–94.

Jung, T. and Blaschke, M. (2004) *Phytophthora* root and collar rot of alders in Bavaria: distribution, modes of spread and possible management strategies. *Plant Pathology* 53, 197–208.

Jung, T. and Burgess, T.I. (2009) Re-evaluation of *Phytophthora citricola* isolates from multiple woody hosts in Europe and North America reveals a new species, *Phytophthora plurivora* sp. nov. *Persoonia* 22, 95–110.

Jung, T. and Nechwatal, J. (2008) *Phytophthora gallica* sp. nov., a new species from rhizosphere soil of declining oak and reed stands in France and Germany. *Mycological Research* 112, 1195–1205.

Jung, T., Cooke, D.E.L., Blaschke, H., Duncan, J.M. and Osswald, W. (1999) *Phytophthora quercina* sp. nov., causing root rot of European oaks. *Mycological Research* 103, 785–798.

Jung, T., Blaschke, H. and Osswald, W. (2000) Involvement of *Phytophthora* species in Central European oak decline and the effect of site factors on the disease. *Plant Pathology* 49, 706–718.

Jung, T., Hansen, E.M., Winton, L., Osswald, W. and Delatour, C. (2002) Three new species of *Phytophthora* from European oak forests. *Mycological Research* 106, 397–411.

Jung, T., Nechwatal, J., Cooke, D.E.L., Hartmann, G., Blaschke, M., Osswald, W.F., Duncan, J.M. and Delatour, C. (2003) *Phytophthora pseudosyringae* sp. nov., a new species causing root and collar rot of deciduous tree species in Europe. *Mycological Research* 107, 772–789.

Jung, T., Hudler, G.W., Jensen-Tracy, S.L., Griffiths, H.M., Fleichmann, F. and Osswald, W. (2005) Involvement of *Phytophthora* species in the decline of European beech in Europe and the USA. *Mycologist* 19, 159–166.

Kaitera, J., Tillman-Sutela, E. and Kauppi, A. (2009) Seasonal fruiting and sporulation of *Thekopsora* and *Chrysomyxa* cone rusts in Norway spruce cones and alternate hosts in Finland. *Canadian Journal of Forest Research* 39, 1630–1646.

Kerling, L.C.P. (1966) The hibernation of the oak mildew. *Acta Botanica Neerlandica* 15, 76–83.

King, M., Reeve, W., Van der Hoek, M.B., Williams, N., McComb, J., O'Brien, P.A. and Hardy, G.E.St.J. (2010) Defining the phosphite-regulated transcriptome of the plant pathogen *Phytophthora cinnamomi*. *Molecular Genetics and Genomics* 284, 425–435.

Klebahn, H. (1904) *Die wirtwechselnden Rostpilze* [*Host-alternating Rust Fungi*]. Borntraeger, Berlin, Germany.

Knudsen, I.M.B., Thomsen, K.A., Jensen, B. and Poulsen, K.M. (2004) Effects of hot water treatment, biocontrol agents, disinfectants and a fungicide on storability of English oak acorns and control of the pathogen, *Ciboria batschiana*. *Forest Pathology* 34, 47–64.

Kohman, K. and Børja, I. (2002) Hot-water treatment for sanitizing forest nursery containers: effects on container microflora and seedling growth. *Scandinavian Journal of Forest Research* 17, 111–117.

Kong, P., Hong, C., Richardson, P.A. and Gallegly, M.E. (2003) Single-strand-conformation polymorphism of ribosomal DNA for rapid species differentiation in genus *Phytophthora*. *Fungal Genetics and Biology* 39, 238–249.

Kox, L.F.F., van Brouwershaven, I.R., van de Vossenberg, B.T.L.H., van den Beld, H.E., Bonants, P.J.M. and de Gruyter, J. (2007) Diagnostic values and utility of immunological, morphological, and molecular methods for *in planta* detection of *Phytophthora ramorum*. *Phytopathology* 97, 1119–1129.

Kurkela, T. (1973) Epiphytology of *Melampsora* rusts of Scots pine (*Pinus sylvestris* L.) and aspen (*Populus tremula* L.). *Communicationes Instituti Forestalis Fenniae* 79, 1–68.

Kurkela, T. (1981) Versosyöpä (*Gremmeniella abietina*) riukuasteen männikössä [Canker and dieback of Scots pine at precommercial stage caused by *Gremmeniella abietina*]. *Folia Forestalia* 485, 1–12 [in Finnish, with English summary].

Kwaśna, H. and Bateman, G.L. (2009) Microbial communities in roots of *Pinus sylvestris* seedlings with damping-off symptoms in two nurseries as determined by ITS1/2 rDNA sequencing. *Forest Pathology* 39, 239–248.

Lane, C.R., Hobden, E., Walker, L., Barton, V.C., Inman, A.J., Hughes, K.J.D., Swan, H., Colyer, A. and Barker, I. (2007) Evaluation of a rapid diagnostic field test kit for identification of *Phytophthora* species, including *P. ramorum* and *P. kernoviae* at the point of inspection. *Plant Pathology* 56, 828–835.

Lecours, N., Toti, L., Sieber, T.N. and Petrini, O. (1994) Pectic enzyme patterns as a taxonomic tool for the characterization of *Gremmeniella* spp. isolates. *Canadian Journal of Botany* 72, 891–896.

Lilja, A. (1994) The occurrence and pathogenicity of uni- and binucleate *Rhizoctonia* and *Pythiaceae* fungi among conifer seedlings in Finnish forest nurseries. *European Journal of Forest Pathology* 24, 181–192.

Lilja, A. and Rikala, R. (2000) Effect of uninucleate *Rhizoctonia* on the survival of outplanted Scots pine and Norway spruce seedlings. *Forest Pathology* 30, 109–115.

Lilja, A., Lilja, S., Poteri, M. and Ziren, L. (1992) Conifer seedling root fungi and root dieback in Finnish nurseries. *Scandinavian Journal of Forest Research* 7, 547–556.

Lilja, A., Hallaksela, A.M. and Heinonen, R. (1995) Fungi colonizing Scots pine cone scales and seeds and their pathogenicity. *European Journal of Forest Pathology* 25, 38–46.

Lilja, A., Heiskanen, J. and Heinonen, R. (1998) Effects of irrigation on uninucleate *Rhizoctonia* on nursery seedlings of *Pinus sylvestris* and *Picea abies* grown in peat growth medium. *Scandinavian Journal of Forest Research* 13, 184–188.

Lilja, A., Poteri, M., Vuorinen, M., Kurkela, T. and Hantula, J. (2005) Cultural and PCR-based identification of the two most common fungi from cankers on container-grown Norway spruce seedlings. *Canadian Journal of Forest Research* 35, 432–439.

Lilja, A., Luoranen, J., Rikala, R. and Heinonen, R. (2007) The effects of calcium on stem lesions of silver birch seedlings. *Forest Pathology* 37, 96–104.

Lilja, S. (1973) Koivun ruoste ja sen torjuminen [Birch rust and the fight against it]. *Metsänviljelyn Koelaitoksen Tiedonantoja* 9, 21–26 [in Finnish].

Linzer, R.E., Rizzo, D.M., Cacciola, S.O. and Garbelotto, M. (2009) AFLPs detect low genetic diversity for *Phytophthora nemorosa* and *P. pseudosyringae* in the US and Europe. *Mycological Research* 113, 298–307.

Liro, J.I. (1906) Kulturversuche mit finnischen Rostpilzen. II [Culture experiments with Finnish rust fungi. II]. *Acta Societas pro Fauna et Flora Fennica* 29.

Liro, J.I. (1908) *Uredineae Fennicae. Finlands Rostsvampar [Finnish rusts]*. Finska Litteratur-Sällskapets Tryckeri, Helsinki, Finland (in Swedish).

Littke, W. (1996) Seed pathogens and seed treatments. In: Landis, T.D. and South, D.B. (eds) *National Proceedings, Forest and Conservation Nursery Associations, 1996*. General Technical Report PNW-GTR-389,US Department of Agriculture, Forest Service, Pacific Northwest Research Station, Portland, Oregon, pp. 187–191.

Longo, N., Moriondo, F. and Naldini Longo, B. (1975) The status of *Melampsora pinitorqua* Rostr. in Italy. *European Journal of Forest Pathology* 5, 147–152.

Marçais, B., Kavkova, M. and Desprez-Loustau, M.-L. (2009) Phenotypic variation in the phenology of ascospore production between European populations of oak powdery mildew. *Annals of Forest Science* 66, 814.

Maresi, G., Luchi, N., Pinzani, P., Pazzagli, M. and Capretti, P. (2007) Detection of *Diplodia pinea* in asymptomatic pine shoots and its relation to the Normalized Insolation index. *Forest Pathology*, 37, 272–280.

McCracken, F.I., Schipper, A.L. and Widin, K.D. (1984) Observations on occurrence of cottonwood leaf rust in central United States. *European Journal of Forest Pathology* 14, 226–233.

Menkis, A. and Burokiene, D. (2012) Distribution and genetic diversity of the root-rot pathogen *Neonectria macrodidyma* in a forest nursery. *Forest Pathology* 42, 79–83.

Menkis, A. and Vasaitis, R. (2010) Fungi in roots of nursery grown *Pinus sylvestris*: ectomycorrhizal colonization, genetic diversity and spatial distribution. *Microbial Ecology* 61, 52–63.

Menkis, A., Vasiliauskas, R., Taylor, A.F.S., Stenström, E., Stenlid, J. and Finlay, R. (2006) Fungi in decayed roots in conifer seedlings in forest nurseries, afforested clear-cuts and abandoned farmland. *Plant Pathology* 55, 117–129.

Migliorini, M., Funghini, L., Marinelli, C., Turchetti, T., Canuti, S. and Zanoni, B. (2010) Study of water curing for the preservation of marrons (*Castanea sativa* Mill., Marrone fiorentino cv). *Postharvest Biology and Technology* 56, 95–100.

Minter, D.W. (1981a) *Lophodermium* on pines. *CMI Mycological Papers* 147, Commonwealth Mycological Institute, Kew UK [now CAB International, Wallingford, UK], pp. 1–54.

Minter, D.W. (1981b) Possible biological control of *Lophodermium seditiosum*. In: Millar, C.S. (ed.) *Current Research on Conifer Needle Diseases. Proceedings of the IUFRO Working Party on Needle Diseases, Sarajevo, 1980*. Aberdeen University, Aberdeen, UK, pp. 75–80.

Minter, D.W. and Millar, C.S. (1980) Ecology and biology of three *Lophodermium* species on secondary needles of *Pinus sylvestris*. *European Journal of Forest Pathology* 10, 169–181.

Mittal, R.K. (1983) Studies on the mycoflora and its control on the seeds of some forest trees. I. *Cedrus deodara*. *Canadian Journal of Botany* 61, 197–201.

Mittal, R.K. and Wang, B.S.P. (1987) Fungi associated with seeds of eastern white pine and white spruce during cone processing and seed extraction. *Canadian Journal of Forest Research* 17, 1026–1034.

Mittal, R.K. and Wang, B.S.P. (1993) Effects of some seed-borne fungi on *Picea glauca* and *Pinus strobus* seeds. *European Journal of Forest Pathology* 23, 138–146.

Mittal, R.K., Singh, P. and Wang, B.S.P. (1987) *Botrytis*: a hazard to reforestation. A literature review. *European Journal of Forest Pathology* 17, 369–384.

Moralejo, E., Pérez-Sierra, A.M., Álvarez, L.A., Belbahri, L., Lefort, F. and Descals, E. (2009) Multiple alien *Phytophthora* taxa discovered on diseased ornamental plants in Spain. *Plant Pathology* 58, 100–110.

Müller, M.M. and Hallaksela, A.-M. (1998) Diversity of Norway spruce needle endophytes in various mixed and pure Norway spruce stands. *Mycological Research* 102, 1183–1189.

Müller, M.M. and Uotila, A. (1997) The diversity of *Gremmeniella abietina* var. *abietina* FAST-profiles. *Mycological Research* 101, 557–564.

Müller, M.M., Hantula, J. and Vuorinen, M. (2009) First observations of *Mycosphaerella pini* on Scots pine in Finland. *Plant Disease* 93, 322.

Munck, I.A. and Stanosz, G.R. (2010) Longevity of inoculum production of *Diplodia pinea* on red pine cones. *Forest Pathology* 40, 58–63.

Munck, I.A., Smith, D.R., Sickley, T. and Stanosz, G.R. (2009) Site-related influences on cone-borne inoculum and asymptomatic persistence of Diplodia shoot blight fungi on or in mature red pines. *Forest Ecology and Management* 257, 812–819.

Munkvold, G.P. (2009) Seed pathology progress in academia and industry. *Annual Review of Phytopathology* 47, 285–311.

Murray, J.S. and Batko, S. (1962) *Dothistroma pini* Hulbary: a new disease on pine in Britain. *Forestry* 3, 57–65.

Nechwatal, J. and Osswald, W. (2001) Comparative studies on the fine root status of healthy and declining spruce and beech trees in the Bavarian Alps and occurrence of *Phytophthora* and *Pythium* species. *Forest Pathology* 31, 257–273.

Nef, L. and Perrin, R. (1999) *Damaging Agents in European Forest Nurseries. Practical Handbook*. European Union Air 2-ct93-1694 Project. Office for Official Publications of the European Communities, Luxembourg.

O'Brien, P.A., Williams, N. and Hardy, G.E.St.J. (2009) Detecting *Phytophthora*. Review article. *Critical Reviews in Microbiology* 35, 169–181.

Orlikowski, L.B., Duda, B. and Szkuta, G. (2004) *Phytophthora citricola* on European beech and silver fir in Polish forest nurseries. *Journal of Plant Protection Research* 44, 57–64.

Oudemans, P., Forster, H. and Coffey, M.D. (1994) Evidence for distinct isozyme subgroups within *Phytophthora citricola* and close relationships with *P. capsici* and *P. citrophthora*. *Mycological Research* 98, 189–199.

Palmer, M.A. and Nicholls, T.H. (1985) Shoot blight and collar rot of *Pinus resinosa* caused by *Sphaeropsis sapinea* in forest tree nurseries. *Plant Disease* 69, 739–740.

Palmer, M.A., McRoberts, R.E. and Nicholls, T.H. (1988) Sources of inoculums of *Sphaeropsis sapinea* in forest tree nurseries. *Phytopathology* 78, 831–835.

Paoletti, E., Danti, R. and Strati, S. (2001) Pre- and post-inoculation water stress affects *Sphaeropsis sapinea* canker length in *Pinus halepensis* seedlings. *Forest Pathology* 31, 209–218.

Patton, R.F., Spear, R.N. and Blenis, P.V. (1984) The mode of infection and early stages of colonization of pines by *Gremmeniella abietina*. *European Journal of Forest Pathology* 14, 193–202.

Pérez-Sierra, A., Landeras, E., León, M., Berbegal, M., García-Jiménez, J. and Armengol, J. (2007) Characterization of *Fusarium circinatum* from *Pinus* spp. in northern Spain. *Mycological Research* 111, 832–839.

Perrin, R. and Sampangi, R. (1986) La fonte des semis en pépinière forestière [Damping off in forest nurseries]. *European Journal of Forest Pathology* 16, 309–321.

Petäistö, R.-L. (1999) Growth phase of bare-root Scots pine seedlings and their susceptibility to *Gremmeniella abietina*. *Silva Fennica* 33, 179–185.

Petäistö, R.-L. (2006) *Botrytis cinerea* and Norway spruce seedlings in cold storage. *Baltic Forestry* 12, 24–33.

Petäistö, R.-L. (2008) Infection of Norway spruce container seedlings by *Gremmeniella abietina*. *Forest Pathology* 38, 1–15.

Petäistö, R.-L. and Heinonen, J. (2003) Conidial dispersal of *Gremmeniella abietina*: climatic and microclimatic factors. *Forest Pathology* 33, 363–373.

Petäistö, R.-L. and Laine, A. (1999) Effects of winter storage temperature and age of *Pinus sylvestris* seedlings on the occurrence of disease induced by *Gremmeniella abietina*. *Scandinavian Journal of Forest Research* 14, 227–233.

Petäistö, R.-L., Heiskanen, J. and Pulkkinen, A. (2004) Susceptibility of Norway spruce seedlings to gray mold in the greenhouse during the first growing season. *Scandinavian Journal of Forest Research* 19, 30–37.

Peterson, G.W. (1967) Dothistroma needle blight of Austrian and ponderosa pines: epidemiology and control. *Phytopathology* 57, 437–441.

Peterson, G.W. (1982) *Dothistroma Needle Blight of Pines*. Forest and Insects Leaflet 143, US Department of Agriculture, Forest Service, Rocky Mountain Forest and Range Experiment Station, Fort Collins, Colorado.

Peterson, G.W. and Read, R.A. (1971) Resistance to *Dothistroma pini* within geographic sources of *Pinus nigra*. *Phytopathology* 61, 149–150.

Poteri, M. (1998) *Screening of Birch,* Betula *spp., for Rust Resistance to* Melampsoridium betulinum. Research Papers 689, Finnish Forest Research Institute, Vantaa, Finland.

Poteri, M. and Rousi, M. (1996) Variation in *Melampsoridium* resistance among European white-birch clones grown in different fertilization treatments. *European Journal of Forest Pathology* 26, 171–181.

Procházková, Z. and Jancarek, V. (1991) Diseases in Czech-Slovak nurseries. In: Sutherland, J.R. and Glover, S.G. (eds) *Proceedings of the First Meeting of IUFRO Working Party S2.07-09. Diseases and Insects in Forest Nurseries, August 23–30, 1990, Victoria, BC*. Information Report BC-X-331, Forestry Canada, Pacific and Yukon Region, Pacific Forestry Centre, Victoria, British Columbia, pp. 37–50.

Procházková, Z., Sikorova, A. and Peskova, V. (2005) Preliminary observations on the occurrence of *Ciboria batschiana* (Zopf) Buchwald in the Czech Republic. *Working Papers of the Finnish Forest Research Institute* 11, 13–18.

Prospero, S., Grünwald, N.J., Winton, L.M. and Hansen, E.M. (1999) Migration patterns of the emerging plant pathogen *Phytophthora ramorum* on the West coast of the United States of America. *Phytopathology* 99, 739–749.

Reglinski, T., Spiers, T.M., Dick, M.A., Taylor, J.T. and Gardner, J. (2009) Management of Phytophthora root rot in radiata pine seedlings. *Plant Pathology* 58, 723–730.

Riedel, M., Julich, S., Kielpinski, M., Möller, R., Fritshche, W., Wagner, S., Henkel, T., Breitenstein, A. and Werres, S. (2010) Chip-based on site diagnosis of *Phytophthora ramorum*. In: Frankel, S.J., Kliejunas, J.T. and Palmieri, K.M. (eds) *Proceedings of the Fourth Sudden Oak Death Science Symposium, Santa Cruz, CA, June 15–18, 2009*. General Technical Report PSW-GTR-229, US Department of Agriculture, Forest Service, Pacific Southwest Research Station, Albany, California, pp. 362–364.

Roll-Hansen, F. (1987) *Phacidium infestans* and *Ph. abies*. Hosts, especially *Abies* species in Norwegian nurseries. *European Journal of Forest Pathology* 17, 311–315.

Roos, J., Hopkins, R., Kvarnheden, A. and Dixelius, C. (2011) The impact of global warming on plant diseases and insect vectors in Sweden. *European Journal of Plant Pathology* 129, 9–19.

Russell, K. (1990) Gray mold. In: Hamm, P.B., Campbell, S.J. and Hansen, E.M. (eds) *Growing Healthy Seedlings. Identification and Management of Pests in Northwest Forest Nurseries*. US Department of Agriculture, Forest Service, Pacific Northwest Region and Forest Research Laboratory, Oregon, pp. 10–13.

Rytkönen, A., Lilja, A., Petäistö, R.-L. and Hantula, J. (2008) Irrigation water and *Phytophthora cactorum* in a forest nursery. *Scandinavian Journal of Forest Research* 23, 404–411.

Sakai, J., Miura, I., Shibata, M., Yonekura, N., Hiyoshi, H., Takagaki, M. and Nagayama, K. (2010) Development of a new fungicide, benthiavalicarb-isopropyl. *Journal of Pesticide Science* 35, 488–489.

Salerno, M.I. and Lori, G.A. (2007) Association of seed-borne *Fusarium* species on *Pinus ponderosa* with germination and seedling viability in Argentina. *Forest Pathology* 37, 263–271.

Santamaría, O., Smith, D.R. and Stanosz, G.R. (2011) Interaction between *Diplodia pinea* and *D. scrobiculata* in red and jack pine seedlings. *Phytopathology* 101, 334–339.

Saunders, J.E., Juzwik, J. and Hutchison, R. (1992) Outplanting survival of Cylindrocladium root rot affected black spruce seedlings. *Canadian Journal of Forest Research* 22, 1204–1207.

Scheuerell, S.J., Sullivan, D.M. and Mahaffee, W.F. (2005) Suppression of seedling damping-off caused by *Pythium ultimum*, *P. irregulare*, and *Rhizoctonia solani* in container media amended with a diverse range of Pacific Northwest compost sources. *Phytopathology* 95, 306–315.

Schröder, T., Kehr, R. and Hüttermann, A. (2002) First report of the seed-pathogen *Geniculodendron pyriforme*, the imperfect state of the ascomycete *Caloscypha fulgens*, on imported conifer seeds in Germany. *Forest Pathology* 32, 225–230.

Schroers, H.-J., Samuels, G.J., Seifert, K.A. and Gams, W. (1999) Classification of the mycoparasite *Gliocladium roseum* in *Clonostachys* as *C. rosea*, its relationship to *Bionectria ochroleuca*, and notes on other *Gliocladium*-like fungi. *Mycologia* 91, 365–385.

Scott, P.M., Burgess, T.I., Barber, P.A., Shearer, B.L., Stukely, M.J.C., Hardy, G.E.St.J. and Jung, T. (2009) *Phytophthora multivora* sp. nov., a new species recovered from declining *Eucalyptus*, *Banksia*, *Agonis* and other plant species in western Australia. *Persoonia* 22, 1–13.

Sieber, T.N. (1988) Endophytische Pilze in Nadeln von gesunden und geschädige Fichten (*Picea abies* (L.) Karsten) [Endophytic fungi in needles of healthy-looking and diseased Norway spruce (*Picea abies* [L.] Karsten)]. *European Journal of Forest Pathology* 18, 321–342.

Sieber, T.N., Rys, J. and Holdenrieder, O. (1999) Mycobiota in symptomless needles of *Pinus mugo* ssp. *uncinata*. *Mycological Research* 103, 306–310.

Sinclair, W.A., Lyon, H.H. and Johnson, W.T. (1987) *Diseases of Trees and Shrubs*. Cornell University Press, Ithaca, New York.

Singh, S.J. and Heather, W.A. (1982) Temperature sensitivity of qualitative race–cultivar interactions in *Melampsora medusa* Thüm. and *Populus* species. *European Journal of Forest Pathology* 12, 123–127.

Smith, H., Wingfield, M.J., Crous, P.W. and Coutinho, T.A. (1996) *Sphaeropsis sapinea* and *Botryosphaeria* endophytic in *Pinus* spp. and *Eucalyptus* in South Africa. *South African Journal of Botany* 62, 86–88.

Stanosz, G.R., Swart, W.J. and Smith, D.R. (1999) RAPD marker and isozyme characterization of *Sphaeropsis sapinea* from diverse coniferous hosts and locations. *Mycological Research* 103, 1193–1202.

Stanosz, G.R., Smith, D.R. and Albers, J.S. (2005) Surveys for asymptomatic persistence of *Sphaeropsis sapinea* on or in stems of red pine seedlings from seven Great Lakes region nurseries. *Forest Pathology* 35, 233–244.

Stenström, E. and Ihrmark, K. (2005) Identification of *Lophodermium seditiosum* and *L. pinastri* in Swedish forest nurseries using species specific PCR primers from the ribosomal ITS region. *Forest Pathology* 35, 163–172.

Sturrock, R.N., Frankel, S.J., Brown, A.V., Hennon, P.E., Kliejunas, J.T., Lewis, K.J., Worrall, J.J. and Woods, A.J. (2011) Climate change and forest diseases. Review. *Plant Pathology* 60, 133–149.

Sutherland, J.R. (1979) The pathogenic fungus *Caloscypha fulgens* in stored conifer seeds in British Columbia and relation of incidence to ground and squirrel-cache collected cones. *Canadian Journal of Forest Research* 9, 129–132.

Sutherland, J.R. (1987) Sirococcus blight. In: Sutherland, J.R., Miller, T. and Quinard, R.S. (eds) *Cone and Seed Diseases of North American Conifers*. NAFC Publication 1, North American Forestry Commission, Victoria, British Columbia, pp. 34–41.

Sutherland, J.R. and Davis, C. (1991) Diseases and insects in forest nurseries in Canada. In: Sutherland, J.R. and Glover, S.G. (eds) *Proceedings of the First Meeting of IUFRO Working Party S2.07-09. Diseases and Insects in Forest Nurseries, August 23–30, 1990, Victoria, BC*. Information Report BC-X-331, Forestry Canada, Pacific and Yukon Region, Pacific Forestry Centre, Victoria, British Columbia, pp. 25–32.

Sutherland, J.R. and Woods, T.A.D. (1978) The fungus *Geniculodendron pyriforme* in stored Sitka spruce seeds: effects of seed extraction and cone collection methods on disease incidence. *Phytopathology* 68, 747–750.

Sutherland, J.R., Lock, W. and Farris, S.H. (1981) Sirococcus blight, a seed-borne disease of container grown spruce seedlings in coastal British Columbia forest nurseries. *Canadian Journal of Botany* 59, 559–562.

Sutherland, J.R., Diekmann, M. and Berjak, P. (2002) *Forest Tree Seed Health for Germplasm Conservation*. IPGRI Technical Bulletin 6, International Plant Genetic Resources Institute, Rome.

Sutton, B.C. (1980) *Coelomycetes*. Commonweath Mycological Institute, Kew, Surrey, UK, pp. 120–121.

Takamatsu, S., Braun, U., Limkaisang, S., Kom-Un, S., Sato, Y. and Cunnington, J.H. (2007) Phylogeny and taxonomy of the oak powdery mildew *Erysiphe alphitoides sensu lato*. *Mycological Research* 11, 809–826.

Talgø, V., Herrero, M.-L., Toppe, B., Brurberg, M.B., Thurston, R. and Stensvand, A. (2009) *Phytophthora plurivora*, ny skadegjerar på tre i Noreg. Bioforsk [*Phytophthora plurivora*, new damage makes three in Norway. Bioresearch]. *Bioforsk FOKUS* 5, 238–239 [in Norwegian].

Talgø, V., Brodal, G., Klemsdal, S.S. and Stensvand, A. (2010) Seed borne fungi on *Abies* spp. *Seed Science and Technology* 38, 477–493.

Terho, M. and Uotila, A. (1999) Virulence of two *Gremmeniella abietina* types (A and B). *European Journal of Forest Pathology* 29, 143–152.

Termorshuizen, A.J., van Rijn, E., van der Gaag, D.J., Alabouvette, C., Chen, Y., Lagerlöf, J., Malandrakis, A.A., Paplomatas, E.J., Rämert, B., Ryckeboer, J., Steinberg, C. and Zmora-Nahum, S. (2006) Suppressiveness of 18 composts against 7 pathosystems: variability in pathogen response. *Soil Biology and Biochemistry* 38, 2461–2477.

Themann, K., Werres, S., Lüttmann, R. and Diener, H.-A. (2002) Observations of *Phytophthora* spp. in water recirculation systems in commercial hardy ornamental nursery stock. *European Journal of Plant Pathology* 108, 337–343.

Tillman-Sutela, E., Kauppi, A., Hilli, A. and Kaitera, J. (2004) Fungal injury to seed tissues of Norway spruce, *Picea abies* (L.) Karst. *Trees – Structure and Function* 18, 151–156.

Tomlinson, J.A., Boonham, N., Hughes, K.J.D., Griffin, R.L. and Barker, I. (2005) On-site DNA extraction and real time PCR for detection of *Phytophthora ramorum* in the field. *Applied and Environmental Microbiology* 71, 6702–6710.

Tomlinson, J.A., Barker, I. and Boonham, N. (2007) Faster, simpler, more specific methods for improved molecular detection of *Phytophthora ramorum* in the field. *Applied and Environmental Microbiology* 73, 4040–4047.

Tomlinson, J.A., Dickson, M., Hobden, E., Robinson, S., Giltrap, P.M. and Boonham, N. (2010a) A five-minute DNA extraction method for expedited detection of *Phytophthora ramorum* following prescreening using *Phytophthora* spp. lateral flow devices. *Journal of Microbiological Methods* 81, 116–120.

Tomlinson, J.A., Dickson, M.J. and Boonham, N. (2010b) Rapid detection of *Phytophthora ramorum* and *P. kernoviae* by two-minute DNA extraction followed by isothermal amplification and amplicon detection by generic lateral flow device. *Phytopathology* 100, 143–149.

Unestam, T., Beyer-Ericson, T. and Strand, M. (1989) Involvement of *Cylindrocarpon destructans* in root death of *Pinus sylvestris* seedlings, pathogenic behaviour and predisposing factors. *Scandinavian Journal of Forest Research* 4, 521–536.

Uotila, A. (1983) Physiological and morphological variation among Finnish *Gremmeniella abietina* isolates. *Communicationes Instituti Forestalis Fenniae* 119, 1–12.

Urošević, B. (1961) The influence of saprophytic and semi-parasitic fungi on the germination of Norway spruce and Scots pine seeds. *Proceedings of International Seed Testing Association* 26, 537–553.

Venn, K. (1979) Effects of benomyl on 2-0 seedlings of *Picea abies* (L.) Karst. *Meddelelser fra Norsk Institutt for Skogforskning* 34, 249–262.

Venn, K., Sandvik, M. and Langerud, B. (1986) Nursery routines, growth media and pathogens affect growth and root dieback in Norway spruce seedlings. *Meddelelser fra Norsk Institutt for Skogforskning* 34, 314–328.

Vercauteren, A., De Dobbelaere, I., Grünwald, N.J., Bonants, P., van Bockstaele, E., Maes, M. and Heungens, K. (2010) Clonal expansion of the Belgian *Phytophthora ramorum* populations based on new microsatellite markers. *Molecular Ecology* 19, 92–107.

Viljoen, A., Wingfield, M.J. and Marasas, W.F.O. (1994) First report of *Fusarium subglutinans* f.sp. *pini* on pine seedlings in South Africa. *Plant Disease* 78, 309–312.

Vuorinen, M. and Kurkela, T. (1993) Concentration of CO_2 under snow cover and the winter activity of the snow blight fungus *Phacidium infestans*. *European Journal of Forest Pathology* 23, 441–447.

Walker, J., Hartigan, D. and Bertus, A.L. (1974) Poplar rusts in Australia with comments on potential conifer rusts. *European Journal of Forest Pathology* 4, 100–118.

Wang, J. and van der Kamp, B.J. (1992) Resistance, tolerance, and yield of western black cottonwood infected by Melampsora rust. *Canadian Journal of Forest Research* 22, 183–192.

Watt, M.S., Kiriticos, D.J., Alcaraz, S., Brown, A.V. and Leriche, A. (2009) The hosts and potential geographic range of Dothistroma needle blight. *Forest Ecology and Management* 257, 1505–1519.

Watt, M.S., Ganley, R.J., Kriticos, D.J. and Manning, L.K. (2011) Dothistroma needle blight and pitch canker: the current and future potential distribution of two important diseases of *Pinus* species. *Canadian Journal of Forest Research* 41, 412–424.

Werres, S., Marwitz, R., Man in't Veld, W.A., De Cock, W.A.M., Bonants, P.J.M., De Weert, K., Ilieva, E. and Baayen, R.P. (2001) *Phytophthora ramorum* sp. nov. a new pathogen on *Rhododendron* and *Viburnum*. *Mycological Research* 105, 1115–1165.

Wickland, A.C., Jensen, C.E. and Rizzo, D.M. (2008) Geographic distribution, disease symptoms and pathogenicity of *Phytophthora nemorosa* and *Phytophthora pseudosyringae* in California, USA. *Forest Pathology* 38, 288–298.

Woods, A., Coates, K.D. and Hamann, A. (2005) Is an unprecedented needle blight epidemic related to climate change? *BioScience* 55, 761–769.

Yakabe, L.E., Blomquist, C.L., Thomas, S.L. and MacDonald, J.D. (2009) Identification and frequency of *Phytophthora* species associated with foliar diseases in California ornamental nurseries. *Plant Disease* 93, 883–890.

Zhang, P.G. (1992) Epidemiology and control of gray mold (*Botrytis cinerea*) of black spruce seedlings. PhD thesis, University of Guelph, Guelph, Ontario.

Zhang, P.G. and Sutton, J.C. (1994) High temperature, darkness, and drought predispose black spruce seedlings to gray mold. *Canadian Journal of Botany* 72, 135–142.

Zhang, P.G., Sutton, J.C. and Hopkin, A.A. (1994) Evaluation of micro-organisms for biocontrol of *Botrytis cinerea* in container-grown black spruce seedlings. *Canadian Journal of Forest Research* 24, 1312–1316.

Zhang, P.G., Sutton, J.C. and Hopkin, A.A. (1995) Low light intensity predisposes black spruce seedlings to infection by *Botrytis cinerea*. *Canadian Journal of Plant Pathology* 17, 13–18.

28 Responding to Diseases Caused by Exotic Tree Pathogens

Marco Pautasso*
Centre d'Ecologie Fonctionnelle et Evolutive (CEFE), National Centre for Scientific Research (CNRS), Montpellier, France

28.1 Introduction

Responding to exotic tree pathogens is a challenge for sustainable land use and rural economies. Many new pathogens are affecting trees in semi-natural ecosystems, causing ecological and/or economic problems in various regions of the world (Liebhold *et al.*, 1995; Rizzo, 2005; Crowl *et al.*, 2008; Loo, 2009; Hunter *et al.*, 2011; Lilja *et al.*, 2011; Parry and Teale, 2011). Examples include:

- *Cronartium ribicola* J.C. Fisch. on pines (*Pinus* spp.) in North America and Europe (Geils *et al.*, 2010)
- *Cryphonectria parasitica* (Murrill) M.E. Barr on chestnuts (*Castanea* spp.) in North America and Europe (Hansen, 1999; Dutech *et al.*, 2010; Prospero and Rigling, 2012)
- *Discula destructiva* Redlin on flowering dogwood (*Cornus florida* L.) in woodlands of the eastern USA (Holzmüller *et al.*, 2010)
- *Erwinia amylovora*, a bacterium affecting trees of the family *Rosaceae* (e.g. *Malus*, *Pyrus*, *Cydonia* and *Crataegus* spp.) throughout the world (Johnson and Stockwell, 1998)
- *Gibberella circinata* Nirenberg & O'Donnell ex Britz, T.A. Cout., M.J. Wingf. & Marasas (anamorph: *Fusarium circinatum* Nirenberg & O'Donnell) causing pine pitch canker worldwide (Garbelotto, 2008; Ganley *et al.*, 2009)
- North American *Heterobasidion irregulare* (Underw.) Garbel. & Otrosina causing root rot of Italian stone pine (*Pinus pinea* L.) in central Italy (Gonthier *et al.*, 2007; Gonthier and Garbelotto, 2011)
- *Hymenoscyphus pseudoalbidus* V. Queloz, C.R. Grünig, R. Berndt, T. Kowalski, T.N. Sieber & O. Holdenrieder (anamorph: *Chalara fraxinea* T. Kowalski) on European ash (*Fraxinus excelsior* L.) in Europe (Kowalski, 2006; Queloz *et al.*, 2011)
- *Mycosphaerella pini* Rostr. ex Munk (anamorph: *Dothistroma septosporum* (Dorog.) M. Morelet) on Monterey pine (*P. radiata* D. Don) plantations, e.g. in Africa (Chungu *et al.*, 2010)
- *Neonectria* spp., the cause of beech bark disease in New England (USA) together with the scale insect *Cryptococcus fagisuga* Lindinger (Lovett *et al.*, 2006)
- *Ophiostoma* spp., the causal agent of Dutch elm disease (Brasier, 2001)

* E-mail: marpauta@gmail.com

- *Phytophthora cinnamomi* Rands, e.g. in the jarrah forest of Western Australia (Cahill *et al.*, 2008)
- *Phytophthora lateralis* Tucker & Milbrath on Port Orford cedar (*Chamaecyparis lawsoniana* (A. Murray bis) Parl.) in Oregon (USA) (Jules *et al.*, 2002)
- *Phytophthora pinifolia* Alv. Durán, Gryzenh. & M.J. Wingf. on Monterey pine in Chile (Durán *et al.*, 2008)
- *Phytophthora ramorum* Werres, De Cock & Man in't Veld in forests and urban areas on the west coast of the USA and in woodland, historic gardens, heathland and tree plantations in the UK (Garbelotto, 2008; Brasier and Webber, 2010)
- *Raffaelea quercivora* Kubono & Shin. Ito on *Fagaceae* in Japan (Kubono and Ito, 2002; Torii *et al.*, 2011)

There is a perception that plant diseases caused by exotic pathogens have been increasing in number and severity over the last few decades owing to the increased magnitude of long distance trade, land-use intensification, urbanization, the already observed climate warming and other global change processes (Anderson *et al.*, 2004; Picco *et al.*, 2011; Stenlid *et al.*, 2011). For a set of European countries for which data on the occurrence of exotic forest pathogens are available, there is a correlation with gross domestic product (GDP; Fig. 28.1), which would suggest that economic activities have a role in introducing tree pathogens – unless another variable correlated with GDP, such as sampling effort, can explain this correlation. Given that global change processes are likely to carry on and intensify in future, in a business as usual scenario it is to be expected that the impacts of many exotic plant pathogens will continue to grow. Devising ways to respond to plant diseases caused by exotic pathogens in a better way than has been the case so far is thus a priority for interdisciplinary research, management and policy.

There are many challenges in fulfilling such an objective. The management of exotic tree pathogens is often made difficult by the limited epidemiological knowledge of the new pathosystem and the temporal lag in recognizing that a new disease has appeared.

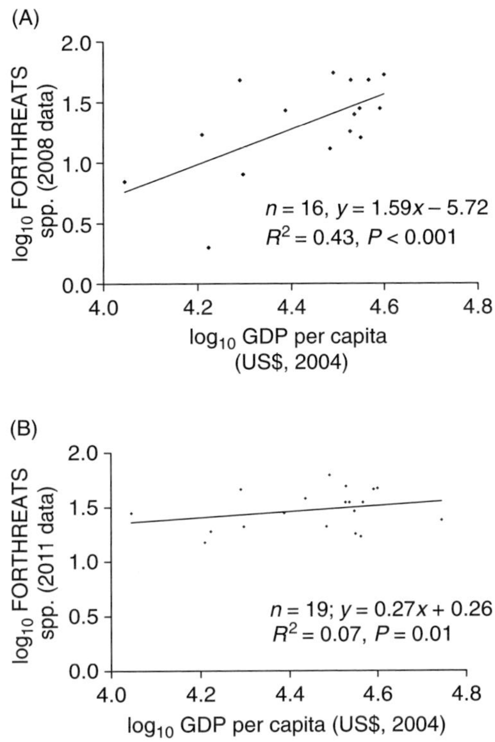

Fig. 28.1. Log–log correlation of numbers of reported exotic forest pathogen species with gross domestic product for 2004 (US$) for European countries for which occurrence data of exotic forest pathogens were available. Panel (A) is based on 2008 data for 16 countries (Austria, Czech Republic, Finland, France, Germany, Greece, Hungary, Italy, Lithuania, Poland, Romania, Slovakia, Spain, Sweden, Switzerland and the UK); panel (B) is based on 2011 data for the same countries with the addition of Belgium, Norway and Slovenia. While the 2008 data show a good correlation between the two variables, the correlation for the updated 2011 data is weak, thus suggesting that sampling effort may play a role in explaining the correlation shown in panel (A). Data from the FORTHREATS project (the European Network on emerging diseases and threats through invasive alien species in forest ecosystems; funded by the European Commission's Sixth Framework Programme); kindly provided by Ottmar Holdenrieder (A) and Alberto Santini (B).

By the time research on the effectiveness of various management options has been undertaken, the spatial extent of the outbreaks may have made such options unrealistic and ineffective (Keller *et al.*, 2011). For generalist pathogens, the number of affected hosts also complicates management efforts. As a rule, for exotic tree pathogens, prevention is better than cure (Krcmar, 2008), but prevention is difficult when the new pathogens are relatively unknown. Further challenges stem from the tendency of scientists to investigate exotic invasive species from the point of view of their own disciplines, without integrating views from other fields and perspectives (Persha *et al.*, 2011; Döring *et al.*, 2012; Lindenmayer and Laurance, 2012). Also, to foster plant health, long-term dialogue, collaboration and trust among scientists, policy makers, economists and the public are required, but these are seldom achieved (Konijnendijk, 2008; Mills *et al.*, 2011; Matyssek *et al.*, 2012).

Management recommendations range from prevention, early detection and eradication, to containment and control, and when these efforts are unsuccessful, from mitigation of the impacts to restoration of the degraded forests. Various such management options are described in the preceding chapters on specific groups of tree pathogens, some of which are exotic and therefore relevant to this chapter. Silvicultural strategies to deal with exotic tree pathogens at the various stages of the invasion process (entry, establishment and spread) and silvicultural cycle have been reviewed extensively by Waring and O'Hara (2005) and Jactel *et al.* (2009). This review attempts to provide an overall framework for dealing with exotic tree pathogens in a changing world.

The rest of the chapter is organized as follows. Section 28.2 covers spatial models, which have great potential to improve our understanding, prediction and management of exotic tree pathogens. This is particularly true if the models are combined with network approaches, given that most exotic pathogens are moved along trade pathways. Hence, Section 28.3 discusses networks. The insights gained from spatial and network models need to be used to improve pathogen management and involve the various stakeholders; and the latter are the subject of Section 28.4. Interaction and engagement, rather than consultation and top-down communication, also make sense in the process of keeping phytosanitary regulations updated with the latest pathogen discoveries, methodological tools and management insights. Our response to exotic tree pathogens is not only governed by stakeholder needs and compromises that result from regulation (Section 28.5), but also by the available resources, a topic that is considered in Section 28.6, which examines economic considerations. Vast amounts of human resources shape how the land is used throughout the planet, with little consideration of how the resulting landscapes will tend to affect outbreaks of exotic tree pathogens, despite the clear influence of land use; both land use and silviculture are covered in Section 28.7. Interdisciplinarity (Section 28.8) is one commendable way to respond to exotic tree pathogens, as it enables a diversity of approaches, broadens the perspectives of researchers and avoids the problems that result from excessive specialization. Climate change and globalized trade, addressed in Section 28.9, are parts of the problem; consequently, strategies to reduce their effects are also a (long-term) response to exotic tree pathogens. The review concludes with some considerations on the effects on our management choices of changing the viewpoint from which we look at novel forest pathosystems (Section 28.10, Changing paradigms).

28.2 Spatial Models

Spatially explicit invasion models are key tools for responding to regional outbreaks of exotic tree pathogens, because they allow prediction of the likely further course of the epidemic, given adequate data on occurrences, absences and rates of spread of the pathogen, as well as the relevant environmental features of the actual and potential outbreak region (Gilbert and Liebhold, 2010; Meentemeyer *et al.*, 2011). Accurate prediction of the likely locations of new invasion foci may, in turn, enable a better allocation of resources in monitoring programmes, as well as improved early detection and thus more likely eradication or control (Lundquist and Klopfenstein, 2001; Harwood *et al.*, 2011). The comparison of monitored and predicted spread rates may help to identify tree populations with genetic resistance against the pathogen, as spread rates will tend to be lower in the presence of resistance, other things being equal (Hatala *et al.*, 2011).

The assumption behind these considerations is that the environmental features of the landscape (e.g. precipitation, soil, aspect, topography, altitude) are more important influences on the development of epidemics than stochastic effects. Even if the initial conditions (i.e. the presence of habitat patches suitable for the pathogen at an appropriate distance from the starting point of the epidemic) are often essential in determining whether an epidemic takes off or not, there is still a role for spatial and landscape variables in shaping where and how severely the epidemic will develop (Jeger, 2004; Holdenrieder *et al.*, 2004).

Spatial autocorrelation analysis is one of the geostatistic tools available for investigating the invasion patterns of exotic pathogens. The presence of short-distance spatial autocorrelation in disease expression can result from an underlying local similarity in environmental features or a dispersal kernel (the probability density function of the dispersal distances for an individual or population) with a preponderantly local transmission (Václavík *et al.*, 2012). Spatial autocorrelation at short distances is also important because, if not accounted for in models of disease incidence as a function of environmental variables, it can lead to inflated degrees of freedom and thereby to potentially misleading parameter estimates and *P* values (Chadfield and Pautasso, 2012).

Together with genetic studies and simulations (González-Varela *et al.*, 2011; Bihon *et al.*, 2012; Hoban *et al.*, 2012), spatial models can deliver useful knowledge about the environmental variables associated with disease expression and the pathways of introduction of exotic pathogens. This approach also suggests which interventions may help to diminish the likelihood of further spread (Woolhouse, 2011). For example, it is often helpful to reduce the connectivity of forest patches from the point of view of a certain pathogen, e.g. by avoiding the carrying of spores on the muddy shoes of tourists along trails (Rizzo *et al.*, 2005). The contribution of people to the dispersal of exotic tree pathogens is well documented, but still rarely included in spatially explicit models of their invasion (Jeger *et al.*, 2007; Yemshanov *et al.*, 2009). A promising refinement of such models is, therefore, the overlaying of plant trade or exchange networks on to spatial simulations of pathogen spread.

28.3 Networks

Networks are sets of nodes (e.g. plant nurseries, wholesalers and retail centres) connected by links (e.g. plants, microbes, insects, customers, or plant pathologists moving among those nodes). Networks are relevant to the management of exotic pathogens of trees because one of the most important pathways of introduction is provided by shipments of plants, seeds, wood and other propagules among regions, countries and continents (Dehnen-Schmutz *et al.*, 2010; Andjic *et al.*, 2011). Fig. 28.2 shows the temporal development of *P. ramorum* interceptions in shipments of ornamental plants among European countries (2002–2010). Epidemiological modelling in directed networks suggests a number of ways to improve our chances of preventing the introduction of new pathogens (Moslonka-Lefebvre *et al.*, 2011):

1. Given that long distance connections between production areas, distribution centres and markets tend to increase the epidemic potential in a given system (Goss *et al.*, 2009; Harwood *et al.*, 2009), control would be more effective if targeted towards these long-distance links.

2. Similarly, given that the presence of super-connected traders markedly reduces the epidemic threshold (the boundary between no epidemic and epidemic development) (Pautasso and Jeger, 2008), and given that the number of links from the starting node of an epidemic is strongly correlated with epidemic final size regardless of the network structure (local, random, small world, or scale free) and level of connectivity (Pautasso *et al.*, 2010b),

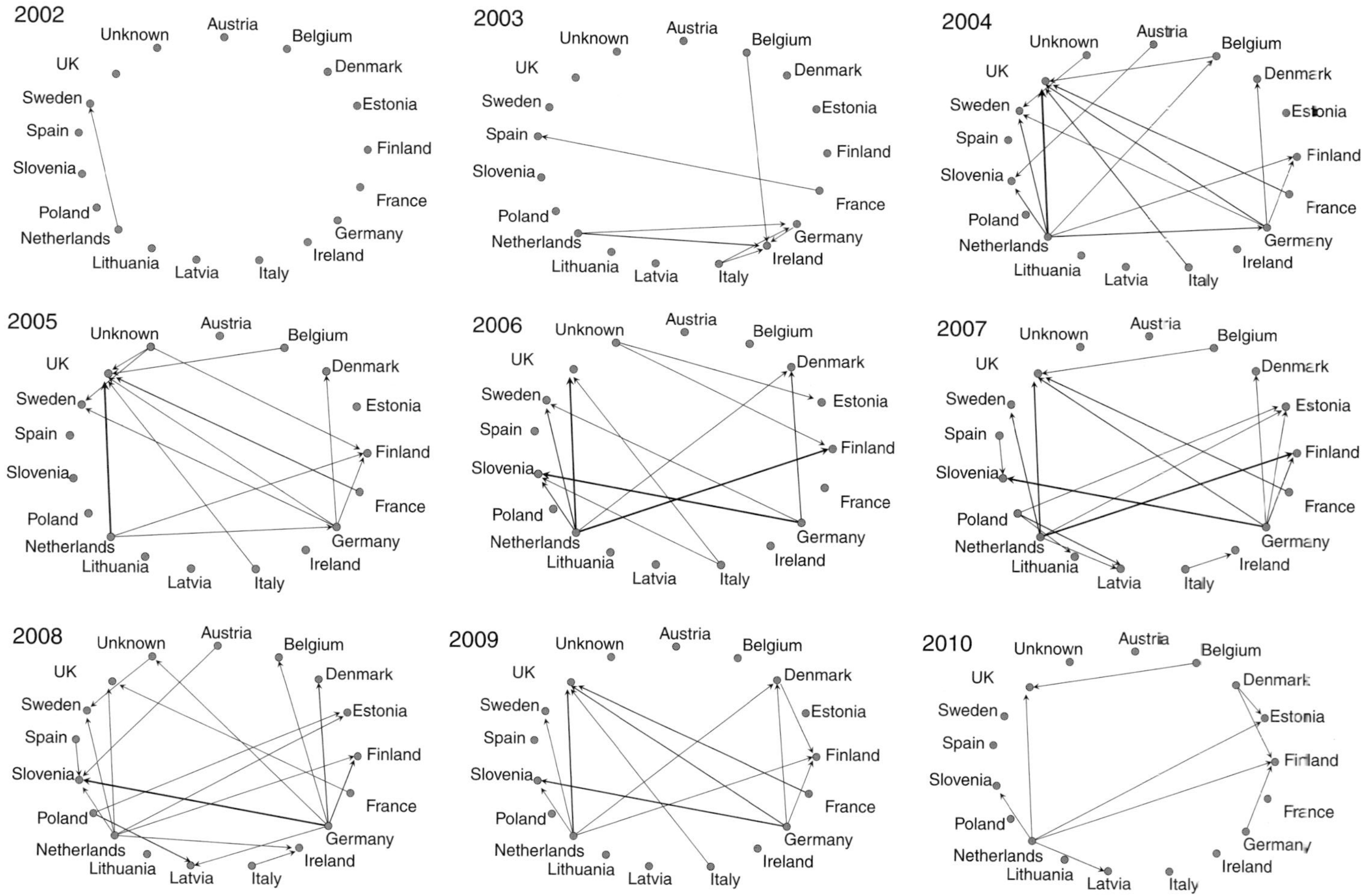

Fig. 28.2. Interceptions of *Phytophthora ramorum* Werres, De Cock & Man in't Veld on shipments of ornamental plants (*Camellia*, *Fagus*, *Hamamelis*, *Kalmia*, *Leucothoe*, *Magnolia*, *Photinia*, *Pieris*, *Rhododendron* and *Viburnum* spp.) among European countries in 2002–2010. Modified from EFSA (2011a), data kindly made available by EUROPHYT, the European Union Notification System for Plant Health Interceptions.

plant health authorities are advised to focus their control efforts towards major traders, i.e. those with many incoming and outgoing connections (Paull *et al.*, 2012).

3. Alternatively, without changing the detection effort, an improvement in the effectiveness of prevention activities can be obtained by modifying the structure of a network, e.g. by moving from a plant trade network with a positive correlation between incoming and outgoing links in a given node to one with a negative correlation (Woolhouse *et al.*, 2005; Moslonka-Lefebvre *et al.*, 2009). This can be achieved by avoiding connections between traders in both incoming and outgoing directions.

4. Given that differences in the proportions of hierarchical categories in plant trade networks can influence the likelihood of an epidemic, incentives, subsidies and/or taxes that favour an increase in the proportion of producers and retailers, and so reduce the presence of wholesalers in a trade network, are likely to result in fewer introductions of exotic pathogens, other things being equal (Pautasso *et al.*, 2010c; Moslonka-Lefebvre *et al.*, 2012). This and the previous intervention, however, may lead to less economically efficient plant trade networks.

Network theory has rarely been considered in the framework of plant pathology and exotic pathogens, but this approach has now delivered clear policy recommendations with the potential to diminish the risk posed by new pathogens, by lowering the probability of introduction rather than acting after this has occurred (Jeger *et al.*, 2007; Chadès *et al.*, 2011; Danon *et al.*, 2011).

28.4 Stakeholders

For reasons of commercial sensitivity and due to the fragmentation of the industry, there is little information with which to assess which network structure (local, random, small world or scale free) is likely to better describe plant trade networks (Mills *et al.*, 2011). Long-distance connections are likely to be common in many countries, and these would confer small world properties to the trade networks. Similarly, preferential attachment mechanisms (new traders may tend to become connected with traders that are already well connected) may result in trade networks with a scale-free connectivity (i.e. heterogeneity in the contact structure: a few nodes with lots of connections and most nodes with just a few connections), which has been observed in many other systems (Moslonka-Lefebvre *et al.*, 2011).

Plant disease management has traditionally occurred without much consideration of risk perception by the various stakeholders (McRoberts *et al.*, 2011; Mills *et al.*, 2011). It is likely that more participatory approaches would have resulted in a higher level of adoption of plant disease management recommendations (Forbes *et al.*, 2009; Prost *et al.*, 2012). It would thus be advisable to try to implement policy recommendations from network analyses that aim to prevent the spread of exotic plant pathogens by involving the many groups with a stake in the health of forests, in the commerce of ornamental plants and in the various ecosystem services provided by both forests and plants for planting.

Given that the use of network models in the prevention of invasions of forest pathogens is a new undertaking, there has been little study of how best to involve stakeholders in the implementation of policy recommendations derived from network epidemiology. Such studies are therefore needed; ideally, they should accompany any implementation of results from network models from the very beginning. For more traditional (i.e. not yet network-based) pest management programmes, there is some evidence that educational and stakeholder involvement in programmes can make those programmes if not more effective, at least better received by the affected communities (Shaw *et al.*, 2011). For example, the overall efficacy of oak wilt management in the USA was improved by educational programmes because they resulted in higher prevention effort and compliance with the recommended management methods (Koch *et al.*, 2010). Similarly, sudden oak death management efforts in California stimulated local involvement with natural resources and improved the preparation of the communities affected by *P. ramorum* for future similar outbreaks (Alexander and Lee, 2010).

28.5 Regulation

The progressive realization that trade is a major contributory factor to plant epidemics has led to the development of phytosanitary regulatory frameworks (MacLeod *et al.*, 2010). Regulation is a way to involve stakeholders, although traditionally in a rather top-down manner. Bottom-up, participatory approaches may result in better understanding and compliance with regulations. There is then a need to make phytosanitary regulation more broadly based, and with interactions and engagement, rather than mere consultation and one-way communication (Mills *et al.*, 2011).

Regulation in an adaptive management setting would itself be subject to tests, so that its effectiveness can be compared with that of other possible measures aiming to diminish the impacts of exotic pathogens. Ideally, the use of experimental approaches would make it easier to establish causality between a policy measure and any decrease, increase or lack of change in the incidence of an exotic tree pathogen in a country/region. In practice, controlled conditions such as those found in laboratory experiments are out of reach for scientists investigating the effectiveness of policy measures against invasive species (Filipe *et al.*, 2012). Still, the political fragmentation of the land masses of our planet – provided that standardized data on the incidence or severity of a certain pathogen are available – may make it possible to compare the effectiveness of various policies, because different countries often adopt different measures to respond to the risk of introduction or further spread of a given pathogen.

An example of an adaptive approach in regulations against exotic tree pathogens is provided by the rapidly developing situation for *P. ramorum* in Europe. The original risk assessment for this pathogen (Sansford *et al.*, 2009) was subsequently reviewed by the European Food Safety Authority (EFSA) so as to take on board new information that had appeared since the publication of the risk assessment report (Mills *et al.*, 2011). The EFSA opinion (EFSA, 2011a) will enable the European Commission to base their decisions on whether to modify the provisional emergency measures against the further spread of *P. ramorum* in European countries on the latest available knowledge.

Phytosanitary regulation would do well to be up to date with the latest pathogen discoveries and scientific developments – from molecular and sequencing tools to models integrating spatial information with trade networks. At the same time, our response to exotic forest pathogens would also profit if modellers were to incorporate the adoption (or absence thereof) of regulations against a given pathogen in their simulations of further potential dispersal and impacts (Harwood *et al.*, 2009). Such an approach, which can also be performed for past epidemics using retrospective modelling (Harwood *et al.*, 2011; Potter *et al.*, 2011), could provide a way to test (*in silico* and *a priori*) the effectiveness of various policy and management options.

28.6 Economics

Stakeholder involvement, adaptive monitoring and testing are approaches that could make phytosanitary regulation more effective, i.e. more likely to achieve a certain level of risk reduction for a given exotic pathogen. However, there is also an issue of efficiency, i.e. of how well resources are employed to achieve a certain outcome. Economic considerations are central to our response to exotic forest pathogens, because resources to study, monitor and manage them are limited. Moreover, most kinds of responses to invasions of exotic pathogens are likely to affect businesses and the rural economy in some way or another (Armsworth *et al.*, 2010). Economic activities, in turn, tend to leave a long-lasting legacy in terms of species introductions (Essl *et al.*, 2011), including invasions of exotic tree pathogens.

When taking decisions on which strategy to adopt against exotic pathogens, there is thus a need to balance the costs inherent in such strategies against those that are likely to be incurred due to an invasion without control measures (Li *et al.*, 2007; Benninga *et al.*, 2012; Vanem, 2012). Soliman *et al.* (2010) provide a review of the main methods for quantitative economic impact assessment of plant pathogens (partial budgeting, partial equilibrium analysis, input/output analysis and computable general equilibrium analysis). Contingent valuation methods have been used to evaluate the willingness of tourists to pay for research on and management of an exotic pathogen (*Seiridium cardinale* (W.W. Wagener) B. Sutton & I.A.S. Gibson) of Italian cypress (*Cupressus sempervirens* L.) in the Lake Garda region of Italy (Notaro and De Salvo, 2010). Such approaches deserve wider application. For an epidemic of *E. amylovora* in Switzerland (Fig. 28.3), considerable sums of money were devoted to destroying infected hosts in orchards (mainly apple and pear trees), with repercussions on the landscape aesthetics of many cantons. There is little knowledge of whether these eradication efforts made a difference to the course of the epidemic, which is still spreading.

Multiple trade-offs are not just present among the various measures (epidemiological studies, monitoring, eradication, mitigation) that can be taken at the different stages of an invasion (Baxter and Possingham, 2011), but also in the choice of which exotic pathogens to focus on among the many already present (or potentially present) in a region (Carrasco *et al.*, 2010). In many cases, the 20/80 rule (that is, 20% of the host population contributes at least 80% of the net transmission potential; Woolhouse *et al.*, 1997) is likely to apply also in relation to the management and economics of exotic forest pathogens (Fig. 28.4; Ricciardi *et al.*, 2011). For example, it can be hypothesized that about 20% of exotic forest pathogens are causing around 80% of damage; similarly, it is possible that an additional 80% of management effort would be needed to achieve the remaining 20% reduction in economic losses due to exotic forest pathogens.

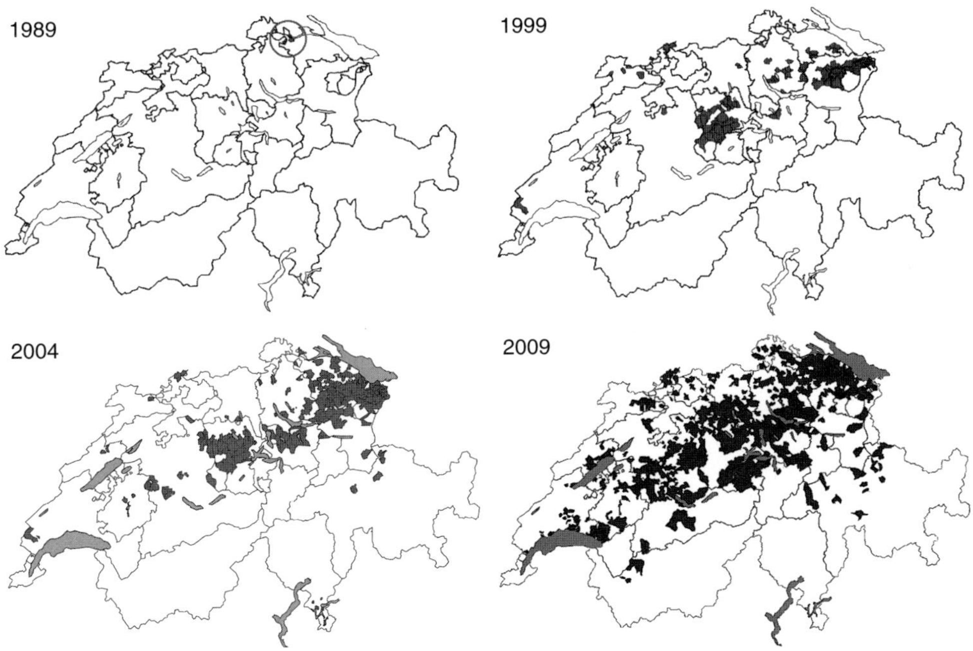

Fig. 28.3. Development of the epidemic of the bacterium *Erwinia amylovora*, the fire blight pathogen, in Switzerland from 1989 to 2009. The pathogen affects tree and shrub species of the family *Rosaceae* (e.g. *Malus*, *Pyrus*, *Cydonia* and *Crataegus* spp.). Maps from Agroscope-Reckenholz – Tänikon Research Station, Switzerland. Available at: http://www.agroscope.admin.ch/feuerbrand/ (accessed October 2011).

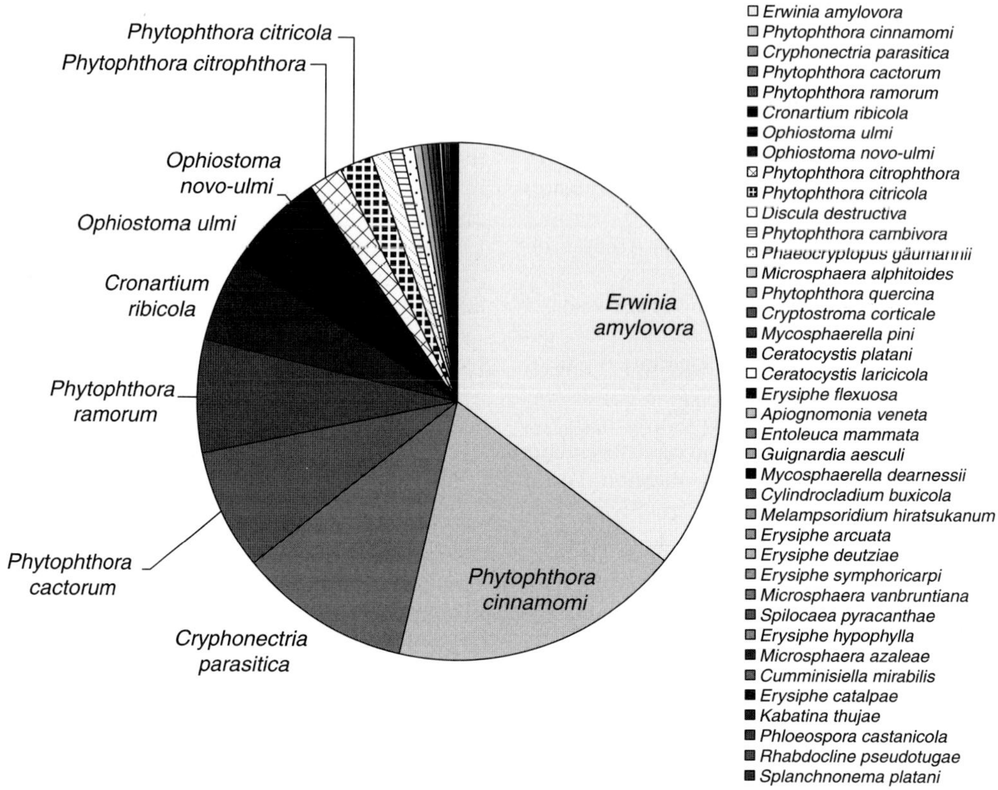

Fig. 28.4. Relative number of publications on the Web of Knowledge (Thomson Reuters, formerly ISI) for the 39 exotic tree pathogens reported in Switzerland up to October 2011. The eight more commonly studied pathogens (approximately 20%; *Erwinia amylovora* to *Ophiostoma novo-ulmi* in the pie chart) cover about 90% of the approximately 5800 publications retrieved (although some publications may be on several pathogens). The search covered all databases and all years available on the Web of Knowledge (up to October 2011), and for each pathogen used its scientific name between inverted commas. Data on the 39 exotic tree pathogens in Switzerland are from the FORTHREATS project (the European Network on emerging diseases and threats through invasive alien species in forest ecosystems; funded by the European Commission's Sixth Framework Programme), and were kindly provided by Ottmar Holdenrieder.

Forest pathologists are starting to include economic considerations in their research (Thor *et al.*, 2006; Holmes *et al.*, 2008; Gonthier *et al.*, 2012). These include, for example: comparative studies of the marginal benefits versus costs of adding a unit of pathogen control; analyses not just of the financial losses due to forest pathogens, but also of their positive effects on wood prices due to the decreased supply; and the socio-economic effects (on the researchers themselves) of outbreaks of diseases such as sudden oak death. Economic perspectives are likely to better inform future attempts to manage exotic tree pathogens, but they need to be aware of the ecological and epidemiological knowledge of the pathogens as well. Economics can also make sense in combination with recent developments in network theory. In the context of trade in plants, the presence of super-connected nodes of trade (in itself a major problem as it will tend to facilitate pathogen dispersal) can be countered by targeting control towards these super-connected trade players, hence increasing control effectiveness and potentially reducing control costs in comparison with systematic surveillance. Identifying super-connected nodes in the absence of detailed information on the structure of trade networks is possible using a technique known as acquaintance immunization: the random choice of a few nodes and the tracing of their contacts will tend to eventually lead to the super-connected nodes, because these, by definition, have more connections than other nodes (Jeger *et al.*, 2007).

28.7 Land Use and Silviculture

Land use is influenced by economic considerations and, in turn, has an influence on regional outbreaks of exotic tree pathogens (Fig. 28.5). None the less, there is little evidence that choices about how to use the land are influenced by whether a particular land use configuration is likely to make it easier to manage exotic pathogens (not just of trees, but also of other plants, animals and even people). Large-scale planting of monocultures has been common not just for agricultural crops, but also for trees, e.g. worldwide plantations of poplar (*Populus* spp.), eucalyptus (*Eucalyptus* spp.), Monterey pine and their associated plant health issues. In towns too, tree avenues have a tendency towards uniformity (in the choice of species and provenance, as well as in the age and management of the trees), thus providing outstanding conditions for the linear dispersal of newly introduced pests and pathogens.

Adding biodiversity to tree plantations and avenues would make sense not just from an insurance perspective against exotic tree pathogens, but also from other points of view, such as landscape aesthetics, the provision of ecosystem services, biodiversity conservation and education, and even tree productivity (Quijas *et al.*, 2010; Isbell *et al.*, 2011). Most of the evidence showing that more diverse ecosystems tend to be more resilient originates from agro-ecosystems, but there are also supportive studies from forests (Pautasso *et al.*, 2005; Keesing *et al.*, 2010; Haas *et al.*, 2011). Some exotic forest pathogens may, if aggressive and polyphagous, challenge the hypothesis that more diverse forests are less susceptible to disease because of the lack of co-evolution between pathogen and hosts, as happened for instance with the introduction of *P. cinnamomi* in the jarrah forest in south-western Australia. However, tree species diversity may still dilute the potential of a generalist exotic forest pathogen to cause damage to a naive ecosystem, as shown for the *P. ramorum* outbreaks in California (Haas *et al.*, 2011).

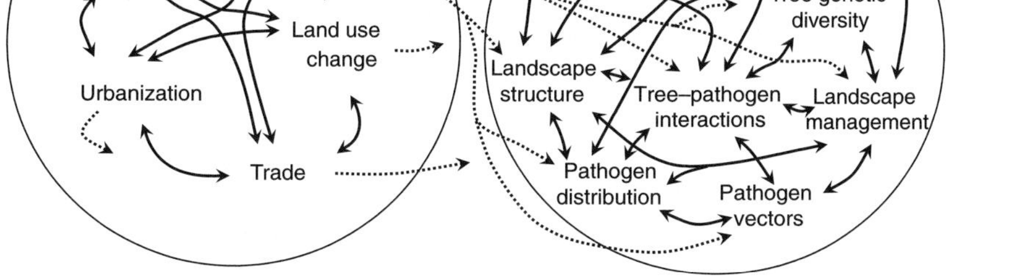

Fig. 28.5. Impacts of global change on forest health. Prediction and management will be complicated by the interactions among global change factors (climate change, increased trade, land use change, pollution, urbanization). All these factors will have direct and indirect impacts on the health of forest ecosystems. Changes in forest health, in turn, will feed back on global change, e.g. due to reduced carbon sequestration (modified from Pautasso *et al.*, 2012).

In addition, the land sparing/sharing alternative has been mostly debated from the point of view of agriculture (Ewers *et al.*, 2009; Clough *et al.*, 2011; Phalan *et al.*, 2011), with less consideration of the issue for forest ecosystems (Fürstenau *et al.*, 2007; Lonsdale *et al.*, 2008; Ranius and Roberge, 2011). We can ask whether wood and non-timber forest products are better reconciled with biodiversity conservation by: (i) intensifying wood production (plantations), and thus preserving a larger area of natural forests; (ii) cultivating forest landscapes in a less intensive way, thereby affecting more natural habitats; or (iii) combining high forest biodiversity with high yields in multi-purpose forests? Which of these scenarios would be preferable when considering the management of risks posed by exotic tree pathogens (both to tree plantations and to semi-natural forests) is a research question that needs to be investigated (Petrokofsky *et al.*, 2010; Grierson *et al.*, 2011; Püttmann, 2011; Pautasso *et al.*, 2012).

Similarly, the issue of habitat connectivity is mainly regarded from the perspective of conservation biology: decreasing the fragmentation of patches of natural habitats would improve the chances of species to track changes in climate. Greater landscape connectivity may, nevertheless, result in a more likely dispersal of invasive tree pathogens across countries (Holdenrieder *et al.*, 2004). The same point applies to assisted migration (managed relocation): it is likely that moving plant species polewards and upwards to enable them to cope with rapid climate change may facilitate the spread and emergence of novel plant health issues, but forest pathologists have rarely been included in the debate about assisted migration (Jeger *et al.*, 2011; Garbelotto and Pautasso, 2012).

28.8 Interdisciplinarity

A necessary condition for a sustainable management of novel diseases of trees in the coming decades is that knowledge about forest pathology needs to reach scientists who are not normally involved in the field. Scientific specialization is probably inevitable, but meeting a lecturer in biology who has not heard of *Cameraria ohridella* Deschka & Dimic (the horse-chestnut leaf miner), or a plant epidemiologist who has no clue about *Hymenoscyphus pseudoalbidus* (the causal fungus of ash dieback), makes it clear that it is time for trees and their health to be taken more seriously in the scientific community. The current widespread budget cuts affecting plant health scientists do not seem a step in the right direction towards improving our long-term response to exotic tree pathogens. We need better ways to encourage young scientists to pursue research on emerging plant diseases (Chungu *et al.*, 2010; Drea, 2011; McDade *et al.*, 2011; Stenlid *et al.*, 2011).

Many approaches are helpful for improving the management of exotic pathogens: from the development of molecular detection tools (Hughes *et al.*, 2011) to the use of meta-analysis in identifying which pathogen traits tend to be associated with invasiveness (Philibert *et al.*, 2011); from the improvement of the best practices for planting nurseries (Ghimire *et al.*, 2011) to breeding for resistance against pathogens such as *G. circinata* and *P. cinnamomi* (González *et al.*, 2011; Vivas *et al.*, 2012); from the use of geographic information systems (GIS) in studying forest health across environmental gradients (Morales *et al.*, 2011) to the assessment of plant species susceptibility to new exotic pathogens (Ireland *et al.*, 2012).

Similarly varied is the potential for interdisciplinary collaborations of forest pathologists, given the repercussions of outbreaks of exotic tree pathogens for hydrology, biodiversity conservation, forest fire regimes, carbon sequestration, sustainability and other key land use issues (Schütt, 1979; Ayres and Lombardero, 2000; Annerstedt, 2010; Pautasso *et al.*, 2010a; Barrett *et al.*, 2011; Olofsson *et al.*, 2011; Ramage *et al.*, 2011; Wilkinson *et al.*, 2011; Adams *et al.*, 2012; Sievanen *et al.*, 2012). Particularly fruitful is likely to be the collaboration with landscape geneticists in terms of the genetic diversity of both trees (so as to identify resistance) and pathogens (so as to reconstruct pathways of dispersal) (Garbelotto, 2008; Pautasso, 2009; Grünwald and Goss, 2011; Lees *et al.*, 2011).

The preventive role of endophytes has much potential against exotic tree pathogens, but is still under studied (Grünig *et al.*, 2011; Martín-García *et al.*, 2011). Moreover, our response to exotic tree pathogens may be improved by biogeographical studies of the first reports of plant diseases, although the bias due to a file-drawer problem may need to be considered in such analyses.

One problem with interdisciplinary studies is that these may require a longer time than specialized investigations. In order to improve the sustainability of plant disease management options, such studies would need to take place at an early stage of the invasion process (entry, establishment, spread, impact). Combining interdisciplinarity with early intervention is thus a challenge that could be met by proactive approaches aiming to prevent the introduction of exotic pathogens, rather than trying to stop their dispersal once they have become established.

28.9 Climate Change and Global Trade

There is a little doubt that climate change will be accompanied by the emergence of many new plant health problems (Baker *et al.*, 2000; Garrett *et al.*, 2006; Shaw, 2009; Jeger *et al.*, 2011; Jactel *et al.*, 2012; Littell *et al.*, 2012). Less appreciated is the potential of our response to climate change to result in plant health challenges, e.g. via the widespread adoption of new biofuel crops, which could facilitate the emergence of new plant pathogens and pests (Stewart and Cromey, 2010).

None the less, actions reducing greenhouse gas emissions (e.g. energy conservation, lower consumption, greater use of public transport, switching to renewable energy sources; MacKay, 2008) can be considered as a way to prevent not just climate change, but also consequences of climate change such as the emergence of new plant health problems. At the same time, rapid climate change, by shuffling the suitability of vast regions to most species, will make it difficult to be certain about the native versus exotic status of newly arrived species, including forest pathogens (Shaw and Osborne, 2011; Webber and Scott, 2012).

In the presence of novel ecosystems and plant pathosystems (Wingfield *et al.*, 2010), research into and the management of plant health problems will have to be creative (Juroszek and von Tiedemann, 2011). A network of sentinel plantings may enable early detection and rapid response (Britton *et al.*, 2010). A similar role may be played by already existing botanic gardens and by urban areas, given their high presence of exotic trees and associated organisms, as well as of observers and traded material (Parmentier and Pautasso, 2010; Tubby and Webber, 2010; Venette *et al.*, 2010; Treseder *et al.*, 2011).

Given that the combination of climate shifts and increased global trade poses a much stronger challenge to plant health than each of these two processes in isolation (Anderson *et al.*, 2004; Webber, 2010; Garrett *et al.*, 2011; Driscoll *et al.*, 2012), a trend towards the consumption of local produce and the planting of locally grown plants and trees could help prevent the emergence of new pathosystems. On its own, climate warming may result in fewer new tree health problems than it would when combined with growing plant trade volumes, shipping distances and areas of plantations of exotic trees (Wingfield *et al.*, 2008, 2011; Moricca and Ragazzi, 2009).

28.10 Changing Paradigms

Tree pathogens (whether exotic or not) are a key factor shaping forest ecosystems (Holdenrieder, 1991; Castello *et al.*, 1995; Hansen, 1999; Teale and Castello, 2011). They are in some cases one of the obstacles in the struggle to ensure that the needs of the growing world population for wood, non-timber forest products, clean water and air, food, recreation and tranquillity are met. However, tree pathogens can also be used as biological control agents to reduce the spread of invasive tree species (Meyer *et al.*, 2012). The combination of rapid climate change and growing long-distance trade is likely to result in declining forest health throughout the planet, regardless of the perspective (e.g. naturalist versus normative, biocentric versus anthropocentric, holistic versus reductionist) used to define forest health (Döring *et al.*, 2012). Furthermore, if future exotic tree pathogens are only going to be a challenge to the ecosystem services essential to humans, and are not going to be a major problem for the continued existence of forests, it does matter from which viewpoint forest health is defined. In this situation, a change of paradigm from anthropocentric to biocentric would rather modify our outlook on whether a particular exotic tree pathogen needs to be managed or not.

Forest pathogens tend to be studied by a relatively small group of scientists, but there is the need for more awareness of tree health issues among the whole scientific community, the public, economists and policy makers. A more holistic approach to risk assessment, regulation and governance of plant diseases could improve scenarios informing national and international responses to emerging plant epidemics (Walshe and Burgman, 2010). Moreover, our understanding of tree diseases due to exotic pathogens is inevitably limited to specific pathosystems and we do not have sufficient knowledge on the myriad potential direct and indirect interactions of a given exotic tree pathogen with all the components of a forest. Holistic approaches to risk assessment are commendable developments, but we may still be far away from a satisfactory understanding of the long-term environmental consequences of pathogen invasions (EFSA, 2011b). It is thus to be welcomed that recent reviews call for an explicit declaration of the degree of uncertainty in the potential development of pathosystems under different scenarios (Hanewinkel *et al.*, 2011).

The growing uniformity of forest landscapes, due to mechanization and the economies of scale deriving from vast monoculture production areas, results in a lower epidemic threshold for many tree diseases. In order to diminish the risk posed by existing and potential exotic forest pathogens (but also for many other reasons), there is today the recognition that we need more diverse woodlands and forests, particularly where past management has contributed to reducing their species and genetic diversity (Pautasso *et al.*, 2012). Then again, the principle that more diverse forest ecosystems are less susceptible to introduced tree diseases is not completely unchallenged (e.g. by generalist *Phytophthora* species). What is more, carried to its extreme consequences, it might end up transforming forests as we know them into a series of extremely diverse, but also rather artificial, species assemblages.

At a time of economic crisis met by austerity, spending money for the management of exotic tree pathogens may appear to make little sense, given that forest enterprises are already struggling to make ends meet without having to cope with new disease outbreaks. Yet this view does not consider the many unsustainable subsidies granted to fossil fuel extraction, fisheries, road construction or mining. Employing more people to respond to exotic tree pathogens may well be a sensible way to react to youth unemployment, thereby reducing social problems deriving from migration into towns due to failing rural economies. Recognizing that a certain presence of exotic tree pathogens may contribute to overall forest health (e.g. by making forests more diverse, by accelerating adaptation to new environmental conditions, and by providing deadwood) is probably also counterintuitive, but could lead to ecological and economic advantages in the long term (Manion, 2003; Ostry and Laflamme, 2009; Carroll, 2011; Geils and Vogler, 2011).

Fungal pathogens do not recognize political boundaries, and the magnitude of trade in plant material makes it impossible to inspect more than a small percentage of potentially infected shipments. However, phytosanitary regulation is one of the tools that we have with which to respond to exotic tree pathogens. Luckily, different countries and cultures will tend to deal with the problem in different ways, a diversity to be welcomed because it could avoid the making of the same mistakes all over the planet. Conversely, complaints that 'the plant health status of the European Union as a whole can only be as good as that of the weakest member state' (MacLeod *et al.*, 2010) call for a greater integration of plant health policies (Engler *et al.*, 2012). Network approaches suggest that marginal nodes pose a lower risk than central and well-connected trade players, unless there is a rapid and uncontrolled link from the former to the latter. There is the need for more intercultural inquiry into how exotic tree pathogens are perceived throughout the world (Kohsaka and Handoh, 2006). Even if, in the past, regulation and management decisions that aimed to avoid further spread of exotic pathogens of trees tended to be reactive, rather than proactive (Bond *et al.*, 2011), new developments in network theory, stakeholder involvement and spatial models may make future management of emerging forest diseases more effective. We should not forget, though, that current global trade and its potential for further introductions of exotic pathogens may dwarf any improvement in management that derives from such developments.

References

Adams, H.D., Luce, C.H., Breshears, D.D., Allen, C.D., Weiler, M., Hale, V.C., Smith, A.M.S. and Huxman, T.E. (2012) Ecohydrological consequences of drought- and infestation-triggered tree die-off: insights and hypotheses. *Ecohydrology* 5, 145–159.

Alexander, J. and Lee, C.A. (2010) Lessons learned from a decade of sudden oak death in California: evaluating local management. *Environmental Management* 46, 315–328.

Anderson, P.K., Cunningham, A.A., Patel, N.G., Morales, F.J., Epstein, P.R. and Daszak, P. (2004) Emerging infectious diseases of plants: pathogen pollution, climate change and agrotechnology drivers. *Trends in Ecology and Evolution* 19, 535–544.

Andjic, V., Dell, B., Barber, P., Hardy, G., Wingfield, M. and Burgess, T. (2011) Plants for planting: indirect evidence for the movement of a serious forest pathogen, *Teratosphaeria destructans*, in Asia. *European Journal of Plant Pathology* 131, 49–58.

Annerstedt, M. (2010) Transdisciplinarity as an inference technique to achieve a better understanding in the health and environmental sciences. *International Journal of Environmental Research and Public Health* 7, 2692–2707.

Armsworth, P.R. *et al.* (2010) The ecological research needs of business. *Journal of Applied Ecology* 47, 235–243.

Ayres, M.P. and Lombardero, M.J. (2000) Assessing the consequences of global change for forest disturbance from herbivores and pathogens. *Science of the Total Environment* 262, 263–286.

Baker, R.H.A., Sansford, C.E., Jarvis, C.H., Cannon, R.J.C., MacLeod, A. and Walters, K.F.A. (2000) The role of climatic mapping in predicting the potential geographical distribution of non-indigenous pests under current and future climates. *Agriculture, Ecosystems and Environment* 82, 57–71.

Barrett, M.A., Bouley, T.A., Stoertz, A.H. and Stoertz, R.W. (2011) Integrating a one health approach in education to address global health and sustainability challenges. *Frontiers in Ecology and the Environment* 9, 239–245.

Baxter, P.W.J. and Possingham, H.P. (2011) Optimizing search strategies for invasive pests: learn before you leap. *Journal of Applied Ecology* 48, 86–95.

Benninga, J., Hennen, W. and Schans, J. (2012) Supply chain risk model for quantifying the cost-effectiveness of phytosanitary measures. *Crop Protection* 32, 64–70.

Bihon, W., Burgess, T., Slippers, B., Wingfield, M.J. and Wingfield, B.D. (2012) High levels of genetic diversity and cryptic recombination is widespread in introduced *Diplodia pinea* populations. *Australasian Plant Pathology* 41, 41–46.

Bond, C.A., Champ, P., Meldrum, J. and Schoettle, A. (2011) Investigating the optimality of proactive management of an invasive forest pest. In: Keane, R.E., Tomback, D.F., Murray, M.P. and Smith, C.M. (eds) *The Future of High-Elevation, Five-Needle White Pines in Western North America: Proceedings of the High Five Symposium, 28–30 June 2010, Missoula, MT*. Proceedings RMRS-P-63, US Department of Agriculture, Forest Service, Rocky Mountain Research Station, Fort Collins, Colorado, pp. 295–302.

Brasier, C.M. (2001) Rapid evolution of introduced plant pathogens via interspecific hybridization. *BioScience* 51, 123–133.

Brasier, C. and Webber, J. (2010) Sudden larch death. *Nature* 466, 824–825.

Britton, K.O., White, P., Kramer, A. and Hudler, G. (2010) A new approach to stopping the spread of invasive insects and pathogens: early detection and rapid response via a global network of sentinel plantings. *New Zealand Journal of Forestry Science* 40, 109–114.

Cahill, D.M., Rookes, J.E., Wilson, B.A., Gibson, L. and McDougall, K.L. (2008) *Phytophthora cinnamomi* and Australia's biodiversity: impacts, predictions and progress towards control. *Australian Journal of Botany* 56, 279–310.
Carrasco, L.R., Mumford, J.D., MacLeod, A., Knight, J.D. and Baker, R.H.A. (2010) Comprehensive bioeconomic modelling of multiple harmful non-indigenous species. *Ecological Economics* 69, 1303–1312.
Carroll, S.P. (2011) Conciliation biology: the eco-evolutionary management of permanently invaded biotic systems. *Evolutionary Applications* 4, 184–199.
Castello, J.D., Leopold, D.J. and Smallidge, P.J. (1995) Pathogens, patterns, and processes in forest ecosystems. *BioScience* 45, 16–24.
Chadès, I., Martin, T.G., Nicol, S., Burgman, M.A., Possingham, H.P. and Buckley, Y.M. (2011) General rules for managing and surveying networks of pests, diseases, and endangered species. *Proceedings of the National Academy of Sciences of the United States of America* 108, 8323–8328.
Chadfield, V. and Pautasso, M. (2012) *Phytophthora ramorum* in England and Wales: which environmental variables predict county disease incidence? *Forest Pathology* 42, 150–159.
Chungu, D., Muimba-Kankolongo, A., Wingfield, M.J. and Roux, J. (2010) Plantation forestry diseases in Zambia: contributing factors and management options. *Annals of Forest Science* 67, 802.
Clough, Y. *et al.* (2011) Combining high biodiversity with high yields in tropical agroforests. *Proceedings of the National Academy of Sciences of the United States of America* 108, 8311–8316.
Crowl, T.A., Crist, T.O., Parmenter, R.R., Belovsky, G. and Lugo, A.E. (2008) The spread of invasive species and infectious disease as drivers of ecosystem change. *Frontiers in Ecology and the Environment* 6, 238–246.
Danon, L., Ford, A.P., House, T., Jewell, C.P., Keeling, M.J., Roberts, G.O., Ross, J.V. and Vernon, M.C. (2011) Networks and the epidemiology of infectious disease. *Interdisciplinary Perspectives on Infectious Diseases* 2011, 284909.
Dehnen-Schmutz, K., Holdenrieder, O., Jeger, M.J. and Pautasso, M. (2010) Structural change in the international horticultural industry: some implications for plant health. *Scientia Horticulturae* 125, 1–15.
Döring, T.F., Pautasso, M., Finckh, M.R. and Wolfe, M.S. (2012) Concepts of plant health – reviewing and challenging the foundations of plant protection. *Plant Pathology* 61, 1–15.
Drea, S. (2011) The end of the botany degree in the UK. *Bioscience Education* 17, 2.
Driscoll, D.A., Felton, A., Gibbons, P., Felton, A.M., Munro, N.T. and Lindenmayer, D.B. (2012) Priorities in policy and management when existing biodiversity stressors interact with climate-change. *Climatic Change* 111, 533–557.
Durán, A., Gryzenhout, M., Slippers, B., Ahumada, R., Rotella, A., Flores, F., Wingfield, B.D. and Wingfield, M.J. (2008) *Phytophthora pinifolia* sp. nov. associated with a serious needle disease of *Pinus radiata* in Chile. *Plant Pathology* 57, 715–727.
Dutech, C., Fabreguettes, O., Capdevielle, X. and Robin, C. (2010) Multiple introductions of divergent genetic lineages in an invasive fungal pathogen, *Cryphonectria parasitica*, in France. *Heredity* 105, 220–228.
EFSA (2011a) Scientific opinion on the pest risk analysis on *Phytophthora ramorum* prepared by the FP6 project RAPRA. *EFSA Journal* 9, 2186.
EFSA (2011b) Guidance on the environmental risk assessment of plant pests. *EFSA Journal* 9, 2460.
Engler, A., Nahuelhual, L., Cofré, G. and Barrena, J. (2012) How far from harmonization are sanitary, phytosanitary and quality-related standards? An exporter's perception approach. *Food Policy* 37, 162–170.
Essl, F. *et al.* (2011) Socioeconomic legacy yields an invasion debt. *Proceedings of the National Academy of Sciences of the United States of America* 108, 203–207.
Ewers, R.M., Scharlemann, J.P.W., Balmford, A.P. and Green, R.E. (2009) Do increases in agricultural yield spare land for nature? *Global Change Biology* 15, 1716–1726.
Filipe, J.A.N., Cobb, R.C., Meentemeyer, R.K., Lee, C.A., Valachovic, Y.S., Cook, A.R., Rizzo, D.M. and Gilligan, C.A. (2012) Landscape epidemiology and control of pathogens with cryptic and long-distance dispersal: sudden oak death in northern Californian forests. *PLoS Computational Biology* 8(1): e1002328. doi: 10.1371/journal.pcbi.1002328.
Forbes, G.A., Mizubuti, E.S.G. and Shtienberg, D. (2009) Plant disease epidemiology and disease management: has science had an impact on practice? In: Peshin, R. and Dhawan, A.K. (eds) *Integrated Pest Management: Innovation-Development Process*. Springer, Berlin, pp. 351–368.
Fürstenau, C., Badeck, F.W., Lasch, P., Lexer, M.J., Lindner, M., Mohr, P. and Suckow, F. (2007) Multiple-use forest management in consideration of climate change and the interests of stakeholder groups. *European Journal of Forest Research* 126, 225–239.

Ganley, R.J., Watt, M.S., Manning, L. and Iturritxa, E. (2009) A global climatic risk assessment of pitch canker disease. *Canadian Journal of Forest Research* 39, 2246–2256.
Garbelotto, M. (2008) Molecular analysis to study invasions by forest pathogens: examples from Mediterranean ecosystems. *Phytopathologia Mediterranea* 47, 183–203.
Garbelotto, M. and Pautasso, M. (2012) Impacts of exotic forest pathogens on Mediterranean ecosystems: four case studies. *European Journal of Plant Pathology* 133, 101–116.
Garrett, K.A., Dendy, S.P., Frank, E.E., Rouse, M.N. and Travers, S.E. (2006) Climate change effects on plant disease: genomes to ecosystems. *Annual Review of Phytopathology* 44, 489–509.
Garrett, K.A. *et al.* (2011) Complexity in climate-change impacts: an analytical framework for effects mediated by plant disease. *Plant Pathology* 60, 15–30.
Geils, B.W. and Vogler, D.R. (2011) A natural history of *Cronartium ribicola*. In: Keane, R.E., Tomback, D.F., Murray, M.P. and Smith, C.M. (eds) *The Future of High-Elevation, Five-Needle White Pines in Western North America: Proceedings of the High Five Symposium, 28–30 June 2010, Missoula, MT*. Proceedings RMRS-P-63, US Department of Agriculture, Forest Service, Rocky Mountain Research Station, Fort Collins, Colorado, pp. 210–217.
Geils, B.W., Hummer, K.E. and Hunt, R.S. (2010) White pines, *Ribes*, and blister rust: a review and synthesis. *Forest Pathology* 40, 147–185.
Ghimire, S.R., Richardson, P.A., Kong, P., Hu, J., Lea-Cox, J.D., Ross, D.S., Moorman, G.W. and Hong, C.X. (2011) Distribution and diversity of *Phytophthora* species in nursery irrigation reservoir adopting water recycling system during winter months. *Journal of Phytopathology* 159, 713–719.
Gilbert, M. and Liebhold, A. (2010) Comparing methods for measuring the rate of spread of invading populations. *Ecography* 33, 809–817.
Gonthier, P. and Garbelotto, M. (2011) Amplified fragment length polymorphism and sequence analyses reveal massive gene introgression from the European fungal pathogen *Heterobasidion annosum* into its introduced congener *H. irregulare*. *Molecular Ecology* 20, 2756–2770.
Gonthier, P., Nicolotti, G., Linzer, R., Guglielmo, F. and Garbelotto, M. (2007) Invasion of European pine stands by a North American forest pathogen and its hybridization with a native interfertile taxon. *Molecular Ecology* 16, 1389–1400.
Gonthier, P., Brun, F., Lione, G. and Nicolotti, G. (2012) Modelling the incidence of *Heterobasidion annosum* butt rots and related economic losses in alpine mixed naturally regenerated forests of northern Italy. *Forest Pathology* 42, 57–68.
González, M.V., Cuenca, B., López, M., Prado, M.J. and Rey, M. (2011) Molecular characterization of chestnut plants selected for putative resistance to *Phytophthora cinnamomi* using SSR markers. *Scientia Horticulturae* 130, 459–467.
González-Varela, G., González, A.J. and Milgroom, M.G. (2011) Clonal population structure and introductions of the chestnut blight fungus, *Cryphonectria parasitica*, in Asturias, northern Spain. *European Journal of Plant Pathology* 131, 67–79.
Goss, E.M., Larsen, M., Chastagner, G.A., Givens, D.R. and Grünwald, N.J. (2009) Population genetic analysis infers migration pathways of *Phytophthora ramorum* in US nurseries. *PLoS Pathogens* 5(9): e1000583. doi: 10.1371/journal.ppat.1000583.
Grierson, C.S. *et al.* (2011) One hundred important questions facing plant science research. *New Phytologist* 192, 6–12.
Grünig, C.R., Queloz, V., and Sieber, T.N. (2011) Structure of diversity in dark septate endophytes: from species to genes. In: Pirttilä, A.M. and Frank, A.C. (eds) *Endophytes of Forest Trees: Biology and Applications*. Springer, Berlin, pp. 3–30.
Grünwald, N.J. and Goss, E. (2011) Evolution and population genetics of exotic and re-emerging pathogens: novel tools and approaches. *Annual Review of Phytopathology* 49, 249–267.
Haas, S.E., Hooten, M.B., Rizzo, D.M. and Meentemeyer, R.K. (2011) Forest species diversity reduces disease risk in a generalist plant pathogen invasion. *Ecology Letters* 14, 1108–1116.
Hanewinkel, M., Hummel, S. and Albrecht, A. (2011) Assessing natural hazards in forestry for risk management: a review. *European Journal of Forest Research* 130, 329–351.
Hansen, E.M. (1999) Disease and diversity in forest ecosystems. *Australasian Plant Pathology* 28, 313–319.
Harwood, T.D., Xu, X.M., Pautasso, M., Jeger, M.J. and Shaw, M. (2009) Epidemiological risk assessment using linked network and grid based modelling: *Phytophthora ramorum* and *Phytophthora kernoviae* in the UK. *Ecological Modelling* 220, 3353–3361.
Harwood, T.D., Tomlinson, I., Potter, C.A. and Knight, J.D. (2011) Dutch elm disease revisited: past, present and future management in Great Britain. *Plant Pathology* 60, 545–555.

Hatala, J.A., Dietze, M.C., Crabtree, R.L., Kendall, K., Six, D. and Moorcroft, P.R. (2011) An ecosystem-scale model for the spread of a host-specific forest pathogen in the Greater Yellowstone Ecosystem. *Ecological Applications* 21, 1138–1153.

Hoban, S., Bertorelle, G., and Gaggiotti, O.E. (2012) Computer simulations: tools for population and evolutionary genetics. *Nature Reviews Genetics* 13, 110–122.

Holdenrieder, O. (1991) Der Forstschutz – Objekte, Probleme, Strategien [Forest protection – objectives, problems, strategies]. *Schweizerische Zeitschrift für Forstwesen* 142, 795–807.

Holdenrieder, O., Pautasso, M., Weisberg, P.J. and Lonsdale, D. (2004) Tree diseases and landscape processes: the challenge of landscape pathology. *Trends in Ecology and Evolution* 19, 446–452.

Holmes, T.P., Bell, K.P., Byrne, B. and Wilson, J.S. (2008) Economic aspects of invasive forest pest management. In: Holmes, T.P., Prestemon, J.P. and Abt, K.L. (eds) *The Economics of Forest Disturbances: Wildfires, Storms, and Invasive Species*. Springer, Berlin, pp. 381–406.

Holzmüller, E.I., Jose, S. and Jenkins, M.A. (2010) Ecological consequences of an exotic fungal disease in eastern U.S. hardwood forests. *Forest Ecology and Management* 259, 1347–1353.

Hughes, K.J.D., Tomlinson, J.A., Giltrap, P.M., Barton, V., Hobden, E., Boonham, N. and Lane, C.R. (2011) Development of a real-time PCR assay for detection of *Phytophthora kernoviae* and comparison of this method with a conventional culturing technique. *European Journal of Plant Pathology* 131, 695–703.

Hunter, G.C., Crous, P.W., Carnegie, A.J., Burgess, T.I. and Wingfield, M.J. (2011) *Mycosphaerella* and *Teratosphaeria* diseases of *Eucalyptus*; easily confused and with serious consequences. *Fungal Diversity* 50, 145–166.

Ireland, K.B., Hüberli, D., Dell, B., Smith, I.W., Rizzo, D.M. and Hardy, G.E.St.J. (2012) Potential susceptibility of Australian native plant species to branch dieback and bole canker diseases caused by *Phytophthora ramorum*. *Plant Pathology* 61, 234–246.

Isbell, F. *et al.* (2011) High plant diversity is needed to maintain ecosystem services. *Nature* 477, 199–202.

Jactel, H. *et al.* (2009) The influences of forest stand management on biotic and abiotic risks of damage. *Annals of Forest Sciences* 66, 701.

Jactel, H., Petit, J., Desprez-Loustau, M.-L., Delzon, S., Piou, D., Battisti, A. and Koricheva, J. (2012) Drought effects on damage by forest insects and pathogens: a meta-analysis. *Global Change Biology* 18, 267–276.

Jeger, M.J. (2004) Analysis of disease progress as a basis for evaluating disease management practices. *Annual Review of Phytopathology* 42, 61–82.

Jeger, M.J., Pautasso, M., Holdenrieder, O. and Shaw, M.W. (2007) Modelling disease spread and control in networks: implications for plant sciences. *New Phytologist* 174, 179–197.

Jeger, M.J., Pautasso, M. and Stack, J. (2011) Climate, globalization and trade: impacts on dispersal and invasion of fungal plant pathogens. In: Olsen, L.A., Relman, D.A., Choffnes, E.R. and Pray, L. (eds) *Fungal Diseases: An Emerging Challenge to Human, Animal and Plant Health*. Institute of Medicine of the National Academies, Washington, DC, pp. 273–295.

Johnson, K.B. and Stockwell, V.O. (1998) Management of fire blight: a case study in microbial ecology. *Annual Review of Phytopathology* 36, 227–248.

Jules, E.S., Kauffman, M.J., Ritts, W.D. and Carroll, A.L. (2002) Spread of an invasive pathogen over a variable landscape: a nonnative root rot on Port Orford cedar. *Ecology* 83, 3167–3181.

Juroszek, P. and von Tiedemann, A. (2011) Potential strategies and future requirements for plant disease management under a changing climate. *Plant Pathology* 60, 100–112.

Keesing, F. *et al.* (2010) Impacts of biodiversity on the emergence and transmission of infectious diseases. *Nature* 468, 647–652.

Keller, R.P., Geist, J., Jeschke, J.M. and Kuehn, I. (2011) Invasive species in Europe: ecology, status, and policy. *Environmental Sciences Europe* 23, 23.

Koch, K.A., Quiram, G.L. and Venette, R.C. (2010) A review of oak wilt management: a summary of treatment options and their efficacy. *Urban Forestry and Urban Greening* 9, 1–8.

Kohsaka, R. and Handoh, I.C. (2006) Perceptions of 'close-to-nature forestry' by German and Japanese groups: inquiry using visual materials of 'cut' and 'dead' wood. *Journal of Forest Research* 11, 11–19.

Konijnendijk, C.C. (2008) *The Forest and the City. The Cultural Landscape of Urban Woodland*. Springer, Berlin, pp. 127–141.

Kowalski, T. (2006) *Chalara fraxinea* sp. nov. associated with dieback of ash (*Fraxinus excelsior*) in Poland. *Forest Pathology* 36, 264–270.

Krcmar, E. (2008) *An Examination of the Threats and Risks to Forests Arising from Invasive Alien Species*. Information Report BC-X-415, Canadian Forest Service, Pacific Forestry Centre, Victoria, British Columbia, Canada.

Kubono, T. and Ito, S. (2002) *Raffaelea quercivora* sp. nov. associated with mass mortality of Japanese oak, and the ambrosia beetle (*Platypus quercivorus*). *Mycoscience* 43, 255–260.

Lees, D.C., Lack, H.W., Rougerie, R., Hernandez-Lopez, A., Raus, T., Avtzis, N.D., Augustin, S. and Lopez-Vaamonde, C. (2011) Tracking origins of invasive herbivores through herbaria and archival DNA: the case of the horse-chestnut leaf miner. *Frontiers in Ecology and the Environment* 9, 322–328.

Li, R., Buongiorno, J., Zhu, S., Turner, J.A. and Prestemon, J. (2007) Potential economic impact of limiting the international trade of timber as a phytosanitary measure. *International Forestry Review* 9, 514–525.

Liebhold, A.M., Macdonald, W.L., Bergdahl, D. and Mastro, V.C. (1995) Invasion by exotic forest pests: a threat to forest ecosystems. *Forest Science Monographs* 30, 1–49.

Lilja, A., Rytköenen, A., Hantula, J., Müller, M., Parikka, P. and Kurkela, T. (2011) Introduced pathogens found on ornamentals, strawberry and trees in Finland over the past 20 years. *Agricultural and Food Science* 20, 74–85.

Lindenmayer, D.B. and Laurance, W.F. (2012) A history of hubris – cautionary lessons in ecologically sustainable forest management. *Biological Conservation* 151, 11–16.

Littell, J.S., Peterson, D.L., Millar, C.I. and O'Halloran, K.A. (2012) U.S. National Forests adapt to climate change through Science–Management partnerships. *Climatic Change* 110, 269–296.

Lonsdale, D., Pautasso, M. and Holdenrieder, O. (2008) Wood-decaying fungi in the forest: conservation needs and management options. *European Journal of Forest Research* 127, 1–22.

Loo, J.A. (2009) Ecological impacts of non-indigenous invasive fungi as forest pathogens. *Biological Invasions* 11, 81–96.

Lovett, G.M., Canham, C.D., Arthur, M.A., Weathers, K.C. and Fitzhugh, R.D. (2006) Forest ecosystem responses to exotic pests and pathogens in Eastern North America. *BioScience* 56, 395–405.

Lundquist, J.E. and Klopfenstein, N.B. (2001) Integrating concepts of landscape ecology with the molecular biology of forest pathogens. *Forest Ecology and Management* 150, 213–222.

MacKay, D.J.C. (2008) *Sustainable Energy – Without the Hot Air*. UIT Cambridge, Cambridge, UK.

MacLeod, A., Pautasso, M., Jeger, M.J. and Haines-Young, R. (2010) Evolution of the international regulation of plant pests and challenges for future plant health. *Food Security* 2, 49–70.

Manion, P.D. (2003) Evolution of concepts in forest pathology. *Phytopathology* 93, 1052–1055.

Martín-García, J., Espiga, E., Pando, V. and Diez, J.J. (2011) Factors influencing endophytic communities in poplar plantations. *Silva Fennica* 45, 169–180.

Matyssek, R. *et al.* (2012) Forests under climate change and air pollution: gaps in understanding and future directions for research. *Environmental Pollution* 160, 57–65.

McDade, L.A., Maddison, D.R., Guralnick, R., Piwowar, H.A., Jameson, M.L., Helgen, K.M., Herendeen, P.S., Hill, A. and Vis, M.L. (2011) Biology needs a modern assessment system for professional productivity. *BioScience* 61, 619–625.

McRoberts, N., Hall, C., Madden, L.V. and Hughes, G. (2011) Perceptions of disease risk: from social construction of subjective judgements to rational decision making. *Phytopathology* 101, 654–665.

Meentemeyer, R.K., Cunniffe, N.J., Cook, A.R., Filipe, J.A.N., Hunter, R.D., Rizzo, D.M. and Gilligan, C.A. (2011) Epidemiological modeling of invasion in heterogeneous landscapes: spread of sudden oak death in California (1990–2030). *Ecosphere* 2, 17.

Meyer, J.-Y., Fourdrigniez, M. and Taputuarai, R. (2012) Restoring habitat for native and endemic plants through the introduction of a fungal pathogen to control the alien invasive tree *Miconia calvescens* in the island of Tahiti. *BioControl* 57, 191–198.

Mills, P., Dehnen-Schmutz, K., Ilbery, B., Jeger, M., Jones, G., Little, R., MacLeod, A., Parker, S., Pautasso, M., Pietravalle, S. and Maye, D. (2011) Integrating natural and social science perspectives on plant disease risk, management and policy formulation. *Philosophical Transactions of the Royal Society B, Biological Sciences* 366, 2035–2044.

Morales, R.M., Idol, T. and Friday, J.B. (2011) Assessment of *Acacia koa* forest health across environmental gradients in Hawaii using fine resolution remote sensing and GIS. *Sensors* 11, 5677–5694.

Moricca, S. and Ragazzi, A. (2009) *Lusus naturae*: cambiamenti climatici ed invasioni di parassiti vegetali modificano il territorio agro-forestale [Freak of nature: climate change and invasions of plant parasites change the agro-forestry territory]. *Rivista Italiana di Agronomia* 3, S13–S17.

Moslonka-Lefebvre, M., Pautasso, M. and Jeger, M.J. (2009) Disease spread in small-size directed networks: epidemic threshold, correlation between links to and from nodes, and clustering. *Journal of Theoretical Biology* 260, 402–411.

Moslonka-Lefebvre, M., Finley, A., Dorigatti, I., Dehnen-Schmutz, K., Harwood, T., Jeger, M.J., Xu, X.M., Holdenrieder, O. and Pautasso, M. (2011) Networks in plant epidemiology: from genes to landscapes, countries, and continents. *Phytopathology* 101, 392–403.

Moslonka-Lefebvre, M., Harwood, T., Jeger, M.J. and Pautasso, M. (2012) SIS along a continuum (SISc) epidemiological modelling and control of diseases on directed trade networks. *Mathematical Biosciences* 236, 44–52.
Notaro, S. and De Salvo, M. (2010) Estimating the economic benefits of the landscape function of ornamental trees in a sub-Mediterranean area. *Urban Forestry and Urban Greening* 9, 71–81.
Olofsson, J., Ericson, L., Torp, M., Stark, S. and Baxter, R. (2011) Carbon balance of arctic tundra under increased snow cover mediated by a plant pathogen. *Nature Climate Change* 1, 220–223.
Ostry, M.E. and Laflamme, G. (2009) Fungi and diseases – natural components of healthy forests. *Botany* 87, 22–25.
Parmentier, I. and Pautasso, M. (2010) Species-richness of the living collections of the world's botanical gardens – patterns within continents. *Kew Bulletin* 65, 519–524.
Parry, D. and Teale, S.A. (2011) Alien invasions: the effects of introduced species on forest structure and function. In: Castello, J.D. and Teale, S.A. (eds) *Forest Health. An Integrated Perspective*. Cambridge University Press, Cambridge, UK, pp. 115–162.
Paull, S.H., Song, S., McClure, K.M., Sackett, L.C., Kilpatrick, A.M. and Johnson, P.T.J. (2012) From superspreaders to disease hotspots: linking transmission across hosts and space. *Frontiers in Ecology and the Environment* 10, 75–82.
Pautasso, M. (2009) Geographical genetics and the conservation of forest trees. *Perspectives in Plant Ecology, Systematics and Evolution* 11, 157–189.
Pautasso, M. and Jeger, M.J. (2008) Epidemic threshold and network structure: the interplay of probability of transmission and of persistence in small-size directed networks. *Ecological Complexity* 5, 1–8.
Pautasso, M., Holdenrieder, O. and Stenlid, J. (2005) Susceptibility to fungal pathogens of forests differing in tree diversity. In: Scherer-Lorenzen, M., Koerner, C. and Schulze, E.D. (eds) *Forest Diversity and Function: Temperate and Boreal Systems*. Springer, Berlin, pp. 263–289.
Pautasso, M., Dehnen-Schmutz, K., Holdenrieder, O., Pietravalle, S., Salama, N., Jeger, M.J., Lange, E. and Hehl-Lange, S. (2010a) Plant health and global change – some implications for landscape management. *Biological Reviews* 85, 729–755.
Pautasso, M., Moslonka-Lefebvre, M. and Jeger, M.J. (2010b) The number of links to and from the starting node as a predictor of epidemic size in small-size directed networks. *Ecological Complexity* 7, 424–432.
Pautasso, M., Xu, X.M., Jeger, M.J., Harwood, T., Moslonka-Lefebvre, M. and Pellis, L. (2010c) Disease spread in small-size directed trade networks: the role of hierarchical categories. *Journal of Applied Ecology* 47, 1300–1309.
Pautasso, M., Döring, T.F., Garbelotto, M., Pellis, L. and Jeger, M.J. (2012) Impacts of climate change on plant diseases – opinions and trends. *European Journal of Plant Pathology* 133, 295–313.
Persha, L., Agrawal, A. and Chhatre, A. (2011) Social and ecological synergy: local rulemaking, forest livelihoods, and biodiversity conservation. *Science* 331, 1606–1608.
Petrokofsky, G. *et al.* (2010) A participatory process for identifying and prioritizing policy-relevant research questions in natural resource management: a case study from the UK forestry sector. *Forestry* 83, 357–367.
Phalan, B., Balmford, A., Green, R.E. and Scharlemann, J.P.W. (2011) Minimising the harm to biodiversity of producing more food globally. *Food Policy* 36, S62–S71.
Philibert, A., Desprez-Loustau, M.-L., Fabre, B., Frey, P., Halkett, F., Husson, C., Lung-Escarmant, B., Marçais, B., Robin, C., Vacher, C. and Makowski, D. (2011) Predicting invasion success of forest pathogenic fungi from species traits. *Journal of Applied Ecology* 48, 1381–1390.
Picco, A.M., Angelini, P., Ciccarone, C., Franceschini, A., Ragazzi, A., Rodolfi, M., Varese, G.C. and Zotti, M. (2011) Biodiversity of emerging pathogenic and invasive fungi in plants, animals and humans in Italy. *Plant Biosystems* 145, 988–996.
Potter, C., Harwood, T., Knight, J. and Tomlinson, I. (2011) Learning from history, predicting the future: the UK Dutch elm disease outbreak in relation to contemporary tree disease threats. *Philosophical Transactions of the Royal Society B, Biological Sciences* 366, 1966–1974.
Prospero, S. and Rigling, D. (2012) Invasion genetics of the chestnut blight fungus *Cryphonectria parasitica* in Switzerland. *Phytopathology* 102, 73–82.
Prost, L., Cerf, M. and Jeuffroy, M.-H. (2012) Lack of consideration for end-users during the design of agronomic models. A review. *Agronomy for Sustainable Development* 32, 581–594.
Püttmann, K.J. (2011) Silvicultural challenges and options in the context of global change: 'simple' fixes and opportunities for new management approaches. *Journal of Forestry* 109, 321–331.
Queloz, V., Grünig, C.R., Berndt, R., Kowalski, T., Sieber, T.N. and Holdenrieder, O. (2011) Cryptic speciation in *Hymenoscyphus albidus*. *Forest Pathology* 41, 133–142.

Quijas, S., Schmid, B. and Balvanera, P. (2010) Plant diversity enhances provision of ecosystem services: a new synthesis. *Basic and Applied Ecology* 11, 582–593.

Ramage, B.S., O'Hara, K.L. and Forrestel, A.B. (2011) Forest transformation resulting from an exotic pathogen: regeneration and tanoak mortality in coast redwood stands affected by sudden oak death. *Canadian Journal of Forest Research* 41, 763–772.

Ranius, T. and Roberge, J.-M. (2011) Effects of intensified forestry on the landscape-scale extinction risk of dead wood dependent species. *Biodiversity and Conservation* 20, 2867–2882.

Ricciardi, A., Palmer, M.E. and Yan, N.D. (2011) Should biological invasions be managed as natural disasters? *BioScience* 61, 312–317.

Rizzo, D.M. (2005) Exotic species and fungi: interactions with fungal, plant, and animal communities. In: Digthon, T., Oudemans, P. and White, J.-F. (eds) *The Fungal Community: Its Organization and Role in the Ecosystem*. CRC, Boca Raton, Florida, pp. 857–877.

Rizzo, D.M., Garbelotto, M. and Hansen, E.M. (2005) *Phytophthora ramorum*: integrative research and management of an emerging pathogen in California and Oregon forests. *Annual Review of Phytopathology* 43, 309–335.

Sansford, C.E. *et al.* (2009) *Report on the Risk of Entry, Establishment, Spread and Socio-economic Loss and Environmental Impact and the Appropriate Level of Management for* Phytophthora ramorum *for the EU*. Deliverable Report D28, EU Sixth Framework Project RAPRA (Risk Analysis of *Phytophthora ramorum*, a Newly Recognised Pathogen Threat to Europe and the Cause of Sudden Oak Death in the USA). Available at: http://rapra.csl.gov.uk/ (accessed 28 December 2012).

Schütt, P. (1979) Ungelöste Probleme der Forstpathologie – 100 Jahre nach Robert Hartig [Unsolved problems of forest pathology – 100 years after Robert Hartig]. *Forstwissenschaftliches Centralblatt* 98, 74–79.

Shaw, D.C., Filip, G.M., Kanaskie, A., Maguire, D.A. and Littke, W.A. (2011) Managing an epidemic of Swiss needle cast in the Douglas-fir region of Oregon: the role of the Swiss needle cast cooperative. *Journal of Forestry* 109, 109–119.

Shaw, M.W. (2009) Preparing for changes in plant diseases due to climate change. *Plant Protection Science* 45, S3–S10.

Shaw, M.W. and Osborne, T.M. (2011) Geographic distribution of plant pathogens in response to climate change. *Plant Pathology* 60, 31–43.

Sievanen, L., Campbell, L.M. and Leslie, H.M. (2012) Challenges to interdisciplinary research in ecosystem-based management. *Conservation Biology* 26, 315–326.

Soliman, T., Mourits, M.C.M., Oude Lansink, A.G.J.M. and van der Werf, W. (2010) Economic impact assessment in pest risk analysis. *Crop Protection* 29, 517–524.

Stenlid, J., Oliva, J., Boberg, J.B. and Hopkins, A.J. (2011) Emerging diseases in European forest ecosystems and responses in society. *Forests* 2, 486–504.

Stewart, A. and Cromey, M. (2010) Identifying disease threats and management practices for bioenergy crops. *Current Opinion in Environmental Sustainability* 3, 75–80.

Teale, S.A. and Castello, J.D. (2011) Regulators and terminators: the importance of biotic factors to a healthy forest. In: Castello, J.D. and Teale, S.A. (eds) *Forest Health. An Integrated Perspective*. Cambridge University Press, Cambridge, UK, pp. 81–114.

Thor, M., Arlinger, J.D. and Stenlid, J. (2006) *Heterobasidion annosum* root rot in *Picea abies*: modelling economic outcomes of stump treatment in Scandinavian coniferous forests. *Scandinavian Journal of Forest Research* 21, 414–423.

Torii, M., Matsuda, Y., Murata, M. and Ito, S. (2011) Spatial distribution of *Raffaelea quercivora* hyphae in transverse sections of seedlings of two Japanese oak species. *Forest Pathology* 41, 293–298.

Treseder, K., Pytel, M., Mappley, M., Griffiths, A. and Pettitt, T. (2011) Evolution of pest management strategies in the rain-forest biome at the Eden project, the first 10 years. *Outlooks on Pest Management* 22, 22–31.

Tubby, K.V. and Webber, J.F. (2010) Pests and diseases threatening urban trees under a changing climate. *Forestry* 83, 451–459.

Václavík, T., Kupfer, J.A. and Meentemeyer, R.K. (2012) Accounting for multi-scale spatial autocorrelation improves performance of invasive species distribution modelling (iSDM). *Journal of Biogeography* 39, 42–55.

Vanem, E. (2012) Ethics and fundamental principles of risk acceptance criteria. *Safety Science* 50, 958–967.

Venette, R.C. *et al.* (2010) Pest risk maps for invasive alien species: a roadmap for improvement. *BioScience* 60, 349–362.

Vivas, M., Zas, R. and Solla, A. (2012) Screening of maritime pine (*Pinus pinaster*) for resistance to *Fusarium circinatum*, the causal agent of pitch canker disease. *Forestry* 85, 185–192.

Walshe, T. and Burgman, M. (2010) A framework for assessing and managing risks posed by emerging diseases. *Risk Analysis* 30, 236–249.

Waring, K.M. and O'Hara, K.L. (2005) Silvicultural strategies in forest ecosystems affected by introduced pests. *Forest Ecology and Management* 209, 27–41.

Webber, B.L. and Scott, J.K. (2012) Rapid global change: implications for defining natives and aliens. *Global Ecology and Biogeography* 21, 305–311.

Webber, J. (2010) Pest risk analysis and invasion pathways for plant pathogens. *New Zealand Journal of Forestry Science* 40, S45–S56.

Wilkinson, K., Grant, W.P., Green, L.E., Hunter, S., Jeger, M.J., Lowe, P., Medley, G.F., Mills, P., Phillipson, J., Poppy, G.M. and Waage, J. (2011) Infectious diseases of animals and plants: an interdisciplinary approach. *Philosophical Transactions of the Royal Society B, Biological Sciences* 366, 1933–1942.

Wingfield, M.J., Slippers, B., Hurley, B.P., Coutinho, T.A., Wingfield, B.D. and Roux, J. (2008) Eucalypt pests and diseases: growing threats to plantation productivity. *Southern Forests* 70, 139–144.

Wingfield, M.J., Slippers, B. and Wingfield, B.D. (2010) Novel associations between pathogens, insects and tree species threaten world forests. *New Zealand Journal of Forestry Science* 40, 95–103.

Wingfield, M.J., Roux, J. and Wingfield, B.D. (2011) Insect pests and pathogens of Australian acacias grown as non-natives – an experiment in biogeography with far-reaching consequences. *Diversity and Distributions* 17, 968–977.

Woolhouse, M.E.J. (2011) How to make predictions about future infectious disease risks. *Philosophical Transactions of the Royal Society B, Biological Sciences* 366, 2045–2054.

Woolhouse, M.E.J. *et al.* (1997) Heterogeneities in the transmission of infectious agents: implications for the design of control programs. *Proceedings of the National Academy of Sciences of the United States of America* 94, 338–342.

Woolhouse, M.E.J., Shaw, D.J., Matthews, L., Liu, W.C., Mellor, D.J. and Thomas, M.R. (2005) Epidemiological implications of the contact network structure for cattle farms and the 20–80 rule. *Biology Letters* 1, 350–352.

Yemshanov, D., McKenney, D.W., Pedlar, J.H., Koch, F.H. and Cook, D. (2009) Towards an integrated approach to modelling the risks and impacts of invasive forest species. *Environmental Reviews* 17, 163–178.

Index

Note: **bold** page numbers refer to figures and tables.

Abies spp. 39, **51**, **98**, 128, 130, 171, 179, 578
- *A. alba* 106, 128, **133**, 136, **139**, **147**, 162, 221, 222, 223, 474
- *A. amabilis* **44**, 475, 573
- *A. balsamea* 422, 474
- *A. borisii-regis* 162, 475
- *A. concolor* **44**, 130, 134, 146, **147**, 162, 573
- *A. grandis* 40, **44**, **147**, 162, 163, 171, 187, 189, 206, 286, 461, 475, 573
- *A. lasiocarpa* **44**, 162, 180, 286, 475
- *A. magnifica* **44**
- *A. pinsapo* 130, 135
- *A. procera* **44**, **140**
- *A. sachalinensis* 135, 137, 422
- branch/tip blights affecting 421, 576
- heart rot in **199**, **201**, **202**, **204**, 205

Acacia spp. 490, 534
- *A. mearnsii* 232

Acer spp. **51**, 65, 82, 205, **493**, **494**, 499
- *A. campestre* **212**, **214**, 215
- *A. rubrum* 503
- *A. saccharum* 231, 243

acholeplasmas 76

ACLSV (*Apple chlorotic leaf spot virus*) **51**, **53**, 61, 67

Adelges geniculatus 475

aerial photography 30, 32, 187, 428

aerial spraying 109, 122, 414, 437, 445–446

Aesculus spp. **51**, **493**, 580
- *A. hippocastanum* 521

afforestation 135, 392, 396, 400, 402, 421

Africa 30, 98, **98**, 107, 121, 232, 287, 352, 422
- *Armillaria* spp. in 161, 163
- *Cryphonectriaceae* in 341, 342–343, 344, 345, 346
- DNB in 436, 437, 449
- foliar diseases of conifers in 460, 467, 468
- rusts in 547

African mistletoe (*Tapinanthus bangwensis*) 107

African oil palm (*Elaeis guineensis*) 116, 117–118, 123

age structure of forests 7–8, 206, 207, 222

Agrilus planipennis 85, 89

air pollution 11, 29, 54, 427, 458, 503

Alaska (USA) 105, 129, 186, **199**, **201**, 205, 207, 407, 521

Alces alces 224, 268

alder *see Alnus* spp.

alder yellows disease **77**, 78–79, 86, 87

AldY phytoplasmas **77**, 78–79, 83, 86

aleurioconidia 231, 236, 239

allelopathy 11

Allygidius atomarius 85

almond (*Prunus dulcis*) 79, **98**, 232, 237

Alnus spp. 78–79, 83, 86, 190, 523, 568
- *A. glutinosa* **77**, 208, **212**, **213**, **214**, 489
- *A. incana* 208, **213**, 489
- *A. rubra* **77**, 190
- *A. viridis* 489
- foliar diseases of 489
- heart rot in **199**, **202**, 208, **214**, 222

Alps 129, 130, 136, **139**, 146, 162, 400, 401, 402

Alternaria spp. 568–569, 570
- *A. alternata* 117
- *A. tenuissima* 506

Amazon region 232

ambrosia beetles 230, 231, 235, 236, 239
- and Ophiostoma diseases 256, 257, 258, 259, 261, 263, 270, 273
- *see also* redbay ambrosia beetle

ambrosia fungi 257, 263
amenity trees 159, 160, 164, 170, 197, 232, 240, 458, 460, 473
American beech (*Fagus grandifolia*) **12**
American chestnut (*Castanea dentata*) 19, **20**, 43, 318–319, 324, 325–326
American palm weevil (*Rhynchophorus palmarum*) 116, 117, 121, 122, 124–125
American sycamore (*Platanus occidentalis*) 80, 241, 242, 245, 247, 509
Amyema spp. 106, 109
  A. miquelli 97
Amylostereum spp. 215, 219, 222
  A. areolatum **202**, 219
  A. chailletii **202**
angiosperms 76, 97, 164, 256, 257, 259, 263, 422, 498, 547
annosus root/butt rots (*Heterobasidion annosum*) 7, 8, 9, 18, 32, **32**, 38, 42, 43, 128–150, 218, 224
  anamorphic stage (*Spiniger meineckellus*) 131
  avoidance strategy for 142
  detection/diagnosis of 129–131, **132**
  disease epidemic models for 21, 22, **22**, 33, 138–141, **139–140**, 149
  dynamics of spread of 134–136
  economic impacts of 129, 149
  epidemiology of 134–141
  eradication strategy for 142–143
  estimating/predicting disease incidence 138–141
  exclusion strategy for 142
  fruiting bodies 130, 136
  host resistance to 134, 146–147
  IDM strategy for 141, 148–150, **148**, **150**
  infection biology of 131–134, **133**
  management of 35, 38, 39, 40, 41, 42, 137, 141–150, 171
  pathogens/significance/distribution 128–129, 162
  protection strategy for 143–146
  as saprophytes 37–38, 129
  species susceptible to **44**, **147**
  spore dispersal range/primary infection patterns 136–137
  and stand characteristics 137–138, **141**
  and stem/root wounds **133**, 134, 142
  stump treatments for 143–145, **145**, 148–150
  taxonomy of 128, 129
  therapy strategy for 147–148
  thinning and 136, 138, 142, 144, 145, 148–149
  trenching for 145
antagonistic fungi 189–190
anthracnose of plane tree (*Apiognomonia veneta*) 247, **494**, 497, 508, **600**
antibiotics 88, 377, 412
aphids 62, 64, 458, 464, 475, 476, 477
APHIS (Animal and Plant Health Inspection Service) 36
Aphrophora canadensis 379
Apiognomonia spp. 490, 495, 498, 500
  A. errabunda 496, 497, **498**
  A. veneta 247, **494**, 497, 508, **600**
Apioplagiostoma populi 490, 495, 502
ApMV (*Apple mosaic virus*) **51**, **53**, 61, 67
apple (*Malus* spp.) **98**, 556, 592
Apple chlorotic leaf spot virus (ACLSV) **51**, **53**, 61, 67
Arabis mosaic virus (ArMV) **51**, **52**, 62, 64, 65
arborvitae *see Thuja* spp.
Arbutus 422
Arceuthobium spp. *see* dwarf mistletoes
  A. campylopodum 31, 37
  A. douglasii 104
  A. tsugense 104–105
Argentina 161, 359
Arizona (USA) 31, 287
Armillaria spp. 2–3, 30, 32, **32**, 33, 35, 37, 38, 159–172, 437
  A. cepistipes 161, 170
  A. ectypa 159
  A. fuscipes 163
  A. gallica 160–161, 162, 165, 170
  A. limonea 161, 163
  A. luteobubalina 161, 163, 170
  A. mellea 16, 37–38, 159, 160, 161, 162, 163, 165
  A. novae-zelandiae 161, 163, 167, 171
  A. ostoyae 6, 11, **17**, 18, 22, 40, 41, 42, 165, 183
    characteristics of 160, 161
    distribution of 162
    management of 35, 37, 38, 43, 143, 170, 171–172
  A. tabescens 159, 160, 162
  avoidance strategy for 168
  biological control of 169–170
  broadleaved trees and 170–171
  chemical control of 169
  diagnosis of 131, 138, **139**, **140**, 163–164
  disease management strategies for 35, 37, 38, 42, 43
  distribution/impacts of 161–163
  epidemiology of 166–167
  eradication strategy for 168–169
  fruiting bodies 159, **160**, 164
  host resistance to 163, 170–171
  IDM for 171–172
  infection biology of 165–166
  inoculum, primary/secondary 166
  long-distance dissemination of 167
  management of 167–172
  in orchards/young stands 166–167
  protection strategy for 169–170
  pruning/fertilization/trenching and 170
  rhizomorphs of 163, 164, 169
  and saprophytic fungi 170
  species susceptibility to **44**
  taxonomy of 159–161

ArMV (*Arabis mosaic virus*) **51**, **52**, 62, 64, 65
Arrhenodes minutus 244
Ascomycota 144, 283, 392, 420, 458, 488, **493**, 495, 498, 508
ascospores 230, 349, 394, 409, 438–439
ash *see Fraxinus* spp.
ash decline 490–491
ash yellows disease (AshY) **77**, 78, 81, 83, 85, 88, 89
Asia 97, 103, 120, 163, 232, 257, 346, 352, 392
  chestnut blight in 318, 325, 333
  Dutch elm disease in 257, 258, 267, 276
  foliar diseases in 460, 461, 464, 467, 468, 502
  heart rot in **199**, **200**, **201**, **204**
  laminated root rot in 178
  rusts in 547
aspen (*Populus* spp.) 21, 208, 209, 217, 221
  canker affecting *see* Hypoxylon canker
  Ceratocystis spp. in 232, 234, 237
  European (*P. tremula*) **77**, **212**, 232, 407, 424, 489, 580
  quaking (*P. tremuloides*) 81, 407, 489, **493**
aster yellows disease **77**, 83, 84, 88
Aulographina eucalypti 506
Aurantiosacculus eucalypti 344
Aurapex penicillata 346, 348, 350
Auratiopycnidiella 347
Aureobasidium **493**, 495, 508
Aurifilum marmelostoma 345, 350
Australia 23, **98**, 168, 231, 324, 341–342, 344, 346, 347, 377, 422, 528, 535, 573, 601
  Armillaria disease in 161, 162–163, 170
  cypress canker in 359, 360, 366
  disease management strategies in 36, 39, 42, 88, 142, 170, 271, 384–385
  DNB in 438, 443, 448, 449, 450–451, 577
  foliar diseases in 461, 464, 479, 502, 506, 508–509, 510
  IDM in 110
  mistletoe management in 104, 105–106, 109, 110
  rusts in 547, 548, 554, 555
  see also New South Wales; Victoria; Western Australia
Austria **56**, **77**, 209, 318, 428, **593**, **596**
avocado (*Persea americana*) **98**, 258, 301
azoxystrobin 368, 371, 463, 466, 480, 576, 578, 579, 581, 582

Bacillus spp. 76, 82, 447
  B. subtilis 40
bacteria 1, 30, 79–80, 369, 447, 458, 561
  and climate change 87–88
  diagnosis of 81–83
  epidemiology of 86–88
  infection biology of 84
  management of 88–89
  symptoms of 81–82, 87
Banksia spp. 524, 529
  bull (*B. grandis*) 42, 524, 534
bark beetles 3, 12, 21, **32**, 33, 42, 162, 180, 232, 237, 238–239
  and cypress canker 362
  elm *see* elm bark beetles
  and heart rots 217
  and *Leptographium* spp. 283, 284, 285, 286, 288, 289–290
  and Ophiostoma diseases 256, 259, 263, 269, 270, 272–273, 274–275, 277
  and pitch canker 378
bark stripping 205, 206, 208–209, **208**, 215, 244
Barriopsis spp. 304
Basidiomycota 128, 159, 420, 459, 488, 508
basidiospores 13, 35, 136, 159, 165, 167, 184
Beauveria bassiana 122
beech (*Fagus* spp.) **51**, **199**, 206, 216, 223, 521
  European (*F. sylvatica*) 146, **212**, **214**, 215, 489, **498**
beetles (Coleoptera) 118, 180, 186, 230, 231, 239, 256, 263, 270, 379, 380
  pine/pine leaf/pine needle 33, 464, 550
  see also ambrosia beetles; bark beetles; weevils
Belgium 402, 422, 560, **596**
benomyl 331, 386, 463, 466, 477, 506, 509, 510
benzimidazoles 331, 368, 371, 386
Berberis thunbergii 65
Betula spp. *see* birch
Bifusella linearis 469
biodiversity 29, 34, 110, 170, 548, 601
bioenergy 37, 143
biological control 7, 38, 39–40, 43, 429
biomass production 8, 29, 437, 490
biotrophs 4, 6, 497
birch (*Betula* spp.) 50, **51**, 54, **55**, 146, 568
  downy (*B. pubescens*) 303
  European white (*B. pendula*) 303
  foliar diseases of 490
  heart rot in **199**, **200**, **202**, **203**, 205, 208, **214**, 220, 221
  paper (*B. papyrifera*) 43
birch rust (*Melampsoridium betulinum*) 476, 581–582
birds 101, 102–103, 232
bitternut hickory (*Carya cordiformis*) 232
Bjerkandera adusta 39, 217, 219–220
black cottonwood *see poplar, P. balsamifera*
black rot (*Ciboria batschiana*) 568, 569
black spruce (*Picea mariana*) 179, 287
black wattle (*Acacia mearnsii*) 232
blackbutt (*Eucalyptus pilularis*) 508–509
blackstain root disease (*Leptographium wageneri*) 41, 43, **44**, 230, 257, 283–291
  diagnosis of 288
  IDM for 291
  infection biology of 289–290
  management of 291
  transmission of 284–285

BLS (bacterial leaf scorch) 79–80, 82
epidemiology of 86–88
blue-stain fungi 119, 231, 256, 258–259, 267, 268, 269, 283
diagnosis of 261–262, 289
management of 269–270, 272–273, 274, 278
BMPs (best management practices) **34**, 35–36, 43
Boddy-Rayner concept 197–198
borax/borates 39, 143, 144
Bordeaux mixture 468, 506, 534
Bosnia-Herzegovina **77**, 318, 326, 327
Botryodiplodia sp. 116
botryosphaeran **306**
Botryosphaeria spp. 233, 234, 299
B. dothidea 299, 300, 303, 304
B. obtusa see Diplodia spp., *D. seriata*
B. stevensii see Diplodia spp., *D. mutila*
Botryosphaeriaceae 422
Botrytis cinerea see grey mould
Brachyramphus marmoratus 104–105
branch dieback 101
branch/tip blights 420–430
chemical/biological control of 428–429
environmental conditions for 420–421
environmental factors affecting 426
epidemiology of 426–428
fungicides for 429, 430
host resistance to 429
human factors in 427
infection biology of 424–426
light factor in 426–427, 428
management of 428–430
molecular identification of 423–424
monitoring for 428, 429
pathogens/significance/distribution 420–423
quarantine for 427–428
stand characteristics and 427
taxonomy of 422, 423, 424
Brassica spp. 40
Brazil 79, 161, 163, 341, 351, 352, 550
Ceratocystis spp. in 233–234, 239, 241, 242, 243
Britain (UK) 39, 104, 461, 519, 521, 522, 567, 593, **593**, **596**
annosus rots in 130, 137, 138, **139**, 143, 148
DNB in 438, 441, 577
Dutch elm disease in 258, 267
larch canker in 392, 402
British Columbia (BC, Canada) 11, **24**, 80, 108, 144, 162, 172, 286, 287, 522, 567
DNB in 436, 437, 438, 445
laminated/tomentosus root rots in 178, 180, 186, 188, 192
rusts in 556, 558
broadleaved trees **55**, 130, 146, 170, **203**, 223, 269, 423
foliar diseases of *see* foliar diseases of broadleaved trees
bronze leaf disease 490
brown felt blight (*Herpotrichia juniperi*) 461, 469–470, 479, 578
brown rot fungus 189–190
brown spot needle blight (*Mycosphaerella dearnessii*) 42, 439, 460, 464, 467, 468–469, **600**
brown spot needle disease (*Scirrhia acicola*) 42
Brunchorstia pinea 421
bull banksia (*Banksia grandis*) 42, 524, 534
Bursaphelenchus spp. 115, 117
B. cocophilus see red ring nematode/disease
B. fraudulentus 117
B. mucronatus 117, 119
B. xylophilus see pine wilt nematode/disease
butt rot 32, 37, 38, 42, 198, 205, 206, 216–217
see also annosus root/butt rots

cacao (*Theobroma cacao*) 166, 232, 241, 522
calcium 138, 426, 529
California (USA) 4, 31, 33, 36, 84, **98**, 107, 160, 178, **201**, 284, 422, 597–598
annosus rots in 130, 134, 135, 145
Ceratocystis spp. in 232, 233, 249
cypress canker in 359, 366, 367, 368, 369–370
oomycete diseases in 520, 523, 524, 527–528, 529, 530, 531–532
pitch canker in 377, 379, 380, 381, 385
California laurel (*Umbellularia californica*) 524, 527–528, 529
California Oak Mortality Task Force 44
Callitropsis spp. 359, 360
Calocedrus spp. **44**, 479
C. decurrens 128, **147**
Calonectria spp.
C. morganii 506
C. quinqueseptata 502
Calophyllum inophyllum 287–288
Caloscypha fulgens 569
Canada 11, 33, 36, 37, 78, 79, 80, 85, 108, 232, 241, 274, 275, 285, 359, 520
annosus rots in 128, **140**, 142, 144
branch/tip blights in 422, 424
DNB in 437, 438, 439, 452, 577
foliar diseases of broadleaved trees in 505, 509
foliar diseases of conifers in 460, 469, 472, 473, 474, 475, 476, 479
Hypoxylon canker in 407
IDM in 110
laminated/tomentosus root rots in 180, 185, 188, 190
larch canker in 392, 398
Leptographium in 285, 286
see also British Columbia; Great Lakes; Ontario; Ottawa

canker rots 197, **198**, 205, 220
canker stain 230, 233–234, 237, 239, 240–241, 242–243
  management of 243, 244, 245, 247, 248–249
cankers/canker fungi 12, 36, 38, 164, 298–310, 340–353, 548
  affecting eucalyptus 341–344
  affecting *Fagales* 340–341
  classification of 298–300
  diagnosis of 299–300, 309
  effects of temperature on 303, 308
  epidemiology of 304, 307
  in fruits 301–302, 307
  genetic analysis of 299
  global movement of 308–309
  histopathology 301–302
  human impact on 301
  infection biology of 298, 299, 300, 302–307
  insect associations of 304
  management of 307–310
  melanization in 304
  moisture/nutrient stress in 302–303, 308
  pathogens/significance/distribution of 298–299, 303, 308–309
  phytotoxin in 304
  potential eradication of 309
  reproduction of 299, 307
  resistance to 309–310
  role of stress in 302–304, 307–308
  symptoms of 298, 299, 301–302, 303
  treatment of 298, 303
canopy gap/dieback 135, 180, 191–192, 378
  see also branch/tip blights
Cape beech (*Rapanea melanophloeos*) 346
carbendazim 331, 466, 477
Caribbean **98**, 116
Carnation Italian ringspot virus (CIRV) 63
Carpinus spp. 50, **51**
Carpophilus sayi 237
Carya spp. 492
  C. cordifomis 232
Cascade Mountains (USA) 187, 190
Castanea spp. 247, 346, 350, 523, 534, **535**, 568
  C. crenata **77**, 318, **494**
  C. dentata 19, **20**, 43, 318–319, 324, 325–326
  C. mollissima 318
  C. sativa 349
Castilleja spp. 550–551, 554, 557
Catharanthus roseus 89
cedar (*Cedrus* spp.) 179, 205, 379, 479, 575
  atlas (*C. atlantica*) 424
  deodar (*C. deodara*) 424
cedar elm (*Ulmus crassifolia*) **77**, 108
cedar leaf blight (*Didymascella thujina*) 460, 479
cedar-apple rusts (*Gymnosporangium* spp.) 552, 556
cedar-of-Goa (*Hesperocyparis lusitanica*) 360
cellulase 364, 497
Celoporthe spp. 344, 348
  C. dispera 344, 345, 347
  C. fontana 344, 345
  C. woodiana 346
Central America 98, **98**, 103, 117, 231, 233, 346, 376, 439, 548
  foliar diseases of conifers in 460, 467, 468
cerambycid beetles (*Monochamus* spp.) 118, 270
Cerambycidae 263
cerato-platinin 237
Ceratocystiopsis spp. 257, 283
Ceratocystis spp. 119, 230–249, 283
  avoidance strategy for 243
  C. albofundus 232, 236
  C. cacaofunesta 232, 241
  C. coerulescens 231
  C. fagacaerum see oak wilt disease
  C. fimbriata 231, 232–234, 235, 236, 239, 241–242
    clades 232
    management of 247–248
  C. larium 232
  C. moniliformis 231
  C. paradoxa 231
  C. platani 231, 232–233, 236, 237, 244
    management of 244, 245, 247, 248, **600**
    transmission of 239, 240–241, 242–243, 244
  C. populicola 232, 236, 241
  C. smalleyi 232, 237
  C. variospora 232, 236, 240, 246
  C. virescens 231, 236
  diagnosis of 234–235
  economic impact of 231, 232
  epidemiology of 238–243
  eradication strategy for 244–245
  exclusion strategy for 244
  fruiting bodies of 231
  fungicides for 243, 246, 248, 406–407
  genetic identification of 235
  host resistance to 247–248
  IDM for 248–249
  infection biology of 230–231, 235–238
  inoculum of 235–236, 242–243
  insects and 230–231, 235, 238, 244
  management of 243–249
  mechanical transmission of 240–241, 243
  odour of 235, 236, 237
  and *Ophiostoma* spp. 259, 262
  pathogens/significance/distribution 230–234
  protection strategy for 246–248
  pruning and 240, 241, 243, 244
  root graft transmission of 242–244
  sawdust/frass and 235, 236, 239, 241, 244, 245
  sporulation/vector activity 238–239
  taxonomy of 230, 231
  therapy strategy for 191, 248–249

Ceratocystis spp. (*continued*)
trenching/physical barriers and 246–247, 248
vegetatively propagated material 241–242
wound colonization/xylem movement 237–238, 243
wound dressings for 246
and wounding/new infection centres 240
Ceratocystis wilt 230, 232, 241, 243, 249
Cercospora needle blight (*Mycosphaerella gibsonii*) 460, 464, 467–468, 479–480
control of 480
cerrado forests (Brazil) 234, 243
Chalara 231
C. fraxinea (teleomorphic stage *Hymenoscyphus pseudoalbidus*) 501, 506, 592
Chamaecyparis spp. 179, 205, 359, 379
C. lawsoniana see Port Orford cedar
C. obtusa 163
chemical control 38–39, 43
regulations for 428–429
cherry (*Prunus* spp.) 166, 489
sweet (*P. avium*) 56, 65, 162
Cherry leaf roll virus (CLRV) **51**, **52**, 54, **55**, 65, 66
chestnut (*Castanea* spp.) 247, 346, 350, 523, 534, **535**, 568
American (*C. dentata*) 19, **20**, 43, 318–319, 324, 325–326
horse (*Aesculus hippocastanum*) 521
Japanese (*C. crenata*) **77**, 318, **494**
chestnut blight (*Cryphonectria parasitica*) 19, **20**, 31, 43, 318–333, 340–341, 346, 347, 348, 349, 350–351, 352
avoidance strategy for 324
diagnosis of 319–320
environmental conditions for 323
epidemiology of 322–324
eradication strategy for 324
exclusion strategy for 324
and fertilizers/soil treatments 325, 331
host resistance to 325–326
host susceptibility to 323–324, 332
hypovirulence and *see* hypoviruses
IDM for 332–333
infection biology of 320–322, **321**
management of 324–333
pathogen/significance/distribution 318–319
reproduction of, sexual/asexual 322
therapy strategy for 326–331
chestnut witches' broom **77**
Chile 161, 359, 376, 377, 384, 437, 461, 569
China 56, **77**, 123, 128, 160, 163, 178, 179, 271, 318, 346, 347
foliar diseases of conifers in 460, 464, 468, 479
larch canker in 392–393
Chinese pinenut (*Pinus koraiensis*) 163, **442**, 471
chlamydospores 4
chloropicrin 38, 189, 191
chlorothalonil 368, 429, 463, 466, 477, 506, 510, 575, 576, 578, 581
chlorotic ringspots 54, **55**, 56–57, **57**, 58
Chondrostereum purpureum **202**, 208, 219, 222, 492
Christmas mistletoe (*Phoradendron tomentosum*) 108
Christmas tree 39, 381, 437, 441, 458, 461, 464, 473, 474, 475, 549, 558
Chrysocrypta corymbiae 344, 346
Chrysomyxa spp. 459, 548, 549, 551, 552
C. pyrolae 548, 551, 556, 568
C. weirii 551, 552
Chrysoporthe spp. 341–342, 346–347, 348, 351
C. austroafricana 342, 346, 347, 350, 351, 352
C. cubensis/deuterocubensis 341–343, 344, 346, 347, 348, 350, 351, 352
C. doradensis 343
C. hodgesiana 343
C. inopina 343, 346
C. syzygiicola 342, 345, 352
C. zambiensis 342, 352
Ciboria batschiana 568, 569
Ciborinia spp. 492, 495, 496, 500
cicada (*Magicicada septendecim*) 415
Cicadellidae 85, 87
CIRV (*Carnation Italian ringspot virus*) 63
Citrus spp. 79, **98**, 160, 164, 236, 301
Citrus tristeza virus (CTV) 54
citrus variegated chlorosis 82
Cladosporium 508
C. fulvum 448–449
clear-cutting/-felling 37–38, 137, 138, 143, 146, 169
Climacocystis borealis **203**, 217
climate 6–7, 12, 31–32, 307, 459–460
modelling 23–25, **24**, 459
climate change 11, 23, 29, 87–88, 298, 303, 427, 443, 460, 501, 502, 567, 594, **601**, 603
CLIMEX model 23, 384, 443–444
clonal colonies *see* genets
Closteroviridae 54
Clostridium 76
clove tree (*Syzygium aromaticum*) 341, 350
CLRV (*Cherry leaf roll virus*) **51**, **52**, 54, **55**, 65, 66
CMV (*Cucumber mosaic virus*) **51**, 65
co-evolution 13, **14**
Coccoloba uvifera 346, 347
coconut palm (*Cocos nucifera*) 116, 117
coffee (*Coffea*) **98**, 236
Coleophora laricella 475
Coleosporium spp. 459, 461, 548
Colletotrichum 488–489, 497
Colombia **77**, 78, 343, 346

Colopterus spp.
C. niger 237
C. semitectus 237
C. truncatus 237, 238
Colorado (USA) 274, 287
columnella 301
composts 534
cone beetle (*Conophthorus radiatae*) 379, 380
cone rusts 551, 556, 568
conidia 3, 7, 65, 122, 131–132, 299, 300, 301, 304, 309
conifer aphid (*Schizolachnus pineti*) 464
conifers
annosus rots in 128
Armillaria disease in 161, 163, 168, 169, 171, 172
foliar diseases of *see* foliar diseases of conifers
heart rot in **199**, **201**, **202**, **203**, **204**, 205
Leptographium spp. in 290
mistletoes and 99, **99**, 101, 104
Ophiostoma diseases in 257, 259, 263, 267, 270
rusts affecting 581
viruses in 50
Coniophora spp. 207
C. puteana **202**, **212**, 216, 221, 222
Conophthorus radiatae 379, 380
constitutive resistance 5
copper compounds 325, 331, 368, 430, 445, 466, 480, 506, 534, 575, 576
coppicing 163, 269
Coprinellus disseminatus 219–220
Cornus sericea 303
Corylus avellana 215, 216, 489
Corymbia spp. 343, 344, 347
Corythucha ciliata 247
Costa Rica 117, 125
cottonwood (*Populus* spp.)
black (*P. balsamifera*) 407, 422, 489, **493**, 559
eastern (*P. deltoides*) 81, 489, **493**, 501, 559, 561
creosote 39
Criptocline taxicola 478–479
Cristulariella leaf spot (*Cristulariella depraedans*) 489, **493**, 499
Croatia **77**, 318, 323, 326, 327
Cronartium spp. 547, 550, 552, 556
C. comandrae 554, 558
C. comptoniae 13, **14**, 554
C. flaccidum 547, 550–551, 554, 558
C. fusiforme 43, 554, 555, 558, 560, 561
C. quercuum 547, 554, 555, 558, 560, 561
C. ribicola see white pine blister rust
crown disease *see* branch/tip blights
Cryphalus trypanus 288
Cryphonectria spp. 320, 340, 346
C. naterciae 320, 348, 349
C. parasitica see chestnut blight
C. radicalis 320, 340–341, 348, 349
pathogens/significance/distribution 340–347
Cryphonectriaceae 340–353
biological control of 350, 352
diagnosis of 347–349
DNA/PCR identification of 348–349
environmental conditions for 350–351
epidemiology of 349–351
eucalyptus and 341–344
Fagales and 340–341
fruiting bodies of 347–348
host resistance to 352
IDM for 352–353
infection biology of 349
management of 351–353
native pathogens and 345–347
therapy strategy for 352
Cryptocephalus pini 464
Cryptomeria japonica 171, 479
Cryptometrion 348
C. aestuescens 344
CSNN (current season needle necrosis) 475
CTV (*Citrus tristeza virus*) 54
Cucumber mosaic virus (CMV) **51**, 65
Cucurbitaria piceae 472
Cupressaceae 160, 479–480
Cupressus spp. *see* cypress
C. sempervirens **305**
Curculionidae 235, 239, 263
Cyclaneusma spp.
C. minus 459, 461, 464, 466
C. niveum 461
Cyclaneusma needle cast (*C. minus*) 458, 459, 460, 461, 464, 466–467, 578, 579
cycloheximide 261, 262, 412, 413–414
cyclopaldic acid 364
Cydia zebeana 395
Cylindrocarpon 38
cypress (*Cupressus* spp.) 79, **305**, 359–360, 364, 469
Italian (*C. sempervirens*) 360, 364, 365–366, 369, 370, 599
Leyland 360, 365
see also Monterey cypress
cypress canker (*Seiridium* spp.) 359–372
avoidance strategy for 366
biological control of 368–369
chemical control of 368
diagnosis of 360–362, **361**
environmental conditions for 365–366
epidemiology of 364–366
eradication strategy for 367–368
exclusion strategy for 366–367
fertilization and 368, 369
genetic variability of 369–370
host defence response to 364, 370
host resistance to 369–371
IDM for 371–372
infection biology of 362–364
insect vectors of 362

cypress canker (*Seiridium* spp.) (*continued*)
management of 366–372
molecular identification of 361–362
patented varieties 370–371
pathogens/significance/distribution 359–360
protection strategy for 368–369
sanitation and 366, 367–368
therapy strategy for 371
toxins produced by 364
Cytospora abietis 475
Czech Republic 318, 402, **593**

Dalbergia sissoo 234
Damping-off disease 570–571
DAPI staining 80–81
Darkera parca 471, 472
date palm (*Phoenix dactylifera*) 116
Davidiella populorum 491, **493**, 495, 497, 501, 502–503, 504–505, 506, 509
de Liocourt's law 11
dead wood 8, 29, 34, **34**, 35, 119, 129, 160, 165, 217
deciduous trees *see* broadleaved trees
Delaware (USA) 79–80
Delphinella abietis 473, 474
Dendroctonus spp. 263
D. frontalis 285, 286
D. ponderosae 287, 550
D. pseudotsugae 180, 186
D. rufipennis 180
D. terebrans 285, 286
Denmark 7–8, 25, **139**, 421, **596**
Diaporthales 318, 340, 347, 459, 498
Diaporthe spp. 347
Dicarpella 495, 498
D. dryina 490, **494**, 495
Dichomera spp. 299
Didymascella spp.
D. tetraspora 479
D. thujina 460, 479
Dienes' stain 80, 81
Diplocarpon rosae 503
Diplodia spp. 304, 437
D. cupressi **305**, 359
D. mutila 304, **306**
D. pinea 6, 9, 41, 298, **305**, 309, 379, 420–421, 422, 423, 424, 471
control of 429–430
infection biology/epidemiology of 425–426, 428
see also Sphaeropsis spp.
D. scrobiculata 43, 300, 422, 423, 424
D. seriata 300, **305**
Diplodina acerina 495, **498**
diplopyrone **306**
Discula spp. 495
D. destructiva 592, **600**
D. quercina 496, **498**
disease cycles 3–5
continuous/discontinuous 4–5
steps in 3–4
disease epidemic models 15–25, 166–167
for annosus rots *see* annosus root/butt rots
climate envelopes/species distribution 23–25, **24**
disease progress curves 16–19, **17**
disease progress curves, application of 18–19
for dispersal/spread of disease 19–21, **20**
dynamic/static 21
economic 33
and forest disease management 31–33
levels/incidence in 15–16
logistic/power-logistic 19–21
regression 31–32
rule-based 33
simulation 31, 32–33, **32**
spatial 21–23, **22**, 594–595
disease syndromes, global 168
disease triangle/pyramid 1–8, **2**, 31
dispersal modes 3
environmental factors in **2**, 5, 6–8
host factors in 5–6
incubation period 4
inoculum potential 2–3
pathogen aggressiveness factor 2
pathogen factors in 1–5, **2**
temporal/human factors in 3, 8
disodium octaborate tetrahydrate (DOT) 143, 144
dispersal of pathogens 30–31
modelling 19–21, **20**
DNA barcoding 551
DNB (Dothistroma needle blight) 3, 5, 8, 9, 23, 25, 33, 436–452, 458, 460, 461, 471
DOT (disodium octaborate tetrahydrate) 143, 144
Dothiora taxicola 478
Dothistroma spp. 3, 5, 8, 9, 23, 25, 33, 460, 467, 468
D. pini 437, 438, 439, 445, 448, 577
D. rhabdoclinis 462
D. septosporum 437, 438, 439, 440, **440**, **441**, 444
control of 447
host resistance to 448–449
in nurseries 577–578
Dothistroma needle blight (DNB) 3, 5, 8, 9, 23, 25, 33, 436–452, 458, 460, 461, 471
avoidance strategy for 444
biological control of 447–448
and brown spot needle blight 439
chemical control (spraying) of 445–447, **446**, **451**, 452
climate/rainfall factors in 441–442, 443–444, **443**, 567
diagnosis of 438–439, 577

distribution of 436, **437**
dothistromin, possible roles of in 440–441
economic impacts of 437, 452
environmental conditions for 442
epidemiology of 441–444, 577
eradication strategy for 444–445
exclusion strategy for 444
genetic diversity of 444
host resistance to 448–449
IDM for 449–452
infection biology of 440–441
insects and 437
management of 444–452
molecular identification of 438, 439, 444
in nurseries 577–578
pine species susceptible to **442**, 459
protection strategy for 445–448
and pruning/thinning 446–447, **447**, 450–451, **451**
taxonomy of 437–438
tree species selection for 449, 452
Douglas-fir (*Pseudotsuga menziesii*) 6, **17**, 31–32, 33, 35, 43, 162, 437, 573
annosus rots in 133, **147**
Armillaria disease in 163, 171, 172
bacterial galls on 80, 83, 84
fertilizers for 41
foliar diseases of 458, 459, 461–464, 471
heart rot in 207
inoculation of 40
laminated root rot in 178, 181, 185, 186, 189
larch canker in 393
Leptographium spp. in 284, 286
pitch canker in 376, 385, 569
Rocky Mountain (*P. menziesii* var. *glauca*) 463
rusts in 549
stump treatments for 38
susceptibility to disease of **44**
Douglas-fir beetle (*Dendroctonus pseudotsugae*) 180, 186
downy birch (*Betula pubescens*) 303
Drepanopeziza spp. 490, 495, 500
D. populi-albae 491, 495, 506, 510
D. punctiformis 490, 491, **493**, 499, 502, 503, 506, 509
drosophilid flies 236
drought **2**, 6, 12, 29, 54, 101, 164, 323, 378, 523
dry twig and cone beetle (*Ernobius punctulatus*) 379
dsRNA analysis 58, 60–61
Dutch elm disease (DED) 3, 5, 13, 31, 230, 256, 267–268, 592
breeding resistance to 88–89
diagnosis of 259–261
disease cycle of **260**
dispersal/transmission of 263, 265, 266
distribution of 257–258
host susceptibility/resistance to 266–267, 276–277
management of 36, 40, 41, 43, 268–269, 270–272, 273–274, 276–278
root graft transmission of 268
dwarf mistletoes (*Arceuthobium* spp.) 31, 33, 36, 37, 38, 43, 97, 211
avoidance strategy for 105
diagnosis of 101
fire and 107
impacts of 98, 99, 110
management of 106, 108, 109
rating system for **98**, 99
seed dispersal by 103–104, 105
wildlife benefits of 104–105
Dynamis borassi 116

eastern cottonwood (*Populus deltoides*) 81, 489, **493**, 501, 559, 561
eastern redcedar (*Juniperus virginiana*) 450
Echinodontium tinctorium **199**, 206, 207, 209, 211, **212**, 220, 221
ecosystem services 29
ecosystem-level management 29–30, 32, 45–46, 170, 180
ectomycorrhizal fungi 40
Ecuador 116, 117, 343, 345
education 248, 249
effector molecules 14–15, 83–84
Elaeis guineensis 116, 117–118, 123
elder/elderberry, black (*Sambucus nigra*) **55**, 215
ELISA tests **59**, 60, 574
elm (*Ulmus* spp.) 3, **53**, 80, 82, 83, 492, **493**, 504
American (*U. americana*) 78, 79, 86, 266, 276
cedar (*U. crassifolia*) **77**, 108
English (*U. procera*) **77**, 85, 266–267, 268
European white (*U. laevis*) **77**, 85, 266, 267, 276–277, 505
field (*U. minor*) 266, 267, 268, 276, 277
foliar diseases of 505, 506
Himalayan (*U. wallichiana*) 257, 276
planting strategies 268–269
Siberian (*U. pumila*) **77**, 85, 267, 276, 277
wych (*U. glabra*) 266, 267, 268, **498**, 505
elm bark beetles (*Scolytus* spp.) 259, 265, 267, 268, 275, 278
elm black spot (*Stegophora ulmea*) 491, **493**, 495, 505, 506
elm yellows disease (EY) **77**, 78, 85–86, 89
Elytroderma needle cast (*Elytroderma deformans*) 458, 470
EMARaV (*European mountain ash ringspot-associate virus*) 50, 54, 56, **56**
emerald ash borer (EAB, *Agrilus planipennis*) 85, 89
endoconidia/aleurioconidia 231, 236
Endocronartium harknessii 379

Endomelanconiopsis spp. 299
reproduction of 299
endophytes 7, 9, 298, 300–301, 302, 304, 307, 309, 448, 459, 498
Endothia spp.
E. gyrosa 340–341, 346, 348, 350, 351
E. singularis 341
Engelmann spruce (*Picea engelmannii*) **44**, 179, 473
engraver beetles (*Ips* spp.) 263, 378, 379
Enterobacter spp. 79, 82
Entoleuca mammata see Hypoxylon canker
entomopathogenic organisms 122
enzymes 183, 184, 364, 497–498
epidemiology of forest diseases 1–25
baseline mortality 11, **12**
co-evolution 13, **14**
density dependence of mortality 10–11
disease cycles *see* disease cycles
disease triangle/pyramid *see* disease triangle/pyramid
forest decline concepts *see* forest decline
and forest disease management 31–33
genetic determinants 13–15
host range 4, 5
in managed forests 8–10, **9**, **10**
modelling in *see* disease epidemic models
in natural forests 10–11
primary/secondary spread 21–22
resting stages/survival structures 4–5
epiphytic phase 5, 300–301
episphaeropsidone **305**
EPPO (European and Mediterranean Plant Protection Organization) 36, 121, 188, 244, 367, 468, 478, 556
EPS (exopolysaccharidal slime) 84
eradication strategy 33, **34**, 36–38
hazard/danger tree management 38
inoculum reduction/sanitation 36–38
Eremolepidaceae 97
Ernobius punctulatus 379
Erwinia spp. 80
E. amylovora 592, 599, **599**, **600**
Erysiphales see powdery mildews
Estonia 444, **596**
ethephon 108–109
ethyl acetate/alcohol 125, 236
Eucalyptus spp. 40, **98**, 161, 168, 529, 601
cankers affecting 341–344, 346, 347, 350–351, 352
Ceratocystis spp. in 233, 236, 239, 241–242, 243, 247–248, 249
E. coccifera 344
E. deglupta 343
E. diversicolor 170
E. doradensis 343
E. globulus 504, 508, 510
E. grandis 248, 343, 550
E. marginata 39, 42, 519, 520, 524, 526, 530–531, **535**, 601
E. nitens 298, 499, 500, 508, 509
E. pilularis 508–509
E. regnans 163
E. urophylla 248
E. wandoo 161
foliar diseases of 490, 499, 503–504
rusts affecting 548, 549, 554
Eurasia 79, 89, 116, 128, **147**, 263
heart rot in **199**, **200**, **201**, **202**, **203**, **204**
Europe 30, 36, 37, **55**, **56**, 83, 89, 98, **98**, 162, 347, 578, 592, 596, **597**, 598
annosus rots in 128, 129, 137, 138, **139**, 142, **147**, 224
Armillaria disease in 160, 161, 162, 163, 171
blackstain root disease in 285
branch/tip blights in 421, 422–423, 424, 427
Ceratocystis spp. in 230, 233, 243, 244, 247, 249
chestnut blight in 318, 319, 323, 324, 325, 326, 327–333
cypress canker in 359, 360
DNB in 436, 437, 438, 577
Dutch elm disease in 257–258, 271, 277
foliar diseases of broadleaved trees in 490, 491, 504–505
foliar diseases of conifers in 460–461, 464, 466, 468, 471, 472, 474, 475, 476, 478, 479
heart rots in **199**, **200**, **201**, **202**, **204**, 205, 207, 208, 219, 223, 224
Hypoxylon canker in 407
larch canker in 392, 399, 400–402
mistletoes in 98, **98**, 106
oomycete diseases in 520, 521, 522, 523
Ophiostoma diseases in 263, 265, 266, 267
phytoplasma diseases in 85–86, 573
pitch canker in 377
rusts in 547, 555, 560
seed diseases in 569
tomentosus root rot in 179
European ash (*Fraxinus excelsior*) **55**, **77**, **212**, **213**
European Union (EU) 138, 249, 324, 366, 368, 428–429, 604
European white birch (*Betula pendula*) 303
European white elm (*Ulmus laevis*) 85
EY phytoplasma group 83

Fagales 340–341
Fagus spp. **51**, **199**, 206, 216, 223, 521, 580
F. grandifolia **12**
F. sylvatica 146, **212**, **214**, 215, 489, **498**
feathery mistletoes (*Misodendrum* spp.) **98**, 103, 105
feeder root diseases 41–42
Fennoscandia 129, 138, **139**, 421

fertilizers 33, 39, 41–42, 107, 170, 369, 387, 412, 427, 503–504, 568
fig (*Ficus* spp.) **98**
Finland 33, 39, 54, **56**, **139**, 144, 146, 149, 163, 271, 407, 579, **596**
  DNB in 442–443, 444, 577
fir (*Abies* spp.) 39, **51**, **98**, 208, 209
  balsam (*A. balsamea*) 422, 474
  foliar diseases of 459, 460, 471, 473–475
  grand *see Abies* spp., *A. grandis*
  Macedonia/Bulgarian (*A. borisii-regis*) 162, 475
  Pacific silver (*A. amabilis*) **44**, 475, 573
  pinsapo (*A. pinsapo*) 130, 135
  silver (*A. alba*) 106, 128, **133**, 136, **139**, **147**, 162, 221, 222, 223, 474
  subalpine (*A. lasiocarpa*) **44**, 162, 180, 286, 475
  Todo (*A. sachalinensis*) 135, 137, 422
  white (*A. concolor*) **44**, 130, 134, 146, **147**, 162, 474–475, 573
fir broom rust (*Melampsorella caryophyllacearum*) 211, 556
fire 11, 29, 44–45, 138, 162, 350, 555
  as disease management strategy 39, 40, 42, 121, 171, 191, 468
  mistletoes and 106–107
firetree (*Morella faya*) 344
firewood 36, 37, 238, 245, 248, 271, 428
Fistulina hepatica **200**, 222
Flavescence Dorée (FD) 78–79, 85, 86
flooding 29
Florida (USA) 80, 344, 346, 376, 548, 550
foliar diseases of broadleaved trees 488–511, 548
  attachment/penetration in 496–497
  avoidance strategy for 504
  and climate change/air quality 501, 502–503
  colonization in 497–498
  diagnosis of 492
  environmental factors in 501–504
  epidemiology of 498–504
  eradication strategy for 505–506
  exclusion strategy for 504–505
  fungicides for 510, 511
  host resistance to 508–510
  host susceptibility to 499–500
  IDM for 510–511
  infection biology of 492–498
  insects and **498**
  management of 504–511
  nutrition/fertilization and 503–504
  overview **493–494**
  pathogen characteristics and 500–501
  pathogens/significance/distribution 488–492
  protection strategy for 506–508
  secondary infection 496
  therapy strategy for 510
foliar diseases of conifers 458–480, 549
  climate factor in 459–460
  control of 39
  needle pathogens 458–460
  needle pathogens, distribution of 460–461
  pathogens/significance 458–461
  of pines *see* pine
  see also DNB; *and see specific conifers*
Foliocryphia eucalypti 344, 346
Fomes fomentarius **199**, 206, 211, **212**, 215, 221
Fomitiporia spp. 220
  F. robusta 197, **198**, **199**, 206, 211, 217, 218–219, 220, 221
Fomitopsis spp.
  F. cajanderi 35, **201**, 207, **212**, 220, 222
  F. officinalis **199**, 207
  F. pinicola 35, 39, **199**, 206, 207, 211, **212**, 215, 216, 217, 219, 221
forest debris/litter 5, 29, 130, 205
  see also dead wood
forest decline 11–13
  arguments against 13
  four types of 12–13
  three factors in 12
forest disease management 29–46, 592–605
  avoidance strategy 33, **34**, 35
  and climate change *see* climate change
  disease detection/surveying/monitoring 30, 45
  economics of 598–601
  epidemiological aspects of 30–33
  eradication strategy 33, **34**, 36–38
  exclusion strategy 33, **34**, 35–36
  global challenge of, overview 592–594
  integrated strategy *see* IDM
  interdisciplinary approach to 602–603
  and land use/silviculture 601–602
  modelling and *see* disease epidemic models
  network approach to 595–597, **596**, 604–605
  passive 34, **34**
  protection strategy 33, **34**, 38–42
  regulation of *see* regulation of forest diseases
  resistance strategy 33, **34**, 42–43
  risk assessments and 36, 271, 384, 551, 604
  stakeholders in 597–598
  strategies, overview of 33–35, **34**, 45–46
  therapy strategy 33, **34**, 43
forest diseases 29–30
  epidemiology of *see* epidemiology of forest diseases
  slow/delayed development of 31
forest management practices 8–10, **9**, 307–309
  modelling 22–23, **22**
forest nurseries 31, **34**, 36, 67, 244, 249, 271, 377, 530
  biological control in 39–40
  control of branch/twig blights in 429, 430
  control of cypress canker in 368, 369, 372

forest nurseries (*continued*)
  DNB in 577–578
  foliar diseases of broadleaved trees in 499
  foliar diseases of conifers in 458, 464, 465, 467, 475
  fumigation in 38
  grey mould in 571–572
  hygiene in 567
  IDM in 45, 567–568
  needle casts in 578–579
  oomycete diseases in 520, 527
  Phytophthora diseases in 572–574
  powdery mildews in 579–580
  rusts in 568, 580–582
  Scleroderris canker in 568, 575–576
  seedling diseases in 570–571
  snow blights in 578
  Sphaeropsis tip blight in 568, 576–577
forest surveys 30
forest tent caterpillar (*Malacosoma disstria*) 415
forest vegetation simulator (FVS) 32, **32**, 33
formosan sweetgum (*Liquidambar formosana*) 350
France 25, **77**, 162, 163, 167, 170, 171, 233, 241, 242, 359, 372, 384, 402, 422, 560, **593**, **596**
  chestnut blight in 318, 326, 327, 328, 329
  DNB in 438, 577
  oomycete diseases in 522, 526
Fraxinus spp. 50, **51–52**, 56, 80–81, 490–491, 580, 592
  F. americana 78, 80
  F. angustifolia 501
  F. anomala **77**, 89
  F. excelsior **55**, **77**, **212**, **213**, 215, 501
  F. pennsylvanica 78, 80
  F. uhdei **77**
  F. velutina **77**, 80, 89
  heart rots in **202**, 205, 208, 216, 221
  phytoplasma diseases of **77**, 78, 85
fruit trees 490
  Armillaria disease in 160
  and bacteria 82
  and mistletoes 98, **98**
  and viruses 50, 54, 67
fumigation 37, 38, 40, 121, 189, 191, 385–386, 398–399
fungal-like organisms 30, 38, 64–65
fungi 1, 30, 301–304, 307–309
  endangered species 205
  inoculum of 2–3
  transmission of viruses by 64–65
fungicides 18, **34**, 39
  aerial spraying of 109, 122, 414, 437, 445–446
  natural 40
  regulations for use of 368
Fusarium spp. 38, 378, 475, 521, 568–569, 570
  F. circinatum see pitch canker
  F. proliferatum 378
  F. solani 235
Fusicoccum spp. 299
fusiform rust (*Cronartium fusiforme*/*C. quercuum*) 43, 549, 554, 555, 558, 560, 561
FVS (Forest Vegetation Simulator) 32, **32**, 33

gall midge 304
galls, bacterial 80, 83, 84
Ganoderma applanatum **204**, **213**, 216
GCPSR (genealogical concordance phylogenetic species recognition) 299–300
GenBank 300
genetic determinants of pathogenicity 13–15
  effector molecules 14–15
  PAMP recognition 14, 15
genetic diversity 9
genetically modified plants 43
genets 133, 134–135, 167, 185, 217, 218
Georgia 327, 384
Georgia (USA) 87, 258
Germany **57**, 63, **77**, 163, 206, 274, 400, 402, 407, 421, 441, 461, 489, 521, **593**, **596**
  chestnut blight in 318, 327
Ghana 341, 342, 345
Gibberella spp.
  G. circinata see pitch canker
  G. subglutinans 377
girdling trees 245
GIS (geographic information systems) 30, 32, 428, 464
Gliocladium spp. 119, 572
  G. virens 40
global plant trade 8, 36, 511, 594, 595–597, **596**, 603
global timber trade *see* timber/timber products industry
Gloeophyllum spp. **213**, 219–220
  G. sepiarium 219
Glomerella spp. 490, 495, 496, 498, 500
  G. cingulata 110, 491, 495, 501
Gmelina arborea 234
Gnomonia spp. 423, 495, 498, 499
  G. leptostyla 502, 503, 509
golden larch (*Pseudolarix* spp.) 392, 398
  Chinese (*P. amabilis*) 399
golf courses 88, 123, 124, 282, 383
GPS (global positioning systems) 30, 31, 428
grafting/budding 54, 62, 301, 308, 310
grape/grapevine (*Vitis*) 62, 78, 79, 82, 86, 87, 89, 169, 300, 308
Great Lakes (USA/Canada) 257, 407, 410, 411, 413
Greece 129, 233, 239, 242, 244, 245, **593**
  chestnut blight in 326, 327, 328, 329
  cypress canker in 359, 360, 365, 366, 367
green ash (*Fraxinus pennsylvanica*) 78, 80
Gremmeniella spp.
  G. abietina see Scleroderris canker
  G. juniperina 421
  G. laricina 421

grey mould (*Botrytis cinerea*) 39, 117, 491, 501, 508, 568, 571–572
control of 572
Grifola frondosa **204**, 216, 222
Grosmannia spp. 257, 258, 283
G. alacris 284, 286, 287
G. huntii 284, 287
G. serpens 286–287
gumhar (*Gmelina arborea*) 234
gummosis 523
gymnosperms 76, 79, 86
Gymnosporangium spp. 552, 556
gypsy moth 33

Haddow-Etheridge concept 197, 198, **199–200**, 211, 215
Hapalopilus croceus **201**, 222, 223
hardwoods 162
decline 33
mistletoe infection of 99, **100**
susceptibility to disease of 41, 43, **44**
Harknessiaceae 347
Hartig concept 197, 198, **200–202**
'hartrot' 118
Hawaii (USA) 98, 341, 344, 548
hazel (*Corylus avellana*) 215, 216, 489
heart rot 5, 8, 38, 129, 197–224, **198**
access to hosts of 217–218
classification of 197–205
diagnosis of 209–210, **210**
four infection concepts of 197–198, 205
fruiting bodies 209, 219, 220–221
habitats of 206
host range and 221–222
host resistance to 222
impact of 206–209
infection biology/epidemiology of 211–222
insect vectors of 217
management of 222–224
molecular detection of 209–210
pathogens 205
root/butt **203–204**, 206–208, 216–217, 222–223
saprotrophs/sap rots and 217, 219–220
stem colonization by 218–220
'true' 198, **199–202**, 206–208, 211–215, **212**, 217–219, 221–222, 222–223
'wound' 198, **202–203**, 205, 206, 208–209, **212–213**, 215–216, **216**, 218, 219, 222, 223–224
Hemiptera 81
hemlock *see Tsuga* spp.
Hendersonia pinicola 470
herbicides 107, 108–109, 122, 169
Hericium spp. 216
H. abietis 35, 205
H. erinaceus 216
Herpotrichia juniperi 461, 469–470, 479, 578
Hesperocyparis spp. 359, 360
H. arizonica **305**, 369
H. lusitanica 360
H. macrocarpa **305**, 359, 360, 366, 368
Heterobasidion spp. 3, **22**, 30, **32**
H. abietinum 128, **132**, 135, 136, 146
H. annosum see annosus root/butt rots
H. irregulare 128, 129, **132**, 136, 137, **140**, 142, 146, 592
H. occidentale 128, 129, 134, **140**, 144, 146
H. parviporum **22**, 128, 129, **132**, 135–136, 137, 146
Heteropyxis canescens 345
Hevea spp. **98**, 164
H. brasiliensis 170, 232, 522
hickory bark beetle (*Scolytus quadrispinosus*) 232
Himalayan region 99, 104, 128, 257, 263
hinoki false cypress (*Chamaecyparis obtusa*) 163
Holocryphia eucalypti 343–344, 346, 347, 348, 350, 351
homothallism 160, 161, 211, 235, 236
Honduras 117, 123, 460
hormones 58, 101, 499, 553
horse chestnut (*Aesculus hippocastanum*) 521
host resistance 5–6, 7, **34**
breeding 30, 42–43
broken 9–10, 15
and genetic uniformity/diversity 7–8, 9–10, **10**, 18, **34**
to nematode diseases 123
to prokaryotes 88–89
systemic induced (SIR) 43, 383, 429
tree species selection 43
human factors in forest diseases 3, 8, 11, 12, 30, 31, 36, 120–121, 427
mechanical damage to trees **202–203**, 206, 223–224, 240
humidity **2**, 7, 41, 135, 137, 365, 396, 402, 426, 440, 507, 572
Hungary **77**, 318, 328, 438, 445, 577, **593**
Hyalorhinocladiella spp. 257, 258, 259, 283
hydrophobins 496–497
hydroxymellein **305**, **306**
hygiene 36, 244, 245, 308, 310, 399, 430, 567
see also sanitation
Hylastes spp. 285, 287, 291
H. angustatus 287
H. nigrinus 284
H. salebrosus 286, 287
H. tenuis 286, 287
Hylobius spp.
H. abietis 224
H. pales 285, 286, 287
Hylurgopinus rufipes 259, 265, 267, 272, 273
Hymenoscyphus pseudoalbidus (anamorphic stage *Chalara fraxinea*) 501, 506, 592

hyperplasia 302
Hypholoma spp. 39
H. australe 170
Hypodermataceae 472
Hypodermella laricis 461, 475, 476
Hypodermina hartigii 472
hypoviruses 326–331, 332, 333
artificial introduction of 328
effectiveness of 328–330, **330**
genetic diversity/distribution of 326, 327
molecular approaches to 330
natural, occurrence of 327, **328**
transmission of 326–327
Hypoxylon canker (*Entoleuca mammata*) 407–415
avoidance strategy for 411
diagnosis of 408
environmental conditions for 409, 410
epidemiology of 409–410
eradication strategy for 411–412
exclusion strategy for 411
host resistance to 412–413, 490–491
IDM for 414–415
infection biology of 408–409
insects and 409, 410, 414, 415
management of 410–415, **600**
pathogen/significance/distribution 407–408
Populus species affected by 407–408
protection strategy for 412
therapy strategy for 413–414
yearly fluctuations of 410

ICTV (International Committee on Taxonomy of Viruses) 50, 57
Idaho (USA) 41, 162, 463
IDM (integrated disease management) 33, **34**, 40, 44–46, **45**
in forest nurseries 45
silviculture and 43
see also specific diseases/pathogens
Illinois (USA) 78, 240, 268
Immersiporthe knoxdaviesiana 346
incense cedar (*Calocedrus decurrens*) 128, **147**
increment borer 131, 138
India **98**, 99, 106, 341, 347, 436, 449, 506
Indian rosewood (*Dalbergia sissoo*) 234
Indonesia **98**, 161, 341, 344, 347, 576
ink disease 325, 333, 519, 520, 521, 534, **535**
inoculum 21, 30, 40, 301, 307, 308
build-up 16, 135
potential 2–3
primary/secondary 166
reduction 18–19, 37, 143, 189
Inonotus spp.
I. andersonii 205
I. dryophilus **201**, 217, 222
I. glomeratus 205
I. hispidus 205
I. leporinus 8, **204**, 207, 217, 222
I. obliquus **199**, 220
insecticides 88, 123, 125, 272, 278
insects 29, 42, 44, 87, 104, 256, 303, 304, 307, 333
and *Ceratocystis* spp. 230–231
defoliation by 12
heart rot transmission by 217, 219
nematode transmission by 115, 118–119
phytoplasma transmission by 83, 84–85
virus transmission by 64
see also aphids; beetles; *and see specific tree species/diseases*
integrated disease management *see* IDM
invasive species 23
Iowa (USA) 85
*i*PhyClassifier 81
IPPC (International Plant Protection Convention) 36
Ips spp. 263, 378, 379, 381
I. pini 287
I. sexdentatus 379
I. typographus 217, 219, 272, 285
Iriartea deltoidea 304
IRPCM (International Research Program on Comparative Mycoplasmology) 76, 78
irrigation 88, 308, 310, 507, 530, 568
isoseiridin 364
ISPM (International Standards for Phytosanitary Measures) 121
Isthmiella faullii 474
Italy 63, **77**, 79, 85, 162, 265, 286, 376, 384, 407, 592, **593**, **596**
annosus rots in 129, 130, 136, 137, 142, 148–149
branch/tip blights in 422, 427
Ceratocystis spp. in 233, 239, 241–242, 243, 244
chestnut blight in 318, 327, 329, 331
cypress canker in 359, 360, 365–366, 368, 369, 371, 372
ITS (internal transcribed spacer) 299, 300
Ivory Coast almond (*Terminalia ivorensis*) 345

Janzen-Connell hypothesis 10–11, 520–521
Japan 115, 118, 120, 121, 122, 123, 124, 128, 137, 178, 346, 360, 422, 593
Armillaria disease in 160
chestnut blight in 318, 323, 325
foliar diseases of conifers in 460, 464, 467, 468, 474, 477
larch canker in 392, 393
Ophiostoma diseases in 257, 258, 259, 263
pitch canker in 376, 377, 568
Japanese cedar (*Cryptomeria japonica*) 171, 479
Japanese pine sawyer (*Monochamus alternatus*) 115, 118, 119, 120, 121, 122, 124
Japanese zelkova (*Zelkova serrata*) **77**, 85, 86

jarrah (*Eucalyptus marginata*) 39, 42, 519, 520, 524, 526, 530–531, **535**, 601
Juglans spp. 60, **77**, **98**, 488, 490, 492, **494**, 504
J. nigra **77**, 503, 509
J. regia 166
juniper witches' broom **77**
Juniperus spp. 128, 130, **147**, 359, 360, 469, 479, 578
branch/tip blights affecting 421, 576
J. occidentalis **77**, 79, 86, **140**
J. virginiana 450

Kabatina spp.
K. abietis 475
K. thujae 359, **600**
karri (*Eucalyptus diversicolor*) 170
Kentucky (USA) 80
Kenya 162, 342, 345, 360, 442, 444, 449
Klebsiella spp. 79, 82
K. oxytoca 86
Koch's postulates 11
Korea 124, 163, 259, 263, 460, 464, 477
Kretzschmaria deusta 205

lacebug (*Corythucha ciliata*) 247
Lachnellula spp. 393
L. arida 393, 394
L. flavovirens 393, 394
L. occidentalis 393, 394–395
L. willkommii see larch canker
Laetiporus spp.
L. conifericola **201**
L. cremeiporus **201**
L. gilbertsonii **201**
L. huroniensis **201**
L. montanus **201**, 207
L. sulphureus **201**, 206, 217–218, 222
L. versisporus **201**
laminated root rot (*Phellinus sulphurascens*) 7, 22, 33, 38, 43, 178–192
antagonistic fungi and 189–190
avoidance strategy for 187–188
and bark beetles 180
diagnosis of 138, **140**, 180–182
ecosystem classification system for 187
epidemiology of 183–187
eradication strategy for 188–190
exclusion strategy for 188
fire and 191
fruiting bodies of 181
host resistance to 179, 183, 190–191
hyphae of 182–183
IDM for 191–192
longevity of 181, 183–184
management of 33, 34, 35, 37–38, 39, 40, 42, 43, 143, 171, 187–192
partial cutting and 192
pathogen/significance/distribution 162, 178–179, 180
and *Porodaedalea pini*, compared 181, 182
protection strategy for 190
saprophytic phase 179, 183, 192
soil conditions and 186–187
species susceptibility to **44**
stand factors in 185–186
survey methods for 187
taxonomy of 178
tree vigour and 186
LAMP (loop-mediated isothermal amplification) 82
larch (*Larix* spp.) 35, **77**, 79, 183, 200, **201**, **204**, 424, 551, 560, 578
branch/tip blights affecting 421, 424
Dahurian (*L. gmelinii*) 398–399
European (*L. decidua*) **139**, 146, **147**, 399, 400–401, 437, 476, 478, 579
foliar diseases of 459, 460, 461, 475–478
hybrid 400, 402
Japanese (*L. kaempferi*) 163, 399–400, 402, 476, 477, 478
tamarack (*L. laricina*) 478
western (*L. occidentalis*) 43, **44**, 171, 190, 399, 579
see also golden larch
larch bark moth (*Cydia zebeana*) 395
larch canker (*Lachnellula willkommii*) 392–402
avoidance strategy for 397–398
diagnosis of 393–395
environmental conditions for 395, 396–397
epidemiology of 396–397
eradication strategy for 399
exclusion strategy for 398–399
fruiting bodies of 394
fumigation for 398–399
host resistance to 399–402
host species susceptibility to 399–400
IDM for 402
infection biology of 395–396
insects and 395
management of 397–402
pathogen/significance/distribution 392–393
protection strategy for 399
therapy strategy for 402
larch dwarfed needle proliferation **77**
larch needle cast (*Meria laricis*) 461, 475, 476, 477, 578, 579
Lasiodiplodia spp. 299, 304
L. theobromae 234, **306**, 309
lasiodiplodin **306**
Latruncellus arorae 345
laurel
Alexandrian (*Calophyllum inophyllum*) 287–288
California (*Umbellularia californica*) 524, 527–528, 529

laurel wilt (LW, *Raffaelea lauricola*) 230, 258, 259, 263, **264**, 269
diagnosis of 261
infection/transmission 265, 266
management of 270, 271, 272, 273, 278
LChD (Little cherry disease) 54–56
leaf blister (*Taphrina* spp.) 488, 489, 492, 495, 496, 497, 500, 506
leaf scorch (*Xylella fastidiosa*) 79–80, 82, 84, 86–88, 233
leafhopper (*Macrosteles quadrilineatus*) 84
leafhoppers (Cicadellidae) 85, 87
Lecanosticta acicola 439
Leptographium spp. 119, 230, 257, 258, 262, 283–291
diagnosis of 288–289
epidemiology of 290–291
infection biology of 289–290
inoculum of 289
L. calophylli 287–288
L. procerum 284, 285, 289
L. terebrantis 284, 286, 290
L. wageneri see blackstain root disease
pathogens/significance/ distribution 283–288
Leptosphaeria sp. 367, 478
Leptostroma austriacum 464–465
lidar (light detection and ranging) 30, 428
lignin 5, 179, 448
lilac (*Syringa* spp.) 81
Liquidambar spp. 303
L. formosana 350
Liriodendron tulipifera 207
Lirula spp.
L. abietis-concoloris 460, 473
L. macrospora 460, 471, 472, 473
L. nervata 473, 474
L. nervisequa 460, 461, 473, 474
Lithocarpus densiflorus 519, 520, 524, 529, 533, 535, **535**
Lithuania **77**, 79, 146, **593**, **596**
Little cherry disease (LChD) 54–56
'little leaf syndrome' 117–118, 123
littleleaf disease *see Phytophthora* spp., *P. cinnamomi*
loblolly pine (*Pinus taeda*) 43, **147**, 163, 171, 284, 285, 286, 287, 468, 469, 549, 560, 570
DNB in **442**, 448
pitch canker in 377, 379, 383, 386
lodgepole pine (*Pinus contorta*) 13, **14**, **44**, 162, 190, 284, 285, 286, 470
DNB in 436, **442**, 448, 577
logging 35, 39, 141–142, 148, 265
and heart rots 205, 206, 208–209, **208**, 215
longhorn beetles (Cerambycidae) 263
Lophodermella spp. 448, 459, 470
L. arcuata 470
L. cerina 470
L. concolor 470
L. conjuncta 461, 470
L. montivaga 470
L. sulcigena 8, 461, 471
Lophodermium spp. 460, 479
L. conigenum 464
L. crassum 460, 472
L. juniperinum 479
L. laricinum 475, 476
L. piceae 471, 472–473
L. pinastri 461, 464, 465
L. seditiosum 458, 459, 460–461, 464–466, 578–579
Lophodermium needle cast (*L. seditiosum*) 458, 459, 460–461, 464–466, 578–579
diagnosis of 464–465
infection biology/epidemiology of 465
management of 465–466
Loranthaceae 97, 98, **98**, 101, 104, 106
Louisiana (USA) 80, 232
lucerne yellows disease 88
Luperus pinicola 464
LVS (Landscape Visualization System) 32–33

Macropsis mendax 85–86
Macrosteles quadrilineatus 84
Magicicada septendecim 415
Maine (USA) 64, 489
mal de machete 232
Malacosoma disstria 415
Malaya 161
Malus spp. **98**, 556, 592
M. domestica **212**, **214**
M. pumila **305**
managed forests 29
epidemiology of diseases in 8–10, **9**, **10**
mancozeb 368, 463, 466, 468, 506, 575, 579
mango (*Mangifera indica*) 234, 241
Manion spiral 13
maple (*Acer* spp.) **51**, 65, 82, 205, 216, 240, 492, 521
foliar diseases of 489, 490
hedge (*A. campestre*) **212**, **214**, 215
red (*A. rubrum*) 503
sugar (*A. saccharum*) 231, 243
maple mosaic virus 50, **51**
marbled murrelet (*Brachyramphus marmoratus*) 104–105
maritime pine (*Pinus pinaster*) 163, 171, 286, 467, 557, 569
Marssonina brunnea 489, **493**
Massachusetts (USA) 286
melaleuca (*Melaleuca* spp.) 303, 548
Melaleuca quinquenervia 303, 550
Melampsora spp. 43, 459, 475–476, 551, 552, 580–581
M. aecidioides 552
M. epitea 547

M. laricis-populina see poplar leaf rust
M. medusae 547, 551, 559, 560, 561, 580, 581
M. occidentalis 547, 549, 559, 580, 581
M. populnea 7, 21, 547–548, 557, 580, 581
Melampsorella spp. 548, 549, 552
M. caryophyllacearum 211, 556
Melampsoridium spp. 548
M. betulinum 476, 581–582
Melastomataceae 346
mellein **305, 306**
Meloderma desmazieri 461, 465, 471
Mensularia radiata **199**, 208, 211, **212**
Meria laricis 461, 475, 476, 477, 578, 579
Merulius tremellosus 219
metalaxyl 532, **532**, 533
Metamasius hemipterus 116
methyl bromide 38, 40, 121, 398
methyl jasmonate 387–388
methylisothiocyanate 189, 191
Mexico **98**, 116, 117, 128, **199**, **201**, 231, 344, 376, 523, 576
pitch canker in 376–377, 569
Michigan (USA) 85, 242, 407, 412, 421
Microstroma spp. 492, 500
M. juglandis 488, 490, **494**, 495
Microthia spp. 348
M. coccolobae 346, 347
M. havanensis 344, 348
mildews 423, 503, 507, 560
powdery *see* powdery mildews
Minnesota (USA) 232, 238, 239, 240, 242, 244, 248, 407, 414
Misodendraceae 97, 101, 103
Misodendrum spp. **98**, 103, 105
Missouri (USA) 171, 232, 237, 238, 240
mistletoes 1, 7, 97–110
avoidance strategy for 105
benefits of 104–105, 110
biological control of 109–110
bird-dispersed 101, 102–103, 105–106
chemical control of 107, 108–109
and Christmas ornament industry 104
diagnosis of 100–101
dwarf *see* dwarf mistletoes
economic impacts of 98–99, 110
endangered species 104
epidemiology of 102–104
eradication strategy for 106–107
exclusion strategy for 105–106
explosive seed discharge by 102, 103–104
fire and 106–107
herbicides for 107, 108–109
host resistance to 97, 107–108, 110
IDM for 110
infection biology of 101–102
loranthaceous 98
management of 104–110
pathogens/significance/distribution 97–99
pollination of 104
protection strategy for 107
rating systems for 99, **99**, **100**
seed characteristics 101–102
taxonomy of 97, **98**
therapy strategies for 108–110
wind-dispersed 102, 103
moisture stress 302–303, 308
molecular analysis 30
see also PCR analysis; RAPD
Monilinia fructicola 117
Monochamus spp. 116, 118–120, **118**, 121
M. alternatus 115, 118, 119, 120, 121, 122, 124
M. carolinensis 118, 120
M. galloprovincialis 118, 120
monocultures 7, 137, 163, 350, 420–421, 500, 508, 601, 604
monocyclic/polycyclic diseases 16
Monterey cypress (*Hesperocyparis macrocarpa*) **305**, 359, 360, 366, 368
Monterey pine (*Pinus radiata*) 5, 8, 23, 24, 39, 41–42, **147**, 286, 422, 592, 601
Armillaria disease in 163, 169, 172
DNB in 436, 440, 442, **442**, 443, 444, 448, 449–451, 452, 578
foliar diseases of 461, 466, 467, 468
oomycete diseases in 522, 534, 535
pitch canker in 377, 379–380, 382–383, 385, 386–387, 388, 569
moose (*Alces alces*) 224, 268
Morella faya 344
moths/moth larvae 475
mountain ash (*Sorbus aucuparia*) **53**, 54, **55**, **56**, 303
mountain hemlock (*Tsuga mertensiana*) 7, 179, 185, 187, 286
Mozambique 352
Mueller-Dombois model 13
mustard (*Brassica* spp.) 40
mycelia 3, 131–133, 159, 301–303, 308, 310
ectotrophic 181, 183, 185, 190
fans of 165–166
homokaryotic/heterokaryotic 135, 211–215
myclobutanil 480, 580, 582
Mycoacia uda 220
mycoherbicides 110
mycorrhizal symbionts 7
mycorrhizas 39, 40
Mycosphaerella spp. 400, 489, 490, 496, 499, 500, 508–509, 510
M. cryptica 501, 503–504, 506
M. dearnessii see brown spot needle blight
M. gibsonii see Cercospora needle blight
M. laricina 461, 475, 476–477
M. laricis-leptolepidis 460, 477, 478
M. pini 7, 438–439, 460, 464, 592, **600**

mycoviruses 148
Myrica gale 13, **14**
Myrtaceae 161, 548
Myrtales 342, 345, 346
myrtle wilt 231

NCBI (the US National Centre for Biotechnology Information) 300
Nebraska (USA) 441, 443, 450
necrophylactic periderms 183, 370
necrotrophic pathogens 4, 6, 14, 15, 497
needle blight 5, 25, 475
needle casts 459–460, 475, 479, 568, 578–579
nematicides 123–124
nematodes 1, 30, 38, 115–125
  avoidance strategy for 120
  diagnosis of 117–118
  epidemiology of 119–120
  eradication strategy for 121–122
  exclusion strategy for 120–121
  extraction techniques for 117
  IDM for 124–125
  infection biology 118–119
  management of 120–125
  pathogens/significance/distribution 115–117
  physical characteristics of 116, 117
  protection strategy for 122–123
  resistance strategy for 123
  taxonomy of 115–116, 117
  therapy strategy for 123–124
  transmission of 115
  transmission of viruses by 64
  see also pine wilt nematode/disease
Nematostoma parasiticum 473, 474
Neodeightonia spp. 304
Neofusicoccum spp. 299
  N. parvum 300, **306**
  N. ribis 300, 303
Neonectria spp. 570, 592
  N. fuckeliana 24
  N. neomacrospora 110
Neopeckia coulteri 460, 469–470, 479
Neoscytalidium spp. 299
Nerium oleander 79
Netherlands 268, 274, **596**
network theory 595–597, **596**, 604–605
New Jersey (USA) 79–80, 240241
New South Wales (Australia) 109
New York (USA) 64, 79, 80, 231, 287, 411, 414, 422
New Zealand 5, 8, 9, 23, 24, 41, 104, 257, 266, 273, 285, 343, 346, 347, 377, 422, 522, 535
  Armillaria disease in 161, 163, 167, 169, 172
  cypress canker in 359, 360, 366
  disease management strategies in 37, 39, 41–42, 142, 271, 384–385
  DNB in 436, 437, 438, 439, 443, 444, 445, 446, 447, 448, 449–450, 451, 452, 577–578
  foliar diseases in 461, 463–464, 467, 495, 504, 506
  rusts in 554, 555
Newfoundland 163, 427–428
NIDRM (National Insect and Disease Risk Map) 33
nitidulids 231, 236–237, 238, 239, 245
nitrogen fertilizers 41, 325, 369, 503
North America 30, 37, 39, 40, 43, **55**, 80, 89, 168, 369, 392, 592
  annosus rots in 128, 138, **140**, 144, 146, **147**
  Armillaria disease in 160, 162, 163, 170
  branch/tip blights in 421, 422–423, 425, 427–428
  cankers in 341, 346, 347
  chestnut blight in 318–319, 323, 324, 327, 329–330, 332, 333
  DNB in 436, 449
  foliar diseases in 460, 461, 464, 466, 468, 469, 470, 471, 474, 501
  heart rot in **199**, **200**, **201**, **202**, **204**, 216, 221, 223, 224
  laminated/tomentosus root rot in 178
  mistletoes in 99, 103, 104
  nematodes in 115, 116, 118
  oomycete diseases in 519, 523
  Ophiostoma diseases in 257, 258, 263, 265, 271, 272, 273, 276, 277
  phytoplasma diseases in **77**, 78, 85, 88
  rusts in 547, 549, 550–551, 555, 557, 559–560
  susceptibility to disease in 43, **44**, 56
North Carolina (USA) 80, 240, 341, 376
Norway 33, 271, 392, 396, 427
Norway spruce (*Picea abies*) 3, 9, **22**, 33, 268, 459, 473, 575, 578
  annosus rots in 7, 128, 129, 130, 133, **133**, 135–136, **139**, **141**, 144, **147**, 149–150, **150**
  Armillaria disease in 163, 170, 171
  branch/tip blights affecting 425, 426, 428
  cone rust in (*Chrysomyxa pyrolae*) 568
  heart rot in **208**, 209, **210**, **212**, **213**, 216, **216**, 219, 221, 222
Nothofagus spp. 103
  N. cunninghamii 231
nutrient cycling 29, 458
nutrient stress 302–303, 308

oak (*Quercus* spp.) 42, **53**, 79–80, 87, **98**, 129, 304, 319, 346, 350, 423, **494**, 580
  Armillaria disease in 160, 161–162
  black (*Q. velutina*) 207
  blackjack (*Q. marilandica*) 232, 238
  bur (*Q. macrocarpa*) 232, 248

California coast live (*Q. agrifolia*) 65, 523–524, 528, 532
chestnut (*Q. prinus*) 207
English (*Q. robur*) **57**, **77**, 162, 197, **212**, **214**, 215, **498**
heart rot in **199**, **200**, **201**, **202**, **203**, 205, 207, 209, 216–217, 220, 223
live (*Q. virginiana*/*Q. fusiformis*) 232, 234, 237, 238, 246
northern pin (*Q. ellipsoidalis*) 232
and Ophiostoma diseases 259
Oregon white (*Q. garryana*) 105
pin (*Q. palustris*) 80, 82, 341
red *see* red oak
rusts affecting 555
seed diseases of 568
white (*Q. alba*) 207, 232, 238, 248, 496, 523
oak decline **77**, 162, 171
oak pinhole borer (*Platypus cylindricus*) 239
oak ringspot virus 50
oak wilt disease (*Ceratocystis fagacearum*) 9, **10**, 36–37, 42, 82, 230, 231–232, 234, 236, 492
colonization by 237–238, 240, 242
diagnosis of 235
host resistance to 247
inoculum of 236
insect vectors of 236–237, 238–239
management of 243, 244–245, 246, 248
odour of 235
and root grafts 242
Oceania 467
ocote chino (*Pinus oocarpa*) 163
Olea 422
oleander (*Nerium oleander*) 79
Oligonychus ununguis 458
olive 54
Olpidium spp. 65
Oncopsis alni 86
Onnia spp.
O. circinata 179, 183
O. tomentosa see tomentosus root rot
O. triquetra 207
Ontario (Canada) 87, 180, 407, 414
ontogenic resistance 5
oomycete diseases 1, 519–536, 569, 570
avoidance strategy for 528–530, **528**
diagnosis of 522–524
epidemiology of 526–528
eradication strategy for 531–532, **532**
exclusion strategy for 529–530
host range and 527–528
host resistance to 534–535, **535**
human dispersal of 526
IDM for 536
infection biology of 524–526
management of 528–536, 574
molecular identification of 521–522, 524, **525**, 534
in nurseries 569, 570, 571
pathogens/significance/distribution 519–522
protection strategy for 532–534
pruning/thinning and 529
sanitation strategy for **530**, 531, 536
therapy strategy for 535
Ophiostoma spp. 36, 119, 230, 256–278, 283, 500, 592
avoidance strategy for 268–270
beetles and 256, 257, 258, 259, **260**, 262–263, 265–266, 267
biological products for 274
and *Ceratocystis* spp. 259, 262
diagnosis of 259–262
disease cycles **260**, **264**
dispersal 262–264
eradication strategy for 271–273
exclusion strategy for 270–271
fungicides for 273, 277, 278
host resistance to 276–277
host susceptibility to 266–267
IDM for 277–278
infection biology/epidemiology 262–268
inoculum 262
insecticides for 272–274, 278
log storage and 268, 270, 273, 275–276
management of 268–278
O. bicolor 258
O. himal-ulmi 257, 263
O. novo-ulmi 3, 5, 13, 18, 21, 230, 257, 258, 259, 262, 266, 268, 492
disease cycle of **260**, 274
insect vectors of 263, 265, 267
management of 270, 271, 273, 277, **600**
O. piceae 258, 262, 268
O. piliferum 258, 274
O. ulmi 230, 257–258, 259, 262, 266, 269
insect vectors of 263
management of 269, 270, **600**
pathogens/significance/distribution 256–259
protection strategy for 273–276
root graft transmission of 268, 274
taxonomy of 256–257
therapy strategy for 277
virulence/variability of 266
and wood moisture content 268
wound dressings for 274–275
and wounds/mechanical transmission 264–265, 268
see also blue-stain fungi; Dutch elm disease; laurel wilt
Ophiostomataceae 283, 289
orchards 37, 79, 166–167, 169, 324, 325, 350, 352
seed **57**, 123, 308, 310, 377, 379, 386, 509
Oregon (USA) 31–33, 39, 40, 41, 42, 44–45, **77**, 86, **140**, 287, 461, 530, 557

Oregon (USA) (*continued*)
  Armillaria disease in 162, 171
  IDM in 44–45, **45**
  laminated root rot in 185, 187
  mistletoes in 104, 105
  oomycete diseases in 522, 526, 528, 532
Oregon white oak (*Quercus garryana*) 105
ornamental trees 36, 38, 39, 54, 109, 233, 347, 350, 360, 370, 372
Orobanchaceae 97, 550–551
Ottawa (Canada) 79, 80
owls 32–33, 103
Oxyporus populinus 205
Ozarks (USA) 171, 232

Pachylobius picivorus 285, 287
Pachysacca spp. 488, 497
Pacific mistletoe (*Phoradendron villosum*) 105
paint treatments 108, 240–241, 243, 246, 248, 308, 310, 534
PAMPs (pathogen associated molecular patterns) 14–15
Pantoea spp. 80, 84, 87
paper birch (*Betula papyrifera*) 43
parasitic plants *see* mistletoes
parenchyma cells 302
PCR analysis 58, 60, 81, 82, 86, 117, 118, 131, **132**, 164, 182, 348, 424, 439, 462, 524, 551
pear thrips (*Taeniothrips inconsequens*) 497
Pedicularis spp. 550, 554, 557
Penicillium spp. 569
  P. glaucoalbidum 471, 473, 475
Peniophora spp.
  P. cinerea 219
  P. limitata **214**, 220
Pennsylvania (USA) 240, 242
Perenniporia spp.
  P. compacta 205
  P. subacida **204**, 222
periderm 302–303
Peridermium spp. 547, 550, 552
periwinkle (*Catharanthus roseus*) 89
Persea spp.
  P. americana **98**, 258, 301
  P. borbonia 258, **264**, 265, 267, 269
Pesotum spp. 257, 258, 260, 261, 262, 283, 289
Pestalotiopsis funerea 359, 475, 479
Phacidium spp.
  P. coniferarum 9
  P. infestans 5, 460, 461, 471, 473, 479, 578
  P. taxicola 478
Phaeobotryosphaeria spp. 299
Phaeocryptopus gaeumannii see Swiss needle cast
Phaeocryptopus needle cast 458
Phaeocryptopus nudus 474
Phaeolus schweinitzii 38, **204**, 207, 217, 218, 222
Phaeoseptoria eucalypti 490, 504
Phellinus spp. 220
  P. chrysoloma 8, **199**, 206, 207, 216, 221
  P. hartigii 8, **199**, 211
  P. himalayensis **200**
  P. ignarius **200**, 206, 221
  P. jezoënsis **200**
  P. nigricans **200**, 221
  P. populicola **200**, 221
  P. sulphurascens see laminated root rot
  P. tremulae **200**, 207–208, 209, 211, **212**, 215, 217, 221, 222, 410
  P. vorax 207
  P. weirii 178, 182
phenolic compounds 5, 304, 448
phenylamides 533, **533**, 535
pheromones 120, 125, 265
Philaenus spumarius 85
Phlebia radiata **214**, 219–220
Phlebiopsis gigantea 7, 18, 39, 143, **214**, 217
Phloeosinus spp. 362
Phoenix dactylifera 116
Pholiota adiposa **202**
Phomopsis spp.
  P. juniperivora 358
  P. occulta 358
phony peach disease 82
Phoradendron spp. **98**, 101, 103, 108, 109
  P. flavescens 108
  P. tomentosum 108
  P. villosum 105
phosphonates 532–533, **533**, 535
phosphorous fertilizers 41, 325, 503
photosynthesis 58, 288, 420, 463, 490
Phyllosticta minima 502
Phytophthora spp. 1, 3, 36, 38, 43, 500, 519–520, 522–530
  diagnosis of 522–524, **525**, 572
  epidemiology of 526–528, 573–574
  infection biology of 524–526
  management of 528–536, 574, 604
  in nurseries 567, 569, 570, 572–574
  P. alni 521, 523, 527, 530, 573
  P. cambivora 521, 522–523, **525**, **535**, **600**
  P. cinnamomi 4, 5, 7, 23, 25, 35, 519, 520, 573, 593
    diagnosis of 522–523, 524, **525**
    infection biology/epidemiology of 525, 526, 527
    management of 36, 39, 41–42, 529–530, 531, 532, 534, **535**, **600**, 601
  P. citricola 521–522, **525**, **600**
  P. cryptogea 531, 573
  P. kernoviae 522, **525**
  P. lateralis 522–523, **525**, 526, 527, 528, 531, 534, **535**, 593
  P. palmivora 522

P. pinifolia 522, 593
P. ramorum 5, 30, 31, 33, 36, 519–520, 522, 593
diagnosis of 523, **525**
and global plant trade 595, **596**
host range of 527–528
infection biology/epidemiology of 523, 526, 527, 573
management of 529, 530, 531, 534, 536, 598, **600**, 601
phytoplasmas 1, 30, 76–79, **77**, 87
classification of 76, **77**, 78
diagnosis of 80–81
and diseases of pine **77**, 458
epidemiology of 84–86
host resistance to 83
infection biology of 83–84
management of 88–89
online resources for 81
symptoms of 80, 83
phytotoxins 237, 364, 497–498
Picea spp. **52**, 128, 130, 179–181, 183, 184, 189, 521, 578
branch/tip blights affecting 421, 422, 423, 424, 576
DNB in 437, 471
foliar diseases of 459, 469, 471–473
heart rot in **199**, **200**, **201**, **202**, **203**, 205, 206, 207, 208, 209
P. abies see Norway spruce
P. engelmannii **44**, 179, 473
P. glauca 179, 286
P. mariana 179, 287
P. pungens 460
P. rubens 63–64
P. sitchensis **44**, 129, 133, 134, 138, **139**, **147**, 223
tomentosus root rot in 178
Picoides pubescens 409
Pierce's disease (PD) 79, 82, 87
Pinaceae spp. **98**, 116, 160, 231, 547
pine (*Pinus* spp.) 3, 7, **22**, 33, 35, 39, 40, **98**, 179, 184, 267, 300, **305**, 529–530, 575, 578
Aleppo (*P. halepensis*) 43, **77**, 79, 387, 422, 424, 427, **442**, 581
annosus rots and 128, 133, 136
Armillaria disease and 168, 171
Austrian (*P. nigra*) 7–8, 171, 421, 422, 436, 443, 449, 464, 577, 578
bishop (*P. muricata*) 380, **442**
branch/tip blights affecting 421, 422, 423, 424, 427, 428, 576
bristlecone (*P. aristata*) 550
Canary Island (*P. canariensis*) 387, **442**, 467
Ceratocystis spp. in 231
Dothistroma needle blight in *see* DNB
eastern white (*P. strobus*) **212**, 284, 286, 287, **442**, 467, 468, 470, 556, 560
foliar diseases of 459, 464–471
brown spot needle blight 439, 468–469
Cercospora needle blight 460, 464, 467–468
Cyclaneusma needle cast *see* Cyclaneusma needle cast
economic impacts of 464
insects and 464
Lophodermium needle cast *see* Lophodermium needle cast
foxtail (*P. balfouriana*) 550
Gregg's (*P. greggii*) 377
heart rot in **200**, **201**, **202**, **203**, **204**, 207, 211, 216, 219, 221
Italian stone (*P. pinea*) 129, **147**, 387, 427, 471, 581, 592
jack (*P. banksiana*) 285, 287
Japanese black (*P. thunbergii*) 124, 286, **442**, 467, 468
Khasi (*P. kesiya*) 460
limber (*P. flexilis*) 550
loblolly *see* loblolly pine
lodgepole *see* lodgepole pine
longleaf (*P. palustris*) 42, 285, 286, 287, 468
luchu/Okinawa (*P. luchuensis*) 377, 467
maritime (*P. pinaster*) 163, 171, 286, 467, 557, 569
Mexican weeping (*P. patula*) 163, 449, 467
Monterey *see* Monterey pine
mountain (*P. uncinata*) 162, 466
mugo (*P. mugo*) 138, **214**, 464, 577
nematodes affecting *see* pine wilt nematode/disease
pitch (*P. rigida*) 377, 467
pitch canker in *see* pitch canker
ponderosa *see* ponderosa pine
red (*P. resinosa*) 142, **147**, 284, 285, 286, 467
rusts affecting 459, 547, 548, 549, 550, 554, 556, 559–560, 561
Scots *see* Scots pine
shortleaf (*P. echinata*) 41, 285
Siberian (*P. sibirica*) 130
singleleaf pinyon (*P. monophylla*) 284
slash *see* slash pine
sugar (*P. lambertiana*) 42, **44**, 559
Swiss stone (*P. cembra*) 130, **139**, 162
twoneedle pinyon (*P. edulis*) 284, 286
Virginia (*P. virginiana*) 387
viruses affecting **52**
western white (*P. monticola*) 7, 43, **44**, 190, 286, 287, 549, 556, 573
whitebark (*P. albicaulis*) 550, 557
see also Pinus spp.
pine beetles 33, 550
pine decline 290–291
pine engravers 33
pine leaf beetle (*Luperus pinicola*) 464
pine needle beetle (*Cryptocephalus pini*) 464

pine stem rusts 554, 555, 556
pine tip moth (*Rhyacionia* spp.) 386
pine twisting rust (*Melampsora populnea*) 7, 21, 547–548, 557, 581
pine weevils 224, 464
pine wilt nematode/disease (*Bursaphelenchus xylophilus*) 3, **116**
biological control of 122
epidemiology of 119–120
infection biology of 117–118, **117**
isolation/diagnosis of 117, 118
management strategies for 120–123, 124
taxonomy of 115–116
pinsapo fir (*Abies pinsapo*) 130, 135
Pinus spp.
P. albicaulis 550, 557
P. aristata 550
P. balfouriana 550
P. banksiana 285, 287
P. canariensis 387, **442**, 467
P. cembra 130, **139**, **147**, 162
P. contorta see lodgepole pine
P. echinata 41, 285
P. edulis 284, 286
P. elliottii see slash pine
P. flexilis 550
P. greggii 377
P. halepensis 43, **77**, 79, 387, 422, 424, 427, **442**, 581
P. jeffreyi 107, **140**, **147**, 284, **442**
P. kesiya 460
P. koraiensis 163, **442**, 471
P. lambertiana 42, **44**, 559
P. luchuensis 377, 467
P. monophylla 284
P. monticola 7, 43, **44**, 190, 286, 287, 549, 556, 573
P. mugo 138, **214**, 464, 577
P. muricata 380, **442**
P. nigra 7–8, 171, 421, 422, 436, 441, 443, 449, 464, 577, 578, 581
P. oocarpa 163
P. palustris 42, 285, 286, 287, 468
P. patula 163, 449, 467
P. pinaster 163, 171, 286, 467, 557, 569
P. pinea 129, **147**, 387, 427, 471, 581, 592
P. ponderosa see ponderosa pine
P. radiata see Monterey pine
P. resinosa 142, **147**, 284, 285, 286, 467
P. rigida 377, 467
P. sibirica 130
P. strobus **212**, 284, 286, 287, **442**, 467, 468, 470, 556, 560
P. sylvestris see Scots pine
P. taeda see loblolly pine
P. thunbergii 124, 286, **442**, 467, 468
P. uncinata 162, 466
P. virginiana 387
Piptoporus betulinus 206, 211–214, **214**, 220
Pisolithus arhizus 40
Pissodes spp.
P. fasciatus 284
P. nemorensis 285, 379
pistachio (*Pistacia vera*) 308
pitch canker (*Gibberella circinata/Fusarium circinatum*) 3, 7, 23, 376–388, 569, 592
canopy dieback caused by 378
diagnosis of 378
disease progress of 382–384
dissemination/epidemics of 380–382
economic impacts of 377–378
epidemiology of 380–384
eradication strategy for 385–386
exclusion strategy for 384–385
fertilization and 387
global risk assessments of 384
host resistance to 387, 388
IDM for 387–388
infection biology of 379–380
insect vectors of 379, 380, 381–382
management of 384–388
pathogen/significance/distribution 376–378
protection strategy for 386–387
SIR and 383, 388
vegetative compatibility groups (VCGs) 381
wound characteristics and 379–380
Pityophthorus spp. 379, 380, 381, 382
P. carmeli 382
P. pubescens 379
P. setosus 382
plane (*Platanus* spp.) 230, 232–235, 237, 242–243, **494**, 504
London (*P. hybrida*) 233, 239, 240–241, 242, 245, 246, 247, 248–249
see also sycamore
plant hoppers (Cixiidae) 85
Platanus spp. 234–235, 242–243, 423, 504
P. hybrida see plane
P. occidentalis 80, 241, 242, 245, 247, 509
P. racemosa 233
Platypodinae 235, 239, 256, 263
Platypus spp.
P. cylindricus 239
P. quercivorus 263, 272
Pleuroceras pseudoplatani 497, 498, 502
Pleurotus spp. 219
P. ostreatus **203**
Ploioderma hedgcockii/P. lethale 471
Plum pox virus (PPV) **52**, 62–63, 67
Poland 41, 77, **78**, 79, 88, 129, 144, 162, 163, 232, 241, 400, 489, **593**, **596**
pollarding 106
polyphenols 5, 61, 81
polysaccharides 61, 81, 84, 165, 496, 497, 521
ponderosa pine (*Pinus ponderosa*) 31, 40, **44**, 107, **147**, 162, 169, 190, 284, 286, 287, 468, 470, 573
DNB in 436, **442**, 443, 450, 578

poplar (*Populus* spp.) 21, 43, **52**, 62–63, 65, **200**, 208, 216, 217, 241, 430, 601
Canadian (*P. canadensis*) 499
foliar diseases of 490, 500, 502–503, 504, 505, 506, 509, 510
Japanese (*P. maximowiczii*) 407, 559
P. balsamifera 407, 422, 489, **493**, 559
P. deltoides 81, 489, **493**, 501, 551, 561
P. tremula **77**, **212**, 232, 407, 424, 489, 500
P. tremuloides 81, 407, 489, **493**
rusts affecting 547–548, 554, 555, 558, 560–561, 580–581
white (*P. alba*) **77**, 407, **493**, 521, 580
see also aspen
poplar leaf rust (*Melampsora laricis-populina*) 547, 551, 553, 554, 555, 580, 581
host resistance to 559, 560–561
Populus spp. *see* aspen *and* poplar
Poria sequoiae 205
Porodaedalea spp. 220
P. pini 5, 8, 35, 181, 182, **200**, 207, 209, 211, **212**, 221
Port Orford cedar (*Chamaecyparis lawsoniana*) 36, 43, **44**, 522, 523, 526, 531, 534, **535**, 593
Portugal 115, 118, 120, 123, 377, 384, 569
chestnut blight in 318, 319, 323, 325, 327, 328
Postia spp. 215, 219, 222
P. caesia/*P. stiptica* **203**
potassium fertilizers 41, 325, 412
powdery mildews (*Erysiphales*) 4, 6, 39, 65, 247, 488, 497, 567, 579–580, 580
PPV (*Plum pox virus*) **52**, 62–63, 67
Pristiphora laricis 475
prokaryotes *see* bacteria; phytoplasmas
propagation techniques 31
propiconazole 43, 246, 248, 273, 466, 480
protection strategy 33, **34**, 38–42
biological control 39–40
chemical control 38–39
fertilization 41–42
prescribed fire 42
pruning 41
thinning 40–41
protein coding genes 299
protozoa 30, 65
pruning 9, **34**, 39, 41, 43
Prunus spp. **52**, 54–56, **98**, 160, 164, 166, 232, 237, 422, 489
P. avium 56, 65
P. dulcis 79, **98**, 232, 237
Pseudocercospora juniperi 479
Pseudofusicoccum spp. 299
Pseudoinonotus dryadeus **204**, 217
Pseudolarix spp. 392, 398
P. amabilis 399
Pseudomonas spp. 82, 86
P. chlororaphis 369
Pseudopityophthorus spp.
P. minutissimus 239, 244
P. pruinosus 244
Pseudotsuga 128, **200**, 576
P. menziesii see Douglas-fir
Psittacanthus 98, **98**
Puccinia psidii 548, 549, 550, 552, 554
Pucciniales see rusts
Pucciniastrum spp. 459
P. areolatum 556, 568
P. epilobii 461, 473
punktree (*Melaleuca quinquenervia*) 303
PWN *see* pine wilt nematode/disease
pycnidia/pycnidial stromata 301, 307
Pyrenees 168
Pyrus spp. **98**, 592, **599**
Pythium spp. 38, 518, 520–521, 524, 525, 529, 534, 570

quarantine **34**, 35–36, 117, 188, 244, 270, 333, 385, 398, 427–428, 477, 504–505, 557
EPPO list of species 324, 468, 469, 556
regulations on 429
Quercus spp. *see* oak
Q. agrifolia 65, 523–524, 528, 532
Q. alba 207, 232, 238, 248, 496, 523
Q. ellipsoidalis 232
Q. fusiformis 232
Q. garryana 105
Q. macrocarpa 232, 248
Q. marilandica 232, 238
Q. palustris 80, 82, 341
Q. prinus 207
Q. robur **57**, **77**, 162, 197, **212**, **214**, 215, **498**
Q. rubra see red oak
Q. suber **306**
Q. velutina 207
Q. virginiana 232

RAB *see* redbay ambrosia beetle
Raffaelea spp. 230, 259, 261, 262, 263
R. lauricola see laurel wilt
R. quercivora 259, 263, 593
R. quercus-mongolicae 259, 263
Rapanea melanophloeos 346
RAPD (random amplification of polymorphic DNA) analysis 135, 184, 369, 395
recreational use of forests 29, 38
red oak (*Quercus rubra*) 9, **10**, 33, 81, 82, 207
Ceratocystis spp. in 232, 234, 237, 238, 242, 246, 247
red osier dogwood (*Cornus sericea*) 303
red ring nematode/disease (*Bursaphelenchus cocophilus*) 116–117
isolation/diagnosis of 117–118
management of 121, 122, 123–125

red spruce (*Picea rubens*) 63–64
redbay (*Persea borbonia*) 258, **264**, 265, 267, 269
redbay ambrosia beetle (RAB, *Xyleborus glabratus*) 258, 261, 263, **264**, 265, 267
  management of 271, 272, 273, 278
redwood (*Sequoia sempervirens*) 205
regulation of forest diseases 36, 121, 244, 324, 333, 368, 428–429, 468, 598
  see also quarantine
resin-top disease (*Cronartium flaccidum*) 547, 550–551, 554, 558
Resinicium bicolor 39, **203**, **213**, 216
resinosis 134, 180, 288
RFLP (restriction fragment length polymorphism) analysis **55**, 76, 78, 81, 300, 326, 348, 361–362, 439, 465
Rhabdocline 463, 476
Rhabdocline needle cast (RNC, *R. pseudotsugae*) 458, 459, 461–462, 463, **600**
  taxonomy of 461
Rhizina undulata 11, 35, 42, 121
Rhizoctonia spp. 21, 38, 569, 570, 571
rhizomorphs 2–3, 159, 160–161
Rhizopogon spp. 40
Rhizosphaera spp.
  R. kalkhoffi 460, 461, 472, 473
  R. macrospora 461, 473, 474
rhododendron (*Rhododendron* spp.) 488, 522, 549
Rhyacionia spp. 386
rhynchophorol 125
Rhynchophorus spp.
  R. ferrugineus 121
  R. palmarum 116, 117, 121, 122, 124–125
Rhytisma spp. 495, 496, 500
  R. acerinum 489, 492, **494**, 497
  R. punctatum 495–496
Ribes spp. 4, 35, 41, 556, 557, 560
ring disease 168
risk assessments 36, 271, 384, 551, 604
Robinia spp. **53**
Rocky Mountains (USA/Canada) 171
root balls 181
root grafts 42, 134, 238, 242, 268, 285
root parasites 97
root pruning 43
root rot 12, 37, 38, 42, 168, 198
  annosus *see* annosus root/butt rots
  Armillaria *see Armillaria* spp.
  control of 570–571
  primary/secondary infections 21–23, **22**
  in seedlings 570–571
  spread of 16
  and thinning 8, **9**, 21, 23, 40, 136, 138, 142, 144, 145, 148–149
Rostraureum tropicale 345, 350
Rotstand disease model 22–23, **22**, 138, **139**, 149
Rotstop treatment 39, 144, 145
rubber tree (*Hevea brasiliensis*) **98**, 170, 232, 522
Russia **56**, 128–129, **140**, 178, 179, 438, 458, 577
rusts (*Pucciniales*) 4, 6, 13, 35, 420, 459, 461, 473, 488, 503, 547–561
  avoidance strategy for 555–556
  cankers caused by 548
  chemical control of 557–558
  cone 551, 556, 568
  diagnosis of 550–551
  DNA barcoding of 551
  economic impact of 549–550
  ecosystem impact of 550
  environmental/landscape conditions for 555
  epidemiology of 554–555
  eradication strategy for 557
  exclusion strategy for 556–557
  foliar damage caused by 548–549
  hazard/risk ratings for 555–556
  heteroecious/autoecious 552, 555
  host resistance to 553, 558–561
  host specificity of 554
  IDM for 561
  infection biology of 552–554
  life cycle of 552–553, 557
  management of 555–561
  in nurseries 568, 580–582
  pathogens/distribution 547–548
  protection strategy for 557–558
  SSP genes of 553–554
  stem 552, 553, 554, 555, 556, 558
  therapy strategy for 561
  thinning/pruning and 558
  types of damage caused by 548–549

Salicaceae 423
salicylic acid 5, 388
Salix spp. *see* willow
  S. babylonica **77**
Sambucus spp. 50, **53**
  S. nigra **55**, 215
sanitation 36–38, 244–245, 269, 271, 272–273, 291, 308, 310, 366, 399, 468–469, 478, 598
Santalaceae 97
sap beetles (Nitidulidae) 231
sap rot **198**, 217, 219–220
sap streak 231, 240, 243
sap-stain *see* blue-stain fungi
saprotrophic fungi 4–5, 8, 37–38, 39
sawdust/frass 235, 236, 239, 241, 244, 245
sawfly (*Pristiphora laricis*) 475
scale insects 464, 592
Scandinavia 35, 39, 144, 162, 179, 422, 471
Scaphoideus spp. 85
  S. luteolus 85

Schizolachnus pineti 464
Schizophyllum commune 219
schweinitzii root/butt rot (*Phaeolus schweinitzii*) 38, 204, 207, 217, 218, 222
Scirrhia acicola 42
Scleroderris canker (*Gremmeniella abietina*) 7–8, 25, 420–422, 424, 427–428, 575–576
  anamorphic stage (*Brunchorstia pinea*) 421
  control of 429, 575–576
  infection biology/epidemiology of 425, 427–428, 575
  races/biotypes 421–422, 424, 575
sclerotia 4, 362, 492, **493**, 495
Scolytinae 235, 256, 263
Scolytus spp. 259, 265, 267, 268, 271, 275, 278
  S. multistriatus 29, 263, 265, 266, 271
  S. pygmaeus 263, 265
  S. quadrispinosus 232
  S. schevyrewi 263
  S. scolytus 263, 265, 267
Scots pine (*Pinus sylvestris*) 5, 7, **22**, 33, 298, 521, 575
  annosus rots in 128, 130, **133**, 134, **139**, **140**, 146, **147**, 149, 150
  Armillaria disease in 162, 163
  branch/tip blights affecting 421, 422
  DNB in 441, **442**, 577
  foliar diseases of 458, 461, 464, 466, 467, 468
  heart rots in **213**, **214**
  Leptographium spp. in 285, 286
  and needle cast disease 8
  phytoplasma diseases of **77**, 79
Scytalidium spp. 39
seagrape (*Coccoloba uvifera*) 346, 347
Sebastiana longicuspis 11
seed orchards **57**, 123, 308, 310, 377, 379, 386, 509
seed rots 520–521
seedling diseases 570–571
seeds 65, 308–310, 430, 568–569
  storage of 569
  treatments for 39
seiridin 364
Seiridium spp. *see* cypress canker
  S. cardinale 359–360, 361, 362–364, **363**, 365–367, 368–371, 599
  S. cupressi 359, 361, 362–363, 364, 367
  S. unicorne 359, 360, 361, 362, 364, 366, 367
Septoria spp. 495
  S. musiva 491
Septotis populiperda 502
Sequoia sempervirens 205
Sequoiadendron spp. 128
  S. giganteum **147**
Serbia **77**, 78, 85, 445
Shigo concept 197–198
shining gum (*Eucalyptus nitens*) 298, 499, 500, 508, 509
Siberian elm (*Ulmus pumila*) **77**, 85, 267, 276, 277
silviculture **2**, 39, 43
single leaf ash (*Fraxinus anomala*) **77**, 89
SIR (systemic induced resistance) 43, 383, 388, 429
siricid woodwasps 219
Sirococcus spp.
  S. conigenus 420, 422–423, 424, 568
    control of 430
    infection biology/epidemiology of 426, 428
  S. piceicola 424
  S. tsugae 424
Sistotrema brinkmannii **203**, **213**, 216, 219
site selection 11, 12, 31, 138, 268, 352, 402, 444, 504
  hazard rating 35, 38, 40, 449, 551
Sitka spruce (*Picea sitchensis*) **44**, 129, 133, 134, 138, **139**, **147**, 223
slash pine (*Pinus elliottii*) 7, **140**, **147**, 163, 285, 449, 467, 470, 549, 560
  pitch canker in 376, 377, 379, 387
Slovakia 329, 402, 478, **593**
Slovenia 318, 326, 327, **593**, **596**
SLRSV (*Strawberry latent ringspot virus*) **51**, **53**, 65
small oak beetle (*Pseudopityophthorus minutissimus*) 239
snow blights 5, 578
sodium hypochlorite 245, 272, 386
sodium nitrate 39
soil 9, 35, 303, 308, 529–530
  and annosus rots 138, **141**, 143
  compaction 189
  disinfection of 169
  nutrient status/pH 7, 8, 40
  pasteurization 38
  viruses transmitted through 63–64
Solanum lycopersicon **306**
Sorbus spp. **53**, 56, 303
  S. aucuparia 54, **55**, **56**, 303
sour-felling 275
South Africa 161, 232, 234, 239, 285, 286, 287, 342, 343, 346, 352, 422, 438, 577
  cypress canker in 359, 360
  foliar diseases in 460, 500, 509
  pitch canker in 376, 377, 569
South America **77**, 78, 98, **98**, 103, 116, 233–234, 342, 343, 346, 350, 352
  foliar diseases of conifers in 460, 461, 471
  rusts in 547, 548, 554
South Carolina (USA) 80, 258
South Korea **77**, 376, 377
Spain **77**, 79, 115, 118, 123, 129, 130, 135, 269, 521, **593**, **596**
  branch/tip blights in 422, 424, 427
  Ceratocystis spp. in 233, 241, 245
  chestnut blight in 318, 319, 328
  pitch canker in 376, 377, 379, 384, 569
Sparassis crispa **204**
spatial resistance 413, 414, 508

species-specific primers 300
Sphaeropsis spp. 299
S. sapinea see Diplodia spp., *D. pinea*
Sphaeropsis tip blight 424, 568, 576–577
Sphaerotheca lanestis 65
spider mites 464, 475
Oligonychus ununguis 458
spiroplasmas 76
spittlebugs (Cercopidae) 85
Aphrophora canadensis 379, 384
Sporothrix spp. 257, 258, 260, 261, 262, 283, 289
spraying 39, 124
aerial 109, 122, 414
spruce *see Picea* spp.
spruce beetles 33
spruce cone rust (*Pucciniastrum areolatum*) 551, 556
SSCP (single strand conformation polymorphism) 184–185
SSPs (small-secreted proteins) 553–554
Stegophora spp. 500
S. ulmea 491, **493**, 495, 502, 504–505, 506
stem cankers 197, 205, 345, 400, 408, 409, 414, 548, 550, 558, 561
stem decay fungi 34–35, 198–205
stem injections **34**, 39, 88
stem rusts 552, 553, 554, 555, 556, 558
stem wound infections **133**, 134, 142
Steremnius carinatus 284
Stereum spp. 215
S. gausapatum **202**, **212**, 215, 222
S. hirsutum **203**, 221
S. sanguinolentum **203**, 206, 217, 218, 219, 221, 222
Stigmina spp.
S. lautii 460
S. thujina 479
Strawberry latent ringspot virus (SLRSV) **51**, **53**, 65
Streptococcus 76
Streptomyces lydicus 40
stress, elastic/plastic 302
Strix occidentalis 104
stump removal 18, **34**, 37, 143, 169, 189, 271
stump temperature 35, 136
stump treatment **22**, 23, **34**, 39, 40, 108, 143–145, **145**
for annosus rots 143–145, **145**, 148–150
stump-top surveys 188
styrax (*Styrax benzoin*) 232
sudden oak death (SOD, *Phytophthora ramorum*) 5, 30, 31, 33, 36, 39, 520, 527
management of 529, 531–532, 533, 597–598
sugarcane weevils 116
Sweden 33, 138, **139**, 143, 144, 146, 149, **150**, 392, 407, 508, **593**, **596**
branch/tip blights in 422, 425, 427
sweet cherry (*Prunus avium*) 56, 65, 162
sweetgale (*Myrica gale*) 13, **14**
sweetgum (*Liquidambar* spp.) 82, 303, 490
Swiss needle cast (SNC, *Phaeocryptopus gaeumannii*) 31, 459, 461, 462–463
management of 463–464
Switzerland **77**, 233, 407, 422, 461, **593**, 599, **600**
chestnut blight in 318, 319, 323, 326, 327, 328, **328**, 329, 330, 333, 407
sycamore (*Platanus* spp.) 82, 490, **494**, 495, 504
American (*P. occidentalis*) 233, 241, 242, 245, 247, 509
California (*P. racemosa*) 233
sycamore anthracnose (*Apiognomonia veneta*) 247, **494**, 497, 508, **600**
Sydowia polyspora 459, 461, 471, 472, 475
synchronous cohort senescence 13
Syringa spp. 81
Syzygium spp. 342, 346
S. aromaticum 341, 350
S. cuminii 344

Taeniothrips inconsequens 497
Taiwan 79, 258, 467, 474, 522
tanoak (*Lithocarpus densiflorus*) 519, 520, 524, 529, 533, 535, **535**
Tanzania 341, 343, 467
Taphrina spp. 488, 489, 492, 495, 496, 497, 500, 506
T. populina **494**, 502
Tapinanthus bangwensis 107
tar spot 489, 492, **494**
Tasmanian bluegum (*Eucalyptus globulus*) 504, 508, 510
TASS (Tree and Stand Simulator) 33
Taxus spp. 478–479
TBSV (*Tomato bushy stunt virus*) **53**, 63
teak (*Tectona grandis*) 162
temperature factor **2**, 35
Terminalia spp. 345
Texas (USA) 80, 232, 237, 238, 240, 242, 245, 246
Theobroma cacao 166, 232, 241, 522
thiabendazole 43, 273, 386
Thielaviopsis 231, 256
T. australis 231, 236, 237
thinning **34**, 35, 39, 40–41
and root rot 8, 21, 23, 40, 136, 138, 142, 144, 145, 148–149
thiophanate-methyl 368, 371, 480, 506, 572, 577
Thuja spp. 179, 359, 379
T. occidentalis 479
T. plicata 43, **44**, **147**, 171, 178, 286
Tiarosporella 459, 474–475
Tibouchina spp. 343, 346–347
T. granulosa 345–346, 350
T. urvilleana 343, 346
timber/timber products industry 29–30, 98–99, 120–121, 142, 233, 239, 244, 266, 366, 398
timberworm (*Arrhenodes minutus*) 244
TMV (*Tobacco mosaic virus*) **52**, **53**, 65

tomentosus root rot (*Onnia tomentosa*) 33, 37, 38, 178–192
avoidance strategy for 187–188
and bark beetles 180
diagnosis of 180–182
economic impacts of 179–180
ecosystem classification system for 187
epidemiology of 183–187
eradication strategy for 188–190
exclusion strategy for 188
fire and 191
fruiting bodies of 181
genotypes of 184–185
host resistance to 183, 190–191
hyphae of 182
IDM for 191–192
infection biology of 182–183
management of 187–192
partial cutting and 192
pathogen/significance/distribution 178, 179–180
protection strategy for 190
saprophytic phase 179, 192
stand factors and 185–186
survey methods for 187–188
taxonomy of 178
therapy strategy for 191
tree vigour and 186
Tomicus piniperda 270, 287
ToMV (*Tomato mosaic virus*) **52**, **53**, 63–64
ToRSV (*Tomato ringspot virus*) **53**, 62
Trametes versicolor **203**, 221
tree density 11, **12**
trenching **34**, 37, 42, 145, 170, 246–247
triadimefon 466, 477, 557–558, 561
triazoles 246, 578, 581
Trichaptum abietinum 217
Trichoderma spp. 35, 39, 170, 447, 526, 569
T. harzianum 40, 508, 510, 572
T. polysporum 508, 572
T. viride 189, 368–369
TRSV (*Tobacco ringspot virus*) **52**, 65
Tsuga spp. **19**, 39, 128, 469, 578
heart rot in **200**, **201**, **202**, **203**, 205
T. heterophylla see western hemlock
T. mertensiana 7, **44**, 179, 185, 187, 286
tulip tree (*Liriodendron tulipifera*) 207
Turkey 129, 233, 318, 327, 328, 329, 349, 445, 461
twig beetle (*Pityophthorus* spp.) 379, 380, 381, 382
Tyromyces amarus 205

Uganda 232, 234, 343
Ukraine **77**, 79, 257, 438
Ulmus spp. 3, **53**, 80, 82, 83, 85–86, **493**
U. alata **77**
U. americana 78, 79, 86, 266, 276
U. chenmoui **77**
U. crassifolia **77**, 108
U. laevis **77**, 85, 266, 267, 276–277
U. minor 266, 267, 268, 276, 277
U. procera **77**, 85, 266–267
U. pumila **77**, 85, 267, 276, 277
U. rubra **77**
U. serotina **77**
U. wallichiana 257, 276
see also Dutch elm disease
Umbellularia californica 524, 527–528, 529
United States (USA) 11, 25, **77**, 103, 179, 192, 325, 592, 593, 597–598
annosus rots in 35, 128, 129, 137, **140**, 142, 144
Armillaria disease in 160, 161–162
bacterial diseases in 79–80
branch/tip blights in 422
cankers in 341
Ceratocystis spp. in 231, 232–233, 236, 238–239, 240–241, 243, 244, 248–249
chestnut blight in 325–326
DNB in 437, 449, 577, 578
foliar diseases in 460, 461, 469, 470, 472, 473, 474, 475, 476, 478, 479, 490
heart rots in 205, 207, 223, 224
Hypoxylon canker in 407, 411, 413–414
larch canker in 392, 398, 399
Leptographium/Grosmannia spp. in 284, 285, 286, 287
mistletoes in 103, 104, 105, 106–107, 108–109
modelling in 31–33
National Institutes of Health 300, 349
oomycete diseases in 520, 522, 526, 527–530
Ophiostoma diseases in 258, 263, 266, 267, 271, 274
pathogen dispersal in 19, **20**
phytoplasma diseases in **77**, 78, 79, 573
pitch canker in 376, 377, 379, 380, 381–383, 569
rusts in 555, 557, 559–560
stump treatment/removal in 37, 39
thinning in 35
see also specific states
urban areas 37, 79, 88, 124, 248, 249, 277–278, 383, 458
urea 39, 41, 143–144, 148
Ursicollum fallax 346
Utah (USA) 85, 130

Valsa spp. 347
V. abietis 461, 473, 475
Vapam 189
vascular stain diseases 256, 258–259, 278
diagnosis of 261–262

vascular wilt diseases 36, 231, 234, 256, 266
see also Dutch elm disease
velvet ash (*Fraxinus velutina*) **77**, 80, 89
Venezuela 117, 122
Venturia shoot blight (*Venturia* spp.) 423, 424, 490, 495, 498, 500
infection biology of 426
V. inaequalis 503
V. inopina 424
V. macularis **423**, 424
V. populina 420, 424, 430, 504
V. saliciperda 490
V. tremulae 424
Verticicladiella spp.
V. alacris 286
V. procera 285
V. wageneri 284
Verticillium spp. 288
V. albo-atrum 274, 492
V. dahliae 260
Verticillium wilt 40, 43, 260, 492
Victoria (Australia) 231, 443, 450
Virgella robusta 460, 473, 475
Virginia (USA) 80, 441
viruses 1, 30, 50–67, 122, 352
assessment of losses from 54
diagnosis/detection of *see specific viruses*
infection biology/epidemiology of 62–65
management of 65–67
mechanical transmission of 62–63
modes of transmission 50
myco- 148
pathogens/significance/distribution of 50–57
size/shape of **51–53**, 54
taxonomy/number of species of 50, 57
transmission through soil/water of 63–64
Viscaceae 97, 101, 104
viscin 102, 104
Viscum spp. 98, **98**, 101
V. album 7, 97, 104
Vitis spp. 62, 78, 79, 82, 86, 87, 89, 169, 300, 308

walnut (*Juglans* spp.) 60, **77**, **98**, 488, 490, 492, **494**, 504
walnut witches' broom **77**
Washington (USA) 40, 41, 42, 44–45, 80, 104, 162, **201**, 287, 557
IDM in 44–45, **45**
water stress 88, 124, 288, 409–410, 423, 503
weather factors 5, 7, 31–32, 38, 303
weed control 88, 241, 291, 398
weevils
American palm (*Rhynchophorus palmarum*) 116, 117, 121, 122, 124–125
and blackstain root disease 284–285, 286
deodar (*Pissodes nemorensis*) 285, 379
and nematodes 116, 119, 120, 121, 122, 123, 124–125
pine 224, 464
West Virginia (USA) 232, 242
Western Australia 36, 39, 161
Phytophthora spp. in 519, 520, 521–522, 524, 526, 530–531, 593
western gall rust (*Endocronartium harknessii*) 379
western hemlock (*Tsuga heterophylla*) 35, **44**, 105, 107–108, 134, **147**, 179, **212**, 286, 424, 573
laminated root rot in 179
western juniper (*Juniperus occidentalis*) **77**, 79, 86
western larch (*Larix occidentalis*) 43, **44**, 171, 190, 286, 399, 579
western redcedar (*Thuja plicata*) 43, **44**, **147**, 171, 178, 286
wetwood, bacterial 79, 81, 81–82, 88
white ash (*Fraxinus americana*) 78, 80
white pine blister rust (*Cronartium ribicola*) 4, 19, **20**, 31, 35, 448, 547
diagnosis of 550
host resistance to 559–560
impacts of 549, 550
infection biology/epidemiology of 554, 555
management of 35, 41, 43, 44, 555–556, 557, 558, 592, **600**
white pine root disease 284, 288, 291
white snow mould (*Phacidium taxicola*) 478
white spruce (*Picea glauca*) 179, 286
white trunk rot *see Phellinus* spp., *P. tremulae*
white-banded elm leafhopper (*Scaphoideus luteolus*) 85
wildfire 40, 106
wildlife 32–33, 45, 104–105, 224, 232
cavity-nesting 34, **34**, 35
see also biodiversity
willow (*Salix* spp.) **53**, 81, **212**, **214**, 215, 490
rusts affecting 547–548
wilt diseases 21
windthrow **32**, 129, 181
Wisconsin (USA) 35, 162, 240, 243, 248, 286, 287, 407
witches' broom **77**, 80, 100, 101, 429, 469, 549
wood decay fungi 8, 129, 144, 217, 218, 220, 221
see also annosus root/butt rots
woodpeckers 34, 35, 260, 409
WRDM (Western Root Disease Model) 22–23, 32, 138, **140**, 171, 192

X-disease **77**
Xanthomonas campestris 84

xylanase 364, 521
Xyleborus glabratus 258
Xylella fastidiosa 79–80, 82, 84, 86–88, 233
xylem feeding insects 87
Xylobolus frustulatus **202**, 222
xylomycetophagous beetles 263

yellows-type disease **77**
yew (*Taxus* spp.) 478–479

Zambia 342, 345, 352
Zeiraphera diniana 475
Zelkova serrata **77**, 85, 86
Zimbabwe 436, 448